Question Bank on
Electrical and Electronics Engineering

with Question Papers from Various Competitive and Recruitment Examinations

for **GATE, UPSC (ESE/IES), Civil Services Examination, NTPC, BHEL, HAL, BEL, CPWD, MTNL, NHPC, DMRC, ECIL, ISRO, DRDO, AIR, Air India, AMIE, IETE, State Services Examination**

Part II

Theory, Concepts, Multiple Choice Questions, and Questions from IES, GATE and Other Competitive/Recruitment Examinations

- ✓ Transformers
- ✓ Rotating Machines—DC Machines
- ✓ Three Phase and Single Phase Induction Motors
- ✓ Alternator and Synchronous Machines
- ✓ Power System—Generation, Transmission, Distribution and Load Flow Studies
- ✓ Switchgear Protection, Stability and High Voltage
- ✓ Power Electronics
- ✓ Electric Traction, Electric Heating, Welding and Braking—Utilization of Electrical Energy and Illumination
- ✓ Analog Electronics and Devices
- ✓ Digital Techniques and Fundamentals of Computer
- ✓ Communication System
- ✓ Computer and Microprocessor
- ✓ Electrical Aircraft, Automotive Electric and Electronic Systems

Contents of Parts I and II

Part I

Part II

Question Bank on
Electrical and Electronics Engineering

with Question Papers from Various Competitive and Recruitment Examinations

for **GATE, UPSC (ESE/IES), Civil Services Examination, NTPC, BHEL, HAL, BEL, CPWD, MTNL, NHPC, DMRC, ECIL, ISRO, DRDO, AIR, Air India, AMIE, IETE, State Services Examination**

Part II

Harish C Rai

PhD (Electrical Engg, IIT, Delhi), FIE (India), FIETE, MISTE, MAeSI

Pro Vice Chancellor
Galgotias University, Greater Noida, UP

Former

Professor, Department of Electrical and Electronics Engineering
Chhotu Ram State College of Engineering
(Presently Deenbandhu Chhotu Ram University of Science and Technology)
Murthal, Haryana 131039

Controller of Examinations
Director, Academic Affairs
Director, Research and Project Monitoring Cell
Director, Organization and Development
GGS Indraprastha University, Delhi

Advisor I, All India Council of Technical Education
(Ministry of Human Resource and Development, Delhi)

CBS Publishers & Distributors Pvt Ltd

New Delhi • Bengaluru • Chennai • Kochi • Kolkata • Mumbai

Bhopal • Bhubaneswar • Hyderabad • Jharkhand • Nagpur • Patna • Pune • Uttarakhand • Dhaka (Bangladesh)

Disclaimer

Science and technology are constantly changing fields. New research and experience broaden the scope of information and knowledge. The author has tried his best in giving information available to him while preparing the material for this book. Although all efforts have been made to ensure optimum accuracy of the material, yet it is quite possible some errors might have been left uncorrected. The publisher, the printer and the author will not be held responsible for any inadvertent errors or inaccuracies.

Question Bank in

Electrical and Electronics Engineering

with Question Papers from Various Competitive and Recruitment Examinations

Part II

ISBN: 978-93-88527-68-2

First Edition: 2019

Published by Satish Kumar Jain and produced by Varun Jain for

CBS Publishers & Distributors Pvt Ltd

4819/XI Prahlad Street, 24 Ansari Road, Daryaganj, New Delhi 110 002, India.

Ph: 23289259, 23266861, 23266867 Fax: 011-23243014 Website: www.cbspd.com

e-mail: delhi@cbspd.com; cbspubs@airtelmail.in

Corporate Office: 204 FIE, Industrial Area, Patparganj, Delhi 110 092

Ph: 4934 4934 Fax: 4934 4935 e-mail: publishing@cbspd.com; publicity@cbspd.com

Branches

- **Bengaluru:** Seema House, 2975, 17th Cross, K.R. Road, Banasankari 2nd Stage, Bengaluru 560 070, Karnataka
  Ph: +91-80-26771678/79 Fax: +91-80-26771680 e-mail: bangalore@cbspd.com
- **Chennai:** 7, Subbaraya Street, Shenoy Nagar, Chennai 600 030, Tamil Nadu
  Ph: +91-44-26680620, 26681266 Fax: +91-44-42032115 e-mail: chennai@cbspd.com
- **Kochi:** 42/1325, 1326, Power House Road, Opposite KSEB Power House, Ernakulam 682 018, Kochi, Kerala
  Ph: +91-484-4059061-65 Fax: +91-484-4059065 e-mail: kochi@cbspd.com
- **Kolkata:** 6/B, Ground Floor, Rameswar Shaw Road, Kolkata-700 014, West Bengal
  Ph: +91-33-22891126, 22891127, 22891128 e-mail: kolkata@cbspd.com
- **Mumbai:** 83-C, Dr E Moses Road, Worli, Mumbai-400018, Maharashtra
  Ph: +91-22-24902340/41 Fax: +91-22-24902342 e-mail: mumbai@cbspd.com

Representatives

• **Bhopal**	0-8319310552	• **Bhubaneswar**	0-9911037372	• **Hyderabad**	0-9885175004	• **Jharkhand**	0-9811541605
• **Nagpur**	0-9021734563	• **Patna**	0-9334159340	• **Pune**	0-9623451994	• **Uttarakhand**	0-9716462459
• **Dhaka (Bangladesh)**	01912-003485						

Printed at: Rashtriya Printers, Dilshad Garden, Delhi, India

to

My loving parents
Late Sh Balraj and Smt Ram Devi
for their everlasting support in everything that
I have achieved in life

My eternally compassionate
wife Sangeeta, loving children
Shivanshu, Himanshu and Shipra,
especially to my pride
grandson Vivaan

Preface

Question Bank on Electrical and Electronics Engineering (in two parts) follows a logical concept building approach rather than only formula based, as offered by the other books. The objective has been to structure a complete examination-oriented reference book covering the fundamental aspects of theory at a glance before proceeding to their relevant questions. This book has a large number of (over 15,500) multiple choice questions from various competitive and recruitment examinations like UPSC (Engineering/Civil Services Examination), GATE, IES, NTPC, BHEL, HAL, AMIE, IETE, CPWD, MTNL, NHPC, DMRC, ECIL, ISRO, DRDO, AIR, Air India, etc. Latest questions (2017 and 2018) from IES with their complete explanations have been given at the end of the text to impart a valuable insight into problem-solving approach.

Uniqueness of this text lies in the fact that it is a harmonious combination of both a textbook and check book, the former being useful to engineering students and latter to the examiner. Multiple choice questions with answers have been given to make students understand the underlying concepts through detailed illustrations and circuit diagrams. The concept has been shared with eminent teachers and their valuable opinion on various aspects has been incorporated while finalizing the text.

This book will be useful to the students not only during their semester examination but more so afterwards, when they have to appear for other competitive and recruitment examinations and to face interview boards for job recruitment of various public and private sector organizations. To the examiner and paper setter, it provides almost an unlimited facility to select questions of his choice for written and oral examinations.

This volume has 13 chapters.

Chapter 7 Transformers

Chapter 8 Rotating Machines—DC Machines

Chapter 9 Three Phase and Single Phase Induction Motors

Chapter 10 Alternator and Synchronous Machines

Chapter 11 Power System—Generation, Transmission, Distribution and Load Flow Studies

Chapter 12 Switchgear Protection, Stability and High Voltage

Chapter 13 Power Electronics

Chapter 14 Electric Traction, Electric Heating, Welding and Braking—Utilization of Electrical Energy and Illumination

Chapter 15 Analog Electronics and Devices

Chapter 16 Digital Techniques and Fundamentals of Computer

Chapter 17 Communication System

Chapter 18 Computer and Microprocessor

Chapter 19 Electrical Aircraft, Automotive Electric and Electronic Systems

At the end of the text, questions from various examinations including IES (2017 and 2018) with detailed explanations are given.

Each chapter is a culmination of theory, MCQs with answers from various competitive and recruitment examinations, supplemented with relevant tables, detailed illustrations and circuit diagrams.

I believe that the systematic coverage of this text will induce proper knowledge to the students which might have been missed by them during their course of study. In addition, the subject centric approach of the book will prepare the reader to crack other parallel examinations like GATE, UPSC (Engineering/Civil Services Examination), NTPC, BHEL, HAL, AMIE, IETE, PWD, MTNL, NHPC, DMRC, ECIL, ISRO, DRDO, AIR, Air India, State Services Examination, etc.

I would like to request my esteemed readers to kindly send me their valuable suggestions for improvement of the book and to notify me of any error they may come across while going through the book.

Harish C Rai

Acknowledgements

I thank all my undergraduate students who suggested that I should write this book and indeed, all those who have encouraged me in this venture. I derive immense pleasure in expressing my sincere thanks to Prof Damodar Acharya, former Director, IIT Kharagpur (former Chairman, AICTE, New Delhi); Prof DP Kothari, former Director, IIT Delhi, and Prof Pritam Singh, former Director, MDI, Gurugram, for the invaluable encouragement throughout this work. I am indebted to their guidance and invaluable suggestions.

I express my gratitude to Prof SS Murthy, former Vice Chancellor, Central University of Karnataka; Prof ZH Zaidi, former Vice Chancellor, MJP Rohilkhand University, Bareilly; Prof BP Singh, Prof Bhim Singh, Department of Electrical Engineering, IIT Delhi, for sparing their valuable time and providing useful guidance on various chapters.

I thank my colleagues, Prof Alok Mittal, Member Secretary, AICTE, New Delhi; Prof Pradeep Kumar, Pro Vice Chancellor, Galgotias University, Greater Noida; Prof AM Agarwal, Pro Vice Chancellor, GLA University, Mathura; Prof JRP Gupta (NSIT, Delhi); Prof DR Bhaskar (Delhi Technological University); Prof SS Inamdar (Vishwaniketan, Mumbai); Prof VK Sharma (NIT, Uttarakhand); Prof Rominder Randhwa, Director, Guru Tegh Bahadur Institute of Technology; Prof SS Tyagi, Director, BSA Institute of Technology, Haryana, and Prof Lajpat Rai, IIT Delhi, with whom I have discussed power electronics while teaching courses on this subject.

My special thanks to Prof VK Mahna, former Director, UPSC, for his valuable suggestions and providing useful material.

I express my gratitude to my brother Dr Mahesh Popli (Income Tax Department), Rajasthan; Dr Vikas Gupta, Director (National Testing Agency); Mr Pankaj Munjal, Director, Training and Development, RVIT, Bijnore, and Mr Ankit Popli, for their immense help and constructive criticism on the manuscript.

I am grateful to Sh Satish Kumar Jain (Mataji), CMD, and Sh Varun Jain, Director, CBS Publishers & Distributors, New Delhi, for their patience, goodwill and cooperation. I express my gratitude to Mr YN Arjuna (Senior Vice President Publishing, Editorial and Publicity); Mrs Ritu Chawla (AGM Production); Mr Kuldeep; Ms Sanjubala Tripathy and Mr Manish Raj, for bringing out the book in the present form.

Harish C Rai

Contents

CHAPTER 7

Transformers

Transformer

Transformer is a static electrical apparatus which converts electrical energy from higher voltage to lower voltage or vice versa at the same supply frequency. There are three types of transformers: (1) Core type (2) Shell type (3) Berry type.

Transformer can be classified based on operation, namely: i. power transformer ii. distribution transformer.

Principle: The energy is transferred at the same frequency from primary winding to secondary winding by means of electromagnetic induction. The flux Φ links not only with the secondary winding but also with the primary winding, thus producing self-induced emf in primary winding which limits the primary current, with magnitude, E_1.

$$E_1 = 4.44\ \Phi_m f N_1\ (\text{V}),$$

rms value of the induced emf in secondary winding is $E_2 = 4.44\ \Phi_m f N_2$.

The laminated core of a transformer is shown in Fig. 7.1.

Transformer must not be connected to a dc source. If the primary winding is connected to dc supply mains, the flux produced will not vary but remain constant in magnitude and therefore no emf will be induced in secondary winding except at the moment of switching on. Also there will be no back emf induced in the primary winding. Therefore, a heavy current will be drawn from the supply which may result in the burning of the winding.

Emf equation: Induced emf in secondary winding of transformer is

$$E_2 = 4.44\ \Phi_m f N_2 \text{ volts}$$

where Φ_m is the max. value of flux in Wb.

Voltage transformation ratio K is defined as:

$$K = \frac{E_2}{E_1} = \frac{N_2}{N_1}$$

In an ideal transformer

$$I_1 N_1 = I_2 N_2$$

or

$$\frac{E_2}{E_1} = \frac{I_1}{I_2} = \frac{N_2}{N_1} = K$$

Induced emfs in primary and secondary winding lag behind the main flux by $\pi/2$ as shown in Fig. 7.2.

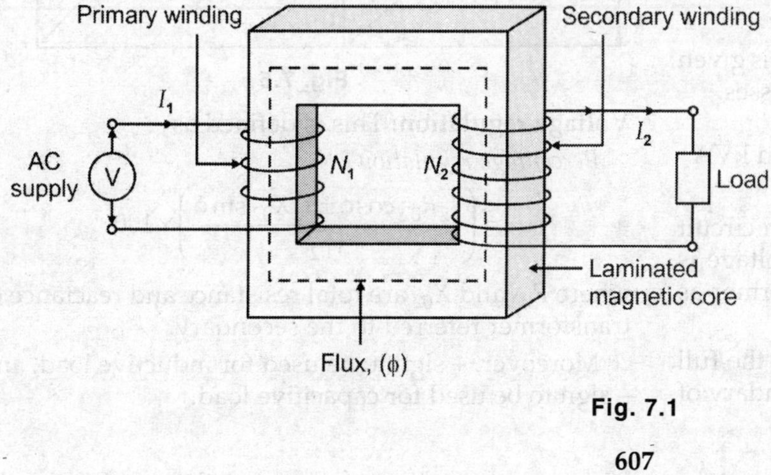

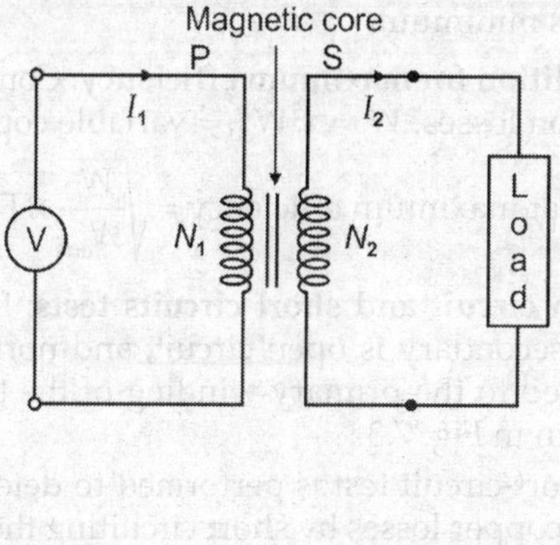

Fig. 7.1

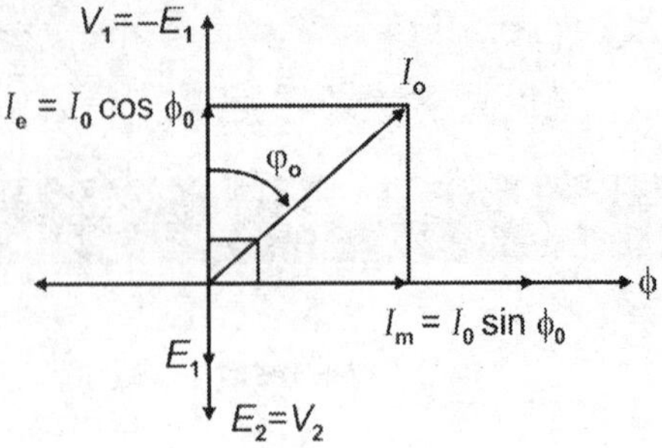

Fig. 7.2

Applied voltage to the primary winding leads the main flux by $\pi/2$ and is in phase opposition to the induced emf in the primary winding.

$$I_e = I_0 \cos \phi_0;\ I_m = I_0 \sin \phi_0;\ \theta = \tan^{-1} I_m/I_e$$

Transformer losses: Losses occurring in the transformer are:

i. Iron losses, W_i: Iron losses are independent of load which occur due to pulsation of flux in the core.

ii. Copper losses, W_c: This loss occurs due to ohmic resistance of the transformer winding.

The purpose of no load test is to determine the iron loss and no load current.

Iron losses has two components:

Hysteresis loss: $W_H \propto \beta_m^{1.6} fv$ (W)

where v is the volume of iron core

n is steinmetz coefficient

Eddy current loss: $W_e \propto \beta_m^2 f^2 t^2$ (W) $= k\beta_m^2 f^2 t^2$,

where t is the thickness of the core.

Efficiency of transformer: Since the transformer is a static device, there is no windage and frictional losses.

$$\text{Efficiency } \eta = \frac{V_2 I_2 \cos\phi_2}{V_2 I_2 \cos\phi_2 + W_i + W_{cu}}$$

$$\text{All day efficiency } \eta_{all} = \frac{\text{Output in kWh}}{\text{Input in kWh}} \text{ (for 24 hours)}$$

Distribution transformers are designed to keep core losses minimum.

Condition for maximum efficiency: Condition is given by iron losses, $W_i = x^2 W_{cu}$ = variable copper losses.

$$\text{kVA at maximum efficiency} = \sqrt{\frac{W_i}{W_{cu}}} \times \text{Full load kVA}$$

Open circuit and short circuits tests: In open-circuit test, secondary is open circuit, and normal voltage is applied to the primary winding of the transformer as shown in Fig. 7.3.

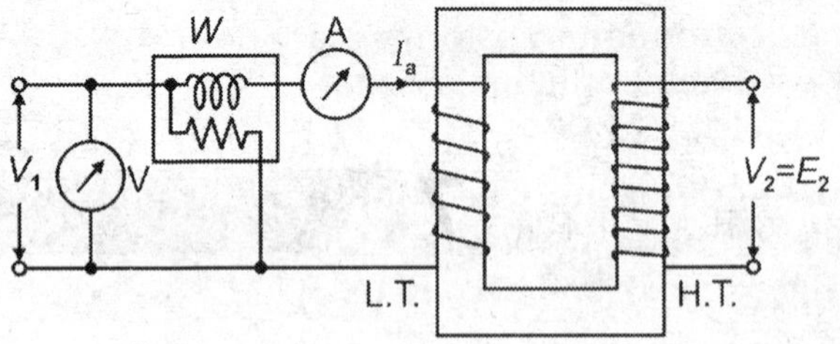

Fig. 7.3

Short-circuit test is performed to determine the full load copper losses by short circuiting the secondary of the transformer. Figure 7.4 shows the short-circuit test on the transformer.

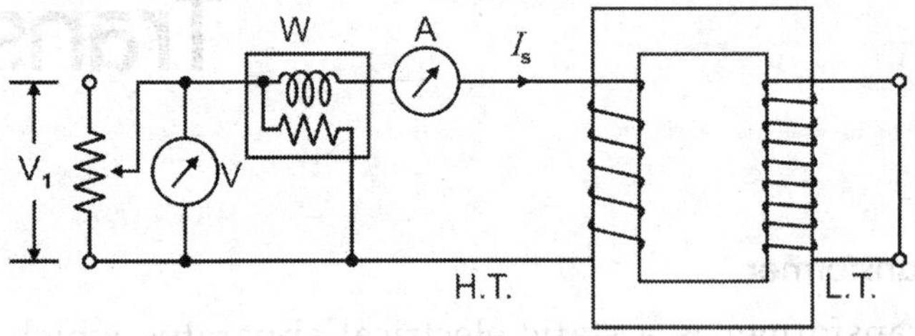

Fig. 7.4

Equivalent resistance and reactance: Equivalent resistance of the transformer as referred to primary R_{01} or as referred to secondary, R_{02} are

$$R_{01} = R_1 + R_2/K^2$$

or $$R_{02} = R_2 + K^2 R_1$$

Equivalent reactance of transformer as referred to primary, X_{01} or as referred to secondary, X_{02} are:

$$X_{01} = X_1 + X_2/K^2$$

$$X_{02} = X_2 + X_1/K^2$$

where R_1, R_2, X_1, X_2 are the resistance and reactance of primary and secondary windings, K being the transformation ratio (Fig. 7.5).

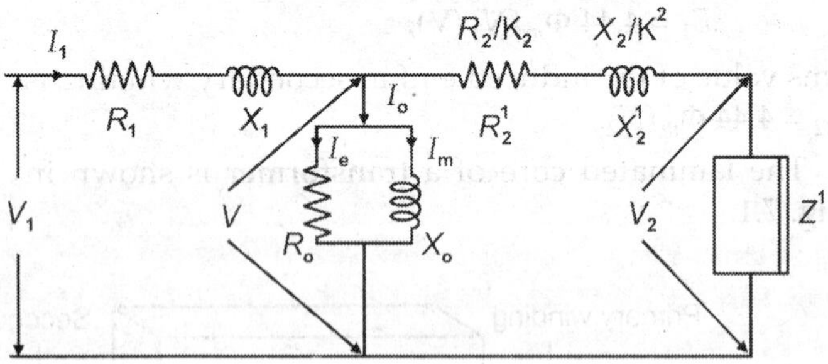

Fig. 7.5

Voltage regulation: This is defined as

Percentage regulation

$$= \left(\frac{I_2 R_{02} \cos\phi \pm I_2 X_{02} \sin\phi}{V_{02}}\right) \times 100$$

where R_{02} and X_{02} are total resistance and reactance of transformer referred to the secondary.

Moreover, + sign to be used for inductive load, and − sign to be used for capacitive load.

The distribution transformers are designed to keep core losses minimum and copper losses are relatively less.

Parallel operation: The conditions for successful parallel operation of the single phase transformers are:

1. The primary and secondary windings should be suitable for supply voltage and frequency.
2. The transformers should be properly connected with regard to polarity.
3. The transformation ratio should be identical.
4. The percentage impedance should be equal.

Pulse transformer: Figure 7.6 shows ideal rectangular pulse input to the pulse transformer, which couples a source of pulses of electrical energy to the load. Its shape and other properties are maintained properly. The size of the transformer is small and have few turns. The inter-winding capacitance and leakage reactance of the windings is small. A high voltage insulation is used between windings and to the ground. Small pulse transformer are used in computers, pulse generators, and so on, whereas large pulse transformers are used in radar systems.

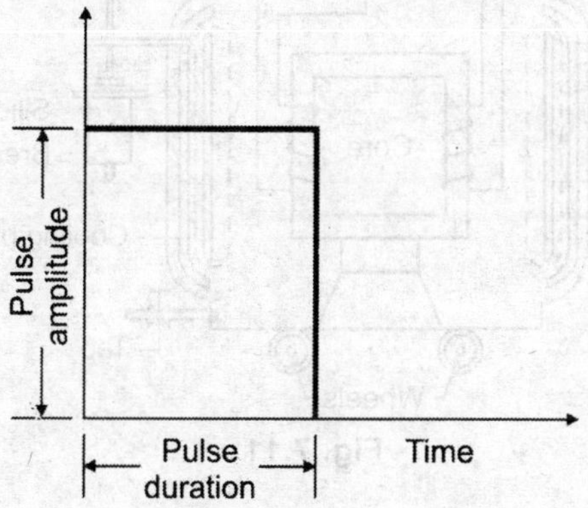

Fig. 7.6

Use of pulse transformers: Pulse transformers are used in:

i. Radar system
ii. Microwave tube circuits
iii. Data-handling circuit
iv. Analogue switching applications
v. Digital signal processing
vi. Fast pulse signal transmission application
vii. Thyristor and switching transistors

Welding transformer: Figure 7.7 shows a diagram of welding transformer having thin primary windings with large number of turns. Secondary transformer has less number of turns having more area of cross-section of winding ensuring less voltage and very high current in the secondary.

Autotransformer: In autotransformer, part of the winding is common to both the primary and secondary circuits. In this, two windings are electrically connected

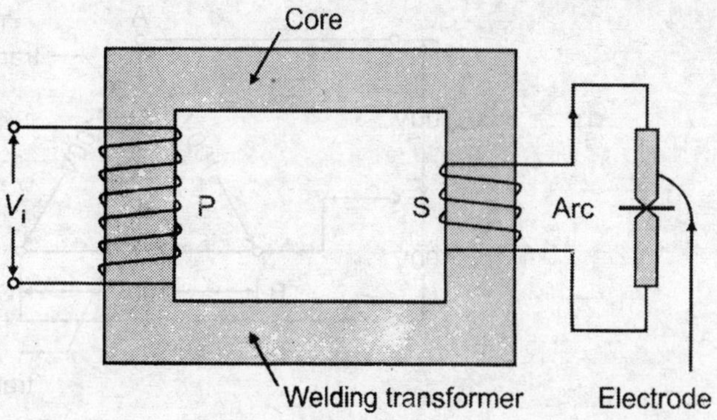

Fig. 7.7

and it works on the principle of induction and conduction.

Application: (i) Autotransformer are used as starter for starting machines such as induction motors and synchronous motor (ii) They can be used as variance (iii) They can be used as furnace transformers (iv) They can be used as boosters.

Current transformer (CT): Current transformers are basically step up transformers, that is, stepping the voltage and hence stepping down the current. The secondary of CT should not be kept open. CT are used for the measurement of heavy currents.

CT is shown in Fig. 7.8.

Potential transformer (PT): This transformer reduces the voltage which is shown in Fig. 7.9.

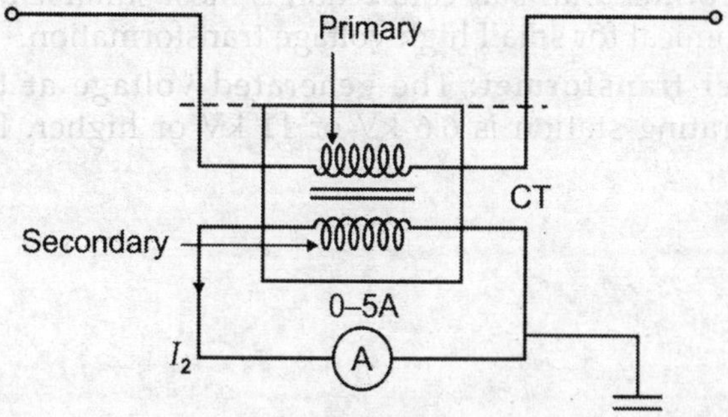

Fig. 7.8

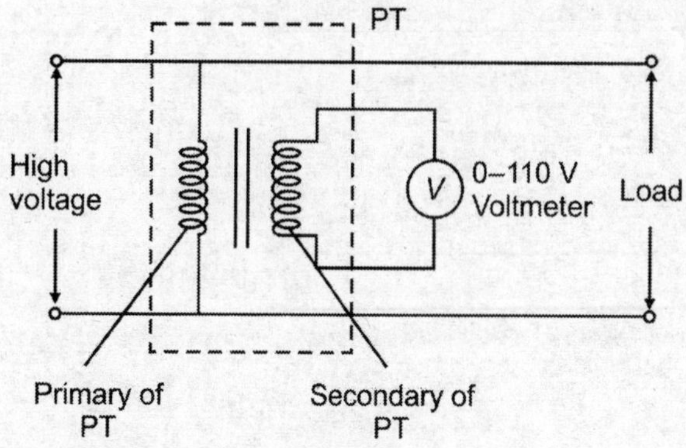

Fig. 7.9

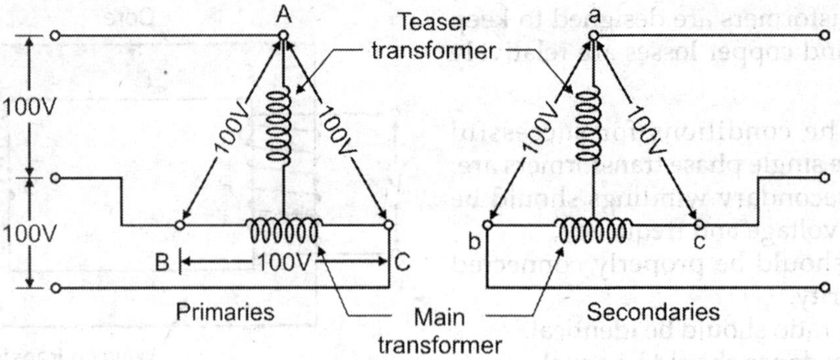

Fig. 7.10

The primary winding PT, consists of a large number of turns and secondary winding consists of small number of turns. The secondary transformer is always grouned for safety purposes.

Three phase transformer connection: The following three phase transformers are the most commonly used in practice: (i) star-star (ii) delta-star (iii) star-delta (iv) delta-delta (v) delta-zigzag connection.

Scott connection or T–T connection: Figure 7.10 shows a scott connection which needs two transformers on each side instead of three transformers and accomplishes three phase to three phase and three phase to two phase transformer.

As shown in Fig. 7.10, one of the transformers known as main transformer, has centre taps both on the primary and secondary windings. The other transformer has a 0.866 tap and is known as teaser transformer. Star-star connection is most suitable and economical for small high voltage transformation.

Power transformer: The generated voltage at the generating station is 6.6 kV or 11 kV or higher. For transmission purposes, it is required to step it up to a voltage of 132 kV or higher. Assembly of power transformer is shown in Fig. 7.11.

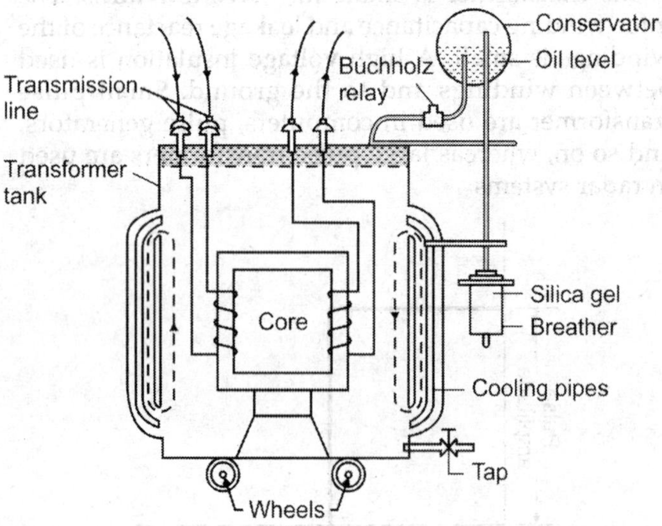

Fig. 7.11

MULTIPLE CHOICE QUESTIONS

1. A transformer transforms
   A. frequency B. voltage
   C. current D. voltage and current
2. An ideal transformer is one which has
   A. a common core for its primary and secondary windings
   B. no losses and magnetic leakage
   C. core of stainless steel and windings of pure copper wire
   D. interleaved primary and secondary windings
3. The main purpose of using core in a transformer is to
   A. decrease iron losses
   B. prevent eddy current losses
   C. eliminate magnetic hysteresis
   D. decrease reluctance of the common magnetic circuit
4. Transformer cores are laminated in order to
   A. simplify its construction
   B. minimise eddy current losses
   C. reduce cost
   D. reduce hysteresis losses
5. A transformer having 1000 primary turns is connected to a 250 V AC supply. For a secondary voltage of 400 V, the number of secondary turns should be
   A. 1600 B. 250
   C. 400 D. 1250
6. In an ideal transformer, the no-load primary current I_0
   A. lags behind V_1 by 90°
   B. is in phase with V_1
   C. leads V_1 by 90° D. lags V_1 by 45°
7. The primary and secondary induced emfs E_1 and E_2 in a two-winding transformer are always
   A. equal in magnitude
   B. antiphase with each other
   C. in-phase with each other
   D. determined by load on transformer secondary
8. A step-up transformer increases
   A. voltage B. current
   C. power D. frequency
9. The voltage transformation ratio of a transformer is equal to the ratio of
   A. secondary induced emf to primary induced emf
   B. secondary terminal voltage to primary applied voltage
   C. primary turns to secondary turns
   D. secondary current to primary current
10. The primary and secondary windings of an ordinary 2-windings transformer always have
    A. different number of turns
    B. same size of copper wire
    C. a common magnetic circuit
    D. separate magnetic circuits
11. In a two-winding transformer, the emf per turn in secondary winding is always .... the induced emf per turn in primary.
    A. equal to K times B. equal to $1/K$ times
    C. equal to D. greater than
12. In relation to a transformer, the ratio 20:1 indicates that
    A. there are 20 turns on primary and one turn on secondary
    B. secondary voltage is 1/20th of primary voltage
    C. primary current is 20 times greater than the secondary current
    D. for every 20 turns on primary, there is one turn on secondary
13. Compared to the secondary of a loaded step-up transformer, the primary has .... voltage and .... current.
    A. lower, higher B. higher, lower
    C. lower, lower D. higher, higher
14. Transformer cores are built up from laminations rather than from solid metal so that
    A. oil penetrates the core more easily
    B. eddy current loss is reduced
    C. insulation required for the windings is reduced
    D. air circulation is improved
15. The equivalent resistance of the primary of a transformer having $K = 5$ and $R_1 = 0.1$ ohm when referred to secondary becomes
    A. 0.5 Ω B. 0.02 Ω
    C. 0.004 Ω D. 2.5 Ω
16. A transformer has negative voltage regulation when its load power factor is
    A. zero B. unity
    C. leading D. lagging
17. The primary reason why open-circuit test is performed on the low-voltage winding of the transformer is that it
    A. draws sufficiently large no-load current for convenient reading
    B. requires least voltage to perform the test
    C. needs minimum power input
    D. involves less core loss
18. No-load test on transformer is carried out to determine

Ans. 1. D 2. B 3. D 4. B 5. A 6. A 7. C 8. A 9. A 10. C 11. C 12. D 13. A 14. B 15. D 16. C 17. A

A. copper loss B. magnetising current
C. magnetising current and no-load loss
D. efficiency of the transformer

19. The main purpose of performing open-circuit test on a transformer is to measure its
A. Cu loss B. core loss
C. total loss D. insulation resistance

20. During short-circuit test, the iron loss of a transformer is negligible because
A. the entire input is just sufficient to meet Cu losses only
B. flux produced is a small fraction of the normal flux
C. iron core becomes fully saturated
D. supply frequency is held constant

21. The iron loss of a transformer at 400 Hz is 10 W. Assuming that eddy current and hysteresis losses vary as the square of flux density, the iron losses of the transformer at rated voltage but at 50 Hz would be
A. 80 W B. 640 W
C. 1.25 W D. 100 W

22. The voltage applied to the h.v side of a transformer during short-circuit test is 2% of its rated voltage. The core loss in terms of the rated core loss will be
A. 4% B. 0.4%
C. 0.25% D. 0.04%

23. Transformers are rated in kVA instead of kW because.
A. load power factor is often not known
B. kVA is fixed whereas kW depends on load pf
C. total transformer loss depends on volt-ampere
D. it has become customary

24. When a 400 Hz transformer is operated at 50 Hz, its kVA rating is
A. reduced to 1/8 B. increased 8 times
C. unaffected D. increased 64 times

25. At relatively light loads, transformer efficiency is low because
A. secondary output is low
B. transformer losses are high
C. fixed loss is high in proportion to the output
D. Copper loss is small

26. It is economical to use an auto-transformer when the transformation ratio is
A. low B. high
C. neither too high nor too low
D. none of these

27. In an autotransformer the input voltage is V_1, input current is I_1, output voltage is V_2 and output current is I_2. The volt-amperes transferred conductively from the input to the output are equal to
A. V_1I_2 B. V_2I_2
C. $V_1(I_1 - I_2)$ D. none of these

28. When a no load test is carried out on a 1000/100 V transformer, with 100 V winding open, the wattmeter reads 110 W. If the test is repeated with 1000 V winding open, the wattmeter will read
A. 11 W B. 110 W
C. 1.1W D. 0.11 W

29. A transformer has iron losses equal to W_1 watts and full load copper losses equal to W_2 watts. If two such transformers are tested in a back to back test and full load current is flowing, the total input will be
A. $2W_1$ B. $2W_2$
C. $2(W_1 + W_2)$ D. $(W_1 + W_2)$

30. The no load test on a transformer is usually carried out by exciting the low voltage winding because
A. the required power input is low
B. no load current becomes significant and can be conveniently measured
C. it is not advisable to work on the high voltage side
D. none of these

31. Transformer works on the principle of
A. self induction B. mutual induction
C. Faraday's law of electromagnetic induction
D. self and mutual induction both

32. During a short circuit test, the iron losses are negligible because
A. the mutual flux is small
B. the power factor is low
C. the current is high D. none of these

33. A 2000/200 V transformer is fed from a 2000 V supply. The transformer is supplying its rated load. The terminal voltage on the secondary side will be
A. 200 V B. about 150 V
C. about 195 V D. none of these

34. Two equal sized transformers A and B are connected in parallel. The impedance of A is more than the impedance of B. Then
A. both will share the load equally
B. A will supply more load than B
C. B will supply more load than A
D. none of these

35. A 500 kVA and a 200 kVA transformer are put in parallel. If it is desired that they should share the load in proportion to their ratings, then their
A. impedances must be equal
B. per unit impedances must be equal
C. iron and copper losses must be equal
D. iron losses should be less than copper losses

36. In a transformer, if in place of sine wave, a peaked voltage is fed to the primary, the
A. iron losses will be less
B. iron losses will be more
C. copper losses will be less
D. noise level will be low

Ans. 18. C 19. B 20. C 21. B 22. D 23. A 24. A 25. C 26. A 27. A 28. B 29. C 30. B
31. B 32. A 33. C 34. C 35. B 36. A

37. Which of the following combinations of 3-phase transformers can successfully operate in parallel
A. Y–Y and Y–Δ B. Δ–Y and Δ–Δ
C. Δ–Y and Δ–Y D. none of these

38. A current transformer is used to measure a 100 A current by 5 A ammeter. The current transformer is a
A. step up transformer B. step down transformer
C. power transformer
D. distribution transformer

39. The following precaution should be observed while using instrument transformers
A. secondary of a CT should never be open circuited
B. secondary of a CT should never be short circuited
C. secondary of a PT should never be open circuited
D. None of these

40. The figure shows the schematic diagram of a transformer. What is the relation between the voltages V_1 and V_2 and the number of turns N_1 and N_2?

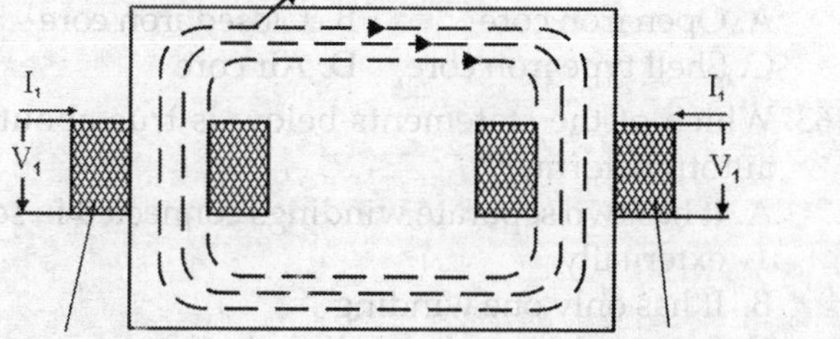

A. $\frac{V_1}{V_2} = \frac{N_1}{N_2}$ B. $\frac{V_1}{V_2} = \frac{N_2}{N_1}$
C. $V_1 \times V_2 = N_1 \times N_2$
D. $V_1 \times N_1 = V_2 \times N_2$

41. The secondary of a PT is generally designed for
A. 220 V B. 110 V
C. 150 V D. 11 V

42. The secondary of a CT is generally designed for
A. 100 A B. 10 A
C. 5 A D. 0.5 A

43. One primary terminal and one secondary terminal of a two winding transformer are connected to an ohm meter. The meter will indicate
A. zero B. infinity
C. a resistance equal to total winding resistance
D. none of these

44. One of the windings of the transformer is made of thin wire and the other of thick wire. Which statement is true?
A. The winding made of thin wire is always the primary winding
B. The winding made of thin wire is always the secondary winding
C. The winding made of thin wire is always the low voltage winding
D. The winding made of thin wire is always the high voltage winding

45. What happens if a too much high voltage is applied to a transformer?
A The transformer draws a very high magnetising current and the primary winding may be destroyed
B. The leakage flux becomes so large that the secondary voltage becomes lower in case the nominal primary voltage is applied
C. The huming of the transformer becomes very loud, but there are no further disadvantages
D. Due to very large hysteresis losses, the iron core heats up and burns

46. The transformer principle is based on
A. self-induction B. mutual-induction
C. electric induction D. the lenz law

47. Which loss does not occur in a transformer?
A. Eddy current loss B. I^2R loss
C. Hysteresis loss D. Friction loss

48. According to the name plate of a small transformer, the secondary nominal voltage is 220 V. Which of the following statements about it is true?
A. The no-load voltage is more than 220 V
B. 220 V is the no-load voltage
C. At a load which draws the rated current the voltage becomes less than 220 V
D. The secondary voltage practically does not change when the transformer is loaded

49. Upon which factor does the short circuit voltage of a transformer mainly depend?
A. On the ohmic resistance of the primary winding.
B. On the ohmic resistance of the secondary winding
C. On the cross-sectional area of the iron core
D. On the magnitude of the leakage flux

50. Which of the following statements about a transformer having a small short circuit voltage is true?
A. The transformation ratio of the transformer is small
B. During operation, the transformer has high iron losses
C. During operation, the transformer has high copper losses
D. A high short circuit current flows through the transformer

51. What is meant by the nominal short circuit voltage VSCN of transformer?

Ans. 37. C 38. A 39. A 40. A 41. B 42. C 43. B 44. D 45. A 46. A 47. D 48. A 49. D 50. D

A. Short circuit voltage
B. After short circuiting two secondary terminals, the voltage appearing across the third and the short circuited terminals
C. The primary voltage when the secondary terminals are short circuited and the rated current flows through the primary winding
D. The voltage across the primary terminals when the secondary terminals are short circuited

52. The efficiency of a power transformer is maximum at
A. nearly full load B. 1/2 full load
C. 1/4 full load D. 1/8 full load

53. A 10,000 V/400 V power transformer has a nominal short circuit voltage Vsc = 4%. Which of the following statements about it is true?
A. A voltage of 400 V appears across the short circuited secondary voltage
B. A voltage of 16 V appears across the short circuited secondary terminals
C. The primary voltage drops to 400 V when the secondary terminals are short circuited
D. When the secondary terminals are short circuited, the rated current flows in the primary side at a primary voltage of 400 V

54. When two identical transformers are available, then the best method to find their efficiency under load conditions is
A. short-circuit test B. back-to-back test
C. open-circuit test D. phase-sequence test

55. Transformers are rated in
A. kilo-watts B. kilo-watt hours
C. kilo-volt-amperes (kVA)
D. kilo-volts

56. What is the phase difference between the low and the high voltage of a Y 5 power transformer?
A. 0° B. 5°
C. 50° D. 150°

57. Which of the following statements is true about a Dy5 power transformer?
A. The high voltage winding is connected in star
B. The high voltage winding is connected in zigzag
C. The low voltage winding is connected in star
D. The low voltage winding is connected in delta

58. What is the advantage of a distribution transformer of group Yz5 as compared to one of group Y?
A. It can be loaded unsymmetrically
B. Its short circuit voltage is lower
C. Its no-load current is lower
D. Its efficiency is high

59. What is the current I_{AC} flowing through the auto-transformer shown in the figure.

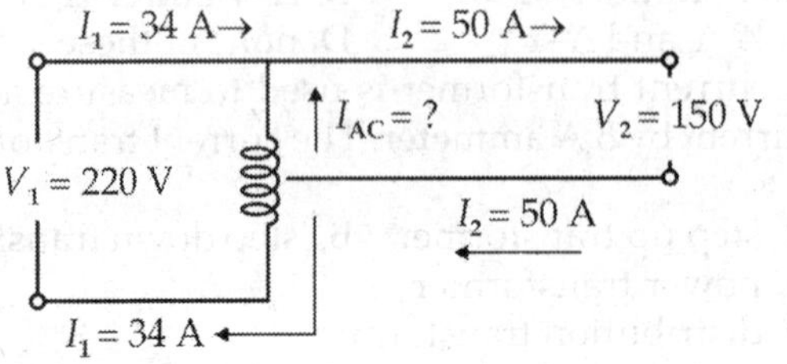

A. $I_{AC} = 50$ A B. $I_{AC} = 34$ A
C. $I_{AC} = 20$ A D. $I_{CA} = 16$ A

60. If in a transformer we double the secondary turns and at the same time reduce the primary voltage by half, then the secondary voltage will
A. be halved B. be four times as high
C. not change D. be doubled

61. Which is the common method of cooling a power transformer?
A. Air-cooling B. Air-blast cooling
C. Oil-cooling D. Natural cooling

62. What type of core is used for a high frequency transformer?
A. Open iron core B. Closed iron core
C. Shell type iron core D. Air core

63. Which of the statements below is true about the autotransformer?
A. It has two separate windings connected in series externally
B. It has only one winding
C. It can only step down the voltage
D. It is most suitable for power transformation

64. If a high voltage is given to the primary and low voltage is taken from the secondary of transformer, it is called
A. step up
B. step down
C. instrument transformer
D. stabiliser

65. Which of the following is an instrument transformer?
A. Bell-transformer
B. Ignition transformer
C. Adjustable transformer
D. Current transformer

66. The self-induced voltage of a coil depends upon the
A. voltage applied to the coil
B. shape of the iron core
C. shape of the bobbin
D. number of turns of the coil

67. For a 100% efficient transformer, the primary winding has 1000 turns and the secondary 100

Ans. 51. C 52. A 53. D 54. B 55. C 56. D 57. A 58. A 59. D 60. C 61. C 62. D 63. B 64. B 65. D 66. D

turns. If the power input to the above transformer is 1000 watts, the power output is
A. 1000 watts B. 100 watts
C. 10 watts D. 10 kW

68. If DC supply is given to a transformer, it will
A. work B. not work
C. give lower voltage than the rated voltage on secondary side
D. burn the winding

69. Why are transformer cores laminated?
A. Laminated cores are cheaper than solid cores
B. There are fewer eddy current losses in laminated cores
C. There are fewer hysteresis losses in laminated cores
D. There is less humming as compared to solid cores

70. A 200 kVA transformer has an iron loss of 1 kW and full-load copper loss of 2 kW. Its load kVA corresponding to maximum efficiency is
A. 100 kVA B. 141.4 kVA
C. 50 kVA D. 200 kVA

71. If copper loss of a transformer at 7/8th full load is 4900 W, then its full-load copper loss would be
A. 5600 kVA B. 6400 kVA
C. 375 kVA D. 429 kVA

72. The ordinary efficiency of a given transformer is maximum when
A. it runs at half-full load
B. it runs at full-load
C. its copper loss equals iron loss
D. it runs slightly overload

73. The output current corresponding to maximum efficiency for a transformer having core loss of 100 W and equivalent resistance referred to secondary of 0.25 Ω is
A. 20 A B. 25 A
C. 5 A D. 400 A

74. The maximum efficiency of a 100 kVA transformer having iron loss of 900 kW and FL copper loss of 1600 W occurs at
A. 56.3 kVA B. 133.3 kVA
C. 75 kVA D. 177.7 kVA

75. The all-day efficiency of a transformer depends primarily on
A. its copper loss B. the amount of load
C. the duration of load
D. Both (B) and (C) are correct

76. The marked increase in kVA capacity produced by connecting a 2-winding transformer as an auto-transformer is due to
A. increase in turn ratio
B. increase in secondary voltage
C. increase in transformer efficiency
D. establishment of conductive link between primary and secondary

77. The kVA rating of an ordinary 2-winding transformer is increased when connected as an auto-transformer because
A. transformation ratio is increased
B. secondary voltage is increased
C. energy is transferred both inductively and conductively
D. secondary current is increased

78. Instrument transformers are used on AC circuits for extending the range of
A. ammeters B. voltmeters
C. wattmeters D. all of these

79. Before removing the ammeter from a current transformer, its secondary must be short-circuited in order to avoid
A. excessive heating of the core
B. high secondary emf
C. increase in iron losses
D. all of these

80. Transformer is used to change the values of
A. voltage B. frequency
C. power D. power factor

81. Transformer is a
A. rotating device B. static device
C. magnetic device D. none of these

82. An ideal transformer is one which has
A. a common core for its primary and secondary windings
B. no losses and magnetic leakage
C. core of stainless steel and windings of pure copper metal
D. interleaved primary and secondary windings

83. Transformer works on
A. AC B. DC
C. AC and DC both D. pulsating DC

84. Transformer works on the principle of
A. self-induction B. mutual induction
C. Faraday's law of electromagnetic induction
D. self and mutual induction both

85. If a higher voltage is given to the primary and low voltage is taken from the secondary of transformer it is called
A. step-up B step-down
C. instrument transformer
D. stabiliser

86. If DC supply is given to a transformer it may
A. work B. not work
C. give lower voltage than the rated voltage on secondary side
D. burn the winding

Ans.													
67. A	68. B	69. B	70. B	71. B	72. C	73. A	74. A	75. D	76. D	77. C	78. D	79. D
80. A	81. B	82. B	83. A	84. D	85. B	86. D						

87. Which of the following is a correct statement about eddy current?
A. Eddy currents improve the efficiency of a motor.
B. Eddy currents do not influence the movement
C. Eddy currents heat up the metal parts
D. Eddy currents are used for arc welding

88. Rating of transformer is given in
A. kVA B. kVAR
C. kW D. watts

89. If lower voltage is given to the primary and higher voltage is taken from the secondary of a transformer, it is called.
A. step up B. step down
C. current transformer D. voltage stabiliser

90. If the input voltage is 100 V and output voltage is 200 V, which side of the transformer windings will be primary?
A. 200 V side B. 100 V side
C. Winding with more turns
D. Winding with less turns

91. The transformation ratio of a transformer is
A. I_2/I_1 B. N_2/N_1
C. N_1/N_2 D. all of these

92. Shell-type transformer is commonly used because it has
A. more magnetic flux leakage
B. less magnetic flux leakage
C. two magnetic paths
D. both (B) and (C) are correct

93. The transformer is commonly used because
A. it has high efficiency
B. frequency remains constant
C. construction cost per kVA is less as compared to other machines
D. all of these

94. In case of transformers Helical coils are very well suited for
A. HV winding of small rating transformers
B. LV winding of large rating transformers
C. HV winding of large rating transformers
D. none of these

95. When the secondary of the transformer is loaded, the flux in the transformer will
A. always constant
B. directly proportional to secondary current
C. directly proportional to the current drawn by primary winding
D. none of these

96. The path of the magnetic flux in a transformer has
A. high reluctance B. low resistance
C. high conductivity D. low reluctance

97. Class-*B* insulation for winding can withstand a maximum temperature of
A. 120°C B. 130°C
C. 105°C D. 135°C

98. What is the phase difference between the low and the high voltage of a YZ5 power transformer?
A. 0° B. 75°
C. 5° D. 150°

99. Electric power is transferred from one coil to the other coil in a transformer the primary induced emf
A. electrically B. electromagnetically
C. magnetically D. physically

100. A transformer operates
A. always at unity power factor
B. has its own power factor
C. at a power factor below a particular value
D. at power factor depending on the power factor of the load

101. The laminations are made from
A. silicon sheet steel
B. nickel alloy steel stamping
C. low carbon steel
D. chrome steel sheets

102. The steel for construction of transformer core is made to have
A. high permeability and low hysteresis loss
B. low permeability and high hysteresis loss
C. low permeability and low hysteresis loss
D. high permeability and high hysteresis loss

103. As compared to an amplifier, a transformer cannot
A. increase the output voltage
B. increase the output current
C. increase output power
D. none of these

104. In an ideal transformer on no-load, the primary applied voltage is balanced by
A. the secondary voltage
B. the drop across resistances and reactance
C. the secondary induced emf
D. the primary induced emf

105. The voltage transformation ratio is
A. E_1/E_2 B. N_1/N_2
C. E_2/E_1 D. None of these

106. A sinusoidal emf
A. lags the flux inducing it by 180°
B. lags the flux inducing it by 90°
C. leads the flux inducing it by 90°
D. leads the flux inducing it by 180°

107. Which of the following statements about a short-circuit power transformer is true?
A. There is a fuse in its primary circuit
B. There is a bimetallic switch in its primary circuit
C. There is a fast magnetic switch in its primary circuit
D. The short-circuit current can continuously flow through it without causing a damage

Ans. 87. C 88. A 89. A 90. B 91. B 92. D 93. D 94. B 95. A 96. D 97. B 98. D 99. C
100. D 101. A 102. A 103. C 104. C 105. C 106. A 107. D

108. The magnitude of mutual flux in a transformer is
A. low at low loads and high at high loads
B. high at low loads and low at high loads
C. same at all loads
D. varies at low loads and constant at high loads

109. At every instant the direction of secondary current in a transformer must be such as to oppose any change in flux. This is in accordance with
A. Faraday's law B. Joule's law
C. Lenz's law D. Coulomb's law

110. Different types of transformers are
A. core type B. shell type
C. berry type D. all of these

111. Transformers are rated in
A. kW B. kV
C. kWh D. None of these

112. If in a transformer the secondary turns are doubled and at the same time the primary voltage is reduced by half, then the secondary voltage will
A. be halved B. be four times
C. not change D. be reduced to a quarter

113. Which is the common method of cooling of a power transformer of 500 kVA?
A. Air-cooling B. Air-blast cooling
C. Oil-cooling D. Natural cooling

114. What type of core is used for a high frequency transformer?
A. Open iron core B. Air core
C. Closed iron core D. None of these

115. Which of the statements given below is true for the auto-transformer?
A. It has to separate windings connected in series externally
B. It has only one winding
C. It can only step down the voltage
D. It is most suitable for power transformation

116. The self-induced voltage of a coil depends upon the
A. voltage applied to the coil
B. shape of the iron core
C. shape of the bobbin
D. number of turns of the coil

117. The insulating materials used for motor windings are classified according to the
A. motor horsepower rating
B. levels of temperature rise
C. overcurrent protection available
D. controller size

118. The use of higher flux density in transformer design
A. increases the weight per kVA
B. decreases the weight per kVA
C. increases the weight per kW
D. decreases the weight per kW

119. The transformer oil used in transformers provides
A. insulation and cooling
B. cooling and lubrication
C. insulation and lubrication
D. insulation, cooling and lubrication

120. The induced emf in the secondary of a transformer will depend on
A. frequency of the supply only
B. number of turns in secondary only
C. maximum flux in the core only
D. frequency, flux and number of turns in the secondary

121. A transformer has a turns ratio of 1 to 10 and a resistance of 5000 ohms is connected across the secondary terminals, the resistance offered to a current flowing in the primary will be
A. 50 ohms B. 500 ohms
C. 5000 ohms D. 50 kilo ohms

122. Two transformers are operating in parallel. They will share the load depending upon their
A. efficiency B. ratings
C. leakage reactance D. per-unit impedance

123. The full load copper-loss and iron-loss of a transformer are 6400 watts and 5000 watts respectively. The above copper loss and iron-loss at half load will be
A. 3200 W, 2500 W respectively
B. 3200 W, 5000 W respectively
C. 1600 W, 1250 W respectively
D. 1600 W, 5000 W respectively

124. Why there is an enamel layer coated over the laminations of the transformer core?
A. To decrease the hum
B. To attain adhesion between the lamination
C. To insulate the laminations against each other
D. To prevent the corrosion of the laminations

125. What is the efficiency of transformer compared with that of electrical motors of the same power?
A. Much smaller B. Somewhat smaller
C. About the same D. Much higher

126. Transformer is a device which transforms the voltage
A. from higher level to lower level
B. from lower level to higher level
C. both (A) and (B) are correct
D. none of these

127. According to the name plate of a small transformer, the secondary normal voltage is 220 V. Which of the following statements about it is true?
A. The no-load voltage is more than 220 V
B. 220 V is the no-load voltage
C. At a load which draws the rated current, the voltage becomes less than 220 V
D. The secondary voltage increases with increasing load

Ans. 108. C 109. C 110. D 111. D 112. C 113. C 114. B 115. B 116. D 117. B 118. B 119. A 120. D
121. A 122. D 123. D 124. C 125. D 126. C 127. A

128. Upon which factor does the short-circuit voltage of a transformer mainly depend. It depends on the
A. ohmic resistance of the primary winding
B. ohmic resistance of the secondary winding
C. cross-sectional area of their core
D. magnitude of the leakage flux

129. Which of the following statements about a transformer having a small short circuit voltage is true?
A. The transformation ratio of the transformer is small
B. During operation the transformer has high iron losses
C. During operation the transformer has high copper losses
D. A high short-circuit current flows through the transformer

130. What is meant by the nominal short circuit voltage V_{SCN} of a transformer?
A. The voltage appearing across the secondary terminals when they are short-circuited
B. After short-circuiting two secondary terminals, the voltage appearing across the primary terminals
C. The primary voltage when the secondary terminals are short-circuited and the rated current flows through the primary winding
D. All of the above

131. What is the typical use of an auto-transformer?
A. Isolating transformer B. Toy transformer
C. Control transformer D. Variable transformer

132. The no load current taken by actual transformer lags the applied voltage by (approximately)
A. 80° B. 55°
C. 40° D. 30°

133. In any transformer, the voltage per turn in primary and secondary remains
A. always same B. always in ratio of K
C. always different D. none of these

134. In any single phase transformer, the primary and secondary induced voltages are
A. 180° out of phase B. 90° out of phase
C. in phase D. None of these

135. The full load copper-loss of any transformer working at unity-power factor is W_C watts. The copper-loss at full load and at 0.6 pf will be
A. W_C watts B. 0.8 W_C watts
C. $W_C/0.6^2$ watts D. 0.8^2 W_C watts

136. In parallel operation of two single-phase transformers, if the impedance triangles of the transformers are not identical in shape and size, then
A. power factors at which the transformers operate will be same but different from power factor of common load
B. power factors at which the transformers operate and power factor of common load-all will be same
C. power factor of one transformer and power factor of common load will be same
D. power factors at which the transformers operate will be different from one another and again these will be different from power factor of common load

137. The full load copper-loss in a transformer is 400 W. At half load, the copper-loss will be
A. 400 W B. 200 W
C. 100 W D. 50 W

138. A transformer is working at its maximum efficiency. Its iron-loss is 500 W. Its copper-loss will be
A. 250 W B. 300 W
C. 400 W D. 500 W

139. The emf induced in the secondary winding depends upon
A. number of turns B. flux
C. supply frequency D. all of these

140. Coupling between the two windings of a transformer can be increased by
A. increasing the number of turns in the two windings
B. increasing the primary voltage
C. interleaving the windings on a common core of low reluctance
D. reducing the insulation of the two windings

141. In an ideal transformer, the no-load primary current I_0
A. lags behind V_1 by 90°
B. is in phase with V_1
C. leads V_1 by 90°
D. lags V_1 by an angle lying between 0° and 90°

142. The primary and secondary induced emfs E_1 and E_2 in a two-winding transformer are always
A. equal in magnitude
B. antiphase with each other
C. in phase with each other
D. determined by load on transformer secondary

143. By definition, a step-down transformer is one in which
A. secondary current is more than primary current
B. secondary turns are less than primary turns
C. secondary winding consists of thinner copper wire
D. secondary power is less than primary power

144. Three number of 10:1 transformers are connected in Y–Δ to supply a load at 230 V. The supply voltage of this transformer must be
A. 2300 V B. 4000 V
C. 23 V D. 40 V

Ans. 128. D 129. D 130. C 131. D 132. A 133. A 134. C 135. C 136. D 137. C 138. D 139. D 140. C 141. A 142. C 143. B 144. C

145. Three units of 1:5 transformers are connected in Δ–Y to supply a 3-phase load from a 400 V, 3-phase source. The line voltage on the load side is
A. 1000 V B. 80 V
C. 3464 V D. 803 V

146. The no-load current of a transformer in terms of full load current is usually
A. 1 to 3% B. 3 to 15%
C. 9 to 12% D. 12 to 20%.

147. When two transformers of different kVA ratings are connected in parallels, they divide the total load in proportion to their respective kVA ratings only when their
A. kVA ratings are identical
B. efficiencies are equal
C. per unit impedances are equal
D. equivalent impedances are equal

148. A transformer is working at its full load and its efficiency is also maximum. Its iron loss is 1000 W. Then, its copper-loss at half of full load will be
A. 250 W B. 300 W
C. 400 W D. 500 W

149. Distribution transformers are designed to have maximum efficiency nearly
A. at full load B. at 50% of FL
C. at no load D. none of these

150. Distribution transformers have good all day efficiency due to
A. low copper-loss
B. low iron-loss
C. low copper as well as iron losses
D. None of these

151. The special silicon steel is used for laminations because
A. eddy current losses are reduced
B. hysteresis losses are reduced
C. both losses are reduced
D. none of these

152. The yoke sections of transformers using hot-rolled laminations is made 15% greater than that of the core so as to
A. reduce the copper-loss
B. increase the size of transformers
C. reduce the iron-loss in yoke and magnetizing current
D. provide better cooling

153. The concentric windings are used in core-type transformers with
A. HT winding placed next to core
B. LT winding placed next to core
C. LT winding on the outer side
D. HT winding on the outer side

154. Cross-over windings are used for
A. high voltage winding of small rating transformers
B. low voltage winding of small rating transformers
C. high voltage winding of large rating transformers
D. none of these

155. Transformer transforms
A. frequency B. voltage
C. current D. power

156. Two transformers when operating in parallel will share the load depending upon their
A. magnetising current
B. leakage reactance
C. per unit impedance
D. efficiency

157. A step-up transformer increases
A. voltage B. current
C. power D. frequency

158. The voltage transformation ratio of a transformer is equal to the ratio of
A. secondary induced emf to primary induced emf
B. secondary terminal voltage to primary applied voltage
C. primary turns to secondary turns
D. secondary current to primary current

159. The primary and secondary windings of an ordinary transformer always have
A. different number of turns
B. same size of copper wire
C. a common magnetic circuit
D. separate magnetic circuits

160. In two-winding transformer, the emf per turn in secondary winding is always
A. equal to K times the induced emf per turn in primary
B. equal to $1/K$ times the induced emf per turn in primary
C. equal to the induced emf per turn in primary
D. greater than induced emf per turn in primary

161. For a given application, the transformation ratio of a transformer
A. is fixed but not constant
B. is constant but not fixed
C. is dependent on secondary load
D. depends on which winding is used as primary

162. If you impress a DC voltage of 230 V to an unloaded 230 V, 50 Hz transformer, the transformer may
A. burn
B. give very high secondary voltage
C. give secondary voltage according to the turn-ratio
D. none of these

Ans. 145. C 146. A 147. C 148. A 149. B 150. B 151. B 152. C 153. B 154. A 155. B 156. C 157. A 158. A 159. C 160. C 161. A 162. A

163. Three single-phase transformers, each with a 10 kVA rating are connected in a closed delta arrangement. If one transformer is taken out, the output capacity of the system will be
A. 20 kVA B. 8.66 kVA
C. 17.32 kVA D. 10 kVA

164. Special precaution to be taken when using a booster transformer is that
A. it should never be left open-circuit
B. there should be no fuse in the high voltage side
C. it should never be left close-circuit
D. none of these

165. The purpose of conservator in transformers is to
A. cool the windings
B. prevent moisture in the transformers
C. take up contraction and expansion of oil
D. none of these

166. When s 440/220 V transformer is connected to 400 V DC supply
A. the output will be 220 V
B. the output will be 0 V
C. the output will be less than 230 V
D. the transformer may burn

167. Ferrite cores are used in high frequency transformer because it has
A. low permeability B. high hysteresis
C. low resistance D. high resistance

168. The maximum temperature permitted for class A insulation is
A. 180° B. 165°C
C. 120° D. 105°C

169. The leakage flux of a transformer is defined as following:
A. it is the flux which is linked with the primary and the secondary winding
B. it is the magnetic flux which is linked either only with the primary or only with the secondary
C. it is the flux whose path is exclusively through the air
D. None of these

170. For a 1.15 kVA, 460/230 V transformers connected to a standard single phase 50 Hz supply, the no load current is likely to be
A. 1.25 A B. 0.1 A
C. 0.5 A D. 2 A

171. A transformer has a core loss of 64 W and copper loss of 144 W, when it is carrying 20% overload current. The load at which this transformer will operate at the maximum efficiency is
A. 80% B. 66%
C. 120% D. 44%

172. Compared to the secondary of a loaded step-up transformer, the primary has
A. lower voltage and higher current
B. higher voltage and lower current
C. lower voltage and lower current
D. higher voltage and higher current

173. Transformer cores are built up from laminations rather than from solid metal so that
A. oil penetrates the core more easily
B. eddy current loss is reduced
C. less insulation is required for the windings
D. turn ratio is higher than voltage ratio

174. The size of transformer core will depend on
A. frequency
B. flux density of the core material
C. area of the core
D. both (A) and (B) are correct

175. The equivalent resistance of a transformer primary winding having $K = 4$ and $R_1 = 0.1$ ohm when referred to secondary, is
A. 0.5 ohm B. 0.02 ohm
C. 0.004 ohm D. 1.6 ohms

176. In performing short-circuit test on a transformer, usually low-voltage side is short-circuited because it has
A. lower terminal voltage and higher current rating
B. more number of turns
C. low insulation D. easy access

177. Iron loss in an actual transformer remains practically constant from no-load to full-load because
A. core flux remains practically constant
B. primary voltage remains constant
C. value of transformation ratio remains constant
D. permeability of transformer core remains constant

178. Which of the following transformers is smallest?
A. 1 kVA, 50 Hz B. 1 kVA, 200 Hz
C. lkVA, 400 Hz D. 1 kVA, 600 Hz

179. If 500-kVA, 400 Hz transformer is operated at 100 Hz, its kVA rating is
A. 125 kVA B. 250 kVA
C. 500 kVA D. 2000 kVA

180. Under operating conditions the secondary of a current transformer is always short circuited because
A. it protects the primary circuits
B. it is safe to human beings
C. it avoids core saturation and high voltage induction
D. none of these

181. Two transformers operating in parallel will share the load depending upon their
A. efficiency ratio B. ratings ratio
C. leakage reactance D. none of these

Ans. 163. C 164. C 165. C 166. D 167. D 168. D 169. B 170. B 171. A 172. A 173. B 174. D 175. D 176. A 177. A 178. D 179. A 180. C 181. B

182. Special silicon steel is used for the laminations of transformer because it has
A. high resistivity and high hysteresis loss
B. high resistivity and low hysteresis loss
C. low resistivity and low hysteresis loss
D. low resistivity and high hysteresis loss

183. In transformer the purpose of breather is to
A. extract moisture from the air
B. take insulating oil from the conservator
C. provide cooling to the winding
D. provide insulation to the winding

184. Conservator consists of
A. an air-tight metal drum fixed at the top of the tank
B. drum placed at the bottom of the tank
C. overload protection D. None of these

185. The nominal short-circuit voltage of the transformer is defined as the percentage of the
A. test voltage
B. rated secondary voltage
C. rated primary voltage
D. nominal transformation ratio

186. The efficiency of transformer is maximum at nearly
A. 70-75% of full load
B. half the full load
C. one third the full load
D. no load

187. For 100% efficiency transformer if the number of turns in the primary and secondary are 1000 and 100 respectively and the input to the transformer is 1000 W, the power output will be
A. 100 W B. 1000 W
C. 10 kW D. 100 kW

188. The primary reason why open-circuit test is performed on the low-voltage winding of the transformer is that it
A. draws sufficiently large no-load current for convenient reading
B. requires least voltage to perform the test
C. needs minimum power input
D. has become a universal custom

189. No-load test on a transformer is carried out to determine
A. copper loss B. full-load current
C. magnetising current
D. efficiency of the transformer

190. During short-circuit test, the iron loss of a transformer is negligible because
A. the entire input is just sufficient to meet copper losses only
B. voltage applied across the hv side is a small fraction of the rated voltage and so is the flux
C. iron core becomes fully saturated
D. supply frequency is held constant

191. A 1000 V/400 V power transformer has a nominal short-circuit voltage V_{sc} = 40%. Which of the following statements about it is true?
A. A voltage of 400 V appears across the short-circuited secondary terminals
B. A voltage of 16 V appears across the short-circuited secondary terminals
C. The primary voltage drops to 400 V when the secondary terminals are short-circuited
D. When the secondary terminals are short-circuited, the rated current flows at the primary side at a primary voltage of 400 V

192. The short-circuit test in a transformer is used to determine
A. iron loss at any load
B. copper loss at any load
C. hysteresis loss D. eddy-current loss

193. An ideal power transformer will have maximum efficiency at a load such that
A. copper loss is less than iron loss
B. copper loss is equal to iron loss
C. copper loss is higher than iron loss
D. None of these

194. When a 400-Hz transformer is operated at 50 Hz, its kVA rating is
A. reduced to 1/8 B. increased 8 times
C. unaffected
D. determined by load on secondary

195. The commercial efficiency of a transformer while on open-circuit is
A. zero B. maximum
C. either zero or 100% D. None of these

196. At relatively light loads, transformer efficiency is low because
A. secondary output is low
B. transformer losses are negligible
C. fixed loss is high in proportion to the output
D. copper losses are small

197. Under heavy loads, transformer efficiency is comparatively low because
A. voltage drops both in primary and secondary become large
B. secondary output is much less as compared to primary input
C. copper loss becomes high in proportion to the output
D. iron loss is increased considerably

198. The iron-losses of a transformer can be calculated by knowing the weights of
A. core only B. yokes only
C. copper winding only D. cores and yokes

199. With the increase in supply frequency of a transformer, there is

Ans. 182. B 183. A 184. A 185. C 186. A 187. B 188. A 189. C 190. B 191. D 192. B 193. B 194. A 195. A 196. C 197. C 198. D

A. decrease in rating B. increase in rating
C. no change in rating D. none of these

200. The main reason for using high voltage for long-distance power transmission is to
A. provide industrial establishments with high voltage
B. enable the national grid system to be used
C. reduce time of transmission
D. reduce transmission losses due to conductor resistance

201. A transformer having 1000 primary turns is connected to a 250 V AC supply for a secondary voltage of 400 V, the number of secondary turns should be
A. 1600 B. 250
C. 400 D. 1250

202. An amplifier works best into an impedance of 500 ohms but is required to drive a speaker of 20 ohms impedance. The primary to secondary turns ratio of the matching transformer should be
A. 1/5 B. 25
C. 5 D. 1/25

203. If a current transformer has a ratio of 100:5 and an ammeter connected to its secondary reads 2.5 A, the actual line current is
A. 12.5 A B. 250 A
C. 50 A D. 0.125 A

204. The ordinary efficiency of a given transformer is maximum when
A. it runs at half full-load
B. it runs at full-load
C. its copper loss equals iron loss
D. it runs overload

205. The output current corresponding to maximum efficiency for a transformer having core loss of 100 W and equivalent resistance referred to secondary of 0.25 ohm is
A. 20 A B. 25 A
C. 40 A D. none of these

206. The saving in copper achieved by converting two winding transformer into an autotransformer is determined by
A. voltage transformation ratio
B. load on the secondary
C. magnetic quality of core material
D. size of the transformer core

207. The kVA rating of an isolation transformer increases when connected to an autotransformer because of
A. decrease in its copper losses
B. increase in secondary current
C. establishment of conductive link between primary and secondary
D. rise in secondary terminal voltage

208. For finding the full-load efficiency of a transformer, its losses must be known. These losses
A. may be found by measuring the input to the primary with secondary open
B. may be found by measuring winding resistances and calculating the I^2R losses
C. cannot be found except by actually loading the transformer fully
D. may be found by performing short-circuit and open-circuit tests

209. A 5-ohm load is connected to the secondary of a 1100/110 V power transformer. The effective load on the 100 V bus-bar is very nearly
A. 50 ohms B. 500 ohms
C. 0.5 ohm D. 0.05 ohm

210. In a transformer, the resistance between its primary and secondary should be
A. infinity B. zero
C. 500 ohms D. nearly 1 K

211. The main purpose of using core in a transformer is to
A. decrease iron losses
B. prevent eddy current loss
C. eliminate magnetic hysteresis
D. decrease reluctance of the common magnetic flux path

212. Transformer cores are laminated in order to
A. simplify its construction
B. minimise eddy current loss
C. reduce cost
D. reduce hysteresis loss

213. In relation to a transformer, the ratio 20:1 indicates that
A. there are 20 turns on primary and one turn on secondary
B. secondary voltage is 1/20th of primary voltage
C. primary current is 20 times greater than secondary current
D. for every 20 turns on primary, there is one turn on secondary

214. While conducting short-circuit test on a transformer the following side is short-circuited
A. HV side B. LV side
C. primary side D. none of these

215. The secondary of a current transformer is always short-circuited under operating conditions because it
A. protects the primary circuits
B. is safer to human beings
C. avoids core saturation and high voltage induction
D. none of these

216. Two transformers operating in parallel will share the load depending upon their

Ans. 199. B 200. D 201. A 202. C 203. C 204. C 205. A 206. A 207. C 208. D 209. B 210. A 211. D
212. B 213. D 214. B 215. C

A. ratings B. leakage reactance
C. efficiency D. PU impedance

217. The iron-losses of a transformer can be calculated by knowing the weights of
A. cores only B. yokes only
C. copper winding only D. cores and yokes

218. For supplying a balanced 3-phase load of 40-kVA rating of each transformer in V-V bank should be
A. 20 kVA B. 23 kVA
C. 34.6 kVA D. 25 kVA

219. The distribution transformers are designed to keep the iron-losses minimum because
A. iron-losses may destroy the insulation
B. iron-losses will heat up the oil of the transformer
C. the primary of distribution transformer is energized for all the twenty four hours
D. iron-losses produce time harmonics

220. If the load pf is 0.886, then the average pf of the V-bank is
A. 0.886 B. 0.75
C. 0.51 D. 0.65

221. The percentage overloading of two 20-kVA transformers when supplying a balanced 3-phase load of 40 kVA in a V-V bank is
A. 15% B. 57.7%
C. 73.2% D. 33.3%

222. If full-load copper loss of a transformer is 1600 W, its copper loss at 75% full-load would be
A. 900 W B. 1200 W
C. 1600 W D. 1800 W

223. Silicon steel is preferred for transformer core because it
A. decreases the permeability of the core
B. reduces both eddy current loss and hysteresis loss
C. reduces eddy current loss
D. decreases the tensile strength

224. Five limb core construction has an advantage over the three limb core construction that the
A. eddy current loss is less
B. magnetic reluctance of the three phases can be balanced
C. hysteresis loss is less
D. permeability is higher

225. The magnetic properties of silicon steel crystal are best along the
A. cube edges B. surface diagonals
C. cubic diagonals D. none of these

226. The magnetic properties of cold rolled grain oriented silicon steel are excellent
A. against the direction of rolling
B. at right angles to the direction of rolling
C. along the direction of rolling
D. none of these

227. For large power transformer the best utilisation of available core space can be made by using
A. rectangular core section
B. square core section
C. stepped core section
D. None of these

228. For cold-rolled grain-oriented silicon steel metred overlap is preferable for the core-yoke joints as it
A. improves mechanical strength
B. makes better use of core space
C. reduces magnetostriction
D. reduces magnetising current and also the core-losses

229. The direction of the central phase winding of a three-phase shell type transformer is reversed with respect to the outer phases to
A. save considerable amount of core material
B. reduce leakage fluxes
C. reduce short circuit forces
D. minimise eddy current loss

230. Low voltage windings are placed nearer to the core in the case of concentric windings because it reduces
A. hysteresis loss
B. eddy current loss
C. insulation requirement
D. leakage fluxes

231. Spiral winding is suitable .... for windings
A. carrying very low current
B. carrying very high current
C. rated for high voltage
D. rated for low voltage

232. Helical winding is suitable for
A. low voltage winding of small transformers
B. high voltage winding of small transformers
C. low voltage winding of large transformers
D. high voltage winding of large transformers

233. A 4-kVA 400/200 V single-phase transformer has resistance of 0.02 pu and reactance of 0.6 pu. What are its resistance and reactance referred to hv side?
A. 0.2 ohm and 0.6 ohm B. 0.8 ohm and 2.4 ohm
C. 0.08 ohm and 0.24 ohm
D. 1 ohm and 3 ohm

234. Cross-over winding is suitable for
A. low voltage winding of small transformers
B. high voltage winding of small transformers
C. low voltage winding of large transformers
D. high voltage winding of large transformers

235. For back-to-back test of two identical transformers
A. both the primaries and the auxiliary transformer have to be fed from voltage sources of rated frequency

Ans. 216. A 217. D 218. B 219. C 220. B 221. A 222. A 223. B 224. B 225. A 226. C 227. C 228. D 229. A 230. C 231. B 232. C 233. A 234. B

B. both the primaries and the auxiliary transformer may be fed from voltage sources of frequencies different from the rated frequency
C. while the primaries have to be fed from a voltage source of rated frequency, the auxiliary transformer can be fed from a voltage source of frequency different from the rated frequency
D. while the primaries can be fed from a voltage source of frequency different from the rated frequency, the auxiliary transformer has to be fed from a voltage source of rated frequency

236. Continuous disc winding is suitable for
A. low voltage winding of small transformers
B. high voltage winding of small transformers
C. low voltage winding of large transformers
D. high voltage transformers

237. Major insulation of a transformer is the insulation between the
A. low-voltage winding and the core only
B. low voltage and the high-voltage winding
C. low-voltage winding and the core and also between the low-voltage and the high-voltage windings
D. turns of the windings

238. Minor insulation of a transformer is the insulation between the
A. low-voltage and the high-voltage windings
B. turns of the windings only
C. layers of the windings only
D. turns of the windings and also between the layers of the windings

239. For distribution transformers of rating less than 50 kVA, it is better to use
A. plain sheet steel tanks only
B. corrugated tanks only
C. either plain sheet steel tanks or corrugated tanks
D. tubed tanks only

240. For voltage above 66 kV, condenser bushings are preferred over non-condenser bushings because
A. the axial stresses can be eliminated
B. the radial stress can be made independent of the radial thickness of the dielectric
C. optimum utilisation of the dielectric can be made due to uniform potential distribution which also reduces both the radial and the axial dimensions of the bushings
D. none of these

241. For off-load tap changing, the best method is to use tap changers
A. inside the tank operated by external selector switches
B. inside the tank but with no selector switch
C. outside the tank operated by selector switches
D. outside the tank but with no selector switch

242. The transformer windings are tapped in the middle because it
A. eliminates radial forces on the windings
B. eliminates axial forces on the windings
C. reduces insulation requirement
D. none of these

243. Tappings are normally provided on the high voltage winding of a transformer
A. only because it has larger number of turns which allows smoother variation of voltage
B. only because it has to handle low currents
C. only because it is easily accessible physically
D. because it has large number of turns, as it has to handle low currents and is also easily accessible physically

244. In general, for all sizes of distribution transformer, it is preferable to use
A. plain sheet steel tanks
B. tubed tanks
C. corrugated tanks
D. radiator tanks

245. With the increase in supply frequency of a transformer there is
A. decrease in rating B. increase in rating
C. no change in rating D. none of these

246. For small power transformers, it is preferable to use
A. tubed tanks B. corrugated tanks
C. radiator tanks
D. tanks with separate coolers

247. For large power transformer, it is preferable to use
A. corrugated tanks B. tubed tanks
C. radiator tanks
D. tanks with separate coolers

248. A transformer may have two or more ratings depending on the type of
A. winding used B. cooling used
C. core used D. insulation used

249. Conservator is used
A. to take up the expansion and contraction of oil due to temperature change
B. to act as an oil storage
C. for better cooling of the transformer
D. none of these

250. Buchholz relay is placed between the
A. conservator and the breather
B. tank and the conservator
C. hv winding and the bushing
D. lv winding and the bushing

251. Buchholz relay is a
A. voltage sensitive device
B. current sensitive device
C. frequency sensitive device
D. gas actuated device

Ans. 235. C 236. D 237. C 238. D 239. C 240. C 241. A 242. B 243. D 244. B 245. B 246. A 247. D 248. B 249. A 250. B 251. D

252. Buchholz relay causes
A. tripping for major fault and alarm for minor faults
B. alarm for major fault and tripping for minor faults
C. tripping for both major and minor faults
D. alarm for both major and minor faults.

253. Transformer oil is used as
A. insulant only
B. coolant only
C. both insulant and coolant
D. inert medium

254. Sludging of transformer oil means
A. continuous expansion and contraction due to heating and cooling
B. formation of semisolid hydrocarbon due to heat and oxidation
C. decomposition of transformer oil under the influence of power arcs
D. evaporation of transformer oil due to heating

255. In a single-phase transformer, if leakage impedance drop is neglected, the flux in the magnetic core and the leakage flux linking a winding are
A. proportional to the applied voltage and proportional to the winding current respectively
B. both proportional to the applied voltage
C. proportional to the applied voltage respectively
D. both proportional to the winding current

256. In a transformer the exciting current has two components–magnetising component and core-loss component. Neglecting leakage impedance drop
A. both of them lag the impressed voltage by 90°
B. both of them are in phase with the impressed voltage
C. the former lags the impressed voltage by 90° while the latter is in phase with impressed voltage
D. the former is in phase with the impressed voltage while the latter lags the impressed voltage by 90°

257. The flux in the transformer core supplied from a constant voltage source
A. slightly decreases with increase in inductive load
B. slightly increases with increase in inductive load
C. remains constant whatever the inductive load may be
D. increases with increase in its inductive load up to its rated capacity and remains constant if the inductive load is increased beyond its rated capacity

258. A transformer has a percentage resistance of 1% and percentage reactance of 4%. Its regulation at power factor 0.8 lagging and 0.8 leading are respectively
A. 3.2% and –1.6%
B. 6% and –4%
C. 4.8% and –3.2%
D. none of these

259. If frequency be f, maximum flux density be B_m and Steinmetz's coefficient be η then eddy current loss in a transformer core is proportional to
A. $B_m f$
B. $B_m^2 f^2$
C. $B_m^2 f$
D. $B_m f^2$

260. If P_i be the iron loss and P_{cu} be the copper loss on full load, the following condition has to be satisfied to obtain maximum efficiency at 3/4 full-load:
A. $P_{cu} = 3/4P_i$
B. $P_{cu} = 4/3P_i$
C. $P_{cu} = 16/9P_i$
D. $P_{cu} = 9/16P_i$

261. For short-circuit and open-circuit test of a transformer, the instruments are connected on
A. the lv side and hv side respectively
B. the hv side and lv side respectively
C. the hv side only
D. the lv side only

262. For short-circuit and open-circuit test of a transformer, the instruments are connected to measure
A. equivalent resistance and leakage reactance
B. equivalent resistance and core loss
C. magnetising current and equivalent leakage reactance
D. magnetising current and core loss

263. Short circuit test was done on a transformer with a certain impressed voltage at rated frequency. If the short circuit is now done on the same transformer with the same magnitude of impressed voltage but at higher frequency. Then
A. both current magnitude and power factor increase
B. both current magnitude and power factor decrease
C. current magnitude decreases and power factor increases
D. current magnitude increases and power factor decreases

264. The utilization factors for transformers connected delta-zigzag and star-zigzag are respectively
A. 0.5 and 0.667
B. 0.667 and 0.75
C. 0.667 and 0.667
D. 0.886 and 0.886

265. Open circuit test was done on a transformer with rated voltage at rated frequency. If the open circuit test was now done on the transformer with double the rated voltage at double the rated frequency
A. both current magnitude and power factor increase
B. both current magnitude and power factor decrease
C. current decreases and power factor increases
D. current increases and power factor decreases

Ans. 252. A 253. C 254. B 255. A 256. C 257. A 258. A 259. B 260. C 261. B 262. D 263. B 264. D 265. A

266. For no-load tap changing, the best method is to use tap changers
A. inside the tank operated by external selector switches.
B. inside the tank but with no selector switch
C. outside the tank operated by selector switch
D. none of these

267. The frequency of the impressed voltage of transformer is increased keeping its voltage magnitude fixed. Its
A. hysteresis loss increases and eddy current loss decreases
B. hysteresis loss decreases and eddy current loss increases
C. hysteresis loss decreases and eddy current remains constant
D. both hysteresis and eddy current losses increase

268. The frequency of the impressed voltage of a transformer is increased keeping its magnitude fixed. Of the two components of the exciting current
A. the magnetising current increases and the core loss component decreases
B. the magnetising current decreases and the core loss component increases
C. both the magnetising and core loss component decrease
D. both the magnetising and core loss components increase

269. The flux involved in the emf equation of a transformer has
A. rms value
B. average value
C. total value
D. maximum value

270. Two transformers A and B with identical ratings, are to be designed with flux densities of 1.2 Wb/m^2 and 1.4 Wb/m^2 respectively. The weight of transformer A per kVA is
A. less than that of transformer B
B. more than that of transformer B
C. equal to that of transformer B
D. none of these

271. Choose the correct statement:
A. Emf per turn in hv winding is more than the emf per turn in lv winding.
B. Emf per turn in both the windings are equal
C. Emf per turn in hv winding is less than the emf per turn in lv winding
D. none of these

272. The no-load current in a transformer lags the applied voltage by nearly
A. 90°
B. about 80°
C. 0°
D. about 115°

273. A 400/200 V transformer has its lv resistance of 0.02 per unit. This resistance when referred to hv side is
A. 0.02 pu
B. 0.04 pu
C. 0.01 pu
D. 0.08 pu

274. The leakage flux in a transformer depends upon the
A. applied voltage
B. frequency
C. load current
D. mutual flux

275. The useful flux of a transformer is 1 Wb. When it is loaded at 0.8 pf lag, then the mutual flux
A. decreases to 0.8 Wb
B. increases to 1.2 Wb
C. remains constant
D. decreases to 0.98 Wb

276. Distribution transformers have core losses
A. greater than copper losses
B. equal to copper losses
C. less than copper losses
D. equal to 1/2 copper losses

277. Good voltage regulation of a transformer means that the difference between its no-load voltage and full-load voltage is
A. very high
B. low
C. very low
D. none of these

278. The voltage regulation of a transformer at full-load 0.8 pf lagging is 4%. Its voltage regulation at full-load 0.8 pf leading
A. will be positive
B. will be negative
C. may be positive
D. may be negative

279. The efficiency of a transformer at full-load 0.8 pf lag is 90%. Its efficiency at full load 0.8 pf lead will be
A. less than 90%
B. more than 90%
C. 90%
D. none of these

280. If P_c and P_{sc} represent core and full-load ohmic losses respectively, then maximum kVA delivered to the load corresponding to maximum efficiency is equal to rated kVA multiplied by
A. P_c/P_{sc}
B. $(P_c/P_{sc})^2$
C. $\sqrt{P_c/P_{sc}}$
D. $(P_{sc}/P_c)^2$

281. Eddy current losses in a transformer depends on
A. both voltage and frequency
B. voltage alone
C. load current alone
D. thickness of core

282. Hysteresis loss in a transformer depends upon
A. frequency
B. supply voltage
C. square of the frequency alone
D. square of the voltage alone

283. The open-circuit test on transformer is conducted to obtain
A. the leakage impedances
B. the ohmic loss
C. hysteresis loss only
D. core losses

284. Short-circuit test on a single-phase transformer gave the following data:30 V at 50 Hz, 20 A, pf = 0.2 lagging. If SC test is performed on 30 V, 25 Hz, then short-circuit current is
A. decreased at pf < 0.2
B. increased at pf < 0.2
C. increased at pf > 0.2
D. decreased at pf > 0.2

Ans. 266. C 267. C 268. C 269. D 270. B 271. B 272. B 273. A 274. C 275. D 276. C 277. D 278. D 279. C 280. C 281. D 282. A 283. D 284. C

285. The short-circuit test on a transformer is conducted to obtain
A. core losses at any load
B. ohmic losses at any load
C. hysteresis losses
D. both core losses and ohmic losses

286. In a transformer
A. both Oc and Sc tests are conducted on lv side
B. Oc test is conducted on hv side and Sc test on lv side
C. Oc test is conducted on lv side and Sc test on hv side
D. both Oc and Sc tests are conducted on hv side

287. In a transformer, tappings are provided on
A. hv side at one end of the winding
B. lv side at one end of the winding
C. lv side at the middle
D. hv side at the middle

288. For successful parallel operation of two single-phase transformers, the essential condition is that their
A. percentage impedances should be equal
B. turns ratios should be exactly equal
C. polarities must be properly connected
D. kVA ratings should be equal

289. Two transformers operating in parallel have different leakage impedances, for a load power factor of 0.8
A. both would operate at pfs less than 0.8
B. both would operate at pfs more than 0.8
C. both would operate at the same pfs
D. one would operate at pf 0.8 and other at pf less than 0.8

290. In a single-phase induction regulator, the output voltage can be varied
A. both in magnitude and phase
B. in magnitude only C. in phase only
D. in magnitude less than the supply voltage

291. Transformer-core laminations are made of
A. cast iron B. wrought iron
C. silicon steel D. cast steel

292. Transformer action requires a
A. constant magnetic flux
B. increasing magnetic flux
C. pulsating magnetic flux
D. alternating electric flux

293. In an oil-filled transformer, oil is provided for
A. cooling B. insulation
C. lubrication
D. both cooling and insulation

294. Transformer cores are laminated to reduce
A. eddy current loss B. hysteresis loss
C. both eddy current and hysteresis
D. loss ohmic loss

295. An air-core transformer, as compared to iron-core transformer, has
A. less magnetic core loss
B. more magnetic core loss
C. no magnetic core loss
D. less ohmic loss

296. One 200 V, 100 W bulb is connected in series with the primary of a 200 V, 10 kVA transformer. If its secondary is left open circuited, then the bulb would have
A. full brightness B. poor brightness
C. a little less than full brightness
D. more than full brightness

297. In case of a power transformer, the no-load current in terms of rated current is
A. 10 to 20% B. 2 to 6%
C. 15 to 30% D. 30 to 50%

298. Transformer zero voltage regulation occurs at
A. unity pf B. leading pf
C. lagging pf D. zero pf leading

299. In transformer, maximum efficiency for a constant load current, occurs at
A. 0.8 pf B. zero pf leading
C. zero pf lagging D. unity power factor

300. Which of the following statement regarding an ideal single-phase transformer having a turn ratio of 1:5 and drawing a current of 4 A from a 200 V AC supply is false?
A. It is step-up transformer
B. Its secondary current is 0.8 A
C. Its secondary current is 4 A
D. Its secondary voltage is 1000 V

301. One transformer has leakage impedance of $(1 + j4)$ ohm and $(3 + j11)$ ohm for its primary and secondary windings respectively. This transformer has
A. hv primary
B. medium voltage primary
C. lv primary D. none of these

302. With core type transformers, the limbs are stepped to
A. reduce the iron material and, therefore, iron loss
B. reduce the conductor material and therefore I^2R
C. provide better cooling
D. provide more mechanical strength to the core

303. For core type power transformers, both primary and secondary windings have circular coil sections, because this section
A. is easier to wind
B. requires minimum copper and, there-fore, less I^2R loss
C. results in less core material and, therefore, less core loss
D. has the strongest mechanical shape

Ans. 285. B 286. C 287. C 288. C 289. C 290. B 291. C 292. C 293. D 294. A 295. C 296. B 297. B 298. B 299. D 300. C 301. C 302. B 303. D

304. If copper loss of a transformer at 7/8th full-load is 4900 W, then its full-load copper would be
A. 5600 W B. 6400 W
C. 373 W D. 429 W

305. The maximum efficiency of a 100-kVA transformer having iron loss of 900 kW and FL. copper loss of 1600 W occurs at
A. 56.3 kVA B. 133.3 kVA
C. 75 kVA D. 177.7 kVA

306. Mark the wrong statement:
A. In a given ideal transformer, volt/turn is constant for a given primary voltage
B. A transformer whose secondary current is greater than primary current is called step-up transformer
C. Hysteresis loss is proportional to supply frequency
D. Eddy current loss is reduced by laminating transformer core

307. The iron-losses of a transformer can be calculated by knowing the weights of
A. cores only B. yokes only
C. copper winding only D. cores and yokes

308. Two transformers A and B are of same type with all linear dimensions in the ratio x:1 and having the same flux density, current density, frequency and window space factor. The outputs of transformer A and B are in the ratio of
A. x:1 B. x^2:1
C. x^3:1 D. x^4:1

309. With the increase in supply frequency of a transformer, there is
A. decrease in rating B. increase in rating
C. no change in rating D. none of these

310. In the above question, copper and iron losses of both the transformers are in the ratio
A. x:1 B. x^2:1
C. x^3:1 D. 1:1

311. The distribution transformers are designed to keep the iron-losses minimum because
A. iron-losses may destroy the insulation
B. iron-losses will heat up the oil of the transformer
C. the primary of distribution transformer is energized for all the twenty four hours
D. iron-losses produce time harmonics

312. Maximum transient currents flow through a transformer winding when this is switched on with
A. secondary side open-circuited and when the input voltage wave is passing through zero value
B. the secondary short-circuited and when the input voltage wave is passing through zero value
C. the secondary open-circuited and when the input voltage wave is passing through maximum value
D. the secondary short-circuited and when the input voltage wave is passing through maximum value

313. For minimum weight of a transformer, the weight of iron should be
A. greater than the weight of copper
B. less than the weight of copper
C. equal to the weight of copper
D. zero

314. The essential condition for parallel operation of two single-phase transformers is that they should have same
A. polarity B. phase sequence
C. voltage ratio D. percentage impedance

315. Two transformers operating in parallel will share the load depending upon their
A. rating B. leakage reactance
C. efficiency D. per-unit impedance

316. 500 kVA, 400 V, 3-phase, 60 Hz transformer is replaced by similar transformer of frequency 50 Hz. Its rating will be
A. 680 kVA B. 500 kVA
C. 416 kVA D. none of these

317. For transformers using cold rolled grain oriented steel, the area of yoke is taken
A. as 15–25% larger than that of core
B. as 15–20% smaller than that of core
C. equal to that of core
D. none of these

318. Oil natural cooling with tubes is used for the transformers of capacity
A. 30 kVA B. 500 kVA
C. 750 kVA D. 3000 kVA

319. Out of the following choices for polyphase transformer connections which one will you select for three-to-two phase conversion?
A. Scott B. Star/star
C. Double scott D. Star/double-delta

320. Air blast cooling is provided for the transformers above a capacity of
A. 500 kVA B. 1000 kVA
C. 5000 kVA D. 10000 kVA

321. Which of the following methods is used for cooling of transformer up to a capacity of 10 MVA?
A. Oil natural cooling B. Air blast cooling
C. Forced oil cooling D. None of these

322. For 3-to-3-phase conversion, which of the following given polyphase connections would you select?
A. Delta/double-delta B. Star/double-star
C. Delta/interconnected-star
D. Delta/double-star

Ans. 304. B 305. C 306. B 307. D 308. D 309. B 310. C 311. C 312. B 313. C 314. A 315. A 316. C 317. C 318. D 319. A 320. D 321. A 322. C

323. The main purpose of conducting no-load test on a transformer is to find out
A. value of flux produced under no-load condition
B. voltage/turn of the low-voltage winding
C. core loss in the transformer at rated voltage and under load conditions
D. its voltage regulation

324. Leakage fluxes of a transformer may be minimised by
A. keeping the magnetising current to the minimum
B. reducing the reluctance of the iron core to the minimum
C. minimizing the number of turns both on primary and secondary
D. sectionalizing and interleaving the primary and secondary windings

325. Of the following statements concerning parallel operation of transformers, the one which is not correct is
A. transformers must have equal voltage ratings
B. transformers must have same ratio of transformation
C. transformers must be operated at the same frequency
D. transformers must have equal kVA ratings

326. The open-circuit test on a transformer is meant for measuring its
A. copper losses B. iron losses
C. insulation resistance D. equivalent resistance

327. A 100 ohms load has a pf of 0.8. A 1 kΩ generator supplies power to the load through a transformer. Best matching will be produced when transformers turn ratio is approximately
A. 3:1 B. 8:1
C. 10:1 D. 2:1

328. Oil is invariably used in large transformers in order to
A. insulate the coils B. insulate the core
C. lubricate the core D. lubricate the coils

329. If full rating of a transformer is 90 kW at a power factor of 0.9, then its kVA rating is
A. 81 B. 100
C. 90 D. 120

330. The secondary winding of a current transformer whose primary is carrying current should
A. not be open-circuited
B. not be short-circuited
C. always be open-circuited
D. not be connected to the current coil of a wattmeter

331. When applying a high-voltage test to the primary winding of a rebuilt transformer
A. secondary winding must be short-circuited
B. secondary transformers case and core must be grounded
C. one end of the primary must be grounded
D. all other parts of the transformer must be grounded

332. Before disconnecting an ammeter from an energised current transformer circuit, the current transformer
A. primary should be opened
B. secondary should be opened
C. primary should be shorted
D. secondary should be shorted

333. An autotransformer is preferred to a conventional 2-winding transformer
A. because it is much safer to use an autotransformer
B. where large number of secondary taps are needed
C. where it is required electrically to isolate the two windings
D. where ratio of transformation is nearly one

334. A single phase transformer has a turn ratio of 10:1. The primary winding has a resistance of 2.1 ohm and the secondary winding has a resistance 0.02 ohm. The total resistance as referred to primary is
A. 2.12 ohm B. 4.1 ohm
C. 0.041 ohm D. 2.3 ohm.

335. The main purpose of using an interphase transformer with a multi-anode rectifier is to
A. permit only one anode to fire at a time
B. cause the anodes to overlap in firing
C. shorten the period of firing of eachanode
D. act as an insulating transformer between the power transformer and load

336. Dust should never be allowed to accumulate on the windings of a dry-type transformer because it
A. may short-circuit the windings
B. tends to corrode metal surface
C. reduces dissipation of heat
D. absorbs oil and grease

337. The relative polarity of the windings of a transformer may be determined by
A. short-circuit test B. open-circuit test
C. phasing out D. back-to-back test

338. The most common cause of contamination by water of oil used in transformers located indoors is
A. condensation of moisture from air in the upper part of the tank
B. decomposition of organic matter in the oil
C. leaky bushings
D. use of filter blotters that have absorbed moisture from the air

Ans. 323. B 324. D 325. D 326. B 327. A 328. A 329. B 330. A 331. B 332. D 333. D 334. B 335. B 336. C 337. C 338. B

339. Part of the transformer which is most subject to damage from overheating is
A. winding insulation B. frame case
C. iron core D. copper winding

340. In general, the most important point to keep under constant watch during the operation of a transformer is its
A. primary voltage B. core loss
C. exciting current D. temperature

341. In a 10:1 step-down 2200/220 V transformer, taps are changed to reduce the number of primary turns by 5%. The secondary voltage will nearly be
A. 209 V B. 231 V
C. 240 V D. 225 V

342. If the secondary of a 10:1 step-down transformer is connected to the primary of a 3:1 step-down transformer, the total step-down of both transformers is
A. 30:1 B. 7:1
C. 13:1 D. 3.3:1

343. Two single-phase transformers with a ratio of 1:2 are to be connected in parallel. The transformers may be connected in parallel by connecting lead A to lead B if voltmeter is connected between them, it reads
A. 110 V B. 220 V
C. 0 V D. 440 V

344. The essential feature of an autotransformer is that it
A. is more suited for single-phase supply
B. is completely automatic in its operation
C. uses only one winding
D. is particularly suited for situations where voltage transformation ratio exceeds unity

345. When secondary of the transformer is loaded, the primary current will
A. not be a effected B. draw more current
C. draw less current than the secondary current
D. be the vector sum of no load current and excessive current drawn due to the secondary current

346. When the secondary of the transformer is loaded, the flux in the transformer will
A. remain constant
B. be directly proportional to secondary current
C. be directly proportional to the excessive current drawn by primary winding
D. none of these

347. The emf induced in the secondary winding depends upon
A. number of turns B. flux
C. supply frequency D. all of these

348. Shell type transformer is commonly used because it has
A. more magnetic flux leakage
B. less magnetic flux leakage
C. two magnetic paths
D. both (B) and (C) are correct

349. If the input voltage is 100 V and output voltage is 200 V, which side of the transformer windings will be primary?
A. 200 V side B. 100 V side
C. Winding with more turns
D. Winding with less turns

350. Which of the following does not change in an ordinary transformer?
A. Voltage B. Current
C. Frequency D. Phase angle

351. The colour of fresh dielectric oil for a transformer is
A. pale yellow B. dark brown
C. white grey D. colourless

352. Core losses in a transformer are usually
A. 1 to 3% B. 10 to 15%
C. 25 to 30% D. 40 to 50%

353. An ideal transformer has secondary winding tapped at point *b*. The number of primary turns is 100 and the number of turns between *a* and *b* is 300 and between *b* and *c* is 2000. The transformer supplies a resistive load connected between a and c and drawing 7.5 kW. Moreover a load impedance of 10 $\angle 45°$ ohm is connected between *a* and *b*. The primary voltage is 1000 V. The primary current will be
A. 435 $\angle -44°$ A B. 435 $\angle 44°$ A
C. 906 $\angle 44°$ A D. 906 $\angle -44°$ A

354. In a 4:1 turn ratio transformer the leakage reactance as referred to the secondary side is 0.05 pu. The leakage reactance as referred to primary is
A. 0.05 ohm B. 0.05 pu
C. 0.8 pu D. 0.2 pu

355. In a 25 kVA 3300/230 V, single-phase transformer the iron and full load copper losses are 350 W and 400 W respectively. The load at which the efficiency will be maximum, is
A. 25 kW B. 21.875 kW
C. 25 kVA D. 21.875 kVA

356. The colour of moist silica gel is
A. pale yellow B. pale pink
C. blue D. green

357. The leakage impedances of primary and secondary windings of a transformer are (3 + *j*4) ohms and (1 + *j*4) ohms respectively. It can be concluded that the transformer has
A. low voltage primary
B. low voltage secondary
C. high voltage secondary
D. no conclusion can be drawn on the basis of information provided

358. The path of the magnetic flux in a transformer should have

Ans.												
339. A	340. D	341. A	342. A	343. C	344. C	345. D	346. A	347. D	348. D	349. B	340. C	351. A
352. A	353. D	354. B	355. D	356. C	357. B							

A. low reluctance B. low resistance
C. high reluctance D. high resistance

359. Losses in a transformer are
I. Hysteresis losses
II. Eddy current losses
III. Copper losses
Open circuit test of a transformer gives
A. I only B. II only
C. II and III only D. I and II only

360. Short circuit (SC) test and open circuit (OC) test on a 110 kVA, 4400/440 V, 50 Hz transformer gave the following results.
OC test: $P = 1200$ W, $I = 2$A, $V = 440$ V
SC test: $P = 2000$ W, $I = 200$ A, $V = 18$ V
Neglecting magnetising losses, if the transformer is supplying the rated current at 0.8 pf lagging, the voltage regulation will be nearly
A. 4.8% B. 9.6
C. 90.4% D. 95.2%

361. Class-A insulations can withstand
A. temperature rise of 150°C
B. maximum temperature of 105°C
C. temperature gradient of 105°C
D. heating rate of 105°C in 24 hours

362. The function of breather in a transformer is to
A. filter the transformer oil
B. provide oxygen to the cooling oil
C. provide cooling air
D. arrest the flow of moisture when outside air enters the transformer

363. Transformer works on
A. AC only B. DC only
C. AC and DC both D. pulsating dc

364. Iron losses in a transformer occur in
A. core B. winding
C. insulating D. main body

365. The efficiency of a transformer depends on
A. current B. voltage
C. power factor D. none of these

366. A 1600 kVA, 200 Hz transformer is operated at 50 Hz. Its kVA rating should be restricted to
A. 6400 B. 800
C. 400 D. 10000

367. In a single-phase core type two winding transformer
A. low voltage winding is put on one limb of the core and the high voltage winding on the other limb
B. one half of each winding is put on each limb of the core
C. each limb carries two-third of one winding and one-third of the other
D. none of these

368. Transformers use laminated core
A. to reduce eddy current loss
B. to reduce eddy current and hysteresis loss
C. to reduce copper loss
D. none of these

369. In a transformer with closed magnetic coupling between primary and secondary windings
A. efficiency will be high
B. all day efficiency will be high
C. regulation will be good
D. none of these

370. The number of turns in the primary winding of a transformer depends on
A. input voltage B. input current
C. both input voltage and current
D. none of these

371. A 10 kVA, 230/1000 V single phase transformer is fed from a 230 V supply. The voltage across the load will be
A. 1000 V B. less than 1000 V
C. more than 1000 V D. 230 V

372. A 5 kVA 200/100 V single phase transformer delivers 50 A at rated voltage. The input current will be
A. 25 A B. less than 25 A
C. more than 25 A D. 50 A

373. Under no load conduction, the power factor of a transformer is
A. zero B. unity
C. about 0.4 lagging D. about 0.7 lagging

374. In a 10 kVA, 230/1000 V single-phase transformer, the no-load current will be about
A. 0.8 A B. 3 A
C. 0.5 A D. 10 A

375. A single-phase transformer has a turn ratio of 4:1. If the secondary winding has a resistance of 1 ohm, the secondary resistance as referred to the primary will be
A. 4 ohms B. 16 ohms
C. 3.5 ohms D. 0.25 ohm

376. A transformer is supplying a unity power, factor load. The power factor at primary terminal will be
A. about 0.95 lagging B. about 0.95 leading
C. about 0.8 lagging D. unity

377. When the secondary winding of a transformer is short circuited, the power factor of the input is
A. unity B. about 0.7 lagging
C. about 0.7 leading D. about 0.2 lagging

378. The voltage regulation of a well designed transformer is of the order of
A. 10% B. 50%
C. 2% D. 0.1%

379. The full-load efficiency of a well designed transformer is around

Ans. 358. A 359. A 360. B 361. D 362. D 363. A 364. A 365. C 366. C 367. B 368. A 369. C 370. A
371. B 372. C 373. C 374. B 375. B 376. A 377. A 378. C

A. 0.98 B. 0.9
C. 0.85 D. 0.7

380. A distribution transformer should be selected on the basis of
A. efficiency B. all day efficiency
C. voltage regulation D. None of these

381. In a transformer the iron losses do not vary with load current because
A. iron losses are equal to copper losses
B. core flux remains constant
C. core area is constant D. None of these

382. In an unloaded transformer having a voltage ratio of 1000/230 V is connected to a 100 V DC supply, the primary current will be
A. very high as compared to rated current
B. much less than the rated current
C. almost the same as the rated current
D. none of these

383. A transformer has a voltage regulation of 3% at full load 0.8 power factor lagging. At full load unity power factor, the regulation will be
A. 3% B. negative
C. less than 3% D. more than 3%

384. A 5 kVA, 1000/100 V two winding transformer is used as an autotransformer by connecting the two windings in the series so as to supply a 1000 A load from an 1100 V supply. The maximum kVA which can be supplied to the load is
A. 10 kVA B. 55 kVA
C. 5.5 kVA D. 550 kVA

385. In case of transformer, would it be easily possible to generally use a frequency of 5 Hz instead of 50 Hz?
A. Yes, because the frequency of 50 Hz was selected arbitrarily. The frequency of 5 Hz would indeed be more favourable because the skin effect would be less
B. Yes, but it is not done due to the high cost of changing the present system
C. No, because in case of 5 Hz the insulation of cables and conductors should be stronger
D. No, because for 5 Hz the number of turns of transformer coils (for the same cross-section of iron core) should be about 10 times more than for 50 Hz

386. Which of the following statements about the humming of transformer is true?
A. The humming is caused by a vibration of the low voltage winding through which the high current flows
B. The humming is caused by the periodic force of attraction between the laminations of the core
C. The frequency of humming tone is 50 Hz
D. The frequency of humming tone is 2 × 50 Hz = 100 Hz

387. The efficiency of a transformer is calculated by using the relation
A. $\eta = \dfrac{\text{output} \times 100}{\text{output} + \text{copper losses}}$
B. $\eta = \dfrac{\text{output} \times 100}{\text{output} + \text{iron losses}}$
C. $\eta = \dfrac{\text{output} \times 100}{\text{input} + \text{copper losses}}$
D. $\eta = \dfrac{\text{output} \times 100}{\text{output} + \text{copper losses} + \text{iron losses}}$

388. What is meant by the leakage flux of transformer?
A. It is the flux which is linked with the primary and the secondary winding
B. It is that part of the magnetic flux which is significant for radio interference
C. It is the magnetic flux which is linked either only with the primary or only with the secondary winding
D. The leakage flux in the transformer is the flux whose path is exclusively through the air

389. Hysteresis loss in a transformer depends upon the
A. applied voltage B. type of core material
C. number of laminations
D. reactance of the winding

390. The loss due to eddy currents in a transformer is
A. directly proportional to the current
B. indirectly proportional to the current
C. indirectly proportional to the resistance of the winding
D. directly proportional to the resistance of the winding

391. Which of the following statements about a short-circuit proof transformer is true?
A. There is a fuse in its primary circuit
B. There is a bimetallic switch in its primary circuit
C. There is a fast magnetic switch in its primary circuit
D. The short circuit current can continuously flow through it without causing a damage

392. Why is there an enamel layer coated over the laminations of the transformer core?
A. To decrease the hum
B. To prevent the friction between the laminations
C. To attain adhesion between the lamination
D. To insulate the laminations against each other

393. What is the efficiency of transformers when compared with that of electrical motors of the same power?
A. Much smaller B. Somewhat smaller
C. About the same D. Much higher

Ans. 379. A 380. B 381. B 382. A 383. C 384. B 385. D 386. D 387. C 388. C 389. B 390. A 391. D 392. D 393. D

394. The rated current flows through the primary winding of the transformer is shown in the figure. What is the voltage V called?

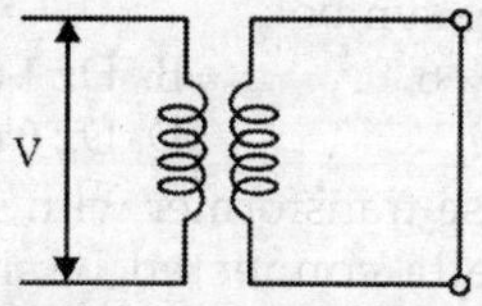

A. Short circuit voltage B. Rated voltage
C. No load voltage D. Operating voltage

395. The tubes welded to the sides of the transformers improve the
A. efficiency of the transformer
B. voltage regulation of the transformer
C. cooling of the transformer
D. none of these

396. The material suitable for the manufacture of transformer and large turbo alternator is
A. hot rolled grain-oriented steel
B. cold rolled grain-oriented steel
C. cast steel D. none of these

397. The saving in copper is achieved by converting a 2-winding transformer into an auto-transformer is determined by
A. voltage transformation ratio
B. load on the secondary
C. magnetic quality of core material
D. size of the transformer core

398. An autotransformer having a transformation ratio of 0.8 supplies a load of 3 kW. The power transferred conductively from primary to secondary is
A. 0.6 kW B. 2.4 kW
C. 1.5 kW D. 0.27 kW

399. The essential condition for parallel operation of two 1-phase tranformers is that they should have the same
A. polarity B. kVA rating
C. voltage ratio D. percentage impedance

400. If the impedance triangles of two transformers operating in parallel are not identical in shape and size, the two transformers will
A. share the load unequally
B. get heated unequally
C. have a circulatory secondary current even when unloaded
D. run with different power factors

401. Two tranformers A and B having equal outputs and voltage ratios but unequal percentage impedances of 4 and 2 are operating in parallel. Transformer A will be running overload by
A. 50% B. 66%
C. 33% D. 25%

402. Which of the following connections is best suited for 3-phase, 4-wire service?
A. Δ–Δ B. Y–Y
C. Δ–Y D. Y–Δ

403. In a three-phase Y-Y transformer connection, neutral is fundamental to the
A. suppression of harmonics
B. passage of unbalanced currents due to unbalanced loads
C. provision of dual electric service
D. balancing of phase voltages with respect to line voltages

404. As compared to Δ–Δ bank, the capacity of the V–V bank of transformers is
A. 57.7% B. 66.7%
C. 50% D. 86.6%

405. If three transformers in a Δ–Δ are delivering their rated load and one transformer is removed, then overload on *each* of the remaining transformers is
A. 66.7% B. 173.2%
C. 73.2% D. 58%

406. When a V–V system is converted into a Δ–Δ system, increase in capacity of the system is
A. 86.6% B. 66.7%
C. 73.2% D. 50%

407. For supplying a balanced 3-phase load of 40 kVA, rating of each transformer in V–V bank should be nearly
A. 20 kVA B. 23 kVA
C. 34.6 kVA D. 25 kVA

408. When a closed Δ bank is converted into an open Δ bank, each of the two remaining transformers supplies of the original load
A. 66.7% B. 57.7%
C. 50% D. 73.2%

409. If the load pf is 0.886, then the average pf of the V–V bank is
A. 0.886 B. 0.75
C. 0.51 D. 0.65

410. A T–T connection has higher ratio of utilization than a V–V connection only when
A. identical transformers are used
B. load power factor is leading
C. load power factor is unity
D. nonidentical transformers are used

411. The biggest advantage of T–T connection over the V–V connection for 3-phase power transformation is that it provides
A. a set of balanced voltages under load
B. a true 3-phase, 4-wire system
C. a higher ratio of utilization
D. more voltages

Ans. 394. D 395. C 396. B 397. A 398. B 399. A 400. D 401. C 402. C 403. A 404. A 405. C 406. B 407. B 408. B 409. B 410. D 411. D

412. Of the following statements concerning parallel operation of transformers, the one which is not correct is
A. transformers must have equal voltage ratings
B. transformers must have same ratio of transformation
C. transformers must be operated at the same frequency
D. transformers must have equal kVA ratings

413. **Statement**: An auto-transformer is more efficient in transferring energy from primary to secondary circuit.
Reason:because it does so both inductively and conductively.
Key
A. Statement is false, reason is correct and relevant
B. Statement is correct, reason is correct but irrelevant
C. Both statement and reason are correct and are connected to each other as cause and effect
D. Both statement and reason are false

414. Out of the following given choices for polyphase transformer connections which one will you select for three-to-two phase conversion?
A. Scott B. Star/star
C. Double Scott D. Star/double-delta

415. A T–T transformer cannot be paralleled with which transformer?
A. V–V B. Y–Δ
C. Y–Y D. Δ–Δ

416. In a three-phase delta transformers, one of the phase burns up. The transformer will supply
A. its full output rating B. zero output
C. 86.6% of its output rating
D. 63% of its output rating

417. In power transformers the tappings are provided on
A. high voltage side B. low voltage side
C. in the middle of both windings
D. LV as well as on HV side

418. The advantage of putting tappings at the phase ends of a transformer is to
A. obtain fine variation of voltage
B. operate with ease
C. reduce the number of bushings
D. obtain better regulation

419. The input power to a three-phase transformer is given by
A. $\sqrt{3} \times V_{ph} I_{ph} \cos\phi$ B. $3 \times V_{ph} I_{ph} \sin\phi$
C. $3 \times V_1 I_c \sin\phi$ D. $3 \times V_{ph} I_{ph} \cos\phi$

420. A three-phase transformer with delta-connected hv side and star-connected lv side gave the following symbol which also indicates phase displacement:
A. Dy 1 or Dy 11 B. Dy 0 or Dy 6
C. Dy 1 only D. Dy 11 only

421. A delta-zigzag three-phase transformer can have the following symbol:
A. Dz 0 or Dz 6 B. Dz 1 or Dz 11
C. Dz 0 only D. Dz 6 only

422. A three-phase transformer with star-connected hv side and delta connected lv side can have the following symbol which also indicates phase displacements:
A. Yd 0 only B. Yd 11 only
C. Yd 1 or Yd 11 D. Yd 0 or Yd 6

423. Consider the following statements:
I. Dy 1 and Yd 11 transformers can operate in parallel
II. Yd 1 and Yd 11 transformers can operate in parallel
III. Yd 1 and Dy 1 transformers can operate in parallel
IV. Yd 1 and Yz 1 can operate in parallel
Use the code below to indicate correct statements
A. II and III only B. I and II only
C. III and IV only D. All of these

424. A star-zig zag three-phase transformer with star connection on hv side can have the following symbol
A. Yz 0 or Yz 6 B. Yz 1 or Yz 11
C. Yz 1 only C. Yz 11 only

425. The utilisation factor for transformers connected in open-delta or double-delta is
A. 0.866 B. 0.75
C. 0.667 D. 0.5

426. The magnitude of each of the voltage phasors represented by sides of a regular hexagon in a 6-phase system produced by double-delta connection of the secondaries of transformers is *k* times the voltage of each secondary winding. The value of k is
A. 1/2 B. 1
C. 3/2 D. $\sqrt{3}$

427. An autotransformer having a transformation ratio of 0.8 supplies a load of 3 kW. The power transferred conductively from primary to secondary is
A. 0.6 kW B. 2.4 kW
C. 1.5 kW D. 0.27 kW

428. The magnitude of each of the voltage phasors represented by sides of a regular hexagon in a 6-phase system produced by double-star connection of the secondaries of transformers is *k* times the voltage of each secondary winding. The value of *k* is
A. 1/2 B. 1
C. 3/2 D. $\sqrt{3}$

Ans. 412. D 413. C 414. A 415. B 416. C 417. A 418. C 419. D 420. A 421. A 422. C 423. C 424. B 425. A 426. D 427. B 428. B

429. In 12-phase star-delta-double-star connection, voltage of each secondary star phase is k times the voltage on each side of the voltage polygon of 12 sides, the value of k being equal to
A. 1/2 sin 30°
B. 1/2 sin 15°
C. 1/2 sin 10°
D. I/sin 30°

430. In 12-phase double-chord connection, voltage of each chord is equal to k times the voltage on each side of the voltage polygon of 12 sides, the value of k being equal to
A. sin 75°/sin 15°
B. sin 60°/sin 12°
C. sin 75°/sin 20°
D. sin 60°/sin 15°

431. A 110 V, 50 Hz single phase supply is needed from a 220 V, 50 Hz source. Determine the ratios of weights of copper needed in a two-winding transformer and an autotransformer employed for the purpose
A. 2:1
B. 1:2
C. 3:1
D. 1:3

432. For short-circuit test on a transformer, the number and duration of tests on a phase are
A. 1 and 0.5 sec
B. 1 and 2 secs ± 10% tolerance
C. 3 and 2 secs ± 10% tolerance
D. 3 and 0.5 sec ± 10% tolerance

433. A two-winding transformer of voltage rating V_1/V_2 has the full-load voltage regulation of r pu at 0.8 pf lagging. What is the full-load regulation at 0.8 pf lagging if the autotransformer formed by joining the two windings in series has the voltage rating $V_1/(V_1+V_2)$?
A. r pu
B. $V_2/(V_1+V_2)\,r$ pu
C. $V_1/(V_1+V_2)\,r$ pu
D. $V_2/V_1\,r$ pu

434. The main purpose of performing open-circuit test at rated voltage is to measure
A. copper loss
B. core loss
C. total loss
D. insulation resistance

435. Two single-phase transformers A and B of identical voltage ratings and equal turn-ratios operate in parallel. The magnitudes of effective impedances of the transformers are equal while resistance reactance ratio of A is higher than that of B. At a certain load
A. the magnitude of I_A is lower than that of I_B and I_C lags I_B
B. the magnitudes of I_A and I_B are equal and I_A is in phase with I_B
C. the magnitudes of I_A and I_B are equal and I_A lags I_B
D. the magnitudes of I A and IB are equal and I_A leads I_B

436. A delta-star transformer with delta connection on the primary side operates in parallel with a star-delta transformer with star connection on the primary side. The ratio (primary turns per phase)/(secondary turns per phase) of the former is x times that of the latter while x is equal to
A. 1/3
B. $1/\sqrt{3}$
C. $\sqrt{3}$
D. 3

437. Consider the following statements concerning the utility of mesh-connected tertiary windings in star-star transformers:
I. It is used to suppress harmonic voltages
II. It is used to allow flow of earth fault current for operation of protective devices
III. It facilitates supply of single-phase loads
IV. It provides low-reactance paths for zero-sequence currents
Use the code below to indicate correct statements
A. All of the above statements are correct
B. Only I and II are correct
C. Only I, II and III are correct
D. Only I, II and IV are correct

438. In star-star transformers distortion in voltage wave shape due to third harmonic flux and voltage unbalance on single-phase loads are minimum if they are
A. three-phase shell type
B. three-phase five-limb core type
C. three-phase three-limb core type
D. banks of single-phase transformers

439. If V is the peak value of lightning impulse striking a terminal of a transformer winding of length l and $\alpha = \sqrt{C_g/C_s}$, where C_g = total earth capacitance of the winding; and C_s = total self capacitance of the winding, then the peak voltage gradient at that terminal of the winding is
A. $(V\alpha/l)\tanh\alpha$ if the other terminal is grounded and $(V\alpha/l)\coth\alpha$ if the other terminal is ungrounded
B. $(V\alpha/l)\coth\alpha$ if the other terminal is grounded and $(V\alpha/l)\tanh\alpha$ if the other terminal is ungrounded
C. $(V\alpha/l)\tanh\alpha$ in both the cases of grounded and ungrounded terminal at the other end
D. none of these

440. CRGO laminations in a transformer are used to minimise
A. eddy current loss
B. hysteresis loss
C. both eddy current and hysteresis loss
D. ohmic loss

441. To avoid breakdown of transformers under surges, any one of the following except one may by adopted.
A. Use of surge absorbers
B. Use of surge diverters
C. Use of metal shields
D. Reinforcement of end-turn insulation

Ans. 429. B 430. A 431. A 432. D 433. B 434. B 435. C 436. D 437. A 438. C 439. B 440. B 441. D

442. Which one of the following statements is incorrect?
A. The thermal ability of a transformer to withstand short circuit is demonstrated by calculation, while its dynamic ability to withstand short circuit is demonstrated by actual short circuit
B. The thermal ability of a transformer to withstand short circuit is demonstrated by calculation, while its dynamic ability of withstand short circuit is demonstrated by calculation
C. Both the thermal ability and the dynamic ability of a transformer to withstand short circuit are demonstrated by calculations
D. Both the thermal ability and dynamic ability of a transformer to withstand short circuit are demonstrated by actual short circuit

443. Winding to earth faults may be caused by
A. sludging of the oil B. surge voltage only
C. current chopping only
D. both surge voltages and current chopping

444. As per the Indian Standards Specifications, the wave-front and wave-tail times with tolerances of impulses for lightning impulse tests are respectively
A. 1 μs ± 30%, 50 μs ± 20%
B. 1.2 μs ± 20%, 50 μs ± 30%
C. 1.2 μs ± 30%, 50 μs ± 20%
D. 1.5 μs ± 30%, 40 μs ± 20%

445. The standard switching impulse as per ISI, has wave-front time and wave-tail time tolerances of
A. 200 μs ± 20%, 2000 μs ± 60% respectively
B. 250 μs ± 20%, 2500 μs ± 60% respectively
C. 250 μs ± 30%, 2500 μs ± 20% respectively
D. 300 μs ± 10%, 3000 μs ± 20% respectively

446. The voltage for induced potential tests on a transformer winding is k times its rated voltage and duration of test, if test frequency does not exceed twice the rated frequency, is s seconds. The values of k and s respectively are
A. 2, 30 B. 2, 60
C. 1.5, 60 D. 2, 15

447. In an autotransformer of voltage ratio V_1/V_2, $V_1 > V_2$, the fraction of power transferred inductively is
A. $V_1/(V_2+V_1)$ B. V_2/V_1
C. $(V_1-V_2)/(V_1+V_2)$ D. $(V_1-V_2)/V_1$

448. Distortion of the whole core and subsequent damage of the coil insulation due to intense local eddy currents may be caused by the failure of
A. the insulation between the laminations only
B. the insulation around the core damping bolts only
C. the insulation between the laminations and insulation around the core damping bolts
D. the insulation of the windings

449. Repeated switching of transformers located nearer the generating station may cause
A. surface flashover of the bushings
B. mechanical distortion of the windings and subsequent turn to turn insulation failure
C. excessive temperature rise
D. intense local eddy currents in the core

450. If the drying out operation of a transformer is unduly shortened and normal voltage is applied, then there is possibility of
A. inter-turn failure
B. winding of earth failure
C. terminal to terminal flashover
D. excessive overheating of the transformer

451. In a transformer where two or more dielectrics, having different permittivities are used in series, incorrect proportioning of their thicknesses may cause
A. insulation failure due to excessive electric stresses
B. very high dielectric loss
C. very high leakage current
D. very poor heat dissipation

452. A 3-phase supply can be converted into a 6-phase supply by joining the six secondaries of the 3-phase transformer in
A. double delta B. double star
C. diametrical D. any of these

453. The most commonly used connection for joining the six secondaries of a transformer used for 3-phase to 6-phase conversion is
A. diametrical B. zigzag
C. double-star D. double-delta

454. Malfunctioning of the Buchholz relay may be caused by
A. excessive overheating of the transformer
B. heavy external short circuit
C. dropping of the oil level below the relay level during operation
D. improper breathing action

455. Surface flashover of the bushing insulators are often caused by
A. moisture present in the transformer oil
B. pollution of the bushing surfaces
C. lining up of the suspended particles present in the oil
D. heavy external short circuit

456. The two windings of a transformer are designated as
A. primary and secondary windings
B. primary and hv windings
C. secondary and lv windings
D. hv and lv windings

Ans. 442. C 443. D 444. C 445. B 446. B 447. D 448. C 449. B 450. A 451. A 452. D 453. A 454. C 455. B 456. D

457. In double-star zig zag connection in the secondary side of transformers for producing 12-phase supply system, the two sets of voltages of the secondary winding are k_1 and k_2 times the voltage of each output terminal neutral. The values of k_1 and k_2 are respectively
A. sin 45°/sin 60°, sin 15°/sin 60°
B. sin 30°/sin 75°, sin 15°/sin 75°
C. sin (45°/2°)/sin 30°, sin (15°/2°)/sin 30°
D. sin 30°/sin 45°, sin 15°/sin 45°

458. In fuel cell, the ..... energy is converted into electrical energy
A. mechanical B. chemical
C. heat D. sound

459. Solar thermal power generation can be achieved by
A. using focusing collector or helistates
B. using flat plate collectors
C. using a solar pond
D. any of the above systems

460. Thermal efficiency of a gas turbine plant as compared to diesel engine plant is
A. higher B. lower
C. same
D. may be higher or lower

461. Mechanical efficiency of a gas turbine as compared to internal combustion reciprocating engine is
A. higher B. lower
C. same D. unpredictable

462. The function of a solar collector is to convert solar energy into
A. electricity B. radiation
C. thermal energy D. light energy

463. Most of the solar radiation received on earth surface lies within the range of
A. 0.2 to 0.4 microns B. 0.38 to 0.78 microns
C. 0 to 0.38 microns D. 0.55 to 0.85 microns

464. Flat plate collector absorbs
A. direct radiation only B. diffuse radiation only
C. both direct and diffuse radiation
D. none of these

465. Main applications of solar energy may be considered in the following categories
A. solar electric applications
B. direct thermal applications
C. fuel from biomass D. all of these

466. Fuel cells have conversion efficiencies of the order of
A. 20% B. 30%
C. 50% D. 70%

467. Fuel cell is a device in which
A. chemical energy is converted into electricity
B. heat energy is first converted into chemical energy
C. heat energy is converted into electricity
D. none of these

468. Temperature attained by a flat-plate collector is of the
A. order of about 90°C
B. range of 100 to 150°C
C. above 150°C D. none of these

469. A pyranometer is used for measurement of
A. direct radiation only B. diffuse radiation only
C. direct as well as diffuse radiation
D. none of these

MULTIPLE CHOICE QUESTIONS FROM VARIOUS COMPETITIVE EXAMINATIONS

1. A single-phase transformer when supplied from 220 V, 50 Hz has eddy current loss of 50 W. If the transformer is connected to a voltage of 330 V, 50 Hz, the eddy current loss will be **(IES 2001)**
A. 168.75 W B. 112.5 W
C. 75 W D. 50 W

2. In case of auto-transformers, which of the following statements are correct? **(IES 2001)**
1. An auto-transformer requires loss copper as compared to a conventional, 2-winding transformer of the same capacity
2. An auto-transformer provides isolation between the primary and secondary windings
3. An auto-transformer has less leakage reactance as compared to the conventional, 2-winding transformer of the same capacity

Select the current answer using the codes given below:
A. 1, 2 and 3 B. 1 and 2
C. 1 and 3 D. 2 and 3

3. If P_c and P_{sc} represent core and full-load ohmic losses respectively, the maximum kVA delivered to load corresponding to maximum efficiency is equal to rated kVA multiplied by **(IES 2001)**
A. P_c/P_{sc} B. $\sqrt{P_c/P_{sc}}$
C. $(P_c/P_{sc})^2$ D. $(P_{sc}/P_c)^2$

4. Generally the no-load losses of an electrical machine is represented in its equivalent circuit by a **(IES 2002)**
A. parallel resistance with a low value
B. series resistance with a low value
C. parallel resistance with a high value
D. series resistance with a high value

5. A two-winding transformer is used as an auto-transformer. The kVA rating of the auto-transformer compared to the two-winding transformer will be **(IES 2002)**
A. 3 times B. 2 times
C. 1.5 times D. same

Ans. 457. A 458. B 459. D 460. D 461. A 462. C 463. A 464. C 465. A 466. D 467. A 468. A 469. C
1. B 2. C 3. B 4. A 5. B

6. A 20 kVA, 2000/200 V, 1-phase transformer has name-plate leakage impedance of 8%. Voltage required to be applied on the high-voltage side to circulate full-load current with the low-voltage winding short-circuite will be (IES 2002)
A. 16 V B. 56.56 V
C. 160 V D. 568.68 V

7. The full-load copper-loss and iron-loss of a transformer are 6400 W and 5000 W respectively. The copper-loss and iron-loss at half load will be, respectively (IES 2002)
A. 3200 W and 2500 W B. 3200 W and 5200 W
C. 1600 W and 1250 W D. 1600 W and 5000 W

8. In a 100 kVA, 1100/220 V, 50 Hz single-phase transformer with 2000 turns on the high-voltage side, the open-circuit test result gives 220 V, 91 A, 5 kW on low-voltage side. The core-loss component of current is, approximately (IES 2002)
A. 9.1 A B. 22.7 A
C. 45.0 A D. 91 A

9. Possible three-to-three phase transformer connection for parallel operation is (IES 2002)
A. $\Delta - Y$ to $\Delta - Y$ B. $D - D$ to $\Delta - Y$
C. $Y - Y$ to $\Delta - Y$ D. $\Delta - Y$ to $Y - \Delta$

10. A 4 kVA, 400/200 V single-phase transformer has resistance of 0.02 pu and reactance of 0.06 pu. Its actual resistance and reactance referred to hv side are, respectively (IES 2002)
A. 0.2 ohm and 0.6 ohm
B. 0.8 ohm and 2.4 ohm
C. 0.08 ohm and 0.24 ohm
D. 2 ohm and 6 ohm

11. A delta/star transformer has a phase-to-phase voltage transformation ratio of K

$$K = \left(\frac{\text{Delta phase voltage}}{\text{Star phase voltage}}\right)$$

The line-to-line voltage ratio of star/delta connection is given by (IES 2002)
A. $K/\sqrt{3}$ B. K
C. $K\sqrt{3}$ D. $\sqrt{3}/K$

12. Two 10 kV/440 V, 1-phase transformers of ratings 600 kVA and 350 kVA are connected in parallel to share a load of 800 kVA. The reactances of the transformers, referred to the secondary side are 0.0198 Ω and 0.0304 Ω respectively (resistances negligible). The load shared by the two transformers will be, respectively (IES 2002)
A. 484.5 kVA and 315.5 kVA
B. 315.5 kVA and 484.5 kVA
C. 533 kVA and 267 kVA
D. 267 kVA and 533 kVA

13. Two transformers, with equal voltage ratio and negligible excitation current, connected in parallel, share load in the ratio of their kVA rating only, if their pu impedances (based on their own kVA) are (IES 2002)
A. equal
B. in the inverse ratio of their ratings
C. in the direct ratio of their ratings
D. purely reactive

14. The use of higher flux density in the transformer design is (IES 2003)
A. reduces the weight per kVA
B. increases the weight per kVA
C. has no relation with the weight of transformer
D. increases the weight per kW

15. The function of oil in a transformer is to provide (IES 2003)
A. insulation and cooling
B. protection against lightning
C. protection against short circuit
D. lubrication

16. Consider the following statements relating to the constructional features of a large power transformer (IES 2003)
1. The conservator is used to maintain the level of oil in the transformer tank
2. The bushing is used to protect trans-former insulation against lightning over-voltages
3. The Buchholz relay is an over current relay
4. Silica gel is used to absorb moisture
Which of these statements are correct?
A. 1, 2, 3 and 4 B. 2 and 3
C. 1 and 4 D. 1, 2 and 4

17. Two transformers operating in parallel will share the load depending upon their (IES 2003)
A. rating B. leakage reactance
C. efficiency D. per unit impedance

18. A 1 kVA, 200/100 V, 50 Hz, single-phase transformer gave the following test results on 50 Hz:
OC (LV side): 100 V, 20 watts
SC (HV side): 5 A, 25 watts
It is assumed that no-load loss components are equally divided. The above tests were then conducted on the same transformer at 40 Hz.
Tests results were:
OC (HV): 160 V, W_1 watts
SC (LV): 10 A, W_2 watts
Neglecting skin effect, W_1 and W_2 will be (IES 2003)
A. $W_1 = 16$ watts, $W_2 = 25$ watts
B. $W_1 = 25$ watts, $W_2 = 31.25$ watts
C. $W_1 = 20$ watts, $W_2 = 20$ watts
D. $W_1 = 14.4$ watts, $W_2 = 25$ watts

Ans. 6. C 7. D 8. B 9. A 10. B 11. D 12. A 13. A 14. A 15. A 16. C 17. D 18. D

19. If per unit impedances of two transformers connected in parallel are not equal, then which one of the following statements is correct? **(IES 2004)**
   A. The power factor of the two transformers will be different from that of the connected load
   B. Transformers will get overloaded
   C. Dead short circuit occurs
   D. The transformer with higher per unit impedance will share more load

20. The exact equivalent circuit of a two-winding transformer is given in the figure given below. For affecting simplification, the parallel magnetising branch, consisting of R_c and X_Q is shifted to the left of the primary leakage impedance $(r_1 + jx_1)$. This simplification introduces the inaccuracy, in the neglect of **(IES 2005)**

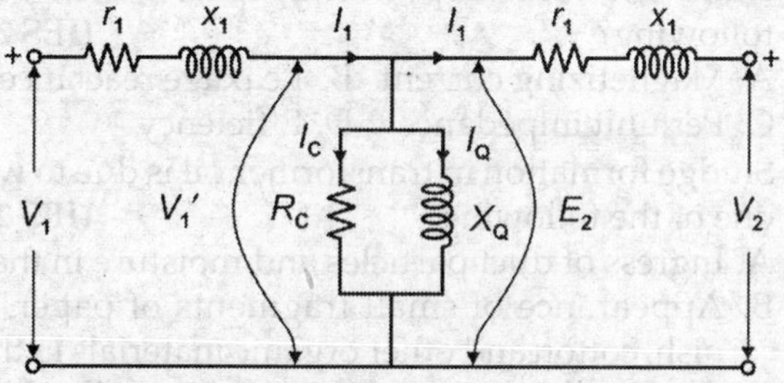

   A. voltage-drop in the primary impedance due to the secondary current
   B. voltage-drop in the primary impedance due to the exciting current
   C. voltage-drop in the secondary impedance due to the exciting current
   D. reduction in values of R_C and R_Q of the exciting circuit

21. A single-phase, 10 kVA, 2000/200 V, 50 Hz transformer is connected to form an auto transformer as shown in the figure given below. What are the values of V_1 and I_2, respectively? **(IES 2005)**

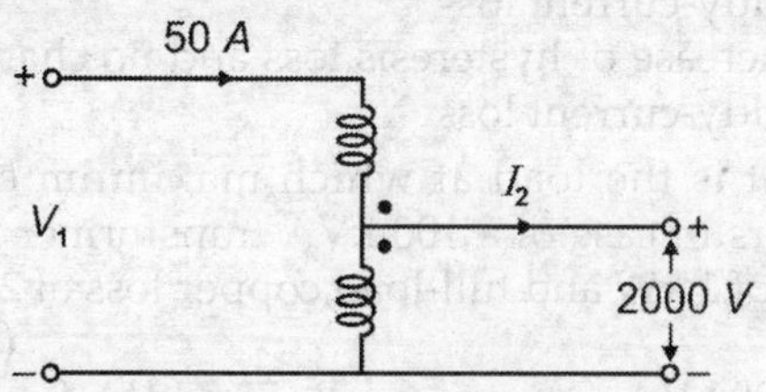

   A. 2200 V, 55 A B. 200 V, 45 A
   C. 2000 V, 45 A D. 1800 V, 45 A

22. Three single phase 11000/220 V transformers are connected to form 3-phase transformer bank. High voltage side is connected in star, and low voltage side is in delta. What are the voltage ratings and turn ratio of 3-phase transformer? **(IES 2005)**
   A. 19052/220 V, 50 B. 19052/220 V, $50\sqrt{3}$
   C. 11000/381 V, $50\sqrt{3}$ D. 11000/220 V, 50

23. Two single phase transformers A and B are connected in parallel, observing all requirements of a parallel operations except that the induced voltage E_a is slightly greater than E_b; Z_{ea} and Z_{eb} being the equivalent impedances of A and B, both referred to the secondary side.

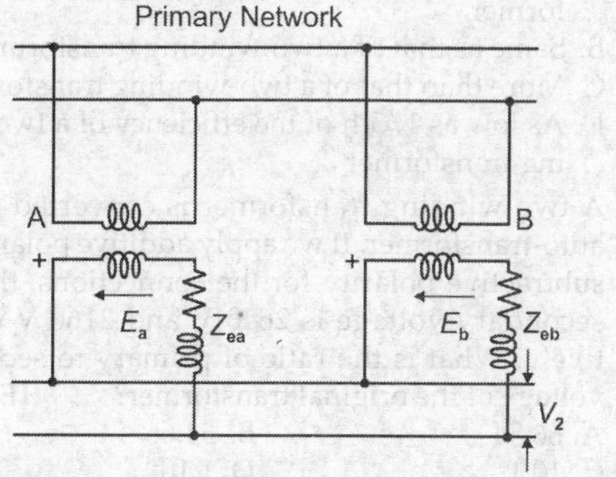

Under this operating condition with the primary bus-bars being energised, a circulating current will flow **(IES 2005)**
   A. only in the secondary windings of A and B
   B. in both the primary and the secondary windings of A and B
   C. in both the primary and the secondary windings of A and B, as well as in the primary side network
   D. in the primary and the secondary windings of A and B and boost the voltages on the secondary side of both A and B

24. A starting torque of 80 Nm is developed in an induction motor by an auto-transformer starter with a tapping of 30%. If the tapping of auto-transformer starter is 60%, then what is the starting torque? **(IES 2005)**
   A. 40 Nm B. 160 Nm
   C. 240 Nm D. 320 Nm

25. What is the phase displacement between primary and secondary voltages for a star-delta, 3-phase transformer connected shown below? **(IES 2005)**

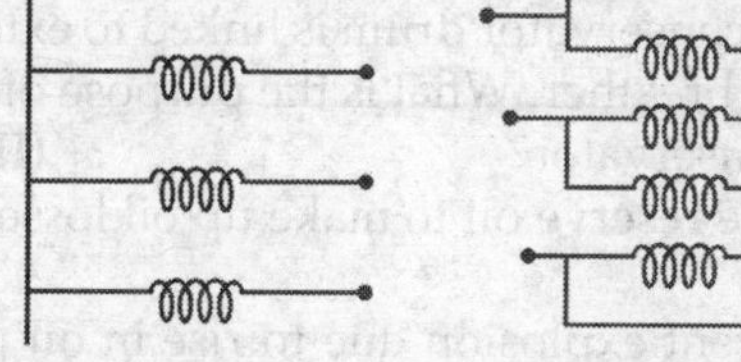

   A. 30° lagging B. 30° leading
   C. 0° D. 180°

26. Which three-phase connection can be used in a transformer to introduce a phase difference of 30° between its output and corresponding input line voltages? **(IES 2006)**

Ans. 19. A 20. B 21. D 22. B 23. B 24. D 25. A

A. Star-delta B. Star-star
C. Delta-delta D. Delta-zigzag

27. What is the efficiency of an auto-transformer in comparison to that of a two-winding transformer of the same rating? **(IES 2006)**
A. Slightly less than that of a two-winding transformer
B. Same as that of a two-winding transformer
C. More than that of a two-winding transformer
D. As low as 1/5th of the efficiency of a two winding transformer

28. A two-winding transformer is converted into an auto-transformer. If we apply additive polarity and subtractive polarity for the connections, then the secondary voltage is 2640 V and 2160 V respectively. What is the ratio of primary to secondary voltage of the original transformer? **(IES 2006)**
A. 66:54 B. 54:66
C. 10:1 D. 1:10

29. If P_1 and P_2 be the iron and copper losses of a transformer at full load, and the maximum efficiency of the transformer is at 75% of the full load, then what is the ratio of P_1 and P_2? **(IES 2006)**
A. 9/16 B. 10/16
C. 3/4 D. 3/16

30. If the iron core of a transformer is replaced by an air core, then the hysteresis losses in the transformer will **(IES 2006)**
A. increase B. decrease
C. remains unchanged D. become zero

31. If the voltage applied to a transformer primary is increased by keeping the V/f ratio fixed, then the magnetizing current and the core loss will, respectively **(IES 2006)**
A. decrease and remain the same
B. increase and decrease
C. remain the same and remain the same
D. remain the same and increase

32. In a large power transformer, a "conservator" drum is provided above the completely oil filled transformer tank and connected to it by a short pipe. The conservator drum is linked to external air through a breather. What is the purpose of providing the conservator? **(IES 2007)**
A. To store reserve oil to make up oil losses due to leakage
B. to prevent explosion due to rise in oil pressure inside the tank during a fault
C. to accommodate change in oil-level during the "load-cycle" of the transformer load
D. to exert additional pressure by the conservator-oil on the oil inside the main tank to prevent disintegration of oil at high temperature

33. A 500 kVA transformer has constant losses of 500 W and copper losses at full load are 2000 W. Then at what load is the efficiency maximum? **(IES 2007)**
A. 250 kVA B. 500 kVA
C. 1000 kVA D. 125 kVA

34. On which of the following factors does hysteresis loss depend? **(IES 2007)**
1. Flux density
2. Frequency
3. Thickness of lamination
4. Time
Select the correct answer using the code given below:
A. 2 and 3 B. 1 and 2
C. 3 and 4 D. 1 and 4

35. Two transformers when operating in parallel will share the load depending upon which of the following? **(IES 2007)**
A. Magnetizing current B. Leakage reactance
C. Per unit impedance D. Efficiency

36. Sludge formation in transformer oil is due to which one of the following? **(IES 2008)**
A. Ingress of duct particles and moisture in the oil
B. Appearance of small fragments of paper, varnish, cotton and other organic materials in the oil
C. Chemical reaction of transformer oil with the insulating materials
D. Oxidation of transformer oil

37. A single-phase transformer rated for 220/440 V, 50 Hz operates at no load at 220 V, 40 Hz. This frequency operation at rated voltage results in which one of the following? **(IES 2008)**
A. Increase of both eddy-current and hysteresis losses
B. Reduction of both eddy-current and hysteresis losses
C. Reduction of hysteresis loss and increase in eddy-current loss
D. Increase of hysteresis loss and no change in the eddy-current loss

38. What is the load at which maximum efficiency occurs in case of a 100 kVA transformer with iron loss of 1 kW and full-load copper loss of 2 kW? **(IES 2008)**
A. 100 kVA B. 70.7 kVA
C. 50.5 kVA D. 25.2 kVA

39. Cores of large power transformers are made from which one of the following? **(IES 2008)**
A. Hot-rolled steel
B. Cold-rolled non-grain oriented steel
C. Cold-rolled grain oriented steel
D. Ferrite

40. A transformer has a percentage resis-tance of 2% and percentage reactance of 4%. What are its

Ans. 26. A 27. C 28. C 29. A 30. D 31. D 32. C 33. A 34. B 35. C 36. D 37. D 38. B 39. C

regulations at power factor 0.8 lagging and 0.8 leading, respectively? **(IES 2008)**

A. 4% and –0.8% B. 3.2% and –1.6%
C. 1.6% and –3.2% D. 4.8 and –0.6%

41. The full-load copper loss and iron loss of a transformer are 6400 W and 5000 W respectively. What are the above copper loss and iron loss, respectively at half-load? **(IES 2008)**

A. 3200 W, 2500 W B. 3200 W, 5000 W
C. 1600 W, 1250 W D. 1600 W, 5000 W

42. **Assertion (A):** Both the efficiency and regulation of a 3-winding ideal transformer are 100%
Reason (R): The flux leakage and the magnetic reluctance of the magnetic core in an ideal transformer are zero. Moreover, losses are absent in ideal transformers. **(IES 2008)**

A. Both A and R are true and R is the correct explanation of A
B. Both A and R are true but R is not the correct explanation of A
C. A is true but R is false
D. A is false but R is true

43. When are eddy-current losses in transformer reduced? **(IES 2008)**

A. If laminations are thick
B. If the number of turns in primary winding is reduced
C. If the number of turns in secondary winding is reduced
D. If laminations are thin

44. Consider the following tests:
1. Load test 2. Short circuit test
3. OC test 4. Retardation test
Which of the above tests are to be conducted for the determination of voltage regulation of a transformer? **(IES 2009)**

A. 1 only B. 2 only
C. 2 and 3 D. 3 and 4

45. What is the power transmitted inductively in an auto-transformer which supplies a load at 161 volts with an applied primary voltage of 230 volts? **(IES 2009)**

A. 35% of the input B. 70% of the input
C. 15% of the input D. 30% of the input

46. What is the power transferred conductively from primary to secondary of an auto-transformer having transformation ratio of 0.8 supplying a load of 3 kW? **(IES 2009)**

A. 0.6 kW B. 2.4 kW
C. 1.5 kW D. 0.27 kW

47. What is the core loss in a high frequency ferrite core transformer used in SMPS power supply? **(IES 2009)**

A. 10% of rated power B. 5% of rated power
C. 2% of rated power D. 1% of rated power

48. At which condition of the transformer, the equivalent circuit will be as shown in the below figure? **(IES 2009)**

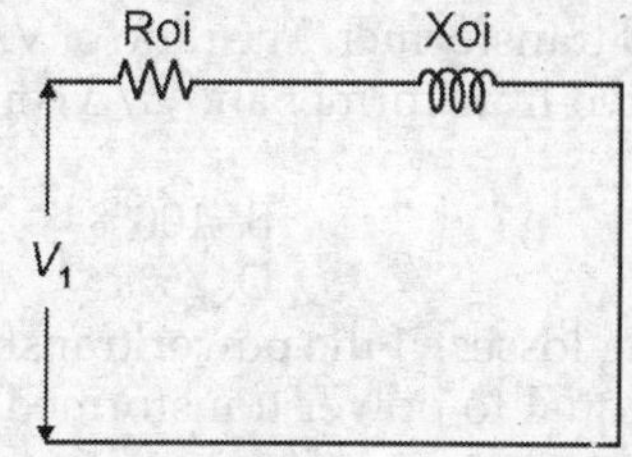

A. Under short circuit B. Under open circuit
C. Under no load D. Under rated load

49. At which condition of the transformer, the equivalent circuit will be as shown below? **(IES 2009)**

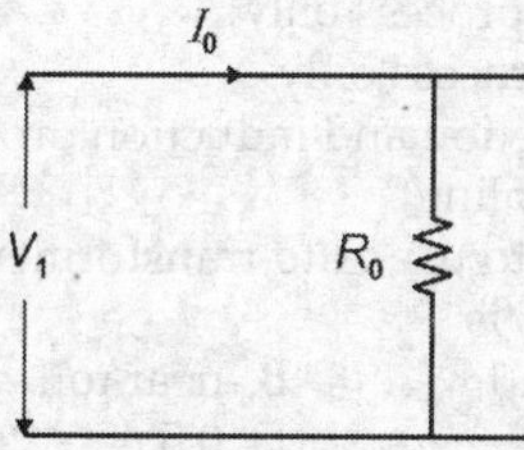

A. Under short circuit B. Under rated load
C. Under open circuit
D. Under load and no load

50. For the parallel operation of transformers, which of the following conditions must be satisfied? **(IES 2010)**

A. Same voltage ratios
B. Must be connected in proper polarities
C. $\frac{R_e}{X_e}$ ratio should be same
D. Same as kVA rating

51. For Δ/Y transformer, connections and terminal markings are shown in the figure below. If 1 and 2 represents positive and negative sequence, then the correct solution in per unit values for Va_1 and Va_2 is **(IES 2010)**

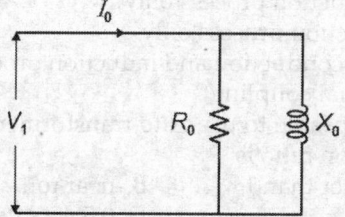

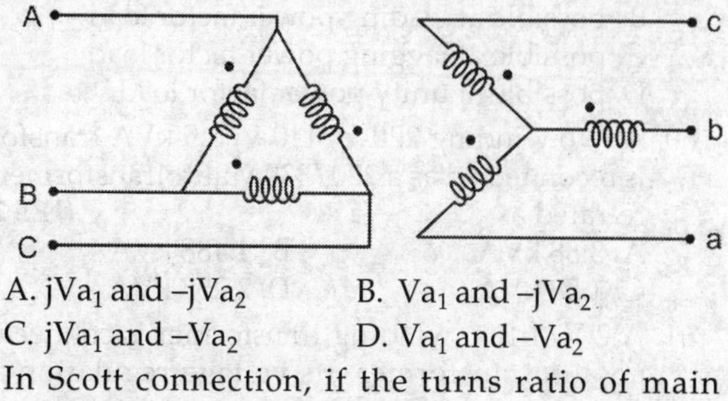

A. jVa_1 and $-jVa_2$ B. Va_1 and $-jVa_2$
C. jVa_1 and $-Va_2$ D. Va_1 and $-Va_2$

52. In Scott connection, if the turns ratio of main transformer is K, then the teaser transformer has turns ratio of **(IES 2010)**

Ans. 40. A 41. D 42. D 43. D 44. C 45. D 46. B 47. D 48. A 49. C 50. B 51. D

A. $\frac{2K}{\sqrt{3}}$ B. $\frac{\sqrt{3}K}{2}$

C. $\frac{K}{\sqrt{3}}$ D. $\frac{K}{2}$

53. A Δ/Δ connected transformer is connected to V/V connected transformer. The ratio of VA rating of V/V connected transformer and Δ/Δ connected transformer is **(IES 2010)**

A. 57.7% B. 100%
C. 50% D. 75%

54. Neglecting losses, if the power transformed inductively is equal to power transformed conductively in an auto-transformer is **(IES 2010)**

A. 0.5 B. 2
C. 1.5 D. 1.25

55. In auto-transformer, power is transferred through **(IES 2010)**

A. conduction process only
B. induction process only
C. both conduction and induction processes
D. mutual coupling

56. It is advisable to use auto-transformer if the transformation ratio is **(IES 2010)**

A. greater than 1 B. near to 1
C. 0.25 D. 0.5

57. A kVA, 400 V/200 V single phase transformer has resistance of 0.02 pu and reactance of 0.06 pu. The resistance and reactance referred to high voltage side are **(IES 2011)**

A. 0.2 Ω and 0.6 Ω B. 0.8 Ω and 2.4 Ω
C. 0.08 Ω and 0.24 Ω D. 1 Ω and 3 Ω

58. The full-load copper loss and iron loss of a transformer are 6400 W and 500 W respectively. The above copper loss and iron loss at half load will be **(IES 2011)**

A. 33200 W and 250 W respectively
B. 3200 W and 500 W respectively
C. 1600 W and 125 W respectively
D. 1600 W and 500 W respectively

59. In a transformer, zero voltage regulation at full load is **(IES 2011)**

A. not possible
B. possible at leading power factor load
C. possible at lagging power factor load
D. possible at unity power factor load

60. A two winding 220 V/110 V, 1.5 kVA transformer is reconnected as a 220/330 V autotransformer. It is re-rated as **(IES 2011)**

A. 3.88 kVA B. 4.488 kVA
C. 1.58 kVA D. 2.258 kVA

61. A 2 V/1 two winding transformer is connected as an autotransformer. Its voltage regulation as an autotransformer compared to the two winding transformer is **(IES 2011)**

A. the same B. 1.5 times
C. 2 times D. 3 times

62. When one transformer is removed from a Δ-Δ bank of 30 kVA transformer, the capacity of the resulting 3-phase transformer corresponding input line voltages? **(IES 2011)**

A. 11.5 kVA B. 17.3 kVA
C. 20 kVA D. 25.9 kVA

63. Which 3-phase connection can be used in a transformer to introduce a phase difference of 30° between its output and corresponding input line voltages? **(IES 2011)**

A. Star-star B. Star-delta
C. Delta-delta D. Delta-zigzag

64. Conditions for parallel operation of transformers are
1. the same voltage ratio
2. the same per unit impedance
3. the same polarity
4. the same phase sequence
5. the relative phase displacement

Which are the absolutely essential conditions? **(IES 2011)**

A. 1, 2, 3, 4 and 5 B. 2, 3 and 4
C. 3, 4 and 5 D. 1, 2 and 3

65. A 200 V/100 V, 50 Hz transformer is to be excited at 40 Hz from 100 V side. For the exciting current to be the same, the applied voltage should be **(IES 2012)**

A. 150 V B. 80 V
C. 100 V D. 125 V

66. A 10 kVA, 2200/220 V transformer gave the following test results: Open circuit test, High voltage side open: V = 220 V, I = 1.5 A, W = 150 W. Short-circuit test. Low voltage side short-circuited: V = 115 V, I = rated, W = 200 W. The half full-load efficiency of the transformer operating at unity power factor is **(IES 2012)**

A. 95% B. 95.5%
C. 96% D. 96.5%

67. Consider the following statements:
The open-circuit test in a transformer can be used to obtain **(IES 2012)**
1. core losses
2. magnitude of exciting current
3. copper losses
4. equivalent series impedance

Correct statements are

A. 1, 2, 3 and 4 B. 1 and 3 only
C. 1 and 2 only D. 2 and 4 only

68. A 100 kVA, 2400 V/240 V, Hz single phase transformer has an exciting current of 0.64 A and core loss 700 W when its high voltage side is energised at rated voltage and frequency. If load current is

Ans.	52. B	53. A	54. A	55. C	56. B	57. B	58. D	59. A	60. B	61. D	62. B	63. D	64. C
	65. B	66. C	67. C										

40 A at 0.8 pf lagging on the LV side, then magnitude of the primary current will be **(IES 2012)**

A. 4.58 A B. 4 A
C. 4.64 A D. 4.85 A

69. In which one of the following sets of 3-phase transformer connections with zero sequence current be present in the transformer windings? **(IES 2012)**
A. Primary in star, neutral grounded; secondary in star, neutral not grounded
B. Primary in star, neutral grounded; secondary in delta
C. Primary in star, neutral not grounded; secondary in star, neutral grounded
D. Primary in star, neutral not grounded; secondary in delta

70. When a transformer is first energised, the transient current during first few cycles is **(IES 2013)**
A. less than full load current
B. equal to full load current
C. equal to no load current
D. much higher than full load current

71. For power transformers of larger ratings, the tapping are located in the middle portion of the winding to **(IES 2013)**
A. increase the bradkdown strength of the winding insulation
B. enable better cooling of the windings
C. enable better distribution of inter-turn voltage
D. reduce the mechanical forces affecting the windings during short-circuits

72. Tapping are normally provided on the high voltage winding of a transformer only, because **(IES 2013)**
A. it has larger number of turns which allows smoother variation of voltage
B. it has to handle low currents
C. it is easily accessible physically
D. it has larger number of turns, has to handle low currents and is also easily accessible physically

73. Short circuit test is performed on a transformer with a certain impressed voltage at rated frequency. If the short circuit test is now performed with the same magnitude of impressed voltage, but at a frequency higher than the rated frequency, then the magnitude of current **(IES 2013)**
A. and power-factor will both increase
B. will decrease but the power factor will increase
C. will increase, but power factor will decrease
D. and power factor will both decrease

74. The impedance of a Δ/Y, 11000 V/400 V, transformer of capacity 100 kVA, on its name plate data base is (0.02 + j 0.07) pu. The ohmic impedance pu phase referred to the primary (11000 V) side is **(IES 2013)**
A. (0.02 + j 0.07) Ω B. (0.55 + j 1.925) Ω
C. (42 + j 147) Ω D. (72.6 + j 254.1) Ω

75. A 2000 kVA, single-phase transformer is in circuit continuously. For 8 hours in a day, the load is 160 kW at 0.8 pf. For 6 hours, the load is 80 kW at unity pf and for the period remaining out of 24 hours, it turns on no-load. If the full-load copper losses are 3.02 kW, the total copper losses in 24 hours are **(IES 2013)**
A. 35.62 kW B. 24.16 kW
C. 11.46 kW D. none of these

76. A 50 kVA transformer has a core loss of 500 W and full-load core loss of 900 W. The load at which the efficiency is maximum is **(IES 2013)**
A. 27.45 kVA B. 37.75 kVA
C. 45.5 kVA D. 47.5 kVA

77. For a two-winding power transformer, with the effects of the no-load current being neglected, the 'Voltage Regulation' can be zero at load, when the load power factor is **(IES 2013)**
A. lagging only B. leading only
C. either lagging or leading, depending upon power rating of the transformer
D. unity

78. The ratio of primary/secondary voltages is 2:1. The saving in terms of weight of copper required if an autotransformer is used instead of a two-winding transformer will be **(IES 2013)**
A. 95% B. 66.7%
C. 50% D. 33.3%

79. Consider the following statements concerning the utility of mesh-connected tertiary windings in star-star transformers:
1. it is used to suppress harmonic volt-ages
2. it is used to allow flow of earth fault current for operation of protective devices
3. it facilitates supply of single-phase loads
4. it provides low-reactances paths for zero sequence currents

Which of these statements are correct? **(IES 2013)**
A. 1, 2, 3 and 4 B. 1, 2 and 3 only
C. 1, 2 and 4 only D. 3 and 4 only

80. The most essential condition for parallel operation of two 1-ϕ transformers is that they should have the same **(IES 2013)**
A. kVA rating B. percentage impedance
C. polarity D. voltage ratio

81. Two-single phase 100 kVA transformer each having different leakage impedance are connected in parallel. When a load of 150 kVA at 0.8 power factor lagging is applied **(IES 2013)**
A. both transformers will operate at power factor more than 0.8 lagging

Ans. 68. A 69. C 70. C 71. D 72. D 73. D 74. D 75. D 76. B 77. B 78. C 79. A 80. C

B. both transformers will operate at power factor less than 0.8 lagging
C. one of the transformers will operate at power factor more than 0.8 lagging
D. both transformers will operate at identical power factors

82. A 200/100 V, 50 Hz transformer is to be excited at 40 Hz from the 100 V side. For the exciting current to remain same, the applied voltage should be **(IES 2014)**

A. 150 V
B. 125 V
C. 100 V
D. 80 V

83. A single-phase two winding transformer is designed to operate at 400/200 V, 50 Hz. If the hv side is now energised from a 400 V, 40 Hz source, the no-load L.v. side voltage would be **(IES 2014)**

A. 300 V
B. 250 V
C. 200 V
D. 150 V

84. A 100 VA, 120/12 V transformer is to be connected so as to form a step-up transformer. A primary voltage of 120 V is applied to the transformer. What is the secondary voltage of the transformer? **(IES 2014)**

A. 1.2 V
B. 12 V
C. 120 V
D. 132 V

85. If a transformer, if the iron losses and copper losses are 40.5 kW and 50 kW respectively, then at what fraction of load will efficiency be maximum? **(IES 2014)**

A. 0.80
B. 0.57
C. 0.70
D. 0.90

86. In a transformer the core loss is 100 Watt at 40 Hz and 72 Watt at 37 Hz, then eddy current and hysteresis losses at 50 Hz respectively are **(IES 2014)**

A. 25 Watt and 105 Watt
B. 20 Watt and 100 Watt
C. 100 Watt and 32 Watt
D. 32 Watt and 100 Watt

87. The voltage regulation of a transformer having 2% resistance and 5% reactance, at full load, 0.8 pf lagging is **(IES 2014)**

A. 4.6%
B. –4.6%
C. –1.4%
D. 6.4%

88. If the percentage impedance of the two transformers working in parallel are different then **(IES 2014)**

A. transformers will be overheated
B. power factor of both the transformers will be same
C. parallel operation will not be possible
D. parallel operation will still be possible

89. In the core-type twowinding transformer, the low voltage winding is placed adjacent to the steel core, in order to **(IES 2014)**

A. facilitate dissipation ofheat during the operation of the transformer
B. minimise the amount of insulation required
C. reduce the chances of axial displacement with respect to the high voltage winding placed outside
D. reduce the mutual radial stress between the two windings

Ans. 81. C 82. D 83. C 84. D 85. D 86. A 87. A 88. D 89. B

CHAPTER 8

Rotating Machines—DC Machines

Rotating Machines

The production of translational motion in an electromagnetic system has been discussed in conventional methods. However, most of the energy converters, particularly the higher power one, produce rotational motion. The essential part of the rotating electromagnetic system is shown in Fig. 8.1. The fixed part of the magnetic system is called stator and the moving part is called the rotor. The latter is mounted on a shaft and is free to rotate between the poles of the stator.

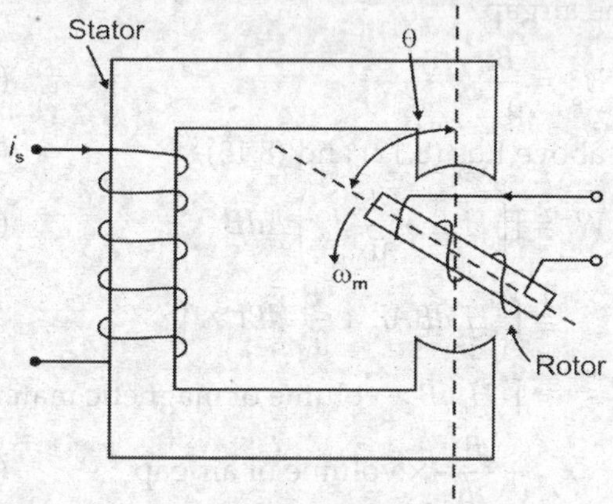

Fig. 8.1

Consider a general case in which both stator and rotor have windings carrying current as shown in Fig. 8.1, the current can be fed into the rotor through fixed brushes and rotor mounted slip rings.

The torque developed in a rotational electromagnetic system in terms of stored field energy W_f is given by

$$T = \frac{\partial W_f'(i, \theta)}{\partial \theta} \qquad [\because i = \text{constant}]$$

In a linear magnetic system, torque developed is given by:

$$T = \frac{1}{2} i_s^2 \frac{dL_{ss}}{d\theta} + \frac{1}{2} i_r^2 \frac{dL_{rr}}{d\theta} + \frac{i_s i_r dL_{sr}}{d\theta}$$

First two terms of the torque equation represents the torques produced in the machines, called *reluctance torque*. The third term represents torque produced by variation of mutual inductance between stator and rotor windings.

Electromagnetic conversion: Three electrical machines (i) dc machine (ii) induction machine (iii) synchronous machine, are used extensively for electromechanical energy conversion. In these machines conversion of energy from electrical to mechanical or vice-versa results from the following two electromagnetic phenomena:

i. When a conductor moves in a magnetic field, voltage is induced in the conductor.
ii. When a current carrying conductor is placed in a magnetic field, the conductor experiences a mechanical force.

These two effects occur simultaneously: Whenever energy conversion takes place from electrical to mechanical or vice-versa.

Electromechanical energy conversion: Various devices can convert electrical energy to mechanical energy and vice-versa. Some devices are used for continuous energy conversion, and these are known as *motors and generators*. Other devices are used to produce translational forces whenever necessary and are known as *actuators*, such as solenoids, relays and electromagnets.

Energy conversion process: There are various methods for calculating the force or torque developed in an energy conversion device based on the principle which states that energy can neither be created nor destroyed, it can only be changed from one form to another. Electromechanical converter system has three essential parts: (a) An electrical system (b) a mechanical system (c) coupling field as shown in Fig 8.2.

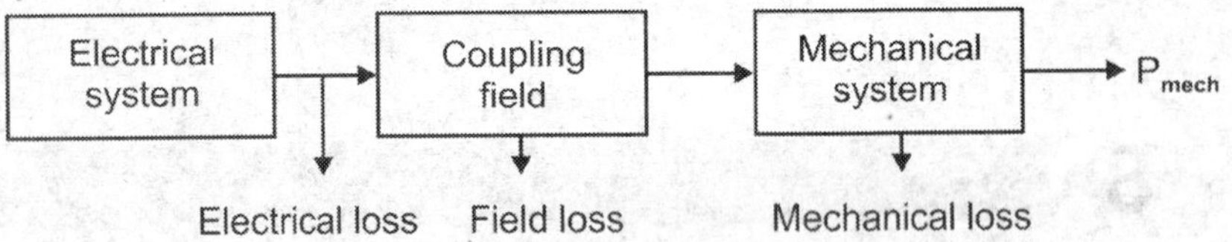

Fig. 8.2: Electromechanical system

All the losses are converted to heat. The energy balance equation can be given by:

Electrical energy input – resistance loss
= Mechanical energy output + windage and frictional loss + Increase in stored field energy
+ core losses (8.1)

Now consider a differential time interval dt during which an increments of electrical energy excluding resistance loss flows into the system. In differential form, Eq. (8.1) can be expressed as

$$dW_e = dW_m + dW_f \tag{8.2}$$

If friction and wintage losses are neglected, then all of dW_m will be available as useful mechanical output. The losses do not contribute to the energy conversion process.

Field energy: Consider the electromechanical system of the Fig. 8.3(a). The movable part can be held in the static equilibrium by spring. Let the movable part is held stationary at same air gap and the current is increased from zero to a value i. Flux will be established in the magnetic system, obviously,

$$dW_m = 0 \tag{8.3}$$

and from Eqs (8.2) and (8.3)

$$dW_c = dW_f \tag{8.4}$$

If the core loss is neglected, all the incremental electrical energy input is stored as incremental field energy.

$$\text{Now} \quad e = \frac{d\lambda}{dt} \tag{8.5}$$

$$dW_e = e \cdot idt \tag{8.6}$$

from Eqs (8.4), (8.5) and (8.6)

$$dW_i = id\lambda \tag{8.7}$$

The relationship between coil flux linkages λ and current i for a particular air gap length is shown in Fig. 8.3(b). The incremental field energy dW_i is shown as the cross hatched area in the figure. When the flux linkages is increased from zero to λ. The energy stored in the field is:

$$W_f = \int_0^{\lambda} i\, d\lambda \tag{8.8}$$

Expression for field energy of the magnetic system can be derived as follows: Let H_c and H_g are the magnetic intensity in the core material and air gap.

$$\text{Thus } Ni = H_c l_c + H_g l_g \tag{8.9}$$

$$\text{Also } \lambda = N\Phi \tag{8.10}$$

$$= NAB \tag{8.11}$$

where A is the cross-sectional area of flux path and B is the flux density, assumed same throughout from Eqs (8.8), (8.9) and (8.11)

$$W_f = \int \frac{H_c l_c + H_g l_g}{N} N A \mathbf{dB} \tag{8.12}$$

For the airgap

$$H_g = \frac{B}{\mu_g} \tag{8.13}$$

From above Eqs (8.12) and (8.13)

$$W_f = \int \left(H_c l_c + \frac{B}{\mu_0} l_g \right) A \mathbf{dB} \tag{8.14}$$

$$= \int \left(H_c dBAl_c + \frac{B}{\mu_0} dBl_g A \right)$$

$$= \int (H_c dB \times \text{volume of magnetic material}$$

$$+ \frac{B^2}{2\mu_0} \times \text{volume of air gap} \tag{8.15}$$

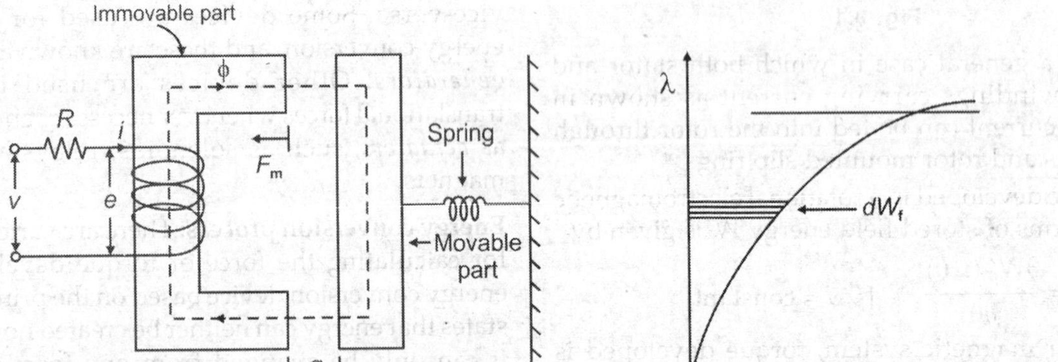

Fig. 8.3: Electromechanical System

$$W_f = w_{fc} \times V_c + w_{fg} \times V_g \quad (8.16)$$
$$= W_{fc} + W_{fg} \quad (8.17)$$

where

$w_{fc} = \int H_c dB_c$ energy density in the magnetic material

$w_{fg} = B^2/2\mu_0$ energy density in the air gap

V_c volume of the magnetic material

V_g volume of the air gap

W_{fc} energy in the magnetic material

W_{fg} energy in the air gap

For a linear magnetic system

$$H_c = B_c/\mu_c \quad (8.18)$$

Therefore

$$w_{fc} = \int \frac{B_c}{\mu_c} dB_c = \frac{B_c^2}{2\mu_c}$$

Normally energy stored in the air gap W_{fg} is much larger than the energy stored in the magnetic material W_{fc}.

Basic Concepts of Rotating Machines

In rotating machines, there are two parts, the stator and the rotor. Rotating electrical machines are also of two types: (i) DC machines (ii) AC machines. The classification of rotating machines is shown in Fig. 8.4.

DC machines: A DC machine have the following parts:

i. Magnetic frame or yok
ii. Pole core or pole shoe
iii. Armature core
iv. Armature winding or conductors
v. Commutator
vi. Brushes and bearings

Types of winding: The following are the classification of windings depending on the connection of their coil ends: (i) lap winding (ii) wave winding.

DC Generator

DC generator operates on the principle of the production of dynamically induced emf. To produce dynamically induced emf, a magnetic field and a conductor moving with respect to the field are essential. The current induced in the coil is collected and conveyed to the external load circuit by connecting the coil terminals to two continuous and insulated rings known as *slip rings*.

DC machine mainly consists of four parts as shown in Fig. 8.5, field magnets, armature, commutator and brushes.

The emf generated within the armature of a dc generator,

$$E_R = \frac{\Phi Z N}{60}\left(\frac{P}{A}\right) \text{Volt}$$

where A = Number of parallel paths in armature.
= 2 for wave winding.
= P for lap winding.
N = Speed in rpm.
Z = Number of conductors in the armature.
F = Flux/pole in Wb.

Condition for Maximum Efficiency

Armature copper loss = Iron loss

The load current corresponding to maximum efficiency is given by

$$I = \sqrt{\frac{W_{Cu}}{R_a}}$$

where W_{Cu} is the copper losses and R_a is armature resistance.

Characteristics of DC Generators

The three important characteristics of a dc machine are shown in Fig. 8.6.

1. **Magnetic open-circuit characteristics:** These show the relation between no-load generated emf in armature and field current at a given fixed speed.

2. **External characteristics:** These curves represent the relationship between the load current and terminal voltage.

3. **Internal or total characteristics:** These characteristics represent the relation between generated emf and armature current.

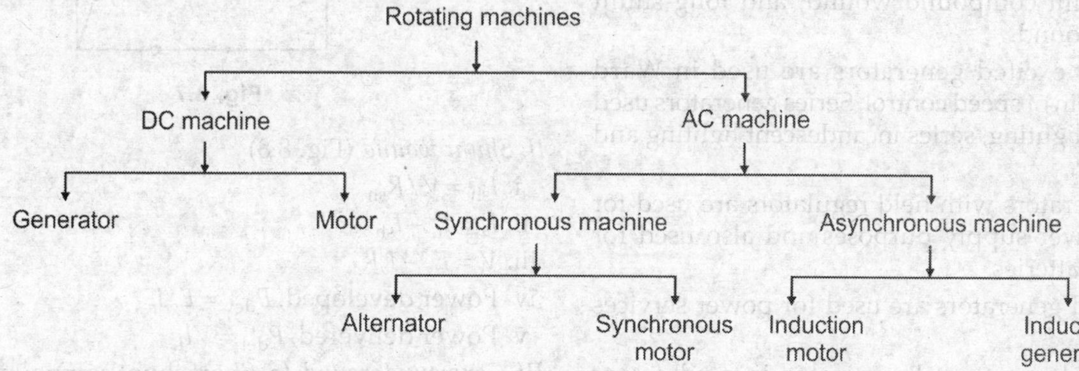

Fig. 8.4

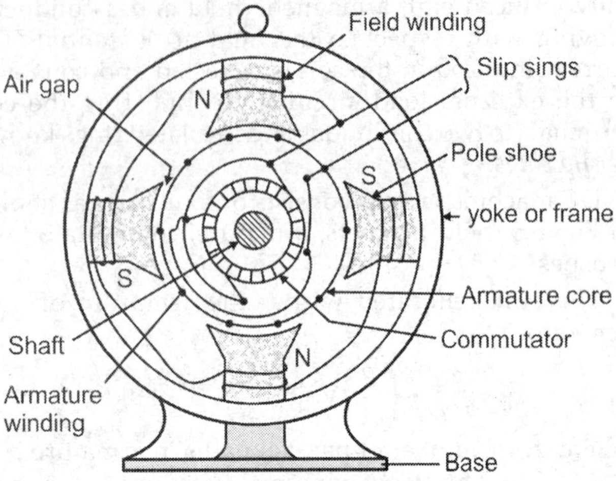

Fig. 8.5

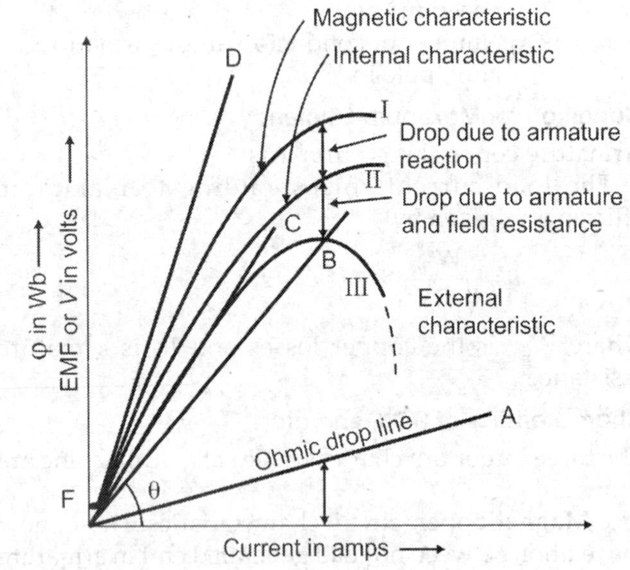

Fig. 8.6

Critical resistance is the maximum resistance with which the generator can be excited.

Connection of different DC generators: Important types of dc generators are separately-excited, series wound, shunt-wound, short shunt compound wound, and long shunt compound wound, and long shunt compound wound.

Separately excited generators are used in Ward Leonard system of speed control. Series generators used for series arc lighting, series incandescent lighting and series booster.

Shunt generators with field regulators are used for light and power supply purposes and also used for charging of batteries.

Compound generators are used for power services and lighting

Differential compound generator is used as arc welding generator.

Reasons for failure of self-excitation: DC generator fails to build-up its voltage due to the following reasons:

1. The residual magnetism may be zero.
2. The residual magnetism may have been reversed in polarity.
3. The terminals of the field winding may have been reversed.

Losses in a generator:

1. Copper losses
2. Iron or magnetic losses
3. Mechanical losses

Magnetic and mechanical losses are collectively known as *stray losses*. In case of shunt generator, the shunt copper loss is practically constant. The condition for maximum efficiency of dc shunt generator is

armature copper loss = iron loss.

The load current corresponding to maximum efficiency is given by

$$I = \sqrt{W_c / R_a}$$

where W_c is the copper losses and R_a is armature resistance.

Connections of different DC generators:

I. Series Wound (Fig. 8.7)

i. $I_a = I_{se} = I_L = I$

ii. $V = E_g - I(R_a + R_{se})$

iii. Power developed $P_{dev} = E_g I$

iv. Power delivered, $P_{del} = VI$

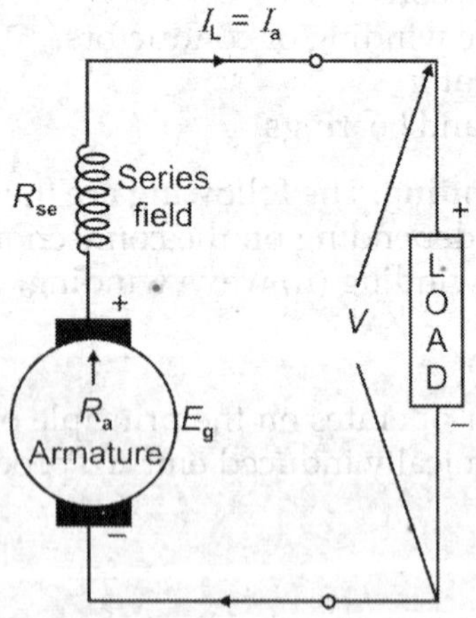

Fig. 8.7

II. Shunt wound (Fig. 8.8)

i. $I_{sh} = V / R_{sh}$

ii. $I_a = I_L + I_{sh}$

iii. $V = E_g - I_a R_a$

iv. Power developed, $P_{dev} = E_g I_a$

v. Power delivered, $P_{del} = VI_L$

III. Compound wound: (a) Short shunt compound wound (b) Long shunt compound wound.

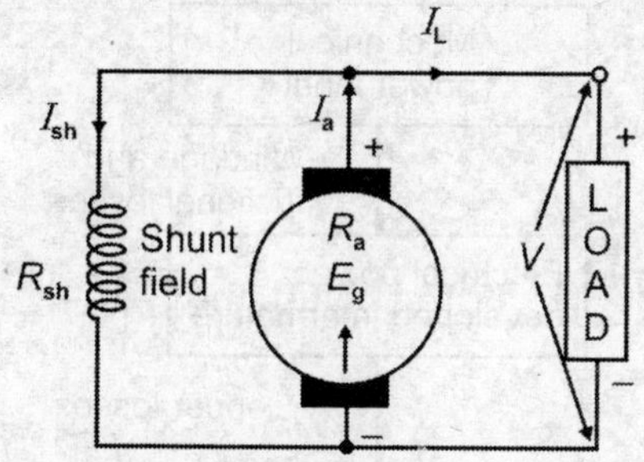

Fig. 8.8

(a) *Short shunt compound wound* (Fig. 8.9)

i. $I_{se} = I_L$

ii. $I_{sh} = \dfrac{V + I_{se}R_{se}}{R_{sh}}$

iii. $I_a = I_L + I_{sh}$

iv. $V = E_g - I_aR_a - I_{se}R_{se}$

v. Power developed $P_{dev} = E_g I_a$

vi. Power delivered $P_{del} = VI_L$

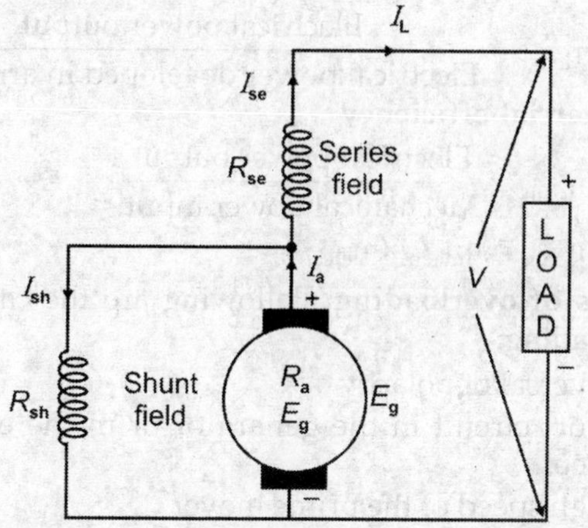

Fig. 8.9

b. *Long shunt compound wound* (Fig. 8.10)

i. $I_{sh} = V/R_{sh}$

ii. $I_a = I_{se} = I_L + I_{sh}$

iii. $V = E_g - I_aR_a - I_{se}R_{se}$

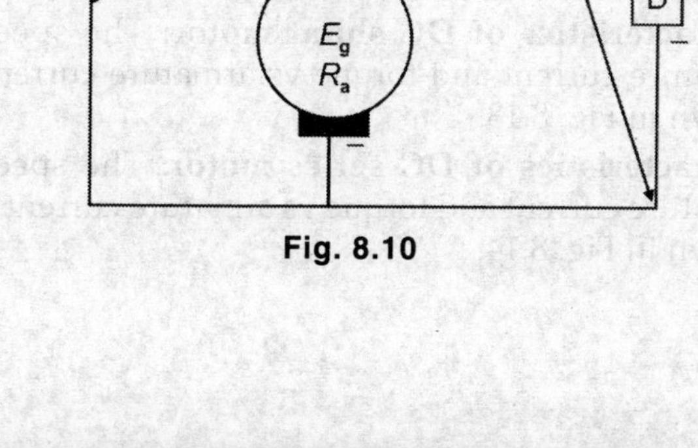

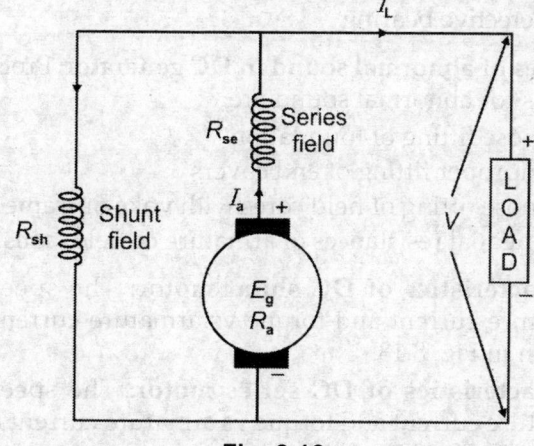

Fig. 8.10

iv. Power developed $P_{dev} = E_g I_a$

v. Power delivered $P_{del} = VI_L$

Armature reaction: It is the effect of the magnetic-field set up by armature current on the distribution flux under the main poles, as shown in Fig. 8.11.

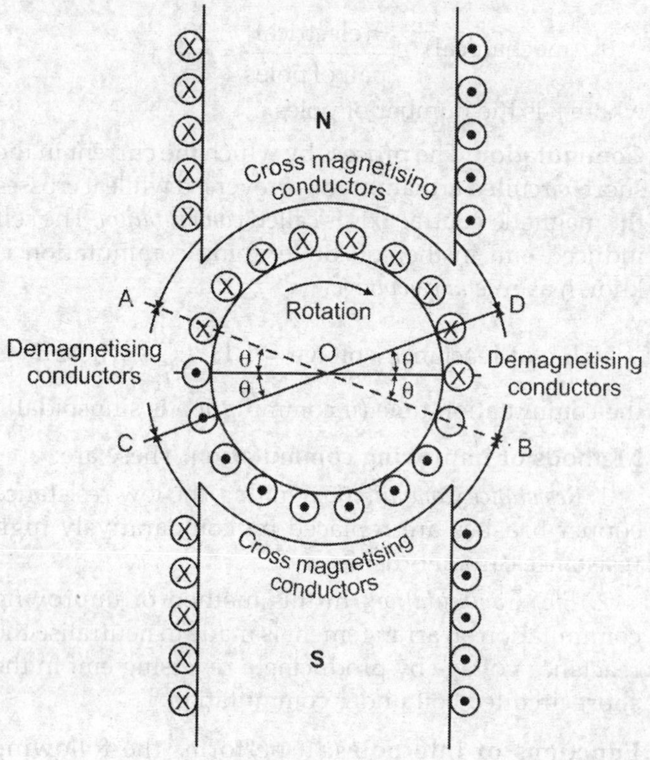

Fig. 8.11

Armature reaction has two effects—it weakens the main flux, and distorts it.

The conductors lying between angles AOC and BOC carry current in such a direction so as to oppose the main mmf. This is known as *demagnetising effect* and thus corresponding conductors are called *demagnetising conductors*.

The conductors lying between angles AOD and BOC carry current is such a direction which causes mmf in the direction perpendicular to that of main mmf. This results in distortion of main field and is called *cross-magnetising effect*.

Demagnetising amp-turns per pole are given by:

$$AT_d/\text{pole} = ZI\frac{\theta}{360}$$

Cross-magnetising amp-turns per pole

$$= ZI\left[\frac{1}{2P} - \frac{\theta}{360}\right]$$

Compensating windings are placed in the pole shoes and their function is to neutralise the cross-magnetising effect of armature reaction.

Armature reaction can be neutralized by

i. increasing the length of air gap

ii. providing the machine with a compensating winding

iii. using commutating poles

iv. reducting the cross-section of pole pieces.

$$\theta_m \text{ (mechanical)} = \frac{\theta(\text{electrical})}{\text{pair of poles}} = \frac{2\theta_e}{p}$$

where p is the number of poles.

Commutation: The process by which the current in the short- circuited armature coil is reversed while it crosses the magnetic neutral axis is called *commutation*. The self induced emf in the coil undergoing commutation is known as *reactance voltage*.

Value of reactance voltage = 1.11 $L\frac{2l}{T_c}$, where T_c is the commutation time (if commutation is sinusoidal).

Methods of improving commutation. These are:

1. *Resistance Commutation.* In this, the low resistance copper brushes are replaced by comparatively high resistance carbon brushes.

2. *Emf Commutation*: In this method of improving commutation an arrangement is made to neutralise the reactance voltage by producing a reversing emf in the short circuited coil under commutation.

Functions of interpoles: It performs the following functions:

1. To make commutation sparkless and ensure automatic neutralisation of reactance voltage
2. To neutralise the cross-magnatising effect of armature reaction.

Equaliser connections: These should be applied to the multi-pole lap windings and to the wave windings with more than two parallel circuits.

Conditions of paralleling DC generators

1. The polarities of the generators must be same
2. The voltages should be nearly identical so that each machine will contribute
3. The change of voltage with change of load should be of the same character.
4. The prime movers that drive the generators should have similar and stable rotational speed characteristics.

Losses in DC generator: The following losses occur in DC generator: (i) Copper losses (ii) Iron losses (iii) Mechanical losses.

Power states: Figure 8.12 shows the various stages in DC generator.

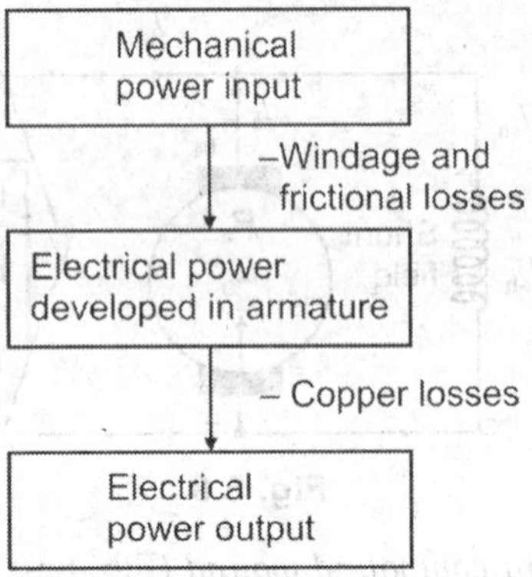

Fig. 8.12

Efficiency of DC generator: DC generator have the following efficiencies due to different power stages.

Mechanical efficiency:

$$\eta_{mech} = \frac{\text{Electrical power developed in armature}}{\text{Mechanical power input}}$$

Electrical efficiency:

$$\eta_{elec} = \frac{\text{Electrical power output}}{\text{Electrical power developed in armature}}$$

Commercial efficiency

$$\eta_{com} = \frac{\text{Electrical power output}}{\text{Mechanical power input}}$$

$$\therefore \quad \eta_{com} = \eta_{mech} \times \eta_{elec}$$

Causes of overloading: Following are the causes of overloading:

i. Reversal of polarity

ii. Short circuit in the generator or in the external circuit

iii. High speed of the prime mover.

Causes of overheating of DC generator on running: The following are the probable reasons of excessive heating of generator on running:

i. Short circuit in the armature and field winding

ii. Weak insulation in lamination

iii. Sparking at the brushes

iv. Defective bearing

Causes of abnormal sound in DC generator: Probable causes for abnormal sound are:

i. Loose fitting of foundation

ii. Improper fitting of end covers

iii. Loose fitting of field cores with yoke or frame

iv. Unequal resistances of armature or field coils.

Characteristics of DC shunt motor: The speed vs armature current and torque vs armature current are shown in Fig. 8.13.

Characteristics of DC series motor: The speed vs armature current and torque vs armature current I_a are shown in Fig. 8.14.

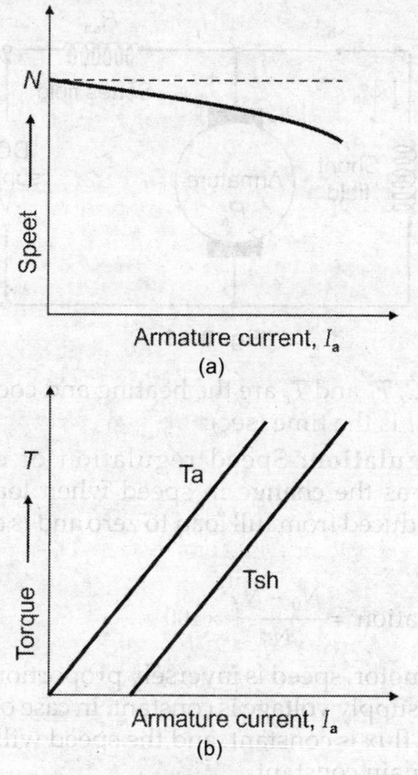

Fig. 8.13

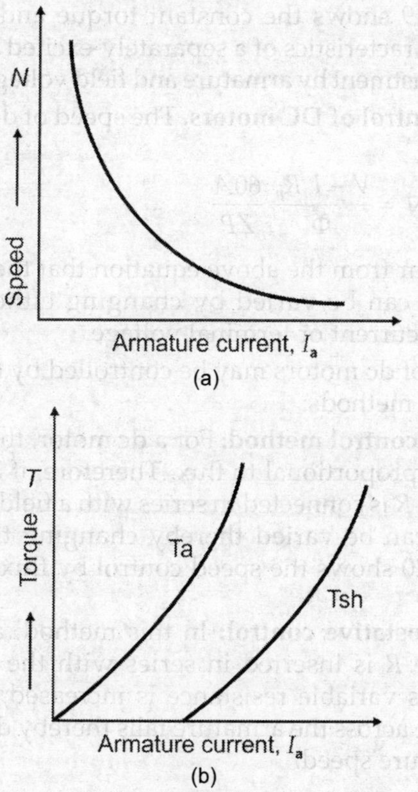

Fig. 8.14

Characteristics of compound motor: The speed vs armature current (I_a) and torque vs armature current I_a are shown in Fig. 8.15.

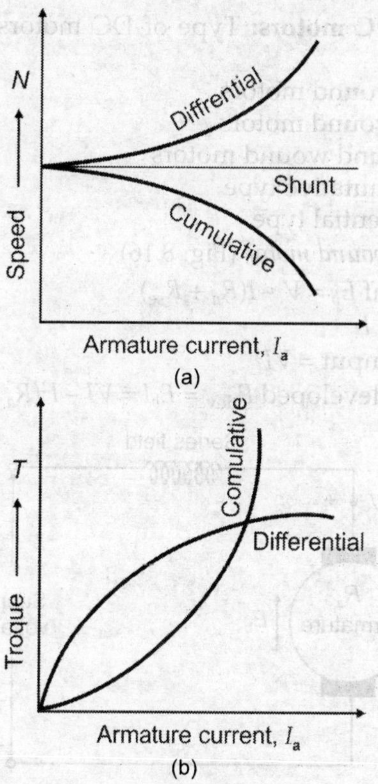

Fig. 8.15

Principle of DC motor: When a conductor carrying current is placed in a magnetic field, emf is procured whose direction is opposite to the main voltage corresponding to Lenz's law. The mechanical force is experienced by the conductor, the direction of which is given by Fleming's left-hand rule.

Armature current of dc motor is, $I = \dfrac{V - E_b}{R_a}$

Voltage equation for shunt motor is

$$V = E_b + I_a R_a + \text{brush drop}$$

where E_b is the back emf, I_a is the armature current, and R_a is the armature resistance.

Condition for maximum power developed in armature is $E_b = V/2$.

Armature torque of a motor is given by

$$T = 0.159\ \Phi\ Z I_a (P/A)\ \text{Nm}$$
$$= 0.0162\ \Phi\ Z I_a\ (P/A)\ \text{kgm}$$

$$\text{Useful torque} = \frac{\text{Output in watt}}{2\pi N}\ \text{Nm}$$

$$\text{Speed of dc motor, } N = \frac{E_b}{\Phi}\left(\frac{A}{ZP}\right) \text{revolution per sec.}$$

Rotation of a motor can be reversed by reversing the current through either the armature winding or the field coils. If the current through both are reversed, the motor will continue to rotate in the same direction.

Types of DC motors: Type of DC motors are listed below:

I. Series wound motors
II. Shunt wound motors
III. Compound wound motors
a. Commutative type
b. Differential type

I. Series wound motor (Fig. 8.16)

(i) Back emf $E_b = V - I(R_a + R_{se})$
(ii) $I_a = I_{se} = I$
(iii) Power Input = VI
(iv) Power developed $P_{dev} = E_b I = VI - I^2(R_a + R_{se})$

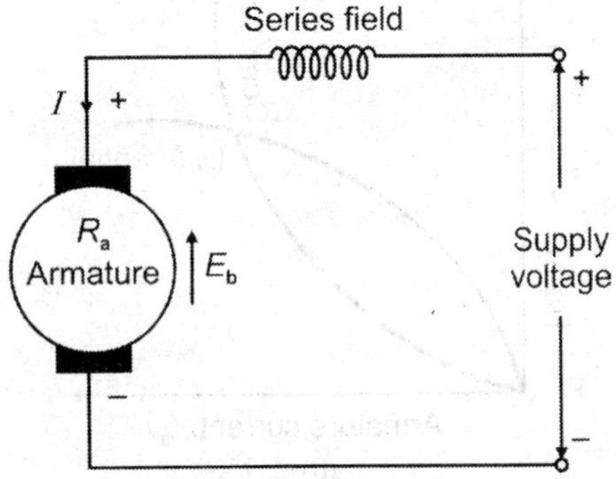

Fig. 8.16

II. Shunt wound motor (Fig. 8.17)

(i) $I_L = I_a + I_{sh}$
(ii) $I_{sh} = V/R_{sh}$
(iii) Back emf $E_b = V - I_a R_a$
(iv) Power Input = VI_L
(v) Power developed $P_{dev} = VI_L - VI_{sh} - I_a^2 R_a$

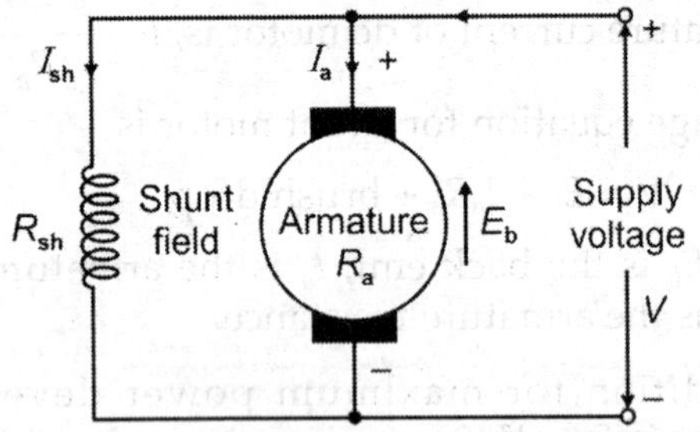

Fig. 8.17

III. Compound wound motors (Fig. 8.18)

(a) Cummulative compound $\Phi = \Phi_{sh} + \Phi_{se}$ (b) Differential compound

$\Phi = \Phi_{se} - \Phi_{sh}$

Heating and Cooling Equations

Equations for DC machines are:

$\theta = \theta_m (1 - e^{-t/Th})$ ...(Heating)

$\theta = \theta_i (1 - e^{-t/Tc})$ ...(Cooling)

where θ_m is the final temperature rise while heating, °C, θ_i is the initial temperature rise over ambient

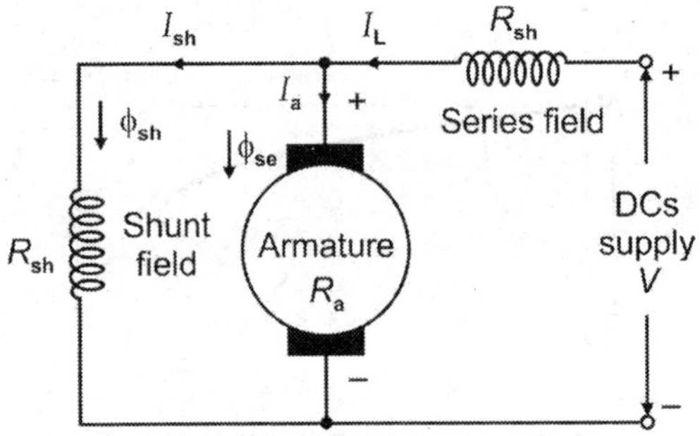

Fig. 8.18

medium °C, T_h and T_c are the heating and cooling time constants, t is the time (sec).

Speed regulation: Speed regulation of dc motor is defined as the change in speed when load on the motor is reduced from full load to zero and is expressed as

$$\% \text{ regulation} = \frac{N_0 - N_f}{N_f} \times 100$$

In a dc motor, speed is inversely proportional to flux per pole if supply voltage is constant. In case of dc shunt motor, the flux is constant, and the speed will approximately remain constant.

Torque and power characteristics of DC motors: Figure 8.19 shows the constant torque and constant power characteristics of a separately-excited motor for speed adjustment by armature and field voltage control.

Speed Control of DC motors. The speed of dc motor is given by

$$N = \frac{V - I_a R_a}{\Phi} \frac{60A}{ZP}$$

It is clear from the above equation that the speed of dc motor can be varied by changing either flux or armature current or terminal voltage.

Speed of dc motors may be controlled by one of the following methods:

a. **Flux control method:** For a dc motor, the speed is inversely proportional to flux. Therefore, if a variable resistance R is connected in series with a field winding, the flux can be varied thereby changing the speed. Figure 8.20 shows the speed control by flux variation method.

b. **Rheostative control:** In this method, a variable resistance R is inserted in series with the armature circuit. As variable resistance is increased, potential difference across the armature falls thereby decreasing the armature speed.

Electric braking: Electric braking of motors can be broadly classified as: (a) Electromechanical (b) Electrical.

a. Electro-mechanical Brakes: Electromechanical or friction brakes are operated by elcctromagnets.

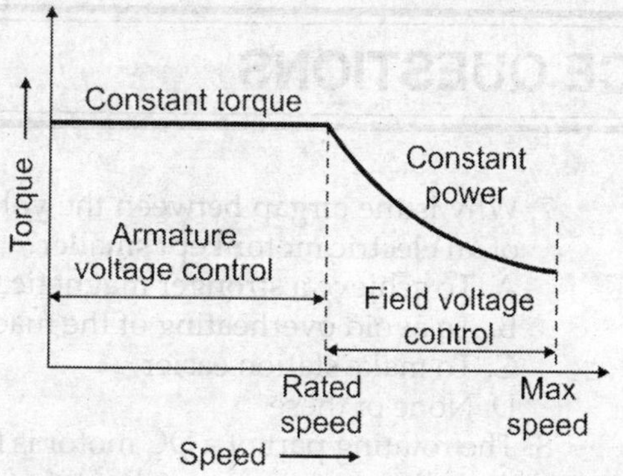

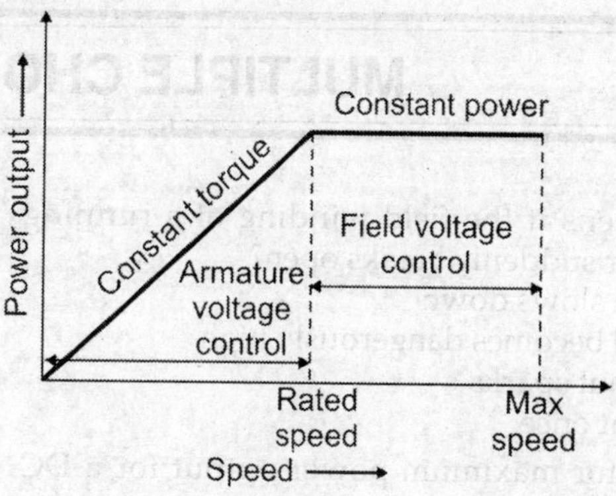

Fig. 8.19

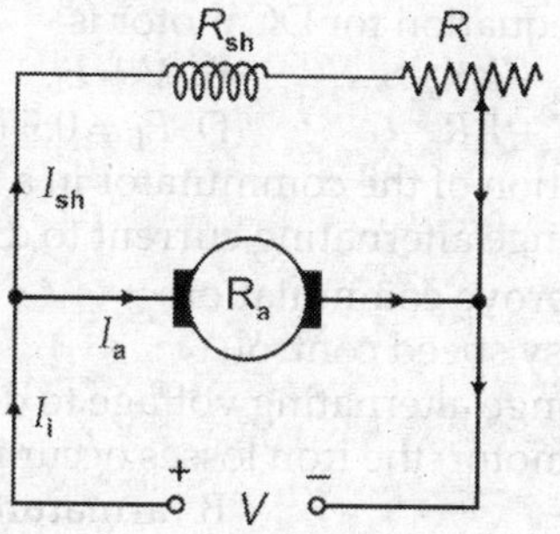

Fig. 8.20

b. Electric Brakes: Electric braking is of three types: (i) Plugging (ii) Rheostatic braking (iii) Regenerative braking.

Efficiency of DC machine. This is defined as :

$$\text{Efficiency} = \frac{\text{Output}}{\text{Input}} = \frac{\text{Output}}{\text{Output + Losses}}$$

Losses in a DC machine may be classified as follows:

1. Electrical losses (copper loss) in
   a. Armature
   b. series field = $I_{se}^2 R_e$
   c. shunt field = (VI_{sh})
   d. Commutative field = $I^2 R_{se}$
2. Rotational losses
   (a) Core loss:
      i. Hysteresis loss
      ii. Eddy current
   (b) Mechanical loss:
      i. Bearings
      ii. Brushes
      iii. Windage

Need of starter for DC motor: When a motor is at rest, there is no back emf in the armature. If it is switched on directly to the mains, its armature would draw a heavy current. This heavy current would damage the armature winding. In order to protect the motor against excessive current at starting, it is necessary that a high resistance be placed in series with the armature winding at the instant of starting, this resistance being gradually cut out in steps as the motor gains speed and develops back emf which can be achieved by starter shown in Fig. 8.21.

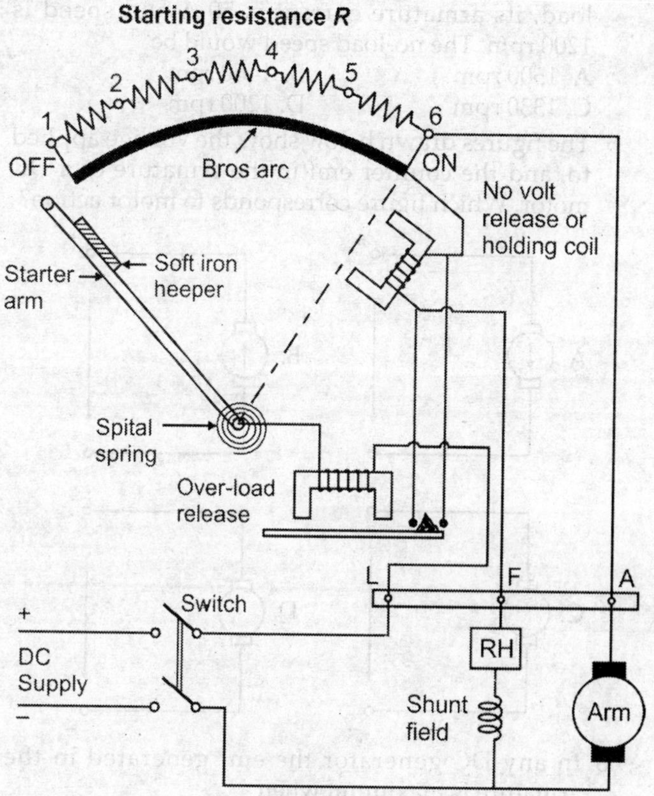

Fig. 8.21

MULTIPLE CHOICE QUESTIONS

1. What happens if the field winding of a running shunt motor suddenly breaks open?
   A. Its speed slows down
   B. Its speed becomes dangerously high
   C. It gives out sparks
   D. It stops at once
2. Condition for maximum power output for a DC motor is
   A. $E_b = V$ B. $E_b = V/2$
   C. $E_b = I_a R_a$ D. $E_b = 0.5\, I_a R_a$
3. The speed of a DC motor is
   A. directly proportional to back emf and inversely proportional to flux
   B. inversely proportional to back emf and directly proportional to flux
   C. directly proportional to emf as well as flux
   D. inversely proportional to emf as well as flux
4. The armature resistance of a 200 V shunt motor is 0.4 ohm and no-load armature current is 2 A. With load, its armature current is 50 A and speed is 1200 rpm. The no-load speed would be
   A. 1500 rpm B. 1460 rpm
   C. 1330 rpm D. 1200 rpm
5. The figures drawn below show the voltage applied to, and the counter emf in the armature of a DC motor. Which figure corresponds to motor action?

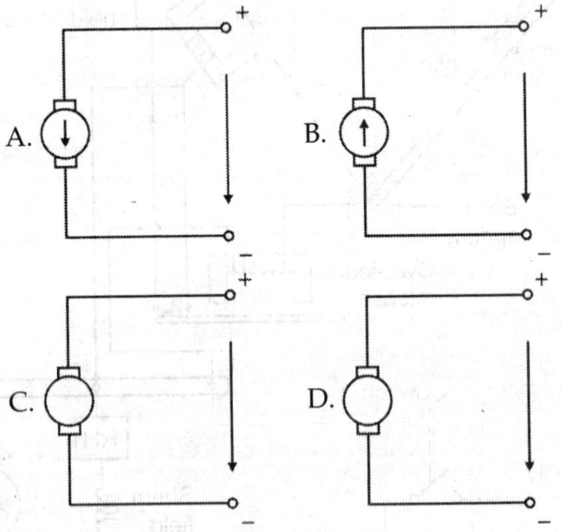

6. In any DC generator the emf generated in the armature is maximum when
   A. rate of change of flux linked is minimum
   B. rate of change of flux linked is maximum
   C. flux linked with conductor is maximum
   D. flux linked with conductor is minimum
7. Why is the airgap between the yoke and armature of an electric motor kept smaller?
   A. To achieve a stronger magnetic field
   B. To avoid overheating of the machine
   C. To make station easier
   D. None of these
8. The rotating part of a DC motor is known as
   A. pole B. stator
   C. armature D. carbon brush
9. Voltage equation for DC motor is
   A. $V = E_b + I_a R_a$ B. $V = E_b - I_a R_a$
   C. $E_b = V + I_a R_a$ D. $E_b = 0.5\, I_a R_a$
10. The function of the commutator in a DC machine is
    A. to change alternating current to direct current
    B. to improve commutation
    C. for easy speed control
    D. to change alternating voltage to direct voltage
11. In a DC motor, the iron losses occur in the
    A. yoke B. armature
    C. field D. none of these
12. Carbon brushes are used in electric motors to
    A. brush off carbon deposits on the commutator
    B. provide a path for flow of current
    C. prevent overheating of armature windings
    D. prevent sparking during commutation
13. A DC voltage is applied to the armature of a DC motor. The current flowing in the armature has the wave form

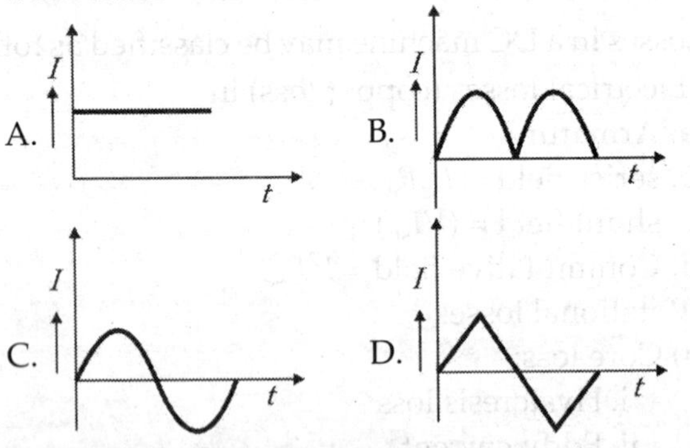

14. DC shunt generator has
    A. slightly drooping characteristics
    B. appreciably rising characteristics
    C. constant voltage characteristics
    D. appreciably falling characteristics
15. Which one of the following generators is used for charging batteries?
    A. Compound generator
    B. Shunt generator
    C. Series generator D. Tacho generator

Ans. 1. B 2. B 3. A 4. C 5. A 6. B 7. A 8. C 9. A 10. D 11. A 12. B 13. C 14. A 15. B

16. The figures drawn below show current drawn by the armature. Which figure corresponds to motor action?

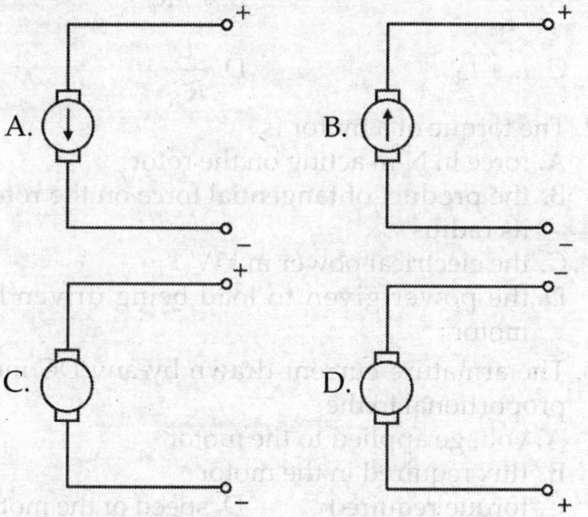

17. What is the effect produced by the electric current in an electric motor?
A. Magnetic effect only
B. Magnetic as well as heating effect
C. Heating effect only
D. Heating as well as chemical effect

18. The highest speed attained by a DC shunt motor is
A. equal to infinity at rated flux
B. higher than no-load speed at rated fluz
C. equal to no-load speed at rated flux
D. lower than no-load speed at rated flux

19. The dummy coil in DC machines is used to
A. eliminate reactance voltage
B. eliminate armature reaction
C. bring about mechanical balance of armature
D. eliminate harmonics developed in the machine

20. A simple method of increasing the voltage of a DC generator is to
A. decrease the air gap flux density
B. increase the speed of rotation
C. decrease the speed of rotation
D. increase the length of the armature

21. The function of the starter in a DC machine is to
A. avoid the excessive current at starting
B. control the speed
C. avoid armature reaction
D. avoid excess heating

22. The emf of a DC generator depends on
A. commutation B. speed
C. frequency D. brush contact drop

23. An external resistance is added in the series with the field of a DC shunt motor. When the motor runs, the effect of resistance is to
A. reduce the speed of the motor
B. increase the speed of the motor
C. reduce the armature current drawn by the motor
D. reduce the losses

24. A conductor is rotating within a magnetic field. At which of the positions shown do the peak voltages occur?

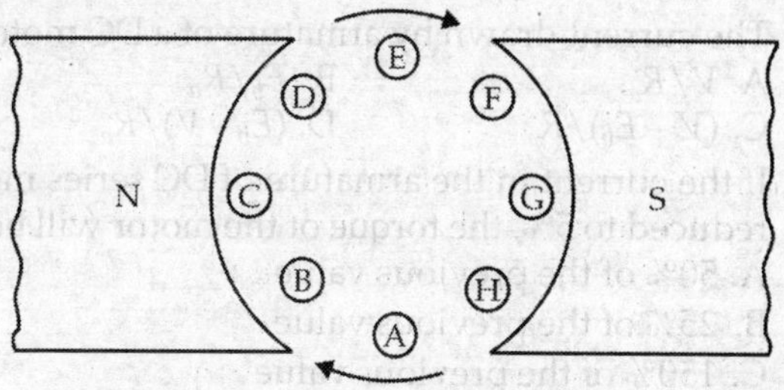

A. Position A B. Positions B and F
C. Positions C and G D. Positions A and E

25. An external resistance added in the field of a DC shunt generator will
A. increase the speed of the generator
B. increase the voltage of the generator
C. decrease the voltage of the generator
D. increase the power delivered.

26. The speed of a DC shunt motor is required to be more than FL speed. This is possible by
A. increasing the armature current
B. decreasing the armature current
C. increasing the excitation current
D. reducing the field current

27. An emf is induced in the windings of an armature of a DC generator when the armature rotates in
A. alternating magnetic flux
B. magnetic field
C. electrostatic field D. electromagnetic flux

28. An external resistance is added in the series with the field of a DC shunt motor. When the motor runs, the effect of resistance is to
A. reduce the speed of the motor
B. increase the speed of the motor
C. reduce the armature current drawn by the motor
D. reduce the losses

29. If speed of a DC shunt motor increases, the back emf
A. increases B. decreases
C. remains constant
D. first decreases and then increases.

30. If the speed of DC generator is increased, the generated emf
A. increases B. decreases
C. remains constant
D. first decreases and then increases.

31. The current flowing in the conductors of a DC motor is
A. AC B. DC
C. AC as well as DC D. transient

Ans. 16. A 17. B 18. C 19. C 20. B 21. A 22. B 23. B 24. D 25. C 26. D 27. B 28. B 29. A 30. A 31. A

32. As the load is increased, the speed of a DC shunt motor will
A. remain constant
B. increase proportionately
C. increase slightly D. reduce slightly

33. The current drawn by armature of a DC motor is
A. V/R_a B. E_b/R_a
C. $(V - E_b)/R_a$ D. $(E_b - V)/R_a$

34. If the current in the armature of DC series motor is reduced to 5%, the torque of the motor will become
A. 50% of the previous value
B. 25% of the previous value
C. 150% of the previous value
D. 125% of the previous value

35. The function of a commutator in a DC generator is to
A. collect current from conductors
B. change DC to AC
C. conduct the current to the brushes
D. change AC to DC

36. A 4-pole lap-wound armature has 480 conductors and a flux per pole of 25 m Wb. The emf generated, when running at 600 rpm, will be
A. 240 V B. 120 V
C. 60 V D. 30 V

37. The shunt current flowing in short shunt compound-wound motor is given by
A. $I_{sh} = \dfrac{V}{R_{sh}}$ B. $I_{sh} = \dfrac{V + I_{sc}R_{se}}{R_{sh}}$
C. $I_{sh} = V - \dfrac{I_{sc}R_{se}}{R_{sh}}$ D. none of these

38. In DC shunt motor, if the terminal voltage is reduced to half and torque remains the same, then
A. speed will be half and armature current also will be half
B. speed will be half but armature current remains the same
C. speed will be half and armature current becomes double
D. speed and armature current will remain the same

39. One DC motor drives another DC motor. The second DC motor when excited and driven, it
A. runs B. runs as a generator
C. does not run as generator
D. also runs as a motor

40. Armature torque of a DC motor is given by
A. $T = 0.159\,\Phi\, ZI_a(P/A)$ kgm
B. $T = 0.01620/2L_a\,(P/A)$ N·m
C. $T = 0.159\dfrac{E_b}{N}I_a$ N·m D. none of these

41. The current in the armature of a dc shunt generator is
A. $\dfrac{E - V}{R_a}$ B. E/R_a
C. $I_L + I_{sh}$ D. $\dfrac{E}{R_{sh}}$

42. The torque of a motor is
A. force in N·m acting on the rotor
B. the product of tangential force on the rotor and its radius
C. the electrical power in kW
D. the power given to load being driven by the motor

43. The armature current drawn by any DC motor is proportional to the
A. voltage applied to the motor
B. flux required in the motor
C. torque required D. speed of the motor

44. The output power of any electrical motor is taken from the
A. armature
B. coupling mounted on the shaft
C. conductors D. poles

45. The power stated on the name plate of any motor is always the
A. output power at the shaft
B. power drawn in kVA
C. power drawn in kW D. gross power

46. The exciting coils of any DC machine are wound
A. in the machine B. in the armature slots
C. around the poles D. separately

47. What is represented by the diagram shown in the figure?

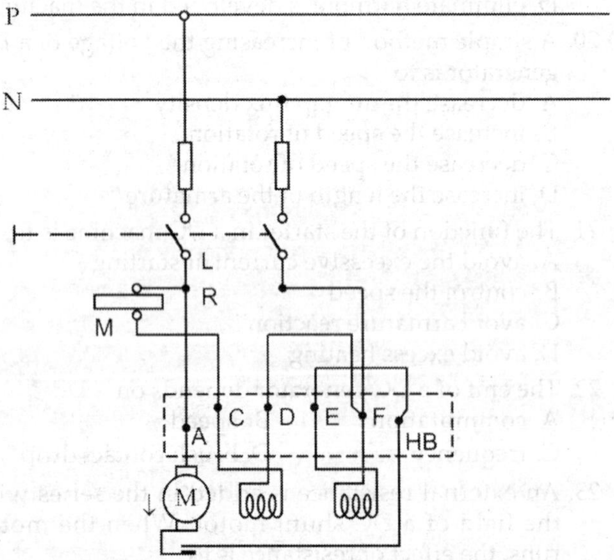

A. A compound generator with auxiliary series winding

Ans.	32. D	33. C	34. B	35. D	36. B	37. B	38. C	39. B	40. C	41. A	42. B	43. A	44. B
	45. A	46. C											

B. A shunt generator with commutating pole winding and compensating winding
C. A shunt motor with commutating pole winding and compensating winding
D. A compound motor with auxiliary series winding and commutating pole winding

48. The efficiency of any electrical machine apparatus will be high, if
A. its losses are minimum
B. its losses are more
C. voltage applied is constant
D. the current drawn is minimum

49. The current delivered by a 4-pole DC shunt generator is 98 A. The shunt-field current is 2 A. The armature is lap wound. The current per parallel path is
A. 100 A B. 25 A
C. 50 A D. 12.5 A

50. The direction of rotation of a DC shunt motor is reversed by
A. reversing armature connections
B. interchanging the armature and field connection
C. adding resistance to the field circuit
D. reversing supply connections

51. A starter is necessary to start a DC motor because it
A. starts the motor
B. limits the speed of the motor
C. limits the back emf to a safe value
D. limits the starting current to a safe value

52. In a properly designed DC generator of 10 kW capacity, total losses may be of the order of
A. 50 W B. 500 W
C. 2500 W D. 5000 W

53. The insulation between the segments of commutator for DC machines is
A. wood B. paper
C. fabric D. mica

54. The DC motor develops a torque of 140 N.m at 1400 rpm. If the motor now runs at 1000 rpm, the torque developed will be
A. 140 N·m B. 2000 N·m
C. 71 N·m D. 51 N·m

55. Usually wide and very sensitive speed control is required in case of
A. Reciprocating pumps B. Colliery winders
C. Centrifugal blowers D. Lathe machines

56. The yoke of a DC machine
A. is laminated B. is not laminated
C. may or may not be laminated
D. none of the above

57. The pole shoes of a DC machine are
A. always laminated B. never laminated
C. sometimes laminated D. none of these

58. The thickness of laminations is about
A. 2 mm B. 5 mm
C. 0.5 mm D. 10 mm

59. The ventilating ducts in most of the machines are
A. radial B. longitudinal
C. radial as well as longitudinal
D. none of these

60. The yoke of DC machine
A. must be made of magnetic material
B. should preferably be made of magnetic material but can be made of non-magnetic material also
C. is always made of non-magnetic material
D. none of these

61. A commutator is made up of
A. iron laminations B. copper segments
C. both iron laminations and copper segments
D. none of these

62. The coil span
A. must be exactly equal to pole pitch
B. may or may not be exactly equal to pole pitch
C. can never be equal to pole pitch
D. none of these

63. The commutator pitch of lap winding is
A. +1 B. –1
C. +1 or –1 D. none of these

64. Equaliser rings are used in
A. lap wound machines
B. wave wound machines
C. both lap and wave wound machines
D. none of these

65. Interpoles are used in
A. lap wound machines
B. wave wound machines
C. both lap and wave wound machines
D. none of these

66. A 4-pole DC machine has lap winding. The winding is removed and then a wave winding is put. The induced emf will
A. increase B. decrease
C. remain the same D. none of these

67. The interpole winding consists of
A. a large number of turns of thin wire
B. a small number of turns of thick wire
C. a large number of turns of thick wire
D. none of these

68. The mmf of an interpole is proportional to
A. armature current B. field current
C. both the armature current and field current
D. none of these

69. The induced emf of a DC machine is
A. directly proportional to speed
B. inversely proportional to speed
C. proportional to square of speed
D. None of these

Ans.													
	47. D	48. A	49. B	50. A	51. D	52. B	53. D	54. A	55. B	56. B	57. A	58. C	59. C
	60. A	61. B	62. B	63. C	64. A	65. C	66. A	67. B	68. A	69. A			

70. For the production of induced emf, field system of an electric machine
A. must be on stator
B. may be on stator or rotor
C. must be on rotor D. none of these

71. A DC generator has 6 poles. A brush shift of 6° actual means a brush shift of
A. 6° electrical B. 18° electrical
C. 30° electrical D. 2° electrical

72. The reactance voltage is
A. directly proportional to armature current and time of commutation
B. indirectly proportional to armature current and directly proportional to the time of commutation
C. directly proportional to armature current and indirectly proportional to time of commutation
D. none of these

73. In a DC generator, the polarity of an interpole is the polarity of the
A. next main pole
B. main pole immediately following the interpole
C. main pole opposite to the interpole
D. none of these

74. For providing proper commutation it is advisable to use
A. emf commutation
B. resistance commutation
C. both emf commutation and resistance commutation
D. none of these

75. In a DC generator emf commutation can be achieved by
A. keeping the brushes in GNP
B. shifting the brushes ahead of GNP
C. shifting the brushes backward
D. none of these

76. The interpoles are connected
A. in series with armature
B. in parallel with armature
C. both in series and parallel with the armature
D. none of these

77. The number of conductors of the compensating winding in a DC machine
A. is always more than the number of armature conductors per pole
B. is always less than the number of armature conductors per pole
C. may be less or more than the number of armature conductors per pole
D. none of these

78. The conductors of the compensating winding are housed
A. in the slots of the armature
B. in the slots in the pole faces
C. partly in armature slots and partly in pole faces
D. none of these

79. If all other quantities are the same, the number of the conductors of compensating winding is
A. less in a lap wound machine than in a wave wound machine
B. more in a lap wound machine than in a wave wound machine
C. the same in the lap wound and wave wound machines
D. none of these

80. Why is the armature of DC motors laminated?
A. To reduce the hysteresis losses
B. To reduce the eddy current losses
C. To reduce the eddy current and hysteresis losses
D. To reduce the inductivity of the armature

81. Which of the following DC motors has the least drop in speed from no load to nominal load?
A. Shunt motor with commutating poles
B. Compound motor without commutating poles
C. Series motor with commutating poles
D. none of these

82. The no load voltage of a generator is 230 V and the rated load voltage is 200 V, then the voltage regulation is
A. 5% B. 10%
C. 15% D. 20%

83. Which of the following is a correct statement about a series motor?
A. Its field winding consists of thicker wire and less turns
B. It can run without load easily
C. It has an almost constant speed
D. It has a poor torque

84. The rotating part of a DC motor is known as
A. pole B. stator
C. armature D. carbon brush

85. A conductor is rotating within a magnetic field. At which of the positions do the zero voltages occur?
A. Along the axis of magnetic field
B. At right angles to the axis of magnetic field
C. Anywhere D. None of these

86. The generator of a power station produces an electric pressure by
A. conversion of heat B. chemical conversion
C. magnetic induction D. conversion of light

87. Why is the air gap between stator and armature of an electric motor kept as small as possible?
A. To get a stronger magnetic field
B. To make the rotation easier
C. To reach at higher speed of rotation
D. To improve the air circulation

88. Which of the following types of generators gives constant voltage output at all loads?

Ans.	70. B	71. B	72. C	73. A	74. C	75. C	76. A	77. B	78. B	79. C	80. B	81. A	82. C
	83. A	84. C	85. A	86. C	87. A								

A. Series generator
B. Shunt generator
C. Short shunt compound generator
D. Level compound generator

89. Interpoles are meant for
A. increasing the speed of the motor
B. increasing counter emf
C. strengthening the main field
D. reducing sparking at the commutator

90. In a Swinburne test on a DC machine
A. all the losses can be determined individually
B. copper losses cannot be determined individually
C. mechanical and iron losses cannot be determined individually
D. none of these

91. Swinburne test cannot be used for
A. series motor B. shunt motor
C. compound motor D. none of these

92. When quick speed reversal is a consideration, the motor preferred is
A. synchronous motor
B. squirrel cage induction motor
C. wound rotor induction motor
D. DC motor

93. DC supply can be obtained from AC supply by the use of
A. motor generator set B. mercury arc rectifier
C. silicon diodes D. any of these

94. The selection of control gear for a particular application is based on the consideration of
A. duty B. starting torque
C. limitations on starting current
D. all of these

95. The basic requirement of a DC armature winding is that it must be
A. a closed circuit B. a lap winding
C. a wave winding D. either B or C

96. The distance between the beginning of two consecutive turns is called
A. back pitch B. front pitch
C. winding pitch D. average pitch

97. For both lap and wave windings, there are as many commutator bars as the number of
A. poles B. armature conductors
C. slots D. winding elements

98. For making coil span equal to a pole pitch in the armature winding of a DC generator, the back pitch of the winding must equal the number of
A. commutator bars per pole
B. winding elements
C. armature conductors per path
D. armature parallel paths

99. The primary reason for making the coil span of a DC, armature winding equal to a pole pitch is to
A. obtain a coil span of 180 (electrical)
B. ensure the addition of emfs of consecutive turns
C. distribute the winding uniformly under different poles
D. obtain a full-pitch winding

100. In a 4-pole, 35-slot dc armature, 180 electrical-degree coil span will be obtained when coils occupy ..... slots.
A. land 10 B. 1 and 9
C. 2 and 11 D. 3 and 12

101. The number of armature parallel paths in a triplex, lap-wound, 12-pole, DC generator is
A. 12 B. 24
C. 36 D. 6

102. The main purpose of a commutator in a DC generator is to
A. increase output voltage
B. reduce sparking at brushes
C. provide smoother output
D. convert the induced AC into DC

103. For a 4-pole, 2-layer. DC, lap-winding with 20 slots and one conductor per layer, the number of commutator bars is
A. 80 B. 20
C. 40 D. 160

104. The commutator pitch of a quadruplex lap winding in a DC generator is
A. 1 B. 2
C. 3 D. 4

105. The only function of dummy coils in a DC armature winding is to
A. increase the induced emf
B. render the average pitch an integer
C. provide mechanical balance to the armature
D. modify its electrical characteristics

106. Lap winding is suitable for ..... current ..... voltage DC generators.
A. high, low B. low, high
C. low, low D. high, high

107. The series field of a short-shunt DC generator is excited by which current?
A. Shunt B. Armature
C. Load D. External

108. In a DC generator, the generated emf is directly proportional to the
A. field current B. pole flux
C. number of armature parallel paths
D. number of dummny coils

109. The field poles and armature of a DC generator are laminated in order to reduce its
A. weight B. speed
C. eddy current loss D. hysteresis loss

Ans. 88. D 89. D 90. B 91. A 92. D 93. D 94. D 95. A 96. C 97. D 98. A 99. B 100. B
101. C 102. D 103. B 104. D 105. C 106. A 107. C 108. B 109. C

110. The commercial efficiency of a shunt generator is maximum when its variable loss equals ..... loss.
A. constant
B. stray
C. iron
D. friction and windage

111. In DC generators, armature reaction is produced actually by
A. its field current
B. armature conductors
C. field pole winding
D. load current in armature

112. In a DC generator, the effect of armature reaction on the main pole flux is to
A. reduce it
B. distort it
C. reverse it
D. both A and B are correct

113. In a clockwise-rotating loaded DC generator, brushes have to be shifted
A. clockwise
B. counterclockwise
C. either (A) or (B)
D. neither (A) nor (B)

114. The primary reason for providing compensating windings in a DC generator is to
A. compensate for decrease in main flux
B. neutralize armature mmf
C. neutralize cross magnetising flux
D. maintain uniform flux distribution

115. Regarding compensating winding which statement is false?
A. It is connected in series with the armature
B. It is embedded in the stationary pole-faces
C. Current flows through it in the same direction as that flowing in armature conductors directly below the pole shoes
D. The number of its conductors is not necessarily equal to the total number of armature conductors

116. In a 10-pole, lap-wound DC generator, the number of active armature conductors per pole is 50. The number of compensating conductors per pole required is
A. 5
B. 50
C. 500
D. 10

117. The commutation process in a DC generator basically involves
A. passage of current from moving armature to a stationary load
B. reversal of current in an armature coil as it crosses MNA
C. conversion of AC to DC
D. suppression of reactance voltage

118. Point out the WRONG statement.
In DC generators, commutation can be improved by
A. using interpoles
B. using carbon brushes in place of copper brushes
C. shifting brush axis in the direction of armature rotation
D. none of these

119. Each of the following statements regarding interpoles is true except
A. they are small yoke-fixed poles spaced in between the main poles
B. they are connected in parallel with the armature so that they carry part of the armature current
C. their polarity, in the case of generators, is the same as that of the main pole ahead
D. they automatically neutralize not only reactance voltage but cross-magnetisation as well

120. Shunt generators are most suited for stable parallel operation because of their .... voltage characteristics.
A. identical
B. drooping
C. linear
D. rising

121. Two parallel shunt generators will divide the total load equally in proportion to their kilowatt output ratings only when they have the same
A. rated voltage
B. voltage regulation
C. internal I_a Ra drops
D. Both (A) and (B) are correct

122. The main function of an equalizer bar is to make the parallel operation of two overcompounded DC generators
A. stable
B. possible
C. regular
D. smooth

123. The essential condition for stable parallel operation of two DC generators having similar characteristics is that they should have
A. same kilowatt output ratings
B. drooping voltage characteristics
C. same percentage regulation
D. same no-load and full-load speeds

124. The main factor which leads to unstable parallel operation of flat and overcompound DC generators is
A. unequal number of turns in their series field windings
B. unequal series field resistances
C. their rising voltage charcteristics
D. unequal speed regulation of their prime movers

125. The simplest way to shift load from one DC shunt generator running in parallel with another is to
A. adjust their field rheostats
B. insert resistance in their armature circuits
C. adjust speed of their prime movers
D. use equalizer connections

126. Which one of the following types of generators does NOT need equalizers for satisfactory parallel operation?
A. Series
B. Overcompound
C. Flat-compound
D. Under-compound

Ans. 110. A 111. D 112. D 113. A 114. C 115. C 116. A 117. B 118. D 119. B 120. B 121. D 122. A 123. B 124. C 125. A 126. D

127. The external characteristic of a shunt generator can be obtained directly from its
A. internal characteristics
B. open-circuit characteristics
C. load-saturation
D. performance

128. Load saturation characteristic of a DC generator gives relation between
A. V and I_a
B. E and I_a
C. E_o and I_f
D. V and I_f

129. The slight curvature at the lower end of the OCC of a self-excited DC generator is due to
A. residual pole flux
B. high armature speed
C. magnetic inertia
D. high field circuit resistance

130. Which condition out of the five given below is not required for the build-up of a shunt generator?
A. There must be sufficient residual magnetism in the poles.
B. For the given direction of rotation, shunt field coils should be correctly connected to the armature
C. If excited on open circuit, the shunt field resistance should be less than the critical resistance
D. Armature speed should be more than the critical speed

131. If a self-excited DC generator after being installed, fails to build up on its first trial run, the first thing to do is to
A. increase the field resistance
B. check armature insulation
C. reverse field connections
D. increase the speed of prime movers

132. If residual magnetism of a shunt generator is destroyed accidentally, it may be restored by connecting its shunt field
A. to earth
B. to an AC source
C. in reverse
D. to a DC source

133. The factor which causes decrease in the terminal voltages of a shunt generator is
A. armature reactance
B. armature resistance
C. armature leakage
D. none of these

134. If field resistance of a DC shunt generator is increased beyond its critical value, the generator
A. output voltage will exceed its name plate rating
B. will not build up
C. may burn out if loaded to its name-plate rating
D. power output may exceed its name-plate rating

135. An ideal DC generator is one that has ..... voltage regulation.
A. low
B. zero
C. positive
D. negative

136. Which generator has poorest voltage regulation
A. series
B. shunt
C. compound
D. overcompound

137. The voltage regulation of an overcompound DC generator is always
A. positive
B. negative
C. zero
D. high

138. Most commercial compound DC generators are normally supplied by the manufacturers as over-compound machines because
A. they are ideally suited for transmission of DC energy to remotely located loads
B. degree of compounding can be adjusted by using a diverter across series field
C. they are more cost-effective than shunt generators
D. they have zero percent regulation

139. In a DC motor, unidirectional torque is produced with the help of
A. brushes
B. commutator
C. plates
D. both (A) and (B) are correct

140. The counter emf of a DC motor
A. often exceeds the supply voltage
B. supply the applied voltage
C. helps in energy conversion
D. regulates its armature voltage

141. Voltage equation of a DC motor is
A. $V = E_b + I_a R_a$
B. $E_b = V + I_a R_a$
C. $V = E_b - I_a R_a$
D. $V = E_b + I_a^2 R_a$

142. The E_b/V ratio of a DC motor is an indication of its
A. efficiency
B. speed regulation
C. starting torque
D. running toque

143. The mechanical power developed by the armature of a DC motor is equal to
A. armature current multiplied by back emf
B. power input minus losses
C. power output multiplied by efficiency
D. power output plus iron losses

144. The no load speed of DC series motor is
A. very small
B. small
C. high
D. very high

145. Which of the following statements is incorrect in DC shunt generators about hysteresis loss?
A. It can be minimised by laminating the armature
B. It is independent of lamination thickness
C. It depends upon the supply frequency
D. None of these

146. Which of the following methods is most economical for finding the no-load losses of a large DC shunt motor?
A. Hopkinson's test
B. Swinburne's test
C. Retardation test
D. None of these

147. An ideal DC generator has a regulation of
A. 0%
B. 50%
C. 75%
D. 100%

Ans. 127. B 128. D 129. C 130. D 131. C 132. D 133. B 134. B 135. B 136. A 137. B 138. B 139. D 140. C 141. A 142. A 143. A 144. D 145. A 146. B 147. A

148. The rated speed of a given DC shunt motor is 1050 rpm. To run this machine at 1200 rpm, which of the following speed control scheme will be used?
A. Armature current resistance control
B. Field resistance control
C. Ward-Leonard control
D. None of these

149. The destroyed residual magnetism of a shunt generator may be restored by connecting its shunt field to
A. earth B. generator
C. a battery D. none of these

150. It is preferable to start a DC series motor with some mechanical load on it because it
A. will not run at no-load
B. acts as a starter to the motor
C. may develop excessive speed and damage itself
D. None of these

151. If the flux of a DC motor approaches zero, its speed will approach
A. infinity B. zero
C. a stable value nearer to the rated speed
D. none of these

152. A 230 V DC series motor is connected to 230 V AC it will
A. run slowly B. not run at all
C. run with less efficiency
D. none of these

153. As the load is increased, the speed of a shunt motor
A. remains constant B. increases slightly
C. reduces slightly D. none of these

154. The main function of a commutator in DC motor is to
A. prevent sparking B. convert AC to DC
C. reduce friction D. convert DC to AC

155. The factor which does not cause decrease in the terminal voltage of a shunt generator is
A. armature resistance B. armature reaction
C. reduction in field current
D. armature reactance

156. A DC generator can be termed as
A. rotating amplifier B. prime mover
C. power pump D. none of these.

157. The function of equalizing connections in a lap wound DC generator is to
A. neutralise the armature reaction effect
B. avoid unequal distribution of currents at brushes
C. avoid short circuit current
D. none of these

158. The function of compensating windings placed in slots in the pole shoes is to
A. neutralise cross magnetising effect
B. neutralise the demagnetising effect
C. neutralise both the effects
D. avoid flash over around the commutator

159. In the case of parallel operation of compound generators, for proper division of load from no-load to full load, it is essential that
A. the regulation of each armature should be the same
B. their series field resistances should be equal
C. their ratings should be identical
D. none of these

160. Which of the following statements is incorrect? If a starter is not used with large DC motor, it will draw a starting current which
A. is many times its full-load current
B. will produce excessive line voltage drop
C. will damage the commutator
D. will produce very low starting torque

161. The simplest way to shift load from one shunt generator to the other running in parallel is to
A. adjust their field rheostats
B. use equalizer connection
C. add resistance in their armature circuit
D. none of these

162. For parallel operation, the polarities of two generators
A. must oppose each other
B. may or may not be the same
C. must be the same D. none of these

163. A 4-pole DC generator is running at 1500 rpm. The frequency of current in the armature winding is
A. 25 Hz B. 50 Hz
C. 0 Hz D. 100 Hz

164. The direction of rotation of a DC shunt motor can be reversed by interchanging
A. the supply terminals
B. the field terminals only
C. the armature terminals only
D. either field or the armature terminals

165. In DC machines, the armature reaction mmf is
A. stationary with respect to armature
B. rotating with respect to stator
C. stationary with respect to stator
D. rotating with respect to brushes

166. The purpose of using interpoles in DC machines is to nullify
A. the demagnetising effect of armature mmf
B. the cross-magnetising effect of armature mmf
C. the reactance voltage
D. both the cross-magnetizing mmf and the reactance voltage

167. In DC generators, the polarity of inter-poles is
A. same as that of main pole behind
B. same as that of main pole ahead

Ans.	148. B	149. D	150. C	151. A	152. C	153. C	154. B	155. C	156. A	157. B	158. D	159. A	160. D
	161. A	162. A	163. B	164. D	165. C	166. D	167. B						

C. opposite to that of the main pole
D. none of these

168. The object of using compensation winding in DC machines is to neutralize the
A. armature reaction in the interpolar zone
B. armature reaction in the commutating zone
C. armature reaction under the pole faces
D. cross-magnetizing armature reaction

169. The emf generated in a DC generator is
A. dynamically induced emf
B. statically induced emf
C. electrostatically induced emf
D. magnetically induced emf

170. The yoke of generator is made of cast iron because
A. it is cheaper
B. it completes the magnetic path
C. it gives mchanical protection to the machine
D. all of these

171. The field of self excited generator is excited by
A. DC B. AC
C. its own current D. AC and DC both

172. The DC generator works on the principle of
A. Faraday's laws of electronmagnetic induction
B. Lenz's law
C. The current carrying conductor placed in the magnetic field an emf is produced
D. induction

173. The electrical efficiency of a generator is
A. $\frac{\text{Output}}{\text{Input}} \times 100$
B. $\frac{\text{Output}}{\text{Input + Stray losses}} \times 100$
C. $\frac{\text{Output}}{\text{Input} + I^2R} \times 100$ D. none of these

174. The carbon brushes are generally used as they
A. are self lubricating B. are soft
C. have negative temperature coefficient
D. possess all the above three properties

175. Armatures are made of laminated steel instead wood because it has
A. low permeability
B. high permeability
C. more mechanical strength
D. more mechanical strength and high permeability

176. A DC generator will not build up the voltage
A. if the field resistance is more than critical resistance
B. due to wrong direction of rotation
C. if it has lost the residual magnetism
D. any one of the above three

177. For battery charging which of the following DC generator is used
A. series B. shunt
C. short shunt compound
D. long shunt compound

178. DC series generator is used for
A. lighting load B. battery charging
C. a booster D. none of these

179. The internal characteristic of generator is the curve between
A. armature current and generated emf
B. load current and terminal voltage
C. field current and no-load voltage
D. armature current and IR loop

180. Emf of generator depends upon
A. number of poles B. flux per pole
C. number of conductors
D. all of these

181. The emf of a generator can be controlled by varying
A. flux B. speed of the generator
C. number of poles D. speed and flux both

182. The DC motor will draw high current at the time of starting because
A. $E_b > V$ B. $E_b < V$
C. $E_b = 0$ D. $E_b = V$

183. In series motor, the number of turns in the field pole is
A. more than shunt field
B. less than shunt field
C. less than armature D. same as in shunt field

184. Poles are made of laminated cores because it
A. increases the iron losses
B. reduces the eddy current losses
C. reduces the copper losses
D. reduces the friction losses

185. The load characteristic of a generator is the curve between
A. load voltage and field current
B. generated emf and armature current
C. load current and terminal voltage
D. load current and voltage drop in the armature winding

186. The resistance of shunt winding is
A. more than series winding
B. more than armature
C. less than series and armature
D. more than series and armature

187. Motor converts
A. mechanical energy into electrical energy
B. electrical energy into mechanical energy
C. chemical energy into electrical energy
D. kinetic energy into mechanical energy

Ans. 168. C 169. A 170. D 171. C 172. A 173. C 174. D 175. D 176. D 177. B 178. C 179. A 180. D 181. D 182. C 183. C 184. B 185. C 186. D 187. B

188. If one generator is to be taken out when two generators are running in parallel, the excitation of
A. first generator is reduced gradually
B. second generator is increased gradually
C. first is reduced and second is increased gradually and simultaneously
D. none of these

189. The type of DC motor control preferred for applications when usually wide and very sensitive speed control is required, will be
A. armature control
B. Ward-Leonard control
C. voltage control D. flux control

190. Generators run in parallel because it
A. keeps stability of supply
B. gives facility of repairing, which results in fewer breakdown
C. gives facility of an additional unit to be installed as and when required
D. all of these

191. The self-excited DC generator is installed. If during the trail run it fails to build up, the remedy lies in
A. reversing the field connections
B. increasing the operational speed
C. increasing the field resistance
D. reducing the field resistance

192. A 220 V DC generator is run at full speed without any excitation. The open circuit voltage will be
A. zero B. about 2 V
C. about 50 V D. 220 V

193. In separately-excited generator supplying rated load, the armature reaction
A. is always absent B. is always present
C. may be sometimes present
D. none of these

194. If residual magnetism is present in a DC generator, the induced emf, at zero speed will be
A. zero B. small
C. the small as rated voltage
D. high

195. A shunt generator has a full load voltage regulation of 10%. If the generator is separately excited and delivers rated load the regulation will be
A. 10% B. more than 10%
C. less than 10% D. zero

196. A separately excited DC generator has an open circuit voltage E_0 for a certain value of the field current. If field current is doubled, the open circuit voltage
A. will be $2E_0$
B. will be less than or more than $2E_0$
C. may be $2E_0$ or less than $2E_0$
D. none of these

197. A series generator can self-excite
A. only if the load current is zero
B. only if the load current is not zero
C. irrespective of the value of load current
D. none of these

198. A shunt generator can self-excite
A. only if the resistance of the field circuit is less than critical value
B. only if the resistance of the field circuit is greater than critical value
C. irrespective of the value of the resistance in the field circuit
D. none of these

199. The open-circuit voltage of a compound generator is 250 V. At full load, the termi-nal voltage
A. will always be less than 250 V
B. will always be 250 V
C. may be greater or less than 250 V
D. none of these

200. When a shunt generator is delivering a certain load current, its terminal voltage
A. will always be less than open circuit voltage
B. will always be greater than open circuit voltage
C. may be greater or less than the open circuit voltage
D. none of these

201. The terminal voltage of a series generator is 150 V when the load current is 5 A. If the load current is increased to 10 A the terminal voltage will be
A. 150 V B. less than 150 V
C. greater than 150 V D. none of these

202. The difference between no load and full load speeds of a DC shunt motor is of the order of
A. 20% B. 1%
C. 10% D. 50%

203. Two shunt generators are operating in parallel. If it is desired to shut down one of the generators
A. its main switch is suddenly open
B. its field current is gradually reduced
C. the input to its prime mover is suddenly reduced to zero
D. none of these

204. Two shunt generators having slightly different external characteristics are operating in parallel and their open circuit emf are equal. Then
A. the machine with more dropping characteristic will supply greater load
B. the machine with more dropping characteristic will supply less load
C. the two machines will supply equal loads
D. none of these

205. The presence of equaliser bar is
A. essential for parallel operation of compound generators

Ans. 188. C 189. B 190. D 191. A 192. B 193. B 194. A 195. C 196. C 197. B 198. A 199. C 200. A 201. C 202. C 203. B 204. B 205. A

B. preferable for parallel operation of compound generators
C. unnecessary for parallel operation of compound generators
D. none of these

206. A series generator is provided with diverter and is delivering its rated current. If the diverter switch is opened the terminal voltage will
A. increase B. decrease
C. remain the same D. none of these

207. Which type of enclosure is preferred for motors to be used in watery dirty atmosphere?
A. Open type B. Screen protected
C. Totally enclosed fan closed
D. Drip proof

208. If the load on an over-compounded DC generator is reduced, the terminal voltage
A. increases B. remains the same
C. decreases D. none of these

209. A shunt generator has a speed of 500 rpm when delivering 100 kW at 220 V. It runs as a motor taking an input of 100 kW from 220 V mains, the speed will be
A. 500 rpm B. less than 500 rpm
C. more than 500 rpm D. none of these

210. In a Hopkinson test on a pair of shunt machines
A. the armature currents of the motor and generator are equal
B. the armature current of the motor is less than that of generator
C. the armature current of the generator is less than that of the motor
D. none of these

211. In a Hopkinson test on two shunt machines
A. the field currents of the two machines are equal
B. the field current of the generator is more than that of the motor
C. the field current of the motor is more than that of the generator
D. none of these

212. A DC machine used as a generator has an efficiency of 90% when the output voltage and current are 220 V and 10 A respectively. If the machine is used as a motor and takes 10 A from 220 A supply the efficiency will be
A. 90% B. less than 90%
C. more than 90% D. none of these

213. In a DC machine the maximum losses occur due to
A. iron losses B. mechanical losses
C. copper losses D. none of these

214. The efficiency of a DC machine is of the order of
A. 90% B. 75%
C. 60% D. 99%

215. The efficiency of a generator is usually expressed as
A. (input – losses)/input
B. output/input
C. output/(output + input)
D. none of these

216. Efficiency of electrical machines should be calculated by measuring
A. output and input B. losses and output
C. losses and input D. losses

217. The efficiency of electrical machines is maximum when
A. constant losses = (variable copper losses)^2
B. constant losses = variable copper losses
C. constant losses = $\sqrt{\text{(variable copper losses)}}$
D. none of these

218. Electrical machines are designed to have maximum efficiency at
A. full load B. 50% of full load
C. near about full load D. no load

219. In 3-phase AC machines, phase spread of 120° is preferred over a phase spread of 60°
A. true B. false
C. depends upon type of machine
D. none of these

220. In aircraft industry, the material used for the conductor is
A. aluminium B. copper
C. silver D. iron

221. The mmf produced by a single phase winding is
A. pulsating and stationary
B. pulsating and rotating
C. constant in amplitude and stationary
D. constant in amplitude and rotating

222. The material used for the magnetic circuit where high value of flux density required is
A. cast iron B. ferro cobalt
C. soft steel D. gray cast iron

223. Rotational losses in electrical machines consist of
A. friction and windage losses
B. stator core, friction and windage losses
C. rotor, core, friction and windage losses
D. stray load losses and friction and windage losses

224. The armature torque of a DC motor is a function of its
A. field flux B. armature current
C. speed
D. Both (A) and (B) are correct

225. Under constant load conditions, the speed of a DC motor is affected by
A. field flux B. armature current
C. back emf D. Both (B) and (C)

Ans. 206. A 227. C 208. C 209. B 210. C 211. B 212. B 213. C 214. A 215. C 216. D 217. B 218. C
219. B 220. A 221. A 222. B 223. C 224. D 225. A

226. It is possible to increase the field flux and, at the same time, increase the speed of a DC motor provided its .... is held constant.
A. applied voltage B. torque
C. armature circuit resistance
D. armature current

227. The current drawn by a 120 V DC motor of armature resistance 0.5 Ω, and back emf 110 V is
A. 20 A B. 240 A
C. 220 A D. 5 A

228. The shaft torque of a dc motor is less than its armature torque because of
A. copper losses B. mechanical losses
C. iron losses D. rotational losses

229. A DC motor develops a torque of 200 N·m at 25 rps. At 20 rps, it will develop a torque of
A. 200 Nm B. 160 Nm
C. 250 Nm D. 128 Nm

230. Neglecting saturation, if current taken by a series motor is increased from 10 A to 12 A, the percentage increase in its torque is
A. 20% B. 44%
C. 20.5% D. 16.6%

231. If load on a DC shunt motor is increased, its speed is decreased due to primarily
A. increase in its flux
B. decreases in back emf
C. increase in armature current
D. increase in brush drop

232. If the load current and flux of a DC motor are held constant and voltage applied across its armature is increased by 10 percent, its speed will
A. decrease by about 10 percent
B. remain unchanged
C. increase by about 10 percent
D. increase by 20 percent

233. If the pole flux of a DC motor approaches zero, its speed will
A. approach zero B. approach infinity
C. not change due to corresponding change in back emf
D. approach a stable value somewhere between zero and infinity

234. If the field circuit of a loaded shunt motor is suddenly opened
A. it would race to almost infinite speed
B. it would draw abnormally high armature current
C. circuit breaker or fuse will open the circuit before too much damage is done to the motor
D. torque developed by the motor would be reduced to zero

235. A large series motor is never started without some mechanical load on it because otherwise it will
A. draw too much current
B. develop excessive speed and damage itself
C. produce vicious sparking at brushes
D. open fuse or circuit breaker

236. As the load is increased, the speed of a DC shunt motor
A. increases proportionately
B. remain constant
C. increase slightly D. reduce slightly

237. Between no-load and full-load, which motor develops the least torque?
A. Series B. Shunt
C. Cumulative-compound
D. Differential-compound

238. The T_a/I_a graph of a DC series motor is a
A. parabola from no-load to overload
B. straight line throughout
C. parabola throughout
D. parabola up to full-load and a straight line at overloads

239. As compared to shunt and compound motors, series motor has the highest torque because of its comparatively ..... at the start.
A. lower armature resistance
B. larger armature current
C. stronger series field D. fewer series turns

240. Unlike a shunt motor, it is difficult for series motor to stall under heavy loading because
A. it develops high overload torque
B. its flux remains constant
C. it slows down considerably
D. its back emf is reduced to almost zero

241. When load is removed, which DC motor will run at the *highest* speed?
A. Shunt motor B. Series motor
C. Differential compound
D. Cumulative-compound

242. A DC series motor is best suited for driving
A. lathes B. cranes and hoists
C. shears and punches D. machine tools

243. The speed of a DC motor can be controlled by varying
A. its flux per pole
B. resistance of armature circuit
C. applied voltage D. all of these

244. The most efficient method of increasing the speed of a 3.75 kW DC shunt motor would be the
A. armature control method
B. flux control method
C. Ward-Leonard method
D. tapped-field control method

245. Which statement about shunt field speed control method for DC, shunt motor is false?

Ans. 226. D 227. A 228. D 229. A 230. B 231. B 232. C 233. B 234. C 235. B 236. D 237. A 238. D 239. C 240. A 241. B 242. B 243. D 244. B

A. It is based on the fact that motor speed is inversely proportional to flux
B. It gives constant horse-power characteristics to the motor
C. It is very wasteful due to high I^2R loss in field rheostat
D. It gives speeds above the basic speed and not below it

246. Regarding Ward-Leonard system of speed control which statement is false?
A. It is usually used where wide and very sensitive speed control is required
B. It is used for motors having ratings from 750 to 4000 kW
C. Capital outlay involved in the system is high since it uses two extra machines
D. It gives a speed range of 10:1 but in one direction only

247. In the rheostatic method of speed control for a DC shunt motor, use of armature divertor makes the method
A. less wasteful B. less expensive
C. unsuitable for changing loads
D. suitable for rapidly changing loads

248. The chief advantage of Ward-Leonard system of DC motor speed control is that it
A. can be used even for small motors
B. has high overall efficiency at all speeds
C. gives smooth, sensitive and wide speed control
D. uses a flywheel to reduce fluctuation in power demand

249. The flux control method using paralleling of field coils when applied to a 4-pole series DC motor can give
A. 2 speeds B. 3 speeds
C. 4 speeds D. 6 speeds

250. The series-parallel system of speed control of series motors widely used in traction work gives a speed range of about
A. 1:2 B. 1:3
C. 1:4 D. 1:6

251. In practice, regenerative braking is used when
A. quick motor reversal is desired
B. load has overhauling characteristics
C. controlling elevators, rolling mills and printing presses, etc.
D. other methods cannot be used

252. A face-plate starter is employed for starting
A. synchronous motor B. universal motor
C. direct current series motor
D. induction motor

253. **Statement 1.** A direct-on-line (DOL) starter is used to start a small DC motor.
Statement 2. It limits initial current drawn by the armature circuit.
A. Both the statements are incorrect
B. Both the statements are correct
C. Statement 1 is correct but 2 is wrong
D. Statement 2 is correct but 1 is wrong

254. Ward-Leonard system of speed control is NOT recommended for
A. frequent motor reversals
B. constant-speed operation
C. wide speed range D. very low speeds

255. The brushes of a DC machine are
A. physically placed in the interpolar axes and electrically connected to the coils in the polar axes
B. physically placed in the interpolar axes and electrically connected to the coils in the interpolar axes
C. physically placed in the polar axes and electrically connected to the coils in the interpolar axes
D. physically placed in the polar axes and electrically connected to the coils in the polar axes.

256. The emf induced in the armature of a DC machine is
A. directly proportional to the flux and inversely proportional to the speed
B. directly proportional to both the flux and the speed
C. inversely proportional to both the flux and the speed
D. none of these

257. The commutator of a DC machine acts as a
A. full-wave rectifier
B. half-wave rectifier
C. controlled full-wave rectifier
D. controlled half-wave rectifier

258. If the armature current of a dc motor is increased keeping the field flux constant, then the developed torque
A. increases proportionally
B. decreases in inverse proportion
C. remains constant
D. increases proportional to square of the current

259. The parts of the armature electric circuit which take active part in emf generation are
A. the coil sides inside the slots
B. the overhangs
C. both the coil sides inside the slots and the overhangs
D. the commutator segments

260. The direction of induced emf in an armature coil of a DC machine is
A. the same as that of the current for both the generator and the motor
B. opposite to that of the current for both generator and the motor

Ans. 245. C 246. D 247. D 248. C 249. B 250. C 251. C 252. C 253. C 254. B 255. C 256. B 257. A 258. A 259. A

C. the same as that of the current for the generator and opposite to that of the current for the motor
D. none of these

261. In a DC machine without any brush shift, the shift of the magnetic neutral axes due to armature reaction is
A. in the direction of rotation for the generator and against the direction of rotation for the motor
B. in the direction of rotation for both the generator and the motor
C. against the direction of rotation for both the generator and the motor
D. against the direction of rotation for the generator and in the direction of rotation for the motor

262. Consider the following statements:
The armature rotation mmf in a DC machine is
I. stationary with respect to the field poles
II. rotating with respect to the field poles
III. rotating with respect to the armature.
State which of the following is correct.
A. only (I) B. only (III)
C. both (I) and (III) D. both (II) and (III)

263. The waveform of the armature mmf in a DC machine is
A. square B. rectangular
C. triangular D. sinusoidal

264. In a DC machine the armature mmf is always directed along the
A. polar axis B. brush axis
C. interpolar axis D. None of these

265. In a DC machine, the armature reaction and the inductance of the commutating coils result in
A. over-commutation B. linear commutation
C. sinusoidal commutation
D. under-commutation

266. Under-commutation gives rise to
A. sparking at the leading edge of the brush
B. sparking at the trailing edge of the brush
C. no sparking at all
D. sparking at the middle of the brush

267. The hysteresis loss per unit weight is
A. proportional to the frequency
B. inversely proportional to the frequency
C. inversely proportional to the flux density
D. proportional to the thickness of laminations

268. For a DC generator if the brushes are given a small amount of forward shift, then the effect of armature reaction is
A. totally demagnetizing
B. totally magnetizing
C. partly demagnetizing and partly cross-magnetizing
D. totally cross-magnetizing

269. In a DC machine, if the brushes are given a backward shift, then commutation is
A. worsened and speed decreases
B. improved and speed decreases
C. unaffected and speed increases
D. improved and speed increases

270. Consider the following statements. The purpose of using interpoles in DC machines is to counteract
I. the demagnetizing effect of armature mmf in the commutating zone
II. the cross-magnetizing effect of armature mmf in the commutating zone
III. the magnetizing effect of armature mmf in the commutating zone.
State which of the following is correct
A. only II B. only III
C. both II and III D. both I and III

271. In DC machines, the polarity of the interpole is
A. same as that of the main pole behind for the generators and that of the main pole ahead for the motors
B. same as that of the main pole ahead for both the generators and the motors
C. same as that of the main pole ahead for the generators and that of the main pole behind for the motor
D. same as that of the main pole behind for both the generators and the motors

272. In a DC machine without interpoles to get improved commutation, the brush shift angle must be
A. varied with change in load
B. kept constant
C. 0 degree D. none of these

273. Consider the following statements:
The purpose of using compensating winding in a DC machine is to counter-act
I. armature reaction mmf under the pole-faces
II. armature reaction mmf in the inter-polar zone.
III. flashover between positive and negative brushes.
State which of the following is correct.
A. both I and III B. only III
C. both I and II D. both II and III

274. A DC machine is provided with both interpole winding (IPW) and compensating winding (CPW) with respect to the armature
A. both IPW and CPW are in parallel
B. both IPW and CPW are in series
C. IPW is in series and CPW is in parallel
D. IPW is in parallel and CPW is in series.

275. Consider the following statements:
Compensating windings are used in a DC motor which are intended to operate
I. with rapidly changing loads of wide range
II. at constant speed over wide range of load

Ans. 260. C 261. A 262. C 263. C 264. B 265. D 266. B 267. A 268. C 269. D 270. C 271. C 272. A 273. A 274. B

III. over wide range of speed by field excitation control

State which of the following is correct

A. only I B. only III
C. both II and III D. both I and III

276. Due to magnetic saturation, the flux per pole in a DC machine without brush shift
A. decreases in the generators and increases in the motors with load
B. decreases in both the generators and the motors with load
C. increases in both the generators and the motors with load
D. increases in the generators and decreases in the motors with load

277. Compared to an uncompensated DC machine, the interpole as required in a compensated DC machine is
A. larger for the generators and smaller for the motors
B. smaller for the generators and larger for the motors
C. smaller for both the generators and the motors
D. larger for both the generators and the motors

278. The core losses in a DC machine occur in
A. the armature only
B. the pole faces only
C. the yoke only
D. both the armature and the pole faces

279. If the thickness of laminations is increased, then the
A. eddy current loss decreases
B. eddy current loss increases
C. hysteresis loss decreases
D. hysteresis loss increases

280. For DC shunt motor, speed control by armature resistance variations is best suited for
A. constant power drive
B. variable power drive
C. constant torque drive
D. variable torque drive

281. The efficiency of a DC machine is maxi-mum when the variable losses are equal to
A. the constant losses
B. the square of the constant losses
C. the square root of the constant losses
D. zero

282. Compared to the air gap under the field-poles, the inter pole air gap is made
A. larger for both the generators and the motors
B. smaller for both the generators and larger for the motors
C. smaller for generators and larger for the motors
D. larger for generators and smaller for the motors

283. The brake test for the determination of efficiency of a DC machine is
A. an indirect method B. a regenerative method
C. a direct method D. none of these

284. In Swinburne's method for the determination of efficiency of a DC machine
A. the no-load losses are calculated and the copper losses are measured
B. the no-load losses are measured and the copper losses are calculated
C. both the no-load losses and the copper losses are measured
D. both the no-load losses and the copper losses are calculated

285. Commutation conditions at full-load for a large DC machine can be checked by
A. the brake test B. the Swinburne's test
C. the Hopkinson's test D. none of these

286. Ohmic losses in a DC machine occur in the
A. armature winding only
B. field winding only
C. brush contact only
D. armature winding, the field winding and also in the brush contact

287. The field efficiency test of DC series motors overcomes the difficulty of obtaining readings relatively for light loads by connecting the
A. series field of the generator in series with the motor armature
B. series field of the motor in series with the generator armature
C. armature of the generator in series with the motor
D. armature of the motor in series with the generator

288. In the Kapp's modification of Hopkinson's efficiency test
A. the power losses in the two machines are supplied mechanically
B. the power output of the generator is dissipated in a resistor
C. the two machines are mechanically decoupled
D. the power losses in the two machines are supplied electrically

289. DC motor starters are used
A. to increase the starting torque
B. to limit the starting current
C. both (A) and (B) are correct
D. none of these

290. For DC shunt motors, the field excitation is kept at maximum value during starting to
A. increase acceleration time
B. decrease starting torque
C. reduce armature heating
D. prevent voltage dip in the supply mains

Ans. 275. D 276. B 277. C 278. D 279. B 280. C 281. A 282. A 283. C 284. B 285. C 286. D 287. A 288. D 289. B 290. C

291. DC motors should be stopped by opening the line switches and not by forcing the starter handle back to the off position because
A. heavy sparking occurs at the brushes
B. heavy sparking occurs at the first stud of the starting resistance steps
C. both (A) and (B) are correct
D. none of these

292. A dc shunt motor has two additional resistances R_1 and R_2 in the field circuit and armature circuit respectively. The starting armature current can be kept to a minimum by keeping.
A. R_1 maximum and R_2 maximum
B. R_1 minimum and R_2 maximum
C. R_1 maximum and R_2 minimum
A. R_1 minimum and R_2 minimum

293. Three point starters of DC shunt motors are not used in applications where speed variation by field flux control is required because the motor may
A. stop at very high speeds
B. stop at very low speeds
C. stop both at very high and at very low speeds
D. run away

294. Direct-on-line starters are not suitable for starting large DC motors because
A. the motor may not start
B. large voltage drop may occur in the supply mains
C. the starting torque becomes very low
D. the motor may run away

295. For DC shunt motor, speed control by the variation of field flux is best suited for
A. constant power drive
B. variable power drive
C. constant torque drive
D. variable torque drive

296. If the field circuit of a DC shunt motor running at rated speed gets open-circuited, then immediately after this the speed of the motor would tend to
A. decrease B. increase
C. remain unchanged
D. oscillate around the rated speed

297. A DC shunt motor is driving a constant torque load without any additional resistance in the armature circuit. If an additional resistance is placed in the armature circuit then the speed of the motor
A. increases B. decreases
C. remains unchanged D. becomes zero

298. A DC shunt motor is running at rated speed with rated excitation, rated voltage and with an additional resistance in the armature circuit. Speeds less than the rated speed can be achieved by
A. reducing the supply voltage and increasing the field excitation
B. increasing the supply voltage and decreasing the armature circuit resistance
C. decreasing the field excitation and increasing the supply voltage
D. decreasing the armature circuit resistance and decreasing the field excitation

299. A DC shunt motor is running at rated speed with rated supply voltage. If the supply voltage is halved, then the speed of the motor becomes
A. half of the rated speed
B. double the rated speed
C. slightly less than the rated speed
D. slightly more than the rated speed

300. In Ward-Leonard method of speed control the direction of rotation of the motor is reversed usually by
A. reversing the connections of the generator field terminals
B. reversing the connection of the generator armature terminals
C. reversing the connections of the motor armature terminals
D. none of these

301. A DC series motor is driving a load with a diverter connected across its armature. If the diverter resistance is decreased, the speed of the motor
A. increases B. decreases
C. remains unchanged D. becomes zero

302. A DC series motor is running with a diverter connected across its field winding. If the diverter resistance is increased then the speed of the motor
A. decreases B. increases
C. remains unchanged
D. becomes excessively high

303. A DC shunt motor is driving a mechanical load at rated voltage and rated excitation. If the load torque becomes double then the speed of the motor
A. increases slightly B. decreases slightly
C. becomes double D. becomes half

304. Two DC series motors connected in series are driving the same mechanical load. If the motors are now connected in parallel the speed becomes slightly
A. less than double B. less than half
C. more than double D. more than half

305. A DC series motor is running at rated speed without any additional resistance in series. If an additional resistance is placed in series, the speed of the motor
A. increases
B. decreases
C. remains unchanged
D. oscillates around the rated speed

Ans. 291. B 292. B 293. A 294. B 295. A 296. B 297. B 298. A 299. C 300. A 301. B 302. A 303. B 304. C 305. B

306. The most economic method of electrical braking is
 A. regenerative braking
 B. dynamic braking with self excitation
 C. dynamic braking with separate excitation
 D. plugging

307. A DC series is running at rated speed with rated excitation. The motor has two resistances R_1 and R_2 connected across the armature and the field respectively. Speeds above the rated speed can be achieved by
 A. decreasing R_1 only B. increasing R_2 only
 C. decreasing R_1 and increasing R_2
 D. increasing R_1 and decreasing R_2

308. Plugging of DC motors is normally executed by
 A. reversing the field polarity
 B. reversing the armature polarity
 C. reversing both armature and field polarity
 D. connecting a resistance across the armature

309. For nonreversing DC drives it is preferable to employ
 A. regenerative braking
 B. dynamic braking with separate excitation
 C. dynamic braking with self excitation
 D. plugging

310. A DC shunt motor is driving a constant torque load with rated excitation. If the field current is halved then the speed of the motor becomes
 A. half B. slightly more than half
 C. double
 D. slightly less than double

311. Consider the following statements. On cross-field generators
 I. to increase amplification factor compensating winding should be provided
 II. the reversal of speed does not reverse the output voltage
 III. the number of brush studs is double that of poles
 Select correct answer from the list given below :
 A. only I and II are true
 B. only I and III are true
 C. only II and III are true
 D. all the statements are true

312. An amplitude has split poles to
 A. provide space for interpoles
 B. increase amplification factor
 C. increase efficiency
 D. damp out mechanical oscillations

313. In DC machines, the axis of armature mmf is along the
 A. direct axis B. interpolar axis
 C. midway between the direct and inter-polar axes
 D. none of these

314. Rosenberg generator is a constant
 A. voltage generator at low speeds and voltage polarity is independent of the direction of rotation
 B. voltage generator at low speeds and voltage polarity changes with the direction of rotation
 C. current generator at high speeds and voltage polarity is independent of the direction of rotation
 D. current generator at high speeds and voltage polarity changes with the direction of rotation

315. If the number of poles in the first stage of a Rototrol is 2, the number of poles in the second stage is
 A. 2 B. 4
 C. 6 D. 8

316. If the number of poles in a cross-field generator is 2, the number of interpoles will be
 A. 2 B. 4
 C. 6 D. 8

317. In DC machines, the armature windings are placed on the rotor because of the necessity for
 A. electromechanical energy conversion
 B. generation of voltage
 C. commutation D. development of torque

318. In DC machines, the space distribution of air-gap flux density wave at no-load is
 A. sinusoidal B. co-sinusoidal
 C. flat-topped D. rectangular

319. A commutator in DC machines provides
 A. half-wave rectification
 B. full-wave rectification
 C. half-wave controlled rectification
 D. full-wave controlled rectification

320. The inductor of a three wire generator should be iron-cored to
 A. increase the current through it
 B. reduce the current through it
 C. reduce the voltage ripple in the output circuit
 D. improve the voltage regulation of the generator

321. A commutator in DC machines can convert
 A. AC to DC
 B. DC to AC
 C. both AC to DC and DC to AC
 D. none of these

322. In DC machines, the space waveform of the air-gap flux distribution affects
 A. torque but not the voltage
 B. voltage but not the torque
 C. neither the voltage nor the torque
 D. both the torque and voltage

323. The magnitude of E_b is given by
 A. $E_b = V - I_a R_a$ B. $E_b = V + I_a R_a$
 C. $E_b = V_b + I_a R_a$ D. $E_b = V_b - I_a R_a$

Ans. 306. A 307. C 308. B 309. B 310. D 311. D 312. A 313. B 314. C 315. C 316. B 317. C 318. C 319. B 320. B 321. C 322. C 323. A

324. The back emf of the DC motor
A. opposes the applied voltage
B. has no effect on the applied voltage
C. favours the applied voltage
D. none of these
325. The speed of a DC shunt motor can be increased more than the rated speed by connecting variable resistance in series with the
A. line B. armature winding
C. field winding D. None of these
326. The construction of DC motor is
A. similar as of DC generator
B. different than DC generator
C. similar but different in frame construction
D. similar in construction and similar in frame construction
327. Motor is called compound when
A. series field flux has an additive effect with shunt field flux
B. series field flux opposes the shunt field flux
C. polarities of series and shunt poles are same
D. it has an additive effect and the polarity of poles are same
328. The function of the starter is to
A. limit the armature current at the time of starting
B. protect the motor from overloading
C. protect the motor from low voltage
D. all of these
329. Maximum efficiency of the motor will occur when
A. Copper losses > Iron losses
B. Copper losses < Iron losses
C. Copper losses = Friction losses
D. Copper losses = Constant losses
330. If a flux of DC motor approaches zero, its speed will be
A. infinity
B. zero
C. between zero and infinity
D. no change in speed
331. The voltage at the terminals of a DC series generator running at rated rpm at no-load will be
A. more than the rated voltage
B. full rated voltage
C. a very small voltage
D. zero
332. A DC shunt motor runs at 200 V supply, if the armature current is 20 A and resistance of the armature is 0.5 ohm, the back emf developed will be
A. 210 V B. 200 V
C. 190 V D. 180 V
333. The load current and flux of a DC motor are kept constant, if the supply voltage is increased by 20%, its speed will
A. remain unchanged B. increase by 20%
C. decrease by 20%
D. depend upon the armature resistance
334. Interpoles are connected in series with the
A. line B. armature winding
C. shunt field winding
D. line and armature winding
335. If the field connection of a loaded shunt motor is suddenly disconnected
A. its speed will become very high
B. it will draw very high current
C. the fuse will be blown off
D. the motor will develop no torque due to zero flux
336. When the field terminals and armature terminals are both interchanged of DC shunt motor, then the
A. motor will not run
B. motor will run in the same direction
C. motor will run in opposite direction
D. speed of motor will change
337. The magnitude of the back emf depends upon
A. flux per pole B. speed of the motor
C. number of the parallel paths in the armature
D. all of these
338. For lifts, which of the following types of DC motor is used?
A. Compound motor
B. Cumulative compound motor
C. Shunt motor D. Series motor
339. In a DC machine, which loss increases rapidly as compared to other, when the number of magnetic reversals is increased?
A. Copper loss B. Eddy current loss
C. Friction loss D. Hyteresis loss
340. If the connections of the armature terminals of DC shunt motor are interchanged then
A. armature will not run
B. armature will rotate in opposite direction
C. armature will rotate in same direction
D. speed of the motor will fall
341. If the connection of the field winding of DC shunt motor is changed
A. the motor will not run
B. it will run in the same direction
C. it will run in the opposite direction
D. motor speed will change
342. Ward–Leonard method for controlling the speed of DC, shunt motor is used
A. where wide range from zero to normal speed is required
B. where speed in both the directions is required
C. where accuracy in speed control is required
D. all of these
343. In a large size machines, compensating winding is used for
A. neutralising the voltage generated in the coil due to high fluctuation of load

Ans. 324. A 325. C 326. C 327. D 328. D 329. C 330. A 331. C 332. A 333. B 334. B 335. C 336. A 337. D 338. D 339. B 340. B 341. C 342. D 343. A

B. compensating the fall in speed due to increase of load
C. compensating the losses
D. all of these

344. If in a four pole lap wound DC machine, if air gap along each pole is not the same, this is likely to result in
A. reduced eddy currents
B. disproportionate increases in windage losses
C. current per parallel path not being the same
D. higher terminal voltage

345. A DC shunt generator produces 450 A at 230 V. The resistances of shunt field and the armature are 50 ohm and 0.025 ohm respectively. The armature voltage drop will be
A. 11.39 V B. 22.7 V
C. 31.6 V D. 38.4 V

346. For parallel operation, the DC generators normally preferred are
A. Series generators B. Shunt generators
C. Compound generators
D. Series, shunt and compound generators

347. Residual magnetism is the essential prerequisite for starting which type of DC generator?
A. Separately excited generator
B. Shunt generator
C. Series generator D. All of these

348. A cumulatively compounded DC generator is supplying 20 A at 200 V. Now if the series field winding is short circuited the terminal voltage will
A. remain unaltered at 200 V
B. rise to 220 V
C. shoot up to a very high value
D. become less than 200 V

349. A 250-V, 50-HP 1000 rpm, DC shunt motor drives a load that requires a constant torque regardless of the speed of operation. The armature circuit resistance is 0.04 ohm when this motor delivers rated power, the armature current is 160 amperes. If the flux is reduced to 70% of its original value, the new value of armature current will be
A. 114.5 A B. 229 A
C. 318 A D. 348 A.

350. The full load current of a 20 HP, 500 V DC motor will be closer to
A. 100 A B. 60 A
C. 35 A D. 15 A

351. The following controls are considered for DC motors:
I. Control of flux
II. Armature resistance control
III. Supply voltage control.
Which of the following controls play significant role in the speed control of DC motors?
A. I only B. II and III only
C. I and III only D. I, II and III

352. If the load current and flux of a DC motor are held constant and voltage applied across its armature is increase by 5%, the speed of motor will
A. increase by about 5%
B. reduce by about 5%
C. remain unaltered
D. depends on other factors

353. The armature voltage control is considered as suitable, in case the DC machine is driven at
A. constant torque B. constant speed
C. constant power D. constant current

354. A 5-HP series motor is provided with a diverter for controlling the speed. For a constant load torque the speed will be minimum when the diverter resistance is
A. zero B. infinite
C. 0.05 ohm D. 2 ohm

355. The value of the diverter resistance for a series motor is of the order of
A. 0.1 ohm B. 25 ohm
C. 100 ohm D. none of these.

356. A shunt motor is fitted with a field regulator for speed control. For a constant load torque the speed will be mininum when the resistance of the regulator is
A. 0 ohm B. infinite
C. about 100 ohms D. none of these

357. The resistance of the field regulator of a DC shunt motor is of the order of
A. 0.05 ohm B. 15 ohm
C. 100 ohm D. none of these

358. For low values of armature current I_a, the torque of a series motor is proportional to
A. I_a B. I_a^2
C. I_a^3 D. $\sqrt{I_a}$

359. The efficiency of a motor is generally expressed as
A. (input – losses)/input
B. output/input
C. output/(output + losses)
D. none of these

360. The torque of a shunt motor is proportional to
A. armature current
B. applied voltage
C. square of the armature current
D. none of these

361. In DC machines
A. torque and induced emf are produced both in motor and generator
B. torque is produced in motor and emf is induced in generator
C. torque is produced in generator and emf is induced in motor
D. none of these

Ans. 344. C 345. A 346. B 347. C 348. D 349. B 350. C 351. D 352. A 363. A 354. B 355. A 356. A 357. C 358. B 359. A 360. A 361. A

362. For high values of armature current, the speed of a DC series motor is
A. proportional to armature current
B. inversely proportional to armature current
C. constant
D. none of these

363. The DC compound motors are generally
A. commutative compound
B. differential compound
C. level compound
D. none of these

364. A 220 V DC motor is supplying 5 metric horse power. If the efficiency is 0.95 the current is
A. 17.6 A
B. 17.85 A
C. 19.2 A
D. 20.1 A

365. A 500 V shunt motor has an armature resistance of 0.5 ohm and a field resistance of 100 ohm. The full load rated current is 10 A, If the motor is directly switched on to the mains, the starting current would be
A. 5 A
B. 1000 A
C. 100 A
D. 10 A

366. The resistance of the starter of a 220 V, 5 hp, DC shunt motor is of the order of
A. 0.1 ohm
B. 1 ohm
C. 10 ohms
D. 0.01 ohm

367. If the field current and armature current of a DC shunt motor are constant and the applied voltage increased by 5%, the speed will
A. increase by 5%
B. decrease by 5%
C. remain constant
D. none of these

368. A series motor should not be started on no-load because
A. the starting current will be high
B. the speed of the motor will be very high
C. excessive sparking will occur at commutator
D. none of these

369. A series motor is driving a load whose torque varies as square of the speed. At a speed of 200 rpm the line current is 10 A. If the speed is 400 rpm the line current would be
A. 5 A
B. 14 A
C. 20 A
D. none of these

370. Two DC machines 500 kW each are tested by Hopkinson test. The power input would be of the order of
A. 500 kW
B. 100 kW
C. 1000 kW
D. none of these

371. Name the stationary part in DC machine
A. The stator of an induction motor
B. The stator of a salient pole synchronous motor
C. The stator of a wound rotor synchronous motor
D. The stator of a DC machine with commutating poles

372. Which of the following winding is connected in series with armature winding of DC machine?
A. Series winding
B. Auxiliary series winding
C. Compensating winding
D. Commutating pole winding

373. In a DC machine how are the commutating pole winding connected?
A. In series with the shunt winding
B. Parallel to the shunt winding
C. In series with the armature winding
D. Parallel to the armature winding

374. The direction of rotation of a DC motor is reversed by
A. reversing armature connections
B. interchanging the armature and field connection
C. adding resistance to the field circuit
D. reversing supply connections

375. The basic function of a rectifier is to
A. change the level of a DC voltage
B. convert DC into AC
C. change frequency of AC voltage
D. convert AC into DC

376. Controlled rectifiers employ
A. diodes
B. thyristors
C. bridge circuits
D. both (A) and (B) are correct

377. Thyristor chopper circuits are employed for
A. lowering the level of a DC voltage
B. rectifying the AC voltage
C. frequency conversion
D. providing commutating circuitry

378. An inverter circuit is employed to convert
A. AC voltage into DC voltage
B. DC voltage into AC voltage
C. high frequency into low frequency
D. low frequency into high frequency

379. One of the main advantages of Swin burne's test is that it
A. is applicable both to shunt and compound motors
B. needs one running test
C. is very economical and convenient
D. ignores any change in iron loss

380. The main disadvantage of Hopkinson's test for finding efficiency of shunt dc motors is that it
A. requires full-load power
B. ignores any change in iron loss
C. needs one motor and one generator
D. requires two identical shunt machines

381. The most economical method of finding no-load losses of a large DC shunt motor?
A. Hopkinson's
B. Swinburne's
C. Retardation
D. Field's

382. Retardation test on a DC shunt motor is used for finding–losses.
A. Stray
B. Copper
C. Friction
D. Iron

Ans. 362. C 363. A 364. A 365. B 366. C 367. A 368. B 369. C 370. B 371. D 372. D 373. C 374. A 375. D 376. D 377. A 378. B 379. C 380. D 381. B 382. A

383. The main thing common between Hopkinson's test and Field's test is that both
A. require two electrically-coupled series motors
B. need two similar mechanically-coupled motors
C. use negligible power
D. are regenerative tests

384. Which test is the usual test for determining the efficiency of a traction motor?
A. Field's B. Retardation
C. Hopkinson's D. Swinburne's

385. The armature of DC motors is laminated to reduce the
A. hysteresis losses B. eddy current losses
C. inductivity of armature
D. mass of the armature

386. Which of the following DC motors has the least drop in speed between no-load to nominal load?
A. Shunt motor with commutating poles
B. Series motor without commutating poles
C. Compound motor without commutating poles
D. Series motor with commutating poles

387. What happens if the field winding of running shunt motor suddenly breaks open?
A. Its speed slows down
B. Its speed becomes dangerously high
C. It gives out sparks D. It stops at once

388. The speed of DC series motor decreases if the flux in the field winding
A. remains constant B. increases
C. decrease D. none of these

389. Which of the following is a correct statement about a series motor?
A. Its field winding consists of thicker wire and less turns
B. It can run easily without load
C. It has an almost constant speed
D. It has poor torque

390. Which of the following motors is used to derive the constant speed line shafting lathes, blowers and fans?
A. DC shunt motor B. DC series motor
C. Commutative compound motor
D. None of these

391. Which of the following motors is used for rolling mills?
A. DC shunt motor B. DC series motor
C. DC commutative compound motor
D. DC differential compound motor

392. The motor used for intermittent, high torque loads is
A. DC shunt motor B. DC series motor
C. differential compound motor
D. commutative compound motor

393. The field flux of a DC motor can be controlled to achieve the speeds
A. lower than rated speed
B. higher than rated speed
C. at rated speed D. none of these

394. If the back emf of a DC motor is absent, then motor will
A. run at very high speed
B. run at very slow speed
C. not run at all D. burn

395. By providing a variable resistance across the series field (diverter) in a DC series motor, speeds above normal can be obtained because
A. armature current decreases
B. line current decreases
C. flux is reduced D. None of these

396. The speed of a series motor at no-load is
A. zero B. 3000 rpm
C. 3600 rpm D. infinity

397. DC series motors are best suited for traction work because
A. torque is proportional to the square of armature current and speed is inversely proportional to torque
B. torque is proportional to the square of armature current and speed is proportional to torque
C. torque and speed are proportional to square of armature current
D. None of these.

398. If a series motor is started without load, the effect is that the
A. torque increases rapidly
B. speed increases rapidly
C. current drawn increases rapidly
D. back emf decreases

399. If the field of a DC shunt motor is opened
A. it will continue to run at its rated speed
B. the speed of the motor will become very high
C. the motor will stop
D. the speed of motor will decrease

400. Armature reaction is attributed to
A. the effect of magnetic field setup by armature current
B. the effect of magnetic field setup by field current
C. copper loss in the armature
D. the effect of magnetic field setup by back emf

401. What will happen if the supply terminals of DC shunt motor are interchanged?
A. Motor will stop
B. Motor will run at its normal speed in the same direction as it was running
C. The direction of rotation will reverse
D. Motor will run at a speed lower than the normal speed in the same direction

402. The direction of rotation of a DC series motor can be reversed by interchanging
A. the supply terminals only
B. the field terminals only

Ans. 383. B 384. A 385. B 386. A 387. B 388. B 389. A 390. A 391. C 392. D 393. B 394. D 395. C 396. D 397. A 398. B 399. B 400. A 401. B 402. B

C. the supply as well as field terminals
D. none of these

403. When the electric train is moving down a hill the DC motor acts as
A. DC series motor B. DC shunt motor
C. DC series generator D. DC shunt generator

404. Which of the following DC motor is suitable for high starting torque?
A. Shunt motor B. Series motor
C. Commutative compound motor
D. Compound motor

405. For which of the following DC motors is the typical field of application mentioned?
A. Shunt motor: electric trains
B. Series motor: machine tools
C. Series motor: belt drive
D. Compound motor: flywheel drive

406. Why is the air gap between stator and armature of an electric motor kept as small as possible?
A. To get stronger magnetic field
B. To make the rotation easier
C. To reach a higher speed of rotation
D. To improve the air circulation

407. Which of the following types of generators gives constant voltage output at all loads?
A. Series generator B. Shunt generator
C. Short shunt compound generator
D. Level compound generator

408. The generator of a power station produces an electric pressure by
A. conversion of heat B. magnetic induction
C. conversion of light D. mechanical pressure

409. Interpoles are meant for
A. increasing the speed'of the motor
B. decreasing counter emf
C. reducing sparking at the commutator
D. converting armature current to DC

410. A DC shunt generator driven at a normal speed in the normal direction fails to build up armature voltage because
A. the resistance of the armature is high
B. there is no residual magnetism
C. the field current is too small
D. none of these

411. Which of the following conditions are necessary for build-up of a shunt generator?
I. There must be some residual magnetism in the poles
II. For the given direction of rotation, the shunt field coils should be correctly connected to the armature
III. If excited on open circuit, its shunt field resistance should be greater than the critical resistance
IV. If excited on load, then its shunt field resistance should be more than a certain minimum value of resistance, which is given by the internal characteristics
A. (I) and (II) B. (II) and (III)
C. (I) and (IV) D. (I), (II) and (IV)

412. The consideration involved in the selection of the type of electric drive for a particular application depends on
A. speed control range and its nature
B. environmental conditions
C. starting torque D. all of these

413. Which of the following is preferred for automatic drives?
A. Squirrel cage induction motor
B. Ward Leonard Controlled dc motors
C. Synchronous motors D. any of these

414. Which type of drive can be used for hoisting machinery?
A. Slip ring induction motor
B. Ward–Leonard controlled DC shunt motor
C. DC compound motor
D. Any of these

415. The motor normaly used for crane is
A. Slip ring induction motor
B. Ward–Leonard controlled DC shunt motor
C. Synchronous motor
D. DC differentially compound motor

416. When smooth and precise speed control over a wide range is desired, the motor preferred is
A. synchronous motor
B. squirrel cage induction motor
C. wound rotor induction motor
D. DC motor

417. In the chopper speed control method for a wound rotor induction motor, the motor speed inversely depends on
A. fixed resistor across the rectifier
B. chopper switching frequency
C. chopper ON time T_{ON}
D. both B and C

418. A steel mill requires a motor having high starting torque, with speed range and precise speed control. Which one of the following motors will you choose?
A. DC shunt motor B. synchronous motor
C. DC series motor
D. slip-ring induction motor

419. Heavy-duty steel-works cranes which have wide load variations are equipped with ..... motor.
A. slip-ring induction B. DC series
C. double squirrel-cage
D. cumulative compound

Ans. 403. C 404. B 405. D 406. A 407. D 408. D 409. C 410. B 411. D 412. D 413. B 414. A 415. A 416. D 417. D 418. C 419. B

MULTIPLE CHOICE QUESTIONS FROM VARIOUS COMPETITIVE EXAMINATIONS

1. Consider the following statements:
   The use of interpoles in DC machines is to counteract the **(IES 2001)**
   1. reactance voltage
   2. demagnetizing effect of armature mmf in the commutating zone
   3. cross-magnetizing effect of armature mmf in the commutating zone

   Which of these statement(s) is/are correct?
   A. 1 and 2 B. 2 and 3
   C. 1 and 3 D. 3 alone
2. The current drawn by a 220 V, DC motor of armature resistance 0.5 Ω and back emf 200 V is **(IES 2001)**
   A. 40 A B. 44 A
   C. 400 A D. 440 A
3. A DC shunt generator when driven without connecting field winding shown an open circuit terminal voltage of 12 V. When field winding is connected and excited, the terminal voltage drops to zero because **(IES 2001)**
   A. field resistance is higher than critical resistance
   B. there is no residual magnetism in the field circuit
   C. field winding has got wrongly connected
   D. there is a fault in armature circuit
4. A DC series motor is accidentally connected to single phase AC supply. The torque produced will be **(IES 2002)**
   A. oscillating B. of zero average value
   C. steady and unidirectional
   D. pulsating and unidirectional
5. For a given torque, reducing the diverter-resistance of a DC series motor **(IES 2002)**
   A. increases its speed but armature current remains the same
   B. increases its speed demanding more armature current
   C. decreases its speed demanding less armature current
   D. decreases its speed but armature current remains the same
6. The direction of rotation of a DC series motor can be reversed **(IES 2003)**
   A. by interchanging supply
   B. by interchanging field terminals
   C. either by interchanging supply terminals or by interchanging field terminals
   D. by interchanging supply terminals as well as field terminals
7. The current drawn by a 120 V DC motor with back emf of 110 V and armature resistance of 0.4 ohm is **(IES 2003)**
   A. 4 A B. 25 A
   C. 274 A D. 300 A
8. Armature torque of a DC motor is a function of which of the following factors? **(IES 2003)**
   1. Speed 2. Field flux
   3. Armature current 4. Residual magnetism

   Select the correct answer using the codes given below:
   A. 2 and 3 B. 1 and 4
   C. 3 and 4 D. 1 and 2
9. The dummy coils in DC machines are useful to **(IES 2003)**
   A. increase the efficiency
   B. improve the commutation
   C. reduce the cost of the machine
   D. maintain mechanical balance of armature
10. The speed of a DC shunt motor may be varied by varying **(IES 2003)**
    1. field current 2. supply voltage
    3. armature circuit

    A. 1, 2 and 3 B. 1 and 2
    C. 1 and 3 D. 2 and 3
11. Four types of DC generators of constant speed are considered List-I. Their external characteristics at constant speed are given in List-II. Match List-I (Type of DC generator) with List-II (External characteristics) and select the correct answer using the code given below: **(IES 2004)**

    List-I
    a. Separately excited b. Series excited
    c. Shunt excited
    d. Over-compound excited

    List-II

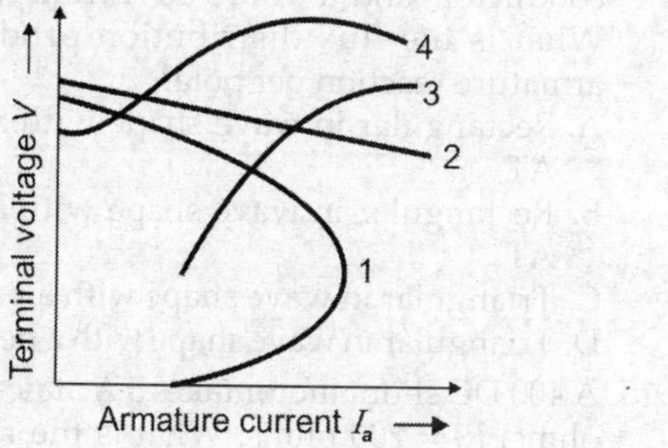

Codes:

	a	b	c	d		a	b	c	d
A.	2	3	1	4	B.	1	4	2	3
C.	1	3	2	4	D.	2	4	1	3

Ans. 1. A 2. A 3. C 4. D 5. B 6. C 7. B 8. A 9. D 10. A 11. A

12. What is the increase in the torque expressed as percentage of initial torque, if the current drawn by a DC series motor is increased from 10 A to 12 A (neglect saturation)? **(IES 2004)**
A. 21% B. 25%
C. 41% D. 44%

13. In a DC machine, for the same value of Φ, Z and N; which one of the following statement is correct? **(IES 2004)**
A. Armature emf is more with wave winding than with lap winding
B. Armature emf is less with wave winding than with lap winding
C. Armature emf depends on whether the machine is running as a motor or a generator
D. Armature emf is the same as long as the flux density in the air gap remains the same

14. Consider the following statements:
In a DC machine, iron loss occurs in
1. armature core 2. yoke
3. pole cores 4. pole shoes
Which of these statements are correct? **(IES 2004)**
A. 1 and 4 B. 1 and 2
C. 1 and 3 D. 2 and 3

15. The armature resistance of a 6-pole lap wound DC machine is 0.05 Ω. If the armature is rewound as a wave-winding, what is the armature resistance? **(IES 2005)**
A. 0.45 Ω B. 0.30 Ω
C. 0.15 Ω D. 0.10 Ω

16. In a DC compound generator, "flat-compound" characteristic, required for certain applications, may be obtained by connecting a variable resistance **(IES 2005)**
A. across the series field
B. in series with the series field
C. in parallel with the shunt field
D. in series wih he shunt field

17. A 6-pole lap wound DC machine armature has 720 conductors and it draws 50 A from supply mains. What is the flux distribution produced by the armature reaction per pole? **(IES 2005)**
A. Rectangular in wave shape with a peak of 500 AT
B. Rectangular in wave shape with a peak of 100 AT
C. Triangular in wave shape with a peak of 500 AT
D. Triangular in wave shape with a peak of 1000 AT

18. A 400 DC shunt motor takes 5 A at no-load. $R_a = 0.5$ ohms, $R_f = 200$ ohms. What is the ratio of speed from full load to no-load, when the DC shunt motor takes 50 A on full load? **(IES 2005)**
A. 0.94 B. 0.8
C. 0.6 D. 0.4

19. A 50 kW DC shunt motor is loaded to draw rated armature current at any given speed. When driven (i) at half the rated speed by armature voltage control and (ii) at 1.5 times the rated speed by field control, what are the approximate output powers delivered by the motor? **(IES 2006)**
A. 25 kW in (i) and 75 kW in (ii)
B. 25 kW in (i) and 50 kW in (ii)
C. 50 kW in (i) and 75 kW in (ii)
D. 50 kW in (i) and 50 kW in (ii)

20. For a constant supply voltage, what are the effects of inserting a series resistance in the field circuit of a DC shunt motor, on its speed and torque? **(IES 2006)**
A. Speed will decrease and the torque will decrease
B. Speed will increase and the torque will increase
C. Speed will increase and the torque will decrease
D. Speed will decrease and the torque will increase

21. A self-excited DC shunt generator, driven by its prime-mover at the rated speed fails to built up voltage across its terminal at no-load. What reason can be assigned for this? **(IES 2006)**
A. The field circuit resistance is higher than the critical resistance
B. The initial shunt field mmf does not assist the residual magnetism
C. One of the inter-pole connections is reversed
D. The brush-axis shifts slightly from the geometrical neutral axis of the machine

22. Wave winding is employed in a DC machine of **(IES 2006)**
A. high current and low voltage rating
B. low current and high voltage rating
C. high current and high voltage rating
D. low current and low voltage rating

23. If the armature current is increased to double its previous value, and the time of commutation is halved, how will the reactance voltage vary? **(IES 2007)**
A. It will be halved B. It will remain the same
C. It will be doubled
D. It will become four times

24. Neglecting all losses, how is the developed torque (*T*) of a DC separately excited motor, operating under constant terminal voltage, related to its output power (*P*)? **(IES 2007)**
A. $T \propto \sqrt{P}$ B. $T \propto P$
C. $T^2 \propto P^3$ D. *T* is independent of *P*

25. A DC shunt generator is supplying a load of 1.8 kW at 200 V. Its armature and field resistance are 0.4 Ω and 200 Ω respectively. What is the generated emf? **(IES 2007)**

Ans. 12. D 13. A 14. B 15. A 16. A 17. C 18. A 19. B 20. C 21. B 22. B 23. D 24. B

A. 190 V B. 196 V
C. 240 V D. 210 V

26. The field flux method of speed control of DC shunt motor is generally used for which one of the following? **(IES 2007)**
A. Constant hp drive below the base speed
B. Constant torque drive above the base speed
C. Constant torque drive below the base speed
D. Constant hp drive above the base speed

27. A 5 kW, 200 V DC shunt motor has armature resistance of 1 ohm and shunt field resistance of 100 ohm. At no load, the motor draws 6 A from 200 V supply and runs at 100 rpm. What is the total copper loss of the machine? **(IES 2007)**
A. 400 W B. 16 W
C. 36 W D. 416 W

28. Torque developed by the armature of a DC motor is proportional to which one of the following? **(IES 2007)**
A. (Back emf) × (Armature current)
B. (Magnetic flux per pole) × (Armature current)
C. (Back emf) × (Magnetic flux per pole)
D. (Back emf)/(Magnetic flux per pole)

29. The graph shown below represents which characteristic of a DC shunt generator? **(IES 2008)**

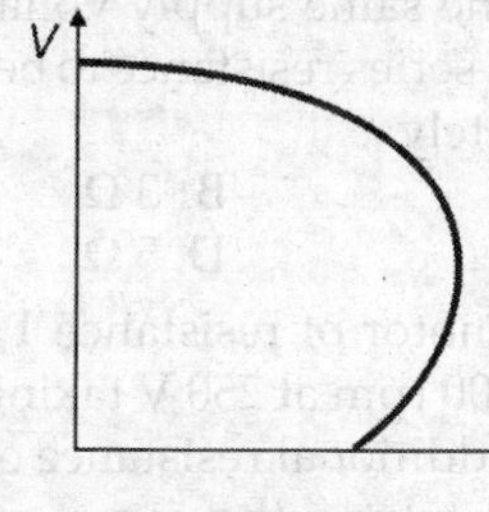

A. Internal characteristic
B. External characteristic
C. Open-circuit characteristic
D. Magnetic characteristic

30. When is the mechanical power developed by a DC motor maximum? **(IES 2008)**
A. Back emf is equal to applied voltage
B. Back emf is equal to zero
C. Back emf is equal to half the applied voltage
D. None of the above

31. A shunt generator has a critical field resistance of 200 Ω at a speed of 800 rpm. If the speed of the generator is increased to 1000 rpm, what is the change in the critical field resistance of the generator? **(IES 2008)**
A. Decrease to 160 Ω B. Increases to 250 Ω
C. Remains the same at 200 Ω
D. Increases to 312.5 Ω

32. The speed of a DC motor is related to the back emf and flux in the following ways: **(IES 2012)**
A. Directly proportional to flux and inversely proportional to back emf
B. Directly proportional to back emf and inversely proportional to flux
C. Inversely proportional to flux and inversely proportional to back emf
D. Directly proportional

33. A 4-pole DC generator is running at 1500 rpm. The frequency of current in the armature winding is **(IES 2012)**
A. 50 Hz B. 150 Hz
C. 25 Hz D. 100 Hz

34. A separately excited DC generator is feeding a DC shunt motor. If the load torque on the motor is reduced to half, then **(IES 2012)**
A. the armature current of both motor and generator are reduced to half
B. the armature current of motor is halved and that of generator remains unchanged
C. the armature current of generator is halved and that of motor remains unchanged
D. the armature current of both machines remains unchanged

35. A DC series motor is running at rated speed and rated voltage, feeding a constant power load. If the speed has to be reduced to 0.25 pu, the supply voltage should be reduced to **(IES 2012)**
A. 0.75 pu B. 0.5 pu
C. 0.25 pu D. 0.075 pu

36. The overall efficiency of a DC shunt generator is maximum when its variable loss equals **(IES 2012)**
A. the stray loss B. the iron loss
C. constant loss D. mechanical loss

37. A triangular mmf wave is produced in the air-gap of an electric machine. Such a wave is produced by **(IES 2013)**
A. stator of an induction machine
B. rotor of a synchronous machine
C. stator of a DC machine
D. rotor of a DC machine

38. At 1200 rpm the reduced emf of a DC machine is 200 V. For an armature current of 15 A the electromagnetic torque produced would be **(IES 2013)**
A. 23.8 Nm B. 238 Nm
C. 2000 Nm D. 3000 Nm

39. A 4-pole lap wound DC generator has a developed power of P watt and voltage of E volt. The adjacent brushes of the machine are removed as they are worn out. If the machine operates with the remaining brushes, the developed voltage and power that can be obtained from the machine are **(IES 2013)**
A. E and P B. $E/2$ and $P/2$
C. E and $P/4$ D. E and $P/2$

Ans.													
	25. C	26. D	27. D	28. B	29. B	30. C	31. B	32. B	33. A	34. A	35. B	36. C	37. D
	38. A	39. C											

40. A 6-pole DC armature has simplex lap-connected 720 conductors, 3 turns per coil and 4 coil-sides per slot. Determine the number of slots in the armature and state whether equalizers can be employed or not. **(IES 2013)**
A. 60 slots and not possible
B. 30 slots and possible
C. 60 slots and possible
D. 30 slots and not possible

41. Determine the flux per pole for 6-pole DC machine having 240 wave connected conductors, which generates an open circuit voltage of 500 volt, which is running at 1000 rpm. **(IES 2013)**
A. 0.129 Wb B. 0.021 Wb
C. 0.042 Wb D. 7 mWb

42. Consider the following statements: **(IES 2013)**
The armature reaction mmf in a DC machine is
1. stationary with respect to the field poles
2. rotating with respect to the field poles
3. rotating with respect to the armature
Which of these statements are correct?
A. 1, 2 and 3 B. 1 and 2 only
C. 1 and 3 only D. 2 and 3 only

43. In case of DC motor, maximum mechanical power is developed when back emf equals **(IES 2013)**
A. the applied voltage
B. half the applied voltage
C. one third of the applied voltage
D. double the applied voltage

44. The induced emf of a DC machine running at 750 rpm is 220 V. The percentage increase in field flux for generating an induced emf of 250 V and 700 rpm would be **(IES 2013)**
A. 7% B. 11.25%
C. 21.7% D. 42.4%

45. The speed of a separately excited DC motor is varied by varying the armature voltage in the range zero the base speed and by varying the field current above the base speed. It is suitable for constant **(IES 2013)**
A. torque drive at all speeds
B. power drive at all speeds
C. torque drive till base speed and constant power drive beyond base speed
D. power drive till base speed and constant torque drive beyond base speed

46. A DC shunt motor is required to drive a constant power load at rated speed while drawing rated armature current. Neglecting saturation and all machine losses, if both the terminal voltage and the field current of the machine are halved then **(IES 2013)**
A. the speed becomes 2 pu but armature current remains at 1 pu
B. the speed remains at 1 pu but armature current becomes 2 pu
C. both speed and armature current becomes 2 pu
D. both speed and armature current remains at 1 pu

47. "A time-varying flux causes an induced electromotive force". What law does this statement represent? **(IES 2014)**
A. Ampere's law
B. Faraday's law
C. Lenz's law
D. Field form of Ohm's law

48. A DC series motor with a resistance between terminals of 1 Ω, runs at 800 rpm from a 200 V supply taking 15 A. If the speed is to be reduced to 475 rpm for the same supply voltage and current, the additional series resistance to be inserted would be approximately **(IES 2014)**
A. 2.5 Ω B. 3 Ω
C. 4.5 Ω D. 5 Ω

49. A DC series motor of resistance 1 Ω across terminals run at 1000 rpm at 250 V taking a current of 20 A. When an additional resistance of 6 Ω is inserted in series and taking the same current, the new speed would be **(IES 2014)**
A. 142.8 rpm B. 166.7 rpm
C. 478.3 rpm D. 956.6 rpm

50. In a DC machine, for the same number of slots and same current in the armature conductor, which one of the following will induce higher emf? **(IES 2014)**
A. Lap winding
B. Wave winding
C. Compensating winding
D. Pole winding

Ans.	40. C	41. C	42. C	43. B	44. C	45. C	46. B	47. B	48. D	49. D	50. B

CHAPTER 9

Three Phase and Single Phase Induction Motors

Induction Motor

The polyphase induction motor is the most widely used AC motor because of its low cost, simple and extremely rugged construction, high efficiency, reasonably good power factor, and simple starting arrangement.

Principle of operation of an induction motor: When a three-phase stator windings are fed by three-phase supply, a magnetic flux of constant magnitude but rotating at synchronous speed is set up. The flux passes through the air gap, and cuts the stationary rotor conductors.

Due to relative speed between the rotating magnetic flux and stationary conductors, an emf is induced in the conductors according to Faraday's law of electromagnetic induction and its direction is given by Fleming's right hand rule. Since the rotor conductors are in a closed circuit, rotor current is produced whose direction is such as to oppose the cause which produces it. And the cause which produces the rotor current is the relative speed between the rotating flux of stator and stationary rotor conductors. Hence to reduce the relative speed, the rotor starts running in the same direction as that of the flux and tries to catch up with the rotating flux.

Induction motor consists of two parts: Stationary part known as stator, and revolving part known as rotor. It is mainly of two types:

1. Squirrelcage type
2. Wound rotor type

Frequency of rotating magnetic field: The phase supply produces a rotating magnetic field which rotates at synchronous speed, given by $N_s = 120f/p$.

An induction motor cannot run at synchronous speed. If it were possible, by some means for the rotor to attain synchronous speed, the rotor would then be standstill with respect to the rotating flux with the result that no emf would be induced in the rotor, no rotor current would flow and therefore there would be no torque developed.

Induction motor relationship: Slip of an induction motor is defined as

$$s = \frac{\text{synchronous speed} - \text{rotor speed}}{\text{synchronous speed}}$$

Rotor current frequency $f' =$ slip × supply frequency

The starting torque is maximum when rotor resistance equals rotor reactance, i.e. $R_2 = X_2$.

The torque under running conditions is maximum at that value of slip s, which makes rotor reactance per phase equal to rotor resistance per phase.

$$s = \frac{R_2}{X_2}$$

The full load torque T_f and maximum torque T_{max} are related by

$$\frac{T_f}{T_{max}} = \frac{2as}{a^2 + s^2}$$

where $a = \dfrac{R_2}{X_2}$

The torque/slip curve is shown in the Fig. 9.1.

The slip of an inductor motors can be measured by actual measurement of motor speed or comparing rotor and stator supply frequencies with dc moving coil millivolt-meter, and stroboscopic method.

Rotor copper loss in induction motor = s × rotor input

The electrical equivalent of mechanical load on motor

$$R_L = R_2\,(1/s - 1)$$

Slip corresponding to maximum gross power output

$$s_m = \frac{R_2 / K^2}{R_2 / K^2 + Z_{01}}$$

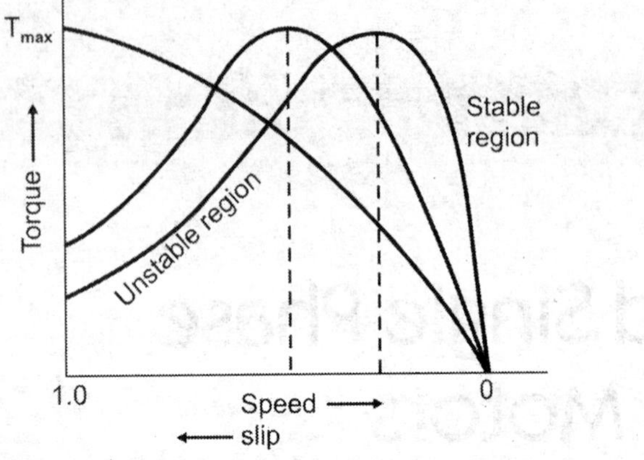

Fig. 9.1

Maximum gross power output:

$$P_g = \frac{3V_1^2}{2(R_{01}+Z_{01})}$$

where, V_1 is the input voltage

$$\text{Gross torque } T_g = \frac{\text{Rotor output in watts}}{2\pi N}$$

Rotor copper loss = $T_g \times 2\pi(N_s - N)$

$$\frac{\text{Rotor gross output}}{\text{Rotor input}} = (1-s) = \frac{N}{N_s}$$

Rotor efficiency = N/N_s

Rotor copper loss/rotor gross output = $s/(1-s)$

Speed of cummulative cascade set

$N_{sc} = 120f/(P_a + P_b)$

where P_a and P_b are the numbers of poles of two machines.

Methods of starting of induction motor:

i. *Torque Developed by direct switching at starting:*

$$T_{st} = T_f\left(\frac{I_{st}}{I_f}\right)^2 \cdot s_f$$

where I_{st} and I_f being the starting and full load currents, T_f and s_f are the full load torque and slip. Starting current is equal to the short-circuit current.

ii. *Torque developed by reduced voltage method:*

$$T_{st} = T_f \cdot k^2\left(\frac{I_{st}}{I_f}\right)^2 \cdot s_f$$

where k is the factor by which voltage is reduced using resistance starter. Starting current = $k\, I_{sc}$, where I_{sc} is the short-circuit current.

iii. *Auto-transformer starting method*

$$T_{st} = T_f\left(\frac{I_{st}}{I_{sc}}\right)^2 \cdot s_f = T_f k^2\left(\frac{I_{st}}{I_f}\right)^2 \cdot s_f$$

= k^2 × torque obtained by DOL

where k is the transformation ratio; supply current = $k^2 I_{sc}$

iv. *Star-delta starting method:*

$$T_{st} = \frac{1}{3}T_f\left(\frac{I_{sc}}{I_f}\right)^2 \cdot s_f$$

Here starting torque is reduced to one third of DOL torque and starting line current is reduced to one-third of line current with DOL starting.

Synchronous wattage of an induction motor is the power transferred across the air-gap to the rotor.

Torque in synchronous watt = rotor input.

The no load current is 40%–50% of full load current.

The electrical equivalent of mechanical load on motor = $R_2(1/s - 1)$.

Power output of an induction motor is maximum when equivalent load resistance is equal to the stand-still leakage impedance of the motor, i.e.

$$R_L = Z_{01}$$

Slip corresponding to maximum gross power output is given by

$$s_m = \frac{R_2'}{R_2' + Z_{01}}$$

Maximum gross power output is given by

$$P_{g\,max} = \frac{3V_1^2}{2(R_{01}+Z_{01})}$$

where V_1 is the input voltage

Gross torque

$$T_s = \frac{\text{Rotor output in watts}}{2\pi N}$$

Rotor copper loss = $T_g \times 2\pi(N_s - N)$

$$\frac{\text{Rotor gross output}}{\text{Rotor input}} = (1-s) = \frac{N}{N_s}$$

$$\text{Rotor efficiency} = \frac{N}{N_s}$$

$$\frac{\text{Rotor copper loss}}{\text{Rotor gross output}} = \frac{s}{1-s}$$

Double squirrelcage motor: A double squirrelcage motor consists of two independent cages on the same rotor one inside the other. The outer cage consists of a high resistance metal whereas the inner cage has low resistance copper bars. It has high starting torque without sacrificing its electrical efficiency under normal running conditions.

Speed control of induction motor: It can be achieved by either of the following methods:

i. Changing the applied voltage
ii. Changing the applied frequency
iii. Changing the number of stator poles

iv. Injecting an emf in the rotor circuit
v. Cascading, and
vi. Rotor rheostat control.

The speed of cumulative cascade set is $N_{sc} = \dfrac{120f}{P_a + P_b}$

where P_a and P_b are the numbers of poles of two machines.

Single phase motors: Single phase supply is mainly used for lighting purposes in offices, houses, hospitals, schools, shops, etc. Domestic applications include mixers, automatic washing machines, electric iron, TV and refrigerators, etc. The power rating of these types of appliances is less than a kilowatt.

Classification of single phase induction machines: Fractional hp motors have the following classifications:

1. Single phase induction motors:
   a. Resistance split type
   b. Capacitor split type
   c. Shaded pole type
2. Single phase synchronous motors:
   a. Hysteresis motor
   b. Reluctance motor
3. Single phase induction motors:
   a. AC series motor
   b. Universal motor
4. Special purpose motor
   a. Stepper motor
   b. Servomotor

Hysteresis motor: The operation of this motor depends on the presence of continuously revolving magnetic field. Hence for the split-phase operation, its stator has two windings which remain connected to the single phase supply continuously both at starting as well as during the running of the motor. Usually shaded pole principle is employed as shown in Fig. 9.2.

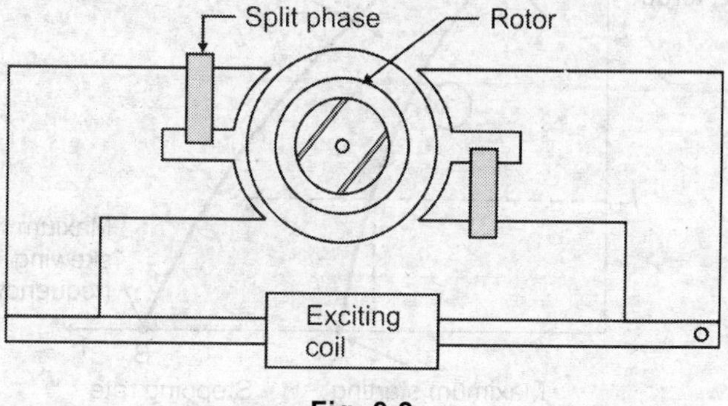

Fig. 9.2

The rotor is a smooth chrome-steel cylinder having high retentivity so that the hysteresis loss is high. It has no winding since the retentivity of the root material is high, it is very difficult to change the magnetic polarities. The rotor revolves synchronously.

The power factor of commercial fractional horse power hysteresis motor is usually low (less than 0.7) and the efficiency is less than 60%, although efficiencies as high as 80% have been claimed for a 1/4 hp motor. The poor efficiency is mainly due to the reduction in developed torque resulting from open stator slots and that caused by various mmf space harmonics.

Reluctance motors: In a single phase reluctance motor, the rotating field can be produced by any of the phase splitting methods. The salient pole structure is given to the rotor by removing some of the teeth of an induction motor rotor. The removing teeth carry short-circuited copper bars to provide the starting induction torque. After starting, the rotor reaches near synchronous speed by induction action and is pulled into synchronism during positive half cycle of sinusoidally varying synchronous torques. This would only be possible if the rotor has low inertia and the load conditions are light (Fig. 9.3).

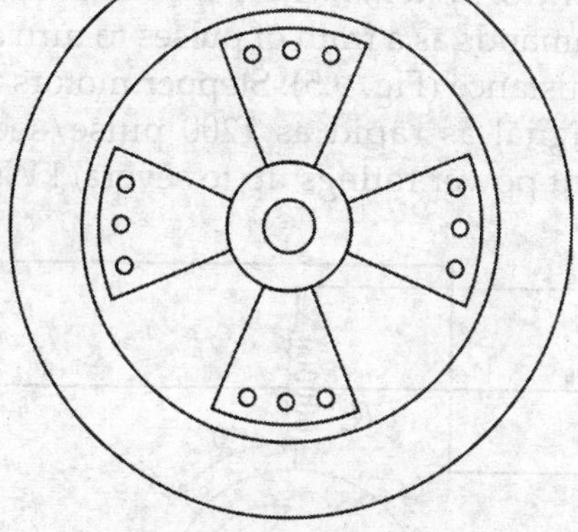

Fig. 9.3

Reluctance motor is similar to a single phase induction motor. This type of motor has variable air gap and DC supply is required to rotor.

Reluctance type motors are used in control apparatus, automatic regulators, clocks and all kinds of timing devices, recording instruments, teleprinters and gramophones.

Reluctance motor has the following advantages:

i. Less maintenance
ii. No requirement of DC supply
iii. Robust construction
iv. Constant speed characteristic

Limitations: i. less efficiency, ii. poor perfection, iii. less capacity to drive load as compared with 3-phase synchronous motors.

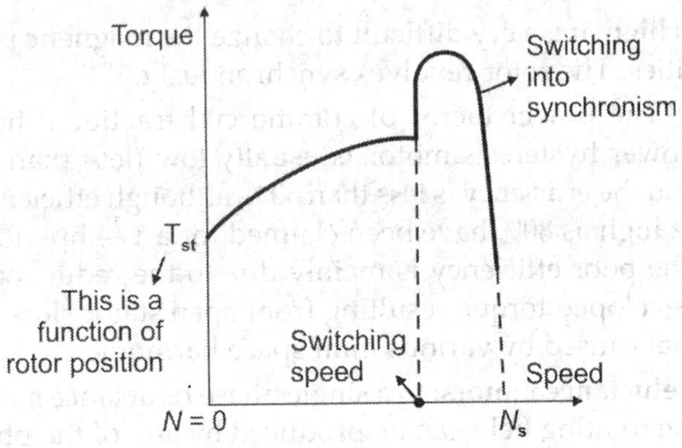

Fig. 9.4

The starting of this type of motor is very high and is solely dependent on the position of the rotor. Figure 9.4 shows the torque speed characteristics of reluctance motor.

Stepper motor: The stepper motor is a form of synchronous motor which is designed to rotate a specific number of degrees for each electrical pulse received by its control unit. Typical steps are 2°, 2.5°, 5°, 7.5° and 15° per pulse. The stepper motor is used in digital control systems where the motor receives open loop commands as a train of pulses to turn a shaft by a specific distance (Fig. 9.5). Stepper motors are built to follow signal as rapid as 1200 pulse/sec and with equivalent power ratings up to several kWs.

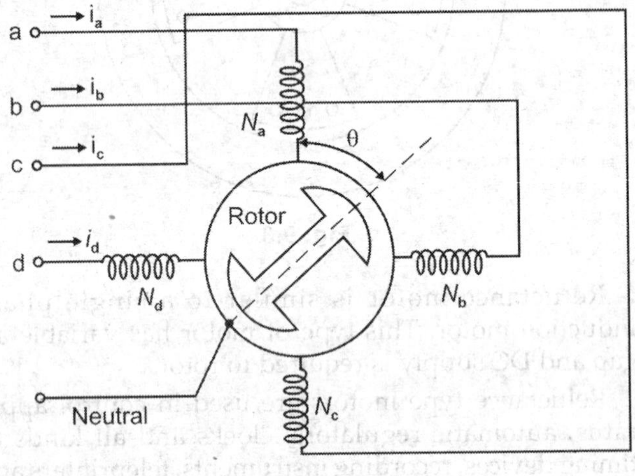

Fig. 9.5

Characteristics of stepper motors: The characteristics of stepper motors are classified as: i. static characteristics ii. dynamic characteristics.

Torque displace characteristics and torque-current characteristic are known as static characteristics shown in Figs 9.6(a) and (b).

Torque stepping rate characteristic is known as dynamic characteristics shown in Fig. 9.6(c).

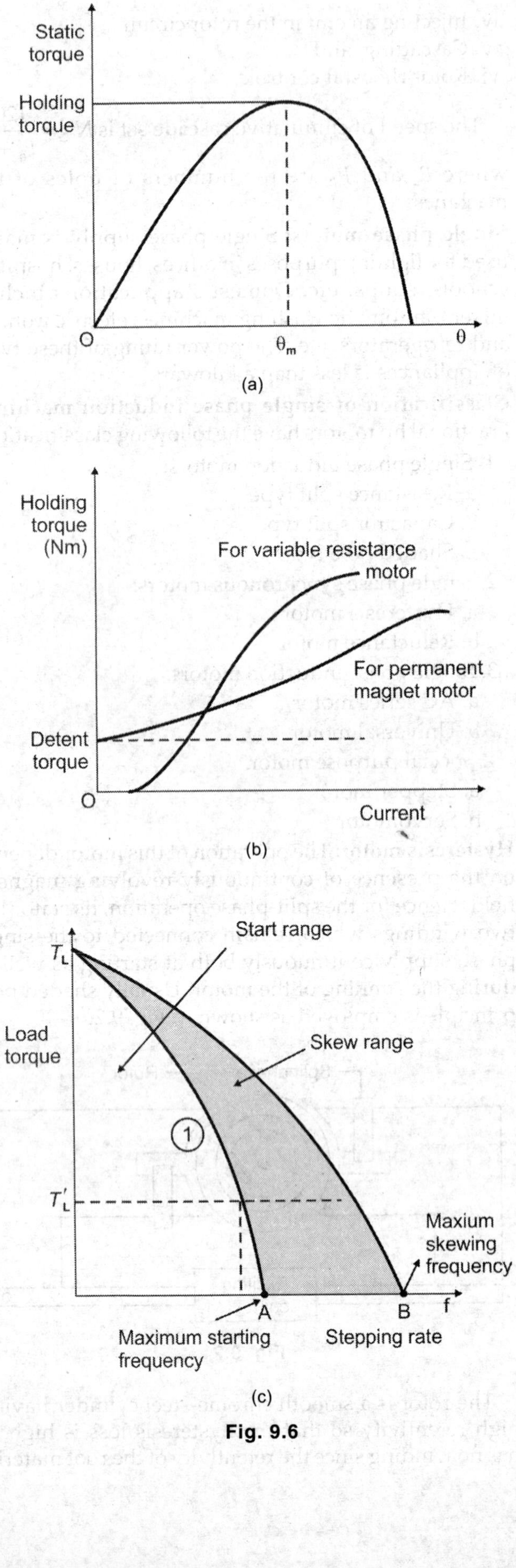

Fig. 9.6

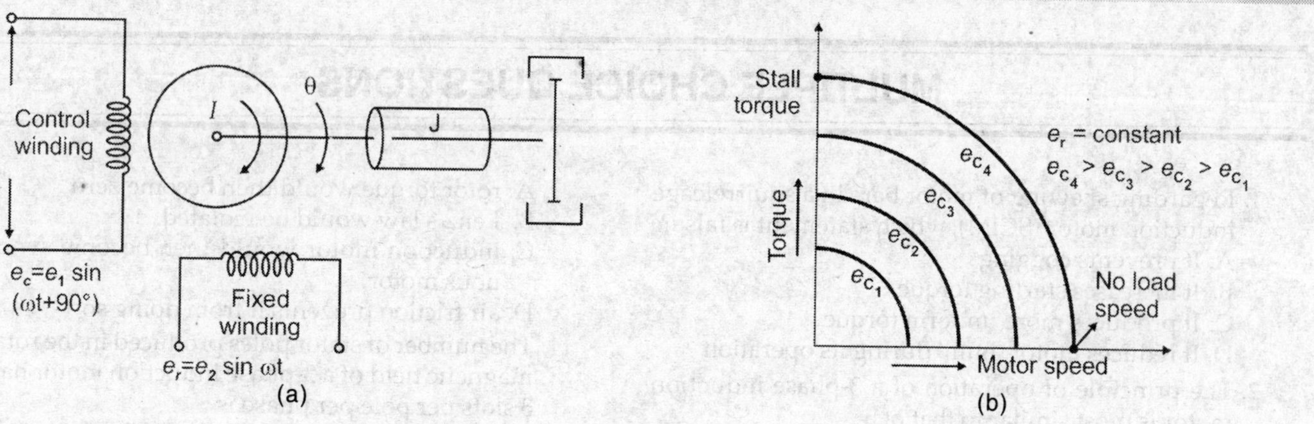

Fig. 9.7

AC Servomotor

An AC servomotor is basically a two-phase induction motor as shown in Fig. 9.7(a). A two phase servomotor differs from a normal induction motor.

The rotor construction is usually squirrelcage or drag-cup type having small diameter to reduce interia.

As in the case of DC motors

$$T_D = J\frac{d^2\theta}{dt^2} + D\frac{d\theta}{dt} = k_m I_e$$

Figure 9.7(b) shows the torque speed characteristic of two phase AC servomotor.

MULTIPLE CHOICE QUESTIONS

1. Regarding skewing of motor bars in a squirrelcage induction motor (SCIM), which statement is false?
   A. It prevents cogging
   B. It increases starting torque
   C. It produces more uniform torque
   D. It reduces motor 'hum' during its operation
2. The principle of operation of a 3-phase induction motor is most similar to that of a
   A. synchronous motor
   B. repulsion-start induction motor
   C. transformer with a shorted secondary
   D. capacitor-start, induction-run motor
3. The two primary parts of a 3-phase induction motor are
   A. rotor and stator B. stator and field
   C. slip-ring and brushes D. rotor and armature
4. The magnetising current drawn by transformers and induction motors is the cause of their .... power factor.
   A. zero B. unity
   C. lagging D. leading
5. The effect of increasing the length of air-gap in an induction motor will be to increase the
   A. power factor B. speed
   C. magnetising current D. air-gap flux
6. The main reason why the rotor of an induction motor runs in the same direction as the rotating stator magnetic flux is to meet the requirement of
   A. Lenz's law B. Faraday's law
   C. Fleming's left-hand rule
   D. Fleming's right-hand rule
7. In a 3-phase induction motor, the relative speed of stator flux is zero with respect to
   A. stator winding B. rotor
   C. rotor flux D. space
8. In a 3-phase induction motor, the rotor field rotates at synchronous speed with respect to
   A. stator B. rotor
   C. stator flux D. none of these
9. Which of the following statements regarding stator flux of an induction motor is false?
   A. It is constant in magnitude
   B. It revolves round the stator at synchronous speed
   C. It induces emf in the motor bars
   D. Its magnitude depends on the motor load
10. The rotor of an induction motor cannot run with synchronous speed because
    A. rotor torque would then become zero
    B. Lenz's law would be violated
    C. induction motor would then become synchronous motor
    D. air friction prevents it from doing so
11. The number of stator poles produced in the rotating magnetic field of a 3-phase induction motor having 3 slots per pole per phase is
    A. 3 B. 6
    C. 2 D. 12
12. Which of the following rotor quantity in a SCIM does NOT depend on its slip?
    A. Reactance B. Speed
    C. Induced emf D. Frequency
13. A 6-pole, 50-Hz, 3-phase induction motor is running at 950 rpm and has rotor copper loss of 5 kW. Its rotor input is
    A. 100 kW B. 10 kW
    C. 95 kW D. 5.3 kW
14. A 6-pole, 50-Hz, 3-phase induction motor has a full-load speed of 950 rpm. At half-load, its speed would be
    A. 475 rpm B. 500 rpm
    C. 975 rpm D. 1000 rpm
15. If rotor input of a SCIM running with a slip of 10% is 100 kW, gross power developed by its rotor is
    A. 10 kW B. 90 kW
    C. 99 kW D. 80 kW
16. Pull-out torque of a SCIM occurs at that value of slip where rotor power factor equals
    A. unity B. 0.707
    C. 0.866 D. 0.5
17. The no load current of 3-phase induction motor is approximately
    A. 45–55% of FL current B. 70–80% of FL current
    C. 1–3% of FL current D. none of these
18. When applied rated voltage per phase is reduced by one-half, the starting torque of a SCIM becomes .... of the starting torque with full voltage.
    A. 1/2 B. 1/4
    C. $1/\sqrt{2}$ D. $\sqrt{3}/2$
19. If maximum torque of an induction motor is 200 kg·m at a slip of 12%, the torque at 6% slip would be
    A. 100 kg·m B. 160 kg·m
    C. 50 kg·m D. 40 kg·m
20. The fractional slip of an induction motor is the ratio
    A. rotor Cu loss/rotor input
    B. stator Cu loss/stator input

Ans.													
	1. B	2. C	3. A	4. C	5. C	6. A	7. C	8. A	9. D	10. A	11. B	12. B	13. A
	14. C	15. B	16. B	17. A	18. B	19. B	20. A						

C. rotor Cu loss/rotor output
D. rotor Cu loss/stator Cu loss

21. The torque developed by a 3-phase induction motor depends on the following three factors
A. speed, frequency, number of poles
B. voltage, current and stator impedance
C. synchronous speed, rotor speed and frequency
D. rotor emf, rotor current and rotor pf

22. If the stator voltage and frequency of an induction motor are reduced proportionately, its
A. locked rotor current is reduced
B. torque developed is increased
C. magnetising current is decreased
D. both (A) and (B) are correct

23. The efficiency and pf of a SCIM increase in proportion to its
A. speed B. mechanical load
C. voltage D. rotor torque

24. A SCIM runs at constant speed only so long as
A. torque developed by it remains constant
B. its supply voltage remains constant
C. its torque exactly equals the mechanical load
D. stator flux remains constant

25. The synchronous speed of a linear induction motor does NOT depend on
A. width of pole pitch B. number of poles
C. supply frequency D. any of these

26. Thrust developed by a linear induction motor depends on
A. synchronous speed B. rotor input
C. number of poles
D. Both (A) and (B) are correct

27. In a circle diagram for a 3-phase induction motor, the diameter of the circle is determined by
A. rotor current B. exciting current
C. total stator current
D. rotor current referred to stator

28. Point out the wrong statement.
Blocked rotor test on a 3-phase induction motor helps to find
A. short-circuit current with normal voltage
B. short-circuit power factor
C. fixed losses
D. motor resistance as referred to stator

29. In the circle diagram of an induction motor, point of maximum input lies on the tangent drawn parallel to
A. output line B. torque line
C. vertical axis D. horizontal axis

30. An induction motor has a short-circuit current 7 times the full-load current and a full-load slip of 4 percent. Its line-starting torque is .... times the full-load torque.
A. 7 B. 1.96
C. 4 D. 49

31. In a SCIM, torque with autostarter is .... times the torque with direct-switching where K is the transformation ratio of the auto starter.
A. K^2 B. K
C. $1/K^2$ D. $1/K$

32. If stator voltage of a SCIM is reduced to 50 percent of its rated value, torque developed is reduced by .... percent of its full-load value.
A. 50 B. 25
C. 75 D. 57.7

33. For the purpose of starting an induction motor, a Y–Δ switch is equivalent to an auto-starter of ratio
A. 33.3% B. 57.7%
C. 73.2% D. 60%

34. A double squirrelcage motor (DSCM) scores over SCIM in the matter of
A. speed regulation under normal operating conditions
B. high efficiency under running conditions
C. starting torque D. all of these

35. In a DSCM, outer cage is made of high resistance metal bars primarily of the purpose of increasing its
A. speed regulation B. starting torque
C. efficiency D. starting current

36. A SCIM with 36-slot stator has two separate windings: one with 3 coil groups/phase/pole and the other with 2 coil groups/phase /pole. The obtainable two motor speeds would be in the ratio of
A. 3:2 B. 2:3
C. 2:1 D. 1:2

37. A 6-pole 3-phase induction motor taking 25 kW from a 50-Hz supply is cumulatively-cascaded to a 4-pole motor. Neglecting all losses, speed of the 4-pole motor would be
A. 1500 rpm B. 1000 rpm
C. 600 rpm D. 3000 rpm

38. Which class of induction motor will be well-suited for large refrigerators?
A. Class E B. Class B
C. Class F D. Class C

39. In a Scharge motor operating at super-synchronous speed, the injected emf and the standstill secondary induced emf
A. are in phase with each other
B. are at 90° in time phase with each other
C. air in phase opposition
D. none of these

40. For starting a Scharge motor, 3-phase supply is connected to
A. stator B. rotor via slip-rings
C. regulating winding
D. secondary winding via brushes

Ans.													
	21. D	22. D	23. B	24. C	25. B	26. D	27. C	28. C	29. D	30. B	31. A	32. C	33. B
	34. D	35. B	36. A	37. C	38. C	39. A	40. B						

41. Two separate induction motors, having 6 poles and 4 poles respectively and their cascade combination from 60 Hz, 3-phase supply can give the following synchronous speeds in rpm
A. 720, 1200, 1500 and 3600
B. 720, 1200, 1800
C. 600, 1000, 1500 D. 720 and 3000
42. Mark the WRONG statement.
A Scharge motor is capable of behaving as a/an
A. inverted induction motor
B. slip ring induction motor
C. shunt motor D. series motor
43. When a stationary 3-phase induction motor is switched on with one phase disconnected
A. it is likely to burn out quickly unless immediately disconnected
B. it will start but very slowly
C. it will make jerky start with loud growling noise
D. remaining intact fuses will be blown out due to heavy inrush of current
44. If single phasing of a 3-phase induction motor occurs under running conditions, it
A. will stall immediately
B. will keep running though with slightly increased slip
C. may either stall or keep running depending on the load carried by it
D. will become noisy while it still keeps running
45. The magnitude of the emf induced in the stator due to revolving flux will depend upon the
A. speed of the motor
B. DC excitation current
C. load on the motor
D. speed and rotor flux
46. Under no-load running condition, the angle between induced voltage and supply voltage will be
A. 180° B. 90°
C. between 90° and 180° D. zero
47. In AC rotating machines, the generated emf
A. is in phase with working flux Φ
B. leads flux Φ by 90°
C. lags flux Φ by 90° D. lags flux Φ by 180°
48. Cogging in induction motor occurs when
A. number of stator teeth and number of rotor teeth = odd number
B. number of stator teeth and number of rotor teeth = even number
C. number of stator teeth and number of rotor teeth = 0
D. number of stator teeth and number of rotor teeth = negative number
49. When the supply voltage for an induction motor is reduced by 10%, maximum running torque will nearly
A. decrease by 10% B. decrease by 20%
C. increase by 10% D. increase by 20%
50. Breakdown torque of a 3-phase induction motor of negligible stator impedance is
A. directly proportional to the rotor resistance
B. inversely proportional to the rotor resistance
C. directly proportional to the reactance
D. inversely proportional to the rotor leakage reactance
51. Consider the following statements:
I. A grid-connected induction generator always supplies leading reactive VARs to the bus
II. An overexcited synchronous motor draws current at a lagging power factor
III. An under excited synchronous generator connected to an infinite bus works at a leading power factor
IV. The load angle of a synchronous machine is the angle between the excitation voltage and the gap voltage
Of these statements
A. I and II are correct B. II and III are correct
C. I, III and IV are correct
D. I, II, III and IV are correct
52. A 6-pole, 50 Hz single-phase induction motor runs at a speed of 95 rpm. The frequency/frequencies of current in the cage rotor will be
A. 5 Hz B. 5 Hz, 55 Hz
C. 5 Hz, 95 Hz D. 55 Hz, 95 Hz
53. In a 3-phase induction machine, motoring, generating and braking operations take place in the range of slip s given by
A. motoring: $1 > s > 0$, generating: $0 > s > -1$, braking: $s > 1$
B. motoring: $s > 1$, generating: $1 > s > 0$, braking: $0 > s > -1$
C. motoring: $s > 1$, generating: $0 > s > -1$, braking: $1 > s > 0$
D. motoring: $0 > s > -1$, generating: $s > 1$, braking: $1 > s > 0$
54. A voltmeter gives 120 oscillations per minute when connected to the rotor of an induction motor. The stator frequency is 50 Hz. The slip of the motor is
A. 2% B. 2.5%
C. 4% D. 5%
55. The relative speed between the stator and the rotor fluxes is equal to
A. n_s rpm B. $(n_s - n)$ rpm
C. $(n_s + n)$ rpm D. zero
56. In all cases of electromagnetic induction, an induced voltage will cause a current to flow in a closed

Ans. 41. A 42. B 43. A 44. C 45. B 46. C 47. C 48. C 49. B 50. D 51. C 52. C 53. A 54. C 55. D

circuit in such a direction that the magnetic field which is caused by that current will oppose the change that produces the current is the original statement of

A. Lenz's law B. Ampere's law
C. Flemming's law of induction
D. Faraday's law of magnetic induction

57. A centre zero ammeter connected in the rotor circuit of a 6-pole 50 Hz induction motor makes 30 oscillations in one minute. The rotor speed in rpm is

A. 970 B. 990
C. 1010 D. 1030

58. If the rotor winding of a wound rotor 3-phase induction motor is connected to balanced 3-phase supply and the stator winding is short-circuited, then the rotor will

A. not run at all
B. run at sub-synchronous speed in the direction of rotating flux
C. run at sub-synchronous speed against the direction of rotating flux
D. run synchronously with rotating flux

59. Consider the following statements :
In a 3-phase induction motor connected to a 3-phase supply if one of the lines suddenly gets disconnected, then the

I. motor will come to standstill
II. motor will continue to run at the same speed with line current unchanged
III. motor will continue to run at a slightly reduced speed with increased line current
IV. rotor currents will have both sf and $(2 - s)f$ component frequencies where s is the slip and f is the supply fre-quency

Of these statements

A. I and IV are correct B. I and II are correct
C. III and IV are correct D. II and III are correct.

60. In a self-excited induction generator, to keep the frequency of generated voltage constant with the increase in load, the speed of the induction machine should be

A. increased B. decreased
C. maintained less than the rated synchronous speed
D. maintained more than the rated synchronous speed

61. Cogging and crawling are phenomena associated with

A. cage induction machines and they are essentially the same
B. squirrelcage induction machines, the former during starting and the latter at a fraction of its rated speed
C. squirrelcage induction machines, the former at a fraction of its rated speed and the latter during starting
D. wound rotor induction machines and they are reduced by skewing, chord-ing and distribution of windings

62. The stator winding of a 3-phase induction motor is connected to a balanced 3-phase supply and the rotor is short-circuited, then the rotor would

A. not run at all
B. run asynchronously in the direction of the rotating flux
C. run synchronously with the rotating flux
D. run against the direction of rotating flux

63. If the speed of an induction motor be synchronous speed, then the relative speed between the rotating flux and the rotor will be

A. maximum and, hence, torque will be maximum
B. maximum and, hence, torque will be zero
C. zero and, hence, torque will be maximum
D. zero and, hence, torque will be zero

64. A 3-phase 50 Hz, 8-pole squirrelcage induction motor will run at a speed

A. less than 750 rpm B. 750 rpm
C. greater than 750 rpm D. 1500 rpm

65. At a slip of 4%, the maximum possible speed of a 3-phase squirrelcage induction motor is

A. 2880 rpm B. 3000 rpm
C. 1500 rpm D. 1440 rpm

66. The relationship between rotor frequency f_2 slip s, and stator frequency f_1 is given by

A. $f_2 = f_1/s$ B. $f_2 = sf_1$
C. $f_2 = (1 - s)f_1$ D. none of these

67. An induction motor, in general, is analogous to

A. an autotransformer
B. a two-winding transformer with its secondary open-circuited
C. a two-winding transformer with its secondary winding short-circuited
D. a three-winding transformer

68. In terms of airgap power, P_g the rotor copper loss and the mechanical power developed are given by

A. sP_g and $(1 - s)P_g$ B. $(1 - s)\ P_g$ and sP_g
C. P_g and P/s D. P_g/s and $P_g(1 - s)$

69. In a 3-phase induction motor, internal developed torque, T_e in terms of supply voltage, V_1 is proportional to

A. V_1 B. $\sqrt{V_1}$
C. V_1^2 D. none of these

70. In a 3-phase induction motor, slip for maximum torque in terms of rotor resistance r_2 is

A. inversely proportional to r_2
B. directly proportional to r_2
C. independent of r_2 D. None of these

Ans. 56. A 57. B 58. C 59. C 60. A 61. B 62. B 63. D 64. A 65. A 66. B 67. C 68. A 69. C 70. B

71. In a 3-phase induction motor, the current is produced in the rotor conductors by
A. giving AC supply B. giving DC supply
C. induction effect D. pulsating DC supply

72. When 3 phase supply is given to the stator of the motor, a
A. revolving field is set up
B. pulsating field is set up
C. revolving field at synchronous speed is set up
D. revolving field at the rotor speed is set up

73. The direction of rotation of 3-phase revolving field can be changed by interchanging
A. R and Y phases only B. B and Y phases only
C. R and B phases only D. any two phases

74. If any two phases of the three-phase supply are interchanged, motor will
A. run in the same direction
B. stop running
C. run in the reverse direction
D. draw high current

75. Different types of a 3-phase induction motors are
A. squirrelcage B. slip-ring
C. commutator D. all of these

76. The rotor of a 3-phase induction motor rotates in the same direction as the direction of revolving field because of
A. Faraday's laws of electromagnetic induction
B. Fleming's right hand rule
C. Lenz's law D. none of these

77. The rotor of a 3-phase induction motor always runs at
A. synchronous speed
B. less than synchronous speed
C. more than synchronous speed
D. none of these

78. The synchronous speed of the revolving field depends upon
A. number of poles B. supply frequency
C. flux
D. both A and B are correct

79. For induction motors, normally
A. the rotor winding is connected to supply and the stator winding is short-circuited
B. the stator winding is connected to supply and the rotor winding is short circuited
C. both the stator and the rotor windings are connected to supply
D. the stator winding is connected to supply and the rotor winding is open-circuited

80. In a 3-phase squirrelcage induction motor, the
A. rotor conductors are kept open
B. rotor conductors are short circuited through end rings
C. ends of the rotor conductors are connected to slip-rings
D. ends of the rotor conductors are short circuited through slip-rings

81. In a 3-phase slip-ring induction motor, the number of poles in the rotor windings are kept
A. same as of the number of stator poles
B. more than the number of stator poles
C. less than the number of stator poles
D. independent of the stator pole

82. At full load, the current induced in the rotor conductor of a 3-phase squirrelcage induction motor is
A. more than rated current
B. less than rated current
C. nearly equal to rated current
D. none of these

83. The frequency of the induced emf in the rotor circuit is
A. maximum at standstill
B. maximum at synchronous speed
C. zero at standstill D. none of these

84. If one of the three phases of a 3-phase induction motor is disconnected while running at light load, it will
A. stop running
B. continue to run at same speed
C. run at low speed D. run at high speed

85. If the rotor terminals of a 3-phase slip-ring induction motor are not short-circuited and the supply is given to the stator, the motor will
A. not start B. start running
C. run at high speed D. run at low speed

86. Generally, a 3-phase slip-ring induction motor is started with
A. DOL starter B. Star-delta starter
C. Autotransformer starter
D. Stator-rotor starter

87. A 3-phase squirrelcage induction motor can be started with
A. Stator rotor starter B. DOL starter
C. Star-Delta starter
D. Both DOL and star delta starters

88. A 3-phase, 4-pole induction motor works on 3-phase, 50 c/s, supply. If the slip of the motor is 4%, the actual speed will be
A. 1500 rpm B. 1460 rpm
C. 1440 rpm D. 720 rpm

89. A 3-phase, 6-pole induction motor operates on 440 V, 50 c/s supply. If the actual speed of the motor is 960 rpm, the slip will be
A. 6% B. 5%
C. 4% D. 0.4%

Ans.												
71. C	72. C	73. D	74. C	75. D	76. C	77. B	78. D	79. B	80. B	81. A	82. C	83. A
84. B	85. A	86. D	87. D	88. C	89. C							

90. When the line voltage of an induction motor is reduced to 75% of its rated value, the starting current will reduced by
A. 75% B. 50%
C. 25%
D. no change in the current

91. If the full load speed of a 4-pole, 3-phase induction motor is 1440 rpm its speed at half load will be
A. 1460 rpm B. 1440 rpm
C. 1400 rpm D. 1000 rpm

92. The advantage of starting 3-phase induction motor with DOL starter in place of ICTP switch is
A. starting current remains same
B. provided with overload protection relay
C. provided with under volt relay
D. both (B) and (C) are correct

93. If a motor of lower speed is purchased in place of high speed motor, its cost will be
A. same B. more
C. less D. slightly less

94. The power drawn by 3-phase induction motor, when first run in star connection and then in delta connection, will be
A. same
B. $\sqrt{3}$ times the power taken in star connection
C. 3 times the power taken in star connection
D. $1/\sqrt{3}$ times the power taken in star connection

95. The slip of an induction motor under full load condition is about
A. 0.1 B. 0.03
C. 0.2 D. 0.8

96. If a 3-phase induction motor runs at 25% supply than the rated voltage, the motor will
A. not start B. start at lower speed
C. run at normal speed
D. get heated and damaged

97. If a 3-phase induction motor runs on 2-phase supply it will
A. run at slow speed B. run with jerks
C. likely to burn D. blow off the fuses

98. Star-delta and autotransformer starters are used with 3-phase induction motors because they
A. have high starting torque
B. may not run in reverse direction
C. draw 5 to 7 times the full load current at the time of starting
D. pick up high speed immediately

99. The speed of the squirrelcage induction motor can be controlled by
A. changing supply frequency
B. changing number of poles
C. reducing supply voltage
D. all of these

100. The speed of the squirrelcage induction motor can be controlled from
A. stator side B. rotor side
C. both stator and rotor sides
D. supply voltage side

101. The speed of a 50 Hz induction motor under full load conditions is 720 rpm The number of poles of the motor are
A. 4 B. 2
C. 8 D. none of these

102. The speed of an induction motor is
A. always less than the synchronous speed
B. can be equal to or less than the synchronous speed
C. always more than the synchronous speed
D. none of these

103. The speed of the rotor field of an induction motor is equal to
A. synchronous speed B. motor speed
C. zero D. none of these

104. Majority of the induction motors have
A. totally enclosed frame
B. drip proof type frame
C. totally enclosed fan closed frame
D. none of these

105. The squirrelcage rotor of a 4-pole induction motor
A. can be used for only 4 pole induction motor
B. can be used for motors having 2, 4 or 8 poles only
C. can be used for motor with any number of poles
D. none of these

106. The number of slip rings in a 3-phase wound motor induction motor is
A. 3 B. 4
C. 6 D. 12

107. The core of the rotor of a 3-phase induction motor is
A. always laminated B. solid
C. sometimes laminated D. none of these

108. In a 50-Hz, 3-phase induction motor, frequency of rotor current is
A. 50 Hz B. about 2 Hz
C. about 10 Hz D. zero

109. In a 3-phase, 50 Hz induction motor the voltage between slip-ring at standstill is 50 V. At full load, the slip is 0.04. The voltage between slip-rings at full load is
A. 50 V B. 2 V
C. 20 V D. 1250 V

110. Under blocked rotor conditions, a 3-phase induction motor is similar to a transformer with secondary
A. open circuited B. underloaded
C. short circuited D. none of these

Ans. 90. C 91. A 92. D 93. B 94. C 95. B 96. D 97. C 98. C 99. D 100. A 101. C 102. A 103. A 104. C 105. C 106. A 107. A 108. B 109. B 110. C

111. The no-load current of 3-phase, 10 hp, 400 V induction motor is about
A. 1 A B. 5 A
C. 10 A D. 20 A

112. In a 3-phase, 50-Hz induction motor rotor core losses are
A. always equal to stator core losses
B. very high as compared to stator core losses
C. always very small as compared to stator core losses
D. none of these

113. At $s = \infty$, the torque of a 3-phase induction motor is
A. zero B. small
C. very high D. high

114. At standstill the power factor of the rotor circuit of an induction motor is 0.2 lagging. At full load the power factor of the rotor circuit will be
A. 0.2 lagging B. about 0.8 lagging
C. almost unity D. none of these

115. An induction motor has a starting torque of 320 N·m when started by direct switching. If started through an auto-transformer with 50% tapping the starting torque will be
A. 0.05 N·m B. 80 N·m
C. 160 N·m D. 640 N·m

116. A wound rotor induction motor runs with a slip of 0.03 when developing full load torque, its rotor resistance is 0.25 ohm per phase. If external resistance of 0.50 ohm per phase is connected across the slip rings, the slip for full load torque will be
A. 0.03 B. 0.06
C. 0.09 D. 0.01

117. A 3-phase induction motor draws a current of 20 A from mains if started by direct switching. If an auto-transformer with 50% tapping is used for starting, the current drawn from the mains will be
A. 20 A B. 10 A
C. 5 A D. 40 A

118. In a double cage 3-phase induction motor
A. the inner cage has lower resistance and higher inductance as compared to the outer cage
B. the inner cage has higher resistance as compared to the outer cage
C. the inner cage has higher resistance and higher inductance as compared to the outer cage
D. None of these

119. An induction motor has a starting torque of 300 N·m when started by direct switching. If a star delta-starter is used for starting, the starting torque will be
A. 300 B. 173.2 N·m
C. 100 N·m D. none of these

120. The change in speed of a 3-phase induction motor from no load to full load is about
A. 2% B. 8%
C. 15% D. 50%

121. The speed of the slip-ring induction motor can be controlled by changing
A. supply frequency B. number of poles
C. resistance of the rotor winding
D. All of these

122. Which of the following motors is represented by the shown characteristic curve?

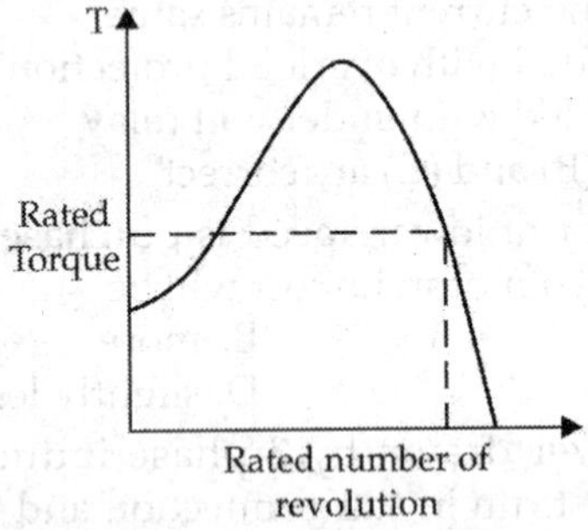

A. DC shunt motor B. DC series motor
C. DC compound motor D. Asynchronous motor

123. What does the data 950 V/1030 A given on the name plate represent?
A. 950 V: The lowest stator voltage
B. 950 V: The voltage across two short-circuited sliprings
C. 950 V: The voltage across two slip-rings when the motor is stationary
D. 1030 A: Current drawn from the circuit at rated load

124. Which of the following motors requires the most complicated speed control arrangement?
A. Three phase squirrelcage induction motor
B. Stator supplied three phase commutator motor
C. Rotor supplied three phase commutator motor
D. DC shunt motor

125. For which of the following fields of application is the three phase induction motor mainly suitable?
A. For running of electric vehicles
B. For running of rolling mills where an exact speed control is required
C. For running of different mechanic tools where one or a few speeds are required
D. For running of paper machines, the paper quality depends upon the exact speed control

126. Which of the following statements about the three phase induction motor is true?
A. The speed of the induction motor cannot be controlled as easily and efficiently as that of a DC shunt motor

Ans. 111. B 112. C 113. A 114. C 115. B 116. C 117. C 118. A 119. C 120. A 121. D 122. D 123. C 124. A 125. C 126. A

B. The induction motor requires more maintenance than a DC motor
C. The price of an induction motor is much higher than that of a DC motor of the same power and speed
D. The induction motor is much more sensitive to overloads than a DC motor

127. Which of the following symbols indicate the correct classes of insulation used for motor windings?
A. A, B, C, D B. A, B, D, E
C. A, B, F, H D. A, B, H, J.

128. A 5-hp motor is started, under no load, by the direct switching. The motor takes 30 seconds to come up to full speed. If started with a star-delta starter, the time taken by the motor to come to full speed will be
A. 30 seconds B. more than 30 seconds
C. less than 30 seconds D. none of these

129. What is the advantage of the slip-ring induction motor over the squirrelcage induction motor?
A. It is suitable for higher speeds
B. Its efficiency is higher
C. Its power factor is higher
D. It can be started with the help of rotor resistances

130. Which motor is widely used?
A. A squirrelcage induction motor
B. A slip-ring induction motor
C. A wound rotor synchronous motor
D. A salient pole synchronous motor

131. Which of the following statements about the squirrel motor is true?
A. The motor can run only in one direction
B. The laminations of the rotor should be properly insulated against one another
C. The cage rotor is made of copper
D. The stator winding produces a rotating magnetic field

132. What is the purpose of the blades in the squirrelcage induction motor?
A. To dynamically balance the rotor
B. To reduce the magnetic resistance of the rotor
C. To reduce the electrical resistance of the rotor cage
D. To cool the rotor

133. An induction motor is running at its rated torque. The applied voltage is reduced from the rated value of 440 V to 320 V. What is the consequence of this?
A. After sometime the motor heats up to an inadmissible extent
B. The current decreases
C. The motor stops
D. The speed reduces considerably

134. Which of the following formula is used to calculate the synchronous speed of an induction motor?
A. $n_s = p \times f$ B. $n_s = \dfrac{60 \times p}{f}$
C. $n_s = \dfrac{60 \times f}{p}$ D. $n_s = \dfrac{f \times p}{60}$

135. What is the disadvantage of starting an induction motor with a star-delta starter?
A. The starting torque increases and the motor runs with jerks
B. The starting torque is one-third of the torque in case of the delta connection
C. By changing over from star to delta connection a voltage peak is produced which endangers the insulation of the motor
D. During starting, high losses are reduced

136. An induction motor has a rated speed of 725 rpm. How many poles has its rotating magnetic field?
A. 2 poles B. 4 poles
C. 6 poles D. 8 poles

137. The ratio of no-load current to full load current of a 3-phase induction motor is about
A. 0.05 B. 0.2
C. 0.5 D. 0.8

138. What is the use of the circuit shown in the diagram?

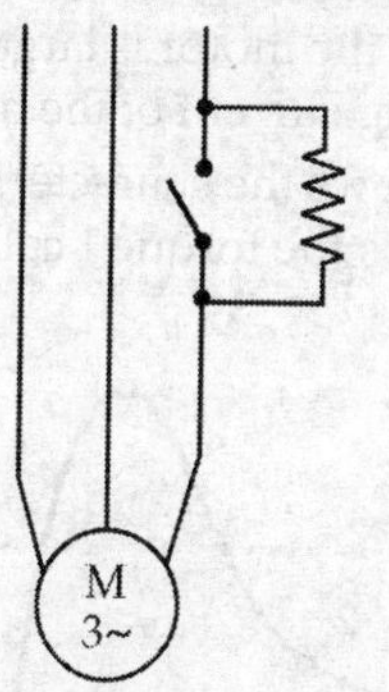

A. To reduce the starting current to a very low
B. To achieve a smooth starting
C. To attain a higher starting torque
D. To attain a higher maximum torque

139. The rated speed of an induction motor is 1410 rpm. What is meant by the statement "the slip s = 0"?
A. The speed of the motor is higher than the synchronous speed
B. The motor runs at a synchronous speed
C. The motor runs at its rated speed
D. The rotor is stationary

140. A machine is driven by an induction motor running at nominal speed. What happens if the counter torque of the machine becomes larger than the maximum torque of the motor?

Ans. 127. C 128. B 129. D 130. A 131. D 132. D 133. A 134. C 135. B 136. D 137. C 138. B 139. B

A. The induction motor heats up to an inadmissible extent
B. The induction motor stops
C. The speed of the induction motor is reduced to less than half of its nominal speed
D. The winding of the motor can be damaged because of high current strength

141. An induction motor is started on the 220/380 V supply mains with the help of a star-delta starter. Which type of induction motor is this?
A. 220/380 V squirrelcage induction motor
B. 220–380 V slip-ring induction motor
C. 380–660 V squirrelcage induction motor
D. 380/660 V slip-ring induction motor

142. What is the advantage of starting a slipring induction motor with the help of rotor resistances as compared to other methods?
A. The starting current is reduced
B. The starting torque increases due to the rotor resistances
C. The starter has to be designed for only a very low current
D. The starter can be built directly into the motor

143. What is the advantage of the double squirrelcage rotor as compared to the round bar cage rotor?
A. The efficiency of the motor is higher
B. The power factor of the motor is higher
C. The slip of the motor is larger
D. The starting current of the motor is lower

144. The figure shows the characteristics of an induction motor. What is the torque 1 called?

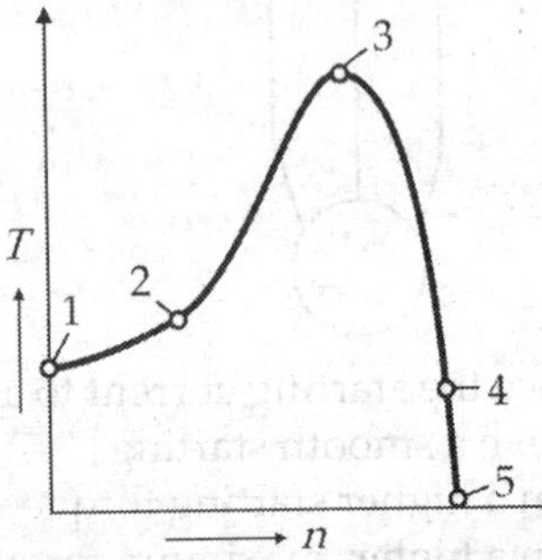

A. Maximum torque B. Starting torque
C. Rated torque D. Frictional torque

145. The above figure shows the characteristics of an induction motor. At which point does the motor operate after starting and accelerating at the rated torque?
A. Point 1 B. Point 2
C. Point 3 D. Point 4

146. The above figure shows the characteristics of an induction motor. What is the torque marked 3 called?
A. Starting torque B. Acceleratory torque
C. Maximum torque D. Rated torque

147. The direction of rotation of a three phase induction motor is reversed by
A. interchanging the connection of any two phases
B. interchanging the connection of all the three phases
C. rewinding the stator
D. adding a capacitor in any phase

148. Which of the following connection is correct for the anticlockwise rotation of an induction motor?

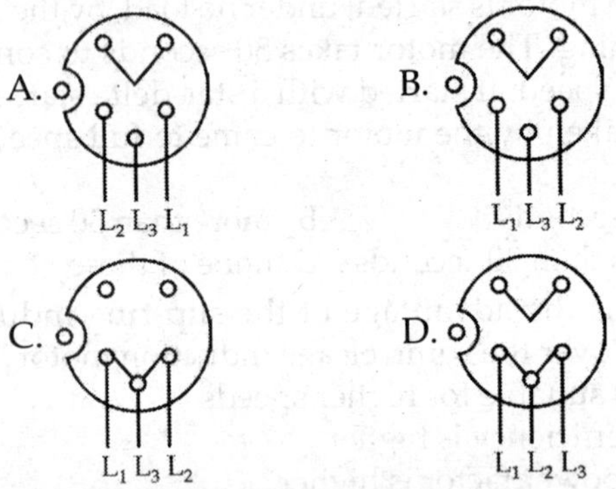

149. What is the disadvantage of the speed control of a slip-ring induction motor with the help of resistances in the rotor circuit?
A. This method of speed control is only applicable to motors having a power of more than 100 kW
B. This method is associated with high losses
C. By using this method, the speed can only be controlled very broadly
D. With reductions in speed the torque decreases considerably

150. Which motor should not be operated on three phase supply?
A. A three phase commutator motor with anti-interference capacitor
B. A three phase induction motor with anti-interference capacitor
C. A three phase induction generator with exciting capacitor
D. A single phase induction motor

151. While starting an induction motor with star-delta starter at the stator position
A. the torque developed is 3 times as high as at the run position
B. the torque developed is $\sqrt{3}$ times as high as at the run position
C. the voltage per phase is $1/\sqrt{3}$ of the supply voltage
D. The current is $1/\sqrt{3}$ of the current at direct online starting

Ans. 140. B 141. C 142. D 143. D 144. B 145. D 146. C 147. A 148. B 149. A 150. D 151. B

152. An induction motor has a rated speed of 715 rpm. How many poles has its rotating magnetic field?
A. 2 poles B. 4 poles
C. 6 poles D. none of these

153. Synchronous speed is defined as
A. the speed of the rotor of an induction motor
B. the speed of a synchronous motor
C. the speed of an induction motor at no-load
D. the natural speed at which a magnetic field rotates

154. Which of the following is not a part of a squirrel-cage induction motor?
A. Rotor B. Stator
C. Carbon brushes D. Shaft

155. What is the purposes of the blades in a squirrelcage induction motor?
A. To dynamically balance the rotor
B. To reduce the magnetic resistance of the rotor
C. To reduce the electrical resistance of the rotor cage
D. To cool the rotor

156. Which of the following statements about induction motor is true?
A. The motor can run only in one direction
B. The laminations of the rotor should be properly insulated against one another
C. The cage rotor is made of copper
D. The stator winding produces a rotating magnetic field

157. In which of the following motors, external resistance can be added to start the motor?
A. A squirrelcage induction motor
B. A slip-ring induction motor
C. A wound rotor synchronous motor
D. A salient pole synchronous motor

158. Which of the following lubricants is the shaft of the motor lubricated?
A. Graphite B. Grease
C. Silicon oil D. Mineral oil

159. An induction motor is running at its rated torque. The applied voltage is reduced from the rated value of 440 to 300 V. What is the consequence of this?
A. After sometime the motor heats up to an inadmissible extent
B. The current decreases
C. The motor stops
D. The speed reduces considerably

160. The rated speed of an induction motor is 1410 rpm. What is meant by the statement "the slip s = 1"?
A. The speed of the motor is higher than the synchronous speed
B. The motor runs at a synchronous speed
C. The motor runs at its rated speed
D. The rotor is stationary

161. A 50 Hz, 10-pole, 3-phase induction motor is used to drive a 60 Hz, 3-phase alternator. The pole number of the alternator should be
A. 10 B. 6
C. 12 D. none of these

162. In a 3-phase induction motor, the frequencies of induced emfs in the stator and the rotor are at 50 Hz and 1.0 Hz, respectively, the slip in this case will be
A. 0.1% B. 2%
C. 0.2% D. 5%

163. A 3-phase, 4 pole induction motor with open circuited rotor is connected to a 400 V, 50 Hz, 3-phase supply. The motor will run at
A. 1500 rpm B. 1480 rpm
C. 1450 rpm D. None of these

164. A three phase induction motor is running at a load of the rated torque. What happens when one of the outer mains is interrupted while the motor is running?
A. The motor stops immediately
B. The motor stops after a few seconds
C. The motor keeps running but draws more current
D. The motor keeps running and the current drawn does not change

165. Two alternators A and *B* are running in parallel and sharing a load. It is desired to shift a part of the load presently shared by alternator A to B. This can be achieved by
A. Increasing the field current of *B* and decreasing it for A
B. Increasing the mechanical power input to the prime-mover of B
C. Decreasing the input mechanical power of the prime-mover of A
D. none of these

166. A 3-phase, 400 V, 50 Hz 4-pole induction motor is fed from a 3-phase, 400 V supply and runs at 1425 rpm. The frequency of the rotor emf is
A. 2.5 Hz B. 50 Hz
C. 48 Hz D. zero

167. A change of 5% in supply voltage to a 3-phase induction motor will produce the approximate change in the torque of
A. 5% B. 7.5%
C. 10% D. 25%

168. The injected emf in the rotor of an induction motor must have
A. the same frequency as the slip frequency
B. the same phase as the motor emf
C. a high value for satisfactory speed control
D. none of these

Ans. 152. D 153. D 154. C 155. D 156. D 157. B 158. B 159. A 160. D 161. C 162. B 163. D 164. C 165. A 166. A 167. C 168. A

169. The stator of an unloaded 3-phase, 6-pole, wound-rotor induction motor is connected to a 60 Hz source and the rotor is connected to a 25 Hz source. The rotor speed at which steady-state motor action will result is given by
A. 1200 rpm B. 700 rpm
C. 500 rpm D. none of these

170. The power factor of an induction motor at low load is low due to
A. high working component of I_0
B. high magnetising component of I_0
C. high speed D. very low speed

171. The rotor output is 15 kW and the corresponding slip is 4%. The rotor copper loss is
A. 700 W B. 0.6 kW
C. 650 W D. 625 W

172. Frequency is 50 Hz and number of poles are 4. If rotor speed is 95%, the actual speed of motor is
A. 1425 rpm B. 1300 rpm
C. 1500 rpm D. none of these

173. An AC voltage of 66.5 volts is applied across an impedance of $(4.11 + j\,8.44)$. The current is
A. $16 + j\,2.36$ A B. $16 + j\,3.5$ A
C. $15 + j\,4$ A D. $16 - j\,2.36$ A

174. The mechanical load in an induction motor can be represented by
A. $(R_s/s - R_2)$ B. $(R_2 - s, R_2)$
C. $1/s(R_2 - 1)$ D. $R_2(1/s - 1)$

175. In the torque/speed characteristics of an induction motor, stable region is

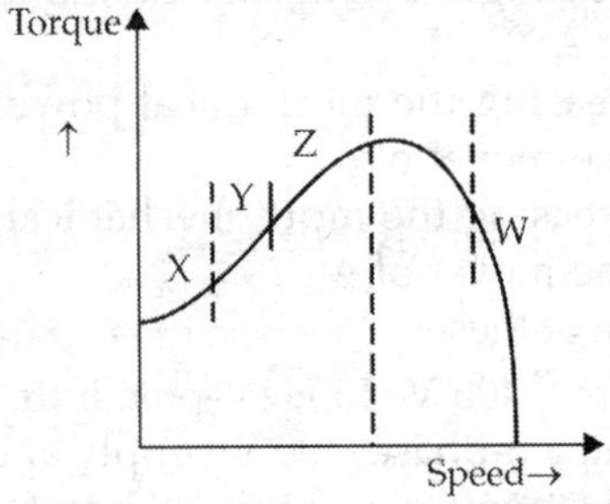

A. X B. Y
C. Z D. W

176. Maximum power is delivered in an induction motor when the load resistance R_L and internal impedance Z are
A. equal B. in 2:1 ratio
C. not equal D. in 1:2 ratio

177. If the input to a table-fan motor is 'a' and its losses are x times 'a', its output will be equal to
A. ax watts B. a/x watts
C. $a(1 - x)$ D. $a - x/a$ watts

178. The reactance per phase as compared to the resistance per phase of an induction motor is
A. quite high B. slightly large
C. almost same D. very small

179. A 500 V, 50 Hz motor takes a full-load current of 40 A at a factor of 0.85 lagging. If a capacitor of 80 μF is connected across the motor terminals, the power factor becomes
A. 0.72 B. 0.85
C. 0.97 D. 1.00

180. In the above problem, the full-load current falls to
A. 21 A B. 35 A
C. 42 A D. 47 A

181. Which of the following motors is used most frequently on AC supply?
A. DC shunt motor B. AC induction motor
C. Three-phase commutator motor
D. Three-phase induction motor

182. A 400 kW, 3-phase, 440 V, 50 Hz induction motor has a speed of 950 rpm on full load. The machine has 6 poles. The slip of the machine will be
A. 0.06 B. 0.05
C. 0.04 D. 0.02.

183. In an induction motor, the ratio of torque exerted at 4% slip to maximum torque, when slip at maximum torque is 12% will be
A.3/5 B. 5/3
C.1/3 D. none of these

184. A change of 5% in supply voltage to an induction motor will produce a change approximately
A. 5% in the rotor torque
B. 8% in the rotor torque
C. 20% in the rotor torque
D. none of these

185. What will happen to rotor copper losses if the frequency of the input to an induction motor increases?
A. Increases B. Decreases.
C. Remains same D. None of these

186. In an induction motor, if the phase sequence of applied voltage is changed
A. no back emf will be developed in rotor
B. no torque will be developed by rotor
C. speed will drop to zero
D. direction of rotation of motor will change

187. If a rotor of induction motor could be theoretically rotated at synchronous speed, the voltage that would be generated in the rotor is
A. zero
B. equal to applied voltage
C. more than applied voltage
D. none of these

188. In an induction motor, if the rotor is locked, then the rotor frequency of induction motor will be
A. more than the supply frequency
B. zero

Ans.													
	169. B	170. B	171. B	172. A	173. D	174. A	175. D	176. A	177. C	178. A	179. C	180. B	181. D
	182. B	183. A	184. D	185. A	186. D	187. A							

C. equal to the supply frequency
D. less than the supply frequency

189. In an induction motor, the rotor frequency is 4.5 Hz when the slip is 7.5%. Then the rotor frequency for the slip 2.5% is assume supply frequency of 50 Hz
A. 13.5 Hz B. 1.5 Hz
C. 0 Hz D. 1 Hz

190. In an induction motor, the rotor input is 600 W and slip is 4%. The rotor copper loss is
A. 700 W B. 0.6 kW
C. 650 W D. 24 W

191. When measuring the no load input to an induction motor using two wattmeter method, it is necessary to subtract wattmeter readings, because the no-load power factor is
A. lagging B. less than 0.5
C. unity D. zero

192. The speed of 25 Hz induction motor is 720 rpm. If speed at no-load is 745 rpm, then the percent slip and percent regulation is
A. 0.667%, 3.47% B. 0.667%, 0.667%
C. 4%, 3.47% D. 4%, 0.667%

193. If the speed is 460 rpm then the number of poles of a 50 Hz induction motor are
A. 4 C. 8
A. 12 C. 18

194. If a 3-phase, 60 Hz, 10-pole induction motor has a full load slip of 0.075, then the speed of rotor relative to revolving field will be
A. 54 rpm B. 1000 rpm
C. zero rpm D. 720 rpm

195. A 3-phase, 440 V, 60 Hz induction motor is to be operated at 25 Hz. The voltage that should be applied to the machine if the air gap flux density is to be maintained at its normal value is
A. 183 V B. 105.7 V
C. 609.7 V D. none of these

196. Which of the following statements is *not true* for an induction motor?
A. If $(1 - s)$ is the power input to the rotor, the power lost in the rotor is $s(1 - s)$
B. Maximum torque of an induction motor is independent of the rotor resistance
C. An induction motor can supply the load stably from $s = 0$ to $s = R_2/X_2$
D. The efficiency of an induction motor is maximum, when $R_2 = X_2$

197. The purpose of skewing of rotor slots in induction motor is to
A. reduce the magnetic hum of motor
B. increase the distribution factor
C. reduce the locking tendency of rotor
D. increase the breadth factor

198. An induction motor has a rotor resistance of 0.02 ohm and a standstill reactance of 0.1 ohm. The value of total resistance of a starter for the rotor circuit for maximum torque to be inserted at starting is
A. 0.002 ohm B. 0.08 ohm
C. 0.8 ohm D. none of these

199. In an induction motor, the ratio of starting torque to maximum torque is 80%. The speed at maximum torque if the synchronous speed is 1500 rpm is
A. 1290 rpm B. 1360 rpm
C. 1440 rpm D. none of these

200. In an induction motor, the short circuit current of the motor at normal voltage is 5 times the full load current and the full load slip is 4%. The starting torque when star-delta switch used is equal to
A. 3 times full load torque
B. 0.33 times full load torque
C. 0.07 times full load torque
D. None of these

201. An induction motor has a slip of 2% at normal voltage. The slip when developing the same torque at 10% below normal voltage is
A. 1.65% B. 2.47%
C. 2.22% D. 1.82%

202. In an induction motor the rotor resistance is equal to standstill reactance then the maximum torque is
A. less than starting torque
B. equal to starting torque
C. more than starting torque
D. none of these

203. The AC line current at slip-rings in a 6-phase, 6-ring rotary converter having 100% efficiency and unity pf is ...... times the DC current.
B. 0.943 B. 0.472
C. 0.236 D. 1.414

204. A 3-phase induction motor with wound rotor has a delta-connected stator winding. For proper operation, the rotor winding
A. must be delta connected
B. must be star connected
C. may be connected in delta or star
D. none of these

205. Under running condition, the rotor circuit of a 3-phase induction motor is
A. always closed B. always open
C. sometimes open and sometimes closed
D. none of these

206. What is the advantage of starting a slip-ring induction motor with the help of the motor resistances as compared to other methods?
A. The starting current is reduced
B. The starting torque increases due to the rotor resistances

Ans. 188. C 189. C 190. D 191. B 192. C 193. A 194. A 195. A 196. D 197. C 198. B 199. C 200. B 201. B 202. B 203. B 204. C 205. A 206. B

C. The starter has to be designed for only a very low current
D. The starter can be built directly into the motor

207. What is the advantage of the double squirrelcage rotor as compared to the round bar cage rotor?
A. The efficiency of the motor is higher
B. The power factor of the motor is higher
C. The slip of the motor is larger
D. The starting current of the motor is lower

208. At three-phase induction motor is running at load of the rated torque. What happens when one of the outer mains is interrupted while the motor is running?
A. The motor stops immediately
B. The motor stops after a few seconds
C. The motor keeps running but draws more current
D. The motor keeps running and the current drawn does not change

209. In a 4 pole, 50 Hz, 3-phase induction motor, blocked rotor reactance per phase is four times the rotor resistance per phase. The speed at which maximum torque developed is
A. 1500 rpm B. 1125 rpm
C. 1000 rpm D. None of these

210. The no-load current of an induction motor is
A. 1% of FL current B. 10% of FL current
C. negligible D. 40–50% of FL current

211. The field of an induction motor rotor rotates with reference to stator at
A. synchronous speed B. slip speed
C. rotor speed D. very low speed

212. Which of the following fields of application is three-phase induction motor mainly suitable in small industry?
A. For running of electric vehicles
B. For running of rolling mills where an exact speed control is required
C. For running of different machine tools where one or a few speeds are required
D. For running of paper machines (the paper quality depends upon the exact speed control)

213. What is the advantage of the slip-ring induction motor over the squirrelcage induction motor?
A. It is suitable for higher speeds
B. Its efficiency is higher
C. Its power factor is higher
D. It can be started with the help of rotor resistances

214. The starting torque of the slip-ring induction motor can be increased by
A. adding resistance to the rotor
B. adding resistance to the stator
C. adding resistance to both stator and rotor
D. none of these

215. The torque of an induction motor in synchronous watt is equal to
A. rotor output B. stator output
C. stator input D. watts at rotor speeds

216. The torque of a rotor in an induction motor under running condition is maximum
A. at unity value of the slip
B. at zero value of slip
C. at the value of slip which makes the rotor reactance half of the rotor resistance
D. at the value of slip which makes the rotor reactance per phase equal to its resistance per phase

217. In three-phase slip-ring induction motor, input is 440 V and the voltage at the slip-rings is 110 V. If an input of 110 V is given to the slip-rings, the stator voltages would be
A. 500 V B. nearly equal to 440 V
C. nearly equal to 110 V D. zero

218. The shape of torque/slip curve of an induction motor is
A. hyperbola B. parabola
C. straight line D. rectangular parabola

219. When the frequency of the rotor of an induction motor is small it can be measured by
A. galvanometer
B. DC moving coil milli-voltmeter
C. DC moving coil ammeter
D. AC voltmeter

220. The stator frame in an induction motor is used
A. as a return path for the flux
B. to hold the armature stampings/stator
C. to protect the whole machine
D. to ventilate the armature

221. A three-phase, 6-pole, 50 Hz induction motor is working with 5% slip. The actual speed of the motor is
A. 1000 rpm B. 950 rpm
C. 900 rpm D. 850 rpm

222. The speed of a three phase cage-rotor induction motor depends on
A. frequency of the supply only
B. number of poles only
C. number of poles and frequency of the supply
D. input voltage

223. In a three-phase induction motor, the power output is maximum when the equivalent load resistance is
A. equal to the standstill resistance of the motor
B. equal to the standstill impedance of the motor
C. equal to the standstill reactance of the motor
D. None of these

Ans. 207. D 208. C 209. B 210. D 211. A 212. C 213. D 214. A 215. B 216. D 217. B 218. D 219. B 220. A 221. B 222. C 223. B

224. In an induction motor the ratio of the rotor output to rotor input is 0.96. Then the percentage slip is
A. 0.04% B. 4%
C. 6% D. 12%

225. A 440 V, 50 Hz, 6-pole, three-phase induction motor has 3% slip. The relative speed between the rotor and stator magnetic field will be
A. zero B. 970 rpm
C. 1,000 rpm D. more than 1000 rpm

226. A three-phase induction motor connected in star takes 5A. When connected in delta, it will take
A. $5/\sqrt{3}$ A B. $5\sqrt{3}$ A
C. 15 A D. 5 A

227. The slip of a 500 HP, 3 ϕ, 400 V, 50 Hz induction motor is 5%. How many complete cycles will the rotor voltage make per minute?
A. 1000 B. 1200
C. 1500 D. 2000

228. A 15 HP, 400 V delta connected, 50 Hz, 4-pole induction motor gives the following test results:

S. No.	Applied voltage	Line current	Input power
1. No-load test	400 V	8 A	1.0 kW
2. Locked rotor test	100 V	25 A	1.75 kW

The short-circuit current with normal voltage will be
A. 8 A B. 25 A
C. 75 A D. 100 A

229. In the same problem, short-circuit power factor angle will be
A. 66.2° B. 78.6°
C. 0.2 D. 0.404

230. The starting current of a three-phase induction motor is 60 A at full load. The current drawn by the same motor at no-load will be
A. 10 A B. 90 A
C. 60 A
D. dependent online voltage

231. Whenever any polyphase induction motor is loaded then induced emf
A. in the rotor increases and its frequency falls
B. in the rotor increases and its frequency rises
C. in the rotor remains constant
D. decreases and frequency increases

232. A three-phase, 8-pole, induction motor runs on 50 Hz supply. The rotor bar current makes 150 complete cycles per minute. Then the rotor moves, at a speed of
A. 710 rpm B. 713.5 rpm
C. 712.5 rpm D. none of these

233. For a three-phase induction motor, input power on no-load is $W_0 = 600$ watts and windage and friction losses is $W_f = 183$ watts. The stator copper-loss will be equal to
A. 783 W B. 600 W
C. 183 W D. 517 W

234. Two induction motors have 12 poles and 8 poles respectively. If the supply frequency is 25 Hz, the possible running speed of the set will be
A. 750 rpm B. 250 rpm
C. 100 rpm D. 150 rpm

235. Imbalance in shaft of the induction motor occurs due to
A. air gap is not uniform
B. overheating of the winding
C. rugged construction D. slip-rings

236. The noise and tooth pulsation losses can be minimised by using
A. large number of open slots in stator
B. small number of open slots in stator
C. large number of narrow slots in stator
D. small number of narrow slots in stator

237. Dispersion coefficient σ is the ratio of
A. magnetising current to supply voltage
B. magnetising current to ideal short circuit current
C. open circuit voltage to short circuit for the same excitation
D. none of these

238. In squirrelcage induction motor, if the rotor is open, the rotor will
A. not run
B. make noise
C. run at very slow speed
D. run at very high speed.

239. In an induction motor with star-delta starter at star position
A. the torque developed is three times as high as at the run position
B. the torque developed is $\sqrt{3}$ times as high as at the run position
C. the current is $1/\sqrt{3}$ times of the current at direct online starting
D. none of these

240. While starting of an induction motor with star-delta starter at the star position
A. the torque developed is three times as high as at the run position
B. the torque developed is 1/3 times that at the run position
C. the current is 1/3 of the current at direct online starting
D. none of these

241. The general area where electrical faults occur most often is in the motor
A. control equipment B. brushes
C. rotor D. all of these

Ans. 224. B 225. A 226. C 227. C 228. A 229. A 230. C 231. A 232. D 233. B 234. D 235. A 236. C 237. B 238. A 239. B 240. C 241. A

242. In two induction motors running at the same speed and having same number of poles, the physical dimensions are in the ratio 3:2. The output of both the motors will be in the ratio of
A. 3:2 B. 9:4
C. 27:8 D. 81:16

243. In the above question, the copper losses will be in the ratio of
A. 3:2 B. 9:4
C. 27:8 D. 81:16

244. If two phases open accidentally of three-phase supply of a running three-phase induction motor, the motor will
A. stop B. run at high speed
C. continue to run depending on load
D. none of these

245. If 3-phase induction motor is switched on with one phase disconnected, it
A. is likely to burn out
B. will run at very low speed
C. will not run at all D. none of these

246. If the input voltage and the frequency of supply is halved, to the three-phase induction motor
A. the maximum torque is halved
B. the maximum torque remains the same
C. the air-gap flux is halved
D. None of these

247. An induction motor crawls at lower speed. This is due to
A. skewing
B. harmonic induction torques
C. high secondary reactance
D. low secondary reactance

248. For good overall design of induction motor, the ratio L/τ for a fixed member of poles should be
A. 1.5 to 2 B. 1.0 to 1.25
C. 1.5 D. 1

249. There are two 3-phase induction motors with 6 poles and 4 poles respectively. Their cascade combination run from 60 Hz, 400 V supply can give one of the following maximum number of synchronous speed combination
A. 720, 1200, 1800 and 3600 rpm
B. 720, 1200, 1800 rpm
C. 3000, 1500, 1000, 600 rpm
D. 3600 and 720 rpm

250. Output coefficient of induction motor is given by
A. output/input
B. active volume of machine/output
C. output/active volume of machine
D. None of these

251. Semi-closed stator slot and not open slot is used in induction motor to
A. increase the pull out torque
B. decrease the slot reactance
C. decrease magnetising current
D. both A and B are correct

252. Efficiency of induction motor can be improved best by choosing L/τ ratio as
A. 2.0 B. 1.5
C. 0.6 D. 1.4

253. An increase in the vlaue of air gap flux density in an induction motor will
A. increase iron-loss B. increase efficiency
C. decrease efficiency
D. both (A) and (C) are correct

254. Peripheral speed of rotor of a normal induction motor should be
A. 10 m/s B. 30 m/s
C. 60 m/s D. 80 m/s

255. In an induction motor, skew of the rotor bar reduces
A. noise B. vibration
C. synchronous cusps D. all of these

256. In an induction motor, the value of average flux density in air gap should be small
A. to get good pf B. to get poor pf
C. to achieve good efficiency
D. for minimum cost

257. The pulsation losses and noise can be reduced in an induction motor by using
A. a large number of narrow slots
B. less number of narrow slots
C. large number of semislots
D. None of these

258. With increase in length of the air gap in an induction motor, the pulsation losses become
A. more B. less
C. unchanged D. None of these

259. Electrical noise cannot be heard with
A. power on B. power off
C. rotor locked D. defective bearings

260. The large number of slots in an induction motor has
A. high over-load capacity
B. low over-load capacity
C. no effect on over-load capacity
D. none of these

261. The capacitor start motor uses
A. paper capacitor B. electrolytic capacitor
C. air capacitor D. none of these

262. A short air gap in an induction motor is necessary to
A. achieve increased overload capacity
B. reduce the pulsation loss
C. reduce the noise
D. achieve a good pf

Ans. 242. D 243. C 244. C 245. A 246. A 247. B 248. D 249. C 250. C 251. C 252. B 253. D 254. C 255. D 256. A 257. A 258. B 259. B 260. A 261. B 262. D

263. The large value of air gap length in an induction motor results in
A. increased overload capacity
B. reducing the pulsation loss
C. providing better cooling
D. reducing the noise

264. In an induction motor, the motor fails to start when number of rotor slots is equal to number of stator slots. This is known as
A. crawling B. cogging
C. synchronous cusps D. none of these

265. The simplest way to eliminate the harmonic induction torques is
A. integral slot winding B. skewing
C. chording D. none of these

266. The motor noise, vibrations cogging and synchronous cusps can be reduced by
A. chording B. integral slot winding
C. skewing
D. increasing the air gap length

267. A 3-phase slip-ring induction motor has 4-pole stator and 2-pole rotor. If its stator is energised from 50 Hz source, the rotor would run at a speed
A. slightly less than 1500 rpm
B. slightly less than 3000 rpm
C. slightly less than 1200 rpm
D. of zero rpm

268. Rotor slot of squirrelcage induction motor are skewed slightly, so as to
A. increase the mechanical strength of rotor
B. make the rotor construction simplex
C. eliminate locking tendency of the rotor and to reduce the noise
D. save the amount of copper required

269. If the rotor of a 3-phase induction motor is assumed purely resistive, then electro-magnetic torque is
A. maximum with load angle of 90°
B. minimum with load angle of 90°
C. optimum with load angle of 90°
D. optimum with load angle of 0°

270. In a 3-phase wound-rotor induction motor, 3-phase balanced supply is given to the rotor and stator winding is short-circuited. The rotor would
A. not run
B. run in the direction of rotating field
C. run against the direction of rotating field
D. run at half synchronous speed

271. The rotor of an induction motor cannot run at synchronous speed. If it did so, then
A. rotor emf would be zero
B. rotor current would be zero
C. rotor torque would be zero
D. all of these

272. A 3-phase induction motor is running at half-full load. If the fuse in one of the phases gets burnt, the motor would
A. come to standstill
B. continue running at the same speed
C. continue running at slightly reduced speed
D. continue running at an increased speed

273. A 3-phase, 50 Hz, 4-pole squirrelcage induction motor has its stator rewound for 6 poles without any alterations in the rotor. The motor would now run at a speed
A. less than 1000 rpm B. less than 1500 rpm
C. less than 3000 rpm D. zero rpm

274. Which of the following motors uses a commutator?
A. Universal motor B. Split phase motor
C. Shaded pole motor D. None of these

275. A 3-phase squirrelcage induction motor, designed to operate with stator in star, requires *W* kg of copper for its stator winding. If this motor is to be designed to operate with stator in delta, then weight of stator copper would be
A. more than W kg B. less than W kg
C. $\sqrt{3}\,W$ kg D. $1/\sqrt{3}\,W$ kg

276. If the rotor of a 3-phase slip-ring induction motor is fed from balanced 3-phase supply with the stator short-circuited, then frequency of stator currents would be
A. sf B. $(2s)f$
C. $(1s)f$ D. $(s1)f$

277. The length of air gap in an induction motor is given by
A. $0.2 + 2\sqrt{DL}$ mm B. $0.2 + \sqrt{2DL}$ mm
C. $0.2 - 2\sqrt{DL}$ mm D. $0.2 + \sqrt{DL}$ mm
where *D* and *L* are standard dimensions of the induction motor.

278. The stator of a 3-phase, 4-pole, slip-ring induction motor is fed from 50 Hz source and its rotor from 30 Hz source. The motor will run at
A. 1500 rpm B. 900 rpm
C. zero rpm D. 2400 rpm

279. If the rotor of an induction motor is assumed purely inductive, then electromagnetic torque is
A. optimum with load angle $\delta = 0°$
B. optimum with $\delta = 90°$
C. maximum with $\delta = 90°$
D. zero

280. If both the supply voltage and frequency are reduced to 90% of their previous values in a 3-phase induction motor, then the new torque T_e in terms of its previous value is
A. 0.90 B. 1.11
C. 2.00 D. 1.00

Ans. 263. A 264. B 265. C 266. C 267. D 268. C 269. C 270. C 271. D 272. C 273. A 274. A 275. D 276. A 277. A 278. D 279. D 280. D

281. The complete circle diagram of a 3-phase induction motor can be drawn with the help of
A. running-light test alone
B. both running-light and blocked-rotor tests
C. running-light, blocked-rotor and stator-resistance tests
D. blocked rotor test alone

282. One unique advantage of employing induction regulator method for controlling the DC output voltage of a rotary converter is that it
A. is extremely simple and relatively cheaper
B. responds instantaneously to changes in load
C. can be used for inverters
D. gives voltage changes in exact jumps

283. In an induction motor the ratio of starting current to full load current is 12.65 and the ratio of starting torque to full load torque is 1.6. Then percentage slip at full load is
A. 0.01% B. 0.5%
C. 1% D. none of these

284. In an induction motor, the short circuit current at normal voltage is 4 times full load current and the full load slip is 3%. The percentage tapping of an auto- transformer required to start the motor against 1/4 of full load torque is
A. 72.2% B. 25%
C. 100% D. none of these

285. What is the speed of rotation of a 5th harmonic component in the stator of an induction motor?
A. 5 times the synchronous speed
B. 1/5 times the synchronous speed
C. 0.5 times the synchronous speed
D. none of these

286. Consider a cascaded set of two motors A and B with 4 poles and 6 poles respectively. The electric power transferred to motors when the input to motor A is 25 kW (neglect losses) is
A. 16.67 kW B. 0.05 kW
C. 1.0 kW D. 10 kW

287. The complete circle diagram of an induction motor can be drawn with the help of data found from
A. no-load test B. blocked rotor test
C. both A and B are correct
D. A, B and stator resistance test

288. The power scale of circle diagram of an induction motor can be found from
A. no-load test only B. short-circuit test only
C. stator resistance test D. None of these

289. Which of the following is not determined by circle diagram?
A. Output B. Efficiency
C. Power factor D. Frequency

290. If a 3-phase, 4-pole, 50 Hz induction motor runs at a speed of 1440 rpm, then the slip is
A. 0.03 B. 0.10
C. 0.04 D. 0.05

291. The rotor of a 3-phase induction motor is running at a speed of n_r rps. The slip-frequency currents in the rotor would produce a field which is
A. stationary with respect to rotor
B. rotating at synchronous speed wrt stator
C. rotating at n_r rps wrt stator
D. rotating at small slip wrt stator

292. In a 3-phase induction motor, the maximum torque
A. is proportional to rotor resistance r_2
B. does not depend on rotor resistance r_2
C. is proportional to $\sqrt{r_2}$
D. none of these

293. In a 3-phase induction motor, if stator impedance is neglected, then the slip at maximum torque is equal to
A. x_2/r_2 B. r_2/x_2
C. $\sqrt{r_2/x_2}$ D. $\sqrt{x_2/r_2}$

294. If the autotransformer tapping K is less than 1, then the starting torque of the cage induction motor with this autotransformer starter is
A. $K \times$ Starting torque T_{st} with direct switching
B. $K^2 \times T_{st}$
C. $\frac{1}{K} \times T_{st}$ D. $\frac{1}{K^2} \times T_{st}$

295. A starting torque of 100 Nm is developed by an autotransformer starter with a tapping of 40%. If the tapping of auto-transformer starter is at 80%, then the starting torque would be
A. 200 Nm B. 50 Nm
C. 400 Nm D. 25 Nm

296. With direct online starter, a 3-phase delta-connected induction motor takes starting line current of 30 A and develops a starting torque of 173.2 Nm. When started by star-delta starter, the starting line-current and starting torque are respectively equal to
A. 10 A, 100 Nm B. 17.32 A, 100 Nm
C. 10 A, 57.73 Nm D. 17.32 A, 57.73 Nm

297. The power factor of a delta-connected 3-phase, 50 kW induction motor is 0.4 when delivering 20% of its rated load. If its stator is reconnected in star, then its pf
A. remains unchanged B. is improved
C. is worsened D. None of these

298. The difference between stator slots and rotor slots in an induction motor should not be equal to p, 2 p, or 5 p to avoid
A. crawling B. noise and vibrations
C. synchronous cusps D. magnetic locking

Ans. 281. C 282. C 283. C 284. A 285. B 286. D 287. D 288. B 289. D 290. C 291. B 292. B 293. B
294. B 295. C 296. C 297. B 298. C

299. A 3-phase induction motor is running at slip *s*. If its two supply leads are inter-changed, then its slip at that instant is equal to
A. 2*s* B. 1 – *s*
C. 2 – *s* D. zero

300. An induction motor in which the stator has the squirrelcage winding is
A. 3-phase squirrelcage motor
B. 1-phase induction motor
C. 1-phase ceiling fan D. 1-phase table fan

301. For satisfactory performance of a 3-phase 420 V, 60 Hz induction motor, the supply voltage at 50 Hz should be equal to
A. 420 V B. 330 V
C. 350 V D. 400 V

302. A 3-phase wound-rotor induction motor, with star-connected rotor, runs clockwise. If the phase sequence of the rotor terminals is changed, then the motor would
A. not run B. run clockwise
C. run anticlockwise D. none of these

303. No-load current of a 3-phase induction motor in terms of its rated current is
A. 10 to 20% B. 2 to 6%
C. 20 to 30% D. 30 to 50%

304. No-load pf of a 3-phase induction motor is
A. 0.2 B. 0.5
C. 0.7 D. 0.85

305. Full-load pf of a 3-phase induction motor is generally
A. 0.2 B. 0.5
C. 0.7 D. 0.85

306. Closed windings are used for
A. AC machines B. DC machines
C. DC commutator machines
D. both AC and DC commutator machines

307. A 3-phase induction motor should have small air-gap length so that it has
A. more starting torque B. more pull-out torque
C. better pf D. improved efficiency

308. Induction motors with open slots have
A. better pf B. improved efficiency
C. more breakdown torque
D. less starting torque

309. Induction motors *A* and *B* have efficiencies of 82% and 92% respectively. Motor 'A' is
A. bigger than *B* B. smaller than *B*
C. equal to *B* D. none of these

310. A 3-phase induction motor runs at 980 rpm. at no-load. If its squirrelcage rotor is replaced by solid cylinder, then its no-load speed would be
A. zero B. 980 rpm.
C. below 980 rpm D. 490 rpm

311. Two induction motors have no-load operating power factors of 0.1 and 0.3. The induction motor with 0.1 pf is
A. bigger than the other motor
B. smaller than the other motor
C. of the same size as the other motor
D. none of these

312. Motor A has deeper and narrow slots, whereas motor *B* has shallow and wide slots. Induction motor A, as compared to motor B, has
A. more starting torque B. more pull-out torque
C. less starting torque D. more operating slip

313. A multi-turn coil has
A. multi-coil-sides B. one-coil-side
C. two-coil-sides D. four coil-sides

314. Closed windings may be
A. single layer B. double layer
C. both single layer and double layer
D. all of these

315. Open windings may be
A. single layer B. double layer
C. both single and double layers
D. none of these

316. For a simplex tap winding, the winding pitch is equal to
A. +2 B. –2
C. ±1 D. ±2

317. For a simplex lap winding, the commutator pitch is equal to
A. + 1 B. ± 1
C. – 1 D. ± 2

318. For *C* coils and *P* poles, the commutator pitch for simplex winding is
A. $\frac{C+1}{P/2}$ B. $\frac{C-1}{P/2}$
C. $\frac{C\pm 2}{P/2}$ D. $\frac{C\pm 1}{P/2}$

319. For C armature coils and P poles, the back pitch for simplex lap winding is
A. $\frac{2}{P}+K$ B. $\frac{2C}{P}-K$
C. $\frac{2C}{P}\pm K$ D. $2CP \pm K$

320. For C coils and *P* poles, the winding pitch for a simplex wave winding is
A. $\frac{2C\pm 1}{P/2}$ B. $\frac{2C+1}{P/2}$
C. $\frac{2C-2}{P/2}$ D. $\frac{2C\pm 1}{P}$

321. For simplex wave and lap windings, the back pitch y_b and front pitch y_f are as follows

Ans.	299. C	300. C	301. C	302. B	303. D	304. A	305. D	306. D	307. C	308. C	309. B	310. C	311. A
	312. C	313. C	314. B	315. C	316. B	317. B	318. D	319. C	320. D				

A. y_b is odd, y_f is even B. y_b is even, y_f is odd
C. both y_b and y_f are odd
D. both y_b and y_f are even

322. A 3-phase, 52-slot, 8-pole machine can have a coil span of
A. 4 B. 7½
C. 6 D. 8

323. A 3-phase machine has 39 slots and 6 poles. This machine can have
A. integral-slot winding
B. integral-slot chorded winding
C. fractional-slot winding
D. none of these

324. The number of parallel paths in an integral-slot winding with P poles are
A. P B. $2P$
C. $P/2$ D. $P/4$

325. Whole-coil and mush windings are similar because their
A. coils are interconnected in the same manner
B. coil-pitches are the same
C. phase spreads are equal
D. slot pitches are equal

326. In a 3-phase machine with 54 slots and 10 poles, the number of parallel circuits
A. 2 B. 5
C. 10 D. 3

327. In single layer winding, the number of coils C and the slot S has the following relation
A. $C = S$ B. $2C = S$
C. $2C = 3S$ D. none of these

328. For the use of mush windings in 3-phase induction motors, the slots should be
A. semi-closed B. open
C. closed D. either (A) or (B)

329. The recommended maximum voltage for mush winding is
A. 230 V B. 440 V
C. 600 V D. 3300 V

330. In most AC machines, it is usually a standard practice to use
A. fractional-slot winding with full-pitch coils
B. fractional-slot winding with chorded coils
C. integral-slot winding with full-pitch coils
D. integral-slot winding with chorded coils

331. If y_b is the back pitch and u is the number of coil-sides per slot, then split coils can be avoided if the following quantity is an integer
A. $\frac{y_b + 1}{u}$ B. $(y_b + 1)$
C. $\frac{y_b - 1}{u}$ D. $y_b - 1$

332. Compensating winding is employed in AC series motor in order to
A. increase the torque
B. reduce sparking at brushes
C. reduce the effects of armature reaction
D. none of these

333. If 220 V DC series motor is operated on 50 Hz AC supply, it will
A. have poor efficiency
B. have poor power factor
C. give spark D. all of these

334. For a 3-phase fractional-slot winding with P poles, the number of parallel paths is
A. P B. $2P$
C. $P/2$ D. None of these

335. The starting winding of a single-phase motor is placed in the
A. rotor B. stator
C. armature D. field

336. One of the characteristics of a single phase motor is that it
A. is self-starting B. is not self-starting
C. requires only one winding
D. can rotate in one direction only

337. After the starting winding of a single-phase induction motor is disconnected from supply, it continues to run only on which winding?
A. Rotor B. Compensating
C. Field D. Running

338. If starting winding of a single-phase induction motor is left open in the circuit, it will
A. draw excessive current and overheat
B. run slower
C. run faster D. spark at light loads

339. The direction of rotation of a single-phase motor can be reversed by
A. reversing connections of both windings
B. reversing connections of starting winding
C. using a reversing switch
D. reversing supply connections

340. If a single-phase induction motor runs slower than normal, the more likely defect is
A. improper fuses B. worn bearings
C. open starting winding
D. shorted running winding

341. The capacitor in a capacitor-start induction-run AC motor is connected in series with which winding?
A. Starting B. Running
C. Squirrelcage D. Compensating

342. A permanent-split single-phase capacitor motor does not have
A. centrifugal switch B. starting winding
C. squirrelcage rotor D. high power factor

Ans. 321. C 322. C 323. D 324. C 325. A 326. A 327. B 328. A 329. B 330. A 331. C 332. B 333. D
334. D 335. B 336. B 337. D 338. A 339. B 340. B 341. A 342. A

343. The starting torque of a capacitor-start induction run motor is directly related to the angle α between its two winding currents by the relation
A. cos α B. sin α
C. tan α D. sin $\alpha/2$

344. In a two-value capacitor motor, the capacitor used for running purposes is a/an
A. dry-type AC electrolytic capacitor
B. paper-spaced oil-filled type
C. air-capacitor D. ceramic type

345. If the centrifugal switch of a two-value capacitor motor using two capacitors, fails to open, then
A. electrolytic capacitor will, in all probability, suffer breakdown
B. motor will not carry the load
C. motor will draw excessively high current
D. motor will not come up to the rated speed

346. Each of the following statements regarding a shaded-pole motor is true except
A. its direction of rotation is from unshaded to shaded portion of the poles
B. it has very poor efficiency
C. it has very poor pf
D. it has high starting torque

347. Compensating winding is employed in an AC series motor in order to
A. compensate for decrease in field flux
B. increase the total torque
C. reduce the sparking at brushes
D. reduce effects of armature reaction

348. A Universal motor is one which
A. is available universally
B. can be marketed internationally
C. can be operated either on DC or AC supply
D. runs at dangerously high speed on no-load

349. In a single-phase series motor, the main purpose of inductively-wound compensating winding is to reduce the
A. reactance emf of commutation
B. rotational emf of commutation
C. transformer emf of commutation
D. None of these

350. A repulsion motor is equipped with
A. a commutator B. slip-rings
C. a repeller D. none of these

351. A repulsion-start induction run single-phase motor runs as an induction motor only when
A. brushes are shifted to neutral plane
B. short-circuiter is disconnected
C. commutator segments are short-circuited
D. stator winding is reversed

352. Which of the following is most economical method for starting single-phase motor?
A. Capacitor-start method
B. Split-phase method
C. Induction-start method
D. Resistance-start method

353. Which of the following motors is most suitable for signalling devices and timer?
A. DC series motor B. DC shunt motor
C. Two phase induction motor
D. Reluctance motor

354. The value of the capacitor in a capacitor start motor controls the
A. starting torque B. speed of the motor
C. efficiency D. none of these

355. Under no-load conditions, the speed of a universal motor is limited by
A. supply frequency B. supply voltage
C. windage and friction D. None of these

356. Which of the following motors does not use a centrifugal switch?
A. Split phase motor
B. Capacitor start capacitor run motor
C. Shaded pole motor D. None of these

357. If a DC series motor is connected to an AC supply, the motor will
A. the motor will not start
B. start but then come to a stop
C. start and run but will have poor performance
D. none of these

358. A capacitor-start single phase induction motor is switched on to supply with its capacitor replaced by an inductor of equivalent reactance value. It will
A. not start at all B. start and run slowly
C. start and run at rated speed
D. start and then stop

359. In a two value capacitor motor, the capacitor used for running purposes is
A. AC electrolytic type
B. paper spaced oil-filled type
C. air capacitor D. ceramic type

360. A wound rotor induction motor is preferred over squirrelcage induction motor when the major consideration involved is
A. high starting torque B. low starting current
C. speed control over limited range
D. all of these

361. As compared to squirrelcage induction motor, a wound rotor induction motor is preferred when the major consideration is
A. high starting torque
B. low windage losses
C. slow speed operation
D. all of these

Ans. 343. B 344. B 345. A 346. D 347. D 348. C 349. D 350. A 351. C 352. A 353. D 354. A 355. C 356. C 357. C 358. A 359. B 360. D 361. A

362. If a DC series motor is operated on AC supply, it will
A. have poor efficiency
B. have poor power factor
C. spark excessively
D. all of these

363. An outstanding feature of a universal motor is its
A. best performance at 50 Hz supply
B. slow speed at all loads
C. excellent performance on DC supply
D. highest output kW/kg ratio

364. The direction of rotation of a hysteresis motor is determined by the
A. retentivity of the rotor material
B. amount of hysteresis loss
C. permeability of rotor material
D. position of shaded pole with respect to the main pole

365. The direction of rotation of a hysteresis motor is determined by
A. position of shaded pole with respect to main pole
B. retentivity of the rotor material
C. interchanging the supply leads
D. None of these

366. Which of the following motors is used in a situation where load increases with speed?
A. Hysteresis motor
B. Three-phase series motor
C. Induction motor D. Scharge motor

367. If all the stator coils of a motor are connected for the same magnetic polarity, they will be formed by an equal number of
A. consequent poles with opposite polarity
B. consequent poles with the same polarity
C. rotor poles with opposite polarity
D. rotor poles with the same polarity

368. Which of the following single-phase induction motors is generally used in time phonographs?
A. Shaded pole B. Capacitor start
C. Resistance start
D. Capacitor start capacitor run

369. The repulsion-start induction-run motor is used because of
A. high efficiency B. high starting torque
C. minimum cost D. good power factor

370. The speed/load characteristics of a universal motor is same as that of
A. DC shunt motor B. DC series motor
C. AC motor D. none of these

371. An induction regulator is used to
A. regulate line voltages
B. regulate speed of induction motor
C. generate AC voltages
D. none of these

372. Which of the following motors is used for unity power factor?
A. Squirrelcage induction motors
B. Slip-ring induction motor
C. Hysteresis motor D. Scharge motor

373. Two-phase win induction voltage regulator is wound for four poles. When the input to it is 100 V/phase, the maximum and minimum output voltage are 120 V and 80 V per phase. If the output is to be changed from 120 V to 90 V, the angle through which the regulator should be turned is
A. 30° B. 45°
C. 60° D. 90°

374. The squirrelcage winding of a single-phase motor is placed in the
A. armature B. stator
C. rotor D. field

375. NEMA standards rate motors according to
A. frame number B. horsepower
C. voltage D. weight

376. A megohmmeter reading of zero or low ohms is indicated between the stator winding and motor frame. The winding is
A. open B. series connected
C. grounded D. cross connected

377. Where or how is the performance data of a motor found?
A. In NEMA Standards
B. By taking electrical measurements
C. By calculation
D. On the motor's nameplate

378. Which of the following is a characteristic of a single-phase motor?
A. It is not self-starting
B. It is self-starting
C. It requires only one winding
D. It can rotate in one direction only

379. A rotating magnetic field is produced by the current in the two windings displaced by 90 electrical degrees. This is the principle of
A. phase timing B. phase splitting
C. phase sequence D. phase relationship

380. Which of the following motors has highest starting torque?
A. Split-phase motor B. Shaded pole motor
C. Capacitor-start motor
D. Repulsion motor

381. The speed of the rotating magnetic field in an induction motor is known as the
A. slip speed B. effective speed
C. shaft speed D. synchronous speed

Ans. 362. D 363. D 364. D 365. A 366. B 367. A 368. A 369. B 370. B 371. A 372. D 373. C 374. C 375. A 376. C 377. D 378. A 379. B 380. C 381. D

382. After the starting winding is disconnected from the circuit, the motor continues to run only on the
A. squirrelcage winding
B. running winding
C. magnetic winding
D. compensating winding

383. The rotating part of an AC motor is called the
A. stator B. rotor
C. running winding D. starting winding

384. Most fractional horsepower motors have what type of bearings?
A. Pair or sleeve bearings
B. Ball or roller bearings
C. Hard and annealed bearings
D. Soft and porous bearings

385. A split-phase motor is started by connecting both the running winding and starting winding across the line in
A. parallel B. series
C. series-parallel D. an inductive coupling

386. If the starting winding is left in the circuit, it will cause the motor to
A. run faster B. run slower
C. draw excessive current and overheat
D. stall on right load

387. How can the direction of rotation of a split-phase be reversed?
A. By using a reversing stater
B. By reversing the connections to both windings
C. By reversing the connections to the starting windings
D. By changing the brush position

388. Instead of using three different windings per pole group in the stator, one large coil is wound called a
A. lap winding B. wave winding
C. complex winding D. skin winding

389. A single-phase motor is made self starting by the addition of a/an
A. running winding B. starting winding
C. electric starter D. autotransformer

390. The stator winding makes an electrical contact with the motor frame. The winding is said to be
A. open B. grounded
C. lap wound D. skein wound

391. An open in the centrifugal switch circuit gives an indication of a/an
A. shorted winding B. shorted switch
C. grounded winding D. open winding

392. A motor runs slower than normal. The defect may be
A. worn bearings B. improper-size fuses
C. open running winding
D. wrong size brushes

393. The armature reaction is reduced in a universal motor by
A. shifting the brushes
B. increasing the number of armature coils
C. using equalizer connections
D. connecting a compensating winding in series with the armature

394. An outstanding feature of a universal motor is its
A. highest horsepower-per-pound ratio
B. slow speed
C. poor performance on DC
D. best performance on 60 Hz power

395. Under no-load conditions, a universal motor tends to run away; its speed is limited only by
A. the supply frequency
B. the number of field poles
C. windage and friction
D. commutation

396. The commutator in a repulsion motor provides a means of connecting the
A. armature windings B. stator windings
C. short-circuiting device
B. starting-winding

398. The recommended brush pressure to use in a universal motor is
A. 0.5 to 1.0 psi B. 1.0 to 1.5 psi
C. 1.5 to 2.5 psi D. 2.5 to 3.5 psi

399. The most common mechanical method of providing speed reduction for universal motors is to use
A. gearing B. brakes
C. chains B. belts

400. The magnetic flux in the shaded part of the stator pole
A. aids the main pole
B. leads the main pole magnetic flux
C. lags the main pole magnetic flux
B. is in phase with the main pole magnetic flux

401. Which of the following is common cause of brush sparking in a universal motor?
A. Shorted armature winding
B. Open armature winding
C. High commutator mica
D. All of these

402. Early breakdown of winding insulation can be caused by
A. low brush pressure
B. dust, oil, or other contaminants
C. High commutator mica
D. All of these

403. The torque-slip characteristic of a polyphase induction motor becomes almost linear at small values of slip, because in this range of slips
A. the effective rotor circuit resistance is very large compared to the rotor reactance

Ans. 382. B 383. B 384. A 385. A 386. C 387. C 388. D 389. B 390. B 391. D 392. A 393. A 394. A
395. C 396. A 397. C 398. C 399. A 400. C 401. D 402. B 403. A

B. the rotor resistance is equal to the stator resistance
C. the rotor resistance is equal to the stator reactance
D. the rotor reactance is equal to the stator reactance

404. The travelling speed of cranes varies from
A. 1 to 2.5 m/s B. 5 to 10 m/s
C. 10 to 20 m/s D. 20 to 40 m/s

405. Which of the following drive can be used for derricks and winches?
A. DC motors
B. Slip-ring induction motor with variable resistance
C. Pole changing squirrelcage motors
D. Any of these

406. Light duty cranes are generally used in
A. automobile workshops
B. pumping stations
C. powerhouses D. all of these

407. In overhead travelling cranes
A. continuous duty motors are preferred
B. slow speed motors are preferred
C. short time rated motors are preferred
D. none of these

408. In motor circuit static frequency changers are used for
A. improved cooling
B. power factor improvement
C. reversal of direction D. speed regulation

409. 15 minute rated motors are suitable for
A. light duty cranes B. medium duty cranes
C. heavy duty cranes D. all of these

410. In case of travelling cranes, the motor preferred for boom hoist is
A. slip-ring induction motor
B. squirrelcage induction motor
C. synchronous motor D. single phase motor

411. The characteristics of drive for crane hoisting is
A. smooth movement B. precise control
C. fast speed control D. all of these

412. The range of horsepower of electric motor drives for rolling mills is of the order of
A. 1 to 10 HP B. 15 to 25 HP
C. 50 to 100 HP D. 100 to 500 HP

413. Motor preferred for rolling mill drive are
A. DC motors
B. AC slip-ring motors with speed control
C. any of these D. none of these

414. Which motors, because of their inherent characteristics, are best suited for the rolling mills?
A. DC motors
B. Slip-ring induction motors
C. Squirrelcage induction motors
D. Single-phase motors

415. The capacity of a crane is expressed in terms of
A. span B. type of drive
C. tonnes D. any of these

416. Heavy duty cranes are used in
A. heavy engineering workshops
B. steel plants
C. ore handling plants D. all of these

417. 1/2-hour rated motors are used for
A. light duty cranes B. medium duty cranes
C. heavy duty cranes D. none of these

418. Belted slip ring induction motor is almost invariably used for
A. centrifugal blowers B. water pumps
C. jaw crushers D. heavy load

419. In jaw crushers, generally a motor has to start against
A. low load B. medium load
C. normal load D. heavy load

420. Motor used for elevators is generally
A. synchronous motor B. induction motor
C. capacitor start single-phase induction motor
D. any of these

421. A pole changing type squirrelcage motor used in derricks has four, eight and twenty four poles. In this, the medium speed is used for
A. lifting B. hoisting
C. lowering D. landing the road

422. The number of the sets used in pole changing type squirrelcage motors for derricks and winches, is
A. 2 B. 3
C. 4 D. 6

423. A pole changing type squirrelcage induction motor used in derricks has four, eight and twenty-four poles. In this, the lowest speed is used for
A. lifting B. hoisting
C. lowering D. landing the road

424. For handling fragile articles in a crane
A. low speed is preferred
B. medium speed is preferred
C. high speed is preferred
D. none of these

425. In case of kiln drives
A. starting torque is almost zero
B. starting torque and running torque are nearly equal
C. starting torque is double of the running torque
D. both torques are zero

426. In case of belt conveyors
A. squirrelcage motors with direct-on-line starters are used
B. single phase induction motors areused

Ans. 404. A 405. D 406. D 407. C 408. D 409. A 410. A 411. D 412. D 413. D 414. A 415. C 416. D
417. B 418. C 419. D 420. B 421. A 422. B 423. D 424. A 425. C 426. A

C. DC shunt motors are used
D. induction motors with star-delta starters are used

427. Which of the following motors is preferred for blowers?
A. Wound rotor induction motor
B. Squirrelcage induction motor
C. DC shunt motor D. DC series motor

428. Motor preferred for kiln drives is usually
A. slip-ring induction motor
B. three phase shunt wound commutator motor
C. cascade controlled AC motor
D. any of these

429. Belt conveyors offer
A. zero starting torque B. low starting torque
C. medium starting torque
D. high starting torque

430. Centrifugal pumps are usually driven by
A. DC shunt motors B. DC series motors
C. squirrelcage induction motors
D. any of these

431. In a centrifugal pump if the liquid to be pumped has density twice that of water, then the horse-power required (as compared to that while pumping water) will be
A. half B. same
C. double D. three times

432. In case of centrifugal pumps the starting torque is generally
A. double the running torque
B. slightly more than running torque
C. same as running torque
D. less than running torque

433. Wound rotor and squirrelcage motors with high slip which develop maximum torque at standstill are used for
A. presses and punches B. machine tools
C. elevators D. all of these

434. In synthetic fibre mills, motor with
A. constant speed is preferred
B. high starting torque is preferred
C. variable speed is preferred
D. low starting torque is preferred

435. A reluctance motor
A. is compact B. has high cost
C. requires starting gear
D. is provided with slip rings

436. In case of contactors the ratio of the in-service period to the entire period, expressed as a percentage is known as
A. duty B. load factor
C. class of contact D. None of these

437. The efficiency of reluctance motor is around
A. 95% B. 80%
C. 75 to 80% D. 60 to 75%

438. Power factor in case of reluctance motor is
A. nearly unity B. always leading
C. 0.8 lagging D. 0.3 to 0.4 leading

439. Which of the following motor is preferred for synthetic fibre mills?
A. DC series motor B. Reluctance motor
C. DC shunt motor D. Synchronous motor

440. Reluctance motor is a
A. self-starting type synchronous motor
B. low torque variable speed motor
C. variable torque motor
D. low noise, slow speed motor

441. A reluctance motor on overload runs as
A. synchronous motor B. induction motor
C. both (A) and (B) are correct
D. none of these

442. The size of an excavator is usually expressed in terms of
A. cubic metres B. angle of swing
C. travel in metres D. crowd motion

443. A contactor having 6-operations per hour is classed as
A. Class 0 B. class I
C. Class II D. class III

444. The number of operations per hour in case of class IV contactor will be around
A. 100 B. 600
C. 900 D. 1200

445. Ward-Leonard controlled DC drives are generally used for
A. light duty excavators
B. medium duty excavators
C. heavy duty excavators
D. all of these

446. In case of contactors, the contacts are generally made of
A. copper B. silver
C. cadmium copper D. any of these

447. Which electromagnet is preferred for noiseless operation?
A. DC operated B. AC operated
C. Any of these D. None of these

448. In case of suitable core reactors, the power gain varies from
A. 1 to 5 B. 5 to 10
C. 5 to 100 D. 100 to 1000

449. The diameter of a 50-SWG wire is closer to
A. 0.0253 mm B. 0.05 mm
C. 0.16 mm D. 0.52 mm

Ans. 427. B 428. D 429. D 430. C 431. C 432. D 433. A 434. A 435. A 436. B 437. D 438. D 439. B 440. A 441. B 442. A 443. A 444. D 445. C 446. D 447. A 448. C 449. A

450. In case of contactors, the duty in which the main contacts remain closed for a period bearing a definite relation to the no-load periods, is known as
A. standard duty B. intermittent duty
C. temporary duty D. un-interrupted duty

451. A class-I contactor should be mechanically sound to withstand
A. 0.05 million times B. 0.25 million times
C. 1.0 million times D. 2.0 million times

452. A saturable core reactor can be used for
A. stepless AC voltage variation
B. plugging of induction motor
C. overload protection of transformers
D. all of these

453. Heat control switches find applications in
A. three-phase induction motors
B. single phase motors
C. transformers D. cooling ranges

454. A rotating magnetic field in a shaded-pole motor is produced by using
A. salient poles B. shading coils
C. a capacitor
D. a high inductance winding

455. What type of construction is used for most stator poles of a shaded-pole motor?
A. Salient pole B. Consequent pole
C. Alternate pole D. Compensated pole

456. The direction of rotation in a conventional shaded-pole motor can be reversed by
A. reversing the line connections
B. using a capacitor
C. reversing the rotor
D. reversing the stator poles 180 degree

457. The direction of rotation in a shaded-pole motor is always toward the
A. main pole B. shaded pole
C. salient pole D. consequent pole

458. Which of the following will determine whether or not the rotor of a synchronous motor locks into step with the rotating magnetic field?
A. Rotor's construction
B. Line frequency
C. The speed of the rotating magnetic field
D. The rate of rotor slip

459. The three types of single-phase synchronous motors are
A. hysteresis, reluctance, repulsion
B. hysteresis, reluctance, inductor
C. reluctance, repulsion, shaded pole
D. repulsion, shaded pole, inductor

460. The stator windings of synchro transmitters and receivers are excited by
A. the differential field flux
B. the AC magnetic field of the rotor
C. a DC source
D. compensation windings

461. What motor is ideal for application, where start/stop timing is required, because of its quick starting and stopping action?
A. Induction motor B. Shaded-pole motor
C. Reluctance motor D. Hysteresis motor

462. The speed of an inductor motor depends online frequency and the
A. stator poles B. number of rotor teeth
C. rotor slip D. shaded poles position

463. A synchro transmitter sends an electrical signal to the
A. synchro receiver B. motor field
C. motor starter D. synchro generator

464. The effective voltage induced in any synchro-transmitter coil depends on the angular position of the
A. phase angle B. stator windings
C. rotor windings
D. coil's axis with respect to the rotor axis

465. Synchro receivers are electrically identical to
A. split-phase motors B. 3-phase motors
C. synchrotransmitters of the same size
D. synchro control transformers of the same size

466. A synchro transmitter is rotated through an angle of 45 degree. The change in angle is called the
A. correspondence B. phase shift
C. signal D. error

467. The stator voltages of synchrotransmitter and receiver approach equality when the receiver
A. sends a strong signal
B. approaches correspondence
C. approaches a critical angle
D. produces a high torque

468. How are the stator and rotor windings connected in a differential synchro transmitter?
A. Series B. Parallel
C. Wye D. Delta

469. The rotor windings of a synchro control transformer are
A. never connected to the AC supply
B. connected to the AC supply
C. connected to the CT taps
D. for 3-phase power input only

470. In a servosystem, the voltage induced in the rotor of a control transformer is the
A. driving voltage
B. opposing the voltage
C. error voltage D. quadrature voltage

471. A system that controls speed, temperature, or position in relation to a reference is called
A. control relay B. servomechanism
C. tachometer C. thermostat.

Ans. 450. B 451. B 452. A 453. D 454. B 455. A 456. D 457. B 458. B 459. B 460. B 461. A 462. B 463. A 464. D 465. C 466. A 467. B 468. C 469. A 470. C 471. B

472. What four characteristics of a servo-system possesses?
A. Fast response, high accuracy, unattended control and remote control operation
B. Good response, good accuracy, attended control and close operation
C. Good regulation, high output, remote control, high power factor
D. Low input, high output, good power factor, good service factor

473. A servocontrol system for DC motors is called a/an
A. series control B. shunt control
C. selsyn D. amplidyne

474. In a servocontrol bridge and positioning loop system, the direction in which the load is driven depends on the
A. polarity of the error voltage
B. polarity of the supply voltage
C. ratio of the bridge arms
D. position of the load shaft

475. An amplidyne is actually a/an
A. DC amplifier B. AC amplifier
C. induction motor D. 3-phase generator.

476. The residual magnetism is removed in an amplidyne by
A. reversing the field windings
B. shifting the brush yoke
C. short-circuiting one set of brushes
D. an AC magneto generator mounted on the amplidyne

477. A servosystem output drive motor should be
A. a universal motor with reversing capabilities
B. easily reversible, have required power, and good speed control
C. easily reversible, highly sensitive, and have constant speed
D. highly sensitive easily controlled and have variable speed

478. The direction or rotation of a hysteresis motor is determined by the
A. hysteresis loss
B. magnetic retentivity of the rotor
C. position of the shaded pole with respect to the main pole
D. magnetic permeability of the stator

479. The principle of the Wheatstone bridge is used in servosystems to
A. measure system resistance
B. position control shafts of mechanical devices
C. amplify error signals
D. reduce equalising signals

480. Servo actuators are designed to produce either
A. constant speed or variable speed
B. rotating motion or linear motion
C. a DC output, or an AC output
D. magnetizing force or demagnetizing force

481. The NEC permits which of the following statements to be applicable for single-phase fractional horsepower motors?
A. One switch can be a disconnecting means and a controller
B. Short-circuit protection and overload protection can be combined
C. A single device can serve more than one function
D. all of these

482. The size of the over current protection for a motor is based on the
A. NEC tables
B. branch circuit conductor size
C. motor name plate current rating
D. horsepower rating

483. The NEC requires a motor to be within sight of the controller, or no farther than
A. 20 ft from the controller
B. 50 ft from the controller
C. 75 ft from the controller
D. 100 ft from the controller

484. For a particular motor, the cooling time-constant is usually
A. smaller than the heating time constant
B. greater than the heating time constant
C. equal to the heating time constant
D. None of these

485. The motor circuit switch, or CB must disconnect the motor and its controller from
A. all grounded conductors
B. all ungrounded conductors
C. the overload relays
D. the branch circuit

486. Motor and controller frames must be grounded to
A. prevent a potential above-ground
B. have a common return path
C. eliminate a grounding conductor
D. isolate the motor from over currents

487. The nearest larger size standard fuse may be used by protection of single-phase circuits under which of the following conditions?
A. Rating of wires do not correspond to a standard size fuse
B. Current rating of motor does not correspond to a standard size fuse
C. Overload relays of the same size are not available
D. all of these

488. When selecting a motor which of the following must be considered?
A. Voltage and horsepower
B. Speed and frame enclosure size

Ans. 472. A 473. D 474. A 475. A 476. D 477. B 478. C 479. B 480. B 481. D 482. C 483. B 484. B 485. B 486. A 487. A

C. Bearings and motor protectors
D. All of these

489. The transformation ratio of a transformer is
A. I_2/I_1 B. V_1/V_2
C. N_2/N_1 D. all of these

490. By pushing and pulling the rotor shaft, a motor can be checked for
A. end play B. side play
C. firm mounting D. free shaft movement

491. A loss in the capacitance of a capacitor start motor results in reduced
A. speed B. voltmeter
C. armature reaction D. starting torque

492. Trouble in the armatures of repulsion and universal motors is usually caused by a/an
A. shorted winding B. shorted commutator
C. open winding D. open commutator

493. The amplifier in a good quality record player would be designed to cover a range of frequencies
A. 500–3000 Hz B. 0–10000 Hz
C. 50–15000 Hz D. 50 kHz–5 kHz

494. According to Indian Electricity rules, extra high voltage implies voltage exceed-ing
A. 440 V B. 66 kV
C. 33 kV D. 110 kV

495. A saturable core reactor is basically a
A. variable resistor
B. step-down transformer
C. thermal relay D. variable impedance

496. A magnetic amplifier can be used for the control of
A. current B. voltage
C. speed D. All of these

497. In case of contactors, the contact chattering may be due to
A. excessive logging B. broken pole shader
C. poor contact in the control pick up circuit
D. Any of these

498. A magnetic amplifier has generally
A. single winding B. two windings
C. three windings D. number of windings

499. In a contactor, overheating of contacts may result from any of the following except
A. excess contact pressure
B. high inductive loads
C. copper oxide on contacts
D. carrying load continuously for a longer time

500. In case of low and medium voltage circuits, the permissible voltage variation is
A. 1% B. 5%
C. 10% D. 15%

501. In case of contactors, the magnet may become noisy due to
A. dirt or rust on magnet faces
B. low voltage
C. broken pole shader D. any of these

502. The failure of a thermal relay may occur due to
A. motor and relay in different ambient temperatures
B. relay previously damaged by short circuit
C. mechanical binding D. any of these

503. Premature blowing of a fuse may occur due to
A. heating at ferrule contacts
B. corrosion or oxidation of ferrules
C. weak contact pressure
D. any of these

504. Which of the following site will be preferred for earthing?
A. Wet mashy ground
B. Clay soil
C. Loam mixed with small quantities of sand
D. Damp and wet sand pit

505. Which of the following is not used as earth continuity conductor?
A. Water pipes B. Gas pipes
C. Structural steel members
D. Any of these

506. Resistivity of earth increases sharply if the moisture falls below
A. 60% B. 50%
C. 40% D. 20%

507. Which of the following is least preferred for earthing?
A. Earth mixed with salt and charcoal
B. Mashy ground containing brine waste
C. Dry earth D. Clay soil

508. Earth electrodes can be in the form of
A. rods and pipes B. strips
C. plates D. any of these

509. Non-metallic conduits for wiring are generally made of
A. rubber B. cork
C. wood D. PVC

510. The maximum horsepower up to which 440 V electric motors are used, is
A. 200 HP B. 50 HP
C. 20 HP D. 10 HP

511. Which of the following mixtures is preferred for filling around the earth for electrode effective earthing?
A. Bone-meat mixture B. Coal-salt mixture
C. Saw-dust sand mixture
D. Lime-sand mixture

512. The distance at which earthing electrode should be situated away from the building whose installation system is being earthed is

Ans. 488. D 489. C 490. A 491. D 492. C 493. C 494. A 495. D 496. A 497. D 498. D 499. A 500. B
501. D 502. D 503. D 504. A 505. D 506. D 507. C 508. D 509. D 510. A 511. B

A. 1.5 m B. 3.5 m
C. 25 m D. 100 m

513. PVC conduits can be hurried on
A. lime B. plaster
C. concrete D. Any of these

514. PVC conduits can be joined by
A. solvent cement B. welding
C. threading D. Any of these

515. Continuous operation of automobile horn will
A. help in charging the battery
B. improve milage
C. damage the operating coil
D. change the tone

516. Miniature lamps on automobiles are used for
A. tail lamp B. dash board lamp
C. side lamp D. All of these

517. Inside the earth or pit, the earthing electrode should be placed
A. vertical B. horizontal
C. inclined at 45°
D. inclined at any angle other than 45°

518. Danger 440 V plates are
A. danger notices B. caution notices
C. advisory notices D. informal notices

519. The earth's potential is taken as
A. infinite B. supply voltage
C. 12 volt D. zero

520. The minimum clearance of any overhead line from the ground should be
A. 20 m B. 10 m
C. 6 m D. 4 m

521. Earthing of electric appliances is done
A. for the safety of human life
B. to reduce line voltage fluctuation
C. for protection of electric equipment
D. All of these

522. Earthing is used as the return conductor for
A. telephone lines B. telegraph lines
C. traction work D. All of these

523. The resistance of earth wire should be
A. infinite B. high
C. reasonable D. very low

524. The earthwire should not be thinner than a
A. 18 SWG wire B. 16 SWG wire
C. 10 SWG wire D. 8 SWG wire

525. The current drawn by a 6-V horn is roughly
A. 0.1 A B. 1 A
C. 2 A D. 20 A

526. In automobiles, the sound produced by horn is due to
A. magnetostriction B. vibrating diaphragm
C. moving coil D. oscillating coil

527. The horns are rated for
A. continuous operation
B. intermittent operation
C. Both A and B are correct
D. None of these

528. Belted wound rotor induction motors are preferred for
A. machine tools B. gyratory crushers
C. belt conveyor D. water pumps

529. For a particular type of motor, the heating time constant
A. increases with increase in size
B. decreases with increase in size
C. same for all sizes D. none of these

530. The heating time constant of a totally enclosed motor is relatively
A. higher B. lower
C. independent of type of enclosure
D. none of these

531. Low frequency operation of AC series motor
A. improves its commutation property but affects adversely the pf and efficiency
B. improves its commutation property, pf and efficiency
C. affects adversely commutation but improves pf and efficiency
D. none of these

532. In a shaded pole motor, shading coils are used to
A. reduce windage losses
B. reduce friction losses
C. produce rotating magnetic field
D. protect against sparking

533. The speed of a universal motor can be controlled by
A. introducing a variable resistance in series with the motor
B. tapping the field at various points
C. centrifugal mechanism
D. All of these

534. Which one of the following capacitor star split-phase induction motors will have the largest value of capacitance?
A. 1/2 HP, 3450 rpm B. 1/4 HP, 1725 rpm
C. 1/2 HP, 1140 rpm D. 3/4 HP, 1140 rpm

535. While a 2-phase ac servomotor is in operation, if the voltage across the control field winding becomes zero, then the motor has a tendency to run as a single-phase induction motor. To prevent this
A. rotor having high mass moment of inertia is used for such a motor
B. drag cup type of light rotor with high rotor resistance is preferred
C. a low rotor resistance is used
D. the number of turns in the control field winding used is less than the main reference winding

Ans. 512. A 513. D 514. D 515. C 516. D 517. A 518. B 519. D 520. C 521. D 522. C 523. D 524. D 525. D 526. A 527. B 528.. B 529. A 530. A 531. B 532. C 533. B 534. D 535. B

536. While operating on variable frequency supplies, the AC motors require variable voltage in order to
A. protect the insulation
B. avoid the effects of saturation
C. improve the commutation capabilities of the inverter
D. protect the thyristors from dv/dt

537. The capacitors used for AC motor starting have no
A. polarity makings B. voltage rating
C. definite capacitance value
D. dielectric rating

538. A capacitor motor has difficulty of starting. What is the probable cause?
A. Reversed magnetic field
B. Shorted commutator segments
C. Shorted windings D. All of these

539. The stator of a 2/4 pole changing cage motor is initially wound for 2-poles. The reconnection of the stator winding to 4-poles through a change-over switch, while the motor is running would result in
A. constant torque drive
B. constant hp drive
C. plugging to standstill
D. regenerative braking to half the original speed

540. The main reason of using a hysteresis motor for high quality tape recorders and record players is that
A. its speed is constant (synchronous)
B. it develops extremely steady torque
C. it requires no centrifugal switch
D. its operation is not affected by mechanical vibrations

541. The capacitor in a capacitor-start motor is connected in series with the
A. running winding
B. starting winding and centrifugal switch
C. compensating winding
D. split-phase winding

542. Before a capacitor-start motor can reverse its direction of rotation, the
A. centrifugal switch must be opened
B. running winding must be opened
C. starting winding must be connected to the line
D. capacitor connections must be changed

543. A dual-voltage single-phase motor can be operated on either
A. 120/240 volts B. 208/240 volts
C. 240/277 volts D. 240/480 volts

544. A permanent-split capacitor motor does not have
A. a starting winding B. a centrifugal switch
C. a squirrelcage rotor D. reversing capabilities

545. The speed of a two-speed capacitor-run motor is decreased by
A. increasing the capacitance value
B. decreasing the capacitance value
C. increasing the strength of the stator magnetic field
D. decreasing the strength of the stator magnetic field

546. A 250 μF, 250 volt capacitor is defective. How many 250 μF, 150 volt capacitors are required and how are they connected to obtain 250 μF at 250 volts?
A. Two connected in series
B. Two connected in parallel
C. Four connected in series parallel
D. Four connected in parallel

547. The starting winding is connected for opposite polarity to reverse the direction of rotation of a motor running at full speed. This is accomplished by short-circuiting the
A. centrifugal switch B. running winding
C. starting winding D. capacitor

548. A repulsion start induction-run motor runs as an induction motor when the
A. commutator segments are short circuited
B. brushes are shifted to a neutral plane
C. shorting devices are disconnected
D. stator connections are reversed

549. The brushes in a repulsion motor are connected
A. to the external supply
B. to the stator winding
C. together by a jumper wire
D. for induction running only

550. In a repulsion-start, induction-run motor, all commutator segments are short circuited by
A. the brushes
B. a short-circuiting device
C. an induction winding
D. equalizer connections

551. The short-circuiting device in a repulsion start induction-run motor operates by
A. circulating currents B. magnetic attraction
C. centripetal force D. centrifugal force

552. The repulsion motor starts and runs as a
A. split-phase motor B. capacitor-start motor
C. repulsion motor D. compound motor

553. Which of the following motors has a squirrelcage winding embedded in the armature under the regular winding?
A. Compound motor
B. Repulsion-start, induction-run motor
C. Repulsion motor
D. Repulsion-induction motor

554. Changing the electrical connections to reverse the direction of rotation of a motor running at full speed is called
A. slugging B. plugging
C. dynamic braking D. brush shifting

Ans. 536. B 537. A 538. C 539. D 540. B 541. B 542. C 543. A 544. B 545. D 546. C 547. A 548. A
549. C 550. B 551. D 552. C 553. D 554. B

555. A DC series motor does not operate satisfactorily on AC for which of thefollowing reasons?
A. Poor efficiency on AC
B. Poor power factor on AC
C. Excessive sparking on AC
D. All of these

556. Cross-connecting the commutator bars in repulsion motors minimises circulating currents caused by
A. unequal airgaps between the stator and armature
B. shorted brushes
C. the short-circuiting device
D. a shorted stator winding

557. Centrifugal short-circuiting mechanisms are used only in
A. split-phase repulsion motors
B. repulsion motors
C. repulsion-induction motors
D. repulsion start, induction-run motors

558. AC to a DC shunt motor field causes a high impedance because of the winding's
A. high resistance B. inductive reactance
C. high induced voltage B. capacitive reactance

559. To overcome the effects of poor power factor, an AC series motor uses
A. laminated iron in the motor frame
B. high resistance windings
C. a brush shifting yoke
B. more field turns than a DC field

MULTIPLE CHOICE QUESTIONS FROM VARIOUS COMPETITIVE EXAMINATIONS

1. The advantage of the double squirrelcage induction motor over single cage rotor is that its **(IES 2001)**
A. efficiency is higher B. power factor is higher
C. slip is larger
D. starting current is lower

2. An induction motor when started on load does not accelerate up to full speed but runs at 1/7th of the rated speed. The motor is said to be **(IES 2001)**
A. locking B. plugging
C. crawling D. cogging

3. Match **List-I** (Tests) with **List-II** (Machines) and select the correct answer: **(IES 2001)**

List-I	**List-II**
a. No-load and blocked rotor test	1. Transformer
b. Sumpner's test	2. Induction motor
c. Swinburne's test	3. Synchronous motor
	4. DC motor

Codes

	a	b	c		a	b	c
A.	1	4	2	B.	2	1	4
C.	3	4	2	D.	3	1	4

4. The starting current of a 3ϕ induction motor is 5 times the rated current, while the rated slip is 4%. The ratio of starting torque to full-load torque is **(IES 2001)**
A. 0.6 B. 0.8
C. 1.0 D. 1.2

5. Which one of the following statements is correct in respect of an induction motor? **(IES 2003)**
A. The maximum torque will depend on rotor resistance
B. Although the maximum torque does not depend on rotor resistance, yet the speed at which maximum torque is produced depends on the rotor resistance
C. The maximum torque will not depend on standstill rotor reactance
D. The slip of induction motor decreases as torque increases

6. The crawling in the induction motor is caused by **(IES 2003)**
A. high loads B. low voltage supply
C. improper design of stator laminations
D. harmonics developed in motor

7. The rotor slots are slightly skewed in squirrelcage induction motor to **(IES 2003)**
A. increase the strengths of rotor bars
B. reduce the magnetic hum and locking tendency of rotor
C. economise on the copper to be used
D. provide ease of fabrication

8. The injected emf in the rotor of induction motor must have **(IES 2003)**
A. the same frequency as the slip frequency
B. the same phase as the rotor emf
C. a high value for satisfactory speed control
D. the same phase as the rotor emf and a high value for satisfactory speed control

9. In an induction motor, when the number of stator slots is equal to an integral multiple of rotor slots **(IES 2003)**
A. there may be a discontinuity in torque-slip characteristics
B. a high starting torque will be available
C. the maximum torque will be high
D. the machine may fail to start

Ans. 555. D 556. A 557. D 558. B 559. A 1. D 2. C 3. A 4. C 5. B 6. D 7. B 8. A 9. D

10. Which one of the following statements is correct? A smaller air gap in a polyphase induction motor helps to (IES 2004)
A. reduct the chances of crawling
B. increase the starting torque
C. reduce the chance of cogging
D. reduce the magnetising current

11. A squirrelcage induction motor having a rated slip of 4% on full load has a starting torque same as the full load torque. Which one of the following statements is correct? (IES 2004)
The starting current is
A. equal to the full load current
B. twice the full load current
C. four times the full load current
D. five times the full load current

12. The stator and the rotor of a 3-phase, 4-pole wound rotor induction motor are excited respectively, from a 50 Hz and a 30 Hz source of appropriate voltage. Neglecting all losses, what is/are the possible no-load speed/speeds at which the motor would run? (IES 2005)
A. 1500 rpm and 900 rpm
B. 2400 rpm and 600 rpm
C. 2400 rpm only
D. 600 rpm only

13. Which of the following methods are suitable for the speed control of squirrelcage induction motors?
1. Voltage control
2. Rotor resistance control
3. Frequency control
4. Pole changing method
Select the correct answer using the codes given below. (IES 2005)
A. 2, 3 and 4
B. 1, 3 and 4
C. 1, 2 and 3
D. 2 and 4

14. What is the shunt resistance component in equivalent circuit obtained by no load test of an induction motor representative of? (IES 2005)
A. Winding and frictional losses only
B. Core losses only
C. Core, windage and frictional losses
D. Coppr losses

15. A wound rotor induction motor runs with a slip of 0.03 when developing full load torque. Its rotor resistance is 0.25 ohm per phase. If an external resistance of 0.50 ohm per phase is connected across the slip rings, what isthe slip for full torque? (IES 2005)
A. 0.03
B. 0.06
C. 0.09
D. 0.1

16. When the applied rated voltage per phase is reduced to one-half, the starting torque of a three-phase squirrelcage induction motor becomes (IES 2006)
A. 1/2 of the initial value
B. 1/4 of the initial value
C. twice of the initial value
D. 4 times of the initial value

17. If the load on an induction motor is increased from no load to full load, its slip and the power factor will respectively (IES 2006)
A. decrease, decrease
B. decrease, increase
C. increase, decrease
D. increase, increase

18. Which of the points on the torque-speed curve of the induction motor represents operation at a slip greater than 1? (IES 2006)

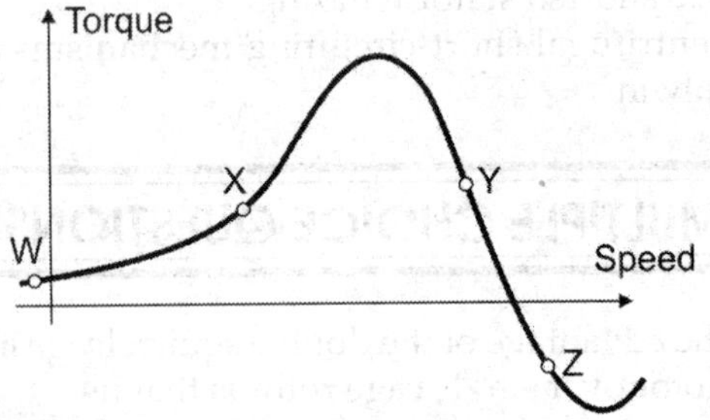

A. W
B. X
C. Y
D. Z

19. The voltage actually used for setting up of the useful flux in the air gap of a 3-phase induction motor is (IES 2006)
A. = applied voltage
B. > applied voltage
C. < applied voltage
D. = rotor induced emf

20. Which of the following parameters in an induction motor influences the magnetising reactance to the maximum extent? (IES 2006)
A. Axial length of the rotor stack
B. Axial length of the stator stack
C. Radial length of air gap
D. Number of slots on the stator

21. Which one of the following is the correct statement? In a 3-phase induction motor, the resultant flux is of a constant nature and is (IES 2007)
A. equal to Φ_m, where Φ_m is maximum flux due to any phase
B. 1.5 time maximum value of flux due to any phase
C. $\sqrt{3}/2$ times maximum value of flux due to any phase
D. $3\Phi_m$

22. In induction motor, air gap power is 10 kW and mechanically developed power is 8 kW. What are the rotor ohmic losses? (IES 2007)
A. 1 kW
B. 3 kW
C. 2 kW
D. 0.5 kW

23. Which one of the following is the correct statement? The output line in an induction motor circle diagram is the line joining the tip of the (IES 2007)

Ans. 10. D 11. D 12. B 13. B 14. B 15. C 16. B 17. D 18. A 19. A 20. C 21. B 22. C

A. no load current phasor to the point corresponding to slip = 0
B. no load current phasor to the point corresponding to slip = 1
C. short circuit current phasor to the point corresponding to slip = 1
D. short circuit current phasor to the point corresponding to slip = 0

24. What is the frequency of rotor current of a 50 Hz induction motor operating at 2% slip? **(IES 2007)**
A. 1 Hz B. 100 Hz
C. 2 Hz D. 50 Hz

25. Consider the following statements:
Skewing of rotor slots in a 3-phase induction motor (cage rotor) may
1. introduce additional leakage reactance
2. eliminate slot harmonics
Which of these statements is/are correct? **(IES 2008)**
A. 1 only B. 2 only
C. Both 1 and 2 D. Neither 1 nor 2

26. In the equivalent circuit of a double-cage induction motor, the two rotor cages can be considered **(IES 2008)**
A. to be in parallel
B. to be in series-parallel
C. to be in series
D. to be in parallel with stator

27. A 3-phase squirrelcage induction motor is started by means of a star/delta switch. What is the starting current of the motor? **(IES 2008)**
A. 3 times the current with direct on line starting
B. 1/3 times the current with direct on line starting
C. $1/\sqrt{3}$ times the current with direct on line starting
D. $\sqrt{3}$ times the current with direct on line starting

28. What is the operating slip of a 400 V, 50 Hz, 6-pole, 3-phase induction motor, while the speed is 936 rpm with a 400 V, 48 Hz, 3-phase supply? **(IES 2008)**
A. 0.036 B. 0.064
C. 0.025 D. 0.075

29. A 3-phase slip-ring induction motor having negligible stator impedance drives a constant torque load. If an additional resistance is included in the rotor circuit, what does the motor experience? **(IES 2008)**
A. Increase in both the stator current and the slip
B. No change in the stator current and increase in the slip
C. Increase in the stator current and no change in the slip
D. Decrease in the stator current and increase in the slip

30. Breakdown torque in a 3-phase induction motor of negligible stator impedance is **(IES 2008)**
A. directly proportional to rotor resistance
B. inversely proportional to rotor resistance
C. directly proportional to rotor leakage reactance
D. inversely proportional to rotor leakage reactance

31. A balanced 3-phase induction motor runs at slip s. If ω_s is its synchronous speed, what is the relative speed between the stator mmf and rotor mmf? **(IES 2008)**
A. $s\omega_s$ B. $(1-s)\omega_s$
C. ω_s D. Zero

32. The starting current and torque of a three phase induction motor on direct line starting is 30 Amp and 300 Nm respectively. What are the corresponding values with star delta starter? **(IES 2009)**
A. 10 A and 100 Nm B. 30 A and 300 Nm
C. 17.32 A and 173.2 Nm
D. 30 A and 173.3 Nm

33. A three phase 6 pole 50 Hz induction motor is running at 5% slip. What is the speed of the motor? **(IES 2009)**
A. 850 rpm B. 900 rpm
C. 950 rpm D. 1000 rpm

34. When will a slip ring induction motor run at super synchronous speed? **(IES 2009)**
A. If a voltage is injected in the rotor circuit in phase opposition to the rotor induced emf
B. If an emf is injected in the rotor circuit in phase with the rotor induced emf
C. If motor is coupled with active load
D. It motor is coupled with passive load

35. The starting current of an induction motor is 5 times the full-load current while the full-load slip is 4%. What is the ratio of starting torque to full-load torque? **(IES 2009)**
A. 0.6 B. 0.8
C. 1.0 D. 1.2

36. What is the rotor copper loss of a 3-phase, 550 Volt, 50 Hz, 6 poles induction motor developing 4.1 kW at the shaft with mechanical loss of 750 W at 970 rpm? **(IES 2009)**
A. 175 W B. 150 W
C. 100 W D. 250 W

37. What is the ratio of starting torque and maximum torque of a 3 phase, 50 Hz, 4 pole induction motor for a maximum torque at 1200 rpm? **(IES 2009)**
A. 0.421 B. 0.384
C. 0.6 D. 0.5

38. The power input to a 415 V, 50 Hz, 6 pole 3-phase induction motor running at 975 rpm is 40 kW. The

Ans. 23. B 24. A 25. C 26. A 27. C 28. B 29. D 30. D 31. D 32. A 33. C 34. A 35. C 36. B 37. B

stator losses are 1 kW and friction and windage losses total 2 kW. What is the efficiency of motor? **(IES 2009)**

A. 92.5% B. 92%
C. 90% D. 88%

39. The MMF produced by the rotor currents of a 3-phase induction motor **(IES 2010)**
A. rotates at the speed of rotor in the air gap
B. is at standstill with respect to stator mmf
C. rotates at synchronous speed with respect to rotor
D. rotates at synchronous speed with respect to rotor

40. A 6-pole, 50 Hz, 3-phase induction motor with a rotor resistance of 0.25 Ω develops a maximum torque of 10 Nm at 875 rpm. The rotor reactance and slip at maximum torque is **(IES 2012)**
A. 2 Ω and 0.125 pu B. 2 Ω and 0.25 pu
C. 1 Ω and 0.25 pu D. 1 Ω and 0.125 pu

41. The rated slip of an induction motor at full-load is 5% while the ratio of starting current to full load current is four. The ratio of the starting torque to full load torque would be **(IES 2012)**
A. 0.6 B. 0.8
C. 1.0 D. 1.1

42. No load test was conducted on a three phase induction motor at different input voltages and the input power obtained was plotted against the input voltage. The intersection of the extrapolated curve on the Y-axis, would give **(IES 2013)**
A. rated core loss
B. windage and friction loss
C. rated copper loss
D. rated core loss and windage and friction loss

43. A 4-pole induction motor (main) and a 6-pole motor (auxiliary) are connected in cumulative cascade. Frequency in the secondary winding of the auxiliary motor is observed to be 1 Hz. For a supply frequency of 50 Hz, the speed of the cascade set is **(IES 2013)**
A. 1485 rpm B. 990 rpm
C. 608 rpm D. 588 rpm

44. In a split phase motor, the running winding should have **(IES 2013)**
A. high resistance and low inductance
B. high resistance as well as high inductance
C. low resistance and high inductance
D. low resistance as well as low inductance

45. A 3-phase induction motor draws 50 kW from a 200 V, 50 Hz mains. The rotor emf makes 100 oscillations minute. If the stator losses are 2 kW, the rotor copper loss would be **(IES 2013)**
A. 0.16 kW B. 0.32 kW
C. 1.6 kW D. 3.2 kW

46. Starting torque can be obtained in the case of a single-phase induction motor with identical main and auxiliary windings by connecting **(IES 2013)**
A. a capacitor across the mains
B. a capacitor in series with the machine
C. a capacitor in series with the auxiliary winding
D. the main and the auxiliary winding in series

47. The ratio of starting to full load current for a 10 kW, 400 V, 3-phase induction motor with star delta starter, given the full load efficiency at 0.86, the full load pf is 0.8 and short circuit current is 30 A at 100 V is **(IES 2013)**
A. 1.9 B. 1.8
C. 2.4 D. 3.2

48. In an electromechanical energy conversion device, the developed torque depends upon **(IES 2001)**
A. stator field strength and torque angle
B. stator field and rotor field strengths
C. stator field and rotor field strengths and the torque angle
D. stator field strength only

49. In a shaded-pole induction motor, the rotor runs from the **(IES 2001)**
A. shaded portion to the unshaded portion of the pole while the flux in the former leads that of the latter
B. shaded portion to the unshaded portion of the pole while the flux in the former lags that in the latter
C. unshaded portion to the shaded portion while the flux in the former leads that in the latter
D. unshaded portion to the shaded portion while the flux in the former lags that in the latter

50. The torque of a reluctance motor can be effectively increased by **(IES 2001)**
A. increasing reluctance of the magnetic circuit along the direct axis
B. decreasing the reluctance of the magnetic circuit along th quadrature axis
C. increasing the ratio of the quadrature axis reluctance to direct axis reluctance
D. decreasing the ratio of quadrature axis reluctance to direct axis reluctance

51. Switched reluctance motor means **(IES 2001)**
A. salient pole synchronous motor without excitation winding
B. a stepper motor with salient poles
C. synchronous motor with salient poles on stator rotor

Ans.	38. C	39. B	40. A	41. B	42. B	43. D	44. C	45. C	46. C	47. A	48. C	49. C	50. C

D. a stepper motor with closed loop control and with rotor position sensor

52. When the excitation of normally operating unloaded salient-pole synchronous motor suddenly gets disconnected, it continues to run as a **(IES 2002)**
A. scharge motor
B. spherical motor
C. switched-reluctance motor
D. variable-reluctance motor

53. Stepper motors are widely used because of **(IES 2002)**
A. wide speed range B. large rating
C. no need for field control
D. compatibility with digital systems

54. A 2-phase servomotor control system uses **(IES 2001)**
A. drag-cup rotor
B. solid cylindrical rotor without slots or windings
C. ordinary squirrelcage rotor
D. slip-ring rotor with inherent low rotor resistance

55. Capacitor in a single-phase induction motor is used for **(IES 2002)**
A. improving the power factor
B. improving the starting torque
C. starting the motor
D. reducing the harmonics

56. A certain RL series combination is connected across a 50 Hz single-phase AC supply. If the instantaneous power drawn was found to be negative for 2 milliseconds in one cycle, the 'power factor angle' of the current must be **(IES 2002)**
A. 9° B. 18°
C. 36° D. 45°

57. Stepper motors are mostly used for **(IES 2002)**
A. high power requirements
B. control system applications
C. very high speed of operation
D. very low speed of operation

58. A coil of 1000 turns is wound on a core. A current of 1 A flowing through the coil creates a core flux of 1 mWb. The energy stored in the magnetic field is **(IES 2003)**
A. 0.25 J B. 0.5 J
C. 1 J D. 2 J

59. Which one of the following types of motors is most suitable for a computer printer drive? **(IES 2004)**
A. Reluctance motor B. Hysteresis motor
C. Shaded pole motor D. Stepper motor

60. A coil of 1000 turns is wound on a core. A current of 1 A flowing through the coil creates a core flux of 1 mWb. What is the energy stored in the magnetic field **(IES 2005)**
A. 1/4 J B. 1/2 J
C. 1 J D. 2 J

61. A capacitor-start single-phase induction motor is used for **(IES 2005)**
A. easy to start loads
B. medium start loads
C. hard to start loads
D. any type of start loads

62. In hand-tool applications, which one of the following single-phase motors is used? **(IES 2005)**
A. Shaded pole motor
B. Capacitor start motor
C. Capacitor run motor
D. AC series motor

63. Which one of the following is the type of single phase induction motor having the highest power factor at full load? **(IES 2006)**
A. Shaded pole type B. Split-phase type
C. Capacitor-start type D. Capacitor-run type

64. A single phase induction motor is running at N rpm. Its synchronous speed is N_s. If its slip with respect to forward field is s, what is the slip with respect to the backward field? **(IES 2006)**
A. s B. $-s$
C. $(1 - s)$ D. $(2 - s)$

65. Which one among the following has the highest numerical value in a stepper motor? **(IES 2006)**
A. Detent torque B. Holding torque
C. Dynamic torque D. Ripple torque

66. Why is a centrifugal switch used in a single-phase induction motor? **(IES 2008)**
A. To protect the motor from overloading
B. To improve the starting performance of the motor
C. To cut off the starting winding at an appropriate instant
D. To cut in the capacitor during running conditions

67. The emf induced in a conductor of machine driven at 6000 rpm, the peak value of flux density is 1.0 Wb/m^2, diameter of machine 2.0 metre and length of machine 0.3 m is **(IES 2011)**
A. 41.83 V B. 29.58 V
C. 9.42 V D. 18.84 V

Ans. 51. D 52. D 53. D 54. A 55. C 56. C 57. B 58. B 59. D 60. B 61. C 62. D 63. D 64. D 65. B 66. C 67. D

Alternator and Synchronous Machines

Alternator

An alternator consists of a stator and a rotor. The stator provides the armature windings whereas rotor provides the rotating magnetic field.

Basic principle: When the rotor is rotated by the prime-mover, the stator winding or conductors are cut by the magnetic flux of the rotor magnetic poles. Hence an emf is induced in the stator conductors. The emf generated in the stator conductors is taken out from three leads connected to the stator winding as shown in Fig. 10.1.

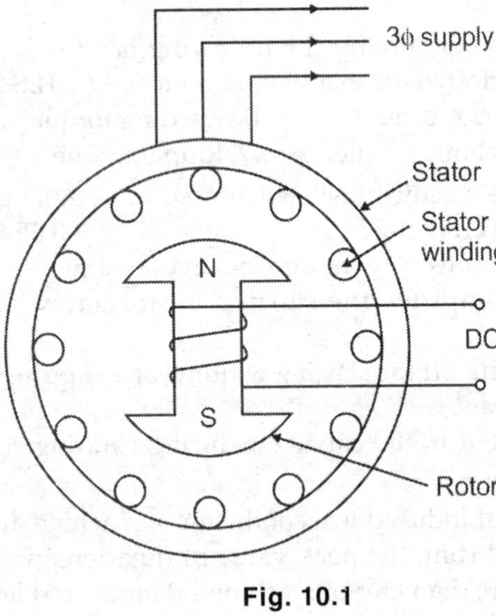

Fig. 10.1

Al alternator consists of the followig two parts: i. stator ii. rotor.

i. *Stator*: The whole structure is held in a frame which may be of cast iron as shown in Fig. 10.2.

ii. *Rotor*: It is a rotating part and carries the field system. Rotor is of two types namely: i. salient pole type ii. nonsalient pole type.

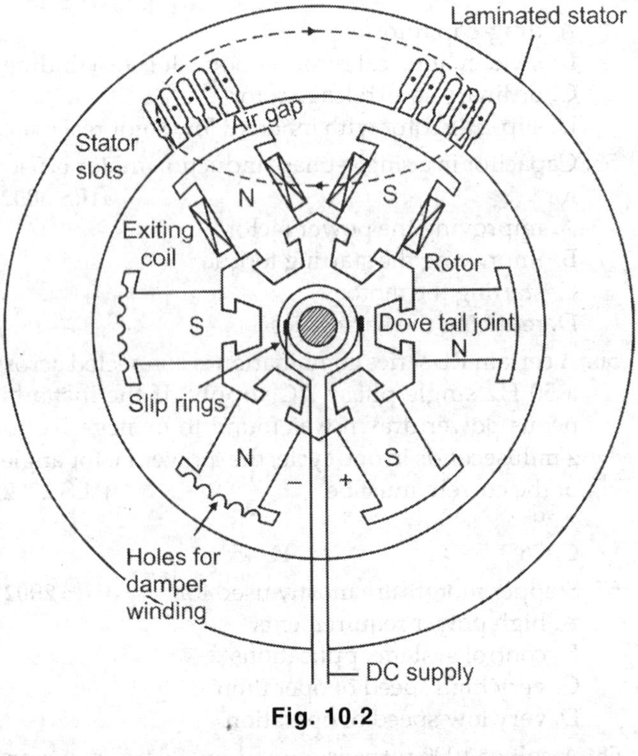

Fig. 10.2

Types of rotors: Rotor of alternator is of two types:

1. *Salient pole type or projecting pole type*: In this type, projecting poles are made of thick laminations, rivetted together and fixed to the rotor by dove tail joint as shown in Fig. 10.3(a). They are used only in low and medium speed alternators.
2. *Nonsalient pole type or cylindrical pole type*: In this type, rotor is made of smooth solid forged-steel radial cylinder having a number of slots along the outer periphery as shown in Fig. 10.3(b). Non-salient pole type rotors are used for high speed alternators which have noise operation.

Emf equation. The emf equation of an alternator is

$$E = 4.44\, K_c K_d \Phi f T \text{ volts/phase}$$

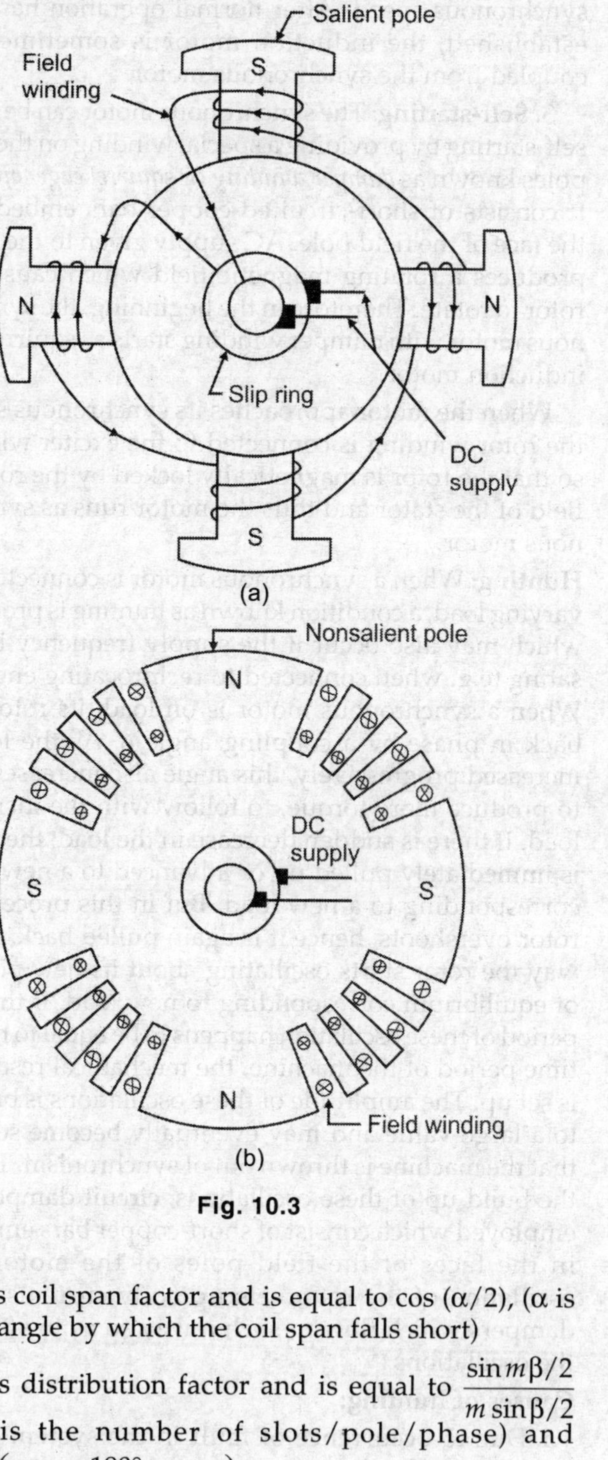

Fig. 10.3

where

K_c is coil span factor and is equal to cos ($\alpha/2$), (α is the angle by which the coil span falls short)

K_d is distribution factor and is equal to $\dfrac{\sin m\beta/2}{m\sin\beta/2}$

(m is the number of slots/pole/phase) and

$$\beta = \left(\frac{180°}{\text{No. of slots/pole}}\right)$$

Synchronous Generator

Parameters of armature winding: An alternator has the following three important parameters:

i. Armature resistance, R_a

ii. Armature leakage reactance, X_L

iii. Reactance (X_{ar}) due to armature reaction

Voltage regulation: The terminal voltage V is less than open-circuit voltage E_o because of

i. armature voltage drops IR_a

ii. synchronous reactance voltage drops IX_s, and

iii. armature reaction voltage drops $I_a X_a$

Therefore, $E = V + IR_a + IX_s + IX_a$

$$\% \textbf{ Voltage Regulation} = \frac{E_o - V}{V} \times 100$$

Voltage regulation can be found by synchronous impedance method, MMF method, and zero power-factor method.

Synchronization: For proper synchronization of alternators, the following conditions must be satisfied:

i. The terminal voltage of the incoming alternator must be the same as bus-bar voltage.

ii. The speed of the incoming machine must be such that its frequency equals bus-bar frequency.

iii. The phase of the alternator voltage must be identical with the phase of the bus-bar voltage.

Synchronizing current, $I_{sy} = \dfrac{E_r}{Z_s}$

Synchronizing power,

$$P_{sy} = \alpha E I_{sy} \text{ watts/phase} = \alpha E^2/X_s \text{ watts/ph,}$$

Load angle

$$\alpha = P/2 \times \text{angle of retardation in mech. degree}$$

Synchronizing torque, $T_{sy} = \dfrac{3 \times 60 \times P_{sy}}{2\pi N_s}$ Nm

Synchronous motor: A synchronous motor is electrically identical with an alternator. Characteristics of synchronous motors:

a. It runs at synchronous speed, $N_s = 120f/P$

b. It is not inherently self-starting,

c. It is capable of being operated under a wide range of power factors both lagging as well as leading.

The mechanical power developed is given by

$$P_m = \frac{E_b V}{Z_s}\cos(\theta - \alpha) - \frac{E_b^2}{Z_s}\cos\theta$$

where α = load angle and θ is internal angle of motor for constant supply voltage V and induced emf E_b

Maximum power developed P_m

$$= \frac{E_b V}{Z_s} - \frac{E_b^2 R_a}{Z_s^2}$$

Maximum power developed, $P_m = \dfrac{V^2}{4R_a}$

Induced emf when pf is lagging,

$$E_b = \sqrt{V^2 + E_r^2 - 2VE_r\cos(\theta - \phi)}$$

Induced emf when pf is leading

$$E_b = \sqrt{V^2 + E_r^2 - 2VE_r\cos(\theta + \phi)}$$

Also $\dfrac{E_r}{\sin\alpha} = \dfrac{E_b}{\sin(\theta+\phi)}$

Effect of excitation: The magnitude of armature current varies with excitation. When overexcited, motor runs with leading pf, and with lagging pf when under-excited.

The torque developed by the motor depends on the coupling angle.

Maximum power developed is given by

$$P_m = \frac{V^2}{4R_a}$$

The *V-curves* present the variation of armature current with excitation as shown in the Fig 10.4.

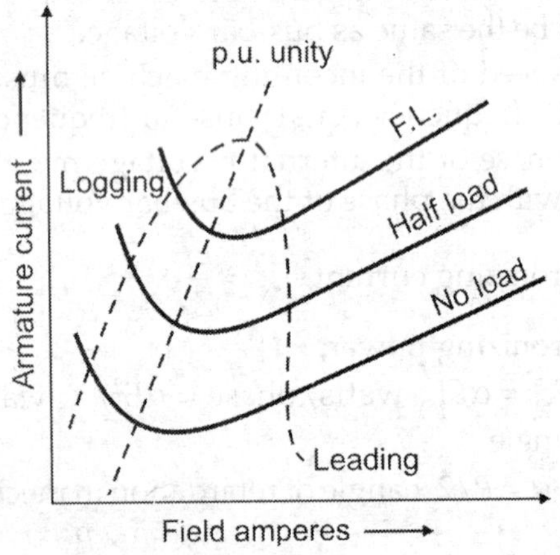

Fig. 10.4

In synchronous motor, minimum current corresponds to unity power factor.

Methods of starting synchronous motors: Since the synchronous motors has no self-starting torque, therefore it is necessary that an external means is to be employed to start the motor. Various methods are described below.

1. **DC source:** The synchronous motor is coupled to a dc compound motor, whose speed is adjusted by a speed regulator. The synchronous motor is then excited and synchronized with the ac supply. At the moment of synchronizing, the synchronous motor is switched on to ac mains and d.c. motor is disconnected from the d.c. supply mains or its field is strengthened until it begins to function as a generator. Otherwise synchronous motor will act as an a.c. motor with d.c. motor as its load.

2. **By means of ac motor:** Synchronous motor with an exciter is coupled with a small induction motor. Before switching the ac supply to the synchronous motor it must be synchronized to the bus-bar. The induction motor used for starting should have lesser number of poles than synchronous motor so that it may be capable of bringing the synchronous motor to the synchronous speed. After normal operation has been established, the induction motor is sometimes uncoupled from the synchronous motor.

3. **Self-starting:** The synchronous motor can be made self-starting by providing a special winding on the rotor poles known as *damper winding* or *squirrel cage winding*. It consists of short-circuited-copper bars embeded in the face of the field pole. AC supply given to the stator produces a rotating magnetic field which causes the rotor to rotate. Therefore in the beginning, the synchronous motor with damper winding starts as squirrelcage induction motor.

When the motor approaches its synchronous speed, the rotor winding is connected to the exciter winding so that the rotor is magnetically locked by the rotating field of the stator and thus the motor runs as synchronous motor.

Hunting: When a synchronous motor is connected to a varying load, a condition known as hunting is produced which may also occur if the supply frequency is pulsating (e.g. when connected to reciprocating engines). When a synchronous motor is on load, its rotor falls back in phase by a coupling angle δ. As the load is increased progressively, this angle also increases so as to produce more torque, to follow with the increased load. If there is sudden decrease in the load, the motor is immediately pulled up or advanced to a new value corresponding to a new load. But in this process, the rotor overshoots, hence it is again pulled back. In this way the rotor starts oscillating about its new position of equilibrium corresponding to new load. If the time period of these oscillation happens to be equal to natural time period of the machine, the mechanical resonance is set up. The amplitude of these oscillations is built up to a large value and may eventually become so great that the machine is thrown out of synchronism. To stop the build up of these oscillations, circuit dampers are employed which consist of short-copper bars embeded in the faces of the field poles of the motor. The oscillation of the rotor sets up eddy currents in the dampers which flow in such a way so as to suppress the oscillations.

Causes of hunting:

i. Due to occurrence of fault in the system that is supplied by the generator
ii. Due to sudden changes of load
iii. Due to sudden changes of field current
iv. Due to cycle variation of load torque

Effects of hunting:

i. Hunting increases the possibility of resonance
ii. Due to hunting, a huge mechanical stress may be developed in the rotor shaft
iii. Hunting causes variation of supply voltage

iv. Hunting increases the chance of loosing synchronism

v. Temperature of the machine rises during hunting

Reduction of hunting: To reduce hunting, the following techniques are used:

i. using damper winding

ii. use of flywheel

iii. synchronous machine can be designed with suitable synchronising power coefficient to reduce hunting

Applications of synchronous motors: Major applications of synchronous motors are:

1. It is used in power houses and substations in parallel with bus-bars to improve the power-factor.
2. In factories having large number of induction motors or other apparatus operating with lagging power factor, these motors are used in order to improve the power factor.
3. It is used as booster to control the voltage at the end of transmission line by varying its excitation.
4. It is used in rubber, textile and cement mills, and other big industries for power application.
5. It is used to derive continuously operating and constant speed equipment such as centrifugal pumps, blowers, and motor generators.

Methods of starting synchronous motors: A synchronous motor has no self-starting torque. Therefore, some external means is required for starting it. The following are the commonly used methods for starting such motors: 1. Induction-motor starting, 2. By means of AC motor.

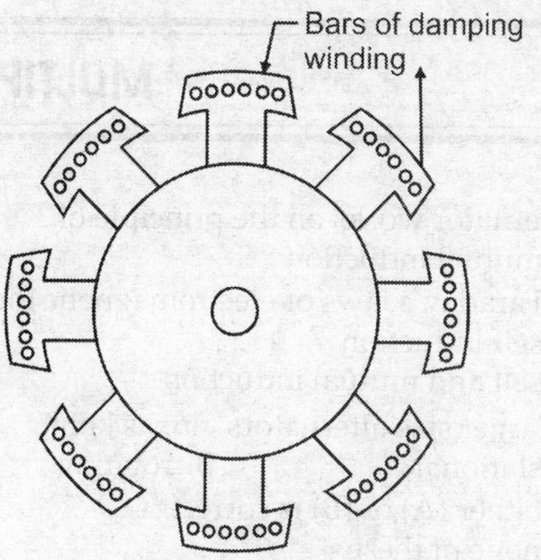

Fig. 10.5

1. **Induction-motor starting:** In this method, a special winding known as damper winding or squirrelcage winding is provided on the rotor. The damper winding consists of copper bars embedded in the pole faces of the motor poles. The bars are short-circuited at the ends to form, in effect, a partial squirrelcage winding. Figure 10.5 shows the salient pole rotor and damper winding.

2. **By means of AC motors:** Synchronous motor with an exciter is coupled with a small induction motor. Before switching the ac supply to the synchronous motor it must be synchronized to the bus-bar. The induction motor used for starting should have lesser number of poles than synchronous motor so that it may be capable of bringing the synchronous motor to the synchronous speed. After normal operation has been established, the induction motor is sometimes uncoupled from the synchronous motor.

MULTIPLE CHOICE QUESTIONS

1. Alternator works on the principle of
   A. mutual induction
   B. Faraday's laws of electromagnetic induction
   C. self induction
   D. self and mutual induction
2. In larger size alternators, flux is kept
   A. stationary B. rotating
   C. either (A) or (B) is correct
   D. none of these
3. The generator which gives DC supply to the rotor is called
   A. convertor B. exciter
   C. invertor D. rectifier
4. The different types of rotor of an alternator are
   A. salient pole type B. cylindrical pole type
   C. both salient pole and cylindrical pole type
   D. none of these
5. The rotor of the alternator has
   A. four slip rings B. three slip rings
   C. two slip rings D. no slip ring
6. Salient pole type rotors are generally used with prime movers of
   A. high speed B. low speed
   C. medium speed D. low and high speed
7. Alternator generates
   A. DC B. AC
   C. DC and AC both D. pulsating DC
8. Cylindrical pole type rotors are generally used with prime movers of
   A. high speed B. low speed
   C. medium speed D. low and high speed
9. The rotor of the alternators requires
   A. DC B. AC
   C. pulsating DC D. none of these
10. The frequency of emf generated depends upon
    A. speed of the alternator
    B. number of poles of the alternator
    C. type of alternator
    D. both (A) and (B) are correct
11. The frequency of a 4 pole alternator running at 1500 rpm will be
    A. 150 Hz B. 100 Hz
    C. 50 Hz D. 25 Hz
12. The salient pole type rotors are
    A. smaller in axial length
    B. larger in axial length
    C. smaller in diameter
    D. larger in diameter and smaller in axial length
13. The advantage of using short pitched winding in an alternator is that it
    A. reduces harmonics in generated emf
    B. saves copper used in the winding
    C. gives the better shape to the waveform
    D. all of these
14. If the speed of an alternator is changed from 3000 to 1500 rpm, the emf generated will become
    A. twice B. half
    C. one fourth D. four times
15. When the load of an alternator is thrown off, the terminal voltage will
    A. increase B. decrease
    C. remain same D. none of these
16. Name the machine which is designed to run without mechanical attachment but with its excitation so controlled as to provide a load taking a purely reactive power is
    A. an induction motor B. a synchronous motor
    C. a synchronous generator
    D. a synchronous compensator
17. An alternator running at 3000 rpm generates voltage at 50 Hz. The number of poles of the alternator will be
    A. 8 poles B. 6 poles
    C. 4 poles D. 2 poles
18. The regulation of an alternator is the ratio of
    A. $\dfrac{V_0 - V}{V_0} \times 100\%$ B. $\dfrac{V_0 - V}{V} \times 100\%$
    C. $\dfrac{V - V_0}{V_0} \times 100\%$ D. $\dfrac{V - V_0}{V} \times 100\%$
19. The conditions for running two alternators in parallel are
    A. terminal voltage should be same
    B. frequency should be same
    C. phase sequence should be same
    D. all the above three
20. The frequency per revolution in an alternator is equal to
    A. number of poles
    B. number of armature conductors
    C. number of pair of poles
    D. None of these
21. What will be the maximum speed at which the field of an alternator can be operated to develop 60 Hz?
    A. 1800 rpm B. 3600 rpm
    C. 7200 rpm D. none of these

Ans.													
	1. B	2. B	3. B	4. C	5. C	6. B	7. B	8. A	9. A	10. D	11. C	12. D	13. D
	14. B	15. A	16. D	17. B	18. B	19. D	20. C	21. B					

22. A 90 slot stator has a three phase, six pole, short coiled winding, each coil of which spans from slot 1 to slot 14. Then the pitch factor is
A. 0.9 B. 0.978
C. 0.407 D. none of these

23. For a full pitch winding, the generated voltages in both coil sides are
A. exactly in phase B. in quadrature
C. exactly 180° out of phase
D. approximately 180° out of phase

24. What will be the maximum speed at which the field of an alternator can be operated to develop 50 Hz?
A. 1800 rpm B. 3000 rpm
C. 7200 rpm D. none of these

25. Distributed winding instead of concentrated winding has the effect of improving the shape of voltage as a sine wave and
A. increasing the speed of machine
B. adding rigidity and mechanical strength to winding
C. reducing armature reaction
D. none of these

26. In an alternator when the load increases then the terminal voltage due to armature reaction
A. drops B. rises
C. either drops or rises
D. does not have any change

27. In a synchronous machine with damper winding, the damping torque at negative slip developed by damper winding acts
A. to reduce the speed B. to increase the speed
C. either to increase or decrease the speed
D. none of these

28. An alternator on open circuit generates 300 V at 50 Hz, when the field current is 3.0 A. Neglecting saturation, when the frequency is 25 Hz and the field current is 2 A, then the open circuit emf is
A. 50 V B. 100 V
C. 200 V D. none of these

29. In a 3-phase, star connected alternator a field current of 40 A produces full load current of 200 A on short circuit and 1160 V on open circuit. If the resistance of the alternator is 0.5 ohm, then the value of synchronous reactance is
A. 5.78 ohms B. 16.5 ohms
C. 3.31 ohms D. 29 ohms

30. The frequency of voltage generated by an alternator having 4-poles and rotating at 1800 rpm is
A. 60 Hz B. 7200 Hz
C. 120 Hz D. 450 Hz

31. A 50-Hz alternator will run at greatest possible speed if it is wound for ...... poles.
A. 8 poles B. 6 poles
C. 4 poles D. 2 poles

32. The main disadvantage of using short-pitch winding in alternators is that it
A. reduces harmonics in the generated voltage
B. reduces the total voltage around the armature coils
C. produces asymmetry in the three phase windings
D. increases copper of end connections

33. Three-phase alternators are invariably Y-connected because
A. magnetic losses are minimised
B. loss turns of wire are required
C. smaller conductors can be used
D. higher terminal voltage is obtained

34. The winding of a 4-pole alternator having 36 slots and a coil span of 1 to 8 is short pitched by
A. 140° B. 20°
C. 20° D. 40°

35. If an alternator winding has a fractional pitch of 5/6, the coil span is
A. 300° B. 150°
C. 30° D. 60°

36. Which harmonic would be totally eliminated from the alternator emf using a fractional pitch of 4/5?
A. Third B. Seventh
C. Fifth D. Ninth

37. For eliminating seventh harmonic from the emf wave of an alternator, the fractional pitch must be
A. 2/3 B. 5/6
C. 7/8 D. 6/7

38. If, in an alternator, chording angle for fundamental flux wave is α, its value for fifth harmonic is
A. 5α B. $\alpha/5$
C. 25α D. $\alpha/25$

39. Regarding distribution factor of an armature winding of an alternator which statement is false?
A. It decreases as the distribution of coils (slots/pole) increases
B. Higher its value, higher the induced emf per phase
C. It is not affected by the type of winding either lap or wave
D. It is not affected by the number of turns per coil

40. When speed of an alternator is changed from 3600 to 1800 rpm, the generated emf/phase will become
A. one-half B. twice
C. four times D. one-fourth

41. The magnitude of the three voltage drops in an alternator due to armature resistance, leakage reactance and armature reactance is solely determined by

Ans.	22. B	23. A	24. B	25. B	26. C	27. A	28. B	29. A	30. A	31. D	32. B	33. D	34. D
	35. B	36. C	37. D	38. A	39. B	40. A							

A. load current, I_a
B. pf of the load
C. whether it is a lagging or leading pf load
D. field construction of the alternator

42. Armature reaction in an alternator primarily affects
A. rotor speed
B. terminal voltage per phase
C. frequency of armature current
D. generated voltage per phase

43. Under no-load condition power drawn, by the prime mover of an alternator goes to
A. produce induced emf in armature winding
B. meet no-load losses
C. produce power in the armature
D. meet copper losses both in armature and rotor windings

44. As load pf of an alternator becomes more leading, the value of generated voltage required to give rated terminal voltage
A. increases
B. remains unchanged
C. decreases
D. varies with rotor speed

45. With a load pf of unity, the effect pf armature reaction on the main-field flux of an alternator is
A. distortional
B. magnetising
C. demagnetising
D. nominal

46. At lagging loads, armature reaction in an alternator is
A. cross-magnetising
B. demagnetising
C. non-effective
D. magnetising

47. At leading pf, the armature flux in an alternator .... the rotor flux.
A. opposes
B. aids
C. distorts
D. does not affect

48. The voltage regulation of an alternator having 0.75 leading pf load, no-load induced emf of 2400 V and rated terminal voltage of 3000 V is
A. 20%
B. –20%
C. 150%
D. –26.7%

49. If, in a 3-phase alternator, a field current of 50 A produces a full-load armature current of 200 A on short-circuit and 1730 V on open circuit, then its synchronous impedance is
A. 8.66 Ω
B. 4 Ω
C. 5Ω
D. 34.6 Ω

50. The power factor of an alternator is determined by its
A. speed
B. load
C. excitation
D. prime mover

51. For proper parallel operation, AC poly phase alternators must have the same
A. speed
B. voltage rating
C. kVA rating
D. excitation

52. Of the following conditions, the one which does NOT have to be met by alternators working in parallel is
A. terminal voltage of each machine must be the same
B. the machines must have the same phase rotation
C. the machines must operate at the same frequency
D. the machines must have equal ratings

53. After wiring up of two 3-phase alternators, you checked their frequency and voltage and found them to be equal. Before connecting them in parallel, you would
A. check turbine speed
B. check phase rotation
C. lubricate everything
D. check steam pressure

54. Zero power factor method of an alternator is used to find its
A. efficiency
B. voltage regulation
C. armature resistance
D. synchronous impedance

55. Some engineers prefer 'lamps bright' synchronization to 'lamps dark' synchronization because
A. brightness of lamps can be judged easily
B. it gives sharper and more accurate synchronization
C. ficker is more pronounced
D. it can be performed quickly

56. It is never advisable to connect a stationary alternator to live bus-bars because it
A. is likely to run as a synchronous motor
B. will get short-circuited
C. will decrease bus-bar voltage though momentarily
D. will disturb generated emfs of other alternators connected in parallel

57. Two identical alternators are running in parallel and carry equal loads. If excitation of one alternator is increased without changing its steam supply, then
A. it will keep supplying almost the same load
B. kVAR supplied decrease
C. its pf will increase
D. kVA supplied by it would decrease

58. Keeping its excitation constant, if steam supply of an alternator running in parallel with another identical alternator is increased, then
A. it would over-run the other alternator
B. its rotor will fall back in phase with respect to the other machine
C. it will supply greater portion of the load
D. its power factor would be decreased

59. The load sharing between two steam-driven alternators operating in parallel may be adjusted by varying the

Ans. 41. A 42. C 43. B 44. C 45. A 46. D 47. B 48. B 49. C 50. B 51. B 52. D 53. B 54. B 55. B 56. B 57. A 58. C

A. field strengths of the alternators
B. power factors of the alternators
C. steam supply to their prime movers
D. speed of the alternators

60. The alternators are operated in parallel because
A. it is convenient and economical in repairing
B. it maintains better stability of supply
C. it is easy to install an additional unit as and when required
D. all the above three

61. Different methods of synchronising the alternators generally used are
A. dark and bright lamp method
B. synchronoscope method only
C. dark lamp method only
D. both (A) and (B) are correct

62. The emf generated in an alternator depends upon
A. coil span factor B. frequency
C. flux per pole D. all the above three

63. When two alternators are running in parallel, if prime mover of one of the alternators is disconnected, that alternator will
A. stop running
B. run as a synchronous motor
C. run as generator D. none of these

64. When two alternators are running in parallel, if one of the alternator is to be disconnected, the
A. load shared by this alternator will be transferred to the other alternator before disconnecting
B. OCB switch will be switched off
C. steam supply of the prime mover is stopped
D. load connected with the bus bar should be reduced

65. A 120 MW turbo-alternator is supplying power to 80 MW load at 0.8 pf lagging. Suddenly the steam supply to the turbine is cut off and the alternator remains connected to the supply network and the field supply also remains on. What will happen to the alternator?
A. The stator winding of the alternator will get burnt
B. The rotor winding of the alternator will get burnt
C. The alternator will continue to run as a synchronous motor rotating in the same direction
D. The alternator will continue to run as a synchronous motor rotating in opposite direction

66. The synchronous impedance method to find voltage regulation of an alternator is not considered an accurate method. This is because the value of synchronous impedance found is always
A. more than its value under normal condition
B. less than its value under normal condition
C. equal to the synchronous reactance
D. none of these

67. Two alternators are running in parallel. If the driving force of both the alternators is changed, this will result in change in
A. frequency B. back emf
C. generated voltage D. all of these

68. What is the experimental data required for Potier method for finding voltage regulation of alternators?
A. No load curve, short circuit test values.
B. No load curve, zero power factor curve
C. Zero power factor curve, short circuit test value
D. none of these

69. A 3-phase, 11 kV, 5 MVA alternator has synchronous impedance of $(1 + j10)$ ohm per phase. Its excitation is such that the generated emf is 14 kV. If the alternator is connected to infinite bus bar, then the maximum output at the given excitation is
A. 15400 kW B. 8000 kW
C. 6200 kW D. 5135 kW

70. The power factor of the input power to a synchronous motor is adjusted by varying
A. number of poles
B. magnitude of excitation
C. magnitude of armature reaction
D. none of these

71. An induction motor is coupled to a synchronous motor. In order to bring up the synchronous motor to synchronous speed at starting, the pairs of pole of induction motor should be
A. less than that of synchronous motor
B. equal to that of synchronous motor
C. greater than that of synchronous motor
D. none of these

72. To be a shunt motor, the counter emf E_b should be always less than applied voltage V. To be a synchronous motor, the counter emf E_b should be, in magnitude
A. always less than the terminal voltage
B. always greater than the terminal voltage
C. always equal to the terminal voltage
D. either equal to, or less than, or greater than the terminal voltage

73. A 10 pole, 25 Hz alternator is directly coupled to, and is driven by a 60 Hz synchronous motor. Then the number of poles in a synchronous motor are
A. 48 poles B. 12 poles
C. 24 poles D. none of these

74. When a synchronous motor delivering rated load is adjusted so that counter emf equals terminal voltage at torque angle 30° then the impedance voltage drop in armature winding as a percentage of terminal voltage is

Ans. 59. C 60. D 61. D 62. D 63. B 64. A 65. C 66. A 67. A 68. B 69. D 70. B 71. A 72. D 73. C

A. 0% B. 51.8%
C. 48.2% D. 100%

75. A synchronous condenser is installed to correct the 2400 kVA, 0.65 power factor load in a plant to unity. Then kilovolt ampere rating of condenser is
A. 27 kVA B. 1560 kVA
C. 1825 kVA D. 3690 kVA

76. The power input to the DC field of a 15 kVA (input) synchronous motor is 6 percent of the rated AC input. Then the direct current that excites the field at 120 volts is
A. 7.5 amps B. 125 amps
C. 15 amps D. none of these

77. At constant load, the magnitude of armature current of a synchronous motor has large values for
A. low values of field excitation only
B. high values of field excitation only
C. both low values and high values of field excitation
D. none of these

78. Chording and distribution of armature windings for AC machines result in reduction in air gap mmf harmonics and
A. reduction in fundamental component of induced voltage
B. increase in effective turns
C. increase in fundamental component of induced voltage
D. None of these

79. The rotors of a high-speed turbo-alternator and a low-speed hydel generator are
A. cylindrical and salient-pole respectively
B. salient-pole and cylindrical respectively
C. both cylindrical D. both salient-pole

80. The salient-pole rotors are not used for high-speed turbo-alternators because of
A. large eddy loss
B. high centrifugal force and windage loss
C. excessive bearing friction
D. harmful mechanical oscillations.

81. Terminal-to-terminal flashover within the oil may be caused
A. only by the moisture present in the oil
B. only by the lining up of the conducting particles present in the oil
C. both by the moisture present in the oil and by the lining up of the conducting particles present in the oil
D. by the rapidly fluctuating loads

82. If diameter and speed of the stator bore of a high-speed turbo-alternator and a low-speed hydel generator of identical ratings are compared, then the turbo-alternator has
A. larger diameter and larger length
B. smaller diameter and smaller axial length
C. larger diameter and smaller axial length
D. smaller diameter and larger axial length

83. To eliminate rth harmonic from the induced emf in a phase of synchronous machine, the pitch of the coils must be
A. $(r-1)/r^{th}$ fraction of full-pitch
B. $(2r-1)/2$ th fraction of full-pitch
C. $r(r+1)$ th fraction of full-pitch
D. $2r(r+1)$ th fraction of full-pitch

84. The fifth space harmonic in the MMF developed by balanced fundamental frequency armature currents rotates at x-times the synchronous speed with respect to the poles while
A. $x = 4/5$ B. $x = 1$
C. $x = 6/5$ D. $x = 7/5$

85. For a uniformly distributed winding with a phase spread of β degrees, the distributed factor at fundamental frequency is
A. $\sin \beta/\beta$ B. $\sin \beta/\beta \times 180/\pi$
C. $(2 \sin \beta/2)/\beta$ D. $(\sin \beta/2)/\beta \times 360/\pi$

86. Which space harmonics are absent in the MMF of a 3-phase synchronous machine produced by balanced sinusoidal currents in the armature?
A. 5, 11, 17, 23, etc B. 3, 9, 15, 21, etc
C. 7, 13, 19, 25, etc
D. 3, 5, 7, 9, 11, 13, 15, 17, 19, etc

87. The seventh space harmonic in the MMF developed by balance fundamental-frequency armature currents rotates at x-times the synchronous speed with-respect to the poles, while
A. $x = 5/7$ B. $x = 6/7$
C. $x = 1$ D. $x = 8/7$

88. The rotor of a high speed turbo-alternator is made up of solid steels forging for
A. high mechanical stress
B. economy
C. reduction of eddy current loss
D. reduction of bearing friction

89. The frequency of voltage generated in an alternator is
A. directly proportional to its number of poles and inversely proportional to its speed of rotation
B. inversely proportional to its number of poles and directly proportional to its speed of rotation
C. directly proportional to its number of poles and speed of rotation
D. inversely proportional to its number of poles and speed of rotation

90. In a synchronous generator the open circuit voltage
A. leads the terminal voltage by an angle known as power angle
B. lags the terminal voltage by an angle known as power angle

Ans. 74. B 75. C 76. A 77. C 78. A 79. A 80. B 81. C 82. D 83. A 84. C 85. D 86. B 87. B 88. A 89. C 90. A

C. leads the terminal voltage by an angle known as overlap angle
D. lags the terminal voltage by an angle known as overlap angle

91. In a synchronous generator the field MMF
A. lags the airgap flux and the airgap flux lags the armature MMF
B. leads the airgap flux and the airgap flux leads the armature MMF
C. leads the airgap flux and the airgap flux lags the armature MMF
D. lags the airgap flux and the airgap flux leads the armature MMF

92. Armature reaction MMF and leakage reactance of a synchronous machine are determined by
A. open circuit test only
B. zero power factor test only
C. open circuit and zero power factor tests
D. open circuit and short circuit tests

93. Consider the following statements regarding synchronous machines:
I. Potier reactance is slightly higher than leakage reactance because of excessive saturation of the field poles during zero power factor test
II. Load characteristic of a synchronous generator represents terminal voltage as a function of field current for constant load current and power factor
III. The regulation curve of a synchronous generator shows the relationship of field current with the load current at constant power factor or with power factor at constant load current
IV. The short-circuit ratio is equal to synchronous impedance in pu

Which of the above statements?
A. All are correct B. I, III and IV are correct
C. I, II and III are correct
D. II and III are correct

94. In calculation of regulation of synchronous generator
A. both EMF and MMF methods give optimistic values
B. both EMF and MMF methods give pessimistic values
C. while EMF method gives optimistic value, MMF method gives pessimistic value
D. while EMF method gives pessimistic value, MMF method gives optimistic value

95. The phasor diagram by ASA method gives
A. reliable results for both regulation and power angle of a salient-pole synchronous generator
B. considerably erroneous results for both regulation and power angle of a salient pole synchronous generator
C. reliable result for regulation but considerably erroneous result for power angle
D. considerably erroneous result for regulation but reliable result for power angle

96. The exciting field coil of an alternator is generally excited by
A. a separate DC generator driven by some source
B. a separate AC generator driven by some source
C. a DC generator coupled directly to the alternator shaft
D. a battery

97. The amortisseur winding in a synchronous motor
A. provides starting torque only
B. provides starting torque and eliminates hunting
C. improves the power factor of the machine
D. none of these

98. Which of the following statements is true?
A. An internal pole dynamo has a rotating magnetic field
B. An internal pole dynamo has a rotating armature
C. An external pole dynamo has a rotating magnetic field
D. Alternators are usually of the external pole type

99. For what purpose the stroboscope instruments are used?
A. To synchronize synchronous motors
B. To synchronize synchronous generators
C. To synchronize induction motors
D. To synchronize induction generators

100. A 3-phase alternator is delivering its rated current at rated terminal voltage. The power factor is 0.8 lagging and the field current is I_f. If the power factor is changed to 0.8 leading, the load current and terminal voltage remaining the same, the new value of field current is
A. I_f B. less than I_f
C. more than I_f D. none of these

101. A four pole turbo-generator supplies a 50 Hz network. What should be the speed of the generator?
A. 6000 rpm B. 3000 rpm
C. 1500 rpm D. 1000 rpm

102. Cylindrical rotor alternators have
A. large length to diameter ratio
B. small length to diameter ratio
C. vertical configuration
D. None of these

103. In a 3-phase alternator the unsaturated synchronous reactance is 3 ohms per phase. The saturated synchronous reactance is
A. 3 ohms B. more than 3 ohms
C. less than 3 ohms D. none of these

Ans. 91. B 92. C 93. C 94. D 95. C 96. C 97. B 98. A 99. B 100. B 101. C 102. A 103. C

104. Under short circuit conditions, the power factor of an alternator is
A. almost zero lagging B. unity
C. about 0.6 lagging D. None of these

105. The synchronous impedance of a larger size modern alternator is about
A. 0.1 pu B. 0.4 pu
C. 1.0 pu D. none of these

106. The ratio of armature leakage reactance to synchronous reactance of a large size modern alternator is about
A. 0.05 B. 0.2
C. 0.6 D. 0.8

107. To ensure proper cooling, cylindrical rotor alternators use
A. radial ducts only B. axial ducts only
C. both radial and axial ducts
D. None of these

108. The rated voltage of alternators used in power stations is usually
A. 11 kV B. 66 kV
C. 132 kV D. 400 kV

109. High speed alternators usually have
A. salient pole rotors B. cylindrical rotors
C. both salient pole and cylindrical rotors
D. none of these

110. Salient pole machines have
A. large number of poles
B. small number of poles
C. small diameters D. long cores

111. The full pitch coil in an alternator has a span of 18 slots. If it is desired to eliminate the third harmonic voltage, the coil span should be
A. 18 slots B. 15 slots
C. 12 slots D. 9 slots

112. The stators of modern alternators are wound for
A. 120° phase groups B. 60° phase groups
C. 240° phase groups D. 180° phase groups

113. An alternator is connected to the bus bars and is supplying load. Its prime mover is suddenly shut-down. The alternator will
A. continue to work as an alternator
B. continue to run as synchronous motor but direction of rotation will reverse
C. continue to run as synchronous motor and the direction of rotation will remain the same
D. none of these

114. Machine A has 120° phase groups and a second similar machine B has 60° phase groups in the armature winding. Then
A. breadth factor of machine *A* is higher than that of *B*
B. breadth factor of machine *B* is higher than that of *A*
C. both the machines have the same breadth factor
D. none of these

115. Modern alternators usually have short pitched winding to
A. improve the voltage wave shape
B. improve the magnitude of the generated voltage
C. increase the machine rating
D. none of these

116. Alternators are usually operated at a power angle of about
A. 10° B. 40°
C. 60° D. 90°

117. If the armature current is leading the generated voltage by 90° the effect of armature reaction will be
A. demagnetising B. magnetising
C. partly magnetising and partly cross magnetising
D. None of these

118. The value of voltage regulation as determined by synchronous impedance method is
A. greater than that found by actual loading
B. less than that found by actual loading
C. the same as that found by actual loading
D. none of these

119. Alternators generally use short pitch coils to remove harmonics from emf wave shape. As compared to full pitch coils, the short pitch coils need
A. more conductor material
B. the same amount of conductor mate-rial
C. less conductor material
D. none of these

120. The biggest size of alternators used in India is
A. 200 MW B. 350 MW
C. 500 MW D. 100 MW

121. The short-circuit characteristic of an alternator is
A. always linear B. always nonlinear
C. sometimes linear and sometimes non-linear
D. None of these

122. In order to extract energy from an electrical system, the coupling magnetic field must react with
A. electrical system B. mechanical system
C. both electrical and mechanical systems
D. None of these

123. Electromagnetic force or torque developed in any physical system tends to
A. increase the magnetic stored energy at constant flux
B. decrease the magnetic stored energy at constant current
C. decrease the reluctance
D. decrease the inductance

124. The pitch-factor, in rotating electrical machines, is defined as the ratio of resultant emf of a
A. full-pitched coil to that of a chorded coil
B. full pitched coil to the phase emf

Ans. 104. A 105. C 106. B 107. C 108. A 109. B 110. A 111. C 112. B 113. C 114. B 115. A 116. B
117. B 118. B 119. C 120. C 121. A 122. A 123. C

C. chorded coil to the phase emf
D. chorded coil to that of a full-pitched coil

125. Electromagnetic force or torque developed in any physical system tends to
A. increase both the field energy and coenergy at constant current
B. decrease both the field energy and coenergy at constant current
C. increase current at constant flux linkages
D. decrease the permeance

126. Reluctance motors are
A. doubly-excited B. singly-excited
C. multi-excited D. none of these

127. Large synchronous machines are constructed with armature winding on the stator because stationary armature winding
A. can be insulated satisfactorily for higher voltages
B. can be cooled more efficiently
C. have reduced slip ring losses
D. all of these

128. The doubly-excited magnetic systems are
A. moving iron instruments
B. electromagnetic relays
C. solenoids D. synchronous motors

129. Reluctance torque in rotating machines is present when
A. airgap is not uniform
B. reluctance seen by stator mmf is constant
C. reluctance seen by rotor mmf varies
D. reluctance seen by rotor mmf is constant

130. The interaction torque depends on
A. stator field strength alone
B. both the stator field and rotor field strengths
C. stator field strength and torque angle
D. stator field strength rotor field strengths and the torque angle

131. If the armature current is in phase with generated emf, the effect of armature reaction is
A. cross magnetising B. magnetising
C. demagnetising D. None of these

132. Electromagnetic torque in rotating machines is present when
A. airgap is uniform
B. stator winding alone carries current
C. rotor winding alone carriers current
D. both stator and rotor windings carry current

133. The torque angle δ is defined as the angle between
A. stator field axis and resultant field axis
B. rotor field axis and resultant field axis
C. stator field axis and rotor field axis
D. stator field axis and mutual, field axis

134. Armature winding is one in which working
A. flux is produced by field current
B. flux is produced by the working current
C. emf is produced by the working flux
D. emf is produced by the leakage flux

135. A pole-pitch in electrical machines is
A. equal to 180° electrical
B. equal to 180° mechanical
C. less than 180° electrical
D. greater than 180° electrical

136. A coil consists of
A. two conductors B. two-coil sides
C. two turns D. four turns

137. One turn consists of
A. two-coil sides B. two conductors
C. four conductors D. four-coil sides

138. The main advantages of distributing the winding in slots is to
A. add mechanical strength to the winding
B. reduce the amount of copper required
C. reduce the harmonics in the generated emf
D. reduce the size of the machine

139. The distribution factor is defined as the ratio of
A. arithmetic sum of coil emfs to phasor sum of coil emfs
B. phasor sum of emf per coil to the arithmetic sum of emf per coil
C. phasor sum of coil emfs 'to the arithmetic sum of coil emfs
D. phasor sum of coil emfs to the per phase voltage

140. The main advantage of using fractional-pitch winding is to reduce
A. amount of copper in the winding
B. size of the machine
C. harmonics in the generated emf
D. cost of the machine

141. For a 3-phase uniformly distributed and narrow-spread winding, the distribution factor is
A. 0.9 B. 0.955
C. 0.965 D. 0.97

142. If F_1 is the constant amplitude of fundamental rotating mmf wave, then for the space harmonics of the order $6k + 1$, the harmonic mmf wave is of
A. constant amplitude and stationary in space
B. constant amplitude and rotates against F_1
C. varying amplitude and rotates along F_1
D. constant amplitude and rotates along F_1

143. When a three-phase synchronous motor is switched on, there exists a rotating magnetic field. The magnitude of this field flux
A. is constant at all loads
B. varies with power factor
C. varies with load D. none of these

Ans. 124. D 125. A 126. B 127. D 128. D 129. C 130. D 131. A 132. D 133. C 134. C 135. A 136. B 137. B 138. C 139. C 140. C 141. B 142. D 143. A

144. The chording angle for eliminating 5th harmonics should be
A. 30° B. 34°
C. 36° D. 35°

145. Two alternators *A* and *B* are sharing an inductive load equally. If the excitation of alternator *A* is increased
A. alternator *B* will deliver more current and alternator A will deliver less current
B. alternator *B* will deliver less current and alternator A will deliver more current
C. both will continue to share load equally
D. both will deliver more current

146. In synchronous alternator, which of the following coils will have emf closer to sine wave form?
A. Concentrated winding in full pitch coils
B. Concentrated winding in short pitch coils
C. Distributed winding in full pitch coils
D. Distributed winding in short pitch coils

147. The windings for an alternator are
I. 36 slots, four poles, span 1 to 3
II. 72 slots, six poles, span 1 to 10
III. 96 slots, six poles, span 1 to 12
The windings having pitch factors of more than 0.9 are
A. I and II only B. II and III only
C. I and III only D. I, II and III

148. Two 3-phase AC generators are such that one has twice the linear dimensions of the other. The field windings of each are excited to give identical sinusoidal air-gap flux density waveform. Both have same number of stator slots and identical winding patterns. The conductor/slot in big generator is *K* times that of the smaller one. The value of *K* to get equal no-load voltage at same frequency is
A. 8 B. 4
C. 1/2 D. 1/4

149. In a synchronous motor, damper winding is provided in order to
A. stabilize rotor motion
B. suppress rotor oscillations
C. develop necessary starting torque
D. both (B) and (C) are correct

150. In a synchronous motor, the magnitude of stator back emf E_b depends on
A. speed of the motor B. load on the motor
C. both the speed and rotor flux
D. DC excitation only

151. An electric motor in which both the rotor and stator fields rotate with the same speed is called a/an
A. DC motor B. charge motor
C. synchronous motor D. universal motor

152. While running, a synchronous motor is compelled to run at synchronous speed because of
A. damper winding in its pole faces
B. magnetic locking between stator and rotor poles
C. induced emf in rotor field winding by stator flux
D. compulsion due to Lenz's law

153. The net armature (stator) voltage of a synchronous motor is equal to the .... of E_b and *V*.
A. vector difference B. arithmetic sum
C. arithmetic difference D. vector sum

154. When running under no-load condition and with normal excitation, armature current I_a drawn by a synchronous motor
A. leads the back emf E_b by a small angle
B. is large in magnitude
C. lags the applied voltage V by a small angle
D. lags the resultant voltage E_R by 90°

155. The angle between the synchronously rotating stator flux and rotor poles of a synchronous motor is called ...... angle.
A. synchronizing angle B. torque angle
C. power factor angle D. slip

156. If load angle of a 4-pole synchronous motor is 8° (elect), its value in mechanical degrees is
A. 4 B. 2
C. 0.5 D. 0.25

157. The maximum value of torque angle α in a synchronous motor is
A. 45° B. 90°
C. between 45° and 90° D. below 60°

158. A synchronous motor running with normal excitation adjusts to load, increases *essentially* by increase in its
A. power factor B. torque angle
C. back emf D. armature current

159. When load on a synchronous motor running with normal excitation is increased, armature current drawn by it increases because
A. back emf E_b becomes less than applied voltage V
B. power factor is decreased
C. net resultant voltage E_R in armature is increased
D. motor speed is reduced

160. When load on a normally-excited synchronous motor is increased, its power factor tends to
A. approach unity
B. become increasingly lagging
C. become increasingly leading
D. remain unchanged

161. The effect of increasing load on a synchronous motor running with normal excitation is to
A. increase both its I_a and pf
B. decrease I_a but increase pf

Ans. 144. C 145. B 146. D 147. A 148. C 149. D 150. D 151. C 152. B 153. A 154. C 155. B 156. A 157. B 158. D 159. C 160. B

C. increase I_a but decrease pf
D. decrease both I_a and pf

162. Ignoring the effects of armature reaction, if excitation of a synchronous motor running with constant load is increased, its torque angle must necessarily
A. decrease B. increase
C. remain constant
D. become twice the no-load value

163. If the field of a synchronous motor is underexcited, the power factor will be
A. lagging B. leading
C. unity D. more than unity

164. Ignoring the effects of armature reaction, if excitation of a synchronous motor running with constant load is decreased from its normal value, it leads to
A. increase in α but decrease in E_b
B. increase in E_b but decrease in I_a
C. increase in both I_a and pf which is lagging
D. increase in both I_a and Φ

165. A synchronous motor connected to infinite bus-bars has at constant full-load, 100% excitation and unity pf. On changing the excitation only, the armature current will have
A. leading pf with under-excitation
B. leading pf with over-excitation
C. lagging pf with over-excitation
D. no change of pf

166. The V-curves of a synchronous motor shows relationship between
A. excitation current and back emf
B. field current and pf
C. DC field current and AC armature current
D. armature current and supply voltage

167. When load on a synchronous motor is increased, its armature current is increased provided it is
A. normally-excited B. over-excited
C. under-excited D. all of these

168. If main field current of a salient-pole synchronous motor fed from an infinite bus and running at no-load is reduced to zero, it would
A. come to a stop
B. continue running at synchronous speed
C. run at sub-synchronous speed
D. run at super-synchronous speed

169. In a synchronous machine when the rotor speed becomes more than the synchronous speed during hunting, the damping bars develop
A. synchronous motor torque
B. DC motor torque
C. induction motor torque
D. induction generator torque

170. In a synchronous motor, the rotor copper losses are met by
A. motor input B. armature input
C. supply lines D. DC source

171. The speed of a motor generator set, consisting of a 6-pole induction motor and a 4-pole DC generator fed from a 3-phase, 50-Hz supply is
A. 1000 rpm B. 600 rpm
C. 1500 rpm D. 3000 rpm

172. A rotary converter generally
A. combines the functions of an induction motor and a DC generator
B. has a set of slip rings at both ends
C. has one armature and two fields
D. is a synchronous motor and a DC generator combined

173. In a synchronous motor, minimum armature current occurs at
A. zero power factor B. unity power factor
C. lagging power factor D. leading power factor

174. A synchronous motor can run at
A. a leading power factor
B. unity power factor
C. lagging or leading or unity power factor
D. zero power factor

175. For a synchronous motor, the inverted "V" curve is the relation between
A. field current and power factor
B. field current and armature current
C. armature current and power factor
D. none of these

176. A coil of 150 degree pitch has third harmonic pitch factor as
A. cos 225° B. sin 225°
C. cos 45° D. sin 45°

177. The function of damper winding in synchronous motor is to provide, starting torque and to
A. reduce speed B. prevent hunting
C. increase speed D. none of these

178. If 4-pole, 50 Hz, 3-phase synchronous motor and a 8-pole, 50 Hz, 3-phase slip ring induction machine are coupled to each other mechanically and operate on the same 3-phase, 50 Hz supply system. The slip rings of the induction machine are left open circuited, the frequency of the voltage across any two slip rings would not be
A. 12.5 Hz B. 25.0 Hz
C. 37.5 Hz D. None of these

179. The operation of a 3-phase synchronous motor working on constant excitation across infinite bus cannot be stable if the following relationship between power angle δ and internal angle γ exists
A. $2\delta > \gamma$ B. $\delta > \gamma$
C. $2\delta < \gamma$ D. $\delta = \gamma$

180. If V be the terminal voltage per phase, E_0 be the open circuit voltage per phase and γ be the internal

Ans. 161. C 162. A 163. A 164. D 165. B 166. C 167. D 168. B 169. D 170. D 171. B 172. D 173. B 174. C 175. A 176. C 177. B 178. D 179. D

angle of a 3-phase synchronous motor, its power factor will always be lagging if

A. $E_0 \cos \gamma < V$ B. $E_0 < V$

C. $E_0 < V \cos \gamma$ D. $E_0 \sec \gamma < V$

181. A 3-phase synchronous motor driving a constant load torque draws power from the infinite bus at a leading power factor. If the excitation increases.
A. the power angle decreases while power factor increases
B. the power angle increases while power factor decreases
C. both power angle and power factor increase
D. both power angle and power factor decreases

182. A 3-phase synchronous motor with constant excitation is driving a certain load drawing electrical power from infinite bus at leading power factor. If the shaft load decreases
A. the power angle decreases while power factor increases
B. the power angle increases while power factor decreases
C. both power angle and power factor increase
D. both power angle and power factor decrease

183. In 3-phase synchronous motor the field MMF
A. lags airgap flux and the airgap flux lags the armature MMF
B. leads the airgap flux and the airgap flux leads the armature MMF
C. leads the airgap flux and the airgap flux lags the armature MMF
D. lags the airgap flux and the airgap flux leads the armature MMF

184. Consider the following statements:
I. A synchonous motor has no starting torque but when started it always runs at a fixed speed.
II. Neglecting residual magnetism, both cylindrical-rotor and salient pole synchronous motor produce some mechanical power even though the field is unexcited.
III. A single-phase reluctance motor is not self-starting even if paths for eddy currents are provided in the rotor
IV. A single-phase hysteresis motor is self-starting.
Which of the above statements
A. All are correct
B. Only I, III and IV are correct
C. Only I and III are correct
D. Only I is correct

185. Which one of the following statements regarding a three-phase hysteresis motor is correct?
A. Both eddy-current and hysteresis torques are effective during starting as well as running
B. Both eddy current and hysteresis torques are effective during starting while only hysteresis torque is effec-tive during running
C. Only eddy current torque is effective during starting while only hysteresis torque is effective during running
D. Only hysteresis torque is effective during starting as well as running

186. For maximum power input and maximum power output of a 3-phase synchronous motor working on constant excitation across infinite bus, its power angle δ is related to its internal angle γ as
A. $\gamma = 180 - \gamma$ and $\delta = \gamma$ respectively
B. $\delta = \gamma$ and $\delta = 180 - \gamma$ respectively
C. $\delta = 180 - 2\gamma$ and $\delta = \gamma/2$ respectively
D. $\delta = \gamma/2$ and $\delta = 180 - 2\gamma$ respectively

187. If one of the 3-phase of synchronous motor is short circuited, motor will
A. start B. not start
C. overheated
D. fail to pull into synchronism.

188. A three-ring synchronous converter supplying a DC 3-wire system must have the secondary of its transformer connected as
A. delta only B. delta or star
C. zigzag only D. star or zigzag

189. If the frequency of supply to a motor converter be f Hz, the number of pole of the induction motor be P_m the number of the poles of the converter be P_g and the DC power generated be P, then the speed of the set and power transferred mecha-nically to the converter from the induc-tion motor are
A. $120f/(P_m + P_g)$ rpm and $p_g/(P_m + P_g)P$ respectively
B. $120f/(P_m + P_g)$ rpm and $P_m/(P_m + P_g)P$ respectively
C. $120f/P_m$ rpm and $P_m/(P_m + P_g)P$ respectively
D. $120f/p_g$ rpm and $p_g/(p_m + p_s)P$ respectively

190. For which of the following cases are synchronous machines used exclusively?
A. For electric trains B. For cranes
C. For machine tools
D. For large three phase generators

191. The synchronous motor runs at
A. less than synchronous speed
B. synchronous speed
C. more than synchronous speed
D. none of these

192. The construction of synchronous motor is similar to
A. DC compound motor
B. slipring induction motor
C. DC shunt generator
D. alternator

193. The synchronous motor runs on
A. 3-phase AC supply

Ans. 180. C 181. D 182. D 183. A 184. C 185. B 186. A 187. B 188. C 189. B 190. B 191. B 192. D

B. 3-phase AC and DC supply
C. DC supply only
D. 3-phase AC and single phase AC

194. If a synchronous motor is switched on to 3-phase supply with its rotor winding short circuited, it will
A. start B. not start
C. start and continue to run as induction motor
D. start and continue to run as synchronous motor

195. Under running condition on load, the angle between the induced voltage and supply voltage will be
A. zero B. 180°
C. 90° D. none of these

196. The standard full load power factor of a synchronous motor is
A. unity B. unity or 0.8 leading
C. unity or 0.8 lagging D. zero or 0.8 leading

197. Net stator voltage of synchronous motor is
A. vector difference of E_b and V
B. vector sum of E_b and V
C. arithmetic difference of E_b and V
D. arithmetic sum of Eb and V

198. If the field of synchronous motor is under-excited, the power factor will be
A. more than unity B. unity
C. lagging D. leading

199. The advantage of synchronous motor over slipring induction motor are that its
A. power factor can be varied
B. speed can be easily varied
C. speed is independent of supply frequency
D. rotor has two slip-rings

200. When a synchronous motor is started, the field winding is initially
A. short circuited B. open circuited
C. excited by a DC source
D. none of these

201. The synchronous motor can be started by
A. coupling with DC compound motor
B. coupling with AC induction motor
C. providing damper winding
D. all of these

202. The advantages of synchronous motor as compared to induction motor are that
A. it runs at constant speed
B. it can run over wide range of power factors both lagging and leading
C. its torque is less sensitive to change in supply voltage
D. all of these

203. Rotor winding of the synchronous motor is excited by
A. induction from stator current
B. AC supply
C. DC supply supplied by the exciter
D. the revolving field

204. The speed of synchronous motor depends upon
A. number of poles B. supply frequency
C. either (A) or (B)
D. both (A) and (B) are correct

205. The voltage supplied to the rotor of the synchronous motor is
A. 400 V AC B. 400V DC
C. 100 to 250 V DC D. $400/\sqrt{3}$V AC

206. The magnitude of the induced emf in the stator of a synchronous motor
A. is equal to the supply voltage
B. is less than the supply voltage
C. can be increased than the supply voltage
D. can be increased or decreased than the supply voltage

207. The damping winding in a synchronous motor is generally used to
A. provide starting torque only
B. reduce noise level
C. reduce eddy currents
D. prevent hunting and provide the starting torque

208. Under full load running condition, the slip of a synchronous motor is
A. zero B. about 0.2
C. about 0.01 D. none of these

209. If the field of synchronous motor is underexcited, the power factor will be
A. power factor angle B. synchronous angle
C. torque angle D. angle of retardation

210. What is the use of the bright lamp circuit?
A. It is used during synchronization to indicate equal voltages of generator and line
B. It is used during synchronization to indicate that generator and line are having the same phase
C. It is used to start synchronous motors
D. It is used to indicate when the power factor of a synchronous motor is exactly 1

211. An unexcited single phase synchronous motor is
A. reluctance motor B. universal motor
C. repulsion motor D. AC series motor

212. A 3-phase 400 V, 50 Hz synchronous motor is working at 50% load. In case an increase in the field current of the motor cause a reduction in the armature current, it can be concluded that
A. the motor is delivering reactive power to the mains
B. the motor is absorbing reactive power from the mains

Ans. 193. B 194. C 195. D 196. B 197. A 198. C 199. A 200. A 201. D 202. D 203. C 204. D 205. C
206. D 207. D 208. A 209. C 210. B 211. A 212. B

C. the motor is neither absorbing nor delivering reactive power
D. none of these

213. A synchronous motor working on leading power factor and not driving any mechanical load is known as
A. synchronous induction motor
B. spinning motor
C. synchronous condenser
D. none of these

214. An overexcited synchronous motor operates at
A. unity power factor B. lagging power factor
C. leading power factor D. none of these

215. When the synchronous motor is to be started, the field system is energised
A. in the very beginning
B. after the motor has attained a speed slightly less than the synchronous speed
C. after the motor has attained synchronous speed and has been loaded
D. none of these

216. V-curves of a synchronous motor show the relation between
A. armature current and field current
B. applied voltage and field current
C. applied voltage and armature current
D. none of these

217. For a given load, the armature current of a synchronous motor is minimum when the power factor is
A. unity
B. slightly less than unity and lagging
C. slightly less than unity and leading
D. none of these

218. As the load on a synchronous motor is increased, the torque angle
A. increases B. decreases
C. remains the same D. none of these

219. Synchronous motors generally have
A. cylindrical rotor B. salient pole rotor
C. both (A) and (B) are correct
D. none of these

220. If the excitation and terminal voltage of a synchronous motor are kept constant and the load is increased then
A. armature current decreases and power factor becomes more leading
B. armature current increases and power factor becomes more leading
C. armature current increases and power factor becomes more lagging
D. none of these

221. A synchronous motor is supplying a certain load and is operating at unity power factor. If field current and terminal voltage are kept constant but the load is increased, the power factor will
A. remain the same B. become leading
C. become lagging D. none of these

222. Synchronous motors are usually operated at
A. unity power factor B. leading power factor
C. lagging power factor D. none of these

223. A salient pole synchronous motor is running at no-load. Its field Current is switched off. The motor will
A. come to stop
B. continue to run at synchronous speed
C. continue to run at a speed slightly less than synchronous speed
D. none of these

224. A synchronous motor always runs at synchronous speed because
A. of the action of amortisseur winding
B. the Lenz's law cannot be violated
C. of magnetic locking between the stator and rotor
D. none of these

225. In a synchronous motor the generated emf
A. must always be less than the terminal voltage
B. must always be more than the terminal voltage
C. can be less than or greater than the terminal voltage
D. none of these

226. An underexcited synchronous motor operates at
A. unity power factor B. leading power factor
C. lagging power factor D. none of these

227. Synchronous motors are used in
A. sizes greater than about 50 HP
B. all sizes
C. generally small sizes D. none of these

228. A synchronous motor is supplying its rated load. If the excitation is increased the power factor becomes
A. more leading
B. more lagging
C. neither leading nor lagging
D. none of these

229. The power output of a synchronous motor is
A. directly proportional to synchronous reactance
B. inversely proportional to synchronous reactance
C. independent of synchronous reactance
D. none of these

230. In a synchronous motor
A. the rotor mmf and stator mmf are stationary with respect to each other
B. the speed of rotor mmf is slightly less than the speed of stator mmf

Ans. 213. C 214. C 215. B 216. A 217. A 218. A 219. B 220. C 221. C 222. B 223. C 224. C 225. C 226. C 227. A 228. A 229. B 230. A

C. the speed of rotor mmf is slightly more than the speed of stator mmf
D. none of these

231. The armature mmf wave of a DC machine is
A. sinusoidal and depends on the speed
B. square and independent of speed
C. sinusoidal and independent of speed
D. triangular and independent of speed

232. When the excitation of synchronous motor is varied,
A. the pf and the armature current vary
B. the pf varies but the armature current does not vary
C. the armature current varies but the pf does not vary
D. neither the pf nor the armature current varies

233. The armature mmf wave for a DC machine is
A. sinusoidal and rotates wrt armature
B. sinusoidal and rotates wrt field
C. triangular and rotates wrt armature
D. triangular and stationary wrt armature

234. Which of the following devices can be used as a phase advancer?
A. 3-phase induction motor squirrelcage type
B. 3-phase induction motor slip-ring type
C. Synchronous motor working at leading power factor
D. Synchronous motor working at lagging power factor

235. The negative phase sequence in a three phase synchronous motor exists when the motor is
A. underloaded B. overloaded
C. supplied with unbalanced voltage
D. hot

236. An inserted V-curve of synchronous motor shows the variation of
A. field current and DC excitation at constant load
B. supply voltage and field current at constant excitation
C. power factor and supply voltage during hunting
D. supply voltage and excitation current at constant load

237. Motor A has shallow and wide slots. Motor B has deeper and narrow slots. If both are 3-phase 400 V, 50 Hz, 1440 rpm induction motors, it can be concluded that
A. motor A has starting torque as compared to motor B
B. motor B has more starting torque as compared to motor A
C. motor A has more pull out torque as compared to motor B
D. motor B has more pull out torque as compared to motor A

238. Two similar synchronous generators are working in parallel to supply a common load demand with identical excitations and steam supplies to their prime movers. Now, if the steam supply to the prime mover of one of the generators is increased as compared to the other with field excitation kept unchanged, then
A. its active power component will remain the same but the reactive power contribution will increase
B. its active power will decrease while the reactive power will increase
C. both active and reactive components of power will increase
D. its active power contribution will increase but reactive power contributions of both will remain unchanged

239. A 3-phase synchronous motor connected to an infinite bus is operating at half full-load with normal excitation. When the load on the synchronous motor is suddenly increased, its speed will
A. first decrease and then become synchronous
B. first increase and then become synchronous
C. fluctuate around synchronous speed and then become synchronous
D. remain unchanged

240. Two mechanically coupled alternators deliver power at 50 Hz and 60 Hz respectively. The highest speed of the alternators is
A. 3600 rpm B. 3000 rpm
C. 600 rpm D. 500 rpm

241. Semiclosed slots or totally closed slots are used in induction motors, essentially to
A. improve starting torque
B. increase pull-out torque
C. increase efficiency
D. reduce magnetising current and improve power factor

242. A synchronous motor is found to be most economical when the load is above
A. 1 kW B. 10 kW
C. 20 kW D. 100 kW

243. The advantage of a synchronous motor in addition to its constant speed is
A. high power factor
B. better efficiency
C. lower cost
D. all of these

244. For low and medium power drives, where speed control is not required, which of the following motor you will recommend?
A. Synchronous motor
B. Squirrel cage induction motor
C. Double cage induction motor
D. Three-phase series motor

Ans. 231. D 232. A 233. C 234. C 235. C 236. A 237. B 238. D 239. C 240. C 241. D 242. D 243. A 244. A

245. The shunt motor, with armature shunted through a resistance as compared to, with aramature not shunted, has no load speed
A. higher B. equal
C. lower D. none of these

246. The number of slip rings of a single-phase and three-phase converters are respectively
A. 1 and 3 B. 2 and 3
C. 2 and 6 D. 1 and 6

247. Which one of the following statements regarding the ratios of line voltages and line currents on the AC side to that on the DC side of a rotary converter is true?
A. Both decrease with increase in number of phases
B. Both increase with increase in number of phases
C. The former increases while the latter decreases with increase in the number of phase
D. The former decreases while the latter increases with increase in the number of phases

248. Which one of the following statements regarding the ratios of line voltages and line currents on the AC side of that on the dc side of a polyphase converter is true?
A. The former is less than unity while the latter is greater than unity
B. The former is greater than unity while the latter is less than unity
C. Both are less than unity
D. Both are greater than unity

249. Which one of the following statements is true about the position of minimum copper loss in a phase of synchronous converter?
A. The position of minimum copper loss is the tapping point whatever may be the power factor
B. The position of minimum copper loss varies with the power factor and this position is displaced from the tapping point by the power factor angle
C. The position of minimum copper loss is the middle of the phase whatever may be the power factor
D. The position of minimum copper loss varies with the power factor and this position is displaced from the middle of the phase by the power factor angle

250. By connecting reactance at the slip ring leads of a rotary converter running at a leading power factor, the output DC voltage can be raised if the power factor is made
A. more leading by increasing the excitation
B. less leading or lagging by increasing the excitation
C. more leading by decreasing the excitation
D. less leading or lagging by decreasing the excitation

251. If H be the ratio of heating in the armature of a synchronous converter running as a converter to that while running as a DC, generator with the same DC output, then the ratio of outputs as a converter to that as a DC generator for the same armature heating is
A. $1/\sqrt{H}$ B. $1/H$
C. H D. $\sqrt{H}$

MULTIPLE CHOICE QUESTIONS FROM VARIOUS COMPETITIVE EXAMINATIONS

1. In the 'V' curve shown in the figure below for a synchronous motor, the parameter of y and x coordinates are, respectively **(IES 2001)**

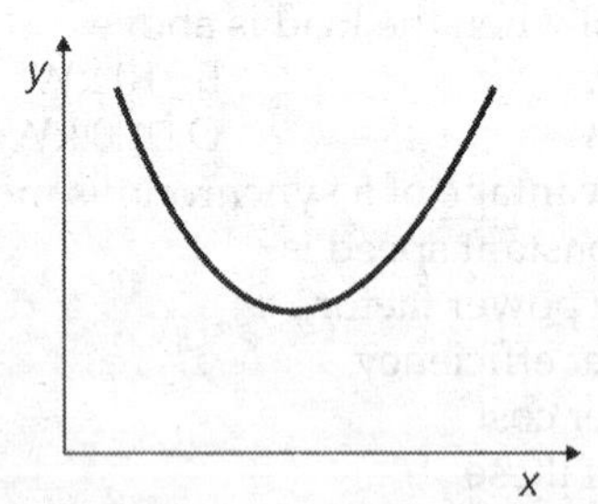

A. armature current and field current
B. power factor and field current
C. armature current and torque
D. torque and field current

2. In an alternator, if m is the number of slots per pole per phase and γ is the slot pitch angle, then the breadth or the distribution factor for the armature winding is **(IES 2001)**

A. $\dfrac{\sin\frac{\gamma}{2}}{m\sin\frac{m\gamma}{2}}$ B. $\dfrac{\sin\left(\frac{m\gamma}{2}\right)}{m\sin\left(\frac{\gamma}{2}\right)}$

C. $\dfrac{m\sin\frac{\gamma}{2}}{\sin\frac{m\gamma}{2}}$ D. $\dfrac{\sin\left(\frac{m\gamma}{2}\right)}{\sin\left(\frac{\gamma}{2}\right)}$

3. Synchronous condenser means **(IES 2001)**
A. a synchronous motor with capacitor connected across stator terminals to improve pf
B. a synchronous motor operating at full load with loading pf
C. an overexcited synchronous motor partially supplying mechanical load, and also improving pf of the system to which it is connected

Ans. 245. C 246. B 247. A 248. C 249. D 250. A 251. A 1. A 2. B

D. an overexcited synchronous motor operating at no-load with leading pf used in large power stations for improvement of pf

4. In a cylindrical rotor synchronous machine, the phasor addition of stator and rotor mmfs is possible because **(IES 2001)**
A. two mmfs are rotating in opposite direction
B. two mmfs are totating in same direction at different speed
C. two mmfs are stationary with respect to each other
D. one mmf is stationary and the other mmf is rotating

5. The 'synchronous-impedance method' of finding the voltage regulation by a cylindrical rotor alternator is generally considered as **(IES 2002)**
A. a pessimistic method because saturation is not considered
B. an optimistic method because saturation is not considered
C. a fairly accurate method even if power factor is not taken into account while determining synchronous impedance
D. a fairly accurate method when power factor is taken into account while determining synchronous impedance

6. The power factor of a synchronous motor **(IES 2002)**
A. improves with increase in excitation and may even become leading at high excitations
B. decreases with increase in excitation
C. is independent of its excitation
D. increases with loading for a given excitation

7. A 6-pole, 3-phase alternator running at 1000 rpm supplies to an 8-pole, 3-phase induction motor which has a rotor current of frequency 2 Hz. The speed at which the motor operates is **(IES 2002)**
A. 1000 rpm B. 960 rpm
C. 750 rpm D. 720 rpm

8. For maximum current during 'slip test' on a synchronous machine, the armature mmf aligns along **(IES 2002)**
A. d-axis B. q-axis
C. 45° to d-axis D. 45° to q-axis

9. A 3-phase 50 MVA 10 kV generator has a reactance of 0.2 ohm per phase. Hence the per-unit value of the reactance on a base of 100 MVA 25 kV will be **(IES 2002)**
A. 1.25 B. 0.625
C. 0.032 D. 0.32

10. The result of a 'slip test' for determining direct-axis (X_d) and quadrature-axis (X_q) reactances of a star-connected, salient-pole alternator are given below: **(IES 2002)**
Phase values: V_{max} = 108 V; V_{min} = 96 V, I_{max} = 12 A, I_{min} = 10 A. Hence the two reactances will be
A. X_d = 10.8 ohms and X_q = 8 ohms
B. X_d = 9 ohms and X_q = 9.6 ohms
C. X_d = 9.6 ohms and X_q = 9 ohms
D. X_d = 8 ohms and X_q = 10.8 ohms

11. If the field of a synchronous motor is underexcited, the power factor will be **(IES 2003)**
A. lagging B. leading
C. unity D. more than unity

12. Which of the following graphs represents the speed-torque characteristic of a synchronous motor? **(IES 2003)**

A.
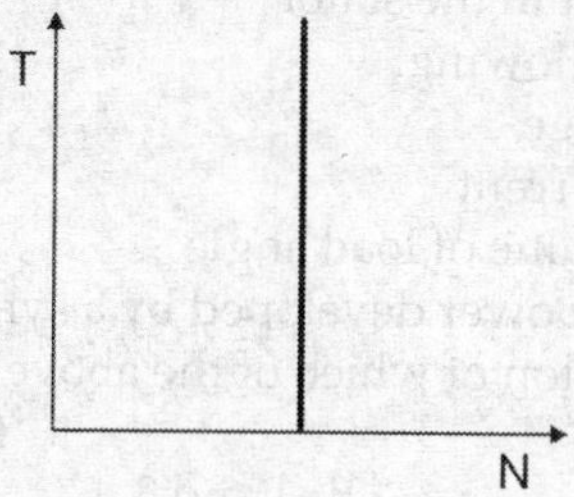

B.
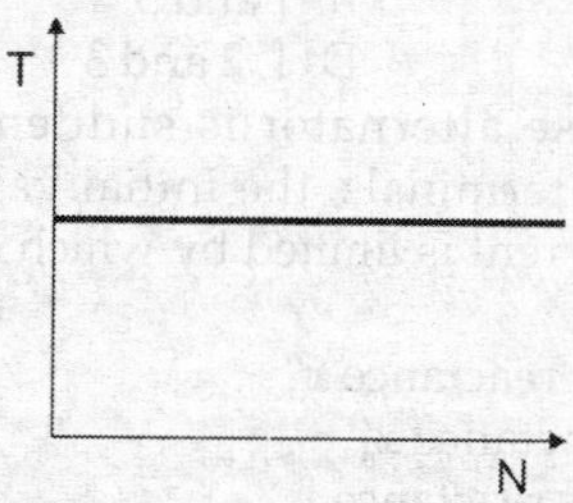

C.
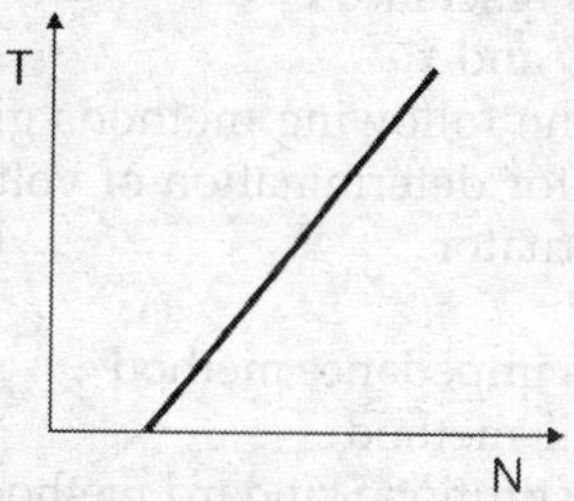

D.
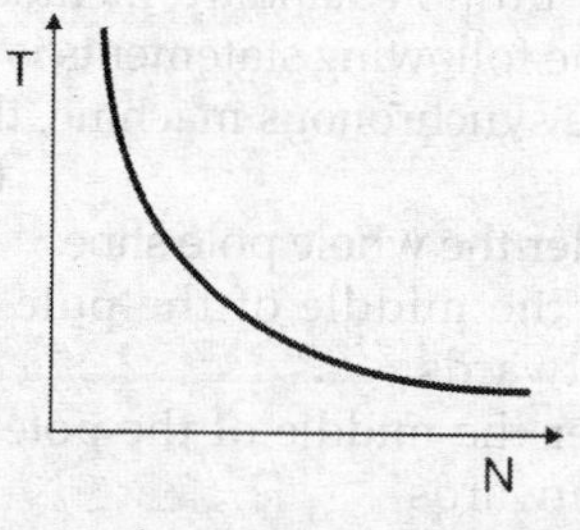

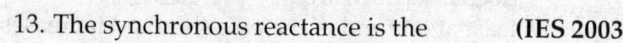

13. The synchronous reactance is the **(IES 2003)**
A. reactance due to armature reaction of the machine
B. reactance due to leakage flux
C. combined reactance due to leakage flux and armature reaction
D. reactance either due to armature reaction or leakage flux

Ans. 3. D 4. C 5. A 6. A 7. D 8. B 9. C 10. A 11. A 12. A 13. C

14. The leakage reactance of a three-phase alternator determined by performing (IES 2003)
A. open circuit and zero power factor tests
B. zero power factor and slip tests
C. open-circuit and short-circuit tests
D. short-circuit and slip tests

15. Which one of the following is the primary reason for placing field on rotor in an alternator? (IES 2004)
A. Small power in field circuit
B. Insulation of high voltage is made easy on stator than on rotor
C. Large power in stator
D. Large current in the stator

16. Consider the following:
1. Supply voltage
2. Excitation current
3. Maximum value of load angle
The maximum power developed by a synchronous motor is a function of which of the above? (IES 2004)
A. 1 and 2 B. 1 and 3
C. 2 and 3 D. 1, 2 and 3

17. When a 3-phase alternator is suddenly short-circuited at its terminals, the initial value of the short-circuit current is limited by which one of the following (IES 2004)
A. Subtransient reactance x_d''
B. Transient reactance x_s'
C. Synchronous reactance x_s
D. Sum of x_d'', x_s' and x_s

18. Which one of the following methods gives more accurate result for determination of voltage regulation of an alternator? (IES 2004)
A. mmf method
B. Synchronous impedance method
C. Potier triangle method
D. American Institution Standard method

19. Which one of the following statements is correct? In a salient pole synchronous machine, the air gap is (IES 2004)
A. uniform under the whole pole shoe
B. least under the middle of the pole shoe and increases outwards
C. largest under the middle of the pole shoe and decreases outwards
D. least at one end of the pole shoe and increases to the maximum value at the other end

20. Which one of the following is not a necessary condition to be satisfied for synchronising an incoming alternator to an already operating alternator? (IES 2004)
A. Same voltage magnitude
B. Same frequency
C. Same prime mover speed
D. Same phase sequence

21. A synchronous motor is operated from a bus voltage of 1.0 pu and is drawing 1.0 pu zero power factor leading current. Its synchronous reactance is 0.5 pu. What is the excitation emf of the motor? (IES 2005)
A. 2.0 B. 1.5
C. 1.0 D. 0.5

22. In the measurement of X_d, X_q (in ohms), following data are obtained by the slip test on a salient pole machine:
$I_{d\,max} = 10$ A $I_{d\,min} = 6.5$ A
$V_{d\,max} = 30$ V $V_{d\,min} = 25$ V
Which one of the following is correct? (IES 2005)
A. $X_d = 3, X_q = 3.86$ B. $X_d = 4.615, X_q = 2.5$
C. $X_d = 3, X_q = 2.5$ D. $X_d = 4.61, X_q = 3.86$

23. Which are the conditions to be satisfied for alternator to be synchronised with an incoming supply?
1. Equal voltage 2. Equal frequency
3. Same power rating 4. Same phase sequence
Select the correct answer using the code given below: (IES 2005)
A. 2, 3 and 4 B. 3 and 4
C. 1, 2 and 3 D. 1, 2 and 4

24. What is the value of the load angle when the power output of a salient pole synchronous generator is maximum? (IES 2006)
A. 0° B. 45°
C. 90° D. None of these

25. Which one of the following statements is not correct in respect of synchronous machines? (IES 2006)
A. In salient pole machines, the direct-axis synchronous reactance is greater than the quadrature axis synchronous reactance
B. The damper bars help the motor to self-start
C. Short circuit ratio is the ratio of field current required to produce the rated voltage on open circuit tothe rated armature current
D. The V-curve of a synchronous motor represents the variation in the arma-ture current with field excitation at a given output power

26. In which one of the following is reluctance power developed? (IES 2006)
A. Salient pole alternator
B. Nonsalient pole alternator
C. Squirrel cage induction motor
D. Transformer

27. A loss-less cylindrical rotor synchronous generator is floating on the system bus-bar after being properly synchronised and is running at no-load. Now, if the prime mover ia acting as a mechanical load on the machine, then (IES 2006)

Ans.	14. A	15. B	16. A	17. A	18. C	19. B	20. C	21. D	22. B	23. D	24. D	25. C	26. A

A. the torque developed by the machine will act in the same direction as the prime-mover torque
B. the torque developed by the machine will act in the opposite direction to that of the prime-mover torque
C. the machine rotor would fall-back and the induced emf would lag behind the terminal voltage
D. the machine rotor would overspeed and the induced emf would lead the terminal voltage

28. The resultant flux density in the air gap of a synchronous generator is the lowest during **(IES 2006)**
A. open circuit B. solid short circuit
C. full load D. half load

29. The equivalent circuit of a transformer has the leakage reactances X_1, X_2' and the magnetising reactance X_m. What is the relationship between their magnitudes? **(IES 2006)**
A. $X_1 >> X_2' >> X_m$ B. $X_1 << X_2' << X_m$
C. $X_1 \approx X_2' >> X_m$ D. $X_1 \approx X_2' << X_m$

30. Consider the following statements:
Assertion (A): Large synchronous machines are constructed with armature winding on stator
Reason (R): Stationary armature winding would have reduced armature reactance **(IES 2007)**
Of these statements
A. Both A and R are true and R is the correct explanation of A
B. Both A and R are true but R is not the correct explanation of A
C. A is true but R is false
D. A is false but R is true

31. When does a synchronous motor operate with leading power factor current? **(IES 2007)**
A. While it is under-excited
B. While it is critically excited
C. While it is over-excited
D. While it is heavily loaded

32. A 10 pole, 25 Hz alternator is directly coupled to and is driven by 60 Hz synchronous motor. What is the number of poles for the synchronous motor? **(IES 2007)**
A. 48 B. 12
C. 24 D. 16

33. What is the angle between the induced voltage and supply voltage of a synchro-nous motor under running condition? **(IES 2007)**
A. Between 90° and 180°
B. Greater than zero but ≤ 90°
C. Zero D. > 180°

34. What does the SCR (short circuit ratio) of a synchronous machine yield? **(IES 2007)**
A. $\frac{1}{X_s \text{ (unsaturated) pu}}$
B. $\frac{1}{X_s \text{ (unsaturated) in Ohm}}$
C. $\frac{1}{X_s \text{ (adjusted) pu}}$ D. $\frac{1}{X_s \text{ (adjusted) in Ohm}}$

35. How can the reactive power delivered by a synchronous generator be controlled? **(IES 2007)**
A. By changing the prime mover input
B. By changing the excitation
C. By changing the direction of rotation
D. By changing the prime mover speed

36. Which one of the following is the correct statement? **(IES 2007)**
A. The armature current upon symmetrical 3-phase short circuit of a synchronous machine (armature resistance is negligible) constitutes q-axis current only
B. The armature current upon symmetrical 3-phase short circuit of a synchro-nous machine (armature resistance is negligible) constitutes d-axis current only
C. The armature current upon symmetrical 3-phase short circuit of a synchronous machine (armature resistance is negligible) has both d-axis and q-axis current only
D. Short circuit current can be divided into d-axis and q-axis components

37. If the excitation of a 3-phase alternator operating on infinite bus bars is changed, which one of the following shall alter? **(IES 2008)**
A. Active power of machine
B. Reactive power of machine
C. Terminal voltage of machine
D. Frequency of machine

38. The stator of a 3-phase, 6-pole AC machine has 45 slots. The stator winding has 45 coils with a coil span of 6 slots. What type of winding will be selected for this machine? **(IES 2008)**
A. Double-layer, fractional slot, short-pitched winding
B. Single-layer, fractional slot, short-pitched winding
C. Single-layer, integral slot, full-pitch winding
D. Double-layer, fractional slot, full-pitch winding

39. A three phase 50 Hz, 11 kV and 37.5 MW at 0.9 pf capacity synchronous generator has its stator bore diameter of 765 cm and an axial core length of 80 cm. For which power plant this generator is suitable? **(IES 2009)**
A. Thermal coal fired plant
B. Hydroelectric power plant

Ans.	27. C	28. B	29. D	30. C	31. C	32. C	33. B	34. C	35. B	36. B	37. B	38. A	39. B

C. Nuclear power plant
D. Pumped storage power plant

40. A salient pole synchronous generator delivering power to an infinite bus through a reactive tie line reaches its steady state stability limit. What is the power angle of the generator relative to the infinite bus voltage reference? **(IES 2009)**
A. Greater than 90° B. Equal to 90°
C. Less than 90° D. Zero

41. Consider the following statements:
EMF induced per phase in an alternator depends on
1. Frequency
2. Number of turns per phase
3. Pitch factor
4. Distribution factor
Which of these statements is/are correct? **(IES 2009)**
A. 1 only B. 1, 2, 3 and 4
C. 2 and 3 only D. 3 and 4 only

42. Transient stability of a 3-phase power systems having more than one synchronous generator is not affected by which one of the following specifications? **(IES 2009)**
A. Initial operating conditions of generators
B. Quantum of large power disturbance
C. Fast fault clearance
D. Small changes in system frequency

43. What is the effect of the field failure of salient pole synchronous motor connected with infinite bus? **(IES 2009)**
A. Reduce motor torque and speed
B. Not charge motor torque and speed
C. Stop the motor
D. Reduce motor torque but motor will continue to run at synchronous speed

44. Armature reaction AT of a synchronous generator supplying power at rated voltage with zero power factor lagging is **(IES 2010)**
A. magnetising B. demagnetising
C. cross-magnetising
D. both magnetising and cross-magnetising

45. Power factor of an alternator driven by constant prime mover input can be changed by changing is **(IES 2010)**
A. speed B. load
C. field excitation D. phase sequence

46. A 3-phase, 11 kV, 5 MVA alternator has synchronous reactance of 10 Ω per phase. Its excitation is such that the generated emf is 14 kV. If the alternator is connected to infinite bus bar, the maximum output at the given excitation is **(IES 2010)**
A. 15400 kW B. 8000 kW
C. 6200 kW D. 5135 kW

47. A 440 V, 3-phase, 10 pole and 50 Hz synchronous motor delivering a torque of 50/π Nm, delivers a power of **(IES 2011)**
A. 50 W B. 500 W
C. 1000 W D. 2000 W

48. A 500 mW, 13.8 kV star connected synchronous generator at 0.8 pf will deliver a full load current of **(IES 2012)**
A. 12.1 kV B. 21.0 kA
C. 26.15 kA D. 46.5 kA

49. A synchronous generator has its effective internal impedance Z_s = 10 Ω and resistance r_a = 1.0 Ω. Its generated voltage E_f and terminal voltage V_t are both 500 V. The maximum power output is **(IES 2012)**
A. 5000 W B. 4550 W
C. 3000 W D. 22500 W

50. Consider the following statement with regard to synchronous machines:
1. When a synchronous motor is overexcited, its back emf is greater than the supply voltage
2. When a synchronous motor is overexcited, its power factor is leading
3. Synchronous motor is used as a capacitor where load is so large that construction of a static capacitor is impractical
Correct statements are **(IES 2012)**
A. 1 and 2 only B. 1 and 3 only
C. 2 and 3 only D. 1, 2 and 3

51. A large synchronous generator is feeding power into an infinite bus at slightly lagging power factor. If a total loss of field occurs and the system can supply sufficient reactive power without a large terminal voltage drop, the unit will **(IES 2012)**
A. continue to run as a synchronous generator and no tripping is necessary
B. get short-circuited and it should be tripped instantaneously
C. run as an induction generator and it should be tripped after a time delay
D. run as a synchronous motor and it should be tripped after a time delay

52. A synchronous generator connected to an infinite bus is supplying electrical power at unity power factor to the bus. If the field current is increased **(IES 2012)**
A. both the active and reactive power output of the machine will remain unchanged
B. the active power of the machine will remain unchanged but the machine will also supply lagging reactive power
C. the active power output of the machine will increase and the machine will draw leading reactive power from the bus

Ans. 40. B 41. B 42. D 43. D 44. B 45. C 46. A 47. C 48. C 49. ? 50. D 51. C 52. B

D. the active power output of the machine will remain unchanged but the machine will also supply leading reactive power

53. A 3-phase synchronous motor with constant excitation is driving a certain load drawing electric power from infinite bus at leading power factor. If the shaft load decreases **(IES 2013)**
A. the power angle decreases while power factor increases
B. the power angle increases while power factor decreases
C. both power angle and power factor increase
D. both power angle and power factor decrease

54. The maximum power delivered by 1500 kW, 3-phase, star-connected, 4 kV, 48 pole, 50 Hz synchronous motor, with synchronous reactance of 4 Ω phase and unity power factor **(IES 2013)**
A. 4271.2 kW B. 3505 kW
C. 1206.1 kW D. 2078 kW

55. Consider the following statements regarding synchronous machines:
1. When a synchronous motor is over-excited, its back emf is greater than the supply voltage
2. When a synchronous motor is overexcited, its power factor is leading
3. Synchronous motor is used as capacitor where load is so large that construction of a static capacitor is impractical

Which of these statements are correct? **(IES 2013)**
A. 1 and 2 only B. 1 and 3 only
C. 2 and 3 only D. 1, 2 and 3

56. In a synchronous machine, the rotor speed becomes more than the synchronous speed during hunting, the damping bars develop **(IES 2013)**
A. synchronous motor torque
B. DC motor torque
C. induction motor torque
D. induction generator torque

57. Damper bars in case of salient pole rotors of hydro-alternators are usually inserted in pole faces to **(IES 2013)**
A. strengthen the excitation current of the poles
B. damp out the rotor oscillations during transient state owing to sudden change in load conditions
C. help improve the power factor of load
D. reduce the no-load current when load is thrown-off

58. In synchronous motor, 'V' curves present the variation of **(IES 2013)**
A. armature current with excitation (field)
B. armature current with maximum power developed
C. field excitation with stalling torque
D. field excitation with minimum power developed

59. The synchronous reactance of a 500 V, 50 kVA alternator having an effective resistance of 0.2 Ω. If an excitation current of 10 A produces 200 A, armature current on short circuit and an emf of 450 volts on open circuit is **(IES 2013)**
A. 2.6 Ω B. 5.2 Ω
C. 2.24 Ω D. 4.5 Ω

60. The main advantage of distributing the winding in slots is to **(IES 2013)**
A. reduce the size of the machine
B. add mechanical strength to the winding
C. reduce the amount of copper required
D. reduce the harmonics in the generated emf

61. When the rotor speed, in a synchronous machine, becomes more than the synchronous speed during hunting, the damper bars develop **(IES 2013)**
A. induction motor torque
B. induction generator torque
C. synchronous motor torque
D. DC motor torque

Ans. 53. D 54. A 55. D 56. D 57. B 58. A 59. C 60. D 61. B

CHAPTER 11

Power System—Generation, Transmission Distribution and Load Flow Studies

Electrical Power

Introduction: Energy is the basic necessity for the economic development of a country. Many functions necessary to present-day living comes to a halt when the supply of electrical energy stops. The modern society is so much dependent upon the use of electrical energy that it has become a part and parcel of our life. The survival of industrial undertakings and our social structures depends upon low cost and uninterrupted supply of electrical energy. In fact, the advancement of a country is measured in terms of per capita consumption of electrical energy.

The most common method of generating energy is by means of electric machines generally called *generator or alternator*. These machines convert mechanical energy into electrical energy when coupled to prime movers.

Electrical energy is produced from energy available in various forms in nature. These sources of energy are: (i) solar (ii) wind (iii) water (iv) fuels (v) nuclear energy.

Out of these sources, the energy due to sun and wind has not been utilised on a larger scale due to a number of limitations. At present, the other three sources, i.e. water, fuels and nuclear energy, are primarily used for generation of electrical energy.

Generation, Transmission and Distribution

Generating stations: The generating stations are classified as:

i. **Hydroelectric station:** Water is a great potential source of energy. The generating stations which obtain energy from water are called *hydroelectric* stations. Water is stored at a suitable place, it possesses potential energy because of the head created. This water energy can be converted into mechanical energy with the help of water turbines. The water turbine drives the alternator which converts mechanical energy into electrical energy. This method of generation of electrical energy has become very popular because it has low production and maintenance cost.

ii. **Steam power station (fuels):** Main sources of energy are fossil fuels, i.e. solid fuel as coal, liquid fuel as oil and gas fuel as natural gas. The heat energy of these fuels is converted into mechanical energy by suitable prime movers such as steam engines, steam turbines and internal combustion engines, etc. The prime movers drive the alternators which convert mechanical energy into electrical energy.

iii. **Nuclear power station.** It was discovered that large amount of heat energy is liberated by the fission of uranium and other fissionable materials. It is estimated that heat produced by 1 kgm of nuclear fuel is equal to that produced by 4500 tons of coal. The heat produced due to nuclear fission can be utilised to raise steam with suitable arrangements. This steam can run the steam turbine which in turn can drive the alternator to produce electrical energy.

Classification of hydroelectric plants: Based on the quantity of water available, plants may be divided into (a) run of river plant without pondage (b) run of river plant with pondage (c) pumped storage plant.

Characteristics of Turbines

Specific speed is defined as the speed of a geometrically similar turbine working under unit head and delivering unit power. The specific speed in British unit N_s is given by the following formula:

$$N_s = \frac{N\sqrt{\text{hp}}}{H^{5/4}} \text{ rpm}$$

where N = rotational speed in rpm;

hp = horse power of the wheel and;

H = head of water in metre.

Further if power is expressed in kW, the above expression becomes

$$N_s = \frac{1.165 N \sqrt{kW}}{H^{5/4}} \text{ rpm}$$

Calculation of hp (Metric) and kW Power

Let Q = Discharge in cu·m/sec

H = Head of water; η = overall efficiency of hydrostation

Since 1 m^3 of water weighs = 1000 kg·m

Work done/sec = 1000 QH kg · m/sec

1 hp (metric) = 75 kg · m/sec.

$$\text{hp developed} = \frac{1000 Q \cdot H \cdot \eta}{75}$$

$$\text{Power (kW)} = \frac{1000 Q \cdot H \cdot \eta \times 735.5}{75 \times 1000}$$

$$P = 9.8\, QH\eta$$

Units of Energy and Relationships

i. Mechanical energy (Joules)
= Force (newton) × distance (m)

ii. Electrical energy (Joules)
= Voltage (V) × current (A) × time (sec)

iii. 1 kilowatt hour (kWh)
= 1 kW × 1 hour
1 kWh = 1000 (W) × 3600 (sec) = 36 × 10^5 watt-sec or Joules
1 kilo calorie = 1000 calories; 1 calorie = 4.18 Joules

Terms Commonly Used in System Operation

Firm power: It is the power intended to be always available.

Cold reserve: It is that reserve generating capacity which is available for service but is not in operation.

Hot reserve: It is that reserve generating capacity which is in operation but is not in service.

Spinning reserve: It is that generating capacity which is connected to bus and is ready to take load.

Connected load: It is the sum of continuous ratings of all the equipment connected to the supply system.

Maximum demand: It is the greatest demand of load on power station during a given period.

Demand factor: It is the ratio of maximum demand on the power station to its connected load, i.e.

$$\text{Demand factor} = \frac{\text{Maximum demand}}{\text{Connected load}}$$

Average load: The average of loads occurring on the power station in a given period is known as average load or average demand.

Daily average load

$$= \frac{\text{No. of units (kWh) generated in a day}}{\text{24 hours}}$$

Monthly average load

$$= \frac{\text{No. of units generated in a month}}{\text{No. of hours in a month}}$$

Yearly average load

$$= \frac{\text{No. of units generated in a year}}{\text{8760 hours}}$$

Load factor: The ratio of average load to the maximum demand during a given period is known as *load factor*.

$$\text{Load factor} = \frac{\text{Average load}}{\text{Maximum demand}}$$

Diversity factor: The ratio of the sum of individual maximum demands to the maximum demand on the power station is known as *diversity factor*.

Diversity factor

$$= \frac{\text{Sum of individual maximum demands}}{\text{Maximum demand on power station}}$$

Capacity factor

$$= \frac{\text{Actual energy produced}}{\text{Maximum energy that could have been produced}}$$

$$= \frac{\text{Average demand}}{\text{Plant capacity}}$$

Reserve capacity

= Plant capacity – maximum demand

Utilization factor (plant use factor): It is the ratio of kWh generated to the product of plant capacity and the number of hours for which the plant was in operation (Table 11.1).

$$\text{Plant use factor} = \frac{\text{Station output in kWh}}{\text{Plant capacity} \times \text{Hours of use}}$$

Table 11.1: Comparison of various systems of transmission

System	*Same maximum voltage to earth*	*Same maximum voltage between conductors*
a. DC System		
i. Two-wire	1	1
ii. Two-wire mid-point earthed	0.25	1
iii. Three-wire	0.3125	1.25
b. Single-Phase System		
i. Two-wire	$2/\cos^2\phi$	$2/\cos^2\phi$
ii. Two-wire with mid-point	$0.5/\cos^2\phi$	$2/\cos^2\phi$
iii. Three-wire	$0.625/\cos^2\phi$	$2.5/\cos^2\phi$
c. Two-Phase System		
i. Two-Phase, four-wire	$0.5/\cos^2\phi$	$2/\cos^2\phi$
ii. Two-Phase, three-wire	$1.457/\cos^2\phi$	$2.194/\cos^2\phi$
d. Three-Phase System		
i. Three-phase, three-wire	$0.5/\cos^2\phi$	$1.5/\cos^2\phi$
ii. Three-phase, four-wire	$0.583/\cos^2\phi$	$1.75/\cos^2\phi$

Kelvin's Law: Economic Choice of Conductor Size

The most economical area of conductor is that for which the total annual cost of transmission line is minimum. This is known as *Kelvin's law*. The total annual cost of transmission line can be divided broadly into two parts, i.e. annual charge on capital outlay and annual cost of energy wasted in the conductor.

Total annual cost = $(P_1 + P_2\, a) + P_3 \neq a$

where P_1, P_2 and P_3 are constants and 'a' is the area of cross-section of the conductor.

The Empirical Formula for Economic Voltage for Transmission Line

$$V = 5.5\sqrt{0.62l + \frac{3P}{150}}$$

where V is line voltage in kV, l is distance of transmission line in km, P is the maximum power in kW/phase to be delivered to a single circuit.

Parameters

Resistance: The DC resistance R of a conductor of length l and cross-sectional area A is

$$R = \rho\frac{l}{a}\ \text{ohm}$$

where ρ is the *resistivity* of the material of the conductor in ohm-metres.

The temperature dependence of resistance is quantified by the relation

$$R_2 = R_1[1 + \alpha(T_2 - T_1)]$$

where R_1 and R_2 are the resistance at temperature T_1 and T_2, respectively, and α is called the *temperature coefficient of resistance*. The resistivities and temperature coefficients of several metals are given in Table 11.2.

Long transmission lines may involve shunt resistance (or conductance) in addition to series resistances.

Table 11.2: Resistance and temperature coefficients of resistance

Material	*Resistivity ρ at 20°C μΩ·cm*	*Temperature coefficient α at 20°C, °C⁻¹*
Aluminium	2.83	0.0039
Brass	6.4–8.4	0.0020
Copper:		
Hard-drawn	1.77	0.00382
Annealed	1.72	0.00393
Iron	10.0	0.0050
Silver	1.59	0.0038
Steel	12–88	0.001–0.005

Inductance of Two-wire, Single-phase Line

The inductance per conductor of a two-wire, single-phase transmission line is given by

$$L_1 = \frac{\mu_0}{8\pi}\left(1 + 4\ln\frac{D}{r}\right)\text{henry/metre}$$

where $\mu_0 = 4\pi \times 10^{-7}$ H/m (the permeability of free space). D is the distance between the centres of the conductors, and r is the radius of the conductor.

The loop, inductance is given by

$$L = 2L_1 = \frac{\mu_0}{4\pi}\left(1 + 4\ln\frac{D}{r}\right)$$

$$= \left(1 + 4\ln\frac{D}{r}\right) \times 10^{-7}\ \text{H/m}$$

Since $\ln e^{1/4} = 1/4$, the above equation may also be written as

$$L = 4 \times 10^{-7} \ln\frac{D}{r'}\ \text{H/m}$$

where $r' = re^{-1/4}$ is known as the geometric mean radius (GMR) of the conductor.

Inductance of Three-Wire, Three-Phase Line

The per-phase (or line-to-neutral) inductance of a three-phase transmission line with equilaterally space conductors is

$$L = \frac{\mu_0}{8\pi}\left(1 + 4\ln\frac{D}{r}\right) = 2\left(\frac{1}{4} + \ln\frac{D}{r}\right) \times 10^{-7}\ \text{H/m}$$

where r is the conductor radius and D is the spacing between conductors.

Inductance of Composite Conductors

For a single-phase line consisting of two composite conductors, X and Y, the inductance L_x of conductor X is

$$L_x = 2 \times 10^{-7} \ln \frac{\sqrt[mn]{(D_{aa'}D_{ab'}D_{ac} \ldots D_{am})(D_{ba'}D_{bb'}D_{bc'} \ldots D_{bm}) \ldots (D_{ma'}D_{nb'}D_{nc'} \ldots D_{nm})}}{\sqrt[n^2]{(D_{aa}D_{ab}D_{ac} \ldots D_{an})(D_{ba}D_{bb}D_{bc} \ldots D_{bn}) \ldots (D_{na}D_{nb}D_{nc} \ldots D_{nm})}}\ (\text{H/m})$$

where $D_{kk} = r_k' = r_k\, e^{-1/4}$ is the geometric mean radius (GMR) of the kth conductor.

The inductance L_Y of conductor Y is similar as L_X. The total line inductance becomes

$$L_X = L_X + L_Y$$

Capacitance

The shunt capacitance per unit length of a single-phase, two-wire transmission line is given by

$$C = \frac{\pi\varepsilon_0}{\ln(D/r)}\ \text{F/m}$$

where ε_0 is the permittivity of free space. For a three-phase line with equilaterally space conductors, the per-phase (or line-to-neutral) capacitance is

$$C = \frac{2\pi\varepsilon_0}{\ln(D/r)}\ \text{F/m}$$

The per-phase capacitance for the double-circuit transmission line is given by

$$C = \frac{4\pi\varepsilon_0}{\ln[\sqrt[3]{2}(D/r)(G/F)^{2/3}}\ \text{F/m}$$

Using the concept of the GMD, we may write the capacitance to neutral of a nonsymmetrical three-phase double-circuit line as

$$C_n = \frac{2\pi\varepsilon_0}{\ln(\text{GMD}/\text{GMR})} = \frac{2\pi\varepsilon_0}{\ln(D_m/D_s)}\ \text{F/m}$$

$$= \frac{10^{-9}}{18\ln(D_m/D_s)}\ \text{F/m}$$

Short-Transmission Line

The short transmission line is represented by the lumped parameters R and L, as shown in Fig. 11.1(a). Notice that R is the resistance per phase and L is the inductance per phase of the *entire line*. The line is shown to have two ends: the sending end designated by the subscript S at the generator, and the receiving end designated R at the load. Quantities of significance here are the voltage regulation and efficiency of transmission for lines of all lengths are defined as follows:

$$\text{Voltage regulation} = \frac{|V_S - V_R|}{|V_R|}\times 100$$

Efficiency of transmission

$$\eta = \frac{\text{power at receiving end}}{\text{power at sending end}} = \frac{P_R}{P_s}$$

where V_R is the receiving-end voltage.

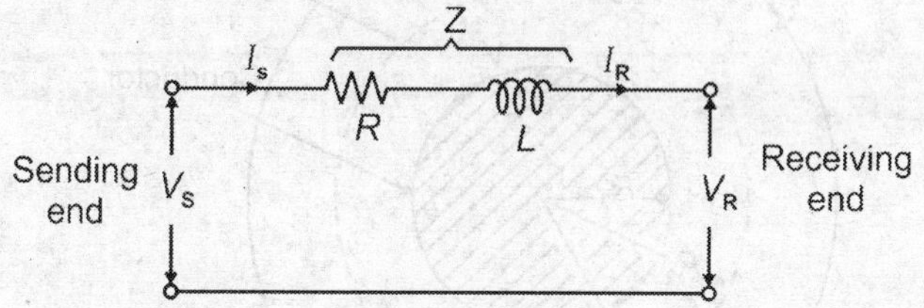

Fig. 11.1(a)

Long Transmission Line

The parameters of a long line are considered to be distributed over the entire length of line L

$$V_S = V_R\cos\gamma L + I_R Z_C \sin\gamma L$$

$$I_S = \frac{V_R \sin\gamma L}{Z_C} + I_R\cos h\gamma L$$

$$\cos\gamma L = 1 + \frac{(\gamma L)^2}{2!} + \frac{(\gamma L)^4}{4!} + \ldots + \cong 1 + \frac{1}{2}yz$$

$$\sin\gamma L = 1 + \frac{(\gamma L)^3}{3!} + \frac{(\gamma L)^5}{5!} + \ldots + \cong \sqrt{yz}\left(1 + \frac{1}{6}yz\right)$$

where γ is known as the propagation constant and $\gamma = \sqrt{yz}$, y is the shunt admittance per unit length of line, z is the series impedance per unit length (Fig. 11.1b).

Transmission Line as a Two-port Network

$$V_S = AV_R + BI_R$$
$$I_S = CV_R + DI_R$$

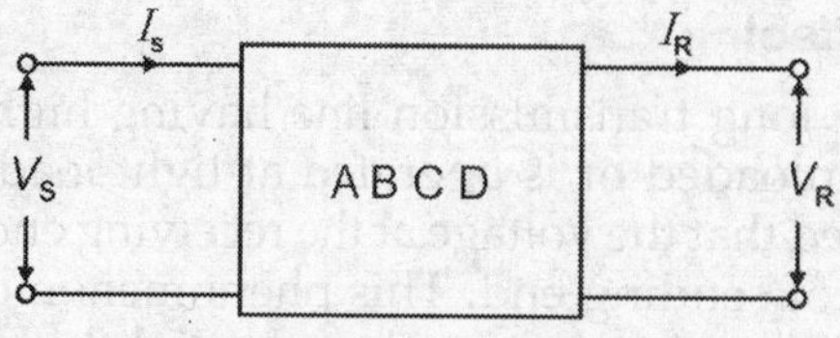

Fig. 11.1(b)

where the constants A, B, C and D are called the *generalized circuit constants* or *ABCD constants* and are complex, in general. By reciprocity they are related to each other as follows:

$$AD - BC = 1$$

ABCD Constants for Transmission Lines

Line length	*Equivalent circuit*	*A*	*B*	*C*	*D*
Short	Series impedance	1	Z	0	1
Medium	Nominal-π	$1+\frac{1}{2}YZ$	Z	$Y\left(1+\frac{1}{4}YZ\right)$	$1+\frac{1}{2}YZ$
	Nominal-T	$1+\frac{1}{2}YZ$	$Z\left(1+\frac{1}{4}YZ\right)$	Y	$1+\frac{1}{2}YZ$
Long	Distributed parameter	$\cos\gamma L$	$Z_c \sin h\gamma L$	$(\sin h\gamma L)/Z_c$	$\cos h\gamma L$

Power flow on transmission lines: The complex power $V_R I_R^*$ at the receiving end is thus given by

$$P_R + jQ_R = \frac{|V_R||V_S|}{|B|}\angle\beta-\delta - \frac{|A||V_R|^2}{|B|}\angle\beta-\alpha$$

$$\text{So that}\quad P_R = \frac{|V_R||V_S|}{|B|}\cos(\beta-\delta) - \frac{|A||V_R|^2}{|B|}\cos(\beta-\alpha)$$

$$Q_R = \frac{|V_R||V_S|}{|B|}\sin(\beta-\delta) - \frac{|A||V_R|^2}{|B|}\sin(\beta-\alpha)$$

Traveling Waves on Transmission Lines

The ratio $\sqrt{L/C}$ has the dimension of ohm and is called the *characteristic impedance* Z_C *of the line.*

$$Z_C = \sqrt{L/C} = R_C \text{ for a lossless line}$$

If only forward-traveling waves exist at the load, then

$$V^+\left(t - \frac{L}{\mu}\right) = R_C I^+\left(t - \frac{L}{u}\right)$$

and if only backward-traveling waves exist at the load, then

$$V^-\left(t - \frac{L}{\mu}\right) = R_C I^-\left(t - \frac{L}{u}\right)$$

Reflection Coefficients

The current reflection coefficients is the negative of the voltage reflection coefficients.

Source voltage reflection coefficient

$$\Gamma_S = \frac{R_s - R_C}{R_s + R_c}$$

Ferranti Effect

When the long transmission line having high capacitance is unloaded or is operated at light loads, it will be observed that the voltage at the receiving end is more than that of sending end. This phenomenon of rise of voltage at the receiving end of the lightly loaded or unloaded line is called as *Ferranti effect*. The rise in voltage is due to the fact that a charging current flowing in the line leads from the receiving end voltage by 90° (assuming that the capacitance of the line is concentrated at the receiving end).

Grading of Cables

The cable in Fig. 11.2(a) is filled with a single layer of a single dielectric, but many cables contain several layers of dielectrics. The dielectric materials in such a cable are chosen and distributed so as to minimize the difference between the maximum and minimum electric field strengths in the cable. This process is known as *grading*. Two types are commonly used in grading: capacitance grading and intersheath grading.

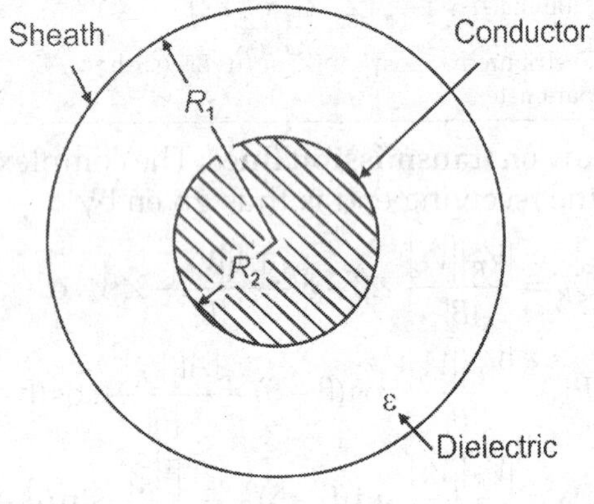

Fig. 11.2(a)

Capacitance grading: In capacitance grading, two or more layers of different dielectric are used to insulate a cable, e.g. two such layers are shown in Fig. 11.2(b). The permittivities of these layers are so chosen that the maximum field strength is the same in both regions. The corresponding variation of the electric field E with radius r is shown in Fig. 11.2(b). For equal maximum field strengths

$$\varepsilon_1 R_1 = \varepsilon_2 R_2$$

Intersheath grading: In intersheath grading, the cable contains several layers of a single dielectric material, separated by coaxial metallic sheaths that are inserted into the dielectric and maintained at predetermined voltages. A cable with one such intersheath is shown in Fig. 11.3.

Cable Inductance

The inductance per unit length of a single conductor is given by

$$L = \frac{\mu_0}{2\pi} \ln \frac{R_1}{R_2} \text{(H/m)}$$

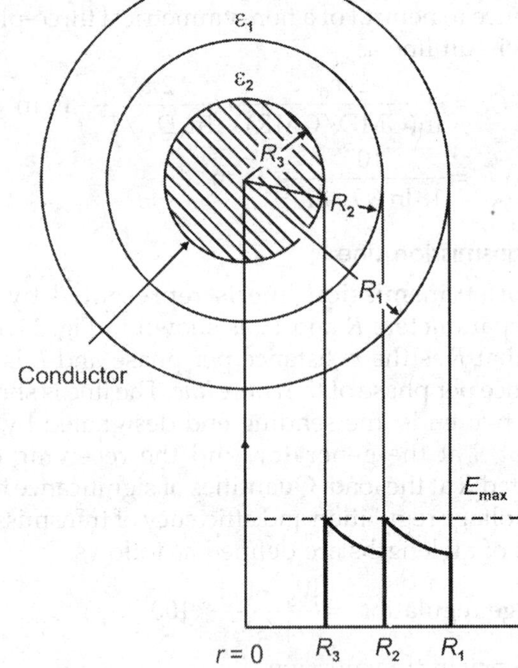

Fig. 11.2(b)

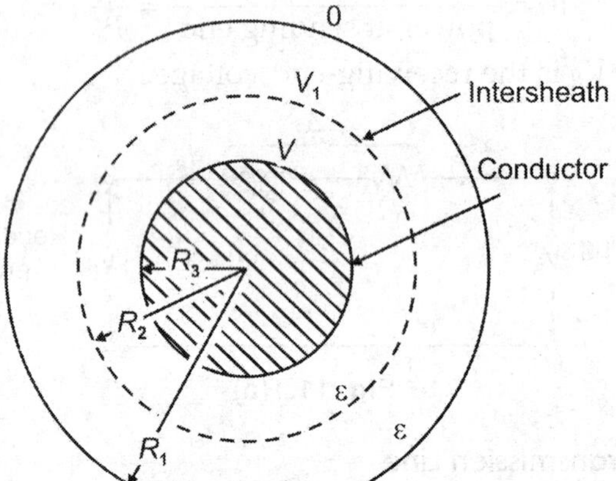

Fig. 11.3

Dielectric loss and heating. In an underground cable, the capacitance of the cable may be considered to be lossy, having a resistance R_i. The loss in R_i is $P = V^2/R_i$

In terms of loss angle δ

$$\tan \delta = \frac{I_{Ri}}{I_C} = \frac{V/R}{\omega CV}$$

$P = \omega CV^2 t$ and $\cong \omega CV^2 \delta$; if δ is small.

Cable Capacitance

The capacitance per unit length of a single conductor cable is given by

$$C = \frac{2\pi\varepsilon_0\varepsilon_r}{\ln(R_1/R_2)} \text{farad/m}$$

$$= \frac{2\pi \times 8.854 \times 10^{-12} \times \varepsilon_r}{2.303 \log_{10}(R_1/R_2)} \text{F/m}$$

$$= \frac{\varepsilon_r \times 10^{-9}}{41.4 \log_{10}(R_1/R_2)} \text{F/m}$$

In a three core cable, the empirical formula for capacitance of each core to neutral C_n is given by

$$\frac{0.03\varepsilon_r}{\log_e\left[1+(T+t/d)(3.84-1.7)(t/T)+0.52(t^2/T^2)\right]} \mu\text{F/km}.$$

where d = diameter of conductor, t = thickness of belt insulation and T = thickness of conductor insulation.

Construction of cables: Figure 11.4 shows the general construction of a 3-conductor cable. The various parts are:

a. *Cores or conductors*: The conductors are made of tinned, copper or aluminum and are usually standed in order to provide flexibility to the cable.

b. *Insulation*: Each conductor is provided with a suitable thickness of insulation, depending upon the voltage to be withstood by the cable. The commonly used materials for insulation are impregnated paper, varnished cambric or rubber mineral compound.

c. *Metallic sheath*: In order to protect the cable from moisture, gases or other damaging liquids (acids or alkalies) in the soil and atmosphere, a metallic sheath of lead or aluminium is provided over the insulation.

d. *Bedding*: It consists of a fibrous material like jute or hessian tape. The purpose of bedding is to protect the metallic sheath against corrosion and from mechanical injury due to armouring.

e. *Armouring*: It consists of one or two layers of galvanised steel wire or steel tape. Its purpose is to protect the cable from mechanical injury while laying it and during the course of handling.

f. *Serving*: In order to protect armouring from atmospheric conditions, a layer of fibrous material (like jute) similar to bedding is provided over the armouring. This is known as serving.

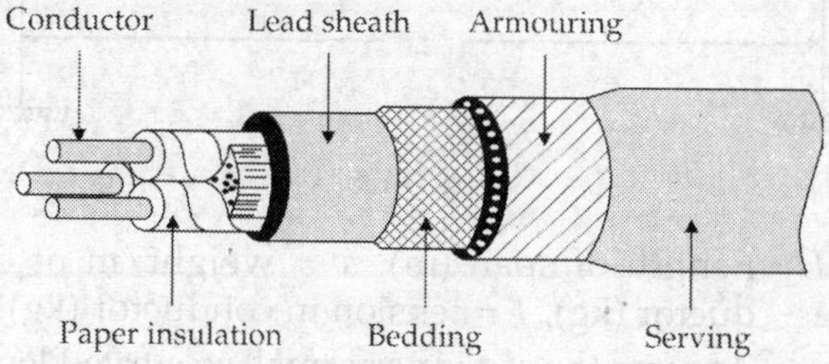

Fig. 11.4

Corona

When an alternating potential difference is applied across two conductors whose spacing is large as compared to their diameters, when the applied voltage exceeds a certain value, and conductors are surrounded by a faint violet glow by hissing sound, with the production of ozone, is called *corona*. The higher the voltage is raised, the larger and higher the luminous envelope becomes, the greater are the sound, the power loss and the radio noise. If the applied voltage is increased to breakdown value, a flash over will occur between the conductors due to the breakdown of air insulation. Thus the phenomenon of violet glow, hissing noise and production of ozone gas in an overhead transmission line is known as *corona*.

If the conductors are polished and smooth, the corona glow will be uniform throughout the length of the conductors otherwise the rough points will appear brighter. With dc voltage, there is difference in the appearance of the two wires. The positive wire has uniform glow around it, while the negative conductor has spotty glow.

Factors affecting corona: The Corona is affected by the following factors:

a. *Atmosphere*: As corona is formed due to ionisation of air surrounding the conductors, therefore, it is affected by the physical state of atmosphere. In the stormy weather, the number of ions is more than normal and as such corona occurs at much less voltage as compared with fair weather.

b. *Conductor*: The corona effect depends upon the shape and conditions of the conductors. The rough and irregular surface will give rise to more corona because unevenness of the surface decreases the value of breakdown voltage. Thus a stranded conductor has irregular surface and hence gives rise to more corona than a solid conductor.

c. *Spacing between conductors*: If the spacing between the conductors is made very large as compared to their diameters, there may not be any corona effect. It is because larger distance between conductors reduces the electrostatic stresses at the conductor surface, thus avoiding corona formation.

d. *Line voltage*: The line voltage greatly affects corona. If it is low, there is no change in the condition of air surrounding the conductors and hence no corona is formed. However, if the line voltage has such a value that electrostatic stresses developed at the conductor surface make the air around the conductor conducting, then corona is formed.

Advantages of corona

i. Due to corona formation, the air surrounding the conductor becomes conducting and hence virtual diameter of the conductor is increased. The increased diameter reduces the electrostatic stresses between the conductors.

ii. Corona reduces the effects of transients produced by surges.

Disadvantages of corona

i. Corona is accompanie by a loss of energy. This affects the transmission efficiency of the line.

ii. Ozone is produced by corona and may cause corrosion of the conductor due to chemical action.
iii. The current drawn by the line due to corona is non-sinusoidal and hence nonsinusoidal voltage drop occurs in the line. This may cause inductive interferes with neighboring communication lines.

Important Corona Terms

Critical disruptive voltage: It is the minimum phase to neutral voltage at which corona occurs. Consider two conductors of radii r cm and spaced d cm apart. If V is the phase to neutral potential, then potential gradient at the conductor surface is given by

$$g_{max} = \frac{V}{r \log_e (d/r)} \text{ volt/cm}$$

In order that corona is formed, the value of g_{max} must be made equal to the breakdown strength of air. The breakdown strength of air at 76 cm pressure and temperature of 25°C is 30 kV/cm (*peak*) or 21.2 kV/cm (rms) and is denoted by g_0. If V_C is the phase-neutral potential required under these conditions, then

$$g_0 = \frac{V_c}{r \log_e d/r}$$

where, g_0 = breakdown strength of air at 76 cm of mercury and 25 °C
= 30 kV/cm (peak) or 21.2 kV/cm (RMS)

$\therefore$ Critical disruptive voltage $V_C = g_0 \, r \log_e \frac{d}{r}$

The above expression for disruptive voltage is under standard conditions, i.e. at 76 cm of Hg and 25 °C. However, if these conditions vary, the air density also changes, thus altering the value of g_0. The value of g_0 is directly proportional to air density. Thus, the breakdown strength of air at a barometric pressure of b cm of mercury and temperature of t °C becomes δg_0, where

$$\delta = \text{air density factor} = \frac{3.92b}{273 + t}$$

Under standard conditions, the value of $\delta = 1$

$\therefore$ Critical disruptive voltage, $V_C = g_0 \delta r \log_e \frac{d}{r}$

Correction must also made for the surface condition of conductor. This is accounted for by multiplying the above expression by irregularity factor m_0.

$\therefore$ Critical disruptive voltage,

$$V_c = m_0 g_0 \delta r \log_e \frac{d}{r} \text{ kV/phase}$$

where, m_0 = 1 for polished conductors
= 0.98 to 0.92 for dirty conductors
= 0.87 to 0.8 for stranded conductors.

Visual Critical Voltage

It is the minimum phase-neutral voltage at which corona glow appears all along the line conductors.

It has been seen that in case of parallel conductors, the corona glow does not begin at the disruptive voltage V_C but at a higher voltage V_V, called *visual critical voltage*. The phase-neutral effective value of visual critical voltage is given by the following empirical formula:

$$V_V = m_v g_0 \delta r \left(1 + \frac{0.3}{\sqrt{\delta r}}\right) \log_e \frac{d}{r} \text{ kV/phase}$$

where m_v is another irregularity factor having a value of 1.0 for polished conductors and 0.72 to 0.82 for rough conductors.

Power Loss Due to Corona

Formation of corona is always accompanied by energy loss which is dissipated in the form of light, heat, sound and chemical action. When disruptive voltage is exceeded, the power loss due to corona is given by:

$$P = 244\left(\frac{f + 25}{\delta}\right)\sqrt{\frac{r}{d}}\,(V - V_c)^2 \times 10^{-5} \text{ kW/km/phase}$$

Caclulation of Sag

Consider a conductor suspended between two equilevel supports A and B in still air as shown in Fig. 11.5, where 0 is the lowest point on the conductor which is the middle of span.

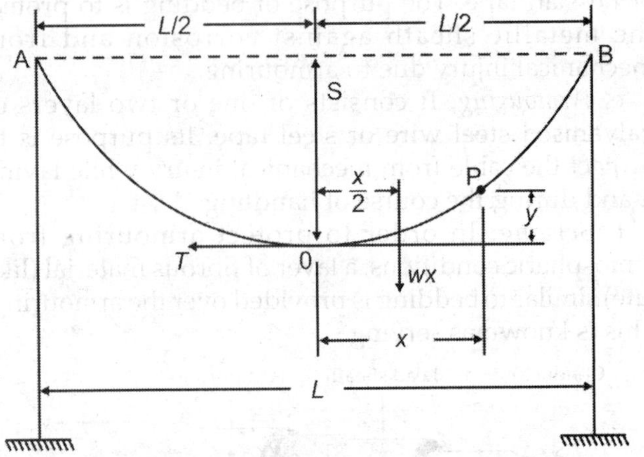

Fig. 11.5

Let L = Length of span (m), w = weight/m of conductor (kg), T = tension in conductor (kg).
Assume the sag is very small so curved length of conductor A, B is equal to length of span. Consider a point P on conductor at a distance x metre from the mid point O. The following two external forces are acting on the portion OP of the conductor.

(i) the weight of conductor of x metres, i.e. $w{\cdot}x$ kg is acting at $x/2$ metres from point P giving anticlockwise moment.

(ii) the tension T acting at point O at y metres from point P giving clockwise moment.

Effect of Wind and Ice Loading

In still air, sag develops due to the weight of conductor only. In actual practice, a conductor may have ice coating and simultaneously subjects to wind pressure. The ice load (w_i) acts vertically downward, i.e. in the same direction as the weight of conductor, whereas, the wind load (w_w) acts horizontally, i.e. perpendicular to the weight of conductor.

Resultant weight of conductor per unit length

$$w_r = \sqrt{(w + w_i) + (w_w)^2}$$

where w = weight of conductor per unit length

w_i = weight of ice per unit length = density of ice × volume of ice per unit length

$$= \text{density of ice} \times \frac{\pi}{4}(D^2 - d^2) \times 1$$

$$= \text{density of ice} \times [(d + 2t)^2 - d^2] \times 1$$

w_w = wind load per unit length

Sag under this condition, $S = \dfrac{w_r L^2}{8T}$

Insulators

The conductors of an overhead transmission or distribution lines are secured to the supporting structures by means of insulating fixtures in order that there is no current leakage to the earth through the supports. Thus the insulators play an important part in successful operation of the lines. The chief requirements for the insulators are:

i. They must be mechanically very strong.
ii. Their dielectric strength must be very high.
iii. They must provide very high insulation resistance to the leakage currents.
iv. They must be free from the internal impurities or flaws.
v. They should not be porous.
vi. They must be impervious to the entrance of gases or liquids into the materials.
vii. They should not be effected with change in temperature.
viii. They must have high ratio of puncture strength of flash over voltage.

The main cause of failure of insulators is due to flash over or puncture. The flash over may occur between the line conductor and the earth, i.e. the pin of the insulator, and due to production of extreme heat produced by the arc, the insulator may puncture.

The material used for insulators of overhead lines is porcelain, glass satellite and special composition materials. The most commonly used material is porcelain whereas the other materials, i.e. glass satellite etc. are analysed to limited extent. This material is preferred over glass since it is mechanically strong, its surface is not affected by dust deposits and is less susceptible to temperature changes. The dielectric strength of a porcelain insulator is 60 kV per cm of thickness and its compressive and tensile strengths are 7000 kg/cm^2 and 500 kg/cm^2 respectively.

Types of insulators: The most commonly used insulator in the overhead lines are: a. Pin type b. Suspension type c. Strain type d. Shackle type e. Egg or Stray type.

a. **Pin type insulators:** Up to 33 kV, pin type insulators are preferred over suspension tape insulators because they are cheaper in cost beyond this voltage they become uneconomical.

b. **Suspension type insulators:** For high voltages more than 33 kV it is usual practice to use suspension type insulators. They consist of a number of protection discs connected in series by metal links in the form of string. Each unit or disc is designed for low voltage say 11 kV.

c. **Strain type insulators:** At the dead ends, on sharp turns, at the river crossings or at the corners, the line is subjected to greater strains. In order to withstand the excessive strain, insulators are used. For high voltage transmission lines, strain insulators consisting of assembly of suspension insulators are used.

d. **Shackle type insulators:** These insulator are mostly used for low voltage distribution lines. The conductor is passed through the place left between the clamp and the insulator and is fixed along the groove with the help of soft binding wires of the same material as the conductor.

e. **Guy or stay type insulators.** Guy or stay wires are used with the poles placed at the dead ends or at sharp ends of low voltage lines. To insulate the lower parts of the guy wire from the pole for the safety of people, stay insulator is placed in between the wire. This insulator is placed in the guy wire at a height of three metres from ground.

String Efficiency

The ratio of voltage across the whole string to the product of number of discs and the voltage across the disc nearest to the conductor is known as string efficiency, i.e.

$$\frac{\text{Voltage across the string}}{n \times \text{Voltage across disc nearest to the conductor}}$$

where n = number of discs in the string.

Methods of Improving String Efficiency

It has been seen above that potential distribution in a string of suspension insulators in not uniform. This necessitates to equalise the potential across the various units of the string to improve the string efficiency. The various methods for this purpose are:

a. *By using longer cross-arms*: The value of string efficiency depends upon the value of K, i.e. ratio of shunt capacitance to mutual capacitance. The lesser the value of K, the greater the string efficiency and more uniform is the voltage distribution. The value of K can be decreased by reducing the shunt capacitance. In order to reduce shunt capacitance, the distance of a conductor from tower must be increased, i.e. longer cross-arms should be used. However, limitations of cost and strength of tower do not allow the use of very long cross-arms, refer to Fig. 11.6. In practice, $K = 0.1$ is the limit that can be achieved by this method.

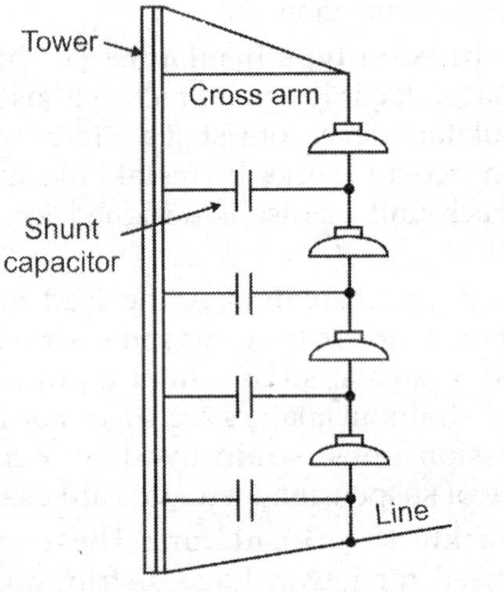

Fig. 11.6

b. *By grading the insulators*: In this method, insulators of different dimensions are so chosen that each has a different capacitance. The insulators are capacitance graded i.e. they are assembled in the string in such a way that the top unit has the minimum capacitance, increasing progressively as the bottom unit (i.e. nearest to conductor) is reached.

c. *By using a guard ring*: The potential across each unit in a string can be equalised by using a guard ring which is a metal ring electrically connected to the conductor and surrounding the bottom insulator as shown in Fig. 11.7. The guard ring introduces capacitance between metal fittings and the line conductor. The guard ring is contoured in such a way that shunt capacitance currents i_1, i_2 etc. are equal to metal fitting—line capacitance currents i_1, i_2 etc. The result is that same charging current I flows through each unit of string. Consequently, there will be uniform potential distribution across the units.

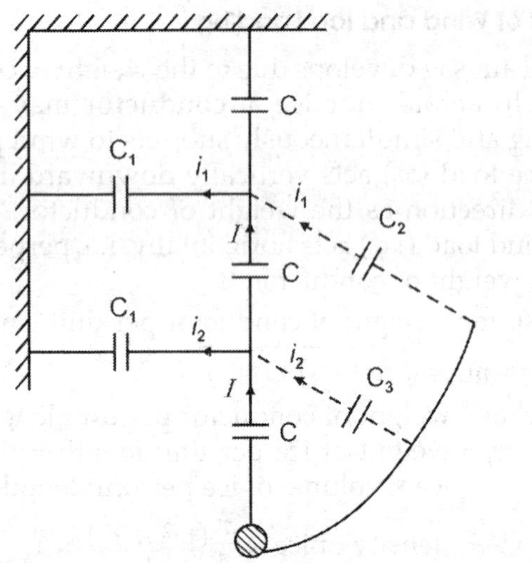

Fig. 11.7

Potential Distribution of Overhead Suspension Insulator String

A string of suspension insulations consists of a number of porcelain discs connected in series through metallic links. Figure 11.8(a) shows 3-disc string of suspension insulators. The porcelain portion of each disc is in between two metal links. Therefore, each disc forms a capacitor C as shown in Fig. 11.8(b). This is known as *mutual capacitance or self capacitance.* If there were mutual capacitance alone, then charging current would have been the same through all the discs.

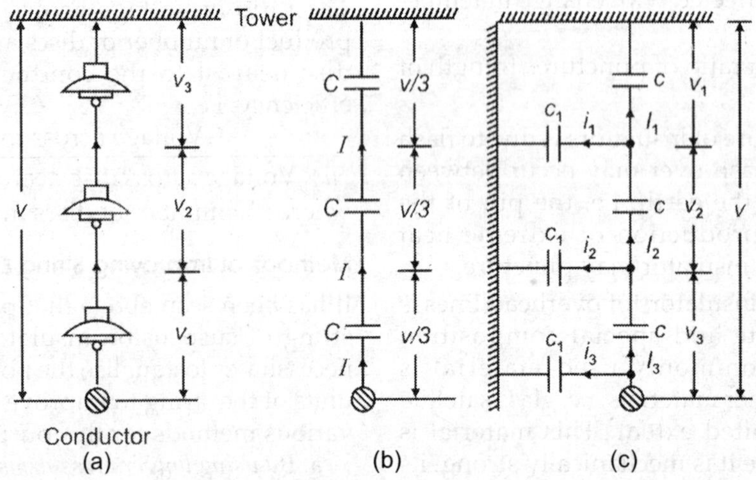

Fig. 11.8

Figure 11.8(c) shows the equivalent circuit for a 3-disc string. Let that self capacitance of each disc is C and shunt capacitance $C_1 = KC$. The voltage across each unit is V_1, V_2 and V_3 are given by

$$V_3 = V_1[1 + 3K + K^2]$$

Voltage between conductor and earth (i.e. tower)

$$V = V_1 + V_2 + V_3$$
$$= V_1 + V_1(1 + K) + V_1(1 + 3K + K^2)$$
$$= V_1(3 + 4K + K^2) = V_1(1 + K)(3 + K)$$

$\therefore$ Voltage across top unit, $V_1 = \dfrac{V}{(1+K)(3+K)}$

String Efficiency

The ratio of voltage across the whole string to the product of number of discs and the voltage across the disc nearest to the conductor is known as *string efficiency*, i.e.

String efficiency

$$= \frac{\text{Voltage across the string}}{n \times \text{voltage across disc nearest to the conductor}}$$

where n = number of discs in the string.

Distribution System

The part of power system by which electric power is distributed among various consumers for their local use in known as *distribution system.*

In general the distribution system is that part of power system which distributes electric power from major substations to the consumers as per their requirement. A low tension distribution system is shown in Fig. 11.9 which comprises *the feeders, distributors* and *service mains.*

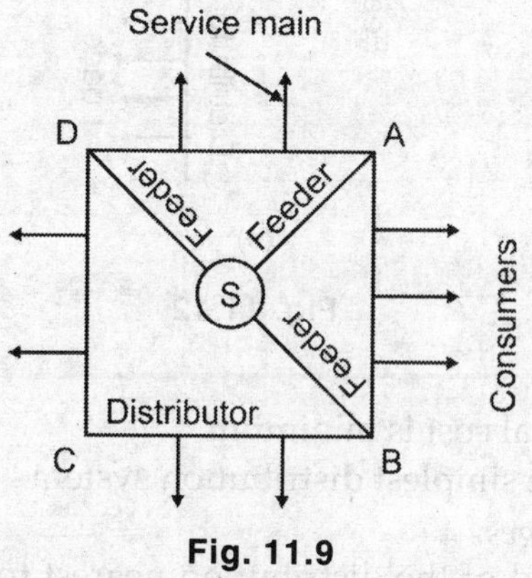

Fig. 11.9

i. *Feeder*: A line or conductor which connects the major station to the distributor is known as *feeder*. It is to feed the electric power (or energy) to the distributor. In Fig. 11.9, SA, SB and SD are the feeders. Since no tapping is taken from the feeder, generally it carries same current throughout its length. The current carrying capacity is the main consideration taken into account while designing a feeder.

ii. *Distributor*: A line or conductor to which various consumers are connected through service mains is known as *distributor.* It is to distribute electric power among various consumers, thus a number of tappings are taken from the distributor. Hence, it carries different currents along its length. In Fig. 11.9, *AB, BCD* and *DA* are the distributors. While designing a distributor, voltage drop along the length of the distributor is the main consideration.

iii. *Service mains*: A line (conductor or cable) which connects the consumer to the distributor is known as *service mains.* It is designed as per the connected load of the consumer.

Classification of distribution system: The distribution system may be classified.

1. **According to nature of current.** According to the nature of current the distribution system may be classified as: a. DC distribution system, b. AC distribution system.

AC distribution system is almost universally employed due to its several advantages over dc distribution system.

2. **According to nature of construction.** According to nature of construction, the distribution system may be classified as: a. Overhead distribution system, b. Underground distribution system.

Overhead system, being cheaper, is mostly employed in our country. However, underground system is employed in densly populated areas where overhead system may not be practicable. It is also employed in big cities because of its better look.

3. **According to the scheme of connections.** According to the scheme of connections, the distribution system may be classified as: a. Radial system b. Ring main system c. Interconnected system.

Interconnected system is preferred because of its better reliability.

DC distribution: Though electric power is almost exclusively generated, transmitted and distributed as ac but for certain applications (e.g. for electrochemical works, for the operation of variable speed machinery, dc motors, etc.) is absolutely necessary. For this purpose, ac is converted into dc at the substation and is then distributed by (i) 2-wire system or (ii) 3-wire system.

i. *Two-wire dc system*: It consists of two-wires one is outgoing (positive) conductor and the other is return (negative) conductor as shown in Fig. 11.10.

ii. *Three-wire dc system*: It consists of three-wire, two outers (positive and negative outer) and a middle or neutral wire which is earthed at the substation. The voltage between any outer and neutral is V, whereas, voltage between the two outers is twice to this value, i.e. 2 V. Thus two voltages are available to the

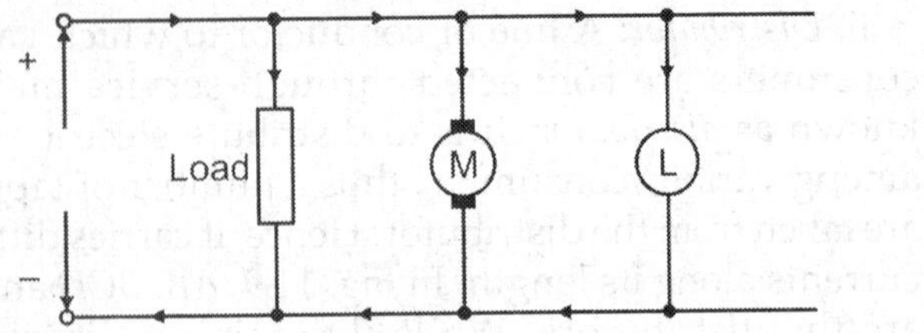

Fig. 11.10

consumers. The lamp loads (low voltage loads) are connected between any outer and neutral, whereas the motor loads (high voltage loads) are connected across the two outer as shown in Fig. 11.11. The current flowing through the neutral is very small (only unbalance current which will be zero when the load is balanced), therefore, generally area of cross-section or neutral is taken half as compared to either of the outer.

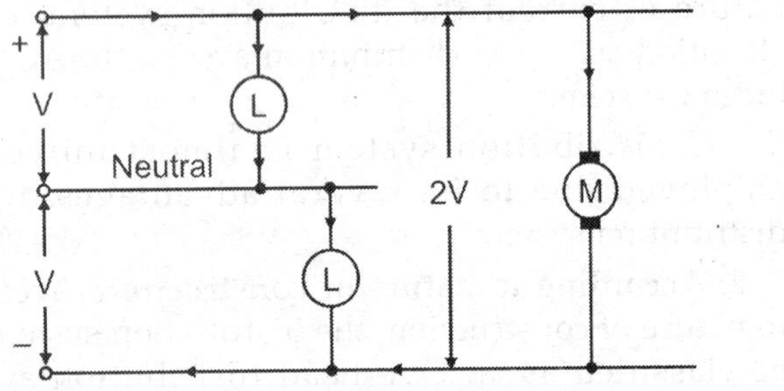

Fig. 11.11

AC distribution: The electric power (or energy) is invariably generated, transmitted and distributed in the form of alternating current. The main reason of adopting AC system for generation, transmission and distribution of electric power is that the alternating voltage can conveniently be changed to any desired value with the help of a transformer. It can be increased to economical value for transmission and can be decreased to a safe value for utilisation of electric power.

Generally, ac distribution system is classified as: (i) primary distribution system, (ii) secondary distribution system.

i. *Primary distribution system*: The bulk consumers are supplied generally at high voltage (11 kV) and system by which they are supplied is *called primary distribution system*. However, the distribution voltage may also be 6.6. kV or 3.3 kV which depends upon the amount of power to be conveyed by the distributor and its length (distance from substation). The consumers install their own transformer to step-down the voltage for utilisation of electric power. The primary distribution is carried out by 3-phase, 3-wire system due to economic distributions.

ii. *Secondary distribution system*: The small consumers are supplied at low voltage 400/230 V and the system by which they are supplied is called secondary distributed system. The secondary distribution employs 400/230 V, 3-phase, 4-wire system, where 400 V is the line voltage and 230 V is the phase voltage.

Connection Schemes of Distribution System

The following connection schemes of distribution system are generally employed.

1. **Radial system:** In radial system, separate feeders radiate from a single substation and feed the distributors at one end only. A single line diagram of a radial system for DC and AC distribution is shown in Fig. 11.12. Here substation (S) supplies power to a distributor (AB) at end (A) through feeder (SC). This system is only employed when power is generated at low voltage and the substation is located at the centre of the load.

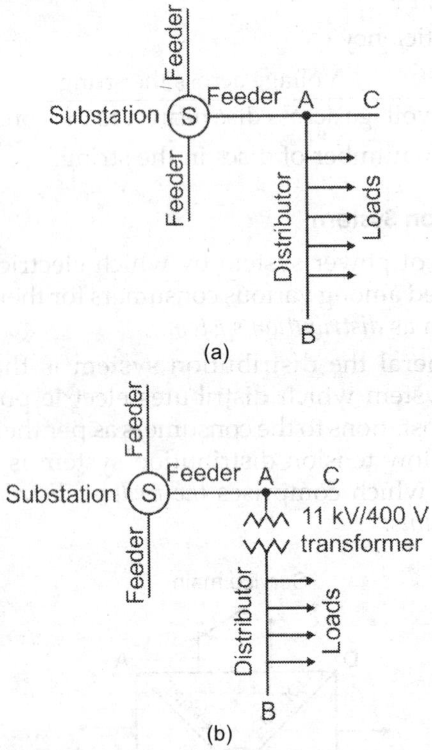

Fig. 11.12

Advantages:

i. Its initial cost is minimum.

ii. It is the simplest distribution system.

Disadvantages:

i. The end of the distribution nearest to the feeding end would be heavily loaded.

ii. The consumers are dependent on single feeder and distribution. Therefore, when fault occurs on the feeder or distributor, the supply is cut off to all the consumers who are on the side of the fault away from the substation.

iii. The consumers at the far end (B) of the distributor would be subjected to serious voltage fluctuations

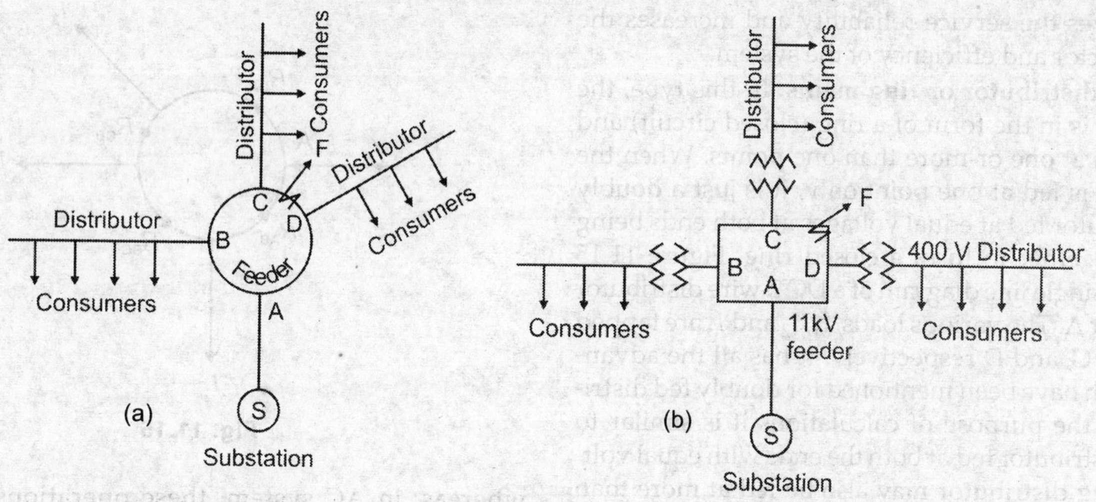

Fig. 11.13

because of the changing load on the distributor. Due to the above limitations this system is used for short distances only.

2. **Ring main system:** In this system, the feeder closes on itself, i.e. it forms a complete ring and hence the name ring main system. Figures 11.13(a) and (b) show respectively DC and AC ring main system, where (ABCDA) is the closed feeder supplied by the substation S at point A. The various distributors are connected to points B, C and D of the feeder as shown.

Advantages:

i. Each distributor is supplied via two feeders, for example the distributor connected at point C is fed via feeder *ABC* and *ADC*. This reduces the voltage fluctuations at the consumers' terminals to some extent.

ii. The system is more reliable since each distributors is fed via two feeders. In the event of fault on any section of feeder, the continuity of supply can be maintained. For instant, suppose fault occurs at any point *F* of section *CD* of the feeder. Then section *CD* for the feeder can be isolated for repairs and at the same time continuity of supply is maintained to all the consumers.

3. **Interconnected system.** In this system, the feeder ring is energized by two or more than two generating stations or substations. Figures 11.14(a) and (b) show the single line diagram of DC and AC interconnected system respectively, where the closed feeder ring ABCA is supplied by two generating or substations S_1 and S_2 at points X and Y respectively. Distributors are connected to the feeder at point A, B and C

Advantages:

i. It increases the reliability of supply.

ii. During overload hours, the area fed from one generating station can be fed from other generating station. Thus it reduces the reserved plant capacity,

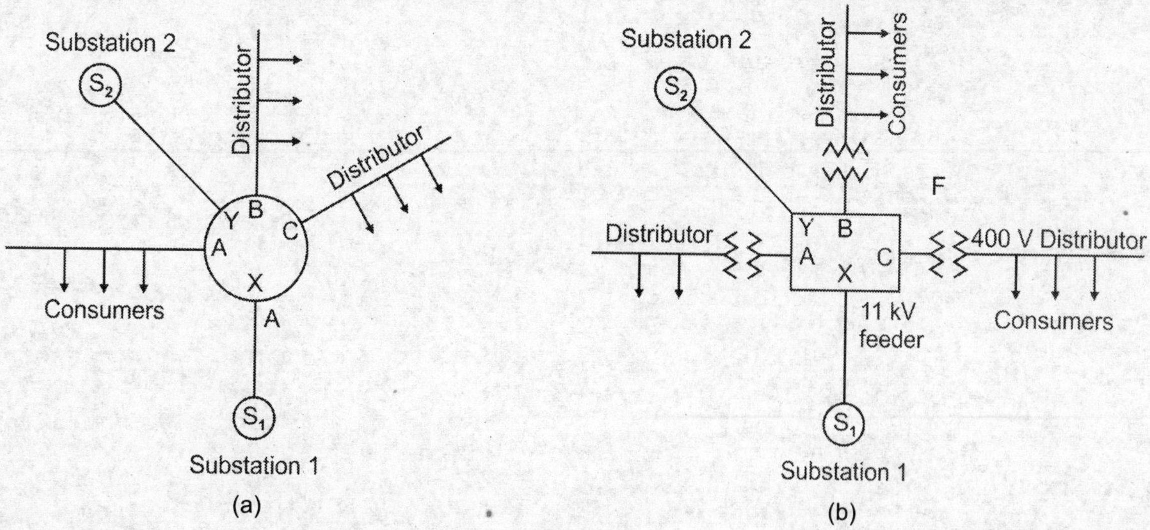

Fig. 11.14

improves the service reliability and increases the load factor and efficiency of the system.

4. **Ring distributor or ring mains:** In this type, the distributor is in the form of a ring (closed circuit) and is supplied at one or more than one points. When the distributor is fed at one point only, it is just a doubly fed distributor fed at equal voltages at both ends being brought together to form a closed ring. Figure 11.15 shows the single line diagram of a DC 2-wire distributor fed at point A. The various loads, I_1, I_2 and I_3 are tapped at point B, C, and D respectively. It has all the advantages which have been mentioned for doubly fed distributor. For the purpose of calculations it is similar to straight distributor fed at both the ends with equal voltages. A ring distributor may also be fed at more than one point.

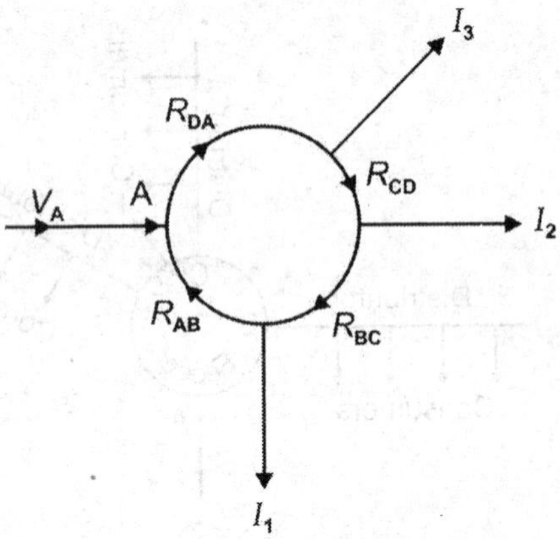

Fig. 11.15

AC distribution: AC distribution differs from the DC distribution in the following respects.

i. In DC system, the voltage drop is due to resistance of the distributor only whereas, in AC system, the voltage drops are due to the combined effects of resistance, inductance and capacitance (capacitance is usually neglected because of smaller lengths).

ii. In DC system, all additions and subtractions of the currents and voltages are done arithmetically whereas, in AC system, these operations are done vectorially. Therefore, all currents and voltages (or voltage drops) are expressed in symbolic notation.

iii. In AC system, the load power factor plays an important role in the calculations, therefore, it has to be taken into account.

iv. The methods used for calculations in DC distributors also hold good for AC distributors but in these cases all the electrical quantities (i.e. impedance, currents and voltages) are expressed symbolically with respect to the reference vector.

MULTIPLE CHOICE QUESTIONS

1. India's first nuclear power plant was installed at
   A. Tarapore B. Kota
   C. Kalpakkam D. Delhi
2. Reflectors of a nuclear reactor are made up of
   A. boron B. cast iron
   C. beryllium D. steel
3. The function of a moderator in a nuclear reactor is to
   A. slow down the fast moving electrons
   B. speed up the slow moving electrons
   C. start the chain reaction
   D. transfer heat produced inside the reactor to a heat exchanger
4. When a nuclear reactor is operating in constant power the multiplication factor is
   A. less than unity B. greater than unity
   C. equal to unity D. none of these
5. The conversion ratio of a breeder reactor is
   A. equal to unity B. more than unity
   C. less than unity D. none of these
6. In the nuclear fission reactions–isotope of uranium is used
   A. U^{223} B. U^{234}
   C. U^{238} D. U^{236}
7. Tarapore nuclear power plant has
   A. pressurised water reactors
   B. boiling water reactors
   C. CANDU type reactors
   D. None of these
8. Critical mass of fuel is the amount required to make the multiplication factor ...... unity.
   A. equal to B. less than
   C. more than D. none of these
9. Fission chain reaction is possible when
   A. fission produces the same number of neutrons which are absorbed
   B. fission produces more neutrons than are absorbed
   C. fission produces less neutrons than are absorbed
   D. none of these
10. In nuclear chain fission reaction, each neutron which causes fission produces
    A. no new neutron B. one new neutron
    C. more than one new neutron
    D. none of these
11. The most commonly used moderator is.
    A. graphite B. sodium
    C. deuterium D. any of these
12. Which of the following are fertile materials?
    A. U^{238} and Th^{239} B. U^{238} and Th^{232}
    C. U^{233} and Pu^{239} D. U^{238} and Pu^{239}
13. In a nuclear reactor, the function of a reflector is to
    A. reduce the speed of the neutrons
    B. stop the chain reaction
    C. reflect the escaping neutrons back into the core
    D. none of these
14. In gas cooled reactor (GCR) the moderator and coolant used respectively are
    A. heavy water and CO_2 B. graphite and air
    C. graphite and CO_2 D. none of these
15. In a pressurised water reactor (PWR)
    A. the coolant water is pressurised to work as moderator
    B. the coolant water boils in the core of the reactor
    C. the coolant water is pressurised to prevent boiling of water in the core
    D. no moderator is used
16. The function of the moderator in a nuclear reactor is to
    A. stop chain reaction
    B. absorb neutrons
    C. reduce the speed of neutrons
    D. reduce temperature
17. Thermal shielding is provided to
    A. protect the walls of the reactor from radiation damage
    B. absorb the fast neutrons
    C. protect the operating personnel from exposure to radiation
    D. both (B) and (C) are correct
18. A CANDU reactor uses
    A. only fertile material
    B. highly enriched uranium (85% U^{235})
    C. natural uranium as fuel and heavy water as moderator and coolant
    D. none of these
19. Fission of U^{235} releases equal to energy
    A. 200 MeV B. 238 MeV
    C. 431 MeV D. none of these
20. In India fast breed reactors are best suited because
    A. of large thorium deposits
    B. of large uranium deposits
    C. of large plutonium deposits
    D. none of these
21. India's first nuclear power plant was started at
    A. Narora B. Tarapore
    C. Kota D. Kalpakkam

Ans.													
	1. A	2. C	3. A	4. C	5. B	6. A	7. B	8. A	9. B	10. C	11. A	12. B	13. C
	14. C	15. C	16. C	17. D	18. C	19. A	20. A	21. B					

22. The function of a surge tank is to
A. supply water at constant pressure
B. produce surges in the pipe line
C. relieve water hammer pressures in the penstock pipe
D. none of these

23. Gross head of a hydropower station is
A. the difference of water level between the level in the storage and tail race
B. the height of the water level in the river where the storage is provided
C. the height of the water level in the river where tail race is provided
D. none of these

24. Location of the surge tank in a hydroelectric station is near to the
A. tailrace B. turbine
C. reservoir D. none of these

25. Pelton wheel turbine is used for minimum of the following heads
A. 40 m B. 20 m
C. 80 m D. 180 m or above

26. Running cost of a hydroelectric power plant is
A. equal to running cost of a steam power plant
B. less than running cost of a steam' power plant
C. more than running cost of a steam power plant
D. none of these

27. Which of the following plants will have the highest capital cost?
A. Nuclear power plant B. Diesel power plant
C. Thermal power plant D. None of these

28. Which of the following plants have the minimum running cost?
A. Nuclear power plant B. Diesel power plant
C. Thermal power plant D. Hydro power plant

29. In a steam power plant, water is used for cooling purposes in
A. condenser B. turbine only
C. boiler tubes
D. boiler tubes and turbine both

30. Water is supplied to a boiler at
A. 140 kg/cm^2 pressure B. atmospheric pressure
C. slightly more than atmospheric pressure
D. more than the steam pressure in the boiler

31. The pH value of water used for boiler is
A. unity B. 7
C. 10 D. slightly more than 7

32. What is the advantage of hydroelectric power stations over thermal power stations?
A. The initial cost of hydroelectric power stations is low
B. Their operating cost is low
C. Hydroelectric power stations can supply the same power throughout the year
D. Hydroelectric power stations can be constructed where electric energy is required

33. Tick out the correct statement
A. The electrostatic precipitator can remove dust particles as small as 0.01 μ and reduce the rating of ID fan required
B. The electrostatic precipitator can remove dust particles as small as 0.001 μ and reduce the rating of ID fan required
C. The electrostatic precipitator can remove dust particles as small as 0.01 μ but increases the rating ID fan required
D. none of these

34. Tick out the correct statement.
A. Radiant type superheater has rising steam output temperature characteristics
B. Convection type has drooping characteristic
C. Radiant type has drooping whereas convection type has rising characteristics
D. Both have flat characteristics

35. Hydrogen when used for cooling in a large alternator, it
A. increases the life of insulation
B. decreases the life of insulation
C. does not affect insulation either way
D. none of these

36. Hydrogen when used for cooling a large alternator, it has a pressure of
A. 0.01 kg/cm^2 B. 0.02 kg/cm^2
C. 0.035 kg/cm^2 D. none of these

37. Tick out the correct statement.
A. The stiffer the shaft is, the higher is its critical speed
B. The stiffer the shaft is, the lower is its critical speed
C. The stiffness of shaft has nothing to do with its critical speed
D. Large size machines have their critical speed higher than their synchronous speed

38. Hydrogen cooling is used for large size
A. turbo alternators only
B. water wheel generators only
C. both turbo alternators and water wheel generators
D. all of these

39. The half-life period of an isotope is 1 h, 15/16 of it radiates out in
A. 8 hrs B. 16 hrs
C. 4 hrs D. 5 hrs

40. In nuclear power plants, it is desirable to use reactor core as
A. cubical or cylindrical
B. cubical or spherical
C. cylindrical or spherical
D. spherical

Ans. 22. C 23. A 24. B 25. D 26. B 27. A 28. D 29. A 30. D 31. D 32. B 33. A 34. C 35. A 36. C 37. A 38. A 39. C 40. A

41. For flue gas flow, tick out the correct sequence.
A. Boiler - air preheater - economiser - ID fan - chimney.
B. Boiler - ID fan - Air preheater - economiser - chimney
C. Boiler - economiser - Air preheater - ID fan - chimney
D. None of these

42. If the reactor is to be used for converting the fertile material into fissionable material
A. the material to be converted should be placed around the core
B. the core should be placed around the fertile material
C. the material to be converted should be inter-mixed with the core
D. the material to be converted can be placed anywhere with respect to the core

43. The nuclear reactor starts generation when
A. the safety rod is inserted into the core
B. the safety rod is withdrawn from the core
C. the slim rod is inserted into the core
D. the slim rod is withdrawn from the core

44. Induced draft fans are located at
A. the top
B. the bottom
C. in the middle part
D. can be anywhere in the cooling tower

45. Under steady state operating condition, the multiplication factor of a nuclear plant is
A. more than unity
B. less than unity
C. equal to unity
D. depends upon the power it is generating

46. The critical mass of a nuclear reactor depends upon
A. the enrichment of the fuel used
B. the moderator material used
C. both the enrichment of fuel and moderater material
D. none of these

47. If natural uranium is used as the fuel, the moderator to be used is
A. ordinary water B. heavy water
C. both (A) and (B) are correct
D. none of these

48. The pressurised water reactors employ
A. light water and natural uranium
B. heavy water and enriched uranium
C. light water and enriched uranium
D. none of these

49. For meeting the changing load of the system, the control rod adjusted is
A. safety rod B. shim rod
C. regulating rod D. none of these

50. The CANDU type reactor can be shut down by controlling
A. safety rod B. shim rod
C. regulating rod D. none of these

51. The windage, friction and iron losses are more when an induction machine is operated as
A. an induction motor
B. an induction generator
C. the losses are same for both the machines
D. none of these

52. Induction generator is more suitable for supplying
A. resistive load B. inductive load
C. capacitive load
D. a combination of A, B and C

53. Which of the following place is not associated with nuclear power plants in India?
A. Tarapur B. Bangalore
C. Kota D. Narora

54. Which of the following plant has least running cost per kWh?
A. Diesel power plant B. Nuclear power plant
C. Thermal power plant
D. Hydroelectric power plant

55. Which of the following materials is used as a moderator?
A. Graphite
B. Boron
C. Sodium potassium liquid
D. Plutonium

56. The capital cost per MW is highest in case of
A. Steam power plants
B. Nuclear power plants
C. Gas turbine power plants
D. Diesel engine power plants

57. A diesel power-plant is best suited as
A. base load plant B. stand by plant
C. general purpose plant
D. any of these

58. In case of an induction generator, for a particular power, the current and pf are both completely defined in terms of
A. resistance and reactance of load
B. constant (parameters) of the induction generator
C. a combination of A and B
D. none of these

59. For self excitation of induction generator, it is desirable economically to connect capacitors in
A. star connection B. delta connection
C. both star and delta connections
D. none of these

60. Which of the following installation at consumer premises usually provides peak load?
A. Air conditioner B. Arc furnace

Ans.													
	41. C	42. A	43. B	44. A	45. C	46. C	47. B	48. C	49. C	50. D	51. B	52. C	53. B
	54. D	55. A	56. B	57. B	58. B	59. B	60. B						

C. Cold storage compressor
D. Air blower running continuously

61. Economisers are used to heat
A. air B. steam
C. feed water D. None of these

62. India's first nuclear power plant was built at
A. Rana Pratap Sagar B. Trombay
C. Tarapur D. Kalpakkam

63. The first nuclear power plant was commissioned in
A. USA B. USSR
C. India D. France

64. The overall thermal efficiency of a thermal power plant lies in the range
A. 25–30% B. 35–40%
C. 45–60% D. 65–80%

65. The cost of fuel transportation is minimum in
A. thermal plant B. nuclear plant
C. hydroelectric plant D. diesel plant

66. The cheapest plant in operation and maintenance is
A. thermal power B. hydro power
C. nuclear power D. diesel power

67. Of all the plants, the minimum quantity of fuel is required in
A. hydro plants B. nuclear power plants
C. thermal power plants
D. none of these

68. Tidal power could not be harnessed in the past as
A. it could not be utilised at the proper position of the load curve
B. lack of technological development
C. it depends upon the seasonal changes
D. none of these

69. Water is supplied to a boiler at
A. atmospheric pressure
B. more than the steam pressure in the boiler
C. at 120 kg/cm^2
D. none of these

70. For low head and high discharge, the hydraulic turbine used is
A. Francis turbine B. Kaplan turbine
C. Pelton wheel D. None of these

71. The fissile material is
A. U^{232} B. thorium
C. plutonium D. none of these

72. The nuclear reactor starts generation when
A. the material to be converted should be placed around the core
B. the core should be placed around the fertile material
C. the material to be converted should be intermixed with core
D. none of these

73. Run away speed of a pelton wheel is
A. full load speed
B. actual speed operating at no load
C. no load speed when governor mechanism fails
D. 80% greater than the normal speed

74. In India, the first tidal power plant is likely to come up in
A. Bay of Bengal B. Korba
C. Singrauli D. Gulf of Kutch

75. The line trap unit employed in carrier current relaying offers
A. high impedance to 50 Hz power frequency signals
B. high impedance to carrier frequency signals
C. low impedance to carrier frequency signals
D. low impedance to carrier frequency signals and high impedance to 50 Hz power frequency signals

76. The suitable turbine used for harnessing tidal power is
A. Francis turbine B. Pelton turbine
C. Kaplan turbine D. None of these

77. Economisers are normally used when boiler pressure exceeds
A. 70 kg/cm^2 B. 50 kg/cm^2
C. 30 kg/cm D. for all pressures

78. The moderator used in fast breeder reactor is
A. heavy water B. graphite
C. ordinary water D. Any of these

79. The enriched uranium consists of approximately
A. 10% of U^{235} and 90% of U^{238}
B. 20% of U^{235} and 80% of U^{238}
C. 30% of U^{235} and 70% of U^{238}
D. none of these

80. Which of the following generating stations has minimum running cost while in operation?
A. Thermal power station
B. Nuclear power station
C. Hydropower station D. None of these

81. In power plants, the condenser normally used is
A. surface type B. jet type
C. both surface and jet type
D. regenerative type

82. The fertile material is
A. U^{235} B. U^{233}
C. plutonium D. U^{238}

83. If natural uranium is tised as the fuel, the moderator to be used is
A. ordinary water B. graphite
C. heavy water D. none of these

84. In water turbines, the run away speed of pelton turbine is
A. 1.8 times the rated speed
B. 2 to 2.2 times the rated speed

Ans.	61. C	62. C	63. B	64. A	65. C	66. B	67. B	68. B	69. B	70. B	71. B	72. C	73. C
	74. D	75. B	76. C	77. A	78. D	79. A	80. C	81. A	82. A	83. B	84. A		

C. 2.5 to 2.8 times the rated speed
D. 3 to 4 times the rated speed

85. For a thermal power plant, which is not the fixed cost?
A. Interest on capital B. Depreciation
C. Cost of fuel D. All of these

86. Which of the following methods of generating electric power from sea water is more advantageous?
A. Wave power B. Ocean currents
C. Tidal power D. None of these

87. Diesel engine power plants usually run on
A. high speed diesel oil B. kerosene
C. light diesel oil D. none of these

88. Nuclear reactor generally employs
A. Fusion B. Fission
C. both fusion and fission
D. none of these

89. Which power plant cannot have single unit of 100 MW?
A. Steam power plant B. Nuclear power plant
C. Hydroelectric power plant
D. Diesel power plant

90. For the same plant size, initial cost of which plant is the highest?
A. Steam power plant B. Diesel engine plant
C. Nuclear power plant D. Gas turbine plant

91. In a steam power plant which component needs maximum maintenance?
A. Condenser B. Boiler
C. Turbine D. Water treatment plant

92. Total input when the plant is generating 35 MW will be
A. 181×10^6 kcal/hr B. 250×10^6 kcal/hr
C. 362.5×10^6 kcal/hr D. 582×12^6 kcal/hr

93. When the load on the generator was 25 MW the heat rate in kcal/MW h was nearly
A. 16×10^6 B. 20×10^6
C. 30×10^6 D. 31.3×10^6

94. In India, tidal power plant is being considered for location
A. in the Gulf of Cambay
B. in the Bay of Bengal
C. near Vishakhapatnam
D. near Goa

95. In a thermal power station of moderate size, the electrical power is generated at a voltage of
A. 11 V B. 230 V
C. 440 V D. 11 kV

96. Which of the following is not a source of power?
A. Solar cell B. Photovoltaic cell
C. Photoelectric cell D. Thermocouple

97. Which of the following has least contribution for power generation in India?
A. Thermal power plants
B. Hydroelectric power plants
C. Gas turbine power plants
D. Nuclear power plants

98. Which of the following organisations is associated with atomic energy in India?
A. ONGC B. NTPC
C. BARC D. NHPC

99. Gas turbine plants can have
I. Multistage compression
II. Heat exchangers
III. Reheating
For maximum efficiency which features should be incorporated in a gas turbine power plant?
A. I only B. I and II only
C. II and III only D. I, II and III

100. Which of the following power plants can generate power at uncontrollable times?
A. Wind power plant B. Tidal power plant
C. Solar power plant D. Any of these

101. In India least share of power is provided by
A. Thermal power plants
B. Hydroelectric power plants
C. Nuclear power plants
D. Diesel power plants

102. For the same cylinder dimensions and speed, which engine will produce least power?
A. Petrol engine B. Diesel engine
C. Super charged engine
D. All will produce the same power

103. The input-output equation of a 0.5 MW power house is given by $I = 30 + 0.8\,L + 0.5\,L^2$, where I is in millions of kcal per hour and L is in megawatts. The input when the plant was running at no load will be
A. 0.5 kcal/hr B. 21.3 kcal/hr
C. 30×10^6 kcal/hr D. zero

104. Which type of plant has the minimum running cost per kWh of energy generated?
A. Hydroelectric plant B. Thermal power plant
C. Nuclear power plant D. Diesel power plant

105. Sun tracking is needed in the case of......
A. flat plate collector
B. cylindrical parabolic and paraboloid collector
C. both A and B are correct
D. none of these

106. In a solar collector, the function of the transparent cover is to
A. transmit solar radiation only
B. protect the collector from dust
C. decrease the heat loss from collector beneath to atmosphere
D. none of these

Ans. 85. C 86. C 87. C 88. B 89. D 90. C 91. B 92. C 93. B 94. A 95. D 96. C 97. C 98. C 99. D 100. D 101. D 102. A 103. C 104. A 105. B 106. C

107. Temperature attained by cylindrical parabolic collector is in the range of
A. 50 to 100°C B. 100 to 150°C
C. 150 to 300°C D. 300 to 500°C
108. Most widely used material of a solar cell is
A. arsenic B. cadmium
C. silicon D. steel
109. Photovoltaic cell or solar cell converts
A. thermal energy into electricity
B. electromagnetic radiation directly into electricity
C. solar radiation into thermal energy
D. none of these
110. Maximum wind energy available is proportional to
A. square of the diameter of rotor
B. air density
C. cube of the wind velocity
D. all of these
111. Which type of wind mill is of simple design
A. Horizontal axis wind mill
B. Vertical axis wind mill
C. both A and B are correct
D. none of these
112. Cost of wind energy generator compared to conventional power plants for the same power output is
A. equal B. lower
C. higher D. none of these
113. The turbine which is used in a tidal power plant for getting continuous power is
A. simple impulse type B. reversible type
C. propeller type D. all of these
114. Largest geothermal plant in operation is in
A. Maxico B. Italy
C. Russia D. California
115. Geothermal plant is suitable for
A. base load power B. peak load power
C. base and peak load power
D. none of these
116. A geothermal field may yield
A. hot water B. wet steam
C. dry steam D. all of these
117. Avagadro's number is the number of
A. atoms in one gram mole
B. molecules in one gram mole
C. both (A) & (B) are correct
D. all of these
118. Geothermal steam and hot water may contain
A. NH_3
B. Na_2S
C. H_2S, NH_3 and radon gas
D. none of these.
119. The nature of the current developed by MHD generator is
A. direct current B. alternating current
C. either direct or alternating in nature
D. none of these
120. In MHD generators the conductor employed is
A. gas B. liquid metal
C. liquid metal or gas D. None of these
121. The thickness of the depletion layer in an PN junction is of the order of
A. 10^{-6} m B. 10^{-10} m
C. 10^{-4} m D. None of these
122. Power output per unit volume of an MHD generator is proportional to
A. square of the magnetic flux density
B. electrical conductivity of the gas
C. square of the fluid velocity
D. all of these
123. Biogas consists of
A. only methane
B. methane and CO_2 with some impurities
C. a special organic gas D. none of these
124. The main byproduct of the biogas plant is
A. biogas B. biomass
C. organic manure D. none of these
125. The commercial sources of energy are
A. solar, wind, biomass
B. fossil fuels, hydropower and nuclear energy
C. wood, animal wastes and agricultural wastes
D. none of these
126. Non-commerical sources of energy are
A. wood, animal wastes and agricultural wastes
B. solar, wind, biomass
C. fossil fuels, hydropower and nuclear power
D. none of these
127. The secondary sources of energy are
A. solar, wind and water
B. coal, oil and uranium
C. none of these
128. In diesel cycle
A. compression ratio and expansion ratio are equal
B. compression ratio is greater than expansion ratio
C. compression ratio is less than expan-sion ratio
D. compression ratio (expansion ratio)^2
129. Compression ratio of an IC engine is the ratio of
A. $\frac{\text{total volume}}{\text{swept volume}}$ B. $\frac{\text{total volume}}{\text{clearance volume}}$
C. either of the above D. none of these
130. In a diesel engine, the heat lost to the cooling water is
A. 10% B. 20%
C. 30% D. 70%

Ans.													
	107. A	108. C	109. B	110. D	111. B	112. C	113. B	114. B	115. A	116. D	117. B	118. C	119. A
	120. C	121. A	122. D	123. B	124. C	125. B	126. A	127. A	128. B	129. B	130. C		

131. The mechanical efficiency of a diesel engine is defined as

A. $\frac{BHP}{IHP}$ B. $\frac{IHP}{BHP}$

C. $\frac{(BHP)^2}{IHP}$ D. BHP × IHP

132. The temperature of cooling water leaving the diesel engine should not be more than

A. 30°C B. 40°C

C. 60°C D. 80°C

133. Total cost of a diesel power plant per kW of installed capacity is less than that of steam power plant by

A. 5 to 10% B. 20 to 30%

C. 40 to 50% D. 70 to 80%

134. The two stroke cycle engine has

A. one suction valve and one exhaust valve operated by one cam

B. one suction valve and one exhaust valve operated by two cams

C. only ports covered and uncovered by piston to affect charging and exhausting

D. none of these

135. The specific fuel consumption of a diesel engine as compared to that of petrol engine is

A. lower B. higher

C. same for same power output

D. same for same speed

136. Compression ratio of pertrol engines is in the range of

A. 2 to 3 B. 7 to 10

C. 16 to 20 D. none of these

137. Compression ratio of diesel engines may have a range

A. 8 to 10 B. 10 to 15

C. 16 to 20 D. none of these

138. The thermal efficiency of good IC engine at the rated load is in the range of

A. 80 to 90% B. 60 to 70%

C. 30 to 35% D. 10 to 20%

139. In case of petrol engine, at starting

A. rich fuel-air ratio is needed

B. weak fuel-air ratio is needed

C. chemically correct fuel-air ratio is needed

D. any fuel-air ratio will do

140. Carburettor is used for

A. SI engines B. Gas engines

C. CI engines D. None of these

141. Fuel injector is used in

A. SI engines B. gas engines

C. CI engines D. None of these

142. Very high speed engines are generally

A. gas engines B. SI engines

C. CI engines D. steam engines

143. In SI engine, to develop high voltage for spark plug

A. battery is installed

B. distributor is installed

C. carburettor is installed

D. ignition coil is installed

144. In SI engine, to obtain required firing order

A. battery is installed

B. distributor is installed

C. carburettor is installed

D. ignition coil is installed

145. For petrol engines, the method of governing employed is

A. quantity governing

B. quality governing

C. hit and miss governing

D. none of these

146. For diesel engines, the method of governing employed is

A. quantity governing

B. quality governing

C. hit and miss governing

D. None of these

147. In a 4-cylinder pertrol engine the standard firing order is

A. 1-2-3-4 B. 1-4-3-2

C. 1-3-2-4 D. 1-3-4-2

148. The torque developed by the engine is maximum

A. at minimum speed of engine

B. at maximum speed of engine

C. at maximum volumetric efficiency speed of engine

D. at maximum power speed of engine

149. Iso-octane content in a fuel for SI engines

A. retards auto-ignition

B. accelerates auto-ignition

C. does not affect auto-ignition

D. none of these

150. The knocking in SI engines increases with increases in

A. inlet air temperature B. compression ratio

C. cooling water temperature

D. all of these

151. The knocking in SI engines gets reduced by

A. increasing the compression ratio

B. retarding the spark advance

C. increasing inlet air temperature

D. increasing the cooling water tempe-rature

152. Increasing the compression ratio in SI engines it

A. increases the tendency for knocking

B. decreases the tendency for knocking

Ans. 131. A 132. C 133. B 134. C 135. A 136. B 137. C 138. C 139. A 140. A 141. C 142. B 143. D 144. B 145. A 146. C 147. D 148. C 149. A 150. D 151. B 152. A

C. does not affect knocking
D. none of these

153. The knocking tendency in petrol engines will increase when
A. speed is decreased
B. speed is increased
C. fuel-air ratio is made rich
D. fuel-air ratio is made lean

154. The ignition quality of fuels for SI engines is determined by
A. acetane number rating
B. octane number rating
C. calorific value rating
D. volatility of the fuel

155. The ratio of piston stroke to bore of cylinder for internal combustion engines varies between
A. 0.9 to 1.9 B. 0.5 to 0.8
C. 0.3 to 0.6 D. 0.1 to 0.2

156. Air rate required for the combustion in diesel engine is about
A. 5:1 B. 10:1
C. 15:1 D. None of these

157. In multicylinder engines a particular sequence in the firing order is necessary to
A. provide the best engine performance
B. obtain uniform turning moment
C. operate the ignition system smoothly
D. obtain non-uniform turning moment

158. Mostly high speed diesel engines work on
A. Diesel cycle B. Carnot cycle
C. Dual combustion cycle
D. Otto cycle

159. In case of diesel engine, the pressure at the end of compression is in the range of
A. 7–8 kgf/cm^2 B. 20–25 kgf/cm^2
C. 35–40 kgf/cm^2 D. 50–60 kgf/cm^2

160. Reciprocating motion of the piston is converted into a rotary one by
A. connecting rod B. crank shaft
C. crankweb D. gudgeon pin

161. Maximum temperature which is developed in the cylinder of a diesel engine is of the order of
A. 1000–1500°C B. 1500–2000°C
C. 2000–2500°C D. 2500–3000°C

162. In a four stroke cycle engine, the four operations namely suction, compression, expansion and exhaust are completed in the number of revolutions of crank shaft equal to
A. four B. three
C. two D. one

163. In a two stroke cycle engine, the operations namely suction, compression, expansion and exhaust are completed in the number of revolutions of crank shaft equal to
A. four B. three
C. two D. one

164. In a four stroke cycle petrol engine, during suction stroke
A. only air is sucked in
B. only petrol is sucked in
C. mixture of petrol and air is sucked in
D. none of these

165. In a four stroke cycle diesel engine, during suction stroke
A. only air is sucked in
B. only fuel is sucked in
C. mixture of fuel and air is sucked in
D. None of these

166. Petrol commercially available in India for Indian passenger cars has octane number in the range
A. 40 to 50 B. 60 to 70
C. 80 to 85 D. 95 to 100

167. Octane number of the fuel used commerically for diesel engine in India is in the range
A. 80 to 90 B. 60 to 80
C. 60 to 70 D. 40 to 45

168. The knocking tendency in CI engines increases with
A. decrease of compression ratio
B. increase of compression ratio
C. increasing the temperature of inlet air
D. increasing cooling water temperature

169. If petrol is uesd in a diesel engine, then
A. low power will be produced
B. efficiency will be low
C. black smoke will be produced
D. higher knocking will occur

170. Voltage developed to strike spark in the spark plug is in the range
A. 6 to 12 volts B. 1000 to 2000 volts
C. 20000 to 25000 volts D. none of these

171. Superheating of steam is desirable for
A. increasing the efficiency of Rankine cycle
B. reducing initial condensation losses
C. avoiding too high moisture in the last stage of turbine
D. all of these

172. Economisers are usually used in boiler plant working above the pressure of
A. 30 kgf/cm^2 B. 50 kgf/cm^2
C. 70 kgf/cm^2 D. 90 kgf/cm^2

173. Which of the following coals has the highest calorific value?
A. Peat B. Lignite
C. Bituminous D. Anthracite

Ans. 153. A 154. B 155. A 156. C 157. A 158. C 159. C 160. A 161. C 162. C 163. D 164. C 165. A 166. C 167. D 168. A 169. D 170. C 171. D 172. C 173. D

174. In a regenerative air preheater, the heat is transferred
A. by direct mixing
B. by extracting some gas from the furnace
C. from heating an intermediate material and then heating the air from this material
D. none of these

175. The height of chimney in a steam power plant is governed by
A. flue gases quantity
B. the draught to be produced
C. control of pollution D. rating of the plant

176. The combined operation of thermal power plant and MHD plant results in efficiency in the range of
A. 40 to 50% B. 50 to 60%
C. 60 to 70% D. 70 to 80%

177. Which of the following power stations is mainly used to cover peak loads on the system?
A. Brown coal power station
B. Pit coal power station
C. Nuclear power station
D. Pump storage water power station

178. In one year there are about
A. 365 full tidal cycles B. 730 full tidal cycles
C. 705 full tidal cycles D. none of these

179. Tick out the incorrect combination.
A. Pelton turbine 'Needle valve'
B. Francis turbine 'Wicket gate'
C. Kaplan turbine 'Needle valve'
D. none of these

180. The oldest geothermal power plant is in
A. France B. Germany
C. Italy D. USA

181. The MHD system generates
A. AC only B. DC only
C. AC and dc both D. either AC or DC

182. A Kaplan turbine is
A. inward flow, impulse turbine
B. outward flow, reaction turbine
C. a high head mixed flow turbine
D. low head axial flow turbine

183. The efficiency of electrostatic precipitator is as high as
A. 99.6 percent B. 90 percent
C. 85 percent D. 80 percent

184. A diesel power plant is best suited for
A. base load plant B. peak load plant
C. general purpose plant
D. none of these

185. A gas turbine power plant usually suits for
A. peak load operation B. base load operation
C. permanent operation D. none of these.

186. Efficiency is a secondary consideration in case of
A. peak load plants
B. base load plants
C. both peak load and base load plants
D. None of these

187. Which of the following plants will take least time in starting from cold conditions to full load operation?
A. Nuclear power plant B. Steam power plants
C. Hydroelectric plants D. Gas turbine plant

188. During load shedding system
A. voltage is reduced B. frequency is reduced
C. some loads are switched off
D. power factor is changed

189. Which of the following can be used as fuel for closed cycle gas turbine plant?
A. Wood B. Furnace oil
C. Any gas D. Any of these

190. Which of the following plant is expected to have the longest expected life?
A. Hydroelectric plant B. Steam plant
C. Diesel plant D. None of these

191. Which of the following plants is almost inevitably used as base load plant?
A. Gas turbine plant B. Diesel engine plant
C. Pumped storage plant
D. Thermal power plant

192. A thermal power station was designed to burn coal containing 12% ash. When the plant actually started operating, coal having 22% ash was made available. Which unit of the plant will need major modifications?
A. Water treatment plant
B. Pulverising unit
C. Ash handling unit D. Cooling towers

193. Anything having heat value can be used as fuel in case of
A. diesel engines B. petrol engines
C. closed cycle turbines D. none of these

194. The useful life of a diesel engine in a power plant is expected to be
A. one year B. five years
C. fifteen years D. fifty years.

195. Hydrogen is used for cooling of large size generators, because it
A. has high thermal conductivity
B. is light
C. offers reduced fire risk
D. all of these

196. Major share of power produced in India is through
A. Thermal power plants
B. Hydroelectric power plants
C. Nuclear power plants
D. Diesel power plants

Ans. 174. B 175. B 176. B 177. D 178. C 179. C 180. C 181. C 182. D 183. A 184. A 185. A 186. A 187. D 188. C 189. D 190. A 191. D 192. C 193. C 194. C 195. D 196. A

107. Which of the following place is not associated with nuclear power plants in India?
A. Delhi B. Tarapur
C. Kota D. Narora

198. Tidal power plant is being installed in India near
A. Gulf of Cambay B. Cochin
C. Vishakhapatnam D. None of these

199. A nuclear power plant is invariably used as a
A. base load plant B. peak load plant
C. standby plant D. spinning reserve plant

200. For underwater movement submarines are given supply by
A. diesel engines B. air motors
C. batteries D. steam engines

201. Ships are generally run by
A. diesel engines B. hydraulic turbines
C. nuclear power plants D. steam engines

202. A turbo jet aeroplane has
A. no blades B. two blades
C. three blades D. four blades

203. An interconnected system has the following plants:
I. Run off river plant
II. Nuclear power plant
III. Steam power plant
IV. Diesel engine plant
Which two plants can be exclusively used for base load?
A. I and II B. II and III
C. II and IV D. II only

204. Air will not be the working substance in a
A. diesel engine B. petrol engine
C. open cycle gas turbine
D. closed cycle gas turbine

205. Which alternator will have more number of poles?
A. Coupled to hydraulic turbine
B. Coupled to steam turbine
C. Coupled to gas turbine
D. None of these

206. An alternator usually runs at slow speed on coupled to
A. gas turbine B. steam turbine
C. hydraulic turbine D. diesel engine

207. Direct conversion of heat into electrical energy is possible through
A. magneto-hydrodynamic generators
B. solar cells
C. fuel cells D. none of these

208. In a fuel cell electricity is produced by
A. combustion of fuel in absence of oxygen
B. oxidation of fuel
C. thermionic D. none of these

209. A diesel power plant is generally used as
A. base load station B. peak load station
C. both (A) and (B) are correct
D. none of these

210. For a gas turbine the pressure ratio may be in the range
A. 2 to 3 B. 3 to 5
C. 16 to 18 D. 18 to 22

211. A closed cycle gas turbine works on
A. Carnot cycle B. Rankine cycle
C. Joule cycle D. Atkinson cycle

212. Thermal efficiency of closed cycle gas turbine plant increases by
A. reheating B. intercooling
C. regenerator D. all of these

213. With the increase in pressure ratio thermal efficiency of a simple gas turbine plant with fixed turbine inlet temperature
A. decreases B. increases
C. first increases and then decreases
D. first decreases and then increases

214. Thermal efficiency of a gas turbine cycle with ideal regenerative heat exhanger is
A. equal to work ratio B. less than work ratio
C. more than work ratio D. unpredictable

215. In a two stage gas turbine plant reheating after first stage
A. decreases thermal efficiency
B. increases thermal efficiency
C. does not affect thermal efficiency
D. none of these

216. In a two stage gas turbine plant reheating after first stage
A. increases work ratio B. decreases work ratio
C. does not affect work ratio
B. None of these

217. In a two stage gas turbine plant, with intercooling and reheating
A. both work ratio and thermal efficiency improve
B. work ratio improves but thermal efficiency decreases
C. thermal efficiency improves but work ratio decreases
D. both work ratio and thermal efficiency decreases

218. For a jet propulsion unit, ideally the compressor work and turbine work are
A. equal
B. unequal
C. not related to each other
D. unpredictable

219. Greater the difference between jet velocity and aeroplane velocity

Ans. 197. A 198. A 199. B 200. C 201. A 202. A 203. A 204. D 205. A 206. C 207. A 208. B 209. B
210. B 211. C 212. D 213. C 214. A 215. D 216. A 217. B 218. A

A. greater is the propulsive efficiency
B. less is the propulsive efficiency
C. unaffected is the propulsive efficiency
D. none of these

220. The blades of the gas turbines rotor are made of
A. carbon steel
B. stainless steel
C. high alloy steel
D. high nickel alloy (Nimic 80)

221. Maximum temperature in a gas turbine is the order of
A. 700°C B. 900°C
C. 1600°C D. 2100°C

222. In gas turbines, high thermal efficiency is obtained in
A. closed cycle B. open cycle
C. in both the cycles D. none of these

223. In a gas turbine plant, a regenerator increases
A. work output B. pressure ratio
C. thermal efficiency D. none of these

224. In boilers, the feed water treatment is done mainly for removing
A. corrosion B. scale formation
C. embrittlement D. all of these

225. Compounding of steam turbine is done for
A. reducing the work done
B. increasing the rotor speed
C. reducing the rotor speed
D. balancing the turbine

226. Topping turbines are
A. low pressure condensing units
B. high pressure non-condensing units
C. low pressure non-condensing units
D. high pressure condensing units

227. Higher calorific value is the heating value of fuel
A. without water vapour which are formed by combustion
B. with water vapour which are formed by combustion
C. either A or B D. None of these

228. Thermal efficiency of the steam plant is of the order of
A. 30% B. 50%
C. 60% D. 70%

229. Ultimate analysis of fuel is the determination of percentage of
A. total carbon by weight
B. total carbon by weight, unit weight of H_2, O_2, N_2, sulphur and ash
C. ash, volatile matter and moisture
D. None of these

230. The proximate analysis of coal gives
A. various chemical constituents, carbon, hydrogen, oxygen and ash
B. fuel constituents as percentage by weight, of moisture, volatile, fixed carbon and ash
C. percentage by weight of moisture, volatile matter, fixed carbon and ash
D. none of these

231. In coal preparation plant, magnetic separators are used to remove
A. dust B. clinkers
C. iron particles D. sand

232. Load carrying capacity of belt conveyor is about
A. 20 to 40 tons/hr B. 50 to 100 tons/hr
C. 100 to 150 tons/hr D. 150 to 200 tons/hr

233. Method which is commonly applied for unloading the coal for small power plant is
A. lift trucks B. coal accelerators
C. tower cranes D. belt conveyor

234. Bucket elevators are used for
A. carrying coal in horizontal direction
B. carrying coal in vertical direction
C. carrying coal in any direction
D. all of these

235. The amount of air which is supplied for complete combustion is called
A. primary air B. secondary air
C. tertiary air D. Either of these

236. Example of overfeed type stoker is
A. chain grate B. spreader
C. travelling grate D. all of these

237. Where unpulverised coal has to be used and boiler capacity is large, the stoker which is used is
A. underfeed stoker B. overfeed stoker
C. either A or B D. none of these

238. Steam pressure in a steam power station, which is usually kept nowadays is of the order of
A. 20 kgf/cm^2 B. 50 kgf/cm^2
C. 100 kgf/cm^2 D. 150 kgf/cm^2

239. Economisers improve boiler efficiency by
A. 1 to 5% B. 4 to 10%
C. 10 to 12% D. 15 to 20%

240. In steam power station, the choice of high temperature steam is for
A. increasing the efficiency of boiler alone
B. increasing the efficiency of turbinealone
C. increasing overall efficiency
D. None of these

241. Capacity of turbine and generator are related as
A. Turbine (kW) $= \dfrac{\text{generator kW}}{\text{generator efficiency}}$

Ans.													
219. B	220. D	221. A	222. A	223. C	224. D	225. C	226. B	227. B	228. A	229. B	230. C	231. C
232. B	233. B	234. B	235. B	236. D	237. B	238. D	239. B	240. D	241. A			

B. Turbine (kW) = generator kW x generator efficieny
C. Turbine (kW) = generator kW
D. Turbine (kW) = (generator kW)^2

242. The capacity of large turbo-generator varies from
A. 20 to 100 MW B. 50 to 300 MW
C. 70 to 400 MW D. 100 to 650 MW

243. Coking coals are those which
A. burn completely
B. burn freely
C. do not form ash
D. form lumps or masses of coke

244. Primary air is that air which is used to
A. reduce the flame length
B. increase the flame length
C. transport and dry the coal
D. provide air round the burners for getting optimum combustion

245. Presence of sulphur in coal will result in
A. corroding air heaters
B. spontaneous combustion during coal storage
C. causing clinkering and slagging
D. all of these

246. Pulverised fuel is used for
A. saving fuel B. better burning
C. obtaining more heat D. all of these

247. Combustible elements in the fuel are
A. carbon and hydrogen
B. carbon, hydrogen and sulphur
C. carbon, hydrogen and nitrogen
D. carbon, hydrogen and ash

248. Calorific value of diesel oil is about
A. 5000 kcal/kg B. 7000 kcal/kg
C. 9000 kcal/kg D. 11000 kcal/kg

249. In a regenerative surface condenser
A. there is one pump to remove air and condensate
B. there are two pumps to remove air and condensate
C. there are three pumps to remove air, vapour and condensate
D. there is no pump, the condensate gets removed by gravity

250. Pipes carrying steam are generally made up of
A. steel B. cast iron
C. copper D. aluminium

251. For the safety of a steam boiler the number of safety valves fitted are
A. four B. three
C. two D. one

252. Steam turbines commonly used in steam power station are
A. condensing type B. non-condensing type
C. either of these D. none of these

253. Belt conveyor can be used to transport coal at inclinations upto
A. 30° B. 60°
C. 80° B. 90°

254. The maximum length of a screw conveyor is about
A. 30 metres B. 40 metres
C. 60 metres D. 100 metres

255. The efficiency of a modern boiler using coal and heat recovery equipment is about
A. 25 to 30% B. 40 to 50%
C. 65 to 70% D. 85 to 90%

256. The average ash content in Indian coals is about
A. 50% B. 10%
C. 15% D. 20%

257. Load centre in a power station is
A. centre of coal fields
B. centre of maximum load of equipments
C. centre of gravity of electrical system
D. None of these

258. In steam turbines the reheat factor
A. increases with the increase in number of stages
B. decreases with the increase in number of stages
C. remains same irrespective of number of stages
D. none of these

259. The thermal efficiency of the engine with condenser as compared to without condenser, for a given pressure and temperature of steam, is
A. higher B. lower
C. same as long as initial pressure and temperature are unchanged
D. none of these

260. In Jet type condensers
A. cooling water passes through tubes and steam surrounds them
B. steam passes through tubes and cooling water surrounds them
C. steam and cooling water mix
D. steam and cooling water do not mix

261. In a shell and tube surface condenser
A. steam and cooling water mix to give the condensate
B. cooling water passes through the tubes the steam surrounds them
C. steam passes through the cooling tubes and cooling water surrounds them
D. all of these varying with situation

262. In a steam power plant the function of condenser is
A. to maintain pressure below atmospheric to increase work output from the prime mover
B. to receive large volumes of steam exhausted from steam prime mover
C. to condense large volumes of steam to water which may be used again in boiler
D. all of these

Ans. 242. B 243. D 244. C 245. D 246. B 247. B 248. D 249. B 250. A 251. C 252. A 253. A 254. A 255. D 256. D 257. C 258. A 259. A 260. C 261. B 262. D

263. Stage efficiency of steam turbine is
A. $blade^{\eta}/\eta$ nozzle B. $nozzle^{\eta}/\eta$ blade
C. $nozzle^{\eta} \times \eta$ blade D. none of these

264. Reheat factor in steam turbines depends on
A. exit pressure only B. stage efficiency only
C. initial pressure and temperature only
D. all of these

265. For multistage steam turbine, reheat factor is defined as
A. stage efficiency × nozzle efficiency
B. cumulative enthalpy drop × η_{nozzle}
C. $\frac{\text{cumulative enthalpy drop}}{\text{isentropic enthalpy drop}}$
D. $\frac{\text{isentropic enthalpy drop}}{\text{cumulative enthalpy drop}}$

266. The value of reheat factor normally varies from
A. 0.5 to 0.6 B. 0.9 to 0.95
C. 1.02 to 1.06 D. 1.2 to 1.6

267. Steam turbines are governed by which of the following methods?
A. Throttle governing
B. Nozzle control govering
C. By pass governing
D. All of these

268. The effect of considering friction losses in steam nozzle for the same pressure ratio leads to
A. increase in exit velocity from the nozzle
B. decrease in exit velocity from the nozzle
C. no change in exit velocity from the nozzle
D. increase or decrease depending upon the exit quality of steam

269. The effect of considering friction in steam nozzles for the same pressure ratio leads to
A. increase in dryness fraction of exit steam
B. decrease in dryness fraction of exit steam
C. no change in the quality of exit steam
D. may decrease or increase of dryness fraction of exit steam depending upon inlet quality

270. In case of impulse steam turbine
A. there is enthalpy drop in fixed and moving blades
B. there is enthalpy drop only in moving blades
C. there is enthalpy drop in nozzles
D. none of these

271. De Laval turbine is
A. pressure compounded impulse turbine
B. velocity compounded impulse turbine
C. simple single wheel impulse turbine
D. simple single wheel reaction turbine

272. In case of reaction steam turbine
A. there is enthalpy drop both in fixed and moving blades
B. there is enthalpy drop only in fixed blades
C. there is enthalpy drop only in moving blades
D. none of these

273. For the induced draught, the fan is located
A. near bottom of chimney
B. near bottom of furnace
C. at the top of the chimney
D. anywhere permissible

274. The pressure at the furnace is minimum in case of
A. forced draught system
B. induced draught system
C. balanced draught system
D. natural draught system

275. For maximum discharge of hot gases through the chimney, the height of hot-gas column producing draught is
A. twice the height of chimney
B. equal to the height of chimney
C. half the height of chimney
D. none of these

276. The efficiency of chimney is approximately
A. 80% B. 40%
C. 20% D. 0.25%

277. In balanced draught system the pressure at force fan inlet
A. is greater than pressure at chimney outlet
B. is less than pressure at chimney outlet
C. approximately same as that at chimney outlet
D. none of these

278. The ratio of exit pressure in inlet pressure for maximum mass flow rate per unit area of steam through a nozzle when steam is initially dry saturated is
A. 0.6 B. 0.578
C. 0.555 D. 0.5457

279. The ratio of exit pressure to inlet pressure for maximum mass flow rate per unit area of steam through nozzle when steam is initially superheated is
A. 0.555 B. 0.578
C. 0.5457 D. 0.6

280. The critical pressure ratio of a convergent nozzle is defined as
A. the ratio of outlet pressure to inlet pressure of nozzle
B. the ratio of inlet pressure to outlet pressure of nozzle
C. the ratio of outlet pressure to inlet pressure only when mass flow rate per unit area is minimum
D. the ratio of outlet pressure to inlet pressure only when mass flow rate per unit is maximum

281. The draught which a chimney produces is called
A. induced draught B. natural draught
C. forced draught D. balanced draught

Ans. 263. C 264. D 265. C 266. C 267. D 268. B 269. A 270. C 271. C 272. A 273. A 274. C 275. B 276. D 277. C 278. B 279. C 280. D 281. B

282. The darught produced by steel chimney as compared to that produced by brick chimney for the same height is
A. less B. more
C. same D. may be more or less

283. In a boiler installation, the natural draught is produced due to the fact that
A. furnace gases being light go through the chimney giving place to cold air from outside to rush in
B. pressure at the grate due to cold column is higher than the pressure at chimney base due to hot column
C. at the chimney top the pressure is more than its environmental pressure
D. all of these

284. The draught produced for a given height of the chimney and given mean temperature of chimney gases
A. decreases with increase in outside temperature
B. increases with increase in outside air temperature
C. remains the same irrespective of outside air temperature
D. may increase or decrease with increase in outside air temperature

285. The draught produced by chimney of given height at given outside temperature
A. decreases if the chimney gas temperature increases
B. increases if the chimney gas temperature increases
C. remains same irrespective of chimney gas temperature
D. may increase or decrease

286. For forced draught systems, the function of chimney is mainly
A. to produce draught to accelerate the combustion of fuel
B. to discharge gases high up in the atmosphere to avoid hazard
C. to reduce the temperature of the hot gases discharged
D. none of these

287. Artificial draught is produced by
A. induced fan B. forced fan
C. induced and forced fans
D. All of these

288. The draught in locomotive boilers is produced by
A. forced fan
B. chimney
C. steam jet
D. only motion of locomotive

289. For the same draught produce the power of induced draught fan as compared to forced draught fan is
A. less B. more
C. same D. not predictable

290. The artificial draught normally is designed to produce
A. less smoke B. more draught
C. less chimney gas temperature
D. All of these

291. Regenerative cycle thermal efficiency
A. is always greater than simple Rankine thermal efficiency
B. is greater than simple Rankine cycle thermal efficiency only when steam is put at particular pressure
C. is same as simple Rankine cycle thermal efficiency
D. is always less than simple Rankine cycle thermal efficiency

292. In a regenerative feed heating cycle, the optimum value of the fraction of steam extracted for feed heating
A. decreases with increase in Rankine cycle efficiency
B. increases with increase in Rankine cycle efficiency
C. is unaffected by increase in Rankine cycle efficiency
D. None of these

293. In a regenerative feed heating cycle, the greatest economy is affected when steam is extracted
A. from only suitable point of steam turbine
B. from several places in different stages of steam turbine
C. only from the last stage of steam turbine
D. only from the first stage turbine

294. The maximum percentage gain in regenerative feed heating cycle thermal efficiency
A. increases with number of feed heaters increasing
B. decreases with number of feed heaters increasing
C. remains same unaffected by number of feed heaters
D. none of these

295. A steam power station requires space
A. equal to diesel power station
B. more than diesel power station
C. less than diesel power station
D. none of these

296. Economiser is used to heat
A. air B. feed water
C. flue gases D. all of these

Ans. 282. B 283. B 284. A 285. B 286. B 287. D 288. C 289. B 290. D 291. A 292. B 293. B 294. A 295. B 296. B

297. The modern steam turbines are
A. impulse turbines B. reaction turbines
C. impulse-reaction turbines
D. none of these
298. Rankine cycle is
A. reversible cycle B. irreversible cycle
C. constant volume cycle
D. none of these
299. The overall efficiency of thermal power plant is equal to
A. Rankine cycle efficiency
B. Carnot cycle efficiency
C. Regenerative cycle efficiency
D. Boiler efficiency × turbine efficiency × generator efficiency
300. Rankine cycle efficiency of a good steam power plant may be in the range of
A. 15 to 20% B. 35 to 45%
C. 70 to 80% D. 90 to 95%
301. Rankine cycle operating on low pressure limit of p_1 and high pressure limit of p_2
A. has higher thermal efficiency than the Carnot cycle operating between same pressure limits
B. has lower thermal efficiency than Carnot cycle operating between same pressure limits
C. has same thermal efficiency as Carnot cycle operating between same pressure limits
D. may be more or less depending upon the magnitude of p_1 and p_2
302. Rankine efficiency of a steam power plant
A. improves in summer as compared to that in winter
B. improves in winter as compared to that in summer
C. is unaffected by climatic conditions
D. none of these
303. Carnot cycle comprises of
A. two isentropic processes and two constant volume processes
B. two isentropic processes and two constant pressure processes
C. two isothermal processes and two constant pressure processes
D. none of these
304. In Rankine cycle the work output from the turbine is given by
A. change of internal energy between inlet and outlet
B. change of enthalpy between inlet and outlet
C. change of entropy between inlet and outlet
D. change of temperature between inlet and outlet
305. Regenerative heating, i.e. bleeding steam to reheat feed water to boiler
A. decreases thermal efficiency of the cycle
B. increases thermal efficiency of the cycle
C. does not affect thermal efficiency of the cycle
D. may increase or decrease thermal efficiency of the cycle depending upon the point of extraction of steam
306. The primary sources of the energy are
A. coal, oil and uranium
B. hydrogen, oxygen and water
C. wind, biomass and geothermal
D. none of these
307. In India largest thermal power station is located at
A. Kota B. Sarni
C. Chandrapur D. Neyveli
308. The %age of O_2 by weight in atmospheric air is
A. 18% B. 23%
C. 77% D. 79%
309. The percentage of O_2 by volume in atmospheric air is
A. 21% B. 23%
C. 77% D. 79%
310. The proper indication of incomplete combustion is
A. high CO content in flue gases at exit
B. high CO_2 content in flue gases at exit
C. high temperature of flue gases
D. the smoking exhaust from chimney
311. The blades of impulse turbine are
A. symmetrically shaped around the centre line
B. asymmetrically shaped around the centre line
C. both A and B D. none of these
312. Function of air pump in condenser is to
A. remove water B. maintain vacuum
C. maintain atmospheric pressure
D. none of these
313. Wet air pump removes
A. air only B. only condensate
C. both air and condensate
D. none of these
314. In high head hydro-power plant the velocity of water in penstock is about
A. 1 m/s B. 4 m/s
C. 7 m/s D. 12 m/s
315. Pelton turbine is suitable for high head and
A. high discharge B. low discharge
C. both low and high discharge
D. none of these
316. In reaction turbine, function of the draft tube is
A. to increase the flow rate
B. to reduce water hammer effect
C. to convert kinetic energy of water to potential energy by a gradual expansion in divergent part
D. none of these
317. Francis turbine is usually used for
A. low head installation up to 30 m
B. medium head installation from 30 to 180 m

Ans. 297. C 298. A 299. D 300. B 301. A 302. B 303. B 304. B 305. B 306. A 307. C 308. B 309. A 310. A 311. A 312. B 313. C 314. C 315. B 316. C 317. B

C. high head installation above 180 m
D. all the heads

318. In Francis turbine runner, the number of blades is generally of the order of
A. 1–2 B. 4–6
C. 6–8 D. 12–16

319. Francis, Kaplan and propeller turbines fall under the category of
A. impulse turbine B. reaction turbine
C. impulse reaction combined
D. axial flow

320. The specific speed (Ns) of the turbine is given by
A. $N_S = N\frac{\sqrt{P}}{H^{5/4}}$ B. $N_S = N\frac{\sqrt{P}}{H^{3/4}}$
C. $N_S = N\frac{\sqrt{P}}{H^{3/2}}$ D. $N_S = N\frac{\sqrt{P}}{H^{2/3}}$

321. The expression for power output (P) in kW, of a hydroelectric station is
A. $\frac{Q\omega H\eta_0}{0.736 \times 75}$ B. $\frac{0.736 Q\omega H}{75 \times \eta_0}$
C. $\frac{0.75 Q\omega H\eta_0}{0.736}$ D. $\frac{0.736 Q\omega H\eta_0}{75}$

322. Pelton turbines are mostly of
A. horizontal type B. vertical type
C. inclined type D. none of these

323. The annual depreciation of a hydro-power plant is about
A. 0.5 to 1.5% B. 10 to 15%
C. 15 to 20% D. 20 to 25%

324. The power output from a hydroelectric power plant depends on three parameters
A. head, type of dam and discharge
B. head, discharge and efficiency of the system
C. efficiency of the system, type of draft tube and type of turbine used
D. type of dam, discharge and type of catchment area

325. The empirical relation for determination of number of buckets (Z) for Pelton turbine in terms of jet ratio (m) is given by
A. Z = 15 m + 0.5 B. Z = 0.5 m + 15
C. $Z = \frac{m}{0.5} + 15$ D. none of these

326. Francis turbine is usually used for
A. high heads B. medium heads
C. low heads D. all of these

327. Water hammer is developed in
A. penstock B. draft tube
C. turbine D. surge tank

328. Operating charges are minimum in the case of ..... for same power output
A. gas turbine plant B. hydel plant
C. thermal plant D. nuclear plant

329. The commonly used material for condenser tubes is
A. aluminium B. cast iron
C. admiralty brass D. mild steel

330. The largest proportion of plane carbon steel is produced by
A. acid bessemer
B. basic bessemer
C. electric furnace process
D. none of these

331. In pumped storage scheme, the generator is also used as
A. induction generator synchronous condenser
B. induction generator or synchronous motor
C. synchronous generator or induction generator
D. synchronous generator or synchronous condenser

332. The economic prospects of nuclear power generation in India depends on a great extent on successful development of
A. fast reactors using enriched uranium as the fuel thus avoiding import of heavy water
B. thermal reactors using natural uranium as the fuel and avoiding import of enriched uranium
C. breeder reactors using thorium as a blanket material and U^{235} as the fuel
D. reactor using plutonium as the fuel, natural uranium as blanket and liquid sodium as the coolant

333. As the operating voltage and consequently the electric stress on the dielectric of solid type cable is increased from a low value, the dielectric power factor cos ϕ remains almost unchanged up to a certain value of the stress beyond which cos ϕ increases very rapidly. This is due to increase in
A. resistivity of the dielectric material
B. ionisation in the voids present in the dielectric
C. core-to-core capacitance of the cable
D. core-to-earth capacitance of the cable

334. The most appropriate operating speed in rpm of generators used in Thermal, Nuclear and Hydro-power plants would respectively be
A. 3000, 300 and 1500 B. 3000, 3000 and 300
C. 1500, 1500 and 3000 D. 1000, 900 and 750

335. Arrange the following in the correct sequences in which the flue gas passes through them after coming out of the boiler in a thermal power station
1. ID fan
2. Air preheater
3. economiser
4. electrostatic precipitator

Ans. 318. D 319. B 320. A 321. D 322. A 323. A 324. B 325. B 326. B 327. B 328. B 329. C 330. B 331. B 332. C 333. B 334. B

Select the correct answer from the codes given below:

Codes:

A. 4, 3, 2, 1 B. 3, 2, 1, 4
C. 2, 1, 4, 3 D. 1, 4, 3, 2

336. The flow-duration curve at a given head of a hydroelectric plant is used to determine the
A. total power available at the site
B. total units of energy available
C. load-factor at the plant
D. diversity-factor for the plant

337. The operation of a nuclear reactor is controlled by controlling the multiplication factor (K), defined as

$$K = \frac{\text{number of neutrons of any one generation}}{\text{number of neutrons of immediately proceeding generation}}$$

The power-level of the reactor can be increased by
A. raising the value of K above 1 and, keeping it at that raised value
B. raising the value of K above 1, but later bringing it back to $K = 1$
C. lowering the value of K below 1 and, keeping it at that lowered value
D. lowering the value of K below 1, but later bringing it back to $K = 1$.

338. In the optimum generator scheduling of different power plants, the minimum fuel cost is obtained when
A. only the incremental fuel cost of each plant is the same
B. the penalty factor of each plant is the same
C. the ratio of the incremental fuel cost to the penalty factor of each plant is the same
D. the incremental fuel cost of each plant multiplied by its penalty factor is the same

339. If a fix amount of power is to be transmitted over a certain length with fixed power loss
A. the weight of the conductor required will be proportional to the voltage
B. the weight of the conductor required will be inversely proportional to the voltage and that of the power factor of the load
C. the weight of the conductor required will be inversely proportional to the square of the voltage and that of power factor of the load
D. the weight of the conductor is proportional to the square of the voltage and directly proportional to the power factor of the load

340. The electrical power being transmitted from generating point is called
A. secondary transmission
B. secondary distribution
C. primary transmission
D. primary distribution

341. The transmission line feeding power on either side of the main transmission line is called
A. secondary distribution
B. secondary transmission
C. primary transmission
D. primary distribution

342. The transmission lines which feed different substations represent
A. primary transmission
B. secondary transmission
C. primary distribution
D. secondary distribution

343. In a power plant, a reserve generating capacity which is not in service but is in operation is known as
A. hot reserve B. cold reserve
C. firm power D. spinning reserve

344. The advantage of transmitting power at high voltage is
A. magnitude of current will be small
B. it will reduce the voltage drop in the line impedance
C. power loss will be less
D. all of these

345. By transmitting power at high voltage it will
A. give poor voltage regulation
B. give better voltage regulation
C. not affect voltage regulation
D. none of these

346. The single phase line at 230 V and 3-phase line at 400 V which feed different domestic circuits and industrial loads represent
A. primary distribution
B. secondary distribution
C. secondary transmission
D. both primary and secondary distribution both

347. Transmitting power at high voltage requires more
A. maintenance and protection of the equipment
B. faster controls for minimising the arcing contacts
C. larger controls for minimising the arcing contacts
D. all of these

348. For transmitting power at high voltage, the ground clearance required is 7.5 m with a sag of 2.5 m, the height above the ground is
A. 5.1 m. B. 9.9 m
C. 7.5 m D. 7.591 m

349. The minimum clearance above the ground of the lowest conductor, for low and medium voltage lines, across a street must be
A. 4 m B. 5.791 m
C. 6.791 m D. 7.591 m

Ans. 335. B 336. A 337. B 338. D 339. C 340. C 341. B 342. C 343. A 344. D 345. B 346. B 347. D 348. B 349. C

350. If the conductors are symmetrically spaced then the capacitance of individual conductor to ground can be determined by

A. $C = \dfrac{\left(\dfrac{2\pi\varepsilon_0}{D-r}\right)}{\log_e \dfrac{D-r}{r}}$ Farads/metre

B. $C = \log_e \dfrac{\left(\dfrac{D-r}{r}\right)}{2\pi\varepsilon_0}$ Farads/metre

C. $C = \dfrac{\varepsilon_0}{\log_e \dfrac{D-r}{r}}$ Farads/metre

D. $C = \dfrac{2\varepsilon_0}{\log_e \dfrac{D-r}{r}}$ Farads/metre

where ε_0 = permittivity of free space: 8.854×10^{-12} F/m; and r = radius of the conductor.

351. If power is transmitted in DC 2-wire system, the volume of the two conductors used will be
A. directly proportional to the square of the length of the conductor
B. inversely proportional to the square of the voltage
C. directly proportional to the power transmitted
D. all of these

352. If power is transmitted in 3-phase, 3-wire system, the volume of copper required will be
A. directly proportional to the square of the length
B. inversely proportional to the square of voltage
C. inversely proportional to the power factor of the load
D. directly proportional to the power transmitted and square of the length of the conductor and inversely proportional to the square of the voltage and square of the power factor of load

353. If same power is transmitted on copper conductor line and on aluminium conductor line, then
A. the weight of the line is more with aluminium
B. the volume of conductor is more with aluminium
C. number of towers required in aluminium will be more
D. All of these

354. The conductors used for transmitting power must have following characteristics
A. it should have low value of specific resistance
B. it should be light in weight and not brittle
C. it should have low cost and high tensile strength
D. all of these

355. Copper conductors are generally used for transmission lines because it
A. has longer life and high conductivity
B. is strong enough to allow long spans
C. requires more support
D. requires more insulators

356. Wooden poles are not used because
A. they easily rot below the ground level and limited to line voltages of 20 kV
B. of light in weight
C. they can easily be fitted and erected
D. of cheap in cost

357. Steel poles are generally used for transmission lines because
A. it has more mechanical strength and life
B. it occupies less space and gives better appearance
C. both A and B are correct
D. it has high cost

358. If the voltage is increased n times taking into consideration the constant efficiency of the transmission, the size of the conductor will be
A. increased n times that of the original
B. decreased to $1/n$ times that of the original
C. decreased to $1/n^2$ times that of the original
D. increased to n^2 times that of the original

359. A single phase line delivers power at 11 kV at a distance of 15 km. first at lagging power factor and then at the same leading power factor. The efficiency of the transmission line will be
A. more at leading pf B. less at lagging pf
C. same at leading and lagging pf
D. more at lagging pf and less at leading pf

360. The most important component required for transmission lines are
A. poles and towers
B. conductors
C. insulators
D. poles, towers, conductors and insulators

361. Different types of line supports used for transmission lines include
A. RCC and PCC poles B. steel towers
C. steel poles
D. wooden poles, steel poles, RCC and PCC poles and steel towers

362. The RCC and PCC poles are commonly employed for medium transmission lines, because they
A. require no maintenance
B. give better mechanical strength and provide insulating property
C. are strong and have long life
D. all of these

363. The steel towers are employed in for transmission lines because they have
A. longer life
B. more mechanical strength
C. better workability D. all of these

Ans. 350. A 351. D 352. D 353. D 354. D 355. D 356. A 357. C 358. C 359. C 360. D 361. D 362. D 363. D

364. Different types of insulators used for transmission lines are
A. pin type only
B. suspension type only
C. strain type only
D. pin type, suspension type and shackle type

365. Sag of conductors between two poles can be determined by
A. WL^2/T B. $WL^2/2T$
C. $WL^2/8T$ D. $WL^2/16T$
where
W = weight per unit length of conductor in kg/m.
L = distance between the two poles
T = tension in conductor in kg·m

366. ACSR stands for
A. All Copper Standard Reinforced
B. Aluminium Conductors Steel Reinforced
C. Aluminium Copper Steel Reinforced
D. All Copper Steel Reinforced Conductors.

367. The basis ingradients of an ac supply consists of
A. generation B. transmission
C. distribution D. All of these

368. In case of high voltage transmission, the control devices must have
A. slow controls B. medium controls
C. faster controls D. all of these

369. The size of the transformer
A. increases with rise of transmitting voltage
B. decreases with rise of transmitting voltage
C. is independent of voltage
D. none of these

370. Wooden poles can be used for a span of maximum up to
A. 100 metres B. 160 metres
C. 300 metres D. 200 metres

371. Wooden poles can be used up to voltage of
A. 11 kV B. 33 kV
C. 66 kV D. 20 kV

372. Wooden poles are not used for long transmission lines because
A. their life is short
B. they require regular inspection
C. their use is limited to 20 kV transmission
D. all are correct

373. When alternating potential difference is applied across a transmission line, the capacitance between two conductors is
A. directly proportional to length of the line
B. inversely proportional to distance between the two conductors
C. directly proportional to length and inversely proportional to distance between two conductors
D. independent of spacing between two conductors

374. The regulation of transmission line is
A. the change in sending end voltage and receiving end voltage
B. the percentage change in voltage at the sending end when full load is thrown off
C. the percentage change in voltage at the receiving end when full load is thrown off, the conditions at sending end remaining constant
D. the percentage change in voltage at the receiving end when full load is not thrown off

375. When potential difference across two conductors is gradually increased whose spacing is large as compared to their diameter, at certain potential, the air gets ionised which results into faint luminous glow of bluish colour accompanied by a hissing sound, this phenomena is called
A. ionisation B. corona
C. breakdown of insulation
D. luminous

376. If the conductors are uniform and smooth, the glow occur due to corona will be
A. more bright at certain points as compared with the remaining length
B. uniform and smooth throughout the length
C. uniform but not smooth throughout the length
D. not uniform but smooth throughout the length

377. The factors which affect the corona are
A. physical state of the atmosphere
B. physical conditions of the line conductors
C. physical state of atmosphere and physical condition of the line conductors
D. None of these

378. The phenomenon of corona reduces
A. the losses of the line
B. the electrostatic stresses between the conductors
C. interference in communication lines
D. the power loss

379. The Corona effect can be minimised by increasing
A. the length of the conductors
B. spacing between conductors
C. diameter of the conductors
D. both spacing between conductors and diameter of the conductor

380. Corona helps in avoiding
A. electrostatic stresses between the conductors
B. effect of lightning
C. effect of surges on the transmission lines
D. all are correct

381. Series capacitors are used for improving line
A. capacitive reactance
B. inductive reactance effect
C. voltage D. regulation

Ans. 364. D 365. C 366. B 367. D 368. C 369. A 370. B 371. D 372. D 373. C 374. C 375. B 376. C 377. B 378. B 379. D 380. D 381. C

382. For a distortionless transmission line (G = shunt conductance between two wires)
A. $RL = GC$ B. $RL - GC = 0$
C. $RC - GL = 0$ D. $G = RLC$

383. Corona effect can be detected by
A. high temperature of conductors
B. vibrations in conductors
C. faint luminous glow of bluish colour
D. emission of smoke

384. For which shape of conductor, the Corona loss will be least?
A. Circular B. Oval
C. Flat
D. It is independent of shape of conductor

385. Guarding transmission line
A. improves power factor
B. reduces earth capacitance of the lowest unit
C. reduces transmission losses
D. improves regulation

386. In case of overhead lines, due to presence of earth
A. capacitance decreases
B. inductance decreases
C. capacitance increases
D. inductance increases

387. What is the disadvantage of connecting phase modifiers?
A. Risk of interruption of supply
B. Decrease of power transmitted
C. Increase of power transmitted
D. Increase of short circuit current of the system

388. A transmission line is normally designed so as to work at a voltage just
A. below visual critical voltage
B. below disruptive critical voltage
C. above visual critical voltage
D. above disruptive critical voltage

389. Stringing chart represents a graph of
A. tension-temperature
B. sag-temperature
C. load-temperature
D. .resistance-temperature

390. Three materials, X, Y and Z having permittivities 2.5, 3.0 and 4.0 are used in a single core cable. The materials on innermost and outer layer respectively should be
A. X and Z B. Y and Z
C. X and Z D. Z and X

391. Phase modifiers in AC transmission lines are
A. Synchronous machines
B. Induction machines
C. DC machines
D. Transformers

392. In comparison with synchronous motors, the phase modifiers are equipped with
A. bigger shafts and bearings
B. smaller shafts and bearings
C. smaller flywheel
D. none of these

393. Ferranti effect on long overhead lines is experienced when it is
A. lightly loaded B. on full load at upf
C. on full load at 0.8 pf lag
D. in all these cases

394. If the supply frequency increases, the skin effect is
A. decreased B. increased
C. remains same D. none of these

395. An overhead transmission line has appreciable inductance because the loop it forms has
A. large cross-sectional area
B. small cross-sectional area
C. zero cross-sectional area
D. none of these

396. The skin effect does not depend on the
A. nature of material B. size of wire
C. supply frequency D. ambient temperature

397. If the conductor diameter decreases, inductance of the line
A. decreases B. increases
C. remains same D. none of these

398. If the capacitance between two conductors of a three-phase line is 4 μF, then the capacitance of each conductor to neutral is
A. 4 μF B. 8 μF
C. 12 μF D. 16 μF

399. If the length of the transmission line is increased, its capacitance
A. increases B. decreases
C. remains same D. none of these

400. For the overhead transmission lines, the self GMD method is used to evaluate
A. inductance B. capacitance
C. inductance and capacitance
D. none of these

401. If the pf of load decreases, the line losses
A. increase B. decrease
C. remain constant D. none of these

402. The nature of line constants in the rigorous solution of transmission lines is
A. distributed parameter
B. lumped
C. both A and B are correct
D. none of these

403. Which voltage determines the insulation strength of an EHV line?
A. Switching over voltage
B. Lightning over voltage

Ans.	381. C	382. C	383. C	384. A	385. B	386. B	387. D	388. B	389. A	390. D	391. A	392. B	393. A
	394. B	395. A	396. D	397. B	398. B	399. A	400. A	401. A	402. A	403. B			

C. Power frequency over voltage
D. None of these

404. The most important cause of power loss in the transmission line is the
A. resistance B. reactance
C. capacitance D. none of these

405. For a given V_R and I_R, the regulation of the transmission line increases with the decrease in pf for
A. lagging loads B. leading loads
C. both (A) and (B) are correct
D. none of these

406. An overhead line has a span of 220 m; conductor weighs 604 kg/km, ultimate tensile strength is 5758 kg. Assuming factor of safety as 2, the sag will be
A. 2 m B. 1.5 m
C. 1.27 m D. 3 m

407. In comparison with the steady state stability limit, transient stability limit is
A. always less B. always more
C. sometimes less D. sometimes more

408. Consider a transmission line transmitting a fixed amount of power. The efficiency of transmission increases, when
A. voltage decreases, power factor remains constant
B. voltage increases and power factor also increases
C. voltage decreases, power factor decreases
D. voltage is constant, power factor decreases

409. In any transmission line, AD - BC is equal to
A. 1 B. 0
C. 2 D. none of these

410. The capacitance of a transmission line is neglected up to about
A. 80 km B. 120 km
C. 180 km D. 200 km

411. The capacitance of a transmission line is a
A. series element B. shunt element
C. neither series nor shunt element
D. none of these

412. The generalised constants A and D of the transmission line have
A. no dimension B. dimension of ohm
C. dimension of mho D. None of these

413. Which of the two generalised constants in a transmission line are equal?
A. B and C B. A and D
C. A and B D. B and D

414. The dimensions of constants B and C respectively are
A. ohm and mho B. mho and ohm
C. ohm and volt D. volt and ohm

415. The line constants of a transmission line are
A. uniformly distributed
B. lumped
C. non-uniformly distributed
D. none of these

416. The shunt admittance of transmission line is 3 micromho. Its complex notation in mho is
A. $3 \times 10^6 \angle 90°$ B. $3 \times 10^6 \angle 0°$
A. 3×10^6 D. none of these

417. If a fixed amount of power is to be transmitted over a given length with certain power loss, it can be said the volume of conductor material required is
A. proportional to voltage
B. inversely proportional to square of voltage
C. inversely proportional to square of voltage and load pf
D. None of these

418. The charging current in the transmission line
A. lags the voltage by 90°
B. leads the voltage by 45°
C. leads the voltage by 90°
D. leads the voltage by 180°

419. The efficiency of a transmission line
A. increases with the decrease in load pf
B. is independent of load pf
C. increases with the increase in load pf
D. decreases with the increase in load pf

420. The sending end voltage of a transmission line will be equal to the receiving end voltage on load when the pf of the load is
A. leading and $IR \cos \phi_R = IX_L \sin \phi_R$
B. lagging and $IR\, \phi_R > IX_L \sin \phi_R$
C. lagging and $IR \cos \phi_R = IX_L \sin \phi_R$
D. leading and $IR \cos \phi_R > IX_L \sin \phi_R$

421. While finding out the relation between V_S and V_R, capacitance is neglected in a
A. short transmission line
B. long transmission line
C. medium transmission line
D. none of these

422. There is no skin effect in DC transmission lines.
A. True B. False

423. The capacitance of the transmission line will be affected if distance between conductors and earth is varied.
A. True B. False

424. String chart is useful in
A. finding the sag in the conductor
B. in the design of tower
C. in the design of insulator string
D. finding the distance between tower

425. The presence of earth wire in case of overhead lines
A. increases the capacitance

Ans. 404. A 405. A 406. C 407. A 408. B 409. A 410. A 411. B 412. A 413. D 414. A 415. A 416. A 417. C 418. C 419. C 420. A 421. A 422. A 423. A 424. A 425. A

B. increases the inductance
C. decreases the capacitance
D. decreases the inductance

426. For an exciting AC transmission line, the string efficiency is 80%. Now if DC voltage is supplied for the same set up, the string efficiency will be
A. 80% B. less than 80%
C. more than 80% D. 100%

427. The sag of a power line with 100 m span is 1 m, what will be the sag if the height of the line is increased by 20%?
A. 1.6 m B. 1.5 m
C. 2 m D. 2.5 m

428. If ACSR conductor has specification as 48/7, which of the following statement is correct for the conductor?
A. The conductor has 48 strands
B. The conductor has 55 strands
C. The conductor has 48 aluminium and 7 steel strands
D. The conductor has 48 steel and 7 aluminium strands

429. The potential across the various discs of suspension string is different because of
A. series capacitance B. shunt capacitance
C. series and shunt capacitance
D. None of these

430. In a string of suspension insulators, maximum voltage appears across the unit
A. nearest to the conductor
B. nearest to the cross-arm
C. maximum voltage is independent of the location
D. none of these

431. The span of transmission line between towers takes the form of
A. hyperbola B. parabola
C. catanary D. none of these

432. In a transmission line, sag depends only on
A. the tension in the conductor
B. height of the tower
C. conductor material D. all of these

433. The coefficient of reflection for current for an open ended line is
A. 1.0 B. 0.5
C. –1.0 D. zero

434. The coefficient of reflection of voltage for a short circulated line is
A. 1.0 B. –1.0
C. 0 D. 2.0.

435. For the same voltage boost, the reactive power capacity is more for a
A. shunt capacitor
B. series capacitor
C. same for both series and shunt
D. none of these

436. If the loading of the line corresponds to the surge impedance loading, the voltage at the receiving end is
A. greater than the sending end
B. less than the sending end
C. equal to the sending end
D. none of these

437. The voltage of a generator and an infinite bus are given as $0.92 \angle 10°$ and $1.0 \angle 0°$ respectively. The generator acts as a
A. shunt coil B. shunt capacitor
C. data is insufficient to find
D. none of these

438. As the transmission voltage increases, the percentage resistance drops
A. increases B. decreases
C. remains constant D. none of these

439. For a lumped inductive load, with increase in supply frequency
A. P and Q increase
B. P increases, Q decreases
C. P decreases, Q increases
D. P and Q decrease

440. An overhead line with surge impedance 400 ohm is terminated through a resistance R. A surge travelling over the line does not suffer any reflection at the junction if the value of R is
A. 20 ohms B. 200 ohms
C. 800 ohms D. none of these

441. For a synchronous phase modifier, the load angle is
A. 0° B. 25°
C. 30° D. none of these

442. The charging reactance of 50 km length of line is 1500 ohms, the charging reactance for 100 km, length of line will be
A. 1500 ohms B. 3000 ohms
C. 750 ohms D. 600 ohms

443. Phase modifier normally installed in the case of
A. short transmission lines
B. medium length lines
C. long length lines D. for all length lines

444. If the capacitance between two conductors of a 3-phase line is 6 μF, the capacitance of each conductor to neutral will be
A. 3 μF B. 12 μF
C. 6 μF D. 1.5 μF

445. The power loss in an overload transmission line is mainly due to
A. impedance of the line conductor
B. resistance of each conductor

Ans. 426. B 427. A 428. C 429. B 430. A 431. B 432. D 433. A 434. B 435. B 436. C 437. A 438. B 439. C 440. D 441. A 442. C 443. C 444. B 445. B

C. inductance of each conductor
D. capacitance of each conductor

446. The economic size of conductor is determined by
A. Kirchhoff's Law B. Ohm's Law
C. Kelvin's Law D. Faraday's Law

447. The economic transmission voltage is one for which the
A. current is restricted to a particular value
B. cost of transmission is minimum
C. total cost of conductor material is minimum
D. height of towers is minimum.

448. The purpose of guarding the transmission line is to
A. reduce the earth capacitance of the lowest unit
B. increase the earth capacitance of the lowest unit
C. reduce the transmission losses
D. none of these

449. The string efficiency of an insulator can be increased by
A. increasing the number of strings in the insulator
B. reducing the number of strings
C. correct grading of insulators of various capacitances
D. None of these

450. If the shunt capacitance is reduced then the string efficiency is
A. increased B. decreased
C. constant
D. independent of shunt capacitance

451. The most commonly used material for insulators of overhead lines is
A. porcelain B. glass
C. mica D. PVC

452. What type of insulator will be used if the direction of the transmission line is to be changed?
A. Pin-type B. Suspension type
C. Strain type D. Shakle type

453. For a given receiving-end voltage in a long transmission line, the sending-end voltage is more than the actual value when calculated by
A. load end capacitance method
B. nominal-T method
C. nominal-π method
D. none of these

454. For a transmission line, the constants A, B, C and D are related as
A. $AD - BC = 1$ B. $BC - AD = 1$
C. $\sqrt{AD - BC} = 0.5$ D. $\sqrt{AD - BC} = 1.2$

455. Transposition of transmission lines is adopted to make
A. reactance uniform throughout the line
B. zero reactance throughout the lint
C. capacitance uniform throughoul the line
D. none of these

456. If the voltage is increased by n times, the size of the conductor would
A. increase by n times B. reduce by $1/n$ times
C. increase by n times D. reduce by $1/n^2$ times

457. The maximum value of power factor can be
A. 1 B. 0.9
C. 0.8 D. 0.7

458. Which method of raising depreciation fund gives heavier changes in the initial years of plant life?
A. Straight line method
B. Sinking fund method
C. Both (A) and (B) are correct
D. None of these

459. Power plant having maximum demand more than installed capacity will have utilisation factor
A. less than 100% B. equal to 100%
C. more than 100% D none of these

460. The annual cost characteristics of two plants are given as
$C_1 = 5\text{ kW}_1 + 0.02\text{ kWhr}$ and
$C_2 = 7\text{ kW}_2 + 0.015\text{ kWhr}$
Which plant can be selected for base load operation?
A. Plant 1 B. Plant 2
C. Any of the two plants
D. None of these

461. In order to have lower cost of electrical generation
A. the load factor and diversity factor should be high
B. the load factor should be low but diversity factor should be high
C. the load factor should be high but diversity factor should be low
D. the load factor and diversity factor should be low

462. It is always economical to improve the pf of an installation to
A. zero
B. unity
C. a little less than unity
D. a little more than unity

463. The choice of number and size of units in a station is governed by best compromise between
A. a plant load factor and plant capacity factor
B. plant capacity factor and plant use factor
C. plant load factor and plant use factor
D. none of these

464. If a plant has zero reserve capacity, the plant load factor always
A. equals plant capacity factor
B. is greater than plant capacity factor
C. is less than plant capacity factor
D. none of these

Ans. 446. C 447. B 448. A 449. C 450. A 451. A 452. B 453. C 454. A 455. A 456. D 457. A 458. A 459. C 460. B 461. A 462. C 463. B 464. B

465. A synchronous machine has higher capacity for
A. leading pf B. lagging pf
C. it does not depend upon the pf of the machine
D. none of these

466. A synchronous phase modifier as compared to synchronous motor used for mechanical loads has
A. larger shaft and higher speed
B. smaller shaft and higher speed
C. larger shaft and smaller speed
D. smaller shaft and smaller speed

467. In a transmission line, if the load is suddenly increased in the receiving end of the system, the phase shift will
A. decrease B. become zero
C. increase D. not be affected

468. In India, the system adopted for transmission of electric power has
A. three-phase three wire system
B. three-phase four wire system
C. both (A) and (B) are correct
D. none of these

469. Tick the correct statement.
A. The greater the power to be transmitted smaller is the economic transmission voltage
B. The greater the power to be transmitted, larger is the economic transmission voltage
C. DC transmission is inferior to AC transmission
D. All are correct

470. A lightning arrester connected to the line and earth in a power system
A. protects the terminal equipment against travelling surges
B. protects the transmission line against direct lightning stroke
C. suppresses high frequency oscillations in the line
D. reflects back the travelling wave approaching it

471. Buses for load flow studies are classified as (i) the load bus, (ii) the generator bus and (iii) the slack bus.
The correct combination of the pair of quantities specified having their usual meaning of different buses is

	Load bus	Generator	Slack bus
A.	P, V	P, Q	P, δ
B.	P, Q	P, V	V, δ
C.	V, Q	P, δ	P, Q
D.	P, δ	Q, V	Q, δ

472. The criterion for selection of size of conductor for a feeder is
A. Voltage drop B. Corona loss
C. Temperature rise D. Radio interference

473. In a high voltage DC transmission scheme, reactive power is needed both for the rectifier at the sending-end and, for the inverter at the receiving-end. During the operation of such a DC link the rectifier receiver
A. lagging reactive power and the inverter supplies leading reactive power
B. leading reactive power and the inverter supplies lagging reactive power
C. lagging reactive power and the inverter supplies lagging reactive power
D. leading reactive power and the inverter supplies leading reactive power

474. The incremental fuel cost for two generating units are given by
$I_{C_1} = 25 + 0.2\,P_{G_1}$
$I_{C_2} = 32 + 0.2\,P_{G_2}$
where P_{G_1} and P_{G_2} are real powers generated by the units.
The economic allocation for a total load of 250 MW, neglecting transmission loss is given by
A. P_{G_1} = 140.25 MW, P_{G_2} = 109.75 MW
B. P_{G_1} = 109.75 MW, P_{G_2} = 140.25 MW
C. $P_{G_1} = P_{G_2}$ = 125 MW
D. P_{G_1} = 100 MW, P_{G_2} = 150 MW

475. The annual cost of a generating station can be expressed in the form of Rs ($A + B \times$ kW + C $\times$ kWh) where A, B, C are constants and kW and kWh represent capacity of the station and energy-generated per year respectively. The choice between the "base-load station" and the "peak-load station" basically depends on the fact that
A. factor B should be less for the peak-load station and factor C should be less for the base-load station
B. factor C should be less for the peak-load station and factor B should be less for the base-load station
C. factor B and C should both be less for the base-load station
D. factors B and C should both be less for the peak-load station

476. In the radial distribution system, the distributor is fed
A. from both ends B. from the centre
C. from one end D. at different points

477. The radial distribution system is adopted when
A. the distributor end nearest to the supply is to be kept heavily loaded
B. the electrical energy is generated at low voltage
C. the power station is situated at the centre of the load
D. both B and C are correct

478. In the ring main distribution system, the distributor is fed

Ans. 465. A 466. B 467. C 468. A 469. B 470. A 471. B 472. D 473. A 474. A 475. A 476. C 477. D

A. by one feeder B. by two feeders
C. by four feeders D. at different points

479. In the interconnected system during peak load hours, any area fed from one generating station
A. can be fed from other generating station
B. cannot be fed from other generating station
C. can be trasferred to other station
D. none of these

480. In the interconnected system, at the time of fault on any part of the feeder, the continuity of supply to all consumers
A. cannot be maintained
B. can be maintained by disconnecting the faulty section
C. can be resorted only after removing the fault
D. none of these

481. In the underground cable system
A. there are less chances of faults to occur
B. it is difficult to locate fault and repair
C. fault does not occur at all
D. fault occurs very rare and difficult to locate and repair

482. The underground cable system for 11 kV can work up to a length of
A. 50 miles B. 75 miles
C. 100 miles D. 400 miles

483. The disruptive critical voltage will
A. decrease with the increase of moisture content in air
B. increase with the increase of moisture content in air
C. increase with the decrease of moisture content in air
D. decrease with the decrease of moisture content in air

484. In a transmission line, if the load is suddenly increased in the receiving end of the system, the phase shift will
A. decrease B. become zero
C. increase D. not be affected

485. What is the network shown in figure called?

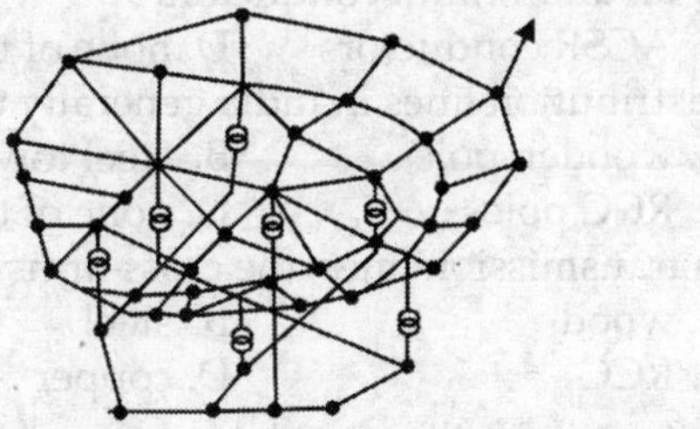

A. Radial system B. Low voltage network
C. Ring system D. Mesh system

486. Ferranti effect on long overhead lines is experienced when the
A. line is lightly loaded B. power factor is unity
C. power factor is leading
D. Corona effect is dominant

487. When is the effective resistance due to skin effect equal to that of ohmic resistance?
A. When the current is uniformly distributed
B. When the frequency is high
C. When the frequency is low
D. None of these

488. If the power transmitted is to be increased, the surge impedance should be
A. decreased B. increased
C. zero D. none of these

489. For a loss-less transmission line, surge impedance when compared with the characteristic impedance is
A. more B. less
C. equal D. none of these

490. In a two-port network of transmission line, if Z_{OC} is the impedance when receiving end open-circuited; and Z_{SC} is the impedance when receiving end short-circuited, then surge impedance would be
A. $Z_C = Z_{SC}/Z_{OC}$ B. $Z_C = Z_{OC}/Z_{SC}$
C. $Z_C = (Z_{OC}/Z_{SC})^2$ D. $Z_C = \sqrt{Z_{OC} \times Z_{SC}}$

491. For which of the reasons mentioned below does the electric supply company want to improve the power factor?
A. To reduce the effective current
B. To reduce the reactive current
C. To reduce the apparent current
D. To reduce the voltage drop at the generating station

492. The inductive interference between power and communication lines can be minimised by
A. transposition of communication lines
B . transposition of power lines
C. increasing the distance between the conductors
D. transposition of power lines and increasing the distance between the conductors

493. Pin type insulators are used for transmission and distribution of electric power at voltage up to
A. 11 kV B. 33 kV
C. 66 kV D. 110 kV

494. The type of insulator that can be used either in horizontal or vertical position is
A. pin-type B. suspension type
C. strain type D. shakle type

495. Ferranti effect on transmission line means
A. rise in receiving end voltage on lagging load
B. rise in receiving end voltage on leading load
C. rise in receiving end voltage on no- load
D. none of these

496. The value of the breakdown stress of atmospheric air is nearly
A. 0.3 kV/cm B. 3 kV/cm
C. 30 kV/cm D. 300 kV/cm

Ans. 478. B 479. B 480. B 481. D 482. C 483. A 484. C 485. D 486. A 487. A 488. A 489. C 490. D
491. B 492. B 493. A 494. D 495. C 496. C

497. What is the main purpose of using bundled conductors for EHV lines?
A. To get reduced reactance
B. To get reduced surge impedance
C. To get reduced voltage gradient
D. To get corona free operation

498. If sag in an overhead line increases, tension in the line
A. increases B. decreases
C. remains same D. none of these

499. By using guard ring, string efficiency is
A. increased B. decreased
C. constant
D. independent of shunt capacitance

500. Suspension type insulators are used for voltages beyond
A. 220 V B. 400 V
C. 11 kV D. 33 kV

501. The discs of the strain insulators are used in
A. vertical plane only B. horizontal plane
C. both the planes D. none of these

502. The AC resistance of conductor is more than DC resistance due to
A. proximity effect B. sking effect
C. both (A) and (B) are correct
D. none of these

503. If the length of the transmitting tower is decreased, the inductance of the line will
A. increase B. decrease
C. remain same D. none of these

504. The skin effect of the conductor increases the effective value of
A. resistance of the conductor
B. reactance of the conductor
C. capacitance of the conductor
D. none of these

505. What will be the surge impedance of a 25 miles long underground cable, if its value for 50 miles long cable is 50 ohms?
A. 25 ohms B. 50 ohms
C. 100 ohms D. none of these

506. Skin effect is
A. proportional tó frequency
B. proportional to square of frequency
C. inversely proportional to frequency
D. independent of frequency

507. Large industrial consumers are supplied electrical energy at
A. 400 V B. 66 kV
C. 11 kV D. 400 kV

508. The voltage drop is the main consideration while designing a
A. feeder B. distributor
C. service main D. none of these

509. In a 3-wire system, the current in the +ve outer is 200 A and in –ve outer is 100 A, the current in the neutral is
A. 0 A B. 100 A
C. 200 A D. 300 A

510. The distribution transformer is generally connected in
A. Delta/Delta B. Star/Delta
C. Delta/Star D. Star/Star

511. Which of the following power distribution system gives the better reliability?
A. Ring-main system B. Radial system
C. DC 3-wire system D. None of these

512. 3-phase, 4-wire ac system of distribution is used for
A. unbalanced load B. balanced load
C. all types of load D. none of these

513. In a balance 3-phase, 4 wire system, the phase sequence is RYB. If the voltage of *R*-phase $230 \angle 0°$ V then for *B*-phase it will be
A. $230 \angle -120°$ V B. $230 \angle 120°$ V
C. $230 \angle 0°$ V D. $230 \angle 90°$V

514. In AC system addition and subtraction of currents are done
A. vectorially B. arithmetically
C. both (A) and (B) are correct
D. none of these

515. A booster is connected in
A. series with the feeder
B. parallel with the feeder
C. both (A) and (B) are correct
D. none of these

516. Most of the high voltage transmission lines in India are
A. underground B. overhead
C. Both (A) and (B) are correct
D. None of these

517. The distributors for residential areas are in
A. single phase B. three phase three wire
C. three phase four wire
D. none of these

518. Overhead lines generally use
A. copper conductors
B. all aluminium conductors
C. ACSR conductors D. none of these

519. Distribution lines in India generally use
A. wooden poles B. steel towers
C. RCC poles D. none of these

520. In transmission lines, the cross-arms are made of
A. wood B. steel
C. RCC D. copper

521. Transmission line insulators are made of
A. glass B. porcelain
C. iron D. PVC

522. High voltage transmission lines use
A. pin insulators B. suspension insulators
C. shakel insulator D. none of these

Ans.													
	497. D	498. B	499. A	500. D	501. A	502. C	503. C	504. A	505. B	506. A	507. B	508. B	509. B
	510. C	511. A	512. B	513. B	514. A	515. A	516. B	517. C	518. C	519. C	520. B	521. B	522. B

523. The material commonly used for sheaths of underground cables is
A. lead B. rubber
C. copper D. iron

524. The material commonly used for insulation in HV cables is
A. lead B. paper
C. rubber D. None of these

525. The purpose of earthing electric appliances is to
A. provide safety against shock
B. ensure that the appliance works properly
C. ensure that the appliance gets full voltage
D. none of these

526. For a fuse wire of diameter *d*, the fusing current is proportional to
A.. $\sqrt{d}$ B. $d^{1.2}$
C. $d^{1.5}$ D. none of these

527. A lighting sub-circuit has 100-W lamps. If the rated current of the fuse in this circuit is 5 A, the maximum number of light points on this circuit without violating regulations and safety should be
A. 8 B. 10
C. 12 D. none of these

528. The insulation resistance to earth of an installation should be about
A. 1 mega ohm B. 0.5 mega ohm
C. 0.2 mega ohm D. none of these

529. The resistance of human body is about
A. 100 ohms B. 500 ohms
C. 1000 ohms D. 10 ohms

530. The loads on distributors systems are generally
A. unbalanced B. balanced
C. either balanced or unbalanced
D. none of these

531. The power factor of industrial loads is generally
A. unity B. lagging
C. leading D. zero

532. Conductor configuration determines
A. inductance B. sag
C. voltage D. current

533. Why are shunt reactors connected at the receiving end of long transmission line system?
A. To increase terminal voltage
B. To compensate voltage rise caused by capacitive charging at light load
C. To improve power factor
D. None of these

534. The highest transmission voltage in India is
A. 400 kV B. 230 kV
C. 220 kV D. 132 kV

535. The most commonly used generation voltage in India is
A. 11 kV B. 6.6 kV
C. 2.2 kV D. 33 kV

536. The voltage of the single phase supply to residential consumers is
A. 400 V B. 230 V
C. 220 V D. 110 V

537. The conductors of the overhead lines are
A. solid B. stranded
C. both solid and stranded
D. none of these

538. The material generally used for armour or high voltage cables is
A. aluminium B. steel
C. brass D. copper

539. The main criterion for the design of a distributor is
A. voltage drop B. corona loss
C. temperature rise D. all of these

540. Ring main system is usually used for
A. urban areas B. rural areas
C. both urban and rural areas
D. none of these

541. The loss load factor is given by (if F is the load factor)
A. $0.25\ F + 0.75\ F^2$ B. $0.25\ F^2 + 0.75\ F$
C. $0.75\ F + 0.25\ F^2$ D. $0.35\ F + 0.7\ F^2$

542. Which of the following is considered as superior quality of coal
A. Peat B. Coke
C. Bituminous coal D. Lignite

543. Multi core cables generally use
A. circular conductors
B. sector-shaped conductors
C. rectangular conductors
D. square conductors

544. The minimum clearance between the ground and a 220 kV line is about
A. 5.5 m B. 10.5 m
C. 7.0 m D. 4.3 m

545. For a particular kW rating of an induction motor, the kVAR rating of the shunt capacitor required is
A. more for high rated speed motor
B. more for lower rated speed motor
C. independent of speed
D. none of these

546. The criterion for selection of size of conductor for a feeder is
A. voltage drop B. corona loss
C. temperature rise D. radio interference

547. The weight of locomotive is 120 tonnes. If the axle load is not to exceed 20 tonnes, the number of axles required is
A. 12 B. 3
C. 7 D. none of these

548. The interest and deprecision charges on the conductor is Rs. $(18a + 25)$ and the annual cost of

Ans. 523. A 524. B 525. A 526. C 527. A 528. A 529. C 530. A 531. B 532. A 533. B 534. A 535. A
536. C 537. B 538. B 539. A 540. A 541. A 542. C 543. B 544. B 545. B 546. C 547. D

energy loss is Rs. $(90/a)$. The area of cross section of the conductor
A. 5 units B. $\sqrt{5}$ units
C. 6.6 units D. none of these

549. The balancer machine connected to the heavily loaded side works as a
A. motor B. generator
C. motor-generator set D. none of these

550. Which of the following distribution system is used for balanced loads?
A. Single phase 2-wire AC system
B. Three-phase, 3-wire AC system
C. Three-phase, 4-wire AC system
D. None of these

551. The life of underground cables is taken as
A. 10 years B. 20 years
C. 25 years D. 40 years

552. For the hydroelectric plant, the life of a RCC dam is taken as
A. 5 years B. 10 years
C. 40 years D. 100 years

553. A 500 kW plant costs Rs. 1000 per kW installed. Fixed charges are estimated at 14% and operating cost is 13 paise per kWh. The plant averages 200 kW for 5000 hours of the year, 450 kW for 1200 hrs and 80 kW for the remaining period. The average cost of production of electricity per kWh will be nearly
A. 5 paise B. 12 paise
C. 17 paise D. 27 paise

554. Two tariffs are offered
(I) Rs. 200 plus 5 paise per unit
(II) A flat rate of 30 paise per unit
From the above it can be concluded that
A. Tariff I will give lower charges for consumption up to 800 kWh.
B. Tariff I will give lower charges for consumption of more than 800 kWh
C. Tariff II will give lower charges for consumption of more than 800 kWh
D. Both will give identical charges beyond 1500 kWh

555. Which of the following distribution system is used for combined power and lightning load?
A. Single phase 2-wire AC system
B. Three-phase, 3-wire AC system
C. Three-phase, 4-wire AC system
D. None of these

556. The total connected load and the number of points on a lightning subcircuit should not exceed
A. 700 W and 8 points respectively
B. 800 W and 10 points respectively
C. 900 W and 10 points respectively
D. 1000 W and 10 points respectively

557. The total connected load and the number of points on a power subcircuit should not exceed
A. 2000 W and 2 points respectively
B. 3000 W and 3 points respectively
C. 4000 W and 2 points respectively
D. 4000 W and 3 points respectively

558. The minimum permissible size of the aluminium cable for lightning circuits is
A. 1.2 sq mm B. 1.5 sq mm
C. 2.2 sq mm D. 3.5 sq mm

559. The minimum permissible size of the aluminium cable for power circuits is
A. 1.5 sq mm B. 2.0 sq mm
C. 2.5 sq mm D. 3.0 sq mm

560. A light controlled by two switches requires
A. single pole single throw switches
B. double pole single throw switches
C. single pole double throw switches
D. None of these

561. The earth continuity conductor is generally made of
A. copper B. aluminium
C. steel D. bronze

562. The minimum permissible size of the earth continuity conductor is
A. 6 sq mm B. 8 sq mm
C. 10 sq mm D. 12 sq mm

563. The length of earthing electrode is about
A. 0.5 m B. 1.0 m
C. 2.5 m D. 5.0 m

564. The resistance of the earth continuity conductor is about
A. 20 ohms B. 2 ohms
C. 6 ohms D. 50 ohms

565. Depreciation cost of a plant is calculated by
A. Straight line method
B. Diminishing value method
C. Sinking fund method
D. Any of these

566. Which method of depreciation charge estimation gives the heaviest charges during early years of plant life?
A. Straight line method
B. Diminishing value method
C. Sinking fund method
D. All give same charges

567. Power generation equipment in a thermal power plant costs Rs 15,75,000 and has a useful life of 25 years. If the salvage value of the plant be interest at 5%, the amount to annual instalment by straight line method will be

Ans.	548. A	549. B	550. B	551. D	552. D	553. C	554. B	555. C	556. B	557. A	558. B	559. C	560. C
	561. C	562. A	563. C	564. B	565. D	566. B							

A. Rs. 30,000 B. Rs. 60,000
C. Rs. 60,500 D. Rs. 80,750

568. In the above question, which of the following could be the amount to be saved annually for replacement of equipment at the end of 25 years, by Sinking Fund Method?
A. Rs. 31,400 B. Rs. 60,000
C. Rs. 66,000 D. Rs. 96,500

569. Which of the following could be the instalment for diminishing value method?
A. Rs. 96,000 B. Rs. 60,000
C. Rs. 48,500 D. Rs. 31,400

570. In the figure shown below which curve represents the variation of cost of generation per kWh with the load factor, for a thermal power plant?

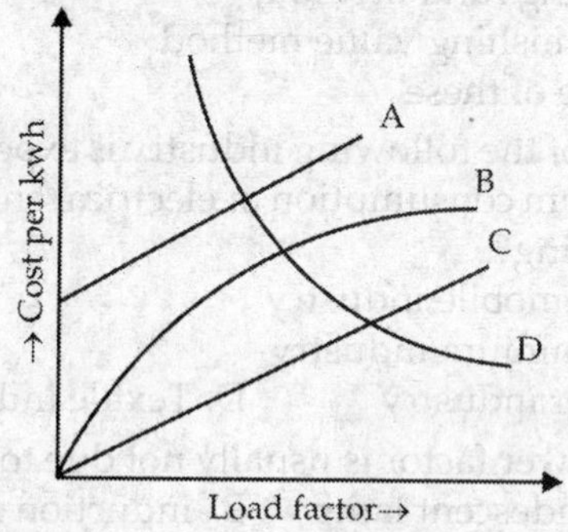

A. Curve A B. Curve B
C. Curve C D. Curve D

571. For a power plant the expenditure on which of the following item is expected to be negligible?
A. Wages B. Taxes
C. Insurance D. Publicity

572. A 120-MW generator is usually
A. air cooled B. hydrogen cooled
C. oxygen cooled D. nitrogen cooled

573. A fuse is inserted in
A. phase wire B. neutral wire
C. both the phase wire and the neutral wire
D. earth continuity conductor

574. Which of the following will not contribute to low power factor?
A. Partially loaded induction motors
B. Replacement of fluorescent lamps with incandescent lamps
C. Use of rectifiers instead of synchro-nous motor-generator sets for DC power supply
D. Increased installation of electronic equipment, air-conditioning units, etc.

575. Which of the following may not be the effect of low plant operating power factor?
A. Overloaded transformers
B. Reduced voltage level
C. Improved illumination from lighting
D. Overloaded cables.

576. Identify the incorrect relation
A. Power factor = kW/kVA
B. kW = kVA × Power factor
C. kVA × kW = Power factor
D. kVA = kW/Power factor

577. The power factor of a system on a 460 V, 3 phase, 60 Hz, in which the ammeter indicates 100 amp and the wattmeter reads 652 kW will be
A. 0.95 B. 0.78
C. 0.65 D. 0.55

578. The simple subtraction of kilowatts from total kVA equals the kVAR when the power factor is
A. unity B. 0.707
C. lagging D. zero

579. The annual peak load on 30 MW power station is 25 MW. The power station supplies loads having maximum demands of 10 MW, 8.5 MW and 4.5 MW. If the annual load factor is 45%, then the average load will be
A. 1025 kW B. 1125 kW
C. 1425 kW D. 1625 kW

580. For the above question, total energy supplied in a year will be
A. 9875000 kWh B. 8345000 kWh
C. 7450000 kWh D. 6395000 kWh

581. Diversity factor for the above question will be
A. 3.80 B. 1.02
C. 1.12 D. 1.22

582. In the same question, demand factor is
A. 0.75 B. 0.83
C. 0.89 D. 0.45

583. Which of the following appliances will offer the least load?
A. Electric Shaver B. Hair Dryer
C. Television D. Vacuum cleaner

584. Which of the following appliances will offer the maximum load?
A. Toaster B. Refrigerator
C. Hot plate D. Electric Bell

585. Which load has the highest load factor?
A. Load A B. Load B
C. Load C D. Load D

586. If all the loads are connected to single source of power, the maximum load on the station will be
A. 9 K/4 B. 2 K
C. 3K D. none of these

587. Four different loads connected to a power plant are shown in the figure below which load has the least value of average load?

Ans.	567. B	568. A	569. A	570. D	571. D	572. B	573. A	574. B	575. C	576. C	577. N	578. A	579. B
	580. A	581. C	582. B	583. A	594. C	585. B	586. C						

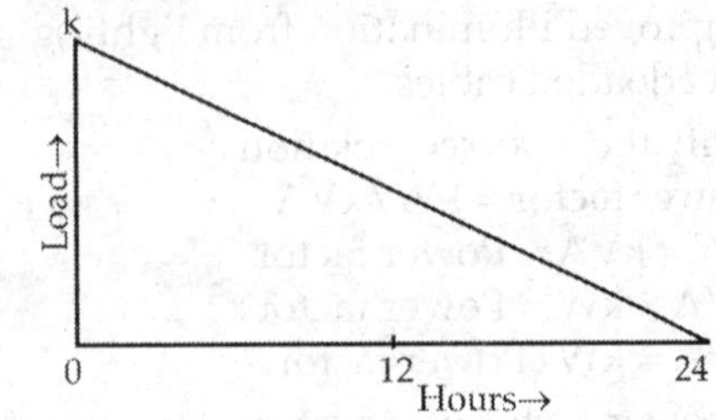

A

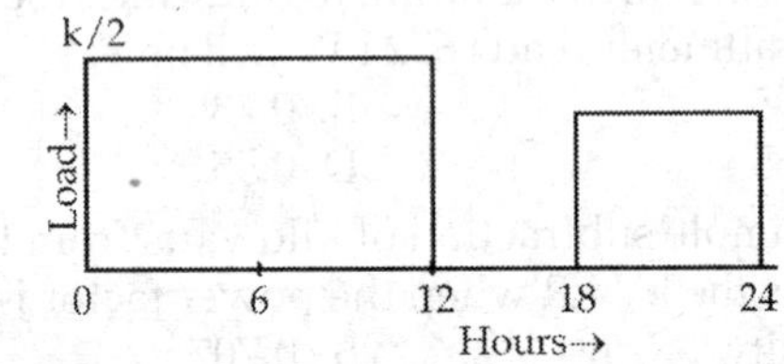

B

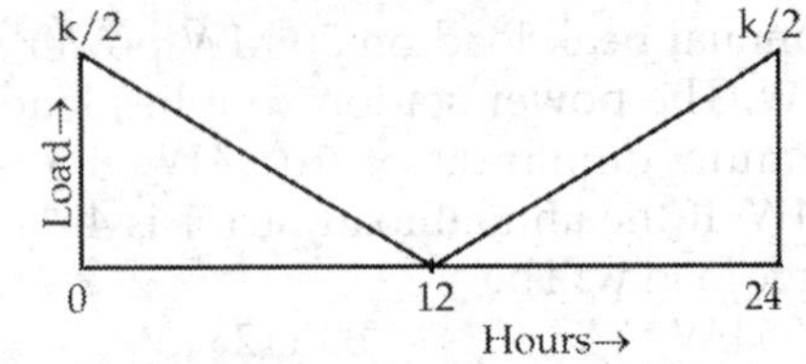

C

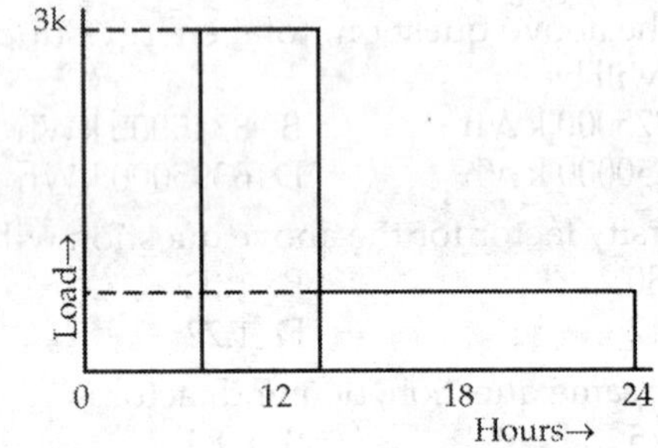

D

A. Load A B. Load B
C. Load C D. Load D

588. The maximum load on the station will occur at
A. 0 h B. 6 h
C. 12 h D. None of these

589. Which of the following industry will consume maximum power per tonne of product?
A. Zinc B. Aluminium
C. Alloy steel D. Cement

590. Which load has the highest value of average load?
A. Load A B. Load B
C. Load C D. Load D

591. Which load has the least load factor?
A. Load A B. Load B
C. Load C D. Load D

592. Maximum span in case of wooden poles is usually restricted to
A. 50 m B. 175 m
C. 100 m D. 200 m

593. A consumer takes a steady load of 200 kW at a power factor of 0.85 lagging for 8 hours per day and 315 days per annum. The annual payment under the tariff of Rs. 80 per annum per kVA plus 10 paise per kWh will be
A. Rs. 5040 B. Rs. 50400
C. Rs. 69200 D. Rs. 84400

594. In the above case load factor of the station will be
A. 0.29 B. 0.31
C. 0.44 D. 0.56

595. A certain plant has fixed cost of Rs. 40,000 and a salvage value of Rs. 4000 at the end of a useful life of 20 years. The depreciated value of the plant at the end of 10 years will be least when calculated by
A. Straight line method
B. Sinking fund method
C. Diminishing value method
D. None of these

596. Which of the following industry is expected to have maximum consumption of electrical energy, during processing?
A. Automobile industry
B. Aluminium industry
C. Paper industry D. Textile industry

597. Low power factor is usually not due to
A. incandescent lamp B. induction motors
C. arc lamps D. fluorescent tubes

598. If some reserve capacity is available in a power plant, its use factor
A. is always greater than its capacity factor
B. equals the capacity factor
C. always less than its capacity factor
D. none of these

599. Diversity factor has direct effect on the
A. fixed cost of unit generated
B. variable cost of unit generated
C. both fixed and variable costs of unit generated
D. none of these

600. More efficient plants are used as
A. base load stations B. peak load stations
C. both (A) and (B) are correct
D. None of these

601. HRC fuses provide best protection against
A. short circuits B. lightning
C. sparking D. fire

602. Sparking occurs when a load is switched off because the circuit has high
A. inductance B. resistance
C. capacitance D. magnetism

603. Zero sequence component always flows through
A. phase wires B. neutral wire
C. earth wire D. any of these

Ans. 587. C 588. B 589. B 590. D 591. D 592. A 593. C 594. D 595. C 596. A 597. A 598. A 599. A 600. A 601. A 602. A 603. B

604. Domestic supply of 220 V, 50 Hz signifies that peak value of voltage waveform is
A. 220 V B. 155 V
C. 185 V D. 310 V

605. The safest value of current which the human body can sustain for more than 3 seconds is
A. 1 A B. 0.5 A
C. 10 mA D. 9 mA

606. Which lightning stroke is most dangerous?
A. Direct stroke on tower top
B. Direct stroke on ground wire
C. Direct stroke on line conductor
D. Indirect stroke on conductor

607. In a synchronous machine, in case the axis of field flux is in line with the armature flux, then the machine
A. is working as synchronous motor
B. is working as synchronous generator
C. is floating D. will not work

608. The resistance of the dry skin of human body, between the tip of the left hand finger and right hand finger, is of the order of
A. 100 ohms B. 200 ohms
C. 10 K D. 100 K

609. The average charge per kWh when the load factor is 20% will be
A. 3 paise B. 5 paise
C. 7 paise
D. cannot be determined on the basis of the information provided.

610. The average charges per kWh for the above question will be least when the plant load factor is
A. 100% B. 75%
B. 50% D. 20%

611. Which statement about the daily load curve is valid?
A. The area under the curve gives the average demand
B. The ratio of the area under the curves to the total area of rectangle in which it is contained gives the load factor for the day
C. The peak of the curve gives the installed capacity of the plant
D. The area of the curves divided by the number of hours gives load factor

612. The effect of electric shock on human body depends on
A. voltage B. current
C. duration of contact D. all of these

613. A current of 10 milliamperes through human body will
A. cause mild sensation B. affect muscle control
C. affect breathing D. cause death

614. Death is almost certain when the current through human body is
A. 10 mA B. 20 mA
C. 40 mA D. 100 mA

615. An equipment purchased for Rs. 10,000 two years ago has a market value of Rs. 13,000 at present. It can be concluded that the value has
A. depreciated according to straight line method
B. depreciated according to diminishing value method
C. depreciated according to sinking fund method
D. appreciated with the time

616. If the resistance of dry skin of human body is 100 K, the resistance of the wet body will be
A. 100 K B. 200 K
C. 5000 K D. 5000 ohms

617. For extinguishing electrical fires
A. water should be used
B. carbon dioxide fire extinguisher should be used
C. foam type fire extinguisher should be used
D. carbon tetrachloride fire extinguisher should be used

618. First aid for electric shock victim is
A. bandage B. cover with blanket
C. pouring water on body
D. artificial respiration

619. Which of the following is the protective device against lightning over voltages?
A. Rod gaps B. Surge absorbers
C. Horn gaps D. All of these

620. A coaxial line is filled with a dielectric of relative permittivity 9. If V denotes the velocity of propagation in free space, the velocity of propagation in the line will be
A. 3 V B. V
C. V/3 D. V/9

621. In case of interconnected system for stable operation the passive element that can be used as interconnecting element is
A. capacitor B. reactor
C. resistor D. any of these

622. Area under the daily load curve divided by 24 gives
A. average load B. maximum demand
C. units generated D. none of these

623. The knowledge of diversity factor helps in determining
A. average load B. units generated
C. plant capacity D. none of these

624. Annual load factor is determined from the load curve based on

Ans. 604. D 605. D 606. C 607. C 608. D 609. C 610. A 611. D 612. D 613. A 614. D 615. D 616. D 617. D 618. D 619. D 620. C 621. B 622. A 623. C

A. daily B. weekly
C. monthly D. annually

625. The connected load of a consumer is 2 kW and his maximum demand is 1.5 kW. The load factor of the consumer is
A. 0.75 B. 0.375
C. 1.33 D. none of these

626. When a power plant is not able to meet the demand of consumers, it will resort to
A. penalising high load consumers by increasing the charges for electricity
B. power factor improvement at the generators
C. efficient plant operation
D. load shedding

627. Load shedding is possible through
A. voltage reduction B. frequency reduction
C. switching off the loads
D. any of these

628. The maximum demand of a consumer is 2 kW and his daily energy consumption is 20 units. Its load factor is
A. 10.15% C. 41.6%
C. 50% D. 60%

629. The loss load factor is given by (if F is the load factor)
A. $(0.25 F + 0.75 F^2)$ B. $(0.25 F^2 + 0.85 F)$
C. $0.75 F + 0.20 F^2)$ D. none of these

630. Industrial heating furnaces such as arc and induction furnaces operate on
A. very low lagging pf B. very low leading pf
C. very high leading pf D. none of these

631. Arc lamp operates at
A. low lagging pf B. high leading pf
C. unity pf D. None of these

632. The annual charge of a transmission line can be expressed as
A. $(P_1 + P_2) \times 1/a$ B. $(P_1 \times P_2 a)$
C. $(aP_1 + P_2)$ D. $(P_1 + P_2/a)$

633. As matter of economy, voltage for power transmission should be
A. low B. high
C. medium D. none of these

634. The economic size of conductor is determined by
A. current carrying capacity
B. transmission voltage
C. kelvin's law
D. none of these

635. In an interconnected grid system, the diversity factor of the whole system
A. decreases B. increases
C. remains same D. none of these

636. The value of demand factor is
A. less than one B. greater than one
C. equal to one D. none of these

637. The value of diversity factor is
A. less than one B. more than one
C. equal to one D. none of these

638. By improving the power factor of the system, the kilowatts delivered by generating stations are
A. decreased B. increased
C. not changed D. none of these

639. An overexcited synchronous motor on no-load is known as
A. synchronous induction generator
B. synchronous condenser
C. alternator D. none of these

640. The smaller the lagging reactive power drawn by a circuit, its pf will be
A. better B. poorer
C. unit D. none of these

641. KVAR is equal to
A. kW tan ϕ B. kW sin ϕ
C. kVA cos ϕ D. none of these

642. For a particular power, the current drawn by the circuit is minimum when the value of power factor is
A. 0.8 lagging B. 0.8 leading
C. unity D. none of these

643. To improve the power factor of 3-phase circuits, the size of each capacitor when connected in delta with respect to when connected in star is
A. 1/6th B. 1/4 th
C. 3 times D. 1/3 rd

644. The power factor improvement equipment is always placed
A. at the generating station
B. near transformer
C. near the apparatus responsible for low pf
D. None of these

645. The most economical power factor for a consumer generally is
A. 0.95 B. unity
C. 0.6 lagging D. 0.6 leading

646. The main reason for low power factor of supply system is due to the use of
A. resistive load B. induction motor
C. synchronous motor D. all are correct

647. Power factor can be improved by installing such a device in parallel with load which takes
A. lagging reactive power
B. leading reactive power
C. apparent power
D. none of these

Ans.	624. D	625. D	626. D	627. D	628. C	629. A	630. A	631. D	632. B	633. B	634. C	635. B	636. A
	637. B	638. B	639. B	640. A	641. A	642. C	643. D	644. C	645. A	646. B	647. B		

648. The main reason for low power factor of supply system is due to the use of
A. inductive load B. synchronous motors
C. charge motors D. none of these

649. If a synchronous machine is under excited, it takes lagging VARs from the system when it is operated as a
A. synchronous motor
B. synchronous generator
C. synchronous motor as well as generator
D. none of these

650. The best location of the power factor correction equipment to be installed in the transmission line is at
A. the sending end B. the receiving end
C. any place D. none of these

651. The tarrif most suitable for large industrial consumers is
A. flat demand rate B. block meter rate
C. two part tariff D. none of these

652. A device required a power of P kW. If a capacitor is put in parallel with the device to improve the power consumed by the device will be
A. P kW B. less than P kW
C. greater than P kW D. none of these

653. The tariff generally used for tubewell loads is
A. flat demand rate B. straight meter rate
C. block meter rate D. none of these

654. The tarriff generally for residential consumers is
A. flat demand rate B. straight meter rate
C. block meter rate D. none of these

655. A device requires a power of P kW. Its power factor is $\cos\theta_1$. A capacitor is put in parallel to improve the power factor to $\cos\theta_2$. The kVA input will decrease by
A. 0 B. $P\left(\frac{1}{\cos\theta_1}-\frac{1}{\cos\theta_2}\right)$
C. $P(\cos\theta_1 - \cos\theta_2)$ D. none of these

656. Which of the following plants has the highest operating costs?
A. Steam plant B. Diesel electric plant
C. Hydroelectric plant D. None of these

657. Which of the following plants has the highest fixed costs?
A. Diesel electric plant B. Steam plant
C. Hydroelectric plant D. None of these

658. If X is the system reactance and R its resistance, the power transferred is maximum when
A. $X = R$ B. $X = \sqrt{2}R$
C. $X = \sqrt{3}R$ D. $X = 2R$

659. In overhead lines, the presence of earth wire causes
A. increase in the capacitance
B. increase in the inductance
C. decrease in the capacitance
D. decrease in the inductance

660. A synchronous motor is operating at a leading power factor. If the applied voltage decreases, (the excitation remains constant) the motor current
A. increases B. decreases
C. remains the same D. none of these

661. A load demands a constant power input of P kW. The amount of capacitor KVAR to improve the power factor from 0.8 lagging to 0.9 lagging is Q KVAR. The capacitor KVAR needed to improve the power factor from 0.9 lagging to unity will be
A. Q B. greater than Q
C. less than Q D. zero.

662. A power system has a rating of s kVA. The improvement of power factor from 0.8 lagging to 0.9 lagging increases active power by 0.1 s. The increase in active power due to improvement of power factor from 0.9 lagging to unity will be
A. 0.1 s B. less than 0.1 s
C. greater than 0.1 s D. zero

663. In an AC transmission line, difference in phase of voltages at the two ends of the line is due to
A. resistance of line B. system voltage
C. reactance of line D. insulators

664. The L/C ratio for 132 kV and 400 kV lines are typically 160×10^2 and 62.5×10^3 respectively. The natural 3-phase loading for the two lines will be respectively
A. 108.9 and 2560 MW B. 44 and 2560 MW
C. 44 and 640 MW D. 640 and 44 MW

665. The inductance of single-phase two-wire power transmission line per kilometre gets doubled when the
A. distance between the wires is doubled
B. distance between the wires is increased four fold
C. distance between the wires is increased as square of original distance
D. radius of the wire is doubled

666. A 3-phase 50 Hz transmission line has the following constants (line to neutral) Resistance = 10 Ω, Inductive reactance = 20 ohms; Capacitive susceptance = 4×10^{-4} ohms. The inductance of the arc suppresser coil to be used in the system is
A. 0.06368 H B. 0.127 H
C. 1.27 H D. 2.654 H

667. In a 400 kV power network, 360 kV is recorded at a 400 kV bus. The reactive power absorbed by a shunt reactor rated for 50 MVAR, 400 kV connected at the bus is
A. 61.73 MVAR B. 55.56 MVAR
C. 45.0 MVAR D. 40.5 MVAR

Ans.	648. A	649. C	650. B	651. C	652. A	653. A	654. C	655. B	656. B	657. C	658. C	659. A	660. A
	661. B	662. A	663. C	664. C	665. C	666. D	667. D						

668. Which one of the following statements is true?
A. Steady-state stability limit is greater than transient stability limit
B. Steady-state stability limit is equal to transient stability limit
C. Steady-state stability limit is less than the transient stability limit
D. No generalisation can be made regarding the equality or otherwise of the steady-state stability limit and transient stability limit

669. The voltages across the various discs of a string of suspension insulators having identical discs is different due to
A. surface leakage currents
B. series capacitance
C. shunt capacitance to ground
D. series and shunt capacitances

670. Compared with a solid conductor of the same radius, corona appears on a stranded conductor at a lower voltage because stranding
A. assists ionisation
B. makes the current flow spirally about the axis of the conductor
C. produces oblique sections to a plane perpendicular to the axis of the conductor
D. produces surfaces of smaller radius

671. Load factor of a power station is defined as
A. maximum demand/average load
B. average load × maximum demand
C. average load/maximum demand
D. none of these

672. Load factor of a power station is generally
A. equal to unity B. less than unity
C. more than unity D. none of these

673. Diversity factor is always
A. equal to unity B. less than unity
C. more than unity D. more than five

674. The load factor for heavy industries may be taken as
A. 10 to 25% B. 25 to 40%
C. 50 to 70% D. 70 to 80%

675. The load factor of domestic load is usually
A. 10 to 15% B. 30 to 40%
C. 50 to 60% D. 60 to 70%

676. Annual depreciation cost is calculated by
A. sinking fund method B. straight line method
C. both (A) and (B) are correct
D. any of these

677. Depreciation charges are high in case of
A. thermal plant B. diesel plant
C. hydroelectric plant D. any of these

678. Demand factor is defined as
A. average load/maximum demand
B. maximum demand/connected load
C. connected load/maximum demand
D. maximum demand × connected load

679. High load factor indicates that
A. cost of generation per unit power is increased
B. total plant capacity is utilised for most of the time
C. total plant capacity is not properly utilised for most of the time
D. none of these

680. A load curve indicates
A. average power used during the period
B. average kWh (kW) energy consumption during the period
C. neither of these D. all of these

681. Approximate estimation of power demand can be made by the method/methods
A. load survey method B. statistical methods
C. economic parameters D. all of these

682. Annual depreciation as per straight line method is calculated by
A. the capital cost divided by number of years of life
B. the capital cost minus the salvage value divided by the number of years of life
C. investing a uniform sum of money per annum at stipulated rate of interest
D. none of these

683. A consumer has to pay lesser fixed charges in
A. flat rate tariff B. two part tariff
C. maximum demand tariff
D. none of these

684. In two part tariff, variation in load factor will affect
A. fixed charges
B. operating or running charges
C. both (A) and (B) are correct
D. none of these

685. In India the tariff for charging the consumers for the consumption of electricity is based on
A. straight meter rate B. block meter rate
C. reverse form of block meter rate
D. two part tariff

686. While calculating the cost of electric power generation, which of the following is NOT considered a fixed cost?
A. interest on capital investment
B. taxes and insurance
C. most of the salaries and wages
D. repair and maintenance

687. Maximum demand of an installation is given by its
A. instantaneous maximum demand
B. greatest average power demand
C. average maximum demand over a definite interval of time during a certain period
D. average power demand during an interval of 1-minute

Ans. 668. D 669. C 670. D 671. C 672. B 673. C 674. D 675. A 676. C 677. A 678. B 679. B 680. B
681. D 682. B 683. C 684. B 685. C 686. D 687. C

688. A diversity factor of 2.5 gives a saving of ...... percent in the generating equipment.
A. 60 B. 50
C. 40 D. 25

689. Mark the WRONG statement. High load factor of a generating equipment
A. leads to lesser charges per kWh
B. implies lower diversity in demand
C. gives more profit to the owner
D. can be obtained by accepting off-peak loads

690. In a generating station, fixed charges at 100% load factor are 6 paise/kWh. With 25% load factor, the charges would become ..... paise/kWh.
A. 1.5 B. 10
C. 24 D. 3

691. When considering the economics of power transmission, Kelvin' law is used for finding the
A. cost of energy loss of bare conductors
B. most economical cross-section of the conductors
C. interest on capital cost of the conductor
D. the maximum voltage drop in feeders

692. A 3-phase balanced system working at 0.9 lagging power factor has a line loss of 3600 kW. If pf is reduced to 0.6, the line loss would become ..... kW.
A. 8100 B. 1600
C. 5400 D. 2400

693. Mark the WRONG statement. While considering power factor improvement, the most economical angle of lag depends on the
A. cost/kVA rating of phase advancer
B. rate of interest on capital outlay
C. rate of depreciation
D. value of original lagging pf angle

694. The method of symmetrical components is very useful for
A. solving unbalanced polyphase circuits
B. analysing the performance of 3-phase electrical machinery
C. calculating currents resulting from unbalanced faults
D. all of these

695. An unbalanced system of 3-phase voltages having *RYB* sequence actually consists of
A. a positive-sequence component
B. a negative-sequence component
C. a zero-sequence component
D. all of these

696. The zero-sequence component of the unbalanced 3-phase system of vectors V_A, V_B and V_C is ..... of their vector sum
A. one-third B. one-half
C. two-third D. one-fourth

697. In the case of an unbalanced star-connected load supplied from an unbalanced 3-phase, 3-wire system, load currents will consist of
A. positive-sequence components
B. negative-sequence components
C. zero-sequence components
D. only (A) and (B) are correct

MULTIPLE CHOICE QUESTIONS FROM VARIOUS COMPETITIVE EXAMINATIONS

1. The ABCD constants of a 3-phase transmission line are
$A = D = 0.8 \angle 1°$
$B = 170 \angle 85° \ \Omega$
$C = 0.002 \angle 90.4°$ mho
The sending end voltage is 400 kV. The receiving end voltage under no-load condition is **(IES 2001)**
A. 400 kV B. 500 kV
C. 320 kV D. 417 kV

2. The surge impedance of a 3-phase, 400 kV transmission line is 400 Ω. The surge impedance loading (SIL) is **(IES 2001)**
A. 400 MW B. 100 MW
C. 1600 MW D. 200 MW

3. For some given transmission line, the expression for voltage regulation is given by $\dfrac{|V_s| - |V_R|}{|V_R|} \times 100\%$.
Hence **(IES 2002)**
A. this must be a 'short' line
B. this may either be a 'medium line' or a 'short line'
C. this expression is true for any line
D. this may either be a 'medium line' or a 'long line'

4. The capacitance of an overhead transmission line increases with **(IES 2002)**
1. increase in mutual geometrical mean distance
2. increase in height of conductors above ground
Select the correct answer from the following:
A. Both 1 and 2 are true B. Both 1 and 2 are false
C. Only 1 is true D. Only 2 is true

5. The component inductance due to the internal flux-linkage of nonmagnetic straight solid circular conductor per metre length has a constant value, and is independent of the conductor-diameter, because **(IES 2003)**

Ans. 688. A 689. B 690. C 691. B 692. A 693. D 694. D 695. D 696. A 697. D 1. B 2. A 3. A 4. B

A. all the internal flux due to a current remains concentrated on the peripheral region of the conductor
B. the internal magnetic flux-density along the radial distance from the centre of the conductor increases proportionately to the current enclosed
C. the entire current is assumed to flow along the conductor-axis and the internal flux is distributed uniformly and concentrically
D. the current in the conductor is assumed to be uniformly distributed throughout the conductor cross-section

6. A 100 km long transmission line is loaded at 110 kV. If the loss of line is 15 MW and the load is 150 MVA, the resistance of the line is **(IES 2003)**
A. 8.06 ohm per phase B. 0.806 ohm per phase
C. 0.0806 ohm per phase D. 80.6 ohms per phase

7. For a fixed receiving end and sending end voltage in a transmission system, what is the locus of the constant power? **(IES 2004)**
A. A straight line B. An ellipse
C. A parabola D. A circle

8. The charging reactance of 50 km length of the line is 1500 Ω. What is the charging reactance for 100 km length of the line? **(IES 2005)**
A. 1500 Ω B. 3000 Ω
C. 750 Ω D. 600 Ω

9. The corona loss on a particular system at 50 Hz is 1 kW/km per phase. What is the corona loss at 60 Hz in kW/km per phase? **(IES 2005)**
A. 0.83 B. 1.0
C. 1.13 D. 1.2

10. What is the approximate value of the surge impedance loading of a 400 kV, 3-phase 50 Hz overhead single circuit transmission line? **(IES 2006)**
A. 230 MW B. 400 MW
C. 1000 MW D. 1600 MW

11. For an extra-high voltage overhead transmission line, four conductors are used per phase (in a bundle) at the corners of a square of side 's' metre. The GMR (geometric mean radius) of each conductor is r'_m metre. **(IES 2008)**
A. $(r'_m \times s^2 \times \sqrt{2}s)^{1/4}$ B. $(r'_m \times s^3)^{1/4}$
C. $(r'_m \times 3s^3)^{1/4}$ D. $[r'_m \times (\sqrt{2}s)^3]^{1/4}$

12. When is the Ferranti effect on long overhead lines experienced? **(IES 2008)**
A. The line is lightly loaded
B. The line is heavily loaded
C. The line is fully loaded
D. The power factor is unity

13. What is the surge impedance loading of a loss less 400 kV, 3-phase, 50 Hz overhead line of average of surge impedance of 400 ohms? **(IES 2008)**
A. 400 MW B. $400\sqrt{3}$ MW
C. $400/\sqrt{3}$ MW D. 400 kW

14. In modelling, the equivalent circuit of a short length overhead transmission line, the line resistance and inductance are only considered because line capacitance to ground is **(IES 2010)**
A. equal to zero B. finite but very small
C. finite but very large D. infinite

15. If a fixed amount of power is to be transmitted over certain length with fixed power loss, it can be said that volume of conductor is **(IES 2010)**
A. inversely proportional to magnitude of the voltage and that of power factor of the load
B. inversely proportional to square of the voltage and square of power factor of the load
C. proportional to square of voltage and that of power factor of the load
D. proportional to magnitude of the voltage only

16. A 10 km long lossless transmission line has a reactance of 0.3 Ω/km and negligible shunt capacitance. The value of $\begin{bmatrix} A & B \\ C & D \end{bmatrix}$ is **(IES 2010)**
A. $\begin{bmatrix} 1 & 0 \\ j3 & 1 \end{bmatrix}$ B. $\begin{bmatrix} 1 & 0 \\ j0.3 & 1 \end{bmatrix}$
C. $\begin{bmatrix} 1 & j3 \\ 0 & 1 \end{bmatrix}$ D. $\begin{bmatrix} j3 & 0 \\ 1 & 1 \end{bmatrix}$

17. Bundled conductors are used for EHV transmission lines primarily for reducing the **(IES 2001)**
A. corona loss
B. surge impedance of the line
C. voltage drop across the line
D. I^2R losses

18. Which one of the following capacitors is suitable for compensation of harmonic and reactive power? **(IES 2004)**
A. Mica capacitor B. Glass capacitor
C. Polypropylene capacitor
D. Electrolytic capacitor

19. An AC capacitor is to be switched in parallel with AC line using back to back connected thyristor. What is the firing angle of thyristor for first switching? **(IES 2009)**
A. 0° B. 180°
C. 90° D. 45°

20. Static VAR controllers are used to provide dynamic voltage regulation. These controllers are primarily of **(IES 2010)**

Ans. 5. B 6. A 7. D 8. C 9. C 10. B 11. A 12. A 13. A 14. B 15. B 16. C 17. A 18. D 19. C

A. thyristor switched inductors
B. thyristor controlled capacitors
C. thyristor switched resistor
D. thyristor switched inductors and thyristor controlled capacitors

21. What will happen if a short circuit fault occur in a switched capacitor controlled reactor? **(IES 2010)**
A. Oscillation B. Capacitor discharge
C. Over voltage D. Noise

22. The main objective of load frequency controller is to apply control of **(IES 2011)**
A. frequency alone
B. frequency and at the same time of real power exchange via the outgoing lines
C. frequency and at the same time of reactive power exchange via the out-going lines
D. frequency and bus voltages

23. The main objective of load frequency control in a power system are **(IES 2011)**
1. to bring the steady state error to zero after load change
2. to maintain the net tie-line flow
3. to maintain voltages on all buses
4. to economise the cost of generation
A. 1 and 2 B. 2 and 3
C. 3 and 4 D. 1, 2, 3 and 4

24. The incremental cost characteristics of two generators delivering 200 MW are as follows:

$$\frac{dF_1}{dP_1} = 2.0 + 0.01P_1, \frac{dF_2}{dP_2} = 1.6 + 0.02P_2$$

For economic operation the generation P_1 and P_2 should be **(IES 2013)**
A. 100 MW and 100 MW B. 80 MW and 120 MW
C. 220 MW and 100 MW D. 120 MW and 80 MW

25. Complete combustion of pulverized coal in a system raising thermal power plant is ensured by what type of an analysis of flue gas going out by the chimney? **(IES 2009)**
A. O_2 content for given air intake
B. CO_2 content for given fuel rate feed
C. CO content
D. All of the above

26. The radiation shield for a nuclear power reactor for biological safety is provided by having the reactor **(IES 2011)**
A. immersed in water pool
B. encased by thick metal walls
C. encased by thick concrete wall
D. isolated from outside world with strong magnetic field

27. A 'Pumped storage hydroelectric plant' consists of **(IES 2011)**
A. a synchronous machine and a multi-stage centrifugal pump in one shaft
B. a synchronous machine, a reaction turbine, and a multistage centrifugal pump all in one shaft
C. an induction generator, a synchronous machine and a reaction turbine, all in one shaft
D. an induction generator, a synchronous machine, and a multistage centrifugal pump, all in one shaft

28. In an interconnected power system, the most suitable power plant to meet the peak load conditions is **(IES 2012)**
A. hydel B. nuclear
C. steam D. pumped storage

29. The advantage of hydro-electric power station over thermal power station is **(IES 2013)**
A. the initial cost of hydroelectric power
B. the operation cost of hydroelectric power station is low
C. hydro-electric power station can supply the power throughout the year
D. hydro-electric power station can be constructed at the place where the energy is required

30. A power generating station has a maximum demand of 100 MW. The annual load factor is 75% and plant capacity factor is 60%. Calculate the reserve capacity. **(IES 2013)**
A. 250 MW B. 500 MW
C. 750 MW D. 1250 MW

31. In order to have lower cost of power generation **(IES 2013)**
A. the load factor and diversity factor should be low
B. the load factor and diversity factor should be high
C. the load factor should be low but diversity factor should be high
D. the load factor should be high but diversity factor should be low

32. The term 'Surg Tank' is associated with which type of power plant? **(IES 2014)**
A. High head hydro B. Low head hydro
C. Medium head hydro D. Thermal

33. In a certain single-phase AC circuit, the instantaneous voltage is given by
$v = V \sin(\omega t + 30°)$ pu and the instanta-neous current is given by $i = I \sin(\omega t - 30°)$ pu. Hence the per unit value of reactive power is **(IES 2002)**
A. 1/4 B. 1/2
C. $\sqrt{3}/4$ D. $\sqrt{3}/2$

Ans. 20. D 21. A 22. B 23. A 24. D 25. D 26. C 27. B 28. D 29. B 30. A 31. B 32. A 33. C

34. The Y_{bus} matrix of a 100-bus interconnected system is 90% sparse. Hence the number of transmission lines in the system must be **(IES 2002)**
A. 450 B. 500
C. 900 D. 1000

35. The bus admittance matrix of a power system is given as

$$\begin{array}{c} \\ 1 \\ 2 \\ 3 \end{array}\begin{array}{c} \begin{array}{ccc} 1 & 2 & 3 \end{array} \\ \begin{bmatrix} -j50 & +j10 & +j5 \\ +j10 & -j30 & +j10 \\ +j5 & +j10 & -j25 \end{bmatrix} \end{array}$$

Theimpedance of line between bus 2 and 3 will be equal to **(IES 2003)**
A. $+j\,0.1$ B. $-j\,0.1$
C. $+j\,0.2$ D. $-j\,0.2$

36. A power system consists of two areas connected via a tie line. While entering the data for load flow, the tie line parameters and its connectivity data were inadvertently left out. If the load flow program is run with this incomplete data, then the load flow calculations will converge only if **(IES 2003)**
A. one slack bus is specified in the first area
B. one slack bus is specified in the second area
C. one slack bus is specified in either of the two areas
D. two slack buses, one in each area are specified

37. In the network as shown below, the marked parameters are pu impedances. The bus admittance matrix of the network is **(IES 2003)**

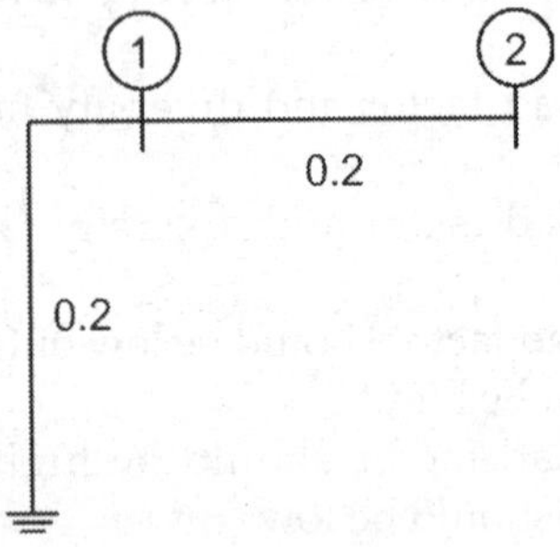

A. $\begin{bmatrix} 10 & -5 \\ -5 & 5 \end{bmatrix}$ B. $\begin{bmatrix} 5 & -5 \\ -5 & 10 \end{bmatrix}$

C. $\begin{bmatrix} -10 & 5 \\ 5 & -5 \end{bmatrix}$ D. $\begin{bmatrix} -5 & 5 \\ 5 & -10 \end{bmatrix}$

38. At slack bus, which one of the following combinations of variables is specified? **(IES 2004)**
(The symbols have their usual meaning)
A. $|V|, \delta$ B. P, Q
C. $P, |V|$ D. $Q|V|$

39. Buses for load flow studies are classified as:
(i) Load bus
(ii) PV bus
(iii) Slack bus
Which one of the following is the correct combination of the pair of quantities specified having their usual meaning for different buses? **(IES 2005)**

	Load bus	PV bus	Slack bus
A.	P, $|v|$	P, Q	P, δ
B.	P, Q	P, $|v|$	$|v|$, δ
C.	$|v|$, Q	P, δ	P, Q
D.	P, δ	Q, $|v|$	Q, δ

40. A sample power system network is shown in the figure given below. The reactances marked are in pu. What is the pu value of Y_{22} of the Bus Admittance matrix (Y_{bus})? **(IES 2005)**

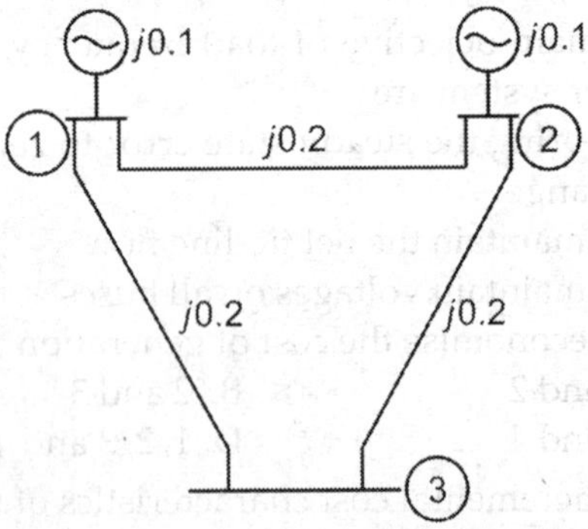

A. $j10.0$ B. $j0.4$
C. $-j0.1$ D. $-j20.0$

41. For a synchronous generator connected to an infinite bus through a transmission line, how are the change of voltage (ΔV) and the change of frequency (Δf) related to the active power (P) and the reactive power (Q)? **(IES 2006)**
A. ΔV is proportional to P and Δf to Q
B. ΔV is proportional to Q and Δf to P
C. Both ΔV and Δf are proportional to P
D. Both ΔV and Δf are proportional to Q

42. For a graph of power system network shown in the figure below, where bus numbers and impedances are marked, assuming equal R/X of impedances, find the bus impedance matrix element Z_{22}. **(IES 2010)**

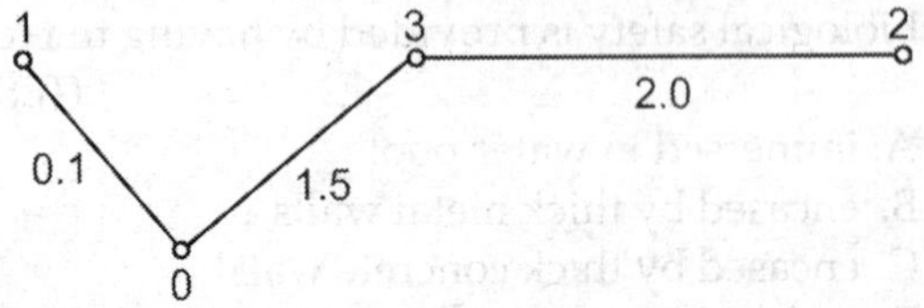

A. 2.0 B. 3.5
C. 0.5 D. 4.5

Ans. 34. A 35. A 36. D 37. A 38. A 39. B 40. D 41. B 42. A

43. Load flow studies must be made on a power system before **(IES 2010)**
A. making short circuit studies but not for transient stability studies on the power system
B. making transient stability studies but not for short circuit studies on the power system
C. making both short circuit and transient stability studies on the power system
D. for neither making short circuit studies nor transient stability studies on the power system

44. In Gauss-Seidel load flow method, the number of iterations may be reduced if the correction in voltage at each bus is multiplied by **(IES 2012)**
A. Gauss constant B. acceleration factor
C. blocking factor D. Lagrange multiplier

45. Consider the following statements regarding convergence of the Newton-Raphson procedure:
1. It does not converge to a root when the second differential coefficient changes sign
2. It is preferred when the graph of (X) is nearly horizontal where it crosses the X-axis
3. It is used to solve algebraic and trans-cendental equations

Which of these statements are correct? **(IES 2013)**
A. 1, 2 and 3 B. 1 and 2 only
C. 2 and 3 only D. 1 and 3 only

46. A surge voltage rising at 100 kV/μs travels along a loss-less open-circuited transmission line. It takes 10 μs to reach the open end. The reflected wave from the open end will be rising at **(IES 2002)**
A. 100 kV/μs B. 200 kV/μs
C. 1000 kV/μs D. 2000 kV/μs

47. The reflection coefficient for the transmission line shown in figure below at point P is **(IES 2002)**

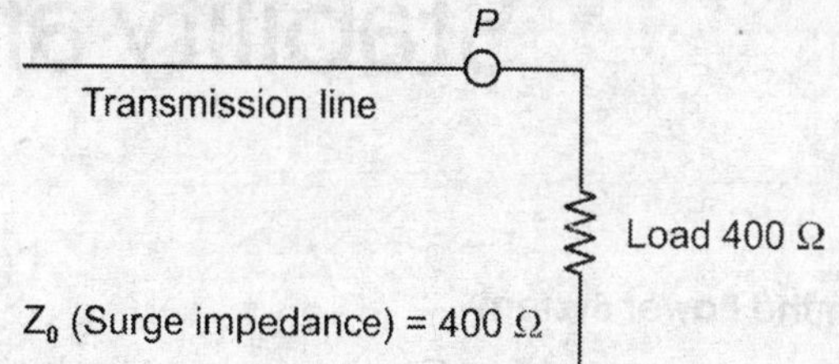

A. +1 B. 0.5
C. 0 D. –1

Ans. 43. C 44. B 45. D 46. C 47. C

CHAPTER 12

Switchgear Protection, Stability and High Voltage

Fault in the Power System

Broadly, faults can be classified as (i) shunt faults (short circuits) (ii) series faults (open conductor). Shunt type of faults involve power conductor or conductor to ground or short circuit between conductors when circuits are controlled by fuses or any other device which does not open all the three phases, one or two phases of the circuit may be opened while the other two phases or phase is closed. These are called *series type of faults* may also occur with one or two broken conductors. Shunt faults are characterised by increase in current and fall in voltage and frequency whereas series faults are characterised by increase in voltage and frequency and fall in current in faulted phases.

The faults can be classified as:

i. single line to ground fault (more than 70%)
ii. double line to ground fault
iii. line to line fault
iv. three phase fault (rarer < 5% of total faults)
v. winding faults
vi. open-circuit fault.

As the faults occurs on the power system, voltage at a faulty point changes. The current changes to many times the normal current. The fall in current varies with time. The region of first few cycles in which the current has very high value, but decreases rapidly is called *subtransient state*. After the first few cycles, the decrease in the current is less rapid. This region of slow decrease in short circuit current is called *transient state*. It covers many cycles after the transient state, the current reaches a *steady state*.

Fault calculations: Knowledge of fault current is useful in determining the relative setting, selecting the circuit breaker of current rating, designing the power system components and their protection system. The fault calculations provide the information about the fault

S. no.	*Causes of faults*	*Total percent*
1. Overhead lines	• Lighting strokes • Iceing in cold countries • Storms • Winds	30–40
2. Underground cables	• Failure of insulation • Failure of joints	8–10
3. Generators	• Stator faults (short-circuit) • Rotor faults • Faults in auxilaries • Faults in terminals	6–8
4. Transformers	• Insulation failure • Faults in bushing • Faults in protection circuit	10–12
5. CTPT switch-gear and other faults		25–30

current and voltage at different points of the system under fault calculation.

Terms related to fault calculation: Per unit impedance of electrical equipment $e_z = IZ/V$ pu

i. The pu impedance is also given by the following expression:

$$\text{Base Z} = \frac{[\text{Base kV}]^2}{[\text{Base kVA}]} \times 1000$$

$$\text{pu Z} = \frac{\text{Ohmic Z}}{\text{Base Z}} = \frac{Z \cdot [\text{Base kVA}]}{[\text{Base kV}]^2 \times 100}$$

$$\% \text{ Z} = \frac{Z \cdot [\text{Base kVA}]}{10[\text{Base kV}]^2}$$

Change of base pu Z_{old}

$$= \text{pu } Z_{\text{old}} \times \frac{[\text{kVA Base new}]}{[\text{kVA Base old}]} \times \left[\frac{\text{kV Base old}}{\text{kV Base new}}\right]^2$$

Percentage impedance = 100 pu impedance

ii. When all the reactance have been referred to the same base kVA, then total impedance or reactance upto the fault point is calculated. This may involve solution of series parallel circuits or star/delta conversion or vice-versa.

iii. Short circuit kVA $= \left[\frac{100}{\%Z} \times \text{base kVA}\right]$

Short circuit kVA (rms) $= \frac{\text{Short circuit kVA}}{\sqrt{3} \times \text{kV (line)}} \cdot \text{amp}$

iv. *Percentage reactance*: It is defined as % reactance

$$= \frac{\text{Reactance} \times [\text{kVA}]}{10\,[\text{kV}]^2}.$$

v. *Percentage reactance at base kVA*: Since different plants in any system have different kVA ratings, it is necessary to convert all reactance to a common kVA rating base.

% Reactance at base kVA

$$= \frac{\text{Base kVA}}{\text{Plant kVA}} \times \text{reactance at plant kVA}$$

Unsymmetrical fault calculations: In the symmetrical fault calculations, attention was confined to the analysis of symmetrical faults, i.e. all the three lines short circuited or all three lines short circuited with a earth connection at the fault. When such a fault occurs, it gives rise to symmetrical fault currents. Although symmetrical faults are the most severe and impose heavy duty on the circuit breakers but the possibility of such a fault in the power system is quite rare. The majority of the faults on the power system are of an unsymmetrical nature; the most common type being a short circuit *from one line to ground*. When such a fault occurs, it gives rise to unsymmetrical currents. The calculation procedure known as *method of symmetrical components* is used to determine the currents and voltages on the occurrence of an unsymmetrical fault. There are three ways in which unsymmetrical faults may occur in a power system (i) single line to ground fault (*L-G*) (ii) line to line fault (*L-L*) (iii) double line to ground fault (*L-L-G*).

The positive sequence system is identical with the actual network of the particular problem. Each synchronous machine in the system is regarded as a voltage source before the fault occurred.

Unsymmetrical currents: The unbalanced phase currents in a three-phase system in terms of symmetrical components are:

$$I_R = I_{R1} + I_{R2} + I_{R0};$$

$$I_Y = I_{Y1} + I_{Y2} + I_{Y0};\; I_B = I_{B1} + I_{B2} + I_{B0}$$

i. **Zero sequence current**

$$\therefore \quad I_0 = \frac{1}{3}(I_R + I_Y + I_B)$$

ii. **Positive sequence current**

$$\therefore \quad I_{R1} = \frac{1}{3}(I_R + aI_Y + a^2 I_B)$$

Omitting the subscript *R*, we have

$$I_1 = \frac{1}{3}(I_R + aI_Y + a^2 I_B)$$

iii. **Negative sequence current**

$$I_{R2} = \frac{1}{3}(I_R + a^2 I_Y + aI_B)$$

Restriking voltage:

$$v(t) = \sqrt{2}\,V[1 - \cos(\omega_0 t)] = V_m\left[1 - \cos\left(\frac{1}{LC}\right)t\right]$$

The restriking voltage (v) can thus rise to a maximum value of $2V_m$, where V_m is the peak value of the system recovery voltage. If a resistance R_s be connected across the contacts of the circuit breaker, the surge will be critically damped when $R_s = \frac{1}{2}\sqrt{L/C}$ and this offers an important method of reducing the severity of transient.

Principles of circuits interruption: When an R-L series circuit is closed with an alternating voltage source, the resulting current consists of two components, namely a dc component and an AC component.

The AC component is superimposed on the dc component. The magnitude of DC component depends on the voltage at the instant of closing the switch, if the switch is closed at voltage zero, the DC component is maximum (Fig. 12.1). But if the switch is closed at voltage maximum, the DC component is zero and the waveform is symmetrical about normal zero axis as shown in Fig. 12.2.

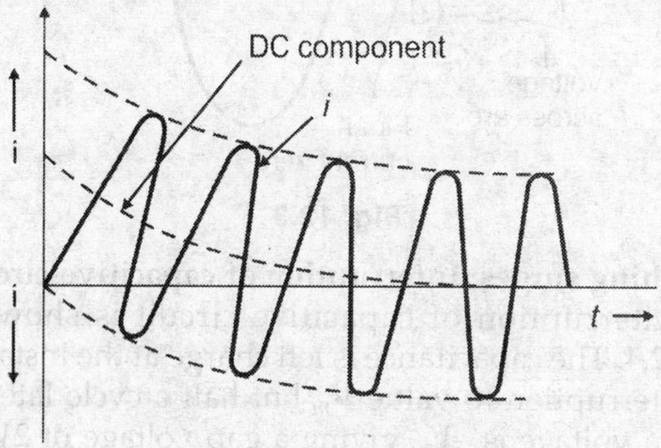

Fig. 12.1

Voltage after final current zero: In alternating current circuit breakers, the current interruption takes place invariably at zero current of the current wave.

At a current zero, the arc gets extinguished if the rate of rise of restriking voltage between contacts is less than the rate of increase of dielectric strength. The voltage appearing across the breaker contacts at the moment of final current zero has a profound influence on the arc extinction process. The voltage appearing

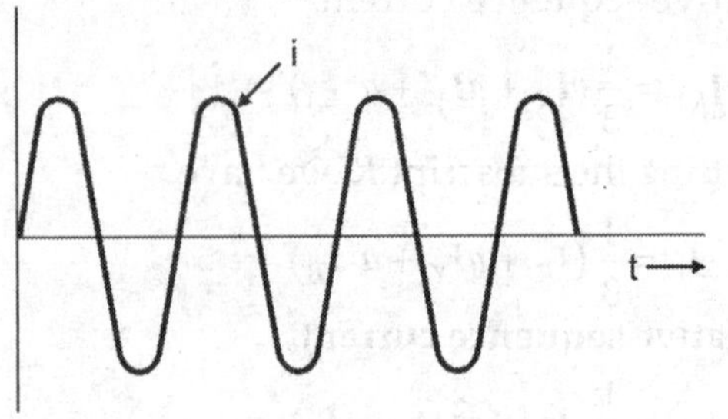

Fig. 12.2

across the contacts at the instant of current zero is a transient voltage of higher natural frequency known as *restriking voltage* and is superimposed on power frequency system voltage known as *recovery voltage,* and the transient component vanishes after a short time of the order of less than half a cycle. After the disappearance of the restriking voltage, the normal frequency voltage is established.

Figure 12.3 illustrates the recovery voltage and restriking voltage. At current zero the voltage appearing across the contacts is composed of transient restriking voltage and power frequency recovery voltage.

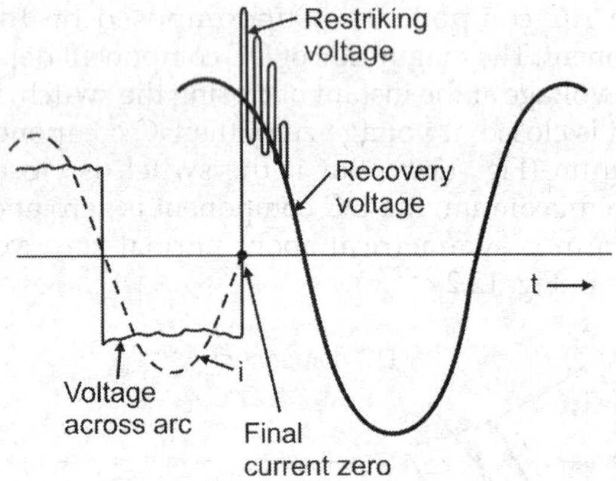

Fig. 12.3

Switching surges: Interruption of capacitive circuits: The interruption of capacitive circuit is shown in Fig. 12.4. The capacitance is left charge at the instant of arc interruption to value V_m but half a cycle later the system voltage is $-V_m$ giving a gap voltage of $2V_m$. If the gap breaks down, an oscillatory transient is set up which can increase the gap voltage still further.

Current chopping: Current chopping arises with air blast circuit breakers which operate on the same air pressure and velocity for all values of interrupted current. Hence on low current interruption the breaker tends to open the circuit before the current natural zero and the electromagnetic energy present is rapidly converted to electrostatic energy.

$$\frac{1}{2}Li_0^2 = \frac{1}{2}Cv^2 \text{ joules} \qquad \therefore v = i_0\sqrt{L/C}$$

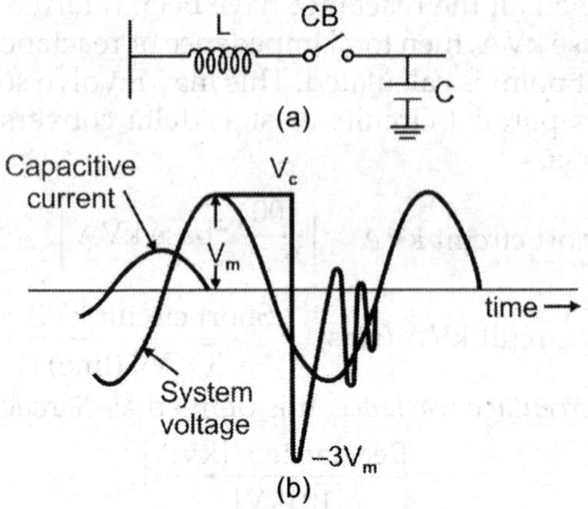

Fig. 12.4

If resistance and time is also included this equation becomes

$$v = i_0\sqrt{L/C}\, e^{-\alpha t} \sin \omega_0 t, \quad \omega_0 = 1/\sqrt{L/C}$$

High transient voltages may be set up on opening a highly inductive current such as transformer on no load.

Resistance switching: A deliberate connection of a resistance in parallel with contact space is called '*resistance switching*. Resistance switching is resorted to circuit breakers having high post zero resistance of contact space (Fig. 12.5).

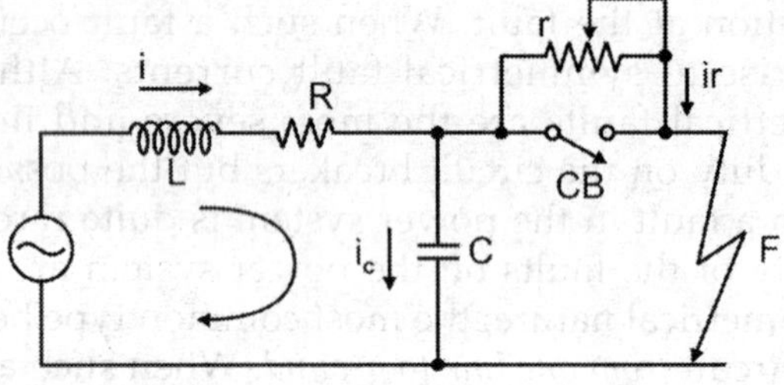

Fig. 12.5

In resistance switching, resistance is connected in shunt with the arc so as to reduce the restriking voltage frequency. The resistance also diverts part of the arc current.

In air-blast circuit breakers, the post zero resistance is high. This may result in severe voltage transients due to current chopping. Hence, the resistance switching is adopted.

The magnitude of resistance for resistance switching is given by

$$r = \frac{1}{2}\sqrt{\frac{L}{C}}$$

Assuming

$$I_{SC} = \frac{E}{\omega L},$$

$$L = \frac{E}{\omega I_{SC}} \qquad \therefore r = \frac{1}{2}\sqrt{\frac{E}{\omega I_{SC} \cdot C}} = K\frac{1}{\sqrt{I_{SC}}}$$

where K is a constant.

Hence value of resistance depends on fault current.

Expression for active recovery voltage: Let V_{ph} be the phase voltage of the system.

The maximum value of the system phase voltage $= \sqrt{2}V_{ph} = V_m$.

The active recovery voltage between phase and neutral $= V_m \sin\phi$.

The active recovery voltage can be represented as

$$V_{AR} = K_1, K_2, K_3, V_m \sin\phi.$$

where K_1 is called the *demagnetizing factor* due to which the recovery voltage will be less than the system voltage.

or $$K_1 = \frac{\text{Recovery voltage}}{\text{System voltage}}$$

K_2 is a condition factor, i.e. it depends on the condition whether the symmetrical fault is grounded or not, its value is either 1 or 1.5 and K_3 factor is equal to 1 if the active recovery voltage between phase and neutral is to be obtained and its value is $\sqrt{3}$ if the active recovery voltage between the two lines is required.

Circuit breaker ratings: A circuit breaker should be capable of

i. opening on the occurrence of a fault and clearing the fault
ii. being closed on to a fault.
iii. carrying fault current for a short time (while another circuit breaker is clearing the fault).

These duties are respectively referred to as

i. Breaking capacity
ii. Making capacity
iii. Short time capacity

Protective relays: These are the devices that detect abnormal conditions in electrical circuits by measuring the electrical quantities which are different under normal and faulty conditions. Due to abnormal conditions, voltage, current, phase angle and frequency may change. After detecting the fault, the relay operates to complete the trip circuit which results in the opening of the circuit breaker and isolating the faulty circuit.

Basic principle: In electromechanical relays, there are one or more coils, movable elements, contact system, etc. The operation of such relays depends on whether the operating torque/force is greater than the restraining torque/force.

The relay operates, if the net force F in equation given below is positive, i.e.

$$F = F_0 - F_R$$

where F is the net force, F_0 is the operating force, and F_R is the restraining force. In other words, the relay operates only if the operating force is greater than the restraining torque. The restraining torque is given by springs.

Basic requirements of protective systems: The efficient protective relaying should possess the following characteristics:

1. *Speed*: Minimum voltage and minimum fault-time to operate.
2. *Selectivity*: Maximum continuity of service by disconnecting the faulty part of the system.
3. *Sensitivity*: Capability of operating reliably under the actual desired condition.

A typical relay circuit is illustrated in Fig. 12.6. When a short circuit occurs at point F on transmission line, the current flowing in the line increases to an enormous value. This results in a heavy current through a relay coil by closing its contacts. This in turn closes the trip coil of the breaker, making the circuit breaker open and isolating the faulty section from the rest of the system.

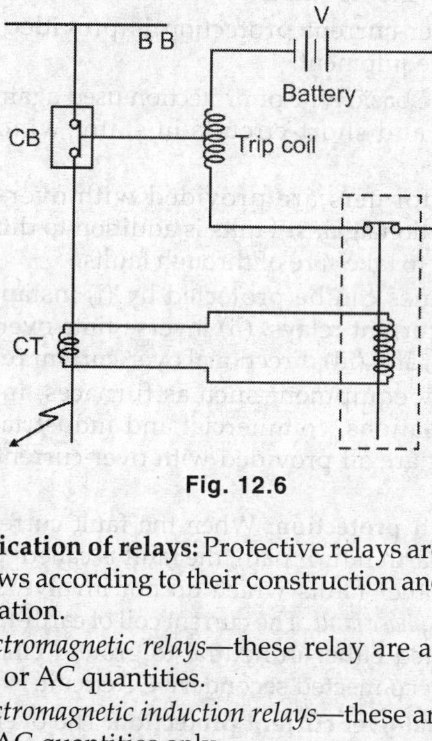

Fig. 12.6

Classification of relays: Protective relays are classified as follows according to their construction and principle of operation.

1. *Electromagnetic relays*—these relay are actuated by DC or AC quantities.
2. *Electromagnetic induction relays*—these are actuated by AC quantities only.
3. *Electrothermal relays*—these are actuated by heat.
4. *Static relays*—these employ transistors or magnetic amplifiers to obtain the operating characteristics.
5. *Electrodynamic relays*—these operate on the same principle as moving coil instrument.

Further, classification depends upon their application:

1. Under-voltage, under-current, and under-power relays.
2. Over-voltage, over-current, and over-power relays.
3. Directional or reverse currents relays.
4. Differential relays.
5. Distance relays.

Another criteria of classification of relays is their time of operation:

1. Definite time-lag relays
2. Inverse time-lag relays
3. Instantaneous relays
4. Inverse-definite minimum time lag-relay or IDMT

Over-current protection: In over-current protection, the relay picks up when the magnitude of current exceeds the pick-up level. It includes the protection from overloads which means machine or equipment is taking more current than its rated value. Hence with overloading, there is an associated temperature rise. Over-current protection for overloads is generally provided by thermal relays. It also includes short-circuit protection. Short-circuits can be phase-faults, earth-faults, or winding-faults.

The over-current protection is provided for the following equipment:

1. It is the basic type of protection used against overloads and short-circuits in stator windings of motors.
2. Transformers are provided with over-current protection against faults is addition to differential relays to take care of through faults.
3. The lines can be protected by (i) instantaneous over-current relays. (ii) inverse time over-current relays, and (iii) directional over-current relays.
4. Utility equipment such as furnaces, industrial installations, commercial and industrial equipments are all provided with over-current protection.

Earth-fault protection: When the fault current flows through earth return path, the fault is called *earth-fault* and the other faults which do not involve earth are known as *phase fault*. The current coil of earth-fault relay is connected either in neutral to ground circuit or in residually connected secondary CT circuit.

Directional over current protection: The over-current protection can be given directional feature by adding directional element in the protection system. Directional over-current protection responds to over current for a particular direction flow. If the power flow is in opposite direction, the directional over-current protection remains inoperative.

Directional over-current protection comprises over-current relay and power directional relay in a single relay casing. The power directional relay does not measure the power but is arranged to respond to the direction of power flow. It is such that it actuates for faults occurring in one direction only. It does not act for faults occurring in other direction. One of the examples of directional protection is that of reverse power protection of a generator.

Differential protection: A differential relay responds to vector difference between two or more similar electrical quantities.

This is perhaps the most important type of protection used for internal phase-to-phase and phase-to-earth faults and is generally applied to transformers having ratings of about 5 MVA and above. Depending upon the importance of the transformer, differential protection is sometimes used to lower capacity transformers as well.

Principle: The principle of operation depends on the two similar actuating quantities, e.g. current. The relay responds to the vector difference between the two currents ($I_1 - I_2$), which includes magnitude or phase angle difference.

Figure 12.7 illustrates the principle of differential protection of generator and transformer where Y is the winding of protected machine. When there is no internal fault, the current entering in Y is equal in phase and magnitude to current leaving winding Y. The CTs are of such a ratio that during normal conditions or for external faults, the secondary currents of CTs are equal. These currents I_1 and I_2 circulate in pilot wires. The polarity connections are such that the currents I_1 and I_2 are in the same direction in pilot wires during normal conditions or external faults. Relay operating coil is connected at the middle of pilot wires. During normal condition and external fault, the protection system is

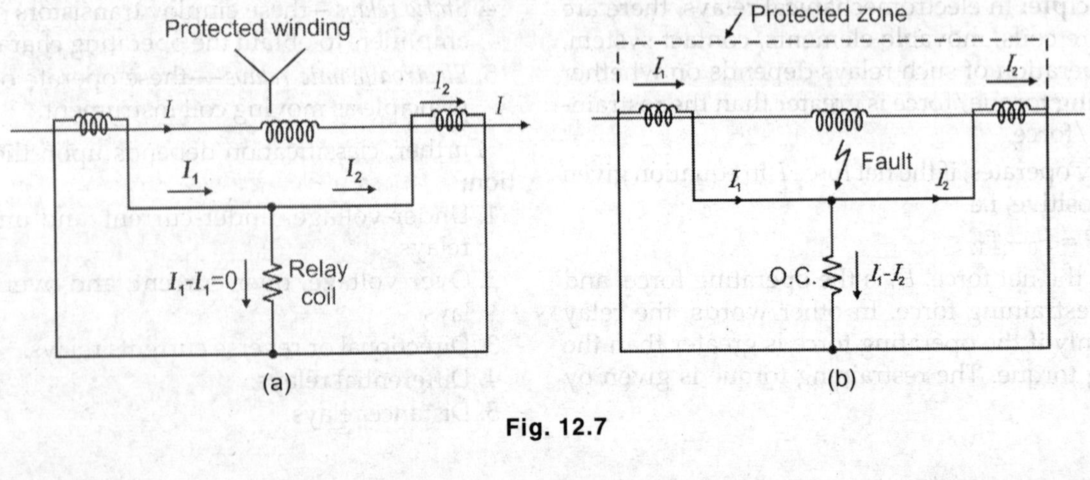

Fig. 12.7

balanced. This balance is disturbed for internal faults, thus the relay operates.

Differential protection in principle is employed in the following systems:

1. Protection of generator.
2. Protection of generator-transformer set.
3. Protection of transformer.
4. Protection of transmission lines.
5. Protection of large motors.
6. Bus-zone protection.

Mertz price protection of power transformers: In protection of transformer, CTs are connected at each end of these lines connected to the transformer as described below.

In both the cases, the current transformer are required at each side of the protected transformer. The connections of CT secondaries are such that during normal conditions and for external faults, no current should flow through the relay operating coils. There is an inherent phase displacement between vectors representing the voltages induced in high voltage windings and low voltage winding having same marking letter and corresponding neutral points.

The CT connections should be such that the resultant currents fed into the pilot wires from either are displaced in phase by an angle equal to the phase shift between the primary and secondary currents. To achieve this, the following rules are followed:

1. Secondaries of CTs on star-connected side of power transformer are connected in delta.
2. Secondaries of CTs connected on delta side of power transformer are connected in star. With such an arrangement, the phase displacement between currents gets cancelled with the phase displacement due to star/delta connections of CT secondaries and the current fed to pilot wires from both sides in phase during normal conditions.
3. The neutrals of CT and power transformer star connections are grounded. Figure 12.8 shows the connections of CT secondaries of delta-star transformer.

Thermal relays: These relays operate on the principle of thermal effect of electric current. Thermal relays sense the current by temperature rise produced by the current. The simplest thermal relay used in motor starters and overload protection devices employs a bimetallic strip mounted above a resistance wound heater coil. The bimetallic strip is being heated by means of a heating coil which is supplied through a current transformer. An insulated arm carrying a contact is pivoted and is held in contact with the strip with the help of spring *S*. The tension of the spring can be varied by rotating the sector shaped plate A, under normal working condition the strip remains straight. But when the strip is heated, it bends and tension of the spring is released; thus the relay contacts are closed which energises the trip circuit. The setting of the relay can be varied by varying the tension of the spring. This arrangement is shown in Fig. 12.9.

These over-current tripping relays are used mostly for motor controls and their heating element is designed

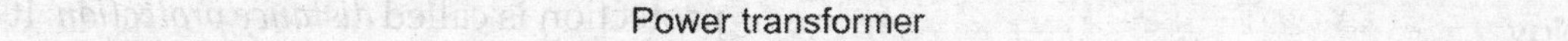

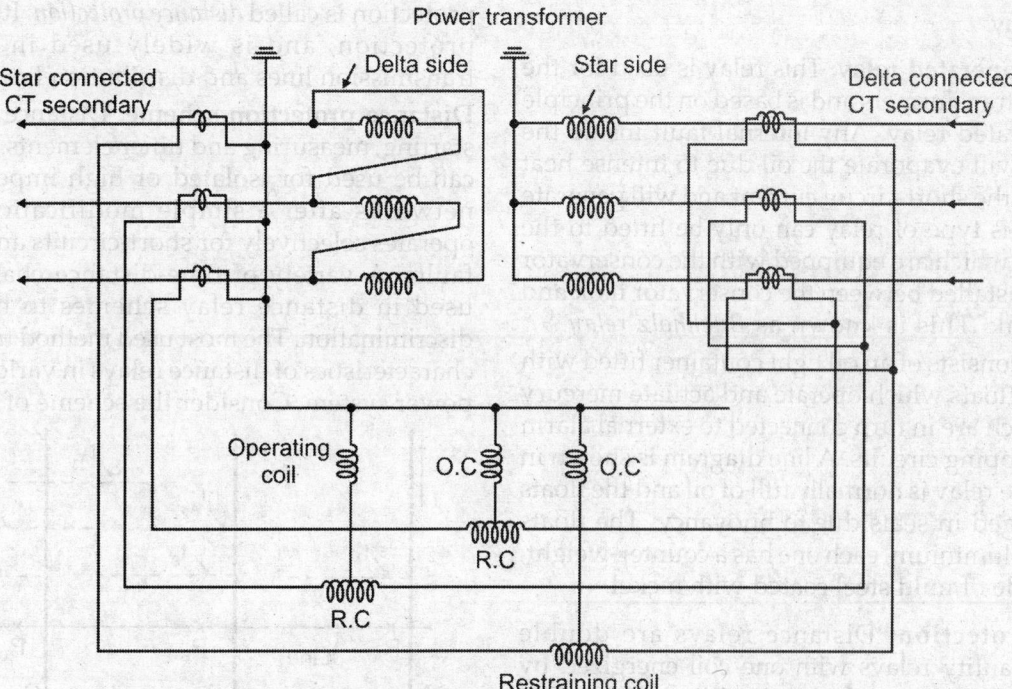

Fig. 12.8

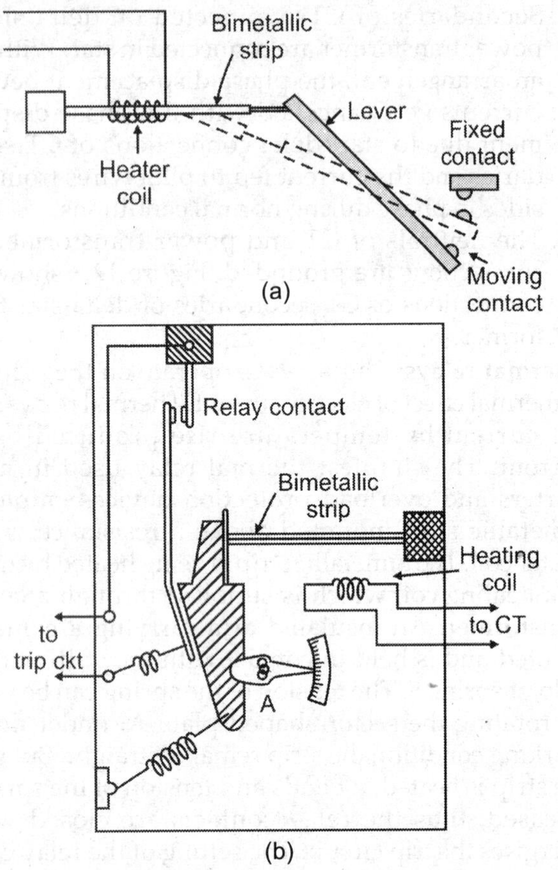

Fig. 12.9

to withstand short-time overload upto 7 times the normal full-load current.

Buchholz Relay

Gas and oil-operated relay: This relay is used for the protection of transformers and is based on the principle of a gas operated relay. Any internal fault inside the transformer will evaporate the oil due to intense heat generated by the short-circuit current and will generate the gases. This type of relay can only be fitted to the transformers which are equipped with the conservator tank as it is installed between the conservator tank and the main tank. This is known as *Buchholz relay*.

The relay consists of an oil tight container fitted with two internal floats which operate and actuate mercury switches which are in turn connected to external alarm and to the tripping circuits. A line diagram is shown in Fig. 12.10. The relay is normally full of oil and the floats remain engaged in seats due to buoyancy. The floats are made of aluminium, each one has a counter-weight, which is made of mild steel coated with nickel.

Distance protection: Distance relays are double actuating quantity relays with one coil energized by voltage and the other coil energized by current. The torque produced is such that when V/I reduces below

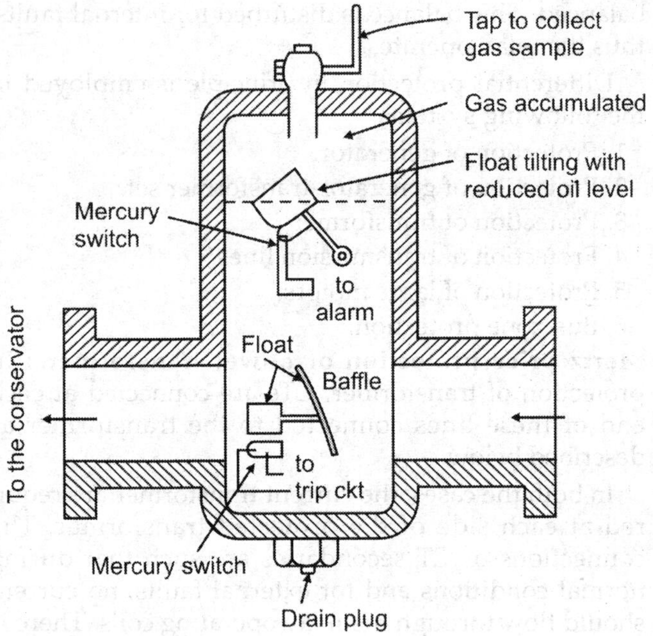

Fig. 12.10

a set value, the relay operates. During a fault on a transmission line, the fault current increases and the voltage at fault point reduces. The ratio V/I is measured at the location of CTs and PTs. The voltage at PT location depends on the distance between the PT and the fault. If the fault is nearer, measured voltage is less. If the fault is farther, measured voltage is more. Hence, assuming constant fault resistance, each value of V/I measured from relay location corresponds to a distance between the relay point and the fault. Hence such protection is called *distance protection*. It is a high speed protection, and is widely used in protection of transmission lines and distribution lines.

Distance protection scheme: Distance protection has starting, measuring and timer elements. The protection can be used for isolated or high impedance earthed networks after a simple modification and then it operates selectively for short-circuits and double earth faults. A variety of time-distance characteristics are used in distance relay schemes to have adequate discrimination. The most used method uses the stepped characteristics of distance relays in various zones of the power system. Consider the scheme of Fig. 12.11.

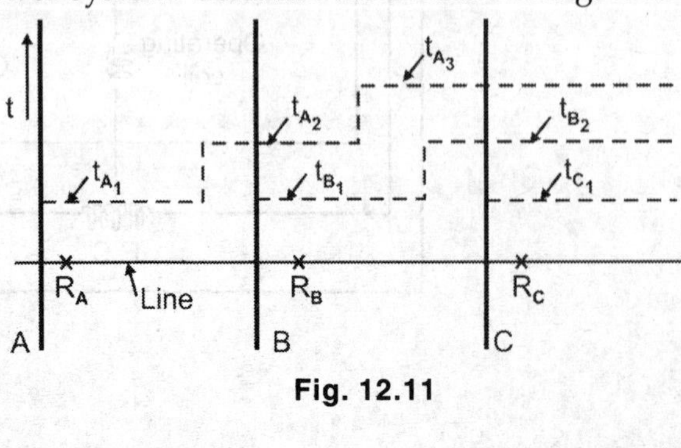

Fig. 12.11

The feeders AB and BC have been divided into three zones for the distance relays located at A, B and C. The relays at A and B see the current from A to B to C. The relay at C sees the current from C to B to A. By such schemes, quick protection can be obtained and back-up of the sections as well as adjoining lines/bus-bars can also be provided. The distance relay R_A, *located at section* A has a 3-step characteristic given by dashed line marked t_{A1}, t_{A2}, t_{A3}. The step t_{A1} is called the first step and covers about 80% of the first line section *AB* and gives instantaneous protection. The section step of relay t_{A2} is at station $A(R_A)$ *and covers the remaining portion of section AB* and about 20% to 50% of the next section (BC). The third step having t_{A3} covers the entire remaining line.

Important terms: Important terms used in connection with relays are:

i. *Pick-up current*: It is the minimum current in the relay coil at which the relay starts to operate. When the relay coil current is equal to or greater than the pick up value, the relay operates to energise the trip coil which opens the circuit breaker.

ii. *Current setting*: It is often desirable to adjust the pick-up current to any required value. This is known as *current setting* and is achieved by the use of tappings on the relay operating coil. This changes the torque on the disc and hence the time of operation of the relay. The values assigned to each tap are expressed in terms of percentage full load rating of CT with which the relay is associated and represents the value above which the disc commences to rotate and finally closes the relay.
Pick-up current = Rated secondary current of CT × current setting.

iii. *Plug setting multiplier (PSM)*: It is the ratio of fault current in relay coil to the pick-up current

$$\text{PSM} = \frac{\text{Fault current in relay coil}}{\text{Pick-up current}}$$

$$= \frac{\text{Fault current in relay coil}}{\text{Rated secondary current of CT} \times \text{current setting}}$$

iv. *Time current multiplier*: A relay is generally provided with control to adjust the time of operation. This adjustment is known as *time setting amplifier*. These figures are the multipliers to be used to convert the time derived from time/PSM curve into the actual operating time.

Power system stability: *Stability* of a power system is the ability of the system to remain in operating equilibrium, or synchronism, while disturbances occur on the system. There are three types of stability (i) steady-state, (ii) dynamic, and (iii) transient.

Steady-state stability relates to the response of a synchronous machine to a gradually increasing load.

Dynamic stability involves the response to small disturbances that occur on the system producing oscillations. If these oscillations are of successively smaller amplitudes, the system is considered dynamically stable. If the oscillations grow in amplitude, the system is dynamically unstable.

Transient stability involves the response to large disturbances, which may cause rather large changes in rotor speech power angles, and power transfers. The system's response to such a disturbance is usually within 1 second.

Inertia constant and swing equation: The angular momentum and inertia constant play an important role in determining the stability of a synchronous machine. The *per-unit constant H* is defined as the kinetic energy stored in the rotating parts of the machine at synchronous speed per unit megavolt ampere (MVA) rating of the machine.

$$H = \frac{\text{Stored energy in megajoules}}{\text{Rating in MVA}}$$

Thus, if *G* is the MVA rating of the machine, then

$$GH = \frac{1}{2} J\omega_s^2 \qquad (12.1)$$

where *J* is the polar moment of inertia of all rotating parts in kilogram-m^2 and ω_s is the angular synchronous velocity in electrical radians per second. If *M* is the corresponding angular momentum, then

$$M = J\omega \qquad (12.2)$$

Since $\omega = 360f$ electrical degrees per second, Eqs (12.1) and (12.2) yield

$$GH = \frac{1}{2} M\omega_s = \frac{1}{2} M(360)f$$

$$M = \frac{GH}{180f} \text{ MJ} \cdot \text{sec/electrical degree}$$

$$= \frac{GH}{\pi f} \text{ Megajoules} \cdot \text{sec/radian}$$

where *f* is the frequency of rotation.

The swing equation

$$\frac{H}{180f} \frac{d^2\delta}{dt^2} = P_i - P_e = P_a \text{ per unit} \qquad (12.3)$$

The swing equation contains information regarding the machine dynamics and stability.

***H* constant on a common MVA base:** An inertia constant H_{mach} based on a machine's own MVA rating may be converted to a value H_{syst} relative to the system base S_{syst} with the formula

$$H_{syst} = H_{mach} \frac{S_{mach}}{S_{syst}}$$

A convenient system base value is 100 MVA.

Critical clearing angle: If a disturbance or fault occurs in a system, δ begins to increase under the influence of positive accelerating power, and the system will become unstable if δ becomes very large. There is a critical angle which the fault must be cleared if the system is to remain stable and the equal-area criterion is to be satisfied. This angle is known as the *critical clearing angle* δ_0.

Equal area criterion: For stability, the area under the graph of accelerating power P_a vs δ is shown in Fig. 12.12 must be zero for some value of δ; that is, the positive area A_1 under the graph must be equal to the negative area A_2. This criterion is therefore known as the *equal area criterion* for stability.

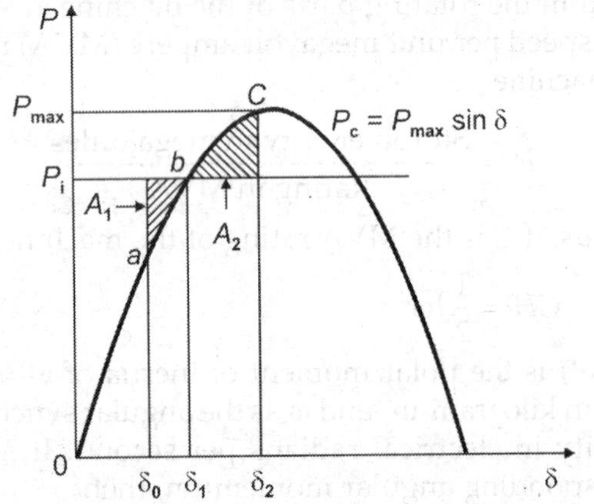

Fig. 12.12

Critical clearing angle: Figure 12.13 shows a power angle curve A before a fault, curve B during the fault, and curve C after the fault such that $A = P_{max} \sin\delta$, before the fault, $B = r_1 P_{max} \sin\delta$ during the fault and $C = r_2 P_{max} \sin\delta$ after the fault with $r_1 < r_2$ for stability.

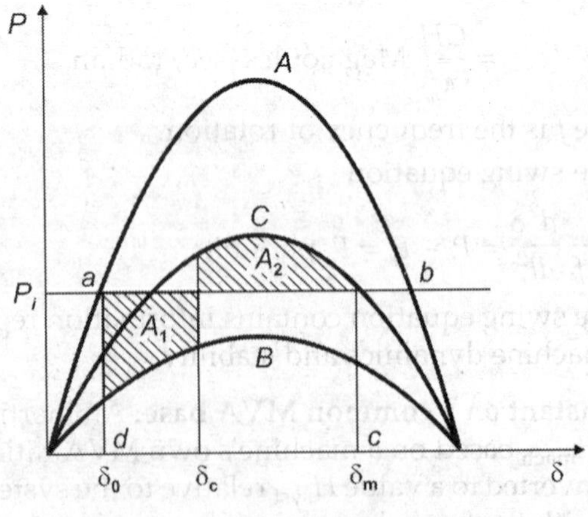

Fig. 12.13

$$\cos\delta_c = \frac{1}{r_2 - r_1}[(\delta_m - \delta_0)\sin\delta_0 - r_1\cos\delta_0 + r_2\cos\delta_m]$$

Two machine system: The swing equation for two machines

$$M_1 \frac{d^2\delta_1}{dt^2} = P_{i1} - P_{e1}$$

$$M_2 \frac{d^2\delta_2}{dt^2} = P_{i2} - P_{e2}$$

For two machines, the above relations are combined and simplified to

$$M\frac{d^2\delta}{dt^2} = P_i - P_e; \text{ (for two machines)}$$

$$\text{where } M = \frac{M_1 M_2}{M_1 + M_2}$$

$$P_i = \frac{M_2 P_{i1} - M_1 P_{i2}}{M_1 + M_2},\ P_e = \frac{M_2 P_{e1} - M_1 P_{e2}}{M_1 + M_2}$$

High Voltage

With ever increasing demand of electrical energy, the power system is growing both in size and complexities. The generating capacities of power plants and transmission voltage are on the increase beause of their inherent advantages. If the transmission voltage is doubled, the power transfer capability of the system becomes four times and the line losses are also relatively reduced. As a result, it becomes a stronger economical system. In India, we already have 400 kV lines in operation and 800 kV lines are being planned. In big cities, the conventional transmission voltages (100 kV, 220 kV etc.) are being used as distribution voltages because of increased demand. A system (transmission, switchgear, etc.) designed for 400 kV and above using conventional insulating materials is both bulky and expensive, and therefore, newer and newer insulating materials are being investigated to bring down both the cost and space requirements.

Test Voltages

For different transmission voltages, the test voltages required are given in Tables 12.1 and 12.2.

Table 12.1: Test voltages for ac equipments

System nominal voltage (rms)	*Power frequency withstand voltage (rms)*	*Impulse withstand voltage*	*Switching surge withstand voltage*
400	520	1425	875
525	670	1800	1100
765	960	2300	1350
1100	1416	2800	1800
1500	1920	3500	2200

If the insulation requirement for a particular operating voltage is required to be studied in a research and development laboratory, the voltage levels required in the laboratory are given in Tables 12.3 and 12.4.

Table 12.2: Test voltages for DC equipments

Normal voltage	*DC withstand voltage kV*	*Impulse withstand voltage kV*	*Switching surge withstand voltage kV*
± 400 kV	800	1350	1000
± 600 kV	1200	1900	1500
± 800 kV	1600	2300	2000

Table 12.3: Test voltages required for different system voltage (AC system)

Nominal voltage kV (rms)	*Power frequency voltage kV (rms)*	*Impulse withstand voltage kV*	*Switching surge voltage kV*
400	800	2400	1150
765	1000	3000	1750
1100	1400	3700	2300
1500	1900	4600	2800

Table 12.4: Test voltages required for different DC system voltage (AC system)

Nominal voltage kV	*DC voltage kV*	*Impulse withstand voltage kV*	*Switching surge voltage kV*
± 400	800	1750	1300
± 600	1200	2500	2000
± 800	1600	3000	2600

Tables 12.5 shows approximate dimension of the testing equipment as an equipment to be tested.

Table 12.5: Test voltages required for different DC system voltage (AC system)

Nominal voltage of equipment kV (rms)	*AC transformer height (m)*	*Impulse generator height (h)*
400	10	4
765	15	8
1100	18	12
1500	22	15

MULTIPLE CHOICE QUESTIONS

1. When all the three phases are short circuited it gives rise to
   A. asymmetrical fault currents
   B. symmetrical fault current
   C. zero current D. none of these
2. What will happen if a short circuit occurs on the system?
   A. Very small current flows through the system
   B. Heavy current flows through the system
   C. No current flows through the system
   D. None of these
3. The most common type of fault on the overhead transmission line is
   A. single line to ground fault
   B, double into to ground fault
   C. three-phase fault D. line to line fault
4. Which fault gives rise to symmetrical fault currents?
   A. Single line to ground fault
   B. Double line to ground fault
   C. Line to line fault D. Three phase fault
5. A 1000 kVA transformer with 5% reactance will have a reactance of 10% at
   A. 1000 kVA base B. 2000 kVA base
   C. 3000 kVA base D. 4000 kVA base
6. If the percentage reactance of the system upto the fault point is 20% and the base kVA is 10,000 then the short circuit kVA is
   A. 10 MVA B. 20 MVA
   C. 50 MVA D. 60 MVA
7. Short circuit kVA is obtained by multiplying the base kVA by
   A. 10/% X B. 20/% X
   C. 50/% X D. 100/% X
8. Reactors are used at various points in the power system to
   A. limit short circuit current
   B. increase short circuit current
   C. avoid short circuit D. none of these
9. The use of reactors permits installation of circuit breakers of the
   A. lower rating B. higher rating
   C. same rating D. none of these
10. The pu impedance value of an alternator corresponding to base values of 13.2 kV and 30 MVA is 0.02 pu The pu value for the base values 13.8 kV and 50 MVA will be
    A. 0.106 pu B. 0.206 pu
    C. 0.306 pu D. 0.318 pu
11. Base impedance of the power system is given by
    A. $\dfrac{(\text{Base kV})^2 \times 1000}{\text{Base kVA}}$ B. $\dfrac{(\text{Base kV})^2 \times 10^3}{\text{Base kV}}$
    C. $\dfrac{(\text{Base kVA})^2}{\text{Base kV}}$ D. none of these
12. If a 5 A fuse is blown frequently in a house lighting circuit, it is required to
    A. insert a fuse of higher current rating
    B. insert a fuse of higher voltage rating
    C. check the load D. none of these
13. HRC fuse provide best protection against
    A. open circuits B. short-circuits
    C. over load D. reverse current
14. Ground wire is used to
    A. give the support to the tower
    B. to connect the circuit conductor or other device to an earth-plate
    C. to avoid overloading
    D. to give good regulation
15. The size of the earth wire is determined by
    A. the ampere capacity of the service wires
    B. the atmospheric conditions
    C. the voltage of the service wires
    D. none of these
16. The grounding is generally done at the
    A. receiving end B. supply end
    C. the middle of receiving and supply end
    D. none of these
17. The most common type of three phase in unsymmetrical fault is
    A. single line to ground
    B. double line to ground
    C. line to line D. three phase
18. In a balanced three phase system, negative and zero phase sequence currents are
    A. equal B. zero
    C. different D. none of these
19. The positive and negative sequence impedance of a transmission line are
    A. equal B. zero
    C. different D. none of these
20. The zero sequence impedance of different elements of power system is generally
    A. equal B. zero
    C. different C. none of these
21. On the occurrence of an unsymmetrical fault which sequence component is always more than the negative sequence component?

Ans.	1. B	2. B	3. A	4. D	5. B	6. C	7. D	8. A	9. A	10. C	11. A	12. C	13. B
	14. B	15. A	16. B	17. A	18. B	19. A	20. C						

A. Zero sequence B. Positive sequence
C. Both are correct D. None of these

22. The zero sequence impedance of an element in a power system is generally different from
A. positive sequence impedance
B. negative sequence impedance
C. both A and B are correct
D. None of these

23. The operator 'a' rotates the vector in the anti-clockwise direction by
A. 90° B. 120°
C. 180° D. none of these

24. Which of the following equation is correct
A. $1 + a + a^2 = 0$ B. $a + a^2 = 1$
C. $1 + a^3 = 0$ D. none of these

25. Which of the following equation is correct
A. $a^2 = 0.5 + j\,0.866$ B. $a^2 = -0.5 + j\,0\text{-}866$
C. $a^2 = -0.5 - j\,0.866$ D. $a^2 = j\,0.866$

26. The positive sequence component of voltage at the point of fault is zero when it is a
A. three phase fault B. L-L fault
C. L-L-G fault D. L-G fault

27. The positive, negative and zero sequence impedances of a solidly grounded system under steady state condition always follow the relations
A. $Z_1 > Z_2 > Z_0$ B. $Z_1 < Z_2 < Z_0$
C. $Z_0 < Z_1 < Z_2$ D. none of these.

28. When a fault occurs on the system a circuit breaker
A. opens automatically B. closes automatically
C. opens manually D. none of these

29. If the length of the arc of circuit breaker increases, its resistance
A. increases B. decreases
C. remains same D. none of these

30. If the dielectric strength of the medium between contacts builds up more rapidly than the restriking voltage, then the arc will
A. be extinguished B. not be extinguished
C. increase D. none of these

31. Current chopping mainly occurs in
A. oil circuit breaker
B. air blast circuit breaker
C. vacuum circuit breaker
D. SF_6 circuit breaker.

32. Capacitive current breaking results in
A. short circuits B. open circuits
C. voltage surges D. none of these

33. The speed of circuit interruption in forced blast circuit breaker is
A. slow B. medium
C. fast D. none of these

34. The property of contact materials, to limit current chopping in vacuum circuit breaker, should have
A. high vapour pressure and high conductivity
B. high vapour pressure and low conductivity
C. low vapour pressure and low conductivity
D. None of these

35. For the interruption of high voltages and low current, the circuit breaker (CB) preferred is
A. air blast CB B. oil CB
C. vacuum CB D. all are correct

36. A fault is more severe from the view point of RRRV if it is a
A. short length line fault
B. medium length line fault
C. long length line fault D. None of these

37. The rate of rise of restriking voltage depends on
A. the type of circuit braker
B. the inductance of the system only
C. the capacitance of the system
D. the inductance and capacitance of the system only

38. A three phase breaker is rated at 2000 MVA, 33 kV, its making current will be:
A. 35 kA B. 70 kA
C. 89 kA D. 160kA

39. The normal practice to specify the making current of a circuit breaker in terms of
A. rms value B. peak value
C. average value D. None of these

40. Resistance switching is normally resorted in case of
A. air blast CB B. bulk oil CB
C. minimum oil CB D. all types of CBs

41. If the inductance and capacitance of a system are 1.0 H and 0.01 μF and the instantaneous value of interrupted current is 10 amps, the voltage across the breaker contacts will be
A. 50 kV B. 100 kV
C. 60 kV D. 57 kV

42. The trip coil of a CB is connected through 40% transformers. If normal line current is 100 amp and circuit breaker opens at 120% normal line current, the trip mechanism should he set to operate when the trip current is
A. 2 amps B. 2.5 amps
C. 3 amps D. 5 amps

43. For remote operation, CB must be equipped with a
A. lime delay trip B. shunt trip
C. inverse time trip D. none of these

44. Which of the following CBs is generally used in Railway applications?
A. Air break CB B. Minimum oil CB
C. Bulk oil CB D. SF_6 CB

45. Which of the following CBs is preferred for EHT applications?
A. Air break CB B. Minimum oil CB
C. Bulk oil CB D. SF_6 CB

Ans.	21. B	22. C	23. B	24. A	25. C	26. A	27. A	28. A	29. A	30. A	31. B	32. C	33. C
	34. B	35. C	36. A	37. D	38. C	39. B	40. A	41. B	42. C	43. B	44. A	45. D	

46. In CB, the arcing contacts are made of
A. porcelain B. copper tungsten alloy
C. aluminum alloy D. electrolytic copper

47. Which statement is correct?
A. SF_6 gas is toxic
B. SF_6 gas is lighter than air
C. SF_6 gas is yellow in colour
D. SF_6 gas has pungent smell

48. A relay performs the function of
A. fault isolation B. fault detection
C. fault prevention D. all are correct

49. The relay operating coil is supplied through
A. fuse B. power transformer
C. instrument transformer
D.nNone of these

50. Choose the correct statement:
A. 1 VA relay is less sensitive than 5 VA relay
B. 1 VA relay is more sensitive than 5 VA relay
C. 1 VA and 5 VA relay have the same sensitiveness
D. None of these.

51. An overcurrent relay having a current setting of 125% is connected to a supply circuit through a current transformer of ratio 400/5. The pick up value is
A. 6.25 amps B. 10 ams
C. 12.5 amps D. 15 amps

52. The pick up value of a relay is 7.5 amps and fault current in relay coil is 30 amps. Its plug setting multiplier is
A. 2 amps B. 4 amps
C. 6 amps D. 8 amps

53. Most of the relays on service on electric power system are
A. electronic relays
B. electromechanical relays
C. thermal relays D. none of these

54. Back up protection is generally employed for protection against
A. short circuit faults only
B. open circuit facults only
C. short and open circuit faults
D. none of these

55. If the fault current is 2000 amps, the relay setting is 50% and CT ratio is 400/5, the plug setting multiplier will be
A. 23 amps B. 50 amps
C. 15 amps D. none of these

56. If the phase angle of voltage coil of a directional relay is 50, the maximum torque angle of relay is
A. 25 B. 100
C. 130 D. none of these

57. A Mho relay is a:
A. voltage restrained directional relay
B. voltage controlled over current relay
C. directional restrained over current relay
D. directional restrained over voltage relay.

58. The Buchholz relay protect a transformer from:
A. a turn to turn fault
B. all types of internal faults
C. winding to winding faults
D. none of these

59. The number of pilot wires required for protecting 3-phase transmission lines using translay system of protection is
A. 6 B. 4
C. 3 D. 2

60. Current transformers for relays and meters usually have
A. 5 amp secondary B. 10 amp secondary
C. 20 amp secondary D. none of these

61. The main function of under voltage protective device generally used with a motor starter is
A. to open supply circuit voltage
B. to control the motor voltage
C. to prevent the opening of the supply circuit
D. none of these

62. Differential relays are used to protect the equipment against
A. internal faults B. reverse current
C. over current D. none of these

63. The least expensive protection for overcurrent in LV system is
A. isolator B. oil circuit breaker
C. rewirable fuse D. air break CB

64. Impedance relay can be used for
A. earth faulty B. phase fault only
C. both earth and phase faults
D. none of these

65. To protect the power transformer with star-delta connection against fault, the current transformer secondaries will have
A. delta-delta connection
B. delta-star connection
C. star-star connection D. star-delta connection

66. In Mertz Price protection, the CTs secondaries of delta-star power transformer will be in
A. star-delta B. star-star
C. delta-star D. delta-delta

67. If the fault occurs near an impedance relay, the V/I ratio will be
A. constant for all distance
B. higher than the value if the fault occurs away from the relay
C. lower than the value if the fault occurs away from the relay
D. none of these

68. A 50 Hz, 4 pole, turbo generator rated at 20 MVA, 13.2 kV has an inertia constant H = 4 kW·sec/kVA

Ans.	46. B	47. A	48. B	49. C	50. B	51. A	52. B	53. B	54. A	55. D	56. D	57. A	58. B
	59. D	60. A	61. A	62. A	63. C	64. C	65. B	66. A	67. C				

The kinetic energy stored in the rotor at synchronous speed is:
A. 20 Megajoules B. 40 Megajoules
C. 80 Kilojoules D. 80 Megajoules

69. The inertia constants of two groups of machines which do not swing together are M_1 and M_2. The equivalent inertia constant of the system is
A. $M_1 + M_2$ B. $\sqrt{M_1 + M_2}$
C. $M_1M_2/M_1 + M_2$ D. $M_1 + M_2/M_1M_2$

70. If the inertia constant H of a machine of 200 MVA is 2 pu, its value corresponding to 400 MVA will be
A. 4 pu B. 3 pu
C. 2 pu D. 1 pu

71. The inertia constant of two groups of machines which do not swing together at M_1 and M_2 such that $M_1 > M_2$. It is proposed to add some inertia to one of the two groups of machine for improving the transient stability of the system. It should be added to
A. M_1 B. M_2
C. $M_1 + M_2$ D. None of tese

72. Transient stability can be improved by
A. increasing the system voltage
B. decreasing the system voltage
C. keeping the system voltage constant
D. none of these

73. Transient stability can be improved by using governors attached to the turbines of
A. low speed B. high speed
C. constant speed D. none ofthese

74. Which of the following methods is the best for improving the transient stability?
A. Using voltage regulators
B. Using high speed circuit breakers
C. Using lightning arrestors
D. None of these

75. An alternator having an induced emf of 1.6 pu is connected to an infinite bus of 1.0 pu. If the bus bar has reactance of 0.6 pu and alternator has reactance of 0.2 pu, the maximum power that can be transferred is given by
A. 2 pu B. 2.67 pu
C. 5 pu D. 6 pu

76. For stability and economic reasons, we operate the transmission line with power angle in the range
A. 10° to 25° B. 30° to 45°
C. 60° to 75° D. 65° to 80°

77. If the torque angle of an alternator increases indefinitely, the system will show
A. instability B. stability
C. none of these

78. If P_m is the maximum power transferred, the loss on the system is
A. $P_m/2$ B. $P_m/3$
C. $P_m/4$ D. None of these

79. The maximum power which can be transferred through the system without loss of stability under sudden disturbances is referred as
A. steady state limit
B. transient stability limit
C. dynamic stability limit
D. none of these

80. Power can be transferred only if reactance is
A. absent B. present
C. zero D. none of these

81. The sending end voltage of a feeder with reactance 0.2 pu is 1.2 pu. If the reactive power supplied at the receiving end of the feeder is 0.3 pu, the approximate drop of volts in the feeder is
A. 0.2 pu B. 0.06 pu
C. 0.05 pu D. 0.072 pu

82. Which of the following statement is correct?
A. Transient stability limit is lower than steady state stability
B. An early fault clearing means poor chance of maintaining system stability
C. A system has better chances of maintaining synchronism if the generators are fitted with acting automatic voltage regulators
D. The use of automatic reclosing circuit breakers improves the system stability

83. Which of the following statement is wrong?
A. A salient pole machine is more stable than a cylindrical rotor machine
B. Present day generators are less stable than the generators made a few years ago
C. An early fault clearing means better chances of maintaining system stability
D. Transient stability is more than steady state limit

84. The relay with inverse time characteristic will operate within
A. 5 sec B. 5 to 10 sec
C. 10 to 15 sec D. 15 to 20 sec

85. Which of the following materials cannot be used for contacts of switching devices?
A. Copper B. Copper alloys
C. Aluminium D. Silver

86. In which of the following circuits there may be zero sequence components?
A. Line voltage of star circuit
B. Phase voltage of star circuit
C. Line current of delta circuit
D. None of these

87. What is the most serious problem in vacuum circuit breaker?

Ans. 68. D 69. C 70. D 71. B 72. A 73. B 74. B 75. A 76. B 77. B 78. A 79. B 80. B
81. C 82. B 83. D 84. C 85. C 86. B

A. Poor arc quenching B. Low thermal stability
C. Current chopping D. All of these

88. In case of power transformer protection by differential relay, why are the CTs connected on secondary star side of transformer is in delta?
A. To avoid response to external fault
B. To avoid response to internal fault
C. To avoid response to CT short circuits
D. None of these

89. When can 2 O/L + 1 E/L scheme be used instead of 3 O/L + 1 E/L scheme to protect a feeder?
A. If the neutral is solidly earthed
B. If the neutral is not earthed
C. If the transformer is star-star connected
D. None of these

90. An over current relay of current rating 5A and setting 150% is connected to the secondary of CT of ratio 300:5. Then the current in the lines for which the relay picks up is
A. 300 A B. 450 A
C. 150 A D. 200 A

91. Which frequency would be lighting arrester for protection of equipment allow to pass?
A. Fundamental frequency component
B. Power frequency component
C. Both fundamental and power frequency component
D. All of these

92. In a 3-phase system, the phasor sum of phase current is zero. It means absence of
A. positive sequence component
B. negative sequence component
C. zero sequence component
D. all of these

93. The positive sequence reactance will be equal to negative sequence reactance in case of
A. transformer B. transmission line
C. generator D. motor

94. Rated making capacity of breaker is equal to
A. $\sqrt{2}\times$ rated symmetrical breaking capacity
B. 2.55 × rated symmetrical breaking capacity
C. 1 × rated symmetrical breaking capacity
D. none of these

95. In load-flow study of a power system, what are the quantities specified at a load bus?
A. $|P|$ and $|V|$ B. P and V
C. P and Q D. P and δ

96. The critical clearing time is related to
A. steady statestability limit
B. transient stability limit
C. short-circuit current limit
D. none of these

97. In what way are long EHV transmission lines in interconnected power systems generally protected?
A. Over current protection
B. Pilot protection
C. Distance protection D. Differential protection

98. The disadvantages of Gauss–Seidel method over Newton's method in load flow programmes are?
A. More memory requirement and less accuracy
B. More programming effort and less accuracy
C. Less accuracy and more iterations
D. None of these

99. For transformer protection, a differential relay is biased to avoid maloperations due to
A. inrush current
B. saturation of current transformers
C. mismatching of current ratio of current transformers
D. current setting multipliers

100. When a circuit breaker interrupt a small inductive current, chopping is most likely to occur when the breaker is
A. an oil circuit breaker
B. an air-blast circuit breaker
C. SF_6 circuit breaker D. all of these

101. In case of protective relays which are energised through PTs and CTs, the relation between the primary and secondary impedances seen by the relay is equal to
A. PT ratio/CT ratio B. CT ratio/PT ratio
C. 1.0 D. $(\text{CT ratio}/\text{PT ratio})^2$

102. The surge impedance of a 100 km long underground cable is 50 ohms. The surge impedance for a 40 km length of similar cable would be
A. 50 ohms B. 60 ohms
C. 70 ohms D. 80 ohms

103. For which line is carrier-current protection scheme normally used?
A. HV transmission lines only
B. HV transmission lines and cables only
C. HV cables D. All of these

104. Mho relay is used to protect
A. long transmission lines
B. medium length lines
C. short length lines D. all of these

105. Reactance relay is used to protect against
A. earth faults only B. phase faults only
C. over voltage fault D. none of these

106. If the fault current is 2000 amps, the relay setting is 50% and the CT ratio is 400/5. The plug setting multiplier will be
A. 5 B. 6
C. 8 D. 10

107. The most economic load on an overhead line is
A. less than the natural load
B. equal to the natural load

Ans. 87. C 88. A 89. A 90. B 91. B 92. C 93. A 94. B 95. C 96. B 97. B 98. C 99. C
100. B 101. B 102. A 103. A 104. A 105. A 106. D

C. greater than the natural load
D. bone of these

108. As compared with circuit breaker, for the same rupturing capacity, the actual current to be interrupted by an HRC fuse is
A. much less B. much more
C. equal D. none of these

109. The stability of arc in vacuum circuit breaker depends on
A. the contact material B. its vapour pressure
C. the circuit parameters
D. all of these

110. The impulse ratio of a rod gap is
A. 0 to 1.0 B. 1.2 to 1.5
C. 1.6 to 1.8 D. 2 to 2.2

111. Shunt compensation in an EHV line is used to improve
A. stability B. fault level
C. voltage profile D. all of these

112. For stability and economic reasons, the transmission line is operated with power angle in the range
A. 10°–25° B. 30°–45°
C. 60°–75° D. 65°–80°

113. The rate of rise of restriking voltage depends upon
A. the type of circuit breaker
B. the inductance of the system only
C. the capacitance of the system only
D. the inductance and capacitance of the system

114. The making current of 3-phase breaker with rating 2000 MVA, 33 kV will be
A. 35 kA B. 50 kA
C. 70 kA D. 89 kA

115. A combination of HRC fuse and circuit breaker is used, the circuit breaker operates for
A. low overload currents
B. under all abnormal current
C. short-circuit currents D. all of these

116. Which of the following is correct statement regarding insulators and lightning arresters?
A. The insulators and lighting arresters should have high impulse ratio
B. The insulators and lightning arresters should have low impulse ratio
C. The insulator should have high impulse ratio and lightning arrester low
D. The lightning arrester should have high impulse ratio but insulator low

117. The leakage resistance of a 50 km long cable is 1 Mohm. For a 100 km long cable, it will be
A. 1 Mohm B. 2 Mohm
C. 0.66 Mohm D. none of these

118. Resistance switching is used in
A. bulk oil circuit breakers
B. minimum oil circuit breakers
C. air-blast circuit breakers
D. all types of breakers

119. Rotating machines are protected against lightning surges by a combination of
A. lightning conductor and capacitor
B. lightning conductor and lightning arrester
C. lightning arrester and capacitor
D. none of these

120. The passive element that is used as interconnecting element for the stable operation of interconnected system is
A. reactor B. resistor
C. capacitor D. resistor and capacitor

121. If two synchronous generators connected in parallel loose synchronism, it will lead to
A. stalling of generators
B. wild fluctuations in current
C. wild fluctuations in current and voltage
D. none of these

122. A generator is connected to a synchronous motor, it is preferable to have from stability point of view,
A. generator neutral reactance grounded and motor neutral resistance grounded
B. generator and motor neutrals resistance grounded
C. generator and motor neutrals reactance grounded
D. generator neutral resistance and motor neutral reactance grounded

123. A 50 Hz, 4 pole turbo alternator rated at 20 MVA, 13.2 kV has an inertia constant H = 4 kW sec/kVA. The KE stored in the rotor at synchronous speed is
A. 80 kilojoules B. 80 megajoules
C. 40 megajoules D. 20 megajoules

124. Which relays are required for complete protection of 3-phase line?
A. Three phase and three-earth fault relays
B. Three-phase and two-earth fault relays
C. Two-phase and two-earth fault relays
D. Two-phase and one-earth fault relays

125. The order of the lightning discharge current is
A. 10 kA B. 100 A
C. 1A D. 1 μA

126. Why does the minimum oil circuit breaker has less volume of oil?
A. There is insulation between contacts
B. The oil between the breaker contacts has greater strength
C. Solid insulation is provided for insulating the contacts from earth
D. None of these

127. Load flow study is used for
A. fault calculations B. stability studies
C. system planning D. all of these

Ans. 107. C 108. A 109. D 110. B 111. C 112. B 113. D 114. D 115. A 116. C 117. D 118. C 119. C 120. A 121. C 122. D 123. B 124. D 125. A 126. C 127. C

128. How is arcing on transmission lines prevented?
A. By circuit breaker B. By protective relay
C. By inductor in the neutral
D. By capacitor in the neutral
129. If in a line, the loading corresponds to the surge impedance loading, the voltage at the receiving end is
A. greater than sending end
B. less than sending end
C. equal to the sending end
D. none of these
130. The making current of a circuit breaker is specified in terms of
A. rms value B. peak value
C. average vale D. either of these
131. Consider two groups of machines which do not swing together and their inertia constants respectively are M_1 and M_2 such that $M_1 > M_2$. If some inertia is to be added to one of the two groups of machines for improving the transient stability of the system. It should be added to
A. M_1 B. M_2
C. either M_1 or M_2
D. the transient stability can not be improved
132. The inertia constants of two groups of machines which do not swing together are M_1 and M_2. The equivalent inertia constant of the system is given by
A. $M_1 + M_2$ B. $M_1 - M_2$ if $M_1 > M_2$
C. $\dfrac{M_1 M_2}{M_1 + M_2}$ D. $\sqrt{M_1 M_2}$
133. The higher the initial load,
A. the larger the critical clearing angle
B. the lower the critical clearing angle
C. the critical clearing angle can have any value
D. none of these
134. The energy in a lightning stroke can be as high as
A. 10 units B. 100 units
C. 250 units D. 1000 units
135. From the view-point of RRRV, a fault is more serious if it is a
A. short-length line fault
B. medium length line fault
C. long line fault D. any of these
136. If the voltage at the two ends of a line are 132 kV and its reactance is 40 ohm, the capacity of the line is
A. 435.6 MW B. 217.8 MW
C. 251.5 MW D. 500 MW
137. In what shape of the conductor, corona loss is less?
A. Circular B. Flat
C. Oval D. Any of these
138. The function of series compensation on EHV lines to
A. improve the stability
B. reduce the fault level
C. improve the voltage profile
D. all of these
139. The stability limit of system is defined as the value of power below which
A. the system is stable and above which also stable
B. the system is unstable and above which it is stable
C. the system is stable and above which it is unstable
D. none of these
140. Swing equation represents
A. the variation of power with respect to time
B. the variation of torque angle with respect to time
C. the variation of input power with respect to time
D. none of these
141. Surge impedance of an overhead line is of the order of
A. tens B. ones
C. hundreds D. thousands
142. An overhead line with surge impedance of 400 ohms is terminated through a resistance *R*. A surge travelling over the line does not suffer any reflection at the junction if the value of *R* is
A. 20 ohms B. 200 ohms
C. 800 ohms D. 400 ohms
143. Merze Price protection is used to protect
A. transmission line and transformers
B. transformers and generators
C. motors only D. transformers only
144. The change in excitation in an alternator connected to an infinite line provides a corresponding change in
A. terminal voltage B. speed of rotor
C. power generated D. power factor
145. In order to improve the voltage profile of the line, compensation is resorted by providing
A. capacitors at the receiving end
B. reactors at the receiving end
C. resistors at the receiving end
D. none of these
146. A surge absorber reduces
A. steepness and magnitude of the wave
B. steepness of the wave tail
C. magnitude of the wave
D. steepness of the wave
147. Switching and protection are undertaken by
A. switch gear B. lightning arrester
C. fuses D. relays
148. The function of Lightning arrester is to
A. limiting the short-circuit fault current
B. provide path to high voltage surge to earth
C. reduce arcing
D. none of these

Ans.													
	128. C	129. C	130. B	131. B	132. C	133. B	134. C	135. A	136. A	137. A	138. A	139. C	140. B
	141. C	142. D	143. B	144. D	145. A	146. A	147. A	148. B					

149. Distance relays operation is dependent upon
A. ratio of current to current
B. ratio of voltage to current
C. ratio of voltage to voltage
D. none of these
150. Which relay is used to detect and protect internal faults of transformer?
A. Buchholz relay B. Directional relay
C. Thermal relay D. Distance relay
151. Impulse insulation breakdown strength of the protected equipment must be
A. greater than that of the protecting device
B. equal to the protecting device
C. less than that of the protecting device
D. none of these
152. Surge protector provides
A. high impedance to normal voltage
B. low impedance to surge
C. both A and B D. none of these
153. Merz-Price Current protection is also called
A. under-voltage protection
B. over-current protection
C. differential protection
D. distance protection
154. In which type of fault, maximum short-circuit current occurs
A. Three-phase faults to ground
B. Double line fault to ground
C. Single line fault to ground
D. None of these
155. Back-up protection is used to protect
A. overvoltages B. overcurrents
C. transient current D. short-circuits
156. In a dynamically stable system, oscillations of synchronous machines following load changes will
A. decrease with time B. increase with time
C. not occur D. none at all
157. When large bulk power is to be transmitted, the reason for using in HV transmission is that
A. line losses are less
B. no technical problem is efficient
C. the line is very sensitive to wind pressure
D. none of these
158. An isolator operates under
A. no load condition B. full load condition
C. 50% load condition D. fault condition
159. The lightning arrester is connected
A. in series with the line
B. between line and earth
C. to a pole near the line
D. none of these
160. In a dynamically unstable system
A. oscillations may increase until synchronism is lost
B. governor action has no effect
C. load system is less than energy input
D. none of these
161. Relays using induction disc principle operate
A. only on DC B. only on AC
C. either on DC and AC D. none of these
162. Which of the following does not influence the strength of an electric shock to a human body?
A. Duration of current flow
B. Frequency
C. Voltage D. Ambient temperature
163. Earthing is necessary to give protection against
A. voltage fluctuation B. overloading
C. danger of electric shock
D. high temperature of the conductors
164. The primary function of a fuse is to
A. protect the appliance B. open the circuit.
C. prevent excessive currents
D. protect the line
165. A thermal protection switch is able to protect against
A. overload B. over voltage
C. temperature D. short circuit
166. In which of the following cases are delay fuses used?
A. For the protection of light circuits
B. For the protection of power outlet circuits
C. For the protection of fluorescent lamps
D. For the protection of motors
167. It is a fatal to touch a live wire because
A. the human body becomes part of the electric circuit
B. the voltage may cause burns to the skin
C. the current may cause burns to the skin or inside the body
D. it causes damage to the heart and nerve system
168. A fuse is able to protect a circuit when
A. it is fitted at the end of the circuit
B. the size of the fuse wire is appropriate for the load
C. the size of the fuse wire is appropriate for the size of the line
D. the fuse is fitted at the beginning of the circuit
169. When extra stop buttons are added to an existing starter control circuit they must be
A. in series with the contactor coil
B. parallel to the overload
C. parallel to the start button
D. in series with the motor
170. A single phase motor protection switch is adjusted according to
A. the maximum starting current
B. the cross-sectional area of the feeding line
C. the rated current
D. the capacity of the fuse

Ans. 149. B 150. A 151. A 152. C 153. C 154. A 155. D 156. A 157. A 158. A 159. B 160. A 161. B 162. D 163. C 164. C 165. A 166. D 167. A 168. D 169. A 170. C

171. An isolating switch is operated under load in an old installation. Which of the following statements is correct?
A. An arc is produced which can destroy the switching section
B. If the switch is operated sufficiently fast, the circuit is interrupted correctly
C. The circuit is interrupted correctly. However, the contacts of the isolating switch must be changed
D. The nominal current can be interrupted five times with the help of an isolating switch

172. How is the isolating switch operated?
A. With revolving crank drive
B. With the help of a switching rod
C. With the help of a compressed air drive
D. With the help of a compressed air motor

173. When is the arc produced by the breaking of an alternating current extinguished?
A. When the product of current and voltage, i.e. the power is the lowest.
B. When the voltage is at its maximum value
C. When the current is at its maximum value
D. When the current reaches its zero value

174. The figure shows the characteristic curve of a manually operated motor protective switch. If the relays are adjusted for a nominal current of 10 A, which of the following statements is true?

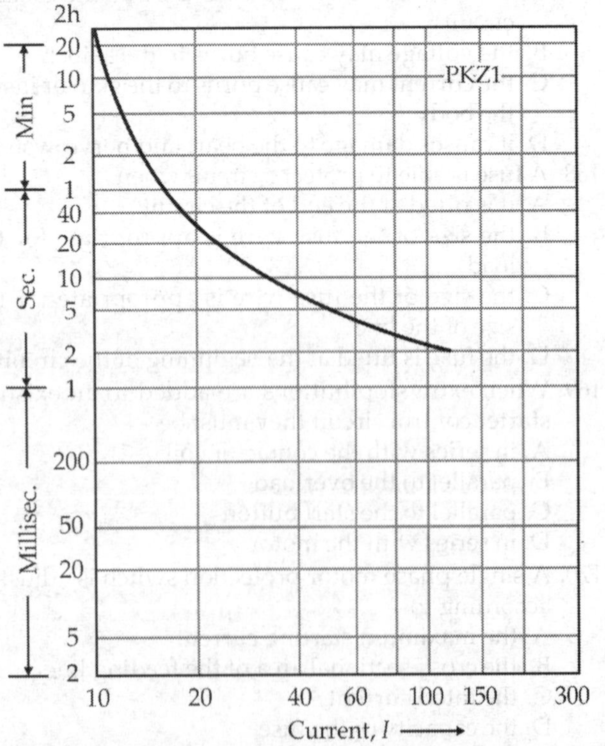

A. The protective switch consists of bimetallic and magnetic relays
B. The switch limits the current in case of short circuits
C. At a current of 50 A the switch operates after 5 seconds
D. A current of 20 A can flow continuously through the switch without operating it

175. In thermal overcurrent delay relays and overcurrent fast relays, which of the following statements is true?
A. The thermal overcurrent delay relays are bimetallic switches
B. The thermal overcurrent delay relays are based upon the magnetic effect of current
C. The overcurrent fast relays are bimetallic switches
D. The thermal overcurrent delay relays as well as the overcurrent fast relays are bimetallic switches

176. The figure shows the characteristic curve of a manually operated motor protective switch. If the relays are adjusted for a nominal current of 10A, which of the following statements is true?

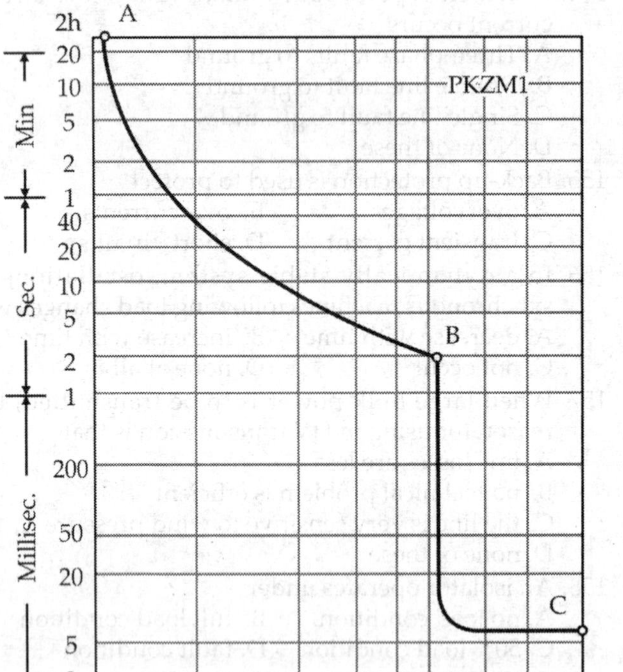

A. The switch contains bimetallic relays only
B. The portion A to B of the curve shows the characteristic of the magnetic fast relays
C. The portion B to C of the curve shows the characteristic of the bimetallic relays

Ans. 171. A 172. B 173. D 174. C 175. A

D. At a current of 150 A the switch operates within 7 ms

177. What can be adjusted by turning screw in the thermal overcurrent delay relays?
A. The short circuit current
B. The operating current of the thermal over-current delay relays
C. The operating current of the over-current fast relays
D. The operating time for the rated current of the switch

178. The turning screw is adjusted at 1.4 in thermal overcurrent delay relays. What current can permanently pass through the switch without operating it?
A. 0.95 A B. 1.1 A
B. 1.25A D. 1.4 A

179. A nominal current of 20*A* is set at the motor protective switch whose characteristic is shown in the figure. After how much time does the switch operate in case of an overcurrent of 30 amp

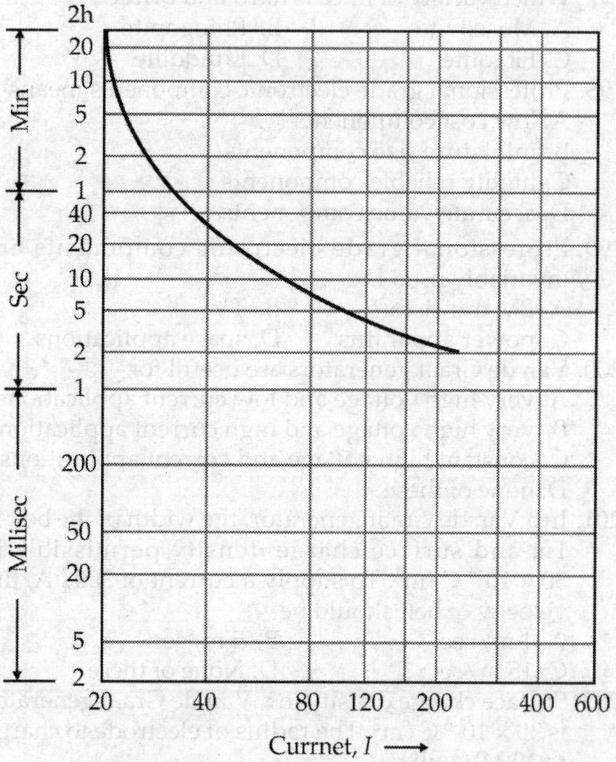

A. after about 2 sec B. after about 5 sec
C. after about 20 sec D. after about 2 minutes

180. The operating voltage of a contactor which operates at 220 V, 50 Hz increases to 250 V due to an inter turn short circuit of the control transformer. What is the effect of this?
A. The contactor is not attracted
B. The contact force is insufficient
C. The attracted contactor is dropped
D. When switching on the armature and the contact start chattering

181. A nominal current of 25 A is set at the motor protective switch whose characteristic is shown in the figure. In which of the following cases does the switch operate?

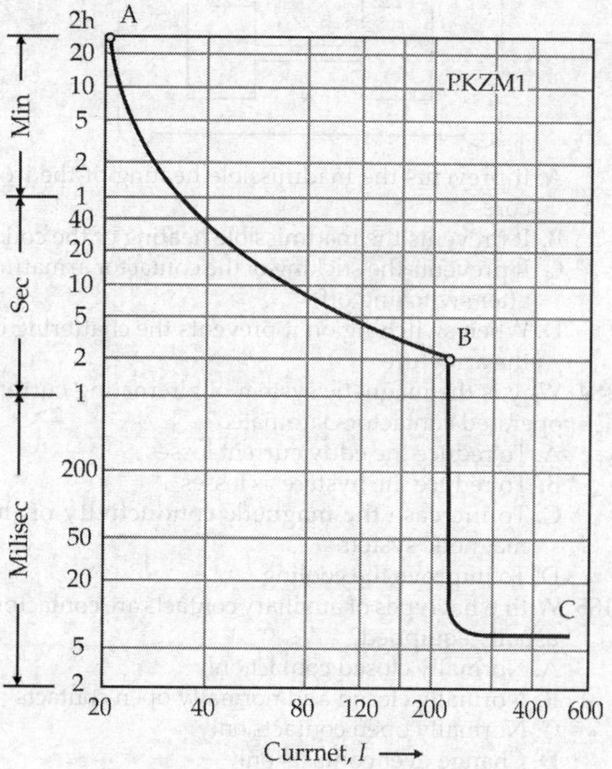

A. When a switching current surge of 245A flows for 5 ms
B. When 250A flow for 1 sec
C. When 150A flow for 5 sec
D. When 50A flow for 20 sec

182. A three-phase contactor is to be used for a 220 V DC motor. Which of the following circuits can be used for this purpose?

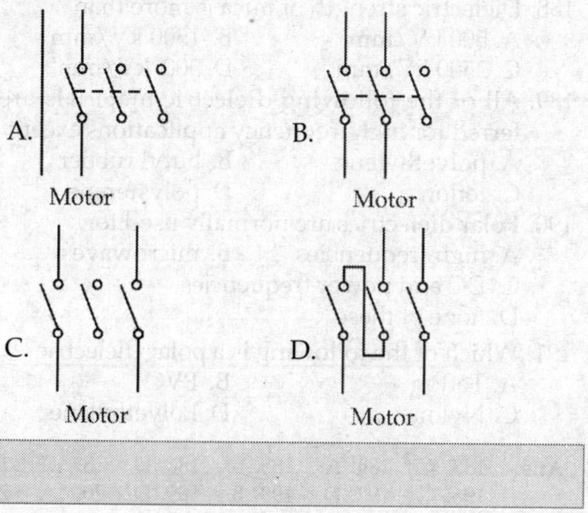

Ans. 176. D 177. B 178. D 179. D 180. D 181. C 182. D

183. The magnetic system of contactors has an air gap (marked X). What is the purpose of this air gap?

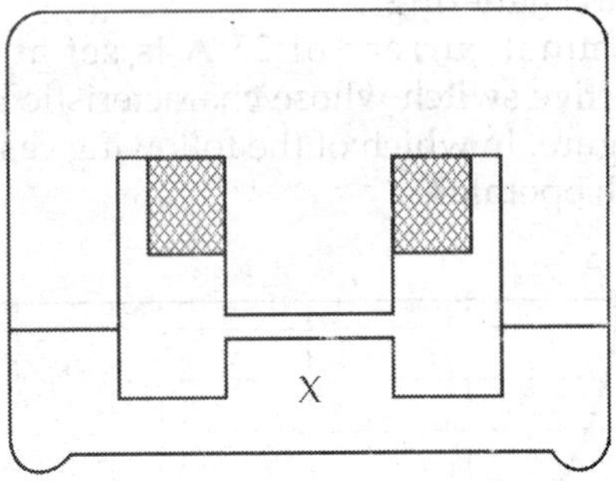

A. It prevents the inadmissible heating of the iron core
B. It prevents the inadmissible heating of the coil
C. It prevents the sticking of the contactor armature after switching off
D. When switching on it prevents the chattering of the armature

184. Why is the magnetic system of alternating current operated contactors laminated?
A. To reduce the eddy current losses
B. To reduce the hysteresis losses
C. To increase the magnetic conductivity of the magnetic system
D. To improve the cooling

185. With what types of auxiliary contacts are contactors usually equipped?
A. Normally closed contact only
B. Normally closed and normally open contacts
C. Normally open contacts only
D. Change over contacts only

186. The process of isonisation is brought about by
A. photons only B. positive ions
C. meta stables only D. all of these

187. The breakdown voltage of a specimen is 65 k V at NTP. The breakdown voltage at 730 mm Hg pressure and 35°C is
A. 69 kV B. 65 kV
C. 69.4 kV D. 70.6 kV

188. Dielectric strength of mica is more than
A. 500 kV/mm B. 1500 kV/mm
C. 2500 kV/mm D. 300 kV/mm

189. All of the following dielectric materials are preferred for high frequency applications except
A. polyethylene B. butyl rubber
C. teflon D. polystyrene

190. Polar dielectrics are normally used for
A. high frequencies B. microwaves
C. DC and power frequencies
D. none of these

191. Which of the following is a polar dielectric?
A. Teflon B. PVC
C. Nylon D. Polyethylene

192. Which of the following is non-polar dielectric?
A. Polysterene B. Phenolic plastics
C. Plasticized cellulose acetate
D. None of these

193. The impurity in liquid dielectric, which has significant effect in reducing the breakdown strength, is
A. dust B. dissolved gases
C. moisture D. none of these

194. The relationship between the breakdown voltage V and gap d is normally
A. $d = kV^2$ B. $d = kV^3$
C. $V = kd$ D. none of these

195. A good dielectric should have all the following properties except
A. high mechanical strength
B. high resistance to thermal deteriora-tion
C. high dielectric loss
D. freedom from gaseous inclusions

196. The variety of paper used for insulation purpose is
A. blotting paper B. rice paper
C. Craft paper D. none of these

197. Which variety of mica is hard and brittle?
A. Muscovite B. Phlogopite
C. Fibiolite D. Lipidolite

198. Professional grade electronic components means
A. low cost components
B. miniature size components
C. highly reliable components
D. accurate value components

199. Professional grade electronic components are invariably used in
A. PM transistors B. TV
C. power apparatus D. space applications

200. Van de Graaf generators are useful for
A. very high voltage and low current applications
B. very high voltage and high current applications
C. constant high voltage and current applications
D. none of these

201. In a Van de Graaf generator, the width of the belt is 1m and surface charge density permissible is 30×10^{-6} C/m^2. To supply a current of 300 μA, the velocity of belt should be
A. 1 m/sec B. 5 m/sec
C. 15 m/sec D. None of these

202. Surface charge density in a Van de Graaf generator is 30×10^{-6} C/m^2. The radius of electrode to charge to 10 MV will be
A. 0.16 m B. 0.295 m
C. 1.234 m D. none of these

203. The radius of the HT electrode is 1 metre in Van de Graaf generator. If the electric field is 21.1 kV/cm, the maximum poten-tial to which the electrode can be charged is
A. 2.1 MV B. 21 MV
C. 4MV D. none of these

Ans. 183. C 184. A 185. B 186. D 187. C 188. A 189. B 190. C 191. C 192. A 193. C 194. D 195. C 196. C 197. D 198. B 199. D 200. A 201. A 202. B 203. A

204. In Van de Graaf generator, output voltage is controlled by
A. controlling the corona source voltage
B. controlling the belt speed
C. controlling the lower spray point
D. all of these

205. A Tesla coil is a
A. cascaded transformer
B. coreless transformer
C. high frequency resonant transformer
D. low impedance transformer

206. Switching surge is
A. high voltage DC
B. high voltage AC
C. short duration transient voltage
D. low impedance transformer,

207. Moles bridge is used to measure
A. properties of dielectric at DC
B. dispersion in insulation
C. high frequency high voltages
D. modulation ratio frequencies

208. Insulators for high voltage applicatiorts are tested for
A. power frequency tests
B. impulse tests
C. both A and B are correct
D. none of these

209. Impulse testing of transformer is done to determine the ability of
A. bushings to withstand vibrations
B. insulation to withstand transient voltages
C. windings to withstand voltage fluctuations
D. all of these

210. Transformers contribute to radio interference due to
A. corona discharges in air
B. internal or partial discharges in insulation
C. sparking D. any of these

211. The accepted value of dielectric strength of transformer oil is
A. 30 kV B. 40 kV
C. 30 kV/cm for one minute
D. none of these

212. While testing transformer oil for dielectric strength the spherical electrodes are placed in
A. venial plane B. horizontal plane
C. diagonal plane D. none of these

213. While testing a speciman for intrinsic dielectric strength, the shape should be so prepared that
A. the electric stress is high at its centre
B. the electric stress is high at its corner
C. the electric stress is same all along the sample
D. none of these

214. The time of application of voltage for determining the intrinsic dielectric strength of the sample is of the order of
A. 10^{-8} sec B. 10^{-6} sec
C. 10^{-4} sec D. none of these

215. During reconditioning of transformer oil it is economical to use electrostatic filters if the water content of oil is
A. greater than 2 ppm B. less than 2 ppm
C. greater than 4 ppm
D. less than 4 ppm

216. The breakdom voltage for air gap of 3mm in a uniform field under standard atmospheric condition is
A. 6.3 kV B. 9.2 kV
C. 10.6 W D. 12.6 W

217. High frequency currents are generally measured using
A. current transformers
B. potential transformers
C. shunts D. all of these

218. Which of the following element has flat frequency response upto 1000 MHz
A. co-axial shunt B. squirrel cage shunt
C. bifilar shunt D. rogowski coil

219. The breakdown voltage is: (assuming all parameters remain same)
A. higher with negative polarity at all pressures
B. lower with negative polarity at all pressures
C. higher with negative polarity at low pressures
D. none of these

220. Liquids with solid impurities
A. have lower dielectric strength as compared to pure liquids
B. have higher dielectric strength
C. large size have higher dielectric strength
D. none of these

221. The electric field of a gas bubble which is immersed in a liquid of permittivity ε_r is
A. lower than that of the field in the liquid
B. higher than that of the field in the liquid
C. same as that in the liquid
D. none of these

222. As compared to air, the relative dielectric strength of sulphur hexafluoride is nearly
A. 1.5 times B. 2.5 times
C. 4.0 times D. 5.0 times

223. The electrical breakdown strength of insulating materials depends on
A. nature of applied voltage
B. imperfections in dielectric material
C. pressure, temperature and humidty
D. all of these

Ans. 204. A 205. C 206. C 207. B 208. C 209. B 210. D 211. D 212. B 213. A 214. A 215. B 216. C 217. A 218. A 219. D 220. A 221. A 222. B 223. D

224. Which of the following gas has been used as insulating medium in electrical appliances?
A. Nitrogen B. Carbon dioxide
C. Sulpher hexafluoride
D. Freon

225. Vacuum insulation is used in all of the following except
A. particle accelerators B. EHT of colour TV
C. field emission tubes D. X-rays

226. Liquids are generally used as insulating materials upto stresses of about
A. 100 MV/cm B. 50 MV/cm
C. 50 kV/cm D. 500 V/cm

227. Electromechanical breakdown of solid insulating materials occurs due to
A. magnetic hum B. vibrations
C. mechanical stresses produced by the electrical, field
D. electrical stresses produced by the voltage fluctuations

228. Surge voltage originate in power system due to
A. lightening
B. switching operations
C. faults D. all of these

229. All the following are the preferred properties of dielectric gas except
A. high dielectric strength
B. physiological inertness
C. low atomic number D. good heat transfer

230. Professional grade electronic components are necessarily used in all of the following except
A. computers B. television
C. defence equipment D. space applications

231. Which of the following techniques is used for the measurements of ac high frequency voltages?
A. Peak voltmeter
B. Series resistance micro ammeter
C. Resistance potential divider
D. All of these

232. Which of the following method or techniques can be used for the measurement of high DC voltages?
A. Generating voltmeter
B. Electrostatic voltmeter
C. Peak voltmeter D. Half effect generators

233. All of the following methods/techniques can be used for the measurement of high ac voltages except
A. potential dividers B. potential transformers
C. electrostatic voltmeters
D. Hall's effect generators

234. Surge diverters are
A. non-linear resistors in series with spark gaps which act as fast switches
B. are quenching devices
C. shunt reactors to limit the voltage rise due to Ferranti effect
D. over voltages of power frequency harmonics.

235. Impulse voltages are characterised by
A. polarity
B. peak value
C. time of half the peak value
D. all of these

236. Paschen's law is associated with
A. breakdown voltage B. ionisation
C. thermal radiations D. none of these

237. The essential condition for the Paschen's law to be valid is that
A. voltage must be dc B. voltage must be ac
C. temperature must be constant
D. humidity must be low

238. If 'p' is the pressure of gas and 'd' the distance between the electrodes, discharge voltage according to Paschen's law is directly proportional to
A. p B. d
C. pd D. None of these

239. If E_r is the radial field due to space charge and E_0 the externally applied field, the transformation of avalanche into a streamer takes place when
A. $E_r < E_0$ B. $E_r > E_0$
C. $E_r \approx E_0$ D. None of these

240. When an avalanche in the air gap reaches a certain critical size the electric spark discharge is mainly due to ionization of gas by
A. electrons impact B. photoionization
C. positive ion barbardment
D. all of these

241. The breakdown voltage in gases depends on
A. distance between the electrodes
B. relative air density
C. humidity
D. all of these

242. At unvarying temperature, breakdown voltage in a uniform field is a function of the product of gas pressure and distance between the electrodes. The above statements is known as
A. electron avalanche B. Paschen's law
C. thermal stability principle
D. breakdown voltage law

243. Large capacity generators are manufactured to generate power at
A. 440 V B. 6.3 to 10.5 kV
C. 132 kV to 220 kV D. 400 kV

244. Klydono graph is used to record impulse voltages
A. greater than 50 kV
B. greater than 50 kV but less than 100 kV
C. greater than 100 kV but less than 10,000 kV
D. None of these

Ans. 224. C 225. B 226. C 227. C 228. D 229. C 230. B 231. A 232. A 233. D 234. A 235. D 236. A 237. C 238. D 239. C 240. D 241. D 242. B 243. B 244. D

245. Klydonograph when used for measurement of voltages provides information on
A. magnitude and polarity
B. polarity and frequency
C. magnitude and frequency
D. none of these

246. Penning effect explains
A. increase in dielectric strength of all mixtures of gases
B. decrease in dielectric strength of any mixtures of gases
C. increase in dielectric strengths of many mixture of gases
D. none of these

247. The onset voltage for Trichel pulses
A. increases with increase in gap length
B. increase with decrease in gap length
C. independent of gap length
D. none of these

248. The frequency of Trichel pulses
A. increases with voltage and gap pressure
B. increases with voltage and decrease with gap pressure
C. decreases with increase in voltage and pressure
D. none of these

249. Trichel pulses are noticed in a point plane gap in air when
A. point is positive
B. point is, negative
C. These pulses are seen in a plane-sphere gap
D. none of these

250. In case of impulse thermal breakdown of solid insulating materials, the critical electric field is proportional to
A. critical absolute temperature
B. square of critical absolute temperature
C. square of T_0
D. none of these

251. In case of impulse thermal breakdown of insulating materials, the critical time to breakdown is
A. proportional to critical electric field
B. inversely proportional to critical electric field
C. proportional to square of critical absolute temperature
D. none of these

252. The ripple voltage in dc output is small if
A. frequency of supply voltage is small
B. the filtering capacitor is small
C. load current is small D. none of these

253. The ratio of voltage drop for an 8 stage to 4 stage cockroft walton circuit approximately is: (Assuming all other parameters remaining same)
A. 2 B. 4
C. 8 D. none of these

254. The voltage drop for a certain cockroft walton circuit is small if
A. load current is small
B. the capacitance of each stage is small
C. the frequency of supply is small
D. none of these

255. In a Cockroft–Walton circuit, what will be the, maximum output voltage and optimum number of stages for maximum output voltage, For input voltage of 400 kV, load current = 100 mA, supply frequency = 100 Hz, each capacitor = 10 μF
A. 133 kV, 1 B. 300 kV, 5
C. 500 kV, 10 D. 1000 kV, 20

256. In order to reduce ripples and /or voltage drop, it is more economical to
A. high frequency and high capacitance
B. high frequency and low capacitance
C. low frequency and low capacitance
D. none of these

257. The short circuit impedance to three stage cascaded transformers as compared to a single unit transformer is
A. 3 to 4 times B. 8 to 9 times
C. 15 to 16 times D. None of these

258. If the quality factor of the HT reactor in ac series resonant circuit is 49 and Q is the reactive power requirement of the test specimen, then the kW rating of the feed transformer is
A. Q/7 B. Q/49
C. 7Q D. 49Q

259. If a single stage system using resonance test voltage is required, it is better to use
A. parallel resonant circuit
B. series resonant circuit
C. series-parallel resonant circuit
D. none of these

260. The impulse ratio for any particular object
A. is a constant
B. depends on the shape of wave
C. depends on the polarity of the impulse wave
D. both B and C are correct

261. The wave front and tail resistances of an impulse generator are 150 and 2000 ohm respectively. The impulse capacitance and load capacitances are 25 and 2 nF. The approximate wave front time is
A. 0.83 μ sec B. 1 μ sec
C. 1.22 μ sec D. none of these

262. The approximate wave tail time for the above question is
A. 20 μ sec B. 40 μ sec
C. 60 μ sec D. 80 μ sec

263. The approximate voltage efficiency of the above impulse generator is

Ans. 245. A 246. B 247. C 248. A 249. B 250. C 251. B 252. D 253. C 254. A 255. A 256. B 257. B 258. D 259. A 260. D 261. A 262. B

A. 85% B. 89.5%
C. 92.6% D. 98.5%

264. The minimum value of impulse capacitance to impulse test for a 3-phase 200 kV 5% impedance, power transformer is
A. 0.45 nF B. 0.826 nF
C. 0.96 nF D. none of these

265. While Impulse testing a power transformer the impulse voltage is applied to
A. one phase, the other two phases grounded, the secondary short circuited
B. one phase, two phases left open, secondary left open
C. one phase two phases left open, secondary short circuited
D. none of these

266. In order that impulse circuit operates consistently, it is essential that
A. the first gap is slightly less than the second one and so on
B. the axes of the gaps should be in the same vertical plane
C. both A and B are correct
D. none of these

267. It is desirable to have wave front control resistance
A. entirely within the impulse circuit
B. entirely outside the impulse circuit
C. partly inside and partly outside the impulse circuit
D. none of these

268. The magnitude of pulse currents depends an
A. voltage applied
B. number of voids
C. both A and B are correct
D. none of these

269. Recurrent surge generator is used to generate
A. current of large magnitude
B. impulse voltage of smaller magnitude
C. current of small magnitude
D. none of these

270. For reducing tower footing resistance, it is better to use
A. chemical and ground road only
B. a chemical and counterpoise only
C. ground rod and counter pose only
D. none of these

271. Impulse patio of a rod-gap is,
A. 1.0 B. between 1.2 to 1.5
C. between 2.0 to 2.5 D. none of these

272. The insulation of the modern EHV lines is designed based on
A. corona B. switching voltages
C. lightning voltage D. all of these

273. High voltage dc testing for HV machines is resorted because
A. the electric stresses do not damage the coil and insulation
B. the electric stress distribution is a representation of the service condition.
C. contain conclusions regarding the continuous aging of an insulation can be drawn
D. none of these

274. Standard impulse testing of power trans-former requires
A. one application of chopped wave followed by one application of full-wave
B. two applications of chopped wave followed by an application of a full-wave
C. one application of chopped wave followed by two application of a full-wave
D. none of these

275. Current ratio arm bridge is used for
A. low voltage high frequency application
B. high voltage low frequency application
C. high voltage high frequency application
D. none of these

276. An inclined straight line on a CRO as a null detector shows
A. magnitudes unbalance
B. magnitude and phase Imbalance
C. phase unbalance D. none of these

277. Partial discharge can be detected by
A. listening to hissing sound
B. optical methods
C. a high $\tan\delta$ D. all of these

278. Partial discharge measurement on a cable gives
A. pf of the cable B. loss angle of the cable
C. location of fault D. none of these

279. If the source is not able to supply charging current required by the test cable, it is desirable to use
A. shunt capacitor B. series capacitor
C. series inductor D. none of these

280. If $\tan\delta_1$, $\tan\delta_2$ and $\tan\delta_3$ are the loss angles of the test capacitors after 16, 24, 48 hrs respectively measured on HV schering bridge, it is desirable that
A. $\tan\delta_1 = \tan\delta_2 = \tan\delta_3$ B. $\tan\delta_1 \le \tan\delta_2 \le \tan\delta_3$
C. $\tan\delta_1 \ge \tan\delta_2 \ge \tan\delta_3$ D. none of these

281. The mechanism responsible for dielectric loss in a dielectric are
A. polarization only B. Ionisationary
C. conduction only D. all of these

282. High voltage Schring bridge is used to measure
A. large capacitance without additional element
B. small capacitance without additional element
C. medium value capacitance
D. none of these

Ans. 263. C 264. B 265. A 266. C 267. C 268. A 269. B 270. C 271. B 272. A 273. C 274. B 275. B
276. A 277. D 278. C 279. C 280. C 281. D 282. B

283. Sphere gap is used for the measurement of
A. AC voltage only
B. DC voltage only
C. both A and B are correct
D. none of these

284. Protective resistance to be connected between the sphere gap and the test equipment is required while measuring
A. power frequency and higher frequency ac voltage
B. power frequency and impulse voltage
C. all kinds of voltages
D. none of these

285. If d is the diameter of the sphere, for better measurements, the gaps s should be such that
A. $0.05\,d \le s \le 0.5\,d$ B. $0.03\,d \le s \le 0.6\,d$
C. $d = s$ D. none of these

286. Tick the Correct answer.
A. The breakdown voltage in sphere gaps decreases with the increase in partial pressure of water vapour
B. The breakdown voltage in sphere gaps increases with increase in partial pressure of water vapour
C. The breakdown voltage in sphere gaps decreases with increase in partial pressure of water vapour
D. None of these

287. A test specimen with specific conductivity 10^{-12} S/m is placed in an electric field of 100 kV/cm. The specific loss in the specimen is
A. 10^{-5} W/m^3 B. 100 W/m^3
C. 10^5 W/m^3 D. none of these

288. Which soil has the maximum specific resistance?
A. Black cotton soil B. Sand
C. 132 kV to 220 kV D. 400 kV

289. In sphere gaps, the sphere are made of
A. aluminium B. brass
C. bronze D. Any of these

290. Using uniform field spark spark gap, the breakdown voltage for 10 mm gap length is
A. 30.3 kV B. 35.3 kV
C. 36.3 kV D. none of these

291. The voltage measurements using uniform spark gap are
A. affected by dust particles
B. affected by polarity of supply
C. affected by temperature
D. none of these

292. A generating voltmeter uses
A. a constant speed motor
B. a variable speed motor
C. Both A and B are correct
D. none of these

293. A generating voltmeter has
A. no-linear scale B. linear-scale
C. no contact with high voltage electrode
D. both B and C are correct

294. A generating voltmeter is used to measure ac voltages if the angular frequency of voltage is
A. equal to the angular frequency of motor
B. twice the angular frequency of motor
C. thrice the angular frequency of motor
D. none of these

295. A potentiometer is normally connected
A. outside the generator circuit towards the load circuit
B. within the generator circuit
C. at a distance of 50 cm from the generator
D. none of these

296. For capacitor dividers, if damping resistor is inserted in series with individual element capacitor divider, the damped capacitive divider acts as
A. resistance divider
B. resistance divider for high frequency
C. capacitance divider only
D. none of these

297. In 'plasma' state a gas
A. loses electrical conductivity
B. becomes perfect insulator
C. conducts electricity D. attracts moisture

298. Which of the following statement about corona is incorrect?
A. Corona gives rise to radio interference.
B. Corona results in loss of power in transmission
C. Corona discharge can be observed as red luminescence
D. Corona is always accompanied by a hissing noise

299. The specific conductivity of a test specimen with $\varepsilon_r = 4.0$ at 50 Hz is 10^{-12} S/m. Its loss angle is
A. 9×10^{-5} B. 9×10^{-8}
C. 9×10^{-12} D. none of these

300. The velocity of a travelling wave through a cable of relative permittivity of 9 is
A. 9×10^8 m/sec B. 3×10^8 m/sec
C. 10^8 m/sec D. none of these

301. Candela is the unit of
A. flux B. luminous intensity
C. illumination D. luminance

302. The unit of illuminance is
A. lumen B. cd/m^2
C. lux D. steradian

303. The illumination at various points on a horizontal surface illuminated by the same source varies as
A. $\cos^2\theta$ B. $\cos\theta$
B. $1/r^2$ D. $\cos^2\theta$

Ans. 283. C 284. A 285. A 286. B 287. B 288. B 289. D 290. A 291. A 292. A 293. D 294. B 295. A
296. B 297. C 298. C 299. A 300. C 301. B 302. C 303. A

304. The MSCP of a lamp which gives out a total luminous flux of 400 p lumen is ..... candela.
A. 200 B. 100
C. 50 D. 400

305. A perfect diffuser surface is one that
A. diffuses all the incident light
B. absorbs all the incident light
C. transmits all the incident light
D. scatters light uniformly in all directions

306. The direct lighting scheme is most efficient but is liable to cause
A. monotony B. glare
C. hard shadows D. both B and C

307. Total flux required in any lighting scheme depends inversely on
A. illumination
B. surface area
C. utilization factor D. space/height ratio

308. Floodlighting is NOT used for ..... purposes.
A. reading B. aesthetic
C. advertising
D. industrial

MULTIPLE CHOICE QUESTIONS FROM VARIOUS COMPETITIVE EXAMINATIONS

1. The below diagram shows the layout of a power station having two generators A and B, connected to the 11 kV buses which are also fed through two transformers C and D from a 132 kV grid. The 11 kV buses are interconnected through a reactor R.

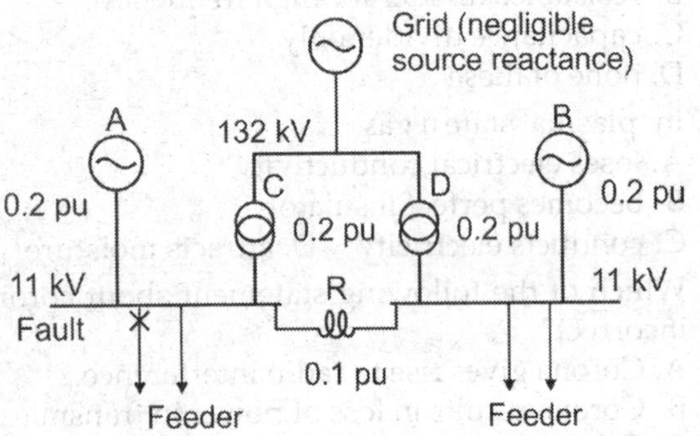

The reactance of A, B, C, D and R in pu on a common MVA- and kV-base. All the generated voltages in A, B and grid are each 1.0 pu and assumed as in phase at the time of fault.
The steady state symmetrical fault-current for a 3-phase fault on the 11 kV feeders is **(IES 2001)**
A. 10 pu B. 15 pu
C. 20 pu D. 25 pu

2. Four identical 100 MVA, 33 kV generators are operating in parallel, as shown below, in two bus-bar sections, inter-connected through a current limiting reactor of x pu reactance on the generator-base. Each generator has a reactance of 0.2 pu.

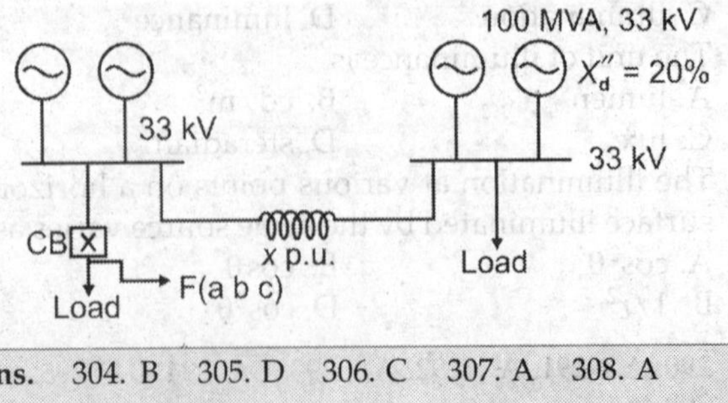

The value of the reactor x to limit a symmetrical short-circuit (a-b-c) current through the circuit breaker to 1500 MVA is **(IES 2003)**
A. 0.05 pu B. 0.10 pu
C. 0.15 pu D. 0.20 pu

3. For the system shown in the diagram given below, what is a line-to-ground fault on the line side of the transformer equivalent to? **(IES 2006)**

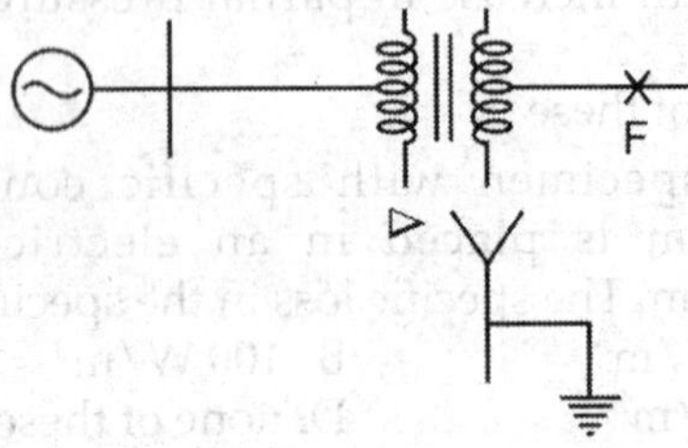

A. A line-to-ground fault on the generator side of the transformer
B. A line-to-line fault on the generator side of the transformer
C. A double line-to-ground fault on the generator side of the transformer
D. A 3-phase fault on the generator side of the transformer

4. How many relays are used to detect inter phase fault of a three-line system? **(IES 2006)**
A. One B. Two
C. Three D. Six

5. A 10 kVA, 400 V/200 V single-phase transformer with 10% impedance, draws a steady short circuit current of **(IES 2010)**
A. 50 A B. 150 A
C. 250 A D. 350 A

6. Fault calculations using computer program are usually done by **(IES 2012)**
A. Y_{bus} method B. Z_{bus} method
C. both of the above D. none of the above

Ans. 304. B 305. D 306. C 307. A 308. A 1. B 2. B 3. B 4. C 5. A 6. A

7. The power stations S_1 and S_2 are interconnected through a transmission line of per unit (pu) reactance of 0.4. Station S_1 has one generator of pu reactance of 0.2 and station S_2 has two generators of pu reactance 0.4 each

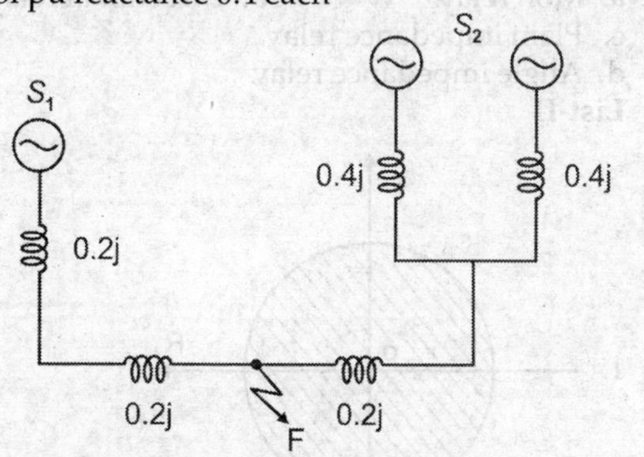

For a 3-phase symmetrical short circuit at the middle of the transmission line (point F), what is the change in the value of the equivalent fault reactance, with one of the generators of S_2 removed? **(IES 2007)**

A. 10% B. 20%
C. 30% D. 40%

8. Figure 2 given below is the "equal-area" criterion diagram for the determination of transient stability limit of the power system shown in figure 1 for a fault on the transmission line.
What is the type of fault, and the time of its clearing, from figure 2? **(IES 2007)**

Figure 1

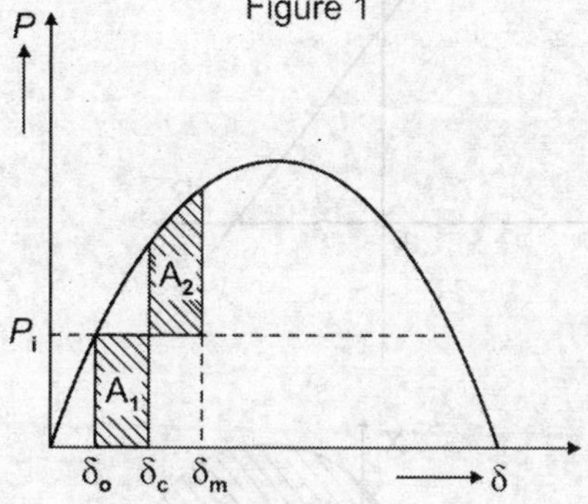

Figure 2

A. Three phase fault with instantaneous clearing
B. Three phase fault with subsequent clearing
C. Single-line to ground fault with instantaneous clearing
D. Single-line to ground fault with subsequent clearing

9. In a transmission line, the mid-point voltage is maintained to V by a compensating device as shown in the circuit below. What is the real power flow through the line? **(IES 2009)**

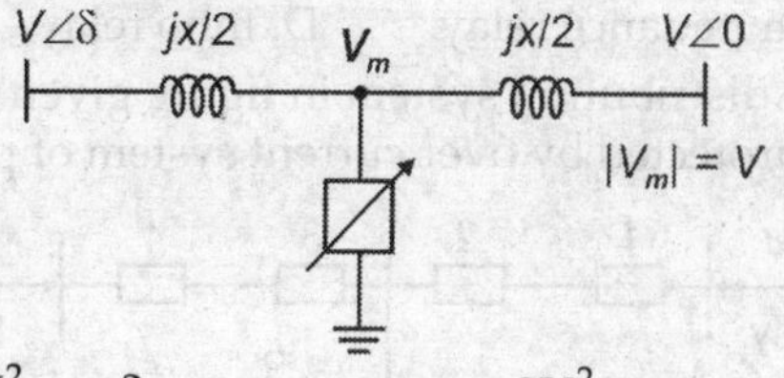

A. $\frac{V^2}{X}\sin\frac{\delta}{2}$ B. $\frac{2V^2}{X}\sin\delta$

C. $\frac{V^2}{X}\sin\delta$ D. $\frac{2V^2}{X}\sin\frac{\delta}{2}$

10. For what value of damping parameter, the transient stability is assured by equal area criterion? **(IES 2010)**

A. Independent of systems damping
B. If only damping is exactly zero
C. For all values of damping parameters
D. If only damping is positive and finite

11. For a fault in a power system, the term critical clearing time is related to **(IES 2012)**

A. reactive power limit
B. transient stability limit
C. short circuit current limit
D. steady state stability limit

12. A 50 MVA, 11 kV, 3-phase generator has a stored energy of 400 MJ. Its inertia constant is **(IES 2013)**

A. 4 B. 8
C. 2 D. 16

13. Equal area criteria in power systems is used in the context of **(IES 2014)**

A. deciding maximum loading for a given excitation
B. stability of a machine connected to infinite bus bar
C. stability of power systems in which many machines are connected to infinite bus bar
D. load distribution between a single machine and load drawn from infinite bus bar

14. The operating characteristic of a distance relay in the R-X plane is shown in the figure below. It represents operating characteristic of a **(IES 2001)**

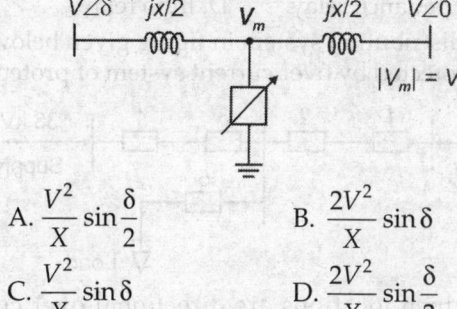

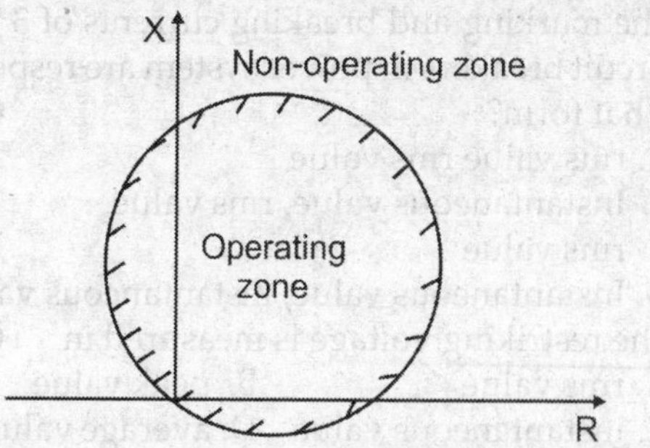

Ans. 7. B 8. B 9. C 10. B 11. B 12. B 13. B

A. reactance relay
B. directional impedance relay
C. impedance relay D. mho relay

15. The distribution system in figure given below is to be protected by over current system of protection.

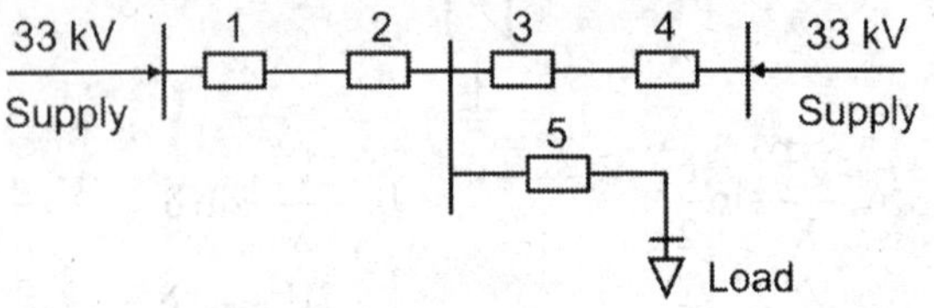

At which locations are directional over current relays required for proper fault discrimination? **(IES 2005)**
A. 1 and 4 B. 2 and 3
C. 1, 4 and 5 D. 2, 3 and 5

16. The figure given below shows a schematic arrangement of distance relay provided with a 'replica impedance' Z_r. The CT ratio = I/i and VT ratio = V/v. **(IES 2008)**

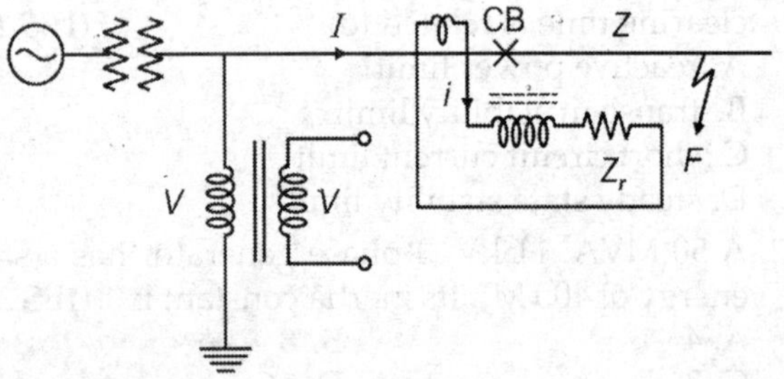

When a fault occurs on the line being protected, when would the relay operate?
A. $Z_r > Z$ B. $Z_r < Z$
C. $Z_r > Z \cdot I/i$ D. $Z_r > Z \cdot V/v$

17. A 50 Hz, 3-phase synchronous generator has inductance per phase of 15 mH. The capacitance of generator and circuit breaker is 0.002 μF. What is its natural frequency of oscillation? **(IES 2008)**
A. 29 kHz B. 2.9 kHz
C. 290 kHz D. 29 MHz

18. Mho relay is usually employed for the protection of **(IES 2008)**
A. short lines only B. medium lines only
C. long lines only D. any line

19. The marking and breaking currents of 3 phase AC circuit breakers in power system are respectively in what form? **(IES 2009)**
A. rms value rms value
B. Instantaneous value, rms value
C. rms value
D. Instantaneous value, instantaneous value

20. The restriking voltage is measured in **(IES 2010)**
A. rms value B. peak value
C. instantaneous value D. average value

21. Match **List-I** with **List-II** and select the correct answer using the code given below the lists: **(IES 2011)**

List-I
a. Moh relay b. Directional relay
c. Plain impedance relay
d. Angle impedance relay

List-II

1. 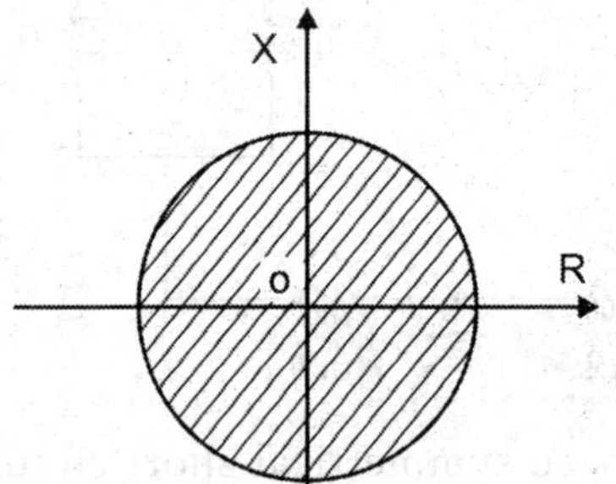

2. 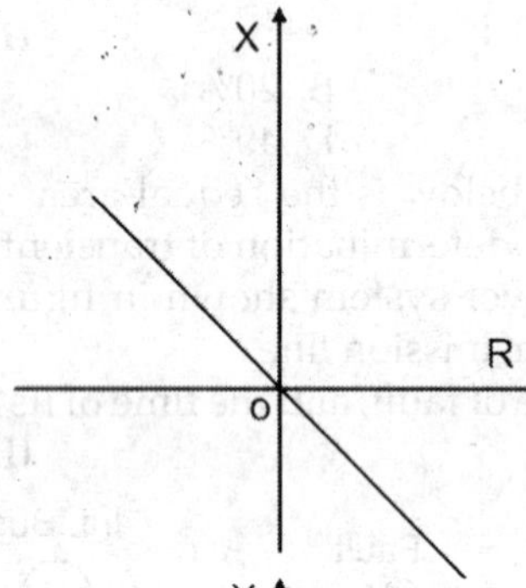

3. 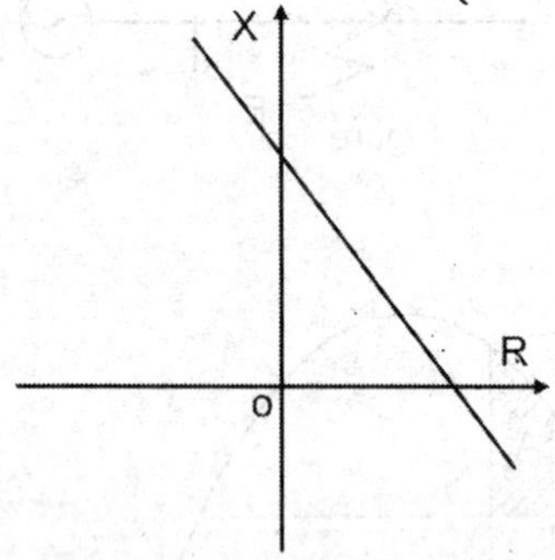

4. 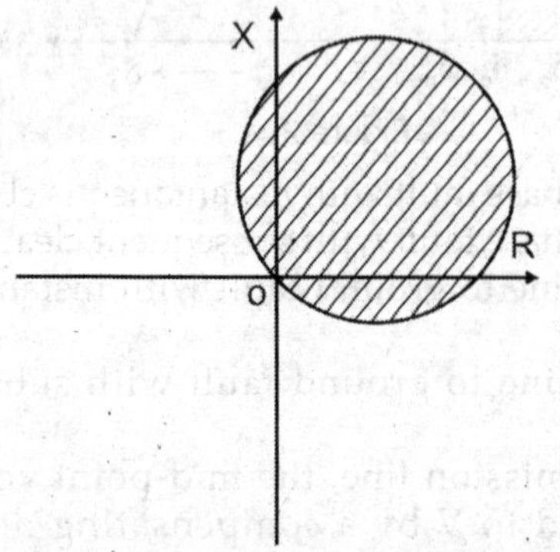

Ans. 14. D 15. B 16. B 17. A 18. C 19. B 20. B

Codes

	a	b	c	d		a	b	c	d
A.	3	1	2	4	B.	4	2	1	3
C.	3	2	1	4	D.	4	1	2	3

22. A relay is connected to a 400/5 A current transformer and set for 150%. The primary fault current of 2400 A needs a plug setting multiplier of **(IES 2014)**
    A. 2 B. 4
    C. 6 D. 8
23. An over current relay is said to over reach when it operates at a current **(IES 2014)**
    A. higher than its setting
    B. equal to its setting
    C. lower than its setting
    D. 2/3rd of its setting
24. Under-voltage relays are mainly used for **(IES 2010)**
    A. motor protection B. transformer protection
    C. transmission line protection
    D. all of the above
25. The use of high speed circuit breakers **(IES 2014)**
    A. reduces the short circuit current
    B. improves the system stability
    C. decreases the system stability
    D. increase short circuit current

Ans. 21. B 22. B 23. C 24. D 25. B

CHAPTER 13

Power Electronics

Power electronics deals with the use of electronics for the control of large power. Power electronics era started with high power tubes like thyratrons, ignitrons and mercury arc rectifiers. With the advent of power semiconductor devices like thyristor, triacs, power MOSFET and power transistors, power electronics has become very important in the control of large power. The major component of power electronics circuit is the thyristor, which is a fast switching semiconductor device whose function is to modulate the power in AC and DC systems. Modulation of power can vary from 10 W to 100 MW by turning the switch on and off in a particular mode.

Applications of power electronics: Power electronics has already found an important role in modern technology. Examples of successful applications are as follows:

1. Rolling mill drives, in which the motor generator set of the conventional Ward–Leonard system is replaced by a controlled rectifier inverter
2. Electric drives for rapid-transit systems, in which wasteful control resistors are eliminated
3. Multiple drive systems for textile and paper mills
4. Machine tool control
5. Aircraft power supplies
6. Uninterruptible power supplies (UPS) for important loads such as computers and space applications
7. Illumination controls for lighting in trains, homes and theatres
8. High voltage direct current (HVDC) systems
9. Frequency converters for motor controls
10. Battery charging and welding systems
11. Vehicle propulsion systems
12. Excitation systems for alternators and synchronous condensers
13. Heat controls involving arc melting, induction heating and electrolysis

Thyristor converter systems:

Advantages

1. Fast response of the thyristor systems as compared to electromechanical converter systems
2. High efficiency due to less losses in thyristors
3. Long life and reliability due to the absence of moving parts
4. Less and easy maintenance
5. Less noise
6. Control flexibility
7. Small size and low weight, require less floor area

Disadvantages

1. Power semiconductor converters have a tendency to introduce current and voltage harmonics into supply systems and controlled systems
2. Harmonics in the supply system causes interference with communication systems and distortion of supply voltage
3. Thyristor controllers have low overload capacity
4. Regeneration of power is difficult
5. Some converters such as controlled rectifier, cycloconverter (AC to AC converters) and the AC voltage controller suffer from a poor power factor, particularly at low output voltages

Power semiconductor devices: Power semiconductor devices can, generally be classified as follows:

1. Power diode
2. Power bipolar junction transistors (BJTs)
3. Power MOSFET
4. Insulated-gate bipolar junction transistors (IGBTs)
5. Thyristor

These devices are operated in the switching mode so that the losses are reduced and conversion efficiency is improved. The disadvantage of switching mode operation is the generation of harmonics and radio frequency interference (RFI). In this section, the external

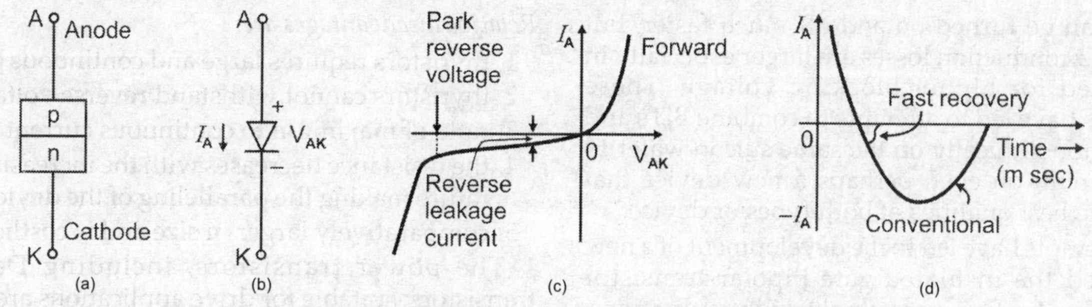

Fig. 13.1: Structure of a power diode and its symbol

electrical characteristics of these devices are discussed briefly.

Power diode: A power diode is a two-terminal p-n junction device and a p-n junction is normally formed by allowing diffusion, and epitaxial growth. The structure of a power diode and its symbol are shown in Figs 13.1(a) and 13.1(b). High power diodes are silicon rectifiers that can operate at high junction temperatures. Power diodes have larger power, voltage, and current handling capabilities than ordinary signal diodes. In addition, the switching frequencies of power diodes are low as compared to signal diodes.

The voltage–current characteristic of a diode is shown in Fig. 13.1(c).

The recovery characteristics of conventional and fast recovery diodes are shown in Fig. 13.1(d).

General purpose diodes: The general purpose rectifier diodes have relatively high reverse recovery time, typically 25 msec, and are used in low-speed applications, where recovery time is not critical. These diodes cover current ratings from less than 1 A to several thousands of amperes with voltage ratings from 50 V to around 5 kV. These diodes are generally manufactured by diffusion. However, alloyed types of rectifiers which are used in welding power supplies are most effective and rugged, and their ratings can go up to 300 A and 1000 V.

Fast recovery diodes: The fast recovery diodes have low recovery time, normally less than 5 msec. They are used in chopper and inverter circuits, where the speed of recovery is often of critical importance. These diodes cover current ratings from less than 1 A to hundreds of amperes, with voltage ratings from 50 V to 3 kV.

For voltage ratings above 400 V, fast recovery diodes are generally made by diffusion and recovery time is controlled by platinum or gold diffusion. For voltage ratings below 400 V, epitaxial diodes provide faster switching speeds than that of diffused diodes. The epitaxial diodes have a narrow base width resulting in fast recovery time of as low as 50 nanosec.

Schottky diodes: The charge storage problem of a p-n junction can be minimised in a Schottky diode. It is accomplished by setting up a 'barrier potential' with a contact between a metal and a semiconductor.

A layer of metal is deposited on a thin epitaxial layer of n-type silicon. The potential barrier simulates the behaviour of p-n junction. However, the rectifying action depends on the majority carriers only and as a result there is no excess minority carriers to recombine. The recovery effect is due to the self-capacitance of the semiconductor junction. The recovered charge of a Schottky diode is much less than that of an equivalent p-n junction diode. It has a relatively low forward voltage drop.

Advantages of bipolar transistor as a switch:

1. High power gain, i.e. low control power requirement. In large collector current levels, the control power requirement is reduced by Darlington connection
2. Low switch-off and switch-on time
3. Fast switching and high frequency periodic switching
4. Long life
5. High reliability
6. High ratio of off-state to on-state resistance
7. Low power dissipation in both the states

Advantages of power MOSFETs over power transistors: Since the power transistor is the power MOSFET's main competitor, it is good to know their relative advantages and disadvantages. The advantages of power MOSFET over the power transistor are:

1. no power switching losses
2. no second breakdown
3. much larger gain and hence simple and cheaper drive circuitry
4. better reliability, ruggedness, and thermal stability
5. higher peak current handling capability
6. easy to parallel due to the positive resistance coefficient
7. relatively more linear transfer characteristics and
8. much higher switching speed

Insulated gate bipolar transistor (IGBT): Bipolar junction transistor (BJT) and MOSFET have characteristics that complement each other in some respects. BJTs have lower conduction losses in the on-state, especially in devices with larger blocking voltages, but have larger switching times, especially at turn off.

MOSFETs can be turned on and off much faster, but their on-state conduction losses are larger especially in devices rated for higher blocking voltage. These observations have led to attempts to combine BJTs and MOSFETs monolithically on the same silicon wafer to achieve a circuit on even perhaps a new device that combines the best qualities of both types of device.

These attempts have led to the development of a new device called the insulated gate bipolar transistor (IGBT) which is finding increasingly wide applications. Other names for this device are:

1. GEMFET
2. COMFET (conductivity modulated field effect transistor)
3. IGT (insulated gate transistor)
4. Bipolar-mode MOSFET or bipolar-MOS transistor

An IGBT combines the advantages of BJTs and MOSFETs. An IGBT has high input impedance like MOSFETs and low on-state conduction losses like BJTs. But there is second breakdown problems like BJTs.

FET controlled thyristors: Research effects are made to improve the devices in respect to increase their blocking voltage capabilities, lowering their on-state losses, and increasing their switching speeds. A recent example of this is the development of the ring emitter BJT by Siemens. The concepts that have not yet found general commercial acceptance or are still in prototype stage, can be termed *emerging devices and circuits*. A list of such emerging devices would include power junction field effect transistors, also termed *static induction transistor*, controlled thyristors (bipolar static induction transistors), MOS-controlled thyristors (MCTs). Some of these devices may became widely used in the future in the industrial application.

Thyristor: The thyristor, also known as silicon controlled rectifier (SCR), is the main workhorse for bulk power conversion and control in industry. A thyristor has a p-n-p structure with three terminals and is fabricated by diffusion process. Light-fired thyristors have recently been developed for high voltage direct current (HVDC) applications. In this application, a number of devices are connected in series-parallel combination, and light firing provides the advantages of electrical isolation between the control and power circuits.

Comparison of transistor with thyristor:

Advantages of the transistor over thyristors are:

1. much higher switching frequency leading to improved and more efficient operation of converters
2. can provide current limit protection by the base drive circuit
3. does not require forced commutation circuit and saves the associated switching losses, cost, weight and volume
4. low conduction drop.

Relative disadvantages are:

1. thyristors requires large and continuous base drive
2. thyristors cannot withstand reverse voltage
3. ratio of maximum to continuous current is low
4. the resistance decreases with the increasing temperature making the paralleling of the device difficult
5. comparatively larger 'n size and is costlier.

The power transistors, including Darlington transistors, suitable for drive applications are available up to the ratings of 120 V, 750 A and 1000 V, 60 A. Because of the high switching frequency and more efficient operation, power transistors have succeeded in replacing thyristors in a number of low and medium power (up to around 200 kW) drives employing inverters and choppers.

Principle of operation of thyristors: Thyristors are the most important type of power semiconductor devices. They are extensively used in power electronic circuits. They are operated as bistable switches from non-conducting state to conducting state thyristor, also known as silicon controlled rectifier (SCR), is the main workhorse for bulk power conversion and control in industry.

The thyristor is a four-layer, three terminal, anode *A*, cathode *K* and gate *G*. The anode and cathode are connected to the main power circuit. The gate terminal carries a low level gate current in the direction from gate to cathode. The thyristor operates in two stable states: ON or OFF.

Volt-ampere characteristics: The static characteristic of a thyristor with no gate current applied is shown by Fig. 13.2(a) with zero gate current ($i_g = 0$), if a forward voltage is applied across the device, i.e. anode positive with respect to cathode, junctions J_1 and J_3 are forward-biased while junction J_2 remains reverse biased, and therefore the anode current is a small leakage current. If the anode to cathode forward voltage reaches a critical limit, called the forward breakover voltage, the device switches into high conduction, thus forward biasing the junction J_2 to turn the thyristor ON. The forward voltage drop then falls to a value between 1 V and 2 V.

The thyristor can be switched to the ON or conducting state by injecting a current into the central p-type layer via the gate terminal. The injection of gate current provides additional holes in the central p-type layer, reducing the forward breakover voltage to a value less than the applied voltage and turning the thyristor ON. These conditions are shown in Fig. 13.2(b) showing the effect of increasing gate current on the level of forward breakover voltage. If the anode current falls below a critical limit, called the holding current, I_H, the device returns to its forward blocking state.

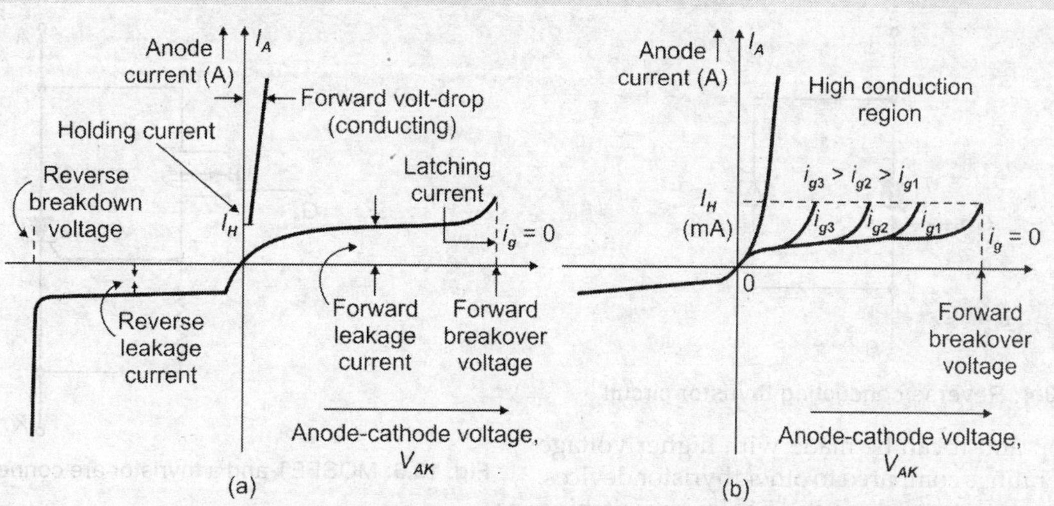

Fig. 13.2: V-I characteristics of thyristor with and without gate current

Phase-control thyristors: The thyristor, also known as silicon-controlled rectifier (SCR) has been widely used in industry for more than two decades for power conversion and control. The device switches on very quickly, the turn on time *t*on typically being 1 to 3 μsec. Mostly, the width of gate pulse is in the range 10 to 50 μsec and its amplitude in the range 10 to 50 μsec and its amplitude in the range 20 to 200 mA. This type of thyristors generally operates at the line frequency and is turned off by natural commutation. The turn off time, t_q is of the order of 50 to 100 ms (Fig. 13.3).

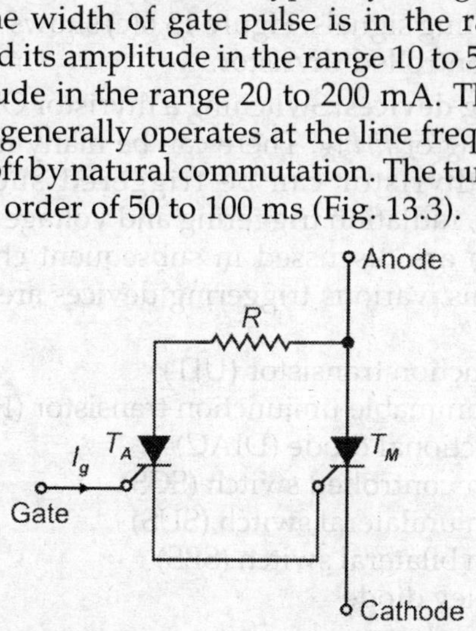

Fig. 13.3: Phase-control thyristor circuit

Fast-switching thyristors/asymmetrical thyristors (ASCRs): An asymmetrical thyristor (ASCR) is fabricated to have limited reverse voltage capability and, as a result, it permits reduction of turn on time, turn off time and conduction drop. These devices are used in high-speed switching applications with forced commutation (e.g. choppers and inverters). A 1500 V, ASCR may typically have a turn off time of 10 to 15 msec compared to 20 to 30 μsec for similar thyristor device. The on-state forward drop varies, approximately as an inverse function of the turn off time, t_q. This type of thyristors is also known as an *inverter thyristor*.

Gate turn off (GTO) thyristor: The conventional thyristor as described earlier has over the years been developed such that two new devices of the thyristor family are now available, the asymmetrical thyristor and gate turn off (GTO) thyristor.

TRIAC: The TRIAC is a bidirectional single gate thyristor device that performs the circuit function of two thyristors connected in inverse parallel. It is used for the control of power in AC circuits. Alternating current power can be efficiently controlled today with a special device that switches conduction on and off during each alternation. Control of this type is accomplished with a special semiconductor device known as a triac. A *triac* is identified as a three-electrode or triode AC semiconductor switch. Conduction is triggered by a gate signal. A triac is the equivalent of two reverse-parallel connected SCRs with one common gate. Conduction can be achieved in either direction with an appropriate gate current. AC can be controlled efficiently and accurately with this device.

Reverse-Conducting Thyristors (RCT): In many choppers and inverter circuits, an antiparallel diode is connected across an SCR in order to allow a reverse current flow due to inductive load and to improve the turn off requirement of commutation circuit. The diode clamps the reverse blocking voltage of the SCR to 1 or 2 V under steady state conditions. However, under transient conditions, the reverse voltage may rise to 30 V due to induced voltage in the circuit stray inductance within the device.

An RCT is a compromise between the device characteristics and circuit requirement, and it may be considered as a thyristor with a built-in antiparallel diode as shown in Fig. 13.4.

Static induction thyristor: The static induction thyristor (SITH) is in many respects similar to the gate turn off thyristor, but is a normally on device. Application of a reverse voltage to the gate with respect to the cathode will turn the SITH off. A SITH is a minority carrier device. As a result, SITH has low on-state resistance or

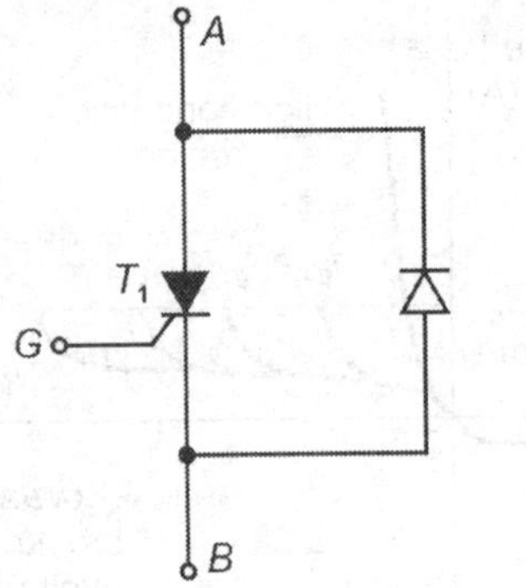

Fig. 13.4: Reverse conducting thyristor circuit

voltage drop and it can be made with higher voltage and current ratings compared to other thyristor devices, the SITH is very fast switching and has lower switching losses.

Light-activated silicon-controlled rectifiers (LASCRs): This device is turned on by direct radiation on the silicon wafer with light. Electron-hole pairs which are created due to the radiation produce triggering current under the influence of electric field. The gate structure is designed to provide sufficient gate sensitivity for triggering from practical light sources.

FET-controlled thyristors (FET-CTHs): A FET-controlled thyristor in which a MOSFET and a thyristor are connected in parallel is shown in Fig. 13.5. If a sufficient voltage is applied to the gate of the MOSFET, typically 3 V, a triggering current for the thyristor is generated internally. It has a high switching speed, high di/dt and high *dv/dt*. This device can be turned on like conventional thyristors, but it cannot be turned off by gate control. This would find applications where optical firing is to be used for providing electrical isolation between the input or control signal and the switching device of the power converter.

MOS -Controlled Thyristors (MCT): The MOS-controlled thyristor or MCT is a new device that is in early stage of development. It is basically a thyristor with one or more MOSFETs built-into the gate structure. The and a thyristor are connected in parallel equivalent circuit of the built-in MOSFETs turn on and turn off

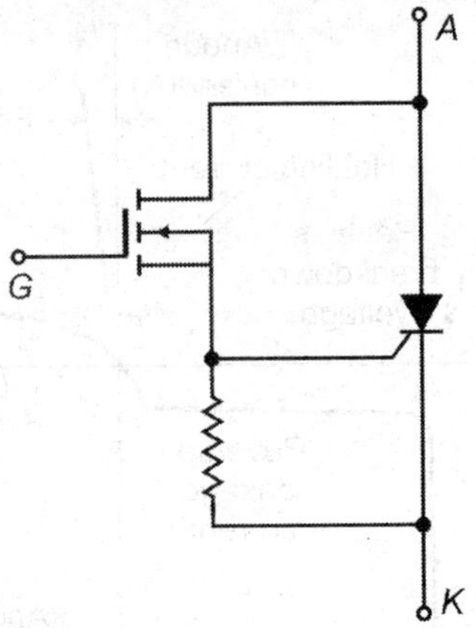

Fig. 13.5: MOSFET and a thyristor are connected in parallel

thyristor is shown in Fig. 13.6(a). Figure 13.6(b) shows the equivalent circuit of the device which is turned on by a gate current like a conventional thyristor and turned off by the built-in MOSFET. Both of these devices would have applications similar to those of GTO but the MCT would have simpler gating requirements, particularly in terms of the magnitude of the needed gating signals. Figure 13.6(c) shows the symbol of MOS-controlled thyristor.

Triggering devices: Switching a thyristor ON, is called its *triggering* or *firing*. There can be many methods by which a thyristor can be triggered such as gate triggering, radiation triggering and voltage triggering, etc. which are discussed in subsequent chapters. To achieve this, various triggering devices are employed like:

1. Unijunction transistor (UJT)
2. Programmable unijunction transistor (PUT)
3. Bidirectional diode (DIAC)
4. Silicon-controlled switch (SCS)
5. Silicon unilateral switch (SUS)
6. Silicon bilateral switch (SBS)
7. Shockley diode
8. Opto-isolator

Unijunction transistor: The most applied method of producing pulses for gate terggering is by means of unjunction transistor (UJT) relaxation oscillator.

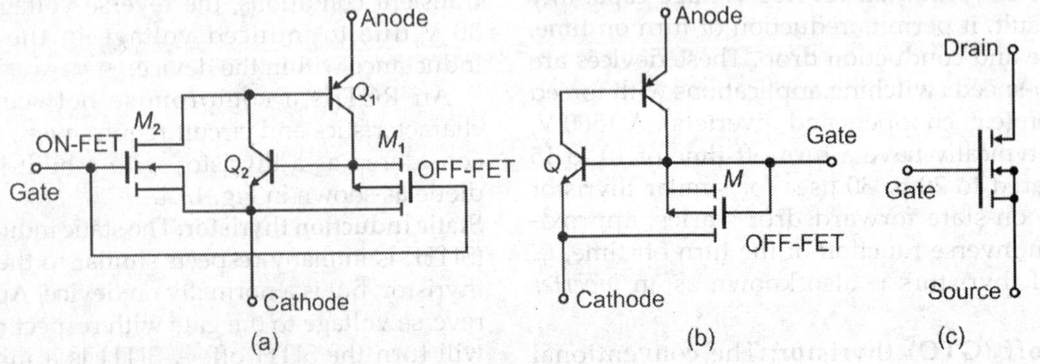

Fig. 13.6: Equivalent circuit of MOS-controlled thyristor (MCT), its symbol

Table 13.1: Symbols and *V-I* characteristics of important power devices

S. No.	*Devices*	*Symbols*	*Characteristics (V-I)*	*No. of terminals*
1.	Diode	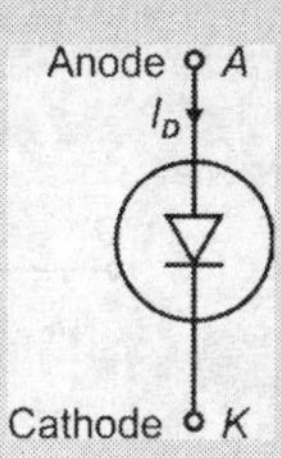	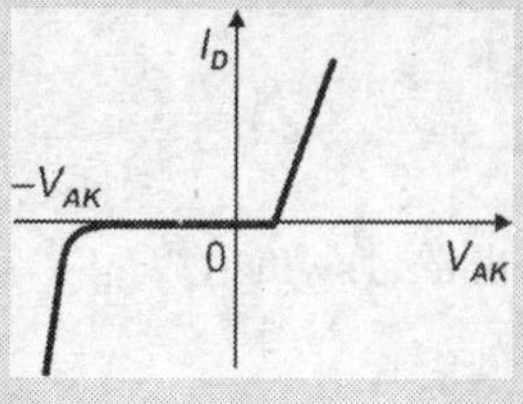	2
2.	Thyristor: (SCR-silicon controlled rectifier)	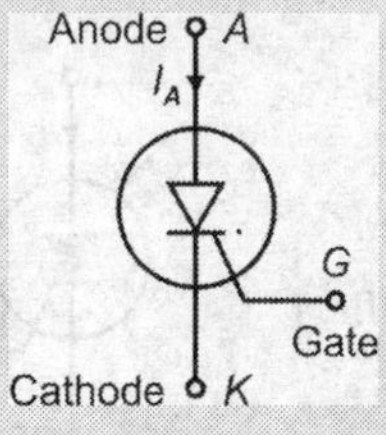	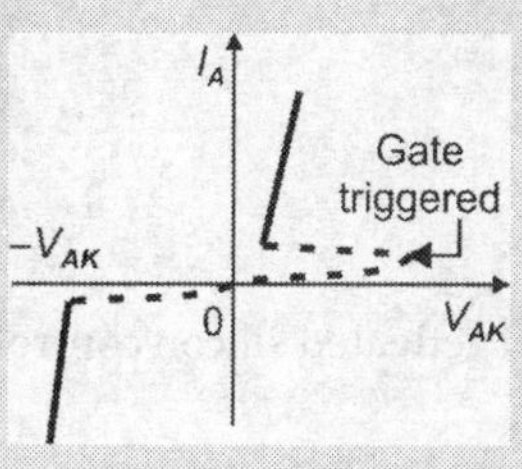	3
3.	TRIAC: (Bidirectional triode thyristor)	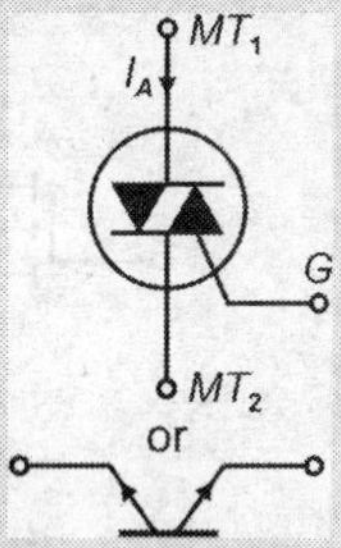	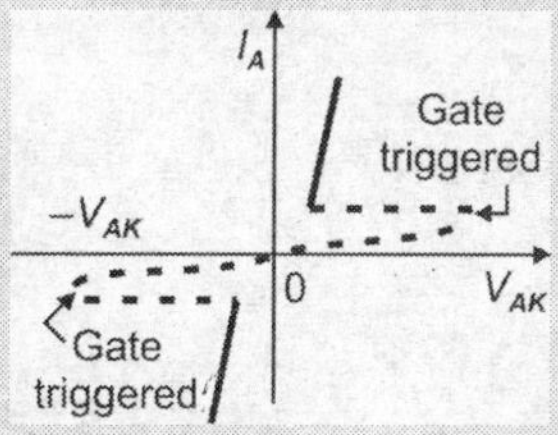	3
4.	DIAC: (Bidirectional trigger diode)		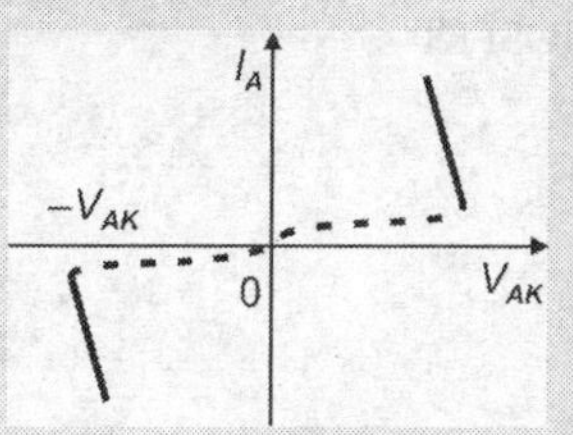	2
5.	GTO: (Gate turn off thyristor)	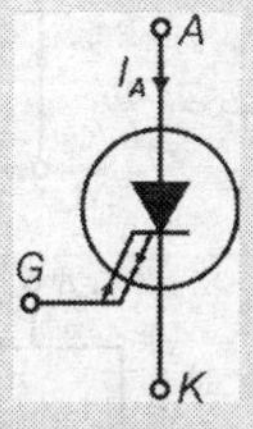	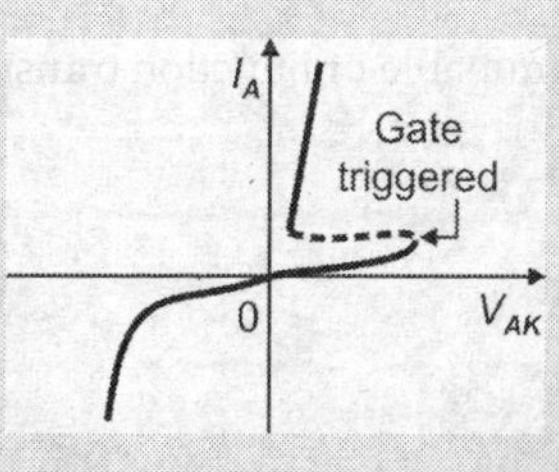	3
6.	SCS: (Silicon-controlled switch)	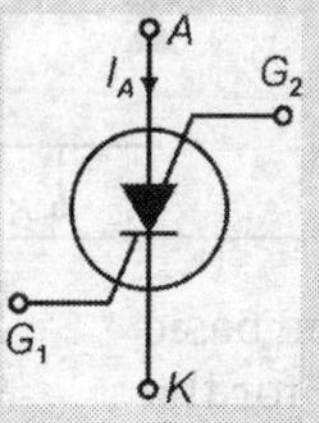	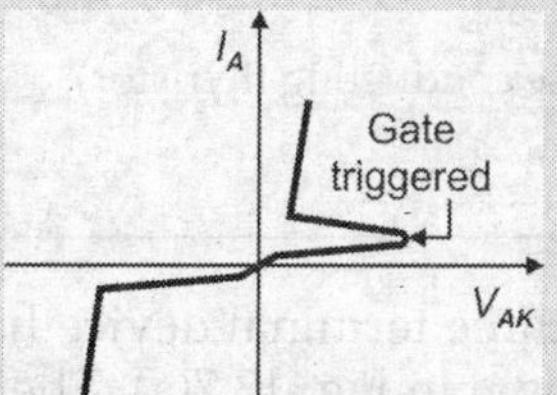	4
7.	SUS: (Silicon unilateral switch)	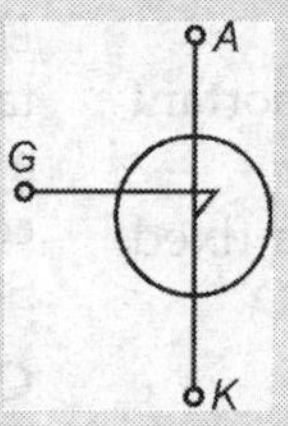	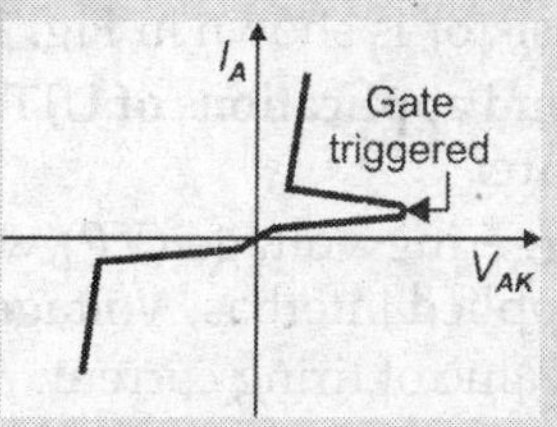	3

(Contd...)

Table 13.1 Symbols and *V-I* characteristics of important power devices (Contd.)

S. No.	*Devices*	*Symbols*	*Characteristics (V-I)*	*No. of terminals*
8.	SBS: (Silicon bilateral switch)	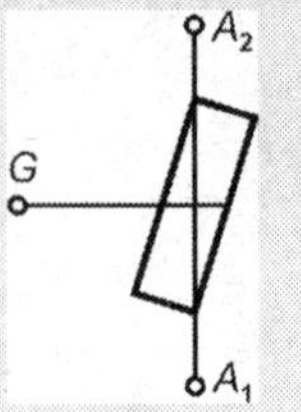	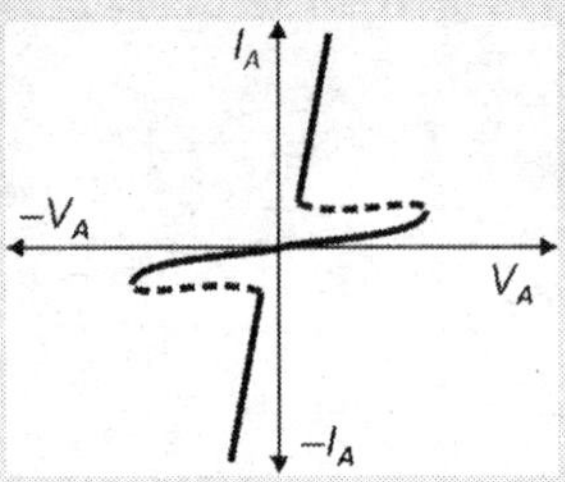	3
9.	LASCR: (Light activated silicon controlled rectifier)	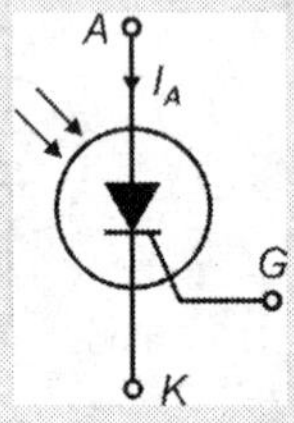	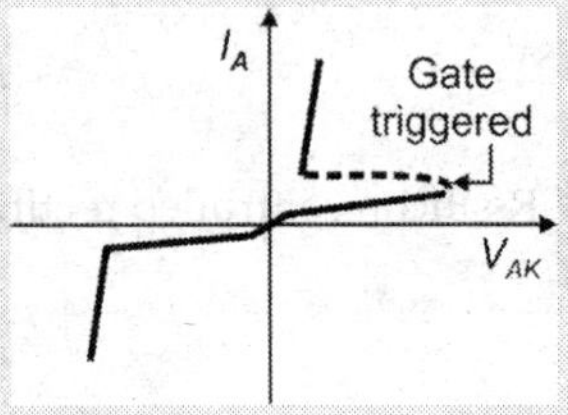	3
10.	N-channel MOSFET	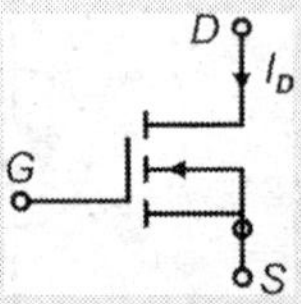	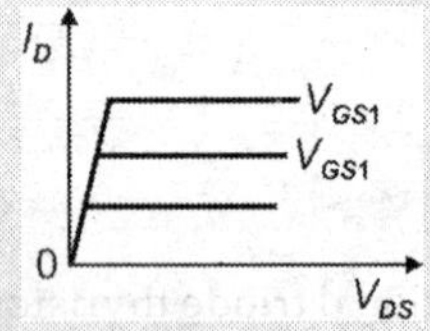	3
11.	P-channel MOSFET	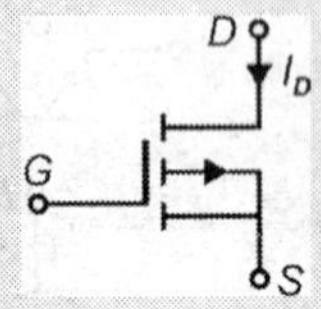	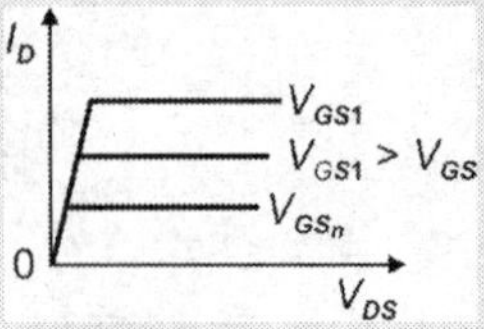	3
12.	PUT: (Programmable unijunction transistor)	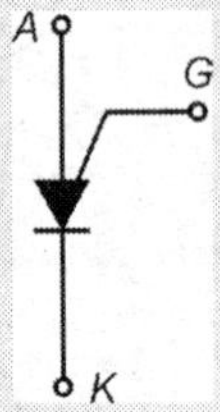	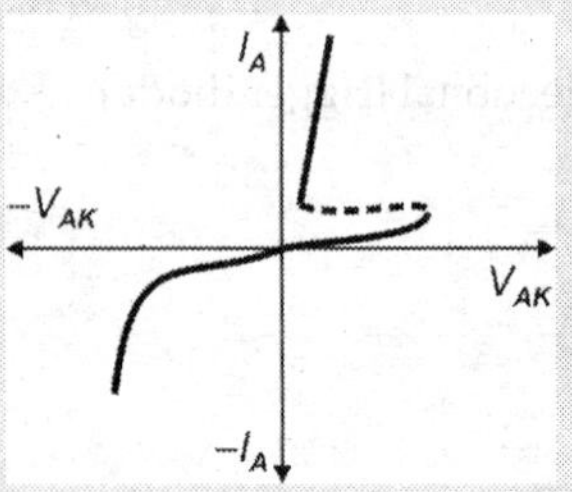	3
13.	RCT: (Reverse conducting thyristor)	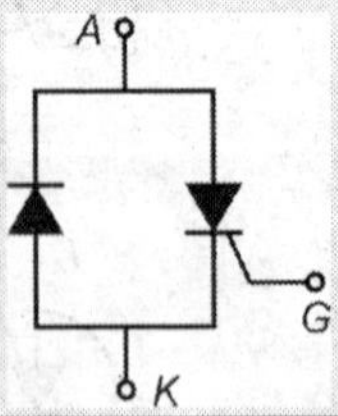	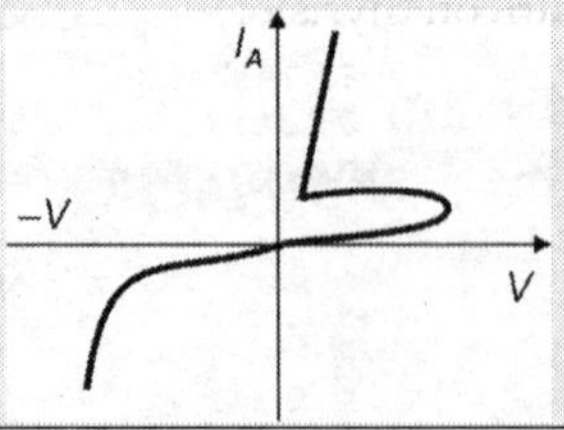	3

The UJT is a three terminal device having the basic construction shown in Fig. 13.7(a). The symbol for the unijunction transistor is shown in Fig. 13.7(b).

Characteristics and applications of UJT: The important features of UJT are:

1. A stable triggering voltage (*VP*) which is a fixed fraction of applied interbase voltage.
2. A very low value of firing current.
3. A negative resistance characteristics.
4. A high pulse current capability.
5. Low cost.

Uses: The above characteristics make the UJT advantageous in oscillators, timing circuits, pulse generators, a trigger circuit or as a saw tooth generator. One common application of the UJT is in the triggering of SCR.

Complementary unijunction transistor (CUJT): CUJT characteristics are like those of a standard UJT except

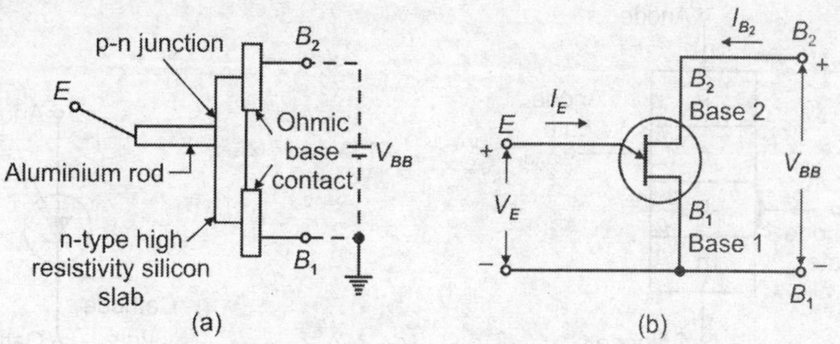

Fig. 13.7: UJT and symbol

that the currents and voltages applied to it are of opposite polarity. The opposite polarity has been chosen so that standard n-p-n planar passivated transistor processing techniques can be used. CUJTs can be used in most applications that use standard UJTs. Their unique stability and uniform properties make them ideal for stable oscillators, timers, and frequency dividers. Figures 13.8(a) and 13.8(b) shows the standard and complementary unijunction transistor.

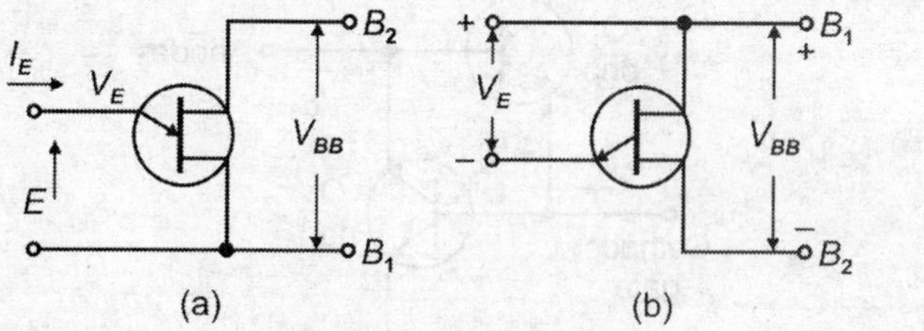

Fig. 13.8: CUJT circuit

Programmable unijunction transistor (PUT): The programmable unijunction transistor (PUT) is similar in operational characteristics to the UJT. It is a four-layer three terminal device designated as anode, gate and cathode. This recent addition to the thyristor family looks and acts like an improved UJT. However, a microscopic examination of the internal structure of the PUT shows it to be another four-layer pn-pn silicon device [Fig. 13.9(a)]. The symbol for the device is shown in Fig. 13.9(b).

Silicon-controlled switch (SCS): Silicon-controlled switch is a four terminal, four-layer pn-pn device used

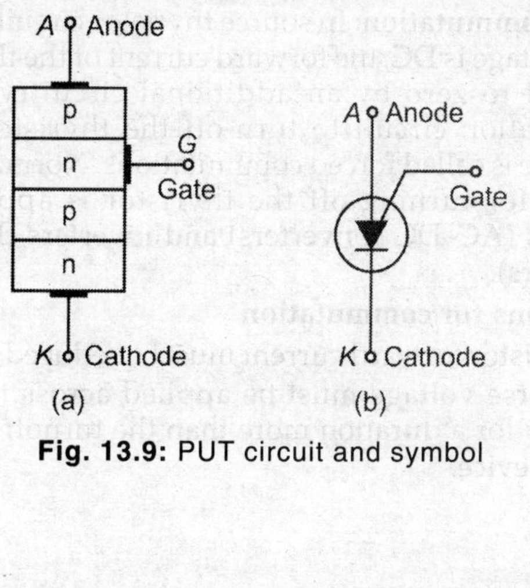

Fig. 13.9: PUT circuit and symbol

in low power switching applications. Similar to the SCR in construction, the SCS has two gate terminals [i.e. anode gate and cathode gate shown in Fig. 13.10(a)]. Each gate can be used to turn on or turn off the main current through the device.

Theory of operation: An equivalent two-transistor switch configuration can be used to describe the operation of the SCS. *V–I* characteristic for an SCS is given in Fig. 13.10(b). The graphic symbol and transistor equivalent circuit are shown in Figs 13.10 (c) and (d).

Requirements for triggering circuits: To turn on the thyristor, these conditions are required to be satisfied:

1. Thyristor should be forward-biased.
2. The load impedance should not be too high so that if the thyristor turn on the current in the thyristor should reach more than the latching current.
3. The gate should be positive with respect to cathode.

Three different types of triggering systems are in use depending on the application, economy and simplicity.

1. **Pulse triggering:** In this process, pulse signals are applied to a forward-biased thyristor. For the application chosen, each pulse is sufficient to trigger the thyristor and the pulse duration is chosen to give a sufficient duration for SCR to latch. Once the thyristor has been latched, no triggering pulse is required. The main advantage of this method is that there is no need of applying continuous signals, thus reducing the gate power losses. Electrical isolation is also provided between the main device supply and its gating signals. The most common method for producing pulses for gate triggering is by a unijunction transistor (UJT) relaxation oscillator.
2. **DC triggering:** In this type of triggering a DC voltage of proper magnitude and polarity is applied between the gate and cathode of the device in such a way that the gate becomes positive w.r.t. the cathode. When the applied voltage is sufficient to produce the required gate current, the device starts conducting. This method is inferior to pulse triggering as there is no isolation between power and control circuits. Another drawback is that a

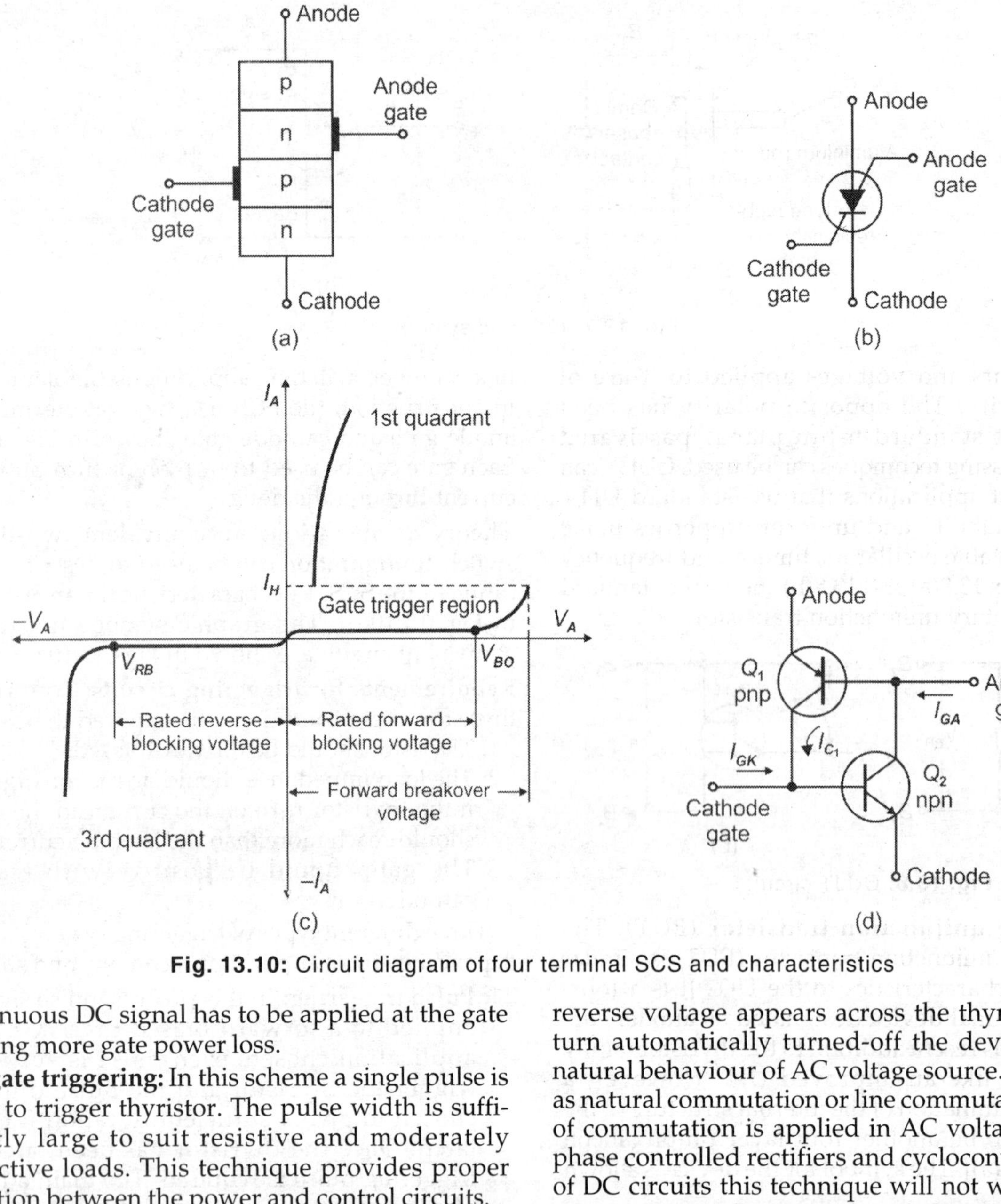

Fig. 13.10: Circuit diagram of four terminal SCS and characteristics

continuous DC signal has to be applied at the gate causing more gate power loss.

3. **AC gate triggering:** In this scheme a single pulse is used to trigger thyristor. The pulse width is sufficiently large to suit resistive and moderately inductive loads. This technique provides proper isolation between the power and control circuits.

Two types of circuits are generally applied for AC triggering:

1. Resistance triggering (*R*-triggering)
2. Resistance–capacitance triggering (*R–C* triggering)

Commutation of a thyristor: The method of switching off a thyristor is known as *commutation of thyristor*. The thyristor can be turned-off:

By reducing the forward current below its holding current or by applying a large reverse voltage across it. There are many techniques to commutate a thyristor. However, these can be broadly classified into *two* types:

1. natural commutation
2. forced commutation.

Natural commutation: If the input voltage is AC, the thyristor current passes through a natural zero, and a reverse voltage appears across the thyristor which in turn automatically turned-off the device due to the natural behaviour of AC voltage source. This is known as natural commutation or line commutation. This type of commutation is applied in AC voltage controllers, phase controlled rectifiers and cycloconverters. In case of DC circuits this technique will not work as the DC current is unidirectional and does not change its direction, thus the reverse polarity voltage does not appear across the thyristor.

Forced commutation: In source thyristor circuits, if the input voltage is DC, the forward current of the thyristor is forced to zero by an additional circuitry called commutation circuit to turn-off the thyristor. This technique is called forced commutations. Normally this method for turning-off the thyristor is applied in choppers (AC–DC converters) and inverters (DC–AC converters).

Conditions for commutation

1. Thyristor forward current must be reduced to zero.
2. Reverse voltage must be applied across the thyristor for a duration more than the turnoff time of the device.

3. Critical rate of rise of voltage of the device should never be exceeded to avoid retriggering of the thyristor.
4. In case of inductive loads, care should be taken against retriggering of the device due to dissipation of stored energy. Generally, a freewheeling diode is used across the inductive load.

Classification of converters: The phase control converter can be classified into *two* types depending on the input supply: 1. single-phase converters 2. three-phase converters

Single-phase converter can be subdivided into:

(i) semiconverter (ii) full converter (iii) dual converter

Semiconverter is a one-quadrant converter and has one polarity of output voltage and current.

Full converter is a two-quadrant converter and the polarity of its output voltage can be either positive or negative. However, the output current of full converter has one polarity only.

A dual converter can operate in four-quadrants; and both the output voltage and current can be either positive or negative. In some applications converters are connected in series to operate at higher voltages and to improve the input power factor.

Industrial application of converters: The class of converters discussed in this chapter uses the phase control principle which include the following applications:

1. Electrochemical processes, such as electroplating, anodising, metal refining and hydrogen production
2. HVDC conversion
3. DC motor speed control
4. DC supply for inverters
5. DC–AC conversion from fuel cells, solar cells and so on
6. Magnet power supply, such as machine excitation, fusion reactor supply, etc.
7. Portable hand tool drives
8. DC traction.

Chopper classification: Depending on the directions of current and voltage, choppers can be classified into following classes: 1. Class A chopper 2. Class B chopper 3. Class C chopper 4. Class D chopper 5. Class E chopper

1. **Class A chopper:** The load current flows into load. Both the load voltage and current are positive as shown in Fig. 13.11(a). This is a single quadrant chopper and is said to be operated as a rectifier.
2. **Class B chopper:** The output load current flows out of the load. The load voltage is positive, but the load current is negative as shown in Fig. 13.11(b). This is a single quadrant chopper but operated in the second quadrant and is said to be operated as inverter.
3. **Class C chopper:** The load current is either positive or negative as shown in Fig. 13.11(c). This is known as two quadrant chopper. Class A and class B choppers can be combined to form class C chopper. Class C chopper can operate either as a rectifier or as an inverter.
4. **Class D chopper:** The load current is always positive. The load voltage is either positive or negative as shown in Fig. 13.11(d). It can also operate either a rectifier or an inverter.
5. **Class E chopper:** The load current is either positive or negative. This is known as *four quadrant chopper*. Two Class C choppers can be combined to form a Class E chopper as shown in Fig. 13.11(e).

Chopper configurations: Various types of chopper configurations and their operation in different quadrants are shown in Fig. 13.12.

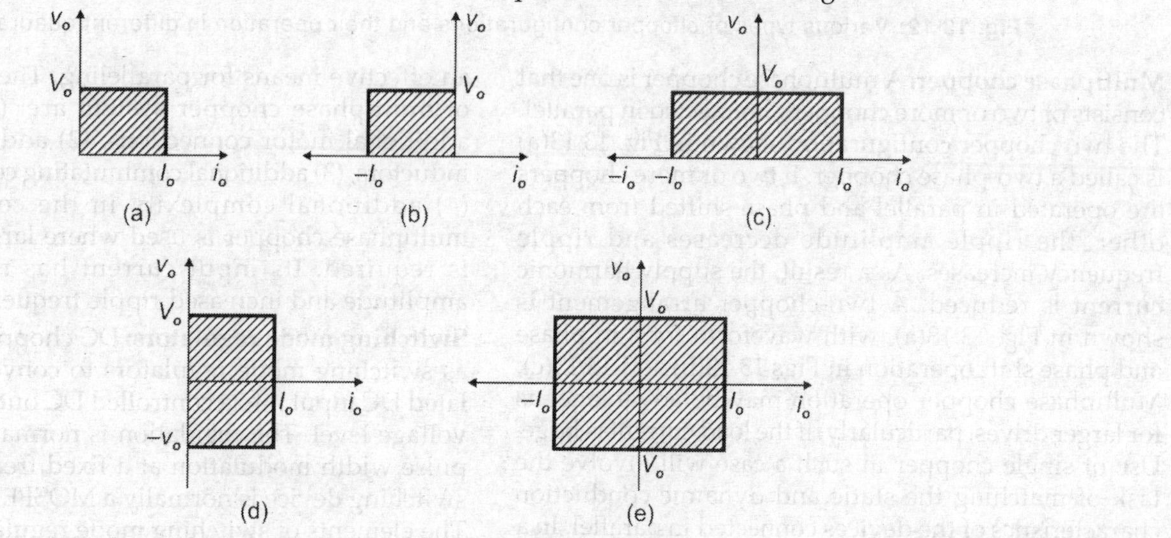

Fig. 13.11 Classification of choppers

Type	Chopper Configuration	Characteristics	Function
First quadrant chopper class A	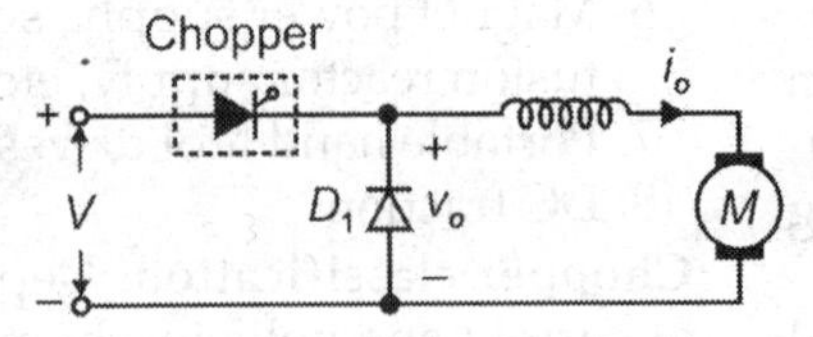	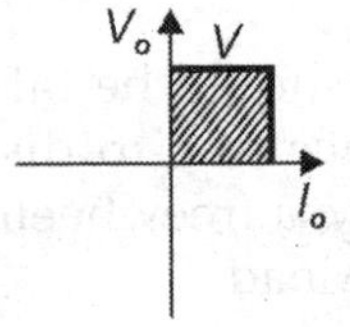	$V_0 = V$, Ch1 ON $= 0$, Ch1 OFF D_1 ON
Second quadrant or regenerative chopper class B	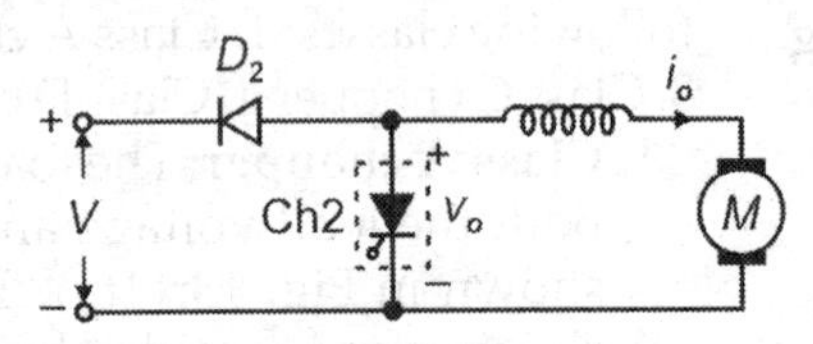	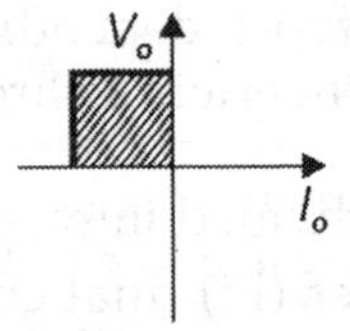	$V_0 = 0$, Ch2 ON $= V$, Ch2 OFF D_2 ON
Two quadrant class C chopper	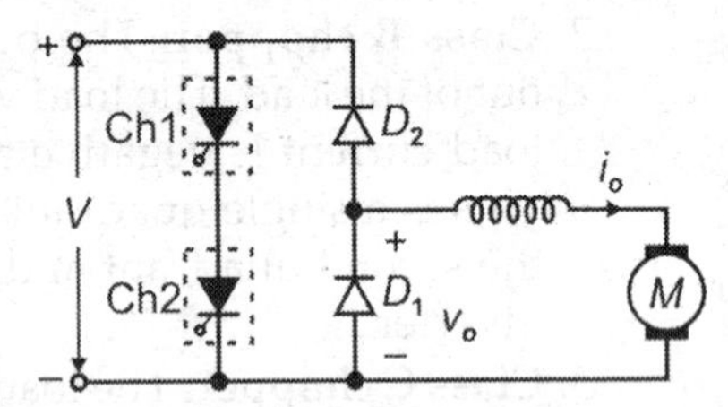	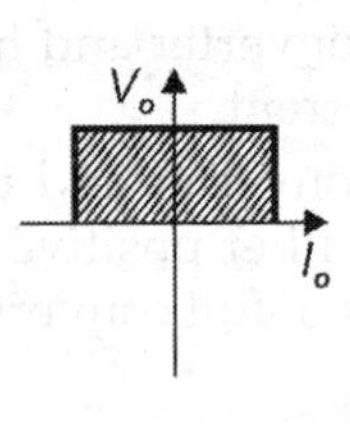	$V_0 = V$, Ch1 ON or D_2 ON $= 0$, Ch2 ON or D_1 ON i_o = Positive Ch1 ON or D_1 ON i_o = Negative Ch2 ON or D_2 ON
Two quadrant class D chopper	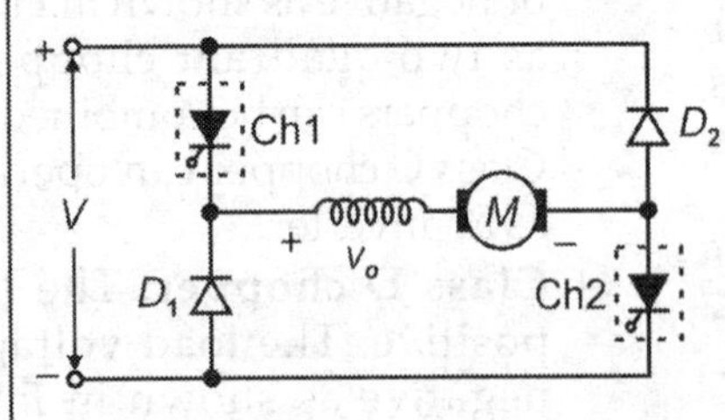	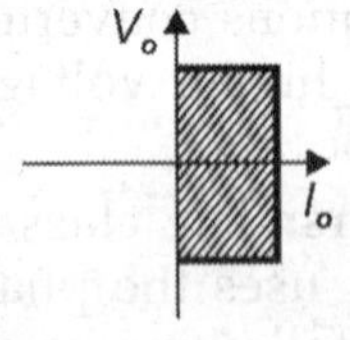	$V_0 = V$ Ch1 ON or D_2 ON $V_0 = -V$ Ch1 and Ch2 OFF D_1 and D_2 ON
Four quadrant chopper class E	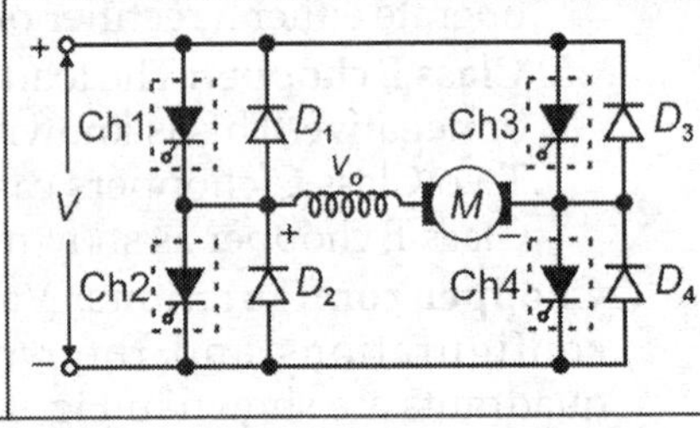	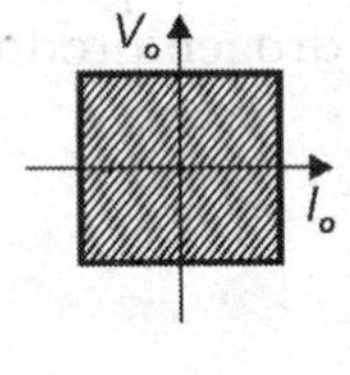	Ch4 ON and Ch3 OFF Ch1 and Ch2 operated V_o = Positive I_o = Reversible Ch2 ON and Ch1 OFF Ch3 and Ch4 operated V_0 = Negative I_0 = Reversible

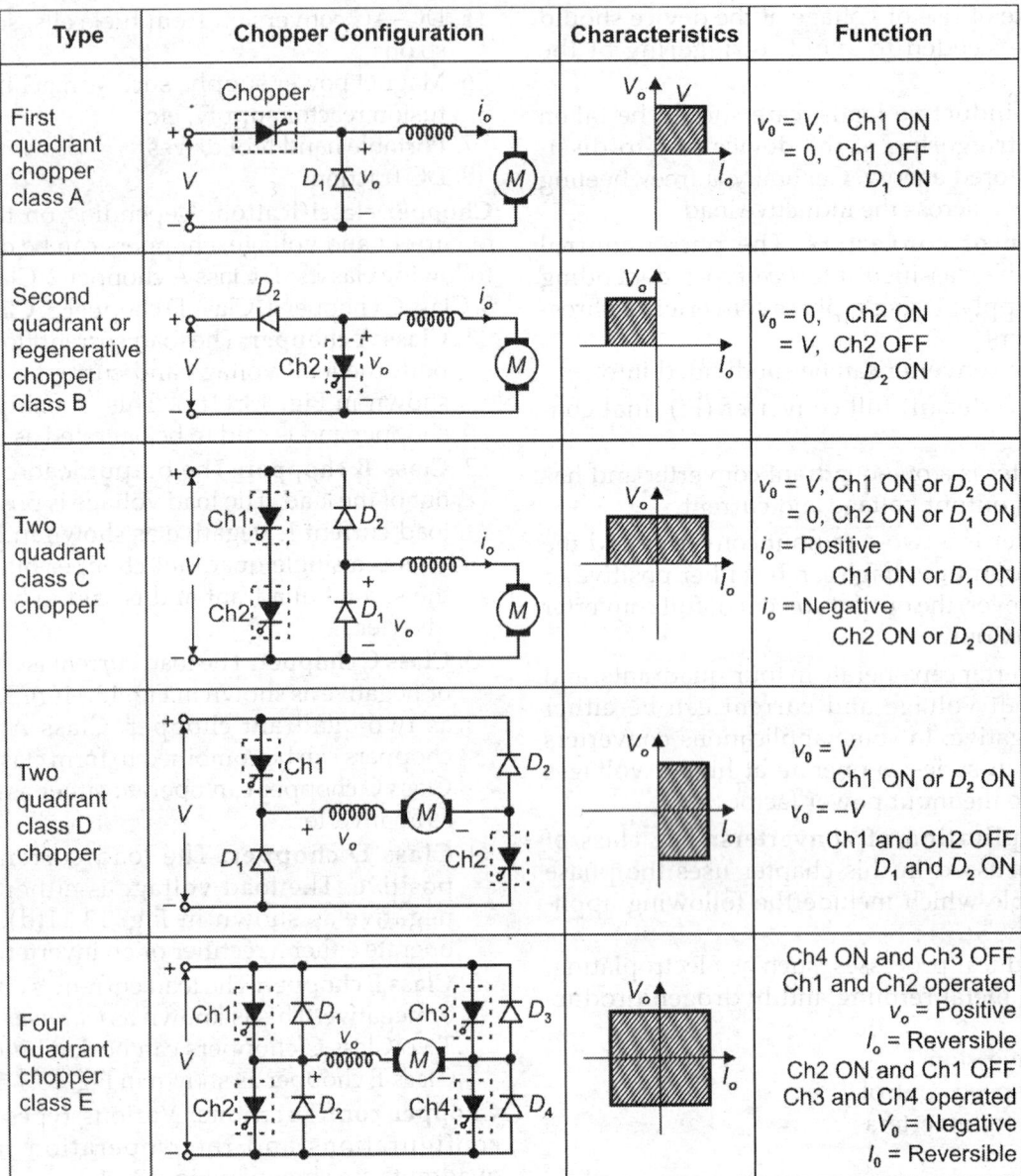

Fig. 13.12: Various types of chopper configurations and their operation in different quadrants

Multiphase chopper: A multiphase chopper is one that consists of two or more choppers, connected in parallel. The two chopper configuration shown in Fig. 13.13(a) is called a two-phase chopper. If two or more choppers are operated in parallel and phase shifted from each other, the ripple amplitude decreases and ripple frequency increases. As a result, the supply harmonic current is reduced. A two-chopper arrangement is shown in Fig. 13.13(a), with waveform for a in phase and phase shift operation in Figs 13.13(b) and 13.13(c). Multiphase chopper operation may be advantageous for larger drives, particularly, if the load current is large. Use of single chopper in such a case will involve the task of matching the static and dynamic conduction characteristics of the devices connected in parallel. In a multiphase chopper system, separate inductors provide an effective means for paralleling. The disadvantages of multiphase chopper system are: (1) the need of additional motor connections, (2) additional external inductors, (3) additional commutating components, and (4) additional complexity in the control logic. A multiphase chopper is used where large load current is required. Its input current has reduced ripple amplitude and increased ripple frequency.

Switching mode regulators: DC choppers can be used as switching mode regulators to convert the unregulated DC input into a controlled DC output at a desired voltage level. The regulation is normally achieved by pulse width modulation at a fixed frequency and the switching device is normally a MOSFET or power BJT. The elements of switching mode regulators are shown in Fig. 13.14(a). Control voltage *vc* is obtained by

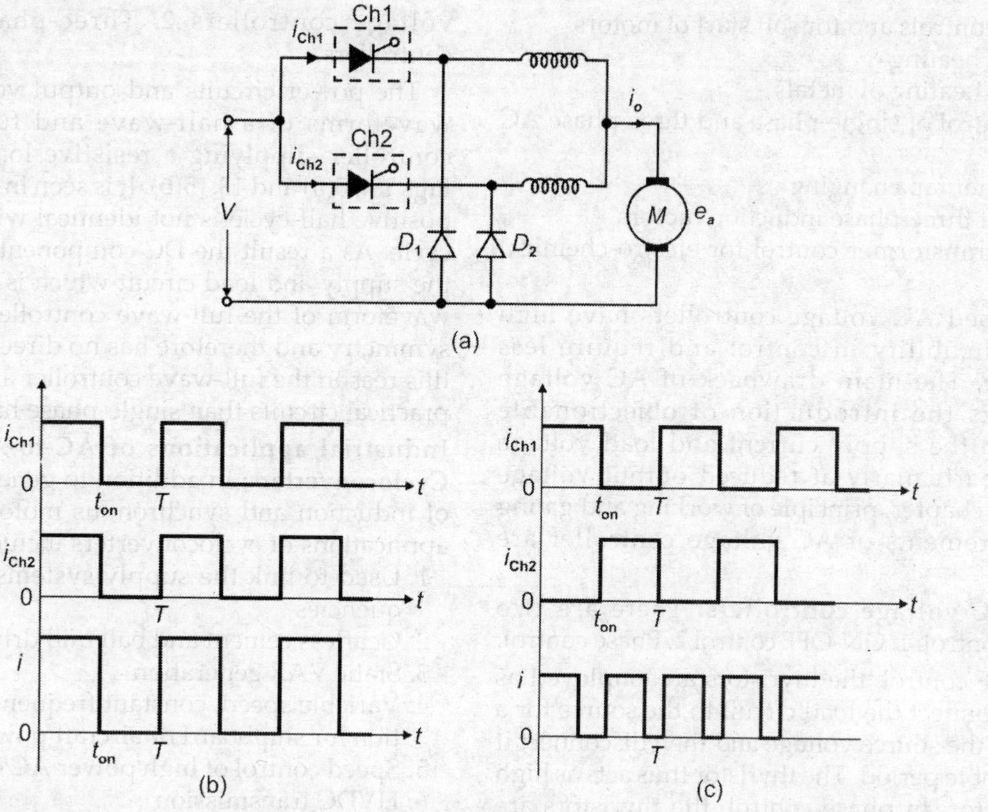

Fig. 13.13: Various types of chopper configurations and their operation in different quadrant

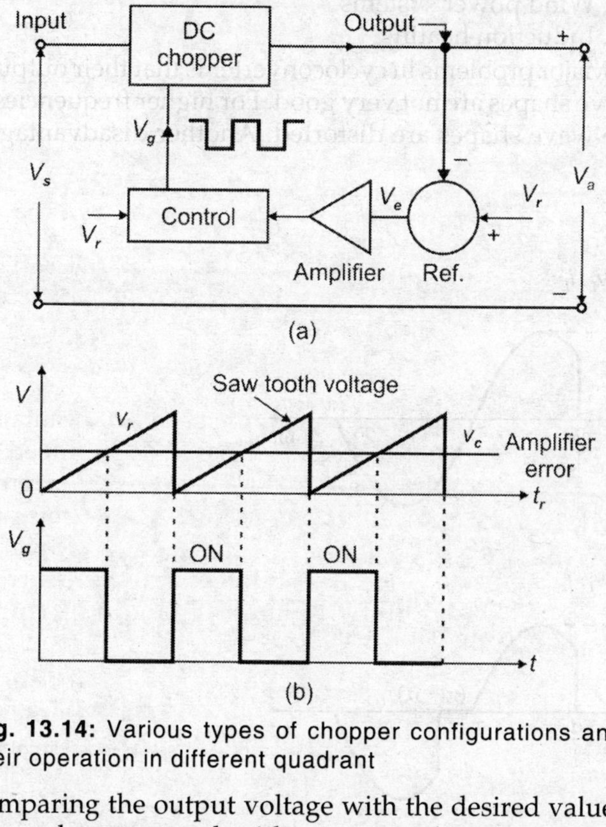

Fig. 13.14: Various types of chopper configurations and their operation in different quadrant

comparing the output voltage with the desired value. *vc* can be compared with a saw tooth voltage *vr* to generate the PWM control signal for the DC chopper. This is shown in Fig. 13.14(b).

Advantages of switched mode regulators: The advantages of switched mode regulators over linear regulators are: (1) very less losses, (2) smaller size filter components, (3) light weight, and (4) noiseless operation.

There are four basic switching regulators:

1. Step-down or buck regulator
2. Step-up or boost regulator
3. Buck–boost regulator
4. Cuk regulator.

Buck–boost regulator: The buck–boost regulator provides output voltage which may be less than or greater than the input voltage, hence the name buck–boost. The output voltage polarity is opposite to that of input voltage. This regulator is known as *inverting regulator*.

A buck–boost converter can be obtained, by cascade connection of the two basic converters: (1) the step-down, and (2) step-up. In steady state, the output to input voltage conversion ratio is the product of conversion ratios of the two converters in cascade assuming that switches in both converters have the same duty ratio.

$$\frac{V_0}{V} = \delta \cdot \frac{1}{1-\delta}$$

Industrial applications of AC voltage controllers: There are some applications of variable AC voltage at line frequency. The applications include:

1. Lighting controls and for soft start of motors
2. Industrial heating
3. Induction heating of metals
4. Speed control of single-phase and three-phase AC drives
5. Transformer tap changing
6. Starting of three-phase induction motors
7. Primary transformer control for electro-chemical processes

Thyristorised AC voltage controllers have high efficiency, flexibility in control and require less maintenance. The main drawback of AC voltage controllers is the introduction of objectionable harmonics in the supply current and load voltage waveforms particularly at reduced output voltage levels. In this chapter, principle of working and gating signal requirements of AC voltage controller are discussed.

Types of AC voltage controllers: There are two methods of control: 1. ON-OFF control 2. Phase control.

In ON-OFF control, the thyristors are employed as switches to connect the load circuit to the source for a few cycles of the source voltage and then disconnect it for a comparable period. The thyristor thus acts as high speed contactor. In phase control, the thyristors are employed as switches to connect the load circuit to the voltage source for a chosen portions of each cycle of the load voltage. Here more attention is concentrated on phase control methods.

AC phase voltage controllers: AC voltage controllers are classified into two categories: 1. Single-phase AC voltage controllers 2. Three-phase AC voltage controllers.

The power circuits and output voltage or current waveforms of a half-wave and fullwave voltage controller supplying a resistive load are shown in Figs 13.15(a) and 13.15(b). It is seen in Fig. 13.15(a), that positive half-cycle is not identical with negative half-cycle. As a result the DC component is introduced in the supply and load circuit which is undesirable. The waveform of the full-wave controller has alternating symmetry and therefore has no direct component. For this reason the full-wave controller are more suited to practical circuits than single-phase half-wave circuits.

Industrial applications of AC-to-AC converters—Cycloconverters: In addition to general speed control of induction and synchronous motors, the industrial applications of cycloconverters include:

1. Used to link the supply systems of different frequencies
2. Gearless cement and ball mill drives
3. Static VAR generation
4. Variable speed, constant frequency power generation for shipboard or aircraft power supply
5. Speed control of high power AC drives
6. HVDC transmission
7. Electric traction
8. Wind power systems
9. Induction heating

Major problems in cycloconverter is that their output wave shapes are not very good. For higher frequencies, the wave shapes are distorted. Another disadvantage

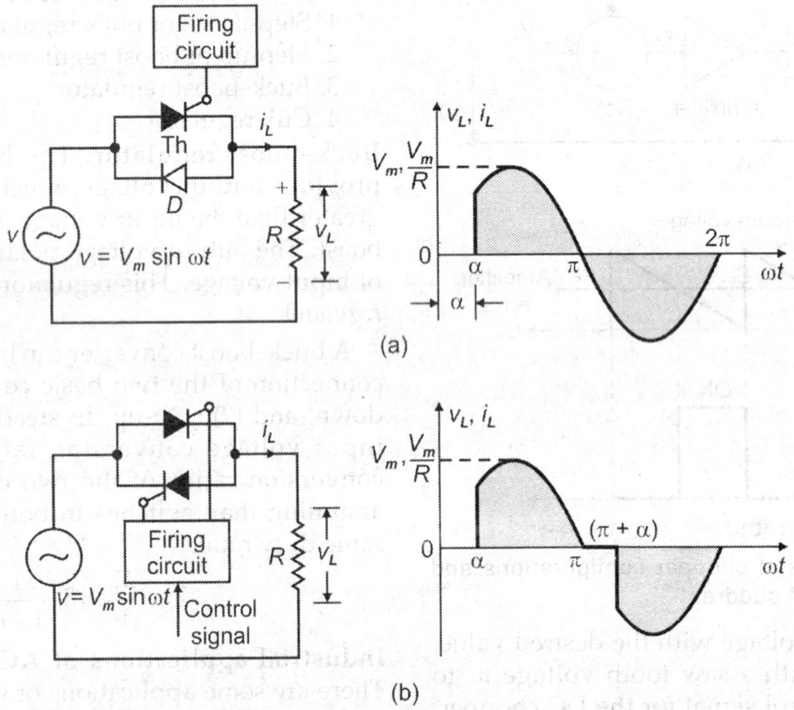

Fig. 13.15: Single-phase AC voltage controllers and waveforms

of using a cycloconverter for motor speed control is that the power factor of the system is reduced.

Types of cycloconverters: Cycloconverter can be categorised as follows: 1. Single-phase cycloconverter 2. Three-phase cycloconverter

Principle of cycloconverter: A cycloconverter can be considered to be composed of two converters connected back to back as shown in Fig. 13.16(a) with a pure resistive load. A positive (P) and negative (N) centre tap converter are connected in parallel so that the voltage and current of either polarity can be supplied to the load. The load waveforms of Fig. 13.16(b) show that the instantaneous power flows in the load fall into one of the four periods. The two periods, when the product of load voltage and current is positive, require power flow into the load dictating a situation where the converter groups rectify, the positive and negative groups conducting respectively during the appropriate positive and negative load current periods.

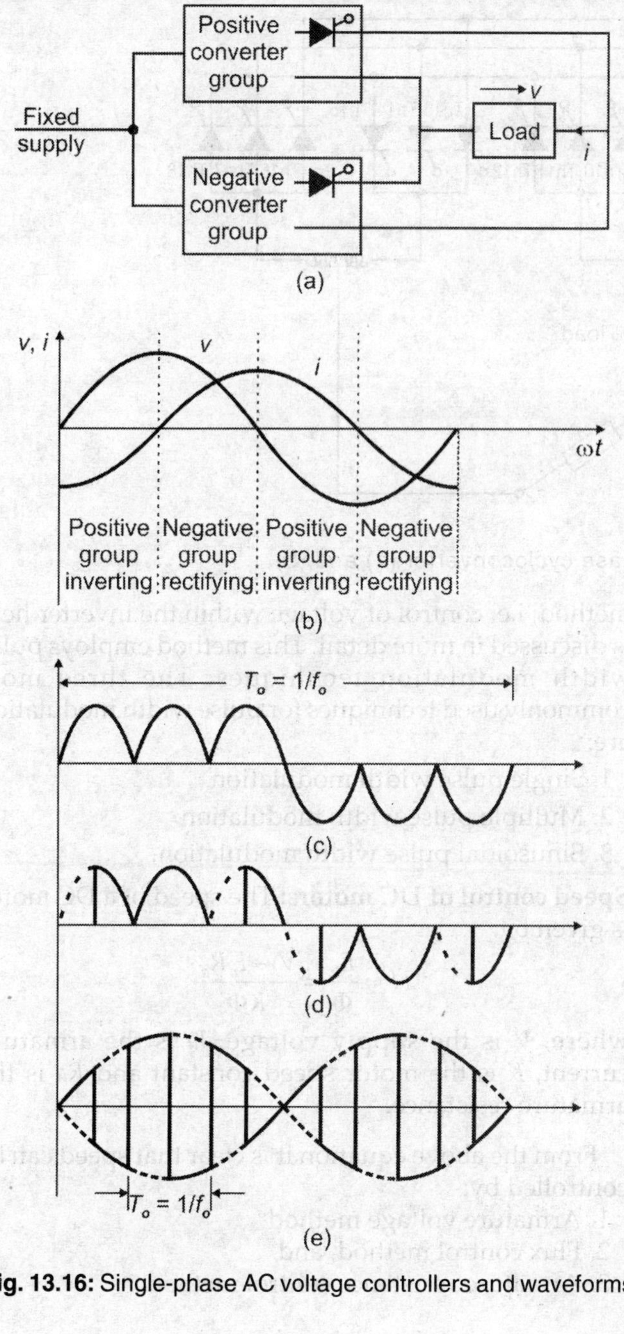

Fig. 13.16: Single-phase AC voltage controllers and waveforms

The other two periods represent times when the product of load voltage and current is negative, hence the power flow is out of the load, demanding that the converter operates in the inverting mode. A positive and a negative centre-tap converter groups are connected in parallel so that voltage and current of either polarity can be supplied to the load. The waveforms are drawn assuming a resistive load. In Fig. 13.16(c) an integral half-cycle output wave is fabricated with fundamental frequency * where n is the number of input half-cycles per half-cycle of the output. The firing angle can be modulated to control the voltages as well as its harmonic content as shown in Fig. 13.16(d). Step up conversion is one in which $fo > fi$ shown in Fig. 13.16(e).

Three-phase to three-phase cycloconverter: When a three-phase output is required, three single-phase cycloconverters with a phase displacement of 120° between their outputs are connected, as shown in Fig. 13.17(a). With a balanced load, the neutral connection is no longer necessary and may be omitted. The simplest arrangement using three-phase half-wave circuits is shown in Fig. 13.17(b). This circuit requires eighteen thyristors, but if each group consists of a three-phase bridge or six-phase circuit or six pulse bridges then thirty-six thyristors are required.

It is considered that the output frequency of the cycloconverter is less than the supply frequency. However, the cycloconverter is also capable of operation as a step-up frequency changer, but the output power is limited and circuit losses are high. In practice, with the basic cycloconverter circuit of Fig. 13.17(b), reasonable power output and efficiency are only obtained in the step-down region with output frequencies from zero to about one-third of the input frequency. As the cycloconverter output frequency approaches the supply frequency, the harmonic distortion in the output voltage increases, since the output voltage waveform is composed of fewer segments of the supply voltage. As a result, the losses in the cycloconverter and in the AC motor become excessive, and there is a drop in overall efficiency. By using more complex rectifier circuits, the output voltage waveform is improved and the maximum useful ratio of output to input frequency is increased to about one-half. The AC motor normally presents a high impedance at the ripple frequency, and, hence, the output current is nearly sinusoidal and no additional filtering is necessary.

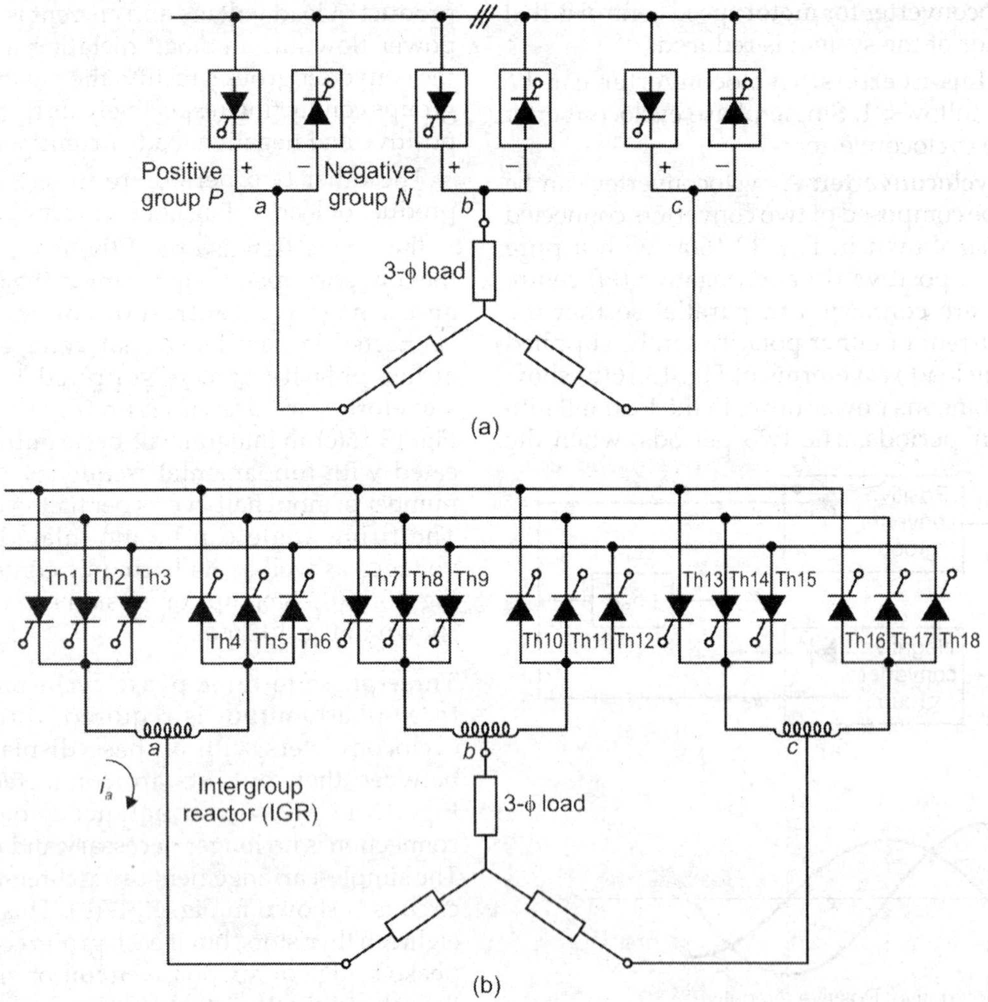

Fig. 13.17: Three-phase to three-phase cycloconverter (a) and (b)

Industrial applications of DC-to-AC converters—intervers: Following are the important industrial applications of inverter, such as:

1. Variable speed AC motor drives
2. High voltage DC transmission lines
3. Induction heating
4. Aircraft power supplies
5. Standby power supplies
6. Uninterruptible power supplies.

The input may be a battery fuel cell, solar cell, etc. Typical single-phase outputs are 220 V, at 50 Hz or 120 V at 60 Hz. For high power three-phase systems, output may be 220/380 V at 50 Hz or 115/200 V at 400 Hz.

Control circuit for square wave inverter: Inverter control can be achieved by analog/digital hardware or by microcomputer software. In this section hardware control square wave inverter is briefly discussed. Figure 13.18 shows the control block diagram of a six-stepped square wave inverter.

Pulse width modulated inverters: Out of the three methods of voltage control discussed above, second method, i.e. control of voltage within the inverter here is discussed in more detail. This method employs pulse width modulation techniques. The three most commonly used techniques for pulse width modulation are:

1. Single pulse width modulation
2. Multiple pulse width modulation
3. Sinusoidal pulse width modulation.

Speed control of DC motors: The speed of a DC motor is given by:

$$N \propto \frac{E_b}{\Phi} = \frac{V - I_a R_a}{K\Phi}$$

where, V is the supply voltage, Ia is the armature current, K is the motor speed constant and Ra is the armature resistance.

From the above equation it is clear that speed can be controlled by:

1. Armature voltage method
2. Flux control method, and
3. Armature resistance control method.

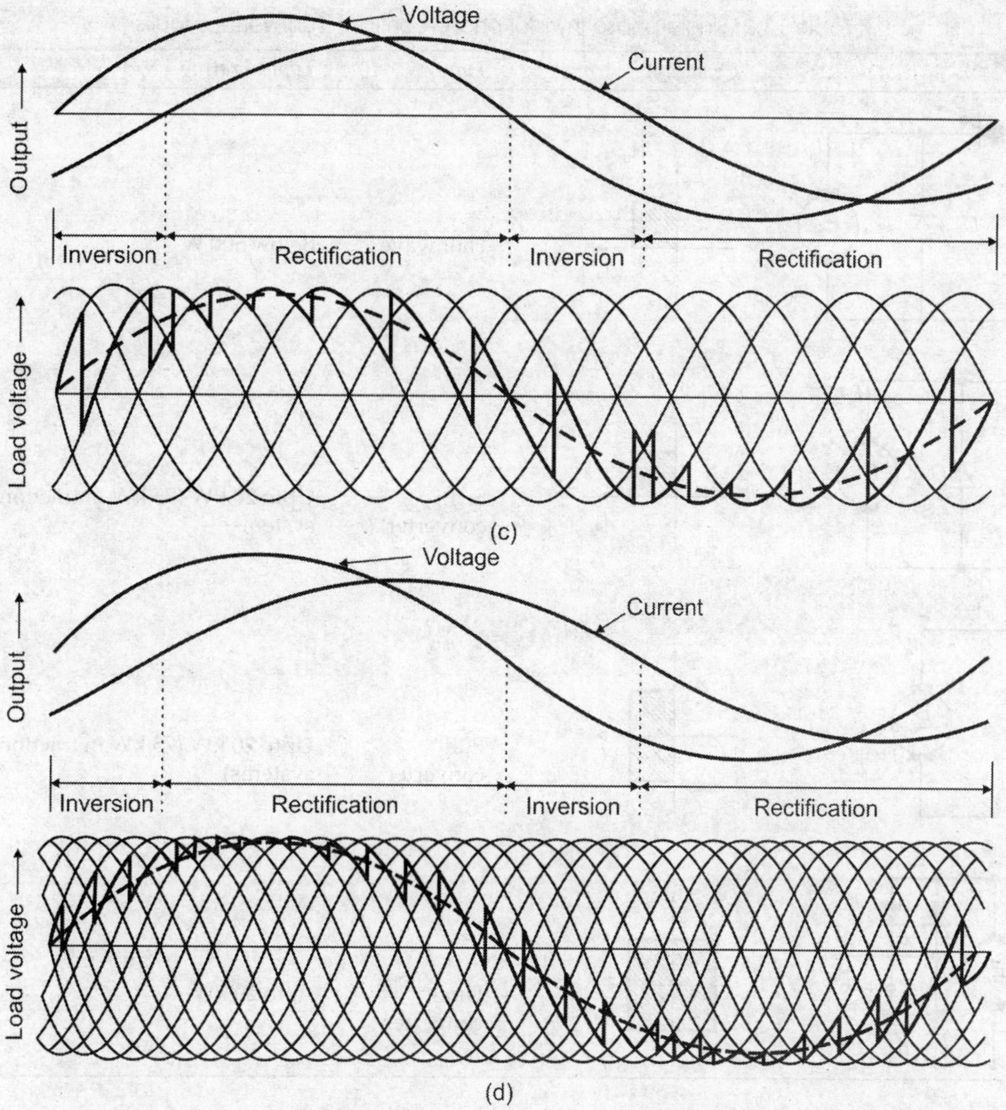

Fig. 13.17: Cycloconverter waveforms for higher pulse conversion (c) and (d)

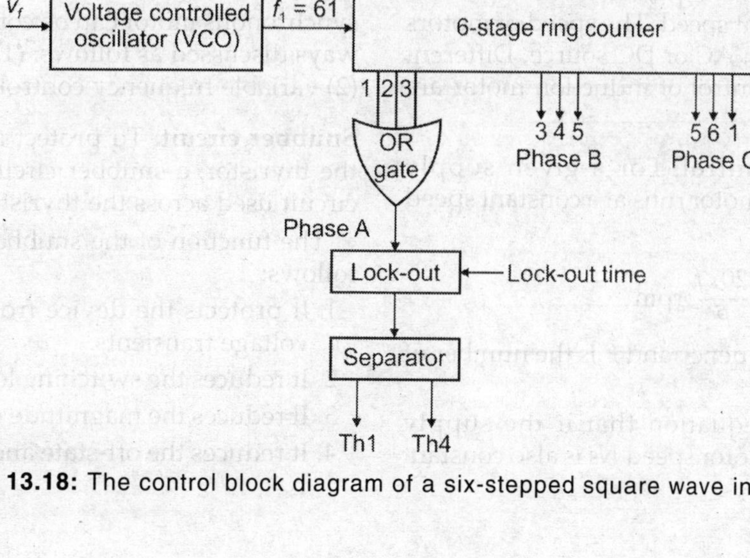

Fig. 13.18: The control block diagram of a six-stepped square wave inverter

Table 13.2: Single-phase thyristor phase-controlled converter circuits

Circuit	*Quadrant*	*Type*	*Typical rating*	*Ripple frequency*
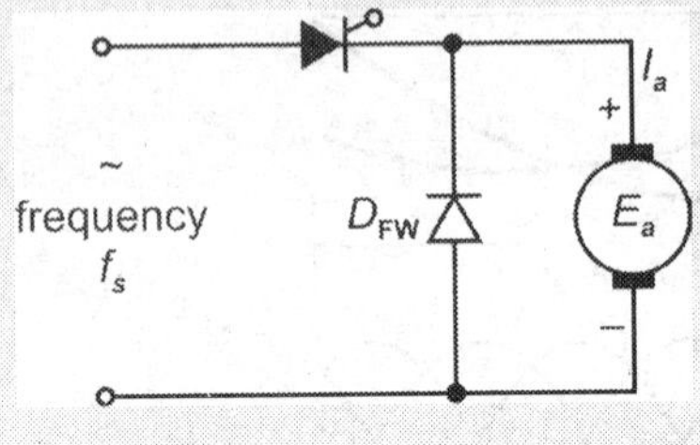	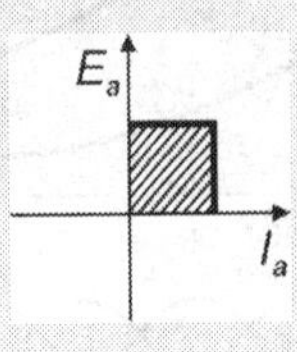	Half wave	Below 500 W	f_s
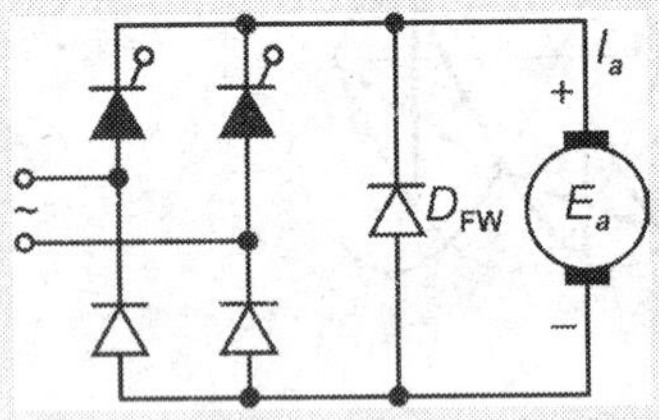	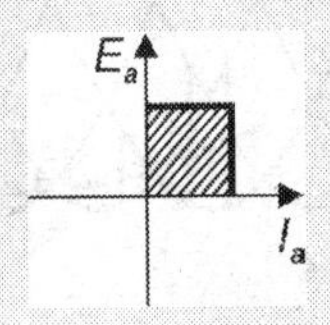	Semi-converter	Upto 20 kW (75 kW in traction systems)	$2f_s$
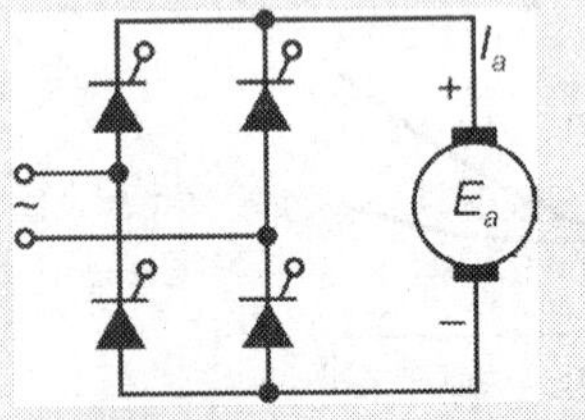	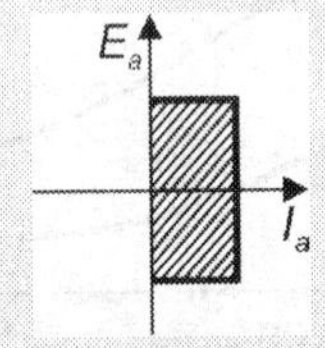	Full converter	Upto 20 kW (75 kW in traction systems)	$2f_s$
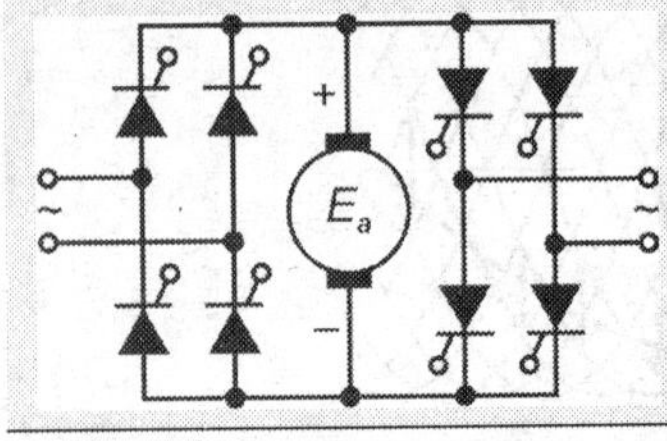	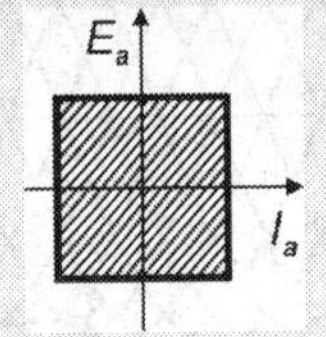	Dual converter	upto 20 kW	$2f_s$

Various schemes of speed control of induction motor: The induction machine is most commonly used in adjustable-speed AC drive systems, for their low cost, less maintenance and simple construction. When operated at constant AC voltages, an induction motor operates at nearly constant speed. The speed of motors can be controlled from an AC or DC source. Different schemes for the speed control of induction motor are shown in Fig. 13.19.

Synchronous motor control: For a given supply frequency, synchronous motor runs at a constant speed given by

$$N_s = \frac{120 \cdot f}{P} \text{ rpm}$$

where f is the supply frequency and P is the number of poles.

It is clear from the equation that if the supply frequency f is constant, motor speed Ns is also constant and no separate control is required to maintain the speed constant, thus it is desired to maintain the frequency constant. The supply frequency may be accurately changed by crystal oscillator. Thus, the same accuracy results in synchronous motor speed. The synchronous motor can be controlled in any of the two ways discussed as follows: (1) load commutation, and (2) variable frequency control.

Snubber circuit: To protect against dv/dt turn on of the thyristor, a snubber circuit, which is a R-C series circuit used across the thyristor.

The function of the snubber can be summarised as follows:

1. It protects the device from supply and load side voltage transients.
2. It reduces the switching losses of devices.
3. It reduces the magnitude of peak recovery voltage.
4. It reduces the off-state and reapplied dv/dt.

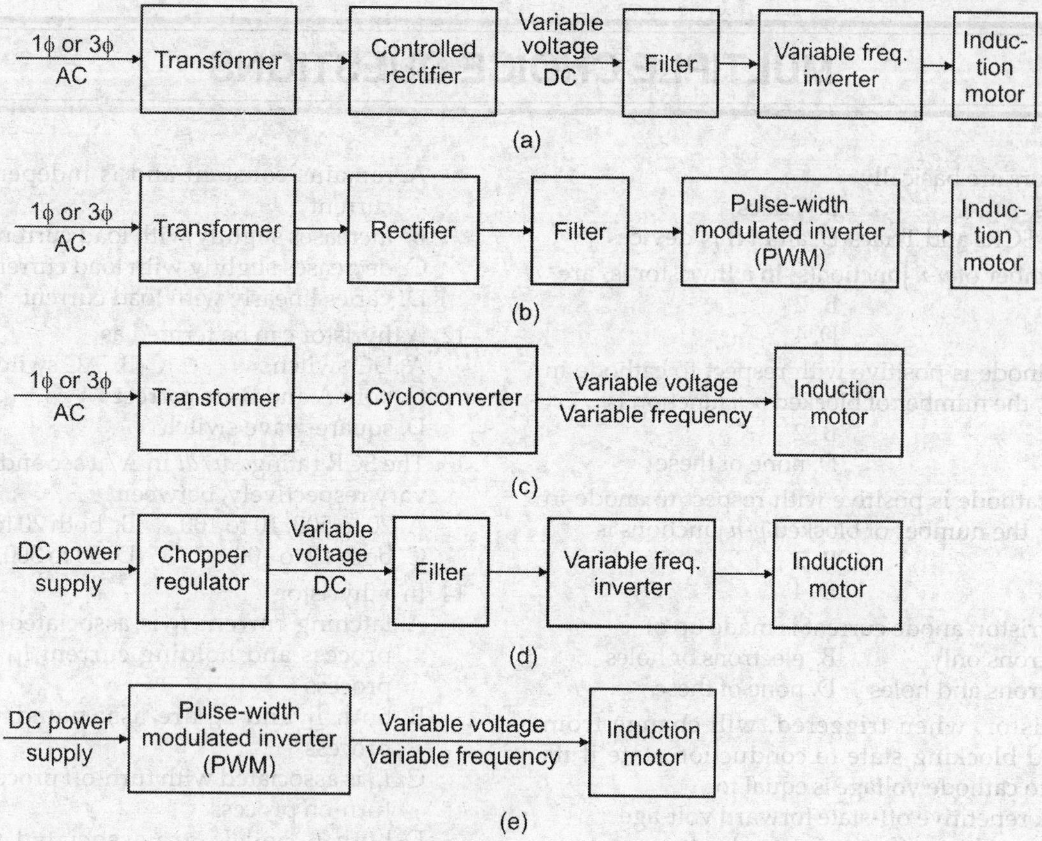

Fig. 13.19: Different schemes for speed control

When the thyristor of Fig. 13.20 is fired, any charge in the capacitor will be discharged into thyristor, giving an excessively high di/dt, but this can be suitably limited by inclusion of the resistor. A diode D may be included to by-pass R for improved dv/dt protection. Typically the values of the protection components will be $C = 0.01$ to $1\ \mu F$, $R = 10$ to $1\ k\Omega$ and $L = 50$ to $100\ \mu H$. The exact values depend on the circuit voltage and stored capacity of the transient source.

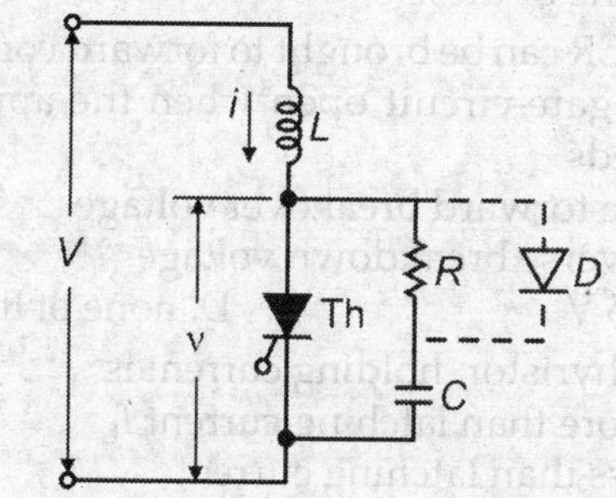

Fig. 13.20: Snubber circuit

MULTIPLE CHOICE QUESTIONS

1. Thyristors are basically
   A. SCRs　　B. Triacs
   C. both SCRs and Triacs　　D. all PNPN devices
2. The number of *p-n* junction(s) in a thyristor is/are
   A. 1　　B. 2
   C. 3　　D. 4
3. When anode is positive with respect to cathode in an SCR, the number of blocked *p-n* junction is
   A. 1　　B. 2
   C. 4　　D. none of these
4. When cathode is positive with respect to anode in an SCR, the number of blocked *p-n* junctions is
   A. 1　　B. 2
   C. 3　　D. 4
5. In a thyristor, anode current is made up of
   A. electrons only　　B. electrons or holes
   C. electrons and holes　　D. none of these
6. A thyristor, when triggered, will change from forward blocking state to conduction state if its anode to cathode voltage is equal to
   A. peak repetitive off-state forward volt-age
   B. peak working off-state forward voltage
   C. peak working off-state reverse voltage
   D. none of these
7. An SCR can be brought to forward conducting state with gate-circuit open when the applied voltage exceeds
   A. the forward breakover voltage
   B. reverse breakdown voltage
   C. 1.5 V　　D. none of these
8. In a thyristor, holding current is
   A. more than latching current I_L
   B. less than latching current
   C. equal to latching current
   D. very small
9. When a thyristor gets turned on, the gate drive
   A. should not be removed as it will turn-off the SCR
   B. may or may not be removed
   C. should be removed
   D. should be removed in order to avoid increased losses and higher junction temperature
10. For normal SCRs, turn-on time is
    A. less than turn-off time t_q
    B. more than turn off time t_q
    C. equal to turn off time t_q
    D. None of these
11. The forward voltage drop during SCR on-state is 1.5 V. This voltage drop
    A. remains constant and is independent of load current
    B. increases slightly with load current
    C. decreases slightly with load current
    D. varies linearly with load current
12. A thyristor can be termed as
    A. DC switch　　B. AC switch
    C. both A and B are correct
    D. square-wave switch
13. The SCR ratings *di*/*dt* in A/μ sec and V/μ sec, may vary respectively, between
    A. 20 to 500, 10 to 100　　B. both 20 to 500
    C. both 10 to 100　　D. 50 to 300, 20 to 500
14. In a thyristor
    A. latching current I_L is associated with turn-off process and holding current I_H with turn-on process
    B. both I_L and I_H are associated with turn-on process
    C. I_H is associated with turn-off process and I_L with turn-on process
    D. both I_L and I_H are associated with turn-on process
15. Once SCR starts conducting a forward current, its gate losses control over
    A. anode circuit voltage only
    B. anode circuit current only
    C. anode circuit voltage and current
    D. none of these
16. On-state voltage drop across a thyristor used in a 250 V supply system is of the order of
    A. 100–110 V　　B. 240–250 V
    C. 1–1.5 V　　D. none of these
17. In a thyristor, ratio of latching current to holding current is
    A. 0.4　　B. 1.0
    C. 2.5　　D. none of these
18. The duration of pulse in a pulse triggering system for thyristors should be at least
    A. 60 μsec　　B. 40 μsec
    C. 20 μsec　　D. 10 μsec
19. For thyristors, pulse triggering is preferred to DC triggering because
    A. gate dissipation is low
    B. pulse system is simpler
    C. triggering signal is required for a very short duration
    D. all of these

Ans. 1. D　2. C　3. A　4. B　5. C　6. B　7. A　8. C　9. D　10. A　11. B　12. A　13. B　14. C　15. C　16. C　17. C　18. D　19. D

20. Which of the following pnpn devices has a terminal for synchronising purpose?
    A. SCS B. Triac
    C. Diac D. SUS
21. In application, a programmable conjunction transistor (PUT) resembles more to
    A. special purpose UJT B. anode-fired SCR
    C. triac D. diac
22. In a normal three-phase SCR rectifier, an SCR cannot be fired during
    A. first 10° of its anode voltage
    B. first 30° of its anode voltage
    C. first 45° of its anode voltage
    D. first 90° of its anode voltage
23. Gate characteristic of a thyristor
    A. is a straight line passing through origin
    B. is of type $V_v = a + bI_v$
    C. is a curve between V_g and I_g
    D. has a spread between two curves of $V_g - I_g$
24. An SCR is used to control the speed of a DC motor. At full speed, the motor is taking 1A at 75 V. The maximum forward surge current rating and maximum forward breakover voltage rating respectively are of the order of
    A. 3 A, 235 V B. 1 A, 294 V
    C. 6 A, 166 V D. 5 A, 210 V
25. In an SCR, anode current flows over a narrow region near the gate during
    A. delay time t_d
    B. rise time t_r and spread time t_p
    C. t_d and t_p D. t_d and t_r
26. Turn-on time for an SCR is 10 μsec. If an inductance is inserted in the anode circuit, then the turn-on time will be
    A. 10 μsec B. less than 10 μsec
    C. more than 10 μsec D. about 10 μsec
27. Turn-off time of an SCR is measured from the instant
    A. anode current becomes zero
    B. anode voltage becomes zero
    C. anode voltage and anode current become zero at the same time
    D. gate current becomes zero
28. A forward voltage can be applied to an SCR after its
    A. anode current reduces to zero
    B. gate recovery time
    C. reverse recovery time
    D. anode voltage reduces to zero
29. For an SCR with turn-on time of 5 micro-second, and ideal trigger pulse should have
    A. short rise time with pulse width = 3 μsec
    B. long rise time with pulse width = 6 μsec
    C. short rise time with pulse width = 6 μsec
    D. long rise time with pulse width = 3 μsec
30. Turn-on time of an SCR in series with R-L circuit can be reduced by
    A. increasing circuit resistance *R*
    B. decreasing *R*
    C. increasing circuit inductance *L*
    D. decreasing *L*
31. Choose the correct answer from the following. High *di/dt* rating of an SCR is required for
    A. rectifier B. inverter
    C. cycloconverter D. none of these
32. Turn-on time of an SCR can be reduced by using a
    A. rectangular pulse of high amplitude and narrow width
    B. rectangular pulse of low amplitude and wide width
    C. triangular pulse D. trapezoidal pulse
33. Static *V-I* characteristics of SCR with different gate pulses applied to the gate are indicated by
    A. $I_{g2} > I_{g1} > I_{g0}$ B. $V_{g2} > V_{g1} > V_{g0}$
    C. $P_{g2} > P_{g1} > P_{g0}$ D. either A or B
34. In a thyristor, the magnitude of anode current will
    A. increase if gate current is increased
    B. decrease if gate current is decreased
    C. increase if gate current is decreased
    D. not change with any variation in gate current
35. The average on-state current for an SCR is 20 A for a conduction angle of 120°. Its average on-state current for 60° conduction angle would be
    A. 20 A B. 10 A
    C. less than 20 A D. none of these
36. The average on-state current for an SCR is 20 A for a resistive load. If an inductance of 5 mH is included in the load, then average on-state current would be
    A. more than 20 A B. less than 20 A
    C. 15 A D. none of these
37. Specification sheet for an SCR gives its maximum rms on-state current as 35 A. This rms rating for a conduction angle of 120° would be
    A. more than 35 A B. less than 35 A
    C. 35 A D. none of these
38. Forward gate current with reverse anode voltage in an SCR will cause
    A. large forward current
    B. large reverse current
    C. low forward current
    D. low reverse current
39. Turn-on gain is defined as the ratio of
    A. rated anode current to minimum gate current to trigger

Ans.													
	20. B	21. B	22. B	23. C	24. A	25. D	26. C	27. A	28. B	29. C	30. D	31. B	32. A
	33. A	34. D	35. B	36. A	37. C	38. B	39. A						

B. rated anode current to maximum gate current
C. rated anode current to latching current
D. rated anode current to holding current

40. Zigzag secondary of a rectifier transformer is required for
A. avoiding DC saturation of core
B. harmonic reduction
C. pf improvement
D. none of these

41. The *VI* characteristics of an ideal diode are

A.
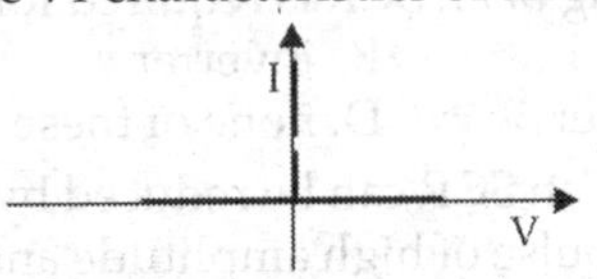

B.
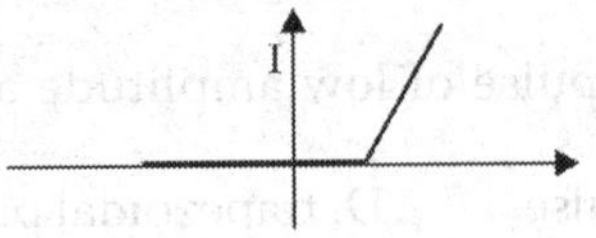

C.
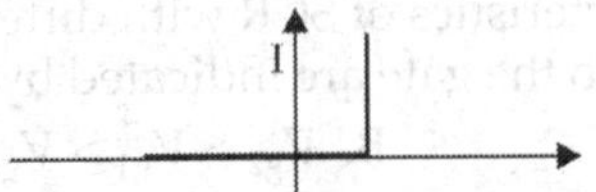

D.
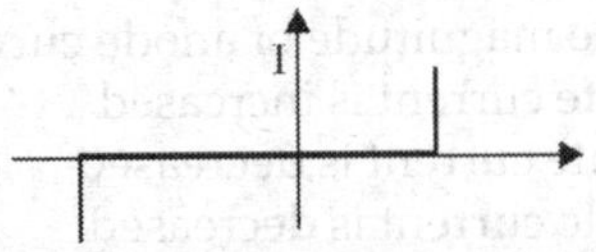

42. Blocking oscillator, triggering of SCR is used for
A. only electrical isolation
B. only lower gate power dissipation
C. successful triggering of inductive loads only
D. all of these

43. So far as generation of triggering pulses is concerned, the inverse cosine method is essentially
A. a closed-loop control
B. an open loop control
C. none of these
D. A and B are correct

44. It is recommended to use h·f pulse train for gate triggering of thyristors to provide
A. reduced *di*/*dt* problem
B. reduced size of pulse transformer
C. reduced *di*/*dt* problem
D. none of these

45. The switching losses for GTOs are much higher than those of normal thyristors. It is
A. false B. true
C. not always D. none of these

46. In gate triggering of thyristor, pulse transformer is usually employed to provide
A. electrical isolation B. hf triggering
C. power frequently triggering
D. none of these

47. A totem-pole arrangement is capable of
A. sourcing large current only
B. both sourcing and sinking large current
C. sinking large current only
D. none of these

48. An SCR is a
A. three-layer three-terminal device
B. three-layer four-terminal device
C. four-layer three-terminal device
D. four-layer four-terminal device

49. The device commonly used for triggering a Triac is
A. diode B. transistor
C. zener diode D. diac

50. How many SCRs are to be connected in series with 800 V rating to be used for a 3 kV circuit using darting factor of 15 percent?
A. 3 B. 4
C. 5 D. 6

51. The work function of metal is generally expressed as
A. joules B. volts
C. electrons D. electro volt

52. The process of conversion AC into DC is called
A. amplification B. modulation
C. rectification D. factorisation

53. Which of the following devices can be used as a programmable UJT?
A. SCR B. LASCR
C. SCS D. SUS

54. The Ge crystal behaves as an insulator at
A. 0° K B. 273° K
C. 5° K D. none of these

55. The addition of impurity atoms to a pure semiconductor makes it an
A. extrinsic semiconductor
B. intrinsic semiconductor
C. both A and B are correct
D. none of these

56. IC timer 555 used for generating signals in an inverter triggering circuit is
A. stable mode B. bistable mode
C. monostable mode D. none of these

57. The majority charge carriers in an N-type semiconductors are
A. holes B. electrons
C. neutrons D. none of these

58. The ripple frequency in case of a full-wave rectifier is
A. twice the supply frequency
B. equal to the supply frequency
C. thrice the supply frequency
D. none of these

59. The semiconductor diodes can be used for
A. half-wave rectifiers
B. full-wave rectifiers

Ans. 40. A 41. A 42. C 43. A 44. B 45. B 46. A 47. B 48. C 49. D 50. C 51. D 52. C
53. D 54. A 55. A 56. A 57. B 58. A

C. both A and B are correct
D. none of these

60. An example of solid state device is a
A. triode B. pentode
C. FET D. thyratron

61. A metal oxide varistor (MOV) is used for protecting
A. gate circuit against overcurrents
B. gate circuit against overvoltages
C. anode circuit against overcurrents
D. anode circuit against overvoltages

62. The functions of connecting a resistor in series with gate-cathode circuit and a zener-diode across gate-cathode circuit are respectively, to protect the gate circuit from
A. overvoltages, overcurrents
B. overcurrents, overvoltages
C. overcurrents, noise signals
D. none of these

63. In the circuit given below, the function of the transistor is to

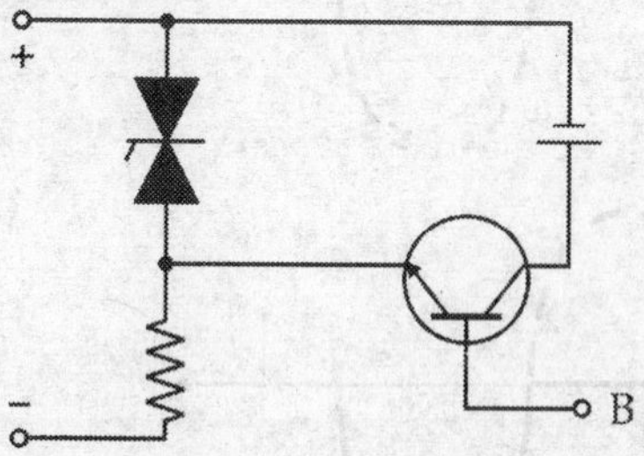

A. provide control signal to trigger SCR
B. make SCR ON
C. make SCR OFF D. amplify anode current

64. In a GTO, anode current begins to fall when gate current
A. is negative peak at time $t = 0$
B. is negative peak at t = storage period t_s
C. just begins to become negative at $t = 0$
D. none of these

65. For a pulse transformer, the material used for its core and the possible turn ratio from primary to secondary are, respectively
A. ferrite; 20:1 B. laminated iron; 1:1
C. ferrite; 1:1 D. powdered iron; 1:1

66. An SCR is a
A. three-layer three terminal device
B. three layer four terminal device
C. four layer three terminal device
D. none of these

67. SCR can be turned on by
A. applying anode voltage at a suffi-ciently fast rate
B. applying sufficiently large anode volt-age
C. increasing the temperature of SCR to a sufficiently large value
D. all of these

68. During forward blocking of two series connected SCRs, a thyristor with
A. higher leakage impedance shares lower voltage
B. high leakage impedance shares higher voltage
C. low leakage impedance shares higher voltage
D. none of these

69. Thyristor A has rated gate current of 2 A and thyristor B, a rated gate current of 100 mA. Then
A. thyristor A is a GTO and B is a conven-tional SCR
B. thyristor B is a GTO and A is a conven-tional SCR
C. thyristor B may operate as a transistor
D. none of these

70. The anode current through a conducting SCR is 10 A. If its gate current is made one-fourth the anode current will become
A. 0 A B. 5 A
C. 10 A D. 20 A

71. A purely AC voltmeter connected anode to cathode across an SCR in a 220 V AC circuit will read
A. 220 V B. about 20 V
C. 220 V or above 200 V D. none of these

72. When two identical SCRs are placed back to back in series with a load and if each is fired at 90°, a DC voltmeter across the load read will be
A. $\frac{2}{\pi} \times$ peak voltage B. zero
C. $\frac{1}{\pi} \times$ peak voltage D. none of these

73. If the gate current of an SCR is increased, the forward breakdown voltage V_{BR} will
A. increase B. decrease
C. not be affected D. infinite

74. An SCR does not conduct for a certain value of load resistance. In order to make it ON, it is necessary to
A. decrease the load resistance
B. increase the resistance
C. increase the gate pulse
D. none of these

75. Most SCRs can be turned off by voltage reversal during negative half cycle of the AC supply for
A. all frequencies
B. frequencies up to 300 Hz
C. frequencies up to 30 KHz
D. frequencies up to 300 KHz

76. In order to make a SCR OFF, it is neces-sary to make
A. anode voltage less than the breakover voltage
B. gate current zero
C. anode current less than the holding current
D. none of these

77. In the circuit shown below the SCR is fired at angle of 60°. The waveform across SCR will be

Ans.	59. C	60. B	61. D	62. B	63. C	64. B	65. C	66. C	67. D	68. B	69. A	70. C	71. D
	72. B	73. B	74. A	75. C	76. C								

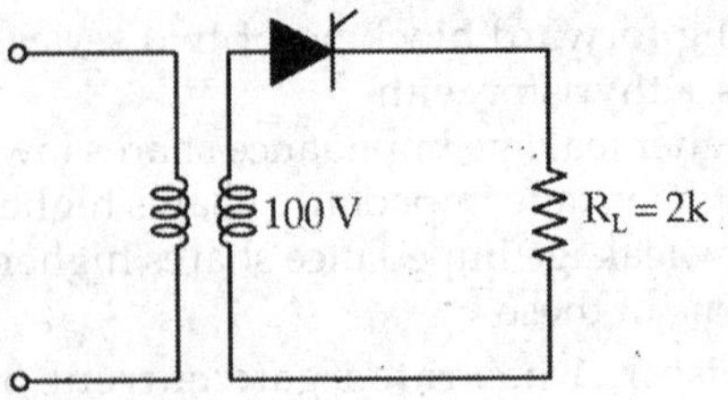

A.
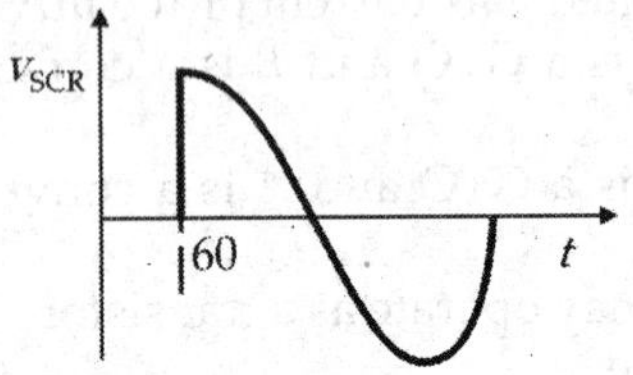

B.
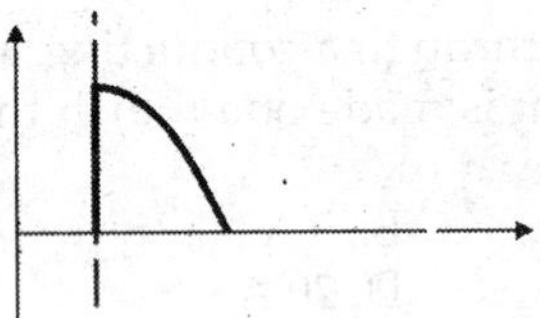

C.
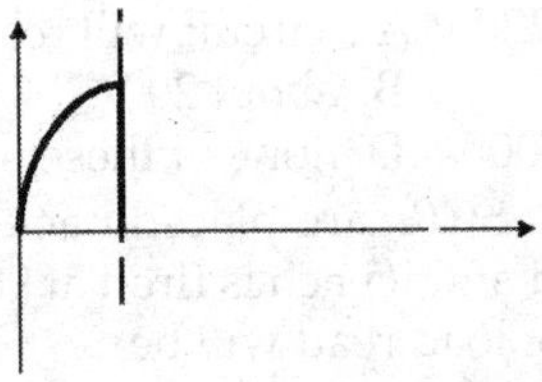

D.
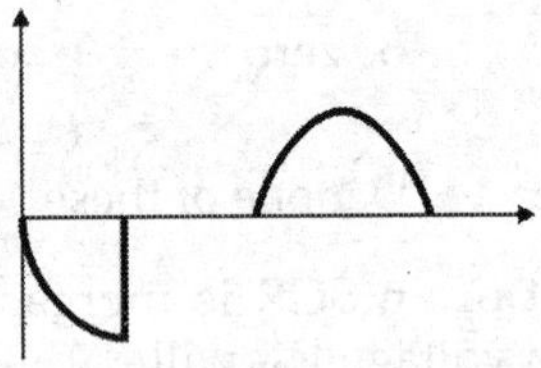

78. In a single-phase one-pulse circuit with *RL* load and a freewheeling diode, extinc-tion angle β is less than π. For a firing angle α, the SCR and freewheeling diode would, respectively, conduct for
 A. (β – α), 0°
 B. (π – α), (π – β)
 C. (α, (β – α))
 D. (β – α), α
79. The three-phase AC to DC converter which required neutrals point connection is
 A. 3-phase semiconverter
 B. 3-phase full converter
 C. 3-phase half-wave converter
 D. 3-phase full-wave converter with diodes
80. Triac is equivalent to
 A. two SCRs connected in parallel
 B. two SCRs connected in antiparallel
 C. one SCR, one diode connected in parallel
 D. none of these
81. In the circuit shown below the SCR is fired at 180°, the waveform across the load will be

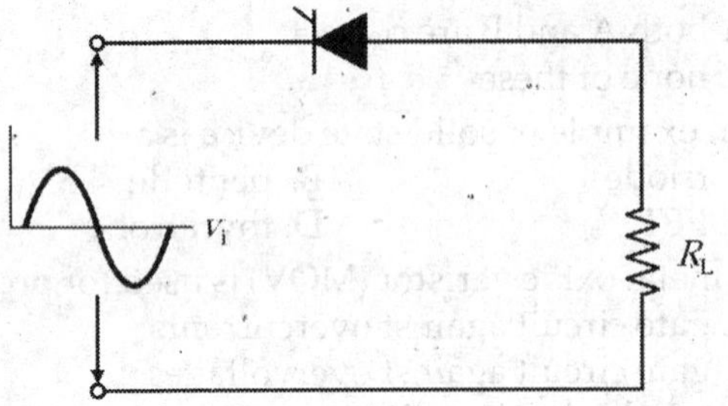

A.
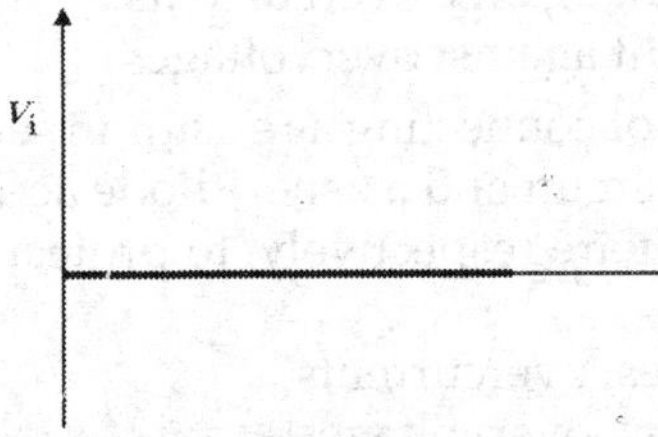

B.
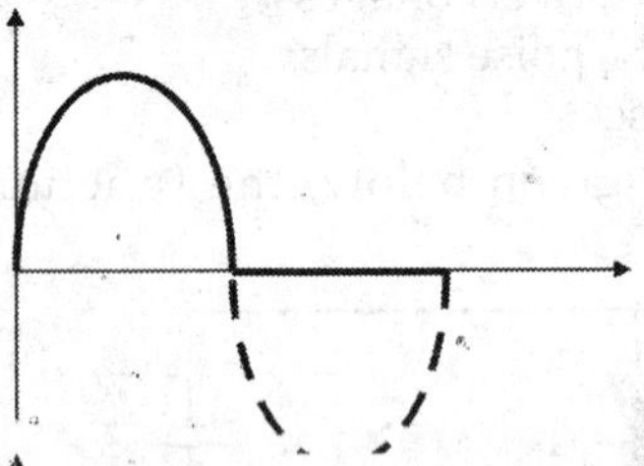

C.
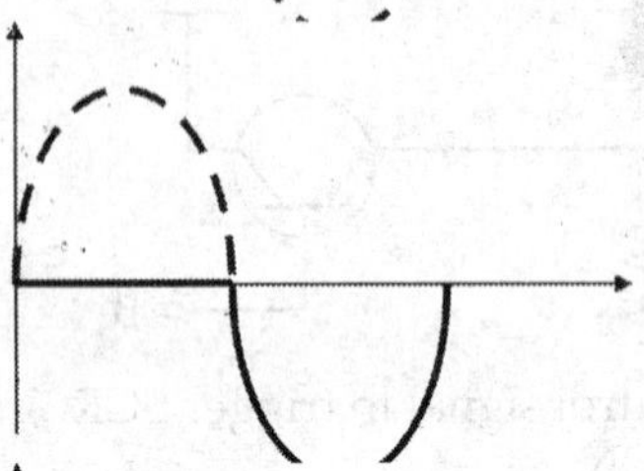

D.
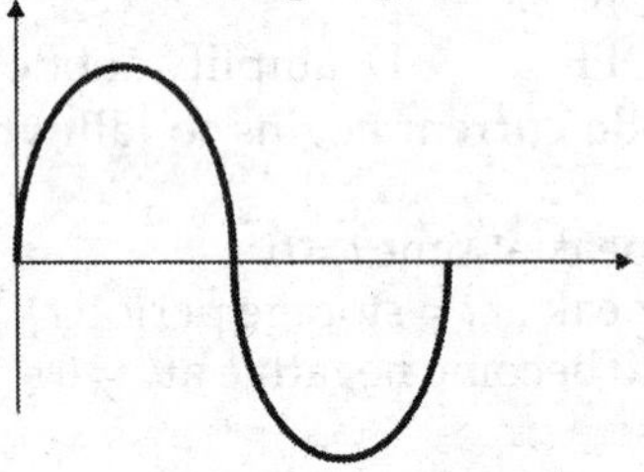

82. The function of diode D_2 is
 A. to dissipate *L di/dt* of the armature
 B. to rectify the armature current
 C. to stop armature current to flow in the capacitor branch and thus charge it
 D. to provide field current
83. If the voltage across a thyrite is increased, the current
 A. decrease to a low value
 B. increases in proportion to voltage increases
 C. increases to a very high value
 D. is not affected
84. Which of the following devices is a three layer device?
 A. SCS B. SUS
 C. TRIAC D. DIAC

Ans. 77. D 78. A 79. C 80. B 81. C 82. A 83. C 84. D

85. Which of the following methods will turn SCS OFF?
A. Applying negative pulse to the anode
B. Applying a positive pulse to the anode gate
C. Applying negative pulse to the cathode gate
D. All of these

86. Which of the following PNPN devices does not have a gate terminal?
A. Triac B. SCS
C. SUS D. Complementary SCR

87. The average current rating as supplied by thyristor manufacturer is based on
A. resistive current B. inductive current
C. capacitive current D. none of these

88. Recovered charge of a Schottky diode is
A. less than that of a junction diode
B. more than that of a junction diode
C. equal to that of a junction diode
D. none of these

89. In circuit given below, in order to make a conducting SCR OFF, it is necessary to

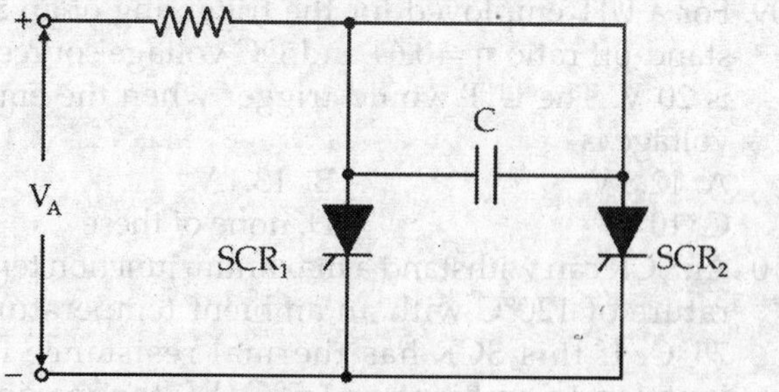

A. make other SCR OFF
B. make other SCR ON
C. reverse the polarity of the applied voltage
D. remove the gate current of conducting SCR

90. In case of thyristors in series
A. only the forward voltages have to be shared
B. only the reverse voltages have to be shared
C. both forward and reverse voltages have to be shared
D. none of these

91. Which of the following statements is true for firing a triac shown in the figure below?

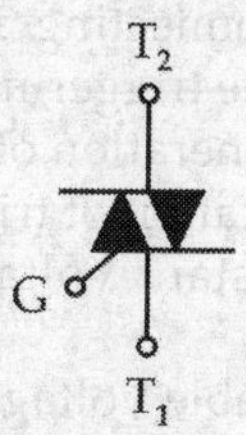

A. **Either** T_1 and T_2 is negative and there is current pulse into the gate
B. T_1 is negative and there is current pulse out of the gate
C. T_2 is negative and there is a current pulse out of the gate
D. Either T_1 or T_2 is positive and there is a current pulse out of the gate

92. Which of the following devices does not have negative resistance characteristics?
A. UJT B. tunnel diode
C. SCR D. FET

93. Which of the following PNPN devices has two gates?
A. Triac B. SCS
C. SUS D. Diac

94. In the triac circuit given below, if the triac fires at 30°, the waveform across the load R_L will be

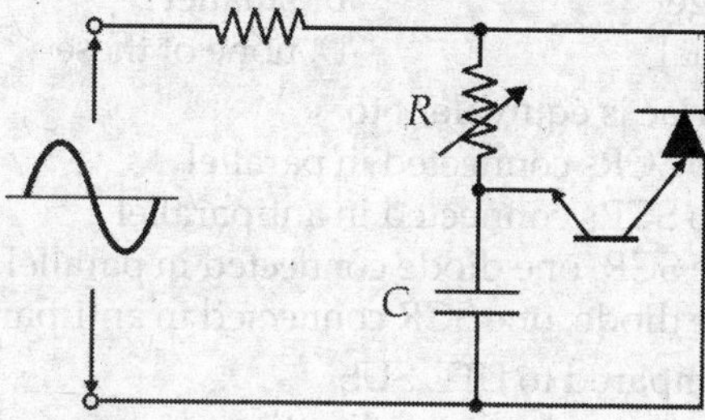

A.
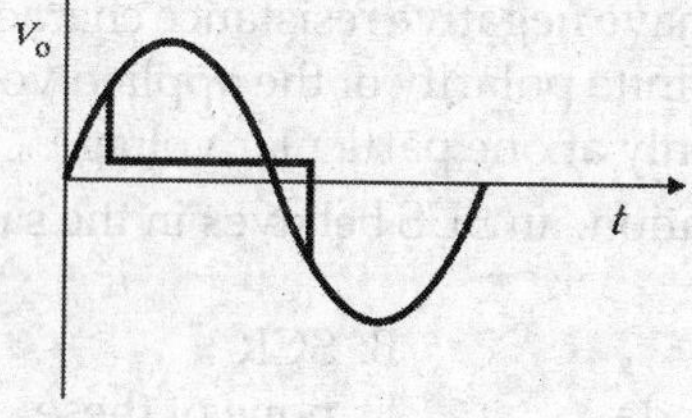

B.
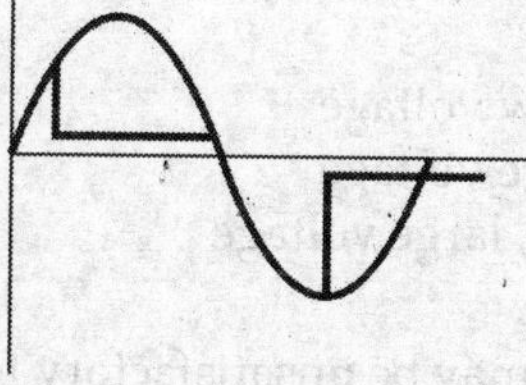

C.
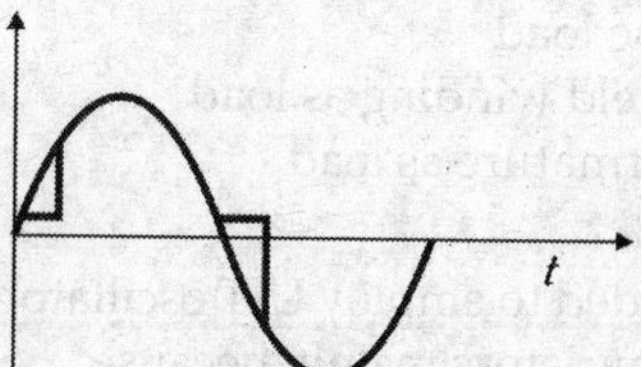

D.
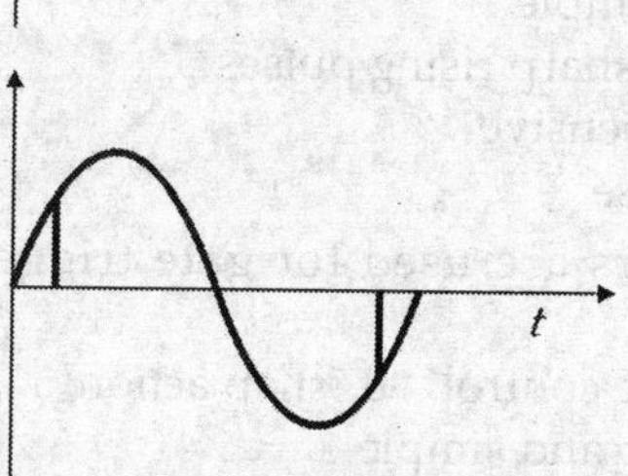

Ans.	85. D	86. C	87. B	88. A	89. B	90. C	91. C	92. D	93. B	94. C

95. Which of the following statements is necessary for a triggering system for thyristors?
A. It must use separate power supply
B. It should provide a train of pulses
C. It should be synchronised with the mains supply
D. None of these

96. Triac are usually operated at
A. high frequency only
B. all frequencies
C. power frequency
D. none of these

97. Compared to thyristor, the size of the heat sink needed by Triac will be
A. larger B. smaller
C. equal D. none of these

98. The Triac is equivalent to
A. two SCRs connected in parallel
B. two SCRs connected in antiparallel
C. one SCR, one diode connected in parallel
D. one diode, one SCR connected in anti-parallel

99. As compared to UJT, SUS
A. triggers only in one direction
B. does not have negative resistance characteristics
C. needs definite polarity of the applied voltage
D. triggers only at one particular voltage

100. In its application, an SUS behaves in the same way as
A. UJT B. SCR
C. tunnel diode D. none of these

101. During forward blocking state, a thyristor is associated with
A. large current, low voltage
B. low current, large voltage
C. medium current, large voltage
D. none of these

102. UJT trigger of SCR may be unsatisfactory for
A. welding type load
B. DC motor field winding as load
C. DC motor armature as load
D. none of these

103. It is recommended to employ UJT oscillator for gate triggering of thyristors mainly because
A. it is fairly simple
B. it provides sharp rising pulses
C. it is less expensive
D. none of these

104. UJT oscillators are used for gate triggering of thyristor for
A. better phase control B. snap action
C. being cheap and simple
D. none of these

105. In a UJT, with V_{BB} as the voltage across two base terminals, the emitter potential at peak point is given by
A. ηV_{BB} B. ηV_D
C. $\eta V_{BB} + V_D$ D. $\eta V_D + V_{BB}$

106. A UJT exhibits negative resistance region
A. before the peak point
B. between peak and valley point
C. after the valley point
D. Both A and B are correct

107. In a UJT, maximum value of charging resistance is associated with
A. peak point B. valley point
C. any point between peak and valley points
D. after the valley point

108. When a UJT is used for triggering of an SCR, the waveshape of the voltage obtained from UJT circuit is a
A. sine wave B. saw-tooth wave
C. trapezoidal wave D. square wave

109. For a UJT employed for the triggering of an SCR, stand-off ratio $\eta = 0.64$ and DC voltage source V_{BB} is 20 V. The UJT would trigger when the emitter voltage is
A. 12.8 V B. 13.1 V
C. 10 V D. none of these

110. An SCR can withstand a maximum junction temperature of 120°C with an ambient temperature of 75°C. If this SCR has thermal resistance from junction to ambient as 1.5°C/W, the maximum internal power dissipation allowed is
A. 30 W B. 60 W
C. 80 W D. 100 W

111. In synchronized UJT triggering of an SCR, voltage V across capacitor reaches UJT threshold voltage thrice in each half cycle so that there are three firing pulses during each half cycle. The firing angle of the SCR can be controlled
A. once in each half cycle
B. thrice in each half cycle
C. twice in each half cycle
D. four times in each half cycle

112. The function of connecting a zener diode in a UJT circuit, used for the triggering of SCRs, is to
A. expedite the generation of triggering pulses
B. delay the generation of triggering pulses
C. provide a constant voltage to UJT to prevent erratic firing
D. provide a variable voltage to UJT as the source voltage changes

113. A flywheel diode across the inductive load
A. helps in the commutation process
B. decreases the range of triggering angle
C. increase voltage ripple factor appreciably

Ans. 95. C 96. C 97. A 98. B 99. D 100. A 101. B 102. B 103. B 104. B 105. C 106. B 107. A 108. B 109. B 110. A 111. A 112. C 113. B

114. The major advantage in the use of a fly-wheel diode in a 3-ph fully controlled bridge rectifier over corresponding half controlled circuit is that
A. the ripple voltage is lower
B. the ripple frequency is lower
C. none of these

115. Compared to an uncontrolled rectifier, the harmonic voltages in phase controlled rectifier feeding resistive load will be
A. less B. more
C. equal D. none of these

116. Freewheeling diodes are mainly used for
A. better utilisation factor
B. *di/dt* protection
C. quick turn on D. none of these

117. Holding current of a thyristor is always
A. more than its latching current
B. less than its latching current
C. equal to its latching current
D. none of these

118. RF dielectric heating is used for
A. highly conducting materials
B. highly insulating materials
C. magnetic materials
D. lossy insulating material

119. Some delay is provided between turn-off and turn-on of thyristors in the same leg of a bridge inverter to
A. prevent short circuit through fault
B. ensure firing of SCRs
C. protect snubber capacitor
D. prevent voltage spikes

120. In an impulse commutation arrangement the current pulse exceeds the load current for an interval T. SCR turn-off time is t_q. For same commutation
A. T should be less than t_q
B. T should be equal to t_q
C. T should be greater than t_q
D. no relation exists between T and t_q

121. During turn-off process of a thyristor, the current does not cause to flow when reaching zero but goes in the negative direction owing to flow of
A. commutation failure B. hole storage effect
C. high reverse recovery loss
D. none of these

122. Natural commutation is considered to a kind of
A. forced commutation B. line commutation
C. commutation failure D. none of these

123. In Forced commutation, electromagnetic noise will be least for
A. thyristor B. GTO
C. triac D. none of these

124. In comparison with the turn on time the turn-off time of a thyristor is
A. lesser B. equal
C. larger D. none of these

125. Commutation overlap in converter will cause
A. improved power factor
B. reduced output voltage
C. higher PIV D. none of these

126. Operating frequency of SMPS is made high for
A. reducing switching losses
B. smaller size filter components
C. high output voltage
D. high output current

127. A 1-phase fully controlled converter is charging a battery from an existing AC mains. It is possible to feedback power to AC supply when firing angle is between
A. 0° and 90° with same battery connection
B. 90° and 180° with same battery connection
C. 0° to 90° with reverse battery connection
D. 90° and 180° with reverse battery connection

128. An ignitron operates in the following mode
A. arc discharge B. townsend discharge
C. glow discharge D. none of these

129. A single-phase AC voltage controller (or regulator) fed from 50 Hz system supplies a load having resistance and inductance of 2.0 ohm and 6.36 mH respectively. The control range of firing angle for this regulator is
A. $0° < \alpha < 180°$ B. $45° < \alpha < 180°$
C. $90° < \alpha < 180°$ D. $0° < \alpha < 45°$

130. For the same AC voltage and load impedance which of the following statements about rectifiers are correct?
A. The average load current in a full wave rectifier is π times that in a half-wave rectifier
B. Half-wave rectifier will have bigger sized transformer compared to full-wave rectifier
C. Half-wave rectifier will have a smaller sized transformer compared to a full-wave rectifier
D. None of these

131. Output voltage expression for controlled rectifier is
A. $\frac{3V_m}{2\pi}(1+\cos\alpha)$ B. $\frac{2V_m}{2\pi}\cos\alpha$
C. $\frac{V_m}{\pi}(1+\cos\alpha)$ D. $\frac{3\sqrt{3}}{2\pi}V_m\cos\alpha$

132. A single-phase voltage controller feeds an induction motor and a heater;
A. in both the loads, fundamental and harmonics are useful
B. in induction motor only fundamental and in heater only harmonics are useful

Ans. 114. A 115. A 116. A 117. B 118. D 119. A 120. C 121. B 122. B 123. A 124. C 125. B 126. A 127. D 128. A 129. B 130. C 131. C

C. in induction motor only fundamental and in heater, harmonics as well as fundamental are useful
D. in induction motor only harmonics and in heater only fundamental are useful

133. A single-phase voltage controller is employed for controlling the power flow from 260 V, 50 Hz source into a load consisting of $R = 5$ ohm and $\omega L = 12$ ohm. The value of maximum rms load current and the firing angle are respectively
A. 20 A, 0°
B. $\frac{260}{10.91}$ A, 0°
C. 20 A, 90°
D. $\frac{260}{10.91}$ A, 60°

134. A load, consisting of $R = 10$ ohm and $\omega L = 10$ ohm is being fed from 230 V, 50 Hz source through a 1-phase voltage controller. For a firing angle delay of 30°, the rms value of load current would be
A. 23 A
B. $\frac{23}{\sqrt{2}}$ A
C. more than $\frac{23}{\sqrt{2}}$ A
D. less than $\frac{23}{\sqrt{2}}$

135. In a single-phase voltage controller with RL load, AC output power can be controlled if
A. firing angle $\alpha > \phi$ (load phase angle) and conduction angle $\gamma = \pi$
B. $\alpha > \phi$ and $\gamma < \pi$
C. $\alpha > \phi$ and $\gamma = \pi$
D. none of these

136. A single-phase voltage controller feeds power to a resistance of 10 ohm. The source voltage is 200 V rms. For a firing angle of 90°, the rms value of thyristor current is
A. 20 A
B. 15 A
C. 10 A
D. 5 A

137. A single-phase voltage controller is connected to a load of resistance 10 ohm and a supply of 200 sin 314t volts. For a firing angle of 90°, the average thyristor current is
A. 10 A
B. $10/\pi$ A
C. $5\sqrt{2}/\pi$ A
D. $5\sqrt{2}$ A

138. A single-phase voltage controller, using two SCRs back to back is found to be operating as a controller rectifier. This is because
A. load is R and pulse gating is used
B. load is R and high-frequency carrier gating is used
C. load is RL and pulse gating is used
D. load is RL and continuous gating is used

139. If the speed of the motor decreases, the SCR will fire
A. earlier
B. later
C. at the same angle
D. according to the anode voltage of SUS

140. A motor armature supplied through phase-controlled SCRs receives a smoo-ther voltage shape at
A. high motor speeds
B. low motor speeds
C. rated motor speeds
D. none of these

141. Free wheeling diode at the output of a rectifier allows
A. SCR to trigger properly
B. turn off SCR at the end of half cycle
C. protection of SCR against over voltage
D. none of these

142. The value of DC load voltage for a single phase semi-converter under continuous current mode is
A. $\frac{V_m}{\pi}(1+\cos\alpha)$
B. $\frac{V_m}{\pi}(1-\cos\alpha)$
C. $\frac{V_m}{2}\pi(1+\cos\alpha)$
D. $\frac{V_m}{2}\pi(1-\cos\alpha)$

143. In the above circuit, the average output voltage is
A. 18.5 V
B. 33.4 V
C. 50 V
D. 62.5 V

144. A single-phase one-pulse controlled circuit has resistance and counter emf load and 400 sin 314t as the source voltage. For a load counter emf of 200 V, the range of firing angle control is
A. 30 to 150°
B. 60 to 180°
C. 60 to 120°
D. none of these

145. In a single-phase half-wave circuit with RL load, and a freewheeling diode across the load, extinction angle β is more than π. For a firing angle α, the SCR and free-wheeling diode would conduct, respec-tively, for
A. $(\pi-\alpha)$, β
B. $(\beta-\alpha)$ $(\pi-\alpha)$
C. $(\pi-\alpha)$, $(\beta-\pi)$
D. $(\pi-\alpha)$, $(\pi-\beta)$

146. A single-phase two-pulse bridge converter has an average output voltage and power output of 500 V and 10 kW respectively. The SCRs used in the two-pulse bridge converter are now re-employed to form a single-phase two-pulse mid-point converter. This new controlled converter wuld give, respectively, an average out-put voltage and power output of
A. 500 V, 10 kW
B. 250 V, 5 kW
C. 250 V, 10 kW
D. 500 V, 5 kW

147. In a single-phase full converter bridge, the average output voltage is given by
A. $\frac{1}{\pi}\int_{\alpha}^{\pi+\alpha} V_m \cos\theta\, d\theta$
B. $\frac{1}{\pi}\int_{0}^{\alpha+\pi} V_m \cos\theta \cdot d\theta$
C. $\frac{1}{\pi}\int_{\alpha-(\pi/2)}^{\alpha+(\pi/2)} V_m \cos\theta \cdot d\theta$
D. $\frac{1}{\pi}\int_{(\pi/2)-\alpha}^{(\pi/2)+\alpha} V_m \cos\theta \cdot d\theta$

148. For continuous conduction, in a single-phase full converter, each pair of SCRs conduct for

Ans. 132. C 133. A 134. B 135. B 136. C 137. B 138. C 139. A 140. A 141. ? 142. A 143. B 144. A 145. C 146. B 147. C

A. $\pi - \alpha$ B. π
C. α D. $\pi + \alpha$

149. For discontinuous load current and extinction and $\beta > (\pi/2)$ in a single-phase full converter each SCR conducts for
A. α B. $\beta - \alpha$
C. β D. $\alpha + \beta$

150. In a single-phase semiconverter with resistive load and for a firing angle α, each SCR and freewheeling diode conduct, respectively, for
A. α, 0° B. $\pi - \alpha$, α
C. $\pi + \alpha$, α D. $\pi - \alpha$, 0°

151. In a single-phase full converter with resistive load and for a firing angle α, the load current is zero and non-zero, respectively, for
A. α, $\pi - \alpha$ B. $\pi - \alpha$, α
C. α, $\pi + \alpha$ D. α, π

152. In a single-phase full converter, if α and β are firing and extinction angles respectively, then the load current is
A. discontinuous if $(\beta - \alpha) < \pi$
B. discontinuous if $(\beta - \alpha) < \pi$
C. discontinuous if $(\beta - \alpha) = \pi$
D. continuous if $(\beta - \alpha) < \pi$

153. With discontinuous conduction and extinction angle $\beta < \pi$, in a single-phase semiconverter, freewheeling diode conducts for
A. α B. $\pi - \beta$
C. $\beta - \pi$ D. zero degree

154. With discontinuous conduction and extinction angle $\beta > \pi$, freewheeling diode conducts for
A. α B. $\beta - \pi$
C. $\pi + \alpha$ D. β

155. For continuous conduction, freewheeling diode in a single-phase semiconverter, conducts for
A. α B. $\pi - \alpha$
C. π D. $\pi + \alpha$

156. For discontinuous conduction and extinction angle $\beta < \pi$, in a single-phase semiconductor, each SCR conducts for
A. $\pi - \alpha$ B. $\beta - \pi$
C. α D. β

157. For discontinuous conduction and extinction angle $\beta > \pi$, in a single-phase semiconverter, each SCR conducts for
A. $\pi - \alpha$ B. $\beta - \pi$
C. α D. β

158. For continuous conduction, in a single-phase semiconverter, each SCR conducts for
A. α B. π
C. $\alpha + \pi$ D. $\pi - \alpha$

159. In controlled rectifiers, the nature of load current
A. does not depend on type of load and firing angle delay
B. depends both on the type of load and firing angle delay
C. depends only on the type of load
D. depends only on the firing angle delay

160. In a single-phase full converter, if output voltage has peak and average values of 325 V and 133 V respectively, then the firing angle is
A. 40° B. 140°
C. 50° D. 130°

161. In a single-phase semiconverter, if output voltage has peak and average values of 325 and 133 V respectively, the firing angle is
A. 40° B. 140°
C. 73.40° D. 80°

162. Each diode of a 3-phase half-wave diode rectifier conducts for
A. 60° B. 120°
C. 180° D. 90°

163. Each diode of a 3-phase, 6-pulse bridge diode rectifier conducts for
A. 60° B. 120°
C. 180° D. 90°

164. In a 3-phase half-wave diode rectifier, if per phase input voltage is 200 V, then the average output voltage is
A. 233.91 V B. 116.95 V
C. 202.56 V D. 101.28 V

165. In a 3-phase, half-wave diode rectifier, the ratio of average output voltage to per phase minimum ac voltage is
A. 0.955 B. 0.827
C. 1.654 D. 1.169

166. In a 3-phase, six-pulse diode rectifier, the average output voltage in terms of maximum value of line voltage V_m is
A. $\frac{3\sqrt{2}}{\pi} V_m$ B. $\frac{3 V_m}{\pi}$
C. $\frac{3\sqrt{3}}{2\pi} V_m$ D. $\frac{3\sqrt{3}}{\pi} V_m$

167. In a 3-phase half-wave rectifier, DC output voltage is 230 V. The peak inverse voltage across each diode is
A. 481.7 V B. 460 V
C. 345 V D. none of these

168. In a 3-phase full-wave diode rectifiers, the peak inverse voltage in terms of average output voltage is
A. 1.571 times B. 0.955 times
C. 1.047 times D. none of these

169. In a 3-phase half-wave diode rectifier, if V_m is the maximum value per phase voltage, then each diode is subjected to a peak inverse voltage of
A. V_m B. $\sqrt{3}\, V_m$
C. $2\, V_m$ D. $3\, V_m$

Ans. 148. B 149. B 150. D 151. A 152. A 153. D 154. B 155. A 156. B 157. A 158. D 159. B 160. C 161. C 162. B 163. B 164. A 165. B 166. B 167. A 168. C 169. B

170. In a 3-phase full-wave diode rectifier, if V_m is the maximum value of line voltage, then each diode is subjected to a peak inverse voltage of
A. V_m B. $\sqrt{3}\,V_m$
C. $2\,V_m$ D. $3\,V_m$

171. In a 3ϕ phase full-wave diode rectifier, if V is the per phase input voltage, then average output voltage is given by
A. 0.955 V B. 1.35 V
C. 2.34 V D. 3 V

172. A converter which can operate in both 3-pulse and 6-pulse modes is a
A. 1-phase full converter
B. 3-phase half-wave converter
C. 3-phase semiconverter
D. 3-phase full converter

173. In a 3-phase semiconverter, for firing angle less than or equal to 60°, each thyristor and diode conduct, respectively, for
A. 60°, 60° B. 90°, 30°
C. 120°, 120° D. none of these

174. In a 3-phase semiconverter, for firing angle less than or equal to 60° freewheel-ing diode conducts for
A. 30° B. 60°
C. 90° D. 0°

175. In a 3-phase semiconverter, for firing angle equal to 90°, and for continuous conduction, each SCR and diode conduct, respectively, for
A. 30°, 60° B. 60°, 30°
C. 60°, 60° D. 30°, 30°

176. In a 3-phase semiconverter, for a firing angle equal to 90° and for continuous conduction, freewheeling diode conducts for
A. 30° B. 60°
C. 90° D. 0°

177. In a 3-phase semiconverter, for firing angle equal to 120° and extinction angle equal to 110°, each SCR and diode conduct, respectively, for
A. 30°, 60° B. 60°, 60°
C. 90°, 30° D. 110°, 30°

178. In a 3-phase semiconverter, for firing angle equal to 120° and extinction angle equal to 110°, freewheel-ing diode conducts for
A. 10° B. 30°
C. 50° D. 110°

179. In a 3-phase semiconverter, for firing angle equal to 120° and extinction angle equal to 100°, none of the bridge elements conduct for
A. 10° B. 20°
C. 30° D. 60°

180. In a 3-phase semiconverter, the three SCRs are triggered at an interval of
A. 60° B. 90°
C. 120° D. 180°

181. In a 3-phase full converter, the six SCRs are fired at an interval of
A. 30° B. 60°
C. 90° D. 120°

182. For a single-phase, phase-controlled rectifier, with a freewheeling diode across the load
A. the instantaneous output voltage v_0 is always positive
B. v_0 may be positive or zero
C. v_0 may be positive, zero or negative
D. none of these

183. The frequency of the ripple in the output voltage of a 3-phase semiconverter depends upon
A. firing angle and load resistance
B. firing angle and load inductance
C. the load circuit parameters
D. firing angle and the supply frequency

184. In a single-phase full converter, if load current is I and ripple free, then average thyristor current is
A. $\frac{1}{2}I$ B. $\frac{1}{3}I$
C. $\frac{1}{4}I$ D. I

185. In a 3-phase full converter, if load current is I and ripple free, then average thyristor current is
A. $\frac{1}{2}I$ B. $\frac{1}{3}I$
C. $\frac{1}{4}I$ D. I

186. In a single-phase voltage controller with RL load, α is the firing angle, ϕ is the load phase angle and β is the extinction angle. For this voltage controller, output power can be controlled if $\alpha > \phi$ and
A. $(\beta - \alpha) = \pi$ B. $(\beta - \alpha) < \pi$
C. $\beta < \pi$ D. none of these

187. The effect of source inductance on the performance of single-phase and three-phase full converters is to
A. reduce the ripples in the load current
B. make discontinuous current as conti-nuous
C. reduce the output voltage
D. increase the load voltage

188. In a single-phase full converter, the number of SCRs conducting during overlap is
A. 1 B. 2
C. 3 D. 4

189. In a single-phase full converter, the output voltage during overlap is equal to
A. zero
B. source voltage
C. source voltage minus the inductance drop
D. none of these

Ans. 170. A 171. C 172. C 173. C 174. D 175. C 176. C 177. B 178. C 179. B 180. C 181. B 182. B
183. D 184. A 185. B 186. B 187. C 188. D 189. A

190. In a 3-phase full converter, the output voltage during overlap is equal to
A. zero B. source voltage
C. source voltage minus the inductance drop
D. average value of the conducting phase voltages

191. The total number of SCRs conducting simultaneously in a 3-phase full conver-ter with overlap considered has the sequence of
A. 3, 3, 2, 2 B. 3, 3, 3, 2
C. 3, 2, 3, 3 D. 2, 2, 2, 3

192. A 3-phase full converter has an average output voltage of 200 V for zero degree firing angle and for resistive load. For a firing angle of 90°, the output voltage would be
A. zero B. 50 V
C. 100 V D. 26.8 V

193. A four quadrant operation requires
A. two full converters connected in series
B. two full converters connected back to back
C. two full converters connected in parallel
D. two semiconverters connected back to back

194. In a 3-phase full converter, the output voltage pulsates at a frequency equal to supply frequency
A. f B. $2f$
C. $3f$ D. $6f$

195. In circulating-current type of dual con-verter, the nature of voltage across reactor is
A. alternating B. pulsating
C. direct D. triangular

196. The peak inverse voltage in AC to DC converter systems is highest in
A. single-phase full-wave mid-point converter
B. single-phase full converter
C. 3-phase bridge converter
D. 3-phase half-wave converter

197. In a 3-phase semiconverter, frequency of the ripple in the output voltage may be
A. 3 times the supply frequency f for firing angle α less than 60°
B. $3f$ for α greater than 60°
C. $6f$ for α less than 60°
D. $6f$ for α greater than 60°

198. In a dual converter, converters 1 and 2 work as under
A. 1 as rectifier, 2 as inverter
B. both as rectifiers
C. both as inverters D. none of these

199. A 3-phase semiconverter can work as
A. converter for $\alpha = 0$ to 180°
B. converter for $\alpha = 0$ to 90°
C. converter for $\alpha = 90$ to 180°
D. converter for $\alpha = 0$ to 90°

200. A single-phase half-wave controlled rectifier has 400 sin 314t as the input voltage and R as the load. For a firing angle of 60° for the SCR, the average out-put voltage is
A. $400/\pi$ B. $300/\pi$
C. $240/\pi$ D. none of these

201. Which of the following statements is true for a dynamotor?
A. It is one type of servomotor
B. It is one type of magnetic amplifier
C. It converts low voltage DC current to high voltage DC current
D. It is a one type of DC generator

202. Fast turn-off SCR should be used for
A. rectifier B. inverter
C. cycloconverter D. none of these

203. PWM technique in inverter is used for
A. harmonic reduction
B. higher output frequency
C. improving efficiency
D. reducing switching losses

204. In a controlled bridge inverter feedback diodes are used for
A. increasing inverter output
B. aiding SCR commutation
C. successful turn on D. none of these

205. The commutation technique in Mc-Murray-Bedford inverter is basically
A. an auxiliary commutation
B. a complementary commutation
C. a class-A commutation
D. none of these

206. Sinusoidal pulse-width modulated inverters will provide
A. reduced lower order harmonics
B. narrow regulation
C. increased distortion
D. none of these

207. In a dual converter operating under circu-lating-current mode
A. one as converter and the other as inverter
B. both operate as converters
C. both operate as inverters
D. none of these

208. If, for a single-phase half-bridge inverter, the amplitude of output voltage is V_s and the output power is P, then their corresponding values for a single-phase full-bridge inverter are
A. V_s, P B. $V_s/2$, $P/2$
C. $2V_s$, $2P$ D. none of these

209. In voltage source inverters
A. load voltage waveform v_0 depends on load impedance Z, whereas load current waveform i_0 does not depend on Z

Ans. 190. D 191. C 192. C 193. B 194. D 195. A 196. A 197. C 198. A 199. A 200. B 201. C 202. B 203. A 204. D 205. A 206. A 207. A 208. C

B. both v_0 and i_0 depend on Z
C. v_0 does not depend on Z whereas i_0 depends on Z
D. none of these

210. A single-phase full bridge inverter can operate in load-communication mode in case load consists of
A. RLC overdamped B. RLC underdamped
C. RLC critically damped
D. none of these

211. A single-phase bridge inverter delivers power to series connected RLC load with $R = 2$ ohm, $\omega L = 8$ ohm. For this inverter-load combination, load commutation is possible in case of magnitude of $1/\omega C$ in ohms is
A. 10 B. 8
C. 6 D. zero

212. In single-pulse modulation of PWM inverters, third harmonic can be eliminated if pulse width is equal to
A. 30° B. 60°
C. 120° D. none of these

213. In single-pulse modulation of PWM inverters, fifth harmonic can be eliminated if pulse width is equal to
A. 30° B. 72°
C. 36° D. 108°

214. In single-pulse modulation of PWM inverters, the pulse width is 120°. For an input voltage of 220 V, DC, the rms value of output voltage is
A. 179.63 V B. 254.04 V
C. 127.02 V D. none of these

215. In single-pulse modulation used in PWM inverters, V_s is the output DC voltage. For eliminating third harmonic, the magnitude of rms value of fundamental component of output voltage and pulse width are respectively
A. $\frac{2\sqrt{2}}{\pi}V_s$, 120° B. $\frac{4V_s}{\pi}$ 60°
C. $\frac{2\sqrt{2}}{\pi}V_s$, 60° D. $\frac{4V_s}{\pi}$ 120°

216. In multiple-pulse modulation used in PWM inverters, the amplitudes of reference square wave and triangular carrier wave are respectively 1 V and 2 V. For generating 5 pulses per half cycle, the pulse width should be
A. 36° B. 24°
C. 18° D. 12°

217. Usually a diode is placed across the base-emitter of a transistor to
A. limit reverse biasing B. provide fast turn-on
C. provide quick turn-off
D. none of these

218. In the inverter circuit shown below, if the SCRs are fired at delayed angles, the frequency of the output waveform will

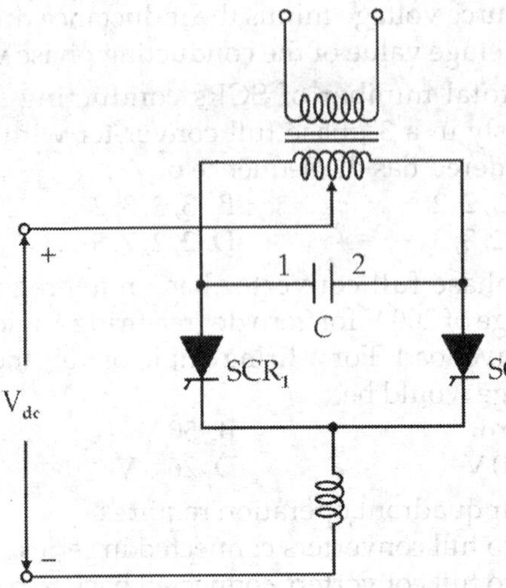

A. increase B. remains the same
C. decrease
D. depend upon which SCR is fired first

219. In the same circuit, the SCR_1 is ON the capacitor C will
A. charge with terminal 2 as positive
B. charge with terminal 1 as positive
C. not charge at all unless SCR_2 is also turned on
D. make SCR_2 ON

220. In multiple-pulse modulation used in PWM inverters, the amplitude and frequency for triangular carrier and square reference signals are respectively 4 V, 6 kHz and 1 V, 1 kHz. The number of pulses per half cycle and pulse width are respectively
A. 6, 90° B. 3, 45°
C. 4, 60° D. 3, 40°

221. Which of the following statements is correct in connection with inverters?
A. Voltage Source Inverter and Current Source Inverter both require feedback diodes
B. Only Current Source Inverter requires feedback diodes
C. GTOs can be used in Current Source Inverter
D. Only VSI requires feedback diodes

222. In sinusoidal-pulse modulation used in PWM inverters, amplitude and frequency for triangular carrier and sinusoidal refe-rence signals are respectively 5 V, 1 kHz and 1 V, 50 Hz. If zeros of the triangular carrier and reference sinusoid coincide, then the modulation index and order of significant harmonics are respectively
A. 0.2, 9 and 11 B. 0.4, 9 and 11
C. 0.2, 17 and 19 D. none of these

Ans. 209. C 210. B 211. A 212. C 213. B 214. A 215. A 216. C 217. ? 218. C 219. A 220. B 221. D 222. C

223. The input of the circuit given below is a 10 kHz sinewave of 200 V peak. The output of the circuit would be

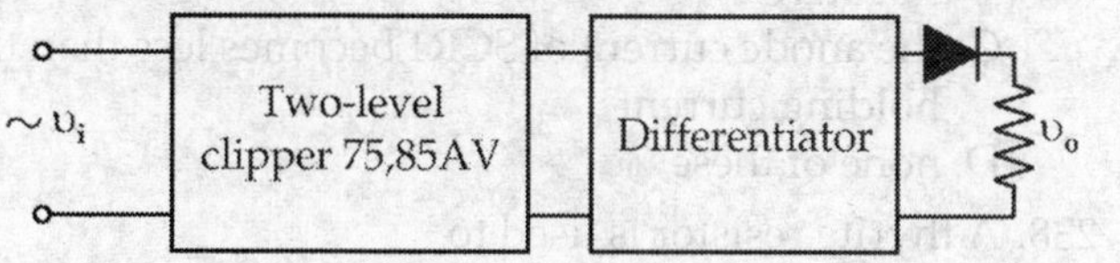

A.
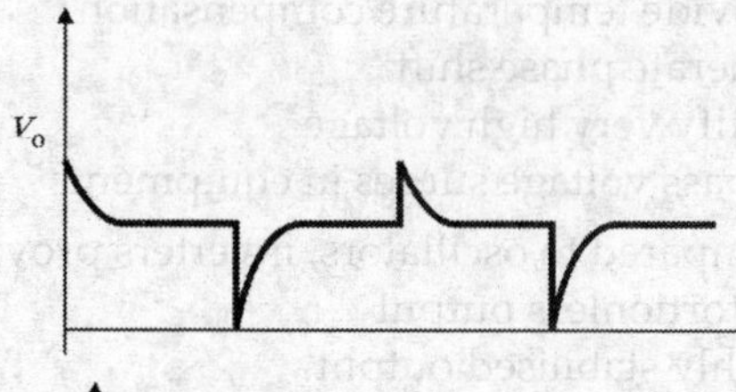

B.
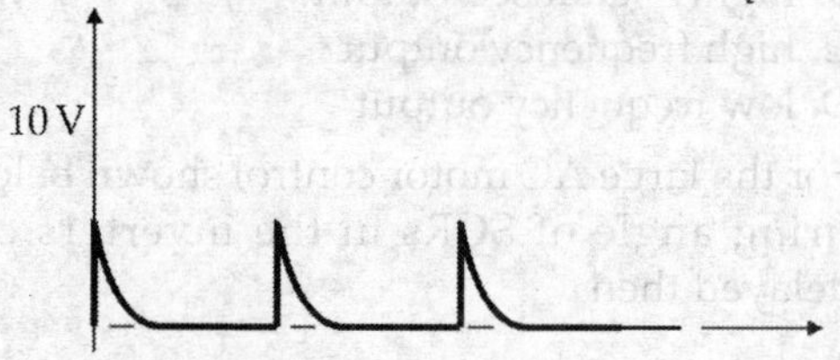

C.
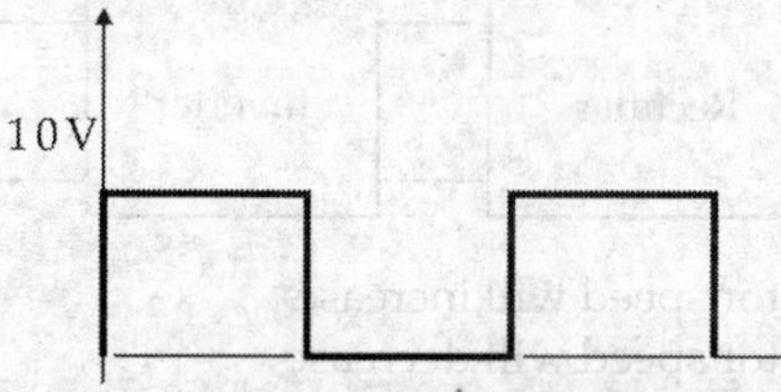

D.
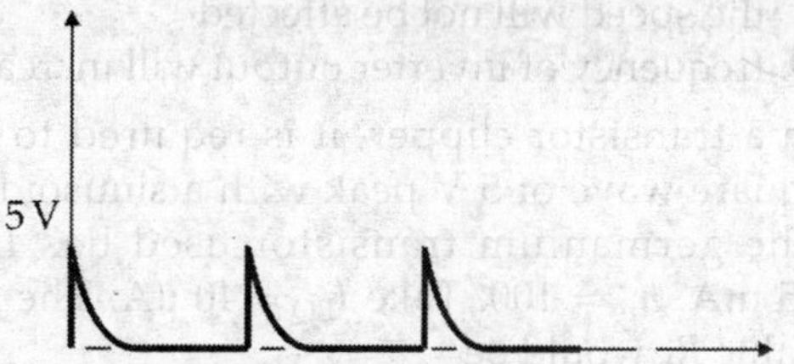

224. In the same circuit, the switching fre-quency of firing the six SCRs should be
A. same as the desired output frequency
B. three times the output frequency
C. five times the output frequency
D. ten times the output frequency

225. In the above circuit, the function tap switching firing sequence of SCRs to obtain positive half cycle is
A. 1–2–3–4–5 B. 5–3–1–3–5
C. 6–4–2–4–6 D. 6–5–4–3–2

226. In a Constant Source Inverter, if frequency of output voltage is f Hz, then frequency of voltage input to Constant Source Inverter is
A. f B. $2f$
C. $3f$ D. $4f$

227. In sinusoidal-pulse modulation used in PWM inverters, amplitude and frequency of triangular carrier and sinusoidal reference signals are respectively 15 V, 1 kHz and 1 V, 50 Hz. If peak of the triangular carrier coincides with the zero of the reference sinusoid, then the modulation index and order of significant harmonics are
A. 0.2, 9 and 11 B. 0.4, 9 and 11
C. 0.2, 17 and 19 D. 0.2, 19 and 21

228. In the SCR tap-switch inverter, when SCR1 is fired

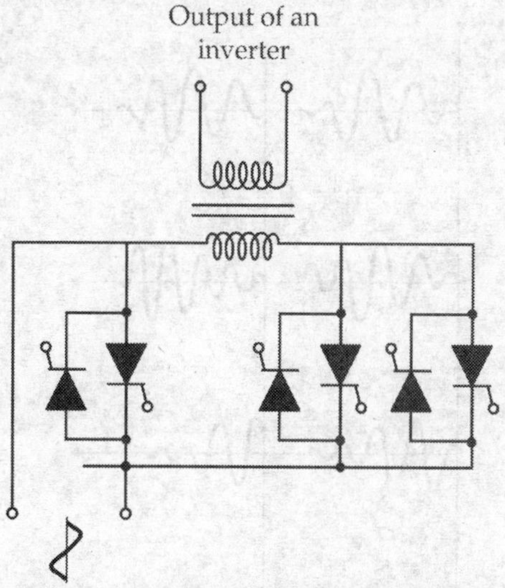

A. positive peak of the AC output is obtained
B. negative peak of the output is obtained
C. two-third to peak value is obtained
D. one-third of peak value is obtained

229. In sinusoidal PWM, there are 'm' cycles of the triangular carrier wave in the half cycle of reference sinusoidal signal. If zero of the reference sinusoid coincides with zero/peak of the triangular carrier waves then number of pulses generated in each half cycle are respectively
A. $(m-1)/m$ B. $(m-1)/(m+1)$
C. $m/(m-1)$ D. none of these

230. In an inverter with fundamental output frequency of 50 Hz, if third harmonic is eliminated, then frequencies of other components in the output voltage wave, in Hz, would be
A. 250, 350, 500, high frequencies
B. 50, 250, 350, 500
C. 50, 50, 350, 550 D. none of these

231. A single-phase CSI has capacitor C as the load. For a constant source current, the voltage across the capacitor is
A. square wave B. triangular wave
C. step function D. none of these

232. A single-phase full bridge VSI has induc-tor L as the load. For a constant source voltage, the current through the inductor is
A. square wave B. triangular wave
C. sine wave D. none of these

Ans. 223. B 224. C 225. B 226. B 227. D 228. B 229. A 230. C 231. B 232. B

233. The output waveform for the circuit shown below, for the given input would be

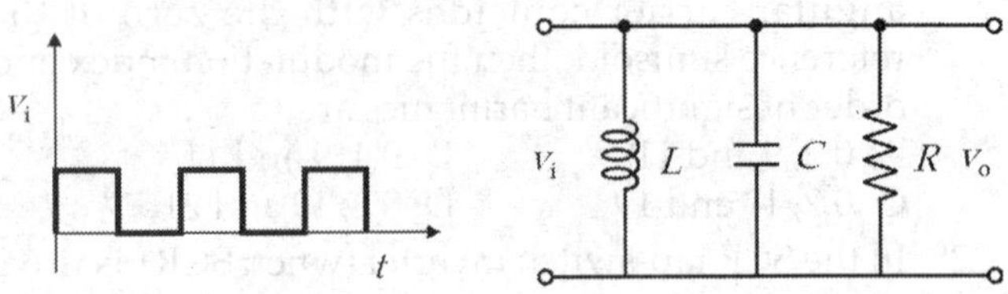

A.
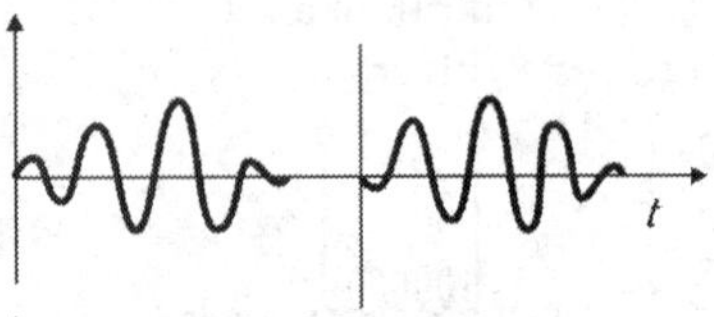

B.
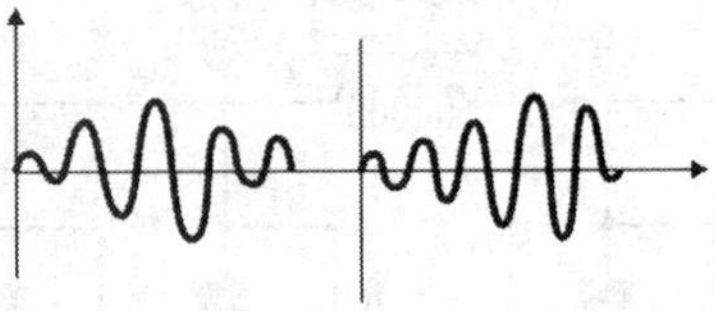

C.
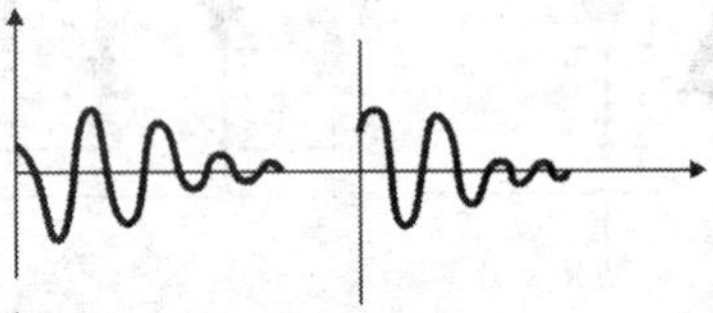

D.
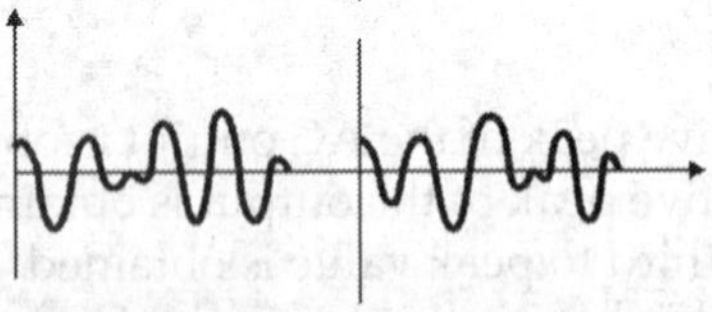

234. In DC choppers, if T_{on} is the on-period and f is the chopping frequency, then output voltage in terms of input voltage V_s is given by

A. $V_s{\cdot}T_{on}/f$ B. $V_s{\cdot}f/T_{on}$
C. $V_s/f{\cdot}T_{on}$ D. $V_s{\cdot}f{\cdot}T_{on}$

235. In DC choppers, the waveforms for input and output voltages are respectively

A. discontinuous, continuous
B. both continuous
C. both discontinuous
D. continuous, discontinuous

236. In PWM method of controlling the aver-age output voltage in a chopper, the on-time is varied but the chopping frequency is

A. varied B. kept constant
C. either of these D. none of these

237. In the circuit given below, SCR2 is turned ON, then

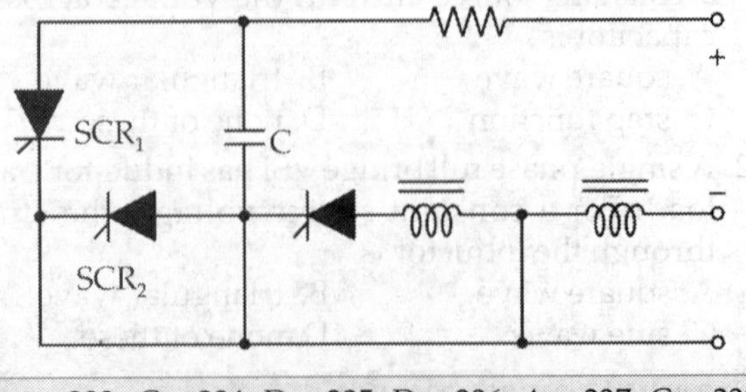

A. SCR1 turns ON
B. the capacitor makes the anode voltage of SCR1 negative
C. the anode current of SCR1 becomes less than the holding current
D. none of these

238. A thyrite resistor is used to

A. provide temperature compensation
B. generate phase shift
C. rectify very high voltage
D. bypass voltage surges in equipments

239. As compared to oscillators, inverters provide

A. distortionless output
B. highly stabilised output
C. high frequency output
D. low frequency output

240. For the large AC motor control shown below, if the firing angle of SCRs in the inverters circuit is delayed then

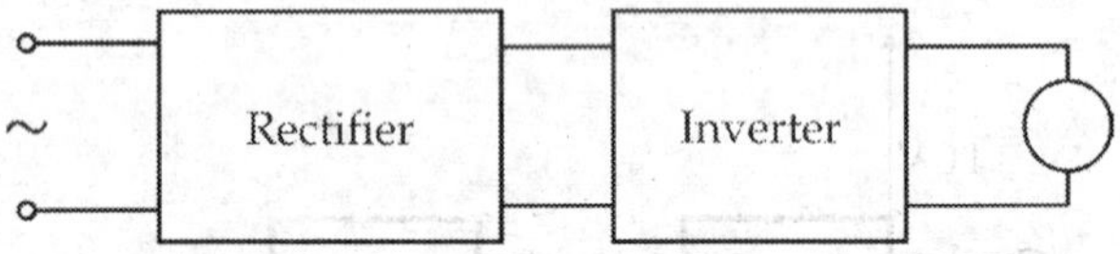

A. motor speed will increase
B. motor speed will decrease
C. the speed will not be affected
D. frequency of inverter output will increase

241. In a transistor clipper, it is required to obtain a square wave of 5 V peak with a sinusoidal input. The germanium transistor used has I_c (sat) = 15 mA, h_{fe} = 100. Take I_{BQ} = 40 μA. The required value R_b would be

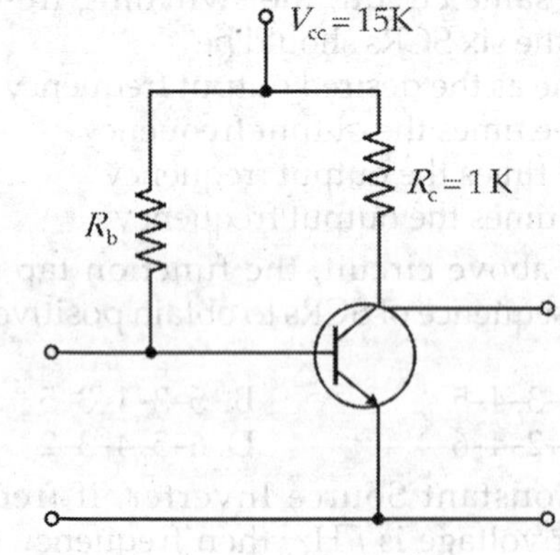

A. 80 K B. 250 K
C. 296 K D. 100 K

242. The Q point in the above circuit would be

A. (5 mA, 10 V) B. (7.5 mA, 7.5 V)
C. (1 mA, 5 V) D. (10 mA, 5 V)

Ans. 233. C 234. D 235. D 236. A 237. C 238. C 239. D 240. A 241. C 242. A

243. If the firing angle of the SCRs of the controlled rectifier is delayed, the motor speed will
A. become high B. become low
C. remain same
D. depend upon firing of inverter

244. The output waveform for the circuit shown below, for the given input would be

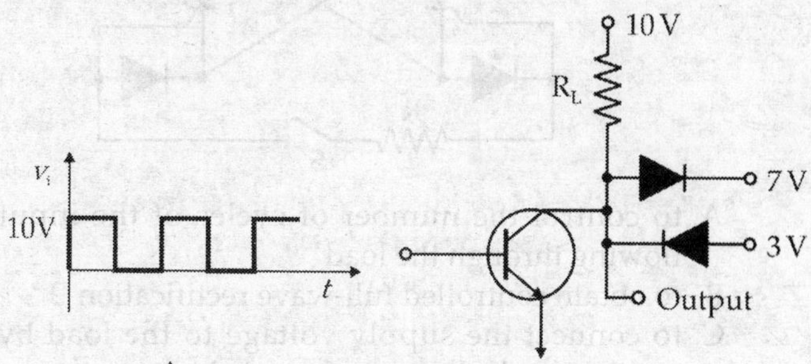

A. 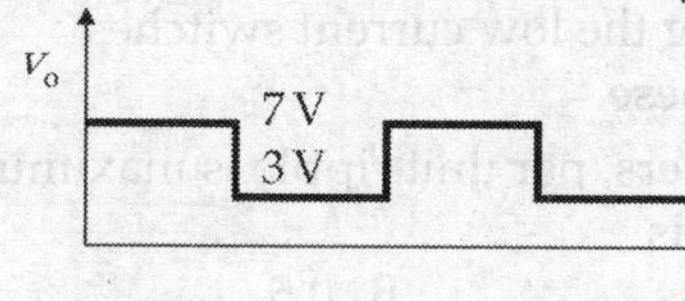

B. 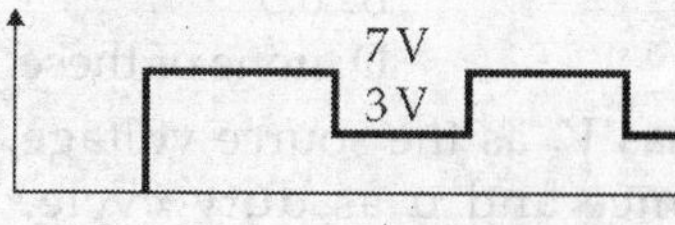

C. 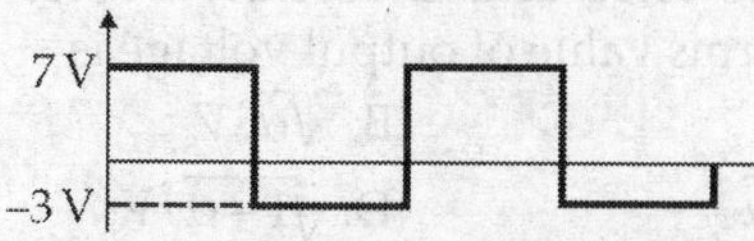

D. 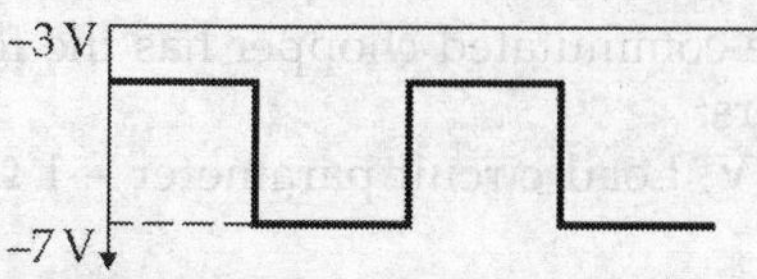

245. Which of the following circuits can be used as a DC transformer?
A. Controlled rectifier using SCRs
B. Magnetic amplifier
C. Inverter D. None of these

246. Which of the following circuit is used for timing purpose?
A. As table multivibrator
B. Monostable multivibrator
C. Schmitt trigger D. Bistable multivibrator

247. In the circuit shown below, if resistance R is increased, the firing angle of the SCR will

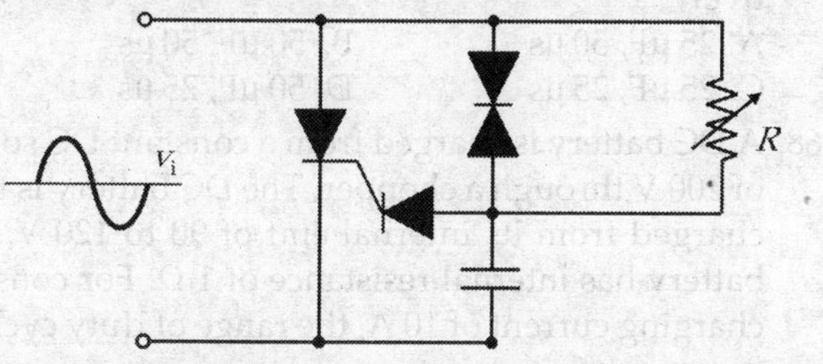

A. increase B. decrease
C. not be affected D. become zero

248. For type-A chopper, V_s is the source voltage, R is the load resistance and α is the duty cycle. The average output voltage for this chopper is
A. αV_s B. $(1-\alpha) V_s$
C. V_s/α D. $V_s/(1-\alpha)$

249. Output frequency of a cycloconverter is significantly lower than the input frequency for
A. keeping commutation failure
B. avoiding commutation failure
C. reducing switching losses
D. none of these

250. From the point of view of low distortion, the ratio of the output to input frequency of a cycloconverter should be
A. low B. high
C. very high D. none of these

251. The cycloconverters require natural or forced commutation as
A. natural commutation in both step-up and step-down cycloconverters
B. forced commutation in both step-up and step-down cycloconverters
C. forced commutation in step-up cycloconverters
D. forced commutation in step-down cycloconverters

252. A load resistance of 10 ohm fed through a 1-phase voltage controller from a volt-age source of 200 sin 314t. For a firing angle delay of 90°, the power delivered to load in kW, is
A. 0.5 B. 0.75
C. 1 D. none of these

253. Three-phase to three-phase cycloconverters employing 18 SCRs and 36 SCRs have the same voltage and current ratings for their component thyristors. The ratio of VA rating of 36-SCR device to that of 18-SCR device is
A. 1/2 B. 1
C. 2 D. none of these

254. Three-phase to 3-phase cycloconverters employing 18 SCRs and 36 SCRs have the same voltage and current ratings for their component thyristors. The ratio of power handled by 36-SCR device to that handled by 18-SCR device is
A. 4 B. 2
C. 1 D. none of these

255. The number of thyristors required for single-phase to single-phase cycloconverter of the three-phase 3-pulse type cycloconverter are respectively
A. 4, 6 B. 4, 4
C. 4, 18 D. none of these

Ans. 243. B 244. B 245. C 246. B 247. A 248. A 249. B 250. A 251. C 252. C 253. C 254. A 255. C

256. A 3-phase to single-phase conversion device employs a 6-pulse bridge cycloconverter. For an input voltage of 200 V per phase, the fundamental rms value of output voltage is
A. $600/\pi$ V
B. $300\sqrt{3}/\pi$ V
C. $300/\pi$ V
D. none of these

257. A cycloconverter is
A. frequency changer from higher to lower frequency with one-stage conversion
B. frequency changer from higher to lower frequency with two-stage conversion
C. frequency changer from lower to high frequency with one-stage conversion
D. either A or B is correct

258. A three-phase to single-phase cycloconverter consists of positive and negative group of converters, in this device, one of the two component converters would operate as a
A. rectifier if the output voltage V_0 and output current I_0 have the same polarity
B. inverter if V_0 and I_0 have the same polarity
C. rectifier if V_0 and I_0 are of opposite polarity
D. none of these

259. Which of the following statements are incorrect for cycloconverters?
A. Step-down cycloconverter works on natural commutation
B. Step-up cycloconverter requires formed commutation
C. Load commutated cycloconverter works on line commutation
D. Load commutated cycloconverter requires a generated emf in the load circuit

260. A 3-phase to 3-phase cycloconverter requires
A. 18 SCRs for 3-pulse device
B. 18 SCRs for 6-pulse device
C. 36 SCRs for 3-pulse device
D. none of these

261. A chopper can be considered as a DC equivalent to a
A. transformer
B. cycloconverter
C. none of these

263. For a chopper, V_s is the source voltage, R is the load resistance and α is the duty cycle. RMS and average values of thyristor currents for this chopper are
A. $\alpha \cdot \frac{V_s}{R}, \sqrt{\alpha} \cdot \frac{V_s}{R}$
B. $\sqrt{\alpha} \cdot \frac{V_s}{R}, \sqrt{\alpha} \cdot \frac{V_s}{R}$
C. $\sqrt{\alpha} \cdot \frac{V_s}{R}, \alpha \cdot \frac{V_s}{R}$
D. none of these

262. The basic function of the circuit shown below is

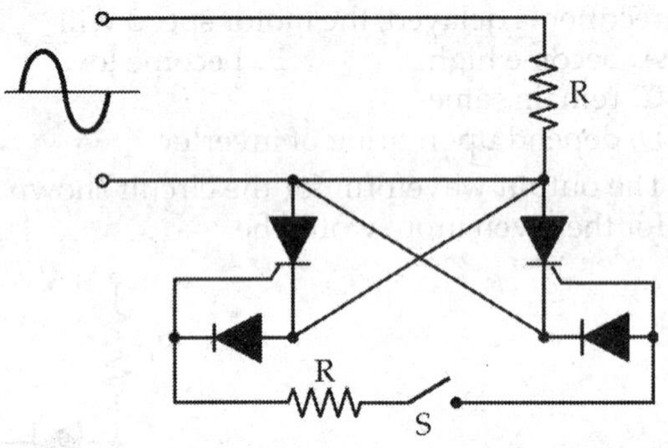

A. to control the number of cycles of the input flowing through the load
B. to obtain controlled full-wave rectification
C. to connect the supply voltage to the load by employing the low current switches
D. none of these

264. In DC choppers, per unit ripple is maximum when duty cycle α is
A. 0.2
B. 0.5
C. 0.7
D. none of these

265. A chopper has V_s as the source voltage, R as the load resistance and α as duty cycle. For this chopper, rms value of output voltage is
A. αV_s
B. $\sqrt{\alpha} \cdot V_s$
C. $V_s / \sqrt{\alpha}$
D. $\sqrt{1-\alpha} \cdot V_s$

266. A voltage commutated chopper has the following parameters:
$V_s = 200$ V, Load circuit parameter – 1 Ω, 2 mH, 50 V
Commutation circuit parameters, $L = 25$ μH, $C = 50$ μF
For a constant load current of 100 A, the effective on period and peak current through the main thyristor respectively are
A. 1000 μs, 200 A
B. 700 μs, 382.8 A
C. 700 μs, 282.8 A
D. none of these

267. A load commutated chopper, fed from 200 V DC source, has a constant load current of 50 A. For a duty cycle of 0.4 and a chopping frequency of 2 kHz, the value of commutating capacitor and the turn-off time for one thyristor pair are respectively
A. 25 μF, 50 μs
B. 50 μF, 50 μs
C. 25 μF, 25 μs
D. 50 μF, 25 μs

268. A DC battery is charged from a constant DC source of 200 V through a chopper. The DC battery is to be charged from its internal emf of 90 to 120 V. The battery has internal resistance of 1 Ω. For constant charging current of 10 A, the range of duty cycle is

Ans. 256. A 257. D 258. A 259. C 260. A 261. A 262. C 263. C 264. B 265. B 266. B 267. A

A. 15 to 65 B. 65 to 0.8
C. 8 to 0.95 D. none of these

269. For type A chopper, V_s, R, I_0 and α are respectively the DC source voltage, load resistance, constant load current and duty cycle. For this chopper, average and rms values of freewheeling diode currents are
A. αI_0, $\sqrt{\alpha} \cdot I_0$ B. $(1-\alpha) I_0$, $\sqrt{1-\alpha} \cdot I_0$
C. $\alpha \cdot V_s / R$, $\sqrt{\alpha} \cdot V_s / R$ D. none of these

270. A step-up chopper has V_s as the source voltage and α as the duty cycle. The output voltage for this chopper is given by
A. $V_s(1+\alpha)$ B. $V_s/(1-\alpha)$
C. $V_s(1-\alpha)$ D. none of these

271. A DC chopper is fed from 100 V DC. Its load voltage consists of rectangular pulses of duration 1 msec in an overall cycle time of 3 msec. The average output voltage and ripple factor for this chopper are respectively
A. 25 V, 1 B. 50 V, 1
C. 33.33 V, 1.5 D. none of these

272. When a series LC circuit is connected to a DC supply of V volts through a thyristor, then the peak current through thyristor is
A. $V \cdot \sqrt{CL}$ B. $V / \sqrt{CL}$
C. $V \cdot \sqrt{C/L}$ D. $V \cdot \sqrt{L/C}$

273. In DC choppers if T is the chopping period, then output voltage can be controlled by PWM by varying
A. T keeping T_{on} constant
B. T_{on} keeping T constant
C. T keeping T_{off} constant
D. none of these

274. In DC choppers, for chopping period T, the output voltage can be controlled by varying
A. T keeping T_{on} constant
B. T_{on} keeping T constant
C. T_{off} keeping T constant
D. none of these

275. For same average output voltage, PIV rating of SCRs will be larger for
A. 1-phase full wave centre tapped rectifier
B. 1-phase full wave bridge rectifier
C. 3-phase full wave bridge circuit
D. same in all cases

276. Transient overvoltage protection of SCRs may be provided by
A. high speed fuse B. crow-bar method
C. snubber circuit D. none of these

277. A low negative gate bias to thyristor will cause
A. increased possibility of false triggering
B. increased formed blocking capability
C. increased possibility of thermal run-way
D. none of these

278. For same average output voltage, the PIV rating of thyristors will be higher for
A. single phase full-wave bridge rectifier
B. three phase full-wave bridge converter
C. single-phase full-wave centre-tapped rectifier
D. none of these

279. A small number of gold atoms are introduced into the silicon wafers in order to reduce
A. on state losses
B. forward voltage drop at full conduction
C. turn-off time D. none of these

280. It is easier to manufacture fast SCR of
A. large power handling capability
B. high PIV
C. low PIV D. none of these

281. So far as harmonic content at the inverter output voltage is concerned, multiple pulse modulation will provide
A. perfect sinusoidal of fundamental frequency
B. more harmonics
C. less harmonics D. none of these

282. Rectifier transformer with zig zag secondary is employed to
A. improve utility factor
B. reduce harmonic content
C. avoid DC saturation
D. none of these

283. A permanent damage to SCR may arise
A. due to high di/dt B. due to high dv/dt
C. none of these

284. Excessive dv/dt to a thyristor may cause
A. maloperation
B. permanent damage
C. better performance
D. none of these

285. Excessive di/dt to a thyristor may cause
A. maloperation B. permanent damage
C. better performance D. none of these

286. For two diodes in series, voltage sharing will be almost equal under
A. forward-biased condition only
B. reverse-biased condition only
C. all conditions D. none of these

287. Snubber circuit is used to provide
A. fast switching
B. freewheeling action
C. dv/dt protection
D. none of these

Ans. 268. A 269. B 270. B 271. C 272. C 273. B 274. A 275. A 276. B 277. C 278. C 279. B 280. C 281. C 282. C 283. A 284. A 285. B 286. A 287. C

288. Surge current rating of an SCR specifies the maximum
A. repetitive current with sine wave
B. non-repetitive current with rectangular wave
C. non-repetitive current with sine wave
D. repetitive current with triangular wave

289. The di/dt rating of an SCR is specified for its
A. decaying anode current
B. decaying gate current
C. rising gate current D. rising anode current

290. For an SCR, dv/dt protection is achieved through the use of
A. R in series with SCR B. RL across SCR
C. L in series with SCR D. none of these

291. For an SCR, di/dt protection is achieved through the use of
A. R in series with SCR B. RL in series with SCR
C. L in series with SCR D. none of these

292. The function of snubber circuit connected across an SCR is to
A. suppress dv/dt B. increase dv/dt
C. decrease dv/dt D. none of these

293. The object of connecting resistance and capacitance across gate circuit is to protect the SCR gate against
A. overvoltages B. dv/dt
C. noise signals D. none of these

294. Thermal resistance in SCRs has the units of °C/W and heat sinks are made from
A. copper B. steel
C. aluminium D. none of these

295. Practical way of obtaining static voltage equalisation in series-connected SCRs is by the use of
A. one resistor across the string
B. resistors of different values across each SCR
C. resistors of the same value across each SCR
D. none of these

296. For series connected SCRs, dynamic equalising circuit consists of
A. resistor R and capacitor C in series but with a diode D across capacitor C
B. series R and D circuit but with C across R
C. series R and C circuit but with D across R
D. none of these

297. For dynamic equalising circuit used for series connected SCRs, the choice of capacitor C is based on
A. reverse recovery characteristic
B. turn-on characteristics
C. turn-off characteristics
D. rise-time characteristics

298. A voltage of controlled induction motor operating at 20% slip will have a limiting efficiency of
A. 10% B. 20%
C. 40% D. 80%

299. Slip-power recovery control is applicable for
A. synchronous motor
B. cage induction motor
C. wound rotor induction motor
D. none of these

300. The advantage of the tachometer speed control method for DC motors is that, it senses
A. back emf B. armature current
C. armature voltage D. speed

301. The ripple factor of a centre tapped full wave rectifier is
A. 0.482 B. 0.707
C. 0.866 D. 1.21

302. Which of the following circuits cannot be operated directly from the mains?
A. Half-wave rectifier
B. Centre-tap full-wave rectifier
C. Bridge rectifier D. None of these

303. The ripple factor (RF) of a waveform is defined as RMS of the ripple/Average output level. By the definition, the ripple factor of the half-wave rectifier is:
A. 0.246 B. 0.48
C. 1.21 D. none of these

304. In a rectifier circuit, the load connected is of low value. For proper filter operation, it is required that:
A. a capacitor is to be included in the circuit
B. a bleeder resistance is to be placed in the circuit
C. an inductor filter is to be included in the circuit
D. all of these

305. The function of bleeder resistance in filter circuit is:
A. to maintain minimum current necessary for optimum inductor filter operation
B. to work as voltage divider in order to provide variable output from the supply
C. to provide discharge path to capacitors so that output becomes zero when the circuit has been de-energised
D. all of these

306. The rectifier, which requires minimum amount of filtering is a:
A. half-wave rectifier B. full-wave rectifier
C. voltage doubler circuit
D. none of these

307. A constant current source supplies a current of 100 mA to a load of 1 kW. When the load is changed to 10 kW, the load current will be
A. 0.1 mA B. 1.0 mA
C. 10 mA D. 100 mA

Ans. 288. C 289. D 290. B 291. C 292. A 293. C 294. A 295. C 296. C 297. A 298. D 299. C 300. D 301. A 302. B 303. C 304. C 305. D 306. B 307. ?

308. A half-wave rectifier circuit with a capacitive filter is connected to 200 V, 50 Hz AC line, the output voltage across the capacitor should be approximately:
A. 100 V B. 180 V
C. 280 V D. 400 V

309. A zener diode:
A. has a high forward voltage rating
B. has a sharp breakdown at low reverse voltage
C. has a negative resistance
D. none of these

310. In a centre-tap full-wave rectifier, V_m is the peak voltage between the centre-tap and one end of the secondary. The maximum voltage across the reverse-biased diode is:
A. V_m B. $2V_m$
C. $3V_m$ D. none of these

311. The important specification of a zener diode is:
A. its breakdown voltage and power dissipation
B. breakdown voltage, dynamic impedance and power dissipation
C. breakdown voltage and dynamic impedance
D. none of these

312. The major advantage of a bridge rectifier is that:
A. centre-tap transformer is not required
B. peak inverse voltage of each diode is double of that for a full-wave rectifier
C. peak inverse voltage of each diode is half of that for a full-wave rectifier
D. the output voltage is more smooth

313. If the ripple factor of the output wave of a rectifier is low, it means that:
A. output voltage will have less ripple
B. output voltage will be low
C. filter circuits may not be required
D. none of these

314. In a three-phase half-wave rectifier, the output voltage is equal to:
A. the most positive input phase voltage at any instant
B. the difference of most positive and most negative input phases at any instant
C. the average value of the three-phases
D. the difference of the two positive phase voltages

315. In a three-phase half-wave rectifier, if the input phase voltage is 200 V, the PIV required for each diode will be:
A. 440 V B. 346 V
C. 370 V D. 220 V

316. The effect of DC saturation in a rectifier transformer is to:
A. decrease the output B. increase the efficiency
C. decrease the AC components of the output
D. increase the DC component

317. For a rectifier output of 100 kW at 400 V an average current rating of 40 A for each diode is sufficient in:
A. single phase full-wave rectifier
B. six-phase half-wave rectifier
C. three-phase half-wave rectifier
D. three-phase full-wave rectifier

318. As the number of phases in a multiphase rectifier are increased, the output will:
A. remains same B. become more smooth
C. decrease
D. the diodes will require high peak inverse voltage

319. Ripple frequency of a six-phase half-wave rectifier for 220 V, 60 Hz input will be:
A. 150 Hz B. 60 Hz
C. 360 Hz D. 100 Hz

320. In a three-phase half-wave rectifier, each diode conducts for a duration of:
A. π B. $\pi/6$
C. $\pi/3$ D. $\pi/4$

321. A capacitor of 100 μF is charged to 10 V through a resistance of 10 kW. It will be fully charged in:
A. 5 sec B. 10 sec
C. 15 sec D. 0.5 sec

322. In the above question, in one second the capacitor will charge to:
A. 3 V B. 6.3 V
C. 12 V D. 15 V

323. A zener diode:
A. has a sharp breakdown at low reverse voltage
B. has a high forward voltage rating
C. has a negative resistance
D. none of these

324. Thyristors are basically:
A. SCRs B. Triacs
C. both SCRs and Triacs D. all pnpn devices

325. The number of p-n junctions in a thyristor is/are:
A. 1 B. 2
C. 3 D. 4

326. When anode is positive with respect to cathode in an SCR, the number of blocked p-n junctions is/are:
A. 1 B. 2
C. 4 D. none of these

327. When cathode is positive with respect to anode in an SCR, the number of blocked p-n junctions is/are:
A. 1 B. 2
C. 3 D. 4

328. In a thyristor, anode current is made up of:
A. electrons only B. electrons or holes
C. electrons and holes D. none of these

329. A thyristor, when triggered, will change from forward blocking state to conductionstate, if its anode to cathode voltage is equal to:
A. peak repetitive off-state forward voltage
B. peak working off-state forward voltage

Ans. 308. C 309. B 310. B 311. B 312. C 313. A 314. A 315. C 316. A 317. B 318. B 319. C 320. C
321. A 322. B 323. A 324. D 325. C 326. A 327. B 328. C 329. B

C. peak working off-state reverse voltage
D. none of these

330. An SCR can be brought to forward conducting state with gate circuit open when the applied voltage exceeds:
A. the forward breakover voltage
B. reverse breakdown voltage
C. 1.5 V D. none of these

331. In a thyristor, holding current is:
A. more than latching current I_L
B. less than latching current
C. equal to latching current
D. very small

332. When a thyristor gets turned on, the gate drive:
A. should not be removed as it will turn off the SCR
B. may or may not be removed
C. should be removed
D. should be removed in order to avoid increased losses and higher junction temperature

333. An SCR is a:
A. two-layer two-junction device
B. three-layer two-junction device
C. four-layer three-junction device
D. four-layer four-junction device

334. An SCR has:
A. one terminal B. two terminals
C. three terminals D. four terminals

335. An SCR can be operated:
A. only on reverse-biased condition
B. only on forward-biased condition
C. both forward and reverse-biased conditions
D. none of these

336. While constructing a thyristor, silicon wafer together with the diffused discrete layers is mounted on a supporting base made of:
A. copper B. molybdenum
C. aluminium D. steel

337. Normally the value of turn on time of commonly used thyristors is approximately:
A. 10 msec B. 100 msec
C. 50 msec D. 200 msec

338. The capacitive current flowing through the junction of a thyristor is:
A. $I_c = C\,dv/dt$ B. $I_C = \frac{1}{C}\cdot\frac{dv}{dt}$
C. $I_c = C{\cdot}di/dt$ D. none of these

339. For normal SCRs, turn on time is:
A. less than turn off time, t_q
B. more than turn off time, t_q
C. equal to turn off time, t_q
D. none of these

340. The forward voltage drop during SCR onstate is 1.5 V. This voltage drop:
A. remains constant and is independent of load current
B. increases slightly with load current
C. decreases slightly with load current
D. varies linearly with load current

341. A thyristor can be termed as:
A. DC switch B. AC switch
C. both A or B are correct
D. square-wave switch

342. The SCR ratings di/dt in A/msec and dv/dt in V/μsec, may vary respectively between:
A. 20 to 500, 10 to 100 B. both 20 to 500
C. both 10 to 100 D. 50 to 300, 20 to 500

343. In a thyristor:
A. latching current I_L is associated withturn off process and holding current I_H with turn on process
B. both I_L and I_H are associated with turn on process
C. I_H is associated with turn off process and I_L with turn on process
D. both I_L and I_H are associated with turn on process

344. Once SCR starts conducting a forward current, its gate looses control over:
A. anode circuit voltage only
B. anode circuit current only
C. anode circuit voltage and current
D. none of these

345. On-state voltage drop across a thyristor used in a 250 V supply system is of the order of:
A. 100–110 V B. 240–250 V
C. 1–1.5 V D. none of these

346. In a thyristor, ratio of latching current to holding current is:
A. 0.4 B. 1.0
C. 2.5 D. none of these

347. During forward blocking state, a thyristor is associated with:
A. large current, low voltage
B. low current, large voltage
C. medium current, large voltage
D. none of these

348. Gate characteristic of a thyristor:
A. is a straight line passing through origin
B. is of the type $V_g = a + bI_g$
C. is a curve between V_g and I_g
D. has a spread between two curves of $V_g - I_g$

349. In an SCR, anode current flows over a narrow region near the gate during:
A. delay time, t_d
B. rise time, tr and spread time, t_p
C. t_d and t_p D. t_d and t_r

Ans.	330. A	331. B	332. D	333. C	334. C	335. B	336. D	337. C	338. A	339. A	340. B	341. A	342. B
	343. C	344. C	345. C	346. C	347. B	348. D	349. D						

350. Turn on time for an SCR is 10 μsec. If an inductance is inserted in the anode circuit, then the turn on time will be:
A. 10 μsec B. less than 10 μsec
C. more than 10 μsec D. about 10 μsec

351. Turn off time of an SCR is measured from the instant:
A. anode current becomes zero
B. anode voltage becomes zero
C. anode voltage and anode current become zero at the same time
D. gate current becomes zero

352. A forward voltage can be applied to an SCR after its:
A. anode current reduces to zero
B. gate recovery time
C. reverse recovery time
D. anode voltage reduces to zero

353. For an SCR with turn on time of 5 microsecond, and ideal trigger pulse should have:
A. short rise time with pulse width = 3 msec
B. long rise time with pulse width = 6 msec
C. short rise time with pulse width = 6 msec
D. long rise time with pulse width = 3 msec

354. Turn on time of an SCR in series with R-L circuit can be reduced by:
A. decreasing *R* B. decreasing *L*
C. increasing circuit resistance *R*
D. increasing circuit inductance *L*

355. Turn on time of an SCR can be reduced by using a:
A. rectangular pulse of high amplitude and narrow width
B. rectangular pulse of low amplitude and wide width
C. triangular pulse D. trapezoidal pulse

356. Static *V–I* characteristics of an SCR with different gate drives applied to the gate are indicated by:
A. $I_{g2} > I_{g1} > I_{g0}$ B. $V_{g2} > V_{g1} > V_{g0}$
C. $P_{g2} > P_{g1} > P_{g0}$ D. either A or B

357. In a thyristor, the magnitude of anode current will:
A. increase, if gate current is increased
B. decrease, if gate current is decreased
C. increase, if gate current is decreased
D. not change with any variation in gate current

358. The average on-state current for an SCR is 20 A for a conduction angle of 120°. Its average on-state current for 60° conduction angle would be:
A. 20 A B. 10 A
C. less than 20 A D. none of these

359. The average on-state current for an SCR is 20 A for a resistive load. If an inductance of 5 mH is included in the load, then average on-state current would be:
A. more than 20 A B. less than 20 A
C. 15 A B. none of these

360. Specification sheet for an SCR gives its maximum rms on-state current as 35 A. This rms rating for a conduction angle of 120° would be:
A. more than 35 A B. less than 35 A
C. 35 A D. none of these

361. Surge current rating of an SCR specifies the maximum:
A. repetitive current with sine wave
B. non-repetitive current with rectangular wave
C. non-repetitive current with sine wave
D. repetitive current with triangular wave

362. The *di*/*dt* rating of an SCR is specified for its:
A. decaying anode current
B. decaying gate current
C. rising gate current D. rising anode current

363. For an SCR, *dv*/*dt* protection is achieved through the use of:
A. *RL* in series with SCR B. *RL* across SCR
C. *L* in series with SCR D. all of these

364. The number of gates in a silicon-controlled switch is:
A. one B. two
C. three D. four

365. A light-activated silicon-controlled rectifier (LASCR) is a:
A. two-layer two-junction device
B. four-layer three-junction device
C. three-layer two-junction devices
D. none of these

366. Diac is:
A. three-layer two-junction device
B. four-layer two-junction device
C. four-layer three-junction device
D. none of these

367. Diac has:
A. one terminal B. two terminals
C. three terminals D. four terminals

368. Diac can be fired by:
A. positive half cycle of the supply only
B. negative half cycle of the supply only
C. positive and negative half cycles of the supply
D. none of these

369. Diac can be used to:
A. trigger a triac B. protect a triac
C. increase the efficiency of a triac
D. none of these

370. Breakover voltage for commonly used diacs is about:
A. 16 V B. 32 V
C. 48 V D. 60 V

Ans.	350. C	351. A	352. B	353. C	354. B	355. A	356. A	357. D	358. C	359. A	360. C	361. C	362. D
	363. B	364. B	365. B	366. B	367. B	368. C	369. A	370. B					

371. TRIAC is a:
A. three-layer two-junction device
B. four-layer three-junction device
C. four-layer four-junction device
D. none of these

372. TRIAC can be considered as:
A. two SCRs connected in antiparallel with a common gate
B. two SCRs connected in parallel with common gate
C. two transistors connected in antiparallel
D. none of these

373. TRIAC has:
A. one terminal
B. two terminals
C. three terminals
D. four terminals

374. TRIAC can be fired by:
A. positive half cycle of the supply only
B. negative half cycle of the supply only
C. positive and negative half cycles of the supply
D. none of these

375. DIAC–TRIAC built-in the same chip is called:
A. ignition
B. thermistor
C. thyristor
D. quadrac

376. TRIAC can be used as a static switch for frequencies up to:
A. 1 kHz
B. 10 kHz
C. 100 kHz
D. 400 kHz

377. LASCR is a:
A. vacuum device
B. unilateral device
C. bilateral device
D. none of these

378. Which of the following devices have similar *v-i* characteristics in shape:
A. SCRs and diacs
B. SCSs and triacs
C. SCRs and UJTs
D. SCRs, SCSs, SUSs and LASCRs

379. In a GTO, anode current begins to fall when gate current:
A. is negative peak at time $t = 0$
B. is negative peak at t = storage time + fall time
C. is negative peak at t = storage time
D. none of these

380. Latching current of thyristor is associated with:
A. turn on process
B. turn off process
C. turn on and turn off proceses
D. none of these

381. Ratio of latching current to holding current of thyristor is generally of the order of:
A. two to three times
B. four to five times
C. five to seven times
D. none of these

382. If a voltage is applied to the thyristor, a displacement of charges occurs. Which of the following movements takes place?
A. The holes to the positive electrode
B. The electrons to the negative electrode
C. The holes to the negative electrode and the electrons to the positive electrode
D. None of these

383. The thyristor is turned off:
A. by switching off the voltage applied to the principal circuit and changing the thyristor into the reverse blocking state
B. by switching off the voltage applied to the gate circuit only
C. by changing the thyristor into forward blocking state
D. none of these

384. The magnitude of anode current can be controlled in the case of the thyristor only by:
A. changing the amplitude of trigger pulse
B. changing the width of the trigger pulse
C. varying the timing of the trigger pulse with respect to the positive half cycle of the applied AC voltage
D. none of these

385. For the transition from the forward blocking to the 'on' state the thyristor must be triggered. This takes place by feeding a gate current into the gate electrode for:
A. the duration of the whole triggering process
B. the duration of the whole triggering and conducting process
C. a fraction of the period of the conducting process
D. none of these

386. From the following which one is an operational state of the thyristor?
A. Triggering
B. Firing
C. Regulating
D. Reverse blocking

387. Two operational states of a thyristor are: when anode is positive with respect to cathode:
A. reverse blocking and conducting
B. reverse blocking and forward blocking
C. conducting and forward blocking
D. none of these

388. In a UJT with V_{BB} as the voltage across two base terminals, the emitter potential, at peak point is given by:
A. ηV_{BB}
B. ηV_D
C. $\eta V_{BB} + V_D$
D. $\eta V_D + V_{BB}$

389. A UJT exhibits negative resistance region:
A. before the peak point
B. between peak and valley points
C. after the valley point
D. both A and B are correct

390. In a UJT, maximum value of charging resistance is associated with:
A. peak point
B. valley point
C. any point between peak and valley points
D. after the valley point

Ans. 371. B 372. A 373. C 374. C 375. D 376. D 377. B 378. D 379. C 380. A 381. A 382. C 383. A 384. B 385. C 386. D 387. C 388. C 389. B 390. A

391. When a UJT is used for triggering of an SCR, the wave shape of the voltage obtained from UJT circuit is a:
A. sine wave B. saw-tooth wave
C. trapezoidal wave D. square wave

392. For a UJT employed for the triggering of an SCR, stand-off ratio $\eta = 0.64$ and DC voltage source V_{BB} is 20 V. The UJT would trigger when the emitter voltage is:
A. 12.8 V B. 13.1 V
C. 10 V D. none of these

393. An SCR can withstand a maximum junction temperature of 120°C with an ambient temperature of 75°C. If this SCR has thermal resistance from junction to ambient as 1.5°C/W, the maximum internal power dissipation allowed is:
A. 30 W B. 60 W
C. 80 W D. 100 W

394. In synchronized UJT triggering of an SCR voltage v_C across capacitor reaches UJT threshold voltage thrice in each half cycle so that there are three firing pulses during each half cycle. The firing angle of the SCR can be controlled:
A. once in each half cycle
B. thrice in each half cycle
C. twice in each half cycle
D. four times in each half cycle

395. The function of connecting a zener diode in a UJT circuit, used for the triggering of SCRs, is to:
A. expedite the generation of triggering pulses
B. delay the generation of triggering pulses
C. provide a constant voltage to UJT to prevent erratic firing
D. provide a variable voltage to UJT as the source voltage changes

396. SCR can be turned on by:
A. applying anode voltage at a sufficiently fast rate
B. applying sufficiently large anode voltage
C. increasing the temperature of SCR to a sufficiently large value
D. all of these

397. Thyristor *A* has rated gate current of 2 A and thyristor *B*, a rated gate current of 100 mA:
A. thyristor *A* is a GTO and *B* is a conventional SCR
B. thyristor *B* is a GTO and *A* is a conventional SCR
C. thyristor *B* may operate as a transistor
D. none of these

398. Which of the following devices does *not* have negative resistance characteristics?
A. UJT B. Tunnel diode
C. SCR D. FET

399. Which of the following statements is true for firing a triac?
A. Either T_1 or T_2 is negative and there is current pulse into the gate
B. T_1 is negative and there is current pulse out of the gate
C. T_2 is negative and there is a current pulse out of the gate
D. Either T_1 or T_2 is positive and there is a current pulse out of the gate

400. As compared to UJT, SUS:
A. triggers only in one direction
B. does not have negative resistance characteristics
C. needs definite polarity of the applied voltage
D. triggers only at one particular voltage

401. In its application, an SUS behaves in the same way as:
A. UJT B. SCR
C. tunnel diode D. none of these

402. Which of the following pnpn devices has two gates?
A. TRIAC B. SCS
C. SUS D. DIAC

403. For thyristors, pulse triggering is preferred to DC triggering because:
A. gate dissipation is low
B. pulse system is simpler
C. triggering signal is required for a very

404. Which of the following pnpn devices has a terminal for synchronising purpose?
A. SCS B. TRIAC
C. DIAC D. SUS

405. Which of the following device is a three layer device?
A. SCS B. SUS
C. TRIAC D. DIAC

406. Which of the following devices cannot be triggered with voltage of either polarity?
A. SCS B. SUS
C. TRIAC D. DIAC

407. Which of the following methods will turn SCS off?
A. Applying negative pulse to the anode
B. Applying a positive pulse to the anode gate
C. Applying negative pulse to the cathode gate
D. All of these

408. In application, a programmable conjunction transistor resembles more to:
A. special purpose UJT B. anode-fired SCR
C. TRIAC D. DIAC

409. Which of the following pnpn devices does not have a gate terminal?
A. TRIAC B. SCS
C. SUS D. Complementary SCR

410. The commonly used method for triggering the thyristor is:
A. gate triggering B. thermal triggering
C. voltage triggering D. all of these

Ans. 391. B 392. B 393. A 394. A 395. C 396. D 397. A 398. D 399. C 400. C 401. A 402. B 403. C 404. D 405. D 406. A 407. D 408. B 409. C 410. A

411. A saw tooth wave generator can be obtained from UJT relaxation oscillator circuit by using a:
A. field effect transistor
B. junction field effect transistor
C. power transistor D. none of these

412. Pulse gate triggering is achieved by:
A. UJT relaxation oscillator
B. rheostatic method
C. R–C method D. none of these

413. R–C triggering is preferred over resistance triggering method because it provides:
A. larger value of firing angle
B. quick firing
C. accurate firing D. all of these

414. In the forward-biased condition of a thyristor:
A. one of three junctions is reverse-biased
B. two out of three junctions are reverse-biased
C. all the three junctions are forward-biased
D. none of these

415. The moment a thyristor starts conducting:
A. it acts like a short circuit device
B. it acts like an open circuit device
C. heavy voltage drop takes place across the device
D. none of these

416. Intrinsic stand-off ratio of a UJT is given by:
A. $R_{B_1} + R_{B_2}$ B. $R_{B_1} \| R_{B_2}$
C. $R_{B_1}/(R_{B_1} + R_{B_2})$ D. none of these

417. The value of forward voltage drop of a UJT at 25°C is:
A. 0.5 V B. 0.25 V
C. 1.0 V D. 1.5 V

418. Series and parallel operations of thyristors require voltage and current sharing networks to protect them under:
A. steady state conditions only
B. transient state conditions only
C. steady state and transient state conditions
D. none of these

419. A pulse train for gating thyristor is used as it:
A. reduces thyristor loss
B. increases thyristor loss
C. reduces harmonics D. none of these

420. Generally, for generating triggering pulses:
A. UJT are used B. PUT are required
C. UJT and PUT are required
D. none of these

421. While the SCR anode is positive and gate current is zero, the voltage appears across just one of its three junction is:
A. nearly all the applied voltage
B. nearly half the applied voltage
C. zero voltage D. none of these

422. For the control purpose of a thyristor:
A. a cathode gate requires a positive supply whereas the anode gate requires a negative supply
B. a cathode and anode gate requires a negative supply
C. a cathode gate as well as anode gate require a positive supply
D. none of these

423. In the reverse-biased condition of an SCR:
A. two out of three junctions are reversebiased
B. one out of three junctions is reversebiased
C. all the three junctions are reverse-biased
D. none of these

424. Method of switching off a thyristor is known as:
A. commutation B. triggering
C. snubbing D. none of these

425. Natural commutation method can be adopted for:
A. dc applications only B. ac applications only
C. both AC and DC applications
D. none of these

426. Forced commutation method is meant for:
A. dc applications only B. ac applications only
C. both DC and AC applications
D. none of these

427. Natural commutation method is applied in:
A. ac voltage controllers
B. choppers
C. inverters
D. none of these

428. Forced commutation method is generally used in:
A. controlled rectifiers B. cycloconverter
C. ac voltage controller D. choppers

429. Auxiliary commutation is a type of:
A. natural commutation method
B. forced commutation method
C. both A and B are correct
D. none of these

430. In case of inductive circuits, retriggering of an SCR is avoided by using a:
A. capacitor B. inductor
C. freewheeling diode D. none of these

431. In class A commutation:
A. the forward current flowing through the device is reduced to less than the level of holding current of the device
B. the current through the device is made maximum
C. the same current is flowing through the device as well as load
D. none of these

432. Class C commutation circuit should be designed such that the circuit recovery time must be:
A. more than the turn-off time of the main SCR
B. equal to the turn-off time of main SCR
C. less than the turn-off time of the main SCR
D. none of these

Ans. 411. B 412. A 413. A 414. A 415. A 416. C 417. ? 418. C 419. A 420. C 421. A 422. A 423. A 424. A 425. B 426. C 427. A 428. D 429. B 430. C 431. A 432. A

433. The main application of class B self commutation process is found in:
A. cycloconverter B. dc chopper circuits
C. power rectifier circuits
D. none of these

434. In class E external pulse commutation method, the external pulse is obtained by:
A. transistor B. zener diode
C. battery D. none of these

435. The condition for underdamped oscillations in a series capacitor method is:
A. R $(4L/C)$ B. $R = (4L/C)$
C. $R_2 < (4L/C)$ D. $R2$ $(4L/C)$

436. Single-phase half-wave controlled bridges use:
A. one thyristor B. two thyristors
C. three thyristors D. none of these

437. A freewheeling diode is used in a controlled rectifier circuit in case of:
A. resistive load B. inductive loads
C. capacitive loads D. none of these

438. A single-phase full-wave fully-controlled bridge uses:
A. two thyristors B. three thyristors
C. four thyristors D. none of these

439. A single-phase half-wave controlled rectifier has 400 sin 314t as the input voltage and R as the load. For a firing angle of 60° for the thyristor, the average output voltage is:
A. $400/\pi$ B. $300/\pi$
C. $240/\pi$ D. none of these

440. A single-phase full-wave half-controlled bridge uses:
A. one SCR B. two SCRs
C. four SCRs D. six SCRs

441. A three-phase semiconverter can work as:
A. converter for $\alpha = 0°$ to 180°
B. converter for $\alpha = 0°$ to 90°
C. inverter for $\alpha = 90°$ to 180°
D. inverter for $\alpha = 0°$ to 90°

442. A four quadrant operation requires:
A. two full converters in series
B. two full converters connected back to back
C. two full converters connected in parallel
D. two converters connected back to back

443. A single-phase one-pulse controlled circuit has resistance and counter emf load and 400 sin 314t as the source voltage. For a load counter emf of 200 V, the range of firing angle control is:
A. 30° to 15° B. 60° to 180°
C. 60° to 120° D. none of these

444. In a single-phase half-wave circuit with R–L load, and a freewheeling diode across the load, extinction angle β is more than π. For a firing angle α, the thyristor and freewheeling diode would conduct respectively, for:
A. $(\pi - \alpha), \beta$ B. $(\beta - \alpha)\ (\pi - \alpha)$
C. $(\pi - \alpha), (\beta - \pi)$ D. $(\pi - \alpha), (\pi - \beta)$

445. In a single-phase one-pulse circuit with R–L load and a freewheeling diode, extinction angle β is less than π. For a firing angle α, the SCR and freewheeling diode would, respectively, conduct for:
A. $(\beta - \alpha), 0°$ B. $(\pi - \alpha), (\pi - \beta)$
C. $\alpha\ (\beta - \alpha)$ D. $(\beta - \alpha), \alpha$

446. A single-phase full-wave midpoint thyristor converter uses a 230/200 V transformer with centre tap on the secondary side. The PIV for each thyristor is:
A. 100 V B. 141.4 V
C. 200 V D. 282.8 V

447. A single-phase two-pulse bridge converter has an average output voltage and power output of 500 V and 10 kW, respectively. The thyristor used in the two-pulse bridge converter is now re-employed to form a single-phase two-pulse midpoint converter. This new controlled converter would give, respectively, an average output voltage and power output of:
A. 500 V, 10 kW B. 250 V, 5 kW
C. 250 V, 10 kW D. 500 V, 5 kW

448. In a single-phase full converter bridge, the average output voltage is given by:
A. $\frac{1}{\pi}\int_{\alpha}^{\pi+\alpha} V_m \cos \omega t\, d\omega t$ B. $\frac{1}{\pi}\int_{\alpha}^{\alpha+\pi} V_m \cos \omega t\, d\omega t$
C. $\frac{1}{\pi}\int_{\alpha-(\pi/2)}^{\alpha+(\pi/2)} V_m \cos \omega t \cdot d\omega t$
D. $\frac{1}{\pi}\int_{\alpha}^{\pi+\alpha} V_m \cos \omega t\, d\omega t$

449. In a single-phase semiconverter, the average output voltage is given by:
A. $\frac{1}{\pi}\int_{\alpha}^{\pi} V_m \cos \omega t \cdot d\omega t$ B. $\frac{1}{\pi}\int_{\alpha}^{\pi+\alpha} V_m \cos \omega t\, d\omega t$
C. $\frac{1}{\pi}\int_{\alpha-(\pi/2)}^{\alpha+(\pi/2)} V_m \cos \omega t \cdot d\omega t$
D. $\frac{1}{\pi}\int_{\alpha-(\pi/2)}^{\pi} V_m \cos \omega t \cdot d\omega t$

450. For continuous conduction, in a single-phase full converter each pair of SCRs conduct for:
A. $\pi - \alpha$ B. π
C. α D. $\pi + \alpha$

451. For discontinuous load current and extinction angle $\beta > \pi$, in a single-phase full converter each thyristor conducts for:

Ans. 433. B 434. A 435. C 436. A 437. B 438. B 439. B 440. B 441. A 442. B 443. A 444. C 445. A 446. D 447. B 448. C 449. C 450. B

A. α B. β – α
C. β D. α + β

452. In a single-phase semiconverter with resistive load and for a firing angle α, each thyristor freewheeling diode conduct, respectively, for:
A. α, 0° B. π – α, α
C. π + α, α D. π – α, 0°

453. In a single-phase full converter with resistive load and for a firing angle α, the load current is zero and non-zero, respectively, for:
A. α, π – α B. π – α, α
C. α, π + α D. α, π

454. In a single-phase full converter, if α and β are the firing and extinction angles, the load current is:
A. discontinuous if (β – α) < π
B. discontinuous if (β – α) > π
C. discontinuous if (β – α) = π
D. continuous if (β – α) < π

455. With discontinuous conduction and extinction angle, β < π, in a single-phase semiconverter, freewheeling diode conducts for:
A. α B. π – β
C. β – π D. zero degree

456. With discontinuous conduction and extinction angle β > π, freewheeling diode conducts for:
A. α B. β – π
C. π + α D. β

457. For a continuous conduction, freewheeling diode, in a single-phase semiconverter, conducts for:
A. α B. π – α
C. π D. π + α

458. For discontinuous conduction and extinction angle β < π, in a single-phase semiconverter, each thyristor conducts for:
A. π – α B. β – α
C. α D. β

459. For discontinuous conduction and extinction angle β > π in a single-phase semiconverter, each thyristor conducts for:
A. π – α B. β – π
C. α D. β

460. For continuous conduction, in a single-phase semiconverter, each thyristor conducts for:
A. α B. π
C. α + π D. π – α

461. In controlled rectifiers, the nature of load current:
A. does not depend on type of load and firing angle delay
B. depends both on the type of load and firing angle delay
C. depends only on the type of load
D. depends only on the firing angle delay

462. In a single-phase full converter, if output voltage has peak and average values of 325 V and 133 V respectively, then the firing angle is:
A. 40° B. 140°
C. 50° D. 130°

463. In a single-phase semiconverter, if output voltage has peak and average values of 325 V and 133 V respectively, the firing angle is:
A. 40° B. 140°
C. 73.49° D. 80°

464. For a single-phase, phase-controlled rectifier, with a freewheeling diode across the load:
A. the instantaneous output voltage v_0 is always positive
B. v_0 may be positive or zero
C. v_0 may be positive, zero or negative
D. none of these

465. In a single-phase full converter, if load current is I, and ripple free, then average thyristor current is:
A. $\frac{1}{2}I$ B. $\frac{1}{3}I$
C. $\frac{1}{4}I$ D. I

466. In a single-phase full converter, the number of SCRs conducting during overlap is:
A. 1 B. 2
C. 3 D. 4

467. In a single-phase full converter, the output voltage during overlap is equal to:
A. zero B. source voltage
C. source voltage minus the inductance drop
D. none of these

468. The frequency of the ripple in the output voltage of a three-phase semiconverter depends upon:
A. firing angle and load resistance
B. firing angle and load inductance
C. the load circuit parameters
D. firing angle and the supply frequency

469. In a three-phase full converter, if load current is I and ripple free, then average thyristor current is:
A. $\frac{1}{2}I$ B. $\frac{1}{3}I$
C. $\frac{1}{4}I$ D. I

470. The effect of source inductance on the performance of single-phase and three-phase full converters is to:
A. reduce the ripples in the load current
B. make discontinuous current as continuous
C. reduce the output voltage
D. increase the load voltage

471. In a three-phase full converter, the output voltage during overlap is equal to:

Ans. 451. B 452. D 453. A 454. A 455. D 456. B 457. A 458. A 459. A 460. D 461. B 462. C 463. C 464. B 465. A 466. D 467. A 468. D 469. B 470. D

A. zero
B. source voltage
C. source voltage minus the inductance drop
D. average value of the conducting-phase voltages

472. The total number of thyristors conducting simultaneously in a three-phase full converter with overlap considered has the sequence of:
A. 3, 3, 2, 2 B. 3, 3, 3, 2
C. 3, 2, 3, 2 D. 2, 2, 2, 3

473. A three-phase full converter has an average output of 200 V for zero degree firing angle and for resistive load. For a firing angle of 90°, the output voltage would be:
A. zero B. 50 V
C. 100 V D. 26.8 V

474. A converter which can operate in both three pulse and six-pulse modes is a:
A. one-phase full converter
B. three-phase half-wave converter
C. three-phase semiconverter
D. three-phase full converter

475. In a three-phase semiconverter, for firing angle less than or equal to 60°, each thyristor and diode conduct, respectively, for:
A. 60°, 60° B. 90°, 30°
C. 120°, 120° D. none of these

476. In a three-phase semiconverter, for firing angle less than or equal to 60°, freewheeling diode conducts for:
A. 30° B. 60°
C. 90° D. zero degree

477. In a three-phase semiconverter, for firing angle equal to 90°, and for continuous conduction, each thyristors and diode conduct, respectively, for:
A. 30°, 60° B. 60°, 30°
C. 60°, 60° D. 30°, 30°

478. In a three-phase semiconverter, for a firing angle equal to 90° and for continuous conduction, freewheeling diode conducts for:
A. 30° B. 60°
C. 90° D. zero degree

479. In a three-phase semiconverter, or firing angle equal to 120° and extinction angle equal to 110°, each thyristor and diode conducts, respectively, for:
A. 30°, 60° B. 60°, 60°
C. 90°, 30° D. 110°, 30°

480. In a three-phase semiconverter, for firing angle equal to 120° and extinction angle equal to 110°, freewheeling diode conducts for:
A. 10° B. 30°
C. 50° D. 110°

481. In a three-phase semiconverter, for firing angle equal to 120° and extinction angle equal to 100°, none of the bridge elements conduct for:
A. 10° B. 20°
C. 30° D. 60°

482. In a three-phase semiconverter, the three thyristors are triggered at an interval of:
A. 60° B. 90°
C. 120° D. 180°

483. In a three-phase full converter, the six thyristors are fired at an interval of:
A. 30° B. 60°
C. 90° D. 120°

484. In a three-phase full converter, the output voltage pulsates at a frequency equal to:
A. supply frequency, f B. $2f$
C. $3f$ D. $6f$

485. The three-phase AC to DC converter which requires neutrals point connection is:
A. three-phase semiconverter
B. three-phase full converter
C. three-phase half-wave converter
D. three-phase full converter with diodes

486. In a three-phase semiconverter, frequency of the ripple in the output voltage may be:
A. 3 times the supply frequency f for firing angle a less than 60°
B. $3f$ for a greater than 60°
C. $6f$ for a less than 60° D. $6f$ for a greater than 60°

487. In circulating-current type of dual converter, the nature of voltage across reactor is:
A. alternating B. pulsating
C. direct D. triangular

488. The peak inverse voltage in AC to DC converter systems is highest in:
A. single-phase full-wave midpoint converter
B. single-phase full converter
C. three-phase bridge converter
D. three-phase half-wave converter

489. In a dual converter, converters 1 and 2 work as under:
A. 1 as rectifier, 2 as inverter
B. both as rectifiers
C. both as inverters D. none of these

490. For the same AC voltage and load impedance, which of the following statements about rectifiers are correct?
A. The average load current in a full-wave rectifier is p times that in a half-wave rectifier
B. Half-wave rectifier will have bigger sized transformer compared to full-wave rectifier
C. Half-wave rectifier will have a smaller sized transformer compared to a fullwave rectifier
D. None of these

Ans. 471. D 472. C 473. C 474. C 475. C 476. D 477. C 478. C 479. B 480. C 481. B 482. C 483. B 484. D 485. C 486. C 487. A 488. A 489. A 490. C

491. Output voltage expression for single-phase
A. $\frac{3V_m}{2\pi}(1+\cos\alpha)$ B. $\frac{2V_m}{\pi}\cos\alpha$
C. $\frac{V_m}{\pi}(1+\cos\alpha)$ D. $\frac{3\sqrt{3}}{2\pi}V_m\cos\alpha$

492. DC chopper converts:
A. DC to DC B. AC to DC
C. DC to AC D. none of these

493. The features of the chopper drives are:
A. smooth control but slow response
B. smooth control but fast response
C. fast response with smooth control but less efficient
D. none of these

494. Choppers can be used in future electric automobiles:
A. for speed control only
B. for speed control and braking
C. for braking only D. none of these

495. Which of the following system is preferred for chopper drives?
A. Constant frequency system
B. Variable frequency system
C. Constant voltage system
D. None of these

496. Which of the following configurations is used for both motoring and regenerative braking?
A. First quadrant configuration
B. Second quadrant configuration
C. Two quadrant configuration
D. Four quadrant configuration

497. Which of the following configurations is used for motoring operation of DC motor load?
A. First quadrant chopper
B. Second quadrant chopper
C. Fourth quadrant chopper
D. None of these

498. If two or more choppers are operated in parallel and shifted from each other, the ripple frequency will:
A. remain same B. increase
C. decrease D. none of these

499. Which of the following systems has the greater possibility of interference with signalling and telephone lines?
A. Constant frequency system
B. Variable frequency system
C. Both are correct D. None of these

500. Constant frequency system gives:
A. low ripple and fast response
B. high ripple and slow response
C. low ripple and slow response
D. none of these

501. Chopper controlled DC motor used in underground traction with regenerative braking, the power consumption will be reduced to:
A. 35–40% B. 50–60%
C. 60–70% D. none of these

502. In DC choppers, if T_{ON} is the on-period and f is the chopping frequency, then output voltage in terms of input voltage V is given by:
A. $V{\cdot}T_{ON}/f$ B. $V{\cdot}f/T_{ON}$
C. $V/f{\cdot}T_{ON}$ D. $V{\cdot}f{\cdot}T_{ON}$

503. In DC choppers, the waveforms for input and output voltages are respectively:
A. discontinuous, continuous
B. both continuous
C. both discontinuous
D. continuous, discontinuous

504. For type A chopper, V is the source voltage, R is the load resistance and d is the duty cycle. The average output voltage for this chopper is:
A. δV B. $(1-\alpha)V$
C. V/α D. $V/(1-\alpha)$

505. A chopper has V as the source voltage, R as the load resistance and d as the duty cycle. For this chopper, rms value of output voltage is:
A. δV B. $\sqrt{\delta}\cdot V$
C. $V/\sqrt{\delta}$ D. $\sqrt{1-\delta}\cdot V$

506. For a chopper, V is the source voltage, R is the load resistance and d is the duty cycle. Rms and average values of thyristor currents for this chopper are:
A. $\delta\cdot\frac{V}{R}, \sqrt{\delta}\cdot\frac{V}{R}$ B. $\sqrt{\delta}\cdot\frac{V}{R}, \sqrt{\delta}\cdot\frac{V}{R}$
C. $\sqrt{\delta}\cdot\frac{V}{R}, \delta\frac{V}{R}$ D. none of these

507. In DC chopper, per unit ripple is maximum when duty cycle 'd' is:
A. 0.2 B. 0.5
C. 0.7 D. none of these

508. A thyristor chopper is preferred over other choppers because it:
A. provides static switching
B. can be easily manufactured
C. is economical D. none of these

509. Duty cycle of chopper circuit is expressed by:
A. $t_{ON}/(t_{ON}+t_{OFF})$ B. t_{ON}/t_{OFF}
C. t_{OFF}/t_{ON} D. T/t_{ON}

510. Chopper frequency is:
A. $1/t_{ON}$ B. $1/(t_{ON}/t_{OFF})$
C. t_{OFF}/t_{ON} D. none of these

511. Load voltage of DC chopper for duty cycle δ, and input voltage V:
A. $\delta\cdot V$ B. V/δ
C. δ/V D. none of these

Ans. 491. C 492. A 493. B 494. B 495. A 496. C 497. A 498. B 499. B 500. A 501. A 502. D 503. D 504. C 505. B 506. C 507. B 508. A 509. A 510. B 511. A

512. In single thyristor chopper circuit, the capacitor discharge current is:
A. oscillating but decaying in nature
B. pulsating in nature
C. alternating in nature D. none of these

513. In DC chopper circuit, the load voltage is governed by:
A. number of thyristors used in the circuit
B. duty cycle of the circuit
C. DC voltage applied to circuit
D. none of these

514. A voltage commutated chopper has the following parameters:
$V = 200$ V, load circuit parameter: 1Ω, 2 mH, 50 V
Commutation circuit parameters, $L = 25$ μH, $C = 50$ μF.
For a constant load current of 100 A, the effective on-period and peak current through the main thyristor are respectively:
A. 1000 μs, 200 A B. 700 μs, 382.8 A
C. 700 μs, 282.8 A D. none of these

515. A load commutated chopper, fed from 200 V DC source, has a constant load current of 50 A. For a duty cycle of 0.4 and a chopping frequency of 2 kHz, the value of commutating capacitor and the turn-off time for one thyristor pair are respectively:
A. 25 μF, 50 μs B. 50 μF, 50 μs
C. 25 μF, 25 μs D. 50 μF, 25 μs

516. A DC battery is charged from a constant DC source of 200 V through a chopper. The DC battery is to be charged from its internal emf of 90 to 120 V. The battery has internal resistance of 1W. For constant charging current of 10 A, the range of duty cycle is:
A. 15 to 65 B. 65 to 0.8
C. 0.8 to 0.95 D. none of these

517. For type *A* chopper V, R, I_0 and δ are respectively the DC source voltage, load resistance, constant load current and duty cycle. For this chopper, average and rms values of freewheeling diode currents are:
A. $\delta I_0, \sqrt{\delta} \cdot I_0$ B. $(1-\delta) I_0, \sqrt{1-\delta} \cdot I_0$
C. $\delta \cdot V/R, \sqrt{\delta} \cdot V/R$ D. none of these

518. A step-up chopper has V as the source voltage and d as the duty cycle. The output voltage for this chopper is given by:
A. $V(1+\delta)$ B. $V/(1-\delta)$
C. $V(1-\delta)$ D. none of these

519. A DC chopper is fed from 100 V DC. Its load voltage consists of rectangular pulses of duration 1 msec in an overall cycle time of 3 msec. The average output voltage and ripple factor for this chopper are respectively:
A. 25 V, 1 B. 50 V, 1
C. 33.33, 1.5 D. none of these

520. When a series LC circuit is connected to a DC supply of V volts through a thyristor, then the peak current through thyristor is:
A. $V \cdot \sqrt{CL}$ B. $V/\sqrt{CL}$
C. $V \cdot \sqrt{C/L}$ D. $V \cdot \sqrt{L/C}$

521. In DC choppers if T is the chopping period, then output voltage can be controlled by PWM by varying:
A. T keeping T_{ON} constant
B. T_{ON} keeping T constant
C. T keeping T_{OFF} constant
D. none of these

522. In DC choppers, for chopping period T, the output voltage can be controlled by FM by varying:
A. T keeping T_{ON} constant
B. T_{ON} keeping T constant
C. T_{OFF} keeping T constant
D. none of these

523. The main part of an inverter is:
A. oscillator circuit B. dc source
C. step up transformer D. filter

524. Now-a-days mostly thyristors are used for developing:
A. medium and high power inverters
B. medium power inverters
C. low power inverters D. none of these

525. An inverter is a device which converts:
A. DC into AC B. AC into DC
C. AC into AC of higher frequency
D. AC into AC of lower frequency

526. Output frequency of inverter circuit depends on:
A. values of the resonant elements
B. transformation ratio of transformer
C. level of DC voltage applied
D. none of these

527. According to the connections of thyristors, inverters can be classified as:
A. series type B. parallel type
C. bridge type D. all of these

528. Oscillator circuit of inverter is build-up of:
A. resistor B. inductor
C. capacitor D. all of these

529. Which of the following statements is incorrect?
A. Bridge inverter does not use any transformer
B. Bridge inverters can be designed for single-phase as well as three-phase circuits
C. In McMurray inverter circuit the commutating pulse is obtained from the main power source
D. Current source inverters are mainly used for high frequency applications

530. In voltage source inverters:
A. load voltage waveform $v0$ depends on load impedance Z_L, whereas, load current waveform i_L does not depend on Z_L

Ans. 512. A 513. B 514. B 515. A 516. A 517. B 518. B 519. C 520. C 521. B 522. A 523. A 524. A 525. A 526. A 527. D 528. D 529. C

B. both v_0 and i_L depend on Z_L
C. v_0 does not depend on ZL, whereas, iL depends on ZL
D. none of these

531. A single-phase full bridge inverter can operate in load commutation mode in case load consists of:
A. RLC overdamped B. RLC underdamped
C. RLC critically damped
D. none of these

532. A single-phase bridge inverter delivers power to series connected RLC load with $R = 2\ \Omega$, $XL = 8\ \Omega$. For this inverter-load combination, load commutation is possible, if the magnitude of $1/\omega C$ is:
A. 10 Ω B. 8 Ω
C. 6 Ω D. zero

533. In single-pulse modulation of PWM inverters, third harmonic can be eliminated if pulse width is equal to:
A. 30° B. 60°
C. 120° D. none of these

534. In single-pulse modulation of PWM inverters, fifth harmonic can be eliminated if pulse width is equal to:
A. 30° B. 72°
C. 90° D. 108°

535. In single-pulse modulation of PWM inverters, the pulse width is 120°. For an input voltage of 220 V dc, the rms value of output voltage is:
A. 179.63 V B. 254.04 V
C. 280 V D. none of these

536. So far as the harmonic content at the inverter output voltage is concerned, multiple pulse modulation will provide:
A. less harmonics B. more harmonics
C. perfect sinusoidal of fundamental frequency
D. none of these

537. In single-pulse modulation used in PWM inverters, V is the input DC voltage. For eliminating third harmonic, the magnitude of rms value of fundamental component of output voltage and pulse width are respectively:
A. $\frac{2\sqrt{2}}{\pi}V, 120°$ B. $\frac{4V}{\pi}, 60°$
C. $\frac{2\sqrt{2}}{\pi}V, 60°$ D. $\frac{4V}{\pi}, 120°$

538. In multiple-pulse modulation used in PWM inverters, the amplitudes of reference square wave and triangular carrier wave are respectively 1 V and 2 V. For generating 5 pulses per half-cycle, the pulse width should be:
A. 36° B. 24°
C. 18° D. 6°

539. In multiple-pulse modulation used in PWM inverters, the amplitudes and frequency for triangular carrier and square reference signals are respectively 4 V, 6 kHz and 1 V, 1 kHz. The number of pulses per half-cycle and pulse width are respectively:
A. 6, 90° B. 3, 45°
C. 4, 60° D. 3, 40°

540. Which of the following statements is correct in connection with inverters?
A. Voltage source inverter and current source inverter both require feedback diodes
B. Only current source inverter requires feedback diodes
C. GTOs can be used in current source inverter
D. Only VSI requires feedback diodes

541. In a constant source inverter, if frequency of output voltage is f Hz, then frequency of voltage input to constant source inverter is:
A. f B. $2f$
C. $3f$ D. $4f$

542. In an inverter with fundamental output frequency of 50 Hz, if third harmonic is eliminated, then frequencies of other components in the output voltage wave in Hz, would be:
A. 250, 350, 500, high frequencies
B. 50, 50, 350, 550
C. 50, 150, 250, 550 D. none of these

543. A single-phase current source inverter has capacitor C as the load. For a constant source current, the voltage across the capacitor is:
A. square wave B. triangular wave
C. step function D. none of these

544. A single-phase full bridge VSI has inductor L as the load. For a constant source voltage, the current through the inductor is:
A. square wave B. triangular wave
C. sine wave D. none of these

545. Natural commutation is considered to a kind of:
A. forced commutation B. line commutation
C. commutation failure D. none of these

546. Which of the following is the limitation of current source inverter?
A. Sluggish dynamic performance
B. Stability problem
C. Unreliable D. None of these

547. In three-phase inverter, leading VAR requirement can be met by a variable VAR generator using:
A. current fed inverter VAR generator
B. rotating synchronous condenser
C. switched capacitor D. all of these

548. Which of the following classes is the most popular among forced commutated current fed inverters?

Ans. 530. C 531. B 532. A 533. C 534. B 535. A 536. A 537. A 538. C 539. B 540. D 541. B 542. B 543. B 544. B 545. B 546. A 547. D

A. Autosequential commutated inverter
B. Auxiliary bridge commutated inverter
C. Third harmonic commutated inverter
D. None of these

549. A cycloconverter is a device which:
A. converters AC into DC
B. converts DC into AC
C. converts AC of one frequency into AC of other frequency
D. none of these

550. Cycloconverter is concerned mostly:
A. with direct conversion of energy to a different frequency
B. with direct conversion of energy at the same frequency
C. with DC source only D. none of these

551. Cycloconverter can be considered to be composed of two:
A. rectifiers connected back to back
B. inverters connected back to back
C. converters connected back to back
D. either of these

552. Whether the load is resistive, inductive, or capacitive, the firing of thyristors of a cycloconverter are:
A. commutated naturally
B. forced commutated
C. either of these D. none of these

553. Cycloconverter is a:
A. frequency converter which has no intermediate DC state
B. device which converts AC to dc
C. device which converts DC to ac
D. none of these

554. Cycloconverter is capable of converting power to:
A. a lower frequency B. a higher frequency
C. the same frequency D. none of these

555. Cycloconverter has:
A. low input power factor
B. high input power factor
C. medium power factor
D. none of these

556. In a single stage conversion of cycloconverter, AC power at one frequency is converted directly:
A. to AC at low frequency
B. to AC at very high frequency
C. to DC only
D. none of these

557. The ruggedness and low weight of solid state cycloconverter makes it attractive for:
A. aircraft electrical system
B. battery operated vehicle
C. electric traction D. all of these

558. The cycloconverter requires:
A. a large number of thyristors but its control circuit is very simple
B. a small number of thyristors and has complicated control circuit
C. a large number of thyristors and its control circuit is more complex
D. none of these

559. Soft start is a:
A. gradual application of power when switching on a load
B. flow of power in one direction
C. switching off D. switching on

560. Cycloconverters are more efficient than DC link as they:
A. do not have any rotating part
B. use very less power
C. produce the output frequency only in one stage
D. none of these

561. A centre tapped transformer configuration of single-phase cycloconverter uses:
A. four thyristors B. six thyristors
C. eight thyristors D. eighteen thyristors

562. To obtain sinusoidal waveform at the output of a centre tapped transformer configuration of a single-phase cycloconverter, the firing angle is of value:
A. 0° B. 45°
C. 90° D. 180°

563. A single-phase bridge type cycloconverter uses:
A. two thyristors B. four thyristors
C. eight thyristors D. sixtyfour thyristors

564. A single-phase bridge type cycloconverter is used for producing an output frequency of:
A. 12.5 Hz B. 25 Hz
C. 50 Hz D. 75 Hz

565. Generally a single-phase centre tapped cycloconverter is used to produce an output frequency of:
A. 12.5 Hz B. 50/3 Hz
C. 25 Hz D. 75 Hz

566. Three-phase three pulse cycloconverter uses:
A. six thyristors B. twelve thyristors
C. eighteen thyristors D. none of these

567. The sum of firing angles of the positive group of thyristors and negative group of thyristors in case of three pulse phase cycloconverter is equal to:
A. 90° B. 180°
C. 360° D. all of these

568. Three-phase six pulse bridge cycloconverter uses:
A. six thyristors B. twelve thyristors
C. eighteen thyristors D. thirty six thyristors

Ans. 548. A 549. C 550. A 551. C 552. A 553. A 554. B 555. A 556. A 557. A 558. C 559. A 560. C 561. A 562. A 563. C 564. B 565. B 566. C 567. B 568. D

569. Cycloconverters are mainly used in:
A. speed control of DC shunt motors
B. speed control of synchronous motors
C. electric traction D. all of these

570. Which of the following statements is correct?
A. Speed reversal of electric motors is not possible with the help of cycloconverter
B. Regenerative braking cannot be achieved with the help of cycloconverter
C. An input transformer is required for a bridge configuration of cycloconverter
D. Minimum firing angle will produce the best possible waveform for a cycloconverter

571. From the point of view of low distortion the ratio of the output to input frequency of a cycloconverter should be:
A. low B. high
C. very high D. none of these

572. A cycloconverter is a:
A. frequency changer from higher to lower frequency with one-stage conversion
B. frequency changer from higher to lower frequency with two-stage conversion
C. frequency changer from lower to high frequency with one-stage conversion
D. either A or B

573. The cycloconverters require natural or forced commutation as under:
A. natural commutation in both step-up and step-down cycloconverters
B. forced commutation in both step-up and step-down cycloconverters
C. forced commutation in step-up cycloconverts
D. forced commutation in step-down cycloconverters

574. Three-phase to three-phase cycloconverters employing 18 SCRs and 36 SCRs have the same voltage and current ratings for their component thyristors. The ratio of VA rating of 36 SCR device to that of 18 SCR device is:
A. 1/2 B. 1
C. 2 D. none of these

575. Three-phase to three-phase cycloconverter employing 18 thyristors and 36 thyristors have the same voltage and current rating for their components thyristors. The ratio of power handled by 36 thyristors to that handled by 18 thyristors is:
A. 4 B. 2
C. 1 D. none of these

576. The number of thyristors required for singlephase to single-phase cycloconverter of the mid-point type and for three-phase to three-phase 3-pulse type cycloconverter are respectively:
A. 4, 6 B. 4, 4
C. 4, 18 D. none of these

577. A three-phase to single-phase conversion device employs a 6-pulse bridge cycloconverter. For an input voltage of 200 V per phase, the fundamental rms value of output voltage is:
A. $600/\pi$ V B. $300\sqrt{3}/\pi$ V
C. $300/\pi$ V D. none of these

578. A three-phase to single-phase cycloconverter consists of positive and negative group of converters in this device, one of the two component converters would operate as a:
A. rectifier, if the output voltage V_o and output current I_o have the same polarity
B. inverter, if Vo and Io have the same polarity
C. rectifier, if Vo and Io are of opposite polarity
D. none of these

579. A three-phase to three-phase cycloconverter requires:
A. 18 SCRs for 3-pulse device
B. 18 SCRs for 6-pulse device
C. 36 SCRs for 3-pulse device
D. none of these

580. Which of the following statements are incorrect for cycloconverters?
A. Step-down cycloconverter works on natural commutation
B. Step-up cycloconverter requires forced commutation
C. Load commutate cycloconverter works on line commutation
D. Load commutated cycloconverter requires a generated emf in the load circuit

581. A single-phase voltage controller feeds an induction motor and a heater:
A. in both the loads, fundamental and harmonics are useful
B. in induction motor only fundamental and in heater only harmonics are useful
C. in induction motor only fundamental and in heater harmonics as well as fundamental are useful
D. none of these

582. Output frequency of a cycloconverter is significantly lower than the input frequency for:
A. keeping commutation failure
B. avoiding commutation failure
C. reducing switching losses
D. none of these

583. In a normal three thyristor phase rectifier, a thyristor can not be fired during:
A. first 10° of its anode voltage
B. first 30° of its anode voltage

Ans. 569. C 570. D 571. A 572. D 573. C 574. C 575. A 576. C 577. A 578. A 579. A 580. C 581. C 582. B 583. B

C. first 50° of its anode voltage
D. first 80° of its anode voltage

584. Armature voltage of a DC motor can be controlled by means of:
A. cycloconverter B. inverters
C. AC-DC converters
D. bridge rectifier circuit with fixed input

585. The speed of DC motor can be increased by:
A. increasing the thyristor firing angle in the armature circuit of a phase controlled converter
B. decreasing the thyristor firing angle in the armature circuit of a phase controlled converter
C. increasing the duty cycle of a chopper
D. none of these

586. A thyristor power converter is said to be in discontinuous conduction when:
A. the load current is zero even though the load voltage is present
B. both the load voltage and load current are zero simultaneously
C. when the load voltage is ripple free
D. none of these

587. A freewheeling diode in a phase controlled converter:
A. increases the chances of discontinuous conduction in the load
B. decreases the chances of discontinuous conduction in the load
C. causes the chances of discontinuous conduction in the load
D. none of these

588. The speed of DC shunt motor above normal speed can be controlled by:
A. armature voltage control method
B. flux control method
C. both the methods D. none of these

589. A freewheeling diode in a phase controlled converter:
A. causes smoothening of load current
B. causes poor voltage regulation
C. enables inverter operation
D. none of these

590. A half-controlled converter:
A. has poor voltage regulation
B. has an improved power factor
C. is capable of inversion also
D. all of these

591. The ripple content of load current of a converter feeding R-L load is decided by:
A. load resistance only
B. load inductance
C. both load resistance and load inductance
D. neither resistance nor inductance

592. For controlling the speed of DC motor of 150 HP rating, the following types of converters are normally used:
A. single-phase full converters
B. single-phase dual converters
C. three-phase full converters
D. three-phase dual converters

593. The overlap of the converter is responsible for:
A. voltage regulation of the converter
B. additional losses of the converter
C. additional harmonics in the load current
D. all of these

594. The advantage of the tachometer speed control method for DC motor is that, it senses:
A. back emf B. armature current
C. armature voltage D. speed

595. In the speed control circuit shown below, if the speed of the motor decreases, the capacitor 'C' will charge (see figure given below):
A. slower
B. faster
C. independent of the speed
D. to the supply voltage

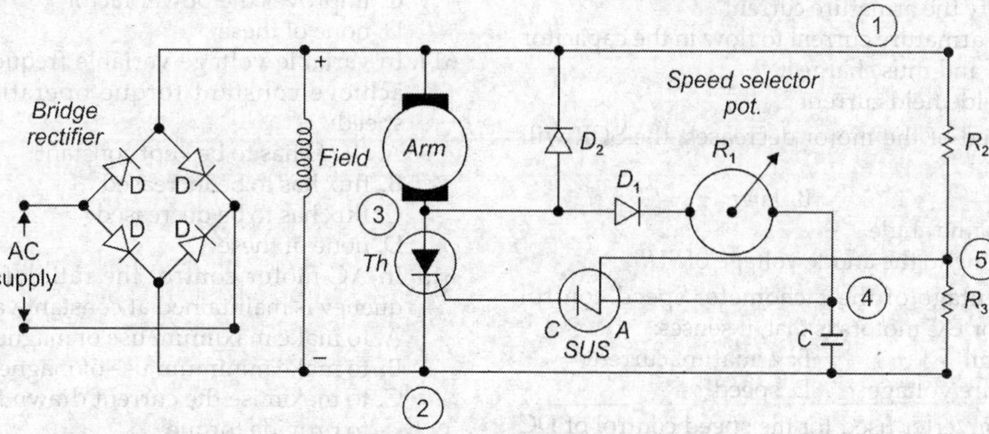

Fig for Q.No. 595

Ans. 584. C 585. C 586. B 587. B 588. B 589. A 590. B 591. C 592. C 593. A 594. D 595. B

596. In the above circuit, if the speed of the motor decreases, the *SUS* will fire at:
A. higher the rated voltage
B. lower the rated voltage
C. the same voltage
D. voltage determined by the gate voltage of *SUS*

597. In the same circuit, if the speed of the motor decreases, the DC voltage across SCR will:
A. remains the same B. decreases
C. increases D. become more smooth

598. In the same circuit, if the motor speed decreases, the rms value of the voltage across SCR will:
A. decreases B. increases
C. remains the same D. become zero

599. In the same circuit, the function of resistances $R_1 - R_2$ is to provide:
A. breakover voltage of *SUS*
B. anode voltage of *SUS*
C. reference voltage of *SUS*
D. none of these

600. In the same circuit, which of the following devices can not replace *SUS*?
A. Diac B. FET
C. UJT D. All of these

601. An SCR is used to control the speed of a DC motor. At full speed, the motor is taking 1 A at 75 V. The maximum forward surge current rating and maximum forward breakover voltage rating respectively are of the order of:
A. 3 A, 235 V B. 1 A, 300 V
C. 5 A, 166 V D. 5 A, 200 V

602. A motor armature supplied through phase controlled SCRs receives a smoother voltage shape at:
A. high motor speed B. low motor speeds
C. rated normal motor speeds
D. none of these

603. The function of diode D_2 is:
A. to dissipate $L\,di/dt$ of the armature
B. to rectify the armature current
C. to stop armature current to flow in the capacitor branch and thus charge it
D. to provide field current

604. If the speed of the motor decreases, the SCR will fire:
A. earlier B. later
C. at the same angle
D. according to the anode voltage of *SUS*

605. The advantage of the tachometer speed control method for DC motors is that, it senses:
A. back emf B. armature current
C. armature voltage D. speed

606. A dual converter used for the speed control of DC motors, will have two bridges, they are:
A. two rectifiers B. two inverters
C. one rectifier and one inverter
D. none of these

607. For DC motors a dual converter is used to obtain:
A. reversible speed control
B. regenerative braking
C. plugging D. all of these

608. A DC chopper circuit controls the average voltage across the DC motor by:
A. controlling the input voltage
B. controlling the field current
C. controlling the line current
D. continuously switching ON and OFF the motor for fixed durations of t_{ON} and t_{OFF} respectively

609. Thyristor switching circuits are used:
A. to reduce the stator voltage
B. to increase the stator voltage
C. to keep the stator voltage constant
D. none of these

610. Variable-speed drives using stator voltage control are normally:
A. open-loop system B. closed loop system
C. both A and B are correct
D. none of these

611. For controlling the speed of a three-phase induction motor, the method generally used is the:
A. fixed voltage fixed frequency method
B. variable voltage variable frequency method
C. fixed voltage variable frequency method
D. none of these

612. Variable voltage fixed frequency supply can be obtained from:
A. three-phase cycloconverter
B. ac chopper
C. three-phase inverter D. none of these

613. The slip power recovery method for the speed control of induction motor (slip ring):
A. increases the efficiency
B. decrease the efficiency
C. improves the power factor
D. none of these

614. In variable voltage variable frequency control, to achieve constant torque operation below base speed:
A. (V/f) has to be kept constant
B. flux has to be increased
C. flux has to be decreased
D. none of these

615. In AC motor control the ratio of voltage to frequency is maintained at constant value:
A. to make maximum use of magnetic circuit
B. to make minimum use of magnetic circuit
C. to maximise the current drawn from the supply to provide torque
D. none of these

Ans. 596. A 597. C 598. A 599. C 600. B 601. A 602. A 603. A 604. A 605. D 606. C 607. D 608. D 609. A 610. B 611. B 612. C 613. A 614. A 615. A

616. Which of the following circuits can be used as a DC transformer?
A. controlled rectifier B. inverter
C. magnetic amplifier D. none of these

617. Power factor of synchronous motor can be made leading by adjusting its:
A. speed B. supply voltage
C. excitation D. supply frequency

618. It is recommended to employ UJT oscillator for gate triggering of thyristors for speed control of IM mainly because:
A. it is fairly simple B. it is less expensive
C. it provides sharp rising pulses
D. none of these

619. The device used for current protection is:
A. the fuse B. R-C network
C. snubber network D. none of these

620. The power semiconductor device requires protection against excessive:
A. voltage B. current
C. certain rate of change D. all of the above

621. A device has a critical *di/dt* rating which, if exceeded, can cause overheating:
A. during turn-on B. during turn-off
C. both are true D. none of these

622. The scheme which can be implemented if two thyristors turn on simultaneously, producing a short circuit on the supply, is known as:
A. protection by ringing
B. gate blocking
C. electronic crowbar D. all the above

623. As soon as fault current is detected, it can be shunted away by turning on a parallel thyristor until the circuit breaker interrupts the fault current. This scheme is known as:
A. electronic crowbar B. protection by ringing
C. gate blocking D. none of these

624. Gates are protected against spurious (or noise) firing by using:
A. a Zener diode across the gate
B. shield cables
C. a series resistance D. all of the above

625. The gate can be protected against over-current by:
A. connecting a series resistance
B. a Zener diode across the gate
C. connecting a series inductor
D. none of these

626. The gate can be protected against overvoltage by:
A. a Zener diode across the gate
B. connecting a series resistance.
C. connecting a series inductor
D. heat sink

627. A suppressor is a device that responds to the rate of change of current or voltage to prevent:
A. a fall below a predetermined level
B. a rise above a predetermined level
C. overloading D. none of these

628. Snubber network is *R-C* series network placed in parallel with a device to protect against:
A. over current transients
B. overvoltage transients
C. thermal losses D. none of these

629. The thyristor can be safeguarded against overvoltages by:
A. thyristor diode B. contactor
C. simple diode D. all of these

630. Over voltages may be generated by:
A. switching of inductive loads
B. variations in the supply voltage
C. bad commutation D. all of the above

631. Heat sink is a mass of metal that is added to a device for the purpose of:
A. absorbing heat B. dissipating heat
C. absorbing and dissipating heat
D. none of these

632. Surge current rating of thyristor specifies the maximum:
A. repetitive current with sine wave
B. non-repetitive current with rectangular wave
C. non-repetitive current with sine wave
D. repetitive current with triangular wave

633. The *di/dt* rating of thyristor is specified for its:
A. decaying anode current
B. decaying gate current
C. rising gate current D. rising anode current

634. For a thyristor, *dv/dt* protection is achieved through the use of:
A. *R-L* in series with thyristor
B. *L* in series with thyristor
C. *RC* across thyristor D. none of these

635. For thyristor *di/dt* protection is achieved through the use of:
A. *R* in series with thyristor
B. *R-L* in series with thyristor
C. *M* in series with thyristor
D. none of these

636. The function of snubber circuit connected across a thyristor is to:
A. suppress *dv/dt* B. increase *dv/dt*
C. decrease *dv/dt* D. none of these

637. The object of connecting resistance and capacitance across gate circuit is to protect the thyristor gate against:
A. overvoltages B. *dv/dt*
C. noise signals D. none of these

Ans. 616. B 617. C 618. C 619. D 620. A 621. A 622. A 623. A 624. B 625. A 626. A 627. B 628. B 629. A 630. D 631. C 632. C 633. D 634. C 635. C 636. A 637. C

638. A microprocessor is also referred to as:
A. the chip that does some calculations for the computer
B. the computer on the chip
C. the chip that is responsible for data transfer
D. none of these

639. 8085 microprocessor is a:
A. a zero address microprocessor
B. one address microprocessor
C. two address microprocessor
D. none of these

640. As compared to 16-bit microprocessor, 8-bit microprocessor are limited in:
A. data handling capability
B. directly addressable memory
C. speed D. all of the above

641. Microprocessors are programmed using:
A. assembly language B. Pascal
C. assembly language or some suitable high level language
D. high level language such as C, C++, Pascal

642. With reference to 8085, ANA R/M is:
A. data transfer instruction
B. an arithmetic instruction
C. a logical instruction D. control instruction

643. The synchronization between microprocessor and memory is done by:
A. ALE signal B. HOLD signal
C. READY signal D. none of these

644. Two operands can be checked for equality using:
A. OR-operation B. AND-operation
C. EX-OR operation D. none of these

645. Microprocessor is called an *n*-bit microprocessor depending on:
A. register's length
B. size of internal data bus
C. size of external data bus
D. none of these

646. Programme counter is used to:
A. store address of next instruction to be executed
B. store temporary data to be used in arithmetic operation
C. store the status of microprocessor
D. none of these

647. Stack memory is used to:
A. provide additional memory to the base memory
B. save return addresses of sub routine
C. save the status of the microprocessor
D. none of these

648. Percentage change in the output voltage for a given change in input voltage is the:
A. line regulation B. load regulation
C. percentage regulation
D. all of these

649. Percentage change in output voltage for a given change in load current is known as:
A. line regulation B. load regulation
C. both are correct D. none of these

650. The fundamental classes of voltage regulators are:
A. linear regulators B. switching regulator
C. both A and B are correct
D. none of these

651. The basic components in a series regulator are:
A. control element B. reference voltage
C. error detector D. all of these

652. Basic configurations of switching regulators are:
A. step up B. step down
C. inverting D. all of these

653. A μA 7805 is a three terminal regulator with output of
A. 78 V B. 5 V
C. 80 V D. 12 V

654. The control element is a transistor in series with the load in case of:
A. series regulator B. shunt regulator
C. series and shunt regulator both
D. none of these

655. The control element is Zener in parallel with the load in case of:
A. series regulator B. shunt regulator
C. neither series nor shunt regulator
D. both A and B are correct

656. Precision voltage regulator belongs to
A. linear regulator B. switching regulator
C. IC regulator D. none of these

657. Three terminal regulator belongs to:
A. switching regulator B. linear regulator
C. IC regulator D. none of these

658. An ordinary power supply contains
A. rectifier only B. filter only
C. rectifier and filter both
D. none of these

659. For a filter, no load voltage is 300 V and full load output voltage is 280 V. The regulation of the circuit is:
A. 9.6% B. 7.1%
C. 6.7% D. 4.2%

660. A typical value of filter capacitor for 50 Hz input is:
A. 100 pF B. 10 PF
C. 50 μF D. 10 μF

661. A typical value of induction filter for 50 Hz input is:
A. 10 H B. 100 H
C. 10 mH D. 100 mH

662. In a rectifier circuit, the load connected is of low value. For proper filter operation, it is required that:
A. a capacitor is to be included in the circuit
B. a bleeder resistance is to be placed in the circuit
C. an inductor filter is to be included in the circuit
D. all of these

Ans.	638. B	639. B	640. D	641. C	642. C	643. C	644. C	645. B	646. A	647. B	648. A	649. B	650. C
	651. D	652. D	653. B	654. A	655. B	656. C	657. C	658. C	659. B	660. C	661. A	662. C	

663. The limitation of the voltage multiplying circuits is that:
A. the output has high ripple content
B. high output voltage is difficult to obtain
C. high output current is difficult to obtain
D. the size of the capacitors becomes very large

664. The function of bleeder resistance in filter circuit is:
A. to maintain minimum current necessary for optimum inductor filter operation
B. to work as voltage divider in order to provide variable output from the supply
C. to provide discharge path to capacitors so that output becomes zero when the circuit has been de-energised
D. all of these

665. A commercial power supply has voltage regulation of:
A. 1% B. 16%
C. 50% D. 100%

666. A Zener voltage regulator is used for:
A. small load current B. heavy load current
C. no load current D. none of these

667. An ideal regulated power supply should have:
A. 100% regulation B. 50% regulation
C. 10% regulation D. 0% regulation

668. An ordinary power supply has:
A. poor regulation B. good regulation
C. very good regulation D. none of these

669. A single-phase voltage controller is connected to resistance of 10 Ω and a supply of 200 sin 314*t* Volts. For a firing angle of 90°, the average thyristor current is:
A. 10 A B. $5\sqrt{2}/\pi$ A
C. $10/\pi$A D. $5\sqrt{2}$ A

670. A single-phase voltage controller, using two back to back thyristors, is found to be operating as a controlled rectifier, because:
A. load is *R* and pulse gating is used
B. load is *R-L* and pulse gating is used
C. load is *R* and high frequency carrier gating is used
D. load is *R-L* and continuous gating is used

671. A single-phase AC voltage controller fed from 50 Hz system supplies a load having resistance and inductance of 2.0 Ω and 6.36 mH respectively. The control range of firing angle for this regulator is:
A. $0° < \alpha < 180°$ B. $45° < \alpha < 180°$
C. $90° < \alpha < 180°$ D. $0° < \alpha < 45°$

672. In a single-phase voltage controller with *R-L* load, with firing angle a the load phase angle ϕ and β is the extinction angle. For this voltage controller, output power can be controlled if $\alpha > \phi$ and
A. $(\beta - \alpha) = \pi$ B. $(\beta - \alpha) < \pi$
C. $\beta < \pi$ D. none of these

673. A single-phase voltage controller feeds power to a resistance of 10 Ω. The source voltage is 200 V. For a firing angle of 90°, the rms value of thyristor current is:
A. 3 A B. 4 A
C. 10 A D. 15 A

674. In a single-phase voltage controller with *R-L* load, AC output power can be controlled if:
A. firing angle $\alpha > \phi$ (load phase angle) and conduction angle $\gamma = \pi$
B. $\alpha > \gamma$ and $\gamma < \pi$
C. $\alpha > \phi$ and $\gamma = \pi$ D. none of these

675. A load, consisting of $R = 10\,\Omega$ and $L = 10\,\Omega$, is being fed from 230 V, 50 Hz source through a single-phase voltage controller. For a firing angle of 30°, the rms value of load current would be:
A. 23 A B. $23/\sqrt{2}$ A
C. more than $23/\sqrt{2}$ A D. less than $23/\sqrt{2}$ A

676. A single-phase voltage controller is employed for controlling the power flow from 260 V, 50 Hz source into a load consisting of $R = 5\,\Omega$ and $X = 12\,\Omega$. The value of maximum rms load current and the firing angle are respectively:
A. 20 A, 0° B. 26 A, 0°
C. 20 A, 90° D. 25 A, 60°

677. A resistance of 10 Ω is fed through a singlephase voltage controller from a voltage source of 200 sin 314*t*. For a firing angle of 90°, the power delivered to load is:
A. 0.5 kW B. 0.75 kW
C. 1 kW D. none of these

678. A single-phase voltage controller feeds an induction motor and a heater:
A. in both the loads, fundamental and harmonics are useful
B. in an induction motor only fundamental and in heater only harmonics are useful
C. in an induction motor only fundamental and in heater harmonics as well as fundamental are useful
D. in an induction motor only harmonics and in heater only fundamental are useful

679. Uninterruptible supply is used in:
A. computers B. communication links
C. essential instrumentation
D. all of the above

680. Integral cycle control does not suffer from the drawbacks given below:
A. the load current takes its fundamental component at a lagging p.f. with harmonic currents
B. the load current takes its fundamental component at leading pf
C. the load current takes its fundamental at upf
D. none of these

Ans. 663. C 664. D 665. A 666. A 667. D 668. A 669. C 670. B 671. A 672. B 673. C 674. B 675. B 676. A 677. C 678. C 679. D 680. A

681. Electricity Supply Authorities normally insist that heating load be controlled by:
A. a single thyristor with diode bridge
B. phase angle control
C. integral cycle control
D. all of the above

682. The majority of heating loads have thermal time constants of:
A. zero second B. 0.2 second
C. several seconds D. none of these

683. A heating load of resistance type can be controlled to different power levels by the use of:
A. triac
B. two thyristors in the inverse parallel connection
C. connection with common cathodes
D. all of the above

684. SCR can be used for protecting equipment from:
A. under voltage B. over voltages
C. overloading D. none of these

685. In Thyristorised Excitation System, the ceiling voltage is decided by:
A. triggering circuit
B. turn ratio of supply transformer
C. thyristor bridge D. all of the above

686. Static excitation to alternators is provided by:
A. brushless excitation system
B. thyristorised excitation
C. both A and B D. none of these

687. Thermal resistance is:
A. the ratio of temperature difference to the heat power flow between the two interfaces
B. the value of resistance at 0°C
C. the value of resistance at room temperature
D. none of these

688. AC load control in which regulation is by alternating continuous whole number of half cycles, on and off, is called:
A. integral cycle
B. inverse parallel connection
C. phase angle control
D. none of these

MULTIPLE CHOICE QUESTIONS FROM VARIOUS COMPETITIVE EXAMINATIONS

1. A chopper operating at a fixed frequency is feeding an *R-L* load. As the duty ratio of the chopper is increased from 25% to 75%, the ripple in the load current: **[GATE 1993]**
A. remains constant.
B. decreases, reaches a minimum at 50% duty ratio and then increases.
C. increase, reaches a maximum at 50% duty ratio and then decreases
D. keeps on increasing as the duty ratio is increased

2. A four quadrant chopper cannot be operated as: **[IES 2001]**
A. one quadrant chopper
B. cycloconverter
C. inverter D. bi-directional rectifier

3. A three-phase wound rotor induction motor is controlled by a chopper-controlled resistance in its rotor circuit. A resistance of 2 Ω is connected in the rotor circuit and a resistance of 4 Ω is additionally connected during OFF periods of the chopper. The OFF period of the chopper is 4 ms. The average resistance in the rotor circuit for the chopper frequency of 200 Hz is: **[IES 2001]**
A. 26/5 Ω B. 24/5 Ω
C. 18/5 Ω D. 16/5 Ω

4. For a step up dc-dc chopper with an input dc voltage of 220 volts, if the output voltage required is 330 volts and the non-conducting time of thyristor-chopper is 100 ms, the ON time of thyristor-chopper would be: **[IES 2002]**
A. 66.6 µs B. 100 µs
C. 150 µs D. 200 µs

5. An ideal chopper operating at a frequency of 500 Hz, supplies a load having resistance of 3 Ω and inductance of 9 mH from a 60 V battery. The mean value of the load voltage for on/off ratio of 4/1 (assuming that load is shunted by a perfect commutating diode and battery is lossless) is: **[IES 2003]**
A. 240 V B. 48 V
C. 15 V D. 4 V

6. A boost-regulator has an input voltage of 5 V and the average output voltage of 15 V. The duty cycle is: **[IES 2003]**
A. 3/2 B. 2/3
C. 5/2 D. 15/2

7. A chopper is employed to charge a battery as shown in figure. The charging current is 5 A. The duty ratio is 0.2. The chopper output voltage is also shown in figure. The peak to peak ripple current in the charging current is: **[GATE 2003]**

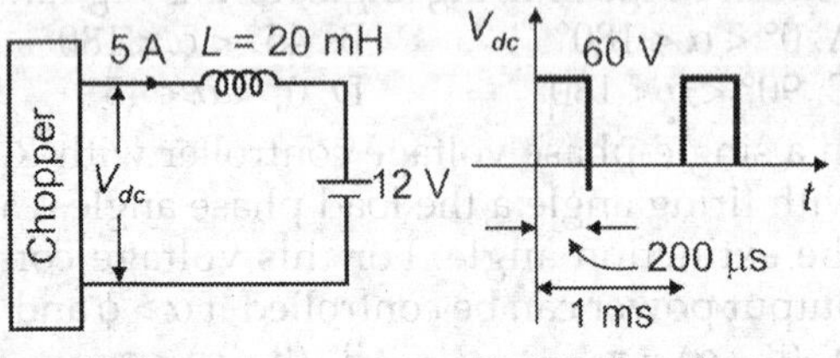

A. 0.48 A B. 1.2 A
C. 2.4 A D. 1 A

Ans. 681. C 682. C 683. D 684. B 685. B 686. C 687. A 688. A 1. C 2. B 3. A 4. D 5. B 6. B 7. A

8. In the buck-boost converter, what is the maximum value of the switch utilization factor? **[IES 2004]**
A. 1.00 B. 0.75
C. 0.50 D. 0.25

9. Figure shows a chopper operating from a 100 V dc input. The duty ratio of the main switch S is 0.8. The load is sufficiently inductive so that the load current is ripple free. The average current through the diode D under steady-state is: **[GATE 2004]**

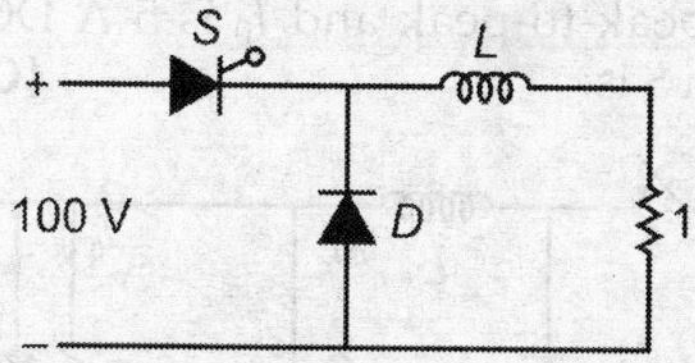

A. 1.6 A B. 6.4 A
C. 8.0 A D. 10.0 A

10. The given figure shows a step-down chopper switched at 1 kHz with a duty ratio $D = 0.5$. The peak-peak ripple in the load current is close to: **[GATE 2005]**

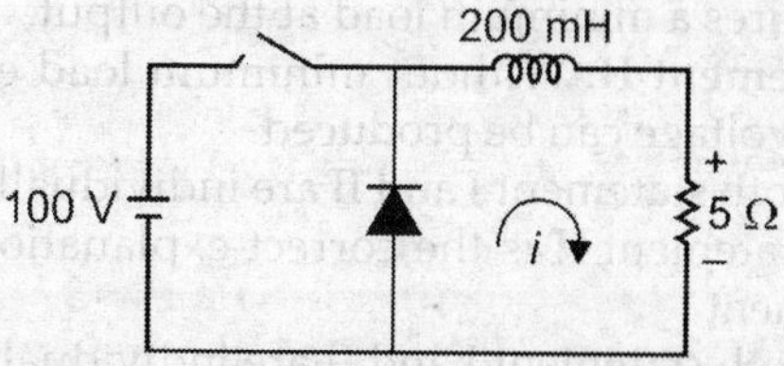

A. 10 A B. 0.5 A
C. 0.125 A D. 0.25 A

Statement for Q. No. 11 and 12

A voltage commutated chopper operating at 1 kHz is used to control the speed of dc motor as shown in figure. The load current is assumed to be constant at 10 A.

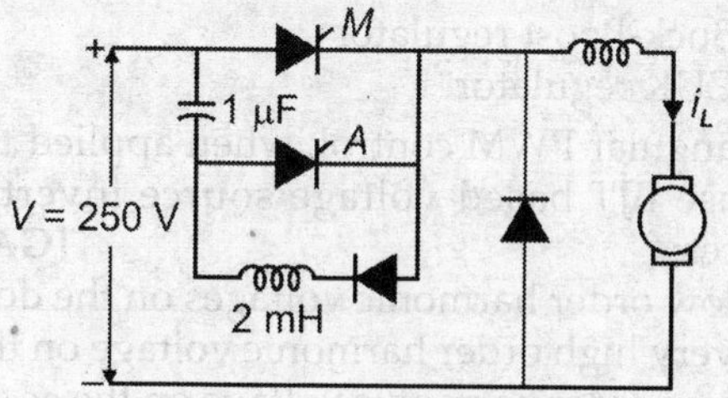

11. The minimum time in m sec for which the SCR M should be ON is: **[GATE 2006]**
A. 280 μs B. 140 μs
C. 70 μs D. 0 μs

12. The average output voltage of the chopper will be: **[GATE 2006]**
A. 70 V B. 47.5 V
C. 35 V D. 0 V

13. The circuit shown in the below figure will work as which one of the following? **[IES 2007]**

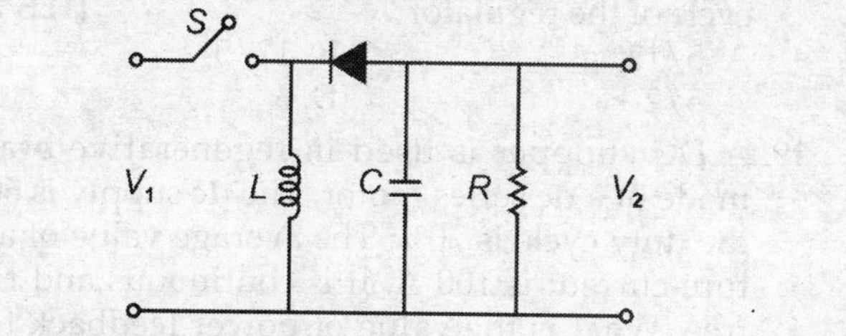

A. Buck-Boost converter
B. Buck converter
C. Boost converter
D. Dual converter

14. For a step-down dc chopper operating with discontinuous load current, what is the expression for the load voltage? (K is duty ratio of chopper) **[IES 2008]**
A. $V_0 = V_{dc} \times K$ B. $V_0 = V_{dc}/K$
C. $V_0 = Vdc/(1 - K)$ D. $V_0 = V_{dc}(1 - K)$

15. An ideal chopper is operating at a frequency of 500 Hz from a 60 V battery input. It is supplying a load having 3 Ω resistance and 9 mH inductance. Assuming the load is shunted by a perfect commutating diode and assuming battery is loss less, what is the mean load current at an on/off ratio of 1/1? **[IES 2008]**
A. 10 A B. 15 A
C. 20 A D. None of these

16. A two-quadrant dc to dc chopper can operate with which of the following load conditions?
1. +ve voltage, +ve current
2. –ve voltage, +ve current
3. –ve voltage, –ve current
4. +ve voltage, –ve current

Select the correct answer using the code given on next page. **[IES 2008]**
A. 1 only B. 1 and 2
C. 1 and 4 D. 3 and 4

17. In the circuit shown in the figure, the switch is operated at a duty cycle of 0.5. A large capacitor is connected across the load. The inductor current is assumed to be continuous. The average voltage across the load and the average current through the diode will respectively be: **[GATE 2008]**

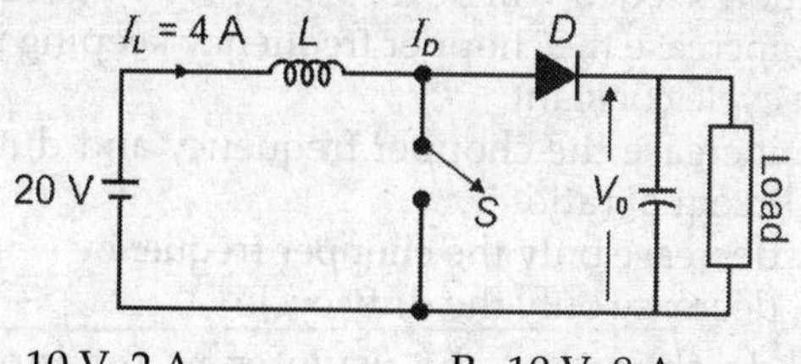

A. 10 V, 2 A B. 10 V, 8 A
C. 40 V, 2 A D. 40 V, 8 A

Ans. 8. B 9. A 10. C 11. B 12. B 13. A 14. A 15. A 16. C 17. C

18. A buck regulator has an input voltage of 12 V and the required output voltage is 5 V. What is the duty cycle of the regulator? **[IES 2008]**
A. 5/12 B. 12/5
C. 5/2 D. 6

19. A DC chopper is used in regenerative braking mode of a dc series motor. The dc supply is 600 V, the duty cycle is 70%. The average value of armature current is 100 A. It is continuous and ripple free. What is the value of power feedback to the supply? **[IES 2009]**
A. 3 kW B. 9 kW
C. 18 kW D. 35 kW

20. If a full wave fully controlled converter is modified as a full wave half controlled converter, what will be the maximum value of active power (*P*) and the maximum value of reactive power demand (*Q*)? **[IES 2009]**

	P	Q
A.	Double	Half
B.	Unchanged	Unchanged
C.	Half	Double
D.	Unchanged	Half

21. For the isolated buck boost converter as shown in the circuit below, the output voltage is to be 35 V at a duty cycle of 30%. The DC input is obtained from a front end rectifier without voltage doubling fed from a 115 V AC. What is the peak forward blocking voltage of the switching element? **[IES 2009]**

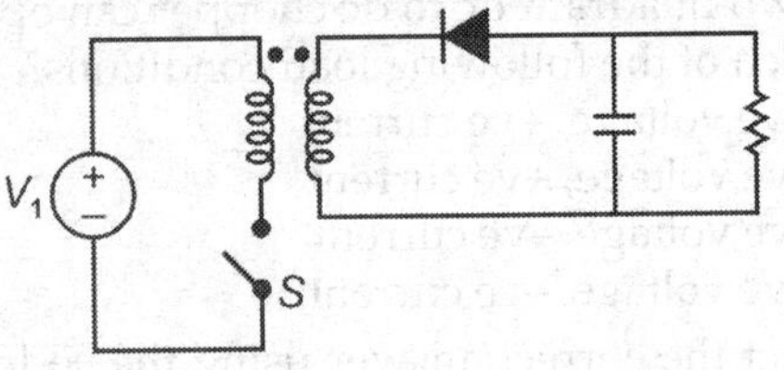

A. 232.3 V B. 69.69 V
C. 162.61 V D. 542 V

22. A DC to DC transistor chopper supplied from a fixed voltage DC source feeds a fixed resistive inductive load and a freewheeling diode. The chopper operates at 1 kHz and 50% duty cycle. Without, changing the value of the average DC current through the load, if it is desired to reduce the ripple content of the load current, the control action needed will be to: **[IES 2010]**
A. increase the chopper frequency keeping its duty cycle constant
B. increase the chopper frequency and duty cycle in equal ratio
C. decrease only the chopper frequency
D. decrease only the duty cycle

23. In dc choppers, the waveforms for input and output voltages are respectively: **[IES 2011]**
A. discontinuous and continuous
B. continuous and discontinuous
C. both discontinuous D. both continuous

24. What is the waveform of the current flowing through the diode in a buck-boost converter? **[IES 2011]**
A. Square wave B. Triangular wave
C. Trapezoidal wave D. Sinusoidal wave

25. In the circuit shown, an ideal switch *S* is operated at 100 kHz with a duty ratio of 50%. Given that Δic is 1.6 A peak-to-peak and I_0 is 5 A DC, the peak current in *S* is: **[GATE 2012]**

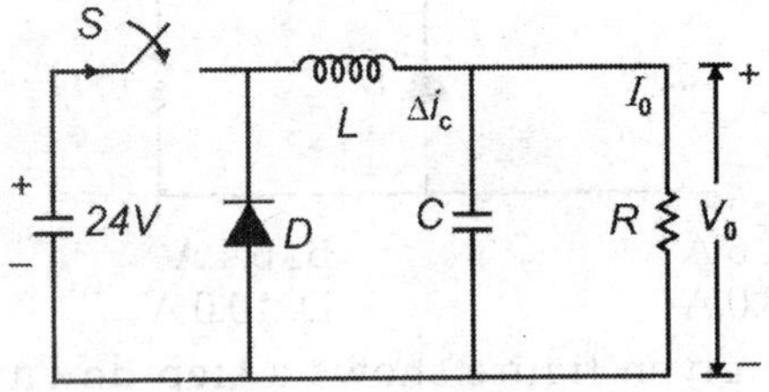

A. 6.6 A B. 5.0 A
C. 5.8 A D. 4.2 A

26. **Statement I:** A forward DC-to-DC converter requires a minimum load at the output.
Statement II: Without minimum load excess output voltage can be produced. **[IES 2013]**
A. both statement I and II are individually true and statement II is the correct explanation of statement I
B. both statement I and II are individually true but statement II is not the correct explanation of statement I
C. statement I is true but statement II is false
D. statement I is false but statement II is true

27. Which of the following regulator provide output voltage polarity reversal without a transformer? **[IES 2014]**
A. Buck regulator B. Boost regulator
C. Buck-Boost regulator
D. CUK regulator

28. Triangular PWM control, when applied to a three-phase BJT based voltage source inverter, introduces: **[GATE 2000]**
A. low order harmonic voltages on the dc side
B. very high order harmonic voltage on the dc side
C. low order harmonic voltage on the ac side
D. very high order harmonic voltage on the ac side

29. PWM switching is preferred in voltage source inverters for the purpose of: **[IES 2001]**
A. controlling output voltage
B. output harmonics
C. reducing filter size
D. controlling output voltage, output harmonics and reducing filter size

Ans.	18. A	19. C	20. B	21. A	22. A	23. B	24. C	25. C	26. A	27. C	28. ?	29. D

30. Match List-I with List-II and select the correct answer: **[IES 2001]**

List-I (Waveforms)

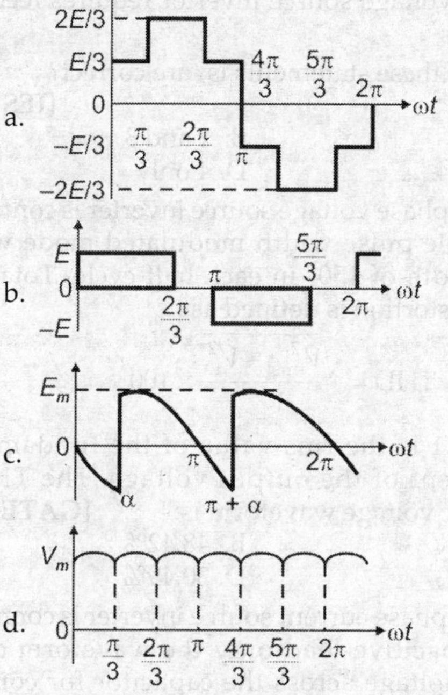

List-II (Descriptions)

1. Single-phase fully controlled AC to DC converter
2. Voltage commutated DC to DC chopper with input DC voltage E
3. Phase voltage of a three-phase inverter with 180° conduction and input DC voltage E
4. Line voltage of a three-phase inverter with 120° conduction and input DC voltage E
5. Three-phase diode bridge rectifier

Codes:

	a	b	c	d		a	b	c	d
A.	3	4	1	5	B.	5	1	4	2
C.	3	1	4	5	D.	5	4	1	2

31. In case of voltage source inverter, freewheeling can be needed for the load of: **[IES 2001]**
A. inductive nature
B. capacitive nature
C. resistive nature
D. back emf nature

32. The operation of an inverter fed induction motor can be shifted from motoring to regenerative braking by: **[IES 2002]**
A. reversing phase sequence
B. reducing inverter voltage
C. decreasing inverter frequency
D. increasing inverter frequency

33. Compared to a single-phase half-bridge inverter, the output power of a single-phase full-bridge inverter is higher by a factor of: **[IES 2002]**
A. 12
B. 8
C. 4
D. 2

34. Figure a shows an inverter circuit with a DC source voltage V_s. The semiconductor switches of the inverter are operated in such a manner that the pole voltage V_{10} and V_{20} are as shown in figure (b), what is the rms value of the pole-to-pole voltage V_{12}? **[GATE 2002]**

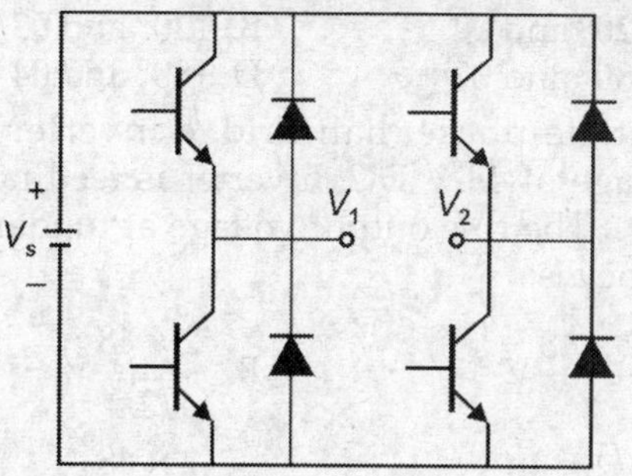

Fig. (a)

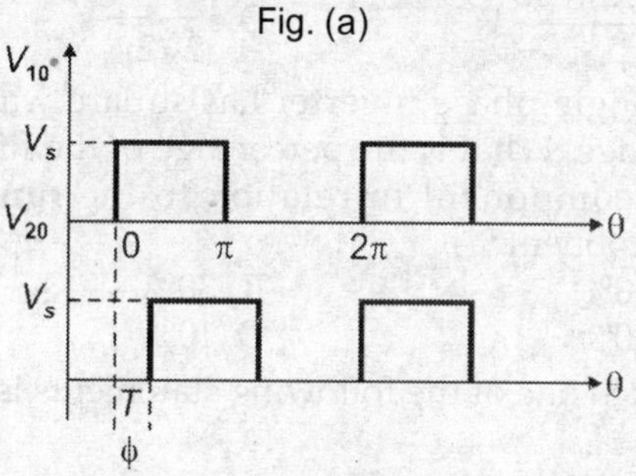

Fig. (b)

A. $\dfrac{V_s\phi}{\pi/2}$
B. $V_s\sqrt{\dfrac{\phi}{\pi}}$
C. $V_s\sqrt{\dfrac{\phi}{2\pi}}$
D. $\dfrac{V_s}{\pi}$

35. In a self-controlled synchronous motor fed from a variable frequency inverter: **[IES 2003]**
A. the rotor poles invariably have damper windings
B. there are stability problems
C. the speed of the rotor decides stator frequency
D. the frequency of the stator decides the rotor speed

36. For a single-phase, full-bridge inverter supplying power to a highly inductive load as shown below, the correct sequence of operation of switches and didoes is: **[IES 2003]**

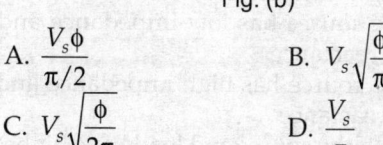

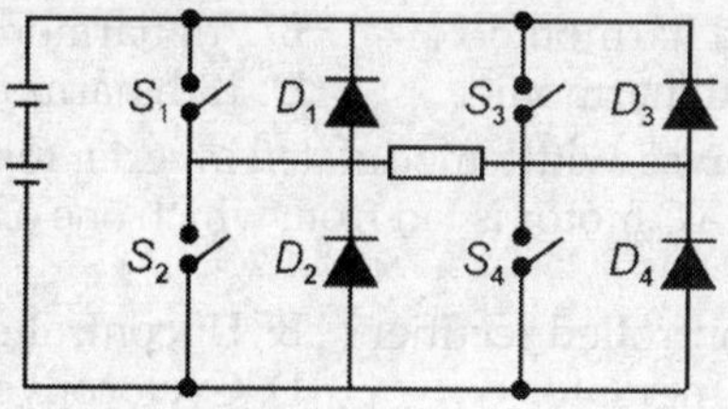

A. $S_1S_4 - S_3S_2 - S_1S_4 - S_3S_2$
B. $S_1S_2 - D_1D_2 - S_3S_4 - D_3D_4$
C. $S_1D_3 - S_1S_4 - S_4D_2 - D_2D_3$
D. $S_2D_4 - D_4D_1 - D_1S_3 - S_3S_2$

Ans. 30. A 31. A 32. C 33. C 34. B 35. D 36. A

37. A DC source is switched in steps to synthesize the three-phase output. The basic three-phase bridge inverter can be controlled. The angle through which each switch conducts, and at any instant, the number of switches conducting simultaneously are, respectively: **[IES 2003]**
A. 120° and 02 B. 120° and 03
C. 180° and 02 D. 180° and 04

38. A single-phase, half-bridge inverter has input voltage of 48 V DC. Inverter is feeding a load of 2.4 Ω. The rms output voltage at fundamental frequency is: **[IES 2003]**
A. $\frac{2\times 48}{\pi}V$ B. $\frac{2\times 48}{\sqrt{2}\pi}V$
C. $\frac{\sqrt{2}\times 48}{\pi}V$ D. $\frac{2\times 48}{2\sqrt{2}\pi}V$

39. A single-phase inverter has square wave output voltage. What is the percentage of the fifth harmonic component in relation to the fundamental component? **[IES 2004]**
A. 40% B. 30%
C. 20% D. 10%

40. Which one of the following statements is correct? **[IES 2005]**
A voltage source inverter is normally employed:
A. when the source has low impedance and load has high reactance
B. when the source has high impedance and load has low reactance
C. when both the source and load have high values of impedance and reactance respectively
D. when both the source and load have low values of impedance and reactance, respectively

41. The output voltage waveform of a three-phase square-wave inverter contains: **[GATE 2005]**
A. only even harmonics
B. both odd and even harmonics
C. only odd harmonics
D. only triple harmonics

42. In a three-phase voltage source inverter operating in square wave mode, the output line voltage is free from: **[IES 2006]**
A. 3rd harmonic B. 7th harmonic
C. 11th harmonic D. 13th harmonic

43. The pulse-width modulated inverter for the control of an AC motor is fed from which one of the following? **[IES 2007]**
A. Controlled rectifier B. Uncontrolled rectifier
C. ac regulator D. Cycloconverter

44. Consider the following statements:
1. Both voltage source inverter and current source inverter require feedback diodes
2. Only current source inverter requires feedback diodes
3. GTOs can not be used in a current source inverter
4. Only voltage source inverter requires feedback diodes

Which of these statements is/are correct? **[IES 2007]**
A. 1 only B. 2 and 3
C. 3 and 4 D. 4 only

45. A single-phase voltage source inverter is controlled in a single pulse-width modulated mode with a pulse width of 150° in each half cycle. Total harmonic distortion is defined as

$$THD = \frac{\sqrt{V_{rms}^2 - V_1^2}}{V_1} \times 100$$

where $V1$ is the rms value of the fundamental component of the output voltage. The THD of output ac voltage waveform is: **[GATE 2007]**
A. 65.65% B. 48.42%
C. 31.83% D. 30.49%

46. A single-phase current source inverter is connected with capacitive load only the waveform of the output voltage across the capacitor for constant source current will be: **[IES 2008]**
A. sine wave B. square wave
C. triangular wave D. step function

47. A single-phase full-bridge inverter is connected to a load of 2.4 Ω. The dc input is 48 V. What is the rms output at fundamental frequency? **[IES 2008]**
A. $\frac{4\times 48}{\sqrt{2}\pi}V$ B. $\frac{2\times 48}{\sqrt{2}\pi}V$
C. $\frac{4\times 48}{\pi}V$ D. $\frac{2\times 48}{\pi}V$

48. A single-phase voltage source inverter is feeding a purely inductive load as shown in the figure.

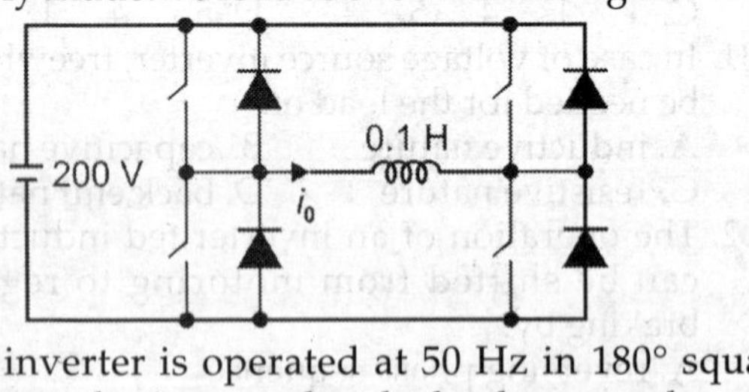

The inverter is operated at 50 Hz in 180° square wave mode. Assume that the load current does not have any dc component. The peak value of the inductor current i_0 **[GATE 2008]**
A. 6.37 A B. 10 A
C. 20 A D. 40 A

49. A three-phase voltage source inverter is operated in 180° conduction mode. Which one of the following statements is true? **[GATE 2008]**

Ans. 37. B 38. B 39. C 40. A 41. C 42. A 43. B 44. C 45. C 46. C 47. A 48. B

A. Both pole-voltage and line-voltage will have 3rd harmonic components
B. Pole-voltage will have 3rd harmonic component but line-voltage will be free from 3rd harmonic
C. Line-voltage will have 3rd harmonic component but pole-voltage will be free from 3rd harmonic
D. Both pole-voltage and line-voltage will be free from 3rd harmonic components

50. Consider the following statements, with respect to the power transistors used in inverters:
1. Maximum collector-emitter voltage V_{CEO}
2. Maximum collector current
3. Maximum power dissipation
4. Maximum current gain at minimum load current
5. Maximum current gain at maximum load current

Which of these statements is/are correct? **[IES 2009]**
A. 1 only B. 1, 2, 3 and 5
C. 2 and 3 only D. 2, 3 and 4

51. For elimination of 5th harmonics from the output of an inverter, what will be the position of pulse in a PWM inverter? **[IES 2009]**
A. 72° B. 36°
C. 60° D. 90°

52. In a single-phase VSI bridge inverter, the load current is $I_0 = 200 \sin(\omega t - 45°)$ mA. The dc supply voltage is 220 V. What is the power drawn from the supply? **[IES 2009]**
A. 9.8 W B. 19.8 W
C. 27.25 W D. 34.03 W

53. What is the effect of blanking time on output voltage in PWM inverter? **[IES 2009]**
A. Distortion in instantaneous voltage at current zero crossing
B. Low order space harmonics in output voltage
C. Distribution in instantaneous voltage at voltage zero crossing
D. High order time harmonics in output voltage

54. In single pulse modulation of PWM inverters, the pulse width is 120°. For an input voltage of 220 V dc, what is the rms value at the fundamental component of the output voltage? **[IES 2009]**
A. 171.5 V B. 254.0 V
C. 127.0 V D. 89.81 V

55. A constant current source inverter supplies 20 A to a load resistance of 1 Ω to a load resistance change to 5 Ω, then the load current: **[IES 2010]**
A. remains same at 20 A and the load voltage changes to 100 V
B. changes to 4 A from 20 A and the load voltage changes to 20 V
C. changes to 4 A from 20 A and the load voltage changes to 80 V
D. load voltage stay at 20 A and 20 V, respectively

56. The below figure shows an inverter circuit with a DC source voltage V_S. The semiconductor switches of the inverter are operated in such a way that the pole voltages of V_{10} and V_{20} are as shown in the figure (b). What is the RMS value of the pole voltage V_{12}? **[IES 2010]**

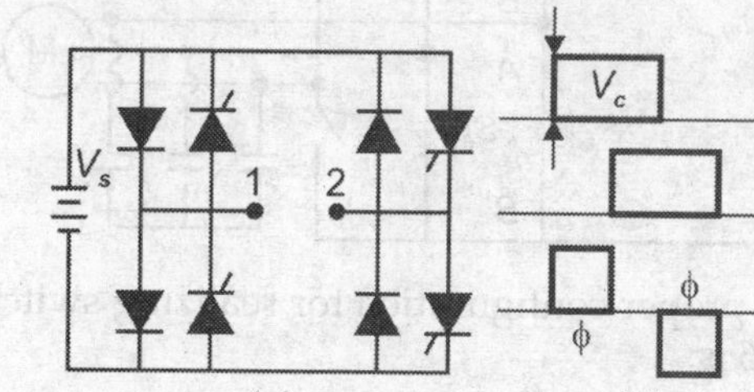

(a) (b)

A. $\frac{V_S\phi}{\sqrt{2\pi}}$ B. $V_S\sqrt{\frac{\phi}{\pi}}$
C. $V_S\sqrt{\frac{\phi}{2\pi}}$ D. $\frac{V_S}{\pi}$

57. **Assertion (A):** The *L* and *C* components of the communication circuit in McMurray inverter are chosen such, that the peak value of resonant current pulse during communication is sufficiently greater than the load current.
Reason (R): A thyristor will successfully turn-off if the current is maintained below 640 W on the holding value for a time greater than the turn-off time of the device. **[IES 2010]**
A. both (A) and (R) are true and (R) is the correct explanation of (A)
B. both (A) and (R) are true but (R) is NOT the correct explanation of (A)
C. (A) is true but (R) is false
D. (A) is false but (R) is true

58. In a PWM inverter, f_0 and f are the frequencies in Hz for the carrier signal and reference signal respectively. Then the number of pulses per half cycle is: **[IES 2010]**
A. $N = f/f_0$ B. $N = f/2f_0$
C. $N = f_0/2f$ D. $N = f_0/f$

59. A voltage source inverter (VSI) is normally employed when: **[IES 2012]**
A. source inductance is large and load inductance is small
B. source inductance is small and load inductance is large
C. both source inductance and load inductance are small
D. both source inductance and load inductance are large

Ans. 49. D 50. B 51. A 52. B 53. D 54. A 55. A 56. B 57. A 58. C 59. B

60. A three-phase current source inverter used for the speed control of an induction motor is to be realized using MOSFET switches as shown below. Switches $S1$ to $S6$ are identical switches. **[GATE 2011]**

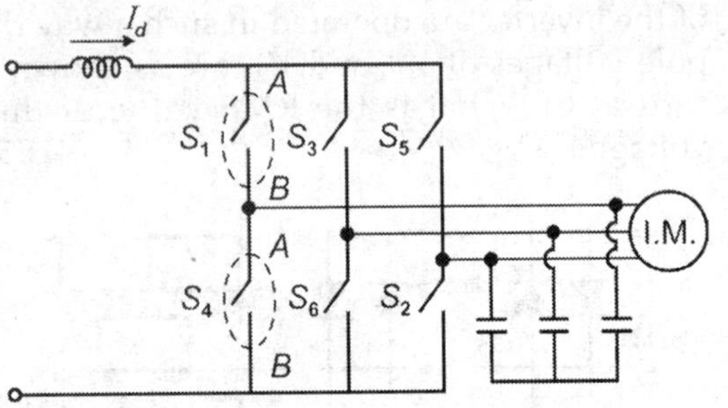

The proper configuration for realizing switches $S1$ to $S6$ is:

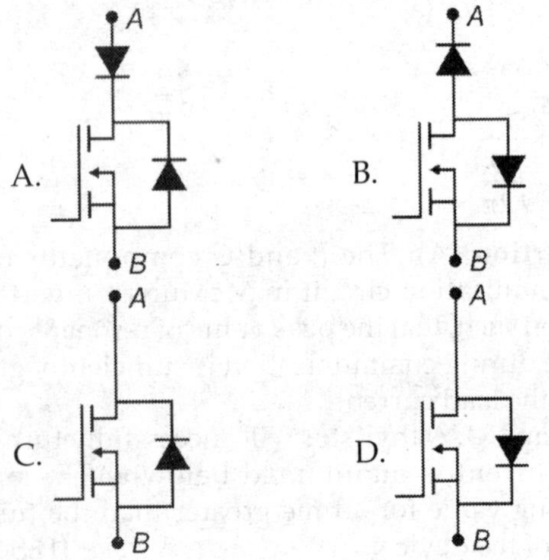

61. A current source inverter is obtained by inserting a large: **[IES 2012]**
A. inductance in series with dc supply
B. capacitance in parallel with dc supply
C. inductance in parallel with dc supply
D. capacitance in series with dc supply

62. **Statement I:** Multiple pulse width modulation is used to reduce the harmonic content in inverters.
Statement II: The high order harmonics can be easily filtered using passive filters. **[IES 2012]**
A. both statement I and II are individually true and statement II is the correct explanation of statement I
B. both statement I and II are individually true but statement II is not the correct explanation of statement I
C. statement I is true but statement II is false
D. statement I is false but statement II is true

63. A PWM switching scheme is used with a three-phase inverter to: **[IES 2013]**
A. reduce the total harmonic distortion with modest filtering
B. minimize the load on the DC side
C. increase the life of the batteries
D. reduce low order harmonics and increase high order harmonics

64. **Statement I:** The output current of a current source inverter remains constant irrespective of load.
Statement II: The load voltage in CSI depends on the load impedance. **[IES 2013]**
A. both statement I and II are individually true and statement II is the correct explanation of statement I
B. both statement I and II are individually true but statement II is not the correct explanation of statement I
C. statement I is true but statement II is false
D. statement I is false but statement II is true

65. A single-phase, voltage source, square wave inverter feeds a pure inductive load. The waveform of the current will be: **[IES 2014]**
A. sinusoidal B. rectangular
C. trapezoidal D. triangular

66. What should be the frequency modulation ratio (m_f) for a three-phase inverter, if the m_fth harmonic and its odd multiples are to be suppressed in the line-to-line voltages? **[IES 2014]**
A. m_f should be odd B. m_f should be even
C. m_f should be an odd multiple of 3
D. m_f should be an even multiple of 3

67. A three-phase semiconverter feeds the armature of separately excited dc motor, supplying a non-zero torque, for steady-state operation, the motor armature current is found to drop to zero at certain instances of time. At such instances, the voltage assumes a value that is: **[GATE 2001]**
A. equal is the instantaneous value of the ac phase voltage
B. equal to the instantaneous value of the motor back emf
C. arbitrary D. zero

68. A single-phase half-controlled rectifier is driving a separately excited dc motor. The dc motor has a back emf constant of 0.25 V/rpm. The armature current is 5 A without any ripple. The armature resistance is 2 Ω. The converter is working from a 230 V, singlephase ac source with a firing angle of 30°. Under this operating condition, the speed of the motor will be: **[GATE 2004]**
A. 339 rpm B. 359 rpm
C. 366 rpm D. 386 rpm

69. A variable speed drive rated for 1500 rpm, 40 Nm is reversing under no load. Figure shows the reversing torque and the speed during the transient. The moment of inertia of the drive is: **[GATE 2004]**

Ans. 60. A 61. A 62. B 63. A 64. A 65. D 66. C 67. B 68. A

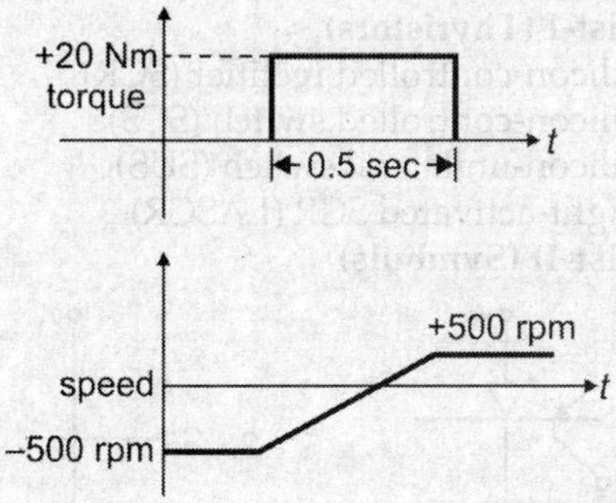

A. 0.048 kg m^2 B. 0.064 kg m^2
C. 0.096 kg m^2 D. 0.128 kg m^2

70. A motor armature supplied through phase controlled SCRs receives a smoother voltage shape at: **[IES 2005]**
A. high motor speed B. low motor speed
C. rated motor speed D. none of these

71. An electric motor, developing a starting torque of 15 Nm, starts with a load torque of 7 Nm on its shaft. If the acceleration at start is 2 rad/sec^2, the moment of inertia of the systems must be (neglecting viscous and coulomb friction) **[GATE 2005]**
A. 0.25 kg m^2 B. 0.25 Nm^2
C. 4 kg m^2 D. 4 Nm^2

72. A solar cell of 350 V is feeding power to an ac supply of 440 V, 50 Hz through a three phase fully controlled bridge converter. A large inductance is connected in the dc circuit to maintain the dc current at 20 A. If the solar cell resistance is 0.5 W, then each thyristor will be reverse biased for a period of: **[GATE 2006]**
A. 125° B. 120°
C. 60° D. 55°

73. Consider the following statements:
Assertion (A): Power electronic converters are extensively used in adjustable speed drives.
Reason (R): Power electronic converters do not produce harmonic distortion.
Of these statements: **[IES 2007]**
A. both (A) and (R) are true and (R) is the correct explanation of (A)
B. both (A) and (R) are true but (R) is not the correct explanation of (A)
C. (A) is true but (R) is false
D. (A) is false but (R) is true

74. A three-phase, 440 V, 50 Hz ac mains fed thyristor bridge is feeding a 440 V dc, 15 kW, 1500 rpm separately excited dc motor with a ripple free continuous current in the dc link under all operating conditions, neglecting the losses, the power factor of the ac mains at half the rated speed is: **[GATE 2007]**
A. 0.354 B. 0.372
C. 0.90 D. 0.955

75. A cycloconverter-fed induction motor drive is most suitable for which one of the following? **[IES 2008]**
A. Compressor drive B. Machine tool drive
C. Paper mill drive D. Cement mill drive

76. A single-phase fully controlled converter bridge is used for electrical breaking of a separately excited dc motor. The dc motor load is respectively by an equivalent circuit as shown in the figure.

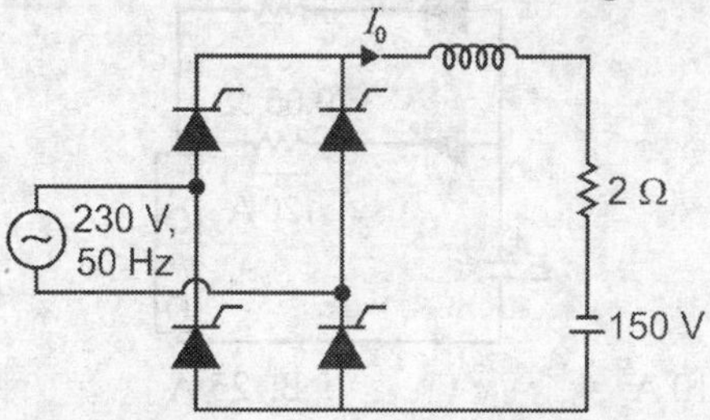

Assume that the load inductance is sufficient to ensure continuous and ripple free load current. The firing angle of the bridge for a load current of I_0 = 10 A will be: **[GATE 2008]**
A. 44° B. 51°
C. 129° D. 136°

77. For low-speed high-power reversible operation, the most suitable drives are: **[IES 2011]**
A. voltage source inverter fed ac drives
B. current source inverter fed ac drives
C. dual converted fed dc drives
D. cycloconverter fed ac drives

78. The separately excited dc motor in the figure below has a rated armature current of 20 A and a rated armature voltage of 150 V. An ideal chopper switching at 5 kHz is used to control the armature voltage. If L_a = 0.1 mH, R_a = 1 Ω, neglecting armature reaction, the duty ratio of the chopper to obtain 50% of the rated torque at the rated speed and the rated field current is: **[GATE 2013]**

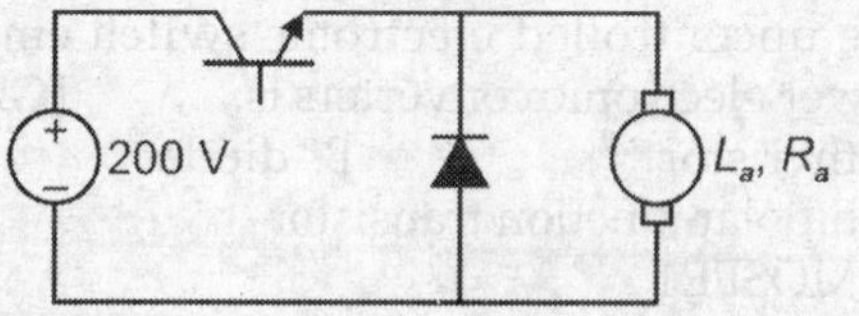

A. 0.4 B. 0.5
C. 0.6 D. 0.7

79. The thermal resistance between the body of a power semiconductor device and the ambient is expressed as: **[Gate 1993]**
A. voltage across the device divided by current through the device
B. average power dissipated in the device divided by the square of the rms current in the device

Ans.	69. A	70. A	71. C	72. D	73. C	74. A	75. B	76. C	77. D	78. D

C. average power dissipated in the device divided by the temperature difference from body to ambient.
D. temperature difference from body to ambient divided by average power dissipated in the device

80. Figure show two thyristors, each rated 500 A (continuous) sharing a load current. Current through thyristor y is 120 A. The current through thyristor x will be nearly: **[GATE 1995]**

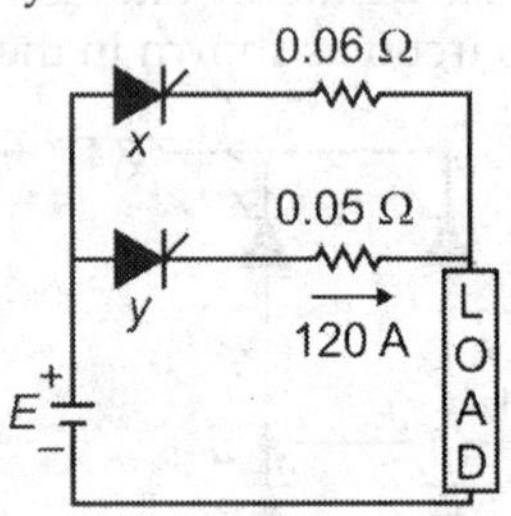

A. 10 A B. 25 A
C. 50 A D. 100 A

81. The Triac can be used in: **[Gate 1996]**
A. inverter B. rectifier
C. multiquadrant chopper
D. ac voltage regulator

82. Which semiconductor power device out of the following is not a current triggered device? **[GATE 1996]**
A. Thyristor B. GTO
C. Triac D. MOSFET

83. Which of the following does not cause permanent damage of an SCR? **[GATE 1996]**
A. High current
B. High rate of rise of current
C. High temperature rise
D. High rate of rise of voltage

84. The MOSFET switch in its ON state may be considered equivalent to: **[GATE 1998]**
A. resistor B. inductor
C. capacitor D. battery

85. The uncontrolled electronic switch employed in power electronic converters is: **[GATE 1998]**
A. thyristor B. diode
C. bipolar junction transistor
D. MOSFET

86. In a commutation circuit employed to turn-off an SCR, satisfactory turn-off is obtained when: **[GATE 1998]**
A. circuit turn-off time < device turn-off time
B. circuit turn-off time > device turn-off time
C. circuit time constant > device turn-off time
D. circuit time constant < device turn-off time

87. Match List-I with List-II and select the correct answer: **[IES 2003]**

List-I (Thyristors)
a. Silicon-controlled rectifier (SCR)
b. Silicon-controlled switch (SCS)
c. Silicon-unilateral switch (SUS)
d. Light-activated SCR (LASCR)

List-II (Symbols)

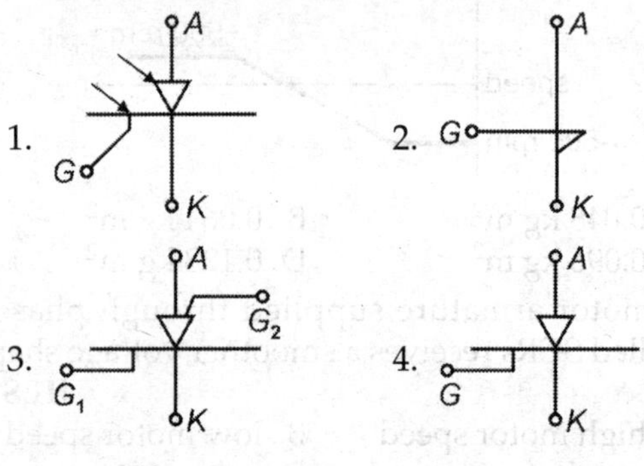

Codes:

	a	b	c	d		a	b	c	d
A.	3	4	1	2	B.	4	3	1	2
C.	3	4	2	1	D.	4	3	2	1

88. Figure shows a MOSFET with an integral body diode. It is employed as a power switching device in the ON and OFF states through appropriate control. The ON and OFF states of the switch are given on the *VDS-IS* plane by: **[GATE 2003]**

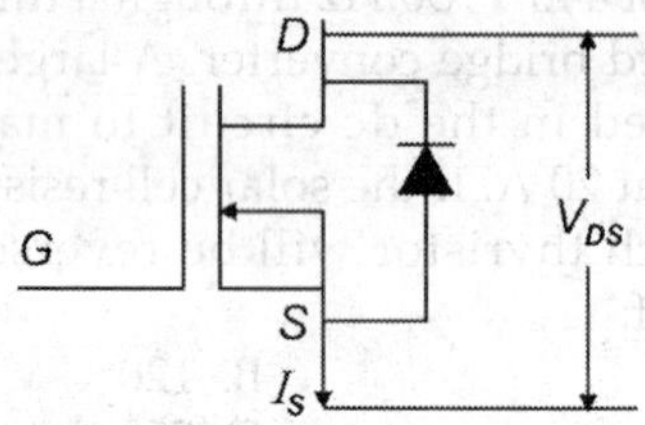

A.
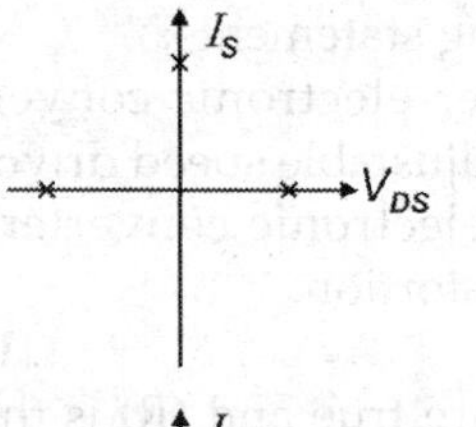

B.
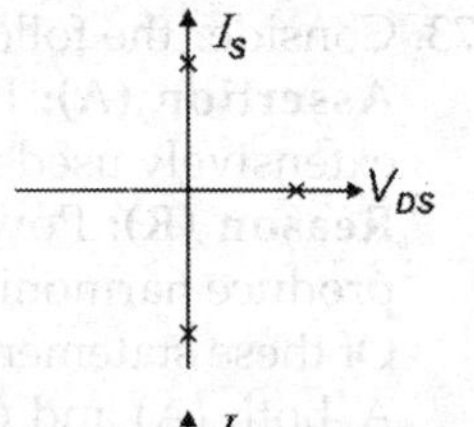

C.
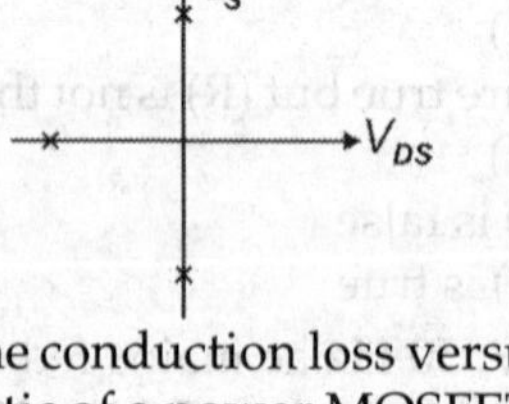

D.
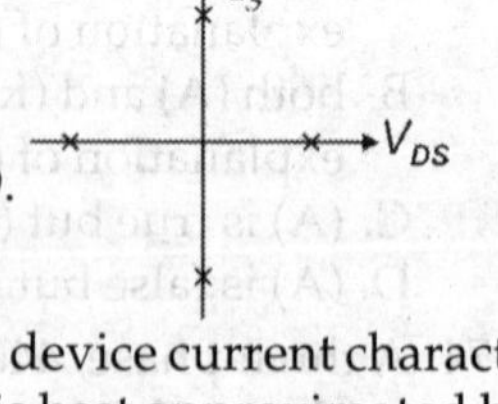

89. The conduction loss versus device current characteristic of a power MOSFET is best approximated by: **[GATE 2005]**
A. a parabola
B. a straight line
C. a rectangular hyperbola
D. an exponentially decaying function

Ans. 79. D 80. D 81. D 82. D 83. D 84. C 85. B 86. ? 87. D 88. B 89. A

90. Figure shows a thyristor with the standard terminations of anode (*A*), cathode (*K*), gate (*G*) and the different junctions named J1, J2 and J3. When the thyristor is turned on and conducting: **[GATE 2003]**

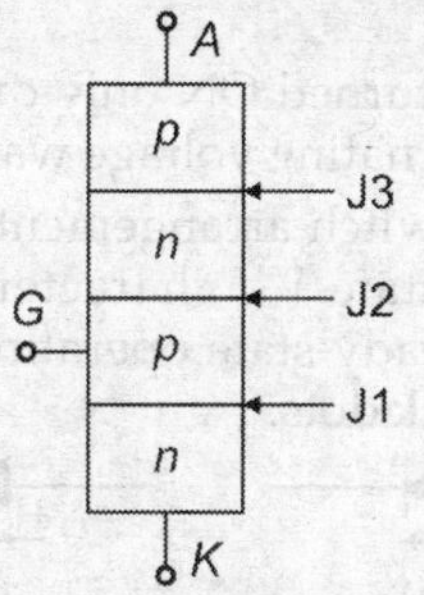

A. J1 and J2 are forward biased and J3 is reversed biased
B. J1 and J3 are forward biased and J2 is reverse biased
C. J1 is forward biased and J2 and J3 are reverse biased
D. J1, J2 and J3 are all forward biased

91. Match List-I with List-II and select the correct answer using the code given below the lists: **[IES 2005]**

List-I (Limiting factor)	**List-II (Safe operating area portion)**
a. The peak voltage limit	1. *PQ*
b. Secondary breakdown limit	2. *QR*
c. Power dissipation limit	3. *RS*
d. Peak current limit	4. *ST*

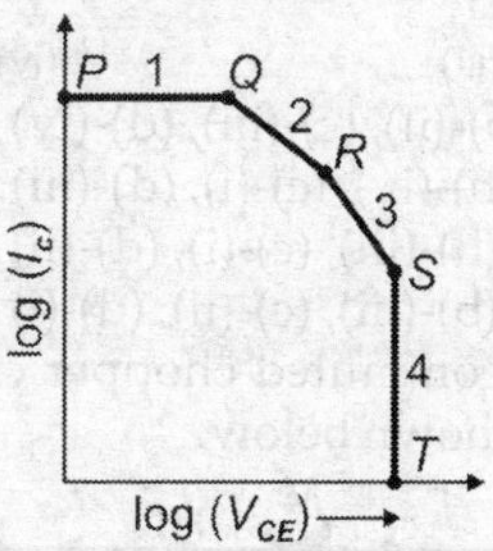

Codes:

	a	b	c	d		a	b	c	d
A.	2	1	4	3	B.	4	3	2	1
C.	2	3	4	1	D.	4	1	2	3

92. Which of the following devices should be used as a switch in a low power switched mode power supply (SMPS)? **[IES 2006]**
A. GTO B. MOSFET
C. TRIAC D. Thyristor

93. The figure shows the voltage across a power semiconductor device and the current through the device during a switching transitions. Whether the transition is a turn ON transition or a turn OFF transition? What is the energy lost during the transition? **[GATE 2005]**

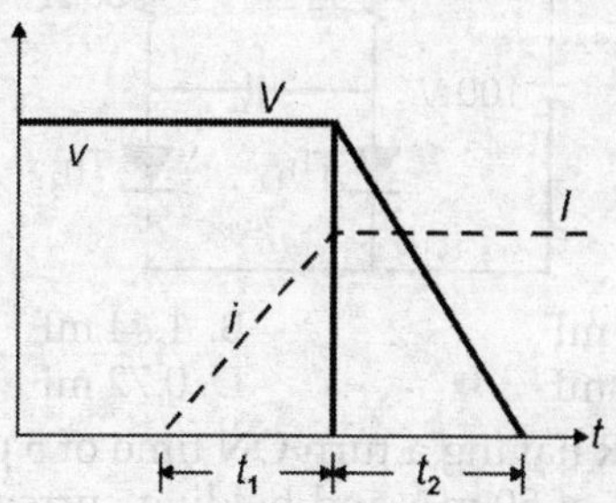

A. Turn ON, $\frac{VI}{2}(t_1 + t_2)$ B. Turn OFF, $VI\,(t_1 + t_2)$
C. Turn ON, $VI\,(t_1 + t_2)$ D. Turn OFF $\frac{VI}{2}(t_1 + t_2)$

94. An electronics switch *S* is required to block voltage of either polarity during its OFF state as shown in the figure (a). This switch is required to conduct in only one direction its ON states as shown in the figure (b).

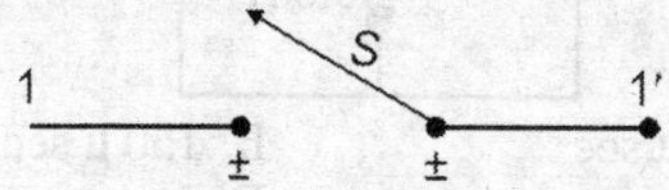

(a)

(b)

Which of the following are valid realizations of the switch *S*? **[GATE 2005]**

1. 

2. 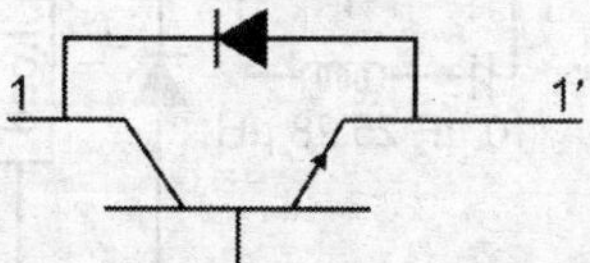

3. 

4. 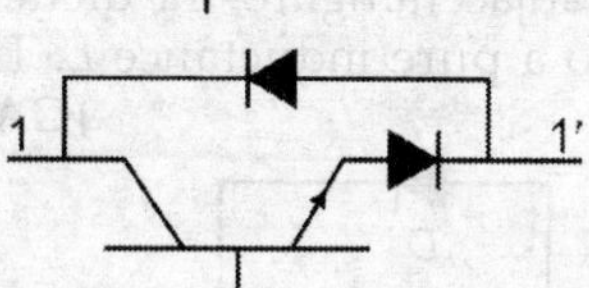

A. Only 1 B. 1 and 2
C. 1 and 3 D. 3 and 4

95. A voltage commutation circuit is shown in figure. If the turn off time of the SCRs is 50 μ sec and a safety margin of 2 is considered, what will be the approximate minimum value of capacitor required for proper commutation? **[GATE 2006]**

Ans. 90. B 91. B 92. B 93. A 94. C

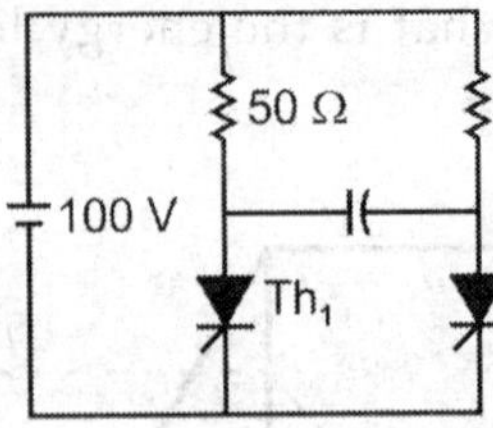

A. 2.88 mF B. 1.44 mF
C. 0.91 mF D. 0.72 mF

96. An SCR having a turn ON time of 5 μ sec, latching current of 50 mA and holding current of 40 mA is triggered by a short duration pulse and is used in the circuit shown in figure. The minimum pulse width required to turn the SCR ON will be: **[GATE 2006]**

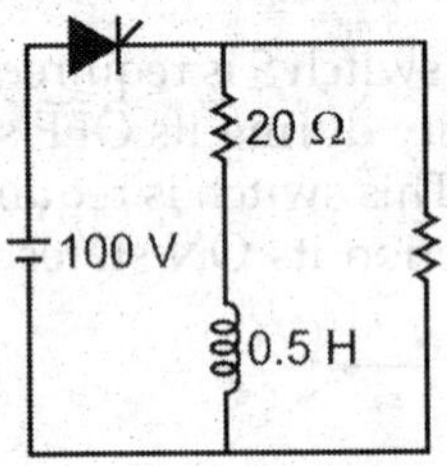

A. 251 μ sec B. 150 μ sec
C. 100 μ sec D. 5 μ sec

97. The circuit in the figure is a current commutated dc-dc chopper where, ThM is the main SCR and Th_{AUX} is the auxiliary SCR. The load current is constant at 10 A. Th_M is ON. Th_{AUX} is triggered at $t = 0$. Th_M is turned OFF between. **[GATE 2007]**

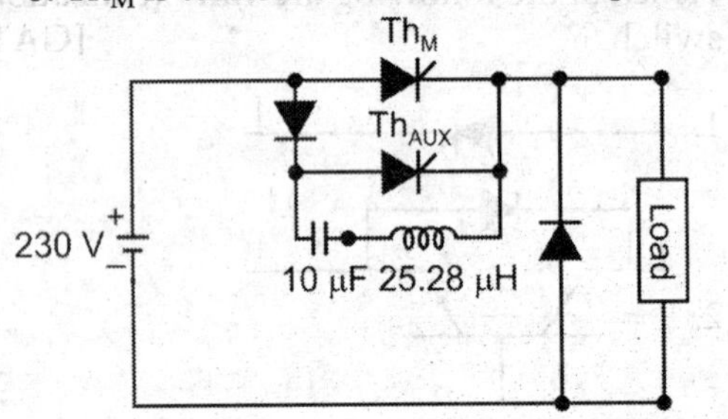

A. $0\ \mu s < t \le 25\ \mu s$ B. $25\ \mu s < t \le 50\ \mu s$
C. $50\ \mu s < t \le 75\ \mu s$ D. $75\ \mu s < t \le 100\ \mu s$

98. In the circuit of adjacent figure, the diode connects the ac source to a pure inductance L. The diode conducts for: **[GATE 2007]**

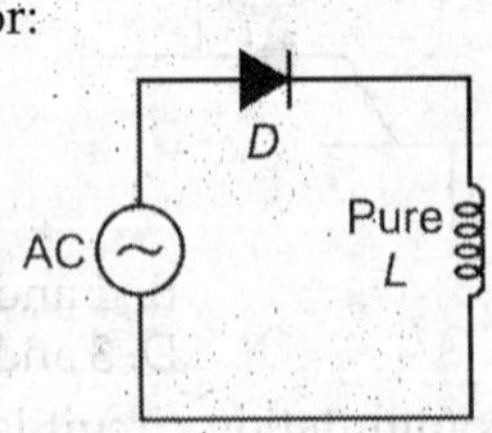

A. 90° B. 180°
C. 270° D. 360°

99. An SCR is considered to be a semicontrolled device because: **[GATE 2009]**

A. it can be turned OFF but not ON with a gate pulse
B. it conducts only during one half-cycle of an alternating current wave
C. it can be turned ON but not OFF with a gate pulse
D. it can be turned ON only during one half-cycle of an alternating voltage wave

100. Match the switch arrangements on the top row to the steady-state *V-I* characteristics on the lower row. The steady-state operating points are shown by large black dots. **[GATE 2009]**

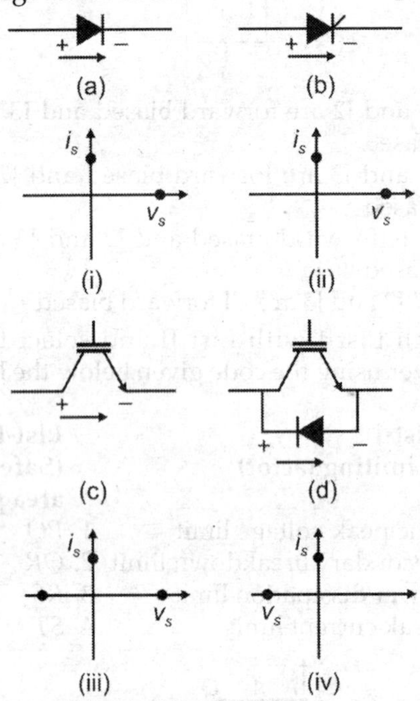

A. (a)-(i), (b)-(ii), (c)-(iii), (d)-(iv)
B. (a)-(ii), (b)-(iv), (c)-(i), (d)-(iii)
C. (a)-(iv), (b)-(iii), (c)-(i), (d)-(ii)
D. (a)-(iv), (b)-(iii), (c)-(ii), (d)-(i)

101. A voltage commuted chopper circuit, operated at 500 Hz, is shown below: **[GATE 2011]**

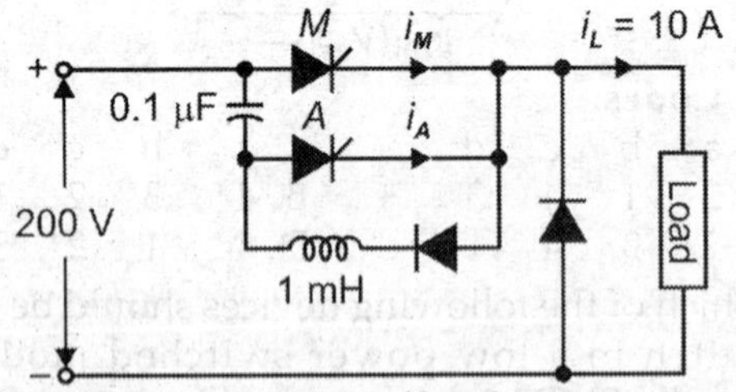

If the maximum value of load currents is 10 A, then the maximum current through the main (*M*) and auxiliary *A*, thyristors will be:

A. $i_{M\,max} = 12$ A and iA max = 10 A
B. $i_{M\,max} = 12$ A and iA max = 2 A
C. $i_{M\,max} = 10$ A and iA max = 12 A
D. $i_{M\,max} = 10$ A and iA max = 8 A

Ans. 95. A 96. B 97. C 98. D 99. C 100. C 101. A

102. Circuit turn-off time of an SCR is defined as the time: **[GATE 2011]**
A. taken by the SCR to turn off
B. required for the SCR current to become zero
C. for which the SCR is reverse biased by the commutation circuit
D. for which the SCR is reverse biased to reduce its current below the holding current

103. Match List-I with List-II and select the correct answer using the code given below the lists: **[IES 2012]**

List-I (Device)	**List-II (Switching time)**
a. TRIAC	1. 5–10 ms
b. SCR	2. 100–400 ms
c. MOSFET	3. 50–100 ms
d. IGBT	4. 200–400 ms

Codes:

	a	b	c	d		a	b	c	d
A.	4	3	2	1	B.	1	2	3	4
C.	4	2	3	1	D.	1	3	2	4

104. A thyristor has a PIV of 650 V. The voltage safety factor is 2. Then the voltage upto which the device can be operated is given by: **[IES 2012]**
A. 1300 V B. 650 V
C. 325 V D. 230 V

105. The typical ratio of latching current to holding current in a 20 thyristor is: **[GATE 2012]**
A. 5.0 B. 2.0
C. 1.0 D. 0.5

106. Figure shows four electronic switches (i), (ii), (iii) and (iv). Which of the switches can block voltages of either polarity (applied between terminals '*a*' and '*b*') when the active device is in the OFF state. **[GATE 2014]**

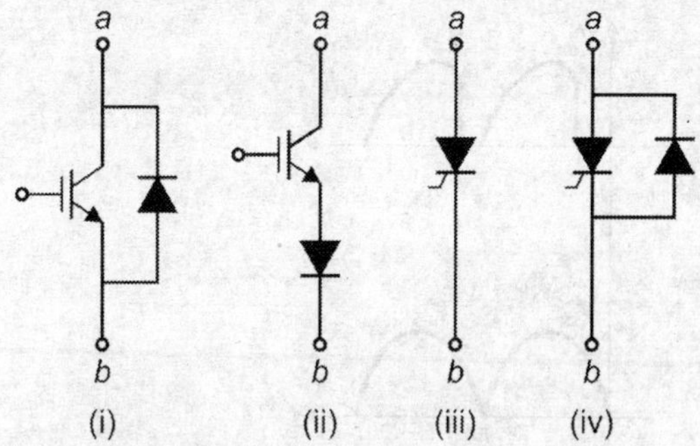

A. (i), (ii) and (iii) B. (ii), (iii) and (iv)
C. (ii) and (iii) D. (i) and (iv)

107. A single-phase diode bridge rectifier supplies a highly inductive load. The load current can be assumed to be ripple free. The ac supply side current waveforms will be: **[GATE 1995]**
A. sinusoidal B. constant dc
C. square D. triangular

108. In the circuit shown in figure, *L* is large and the average value of '*i*' is 100 A. The thyristor is gated in the half cycle of '*e*' at a delay angle is equal to ...... where $e(t) = 200\sqrt{2}\sin 314t$. **[GATE 1992]**

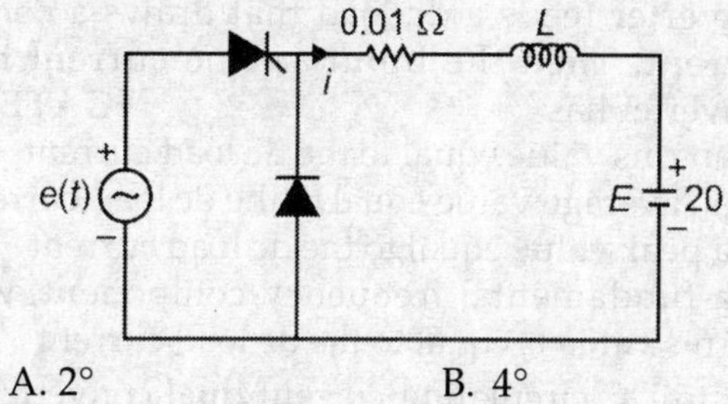

A. 2° B. 4°
C. 6° D. 8°

109. Referring to the figure, the type of load is: **[GATE 1994]**

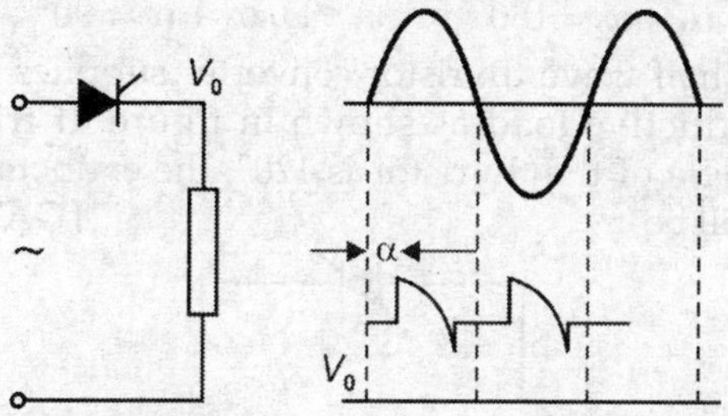

A. inductive load B. resistive load
C. dc motor D. capacitive load

110. In a three-phase controlled bridge rectifier, with an increase of overlap angle the output dc voltage: **[GATE 1996]**
A. decreases B. increases
C. does not change
D. depends upon load inductance

111. In a dual converter, the circulating current: **[GATE 1997]**
A. allows smooth reversal of load current, but increases the response time
B. does not allow smooth reversal of load current, but reduces the response time
C. allows smooth reversal of load current with improved speed of response
D. flows only if there is not interconnecting inductor

112. A three-phase, fully controlled, converter is feeding power into a dc load at a constant current of 150 A. The rms current through each thyristor of the converter is: **[GATE 1998]**
A. 50 A B. 100 A
C. $\frac{150\sqrt{2}}{\sqrt{3}}A$ D. $\frac{150}{\sqrt{3}}A$

113. When the firing angle α of a single-phase fully controlled rectifier feeding a constant dc current into a load is 30°, the displacement power factor of the rectifier is: **[GATE 1998]**

Ans. 102. C 103. C 104. D 105. B 106. C 107. C 108. B 109. C 110. A 111. C 112. D

A. 1 B. 0.5
C. $1/\sqrt{3}$ D. $\sqrt{3}/2$

114. A thyristor based, three-phase, fully controlled converter feeds a dc load that draws a constant current. Then the input ac line current to the converter has: **[GATE 2000]**
A. an rms value equal to the dc load current
B. an average value equal to the dc load current
C. a peak value equal to the dc load current
D. a fundamental frequency component, whose rms value is equal to the dc load current

115. AC to DC circulating current dual converters are operated with the following relationship between their triggering angles (α_1 and α_2). **[GATE 2001]**
A. $\alpha_1 + \alpha_2 = 180°$ B. $\alpha_1 + \alpha_2 = 360°$
C. $\alpha_1 - \alpha_2 = 180°$ D. $\alpha_1 + \alpha_2 = 90°$

116. A half wave thyristor converter supplies a purely inductive load as shown in figure. If triggering angle of the thyristor is 120°, the extinction angle will be **[GATE 2001]**

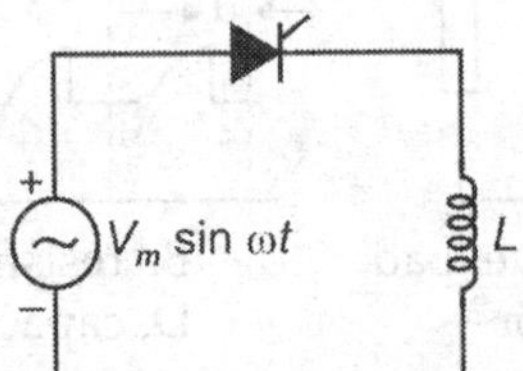

A. 240° B. 180°
C. 200° D. 120°

117. When fed from a fully controlled rectifier, a dc motor, driving an active load, can operate in: **[IES 2002]**
A. forward motoring and reverse braking mode
B. forward motoring and forward braking mode
C. reverse motoring and reverse braking mode
D. reverse motoring and forward braking mode

118. In the single-phase, diode bridge rectifier shown in figure, the load resistor is $R = 50\ \Omega$. The source voltage is $V = 200 \sin \omega t$, where $\omega = 2\pi \times 50$ radians per second. The power dissipated in the load resistor R is: **[GATE 2002]**

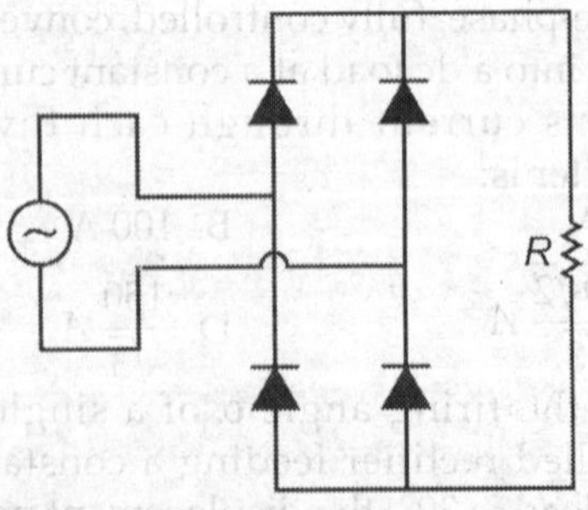

A. $3200/\pi$ W B. $400/\pi$ W
C. 400 W D. 800 W

119. A six pulse thyristor rectifier bridge is connected to a balanced 50 Hz three-phase ac source. Assuming that the dc output current of the rectifier is constant. The lowest frequency harmonics component of the ac source line current is: **[GATE 2002]**
A. 100 Hz B. 150 Hz
C. 250 Hz D. 300 Hz

120. A fully controlled natural commutated threephase bridge rectifier is operating with a firing angle $\alpha = 30°$. The peak to peak voltage ripple expressed as a ratio of the peak output dc voltage at the output of the converter bridge is: **[GATE 2003]**
A. 0.5 B. 3/2
C. $\left(1 - \dfrac{\sqrt{3}}{2}\right)$ D. $\sqrt{3} - 1$

121. A phase-controlled half-controlled singlephase converter is shown in figure. The control angle $\alpha = 30°$.

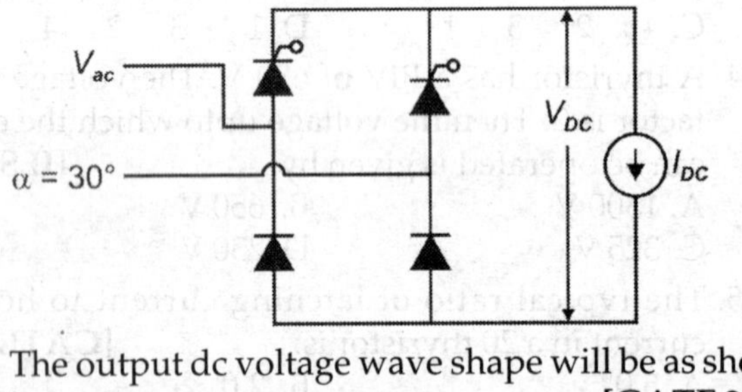

The output dc voltage wave shape will be as shown in: **[GATE 2003]**

A.
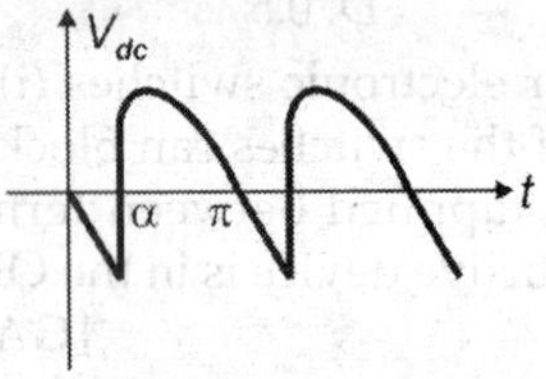

B.
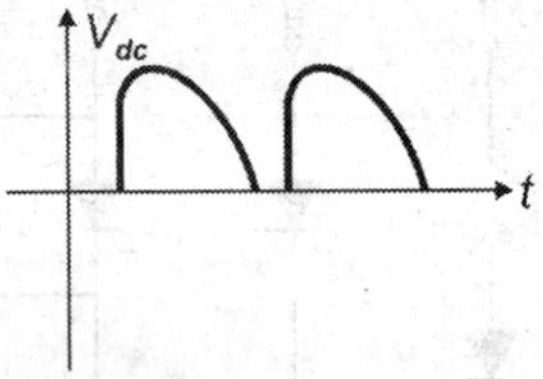

C.
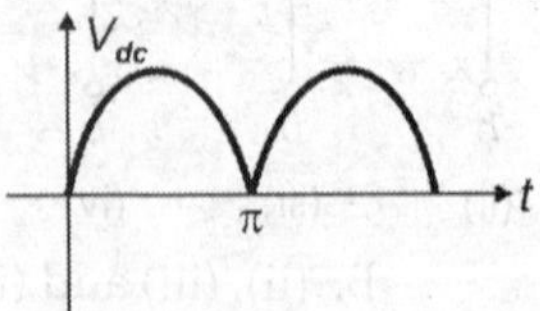

D.
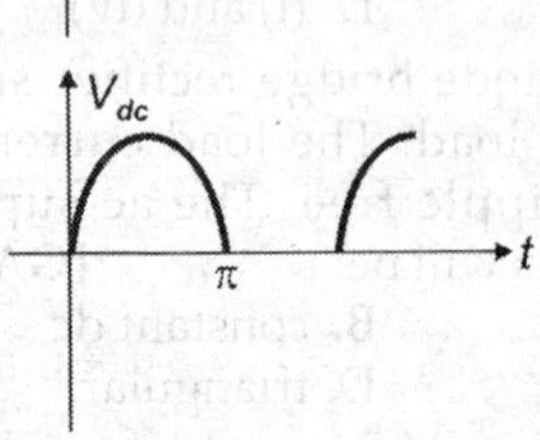

Ans. 113. D 114. C 115. A 116. A 117. B 118. C 119. C 120. A 121. A

122. The triggering circuit of a thyristor is shown in figure. The thyristor requires a gate current of 10 mA, for guaranteed turn-on. The value of R required for the thyristor to turn-on reliably under all conditions of V_b variation is: **[GATE 2004]**

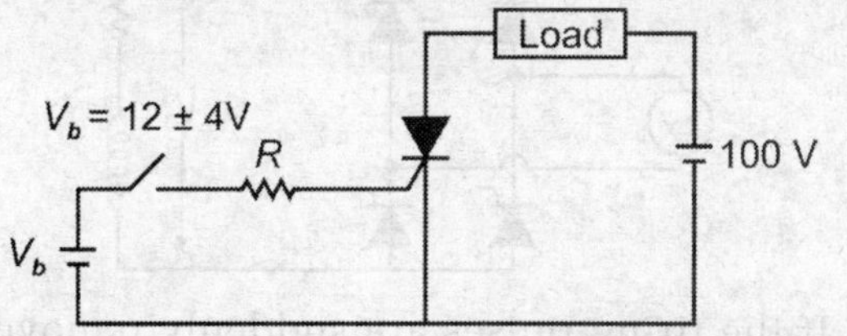

A. 10000 Ω B. 1600 Ω
B. 1200 Ω D. 800 Ω

123. The circuit in figure shows a full-wave rectifier. The input voltage is 230 V (rms) single-phase ac. The peak reverse voltage across the diodes D_1 and D_2 is: **[GATE 2004]**

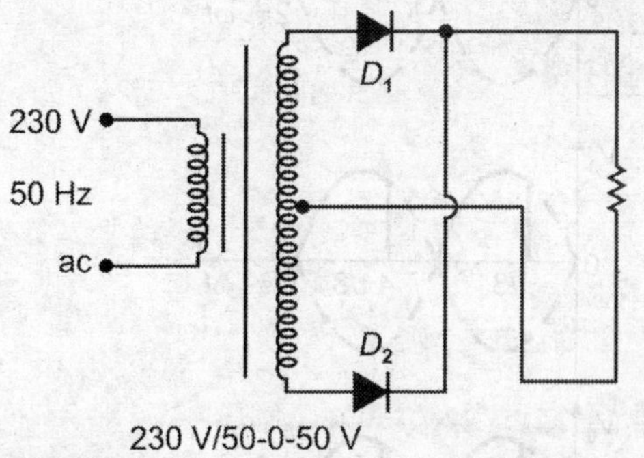

A. $100\sqrt{2}V$ B. 100 V
C. $50\sqrt{2}V$ D. 50 V

124. The circuit in figure shows a three-phase halfwave rectifier. The source is a symmetrical, three-phase four-wire system. The line-to-line voltage of the source is 100 V. The supply frequency is 400 Hz. The ripple frequency at the output is: **[GATE 2004]**

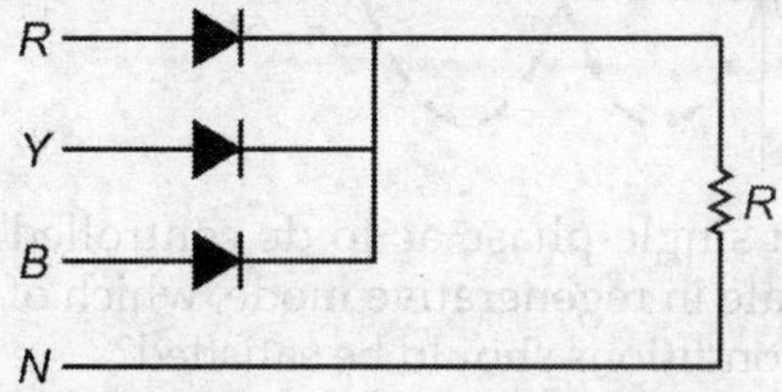

A. 400 Hz B. 800 Hz
C. 1200 Hz D. 2400 Hz

125. A three-phase diode bridge rectifier is fed from a 400 V rms, 50 Hz, three-phase AC source. If the load is purely resistive, then peak instantaneous output voltage is equal to: **[GATE 2005]**

A. 400 V B. $400\sqrt{2}$ V
C. $400\sqrt{2/3}$ V D. $400\sqrt{3}$ V

126. Consider a phase-controlled converter shown in the figure. The thyristor is fired at an angle α in every positive half cycle of the input voltage. If the peak value of the instantaneous output voltage equals 230 V, the firing angle a is close to: **[GATE 2005]**

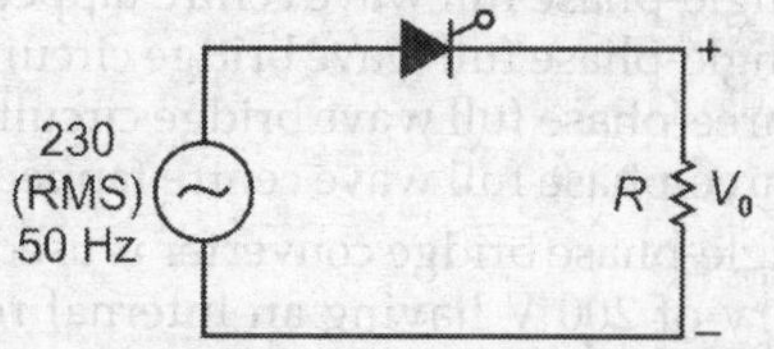

A. 45° B. 135°
C. 90° D. 83.6°

127. In a dual converter, the circulating current: **[IES 2006]**

A. allows smooth reversal of load current, but increases the response time
B. allows smooth reversal of load current with improved speed of response
C. does not allow smooth reversal of load current, but reduces the response time
D. flows if there is no interconnecting inductor

128. When the firing angle a of a single-phase fully controlled rectifier feeding constant dc current into the load is 30°, what is the displacement factor of the rectifier? **[IES 2006]**

A. 1 B. 0.5
C. $\sqrt{3}$ D. $\sqrt{3}/2$

129. A single-phase half-wave uncontrolled converter circuit is shown in figure. A 2-winding transformer is used at the input for isolation. Assuming the load current to be constant and $v = V_m \sin \omega t$, the current waveform through diode D_2 will be: **[GATE 2006]**

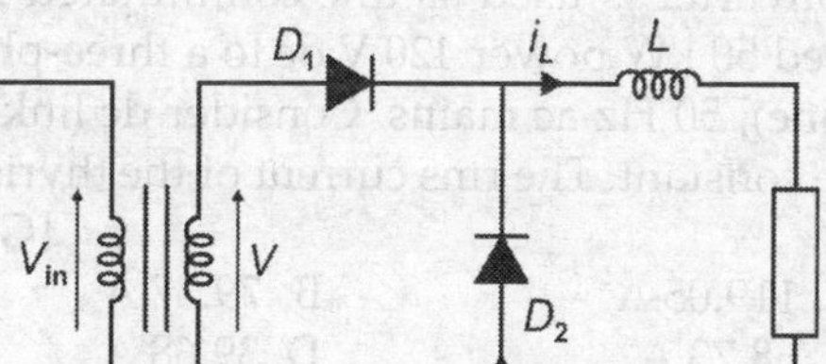

A. 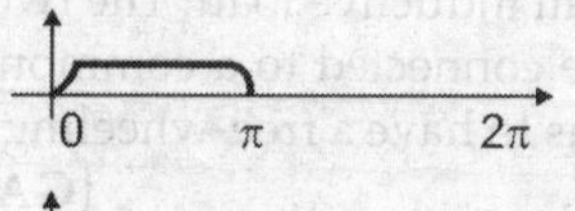

B. 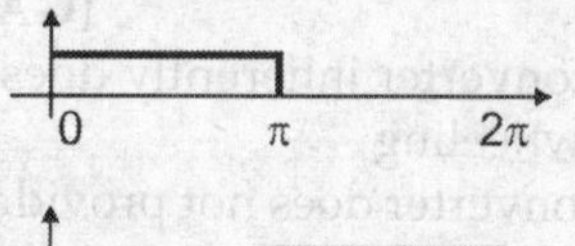

C. 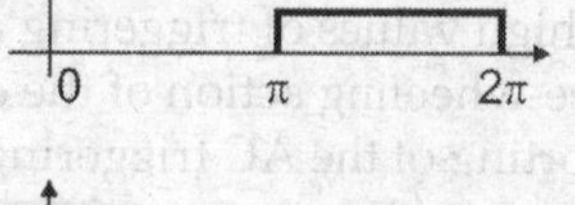

D. 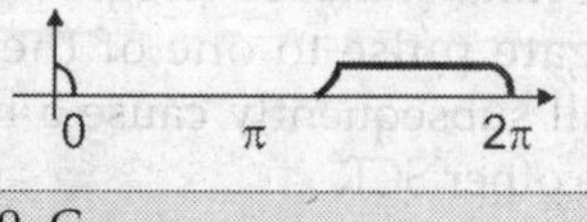

Ans. 122. D 123. A 124. C 125. B 126. B 127. B 128. D 129. C

130. For the same voltage output, which one of the following has larger peak inverse voltage of the thyristor? **[IES 2006]**
A. Single-phase full wave centre tapped circuit
B. Single-phase full wave bridge circuit
C. Three-phase full wave bridge circuit
D. Three-phase full wave centre tapped circuit

131. A single-phase bridge converter is used to charge a battery of 200 V having an internal resistance of 2 Ω as shown in figure. The SCRs are triggered by a constant dc signal. If SCR 2 gets open circuited, what will be the average charging current? **[GATE 2006]**

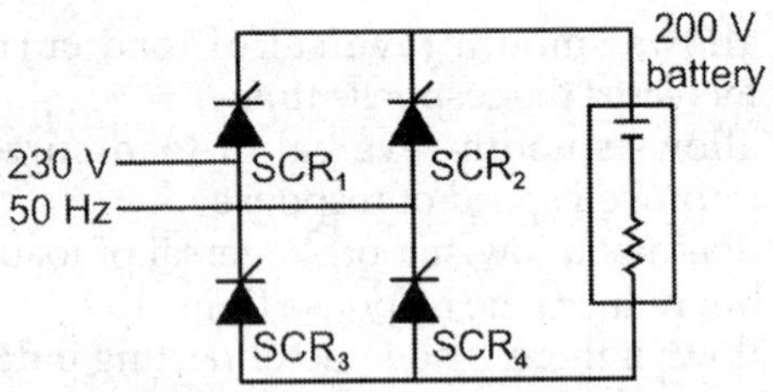

A. 23.8 A B. 15 A
C. 11.9 A D. 3.54 A

132. A single-phase fully controlled, the thyristor bridge ac-dc converter is operating at a firing angle of 25° and on overlap angle of 10° constant dc output current of 20 A. The fundamental power factor (displacement factor) at input ac mains is: **[GATE 2007]**
A. 0.78 B. 0.827
C. 0.866 D. 0.9

133. A three-phase fully-controlled thyristor bridge converter is used as line commutated inverter to feed 50 kW power 420 V dc to a three-phase, 415 V (line), 50 Hz ac mains. Consider dc link current to be constant. The rms current of the thyristor is: **[GATE 2007]**
A. 119.05 A B. 79.37 A
C. 68.73 A D. 39.68 A

134. A single-phase full-wave half-controlled bridge converter feeds an inductive load. The two SCRs in the converter are connected to a common DC bus. The converter has to have a free-wheeling diode: **[GATE 2007]**
A. because the converter inherently does not provide for free-wheeling
B. because the converter does not provide for free-wheeling for high values of triggering angles
C. or else the free-wheeling action of the converter will cause shorting of the AC triggering angles
D. or else if a gate pulse to one of the SCRs is missed, it will subsequently cause a high load current in the other SCR.

135. A single-phase half controlled converter shown in the figure feeding power to highly inductive load. The converter is operating at a firing angle of 60°.

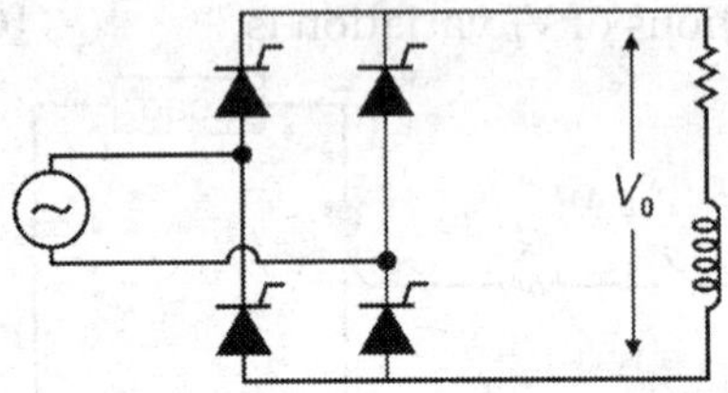

If the firing pulses are suddenly removed, the steady-state voltage (V_0) waveform of the converter will become: **[GATE 2008]**

A. 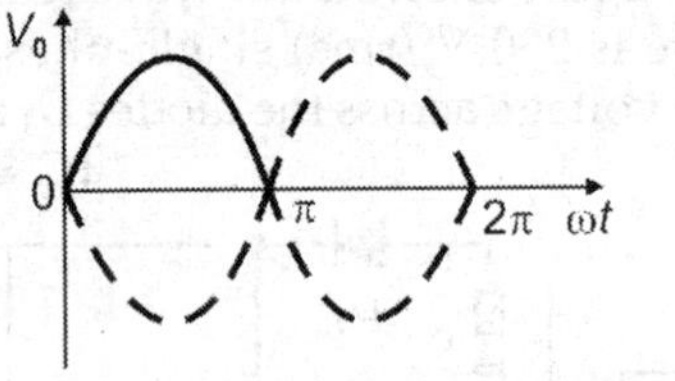

B. 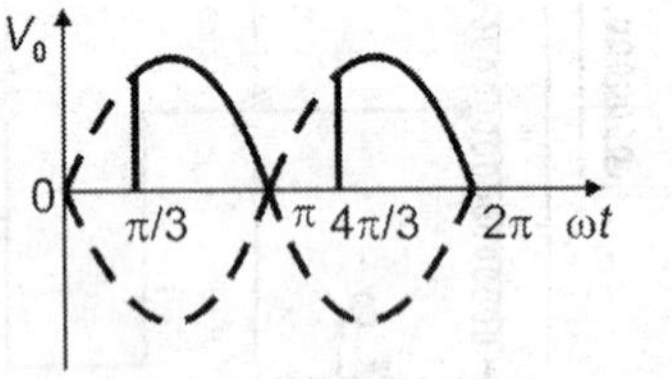

C. 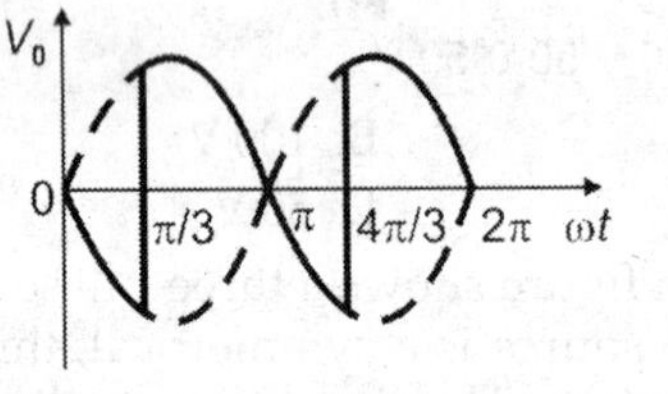

D. 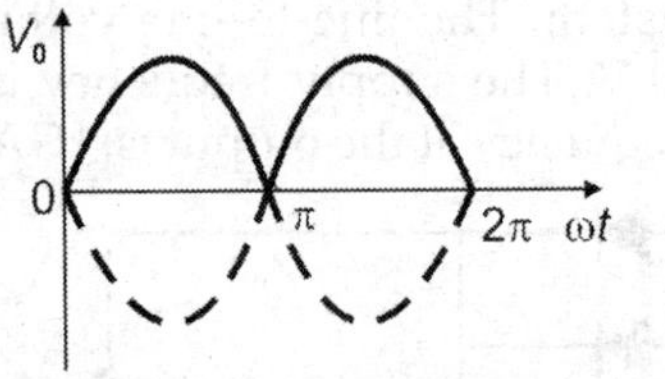

136. For a single-phase ac to dc controlled rectifier to operate in regenerative mode, which of the following conditions should be satisfied? **[IES 2008]**
A. Half-controlled bridge, α < 90°, source of emf in load
B. Half-controlled bridge, α > 90°, source of emf in load
C. Full-controlled bridge, α > 90°, source of emf in load
D. Full-controlled bridge, a < 90°, source of emf in load

137. The fully controlled thyristor converter in the figure is fed from a single-phase source. When the firing angle is 0°, the dc output voltage of the converter is 300 V. What will be the output voltage

Ans. 130. D 131. C 132. C 133. C 134. C 135. A 136. C

for a firing angle of 60°, assuming continuous conduction? **[GATE 2010]**

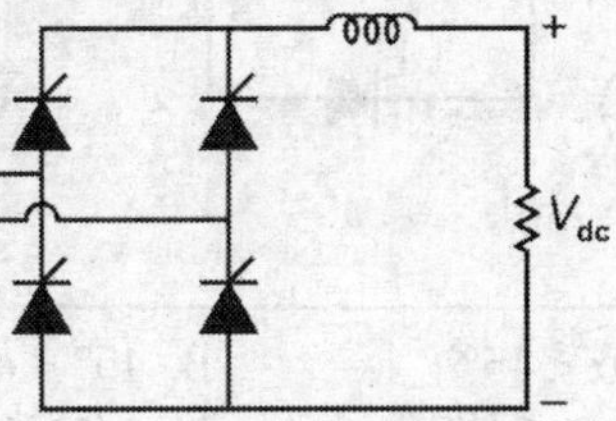

A. 150 V B. 210 V
C. 300 V D. 100p V

138. A half-controlled single-phase bridge rectifier is supplying an *R-L* load. It is operating at a firing angle α and the load current is continuous. The fraction of cycle that the free-wheeling diode conduct is: **[GATE 2012]**

A. 1/2 B. $\left(1-\frac{\alpha}{\pi}\right)$
C. $1/2\pi$ D. α/π

139. Thyristor *T* in the figure given is initially off and is triggered with a single-pulse of width 10 μs. It is given that $L = \left(\frac{100}{\pi}\right)\mu H$ and $C = \left(\frac{100}{\pi}\right)\mu F$. Assuming latching and holding currents of the thyristor are both zero and the initial charge on C is zero, *T* conducts for: **[GATE 2013]**

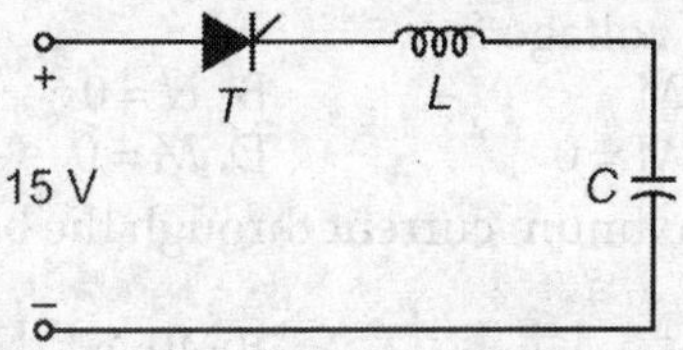

A. 10 μs B. 50 μs
C. 100 μs D. 200 μs

140. **Statement I:** For the same voltage output, the power factor of a single-phase semiconverter is better than a full converter.
Statement II: The single-phase semiconverter uses two diodes and two controlled switches.
Codes:
A. both statement I and statement II are individually true and statement II is the correct explanation of statement I
B. both statement I and statement II are individually true but statement II is not the correct explanation of statement I
C. statement I is true but statement II is false
D. statement I is false but statement II is true

141. In the following circuit, the input voltage V_{in} is 100 sin (100π*t*). For 100π*RC* = 50, the average voltage across *R* (in volts) under stead-state is nearest to: **[GATE 2015]**

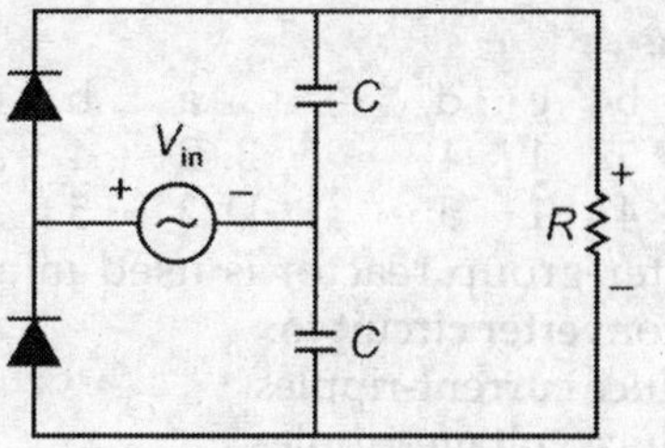

A. 100 B. 31.8
C. 200 D. 63.6

142. In the given rectifier, the delay angle of the thyristor T_1 measured from the positive going zero crossing of *Vs* is 30°. If the input voltage *Vs* is 100 sin (100π*t*) *V*, the average voltage across *R* (in volt) under steady-state is .... **[GATE 2015]**

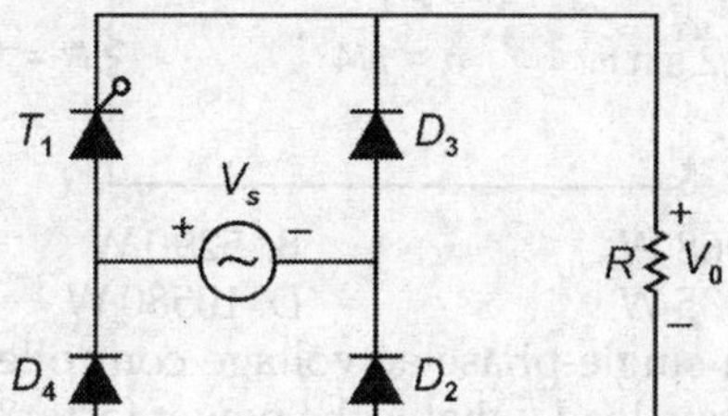

A. 31.5 V B. 41.5 V
C. 51.5 V D. 61.5 V

143. If a diode is connected in anti-parallel with a thyristor, then: **[GATE 1997]**
A. both turn-off power loss and turn-off time decreases
B. turn-off power loss decreases but turn-off time increases
C. turn-off power loss increases, but turn-off time decreases
D. none of these

144. Resonant converter's are basically used to: **[GATE 1999]**
A. generate large peaky voltage
B. reduce the switching losses
C. eliminate harmonics
D. convert a square wave into a sine wave

145. A cycloconverter is operating on a 50 Hz supply. The range of output frequency that can be obtained with acceptable quality, is: **[IES 2001]**
A. 0–16 Hz B. 0–32 Hz
C. 0–64 Hz D. 0–128 Hz

146. Match List-I with List-II and select the correct answer: **[IES 2002]**

List-I	List-II
a. Commutation	1. Inductive load
b. *V*-curves	2. Capacitive load
c. Free-wheeling diode	3. Interpole
d. Overlap	4. Source inductance
	5. Synchronous motor

Ans. 137. A 138. D 139. C 140. B 141. C 142. D 143. ? 144. B 145. D

Codes:

	a	b	c	d		a	b	c	d
A.	3	5	1	4	B.	2	4	3	5
C.	3	4	1	5	D.	2	5	3	4

147. An inter-group reactor is used in a singlephase cycloconverter circuit to: **[IES 2003]**
A. reduce current-ripples
B. reduce voltage-ripples
C. limit circulating current
D. limit di/dt in the semiconductor switch

148. The triac circuit shown in figure controls the ac output power to the resistive load. The peak power dissipation in the load is: **[GATE 2004]**

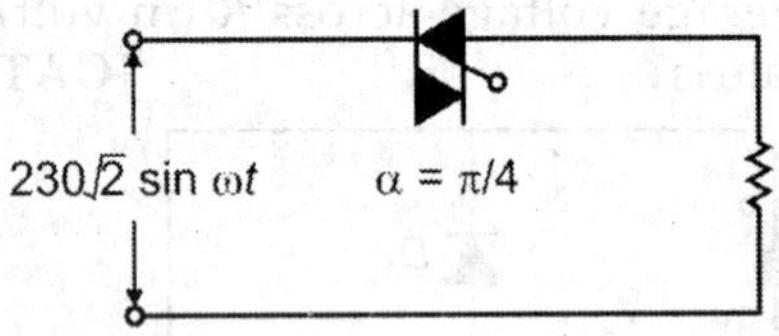

A. 3968 W
B. 5290 W
C. 7935 W
D. 10580 W

149. For a single-phase ac voltage controller feeding a resistive load, what is the power factor? **[IES 2005]**
A. Unity for all values of firing angle
B. $\left[\frac{1}{\pi}\left\{(\pi-\alpha)+\frac{1}{2}\sin 2\alpha\right\}\right]^{1/2}$
C. $\left[\frac{1}{\pi}\left\{(\pi+\alpha)+\frac{1}{2}\sin 2\alpha\right\}\right]^{1/2}$
D. $\left[\frac{1}{\pi}\left\{(\pi+\alpha)-\frac{1}{2}\sin 2\alpha\right\}\right]^{1/2}$
where a is firing angle measured from voltage zero.

150. Match List-I with List-II and select the correct answer using the code given below the lists: **[IES 2008]**

List-I
A. Chopper controlled resistance in the rotor circuit of an induction motor
B. Sub-synchronous convertercascade in the rotor circuit of an induction motor
C. 3-phase ac voltage controller
D. Cycloconverter

List-II
1. Very low speed, high-power reversible drive
2. Centrifuges in sugar industry
3. Blowers and compressors
4. Loads requiring good starting performance

	a	b	c	d		a	b	c	d
A.	3	4	2	1	B.	3	4	1	2
C.	4	3	1	2	D.	4	3	2	1

151. In the single-phase voltage controller circuit shown in the figure, for what range of triggering angle (α), the output voltage (V_0) is not controllable? **[GATE 2008]**

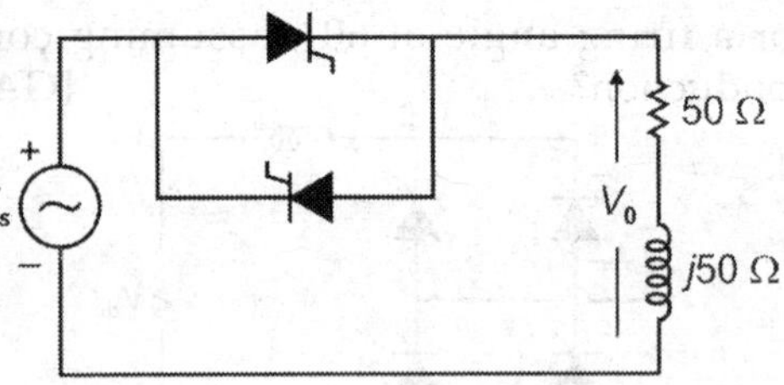

A. 0° < α < 45°
B. 45° < α < 135°
C. 90° < α < 180°
D. 135° < α < 180°

152. The power electronic converter shown in the figure has a single-pole double-throw switch. The pole *P* of the switch is connected alternately to throws *A* and *B*. The converter shown is a: **[GATE 2010]**

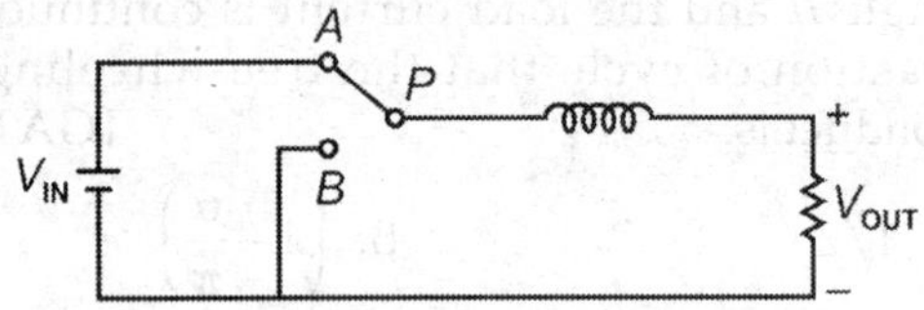

A. step-down chopper (buck converter)
B. step-up chopper (boost converter)
C. half-wave rectifier
D. full-wave rectifier

153. An integral cycle ac voltage controller is feeding a purely resistive circuit from a singlephase ac voltage source. The current waveform consists alternately burst of *N*-complete cycle of conduction followed by *M*-complete of extinction. The rms value of the load voltage equals the rms value of supply voltage for: **[IES 2011]**
A. $N = M$
B. $N = 0$
C. $N = M = 0$
D. $M = 0$

154. The maximum current through the battery will be: **[GATE 2011]**
A. 14 A
B. 40 A
C. 80 A
D. 94 A

155. The kVA rating of the input transformer is: **[GATE 2011]**
A. 53.2 kVA
B. 46.0 kVA
C. 22.6 kVA
D. 19.6 kVA

156. The SCR in the circuit shown has a latching current of 40 mA. A gate pulse of 50 μs is applied to the SCR. The maximum value of R in Ω to ensure successful firing of the SCR is .... **[Gate 2014]**

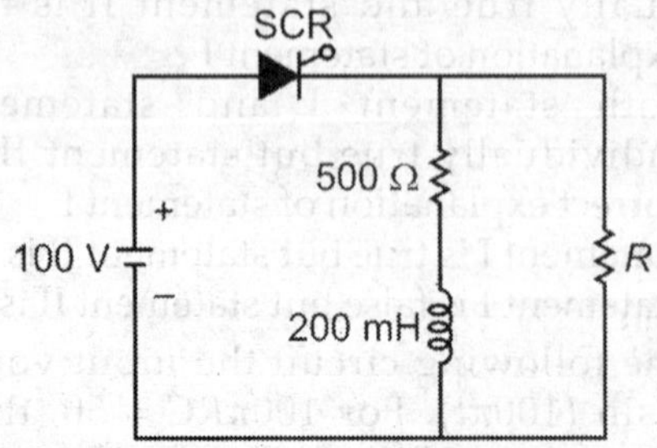

A. 2 kΩ
B. 4 kΩ
C. 6 kΩ
D. 10 kΩ

Ans. 146. A 147. C 148. D 149. A 150. A 151. A 152. A 153. A 154. B 155. C 156. D

157. A single-phase AC regulator fed from 50 Hz supply feeds a load having 4 Ω resistance and 12.73 mH inductance. The control range of firing angle will be: **[IES 2012]**
A. 0° to 180° B. 45° to 180°
C. 90° to 180° D. 0° to 45°

158. The latching current in the below circuit is 4 mA. The minimum width of the gate pulse required to turn on the thyristor is **(IES 2001)**

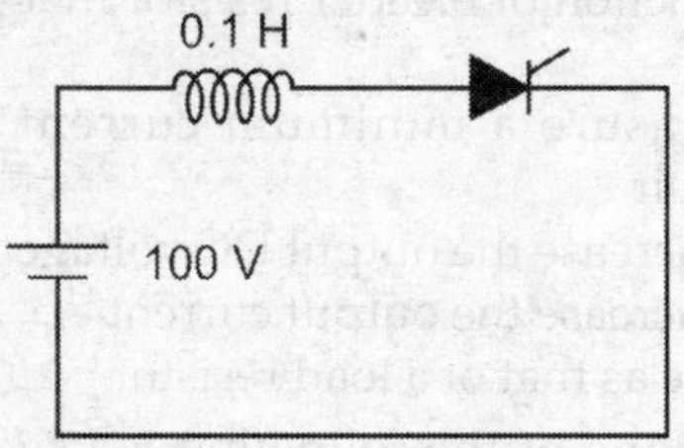

A. 6 μs B. 4 μs
C. 2 μs D. 1 μs

159. Triac cannot be used in **(IES 2001)**
A. AC voltage regulators
B. cycloconverters
C. solid state type of switch
D. inverter

160. The snubber circuit is used in thyristor circuits for **(IES 2001)**
A. triggering B. dv/dt protection
C. di/dt protection D. phase shifting

161. It is preferable to use a train of pulse of high frequency for gate triggering of SCR in order to reduce **(IES 2001)**
A. dv/dt problem
B. di/dt problem
C. the size ofthe pulse transformer
D. the complexity of the firing circuit

162. Which one of the following is not the advantage of solid state switching of AC capacitors into AC supply over relay-based switching? **(IES 2001)**
A. low transients B. low losses
C. fast response D. long life

163. The most suitable device for high frequency inversion in SMPS is **(IES 2001)**
A. BJT B. IGBT
C. MOSFET D. GTO

164. When cathode of a thyristor is made more positive than its anode **(IES 2002)**
A. all the junctions are reverse biased
B. outer junctions are reverse biased and central one is forward biased
C. outer junctions are forward biased and central one is reverse biased
D. all the junctions are forward biased

165. The sharing of the voltage between thyristors operating in series is influenced by the **(IES 2002)**
A. di/dt capabilities
B. dv/dt capabilities
C. junction temperatures
D. static V-I characteristics and leakage currents

166. RC snubber is used in parallel with the thyristor to **(IES 2002)**
A. reduce dv/dt across it
B. reduce di/dt through it
C. limit current through the thyristor
D. ensure its conduction after gate signal is removed

167. A thyristor controlled reactor is used to get **(IES 2002)**
A. variable resistance B. variable capacitance
C. variable inductance
D. improved reactor power factor

168. In a switched-mode power supply (SMPS), after conversion of AC supply to a highly filtered DC voltage, a switching transistor is switched ON and OFF at a very high speed by a pulse width modulator (PWM) which generates very high frequency square pulses. The frequency of the pulses is typically in the range of **(IES 2002)**
A. 100 Hz–200 Hz B. 500 Hz–1 kHz
C. 2 kHz–5 kHz D. 20 kHz–50 kHz

169. Match **List-I** (Thyristors) with **List-II** (Symbols) and select the correct answer: **(IES 2003)**

List-I
a. Silicon-controlled rectifier (SCR)
b. Silicon-controlled switch (SCS)
c. Silicon-unilateral switch (SUS)
d. Light-activated SCR (LASCR)

List-II

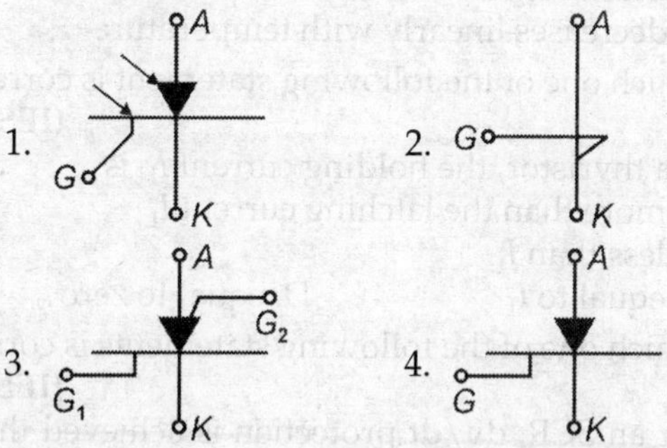

Codes

	a	b	c	d
A.	3	4	1	2
B.	4	3	1	2
C.	3	4	2	1
D.	4	3	2	1

170. Turn-on and turn-off times of transistor depend on **(IES 2003)**
A. static characteristic
B. junction capacitances
C. current gain
D. none of the above

Ans. 157. B 158. B 159. C 160. B 161. C 162. A 163. C 164. B 165. D 166. D 167. C 168. D 169. D 170. B

171. Which one of the following statements is correct? **(IES 2004)**

A triac is a
A. 2 terminal switch
B. 2 terminal bilateral switch
C. 3 terminal unilateral switch
D. 3 terminal bidirectional switch

172. Which one of the following is the most suitable device for a DC-DC converter? **(IES 2004)**
A. BJT B. GTO
C. MOSFET D. Thyristor

173. Which one of the following shows current fold back characteristics curve for an SCR controlled shunt regulated power supply? **(IES 2004)**

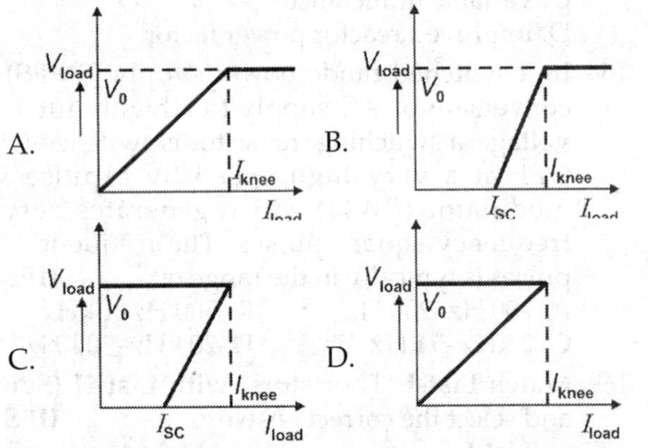

174. Which one of the following statements is correct? **(IES 2004)**
In a transistor, the reverse saturation current I_{CO}
A. doubles for every 10°C rise in temperature
B. double for every 1°C rise in temperature
C. increases linearly with temperature
D. decreases linearly with temperature

175. Which one of the following statement is correct? **(IES 2004)**
In a thyristor, the holding current I_H is
A. more than the latching current I_L
B. less than I_L
C. equal to I_L D. equal to zero

176. Which one of the following statements is correct? **(IES 2004)**
For an SCR, dv/dt protection is achieved through the use of
A. RC across SCR B. RL in series with SCR
C. L in series with SCR D. RC in series with SCR

177. A power diode is in the forward conduction mode and the forward current is now decreased. The reverse recovery time of the diode is t_r and the rate of fall of the diode current is di/dt. What is the stored charge? **(IES 2004)**
A. $(di/dt) \cdot f_r$ B. $1/2(di/dt) \cdot t_r^2$
C. $(di/dt) \cdot t_r^2$ D. $1/2(di/dt) \cdot t_r$

178. Which one of the following statements is correct? **(IES 2004)**
In order to get best results per unit cost, the heat sinks on which the thyristors are mounted, and made of
A. aluminium B. copper
C. nickel D. stainless steel

179. Which one of the following statements is correct? The function of bleeder resistor in a power supply is **(IES 2004)**
A. to ensure a minimum current drain in the circuit
B. to increase the output DC voltage
C. to increase the output current
D. same as that of a load resistor

180. Which one of the following is used as the main switching element in a switched mode power supply operating in 20 kHz to 100 kHz range? **(IES 2004)**
A. Thyristor B. MOSFET
C. Triac D. UJT

181. A gate turn off (GTO) thyristor has capacity to **(IES 2005)**
A. amplify the gate-current
B. turn-off when positive current pulse is given at the gate
C. turn-off when a gate-pulse is given at the gate even though it is reverse biased
D. turn-off when a negative current pulse is given at the gate

182. Power electronic device with poor turn-off gain is **(IES 2005)**
A. a symmetrical thyristor
B. a conventional thyristor
C. power bipolar junction transistor
D. gate turn-off thyristor

183. Carrier frequency gate drive is used for turn-on of a thyristor to reduce **(IES 2005)**
A. di/dt B. turn-on time
C. dv/dt
D. size of pulse transformer

184. Which one of the following statements is not correct? **(IES 2005)**
A. Power MOSFETs are so constructed as to avoid punch through
B. In a power MOSFET, the channel length is relatively large and channel width is relatively small
C. Power MOSFETs do not experience any minority carrier storage
D. Power MOSFETs can be put in parallel to handle large currents

Ans. 171. D 172. B 173. C 174. A 175. C 176. A 177. B 178. A 179. A 180. B 181. D 182. D 183. D 184. B

185. Match **List-I** (Limiting factor) with **List-II** (Safe Operating Area Portion) and select the correct answer using the code given below the lists: **(IES 2005)**

List-I
a. The peak voltage limit
b. Secondary breakdown limit
c. Power dissipation limit
d. Peak current limit

List-II
1. PQ
2. QR
3. RS
4. ST

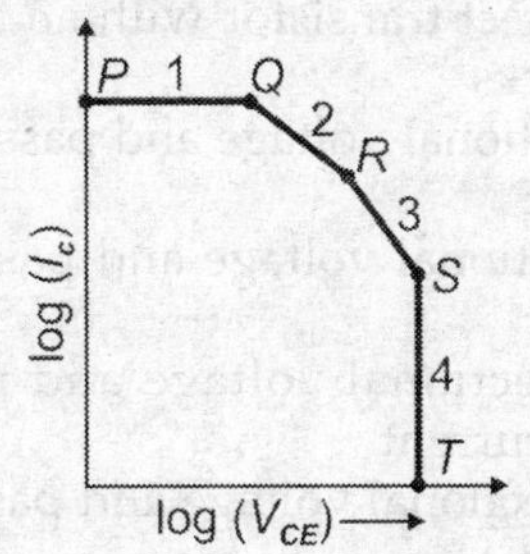

Codes

	a	b	c	d		a	b	c	d
A.	2	1	4	3	B.	4	3	2	1
C.	2	3	4	1	D.	4	1	2	3

186. In a thyristor, di/dt protection is achieved by the use of **(IES 2005)**
A. an inductance L in series withthe thyristor
B. A resistor in series with the thyristor
C. RC in series with the thyristor
D. none of these

187. Triacs cannot be used in AC voltage regulator for a **(IES 2005)**
A. resistive load B. back emf load
C. inductive load
D. all of these

188. A resistor connected across the gate and cathode of an SCR **(IES 2006)**
A. increases dv/dt rating of SCR
B. increases holding current of SCR
C. increases noise immunity of SCR
D. increases turn-off time of SCR

189. Turn-on of a thyristor takes place when **(IES 2006)**
A. anode to cathode voltage is positive
B. anode to cathode voltage is negative
C. there is a positive current pulse at the gate
D. the anode to cathode voltage is positive and there is a positive current pulse at the gate

190. Which is the most suitable power device for high frequency (> 100 kHz) switching application? **(IES 2006)**
A. Power MOSFET
B. Bipolar junction transistor
C. Schotty diode D. Microwave transistor

191. Which of the following devices should be used as a switch in a low power switched mode power supply (SMPS)? **(IES 2006)**
A. GTO B. MOSFET
C. Triac D. Thyristor

192. Which one of the following is the main advantage of SMPS over linear power supply? **(IES 2006)**
A. No transformer is required
B. Only one stage of conversion
C. No filter is required
D. Low power dissipation

193. Snubber circuits are used to protect thyristor from which ofthe following? **(IES 2007)**
A. High di/dt and low dv/dt
B. High dv/dt and low di/dt
C. Low dv/dt and low di/dt
D. High dv/dt and high di/dt

194. Consider the following statements in respect of IGBT:
1. It combines the attributes of MOSFET and BJT
2. It has low forward voltage drop
3. Its switching speed is very much lower than that of MOSFET
4. It has high input impedance

Which of these statements are correct? **(IES 2007)**
A. 1, 2, 3 and 4 B. 1, 2 and 4
C. 1, 2 and 3 D. 3 and 4

195. A modern power semiconductor device that combines the characteristics of BJT and MOSFET is **(IES 2008)**
A. GTO B. FCT
C. IGBT D. MCT

196. The anode current through a conducting SCR is 10 A. If its gate current is made one-fourth, then what will be the anode current? **(IES 2009)**
A. 0 A B. 5 A
C. 10 A D. 20 A

197. In a power circuit of 3 kV, four thyristor each of rating 800 V are connected in series. What is the percentage series derating factor? **(IES 2009)**
A. 30.5 B. 25.5
C. 12.5 D. 6.25

198. For an SCR, the gate cathode characteristic has a straight line slope of 140. For trigger source voltage of 20 V and allowable gate power dissipation of 0.5 watts, what is the gate source resistance? **(IES 2009)**
A. 200 Ω B. 255 Ω
C. 195 Ω D. 185 Ω

199. An SCR is rated for 650 V, PIV. What is the voltage for which the device can be operated if the voltage safety factor is 2? **(IES 2009)**
A. 325 V_{rms} B. 230 V_{rms}
C. 459 V_{rms} D. 650 V_{rms}

200. In an LC series circuit connected to a DC supply of E volts via a thyristor when it turns off, the voltage that appears across the thyristor is **(IES 2010)**

Ans. 185. B 186. A 187. C 188. C 189. D 190. A 191. B 192. D 193. D 194. A 195. C 196. C 197. D 198. C 199. B

A. +E B. +2E
C. –E D. –2E

201. An SCR is in conducting state, a reverse voltage is applied between anode and cathode, but it fails to turn off. What could be reason? **(IES 2011)**
A. Positive voltage is applied to the gate
B. The reverse voltage is small
C. The anode current is more than the holding current
D. Turn off time of SCR is alrge

202. A reverse conducting thyristor (RCT) normally replaces **(IES 2011)**
A. a pair of anti-parallel thyristors in a circuit
B. a combination of a thyristor and an anti-parallel diode in a circuit
C. a thyristor in situation where it is not required to have reversed blocking capability at all
D. conventional conversion grade thyristors having large turn off time

203. A thyristor can be switched from a non-conducting state to a conducting state by applying: **(IES 2011)**
1. voltage more than forward break over voltage
2. a voltage with high dv/dt
3. positive gate current with positive anode voltage
4. negative gate current with positive anode voltage
A. 1, 2, 3 and 4 are correct
B. 1, 2 and 4 are correct
C. 1, 2 and 3 are correct D. 2, 3 and 4 are correct

204. A structure obtained by lightly doped in drift region between the layers of a pn-junction a PIN diode is obtained. This structure is effective in **(IES 2011)**
A. making the diode support large reverse blocking voltages
B. making reverse recovery process slow
C. making the diode have high on-state voltage drop
D. reducing the voltage spike during turn off due to stray inductance

205. Which one of the following statements is not correct for a MOSFET? **(IES 2011)**
A. Are easy to parallel for higher current
B. Leakage current is relatively high
C. Have more linear characteristic
D. Overload and peak current handling capability are high

206. In a GTO, anode current begins to fall when the gate current **(IES 2011)**
A. is negative peak at time $t = 0$
B. is negative peak at t = storage period t
C. just begins to become negative at $t = 0$
D. just begins to become positive at $t = 0$

207. Consider the following statements:
1. A thyristor requires turn off circuit while transistor does not
2. The voltage drop of a thyristor is less than that of a transistor
3. A thyristor requires a continuous gate current
4. A transistor draws continuous base current
Which of these statements are correct? **(IES 2011)**
A. 1, 2, 3 and 4 B. 1 and 2
C. 2 and 4 D. 1 and 4

208. A field effect transistor with an anti-parallel body diode blocks **(IES 2011)**
A. bidirectional voltage and passes uni-directional current
B. bidirectional voltage and passes bi-directional current
C. unidirectional voltage and passes uni-directional current
D. unidirectional voltage and passes bi-directional current

209. The following is a unipolar device. **(IES 2012)**
A. BJT B. IGBT
C. GTO D. MOSFET

210. A thyristor has a PIV of 650 V. The voltage safety factor is 2. Then the voltage upto which the device can be operated is given by **(IES 2012)**
A. 1300 V B. 650 V
C. 325 V D. 230 V

211. When a thyristor in the forward blocking state, then **(IES 2012)**
A. all 3 junctions are reverse biased
B. anode and cathode junctions are forward biased but gate junction is reverse biased
C. anode junction if forward biased but other two are reverse biased
D. anode and gate junctions are forward biased but cathode is reverse biased

212. In forward-bias portion of the thyristor i-v characteristic, the number of stable operating regions is **(IES 2012)**
A. one B. two
C. three D. none

213. A DC source of 100 volts supplies a purely inductive load of 0.1 H; the controller is an SCR in series with source and load. If the specified latching current is 100 mA, then the minimum width of the gating pulse to ensure turn-on of SCR would be **(IES 2012)**
A. 10 μs B. 50 μs
C. 100 μs D. 1 μs

214. A single-phase two pulse converter feeds an R-L load with insufficient smoothing but the conduction is continuous. If the resistance of the load circuit is increased, then **(IES 2012)**

Ans. 200. C 201. C 202. B 203. C 204. B 205. B 206. B 207. D 208. D 209. D 210. D 211. B 212. B 213. C

A. the ripple content ofthe load current will remain the same
B. the ripple content of the load current will decrease
C. the ripple content of the load current will increase
D. there is possibility of discontinuous conduction due to an increase in the ripple content

215. A thyristor has internal power dissipation of 40 W and is operated at an ambient temperature of 20°C. If thermal resistance is 1.6° C/W, the junction temperature is **(IES 2013)**
A. 114°C B. 164°C
C. 90°C D. 84°C

216. Which one of the following power semiconductor device has bidirectional current capability? **(IES 2014)**
A. SCR B. TRIAC
C. IGBT D. MOSFET

217. Turn on time of an SCR can be reduced by using a **(IES 2014)**
A. rectangular pulse of high amplitude and narrow width
B. rectangular pulse of low amplitude and wide width
C. triangular pulse
D. trapezoidal pulse

218. Which of the following is the fastest switching device? **(IES 2014)**
A. JFET B. BJT
C. MOSFET D. Triode

219. Which of the following does not cause damage of an SCR? **(IES 2014)**
A. High current
B. High rate of rise of current
C. High temperature rise
D. High rate of rise of voltage

220. Which of the following transistors is symmetrical in the same manner that emitter and collector or source and drain terminals can be interchanged? **(IES 2014)**
A. JFET B. MOSFET
C. NPN transistor D. PNP transistor

221. The snubber circuit used to shape the turn-on switching trajectory of thyristor and/or to limit di/dt during turn-on is **(IES 2014)**
A. L-R snubber polarised
B. R-C snubber polarised
C. R-C snubber unpolarised
D. L-R snubber unpolarised

222. The total harmonic distortion (THD) of AC supply input current of rectifiers is maximum for **(IES 2001)**
A. single-phase diode rectifier with DC inductive filter
B. 3-phase diode rectifier with DC inductive filter
C. 3-phase thyristor rectifier with inductive filter
D. Single-phase diode rectifier with capacitive filter

223. In a thyristor-controlled rectifier, the firing angle of thyristor is to be controlled in the range of **(IES 2001)**
A. 0° to 90° B. 0° to 180°
C. 90° to 180° D. 90° to 270°

224. A single-phase full-bridge converter with a free-wheeling diode feeds an inductive load. The load resistance is 15.53 Ω and it has a large inductance providing constant and ripple free DC current. Input to converter is from an ideal 230 V, 50 Hz single phase source. For a firing delay angle of 60°, the average value of diode current is **(IES 2002)**
A. 10 A B. 8.165 A
C. 5.774 A D. 3.33 A

225. In a three-phase full wave AC to DC converter, the ratio of output ripple-frequency to the supply-voltage frequency is **(IES 2002)**
A. 2 B. 3
C. 6 D. 12

226. A 6-phase bridge-converter feeds a purely resistive load. The delay angle α is measured from the point of natural-commutation. The effective control of voltage can be obtained when α lies in the range **(IES 2002)**
A. $0 \le \alpha \le 105°$ B. $0 \le \alpha \le 120°$
C. $0 \le \alpha \le 150°$ D. $0 \le \alpha \le 180°$

227. When fed from a fully controlled rectifier, a DC motor, driving an active load, can operate in **(IES 2002)**
A. forward motoring and reverse braking mode
B. forward motoring and forward braking mode
C. reverse motoring and reverse braking mode
D. reverse motoring and forward braking mode

228. The characteristic features of discontinuous conduction compared to continuous conduction in a four-pulse, single-phase bridge converter are **(IES 2003)**
A. larger average value of load voltage and larger ripple-content
B. larger average value of load voltage and smaller ripple-content
C. smaller average value of load voltage and smaller ripple-content
D. smaller average value of load voltage and larger ripple-content

229. If the rms source voltage is V volts, the minimum and maximum values of firing angles for a single-

Ans. 214. D 215. D 216. B 217. A 218. C 219. D 220. A 221. A 222. D 223. B 224. D 225. C 226. B 227. B 228. D

phase, half-wave controlled rectifier, supplying a load with a back emf of 40 volts are: **(IES 2003)**
A. 0° to 180°
B. $\alpha = \sin^{-1}(40/\sqrt{2}\,V)$ and 180°
C. $\alpha = \sin^{-1}(40/\sqrt{2}\,V)$ and $[\pi - \sin^{-1}(40/\sqrt{2}\,V)]$
D. 0° and $[\pi - \sin^{-1}(40/\sqrt{2}\,V)]$

230. A converter which can operate both in 3-pulse and 6-pulse modes is a **(IES 2005)**
A. 1-phase full converter
B. 3-phase half-wave converter
C. 3-phase semi-converter
D. 3-phase full converter

231. The circulating current inductor is required in a dual converter to **(IES 2005)**
A. improve the pf
B. smoothen the waveform of circulating current
C. limit the circulating current
D. increase the circulating current

232. In a single-phase semiconvecter with discontinuous conduction and extinction angle $\beta < \pi$, freewheeling action takes place for **(IES 2005)**
A. α B. $\alpha - \beta$
C. $\beta - \pi$ D. 0°

233. A three pulse converter is feeding a purely resistive load. What is the value of firing delay angle α, which dictates the boundary between continuous and discontinuous mode of current conduction? **(IES 2005)**
A. $\alpha = 0°$ B. $\alpha = 30°$
C. $\alpha = 60°$ D. $\alpha = 150°$

234. In a dual converter, the circulating current **(IES 2006)**
A. allows smooth reversal of load current, but increases the response time
B. allows smooth reversal of load current with improved speed of response
C. does not allow smooth reversal of load current, but reduces the response time
D. flows it there is no interconnecting inductor

235. When the firing angle α of a single phase fully controlled rectifier feeding constant DC current into the load is 30°, what is the displacement factor of the rectifier? **(IES 2006)**
A. 1 B. 0.5
C. $\sqrt{3}$ D. $\sqrt{3/2}$

236. For the same voltage output, which one of the following has larger peak inverse voltage of the thyristor? **(IES 2006)**
A. Single phase full wave centre tapped circuit
B. Single phase full wave bridge circuit
C. Three phase full wave bridge circuit
D. Three phase full wave centre tapped circuit

237. A fully controlled line commutated converter functions as an inverter when firing angle (α) is in the range **(IES 2006)**
A. 0°–90° B. 90°–180°
C. 90°–180° only when there is a suitable DC source in the load
D. 90°–180° only when it supplies a back emf load

238. Which one of the following is the correct statement? **(IES 2007)**
In a two quadrant converter working in the 1st and 2nd quadrants
A. load current and load voltage are always positive
B. load current is always negative
C. load current can be positive or negative
D. load current and load voltage are always negative

239. What is the maximum output voltage of a 3-ph bridge rectifier supplied with line voltage of 440 V? **(IES 2007)**
A. 528 V B. 396 V
C. 594 V D. 616 V

240. In a single phase full wave controlled bridge rectifier, minimum output voltage and maximum output voltage are obtained at which conduction angles? **(IES 2007)**
A. 0°, 180° respectively B. 180°, 0° respectively
C. 0°, 0° respectively D. 180°, 180° respectively

241. For a single-phase AC to DC controlled rectifier to operate in regenerative mode, which of the following conditions should be satisfied? **(IES 2008)**
A. Half-controlled bridge, $\alpha < 90°$, source of emf in load
B. Half controlled bridge, $\alpha > 90°$, source of emf in load
C. Full-controlled bridge, $\alpha > 90°$, source of emf in load
D. Full-controlle bridge, $\alpha < 90°$, source of emf in load

242. A half-controlled bridge converter is operating from an rms input voltage of 120 V. Neglecting the voltage drops, what are the mean load voltage at a firing delay angle 0° and 180°, respectively? **(IES 2008)**
A. $\frac{120 \times 2\sqrt{2}}{\pi}$ V and 0 B. 0 and $\frac{120 \times 2\sqrt{2}}{\pi}$ V
C. $\frac{120 \times \sqrt{2}}{\pi}$ V and 0 D. 0 and $\frac{120 \times \sqrt{2}}{\pi}$ V

243. A large DC motor is required to control the speed of blower from a 3-phase ac source. What is the most suitable AC to DC converter? **(IES 2008)**

Ans. 229. C 230. C 231. C 232. D 233. B 234. B 235. D 236. D 237. D 238. C 239. C 240. A 241. C 242. A

A. 3-phase fully controlled bridge converter
B. 3-phase fully controlled bridge converter with free wheeling diode
C. 3-phase half-controlled bridge converter
D. A pair of 3-phase converters in sequence control

244. The input voltage for the given full-wave rectifier circuit is 230 V AC, then what is the peak inverse voltage across diodes D_1 and D_2? **(IES 2009)**

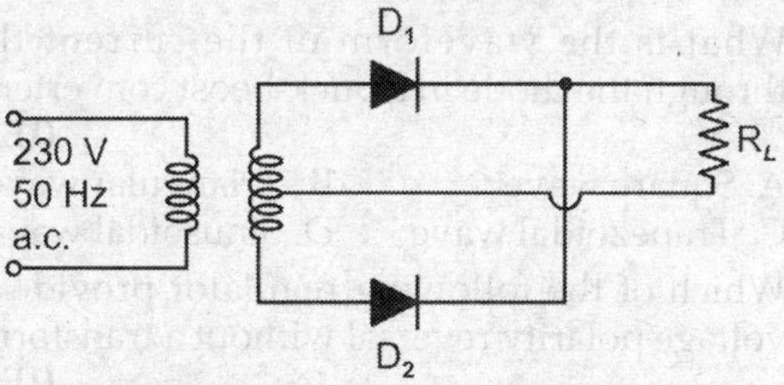

A. $100\sqrt{2}$ Volts B. 100 Volts
C. $50\sqrt{2}$ Volts D. 50 Volts

245. In a 3-phase semiconverter, for firing angle less than or equal to 60°, free wheeling diode conducts for **(IES 2011)**
A. 90° B. 60°
C. 30° D. 0°

246. In a three-phase semiconverter, if firing angle is less than or equal to 60°, then the duration of conduction of each thyristor and diode would be respectively **(IES 2012)**
A. 60° and 60° B. 90° and 30°
C. 120° and 120° D. 180° and 180°

247. The average output of a semiconverter connected to a 120 V, 50 Hz supply and firing angle of $\pi/2$ is **(IES 2013)**
A. 54.02 V B. 56.02 V
C. 108.04 V D. zero

248. In a 3ϕ controller bridge rectifier, the maximum conduction of each thyristor is **(IES 2014)**
A. 60° B. 90°
C. 120° D. 150°

249. A four quadrant chopper cannot be operated as **(IES 2001)**
A. inverter B. cycloconverter
C. one quadrant chopper
D. bi-directional rectifier

250. A 3-phase wound rotor induction motor is controlled by a chopper-controlled resistance in its rotor circuit. A resistance of 2 Ω is connected in the rotor circuit and a resistance of 4 Ω is additionally connected during OFF periods of the chopper. The OFF period of the chopper is 4 ms. The average resistance in the rotor circuit for the chopper frequency of 200 Hz is **(IES 2001)**
A. 26/5 Ω B. 24/5 Ω
C. 18/5 Ω D. 16/5 Ω

251. For a step up DC–DC chopper with an input DC voltage of 220 volts, if the output voltage required is 330 volts and the non-conducting time of thyristor-chopper is 100 μs, the ON time of thyristor-chopper would be **(IES 2002)**
A. 66.6 μs B. 100 μs
C. 150 μs D. 200 μs

252. An ideal chopper operating at a fre-quency of 500 Hz, supplies a load having resistance of 3 ohms and inductance of 9 mH from a 60 V battery. The mean value of the load voltage for on/off ratio of 4/1 (assuming that load is shunted by a perfect commutating diode and battery is loss-less), is **(IES 2003)**
A. 240 V B. 48 V
C. 15 V D. 4 V

253. A boost-regulator has an input voltage of 5 V and the average output voltage of 15 V. The duty cycle is **(IES 2003)**
A. 3/2 B. 2/3
C. 5/2 D. 15/2

254. In the buck-boost converter, what is the maximum value of the switch utilisation factor? **(IES 2004)**
A. 1.00 B. 0.75
C. 0.50 D. 0.25

255. The circuit shown in the below figure will work as which one of the following? **(IES 2007)**

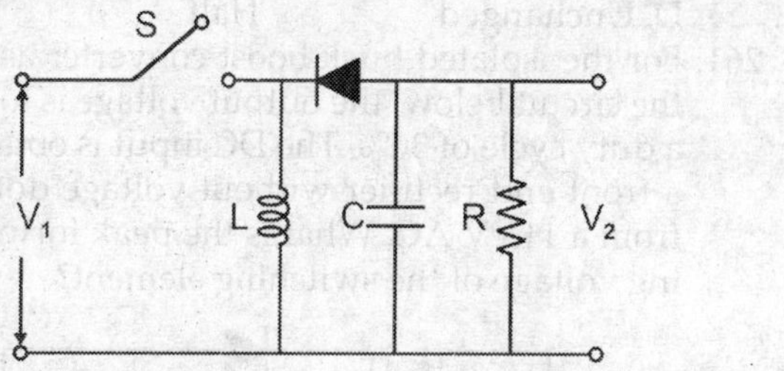

A. Buck-Boost converter
B. Buck converter
C. Boost converter D. Dual converter

256. An ideal chopper is operating at a frequency of 500 Hz from a 60 V battery input. It is supplying a load having 3 Ω resistance and 9 mH inductance. Assuming the load is shunted by a perfect commutating diode and assuming battery is loss less, what is the mean load current at an on/off ratio of 1/1? **(IES 2008)**
A. 10 A B. 15 A
C. 20 A D. None of these

257. A two-quadrant DC to DC chopper can operate with which of the following load conditions? **(IES 2008)**
1. +ve voltage, +ve current
2. –ve voltage, +ve current
3. –ve voltage, –ve current
4. +ve voltage, –ve current

Ans. 243. C 244. A 245. D 246. C 247. A 248. C 249. B 250. A 251. D 252. B 253. B 254. B 255. A 256. A

Select the correct answer using the code given below

Code:

A. 1 only B. 1 and 2
C. 1 and 4 D. 3 and 4

258. A buck regulator has an input voltage of 12 V and the required output voltage is 5 V. What is the duty cycle of the regulator? **(IES 2008)**
A. 5/12 B. 12/5
C. 5/2 D. 6

259. A DC chopper is used in regenerative braking mode of a DC series motor. The DC supply is 600 V, the duty cycle is 70%. The average value of armature current is 100 A. It is continuous and ripple free. What is the value of power feedback to the supply? **(IES 2009)**
A. 3 kW B. 9 kW
C. 18 kW D. 35 kW

260. If a full wave fully controlled converter is modified as a full wave half controlled converter, what will be the maximum value of active power (P) and the maximum value of reactive power demand (Q)? **(IES 2009)**

	P	Q
A.	Double	Half
B.	Unchanged	Unchanged
C.	Half	Double
D.	Unchanged	Half

261. For the isolated buck boost converter as shown in the circuit below, the output voltage is to be 35 V at a duty cycle of 30%. The DC input is obtained from a front end rectifier without voltage doubling fed from a 115 V AC. What is the peak forward blocking voltage of the switching element? **(IES 2009)**

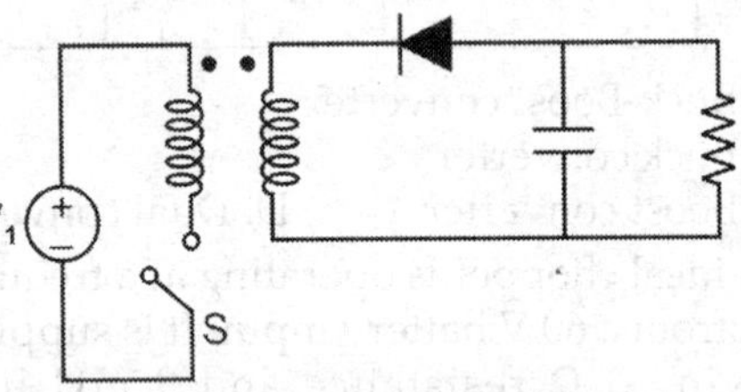

A. 232.3 V B. 69.69 V
C. 162.61 V D. 542 V

262. A DC to DC transistor chopper supplied from a fixed voltage DC source feeds a fixed resistive inductive load and a free wheeling diode. The chopper operates at 1 kHz and 50% duty cycle. Without changing the value ofthe average DC current through the load, if it is desired to reduce the ripple content of the load current, the control action needed will be to **(IES 2010)**
A. increase the chopper frequency keeping its duty cycle constant
B. increase the chopper frequency and duty cycle in equal ratio
C. decrease only the chopper frequency
D. decrease only the duty cycle

263. In DC choppers, the waveforms for input and output voltages are respectively **(IES 2011)**
A. discontinuous and continuous
B. both continuous
C. both discontinuous
D. continuous and discontinuous

264. What is the waveform of the current flowing through the diode in a buck-boost converter? **(IES 2011)**
A. Square wave B. Triangular wave
C. Trapezoidal wave D. Sinusoidal wave

265. Which of the following regulator provide output voltage polarity reversal without a transformer? **(IES 2014)**
A. Buck regulator B. Boost regulator
C. Buck-Boost regulator D. CUK regulator

266. In case of voltage source inverter, free-wheeling can be needed for the load of **(IES 2001)**
A. inductive nature B. capacitive nature
C. resistive nature D. back emf nature

267. PWM switching is preferred in voltage source inverters for the purpose of **(IES 2001)**
A. controlling output voltage
B. output harmonics
C. reducing filter size
D. controlling output voltage, output harmonics and reducing filter size

268. The operation of an inverter fed induction motor can be shifted from motoring to regenerative braking by **(IES 2002)**
A. reversing phase sequence
B. reducing inverter voltage
C. decreasing inverter frequency
D. increasing inverter frequency

269. Compared to a single-phase half-bridge inverter, the output power of a single-phase full-bridge inverter is higher by a factor of **(IES 2002)**
A. 12 B. 8
C. 4 D. 2

270. In a self-controlled synchronous motor fed from a variable frequency inverter **(IES 2003)**
A. the rotor poles invariably have damper windings
B. there are stability problems
C. the speed of the rotor decides stator frequency
D. the frequency of the stator decides the rotor speed

271. A DC source is switched in steps to syn-thesize the three-phase output. The basic three-phase bridge inverter can be controlled. The angle through which each switch conducts, and at any instant the

Ans. 257. C 258. A 259. C 260. B 261. A 262. A 263. D 264. C 265. C 266. A 267. D 268. C 269. C 270. D

number of switches conducting simultaneously are, respectively (IES 2003)

A. 120° and 02 B. 120° and 03
C. 180° and 02 D. 180° and 04

272. For a single-phase, full-bridge inverter supplying power to a highly inductive load as shown below, the correct sequence of operation of switches and diodes is (IES 2003)

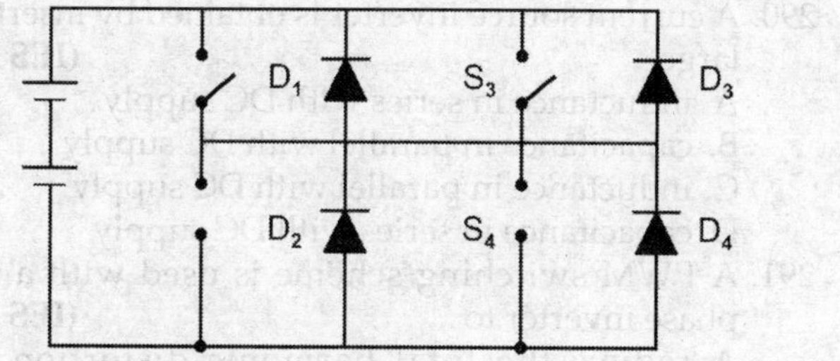

A. $S_1S_4 - S_3S_2 - S_1S_4 - S_3S_2$
B. $S_1S_2 - D_1D_2 - S_3S_4 - D_3D_4$
C. $S_1D_3 - S_1S_4 - S_4D_2 - D_2D_3$
D. $S_2D_4 - D_4D_1 - D_1S_3 - S_3S_2$

273. A single-phase, half-bridge inverter has input voltage of 48 V DC. Inverter is feeding at a load of 2.4 Ω. The rms outout voltage at fundamental frequency is (IES 2003)

A. $\frac{2\times 48}{\pi}$ V B. $\frac{2\times 48}{\sqrt{2}\pi}$ V
C. $\frac{\sqrt{2}\times 48}{\pi}$ V D. $\frac{2\times 48}{2\sqrt{2}\pi}$ V

274. A single-phase inverter has square wave output voltage. What is the percentage of the fifth harmonic component in relation to the fundamental component? (IES 2004)

A. 40% B. 30%
C. 20% D. 10%

275. Which one of the following statements is correct? (IES 2005)

A voltage source inverter is normally employed
A. when the source has low impedance and load has high reactance
B. when the source has high impedance and load has low reactance
C. when both the source and load have high values of impedance and reactance respectively
D. when both the source and load have low values of impedance and reactance respectively

276. In a three phase voltage source inverter operating in square wave mode, the output line voltage is free from (IES 2006)

A. 3rd harmonic B. 7th harmonic
C. 11th harmonic D. 13th harmonic

277. The pulse-width modulated inverter for the control of an AC motor is fed from which one of the following? (IES 2007)

A. Controlled rectifier B. Uncontrolled rectifier
C. AC regulator D. Cycloconverter

278. Consider the following statements:
1. Both voltage source inverter and current source inverter require feed-back diodes
2. Only current source inverter requires feedback diodes
3. GTOs can not be used in a current source inverter
4. Only voltage source inverter requires feedback diodes

Which of these statements is/are correct? (IES 2007)

A. 1 only B. 2 and 3
C. 3 and 4 D. 4 only

279. A single-phase curren source inverter is connected with capacitive load only the waveform of the output voltage across the capacitor for constant source current will be (IES 2008)

A. sine wave B. square wave
C. triangular wave D. step function

280. A single-phase full-bridge inverter is connected to a load of 2.4 Ω. The DC input voltage is 48 V. What is the rms output at fundamental frequency? (IES 2008)

A. $\frac{4\times 48}{\sqrt{2}\pi}$ V B. $\frac{2\times 48}{\sqrt{2}\pi}$ V
C. $\frac{4\times 48}{\pi}$ V D. $\frac{2\times 48}{\pi}$ V

281. Consider the following statements with respect to the power transistors used in inverters:
1. Maximum collector-emitter voltage V_{CEO}
2. Maximum collector current
3. Maximum power dissipation
4. Maximum current gain at minimum load current
5. Maximum current gain at maximum load current

Which of these statements is/are correct? (IES 2008)

A. 1 only B. 1, 2 3 and 5
C. 2 and 3 only D. 2, 3 and 4

282. For elimination of 5th harmonics from the output of an inverter, what will be the position of pulse in a PWM inverter? (IES 2009)

A. 72° B. 36°
C. 60° D. 90°

283. In a single phase VSI bridge inverter, the load current is $I_0 = 200 \sin(\omega t - 45°)$ mA. The DC supply voltage is 220 V. What is the power drawn from the supply? (IES 2009)

A. 9.8 W B. 19.8 W
C. 27.25 W D. 34.03 W

Ans. 271. B 272. A 273. B 274. C 275. A 276. A 277. B 278. C 279. C 280. A 281. B 282. A 283. B

284 What is the effect of blanking time on output voltage in PWM inverter? **(IES 2009)**
A. Distortion in instantaneous voltage at current zero crossing
B. Low order space harmonics in output voltage
C. Distribution in instantaneous voltage
D. High order time harmonics in output voltage

285. In single pulse modulation of PWM inverters, the pulse width is 120°. For an input voltage of 220 V DC, what is the rms value at the fundamental component of the output voltage? **(IES 2009)**
A. 171.5 V B. 254.0 V
C. 127.0 V D. 89.81 V

286. A constant current source inverter supplies 20 A to a load resistance of 1 Ω to a load resistance change to 5 Ω, then the load current **(IES 2010)**
A. remains same at 20 A and the load voltage changes to 100 V
B. changes to 4 A from 20 A and the load voltage changes to 20 V
C. changes to 4 A from 20 A and the load voltage changes to 80 V
D. and load voltage stay at 20 A and 20 V respectively

287. The below figure shows an inverter circuit with a DC source voltage V_S. The semiconductor switches of the inverter are operated in such a way that the pole voltages of V_{10} and V_{20} are shown in the figure (b). What is the RMS value of the pole voltage V_{12}? **(IES 2010)**

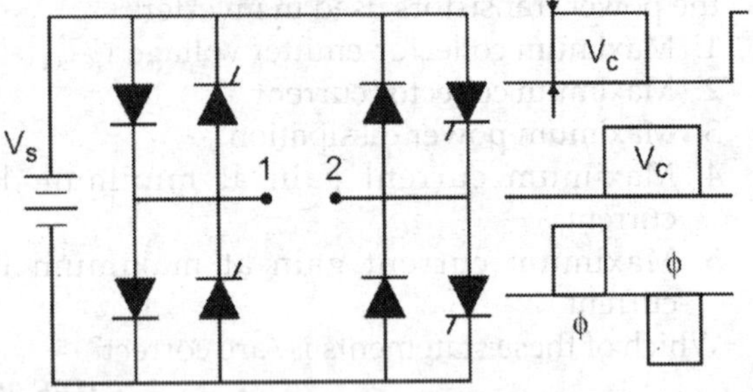

A. $\frac{V_S\phi}{\sqrt{2\pi}}$ B. $V_S\sqrt{\frac{\phi}{\pi}}$
C. $V_S\sqrt{\frac{\phi}{2\pi}}$ D. $\frac{V_S}{\pi}$

288. In a PWM inverter, f_0 and f are the frequencies in Hz for the carrier signal and reference signal respectively. Then the number of pulses per half cycle is **(IES 2010)**
A. $N = f/f_0$ B. $N = f/2f_0$
C. $N = f_0/2f$ D. $N = f_0/f$

289. A Voltage Source Inverter (VSI) is normally employed when **(IES 2012)**
A. source inductance is large and load inductance is small
B. source inductance is small and load inductance is large
C. both source inductance and load inductance are small
D. both source inductance and load inductance are large

290. A current source inverter is obtained by inserting a large **(IES 2012)**
A. inductance in series with DC supply
B. capacitance in parallel with DC supply
C. inductance in parallel with DC supply
D. capacitance in series with DC supply

291. A PWM switching scheme is used with a three phase inverter to **(IES 2013)**
A. reduce the total harmonic distortion with modest filtering
B. minimise the load on the DC side
C. increase the life of the batteries
D. reduce low order harmonics and increase high order harmonics

292. A motor armature supplied through phase controlled SCRs receives a smoother voltage shape at **(IES 2005)**
A. high motor speed B. low motor speed
C. rated motor speed D. none of the above

293. A cycloconverter-fed induction motor drive is most suitable for which one of the following? **(IES 2008)**
A. Compressor drive B. Machine tool drive
C. Paper mill drive D. Cement mill drive

294. For low-speed high-power reversible operation, the most suitable drives are **(IES 2011)**
A. voltage source inverter bed AC drives
B. current source inverter bed AC drives
C. dual converted bed DC drives
D. cycloconverter bed AC drives

295. A single phase AC voltage controller feeding a pure resistance load has a load voltage of 200 V(rms) when fed from a source of 250 V(rms). The input power factor of the controller is **(IES 2001)**
A. 0.64 B. 0.8
C. 0.894
D. difficult to estimate because of insuffi-ciency of data

296. The most suitable solid state converter for controlling the speed of the three-phase cage motor at 25 Hz is **(IES 2001)**
A. cycloconverter
B. current source inverter
C. voltage source inverter
D. load commutated inverter

297. The quality of output AC voltage of a cycloconverter is improved with **(IES 2001)**

Ans. 284. D 285. A 286. A 287. B 288. C 289. B 290. A 291. A 292. A 293. B 294. D 295. B 296. A

A. increase in output voltage at reduced frequency
B. increase in output voltage at increased frequency
C. decrease in output voltage at reduced frequency
D. decrease in output voltage at increased frequency

298. A cycloconverter is operating on a 50 Hz supply. The range of output frequency that can be obtained with acceptable quality is **(IES 2001)**
A. 0–16 Hz B. 0–32 Hz
C. 0–64 Hz D. 0–128 Hz

299. AC voltage regulators are widely used in **(IES 2002)**
A. traction drives B. fan drives
C. synchronous motor drives
D. slip power recovery scheme of slip-ring induction motor

300. How many switches are used to construct a three-phase cycloconverter? **(IES 2002)**
A. 3 B. 6
C. 12 D. 18

301. A 3-phase cycloconverter is used to obtain a variable-frequency single-phase AC output. The single phase AC load is 220 V, 60 A at a power factor of 0.6 lagging. The rms value of input voltage per phase required is **(IES 2002)**
A. 376.2 V B. 311.12 V
C. 266 V D. 220 V

302. An AC voltage-regulator using back-to-back connected SCRs is feeding an R-L load. The SCR firing angle $\alpha < \phi$ (ϕ is power factor angle of the load). If SCRs are fired using short-duration gate pulses, the output load-voltage wave-form will be **(IES 2003)**
A. symmetrical chopped AC voltage
B. half-wave rectified
C. full-wave rectified D. sinusoidal

303. An inter-group reactor is used in a single-phase cycloconverter circuit to **(IES 2003)**
A. reduce current-ripples
B. reduce voltage-ripples
C. limit circulating current
D. limit di/dt in the semiconductor switch

304. For a single phase AC voltage controller feeding a resistive load, what is the power factor where α is firing angle measured from voltage zero? **(IES 2005)**

A. Unity for all values of firing angle

B. $\left[\dfrac{1}{\pi}\left\{(\pi - \alpha) + \dfrac{1}{2}\sin 2\alpha\right\}\right]^{1/2}$

C. $\left[\dfrac{1}{\pi}\left\{(\pi + \alpha) + \dfrac{1}{2}\sin 2\alpha\right\}\right]^{1/2}$

D. $\left[\dfrac{1}{\pi}\left\{(\pi + \alpha) - \dfrac{1}{2}\sin 2\alpha\right\}\right]^{1/2}$

305. In the AC regulator for Fig. 1, the supply voltage and gate currents waveforms are as in Fig. 2, what is the load voltage waveform for $R = 0$? **(IES 2009)**

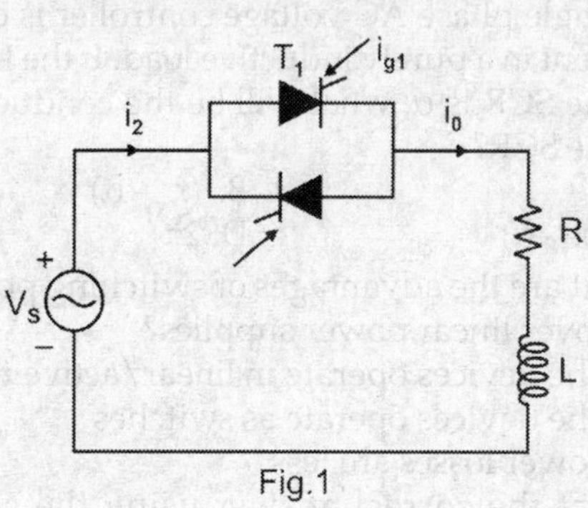

Fig.1

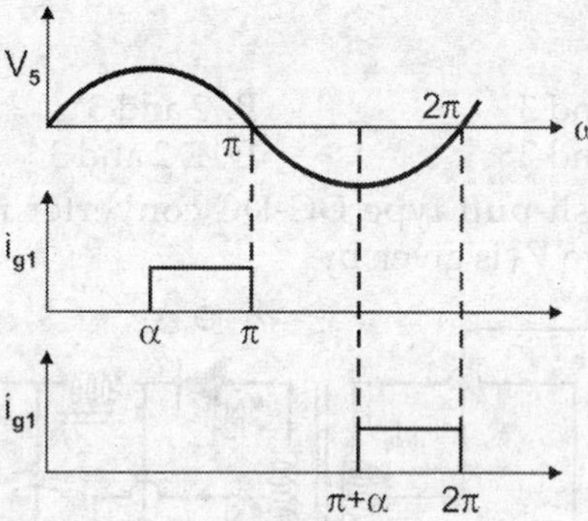

Fig.2

A. 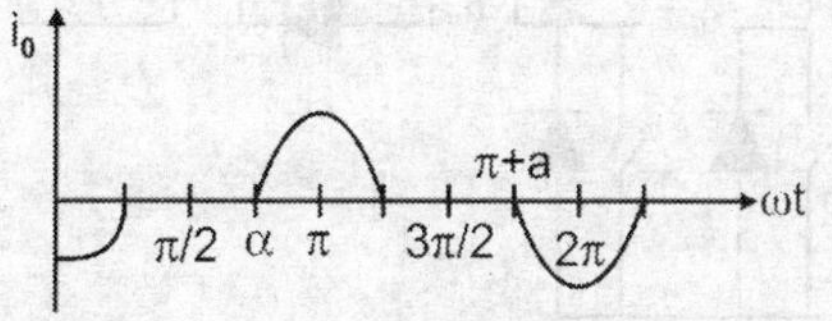

B. 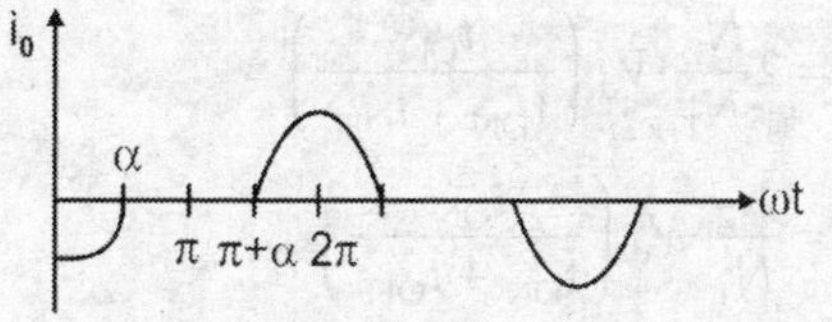

C. 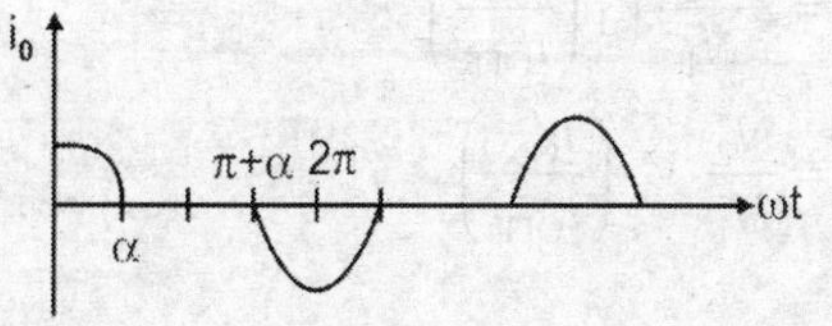

D. 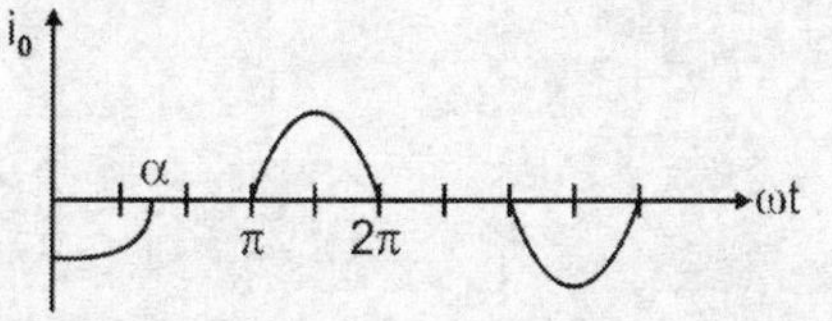

Ans. 297. B 298. D 299. B 300. D 301. C 302. D 303. C 304. B 305. A

306. What is the power factor of a single phase AC regulator feeding a resistive load? **(IES 2006)**
A. $(\text{Per unit power})^2$ B. $(\text{Per unit power})^{1/2}$
C. $\frac{(\text{Per unit power})^2}{\sqrt{2}}$ D. $\frac{(\text{Per unit power})^{1/2}}{2}$

307. A single phase AC voltage controller is controlling current in a purely inductive load. If the firing angle of the SCR is α, what will be the conduction angle of the SCR? **(IES 2007)**
A. π B. $(\pi - \alpha)$
C. $(2\pi - \alpha)$ D. 2π

308. What are the advantages of switching power supplies over linear power supplies? **(IES 2008)**
1. The devices operate in linear/active region
2. The devices operate as switches
3. Power losses are less
Select the correct answer using the code given below:
Code:
A. 1 and 3 B. 2 and 3
C. 1 and 2 D. 1, 2 and 3

309. In push-pull type DC-DC converter the output voltage V_0 is given by **(IES 2010)**

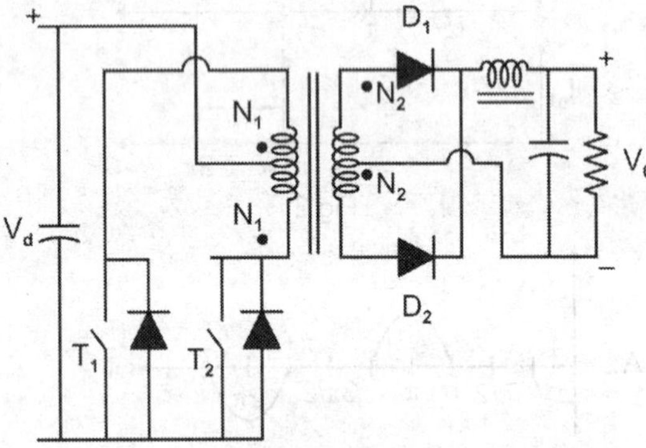

A. $V_0 = 2\frac{N_2}{N_1} \cdot V_d \left(\frac{t_{ON}}{t_{ON} + t_{OFF}}\right)$
B. $V_0 = \frac{N_2}{N_1} \cdot V_d \left(\frac{t_{ON}}{t_{ON} + t_{OFF}}\right)$
C. $V_0 = 2\frac{N_2}{N_1} \cdot V_d \left(\frac{t_{ON}}{t_{OFF}}\right)$
D. $V_0 = \frac{N_2}{N_1} \cdot V_d \left(\frac{t_{ON}}{t_{OFF}}\right)$

310. An advantage of a cycloconverter is **(IES 2011)**
A. very good power factor
B. requires few number of thyristors
C. commutation failure does not short circuit the source
D. load commutation is possible

311. An integral cycle AC voltage controller is feeding a purely resistive circuit from a single-phase AC voltage source. The current waveform consists alternately burst of *N*-complete cycle of conduction followed by *M*-complete of extinction. The rms value of the load voltage equals the rms value of supply voltage for **(IES 2011)**
A. $N = M$ B. $N = 0$
C. $N = M = 0$ D. $M = 0$

312. A single-phase AC regulator fed from 50 Hz supply feeds a load having 4 Ω resistance and 12.73 mH inductance. The control range of firing angle will be **(IES 2012)**
A. 0° to 180° B. 45° to 180°
C. 90° to 180° D. 0° to 45°

313. Consider the following statements:
Switched mode power supplies are preferred over the continuous types, because they are **(IES 2013)**
1. suitable for use in both AC and DC
2. more efficient
3. suitable for low-power circuits
4. suitable for high-power circuits
Which of these statements are correct?
A. 1 and 2 B. 1 and 3
C. 2 and 3 D. 3 and 4

314. In a forward converter, a tertiary winding is used. What is the reason? **(IES 2014)**
A. To provide di/dt protection to the switching device
B. To provide dv/dt protection to the switching device
C. To provide electrical isolation between the input and output
D. To demagnetize the core before the application of the next switching

315. The device used for switching in a switched mode power supply is **(IES 2014)**
A. diode B. thyristor
C. GTO D. MOSFET

Ans. 306. B 307. B 308. B 309. C 310. D 311. A 312. B 313. C 314. D 315. D

CHAPTER 14

Electric Traction, Electric Heating, Welding and Braking—Utilization of Electrical Energy and Illumination

Electric Traction

Advantages: Advantages of electric traction over steam traction are:

1. *Cleanliness*: Electric traction is free from smoke and flue gases, etc.
2. *Maintenance cost*: The maintenance cost of electric traction system is 50% of that of steam traction system.
3. *Starting time*: Electric locomotive can be started without any loss of time.
4. *High starting torque*: The motor used in electric traction is DC series/AC series motor which has a very high starting torque.
5. *Higher acceleration rate*: It makes the traffic handling capacity almost double.
6. *Braking*: Regenerative braking is used in electric traction which feeds back 80% of the energy required during current back during decent.
7. *Centre of gravity*: Centre of gravity of electric locomotive is quite low which makes it quite stable and as a result it is able to take curves even at higher speeds quite safely.
8. *Absence of unbalanced forces*: Due to absence of unbalanced forces produced by reciprocating, *coefficient of adhesion* is more in case of electric traction. This reduces the wight to HP ratio of the locomotive

Disadvantages:

1. The capital cost of overall electric traction system is quite high.
2. Failure of power disturbs the traction.
3. Communication lines which run parallel to the power lines face interference.
4. Electric traction can be used only at places which are electrified.

Systems of traction electrification: There are four types of track electrification systems normally used:

i. Direct current system
ii. Single phase low frequency system
iii. Three phase AC system
iv. Composite System

Railway service: Railway service may be classified into following categories:

i. *Main Service*: This involves operation over a long distance within frequent stops.
ii. *City Service*: This involves operation over stations where the distance between the stops is of the order of a kilometre.
iii. *Suburban Service*: In this case the distance between the stops is larger, upto about 6 kilometres. Here, high values of acceleration and braking are required.

Speed-time curves of suburban run: A typical speed-time curve is shown in Fig. 14.1.

i. *Acceleration while notching up or constant acceleration*: During this period of run (0 to t_1), starting resistance is gradually cut off so that the motor current is limited to a certain value and the voltage across the motor is gradually increased.

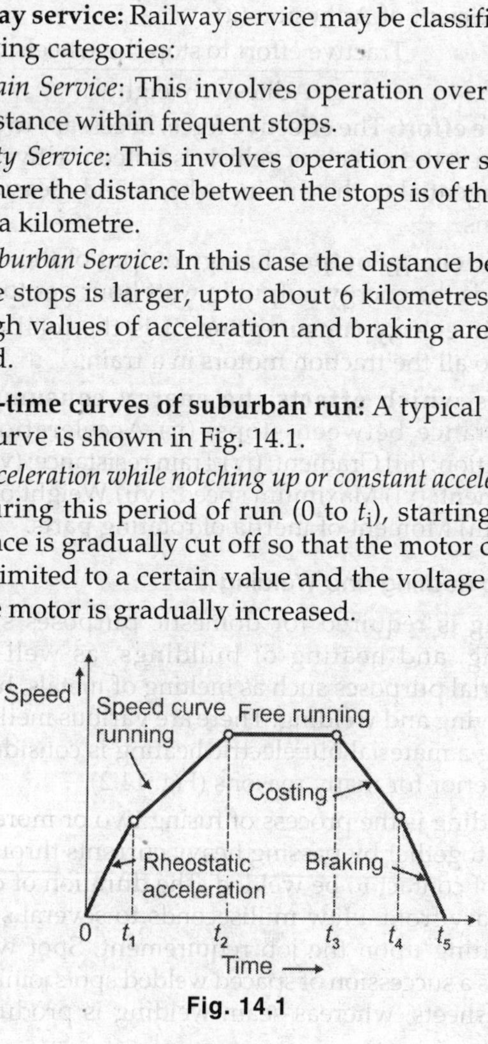

Fig. 14.1

ii. *Speed curve running:* During speed curve running (t_1 to t_2), the current starts decreasing with the increase in speed.
iii. *Free running (t_2 to t_3):* This period occurs, on level track, when the power output from the driving axels balances the rate at which energy is expended against the resistance to motion.
iv. *Coasting period (t_3 to t_4)*: At the end of free running period, supply to the motors is cut off and train is allowed to run under its own momentum
v. *Braking period (t_4 to t_5)*: At the end of coasting period, brakes are applied to bring the train to stop.

Limiting factors:

1. *Crest speed*: The maximum speed attained by the train during the run is known as crest speed.
2. Average speed $= \dfrac{\text{Total distance}}{\text{Running time (excluding stops)}}$
3. Scheduled speed $= \dfrac{\text{Total distance}}{\text{Running time (including stops)}}$

Adhesive weight: This is total weight to be carried on the driving wheels is called adhesive weight.

Coefficient of Adhesion (μ)

$$= \frac{\text{Tractive effort to stop the wheels}}{\text{Adhesive weight}}$$

Tractive effort: The effective force, necessary to propel the train at the wheels of vehicle is called tractive effort. It is tangential to the driving wheels and measured in Newtons.

Specific energy output: Energy output of the driving wheels is the energy output in watt hour per ton km.

Specific energy consumption: It is the total energy input to all the traction motors in a train.

Factors which effects the energy consumption: (i) Distance between stops; (ii) Acceleration and retardation; (iii) Gradient; (iv) Train resistance; (v) Train equipment; (vi) Maximum speed; (vii) Weight of train; and (viii) Moment of inertia of rotating parts.

Electric Heating and Welding

Heating is required for domestic purposes such as cooking, and heating of buildings, as well as for industrial purposes such as melting of metals, hardening, drying and welding. There are various methods of heating a material but electric heating is considered to be superior for many reasons (Fig. 14.2).

Welding is the process of fusing two or more metal pieces together by passing heavy currents through the areas of contact to be welded. The duration of current may vary from a few milliseconds to several seconds depending upon the job requirement. Spot welding implies a succession of spaced welded spots joining two metal sheets, whereas seam welding is produced by succession of overlapping and limit spots, using water-cooled roller electrodes. Electronic control of the duration and magnitude of the welding current has made process welding possible, and the materials such as aluminum and stainless steel can be satisfactorily welded.

Advantages of electric heating: These are as follows:

1. *Cleanliness*: In the absence of dust and ash of the fuel, charge never gets contaminated. So it is a clean system.

2. *Economical*: Electric heating is economical as electric furnaces are cheaper in initial cost as well as maintenance cost. Electrical energy is also very cheap as it is being produced on larger scale.

3. *Absence of flue gases*: Since no flue gas is produced in this system so there is no risk of atmosphere to be polluted. Thus the operation is hygienic clean.

4. *Ease of control*: Simple, accurate and reliable temperature control can be achieved either by hand operated or by fully automatic switches. Desired temperature or temperature cycle can be obtained accurately in electric heating system which is not convenient in other heating systems.

5. *Automatic*: Automatic protection against over current or over heating can be provided through suitable switch gears in the electric heating.

6. *Uniform heating*: Certain requirements of heating such as uniform heating of a material or heating with no oxidation can be met only with electric heating system.

7. *Heating of nonconducting materials*: It is possible only with electric heating, to get nonconducting materials heated uniformly throughout the section. This is possible because heat is generated inside the material itself.

8. *Efficiency*: The overall efficiency of electric heating is comparatively higher since in this system of heating, the source can be brought directly to the point where heat is required, thereby reducing the losses. Further there is no products of combustion in which heat losses are involved.

9. *Better working conditions*: Electric heating system produces no irritating noise and also the radiating losses are low. Thus working with electric furnaces is convenient and cool.

10. *Safety*: Electric heating is quite safe and responds quickly.

Resistance heating: When electric current passes through a resistance, power loss takes place therein, which appears in the form of heat.

$$\text{Power loss} = I^2R = IV = V^2/R \text{ watt}$$

where R is the effective resistance of the element.

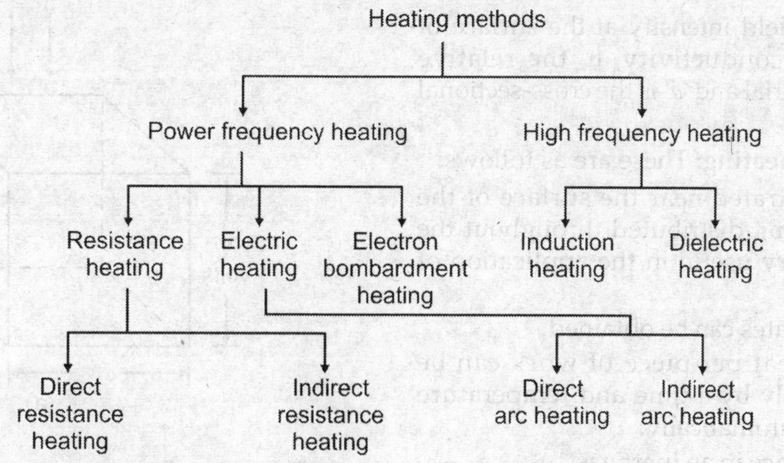

Fig. 14.2: Classification of heating methods

Table 14.1: Different types of heating materials

Types of alloys	*Composition*	*Max temp. of operation*	*Sp. resistance at room temp.*	*Sp. gravity*
Nickel–Copper	45% Ni 55% Cu	400°C	49 μ ohm/cm³	8.88
Nickel–Chromium–Iron	60% Ni, 16% Cr, 24% Fe	950°C	10 μ ohm/cm³	8.28
Nickel–Chromium	80% Ni, 20% Cr	1150°C	109 μ ohm/cm³	8.36
Iron–Chromium–Aluminium	65–75% Fe, 20–30% Cr, 5% Al	1150–1350°C	140 μ ohm/cm³	7.2

Design of resistance heating elements: When the element has reached a steady temperature, all the electrical input is given out as heat. The heat produced according to Stefan's law is

$$H = 5.72 \times 10^4 ke\left[\left(\frac{T_1}{1000}\right)^4 - \left(\frac{T_2}{1000}\right)^4\right] \frac{\text{watts}}{\text{m}^2}$$

The length of wire l, diameter of wire d of specific resistance ρ are given by:

$$\frac{d}{l^2} = \frac{4\rho H}{V^2}$$

$$\frac{l}{d^2} = \frac{V^2 \cdot \pi}{4\rho P}$$

Induction heating: Induction heating is employed for electrically conductive material and uses the magnetic field set up by an alternating current. The material to be heated known as the *work* is placed within or adjacent to a coil which carries current. This is the work coil. Fig. 14.3 shows the circuit for induction heating. The varying magnetic field induces a voltage in the work piece. This voltage, in turn, causes a current i_0 to flow in the surface of the work.

The resulting ac eddy current flows in the conductor. The action is essentially the same as that of a transformer, and the work coil can be considered as a primary, while the work itself serves as a single turn secondary. Because of resistance losses, the resulting current in the work coil develops the desired heat, which is confined mainly to the surface layer.

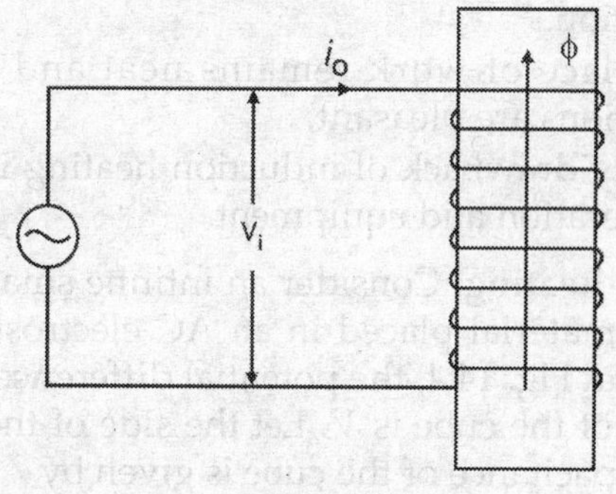

Fig. 14.3

Thus *depth of penetration* is defined as the depth of surface inside the cylinder at which the current density becomes equal to 37% of its value at the surface. In terms of frequency, it is given by

$$\delta = \frac{1}{2\pi\sqrt{10^{-9}\mu_r \sigma f}} \text{ cm}$$

where μ_r = relative magnetic permeability of the material

σ = conductivity of the material in mhos per cm³

f = frequency of input current in Hz.

This equation shows that the depth of penetration can be decreased by increasing either permeability μ_r, conductivity σ or the oscillator frequency f.

For induction heating of cylindrical bars, the surface power density is given by

$$P = \frac{4H_0}{\sigma\delta} \text{ watt/cm}^2$$

The desirable frequency for a given work piece should be more than the critical frequency given by

$$f_c \cong \frac{2}{u_r \sigma a^2}$$

where H_0 is magnetic field intensity at the surface of piece of metal, σ the conductivity, μ_r the relative permeability of the material and 'a' is the cross-sectional area.

Features of induction heating: These are as follows:

1. The heat is concentrated near the surface of the work instead of being distributed throughout the work. Thus it is very useful in the application of case hardening.
2. Very high heating rates can be obtained.
3. Total electronic heat per piece of work can be controlled accurately by a time and temperature can be controlled automatically.
4. Heating can take place in an inert gas.
5. Relatively unskilled labour may be used for its operation.
6. The place of work remains neat and working conditions are pleasant.

The only drawback of induction heating is its high cost of operation and equipment.

Dielectric heating: Consider an infinite small cube of dielectric material placed in an AC electrostatic field as shown in Fig. 14.4, the potential difference between two faces of the cube is V. Let the side of the cube be 'a'. The capacitance of the cube is given by

$$C = \varepsilon_0 \varepsilon_r \frac{\text{Area of the face}}{\text{Thickness}}$$

$$= \varepsilon_0 \varepsilon_r \frac{a^2}{\text{a}} = \varepsilon_0 \varepsilon_r a \text{ farad}$$

where ε_r is the relative permittivity of the material. The resulting current makes an angle θ with the voltage V across the capacitor. θ is slightly less than 90°. The small in-phase component I is given by

$$I \cos\theta = \frac{|V|}{Z_C} \cos\theta$$

where Z_C is the impedance of elementary capacitor and nearly equals to $1/2\pi fC$.

The power dissipated in the cube

$$= 2\pi f \cdot \varepsilon_0 \varepsilon_r a \cdot V^2 \cdot (pf)$$

Power per unit volume

$$P_v = 2\pi f \cdot \varepsilon_0 \varepsilon_r (pf)\, v^2 \text{ watt/m}^3$$

where v is the applied voltage per unit length. The product of dielectric constant ε_r and power factor (pf), i.e. ($\varepsilon_r \cdot pf$) is called the *loss factor*

$$f_c = \frac{14.31}{a\sqrt{\varepsilon_r}} \text{ MHz}$$

where a is the radius of material.

Thermal losses in dielectric heating: The total power P_i delivered to a dielectric piece is given by

$$P_i = P_s + P_{cd} + P_{cv} + P_{cp} + P_{cr} \text{ watts}$$

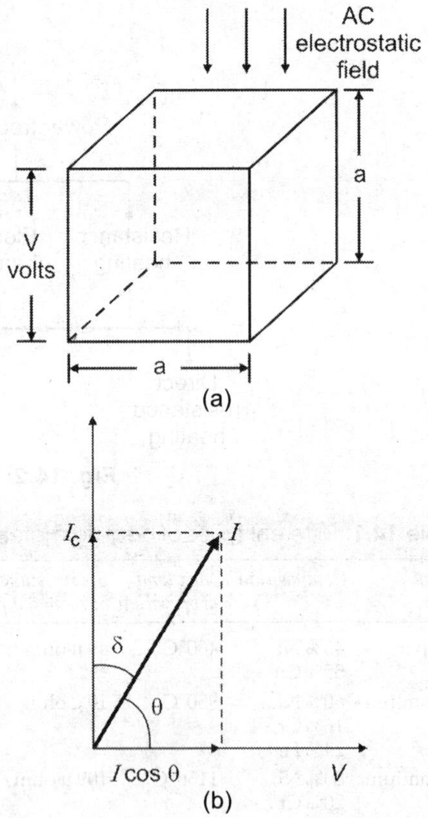

Fig. 14.4

where

P_s is the specific heat power in watts or thermal power

P_{cd} is the conduction loss in watts

P_{cv} is the convection loss in watts

P_{cp} is the power in watts required to change the physical state of material if needed

P_{cr} is the radiation loss in watts

Thermal power P_s: This is the power required to raise the temperature of a given dielectric material to the desired final temperature within a specified heating time.

Specific heat power: $P_s = 17.6\, mc\Delta T$, where m is the mass, c is the specific heat and ΔT is the temperature rise.

Conduction loss P_{cd}: The flow of heat along a substance or object depends upon the temperature gradient. Each molecule of the substance gets heated and transfers the heat to the adjacent one, thus making heat travel from one point to another. The conduction loss does not remain constant during the entire heating interval.

Convection loss P_{cv}: Most common example of heat transfer by this method is the heating of water by an immersion heater where convection currents are set-up and the water gets heated by these. For the heating periods and temperatures involved in the dielectric heating, the convection loss generally is insignificant.

Radiation loss P_{cr}: Heat reaches the object from the source without heating the medium in between, (Stefan's law of heat radiation).

Heat dissipated

$$= 5.72 \times 10^4 ke\left[\left(\frac{T_1}{1000}\right)^4 - \left(\frac{T_2}{1000}\right)^4\right] \text{ watts/m}^2$$

where

k is a constant called *radiating efficiency* whose value is 1 for single element and 0.5 to 0.8 for several elements placed side by side

e is emissivity equal to 1 for black body and 0.9 for resistance heating elements

T_1 is the temperature of the source in °K observed

T_2 is the temperature of object absorbing the heat in °K

Resistance welding: Resistance welding is the process of fusing two or more metal pieces together by passing heavy currents through the areas of contact to be welded. The duration of the current may vary from a few milliseconds to several seconds depending upon the job requirement. Spot welding implies a succession of spaced welded spots joining two metal sheets, whereas seam welding is produced by succession of overlapping and timed spots, using water-cooled roller electrodes. Pulsation welding applies a number of successive current pluses to the same weld, used mainly to prevent electrode overheating in the welding of thick plate. Electronic control of the duration and magnitude of the welding current has made precise welding possible, and the materials such as aluminium and stainless steel can be satisfactorily welded in production.

This process is based on the resistance to a current flow between the two metal pieces where weld heat is produced. The welding heat is given by

$$H = \int_0^{t_w} i^2 r dt$$

where r is the resistance between the pieces to be welded and t_w, time during which weld current flows. Either alternating current or short pulses of unidirectional current may be used.

Types of resistance welding: Following are the principal types of resistance welding processes.

(a) *Spot welding*: Spot welding is done by clamping two or more pieces of metal in the form of sheets together between two pointed electrodes. The electrode tips are made of copper or copper alloy and are water-cooled. The welding current depends on the thickness and the composition of the plates. Fig. 14.5(a) shows the arrangement.

(b) *Projection welding*: In projection welding, instead of using pointed electrode for concentration of heat, use is made of either an embossing on or both of the pieces at a location where welding is desired. Figure 14.5(b) shows the general arrangement.

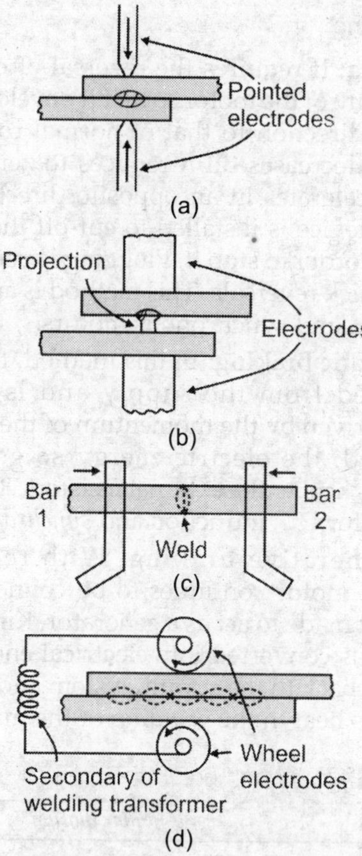

Fig. 14.5

(c) *Butt Welding*: In butt welding, one bar is held in a fixed clamp and the other is held in a moving clamp. The two bars are brought together as shown in Fig. 14.5(c). A heavy current is passed from the welding transformer or raise the temperature when the desired temperature is reached, the two pieces are ramped together either manually or automatically resulting in a weld. Rods, pipes and wires are welded in this way.

(d) *Seam welding*. In seam welding, two overlapping sheets of metal are welded together along a continuous line by moving them between wheel-shaped electrodes through which the welding current flows at short intervals. Fig. 14.5d illustrates the principle. This process can produce continuous pressure tight seams at high speeds.

(e) *Pulsation Welding*. In pulsation welding, the welding current flows intermittently unlike spot welding in which current flows continuously until the weld is formed. This intermittent flow of current permits the water cooled electrodes to remove heat from the surface of the materials and thereby causes temperature gradient near the weld. This method is used for pieces thicker than 1 cm.

Electric Braking

i. **Plugging:** It requires the reversal of connections to the armature of the motor so that it develops a torque in opposite direction to that of normal rotation. The rotor speed decreases till it reduces to zero and then the rotor accelerates in an opposite direction, unless the special device is installed to cut-off the supply to the motor in order to stop it. Plugging is used for rapid stops and quick reversals. This method is applicable to DC induction and synchronous motors.

ii. **Rheostatic braking:** In this method, the motor is disconnected from the supply and is used as a generator, driven by the momentum of the equipment to be braked, the electric energy so generated is dissipated as heat to the external resistors. This method can be used for DC, induction and *synchronous motors.*

iii. **Regenerative braking:** With regenerative braking, the motor continues to be connected to the supply but it made to act as a generator. Kinetic energy of the drive is converted into electrical energy part of which is fed back to the supply system and part of it is dissipated as heat in the winding of the machine.

Table 14.2: Fundamental formulae in dynamics

Linear motion	*Angular motion*
Mass = W/g kg	Moment of inertia J in kg·m^2
Linear acceleration	Angular acceleration
$a = f_t \times \frac{g}{W}$ m/sec	$\alpha = \frac{d\omega}{dt}$ radians/sec
Linear velocity, $v = \frac{ds}{dt}$ m/sec	Angular velocity, $\omega = 2\pi \frac{dn}{dt}$ radians/sec or $n = 1/2\pi \int \omega . dt$
Linear distance, $s = \int v \cdot dt$ m	Angular motion, $n = \int \frac{\omega}{2\pi} d\theta$
Tractive force $f_t = \frac{W}{g} a$ kg	Torque, $T_m = \frac{J}{g} \alpha$ kg · m

Time calculations: Angular retardation

$$\alpha_r = \frac{T_B g}{J}$$

$$t = \frac{J}{Kg} \log_e \frac{K\omega_i + T_M}{T_M}$$

Speed: $$n = \frac{1}{2\pi K}\left[\frac{J}{gK}(K\omega_i + T_M)(1 - e^{\frac{-gK}{j}t}) - T_M \times t\right]$$

Standard ratings for motors: The ratings of a motor may either be continuous, continuous maximum or short time. ISI specification has defined these ratings as given below:

i. *Continuous rating*: This is a rating or the output of a motor which is capable of delivering continuously without exceeding the maximum permissible temperature rise. This term also indicates that the motor is capable of delivering 25% overload for 2 hours.

ii. *Continuous maximum rating*: This term is similar to the continuous rating with the only difference is that the motor is not capable of running on overload. The term is employed for motors having capacity more than 2.5 hp per rpm.

iii. *Short time rating*: This term denotes the output of a motor for a short and specified period but without exceeding the specified temperature rise.

Choice of motor: Assuming heating to be proportional to load,

rating of the motor

$$= \text{(rms) HP}$$

$$= \sqrt{\frac{1}{\text{time for 1 cycle}} \int (\text{hp})^2\, dt}$$

$$= \sqrt{\frac{\sum (\text{hp})^2 \times \text{time}}{\text{time for 1 cycle}}}$$

Illumination Engineering

Terms used in illumination are described below:

Light: Light is defined as the radiant energy from a hot body causing visual sensation on the human eye.

Flux: It is also known as luminous flux. It is defined as the total quantity of light energy radiated or emitted per second from a luminous body in the form of light waves. It is measured in lumens and is denoted by F or Φ. It is defined as the rate of luminous energy.

If Q lumen-hour is the luminous output of source of light and energy radiated for t hours, then flux is $\Phi = Q/t$ lumens.

Light energy: It is the energy contained in visual radiations in a given time and is expressed in lumen-hour and is denoted by Q.

Luminous efficiency or radiant efficiency: It is defined as the output in lumens per watt of the power consumed by the source of light. It is measured in lumens per watt. If a bulb is rated at 500 watts and 250 volts, has an efficiency of 13 lumens/watt. Then total flux produced by bulb = 500 × 13 = 6500 lumens.

Plane angle: Plane angle is subtended in a plane by two converging lines. This angle is measured in radians or degrees.

One radian is the angle subtended at the centre of a circle by an arc whose value is equal to the radians of the circle.

$$\theta = \frac{\text{Arc}}{\text{radians}}$$

and one radian = $180/\pi$ degree.

The largest plane angle subtended at a point may be 2π radians.

Solid angle: Solid angle is the angle generated by the line passing through the point in space and the periphery of the area. In other words, a solid angle encloses a volume by an infinite number of lines lying on a surface and meeting at a point. It is measured in steradians and is denoted by ω and

$$\omega = \frac{\text{Area}}{(\text{Radius})^2} \text{ steradians}$$

Luminous intensity: Luminous intensity in any particular direction is the luminous flux emitted per unit solid angle by a point source and is denoted by *I*.

$$I = \frac{F}{\omega} = \frac{\Phi}{\omega} \text{ lumen/steradian or candela}$$

It is the ratio of the brightness of a source of light to that of standard candle. One candle gives out luminous flux of 4π lumens in space. Thus lumens emitted by one candle source of light is one lumen/steradian.

In scientific terms, candle is defined as the luminous intensity in the perpendicular direction of a surface of 1/600,000 sq metre of a full radiator at the temperature of freezing platinum under a pressure of 101,325 N/sq. metre.

Lumen: Lumen is defined as the amount of luminous flux given out in space represented by one unit of solid angle by a source having an intensity of one candle power in all directions.

$$\text{Lumen} = \text{Candle power} \times \text{Solid angle} = \text{CP} \times \omega$$

Candle Power (*CP*): Candle power is the light rendering capacity of a source in a given direction and is defined as number of lumens given out by the source in a unit solid angle in a given direction. It is denoted by symbol CP i.e.

$$\text{CP} = \frac{\text{Lumen}}{\omega}$$

Illumination: When light falls upon any surface, the phenomenon is called the illumination. It is defined as the number of lumens, falling on the surface per unit area. It is denoted by symbol *E* and is measured in lumens per square metre or lux or metre candle.

If a flux of *F* lumens falls on a surface of area *A*, then the illumination of that surface is $E = F/A$ lumens/m^2 or lux.

Bigger unit of illumination is phot:

One phot = 10^4 lux.

Metre candle: It is also known as lux. It is the unit of illumination and is defined as the luminous flux falling per sq metre on the surface which is everywhere, perpendicular to the rays of light from a source of one candle power and one metre away from it or it is the illumination produced by a source of one candle power on the inner surface of a sphere of one metre radius.

Centimetre candle: It is also a unit of illumination and is defined as the luminous flux falling per square centimetre on the surface which everywhere is perpendicular to the rays of light from a source of one CP and one cm away from it.

Mean spherical candle power (*MSCP*): It is defined as average of candle powers in all directions and in all plans from the source of light, i.e.

$$\text{MSCP} = \frac{\text{Total flux in lumens}}{4\pi}$$

Mean hemispherical candle power (*MHSCP*): It is defined as average of candle powers in all directions above or below the horizontal plane passing through the source of light.

Mean horizontal candle power (*MHCP*): It is average of all the candle power in all directions in the horizontal plane containing the source of light.

Reduction factor: Reduction factor of a source of light is defined as the ratio of its mean spherical candle power to its mean horizontal candle power.

$$\text{Reduction factor} = \frac{\text{MSCP}}{\text{MHCP}}$$

Reflection factor: It is defined as the ratio of reflected light to incident light. It is always less then unity

$$\text{Reflection factor} = \frac{\text{Reflected}}{\text{Incident light}}$$

A polished surface like mirror reflects light incident over it. The part of light absorbed by such a surface is too small in comparison with the reflected part. Since all the incident light is not reflected, reflection factor is always less than one. Following are some typical values of RF.

Source	*RF*
Mirror	0.80 to 0.85
Steel stainless	0.60 to 0.65
Chromium plating	0.55 to 0.60

Lamp efficiency: It is defined as the ratio of luminous flux to the power input. It is expressed as lumens/watt or watts/candle power.

$$\text{Lamp efficiency} = \frac{\text{Lumens emitted to source}}{\text{Wattage of source of light}}$$

Space height ratio: It is defined as the ratio of horizontal distance between two adjacent lamps and height of lamps from the working plane.

$$\text{Space height ratio} = \frac{\text{Horizontal distance between two adjacent lamps}}{\text{Height of lamps above working plane}}$$

Depreciation factor: Due to accumulation of dirt, dust and smoke, lamps emit less light as compare to they emit when they are new ones. Similarly, ceilings and walls, etc. after being covered by dust, etc. reflect less light as compare to that when they are new.

Therefore, the ratio of illumination under everything is perfectly clean to the illumination under normal working condition is known as *depreciation factor*.

Depreciation factor

$$= \frac{\text{Illumination when everything is clean}}{\text{Illumination under normal working condition}}$$

It is more then unity.

Maintenance factor: It is merely reverse of the depreciation factor. So it is always less than unity.

Specific consumption: It is the ratio of power input to the source of light to its luminous intensity. Its unit is watt per candela.

Absorption factor: In case of factories or places where atmosphere is full of smoke fumes, there is a possibility of absorption of light energy. The absorption factor is defined as

Absorption factor

$$= \frac{\text{Net lumens available after absorption}}{\text{Total lumens emitted by source of light}}$$

Coefficient of utilisation: All the light emitted by the source does not reach the surface to be illuminated. Some portion of light directly falls upon surface and some light falls upon surface after reflection from wall and ceilings. All the light falling upon the walls and ceiling does not reach the surface because some light is absorbed.

Therefore, coeffcient of utilisation is the ratio of lumens reaching the working plane to the total lumens given out by lamp.

Coefficient of utilisation

$$n = \frac{\text{Lumens reacting on working plane}}{\text{Total lumens emitted by source of light}}$$

Waste light factor: In flood lighting scheme, a number of projector lamps is employed so the light of each projector not only overlaps that of its adjacent projector but at the edges of the area, some light falls outside. In order to compute total lumens required, theoretical lumens required are multiplied by waste light factor. Value of this factor is about 1.2 for rectangular areas and 1.5 for irregular objects such as statues, monuments, etc.

Beam factor: It is used in turns of projector lamps. It is the ratio of lumens in the beam of a projector to lamp lumens. This factor takes into account the absorption of light by reflector and front glass of the projector lamp. Its value lies between 0.3 to 0.6.

Laws of Illumination: There are two laws of illumination:

i. **First law:** It is also known as *law of inverse square*.

According to this law, the illumination of a surface is inversely proportional to square of distance between the source and surface provided that the distance between the surface and the source is sufficiently large so that the source can be regarded as a point source.

ii. **Lambert's cosine law:** According to this law, illumination of a surface varies directly as the cosine of the angle between the normal to the surface and direction of incident light.

Table 14.3: Illumination parameters, relationships and units in lighting engineering

Physical quantity	*Symbol*	*Relationship (simplified)*	*SI unit*	*Explanation*
Luminous flux	Φ		Lumen or Lm	Photometrically evaluated, radiant flux (light output)
Luminous intensity	I	$I = \frac{\Phi}{\Omega}$	Candela, cd	Quotient of luminous flux Φ emitted from a light source in a specific direction and the solid angle Ω
Illuminance or	E	$E = \frac{\Phi}{A}$	Lumen per square metre or	Quotient of luminous flux Φ hitting surface A and the illuminated surface A.
Illumination		$E = \frac{I_e}{r^2}\cos^5 \varepsilon$ $E = \frac{I_e}{h^2}\cos^5 \varepsilon$	Lux	A surface in m^2, r distance in m, h distance of light source to surface in m, ε radiation angle of lamp or incidence angle on surface measured to vertical I_e, luminous intensity at radiation angle
Luminance	L	$L = \frac{\Phi}{A\cos\varepsilon \cdot \Omega}$ $L = \frac{I_e}{A \cdot \cos\varepsilon}$	Candela per square metre cd m^{-2}	Quotient of luminous flux Φ passing through surface A in a specific direction ε and the product of the solid angle Ω and the projection of surface A × cos ε on a plane vertical to viewed direction

Table 14.4: Lighting calculation for indoor working spaces according to the efficiency method

	n	*Number of lamps required*
$n = \dfrac{1.25 \cdot E \cdot A}{\Phi_L \cdot \eta_B}$	1.25	The factor by which the design value should be chosen to exceed the nominal illuminance E_n. Design values for new installations: 1.25 E_n
	A	Working surface in m^2
	Φ_L	Luminous flux of lamp in 1m
	η_B	Utilization factor

Table 14.5: Utilization factors

Type of lighting	η_B*	*Type of lighting*	η_B*
Direct	0.6 to 0.45	Semi-indirect	0.45 to 0.35
Semi-direct	0.55 to 0.45	Indirect	0.35 to 0.25
Direct-indirect	0.5 to 0.35	Indirect cornice lighting	0.2 to 0.1 5

* According to room shape, as well as reflection factors of walls, ceilings and floors.

Table 14.6: Commonly used light sources general purpose lamps (220 to 230 V)

Wattage W	*Luminous flux at 225 V lm*	*Base*	*Wattage* W	*Luminous flux at 225 V lm*	*Base*
25	230	E27	300	5000	E40
40	430	E27	500	8400	E40
60	730	E27	1000	18800	E40
75	960	E27			
100	1380	E27			

Table 14.7: Fluorescent lamps OSRAM Lamilux (with fully electronic control gear)

Wattage without ballast W	*With ballast* W	*Rated current* A	*Length of lamp* mm	*Luminous flux acc. to type of lamp and light colour* lm	
16 (L18W)	19	0.095	590	1350	
32 (L 36 W)	36	0.18	1200	3200	
34 (L 38 W)	38	0.19	1047	3150	
50 (L 38 W)	55	0.25	1500	5200	
Fluorescent lamps (with ballast)				*Standard*	*Lumilux*
8	14	0.145	288	330	430
15	25(19.5)[1]	0.33	438	720	1000
18	30(23)[1]	0.37	590	1150	1450
20	32(26)[1]	0.37	590	1150	–
30	40	0.365	895	1800	2400
36	46	0.43	1200	3000	3450
38	50[2]	0.43	1047	—	3200[2])
40	50	0.43	1200	3000	
58	71	0.67	1500	4800	5400
65	78	0.67	1500	4800	

1. For each lamp with series connection of two lamps to ~220 V and one ballast.
2. With 40 W ballast.

Table 14.8: Recommended levels of illumination

Premises	*Illumination level (lumens per square metre or Lux.)*
Residential	
Living room, kitchen, bedroom	150
dinning room	150
study room	300
corridor, toilet, stair-case	50–100
Offices	
General	200–300
Drawing, architecture, typing	400–1500
Factories	
Assembly, rough and medium	150–250
Fine and very fine assembly of components, precious stone cutting, optics and watchmaking	600–1500
Machine shop	200–800
Printing General	
Composing	500
Colour print inspection	800
Textile carding, spinning	150–250
Weaving coarse	350
Weaving fine	800
Wood working	200–800
Welding general precision	250–350
Inspection medium and fine	350–800
Very fine and minute	1500–2500
Schools	
Classrooms	250–350
Classrooms for chemistry, physics, works, arts	500
Shops	
General	500–1000
Show windows	1000–2000
Hospitals	
General	150–200
Operation theatre	500–1000
Operation table	Over 5000
Gardens	5–30
Railway yards	5–20
Canteens	150–250
Car parks	5–10
Traffic, routes acc. to traffic density	50–100

MULTIPLE CHOICE QUESTIONS

1. Overall efficiency of streams locomotive system is close to
   A. 5 to 10% B. 25 to 30%
   C. 55 to 60% D. 75 to 80%
2. In a steam locomotive, electric power is provided through
   A. battery system
   B. diesel engine generator
   C. overhead wire D. small turbo generator
3. Maximum horse power of steam locomotive is
   A. 100 hp B. 500 hp
   C. 1500 hp D. 2500 hp
4. The pressure of steam in a locomotive is
   A. 10–15 kg/cm^2 B. 20–30 kg/cm^2
   C. 40–50 kg/cm^2 D. 80–90 kg/cm^2
5. The efficiency of diesel locomotives is nearly
   A. 20–25% B. 35–40%
   C. 50–55% D. 70–75%
6. Diesel locomotives are manufactured in India at
   A. Bangalore B. Ajmer
   C. Jamalpur D. Varanasi
7. Suburban railways use
   A. 1500 V, DC B. 440 V, three-phase AC
   C. 660 V, three-phase AC
   D. 3.3 V, three-phase AC
8. Long distance railways use
   A. 220 V, DC
   B. 25 kV, single-phase AC
   C. 25 kV, two-phase AC
   D. 25 kV, three-phase AC
9. The range of horsepower for diesel locomotives is
   A. 100 to 500 hp B. 500 to 1000 hp
   C. 1500 to 2500 hp D. 4000 to 5500 hp
10. The maximum number of passenger coaches that can be attached to diesel locomotives on broad gauge is
    A. 18 to 20 B. 25 to 30
    C. 32 to 35 D. 36 to 40
11. A submarine while moving under water, is provided driving power through
    A. diesel engines B. steam turbine
    C. gas turbine D. batteries
12. Overload capacity of diesel engines is usually restricted to
    A. 1% B. 10%
    C. 25% D. 50%
13. Which locomotive has the highest operational duty?
    A. Diesel B. Electric
    C. Steam D. All of these
14. Which motor is used in tramways?
    A. AC single-phase capacitor-start motor
    B. AC three-phase motor
    C. DC series motor D. DC shunt motor
15. A drive suitable for mines where explosive gas exist, is
    A. Diesel engine B. Steam engine
    C. Battery locomotive D. Any of these.
16. The advantage of electric braking is
    A. it is instantaneous
    B. more heat is generated during breaking
    C. it avoids wear of track
    D. motor continue to remain loaded during braking
17. Which braking system on the locomotives is costly?
    A. Vacuum braking on steam locomotives
    B. Vacuum braking on diesel locomotives
    C. Regenerative braking on electric locomotives
    D. All of these
18. The acceleration rate of trains on suburban services is generally
    A. 0.1 to 0.4 km-phps B. 0.8 to 1 km-phps
    C. 0.4 to 6.5 km-phps D. 10 to 26 km-phps
19. The coasting retardation on trains is approximately
    A. 0.16 km-phps B. 1.6 km-phps
    C. 16 km-phps D. 25 km-phps
20. The coefficient of adhesion is
    A. same on ACand DC traction systems.
    B. low in case of ACtraction and high in DC traction
    C. high in case of DC traction and low in AC traction
    D. None of these
21. Braking retardation on suburban trains is
    A. 0.3 to 0.5 km-phps B. 0.5 to 1 km-phps
    C. 3 to 5 km-phps D. 30 to 40 km-phps
22. Power supply frequency for 25 kVsingle-phase system
    A. 50/3 Hz B. 25 Hz
    C. 50 Hz D. 60 Hz
23. For supply of 25 kV, 50 Hz single-phase, suitable motor for electric traction is
    A. ac single-phase split phase motor
    B. ac single-phase universal motor
    C. dc shunt motor D. dc series motor
24. Method of speed control used on 25 kV, 50 Hz single-phase traction is

Ans. 1. A 2. D 3. C 4. A 5. A 6. D 7. A 8. B 9. C 10. B 11. D 12. B 13. A 14. C 15. C 16. C 17. C 18. C 19. A 20. B 21. C 22. A 23. D

A. Tap changing control of transformer
B. Reduced current method
C. Series parallel operation of motors
D. Any of these.

25. The coefficient of adhesion is highest when
A. the rails are dry B. the rails are oiled
C. the rails are wet with dew
D. the rails are dusty.

26. When the speed of the train is estimated taking into account the time of stop at a station in addition to the actual running time between stops, is known as
A. Average speed B. Schedule speed
C. Notching speed D. Free running speed

27. A schedule speed of 45 km, per hour is required between two stops 1.5 km apart. The duration of stop is 20 seconds. The acceleration is 2.4 km phps and retardation is 3.2 km phps. For a simplified trapezoidal curve, the maximum speed over the run will be
A. 40 km per hour B. 48 km per hour
C. 74 km per hour D. 90 km per hour

28. A three tier sleeper bogie on broad gauge has
A. 40 berths B. 50 berths
C. 75 to 80 berths D. 96 to 100 berths

29. The specific energy consumption for suburban services is usually
A. 18 to 25 watt-hours per tonne km
B. 50 to 75 watt-hours per tonne km
C. 125 to 150 watt-hours per tonne km
D. 155 to 200 watt-hours per tonne km

30. If the specific energy consumption for suburban services is 50 to 75 watts hours per tonne km, which of the following could be a representative figure for energy consumption on main line service?
A. 150 to 200 watt-hours per tonne km
B. 100 to 125 watt-hours per tonne km
C. 50 to 75 watt-hours per tonne km
D. 20 to 30 watt-hours per tonne km

31. Specific energy consumption is least in
A. urban service
B. sub-urban service
C. main-line service
D. equal for all types of services.

32. Bearings used to support axles of rolling stock are
A. Bush bearings B. Journal bearings
C. Ball bearings D. Roller bearings

33. If the coefficient of adhesion on dry rails is 0.25, which of the following could be the value for wet rails?
A. 0.32 B. 0.25
C. 0.245 D. 0.15

34. A train has a schedule speed of 36 km per hour on a level track. If the distance between the stations is 2 km. and the stoppage is 30 seconds, the actual time of run will be
A. 260 sec B. 170 sec
C. 200 sec D. 230 sec

35. B_0B_0 locomotives have two bogies with
A. four driving axles each with individual driving motors
B. three driving axles with group drives
C. two driving axles with group drives
D. two driving axles with individual drive motor

36. A locomotive exerts a tractive effort of 30000 Newtons in pulling a train at 50 km per hour on the level track. It is to haul the same train at the same speed on a gradient and the tractive effort required is 45000 Newtons. The horse power delivered by the motor will be more if it is driven by
A. dc series motors B. induction motor
C. Same in both cases D. None of these

37. A composite system consists of
A. a combination of diesel engine and DC series motor
B. a combination of diesel engine and AC single-phase motor
C. single-phase power received is converted into DC or three-phase power AC system
D. use of combination of DC and AC motors on the same locomotive

38. Which city in India was first to adopt electric traction?
A. Bombay B. Madras
C. Calcutta D. Delhi

39. Which of the following route has electric traction?
A. Delhi–Madras B. Delhi–Ahmedabad
C. Delhi–Howrah D. Delhi–Bangalore

40. Locomotives with manometer bogie have
A. lot of skidding
B. low coefficient of adhesion
C. uneaten distribution of tractive
D. suitability for passenger as well as freight service

41. The speed time curve for a local train is shown in figure below. In this time, AB represents:

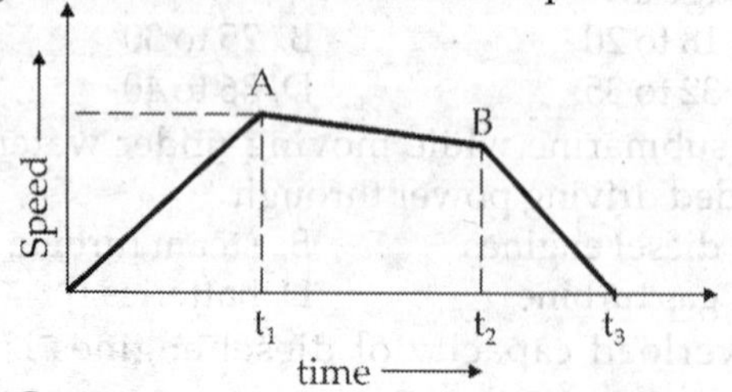

A. Coasting B. Acceleration
C. Braking D. Regeneration

42. The duration for braking is represented by the time
A. $0 - t_1$ B. $0 - t_2$
C. $t_1 - t_2$ D. $t_2 - t_3$

Ans.	24. A	25. A	26. B	27. C	28. C	29. B	30. C	31. C	32. B	33. D	34. D	35. D	36. B
	37. C	38. A	39. C	40. D	41. A	42. D							

43. Area under the curve represents
A. average speed B. distance travelled
C. net acceleration D. average acceleration

44. From the above figure, it can be concluded that
A. rate of acceleration is the same as the rate of retardation during braking
B. average acceleration is zero
C. time taken during costing is equal to the time during acceleration and braking
D. during coasting the acceleration is negative

45. For tram ways, the return circuit is
A. through cables B. through rails
C. through neutral wire
D. through common earthing

46. For 600 V DC line for tram cars
A. track is connected to negative of the supply
B. track is connected to positive of the supply
C. track is connected to mid voltage of 300 V
D. None of these

47. Over-head lines for power supply to tram cars are at a minimum height of
A. 2 m B. 6 m
C. 10 m D. 15 m

48. Which of the following traction systems is latest used in the world?
A. Three-phase 3.7 kV
B. 20 kV, 50 Hz, single-phase
C. 600 V, DC D. 3 kV, DC

49. Which of the following frequencies is not common in low frequency traction system?
A. 50/3 Hz. B. 25 Hz
C. 40 Hz D. 60 Hz

50. In a long distance electric train, power for lighting in passenger coach is provided
A. through locomotive
B. directly through overhead electric line.
C. through individual generator of bogie and batteries
D. through rails

51. In Kando system
A. single-phase supply is converted into three-phase system
B. single-phase ac is converted into dc
C. three-phase AC is converted into dc
D. dc supply is to run dc motors.

52. Free running and coasting periods are generally long in case of
A. urban service B. sub-urban service
C. main-line service D. All of these.

53. Which of the following factors affects specific energy consumption?
A. Distance between stops
B. Gradient
C. Retardation and acceleration values
D. All of these

54. A train runs at an average speed ol 50 kmph between stations situated 2.5 km apart. The train accelerates at 2 kmphs and retards at 3 kmphs. Speed-time curve may be assumed to be trapezoidal. The maximum speed is
A. 27.75 kmph B. 38.50 kmph
C. 44.25 kmph D. 57.75 kmph

55. In the above question, the distance travelled before the brakes are applied is
A. 0.75 km B. 1.35 km
C. 2.0 km D. 2.35 km

56. The draw bar horsepower per tonne in case of steam locomotives is
A. 7 to 8 B. 15 to 17
C. 20 to 25 D. 30 to 35

57. At an average the coal consumption per km in case of steam engine is nearly
A. 28 to 30 kg B. 80 to 100 kg
C. 150 to 160 kg D. 200 to 250 kg

58. Electrification of railway track in India, was done for the first time in
A. 1857–1863 B. 1887–1893
C. 1925–1932 D. 1940–1947

59. Unbalanced forces are maximum in case of
A. electric locomotive B. diesel locomotives
C. diesel and electric D. steam locomotives

60. Maintenance requirements are least in case of
A. electric locomotives B. diesel locomotives
C. steam locomotives D. none of these

61. Suri transmission is
A. hydro-mechanical B. hydro-pneumatic
C. mechanical-electrical D. electrical-pneumatic

62. If the resistance to electric train is given by: $F_r = a + bv + cv^2$. In this equation, constant c is likely to cover
A. air resistance b. frictional resistance
C. flange resistance D. track resistance

63. A train is required to run between two stations 16 km apart at an average speed of 43 kmph. The run is to be made to a simplified quadrilateral speed-time curve. The maximum speed is to be limited to 64 kmph, acceleration to 2 kmphs and coasting and braking retardations to 0.16 and 3.2 kmphs respectively. The duration of acceleration is
A. 32 sec B. 20 sec
C. 15 sec D. 10 sec

64. In the above question, the duration of costing is
A. 48.4 sec B. 96.8 sec
C. 144.2 sec D. 15.15 sec

65. In the same question, the braking period is
A. 1.0 sec B. 5.15 sec
C. 12.35 sec D. 15.15 sec

Ans.													
	43. B	44. D	45. B	46. A	47. C	48. A	49. C	50. C	51. A	52. C	53. D	54. D	55. D
	56. B	57. A	58. C	59. D	60. A	61. A	62. A	63. A	64. B	65. D			

66. When locomotive for Indian Railways is designated as WAM1, in this, the letter W indicates that
A. the locomotive is to run on broad gauge track
B. the locomotive is to run on metre gauge track
C. the locomotive is for shunting duty
D. the locomotive is for goods trains only

67. An ideal traction system should have
A. high starting tractive effort
B. the locomotive is to run on metre gauge track
C. the locomotive is for shunting duty
D. the locomotives is for goods trains only

68. A train runs at an average speed of 45 kmph between stations 2.5 km apart. The train accelerates at 2 kmph and retards at 3 kmph speed-time curve may be assumed to be trapezoidal. The maximum speed attained will be nearly
A. 80 kmph B. 60 kmph
C. 50 kmph D. 35 kmph

69. In the above question, the distance traveled before the brakes are applied is
A. 2.383 km B. 2.103 km
C. 1.887 km D. 1.535 km

70. The main difference between speed-time curves of main line service as compared to suburban services lies in
A. longer free running periods
B. longer coasting periods
C. shorter accelerating and braking periods
D. All of these

71. An electric train is to have braking retardation 3.2 kmph. The ratio of maximum speed to average speed is 1.3, the time for stop is 26 seconds and acceleration is 0.8 kmph. The run is 1.5 km. Actual time of run is
A. 77 sec B. 101 sec
C. 154 sec D. 231 sec

72. In the above question, the schedule time is
A. 154 sec B. 180 sec
C. 210 sec D. 240 sec

73. In the question, the schedule speed is
A. 25 kmph B. 30 kmph
C. 45 kmph D. 60 kmph

74. Energy consumption in propelling the train is required for
A. acceleration
B. work against gravity while moving up the gradient
C. work against the resistance to motion
D. all of these

75. Quadrilateral speed-time curve is the closer approximation for
A. main line service B. suburban service
C. urban service
D. urban and suburban service

76. As compared to suburban service, in main line service
A. distance between stops is more
B. higher maximum speeds can be achieved
C. acceleration and retardation rates are low
D. All of these

77. The train is moving a uniform upgrading of 1 in 100. The train resistance is 40 N/tonne, rotational inertia effect 10% of dead weight and duration of stop 30 seconds. Braking retardation is
A. 0.87 kmphs B. 1.27 kmphs
C. 1.87 kmphs D. 2 kmphs

78. In the above question, the distance travelled is
A. 1.03 km B. 1.53 km
C. 2.03 cm D..2.53 cm

79. In the same question, schedule time is
A. 100 sec B. 110 sec
C. 120 sec D. 130 sec

80. The electric locomotives run faster at curved routes as compared to steam locomotives as:
A. the centre of gravity of electric locomotive is higher than that of steam locomotive
B. the centre of gravity of electric locomotive is lower than that of steam locomotive
C. the speed at curved routes is independent of location of centre of gravity
D. none of these

81. Low frequency operation of AC series motor
A. improves its commutation property but affects adversely the pf and efficiency
B. improves its commutation property, pf and efficiency
C. affects adversely commutation but improves pf and efficiency
D. none of these

82. Low frequency operation of overhead line in traction system
A. reduces the spacing between sub-stations
B. increases the spacing between sub-stations
C. spacing is independent of frequency of supply
D. none of these

83. At low frequency of the order of 1/2 Hz to 10 Hz, the induction motors develop
A. high starting torque with excessive starting current
B. high starting torque without excessive starting current
C. low starting torque with excessive starting current
D. low starting torque without excessing starting current

84. Typical values of acceleration and deceleration for urban services are

Ans.	66. B	67. D	68. C	69. A	70. D	71. C	72. B	73. B	74. D	75. D	76. D	77. C	78. A
	79. D	80. B	81. B	82. B	83. B								

A. 2 km php and 3 km phps
B. 3 km php and 2 km phps
C. 1.5 km php and 4 km phps
D. none of these

85. Typical value of acceleration and deceleration for suburban services are
A. 4 km php and 6 km phps
B. 6 km php and 4 km phps
C. 2.5 km php and 4 km phps
D. none of these

86. In suburban services as compared with urban services.
A. the coasting period is longer
B. the coasting period is smaller but free running period is longer
C. the coasting and free running periods are smaller
D. none of these

87. The distance between two stations is 1 km. and the schedule speed is 30 km per/hour. Station stopping time 20 sec. The average speed is
A. 25.7 km per hr B. 36 km per hr
C. 45 km per hr D. 54 km per hr

88. The distance between two stations is 1 km and the average and schedule speeds of the train are 36 kmph and 30 kmph respectively. The stations stopping time is
A. 20 sec B. 40 sec
C. 10 sec D. 16 sec

89. For a given run and a given schedule speed
A. the specific energy consumption is lower, the higher the acceleration and retardation
B. the specific energy consumption is higher, the higher the acceleration and retardation
C. the specific energy consumption is lower, the higher the acceleration and higher the deceleration
D. none of these

90. The specific energy consumption.
A. increases with increase in maximum speed
B. decreases with increase in maximum speed
C. is independent of maximum speed
D. none of these

91. If the speed-time curves are similar (not identical), the specific energy consumption of the curve having higher maximum speed is
A. higher than that of lower maximum speed
B. lower than that of lower maximum speed
C. equal to that of lower maximum speed
D. None of these

92. Coefficient of adhesion is higher if
A. the speed is low and the rails are sandy and dry
B. the speed is high and the rails are sandy and dry
C. The speed is high and the rails are wetty
D. the rails are wet and greasy

93. Two shunt motors having identical characteristics are used to drive a car with unequal wheel diameters. The two motors share load equally if they are connected in
A. series B. parallel
C. the loading will always be unequal
D. none of these

94. Two locomotives having identical tractive-effort-slip characteristic haul a heavy train. The loco with larger diameter shares
A. larger tractive effort B. smaller tractive effort
C. equal tractive effort. D. none of these.

95. For rheostatic braking of two series motors connected in parallel
A. equaliser connection is better
B. cross-connection is better
C. both are equally good
D. none of these

96. For braking of electric trains, normally:
A. regenerative braking is used to bring the speed down to zero
B. regenerative breaking is used down to a speed of 4 kmph followed by mechanical brakes
C. regenerative braking is used down to a speed of 16 kmph and finally mechanical brakes
D. regnerative braking is used down to a speed of 16 kmph followed by rheostalic braking till zero speed

97. Run away speed of a Pelton wheel is
A. full load speed
B. actual speed operating at no load
C. no load speed when governor mechanism fails
D. 80% greater than the normal speed

98. For regenerative braking, the regenerated power should have
A. the same frequency as that of the main supply
B. frequency 1/3 that of the main supply
C. frequency 1/2 of that of the main supply
D. any frequency

99. A train having inertial weight of 220 tonnes uses regenerative braking when the speed is changed from 60 kmph to 20 kmph, the electric energy returned to supply is (assume 100% efficiency)
A. 7.546 kW hr B. 8.32 kW hr
C. 5.6595 kW hr D. none of these

100. In case of bow and pantograph current collectors,
A. the bow collector is used for large current at higher speeds
B. the pantograph collector is used for large currents at higher speeds
C. the bow collector is used for large currents at lower speeds
D. the pantograph collector is used for large currents at lower speeds

Ans. 84. D 85. C 86. A 87. B 88. A 89. A 90. A 91. C 92. B 93. A 94. A 95. B 86. C 97. C 98. A 99. A 100. D

101. Contact with wire is maintained
   A. by springs in both bow and pantograph collectors
   B. by air pressure in both bow and pantograph collectors
   C. by spring in case of bow and air pressure in case of pantograph
   D. the spring in case of pantograph and air pressure in case of bow collectors
102. The coefficient of adhesion is
   A. same on AC and DC traction
   B. high in DC and low in AC
   C. high in AC and low in DC
   D. none of these
103. Tick out the correct statement:
   A. Smaller size have their critical speeds above their synchronous speeds
   B. Smaller size machines have their critical speeds below their synchronous speeds
   C. Critical speed is always greater than synchronous speed irrespective of size of machine
   D. Critical speed is always less than synchronous speed irrespective of the size of the machine
104. The coefficient of adhesion is highest when
   A. the rails are dusty B. the rails are wet
   C. the rails are dry D. the rails are oiled
105. Specific energy consumption is minimum in
   A. main-line service B. sub-urban service
   C. urban service D. equal for all types
106. Quadrilateral speed time curve is used for
   A. sub-urban service B. urban service
   C. main line service
   D. urban and sub-urban service
107. Acceleration rate of trains on suburban services is
   A. 0.2 to 0.4 kmphps B. 0.8 to 2 kmphps
   C. 0.4 to 6.5 kmphps D. None of these
108. Overall efficiency of streams locomotive system is close to
   A. 7 to 10% B. 25 to 30%
   C. 55 to 60% D. 75 to 80%
109. Long distance railways use
   A. 220 kV, DC
   B. 25 kV, single-phase AC
   C. 25 kV, two-phase AC
   D. 25 kV, three-phase AC
110. Diesel locomotives are manufactured in India at
   A. Bangalore B. Jaipur
   C. Bharatpur D. Varanasi
111. The range of horsepower for diesel locomotives is
   A. 100 to 500 hp B. 500 to 1000 hp
   C. 1500 to 2500 hp D. 4000 to 5500 hp
112. Suburban railways use
   A. 1500V, DC B. 440 V, three-phase AC
   C. 660 V, three-phase AC
   D. 12 kV, three-phase AC
113. Which locomotive has the highest operational duty?
   A. Diesel B. Electric
   C. Steam D. All of these
114. A drive suitable for mines where explosive gas exists, is
   A. diesel engine B. steam engine
   C. battery locomotive D. Any of these
115. Which motor is used in tramways?
   A. AC single-phase capacitor-start motor
   B. AC three-phase motor
   C. DC series motor D. DC shunt motor
116. The advantage of electric braking is that
   A. it is instantaneous
   B. more heat is generated during breaking
   C. it avoids wear and tear of track
   D. motor continues to remain loaded during braking
117. A three tier sleeper bogie on broad gauge has
   A. 50 berths B. 60 berths
   C. 75 to 80 berths D. 96 to 100 berths
118. The specific energy consumption for suburban services is usually
   A. 18 to 25 watt-hours per tonne km
   B. 50 to 75 watt-hours per tonne km
   C. 125 to 150 watt-hours per tonne km
   D. 155 to 200 watt-hours per tonne km
119. If the specific energy consumption for suburban services is 50 to 75 watt hours per tonne km, which of the following could be a representative figure for energy consumption on main line service?
   A. 150 to 200 watt-hours per tonne km
   B. 100 to 125 watt-hours per tonne km
   C. 50 to 75 watt-hours per tonne km
   D. 20 to 30 watt-hours per tonne km
120. The acceleration rate of trains on suburban services is generally
   A. 0.1 to 0.4 km/hr/s B. 0.8 to 1 km/hr/s
   C. 0.4 to 6.5 km/hr/s D. 10 to 26 km/hr/s
121. The coasting retardation on trains is approximately
   A. 0.16 km/hr/s B. 1.6 km/hr/s
   C. 16 km/hr/s D. 25 km/hr/s
122. Which braking system on the locomotives is costly?
   A. Vacuum braking on steam locomotives
   B. Vacuum braking on diesel locomotives
   C. Regenerative braking on electric locomotives
   D. All of these
123. The coefficient of adhesion is
   A. same on AC and DC traction systems
   B. high in case of DC traction and low in AC traction

Ans. 101. A 102. C 103. A 104. C 105. A 106. D 107. C 108. A 109. B 110. D 111. C 112. A 113. B 114. C 115. C 116. C 117. C 118. B 119. D 120. C 121. A 122. C

C. low in case of ac traction and high in DC traction
D. none of these

124. For supply of 25 kV, 50 Hz single-phase, suitable motor for electric traction is
A. AC single-phase split phase motor
B. AC single-phase universal motor
C. DC shunt motor
D. DC series motor

125. Braking retardation on suburban trains is
A. 0.3 to 0.4 km/hr/s B. 0.5 to 1 km/hr/s
C. 3 to 5 km/hr/s D. 30 to 40 km/hr/s

126. Power supply frequency for 25 kV single-phase system is
A. 50/3 Hz B. 25 Hz
C. 50 Hz D. 60 Hz

127. Method of speed control used on 25 kV, 50 Hz single-phase traction is
A. tap changing control of transformer
B. reduced current method
C. series parallel operation of motors
D. any of these

128. When the speed of the train is estimated taking into account the time of stop at a station in addition to the actual running time between stops, is known as
A. average speed B. schedule speed
C. notching speed D. free running speed

129. A schedule speed of 45 km, per hour is required between two stops 1.5 km. apart. The duration of stop is 20 seconds. The acceleration is 2.4 km/hr/s and retardation is 3.2 km/hr/s. For a simplified trapezoidal curve, the maximum speed over the run will be
A. 40 km per hour B. 48 km per hour
C. 74 km per hour D. 90 km per hour

130. The coefficient of adhesion is highest when the rails are
A. dry B. oiled
C. wet with dew D. dusty

131. Specific energy consumption is least in
A. urban service B. suburban service
C. main-line service
D. equal for all types of service

132. If the coefficient of adhesion on dry rails is 0.25, which of the following could be the value for wet rails?
A. 0.35 B. 0.25
C. 0.245 D. 0.15

133. A locomotive exerts a tractive effort of 30,000 newtons in pulling a train at 50 km per hour on the level track. It is to haul the same train at the same speed on a gradient and the tractive effort required is 45000 newtons. The horsepower delivered by the motor will be more if it is driven by
A. DC series motor B. induction motor
C. same in both cases D. none of these

134. A composite system consists of
A. a combination of diesel engine and DC series motor
B. a combination of diesel engine and AC single-phase motor
C. single-phase power received is converted into DC or three-phase power AC system
D. use of combination of DC and AC motors on the same locomotive

135. A train has a schedule speed of 36 km per hour on a level track. If the distance between the stations is 2 km and the stoppage is 30 seconds, the actual time of run will be
A. 260 sec B. 210 sec
C. 200 sec D. 170 sec

136. Bearings used to support axles of rolling stock are
A. bush bearings B. journal bearings
C. ball bearings D. roller bearing

137. B_0B_0 locomotives have two bogies with
A. four driving axles each with individual driving motors
B. three driving axles with group drives
C. two driving axles with group drives
D. two driving axles with individual drive motor

138. Which city in India was first to adopt electric traction?
A. Bombay B. Madras
C. Calcutta D. Bangalore

139. Which one of the following is correct?
A. Rate of acceleration is the same as the rate of retardation during braking
B. Average acceleration is zero
C. Time taken during coasting is equal to the time during acceleration and braking
D. During coasting the acceleration is negative

140. Locomotives with manometer bogie have
A. lot of skidding
B. low coefficient of adhesion
C. uneaten distribution of tractive
D. suitability for passenger as well as freight service

141. Which of the following route has electric traction?
A. Delhi-Madras B. Delhi-Ahmedabad
C. Delhi-Howrah D. Delhi-Bangalore

142. For 600 V DC line for tram cars track is connected to
A. negative of the supply
B. positive of the supply
C. mid voltage of 300 V
D. none of these

Ans. 123. C 124. D 125. C 126. A 127. A 128. B 129. C 130. A 131. C 132. D 133. B 134. C 135. D 136. D 137. D 138. A 139. D 140. D 141. C 142. A

143. The speed time curve for a local train is shown in the figure below. In this time, OA represents:

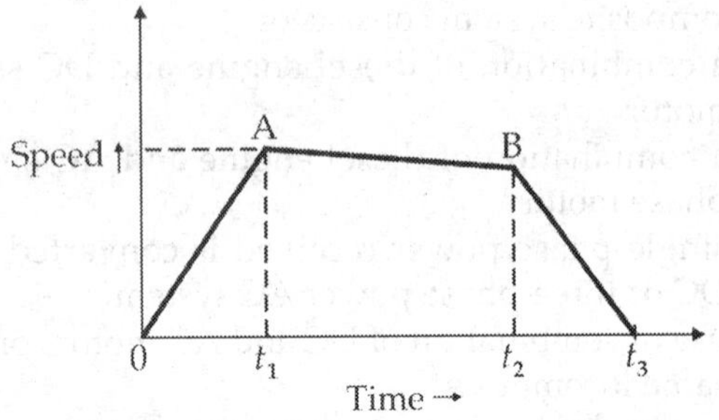

A. coasting B. acceleration
C. braking D. regeneration.

144. The duration for braking is represented by the time
A. $0-t_1$ B. $0-t_2$
C. t_1-t_2 D. t_3-t_2

145. Area under the curve represents
A. average speed B. average acceleration
C. net acceleration D. distance travelled

146. Which of the following frequencies is not common in low frequency electric traction system?
A. 50/3 Hz B. 25 Hz
C. 40 Hz D. 60 Hz

147. In Kando system
A. single-phase supply is converted into three-phase supply system
B. single-phase AC is converted into DC supply
C. three-phase AC is converted into DC supply
D. DC supply is to run DC motors

148. In a long distance electric train, power for lighting in passenger coach is provided
A. through locomotive
B. directly through overhead electric line.
C. through individual generator of bogie and batteries
D. through power supply

149. Free running and coasting periods are generally long in case of
A. urban service B. suburban service
C. main-line service D. all of these

150. A train runs at an average speed of 50 kmph between stations situated 2.5 km apart. The train accelerates at 2 kmphs and retards at 3 kmphs. Speed-time curve may be assumed to be trapezoidal. The maximum speed is
A. 27.75 kmph B. 40.50 kmph
C. 44.25 kmph D. 57.75 kmph

151. Which of the following factors affects specific energy consumption?
A. Distance between stops
B. Retardation and acceleration values
C. Gradient D. All of these

152. Armature short circuits can be detected and identified by which of the following tests?
A. Bar-to-bar B. Voltage drop
C. Growler D. All of these

153. When locomotive for Indian Railways is designated as WAM1, in this the letter W indicates that the locomotive is
A. to run on broad gauge track
B. to run on metre gauge track
C. for shunting duty
D. for goods trains only

154. An ideal traction system should have
A. high starting tractive effort
B. the locomotive is to run on metre gauge track
C. the locomotive is for shunting duty
D. the locomotive is for goods trains only

155. A train runs at an average speed of 45 kmph between stations 2.5 km apart. The train accelerates at 2 kmph and retards at 3 kmph speed-time curve may be assumed to be trapezoidal. The maxi-mum speed attained will be nearly
A. 80 kmph B. 60 kmph
C. 50 kmph D. 35 kmph

156. At an average, the coal consumption per km in case of steam engine is nearly
A. 28 to 30 kg B. 40 to 60 kg
C. 150 to 160 kg D. 200 to 250 kg

157. In the above question, the distance travelled before the brakes are applied is
A. 0.75 km B. 1.35 km
C. 2.0 km D. 2.35 km

158. The draw-bar horsepower per tonne in case of steam locomotives is
A. 7 to 8 B. 15 to 17
C. 20 to 25 D. 30 to 35

159. The distance between two stations is 1 km. and the schedule speed is 30 km per/hour. Station stopping time 20 sec. The average speed is
A. 26 km per hr B. 36 km per hr
C. 50 km per hr D. 54 km per hr

160. The distance between two stations is 1 km and the average and schedule speeds of the train are 36 kmph and 30 kmph respectively. The stations stopping time is
A. 20 sec B. 30 sec
C. 40 sec D. None of these

161. For a given run and a given schedule speed
A. the specific energy consumption is higher, the lower the acceleration and retardation
B. the specific energy consumption is higher, the higher the acceleration and lower retardation
C. the specific energy consumption is lower, the higher the acceleration and higher the deceleration
D. none of these

Ans.	143. A	144. D	145. D	146. C	147. A	148. C	149. C	150. D	151. D	152. D	153. A	154. D	155. C
	156. A	157. D	158. B	159. B	160. A	161. A							

162. The electric locomotives run faster at curved routes as compared to steam locomotives as
A. the centre of gravity of electric locomotive is higher than that of steam locomotive
B. the centre of gravity of electric locomotive is lower than that of steam locomotive
C. the speed at curved routes is independent of location of centre of gravity
D. none of these

163. A train is required to run between two stations 16 km apart at an average speed of 43 kmph. The run is to be made to a simplified quadrilateral speed-time curve. The maximum speed is to be limited to 64 kmph, acceleration to 2 kmphs and coasting and braking retardations to 0.16 and 3.2 kmphs respectively. The duration of acceleration is
A. 32 sec B. 20 sec
C. 15 sec D. 10 sec

164. If the resistance to electric train is given by: $F_r = a + bv + cv^2$. In this equation, constant c is likely to cover
A. air resistance B. frictional resistance
C. flange resistance D. track resistance

165. An electric train is to have braking retardation 3.2 kmphs. The ratio of maximum speed to average speed is 1:3, the time for stop is 26 seconds and acceleration is 0.8 kmphs. The run is 1.5 km. Actual time of run is
A. 77 sec B. 101 sec
C. 154 sec D. 200 sec

166. The main difference between speed-time curves of main line service as compared to suburban services lies in
A. longer free running periods
B. longer coasting periods
C. shorter accelerating and braking periods
D. all of these

167. Quadrilateral speed-time curve is the closer approximation for
A. main line service B. suburban service
C. urban service
D. urban and suburban services

168. Energy consumption in propelling the train is required for
A. acceleration
B. work against gravity while moving up the gradient
C. work against resistance to motion
D. all of these

169. As compared to suburban service, in main line service
A. distance between stops is more
B. higher maximum speeds can be achieved
C. acceleration and retardation rates are low
D. all of these

170. In case of bow and pantograph current collectors,
A. the bow collector is used for large current at higher speeds
B. the pantograph collector is used for large currents at higher speeds
C. the bow collector is used for large currents at lower speeds
D. the pantograph collector is used for large currents at lower speeds

171. Contact with wire is maintained
A. by springs in both bow and pantograph collectors
B. by air pressure in both bow and pantograph collectors
C. by spring in case of bow and air pressure in case of pantograph
D. the spring in case of pantograph and air pressure in case of bow collectors

172. The coefficient of adhesion is
A. same on AC and DC traction
B. high in DC and low in AC
C. high in AC and low in DC
D. none of these

173. Tick out the correct statement:
A. Smaller size have their critical speeds above their synchronous speeds
B. Smaller size machines have their critical speeds below their synchronous speeds
C. Critical speed is always greater than synchronous speed irrespective of size of machine
D. Critical speed is always less than synchronous speed irrespective of the size of the machine

174. Quadrilateral speed time curve is used for
A. suburban service B. urban service
C. main line service
D. Both A and B are correct

175. The train is moving a uniform upgrading of 1 in 100. The train resistance is 40 N/tonne, rotational inertia effect 10% of dead weight and duration of stop 30 seconds. Braking retardation is
A. 0.87 kmphs B. 1.27 kmphs
C. 1.87 kmphs D. 2 kmphs

176. Low frequency operation of overhead line in traction system
A. reduces the spacing between substations
B. increases the spacing between substations
C. spacing is independent of frequency of supply
D. All of these

177. Coefficient of adhesion is higher if
A. the speed is low and the rails are sandy and dry

Ans. 162. B 163. A 164. A 165. C 166. D 167. D 168. D 169. D 170. D 171. A 172. B 173. A 174. D 175. C 176. B

B. the speed is high and the rails are sandy and dry
C. the speed is high and the rails are wetty
D. the rails are wet and greasy

178. In suburban services as compared with urban services
A. the coasting period is longer
B. the coasting period is smaller but free running period is longer
C. the coasting and free running periods are smaller
D. none of these

179. The specific energy consumption
A. increases with increase in maximum speed
B. decreases with increase in maximum speed
C. is independent of maximum speed
D. none of these

180. Run away speed of a Pelton wheel is
A. full load speed
B. actual speed operating at no load
C. no load speed when governor mechanism fails
D. 80% greater than the normal speed

181. If the speed-time curves are similar (not identical), the specific energy consumption of the curve having higher maximum speed is
A. higher than that of lower maximum speed
B. lower than that of lower maximum speed
C. equal to that of lower maximum speed
D. none of these

182. Two shunt motors having identical characteristics are used to drive a car with unequal wheel diameters. The two motors share load equally if they are connected in
A. series B. parallel
C. the loading will always be unequal
D. none of these

183. Two locomotives having identical tractive-effort-siip characteristic haul a heavy train. The loco with larger diameter shares
A. larger tractive effort B. smaller tractive effort
C. equal tractive effort D. none of these

184. For braking of electric trains, normally regenerative braking is used
A. to bring the speed down to zero
B. down to a speed of 4 kmph followed by mechanical brakes
C. down to a speed of 16 kmph and finally mechanical brakes
D. down to a speed of 16 kmph followed by rheostatic braking till zero speed

185. For rheostatic braking of two series motors connected in parallel
A. equaliser connection is better
B. cross-connection is better
C. both are equally good
D. none of these

186. At low frequency of the order of 1/2 Hz to 10 Hz, the induction motors develop
A. high starting torque with excessive starting current
B. high starting torque without excessive starting current
C. low starting torque with excessive starting current
D. low starting torque without excessive starting current

187. Specific energy consumption is minimum in
A. main-line service B. suburban service
C. urban service D. equal for all types

188. For regenerative braking, the regenerated power should have
A. the same frequency as that of the main supply
B. frequency 1/3 that of the main supply
C. frequency 1/2 that of the main supply
D. any frequency

189. Acceleration rate of trains on suburban services is
A. 0.2 to 0.4 kmphs B. 0.8 to 2 kmphs
C. 0.4 to 6.5 kmphs D. None of these

190. A train having inertial weight of 220 tonnes uses regenerative braking when the speed is changed from 60 kmph to 20 kmph, the electric energy returned to supply is (assume 100% efficiency)
A. 7.546 kWhr B. 8.32 kWhr
C. 5.6595 kWhr D. none of these

191. The coefficient of adhesion is highest when the rails are
A. dusty B. wet
C. dry D. oiled

192. Which type of motor is recommended for locomotive drive?
A. DC series motor
B. DC shunt motor
C. DC compound motor
D. Synchronous motor

193. For the same series armature resistance, the speed torque characteristic of a shunted armature DC shunt motor as compared to when it is not shunted, has a slope
A. higher B. smaller
C. equal D. none of these

194. With V/f speed control of 3-phase induction motor, the maximum torque and slip at which maximum torque occurs respectively
A. increases and decreases
B. increases and increases

Ans. 177. B 178. A 179. A 180. C 181. C 182. A 183. A 184. C 185. B 186. B 187. A 188. A 189. C 190. A 191. C 192. A 193. B 194. A

C. decreases and increases
D. decreases and decreases

195. In a steam locomotive, electric power is provided through
A. battery system
B. diesel engine generator
C. overhead wire
D. small turbo generator

196. The maximum number of passenger coaches that can be attached to diesel locomotives on broad gauge is
A. 18 to 20
B. 25 to 30
C. 32 to 35
D. 36 to 40

197. A submarine while moving under water is provided driving power through
A. diesel engines
B. steam turbine
C. gas turbine
D. batteries

198. Overload capacity of diesel engines is usually restricted to
A. 1%
B. 10%
C. 25%
D. 50%

199. Maximum horsepower of steam locomotive is
A. 100 hp
B. 500 hp
C. 1500 hp
D. 2500 hp

200. The pressure of steam in a locomotive is
A. 10–15 kg/cm^2
B. 20–30 kg/cm^2
C. 40–50 kg/cm^2
D. 80–90 kg/cm^2

201. The efficiency of diesel locomotives is nearly
A. 20–25%
B. 35–40%
C. 50–55%
D. 70–75%

202. A 3-phase delta connected squirrel cage induction motor when started with DOL stater has a starting torque of 600 NM. Its starting torque when star-delta is used, is
A. 600 NM
B. 200 NM
C. 300 NM
D. none of these

203. In case of 3-phase slip-ring induction motor, as the rotor resistance is increased, the starting torque
A. increases
B. decreases
C. increases up to certain value of resistance and remains constant
D. increases up to certain value of resistance and then decreases

204. In case of voltage injection method of speed control, the injected emf should be of
A. supply frequency
B. slip frequency
C. $(1 - s)f$
D. $(2 - s)f$

205. If in the above question, auto transformer with 70.7% tap setting is used for start-ing, the starting torque is
A. 600 NM
B. 200 NM
C. 300 NM
D. none of these

206. During starting, the energy consumed by the armature of a DC series motor
A. equals the KE stored by the rotor and increases with increase in armature resistance
B. equals the KE stored by the rotor and decreases with increase in armature resistance
C. equals the KE stored by the rotor and is independent of armature resistance
D. None of these

207. During starting, the energy consumed by the armature of a DC shunt motor
A. equals the KE stored by the rotor and increases with increase in armature resistance
B. equals the KE stored by the rotor and decreases with increase in armature resistance
C. equals the KE stored by the rotor and is independent of armature resistance
D. none of these

208. The consideration involved in the selection of the type of electric drive for a particular application depends on
A. speed control range and its nature
B. starting torque
C. environmental conditions
D. all of these

209. Which of the following is preferred for automatic drives?
A. Synchronous motors
B. Squirrel cage induction motor
C. Ward-Leonard Controlled DC motors
D. Any of these

210. Which type of drive can be used for hoisting machinery?
A. Slip ring induction motor
B. Ward Leonard Controlled DC shunt motor
C. DC compound motor
D. Any of these

211. The motor normaly used for crane is
A. Slip ring induction motor
B. Ward Leonard Controlled DC shunt motor
C. Synchronous motor
D. DC differentially compound motor

212. A wound rotor induction motor is preferred over squirrel cage induction motor when the major consideration involved is
A. high starting torque
B. low starting current
C. speed control over limited range
D. all of these

213. When smooth and precise speed control over a wide range is desired, the motor preferred is
A. wound rotor induction motor
B. squirrel cage induction motor
C. synchronous motor
D. DC motor

214. When quick speed reversal is a consideration, the motor preferred is

Ans. 195. D 196. B 197. D 198. B 199. C 200. A 201. A 202. B 203. D 204. B 205. C 206. A 207. C 208. D 209. C 210. D 211. A 212. D 213. D

A. synchronous motor
B. squirrel cage induction motor
C. wound rotor induction motor
D. DC motor

215. DC supply can be obtained from AC supply by the use of
A. motor generator set B. mercury arc rectifier
C. silicon diodes D. Any of these

216. The selection of control gear for a particular application is based on the consideration of
A. duty B. starting torque
C. limitations on starting current
D. All of these

217. As compared to squirrel cage induction motor, a wound rotor induction motor is preferred when the major consideration is
A. high starting torque B. low windage losses
C. slow speed operation D. All of these

218. A synchronous motor is found to be most economical when the load is above
A. 1 kW B. 10 kW
C. 20 kW D. 100 kW

219. The advantage of a synchronous motor in addition to its constant speed is
A. high power factor B. better efficiency
C. lower cost D. All of these

220. In motor circuit static frequency changers are used for
A. improved cooling
B. power factor improvement
C. reversal of direction D. speed regulation

221. In case of travelling cranes, the motor preferred for boom hoist is
A. slip ring induction motor
B. squirrel cage induction motor
C. synchronous motor D. single phase motor

222. The characteristics of drive for crane hoisting is
A. smooth movement B. precise control
C. fast speed control D. All of these

223. The capacity of a crane is expressed in terms of
A. Span B. Type of drive
C. Tonnes D. Any of these

224. The travelling speed of cranes varies from
A. 1 to 2.5 m/s B. 5 to 10 m/s
C. 10 to 20 m/s D. 20 to 40 m/s

225. In overhead travelling cranes
A. continuous duty motors are preferred
B. slow speed motors are preferred
C. short time rated motors are preferred
D. None of these

226. 15 minute rated motors are suitable for
A. light duty cranes B. medium duty cranes
C. heavy duty cranes D. all of these

227. Light duty cranes are generally used in
A. automobile workshops
B. pumping stations
C. power houses D. all of these

228. Heavy duty crenes are used in
A. Heavy engineering workshops
B. Steel plants
C. Ore handling plants D. all of these

229. 1/2-hour rated motors are used for
A. light duty cranes B. medium duty cranes
C. heavy duty cranes D. none of these

230. Which of the following drive can be used for derricks and winches?
A. Pole changing squirrel cage motors
B. Slip-ring induction motor with variable resistance
C. DC motors D. Any of these

231. The number of the sets used in pole changing type squirrel cage motors for derricks and winches, is
A. 2 B. 3
C. 4. D. 6

232. A pole changing type squirrel cage motor used in derricks has four, eight and twenty four poles. In this, the medium speed is used for
A. lifting B. hoisting
C. lowering D. landing the road

233. A pole changing type squirrel cage induction motor used in derricks has four, eight and twenty four poles. In this the lowest speed is used for
A. lifting B. hoisting
C. lowering D. landing the road

234. For handing fragile articles in a crane
A. low speed is preferred
B. medium speed is preferred
C. high speed is preferred
D. None of these

235. The range of horsepower of electric motor drives for rolling mills is of the order of
A. 1 to 10 HP B. 15 to 25 HP
C. 50 to 100 HP D. 100 to 500 HP

236. Motors preferred for rolling mill drive is
A. DC motors
B. AC slip ring motors with speed control
C. any of the above D. none of the above

237. Which motors, because of their inherent characteristics, are best suited for the roling mills?
A. DC motors
B. Slip ring induction motors
C. Squirrel cage induction motors
D. Single-phase motors

Ans.	214. D	215. D	216. D	217. A	218. D	219. A	220. D	221. A	222. D	223. C	224. A	225. C	226. A
	227. D	228. D	229. B	230. D	231. B	232. A	233. D	234. A	235. D	236. D	237. A		

238. In case of kiln drives
A. starting torque is almost zero
B. starting torque and running torque are nearly equal
C. starting torque is double of the running torque
D. Both torques are zero

239. Motor preferred for kiln drives is usually
A. slip-ring induction motor
B. three phase shunt wound commutator motor
C. cascade controlled ACmotor
D. Any of the above

240. Belt conveyors offer
A. zero starting torque B. low starting torque
C. medium starting torque
D. high starting torque

241. In case belt conveyors
A. squirrel cage motors with direct-on-line starters are used
B. single-phase induction motors are used
C. DC shunt motors are used
D. induction motors with star-delta starters are used

242. Which of the following motor is preferred for blowers?
A. Wound rotor induction motor
B. DC shunt motor
C. Squirrel cage induction motor
D. DC series motor

243. Centrifugal pumps are usually driven by
A.DC shunt motors B. DC series motors
C. squirrel cage induction motors
D. any of these

244. In case of centrifugal pumps the starting torque is generally
A. double the running torque
B. slightly more than running torque
C. same as running torque
D. less than running torque

245. In a centrifugal pump if the liquid to be pumped has density twice that of water, then the horsepower required (as compared to that while pumping water) will be
A. half B. same
C. double D. three time

246. Wound rotor and squirrel-cage motors with high slip which develop maximum torque at stand still are used for
A. machine tools B. presses and punches
C. elevators D. All of these

247. Belted slip ring induction motor is almost invariably used for
A. centrifugal blowers B. water pumps
C. jaw crushers D. heavy load

248. In jaw crushers, generally a motor has to start against
A. low load B. medium load
C. normal load D. heavy load

249. Motor used for elevators is generally
A. synchronous motor B. induction motor
C. capacitor start single-phase induction motor
D. Any of these.

250. In synthetic fibre mills, motor with
A. constant speed is preferred
B. high starting torque are preferred
C. variable speed are preferred
D. low starting torque are preferred

251. Which of the following motor is preferred for synthetic fibre mills?
A. DC Series motor B. Reluctance motor
C. DC Shunt motor D. Synchronous motor

252. Reluctance motor is a
A. self-starting type synchronous motors
B. low torque variable speed motor
C. variable torque motor
D. low noise, slow speed motor.

253. A reluctance motor
A. is compact B. has high cost
C. requires starting gear
D. is provided with slip rings

254. Power factor in case of reluctance motor is
A. nearly unity B. always leading
C. 0.8 lagging D. 0.3 to 0.4 leading

255. The efficiency of reluctance motor is around
A. 95% B. 80%
C. 75 to 85% D. 60 to 75%

256. A reluctance motor on over load runs as
A. synchronous motor B. induction motor
C. both (A) and (B) are correct
D. none of these

257. The size of an excavator is usually expressed in terms of
A. cubic metres B. angle of swing
C. travel in metres D. crowd motion

258. Ward-Leonard controlled DC drives are generally used for
A. heavy duty excavators
B. medium duty excavators
C. light duty excavators D. All of these

259. In case of contactors, the contacts are generally made of
A. copper B. silver
C. cadmium copper D. Any of these

260. Which electromagnet is preferred for noiseless operation?
A. DC operated B. AC operated
C. Any of these D. None of these

Ans. 238. C 239. D 240. D 241. A 242. C 243. C 244. D 245. C 246. B 247. C 248. D 249. B 250. A
251. B 252. A 253. A 254. D 255. D 256. B 257. A 258. A 259. D 260. A

261. A contractor having 6-operations per hour is classed as
A. class O B. class I
C. class II D. class III
262. The number of operations per hour in case of class IV contactor will be around
A. 100 B. 600
C. 900 D. 1200
263. In case of contactors, the duty in which the main contacts remain closed for a period bearing a definate relation to the no-load periods, is known as
A. standard duty B. intermittent duty
C. temporary duty D. uninterrupted duty
264. In case of contactors the ratio of the in-service period to the entire period, expressed as a percentage is known as
A. duty B. load factor
C. class of contact D. None of these
265. A class I contactor should be mechanically sound to withstand
A. 0.05 million times B. 0.25 million times
C. 1.0 million times D. 2.0 million times
266. Heat control switches find applications in
A. three-phase induction motors
B. single-phase motors
C. transformers D. cooling ranges
267. A saturable core reactor is basically a
A. variable resistor
B. step-down transformer
C. thermal relay D. variable impedance
268. A saturable core reactor can be used for
A. stepless ACvoltage variation
B. plugging of induction motor
C. overload protection of transformers
D. All of these
269. In case of suitable core reactors, the power gain varies from
A. 1 to 5 B. 5 to 10
C. 5 to 100 D. 100 to 1000
270. A magnetic amplifier can be used for the control of
A. current B. voltage
C. speed D. All of these
271. A magnetic amplifier has generally
A. single winding B. two windings
C. three windings D. number of windings
272. In case of contactors, the contact chattering may be due to
A. excessive jogging B. broken pole shader
C. poor contact in the control pick up circuit
D. Any of these
273. In a contactor, overheating of contacts may result from any of the following except
A. excess contact pressure
B. high inductive loads
C. copper oxide on contacts
D. carrying load continuously for a longer time
274. In case of contactors, the magnet may become noisy due to
A. dirt or rust on magnet faces
B. low voltage
C. broken pole shader D. Any of these
275. The failure of a thermal relay may occur due to
A. motor and relay in different ambient temperatures
B. relay previously damaged by short circuit
C. mechanical binding D. Any of these
276. Premature blowing of a fuse may occur due to
A. heating at ferrule contacts
B. corrosion or oxidation of ferrules
C. weak contact pressure
D. Any of these
277. According to Indian Electricity rules, extra high voltage implies voltage exceeding
A. 440 V B. 66 kV
C. 33 kV D. 110 kV
278. In case of low and medium voltage circuits, the permissible voltage variation is
A. 1% B. 5%
C. 10% D. 15%
279. Which of the following site will be preferred for earthing?
A. wet mashy ground B. clay soil
C. loam mixed with small quantities of sand
D. damp and wet sand pit
280. Resistivity of earth increases sharply if the moisture falls below
A. 60% B. 50%
C. 40% D. 20%
281. Which of the following is least preferred for earthing?
A. earth mixed with salt and charcoal
B. dry earth
C. mashy ground containing brine waste
D. clay soil
282. Earth electrodes can be in the form of
A. rods and pipes B. strips
C. plates D. Any of these
283. Which of the following is not used as earth continuity conductor?
A. water pipes B. gas pipes
C. structural steel members
D. Any of these
284. The diameter of a 50-SWG wire is closer to
A. 0.0253 mm B. 0.05 mm
C. 0.16 mm D. 0.52 mm.
285. Non metallic conduits for wiring are generally made of

Ans. 261. A 262. D 263. B 264. B 265. B 266. D 267. D 268. A 269. C 270. A 271. D 272. D 273. A 274. D 275. A 276. D 277. A 278. B 279. A 280. D 281. B 282. D 283. D 284. A

A. rubber B. cork
C. wood D. PVC

286. PVC conduits can be buried on
A. lime B. plaster
C. concrete D. Any of these

287. PVC conduits can be joined by
A. solvent cement B. welding
C. threading D. Any of these

288. The maximum horsepower upto which 440 V electric motors are used, is
A. 200 HP B. 50 HP
C. 20 HP D. 10 HP

289. The distance at which earthing electrode should be situated away from the building whose installation system is being earthed is
A. 1.5 m B. 3.5 m
C. 25 m D. 100 m

290. Which of the following mixtures is preferred for filling around the earth for electrode effective earthing?
A. bone-meat mixture B. coal-salt mixture
C. saw-dust sand mixture
D. lime-sand mixture

291. Inside the earth or pit, the earthing electrode should be placed
A. vertical B. horizontal
C. inclined at 45°
D. inclined at any angle other than 45°

292. Danger 440 V plates are
A. danger notices B. caution notices
C. advisory notices D. informal notices.

293. The minimum clearance of any overhead line from the ground should be
A. 20 m B. 10 m
C. 6 m D. 4 m

294. Earthing of electric appliances is done
A. for the safety of human life
B. to reduce line voltage fluctuation
C. for protection of electric equipment
D. All of these

295. Earthing is used as the return conductor for
A. telephone lines B. telegraph lines
C. traction work D. All of these

296. The resistance of earth wire should be
A. infinite B. high
C. reasonable D. very low

297. The earth's potential is taken as
A. infinite B. supply voltage
C. 1 volt D. zero

298. The earthwire should not be thinner than a
A. 18 SWC wire B. 16 SWG wire
C. 10 SWC wirc D. 8 SWG wire

299. In automobiles, the sound produced by horn is due to
A. magnetostriction B. vibrating diaphragm
C. moving coil D. oscillating coil

300. The current drawn by a 6-V horn is roughly
A. 0.1 A B. 1 A
C. 2 A D. 20 A

301. The horns are rated for
A. continous operation
B. intermittent operation
C. Both A and B are correct
D. None of these

302. Continuous operation of automobile horn will
A. help in charging the battery
B. improve mileage
C. damage the operating coil
D. change the tone

303. Miniature lamps on automobiles are used for
A. tail lamp B. dash board lamp
C. side lamp D. All of these

304. Which of the following pair is used for frequency converters?
A. Squirrel cage induction motor and synchronous motor
B. Wound-rotor induction motor and synchronous motor
C. Wound rotor induction motor and squirrel cage induction motor
D. Any of these.

305. Belted wound rotor induction motors are preferred for
A. machine tools B. gyratory crushers
C. belt conveyor D. water pumps

306. Electric braking is preferred because it is
A. smooth
B. maintenence cost is less
C. energy is saved in regenerating braking
D. all of these

307. The various types of electric braking are
A. plugging B. rheostatic braking
C. regenerative braking D. All of these

308. For low and medium power drives, where speed control is not required, which of the following motor you will recommend?
A. Synchronous motor
B. Squirrel cage induction motor
C. Double cage induction motor
D. Three-phase series motor

309. Which type of motor is recommended for locomotive drive?
A. DC series motor B. DC shunt motor
C. DC compound motor
D. Synchronous motor

Ans.													
	285. D	286. D	287. D	288. A	289. A	290. B	291. A	292. B	293. C	294. D	295. C	296. D	297. D
	298. D	299. B	300. D	301. B	302. C	303. D	304. A	305. B	306. D	307. D	308. A	309. A	

310. The shunt motor with armature shunted through a resistance as compared to with aramature not shunted, has no load speed
A. higher B. lower
C. equal D. none of these

311. For the same series armature resistance, the speed torque characteristic of a shunted armature DC shunt motor as compared to when it is not shunted, has a slope
A. higher B. smaller
C. equal D. none of these

312. In case of DC series motor it is possible to have finite no load speed if a resistance is connected across its
A. field terminals B. armature terminals
C. field and armature together
D. it is always very high at no load

313. With V/f speed control of three-phase induction motor, the maximum torque and slip at which maximum torque occurs respectively
A. increases and decreases
B. increases and increases
C. decreases and increases
D. decreases and decreases

314. In case of three-phase slip ring induction motor, as the rotor resistance is increased, the starting torque
A. increases B. decreases
C. increases upto certain value of resistance and remains constant
D. increases upto certain value of resistance and then decreases

315. In case of voltage injection method of speed control, the injected emf should be of
A. supply frequency (f) B. slip frequency (sf)
C. $(1-s)f$ D. $(2-s)f$

316. A 3-phase delta connected squirrel cage induction motor when started with DOL stater has a starting torque of 600 NM. Its starting torque when star-delta is used, is
A. 600 NM B. 200 NM
C. 300 NM D. none of these

317. If in the above question, auto transformer with 70.7% tap setting is used for starting, the starting torque is
A. 600 NM B. 200 NM
C. 300 NM D. none of these

318. During starting, the energy consumed by the armature of a DC shunt motor
A. equals the K.E. stored by the rotor and increases with increase in armature resistance
B. equals the KE stored by the rotor and decreases with increase in armature resistance
C. equals the KE stored by the rotor and is independent of armature resistance
D. none of these

319. During starting, the energy consumed by the armature of a DC series motor
A. equals the KE stored by the rotor and increases with increase in armature resistance
B. equals the KE stored by the rotor and decreases with increase in armature resistance
C. equals the KE stored by the rotor and is independent of armature resistance
D. none of these

320. For a particular type of motor, the heating time constant
A. increases with increase in size
B. decreases with increase in size
C. same for all size
D. none of these

321. The heating time constant of a totally enclosed motor is relatively
A. higher B. lower
C. independent of type of enclosure.
D. none of these

322. For a particular motor, the cooling time-constant is usually
A. smaller than the heating time constant
B. greater than the heating time constant
C. equal to the heating time constant
D. none of these

323. Electric braking is preferred because it is
A. smooth
B. maintenance cost is less
C. energy is saved in regenerating braking
D. all of these

324. The various types of electric braking are
A. plugging B. rheostatic braking
C. regenerative braking D. all of these

325. In case of DC series motor it is possible to have finite no-load speed if a resistance is connected across its
A. field terminals B. armature terminals
C. field and armature together
D. it is always very high at no load

326. Which of the following pair is used for frequency converters?
A. Squirrel cage induction motor and synchronous motor
B. Wound-rotor induction motor and synchronous motor
C. Wound rotor induction motor and squirrel cage induction, motor
D. Any of these

327. The basic electrical requirement in arc welding is that there should be
A. coated electrodes
B. high open-circuit voltage
C. no arc blow D. DC power supply

Ans. 310. B 311. B 312. B 313. A 314. D 315. B 316. B 317. C 318. C 319. A 320. A 321. A 322. B 323. D 324. D 325. B 326. A 327. B

328. Welding is not done directly from the supply mains because
A. it is customary to use welding machines
B. its voltages is too high
C. its voltage keeps fluctuating
D. it is impracticable to draw heavy currents

329. AC welding machine cannot be used for ..... welding.
A. MIG B. atomic hydrogen
C. resistance D. submerged arc

330. In electric welding, arc blow can be avoided by
A. using bare electrodes
B. welding away from earth ground connection
C. using AC welding machines
D. increasing arc length

331. In DCSP welding
A. electrode is the hottest
B. workpiece is relatively cooler
C. base metal penetration is deep
D. heavily-coated electrodes are used

332. Overhead welding position is assumed to be the most
A. hazardous B. difficult
C. economical D. useful

333. The ultimate aim of using electrode coating is to
A. provide shielding to weld pool
B. prevent atmospheric contamination
C. improve bead quality
D. cleanse the base metal

334. In electrode-positive welding ..... of the total heat is produced at the electrode.
A. one-third B. two-third
C. one-half D. one-fourth

335. Submerged arc process is characterised by
A. deep penetration B. high welding current
C. exceptionally smooth beads
D. all of the above

336. The major disadvantage of carbon arc welding is that
A. there is occurrence of blow holes
B. electrodes are consumed fast
C. separate filler rod is needed
D. bare electrodes are necessary

337. In atomic hydrogen welding, electrodes are long-lived because
A. two are used at a time
B. electrode arc in the shape of a fan
C. of ac supply
D. it is a nonpressure process

338. Unlike TIG welding, MIG welding
A. requires no flux
B. uses consumable electrodes
C. provides complete protection from atmospheric contamination
D. requires no post-weld cleansing

339. The major power supply used in MIG welding is
A. AC supply B. DCSP
C. electrode-negative D. DCRP

340. MIG welding process is becoming increasingly popular in welding industry mainly because of
A. its easy operation
B. its high metal deposit rate
C. its use in both ferrous and non-ferrous metals
D. both A and B are correct

341. A weld bead of wineglass design is produced in ..... welding.
A. plasma arc B. electron beam
C. laser D. MAG

342. Spot welding process basically depends on
A. ohmic resistance B. generation of heat
C. application of forging pressure
D. both B and C are correct

343. Electric resistance seam welding uses ..... electrodes.
A. pointed B. disc
C. flat D. domed

344. Projection welding can be regarded as a mass production form of ..... welding.
A. seam B. butt
C. spot D. upset

345. In the process of electroslag welding, theoretically there is no upper limit to the
A. thickness of weld bead
B. rate of metal deposit
C. slag bath temperature
D. rate of slag consumption

346. High temperature metals like Columbium can be easily welded by ..... welding.
A. flash B. MIG
C. TIG D. electron beam

347. Motors used in wood-working industry have ..... enclosure.
A. screen protected B. drip proof
C. TFFC D. TE

348. Major disadvantage of using three sets of SCRs for variable-voltage speed control of a squirrel cage induction motors is the
A. considerable I^2R loss
B. poor power factor
C. long delay of thyristor firing pulses
D. necessity of using a processor

349. While plugging DC motors, ... connections are reversed
A. supply B. armature
C. field
D. both armature and field

350. In arc welding, the temperature of the arc is of the order of
A. 100° C B. 1000° C
C. 3500°C D. 35000°C

Ans. 328. D 329. A 330. C 331. C 332. A 333. C 334. B 335. D 336. A 337. C 338. B 339. D 340. D 341. A 342. D 343. B 344. C 345. A 346. D 347. C 348. A 349. B 350. C

351. The arc has
A. linear resistance characteristics
B. positive resistance characteristics
C. negative resistance characteristics
D. highly inductive characteristics
352. Arc can be produced by
A. AC current only B. DC current only
C. Either AC or DC current
D. None of these.
353. The resistance of arc
A. decreases with increase of the current
B. increases with increase of the current
C. does not depends on current
D. None of these
354. In arc welding, the voltage on ac supply system is in the range
A. 1000–1220 V B. 400–500 V
C. 200–250 V D. 70–100 V
355. In arc welding by DC supply, the voltage required is
A. 10 to 20 V B. 50 to 60 V
C. 100 to 120 V D. 200 to 250 V
356. In arc welding, once the arc is struck, the voltage required to maintain it will be
A. 20–30 V B. 100–120 V
C. 200–220 V D. 500–1000 V
357. A DC generator used for ACwelding should have
A. rising characteristics
B. drooping characteristics
C. straight characteristics
D. None of these
358. The transformer used for ac, welding sets is
A.booster type B. step-up transformer
C. step-down transformer
D. none of these
359. As the thickness of the part to be welded increases, which of the following parameter (s) for AC, welding should also increase?
A. Voltage B. Current
C. Frequency D. All of these
360. In Argon arc welding, the electrode is made of
A. carbon B. steel
C. tungsten D. no electrode is needed
361. In argon arc welding the purpose of using argon is
A. to prevent oxidation of metal by coming in contact with oxygen of air
B. to creat inert atmosphere around the job to be welded
C. to obviate the necessity for using flux
D. all of these
362. Steel rails are welded by
A. resistance welding B. thermit welding
C. argon arc welding D. gas welding
363. In gas welding, the gases used are
A. oxygen and nitrogen
B. argon and helium
C. helium and carbondioxide
D. acetylene and oxygen
364. Steel pipes are manufactured by
A. arc welding B. thermit welding
C. resistance welding D. argon arc welding
365. Which of the following is different trom the remaining?
A. Spot welding B. Seam welding
c. Butt welding D. Argon arc welding
366. Two 3 mm thick mild steel sheets are to be welded. The electrodes of No. 18, 16, 12 and 10 are available. Which one would you select?
A. No. 10 B. No. 12
C. No. 16 D. No. 18
367. Two 12 mm steel plates are to be welded using arc welding. Electrode of No, 8 is to be used. What will be current requirement?
A. 10 A B. 50 A
C. 100A D. 150A
368. Grey iron is usually welded by
A. arc welding B. gas welding
C. TIG welding D. MIG welding
369. In ultrasonic welding, the frequency range is generally
A. 100–4000 Hz B. 4000–20000 Hz
C. 20000–80000 Hz D. 80000–200000 Hz
370. Arc blow is a welding defect which is encountered in
A. arc welding using dc current
B. arc welding using ac current
C. gas welding D. thermit welding
371. A rectifier for welding has voltage/current characteristics as
A. drooping B. rising
C. static D. variable
372. The efficiency of a welding motor generator is usually in the rahge of
A. 90 to 95% B. 80 to 90%
C. 50 to 60% D. 30 to 45%
373. For welding duty the rectifiers commonly used are
A. Mercury arc rectifiers
B. Selenium metal rectifiers
C. Any of the above D. none of these
374. Which of the following conditions determines the selection of an engine-driven welding set?
A. The set must have speed generator
B. The flywheel must be capable of limiting the fall in speed when the welding load is applied
C. The engine power must be enough to provide full generator load as well as auxiliaries
D. All of these

Ans.													
Ans.	351. C	352. C	353. A	354. D	355. B	356. A	357. B	358. C	359. B	360. C	361. D	362. B	363. C
	364. C	365. D	366. A	367. A	368. B	369. B	370. B	371. A	372. C	373. B	374. D		

375. Which of the following is not a welding accessory?
A. Electrode holder B. Hand screen
C. Cable D. Gloves

376. Chipping hammers are used
A. to remove slag from weldings
B. to align the pieces to be welded
C. for tag welding
D. for marking spots to be welded

377. The welding electric circuit is
A. always earthed B. never earthed
C. connected through cables only
D. None of these

378. The eyes of welding operator must be protected against
A. ultra violet radiations
B. infrared radiations
C. Both A and B D. solar radiations

379. The danger of electric shock is maximum
A. during arcing B. after arcing
C. before welding
D. while inserting electrode into the holder

380. A 10-SWG electrode has approximate diameter of
A. 0.9 mm B. 1.2 mm
C. 3.3 mm D. 10 mm

381. Which of the following electrodes will have least diameter?
A. 20 SWG B. 14 SWG
C. 8 SWG D. 4SWG

382. Which of the following electrodes will have maximum diameter?
A. 8 SWG B. 10 SWG
C. 16 SWG D. 20 SWG

383. A 10-SWG electrode usually operates in the current range
A. 25 to 45 A B. 50 to 75 A
C. 95 to 135 A D. 400 to 600 A

384. TIG welding is
A. temperature insulated gas welding
B. tungsten inert gas welding
C. thermally induced gas welding
D. thorium iodine gas welding

385. Electrode is not consumed in case of
A. DC arc welding B. AC arc welding
C. gas welding D. TIG welding

386. Flux used in TIG welding is
A. ammonium chloride B. borax
C. ash D. none of these

387. Which method would you recommend for the welding of aluminium alloys?
A. DC arc welding B. AC arc welding
C. Tungsten arc welding
D. Acetylene-oxygen gas welding

388. Argon is
A. inactive gas B. inert gas
C. rare gas D. oxidising agent

389. With manual arc welding in mild steel, the metal deposition rate will be nearly
A. 2 to 5 kg per hour B. 5 to 10 kg per hour
C. 10 to 20 kg per hour D. 20 to 50 kg per hour

390. Which of the following automatic welding processes is likely to give maximum rate of metal deposition
A. gas shielded bare wire
B. submerged arc (single wire)
C. multiple power submerged arc
D. continuous flux convered electrode

391. The load taken by a welding transformer is
A. purity resistive B. noninductive
C. highly inductive D. none of these

392. The power factor of load using welding transformer is usually
A. 1.0
B. nearly unity lagging
C. nearly unity leading
D. very low of the order of 0.3 to 0.5 lagging

393. The power factor of load using welding transformer least depends on
A. arc length B. type of electrode
C. number of operation D. material to be welded

394. For power factor correction of welding transformer, a capacitor is usually connected on
A. primary side B. secondary side
C. parallel to arcing electrodes
D. parallel to mains

395. The welding load is always
A. continuous and constant
B. continuous but varying
C. intermittent D. none of these

396. MIG welding is
A. mildsteel inert gas welding
B. medium inert gas welding
C. maximum inner depth gas welding
D. metal inert gas welding

397. The advantage of resistance welding is
A. less skill required B. reduced distortion
C. higher production rates
D. all of these

398. Which of the following is not resistance welding?
A. Projection welding B. MIG welding
C. Seam welding D. Flash-but welding

399. When t is the thickness of the sheet, the tip diameter for spot welding is usually
A. $2t$ B. t
C. $\sqrt{t}$ D. $1/t$

400. Heat is transferred simultaneously by conduction, convection and radiation
A. through the surface of the insulated pipe carrying steam

Ans.	375. C	376. A	377. A	378. C	379. D	380. C	381. A	382. A	383. C	384. B	385. D	386. D	387. C
	388. B	389. A	390. C	391. C	392. D	393. D	394. A	395. C	396. C	397. D	398. B	399. C	

B. during melting
C. inside boiler furnaces
D. from refrigerator coils to freezer of a refrigerator

401. Essential requirement for the transmission of heat from one body to another is
A. both bodies must be solids
B. both bodies must be solids in contact
C. atleast one of the bodies must have some source of heating
D. temperature of the two bodies must be different

402. Which of the following is the unit for thermal conductivity?
A. $W/m°K$ B. $W/m^2°C$
C. $W^2/m°K$ D. $W^2/m^2°K$

403. The rate of heat flow through a 50 mm thick wall of a material having thermal conductivity of 40 $W/m°K$ for a temperature difference of 10°C will be
A. 80 W/m^2 hr B. 800 W/m^2 hr
C. 8000 W/m^2 hr D. 80000 W/m^2 hr

404. The ratio of heat flow $Q_A//Q_B$, from two walls of same thickness having thermal conductivity $K_A = 2 K_B$ for the same temperature difference will be?
A. 1 B. 0.5
C. 2 D. 4

405. The highest value of thermal conductivity is for
A. solid ice B. melting ice
C. water D. steam

406. Which insulating material is suitable for low temperature applications?
A. Cork B. Asbestos paper
C. Diatomaceous earth D. 85% magnesia

407. Which concrete is expected to have highest value of thermal conductivity?
A. Dry concrete
B. Dry concrete having 0.4% reinforcement
C. Concrete having 10% moisture by volume
D. Concrete having 0.4% reinforcement and 10% moisture by volume

408. The quantity of heat required to change the temperature of 1 gm of ice from, –6°C to –5°C is known as
A. latent heat of freezing
B. temperature coefficient
C. specific heat D. freezing heat

409. A body at temperature T°K radiates heat in proportion to
A. T^2 B. $1/T^2$
C. T^4 D. $1/T^4$

410. If a body reflects entire radiation incident on it, then it is known as
A. black body B. grey body
C. white body D. transparent body

411. A body cools from 70°C to 60°C in 10 minutes. Under the same external conditions to cool from 60°C to 50°C, the body will take
A. more than 10 minutes
B. 10 minutes
C. less than 10 minutes
D. none of these

412. All temperatures above the freezing point of gold are usually determined by
A. thermocouple B. gas thermometer
C. optical pyrometer D. resistance bridge

413. Which of the following is of high importance in case of induction heating?
A. Voltage B. Current
C. Frequency D. All of these

414. Induction heating process is based on
A. electromagnetic induction principle
B. resistance heating principle
C. thermal-ion release principle
D. nuclear heating principle

415. In case of surface hardening by induction heating,
A. heating occurs uniformly in the part to be heated
B. heating occurs in the core of the part to be heated
C. heating does not occur in the core of the part to be heated
D. heating is more at the core and less on the surface of the part to be heated.

416. In induction heating, the depth upto which the current will penetrate is proportional to
A. frequency B. $(\text{frequency})^2$
C. 1/(frequency) D. $1/\sqrt{\text{frequency}}$

417. The method of heating used in an electric room heat convector is
A. resistance heating B. induction heating
C. dielectric heating D. arc-heating

418. Highest power factor can be expected in which method of heating?
A. Electric arc heating B. Dielectric heating
C. Induction heating D. Resistance heating

419. Which of the following heating elements can give highest temperature in resistance heating?
A. Copper B. Nickel copper
C. Nichrome D. Silicon carbide

420. If the temperature coefficient of silver is 0.0338, the temperature coefficient for Manganin may be expected to be
A. 0.001 B. 0.0001
C. 0.0038 D. 0.00001

421. Nichrome wires can be rafely used for heating up to
A. 2000°C B. 1600°C
C. 1450°C D. 1050°C

Ans. 400. C 401. D 402. A 403. C 404. C 405. A 406. C 407. D 408. C 409. C 410. C 411. A 412. C 413. C 414. A 415. B 416. D 417. A 418. D 419. D 420. D 421. C

422. Nichrome wires can be safely used for heating to temperature around 1500°C. Which of the following materials for heating element should be selected?
A. Eureka B. Kanthal
C. Platinum molybdenum, carbon compound
D. Nichrome

423. Which of the following element will have the least range of temperature?
A. Eureka B. Nichrome
C. Silicon carbon D. Kanthal

424. Which of the following devices is necessarily required for automatic temperature control in a furnace?
A. Thermocouple B. Thermostat
C. Auto-transformer
D. Heating elements of variable resistance material

425. Furnaces used for cremation use is of
A. resistance heating B. induction heating
C. dielectric heating D. arc heating

426. Induction hardening is possible
A. on ferrous materials only
B. on magnetic materials only
C. on dc supply only
D. on ac supply only

427. A piece of steel is to be heated to a predetermined temperature, in which of the following furnace it will attain the desired temperature in the shortest possible time?
A. Oil fired furnace with heat exchanger
B. Electric resistance furnace
C. Induction heating furnace
D. Electric arc furnace

428. In a domestic cake baking oven, the temperature is controlled by
A. series parallel operation
B. auto transformer
C. thermostat D. voltage variation

429. If the supply frequency is reduced from 50 Hz to 1 Hz, which of the following methods of heating will be least affected?
A. Resistance heating B. Dielectric heating
C. Induction heating D. None of these

430. Thermal conductivity is least for
A. air B. water
C. glass D. copper

431. Which method of heating is likely to give leading power factor?
A. Electric arc B. Induction heating
C. Dielectric heating D. Resistance heating

432. In an electric press, mica is used
A. for dielectric heating
B. for induction heating
C. as an insulator
D. as a device for power factor improvement

433. Which of the following methods is suitable for the heating of conducting medium?
A. Radiant heating B. Eddy current heating
C. Induction heating D. Indirect arc heating

434. In dielectric heating, current flows through
A. metallic conductor
B. ionic discharge between dielectric medium and metallic conductor
C. dielectric D. air

435. Which of the following is the desirable property of resistance heating element materials?
A. High resistivity B. High melting point
C. Low temperature coefficient
D. All of these

436. Which of the following is the ideal method of heating plastics?
A. Oil fired furnace
B. Resistance furnace heating
C. Dielectric heating D. Coal fired furnace

437. If 'f' be the frequency, then dielectric loss is proportional to
A. f B. f^2
C. $1/f$ D. $1/f^2$

438. If E is the voltage impressed on a dielectric, the dielectric loss will be proportional to
A. E B. E^2
C. $1/E$ D. $1/E^2$

439. Radiant heating is used for
A. melting of ferrous metals
B. drying of paints and varnishes
C. heating of liquids in electric kettle
D. annealing of metals

440. Which method is appropriate for heating non-ferrous metals?
A. Dielectric heating B. Radiant heating
C. Indirect arc heating
D. Indirect resistance heating

441. For arc heating? the electrodes are made of
A. Copper B. Aluminium
C. Graphite D. ACSR conductor

442. A plywood board is to be heated through 100°C. Which method will be suitable for this purpose?
A. Induction heating B. Resistance heating
C. Arc heating D. Any of these

443. For heating of plywood, the frequency should be
A. 100 Hz B. 1000 Hz
C. 10–20 kHZ D. 1–2 MHz

444. High frequency for induction heating can be generated by
A. motor generator set B. spark gap oscillator
C. vacuum tube oscillator
D. Any of these

Ans. 422. C 423. A 424. B 425. A 426. D 427. C 428. C 429. A 430. A 431. C 432. C 433. C 434. C
435. D 436. C 437. A 438. B 439. B 440. C 441. C 442. A 443. D 444. B

445. Steel rails are welded by
A. argon arc welding B. thermit welding
C. gas welding
D. resistance welding

446. In arc welding, the voltage on a AC supply system is in the range
A. 1000–1220 V B. 400–500 V
C. 200–250 V D. 70–100 V

447. Which of the following is not a welding accessory?
A. Electrode holder B. Hand screen
C. Cable D. Gloves

448. Arc can be produced by
A. AC current only B. DC current only
C. either AC or DC current
D. none of these

449. The arc has
A. linear resistance characteristics
B. positive resistance characteristics
C. negative resistance characteristics
D. highly inductive characteristics

450. In arc welding, the temperature of the arc is of the order of
A. 100°C B. 1000°C
C. 3500°C D. 35000°C

451. In arc welding by DC supply, the voltage required is
A. 10 to 20 V B. 50 to 60 V
C. 100 to 120 V D. 200 to 250 V

452. In arc welding, once the arc is struck, the voltage required to maintain it will be
A. 20–30 V B. 100–120 V
C. 200–220 V D. 500–1000 V

453. The resistance of arc
A. decreases with increase of the current
B. increases with increase of the current
C. does not depend on current
D. none of these

454. In argon arc welding, the electrode is made of
A. carbon B. steel
C. tungsten D. no electrode is needed

455. A DC generator used for AC welding should have
A. rising characteristics
B. drooping characteristics
C. straight characteristics
D. none of these

456. The transformer used for AC welding sets is
A. booster type B. step-up transformer
C. step-down transformer
D. none of these

457. In argon arc welding, the purpose of using argon is
A. to prevent oxidation of metal by coming in contact with oxygen or air
B. to create inert atmosphere around the job to be welded
C. to obviate the necessity for using flux
D. all of these

458. As the thickness of the part to be welded increases, which of the following parameter(s) for AC welding should also increase?
A. Voltage B. Current
C. Frequency D. All of these

459. Steel pipes are manufactured by
A. arc welding B. thermit welding
C. resistance welding D. argon arc welding

460. In gas welding, the gases used are
A. oxygen and nitrogen B. argon and helium
C. helium and carbon dioxide
D. acetylene and oxygen

461. Arc blow is a welding defect which is encountered in
A. arc welding using DC current
B. arc welding using AC current
C. gas welding D. thermit welding

462. Steel rails are welded by
A. resistance welding B. thermit welding
C. argon arc welding D. gas welding

463. A rectifier for welding has voltage/current characteristics as
A. drooping B. rising
C. static D. variable

464. For welding duty, the rectifiers common-ly used are
A. mercury arc rectifiers
B. selenium metal rectifiers
C. any of these D. none of these

465. The eyes of welding operator must be protected against
A. ultraviolet radiations B. infrared radiations
C. both A and B D. solar radiations

466. Which of the following conditions deter-mines the selection of an engine-driven welding set?
A. The set must have speed generator
B. The flywheel must be capable of limit-ing the fall in speed when the welding load is applied
C. The engine power must be enough to provide full generator load as well as auxiliaries
D. All of these

467. A 10-SWG electrode has approximate diameter of
A. 0.9 mm B. 1.2 mm
C. 3.3 mm D. 10 mm

468. The danger of electric shock is maximum
A. during arcing
B. after arcing
C. before welding
D. while inserting electrode into the holder

Ans. 445. B 446. D 447. C 448. C 449. C 450. C 451. B 452. A 453. A 454. C 455. B 456. C 457. D 458. B 459. C 460. D 461. B 462. B 463. A 464. B 465. C 466. D 467. C 468. D

469. Two 12 mm steel plates are to be welded using arc welding. Electrode of No. 8 is to be used. What will be the current requirement?
A. 10 A B. 50 A
C. 100 A D. 150 A

470. Which of the following electrodes will have least diameter?
A. 20 SWG B. 14 SWG
C. 8 SWG D. 4 SWG

471. Grey iron is usually welded by
A. arc welding B. gas welding
C. TIG welding D. MIG welding

472. Which of the following is different from the remaining?
A. Spot welding B. Seam welding
C. Butt welding D. Argon arc welding

473. All temperatures above the freezing point of gold are usually determined by
A. thermocouple B. gas thermometer
C. optical pyrometer D. resistance bridge

474. Which of the following is of high importance in case of induction heating?
A. Voltage B. Current
C. Frequency D. All of these

475. Induction heating process is based on
A. electromagnetic induction principle
B. resistance heating principle
C. thermal-ion release principle
D. nuclear heating principle

476. In case of surface hardening by induction heating,
A. heating occurs uniformly in the part to be heated
B. heating occurs in the core of the part to be heated
C. heating does not occur in the core of the part to be heated
D. heating is more at the core and less on the surface of the part to be heated

477. In induction heating, the depth up to which the current will penetrate is proportional to
A. frequency B. $(\text{frequency})^2$
C. 1/(frequency) D. $1/\sqrt{\text{frequency}}$

478. The method of heating used in an electric room heat convector is
A. resistance heating B. induction heating
C. dielectric heating D. arc-heating

479. Highest power factor can be expected in which method of heating?
A. Electric arc heating B. Dielectric heating
C. Induction heating D. Resistance heating

480. Two 3 mm thick mild steel sheets are to be welded. The electrodes of No. 18, 16, 12 and 10 are available. Which one would you select?
A. No. 10 B. No. 12
C. No. 16 D. No. 18

481. In ultrasonic welding, the frequency range is generally
A. 100–4000 Hz B. 4000–20000 Hz
C. 20000–80000 Hz D. 80000–200000 Hz

482. The efficiency of a welding motor generator is usually in the range of
A. 90 to 95% B. 80 to 90%
C. 50 to 60% D. 30 to 45%

483. A 10-SWG electrode usually operates in the current range
A. 25 to 45 A B. 50 to 75 A
C. 95 to 135 A D. 400 to 600 A

484. TIG welding is
A. temperature insulated gas welding
B. tungsten inert gas welding
C. thermally induced gas welding
D. thorium iodine gas welding

485. Which of the following electrodes will have maximum diameter?
A. 8 SWG B. 10 SWG
C. 16 SWG D. 20 SWG

486. Electrode is not consumed in case of
A. DC arc welding B. AC arc welding
C. gas welding D. TIG welding

487. Which method would you recommend for the welding of aluminium alloys?
A. DC arc welding B. AC arc welding
C. Tungsten arc welding
D. Acetylene-oxygen gas welding

488. With manual arc welding in mild steel, the metal deposition rate will be nearly
A. 2 to 5 kg per hour B. 5 to 10 kg per hour
C. 10 to 20 kg per hour D. 20 to 50 kg per hour

489. Argon is
A. inactive gas B. inert gas
C. rare gas D. oxidising agent

490. Flux used in TIG welding is
A. ammonium chloride B. borax
C. ash D. none of these

491. The load taken by a welding transformer is
A. purely resistive B. non-inductive
C. highly inductive D. none of these

492. Which of the following automatic welding processes is likely to give maximum rate of metal deposition?
A. Gas shielded bare wire
B. Submerged arc (single wire)
C. Multiple power submerged arc
D. Continuous flux converted electrode

493. The power factor of load using welding transformer is usually
A. 1.0 B. nearly unity lagging

Ans. 469. A 470. A 471. B 472. D 473. C 474. C 475. A 476. B 477. D 478. A 479. D 480. A 481. B 482. C 483. C 484. B 485. A 486. D 487. C 488. A 489. B 490. D 491, C 492. C

C. nearly unity leading
D. very low of the order of 0.3 to 0.5 lagging

494. Which of the following is not resistance welding?
A. Projection welding B. MIG welding
C. Seam welding D. Flash-butt welding

495. Essential requirement for the transmission of heat from one body to another is
A. both bodies must be solids
B. both bodies must be solids in contact
C. at least one of the bodies must have some source of heating
D. temperature of the two bodies must be different

496. The rate of heat flow through a 50 mm thick wall of a material having thermal conductivity of 40 W/m °K for a temperature difference of 10°C will be
A. 80 W/m^2 hr B. 800 W/m^2 hr
C. 8000 W/m^2 hr D. 80000 W/m^2 hr

497. Which of the following is the unit for thermal conductivity?
A. W/m °K B. W/m^2 °K°C
C. W^2/m °K D. W^2/m^2 °K

498. The ratio of heat flow Q_A/Q_B, from two walls of same thickness having thermal conductivity $K_A = 2K_B$ for the same temperature difference will be
A. 1 B. 0.5
C. 2 D. 4

499. Which insulating material is suitable for low temperature applications?
A. Cork B. Asbestos paper
C. Diatomaceous earth D. 85% magnesia

500. The highest value of thermal conductivity is for
A. solid ice B. melting ice
C. water D. steam

501. Which concrete is expected to have highest value of thermal conductivity?
A. Dry concrete
B. Dry concrete having 0.4% reinforcement
C. Concrete having 10% moisture by volume
D. Concrete having 0.4% reinforcement and 10% moisture by volume

502. For power factor correction of welding transformer, a capacitor is usually connected on
A. primary side
B. secondary side
C. parallel to arcing electrodes
D. parallel to mains

503. The power factor of load using welding transformer depends on
A. arc length B. type of electrode
C. number of operation D. material to be welded

504. The welding load is always
A. continuous and constant
B. continuous but varying
C. intermittent D. none of these

505. When t is the thickness of the sheet, the tip diameter for spot welding is usually
A. $2t$ B. t
C. $\sqrt{t}$ D. $1/t$

506. Heat is transferred simultaneously by conduction, convection and radiation
A. through the surface of the insulated pipe carrying steam
B. during melting
C. inside boiler furnaces
D. from refrigerator coils to freezer of a refrigerator

507. MIG welding is
A. mild steel inert gas welding
B. medium inert gas welding
C. maximum inner depth gas welding
D. metal inert gas welding

508. The advantage of resistance welding is
A. less skill required B. reduced distortion
C. higher production rates
D. all of these

509. A body at temperature T °K radiates heat in proportion to
A. T^2 B. $1/T^2$
C. T^4 D. $1/T^4$

510. A body cools from 70 to 60°C in 10 minutes. Under the same external conditions to cool from 60 to 50°C, the body will take
A. more than 10 minutes
B. 10 minutes
C. less than 10 minutes D. none of these

511. Furnaces used for cremation use is of
A. resistance heating B. induction heating
C. dielectric heating D. arc heating

512. Which of the following heating elements can give highest temperature in resistance heating?
A. Copper B. Nickel copper
C. Nichrome D. Silicon carbide

513. If the temperature coefficient of silver is 0.0338, the temperature coefficient for manganin may be expected to be
A. 0.001 B. 0.0001
C. 0.0038 D. 0.00001

514. Induction hardening is possible on
A. ferrous materials only
B. magnetic materials only
C. DC supply only D. AC supply only

515. A piece of steel is to be heated to a predetermined temperature, in which of the following furnace it will attain the desired temperature in the shortest possible time?
A. Oil fired furnace with heat exchanger
B. Electric resistance furnace
C. Induction heating furnace
D. Electric arc furnace

Ans.													
	493. D	494. B	495. D	496. C	497. A	498. C	499. C	500. A	501. D	502. A	503. D	504. C	505. C
	506. C	507. C	508. D	509. C	510. A	511. A	512. D	513. D	514. D	515. C			

516. What is the use of the white matter coated inside the fluorescent lamp?
A. It reduces the brightness
B. It provides a proper exterior to the tube
C. It converts the ultraviolet radiation into visible light
D. It provides the ions necessary for the gas discharge

517. In a domestic cake baking oven, the temperature is controlled by
A. series parallel operation
B. auto transformer
C. thermostat D. voltage variation

518. What is the function of the capacitor marked in the circuit shown?
A. To eliminate interference of the fluorescent tube
B. To producethe starting voltage
C. To limit the operating current
D. To compensate for the phase difference caused by the choking coil

519. Which of the following methods is suitable for the heating of conducting medium?
A. Radiant heating B. Eddy current heating
C. Induction heating D. Indirect arc heating

520. Nichrome wires can be safely used for heating up to
A. 2000°C B. 1600°C
C. 1450°C D. 1050°C

521. Nichrome wires can be safely used for heating of temperature around 1500°C. Which of the following materials for heating element should be selected?
A. Eureka B. Kanthal
C. Platinum molybdenum, carbon compound
D. Nichrome

522. Which of the following element will have the least range of temperature?
A. Eureka B. Nichrome
C. Silicon carbon D. Kanthal

523. Which of the following devices is necessarily required for automatic temperature control in a furnace?
A. Thermocouple B. Thermostat
C. Autotransformer
D. Heating elements of variable resistance material

524. If the supply frequency is reduced from 50 Hz to 1 Hz, which of the following methods of heating will be least affected?
A. Resistance heating B. Dielectric heating
C. Induction heating D. None of these

525. Thermal conductivity is least for
A. air B. water
C. glass D. copper

526. If a body reflects entire radiation incident on it, then it is known as
A. Black body B. Grey body
C. White body D. Transparent body

527. Which method of heating is likely to give leading power factor?
A. Electric arc B. Induction heating
C. Dielectric heating D. Resistance heating

528. What is the function of the capacitor in the circuit?
A. To increase the heating current
B. To reduce the heating current
C. To compensate for the phase difference caused by the choke
D. To eliminate radio-interference caused by the starter

529. Which of the following is more efficient?
A. Sodium vapour lamp
B. Mercury vapour lamp
C. Zenon lamp D. Halogen lamp

530. Which of the following is the desirable property of resistance heating element materials?
A. High resistivity B. High melting point
C. Low temperature coefficient
D. All of these

531. In an electric press, mica is used
A. for dielectric heating B. for induction heating
C. as an insulator
D. as a device for power factor improvement

532. A plywood board is to be heated through 100°C. Which method will be suitable for this purpose?
A. Induction heating B. Resistance heating
C. Arc heating D. Any of these

533. In dielectric heating, current flows through
A. metallic conductor
B. ionic discharge between dielectric medium and metallic conductor
C. dielectric D. air

534. Which of the following is the ideal method of heating plastics?
A. Oil fired furnace
B. Resistance furnace heating
C. Dielectric heating D. Coal fired furnace

535. If 'f' be the frequency, then dielectric loss is proportional to
A. f B. f^2
C. $1/f$ D. $1/f^2$

536. If E is the voltage impressed on a dielectric, the dielectric loss will be proportional to
A. E B. E^2
C. $1/E$ D. $1/E^2$

537. Radiant heating is used for
A. melting of ferrous metals
B. drying of paints and varnishes
C. heating of liquids in electric kettle
D. annealing of metals

538. Which method is appropriate for heating non-ferrous metals?

Ans. 516. C 517. C 518. D 519. C 520. C 521. C 522. A 523. B 524. A 525. A 526. C 527. C 528. D 529. B 530. D 531. C 532. A 533. C 534. C 535. A 536. B 537. B

A. Dielectric heating B. Radiant heating
C. Indirect arc heating
D. Indirect resistance heating

539. For arc heating, the electrodes are made of
A. Copper B. Aluminium
C. Graphite D. ACSR conductor

540. Steel rails are welded by
A. argon arc welding B. thermit welding
C. gas welding D. resistance welding

541. Which of the following comparisons between the filament lamp and the fluorescent lamp is correct?
A. The fluorescent lamp has a higher dazzle
B. The fluorescent lamp produces sharper shadows
C. The fluorescent lamp produces greater brightness
D. The average life of the fluorescent lamp is five to seven times higher

542. For heating of plywood, the frequency should be
A. 100 Hz B. 1000 Hz
C. 10–20 kHz D. 1–2 MHz

543. High frequency for induction heating can be generated by
A. motor generator set B. spark gas oscillator
C. vacuum tube oscillator
D. any of these

544. For which of the following work is highest illumination required?
A. For office work
B. For mounting of wrist watches
C. For sorting work in the stores
D. All of these

545. What is attained by heating the electrodes of the fluorescent lamp?
A. The thermionic emission of electrons
B. The photo emission of electrons
C. The heating up of the gas filling and thus the creation of ions
D. An increase in the voltage across the electrodes

546. In dielectric heating, if the capacitor is loss-free, the heat produced will be
A. zero B. infinity
C. porportional to the value of capacitance
D. proportional to the frequency

547. The power density generated in dielectric heating is not proportional to
A. voltage gradient B. frequency
C. power factor D. capacitance

548. The frequency used in dielectric heating lies in the range of
A. 500–1000 Hz B. 1–50 MHz
C. 600 Hz–10 KHz D. 10–200 KHz

549. In dielectric heating, non-uniform heating
A. occurs for higher frequencies
B. occurs for lower frequencies
C. is independent of frequency
D. occurs for higher power factors

550. In dielectric heating, the rate of heating cannot be increased by increasing the potential gradient because
A. coupling problems become highly pronounced
B. very high voltages are not easily avail-able
C. heating becomes non-uniform
D. corona takes place

551. In resistance welding, aluminium as compared to steel, requires
A. larger welding time B. smaller welding time
C. equal welding time
D. welding time depending upon the value of weld current

552. In resistance welding, the SCR contractor will close during
A. squeeze time B. hold time
C. weld time
D. weld as well as off time

553. In a synchronous welding control, the welding current begins at
A. later than the power-factor angle
B. before the power-factor angle
C. the power-factor angle
D. a phase delay determined by number of weld cycles

554. In a resistance welder, pneumatic pressure is applied during
A. squeeze time B. weld time
C. squeeze and weld time
D. squeeze, weld and hold time

555. The welding transformer used in resistance welding will
A. step up current B. step down current
C. step up voltage D. step up power

556. A digital timer in a resistance welding machine provides
A. accurate timings B. identical welds
C. synchronous operation
D. all of these

557. The primary reason for using flux when welding aluminum is to
A. prevent oxides from forming ahead of the weld.
B. promote better fusion of the base metal at a lower temperature
C. prevent molten metal from flowing too widely
D. clean the base metal ahead of the weld

558. Magnesium aircraft structural members are usually heat treated. Which statement is true concerning the welding of heat treated magnesium?
A. The welded section can never have the strength of the original metal

Ans.	538. C	539. C	540. B	541. D	542. D	543. B	544. B	545. A	546. A	547. D	548. B	549. A	550. D
	551. B	552. C	553. C	554. D	555. A	556. D	557. B	558. A					

B. Flux should not be used as it causes corrosion to commence
C. Use an oxidizing flame held at a flat angle to the work
D. Magnesium cannot be repaired by fusion welding

559. The oxyacelylene flame for silver solder-ing should be
A. harsh B. oxidizing
C. neutral D. carburizing

560. It is necessary to use flux in all silver soldering operations so as to
A. chemically clean the base metal of oxide film
B. prevent overheating of the base metal
C. increase the strength of the joint and save the expensive silver solder
D. increase heat conductivity

561. Engine mount members should preferably be repaired by using
A. a larger dimeter tube with fishmouth and no rosette fields
B. a larger diameter tube with fishmouth and rosette welds
C. a smaller diameter tube with fish-mouth and rosette welds
D. a larger diameter tube with 30° cuts and rosette welds

562. The method of repair is recommended for a steel tube long on dented at a cluster is
A. Welded split sleeve B. Welded outer sleeve
C. Welded patch plate D. Welded inner sleeve

563. In aircraft welding, the usual practice in the selection of a welding tip is to use
A. the type of material to be welded as the primary basis for selection
B. a tip with a hole size equal to the diameter of the welding rod used
C. as small as tip as possible with the tip adjusted to the maximum operating range
D. as large as tip is possible with the tip adjusted to the maximum operating range

564. Carburizing flame is avoided when welding steel because
A. It removes the carbon content
B. It hardness the surface
C. It causes excessive sparking.
D. A cold weld will result.

565. When welding a tank fabricated from 710 sheet aluminum, a welding tip slightly larger than that used for welding steel of the same thickness should be used to obtain sufficient heat to melt the base metal. The flame used for aluminum welding should.
A. be neutral and soft B. be slightly oxidizing
C. contain and excess of acetylene and leave the tip at a relatively low speed
D. contain an excess of acetylene and leave the tip at a relatively high speed

566. New or resurfaced soldering copper tip should be tinned or coated with solder so as to
A. aid in the transfer of heat from the soldering tip to the joint to be soldered
B. prevent the soldering tip from overheating the joint to be soldered
C. prevent excessive heat from being radiated from the tip
D. none of these

567. A very thin and pointed tip on a soldering copper is undesirable because it will
A. burn the alloys out of the solder
B. be very difficult to tin
C. cool too rapidly
D. punch holes in the metal being soldered

568. The characteristics of a gas welding which is successfully completed will be
A. the finish weld should have a rough seam and be non-uniform in thickness
B. the weld metal should be tapered smoothly into the base metal
C. oxide should be formed on the base metal close to the weld
D. The base metal should show signs of pits

569. An acceptable line pressure for acetylene is
A. 5 psi B. 18 psi
C. 22 psi
D. the same as the oxygen pressure

570. Cylinders used to transport and store acetelyne
A. are purged after each use
B. are pressure tested to 2,000 psi
C. are green in color D. contain acetone

571. Which of the following statements concerning a welding process is true?
A. The inert-arc welding process uses an inert gas to protect the weld zone from the atmosphere
B. In the metallic-arc welding process, filler material if needed, is provided by a separate metal rod of the proper material held in the arc
C. In the carbon-arc welding process, filler material, if needed, is provided by the arc
D. In the oxyacetylene welding process, the filler rod used for steel is covered with a thin coating of flux

572. Flux is applied when oxyacetyene welding aluminum in the form
A. Paint only on the surface to be welded
B. Paint on the surface to be welded and applied to the welding rod
C. Apply only to the welding rod
D. Paint on the surface to be welded and applied to the welding rod after tack weld

573. Oxides form very rapidly when alloys or metals are hot. It is important, therefore, when welding aluminum is to use a

Ans. 559. C 560. A 561. B 562. C 563. A 564. B 565. A 566. A 567. C 568. B 569. A 570. D 571. A 572. B

A. solvent B. float
C. filler D. flux

574. In gas welding, the amount of heat applied to the material being welded is controlled by
A. the amount of gas pressure used
B. the size of the tip opening
C. the distance the tip is held from the work
D. the temperature of the flame

575. Oxygen and acetylene cylinders are made of
A. heat-treated seamless copper
B. seamless aluminum
C. steel D. bronze

576. When inspecting a butt welded joint by visual means.
A. the penetration should be 25 percent to 50 percent of the thickness of the base metal.
B. the width of the bead should be twice the thickness of the base metal.
C. the penetration should be 100 percent of the thickness of the base metal
D. look for evidence of excessive heat in the form of a very high bead.

577. What is the advantage of annealing of aluminum?
A. It increases the tensile strength.
B. It makes the material brittel.
C. It removes stresses caused by forming.
D. It makes the material hard

578. Which of the following quantities has the unit "lux"?
A. Utilization factor B. Luminous flux
C. Luminous intensity D. Illumination

579. Which of the following statements about illumination is true?
A. If the distance from the source doubles, the illumination becomes half
B. If the distance from the source doubles, the illumination reduces to one fourth
C. The greater the illumination, the better one sees
D. The finer the work, the less the required illumination

580. For which of the following work is the highest illumination required?
A. For office work B. For painting work.
C. For mounting of wrist watches
D. For sorting work in the stores

581. What is the average life of a normal filament lamp?

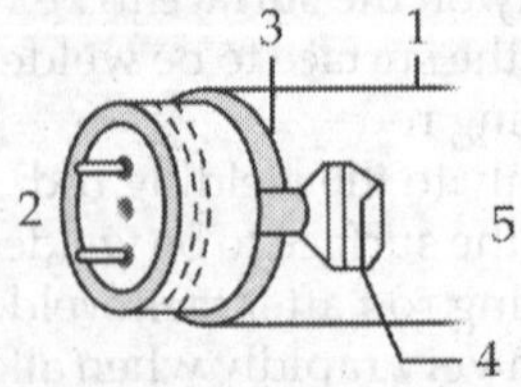

A. 10000 hours B. 5000 hours
C. 1000 hours D. 500 hours

582. Which of the following comparisons between the filament lamp and the fluorescent lamp is correct?
A. The fluorescent lamp has a higher dazzle
B. The fluorescent lamp produces sharper shadows
C. The fluorescent lamp produces greater brightness
D. The average life of the fluorescent lamp is five to seven times higher

583. What is the function of the choking coil in the circuit of the figure shown

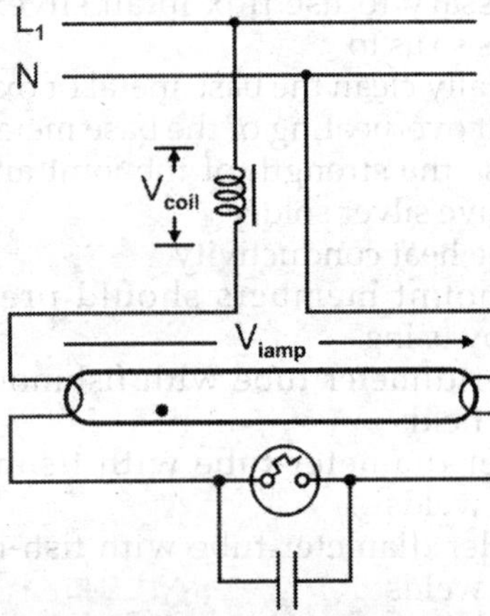

A. Only to limit the operating current
B. Only to produce the starting voltage impulse
C. To produce the starting voltage impulse and to limit the operating current
D. To limit the heating current

584. What is the approximate power factor of the fluorescent lamp (including starting device) shown in the figure

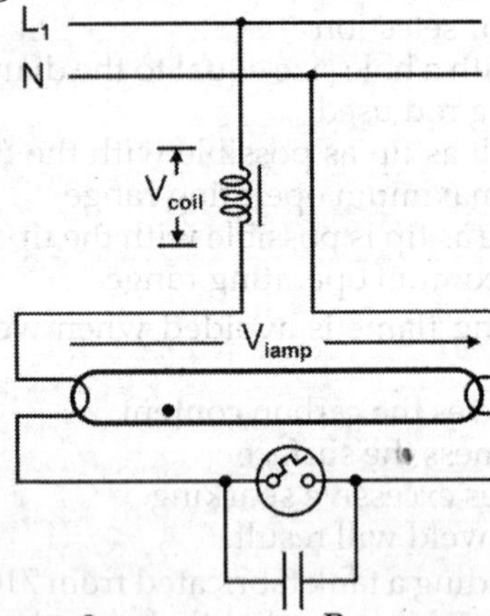

A. $\cos\varphi = 0$ cap B. $\cos\varphi = 0.5$ cap
C. $\cos\varphi = 1$ D. $\cos\varphi = 0$ ind

585. What is attained by heating the electrodes of the fluorescent lamp?
A. The thermionic emission of electrons
B. The photo emission of electrons
C. The heating up of the gas filling and thus the creation of ions
D. An increase in the voltage across the electrodes

Ans. 573. D 574. B 575. C 576. C 577. C 578. D 579. B 580. C 581. C 582. D 583. C 584. D 585. A

586. What is the process taking place in a fluorescent tube called?
A. Gaseous discharge B. Secondary emission
C. Phosphorescence D. Thermionic emission

587. What is the difference between fluorescent lamps that produce different coloured light?
A. The colour of the glass
B. The composition of the fluorescent material
C. The pressure of the filled gas
D. The composition of the filled gas

588. What is the function of the capacitor marked 1 in the circuit shown?

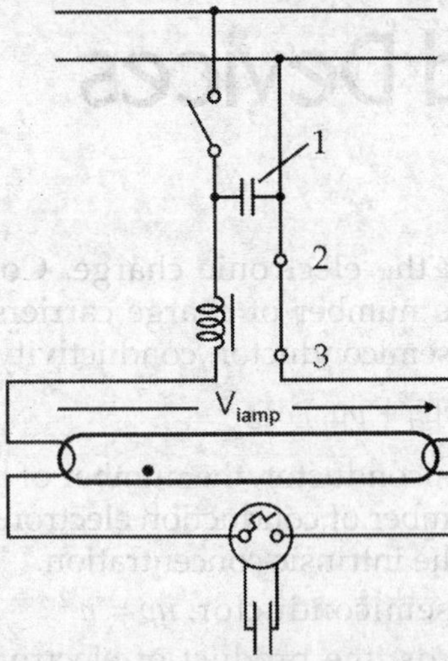

A. To eliminate interference of the fluorescent tube
B. To produce the starting voltage
C. To limit the operating current
D. To compensate for the phase difference caused by the choking coil

589. What is the advantage of the circuit shown in the figure

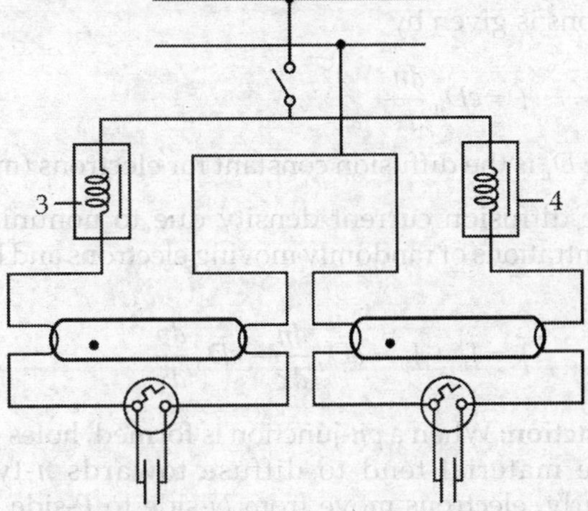

A. The tube power consumed in the circuit is almost zero
B. The power factor of the circuit is 0.5 leadng
C. A very small current flows in both the branches
D. The power factor is 1 and the light does not flicker

590. What is the use of the white matter coated inside the fluorescent lamp?
A. It reduces the brightness
B. It provides a proper exterior to the tube
C. It converts the ultraviolet radiation into visible light
D. It provides the ions necessary for the gas discharge

591. A fluorescent lamp flickers. Why should the lamp be immediately switched off?
A. The choking coil or the starter can be damaged
B. A strong interference is produced in radio receivers
C. Due to voltage peaks other lamps might also be damaged
D. The lamp can cause a short circuit

592. What is the function of the capacitor in the circuit given?

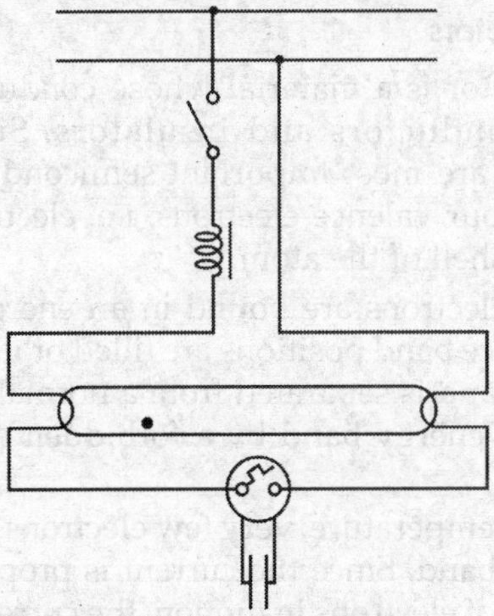

A. To increase the heating current
B. To reduce the heating current
C. To compensate for the phase difference caused by the choke
D. To eliminate radio-interference caused by the starter

593. What does the figure show?

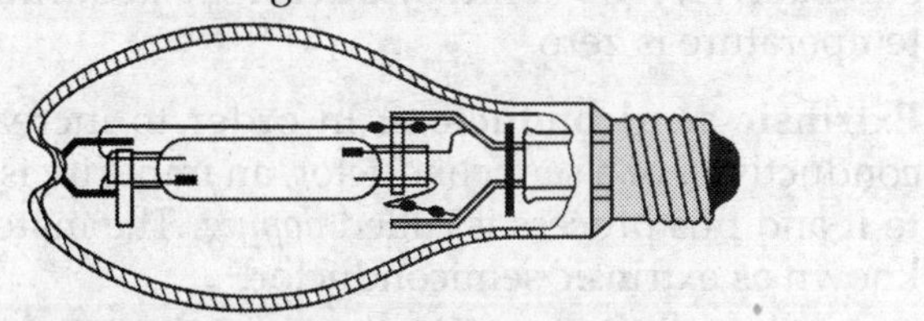

A. Sodium vapour lamp B. Xenon lamp
C. Mercury vapour lamp
D. Halogen lamp

594. The choking coil of an operating fluorescent lamp is short circuited. What is the consequence?
A. The fluorescent lamp goes out immediately
B. The lamp becomes brighter
C. The lamp becomes less bright
D. The current becomes so large that it damages the fluorescent lamp

Ans. 586. A 587. B 588. D 589. D 590. C 591. A 592. D 593. C 594. D

CHAPTER 15

Analog Electronics and Devices

Semiconductors

Semiconductor is a material whose conductivity lies between conductors and insulators. Silicon and germanium are most important semiconductors and both have four valence electrons, i.e. electrons in the outer most shell of the atom.

Valence electrons are bound in an energy band in which valence band positions are filled or unoccupied. This filled band is separated from a normally unoccupied higher energy band by a forbidden energy gap ΔE.

At room temperature, very few electrons exist in the conduction band. Since the current is proportional to the number of electrons in motion, the current is small, and hence the material has a higher resistance.

When an energy band is completely filled, its electrons do not contribute to electrical conductivity as there are no open energy level to which they can move after absorbing energy from an electric field. Therefore, conductivity of semiconductor at absolute zero temperature is zero.

Extrinsic semiconductors: In order to increase the conductivity of a semiconductor, an impurity is added to it and this process is called *doping.* The materials is known as extrinsic semiconductor.

An acceptor material accepts electrons from the valence band of silicon since it has only three valence electrons. Example of donor impurity are boron, indium, gallium, and aluminium.

Current in semiconductor: In a semiconductor, both holes and electrons contribute to electrical conduction. With an applied electric field E, the expression for current density is

$$J = (n\mu_n + p\mu_p)\, eE = \sigma E$$

where n and p are the concentrations of electrons and holes (number/m^3) and μ_n and μ_p are the corresponding mobilities, e is the electronic charge. Conductivity depends on the number of charge carriers and their mobility; for a semiconductor, conductivity is

$$\sigma = (n\mu_n + p\mu_p)\, e$$

In a pure semiconductor, the number of holes is just equal to the number of conduction electrons or $n = p = n_i$, where n_i is the intrinsic concentration.

In a doped semiconductor, $np = n_i^2$.

In other words, the product of electron and hole concentrations is a constant; if one is increased (by doping), the other must decrease. If the doping concentration is nonuniform, and it is possible to have charge motion by the mechanism called *diffusion.* The *diffusion current* is proportional to the concentration gradient *dn/dx.* the diffusion current density due to electrons is given by

$$J = eD_n \frac{dn}{dx}$$

where D_n is the diffusion constant for electrons (m^2/s).

The diffusion current density due to nonuniform concentrations of randomly moving electrons and holes is

$$J = J_n + J_p = eD_n \frac{dn}{dx} - eD_p \frac{dp}{dx}$$

***pn*-junction:** When a *pn*-junction is formed, holes from *p*-type material tend to diffuse towards *n*-types. Similarly, electrons move from *N*-side to *P*-side. Due to this phenomena, most of holes and electrons near the junction get neutralised and a small region On both side of the junction gets devoid of any mobile carriers with the result that impurity ions in this region become electrically un-neutralised. This region, called *depletion region* or *layer,* now acts like battery or potential barrier V_0 which opposes the further diffusion of majority carriers.

When an external voltage V_f is applied across the junction with *P*-side positive and *N*-side negative so as to reduce the potential barrier V_0, the junction is said to be *forward biased* and the majority carrier of each side can go over to other side more easily leading to external current as shown in Fig 15.1.

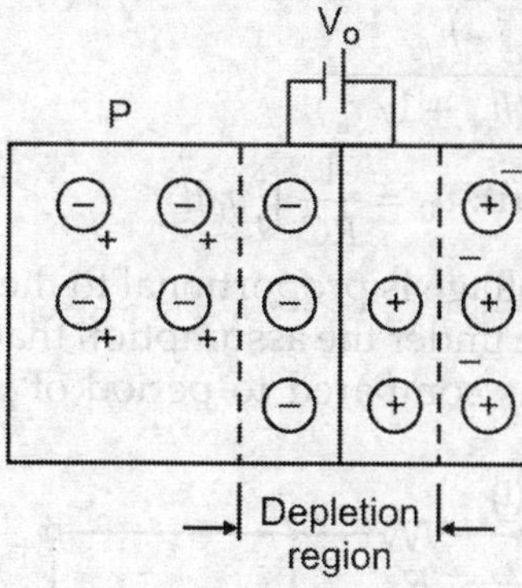

Fig. 15.1

When an external voltage V_R is connected in the reverse way, the potential barrier increases and the majority carriers find it difficult to cross to the other side. Only a vary small reverse current I_0 flows in this reverse biased *pn*-junction. This current is not affected by potential barrier and depends on temperature, material and construction of the junction. Therefore *PN*-junction allows large current only in forward bias direction and only small reverse current I_0 in the reverse bias direction.

Semiconductor diodes: A semiconductor diode conducts forward current with a small forward voltage drop across the device, simulating a closed switch. The relationship between the forward current and forward voltage is a good approximation given by the Shockely diode equation,

$$i = I_s [e^x - 1],$$

where $x = \frac{eV}{kT}$ and I_s is the leakage current through the diode, *e* is electronic charge, *k* is Boltzman's constant, *T* is the temperature of diode, and *V* is the voltage across the diode.

Bipolar junction transistor (BJT): A bipolar junction transistor consists of two *pn* junctions in close proximity normally, the emitter junction is forward biased, the collector is reverse biased. In common-base operation, the collector current

$$i_C = -\alpha i_E + I_{CBO}, \quad \text{where } \alpha \cong 1$$

where I_{CBO} is the collector cut-off current and α is the forward current–transfer ratio. In common emitter operation, a small base current controls the relatively larger collector current to achieve current amplification.

$$i_C = \beta i_B + I_{CEO}, \text{ where } \beta = \frac{\alpha}{1-\alpha}$$

where i_B is the base current and I_{CEO} is the collector cut-off current in the common emitter configuration.

Transistor connectors: (a) Common base configuration, (b) Common emitter configuration, (c) Common collector configuration.

(a) **Common base configuration (CB):** *Current amplification factor* (α): It is the ratio of change in collector current to the change in emitter current at a constant collector base voltage V_{CB}, i.e.

$$\alpha = \frac{\Delta t_c}{\Delta I_E} \text{ at constant } V_{CB}$$

Collector current: $I_C = \alpha I_E + I_{CBO}$, where I_{CBO} is the leakage collector current.

Input resistance: It is the ratio of change in emitter base voltage (ΔV_{EB} to the resulting change in emitter current (ΔI_E) at constant collector base voltage (V_{CB}), *i.e.* input resistance $r_i = v_{EB}/I_E$ at constant (V_{CB}).

Output resistance: It is the ratio of change in collector base voltage (ΔV_{CB}) to the change in collector current (ΔI_c) at constant emitter current, *i.e.* output resistance, or, $r_o = \Delta V_{CB}/\Delta I_C$ at constant I_E.

(b) **Common emitter configuration (CE):** *Base current amplification factor* (β): It is the ratio of change in collector current (ΔI_C) to the change in base current (ΔI_B)

i.e. $$\beta = \frac{\Delta I_C}{\Delta I_B}$$

Relation between β and α: $\beta = \alpha/(1 - \alpha)$

Collector current: $I_C = \beta I_B + I_{CEO}$ where I_{CEO} is the collector emitter current with base open.

Emitter current: $I_E = (\beta + I)\, I_B + I_{CEO}$

Input resistance: It is the ratio of change in base emitter voltage (ΔV_{BE}) to the change in base current (ΔI_B) at constant V_{CE}.

Input resistance $r_i = \Delta V_{BE}/\Delta I_B$ at constant V_{CE}.

Table 15.1: Comparison of transistor configurations

S. no.	*Emitter*	*Work function*	*Operating temperature*	*Emission efficiency*
1.	Input resistance	Low (about 100 ohm)	Low (about 750 ohm)	Very high (about 750 k ohm)
2.	Output resistance	Very high (about 450 k ohm)	High (about 45 kohm)	Low (about 50 ohm)
3.	Voltage gain	About 150	About 500	Less than 1
4.	For frequency applications	For high frequency applications	For audio frequency applications	For impedance matching

(c) **Common collector configuration (CC):** *Current amplification factor* (γ): It is the ratio of change in emitter current (ΔI_E) to the change in base current (ΔI_B), i.e. $\gamma = \Delta I_E/\Delta I_B$.

Relation between γ and α:

$$\gamma = 1/(1-\alpha)$$

Collector current $I_C = \alpha I_E + I_{CBO}$

Emitter current

$$I_E = (\beta+1)\,I_B + (\beta+1)\,I_{CBO}$$
$$= I_B/(1-\alpha) + I_{CBO}/(1-\alpha)$$

Ebers–Moll equations:

$$i_E = I_{ES}\,(e^{qV_{EB}/KT} - 1) - \alpha_R I_{CS}\,(e^{qV_{EB}/KT} - 1)$$
$$i_C = \alpha I_{ES}\,(e^{qV_{EB}/KT} - 1) - I_{CS}\,(e^{qV_{CB}/KT} - 1)$$

Hybrid model: The pair of equations for the two-port network is shown in Fig. 15.2.

$$v_1 = h_{11}i_1 + h_{12}v_2$$
$$i_1 = h_{21}i_1 + h_{22}v_2$$

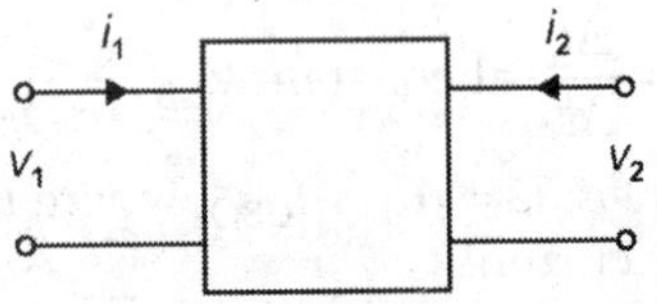

Fig. 15.2

For a transistor connected in Common–Emitter (CE) configuration, base is the input terminal and collector is the output terminal with respect to the terminal.

Therefore, $v_{be} = h_{11}i_b + h_{12}v_{ce}$

$$i_c = h_{21}i_b + h_{22}v_{ce}$$

From these equations, the h-parameters can be defined as follows:

$$h_{11} = h_{ie} = \frac{v_{be}}{i_b}\Big|_{v_{ce}=0} = \text{short-circuit input impedance}$$

$$h_{12} = h_{re} = \frac{v_{be}}{v_{ce}}\Big|_{i_b=0} = \text{open-circuit reverse voltage gain}$$

$$h_{21} = h_{fe} = \frac{i_c}{i_b}\Big|_{v_{ce}=0} = \text{short-circuit forward current gain}$$

$$h_{22} = h_{oe} = \frac{i_c}{v_{ce}}\Big|_{i_b=0} = \text{open-circuit output admittance}$$

Thus, the equations become

$$v_{be} = h_{ie}\,i_b + h_{re}\,v_{ce}$$
$$i_c = h_{fe}i_b + h_{oe}v_{ce}$$

Performance of linear circuit in h-parameters and for Common emitter configuration:

i. **Input impedance**

$$Z_m = h_{11} - h_{12}h_{21}\Big/\left(h_{22} + \frac{1}{r_2}\right)$$

$$Z_i = h_{11} \text{ (if } h_{12} \text{ or } r_L \text{ is very small)}$$

ii. **Current gain**

$$A = \frac{h_{21}}{1 + h_{22} + r_L} \approx h_{21}\ (h_{22}\cdot r_L \ll 1)$$

$$A_i = \frac{h_{fe}}{1 + h_{oe}r_L}$$

iii. **Voltage gain**

$$A_v = \frac{-h_{21}}{Z_i\,(h_{22} + 1/r_L)}$$

or

$$A_v = \frac{-h_{fe}}{Z_i\,(h_{oe} + 1/r_L)}$$

Integrating circuit: $v_0 = \dfrac{1}{RC}\int_0^t v_i\,dt$

The output voltage is proportional to the integral of the input voltage under the assumption that R_{Ci} is very large, the time as compared to period of input wave (Fig. 15.3).

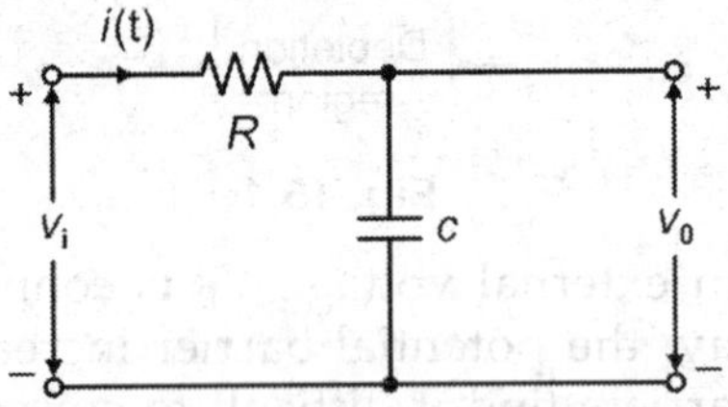

Fig. 15.3

Differentiator: By making the RC, time constant small with respect to the interval of any applied voltage, the output voltage

$$v_0 = iR \cong RC\frac{dv_i}{dt}$$

Thus, with small value of time constant RC, the output voltage will be proportional to the derivative of the input voltage and the circuit is known as **differentiating circuit** (Fig. 15.4).

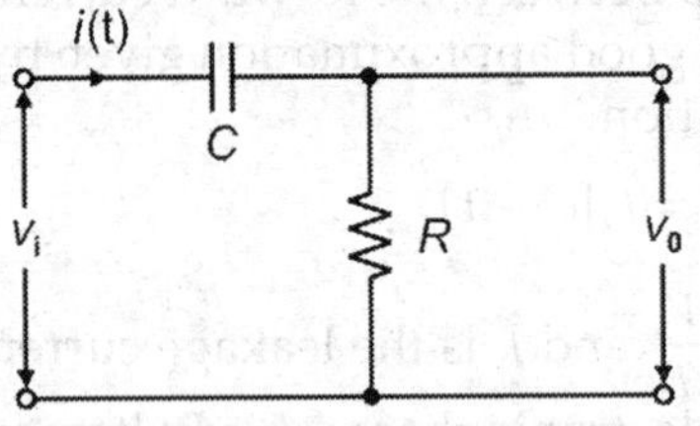

Fig. 15.4

Field effect transistors: In a *junction field effect transistor* (JFET), the width of depletion layers controls conductance. For the JFET, the drain current in the constant-current–region is

$$i_{DS} = I_{DSS}(1 - V_{GS}/V_p)^2$$

where i_{DS} is the drain current in the constant–current region, I_{DSS} is the value of i_{DS} with gate shorted to source, and V_p is the pinch-off voltage.

For an enhancement MOSFET, the transfer characteristic is

$$i_{DS} = K(V_{GS} - V_T)^2$$

where K is a device parameter and V_T is the turn-on or threshold voltage.

Differential amplifier: The function of a differential amplifier is to amplify the difference between two signals. Figure represents a linear active device with two signals v_1, v_2 and one output voltage v_0 measured with respect to ground. In ideal differential amplifier, having A_d, the gain of differential amplifier (Fig. 15.5).

$$v_0 = A_d\,(v_1 - v_2)$$

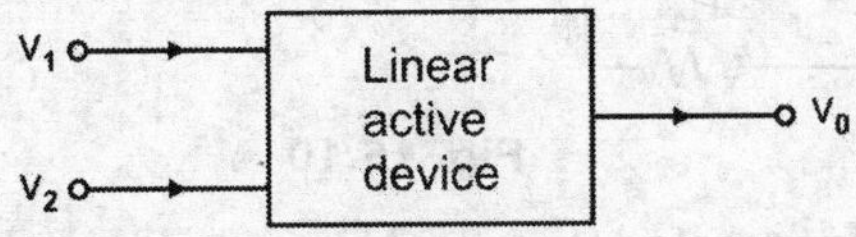

Fig. 15.5

Differential-mode operation: When a differential input signal is applied, the input voltages v_{i1} and v_{i2} will be equal to amplitude and opposite in phase. Therefore, if I_{c1} increases, I_{c2} will decrease by an equal amount. Since V_{c1} decreases and V_{c2} increases simultaneously, the output voltage (= $V_{c1} - V_{c2}$) and hence the gain will be very high as compared to a circuit using single transistor. Differential mode signals are not affected by R_e because as I_{c1} increases, I_{c2} decreases, and as a result the total emitter current (= $I_{c1} + I_{c2}$) remains constant.

In order to amplify differential signals and suppress common-mode signals, R_e should be as large as possible so that a small change in emitter voltage does not lead to any change in the emitter current.

A differential amplifier is characterised by a parameter called *common-mode rejection ratio* (CMRR) which is defined as

$$\text{CMRR} = \frac{\text{Differential-mode gain}}{\text{Common-mode gain}} = \frac{A_d}{A_c}.$$

Obviously, CMRR should be as high as possible. Typical values of CMMR is 40–120 dB.

Other specifications-common-mode voltages range V_{CMR}: It is the input voltage range over which the transistor operate in the linear region and not get saturated. It is usually specified for a given output distortion or a given CMRR reduction (Fig. 15.6).

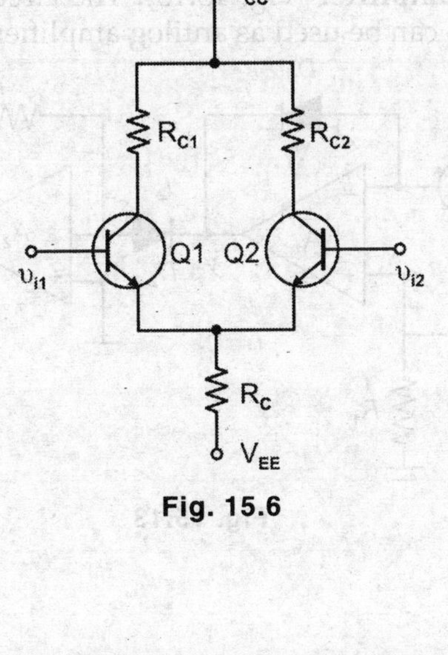

Fig. 15.6

Output-offset voltage V_{OO}: It is the dc voltage difference between the collectors when base inputs are equal and thus a measure of unbalance of the amplifier.

Input-offset voltage V_{IO}: It is the input difference that must be applied between the bases to make the two collector voltages equal. Thus it is a measure of mismatching of V_{BES} of the transistors. Its value is of the order of 1 mV. It is a more useful specification because it in independent of the gain.

Input bias-current offset I_{IO}: It is a measure of h_{fe} mismatch of the transistor. It is the maximum difference in input bias currents that must be supplied to make the dc voltage difference between collectors zero. It is about 5–10% of the dc input current.

BW of differential amplifier: The bandwidth of a differential amplifier is typically nearly double of the bandwidth obtainable from a single-stage *RC* coupled amplifier. It is comparable with that of a cascade amplifier. Sometimes only one R_c is used to improve the *BW*. Due to high *BW* and CMRR, it is an important configuration in integrated circuits. Different stages can be cascaded by direct coupling.

Direct-coupled amplifiers: If the amplifier is to respond to low frequency alternating signals, a direct connection is made from the collector of Q_1 to the base of Q_2 in the two-stage direct-coupled amplifier shown in Fig. 15.7. If a small increase in signal raises the base of Q_1, this increases the collector current of Q_1 causing larger voltage to appear across R_{c1}. This lowers the collector of Q_1, decreasing the base current of Q_2, and thus decreasing its collector current. Due to this, the output voltage at the collector of Q_2 increases.

These amplifiers are very important in integrated circuit as the coupling capacitors which require lot of space proportional to their values and take much larger area that needed to form a transistors are not present. Moreover in designs where size is a crucial factor (e.g. in case of hearing aids), DC-coupled amplifiers are very common.

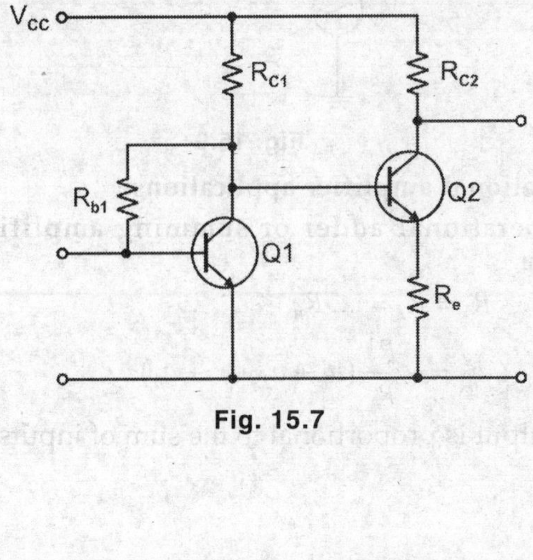

Fig. 15.7

Operational amplifiers

Ideal OP amplifiers: An ideal operational amplifier has the following characteristics:

i. Infinite input resistance, $R_i = \infty$
ii. Zero output resistance, $R_0 = 0$
iii. Infinite voltage gain, $A_v = -\infty$
iv. Infinite bandwidth
v. Zero drift
vi. Perfet balance

1. **Inverting amplifier** (Fig. 15.8): Voltage gain

$$Av = \frac{V_0}{V_{in}} = -\frac{Z_2}{Z_1} = -\frac{R_2}{R_1}$$

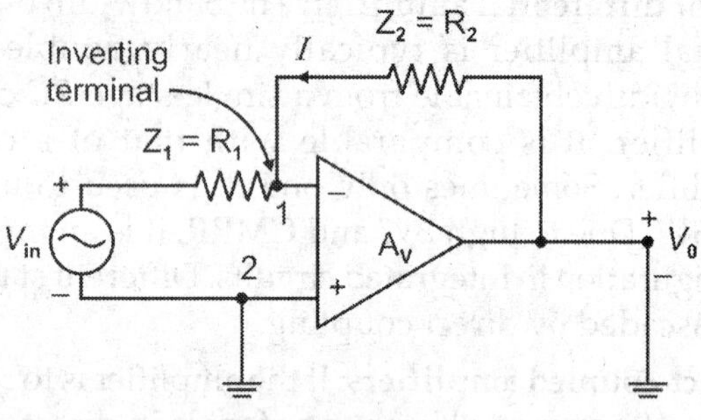

Fig. 15.8

2. **Non-inverting amplifier** (Fig. 15.9): Voltage gain

$$A_v = \frac{V_0}{V_{in}} = \left(1 + \frac{R_2}{R_1}\right)$$

Hence, the closed loop gain is always greater than unity.

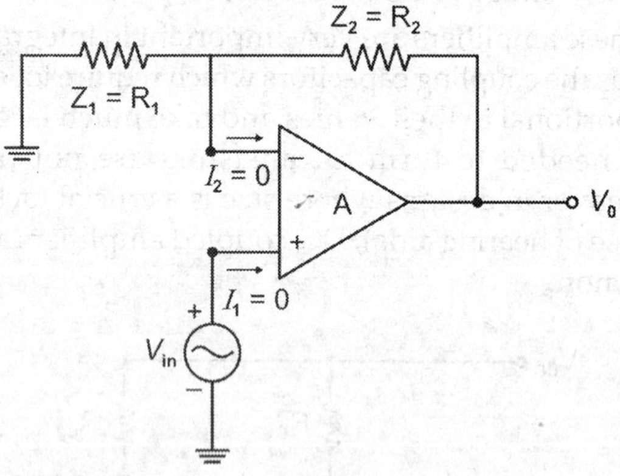

Fig. 15.9

Operational amplifier applications:

1. **Operational adder or summing amplifier** (Fig. 15.10):

If $R_1 = R_2 = ..., R_n$

then $v_0 = -\frac{R^1}{R_1}(v_1 + v_2 + ... + v_n)$

Output is proportional to the sum of inputs.

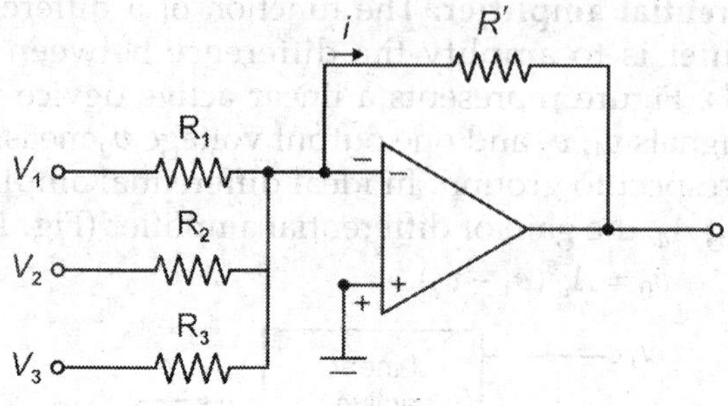

Fig. 15.10

Analog integrator (Fig. 15.11):

$$v_0 = -\frac{1}{C}\int i\,dt = -\frac{1}{RC}\int v\,dt$$

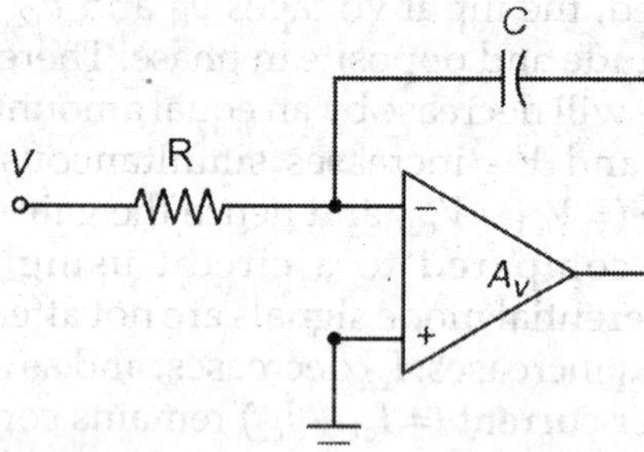

Fig. 15.11

Logarithmic amplifier (Fig. 15.12): Output voltage is proportional to the logarithm of input voltage

$$v_0 = -V_f - \eta V_T\left(\ln\frac{V_s}{R} - \ln I_0\right)$$

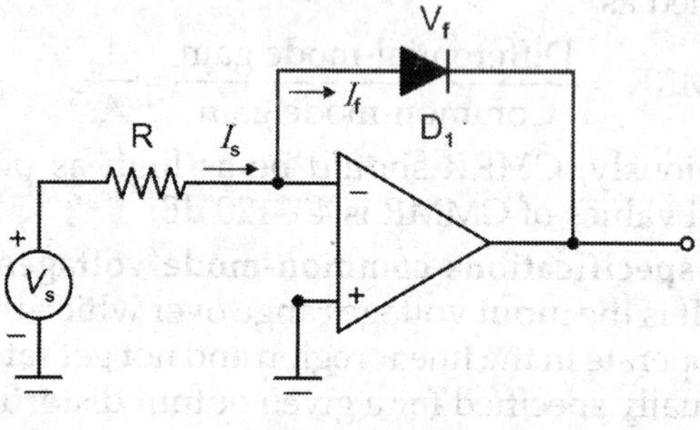

Fig. 15.12

Antilog amplifier (Fig. 15.13): The circuit shown in Fig. 15.13 can be used as antilog amplifier

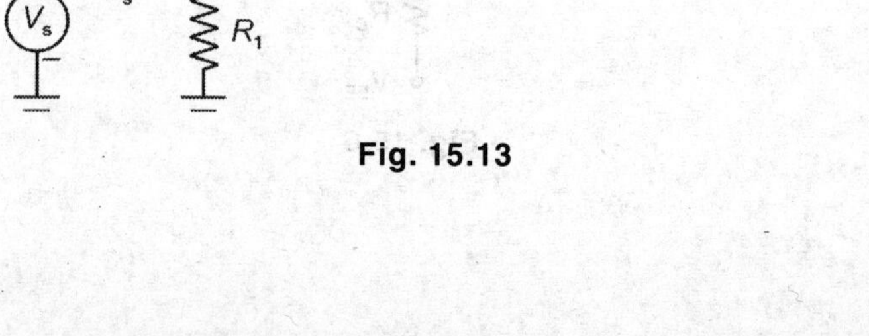

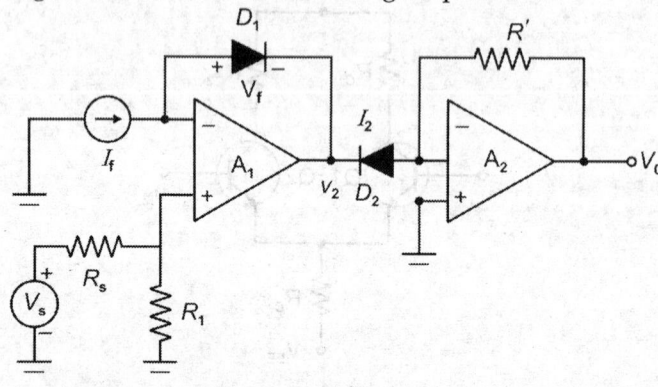

Fig. 15.13

$$V_0 = R'I_f \ln^{-1}\left[-V_s\left(\frac{R_1}{R_1+R_2}\cdot\frac{1}{\eta V_T}\right)\right]$$

Logarithmic multiplier (Fig. 15.14): The log and antilog amplifier can be used for the multiplication or division of two analog signals V_{s1} and V_{s2}.

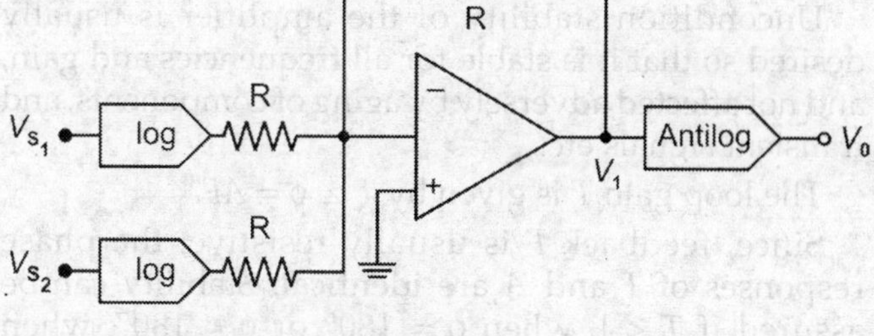

Fig. 15.14

Logarithmic multiplier of two analog signals

$$V_0 = KVs_1V_{s2}$$

Feedback amplifiers: Figure 15.15 shows general flow diagram of a feedback amplifier. The gain of the amplifier without feedback is *A* and feedback from output to input occurs through the feedback network *F*.

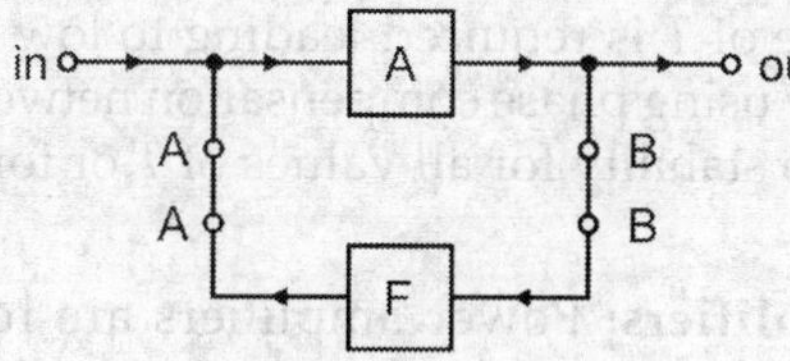

Fig. 15.15

Basic principle: Figure 15.16 shows four feedback topologies due to two ways in which the output and feedback signals can be combined and two ways in which output connection can be formed. If the output voltage is also the input voltage to *F*, the parallel output connection is called *shunt or voltage feedback.* If the feedback networks is in series with the output, it samples the output current and called *series or current feedback.*

Consider Fig. 15.16(a). Let *A*(*s*) be the gain of the amplifier and the feedback be *F*(*s*). The variables s shows that these gains ar frequency dependent.

Now, $A(s) = V_0(s)/V_1(s)$

$$F(s) = V_2(s)/V_0(s)$$

$$V_i(s) = V_1(s) - V_2(s)$$

From these three equations, the overall gain of the feedback amplifier is

$$A(s) = \frac{V_0(s)}{V_i(s)} = \frac{A(s)}{1 - A(s)F(s)}$$

The gain $A_f(s)$ is also called *closed-loop gam* and represents the actual gain of the amplifier with feedback. On the other hand. *A*(*s*) is called *open-loop gain.* The term *A*(*s*)*F*(*s*) is referred as *return-ratio* or *loop gain T*(*s*). It denotes the amount of the input required to *A* is referred as *return-ratio* or *loop gain T*(*s*). It denotes the amount of the output signal of *A* that is transmitted to returned through *F*.

$$T(s) = A(s)\ T(s)$$

and $$A_f(s) = \frac{A(s)}{1 - T(s)}$$

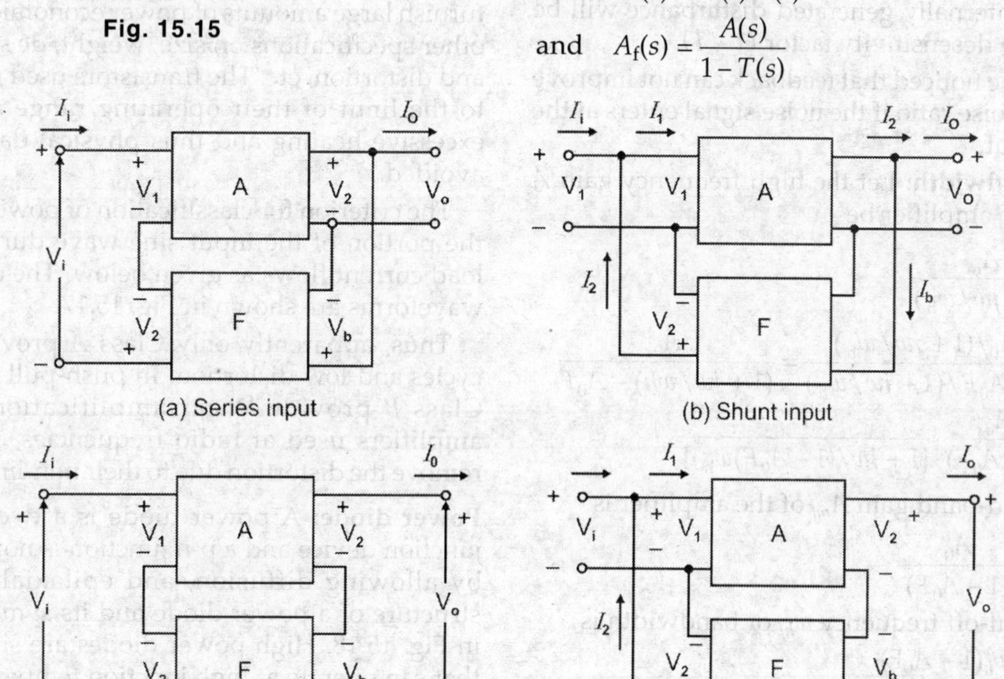

Fig. 15.16

If $T(s) > 0$, then $V_i(s)$ has the same sign as $V_1(s)$ and reinforces the original signal $V_1(s)$ present al the input to A This results in increase in $V_0(s)$ leading to decreased input to F. This cycle continues until the amplifier saturates. This is called *regenerative or positive feedback.* If $T(s) < 0$, the feedback signal now opposes the initial output signal and hence initial output change.

Desensitivity: Consider the feedback formula,

$$A_f(s) = \frac{A(s)}{1 - A(s)F(s)}$$

Differentiating this equation with respects to A

$$\frac{dA_f}{dA} = \frac{1}{(1-AF)^2} = \frac{A_f}{A(1-AF)}$$

or $$\frac{dA_f}{A_f} = \frac{1}{(1-AF)} \cdot \frac{dA}{A}.$$

The term $(1 - AF) = (1 - T)$ is called *desensitivity factor.* Thus, larger the loop gain, the more insensitive or stable the amplifier to parameter variations.

Noise reduction: Let a noise voltage V_n appears at the output of the amplifier A due to noise, distortion or any other nonlinear, disturbance is generated. If V_{nf} is noise voltage present at the output of the complete amplifier, then a voltage of AFV_{nf} will be present at the output. Thus, the total amount of noise voltage will be

$$V_{nf} = V_n + AFV_{nf}$$

or $$V_{nf} = \frac{V_n}{1-AF} = \frac{V_n}{1-T}$$

Thus, any internally generated disturbance will be reduced by the desensitivity factor $(1 - T)$.

But it must be noticed that feedback can not improve tfie signal-to-noise ratio if the noise signal enters at the amplifiers input.

Increased Bandwidth: Let the high frequency gain $A(\omega)$ of the basic amplifier be

$$A(w) = \frac{A_m}{(1 + jw/w_h)}$$

$$A_f(w) = \frac{A_0/(1 + jw/w_h)}{1 - A_0F/(1 + jw/w_h)} = \frac{A_0}{(1 + jw/wh) - A_0F}$$

$$= \frac{A_0}{(1 - A_0F)} \cdot \frac{1}{[1 + jw/(1 - A_0F)w_h)]}$$

The new mid-band gain A_{mf} of the amplifier is

$$A_{mf} = \frac{A_0}{(1 - A_0F)}$$

and its high cut-off frequency w_{hf} or bandwidth is

$$w_{hf} = w_h(1 - A_0F).$$

Thus, the gain decreases, but bandwidth of the amplifier increases by the same factor.

Feedback effects on Z_i and Z_0: By employing feedback, impedance level control can be obtained which provides match impedance levels and thus maximising the gain.

Frequency response of amplifiers: Generally speaking, A, F and hence T can vary with frequency. The loop gain T may.be positive in one frequency range and negative in another with the result that if $T(\omega) \to +1$, $A_f(\omega) \to \infty$ and the amplifier becomes highly unstable.

Uncondition stability of the amplifier is usually desired so that it is stable for all frequencies and gain, and not affected adversely by aging of components, and transient signals etc.

The loop gain T is given by $T \angle \phi = AF$.

Since, feedback F is usually resistive, the phase responses of T and A are identical. Stability can be assured if $T \leq 1$, when $\phi = 180°$ or $\phi \leq 180°$, when $T = 1$.

Amplitude or gain margin is defined as the amount by which T is less than 1 when $\phi = 180°$.

Phase Margin is the amount by which ϕ is less than 180° when $T = 1$. Generally a $PM = 45°$ is used to ensure stability which means that T should be ≤ 1 when its phase angle is $-(180° - 45°)$ or $-135°$.

For larger bandwidths or greater desensitivity, a larger value of T is required leading to low A_f. This is achieved by using phase compensation networks which may lead to stability for all values of T or for specified ranges of T.

Power amplifiers: Power amplifiers are required to furnish large amounts of power economically and meet other specifications on size, weight, dc supply voltage and distortion, etc. The transistors used may be driven to the limit of their operating range and therefore excessive heating and thus physical damage is to be avoided.

The criterion for classification of power amplifiers is the portion of the input sine wave during which the load current flows as given below. The corresponding waveforms are shown in Fig. 15.17.

Thus, apparently only Class A provides complete cycles and low distortion. In push-pull configuration, Class B provide linear amplification. In Class C amplifiers used at radio frequencies, tuned circuits remove the distortion due to their non-linear operation.

Power diode: A power diode is a two-terminal p-n junction device and a p-n junction is normally formed by allowing diffusion, and epitaxial growth. The structuie of a power diode and its symbol are shown in Fig. 15.18. High power diodes are silicon rectifiers that can operate at high junction temperatures. Power diodes have larger power, voltage, and current handling capabilities than ordinary signal diodes. In addition, the switching frequencies of power diodes are low as compared to signal diodes.

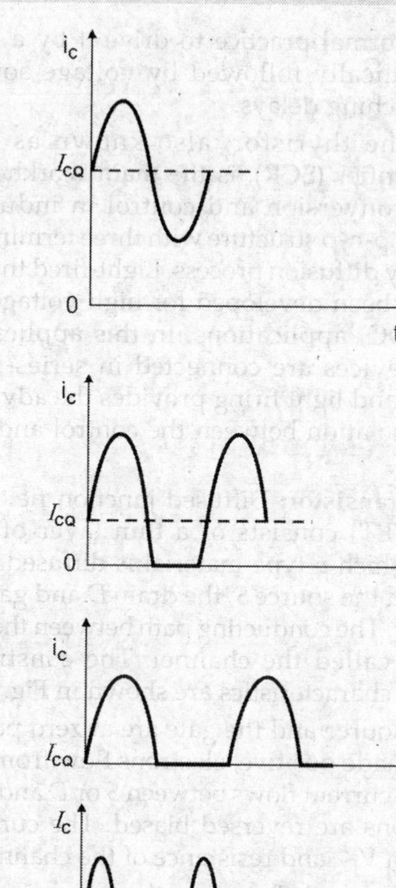

Fig. 15.17

General purpose diodes: The general purpose rectifier diodes have relatively high reverse recovery time, typically 25 μsec, and are used in low-speed applications, where recovery time is not critical. These diodes cover current ratings from less than 1A to several thousands of amperes with voltage ratings from 50 V to around 5 kV. These diodes are generally manufactured by diffusion. However, alloyed types of rectifiers which are used in welding power supplies are most effective and rugged, and their ratings can go up to 300 A and 1000 V.

Fast recovery diodes: The fast recovery diodes have low recovery time, normally less than 5 μsec. They are used in chopper and inverter circuits, where the speed of recovery is often of critical importance. These diodes cover current ratings from less than 1 A to hundreds to amperes, with voltage ratings from 50 V to 3 kV.

For voltage ratings above 400 V, fast recovery diodes are generally made by diffusion and recovery time is controlled by platinum or gold diffusion. For voltage ratings below 400 V, epitaxial diodes provide faster switching speeds than that of diffused diodes. The epitaxial diodes have a narrow base width resulting in fast recovery time of as low as 20 nanosec.

Schottky diodes: The charge storage problem of a p-n junction can be minimised in a Schottky diode. It is accomplished by setting up a 'barrier potential' with a contact between a metal and a semiconductor.

A layer of metal is deposited on a thin epitaxial layer of n-type silicon. The potential barrier simulates the behaviour of p-n junction. However, the rectifying action depends on the majority carriers only and as a result there is no excess minority carriers to recombine. The recovery effect is due to the self-capacitance of the semiconductor junction. The recovered charge of a Schottky diode is much less than that of an equivalent p-n junction diode. It has a relatively low forward voltage.

Power MOSFET (metal oxide semiconductor field effect transistor): It is a very fast switching transistor that has shown great promise for low power (up to a few kWs) involving frequency (up to 1 MHz) applications. Originally, the devices employed surface-groove technology, called VMOS but today planner DMOS structure is used for high voltage devices.

The circuit symbol of MOSFET is shown in Fig. 15.19. The three terminals are called drain (D), source (S) and gate (G). The current flow is from drain to source. The device has no reverse-voltage blocking capability and it always comes with an integrated reverse rectifier.

Unlike a bipolar transistor (which is a current driven device), a MOSFET is a voltage controlled majority carrier device. With positive voltage applied to the gate

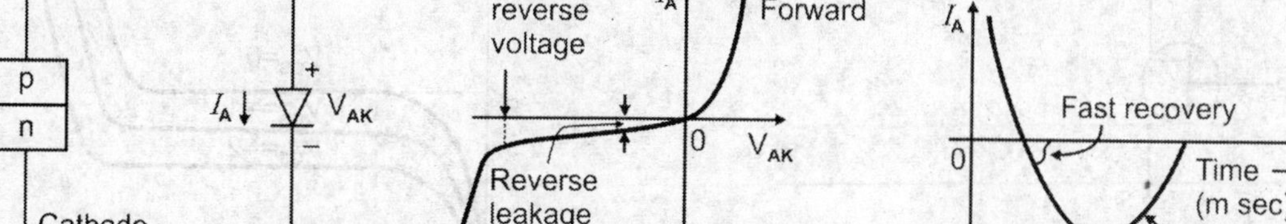

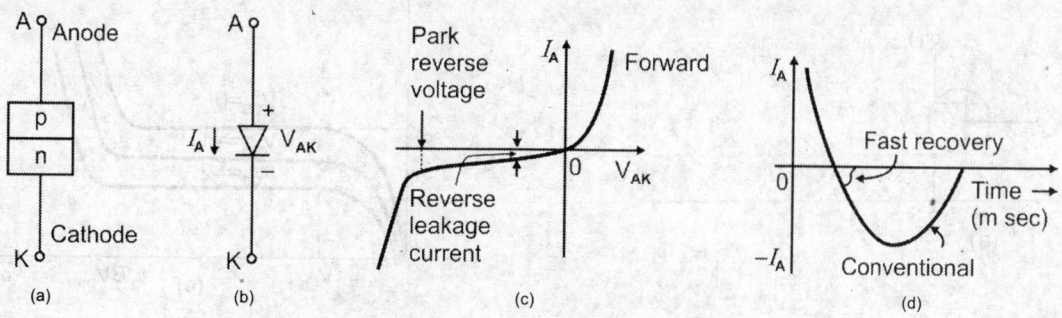

Fig. 15.18: Structure of a power diode and its symbol

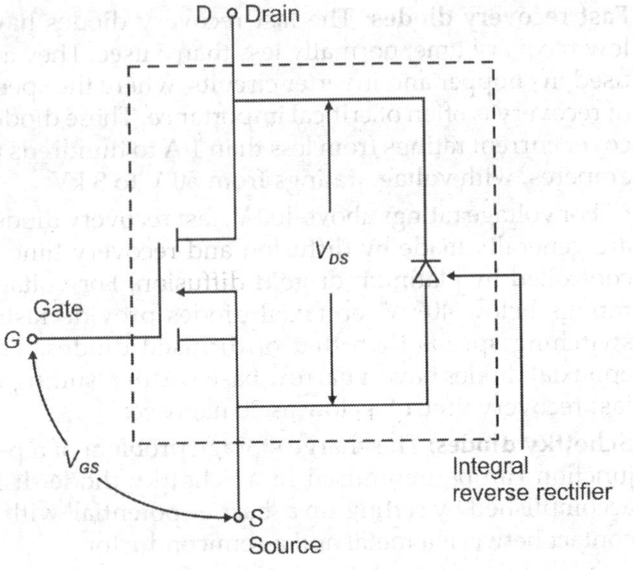

Fig. 15.19

(i.e. V_{GS} positive), the transistor switches on. The gate is isolated by a silicon oxide SiO_2 layer, and therefore the gate circuit circuit input impedance is extremely high. This feature allows a MOSFET to be driven directly from CMOS or TIL logic. The gate drive current is therefore very low, it can be less than a milliampere.

The conduction loss of a high voltage MOSFET is very high, the switching loss is almost negligible. The device does not have minority-carrier delay time as in a bipolar device, and its switching times are determined essentially by the ability of the drive to charge and discharge a tiny input capacitance, C_{ISS} defined as $C_{GS} + C_{GD}$ with C_{DS} shorted, where C_{GS} is the gate to source capacitance C_{GD}, the gate to drain capacitance, and C_{DS} the drain to source capacitance. Although MOSFET can be controlled statistically by the voltage source, it is normal practice to drive it by a current source dynamically followed by voltage source to minimise switching delays.

Thyristor: The thyristor, also known as silicon controlled rectifier (SCR), is the main workhorse for bulk power conversion and control in industry. A thyristor has a p-n-p structure with three terminals and is fabricated by diffusion process. Light-fired thyristors have recently been developed for high voltage direct current (HVDC) applications. In this application, a number of devices are connected in series-parallel combination, and light firing provides the advantages of electrical isolation between the control and power circuits.

Field effect transistor: Diffused junction field effect, transistor (JFET) consists of a thin layer of *n*-type material on which *p*-type material is diffused. Ohmic contacts called the source *S*, the drain *D* and gate *G* are then taken out. The conducting path between the source and drain is called the channel. The construction, principle and characteristics are shown in Fig. 15.20.

When the source and the gate are at zero potential, and drain is made positive, electrons flow from *S* to *D*, but negligible current flows between *S* or *D* and *G* since the *pn*-junctions are reversed biased. The current I_{DS} depends upon V_{DS} and resistance of the channel.

As V_{DS} is increased, the *pn*-junctions are more heavily reverse biased and the area of depletion region increases leading to narrowing of the channel. This leads to increase in channel resistance. This process continues until at some voltage, V_{PO} called the *pinch-off* voltage since it cuts off the channel between *S* and *D*. The drain current I_D remains constant because potential of point *A* essentially remains equal to V_{PO}. When $V_{DS} > V_{PO}$, the current remains I_{PO}, and then rises slowly. When

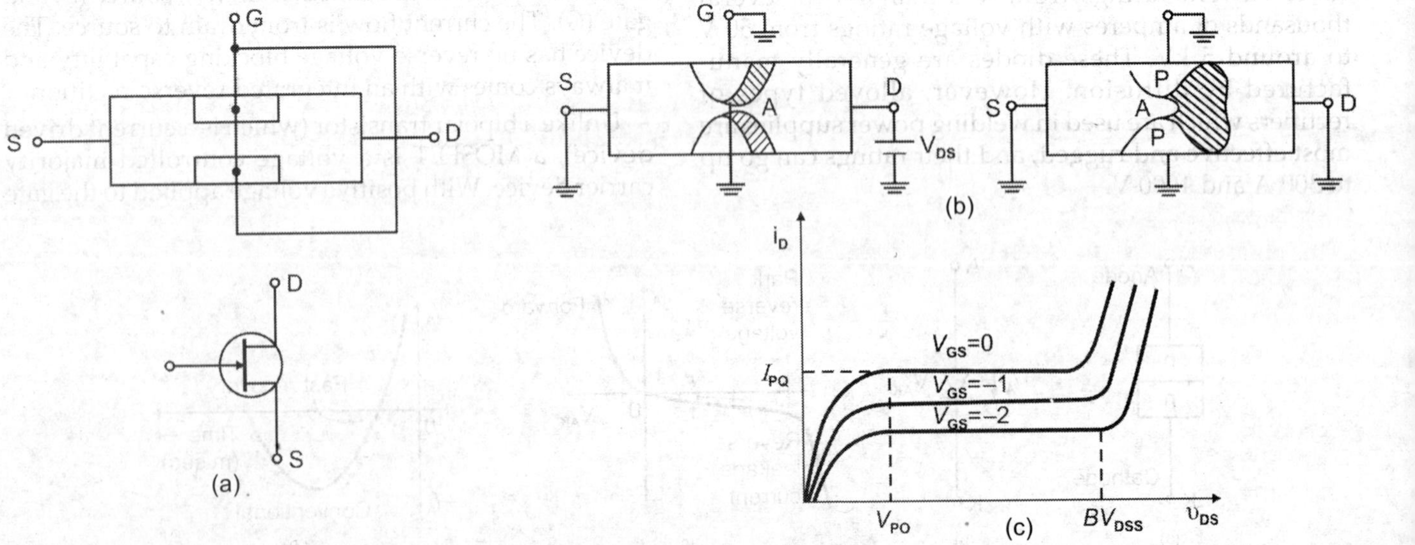

Fig. 15.20

$V_{DS} = BV_{DSS}$, the breakdown voltage, an avalanche break-down occurs and i_D rises very fast.

When V_{DS} is held constant and V_{GS} is made negative, the *pn*-junction is reverse biased with the result that depletion regions spread. As a consequence, channel resistance increases and i_D decreases. When V_{GS} is made positive, the depletion region decreases and channel opens as *pn*-junction get forward biased and I_{GS} starts flowing. Therefore, in order to avoid I_{GS}, *n*-type JFET is usually operated with V_{GS} either negative or slightly positive.

Thus in a JFET, varying the gate voltage varies the channel width, resistance and current i_D. This means that FET is a voltage-sensitive device in contrast to junction transistor which is a current sensitive device.

In the saturation region (between V_{PO} and BV_{DSS}),

$$i_D = I_{PO}\left[1 + \frac{3V_{GS}}{V_{PO}} + 2\left(-\frac{V_{GS}}{V_{PO}}\right)^{1/2}\right] \cdot V_{GS} < 0$$

Fig. 15.20(b) shows a typical biasing circuit. The dc load line for the FET is

$$V_{DD} = V_{DS} + i_D\,(R_a + R_s)$$

Since gate current is very small,

$$V_{GS} = -\,i_D R_S$$

Therefore, V_{GS} is developed directly across R_S by the drain current i.e., JFET is self-biased. There is no need for external bias circuit in contrast to a junction transistor.

IGFET or MOSFET: Insulated Gate FET consists of definite layer of insulating silicon dioxide formed between gate and channel. Thus, it does not has a junction and consequently, the disadvantage of leakage current and the need for reverse bias are eliminated. This insulated gate can be forward-biased without any flow in the gate.

In Fig. 15.21(a), a channel of n-material has been diffused on a *p*-type semiconductor and enlarged ends of the *n*-channel form the source and drain. The gate is only a layer of conducting metal. It is insulated from the channel by a layer of silicon dioxide. If the gate is made more negative, its voltage produces a field effect. This enlarges the depletion region in the *n*-channel with the result that the flow of electrons decreases through the channel. This mode of operation of MOSFET is called *depletion mode.*

In the enhancement mode of operation, the MOSFET is usually operated with a positive gate-source potential which leads to increase in drain current.

In this saturation region, the drain current is given by $i_D = I_{PO}\left(1 + \frac{V_{GS}}{V_{PO}}\right)^2$.

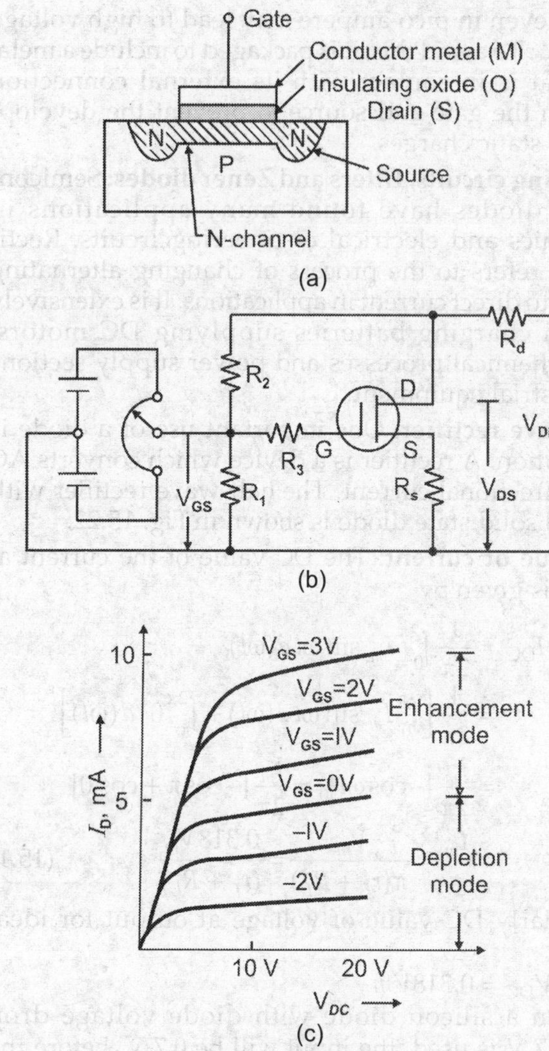

Fig. 15.21

The MOSFET is symmetrical in the sense that its drain and source connections can be interchanged with only minor changes in its performance.

A basic amplifier and the characteristics of MOSFET are shown in Figs 15.21(b) and 15.21(c). In IGFET, the gate is most often forward biased. The *dc* load line equation is

$$V_{DD} = V_{DS} + i_D(R_d + R_S)$$

and $$V_{GS} = \left(\frac{R_1}{R_1 + R_2}\right) \cdot V_{DD} - i_D R_S = V_{GG} - i_D R_S$$

The resistance R_3 does not affect the dc bias since the gate current is negligible. Its function is to make ac input impedance high.

The main disadvantage of MOSFET is that it is easily damaged by static charges. With very small gate capacitance and gate current, a small amount of charge is sufficient to develop a gate voltage of 35 to 50 V, large enough to cause a destructive avalanche breakdown of the insulating layer which gets punctured. Static current even in pico-ampere may lead to high voltage across r_{gs}. To avoid this, it is packaged to include a metal ring that is in contact with its external connection between the gate and source to prevent the development of static charges.

Rectifying circuits, filters and Zener diodes: Semiconductor diodes have found many applications in electronics and electrical engineeringcircuits. Rectification refers to the process of changing alternating current to direct current in applications. It is extensively used in charging batteries supplying DC motors, electrochemical processes and power supply sections of industrial equipment.

Half wave rectifier: One important use of a diode is rectification. A rectifier is a device which converts AC to unidirectional current. The half-wave rectifier with an ideal solid state diode is shown in Fig. 15.22.

DC value of current: The DC value of the current at output is given by

$$I_{DC} = \frac{1}{2\pi}\int_0^{2\pi} I_m \sin\omega t\, d(\omega t)$$

$$= \frac{1}{2\pi}\left[\int_0^{\pi} I_m \sin\omega t\, d(\omega t) + \int_{\pi}^{2\pi} 0 \cdot d(\omega t)\right]$$

$$= \frac{I_m}{2\pi}[-\cos\omega t]_0^{\pi} = \frac{I_m}{2\pi}[-\cos\pi + \cos 0]$$

$$= \frac{I_m}{\pi} = \frac{V_m}{\pi(r_f + R_L)} = \frac{0.318\,V_m}{(r_f + R_L)} \quad (15.1)$$

Similarly DC value of voltage at output for ideal diode,

$$V_{DC} = 0.318Vm.$$

When a silicon diode with diode voltage drop $VT = 0.7$ V is used, the input will be 0.7 V, before the diode conducts, resulting in a zero output level until the transition occurs. When conducting, the difference between vo and vi is a fixed level of $VT = 0.7$ V and $v_o = v_i - VT$, which reduces the resulting DC voltage level. If Vm is closer to VT then

$$V_{DC} = 0.318\,(Vm - VT) \quad (15.2)$$

RMS value of current: This is given by the expression

$$I_{rms} = \sqrt{\frac{1}{2\pi}\int_0^{2\pi}(I_m \sin\omega t)^2 d(\omega t)}$$

$$= \sqrt{\frac{1}{2\pi}\left[\int_0^{\pi}(I_m \sin\omega t)^2 d(\omega t) + \int_{\pi}^{2\pi} 0 \cdot d(\omega t)\right]}$$

$$= \frac{I_m}{2} = \frac{V_m}{2(r_f + R_L)} = \frac{0.5V_m}{(r_f + R_L} \quad (15.3)$$

DC voltage across the load

$$V_{DC} = I_{DC} \cdot R_L = \frac{V_m}{\pi(r_f + R_L)} \cdot R_L \quad (15.4)$$

DC output power: P DC is given by

$$P_{DC} = I_{DC}^2 \cdot R_L = \left(\frac{I_m}{\pi}\right)^2 \cdot R_L = V_{DC} \cdot I_{DC}$$

$$= V_{DC} \cdot \frac{V_{DC}}{(R_L + r_f)} = \frac{(0.318V_m)^2}{(R_L + r_f)} \quad (15.5)$$

AC power input to the rectifier

$$P_{AC} = (I_{rms})^2 \cdot (r_f + R_L)$$

$$= \left(\frac{I_m}{2}\right)^2 (r_f + R_L) = (0.5I_m)^2 (r_f + R_L) \quad (15.6)$$

Ripple factor γ: The lack of smoothness of the waveform is given by ripple factor.

$$\gamma = \frac{\text{Effective value of AC component of current}}{\text{DC component of current}}$$

The total load current

$$I_m = \sqrt{I_{DC}^2 + I_{AC}^2}$$

$$I_{AC} = \sqrt{I_{rms}^2 + I_{DC}^2}$$

$$\gamma = \frac{\sqrt{I_{rms}^2 + I_{DC}^2}}{I_{DC}}$$

$$= \sqrt{\left(\frac{I_{rms}}{I_{DC}}\right)^2 - 1} = \sqrt{\frac{I_m/2}{I_m/\pi}} - 1 = 1.21 \quad (15.7)$$

Peak inverse voltage (PIV): The peak inverse voltage is another factor in rectifier design. It is the maximum instantaneous voltage which the diode has to withstand without destroying the junction in the blocking state. In a half wave rectifier, the reverse voltage across the diode during nonconduction period equals the secondary voltage. Thus, PIV = $V_m = \sqrt{2}V = 1.414$ V.

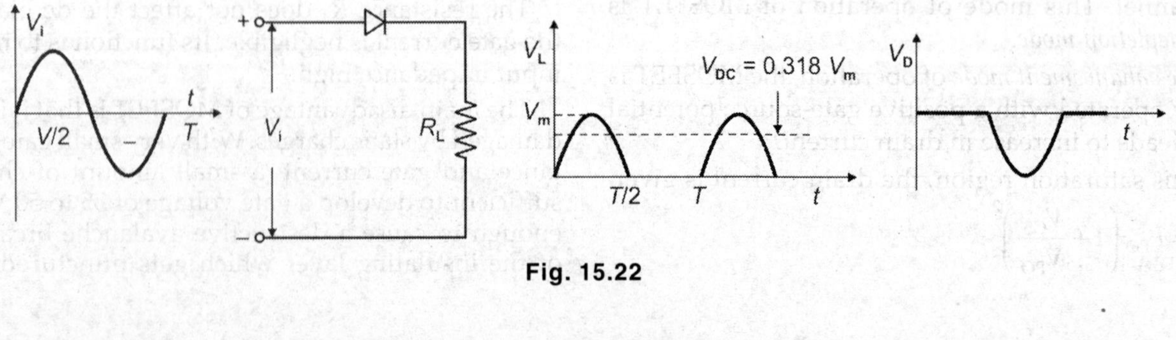

Fig. 15.22

Rectifier efficiency, η: It is denoted by η

$$= \frac{\text{DC load power}}{\text{AC power input from transformer}}$$

$$\eta\% = \frac{P_{DC}}{P_{AC}} = \frac{(I_L/\pi)^2 \cdot R_L}{(I_m/2)^2 (r_f + R_L)} \times 100$$

$$= \frac{0.406}{1 + r_f/R_L} \times 100 \qquad (15.8)$$

The efficiency will be maximum if r_f is negligible as compared to R_L,

i.e. $\eta_{max} \cong 40.6\%$

Transformer utilisation factor

$$\text{TUF} = \frac{\text{DC load power}}{\text{AC rating of transformer secondary}} = \frac{P_{DC}}{V_s I_s}$$

$$= \frac{P_{DC}}{P_{AC}\text{ (rated)}} = \frac{(I_m/\pi)^2 \cdot R_L}{(V_m/\sqrt{2})(I_m/2)}$$

As $V_m = I_m (r_f + R_L)$

$$\therefore \text{TUF} = \frac{0.286 R_L}{(r_L + R_L)} = \frac{0.286}{\left(1 + \frac{r_f}{R_L}\right)} \qquad (15.9)$$

If $r_f << R_L$ then TUF = 0.286 ≈ 0.29.

Displacement factor: The angle between the fundamental components of the input current and voltage is called the displacement angle. If a represents the displacement angle, then cos α is called the displacement factor, i.e.

$$\text{DF} = \cos\alpha \qquad (15.10)$$

Harmonic factor:

$$\text{HF} = \left[\frac{I_s^2 - I_{sf}^2}{I_{sf}}\right]^{1/2} = \left[\left(\frac{I_s}{I_{sf}}\right)^2 - 1\right]^{1/2} \qquad (15.11)$$

where I_{sf} is the fundamental component of input current I_s.

Input power factor: It is defined as

$$\text{PF} = \frac{V_s I_{sf}}{V_s I_s} \cos\alpha = \frac{I_{sf}}{I_s} \cdot \cos\alpha \qquad (15.12)$$

Full-wave rectifier: In full-wave rectification, current flows through load in the same direction for both half cycles of input AC voltage. There are two types of circuits used for full-wave rectifiers, namely (1) centre-tap circuit, and (2) bridge circuit.

Centre-tap rectifier with resistive load: The circuit employs two-diodes D_1 and D_2 and a centre-tapped secondary winding. The diode D_1 is forward-biased and D_2 reverse-biased in the positive half cycle and current flows from point 2 to point 1 in the load *RL*. Whereas in the negative half cycle D_1 is reverse-biased and D_2 is forward-biased and again current flows from point 2 to point 1. Hence, DC output is obtained across *RL*, the corresponding waveforms are shown in Fig. 15.23).

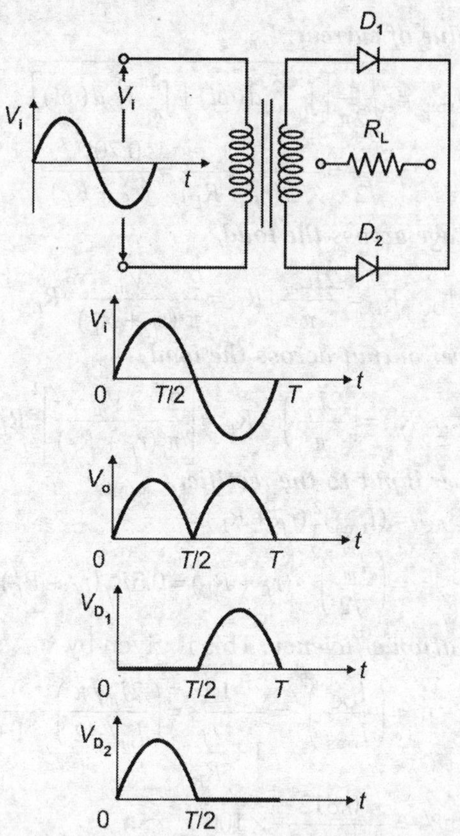

Fig. 15.23

Analysis of full-wave rectifier: Let the forward and load resistance be *rf* and *RL* and the AC input voltage be

$$v_i = \sqrt{2}V \sin\omega t = V_m \sin\omega t$$

Current due to diodes D_1, D_2

$$\left.\begin{aligned} i_{d1} &= \frac{V_m}{r_f + R_L}\sin\omega t \\ i_{d2} &= 0 \\ i_{d1} &= 1 \end{aligned}\right\} \quad 0 \le \omega t \le \pi$$

$$\left.\begin{aligned} i_{d1} &= 0 \\ i_{d2} &= \frac{V_m}{r_f + R_L}\sin\omega t \\ &= I_m \sin\omega t \end{aligned}\right\} \quad \pi \le \omega t \le 2\pi$$

where $I_m = \frac{V_m}{(r_f + R_L)}$

Diode currents, i_{d1}, i_{d2} are however, the same as both diodes are similar.

DC value of current

$$I_{DC} = \frac{1}{\pi}\int_0^{\pi} I_m \sin\omega t\,(\omega t) = \frac{1}{\pi}[-I_m\cos\omega t]_0^{\pi} = \frac{2I_m}{\pi}$$

$$= \frac{2V_m}{\pi(r_f + R_L)} = \frac{0.63 V_m}{(r_f + R_L)} \qquad (15.13)$$

RMS value of current, I_{rms}

$$I_{\text{rms}} = \sqrt{\frac{1}{2\pi}\left[\int_0^{\pi} i_{d1}^2 d(\omega t) + \int_{\pi}^{2\pi} i_{d2}^2 d(\omega t)\right]}$$

$$= \frac{I_m}{\sqrt{2}} = \frac{V}{\sqrt{2}(r_f + R_L)} = \frac{0.70V_m}{(r_f + R_L)} \quad (15.14)$$

DC voltage across the load,

$$V_{DC} = I_{DC} \cdot R_L = \frac{2IL_m}{\pi} \cdot R_L = \frac{2}{\pi}\frac{V_m}{(r_f + R_L)} \cdot R_L \quad (15.15)$$

DC power output across the load

$$P_{DC} = I_{DC}^2 \cdot R_L = \left(\frac{2I_m}{\pi}\right)^2 \cdot R_L = \left[\frac{2}{\pi}\frac{V_m}{(r_f + R_L)}\right]^2 R_L \quad (15.16)$$

AC power input to the rectifier,

$$P_{AC} = (I_{\text{rms}})^2 (r_f + R_L)$$

$$= \left(\frac{I_m}{\sqrt{2}}\right)^2 \cdot (r_f + R_L) = 0.5I_m^2 (r_f + R_L) \quad (15.17)$$

***Rectification efficiency*:** This is given by

$$\eta = \left(\frac{I_{DC}}{I_{\text{rms}}}\right)^2 = \frac{1}{1 + \dfrac{r_f}{R_L}} = \left(\frac{2I_m/\pi}{I_m/\sqrt{2}}\right)^2 \cdot \frac{I}{1 + r_j/R_L}$$

$$\eta\% = \frac{0.812}{1 + r_f/R_L} \times 100 \quad (15.18)$$

Ripple factor,

$$\gamma = \sqrt{\left(\frac{I_{\text{rms}}}{I_{DC}}\right)^2 - 1} = \sqrt{\frac{I_m/\sqrt{2}}{2I_m/\pi}} - 1 = 0.483 \quad (15.19)$$

***Peak inverse voltage, PIV*:** The peak inverse voltage of a full-wave rectifier, from Fig. 15.23.

$$= V_{\text{sec}} + V_R = V_m + V_m = 2V_m \quad (15.20)$$

***Voltage regulation*:** It is a measure of the ability of a rectifier to maintain a specified output voltage irrespective of the variation in load resistance. It is defined as

% voltage regulation

$$= \frac{\text{Output voltage at no load} - \text{voltage at full load}}{\text{Output voltage at full load}} \times 100 \quad (15.21)$$

Average of the load voltage is given by

$$V_{DC} = \frac{2V_m}{\pi} \quad (15.22)$$

RMS value of the load voltage

$$V_{\text{rms}} = \frac{V_m}{\sqrt{2}} \quad (15.23)$$

***Transformer utilization factor*:** For full-wave rectifier, it has a value equal to

$$\text{TUF} = 0.67 \quad (15.24)$$

Zener diodes: A zener diode is like an ordinary silicon p-n diode. The only difference between zener diode and ordinary crystal diode is that zener diode has a sharply defined knee in its reverse characteristics.

The breakdown phenomenon in case of a zener diode is the result of electrons breaking their covalent bonds due to existence of strong electric field at the junction. The new holeelectron pair created, increases the reverse current. It does not involve collisions of carriers with the lattice atoms. A zener break down phenomenon occurs for heavily doped diodes having a narrow depletion-region width and high field intensity. They have breakdown voltages below 6 V with increase in temperature, the energy of the valence electrons increases, making it easier for those electrons to break through the covalent bonds and hence the breakdown voltage decreases. Hence, zener diodes have negative temperature coefficient of breakdown voltage. If the applied reverse voltage exceeds the breakdown voltages, the zener diode acts like a constant voltage source.

The circuit symbol of a zener diode is shown in Fig. 15.24.

A zener diode is specified by its breakdown voltage and the maximum power dissipation.

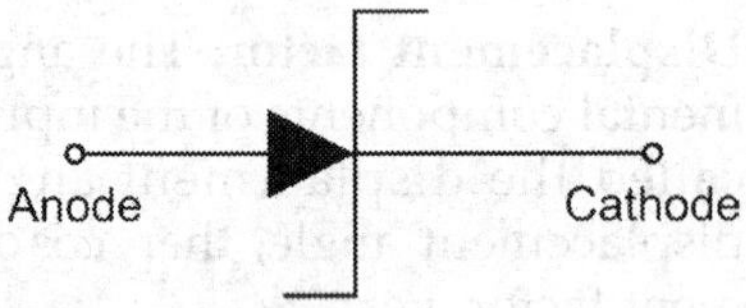

Fig. 15.24

The most common application of a zener diode is in the voltage stabilizing or regulator circuits.

Voltage stabilizer: After the ripples have been filtered from the rectifier output, a sufficiently steady state DC output voltage is obtained. But in many applications, this DC output does not serve the purposes because this DC output voltage reduces as the load current connected to it is increased. Secondly, the DC output voltage varies with the change in the AC input voltage. To obtain constant DC output voltage as the load and/or the AC input voltage vary, a voltage stabilizer circuit is used.

The output of the rectifier goes to filter and then to a voltage stabilizer circuit, the block diagram of power supply is shown in Fig. 15.25. The rectifier circuit is the heart of the power supply.

The stabilizer circuit is connected between the output of the filter and the load.

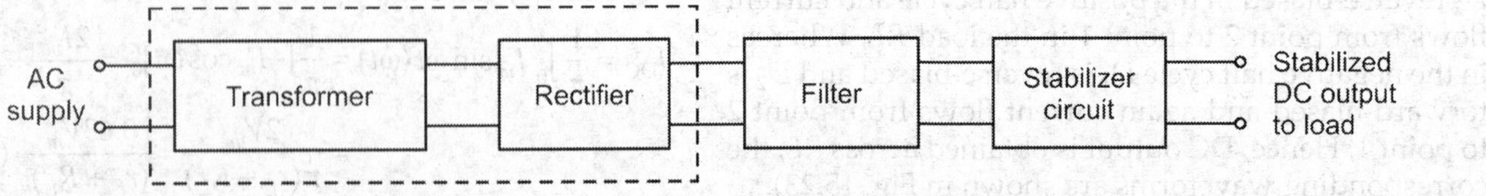

Fig. 15.25: Block diagram of power supply

MULTIPLE CHOICE QUESTIONS

1. The total energy of a revolving electron in an atom can
   A. have any value above zero
   B. never be positive
   C. never be negative D. not be calculated
2. The maximum number of electrons which the M-shell of an atom can contain is
   A. 32 B. 8
   C. 18 D. 50
3. Semiconductor materials have ...... bonds.
   A. ionic B. covalent
   C. mutual D. metallic
4. The maximum number of electrons which the valence shell of an atom can have is
   A. 6 B. 8
   C. 18 D. 2
5. Silicon has Z = 14. Its outermost orbit is
   A. partially filled B. half filled
   C. completely occupied D. empty
6. Major part of the current in an intrinsic semiconductor is due to
   A. conduction-band electrons
   B. valence-band electrons
   C. holes in the valence band
   D. thermally-generated electron.
7. Conduction electrons have more mobility than holes because they
   A. are lighter
   B. experience collisions less frequently
   C. have negative charge
   D. need less energy to move them
8. Doping materials are called impurities because they
   A. decrease the number of charge carriers
   B. change the chemical properties of semiconductors
   C. make semiconductors less than 100 percent pure
   D. alter the crystal structure of the pure semiconductors
9. Current flow in a semiconductor depends on the phenomenon of
   A. drift B. diffusion
   C. recombination D. All of these
10. The width of depletion layer of a P-N junction
    A. decreases with light doping
    B. increases with heavy doping
    C. is independent of applied voltage
    D. is increased under reverse bias
11. At room temperature of 25°C, the barrier potential for silicon is 0.7 V. Its value at 125°C is
    A. 0.5 V B. 0.3 V
    C. 0.9 V D. 0.7 V
12. Junction breakdown occurs
    A. under reverse bias B. with forward bias
    C. under high temperature condition
    D. because of manufacturing defect
13. Avalanche breakdown is primarily dependent on the phenomenon of
    A. collision B. doping
    C. ionization D. recombination
14. Reverse current in a silicon junction nearly doubles for every .... °C rise in temperature.
    A. 10 B. 2
    C. 6 D. 5
15. In the forward region of its characteristic, a diode appears as a/an
    A. OFF switch B. high resistance
    C. capacitor D. ON switch
16. The approximate value of v_0 across the diode shown below

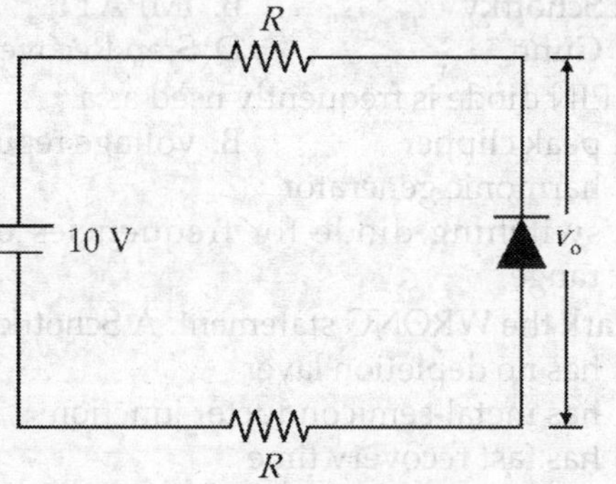

    A. zero B. 10 V
    C. 5 V
    D. dependent on value of R
17. Without a DC source, a clipper acts like a
    A. rectifier B. clamper
    C. demodulator D. chopper
18. The primary function of a clamper circuit is to
    A. suppress variations in signal voltage
    B. raise positive half-cycle of the signal
    C. lower negative half-cycle of the signal
    D. introduce a DC fuel into an AC signal
19. Silicon is preferred for manufacturing Zener diodes because it
    A. is relatively cheap
    B. needs lower doping level

Ans.	1. B	2. C	3. B	4. B	5. B	6. A	7. D	8. D	9. D	10. D	11. C	12. A	13. A
	14. C	15. D	16. B	17. A	18. D								

C. has higher temperature and current capacity
D. has lower breakdown voltage

20. When used in a circuit, a Zener diode is always
A. forward-biased B. connected in series
C. troubled by overheating
D. reverse-biased

21. The main reason why electrons can tunnel through a PN junction is that
A. they have high energy
B. barrier potential is very low
C. depletion layer is extremely thin
D. impurity level is low

22. The I_P/I_V ratio of a tunnel diode is of primary importance in
A. determining tunneling speed of electrons
B. the design of an oscillator
C. amplifier designing
D. computer applications

23. Mark the *incorrect* statement.
A varactor diode
A. has variable capacitance
B. utilizes transition capacitance of a junction
C. has always a uniform doping profile
D. is often used as an automatic frequency control device

24. The microwave device used as an oscillator within the frequency range 10–1000 GHz is .... diode.
A. Schottky B. IMPATT
C. Gunn D. Step Recovery

25. A PIN diode is frequently used as a
A. peak clipper B. voltage regulator
C. harmonic generator
D. switching diode for frequencies up to GHz range

26. Mark the WRONG statement. A Schottky diode
A. has no depletion layer
B. has metal-semiconductor junction
C. has fast recovery time
D. is a bipolar device

27. A Step-recovery diode
A. has an extremely short recovery time
B. conducts equally well in both directions
C. is mainly used as a harmonic generator
D. is an ideal rectifier of high-frequency signals

28. An electron rising through a potential of 500 V, will acquire an energy of
A. 800×10^{-22} joules B. 800 eV
C. 500 joules D. 500 eV

29. A *P-N* junction is
A. a rectifier B. an amplifier
C. an insulator D. none of these

30. For pure germanium, the density of the charge carriers is $2.5 \times 10^{18}/\text{cm}^2$ and electron and hole mobilities at 300°K are 3600 cm^2/V-sec and 1700 cm^2/V-sec, respectively. Its conductivity would be
A. 0.02 mho/cm B. 0.5 mho/cm
C. 0.05 ohm·cm D. 0.2 ohm·cm

31. In the depletion region of a *P-N* junction, there are
A. no charges B. no mobile charges
C. no currents D. all of these

32. In a *P*-type semiconductor with acceptor atom concentration of N_a, the electron concentration (n) in terms of intrinsic carrier concentration (N_i) is given by
B. $N_i \cdot N_a$ B. N_i
C. N_i^2/N_a D. N_a

33. For silicon the intrinsic concentration is approximately 10^{10} carriers/cm^3 respectively. If an impurity concentration of 10^{14} donor atoms/cm^2, the electron concentration will be
A. 10^6 B. 10^{10}
C. 10^{16} D. 10^{20}

34. In the same question, the hole concentration is
A. 10^6 B. 10^{10}
C. 10^{14} D. 10^{20}

35. A specimen of intrinsic germanium, with the density of charge carriers of $2.5 \times 10^{13}/\text{cm}^3$, is doped with impurity atoms such that there is one impurity atom for every 10^6 germanium atoms. The density of germanium atoms is $4.4 \times 10^{22}/\text{cm}^2$. The hole density would be
A. $4.4 \times 10^{13}/\text{cm}^3$ B. $1.4 \times 10^{10}/\text{cm}^3$
C. $4.4 \times 10^{10}/\text{cm}^3$ D. 1.4×10^4/cm/sec

36. The electron and hole concentration in an *N*-type semiconductor is $10^{16}/\text{cm}^2$ and $10^6/\text{cm}^2$. If the mobilities of electrons and holes are 400 and 600 cm^2/C-sec respectively, the conductivity of the material approximately is
A. $10^{32} \times e$ B. $10^{19} \times e$
C. $4 \times 10^{18} \times e$ D. none of these

37. Which of the following statements is *not* true for a hole?
A. Holes may constitute an electric current
B. Holes can be considered as a net positive charge
C. Holes can exist in any material including conductors
D. Holes can exist in certain semiconductor materials only

38. The potential-barrier across a *P-N* junction corresponds
A. height of the barrier B. width of the barrier
C. forward bias of the junction
D. reverse bias of the junction

39. When acceptor type impurity is added to a semiconductor material
A. electrons are generated and material is called N-type

Ans.	19. C	20. D	21. B	22. C	23. C	24. B	25. D	26. D	27. C	28. D	29. A	30. A	31. B
	32. C	33. B	34. A	35. B	36. C	37. C	38. A						

B. electrons are generated and material is called *P*-type
C. holes are generated and material is called *N*-type
D. holes are generated and material is *P*-type

40. The potential barrier existing across a *pn* junction
A. facilitates recombination of electrons and holes
B. prevents flow of minority carriers
C. prevents total recombination of holes and electrons
D. prevents neutralisation of acceptor and donor ions

41. In a depletion region of a *pn* junction,
A. *P*-side is positively charged and *N*-side is negatively charged
B. *N*-side is positively charged and *P*-side is negatively charged
C. there is hole concentration on *P*-side and electron concentration on *N*-side
D. none of these

42. The nature of atomic bond found in diamond is
A. ionic B. covalent
C. metallic D. none of these

43. In Hall's Effect, the output voltage produced across the crystal is due to
A. induced voltage by the applied magnetic field
B. drop across the crystal due to the current passed through it
C. movement of charge carriers towards one end
D. all of these

44. In correctly biased NPN transistor
A. the base is positive with respect to collector
B. the base is positive with respect to emitter
C. the base is positive with respect to both emitter and collector
D. none of these

45. In transistor, if electrons flow into the emitter,
A. holes flow out of the emitter
B. electrons flow into the collector
C. electrons flow out of the collector
D. holes flow out of the collector

46. In a metal, there are 1028 electrons per cubic metre and mean free time is 10^{-14} sec. When a field of 10^4 V/m is applied, the current density will be
B. 2.8×10^6 A/m^2 B. 2.8×10^6 A/m^2
C. 2.8 A/m^2 D. none of these

47. The resistivity of all metals as temperature is lowered
A. remains constant B. tends to zero
C. decreases first and then starts increasing
D. None of these

48. Fermi level is
A. the highest occupied level at 0°K
B. the highest available level at 0°K
C. the highest velocity of electron of 0°K
D. none of these

49. Halls's Effect can be used to measure
A. electric field intensity
B. magnetic field intensity
C. carrier concentration D. none of these

50. If the reverse-bias voltage of a *P-N* junction is increased within limits, the reverse saturation current will
A. decrease B. increase
C. remain unaffected D. none of these

51. The conductivity of a pure semiconductor
A. is proportional to temperature
B. increase exponentially with temperature
C. decreases exponentially with increasing temperature
D. all of the above

52. Which of the following has the highest mobility?
A. An electron B. A negative ion
C. A positive ion D. A hole

53. The α and β of a transistor are related as, $B =$ ......
A. $\alpha/(1+\alpha)$ B. $\alpha/(1-\alpha)$
C. $(1+\alpha)/\alpha$ D. $1-\alpha$

54. The cut-in voltage, V_γ for a silicon diode is about
A. 0.3 V B. 0.4 V
C. 0.5 V D. 0.7 V

55. At half power frequencies, an amplifier voltage gain is lower by
A. 6 dB B. 3 dB
C. 2 dB D. 0.5 dB

56. The AC input to a half wave rectifier is 28.3 V peak. Neglecting the drop across the diode, the DC voltage across the load will be about
B. 28.3 V B. 20 V
C. 14.15 V D. 9.0 V

57. Electronics is the branch of engineering which deals with flow of electrons through
A. semiconductor B. gas
C. vaccum D. all of these

58. The liberation of electrons from the surface of a substance is called
A. bombardment of electrons
B. movement of free electrons
C. electronic emission D. ions emission

59. The circuit shown in the figure below represents a feedback amplifiers. For this circuit, the current and voltage gains would respectively be

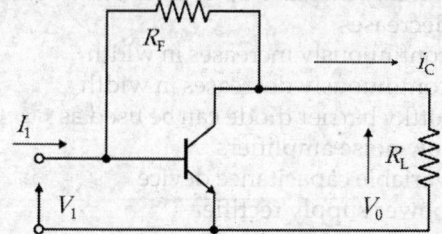

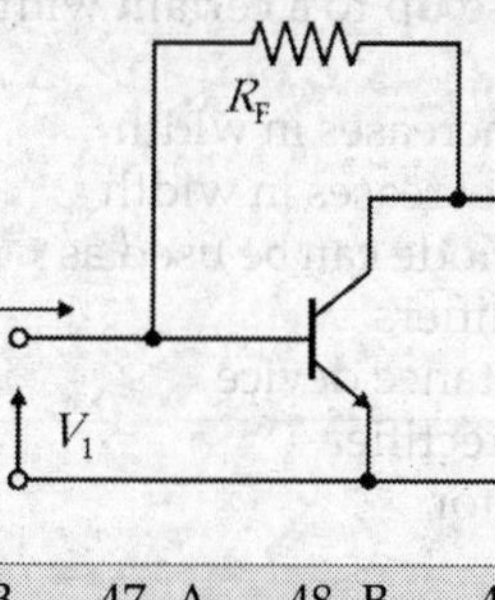

Ans.													
	39. D	40. C	41. B	42. B	43. C	44. B	45. C	46. B	47. A	48. B	49. A	50. C	51. B
	52. A	53. B	54. C	55. B	56. D	57. D	58. C						

A. h_{fe} $h_{fe}R_L/h_{ie}$ B. R_f/R_L $h_{fe}R_L/h_{ie}$
C. h_{fe} R_L/R_F D. R_F/R_L R_L/R_F

60. The metal from which the electrons are emitted is called
A. an emitter B. a cathode
C. an emitter or a cathode
D. none of these

61. The semiconductor are replacing electron tube because they are
A. light in weight B. smaller in size
C. longer in life and greater in ruggedness
D. all of these

62. Oxide-coated cathode is used in
A. low power tubes B. medium power tubes
C. high power tubes
D. medium and high power tubes

63. The emitter must have the following characteristics:
A. immunity to reaction from residual gasses and water vapours
B. low value of work function
C. high melting point D. all of these

64. In an indirectly heated cathodes
A. filament is not used
B. filament heater is used
C. filament heater may or may not be used
D. high voltage is used

65. An electron tube can only operate when
A. anode is negative with respect to cathode
B. anode is positive with respect to cathode
C. anode must be at zero potential with respect to cathode
D. anode must be earthed

66. Electrical conductivity of copper reduces with addition of a small amount of nickel because
A. the mobility decreases due to creation of new scattering centers for electrons
B. nickel is ferromagnetic
C. less free electrons are available for conduction
D. the electrons of copper become bound to nickel atoms and are not free to move

67. When the reverse voltage across *p-n* junction diode is gradually decreased, the depletion region inside the diode
A. does not change in width
B. initially increase up to a certain width and then decreases
C. continuously increases in width
D. continuously decreases in width

68. Schottky barrier diode can be used as
A. low noise amplifiers
B. variable capacitance device
C. power supply rectifier
D. low level detector

69. Which of the following statements is *not true* for an intrinsic semiconductor?
A. The number of holes is less than the number of electrons
B. The product of hole concentration and electron concentration is a constant
C. The net charge density of the material is zero
D. All of these

70. In an intrinsic semiconductor
A. there are no holes in the material
B. the number of holes is too small
C. electrons in the material are neutralised by holes
D. there are no electrons in the material

71. An element used in semiconductors whose atoms have three valence electrons is
A. germanium B. a donor
C. an acceptor D. silicon

72. The impurity atoms in semiconductors
A. reduce the energy gap
B. increase the kinetic energy of valence electrons
C. inject more charge carriers
D. all of these

73. With increase in temperature, the conductivity of an extrinsic semiconductor
A. decreases B. increases
C. remains constant D. all of these

74. The donor atoms in an *N*-type semiconductor at normal temperature
A. carry a positive charge
B. carry a negative charge
C. are neutral D. none of these

75. When a semiconductor is doped with a *P*-type impurity, each impurity atom will
A. acquire negative charge
B. acquire positive charge
C. remain electrically neutral
D. give away one electron

76. The minority carrier concentration is largely a function of
A. the amount of doping
B. temperature
C. forward biasing voltage
D. reverse biasing voltage

77. In a semiconductor, the movement of holes is due to
A. movement of holes in conduction band
B. movement of electrons in conduction band
C. movement of electrons in valence band
D. movement of holes in valence band

78. It is possible to obtain *P*-type or *N*-type semiconductor from a single compound by adding Ge. That compound is

Ans.													
	59. A	60. C	61. D	62. A	63. D	64. B	65. B	66. A	67. D	68. D	69. A	70. B	71. C
	72. A	73. B	74. B	75. A	76. A	77. B							

A. InSb B. GaP
C. GaAs D. none of these

79. Fermi energy is the amount of energy which
A. a valence electron can have at room temperature
B. must be given to an electron to move it to conduction band
C. must be given to a hole to move it to valence band
D. a hole can have at room temperature

80. In the energy band diagram of a *P*-type semiconductor
A. the acceptor band is near to the conduction band
B. the acceptor band is near to valence band
C. the donor band is near to valence band
D. the donor band is near to conduction band

81. Transformer coupling is used to cascade two such stages of which the first has
A. high collector impedance and succeeding has low input impedance
B. low collector resistance and succeeding has high input impedance
C. low collector impedance and succeeding has high input impedance
D. high collector impedance and succeeding has high input impedance

82. As compared to a LED, an LCD has the distinct advantage of
A. extremely low power requirement
B. providing a silver display
C. being extremely thin
D. giving two types of displays

83. The emitter of a transistor is generally doped the heaviest because it
A. has to dissipate maximum power
B. has to supply the charge carriers
C. is the first region of the transistor
D. must possess low resistance

84. For correct working of an NPN bipolar junction transistor, the different electrodes should have the following polarities with respect to emitter
A. collector +ve, base –ve
B. collector –ve, base +ve
C. collector –ve, bese –ve
D. collector +ve, base +ve

85. Select the correct alternative. In a bipolar transistor
A. emitter region is of low/high resistivity material which is lightly/heavily-doped
B. collector region is of lower/higher conductivity than emitter region
C. base region is of high/low resistivity material which is only lightly/heavily-doped
D. none of these

86. In a properly-biased NPN transistor, most of the electrons from the emitter
A. recombine with holes in the base
B. recombine in the emitter itself
C. pass through the base to the collector
D. are stopped by the junction barrier

87. The following relationships between α and β are correct except
A. $\beta = \dfrac{\alpha}{1-\alpha}$ B. $\alpha = \dfrac{\beta}{1-\beta}$
C. $\alpha = \dfrac{\beta}{1+\beta}$ D. $1-\alpha = \dfrac{1}{1+\beta}$

88. The value of total collector current in a CB circuit is
A. $I_C = \alpha I_E$ B. $I_C = \alpha I_E + I_{CO}$
C. $I_C = \alpha I_E - I_{CO}$ D. $I_C = \beta I_E$

89. In the case of a bipolar transistor, α is
A. positive and –1 B. positive and 1
C. negative and –1 D. negative and 1

90. If we neglect V_{EB} shown below, value of R_E required to make V_{BC} = 10 V is

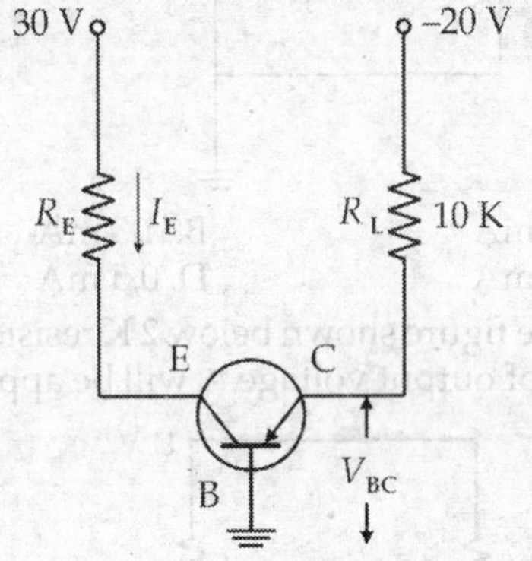

A. 30 kΩ B. 20 kΩ
C. 15 kΩ D. 5 kΩ

91. In the ideal transistor shown below, the value of R_B required for making V_{CE} = 10 V is

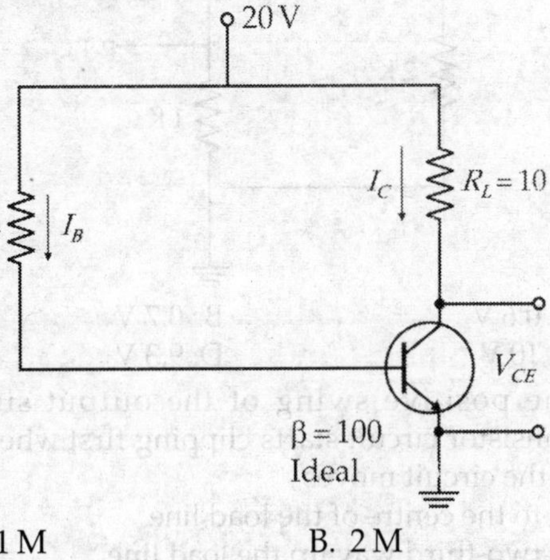

A. 1 M B. 2 M
C. 20 K D. 10 K

92. The Emitter based junction of a given transistor is forward-biased and its CBJ reverse-biased. If the base current is increased, then its

Ans.	78. C	79. C	80. B	81. A	82. A	83. A	84. D	85. A	86. C	87. B	88. B	89. D	90. A
	91. B												

A. I_C will decrease B. V_{CE} will increase
C. I_C will increase D. V_{CG} will increase

93. The collector characteristics of CE-connected transistor may be used to find its
A. input resistance B. base current
C. output resistance D. voltage gain

94. Taking V_{BE} = 0.7 V, approximate collector current of the transistor shown below

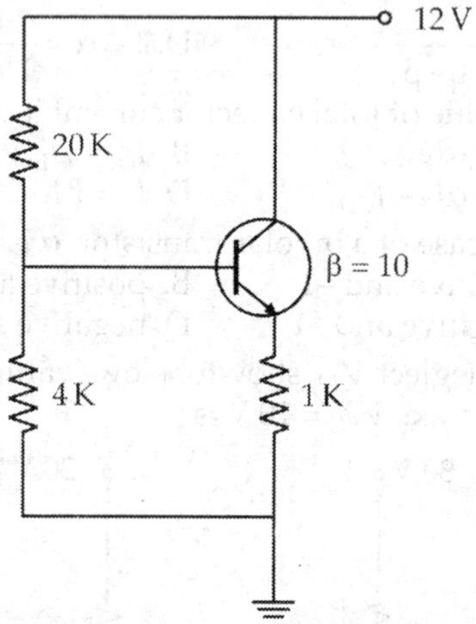

A. 0.6 mA B. 1.3 mA
C. 1.2 mA D. 0.5 mA

95. If in the figure shown below 2 K resistor opens, then value of output voltage v_0 will be approximately

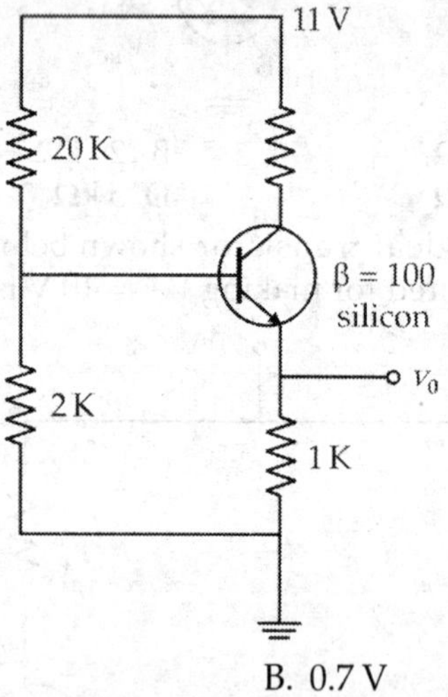

A. 0.6 V B. 0.7 V
C. 10 V D. 9.3 V

96. The positive swing of the output signal in a transistor circuit starts clipping first when *Q*-point of the circuit moves
A. to the centre of the load line
B. two-third way up the load line
C. towards the saturation point
D. towards the cut-off point

97. In the silicon transistor circuit shown below, voltage of point *A* relative to ground is

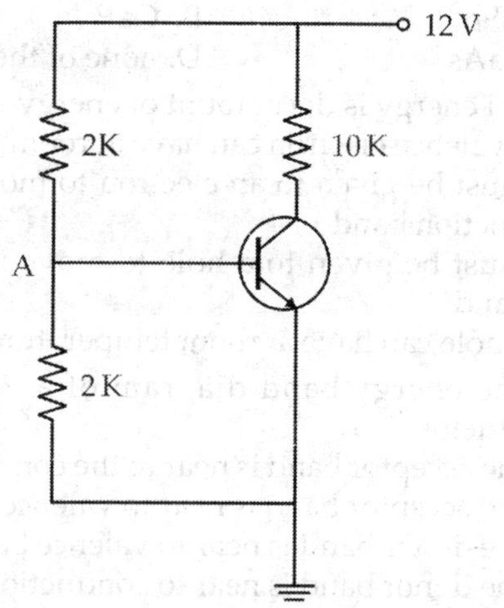

A. 6 V B. 12 V
C. 0.7 V D. 3 V

98. The alue of I_c in the figure is

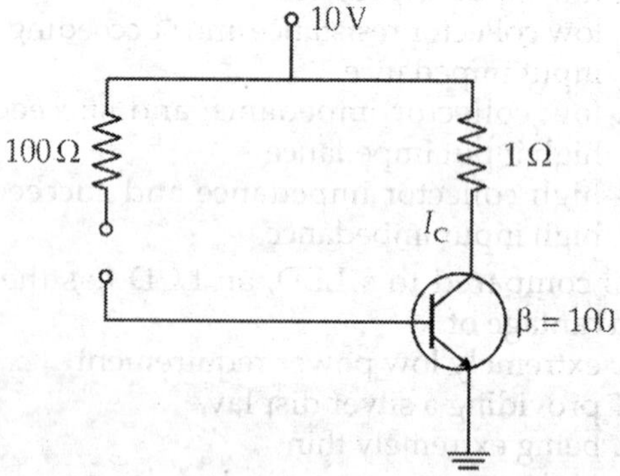

A. 10 A B. 0 A
C. 0.1 A D. 1000 A

99. The value of V_{CE} in the figure is

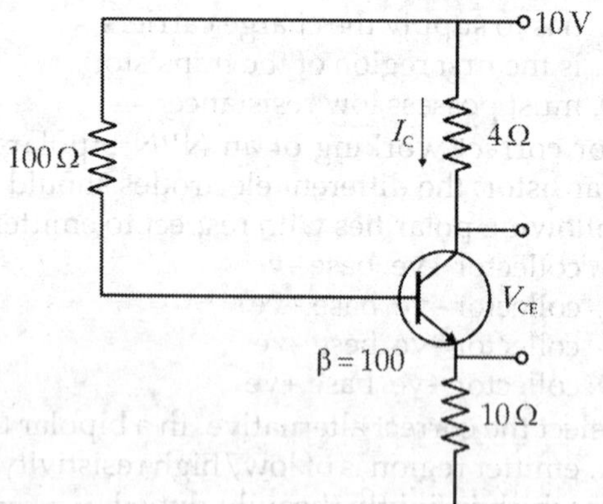

A. 8.6 V B. 3 V
C. 4.17 V D. –3 V

100. The DC load line of a transistor circuit
A. has a negative slope B. is a curved line
C. gives graphic relation between I_C and I_B
D. does not contain the *Q*-point

Ans. 92. C 93. C 94. B 95. D 96. A 97. C 98. B 99. B 100. A

101. A transistor circuit employing base bias with collector feedback has greater stability than the one without feedback because
A. I_C decreases in magnitude
B. V_{BE} is decreased
C. of negative feedback effect
D. I_C becomes independent of β

102. The universal bias stabilization circuit is most popular because
A. I_C does not depend on transistor characteristics
B. its β-sensitivity is high
C. voltage divider is heavily loaded by transistor base
D. I_C equals I_E

103. Improper biasing of a transistor circuit leads to
A. excessive heat production at collector terminal
B. distortion in output signal
C. faulty location of load line
D. heavy loading of emitter terminal

104. Which reference to the voltage divider bias circuit of figure below?

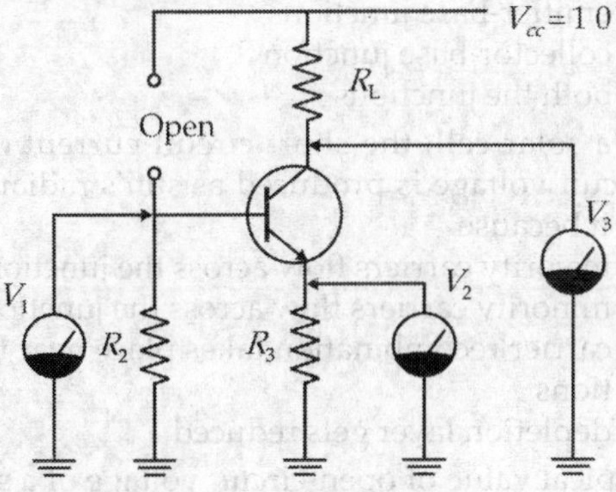

A. V_1 will read......... volt.
(i) 0 (ii) 10 (iii) 5 (iv) 0.7
B. V_2 will read......volt.
(i) 10 (ii) 0 (iii) 0.7 (iv) 9.3
C. V_3 will read.......volt.
(i) 0 (ii) 0.7 (iii) 5 (iv) 10

105. In the silicon transistor circuit of figure below

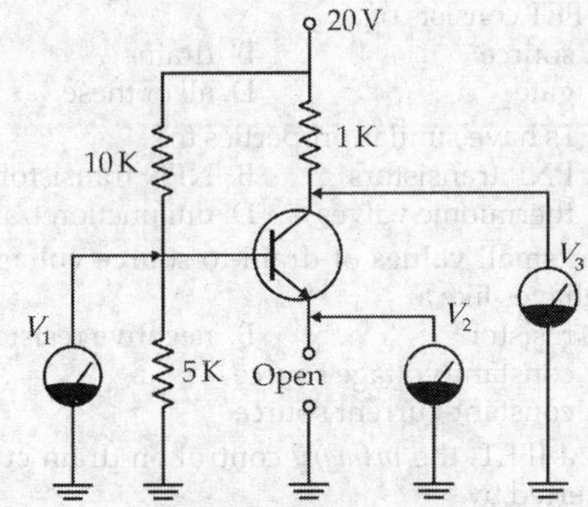

a. V_1 will read.......volt.
A. 0 B. 10 C. 6.67 D. 5
b. V_2 will read.......volt.
A. 5.97 B. 0 C. 20 D. 10
c. V_3 will read .... volt.
A. 13.33 B. 10 C. 0 D. 20

106. The negative output swing in a transistor circuit starts clipping first when Q-point
A. has optimum value
B. is near saturation point
C. is near cut-off point
D. is in the active region of the load line

107. The primary switching delay in an overdriven dipolar junction transistor is contributed by which one of the following characteristics of the transistor?
A. Storage time B. Fall time
C. Rise time D. Charging time

108. Consider the following statement: The energy band gap of GaAs is
1. indirect 2. direct
3. 1, 43 CV
4. corresponds to wavelength of 0.89 μm. (Given $h = 6.63 \times 10^{34}$ Joule-sec)
Of these statements
A. 1 alone is correct B. 2, 3 and 4 are correct
C. 2 and 3 are correct D. 1 and 4 are correct

109. Consider a single crystal of an intrinsic semiconductor. The number of free carriers at the Fermi level at room temperature is
A. half the total number of electrons in the crystal
B. half the number of free electrons in the crystal
C. half the number of atoms in the crystal
D. zero

110. If the molecular structure and density of the material are known, then which of the following material parameters may be determined from the ε_r versus $1/T$ plot?
A. Sum of electronic and ionic polarizabilities
B. Lattice constant
C. Breakdown voltage D. Loss tangent

111. The electron relaxation time of metal A is 2.7×10^{-4} sec, that of B is 1.35×10^{-4} sec. The ratio of resistivity of B to resistivity of A will be
A. 4.0 B. 2.0
C. 0.5 D. 0.25

112. The threshold wavelength in the case a Caesium surface for which the work function 1.8 eV is
A. 6890 Å B. 8000 Å
C. 4000 Å D. 5240 Å

113. A surface having a work-function of 4.3 eV is radiated with mercury line of 2537 Å. The maximum velocity of the emitted photoelectrons will be

Ans. 101. C 102. A 103. B 104. D 105a. C 105b. A 105c. D 106. B 107. A 108. C 109. B 110. A 111. B 112. A

A. 1.55×10^5 m/sec B. 2.45×10^5 m/sec
C. 4.56×10^5 m/sec D. 8.35×10^5 m/sec

114. The given figure represents the AC equivalent circuit of a parallel voltage feedback amplifiers stage. h_{fe}, h_{fc} are h parameters in Common Emitter and Common Collector operations respectively of Q. The current gain ($= i_o/i_i$) of this stage is

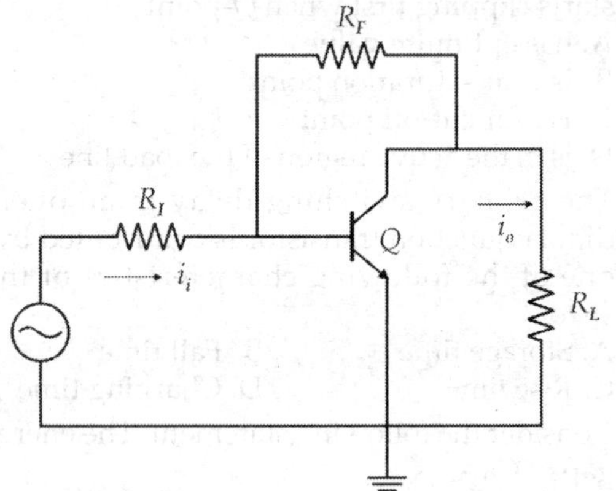

A. R_F/R_L B. h_{fe}
C. R_L/R_F D. h_{fc}

115. For photoelectric emission to over the whole visible region is 4000°A, the work function of the photo-sensitive must be
B. more than 2.5 eV B. less than 1.55 eV
C. equal to 3.5 eV D. equal to 1.2 eV

116. The sensitivity of a photo multiplier of six stages with gain of each stage 5 and cathode sensitivity 24 μA/lumen, will be
A. 375 mA/lumen B. 220 mA/lumen
C. 310 mA/lumen D. 460 mA/lumen

117. In the above problem, if maximum safe output current is 3 mA, the maximum safe illumination is
A. 2 m lumens B. 4 m lumens
C. 6 m lumens D. 8 m lumens

118. The differential mode voltage gain of the amplifier is $\left(\dfrac{v_o}{v_{i1} - v_{i2}}\right)$ given in the figure is

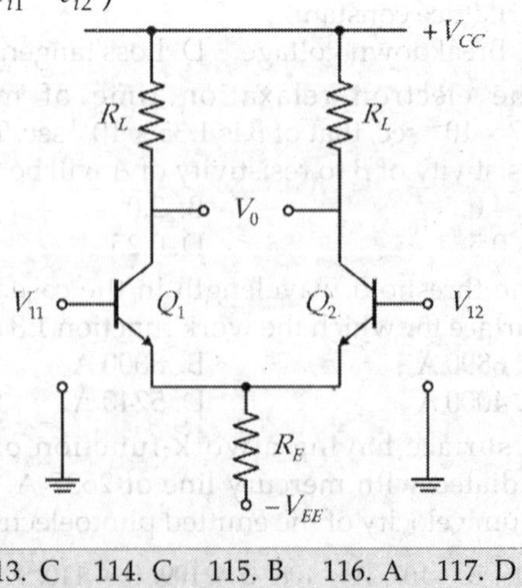

A. $h_{fe}R_L/h_{ie}$ B. $R_L/2R_E$
C. R_L/R_E D. $h_{fe}R_L/R_E$

119. In photoconductive effect, radiations of wavelength greater than critical wavelength will
A. produce free electrons
B. not produce free electrons
C. emit electrons from the surface
D. produce protons

120. Photodiode is used in reverse biased because
A. majority swept are reversed across the junction
B. only one side is illuminated
C. reverse current is small as compared to photocurrent
D. reverse current is large as compared to photocurrent

121. Photo-multipliers are based on the principle of
A. pyroelectric effect B. photovoltaic effect
C. photoconduction D. secondary emission

122. In a phototransistor, the photocurrent is generated at
A. either of the junctions
B. emitter-base junctions
C. collector-base junction
D. both the junctions

123. In a solar cell, the short circuit current or open-circuit voltage is produced as sun's radiation falls on it because
A. majority carriers flow across the junction
B. minority carriers flow across the junction
C. carrier recombination takes place near the junctions
D. depletion layer gets reduced

124. Typical value of open-circuit voltage of a solar cell is
A. 10 mV B. 100 mV
C. 0.5 V D. 1 V

125. Which of the following rays is used to photograph thick metal objects?
A. infrared rays B. X-rays
C. cosmic rays D. ultraviolet

126. A FET consists of a
A. source B. drain
C. gate D. all of these

127. FETs have similar properties to
A. PNP transistors B. NPN transistors
C. thermionic valves D. unijunction transistors

128. For small values of drain-to-source voltage, JFET behaves like a
A. resistor B. negative resistance
C. constant-voltage source
D. constant-current source

129. In a JFET, the *primary* control on drain current is exerted by

Ans. 113. C 114. C 115. B 116. A 117. D 118. A 119. B 120. C 121. D 122. C 123. B 124. C 125. B 126. D 127. B 128. A

A. channel resistance B. gate reverse bias
C. voltage drop across channel
D. size of depletion regions

130. After V_{DS} reaches pinch-off value V_P in a JFET, drain current I_D becomes
A. zero B. low
C. saturated D. reversed

131. In a JFET, as external bias applied to the gate is increased
A. channel resistance is decreased
B. drain current is increased
C. pinch-off voltage is reached at lower values of I_D
D. size of depletion regions is reduced

132. In a JFET, drain current is maximum when V_{GS} is
A. zero B. negative
C. positive D. equal to V_P

133. The voltage gain of a given common-source JFET amplifier depends on its
A. input impedance B. amplification factor
C. dynamic drain resistance
D. drain load resistance

134. A JFET has the disadvantage of
A. being noisy
B. having small gain-bandwidth product
C. possessing positive temperature coefficient
D. having low input impedance

135. A JFET can be cut-off with the help of
A. V_{GS} B. V_{DS}
C. V_{DG} D. V_{DD}

136. DE-MOSFET differs from a JFET in the sense that it has no
A. channel B. gate
C. *PN* junctions D. substrate

137. For the operation of enhancement only *N*-channel MOSFET, value of gate voltage has to be
A. high positive B. high negative
C. low positive D. zero

138. The extremely high input impedance of a MOSFET is primarily due to the
A. absence of its channel
B. negative gate-source voltage
C. depletion of current carriers
D. extremely small leakage current of its gate capacitor

139. The main factor which makes a MOSFET likely to breakdown during normal handling is its
A. very low gate capacitance
B. high leakage current
C. high input resistance
D. both A and C are correct

140. The main factor which differentiates a DE-MOSFET an E only MOSFET is the absence of
A. insulated gate B. electrons
C. channel D. *PN* junctions

141. The polarity of V_{GS} for *E* only MOSFET is
A. positive B. negative
C. zero
D. depends on *P*- or *N*-channel

142. A unijunction transistor has
A. anode, cathode and a gate
B. two bases and one emitter
C. two anodes and one gate
D. anode, cathode and two gates

143. Which semiconductor device acts like a diode and two resistors?
A. SCR B. triac
C. diac D. UJT

144. A UJT has R_{BB} = 10 K and R_{B2} = 4 K. Its intrinsic stand-off ratio is
A. 0.6 B. 0.4
C. 2.5 D. 5/3

145. A source follower using an FET usually has voltage gain which is
A. greater than +100 B. about –10
C. exactly unity but negative
D. slightly less than unity but positive

146. Consider the following four common types of transistors:
1. Point Contact Transistor
2. Bipolar Junction Transistor
3. MOS Field Effect Transistor
4. Junction Field Effect Transistor
The correct arrangement of these transistors in the increasing of input impedance
A. 1, 2, 4, 3 B. 1, 2, 3, 4
C. 2, 1, 3, 4 D. 2, 1, 4, 3

147. A FET has three terminals namely, source, drain and ......
A. grid B. substrate
C. ground D. gate

148. From the output characteristics, FET is considered to be a solid-state device equivalent to a
A. diode valve B. tetrode valve
C. pentode valve D. none of these

149. A FET is essentially a
A. current driven device B. voltage driven device
C. power driven device D. none of these

150. A FET is a
A. unipolar transistor B. bipolar transistor
C. tripolar transistor D. none of these

151. The input impedance of a FET is of the order of
A. 10^2 ohm
B. hundreds of mega ohms
C. hundred ohms D. a few ohms

Ans. 129. B 130. C 131. C 132. A 133. D 134. B 135. A 136. C 137. A 138. D 139. D 140. C 141. D 142. B 143. D 144. A 145. B 146. D 147. D 148. C 149. B 150. A 151. B

152. A FET is characterised by
A. current gain B. voltage gain
C. power gain D. none of these

153. Before connecting FET in the circuit
A. its source and drain terminals are interchangeable
B. its source and drain terminals are not interchangeable
C. its drain terminal is marked
D. none of these

154. The noise level in FET is
A. more than BJT B. negligibly small
C. slightly less than BJT D. none of these

155. Input impedance of MOSFET is
A. less than that of FET but more than BJT
B. more than that of FET and BJT
C. more than that of FET but less than BJT
D. less than that of FET and BJT

156. MOSFET uses the electric field of
A. gate capacitance to control the channel current
B. barrier potential *pn*-junction to control the channel current
C. both A and B D. none of these

157. In an AC amplifier, larger the internal resistance of the AC signal source
A. greater the overall voltage gain
B. greater the input impedance
C. smaller the current gain
D. smaller the circuit voltage gain

158. The main use of an emitter follower is a
A. power amplifier
B. impedance matching device
C. low-input impedance circuit
D. follower of base signal

159. In the *CE* amplifier circuit shown below, if R_E went short-circuit

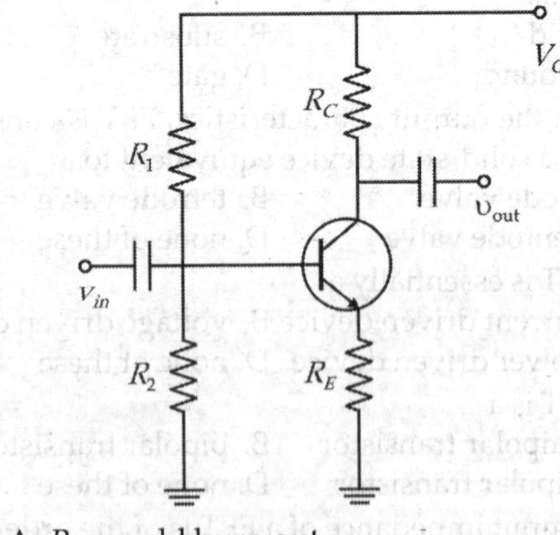

A. R_E would burn out
B. amplifier output signal would become zero
C. collector current would fall to zero
D. transistor will be overloaded

160. An ideal amplifier is one which
A. has infinite voltage gain
B. responds only to signals at its input terminals
C. has positive feedback
D. gives uniform frequency response

161. The smallest of the four *h*-parameters of a transistor is
A. h_i B. h_r
C. h_o D. h_f

162. The voltage gain of a single-stage *CE* amplifier is increased when
A. its AC load is decreased
B. resistance of signal source is increased
C. emitter resistance R_E is increased
D. AC load resistance is increased

163. A *CB* amplifier has very low input resistance because
A. low emitter AC resistance r_e shunts all other resistances
B. it handles small input signals
C. emitter bulk resistance is small
D. its base is at AC ground

164. *CE* amplifier is characterised by
A. low voltage gain B. moderate power gain
C. signal phase reversal
D. very high output impedance

165. A CC amplifier has the highest
A. voltage gain B. current gain
C. power gain D. output impedance

166. In a CC amplifier, voltage gain
A. cannot exceed unity
B. depends on output impedance
C. is dependent on input signal
D. is always constant

167. In a class-A amplifier, conduction extends over 360° because *Q*-point is
A. located on load line
B. located near saturation point
C. centred on load line
D. located at or near cut-off point

168. The circuit efficiency of a class-A amplifier can be increased by using
A. low DC power input B. direct-coupled load
C. low-rating transistor
D. transformer-coupled load

169. In a class-A amplifier, worst-case condition occurs with
A. zero signal input B. maximum signal input
C. high load resistance D. transformer coupling

170. The output of a class-B amplifier
A. is distortion-free
B. consists of positive half-cycles only
C. is like the output of a full-wave rectifier
D. comprises short-duration current pulses

Ans.	152. B	153. A	154. B	155. B	156. A	157. D	158. B	159. B	160. B	161. C	162. D	163. A	164. C
	165. B	166. A	167. C	168. D	169. A	170. B							

171. The maximum overall efficiency of a transformer-coupled class-A amplifier is
A. 78.5% B. 25%
C. 50% D. 85%

172. A transistor audio amplifier is found to have an overall efficiency of 70 percent. Most probably, it is a........ amplifier.
A. class-B push-pull B. single-stage class-C
C. transformer-coupled class-A
D. direct-coupled class-A

173. The main purpose of using transformer coupling in a class-A amplifier is to make it more
A. distortion-free B. bulky
C. costly D. efficient

174. A class-B push-pull amplifier has the main advantage of being from
A. any circuit imbalances
B. unwanted noise
C. even-order harmonic distortion
D. DC magnetic saturation effects

175. Crossover distortion occurs in .... amplifiers
A. push-pull B. class-A
C. class-B D. class-AB

176. The maximum overall efficiency of a class-B push-pull amplifier cannot exceed
A. 100% B. 78.5%
C. 50% D. 85%

177. The main use of a class-C amplifier is
A. as an RF amplifier B. as stereo amplifier
C. in communication sound equipment
D. as distortion generator

178. The primary cause of linear distortion in amplifiers is
A. change of gain with frequency
B. unequal phase shift in component frequencies
C. reactances associated with the circuit and active amplifying element
D. inherent limitations of the active device

179. An amplifier is said to suffer from distortion when its output is
A. low
B. different from its input
C. noisy D. larger than its input

180. While discussing amplifier performance, noise is defined as any kind of unwanted signal in the output which is
A. unrelated to the input signal
B. derived from the input signal
C. not generated by the amplifier
D. due to associated circuit

181. An ideal amplifier has
A. noise figure of less than 1 dB
B. noise factor of unity
C. output S/N more than input S/N
D. noise figure of more than 0 dB

182. The decibel is a measure of
A. power B. voltage
C. current D. power level

183. When power output of an amplifier doubles, the increase in its power level is
A. 2 dB B. 20 dB
C. 3 dB D. 10 dB

184. When output power level of a radio receiver increases by 3 dB, its absolute power changes by a factor of
A. 2 B. 10
C. 1/2 D. 3

185. Zero watt cannot be chosen as zero decibel level because
A. it is impossible to measure zero watt
B. it is too small
C. every power compared with it would be zero
D. it would be impossible to define a decibel

186. A minus 3 dB point on the gain versus frequency curve of an amplifier is that point where
A. signal frequency drops to half the midband frequency
B. voltage amplification becomes half of its maximum value
C. power falls to half its maximum value
D. upper cut-off frequency becomes twice the lower cut-off frequency

187. The bandwidth of an amplifier may be increased by
A. decreasing the capacitance of its bypass capacitors
B. minimizing its stray capacitances
C. increasing input signal frequency
D. cascading it

188. Low cut-off frequency of an amplifier is primarily determined by the
A. internal capacitances of the active device used
B. stray capacitances between its wiring and ground
C. β value of its active devices
D. capacitances of coupling and bypass capacitors

189. The main reason for the variation of amplifier gain with frequency is
A. the presence of capacitances, both external and internal
B. due to interstage transformers
C. the logarithmic increase in its output power
D. the Miller effect

190. The gain-bandwidth product of an amplifier is given by
A. $f_2 - f_1$ B. $f_\alpha - f_\beta$
C. f_T D. βf_α

Ans. 171. C 172. A 173. D 174. C 175. A 176. B 177. A 178. C 179. B 180. A 181. B 182. D 183. C 184. A 185. D 186. C 187. B 188. D 189. A 190. A

191. The decibel gain of a cascaded amplifier equals the
A. product of individual gains
B. sum of individual gains
C. ratio of stage gains
D. product of voltage and current gains

192. If two stages of a cascaded amplifier have decibel gains of 60 and 30, then overall gain is
A. 90 dB B. 1800 dB
C. 2 dB D. 0.5 dB

193. *RC* coupling is popular in low-level audio amplifiers because it
A. has better low frequency response
B. is inexpensive and needs no adjustments
C. provides an output signal in phase with the input signal
D. needs low voltage battery for collector supply

194. The most desirable feature of transformer coupling is its
A. higher voltage gain B. wide frequency range
C. ability to provide impedance matching between stages
D. ability to eliminate hum from the output

195. In multistage amplifiers, direct coupling is especially suited for amplifying
B. high frequency AC signals
B. changes in DC voltages
C. high-level voltage D. sinusoidal signals

196. The outstanding characteristic of a direct-coupled amplifier is its
A. utmost economy B. temperature stability
C. avoidance of frequency-sensitive components
D. ability to amplify direct current and low-frequency signals

197. Darlington pairs are frequently used in linear ICs because they
A. do not require any capacitors or inductors
B. have enormous impedance transformation capability
C. can be readily formed from two adjacent transistors
D. resemble emitter followers

198. When same input signal is applied to both the inputs of an ideal differential amp., the output
A. is zero B. depends on its CMMR
C. depends on its voltage gain
D. is determined by its symmetry

199. The common-mode rejection ratio (CMRR) of an Ideal differential Amp is
A. zero B. infinity
C. less-than unity D. greater than unity

200. Feedback in an amplifier always helps to
A. control its output B. increase its gain
C. decrease its input impedance
D. stabilize its gain

201. Closed-loop gain of a feedback amplifier is the gain obtained when
A. its output terminals are closed
B. negative feedback is applied
C. feedback loop is closed
D. feedback factor exceeds unity

202. A large sacrifice factor in a negative feedback amplifier leads to
A. inferior performance
B. increased output impedance
C. characteristics impossible to achieve without feedback
D. precise control over output

203. Negative feedback in an amplifier
A. lowers its lower 3-dB frequency
B. raises its upper 3-dB frequency
C. increases its bandwidth
D. All of these

204. Regarding negative feedback in amplifiers which statement is wrong?
A. It widens the separation between 3 dB frequencies
B. It increases the gain-bandwidth product
C. It improves gain stability
D. It reduces distortion

205. Negative feedback reduces distortion in an amplifier only when it
A. comes as part of input signal
B. is part of its output
C. is generated within it
D. exceeds a certain safe level

206. An amplifier with no feedback has a gain-bandwidth product of 4 MHz. Its closed-loop gain is 40. The new bandwidth is
B. 100 kHz B. 160 MHz
C. 10 MHz D. 20 kHz

207. An electronic oscillator is
A. just like an alternator
B. nothing but an amplifier
C. an amplifier with feedback
D. a converter of AC to AC energy

208. The frequency of oscillation of an elementary *LC* oscillatory circuit depends on
A. coil resistance B. coil inductance
C. capacitance
D. both B and C are correct

209. For sustaining oscillations in an oscillator
A. feedback factor should be unity
B. phase shift should be 0°
C. feedback should be negative
D. both A and B are correct

210. If Barkhausen criterion is not fulfilled by an oscillator circuit, it will

Ans.	191. B	192. A	193. B	194. C	195. B	196. D	197. C	198. A	199. B	200. A	201. C	202. C	203. D
	204. B	205. C	206. A	207. C	208. D	209. D							

A. stop oscillating
B. produce damped wave continuously
C. become an amplifier
D. produce high-frequency whistles

211. In a transistor Hartley oscillator
A. inductive feedback is used
B. untapped coil is used
C. entire coil is in the output circuit
D. no capacitor is used

212. A Colpits oscillator uses
A. tapped coil B. inductive feedback
C. tapped capacitance D. no tuned LC circuit

213. In RC phase-shift oscillator circuits
A. there is no need for feedback
B. feedback factor is less than unity
C. pure sine wave output is possible
D. transistor parameters determine oscillation frequency

214. When bridge oscillator is most often used whenever
A. wide range of high purity sine waves is to be generated
B. high feedback ratio is needed
C. square output waves are required
D. extremely high resonant frequencies are required

215. Non-sinusoidal wave forms
A. are departures from sine waveform
B. have low mark to space ratio
C. are much easier to generate
D. are unit for digital operation

216. A relaxation oscillator is one which
A. has two stable states B. relaxes indefinitely
C. produces non-sinusoidal output
D. oscillates continuously

217. A square pulse has a mark-to-space ratio of
A. 1:1 B. 1:2
C. 2:1 D. 1:4

218. Apart from a DC power source, the essential requirements of a sawtooth generator are
A. a resistor B. a capacitor
C. a switching device D. all of these

219. Which of the following statement is WRONG?
A. Output is available continuously
B. Feedback between two stages is 100%
C. Positive feedback is employed
D. When one transistor is ON, the other is OFF

220. An MMV circuit
A. has no stable state
B. gives two output pulses for one input trigger pulse
C. returns to its standby states automatically
D. has no energy-storage element

221. A BMV circuit
A. has two unstable states
B. has one energy-storage element
C. switches between its two states automatically
D. is not an oscillator

222. The frequency of oscillation of an AMV depends mainly on the
A. value of collector load resistors
B. RC values of the circuit
C. value of transistor β
D. width of the input pulse

223. An MMV is frequently used
A. in memory and timing circuits
B. for producing triangular waves
C. in counting circuits
D. for regeneration of distorted waves

224. A BMV is a ...... oscillator
A. triggered B. free-running
C. sine-wave D. sawtooth

225. A blocking oscillator
A. is a triggered oscillator
B. is an amplifier with negative feedback
C. generates sinusoidal waves
D. produces sharp and narrow pulses

226. In an astable multivibrator
A. $\beta = 1$ B. $\beta A = 1$
C. $\beta > 1$ D. $\beta < 1$

227. Transformer coupling is used to cascade two such stages of which the first has
A. high collector impedance and succeeding has low input impedance
B. low collector resistance and succeeding has high input impedance
C. low collector impedance and succeeding has high input impedance
D. high collector impedance and succeeding has high input impedance

228. Which material is generally used for LED?
A. Compounds of silica
B. Compounds of gallium
C. Compounds of phosphorus
D. Compounds of sulphur

229. An NPN transistor operated at $I_c = 2.5$ mA has a collector load resistor of 1 kilo ohms and its emitter is grounded. The voltage gain of this stage is about
A. 25 B. 100
C. 250 D. 1000

230. If an NPN silicon transistor is operated at $V_{CE} = 5$ V and $I_c = 100$ microamperes and has a current gain of 100 in the CE connection, then the input resistance of this circuit will be
A. 250 Ω B. 2500 Ω
C. 25 kΩ D. 250 kΩ

Ans. 210. A 211. A 212. C 213. C 214. A 215. A 216. C 217. A 218. D 219. A 220. C 221. D 222. B 223. D 224. A 225. D 226. A 227. A 228. B 229. B 230. B

231. The most preferred high power output stage for hi-fi AF amplifiers is of the
A. Class-B push-type
B. Class-B single ended type
C. Class-A type D. Class-C type

232. The circuit schematic of a source follower having $A_v = v_o/v_i$ is shown in the figure. The input resistance of this circuit is given by

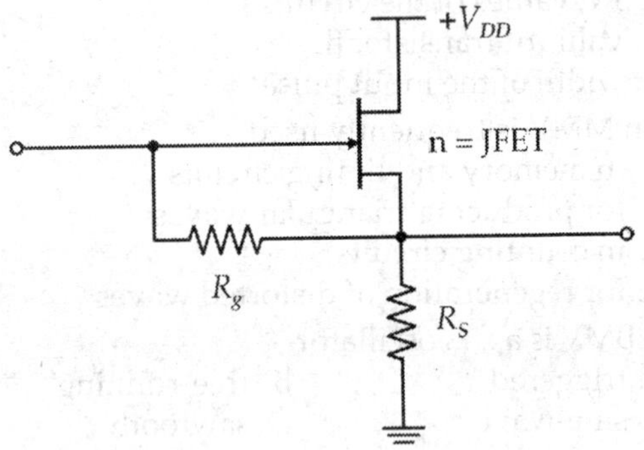

A. $R_g(1 + A_v)$ B. $R_g/(1 + A_v)$
C. $R_g(1 - A_v)$ D. $R_g/(1 - A_v)$

233. Consider the following statements : In a CMOS inverter
1. one transistor is N-channel and the other is P-channel
2. both the transistors are enhancement type
3. one transistor is enhancement type and the other is depletion type
4. both the transistors are N-channel

Of these statements
A. 1 and 2 are correct B. 2 and 4 are correct
C. 1 and 3 are correct D. 3 and 4 are correct

234. Circuit of a feedback amplifiers having series type of feedback is shown in the given figure: The β of feedback is determined by

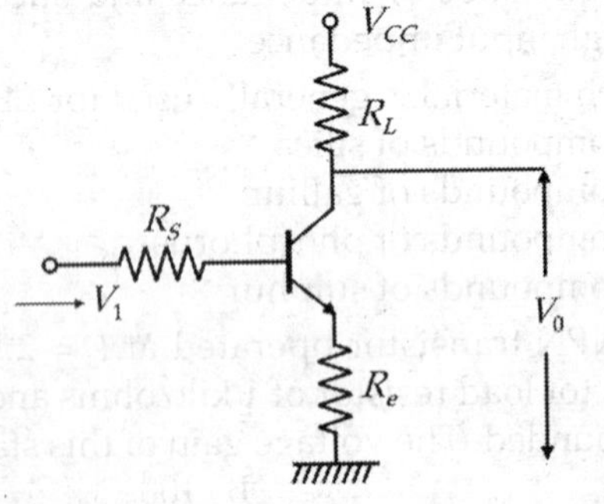

A. $-\frac{R_e}{R_L}$ B. $-R_e$
C. $-\frac{R_L}{R_e}$ D. $-\left(\frac{R_e}{R_L}\right)^2$

235. Consider the following statements.
Negative feedback in amplifiers results in
1. reduced voltage gain
2. reduced bandwidth
3. increased signal to noise ratio
4. reduced distortion

Of these statements
A. 1 and 2 are correct B. 1, 3 and 4 are correct
C. 2, 3 and 4 are correct D. 1 and 4 are correct

236. Consider the self-biased circuit given in the figure:
If driven by a current source

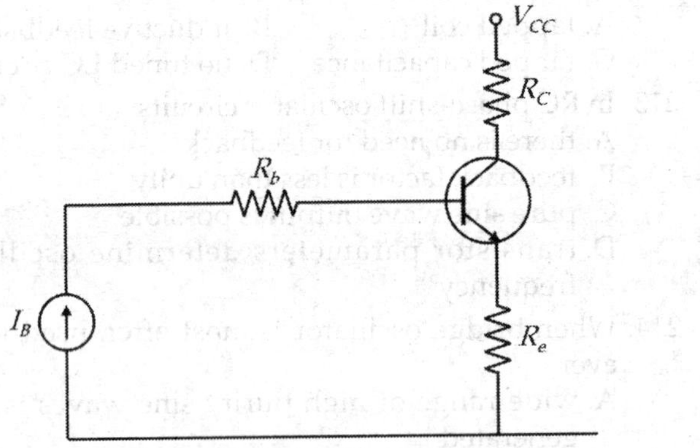

A. it will not work as self-biased circuit
B. circuit would stabilize operating point as in the case when driven by voltage source
C. stability can be increased with small value of R_e
D. stability can be increased with the increase of α of the transistor

237. In the circuit shown in the given figure, the transistor is in the active mode. V_{BE} = 0.7 volt. The best approximation without knowing h_{fe}/V as marked in the figure is:

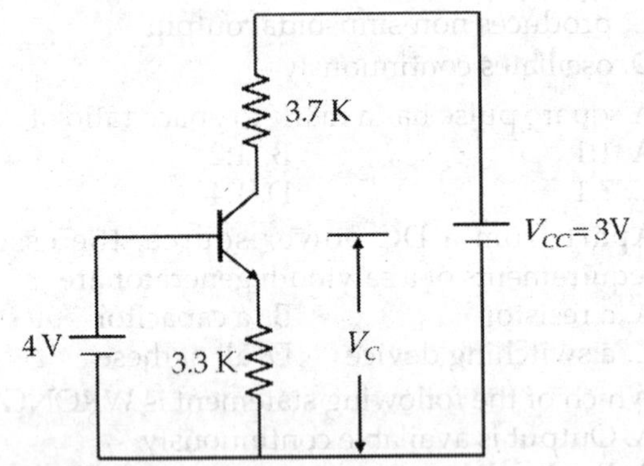

A. 1.0 V B. 3.3 V
C. 4.3 V D. 4.7 V

238. Which of the following rays is used to photograph thick metal objects?
A. infrared rays B. X-rays
C. Cosmic rays D. ultraviolet

239. In the circuit shown below, the output will be of the form

Ans. 231. D 232. C 233. A 234. A 235. B 236. A 237. C 238. B

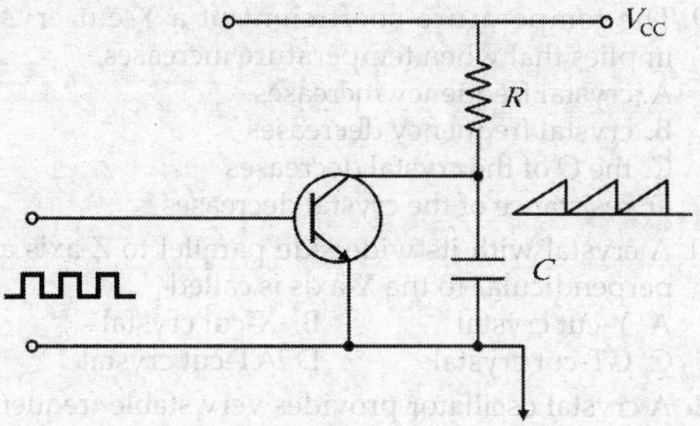

A. 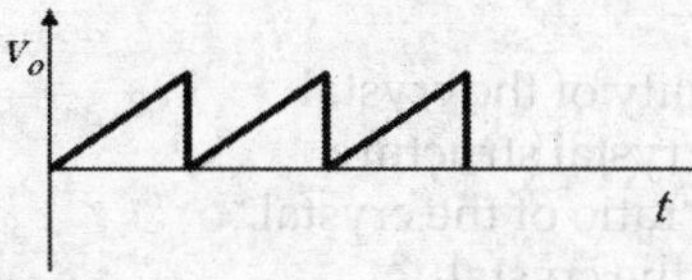

B. 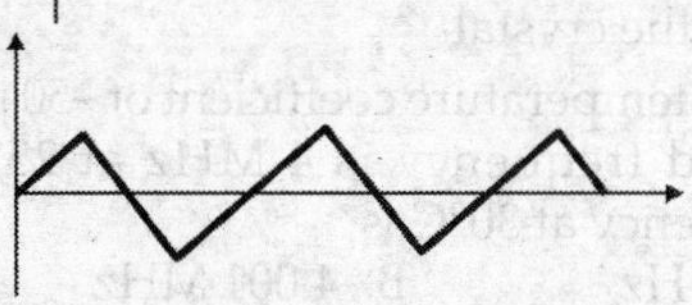

C. 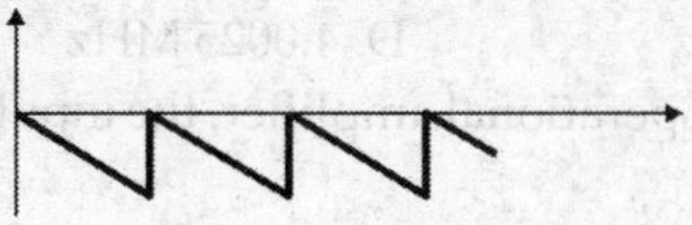

D. 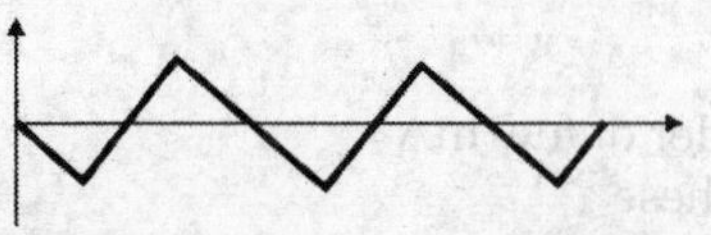

240. Cascade amplifiers are used on
A. video amplifiers B. voltage amplifiers
C. power amplifier D. tuned amplifier design

241. The advantages of Bootstrap circuit is that
A. the flyback shape of the output is very sharp
B. the charging circuit of the capacitor is constant
C. it provides high output voltage
D. all of these

242. To eliminate the crossover distortion, the transistors are given at their bases
A. small forward bias B. self bias
C. larger current D. smaller current

243. In an oscillator, the greatest amount of frequency instability is due to variations in
A. Q point B. $C_{b'e}$
C. C_e D. temperature

244. In an oscillator employing *CE* configuration, the voltage feedback is
A. phase shifted by 0° B. phase shifted by 90°
C. phase shifted by 180°
D. dependent upon the type of the oscillator

245. In a certain wien-bridge type of oscillator, the frequency selective network employs 120 K resistor and 0.001 μF capacitor. Its frequency of oscillation will be
A. 1985 Hz B. 1326 Hz
C. 1100 Hz D. 955 Hz

246. The output of the circuit shown below for the given input will be

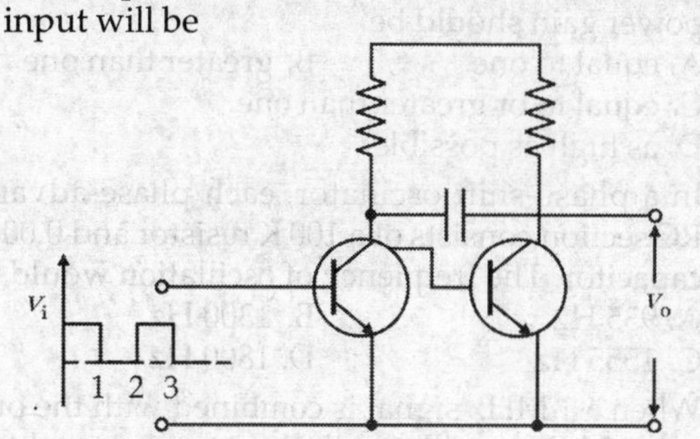

A. 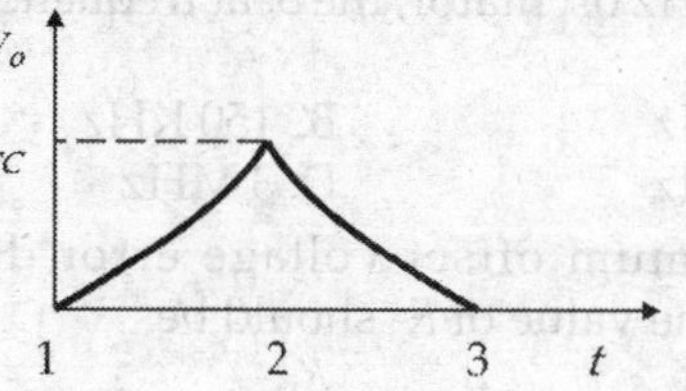

B. 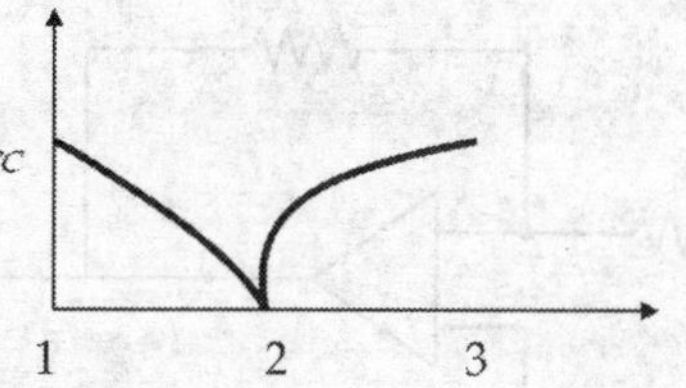

C. 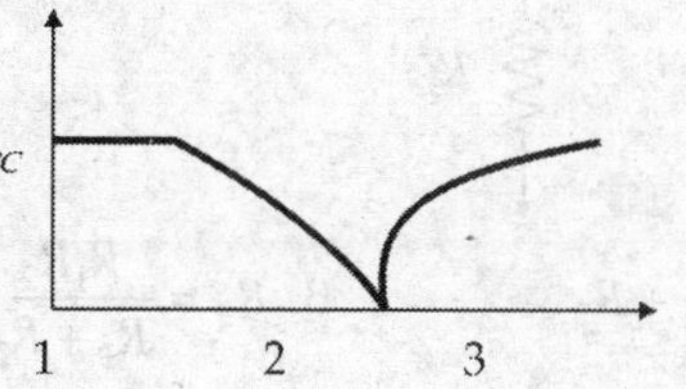

D. 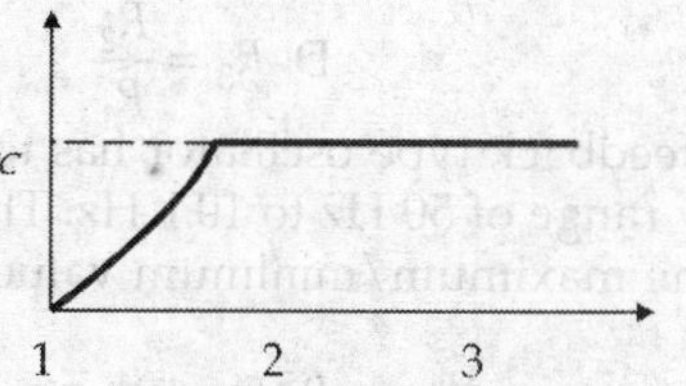

247. In a phase-shift oscillator, the minimum number of *RC* circuits required to generate a phase shift of 180° between input and output is
A. 1 B. 2
C. 3 D. 4

248. In an oscillator, the feedback signal is
A. in phase with the input signal
B. in phase with the output signal
C. out of phase with input signal
D. out of phase with the output signal

249. The most efficient oscillator operate as
A. Class A B. Class AB
C. Class B D. Class C

250. In the Hartley and Colpitts oscillators, the feedback is provided respectively through
A. a coil and a capacitor B. a capacitor and a coil
C. a capacitor in both D. a coil and a resistor

Ans. 239. C 240. A 241. B 242. A 243. B 244. A 245. B 246. C 247. C 248. A 249. D 250. A

251. If an amplifier has to sustain oscillations, then its power gain should be
A. equal to one B. greater than one
C. equal to or greater than one
D. as high as possible

252. In a phase-shift oscillator, each phase-advancing RC section consists of a 100 K resistor and 0.0005 μF capacitor. The frequency of oscillation would be
A. 935 Hz B. 1300 Hz
C. 1555 Hz D. 1800 Hz

253. When a 1 MHz signal is combined with the output of a 1.3 MHz oscillator, the beat frequency obtained will be
A. 100 KHz B. 150 KHz
C. 300 KHz D. 3 MHz

254. For minimum offset voltage error due to bias current, the value of R_3 should be

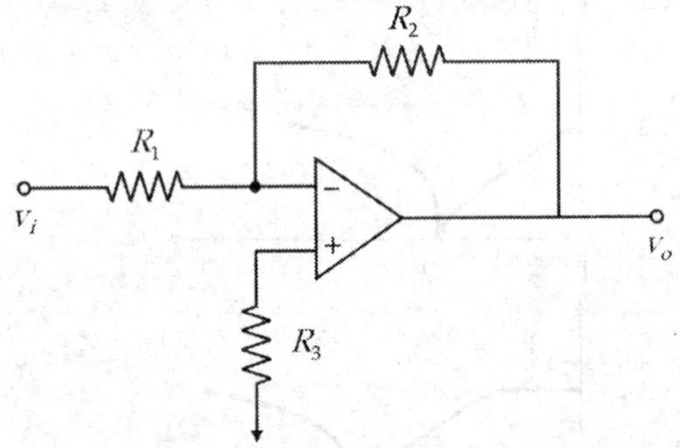

A. $R_3 = R_1 + R_2$ B. $R_3 = \frac{R_1 R_2}{R_1 + R_2}$
C. $R_3 = R_1$ D. $R_3 = \frac{R_2}{R_1}$

255. A single feedback type oscillator has to provide a frequency range of 50 Hz to 10 KHz. The required ratio of the maximum/minimum value of capacitance is
A. 4×10^4 B. 5×10^4
C. 1×10^5 D. 4×10^5

256. A blocking oscillator employs
A. degenerative feedback
B. feedback through a couple transformer
C. pulse type feedback D. none of these

257. In a tunnel diode oscillator, the function of the tunnel diode is to provide
A. positive feedback B. negative feedback
C. negative resistance to maintain oscillations in the resonant circuit
D. zero phase difference

258. For a crystal, its equivalent circuit capacitance = 0.042 pF, inductance = 3.3 H, and resistance = 390 ohms. The crystal frequency and Q are given by
A. 950 KHz, 26 B. 16 MHz, 2000
C. 670 KHz, 2700 D. 430 KHz, 23000

259. The temperature coefficient of a Y-cut crystal implies that when temperature increases,
A. crystal frequency increases
B. crystal frequency decreases
C. the Q of the crystal decreases
D. resistance of the crystal decreases

260. A crystal with its wide side parallel to Z-axis and perpendicular to the X-axis is called
A. Y-cut crystal B. X-cut crystal
C. GT-cut crystal D. AT-cut crystal

261. A crystal oscillator provides very stable frequency because of
A. high stability of the crystal
B. the rigid crystal structure
C. low X_L/R ratio of the crystal
D. high Q of the crystal

262. A crystal has temperature coefficient of –50 ppm/°C and its rated frequency is 4 MHz at 25°C. The crystal frequency at 30°C is
A. 3.90075 MHz B. 4.001 MHz
C. 3.999 MHz D. 4.0025 MHz

263. For an ideal operational amplifier, the input current is
A. infinite
B. zero
C. of the order of few mA
D. none of these

264. A differentiator is rarely used in analog computers because
A. it reduces the gain
B. it decreases the output of the amplifier
C. it amplifies noise, drift and other unwanted disturbances
D. none of these

265. Which one of the following statements is true of phase-shift type and Wain-bridge type RC oscillators?
A. Both use positive feedback
B. The former uses positive feedback only whereas the latter uses both positive and negative feedback
C. The former uses both positive and negative feedback, whereas the latter uses positive feedback only
D. Both use negative feedback

266. An OP-AMP can be classified as.......... amplifier
A. linear B. nonlinear
C. positive feedback D. RC-coupled

267. An ideal OP-AMP has
A. infinite A_v B. infinite R_i
C. zero R_o D. All of these

268. OP-AMPs have become very popular in industry mainly because

Ans. 251. C 252. B 253. C 254. B 255. A 256. B 257. C 258. D 259. A 260. B 261. D 262. C 263. B 264. C 265. B 266. A 267. D

A. they are very cheap
B. their external characteristics can be changed to suit any application
C. of their extremely small size
D. they are available in different packages

269. Since input resistance of an ideal OP-AMP is infinite
A. its output resistance is zero
B. its output voltage becomes independent of load resistance
C. its input current is zero
D. it becomes a current-controlled device

270. The gain of an actual OP-AMP is around
A. 1,000,000 B. 1000
C. 100 D. 10,000

271. When an input voltage of 1 V is applied to an OP-AMP having $A_v = 10^6$ and bias supply of 15 V, the output voltage available is
A. 15×10^6 V B. 10^6 V
C. 15 μV D. 15 V

272. Potentiometers are extensively used in analog computers to
A. multiply currents by a constant less unity
B. multiply currents by a constant greater than unity
C. multiply voltages by a constant less than unity
D. multiply voltages by a constant greater than unity

273. In the circuit shown in the figure, the expression for I_o/I_i is

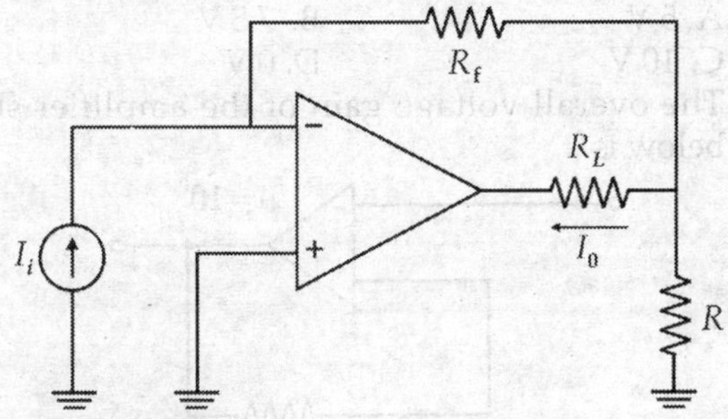

A. $R_1/(R_L + R)$ B. R_f/R_L
C. $(1 + R_f/R)$ D. $(1 - R_f/R)$

274. The circuit shown below is used for

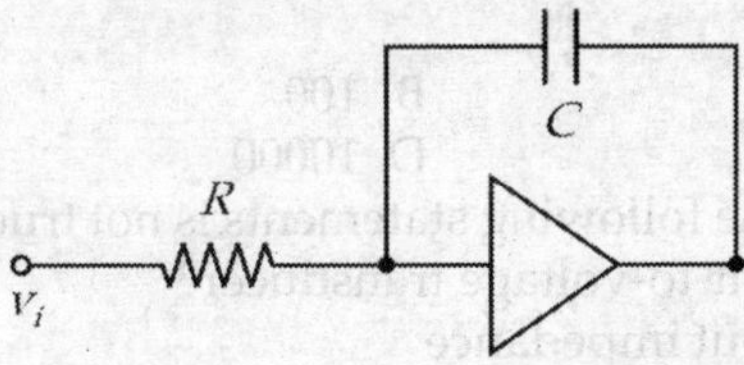

A. summation B. subtraction
C. integration D. differentiation

275. The output impedance of an ideal operational amplifier is
A. zero B. infinite
C. few K ohms D. few ohms

276. The following circuit is used for

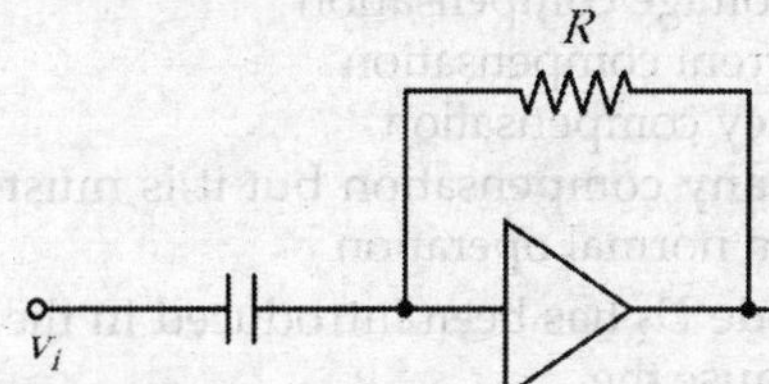

A. summation B. subtraction
C. integration D. differentiation

277. The bandwidth, over which the operational amplifier has constant gain is from
A. 0 to a few Hz B. 20 KHz to 200 KHz
C. 1 MHz to 10 MHz D. none of these

278. The cut-off frequency, when the OP-AMP integrator circuit will act as low pass filter is
A. $f_c = 1/2\pi R_2C_1$ B. $f_c = 1/2\pi R_1C_1$
C. $f_c = R/9\pi R_3C_1$ D. none of these

279. The differentiator Op amp circuit is subjected to damage from fast rising input signals because
A. it can not differentiate them
B. it is susceptiable to high frequency instability
C. it will puncture the OP-AMP IC because of increased gain at high frequency
D. none of these

280. The feedback path of zero-crossing detector should have
A. capacitive circuit B. limiting circuit
C. open circuit D. short circuit

281. The circuit shown below is a

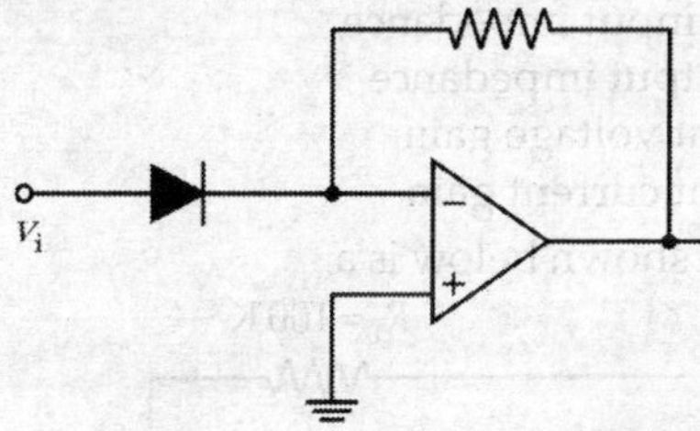

A. log amplifier B. antilog amplifier
C. clipper D. clamping circuit

282. The circuit shown below is a

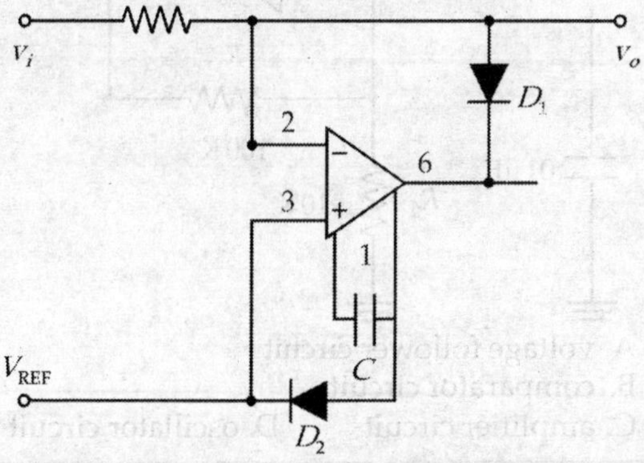

Ans. 268. B 269. C 270. A 271. D 272. C 273. C 274. C 275. A 276. D 277. A 278. A 279. B 280. B 281. B

A. log amplifier B. antilog amplifier
C. clamping circuit D. rectifier circuit

283. In the above circuit, the capacitor C is used for
A. offset-voltage compensation
B. bias current compensation
C. frequency compensation
D. not for any compensation but it is must in the circuit in normal operation

284. Clamp diode D_2 has been introduced in the above circuit because the
A. clamp diode prevents op amp from saturating when D_1 is reverse biased
B. clamp diode prevents op amp from saturating when D_1 is forward biased
C. clamp diode helps in offset voltage compensation
D. none of these

285. The circuit shown below is a

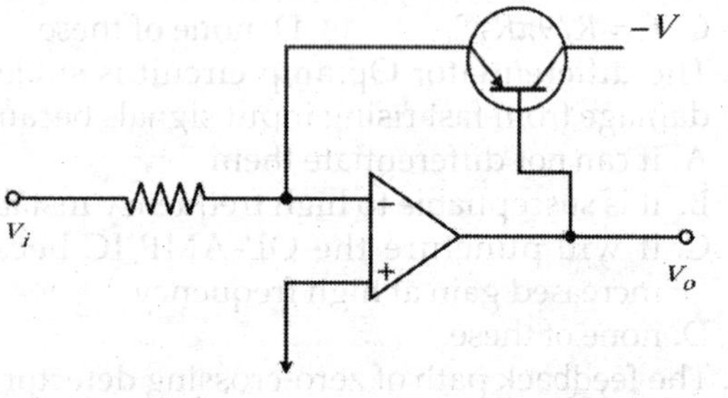

A. log amplifier B. antilog amplifier
C. clipper D. clamping circuit

286. Which of the following statements is not true for an ideal voltage amplifier?
A. Infinite input impedance
B. Zero output impedance
C. Constant voltage gain
D. Constant current gain

287. The circuit shown below is a

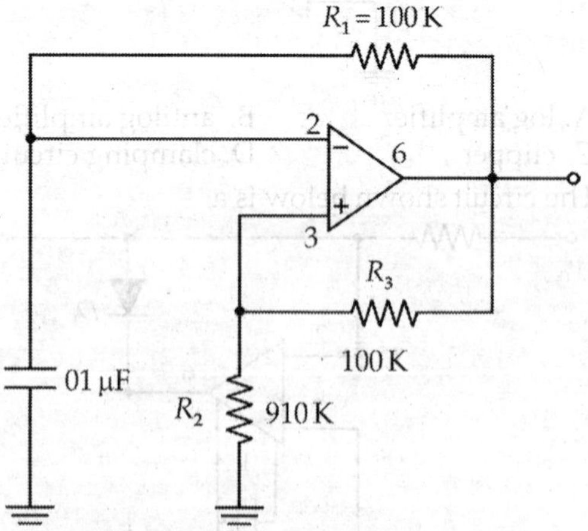

A. voltage follower circuit
B. comparator circuit
C. amplifier circuit D. oscillator circuit

288. For the above circuit, which of the following statement is correct?
A. It is a free running multivibrator
B. It gives triangular output
C. It gives sinusoidal waveform output
D. None of these

289. The frequency of oscillation for the same circuit is given by

A. $f = 2R_2C\ln\left(\dfrac{2R_2}{R_2}\right)$ B. $f = \dfrac{1}{2R_2C\ln\left(\dfrac{2R_2}{R_3}\right)}$

C. $f = \dfrac{1}{2R_2\ln\left(\dfrac{R_2}{R_3}\right)}$ D. $f = \dfrac{1}{2R_2C\ln\left(\dfrac{R_2}{R_3}\right)}$

290. In the differencial amplifier circuit shown below, the collector to ground voltage is

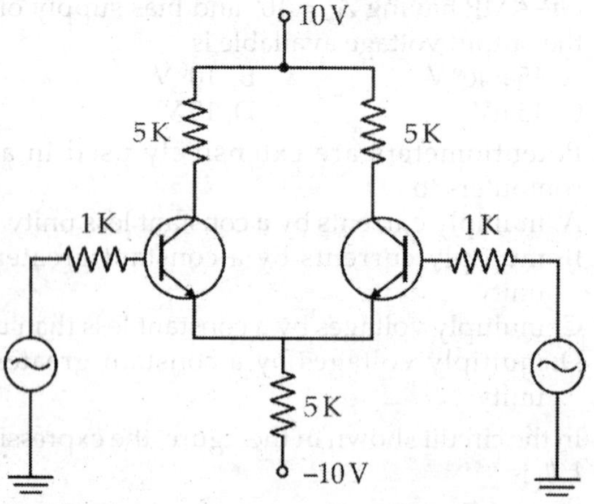

A. 5 V B. 7.5 V
C. 10 V D. 0 V

291. The overall voltage gain of the amplifier shown below is

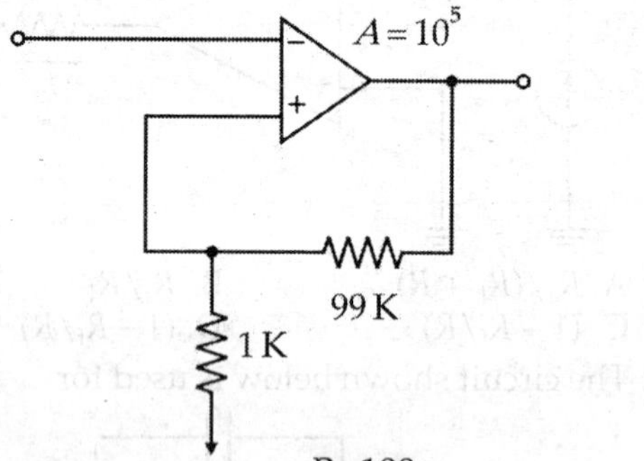

A. 50 B. 100
C. 1000 D. 10000

292. Which of the following statements is not true for an ideal current-to-voltage transducer?
A. Zero input impedance
B. Infinite input impedance
C. Zero output impedance
D. A fixed relation between input current and output

Ans. 282. C 283. C 284. A 285. A 286. D 287. D 288. A 289. B 290. A 291. B 292. B

293. Which of the following statements is not true for an ideal voltage-to-current transducer?
A. Infinite input impedance
B. Zero output impedance
C. A fixed relationship between input current and output voltage
D. A fixed relationship between current output and input voltage

294. In the circuit shown below, an input current of 1 μA produces an output of 10 mV. The value of resistance *R* is

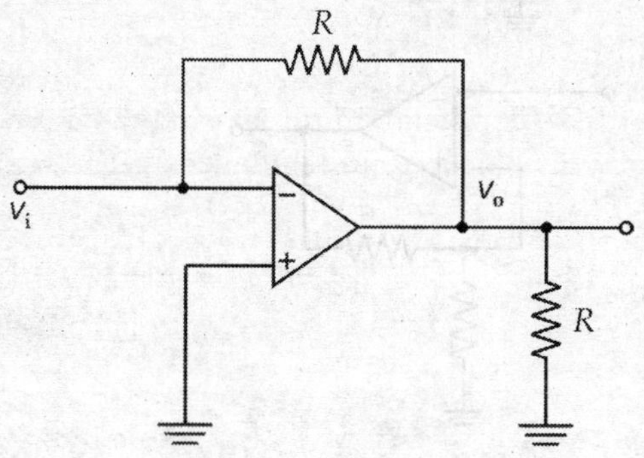

A. 0.1 K B. 1 K
C. 10 K D. 100 K

295. The overall voltage gain of the amplifier shown below is

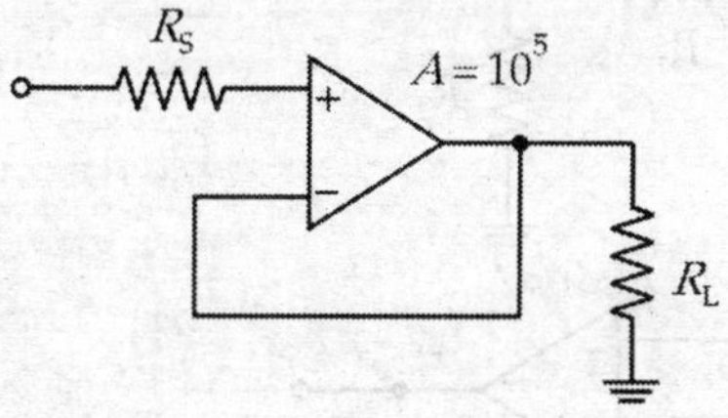

A. zero B. 1
C. 100 D. 100000

296. In the circuit shown below, output current is 100 μA for input voltage of 1 mV, the value of resistance *R* should be

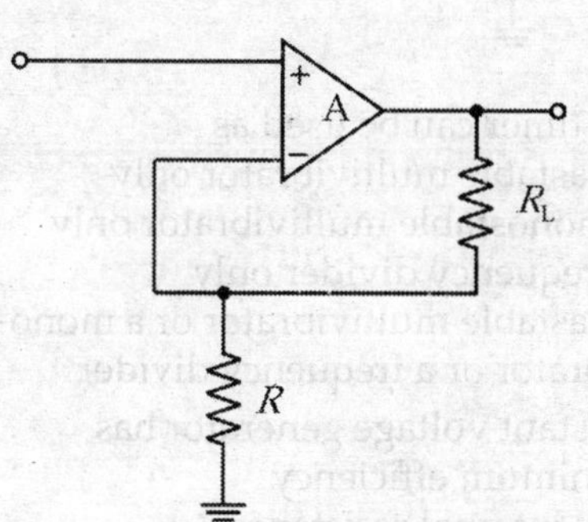

A. 10 K B. 1 K
C. 500 ohms D. 10 ohms

297. In the above problem, the output impedance is
A. 10 K B. 500 K
C. 0.1 M ohms D. 1 M ohms

298. The bandwidth of a feedback amplifier is
A. larger than that of internal amplifier
B. smaller than that of internal amplifier
C. equal to that of internal amplifier
D. independent of feedback

299. The 741 chip used in the feedback amplifier has mid-band gain of 120 dB and break frequency of 10 Hz. The bandwidth of the amplifier will be

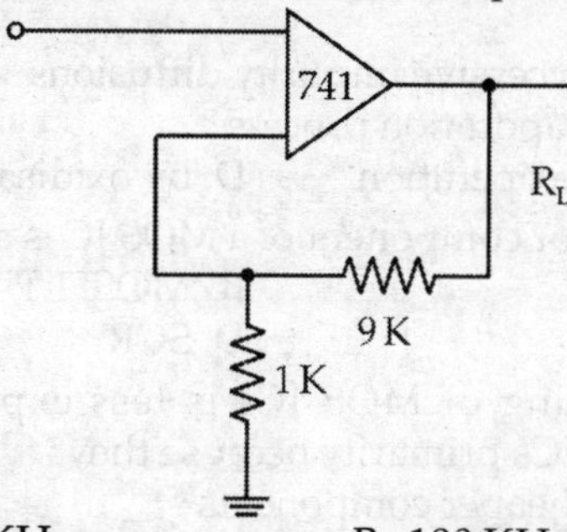

A. 10 KHz B. 100 KHz
C. 1 MHz D. 10 MHz

300. In ICs, Cascode amplifier can be obtained from
A. CA 3085 B. CA 3028A
C. CA 723 D. CA 741

301. Which of the following ICs is a voltage regulator?
A. CA 723 B. CA 3028A
C. CA 3065 D. CA 741

302. First integrated circuit chip was developed by
A. CV Raman B. WH Brattain
C. JS Kilby D. Robert Noyce

303. An integrated electronic circuit is
A. a complicated circuit
B. an integrating device
C. much costlier than a single transistor
D. farbricated on a tiny silicon chip

304. Most important advantage of an IC is its
A. easy replacement in case of circuit
B. extremely high reliability
C. reduced cost
D. low power consumption

305. In monolithic ICs, all components are fabricated by
A. evaporation B. sputtering
C. diffusion D. oxidisation

306. Monolithic ICs are fabricated within a
A. soft stone B. single stone
C. silicon layer D. ceramic base

307. As compared to monolithic ICs, film ICs have the advantage of
A. better high-frequency response
B. less flexibility in circuit design
C. smaller size D. much reduced cost

Ans. 293. B 294. C 295. B 296. D 297. C 298. A 299. C 300. B 301. A 302. C 303. D 304. B 305. C 306. C 307. A

308. Monolithic technique is ideally suited for fabricating....... ICs.
A. thin-film B. digital
C. linear D. thick-film

309. In the context of IC fabrication, metallisation means
A. connecting metallic wires
B. forming interconnecting conduction pattern and bonding pads
C. depositing SiO_2 layer
D. covering with a metallic cap

310. Monolithic transistors are formed in the epitaxial *N*-layer
A. by successive impurity diffusions
B. by evaporation process
C. in one operation D. by oxidisation

311. The major component of a MOS IC is a/an
A. FET B. MOSFET
C. BJT D. SCR

312. Processing of MOS ICs is less expensive than bipolar ICs primarily because they
A. use cheaper components
B. need no component isolation
C. require much less diffusion steps
D. have very high packing density

313. Most of the linear ICs are based on the two-transistor differential amplifiers because of its
A. input voltage dependent linear transfer characteristic
B. high voltage gain
C. high input resistance D. high CMRR

314. Ultraviolet radiation is used in IC fabrication process for
A. diffusion B. masking
C. isolation D. metallisation

315. Which of the following statement is not correct?
A. A beam of light may be modulated to carry speech
B. Blinking light signals from ship to ship use modulated light beams
C. Short-wave radio uses shorter wave-lengths than that exist in light beams
D. Rays of 1A wavelength pass through glass more easily than rays of 1000A

316. In a X-ray tube, if the anode to cathode DC voltage is increased, the wavelength of X-rays will
A. increase B. decrease
C. remains unaffected D. None of these

317. Which of the following statement is correct?
A. A lamp used for infrared heating has a hotter filament than a lamp for lighting
B. A phototube that 'sees' very short wavelength, can operate a relay faster than a phototube that 'sees' only longer wavelengths
C. A fluorescent lamp at 3500°K feels warmer than a tungsten lamp at 2800°K
D. None of these

318. Which of the following circuits is an ideal current amplifier?

A. 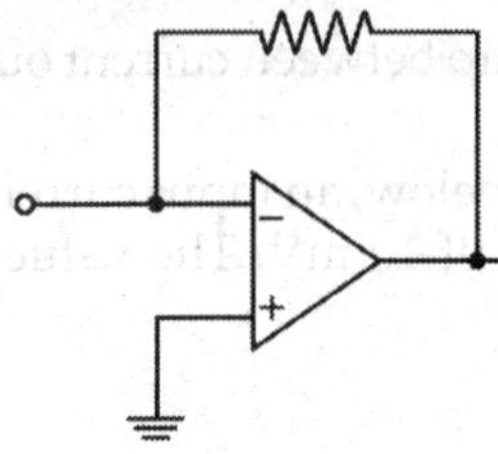

B. 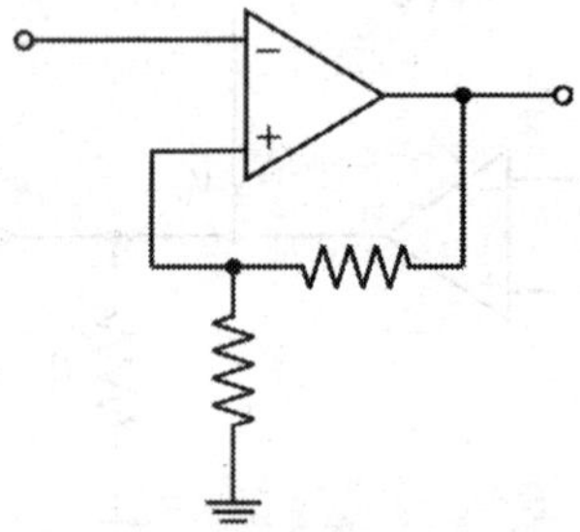

C. 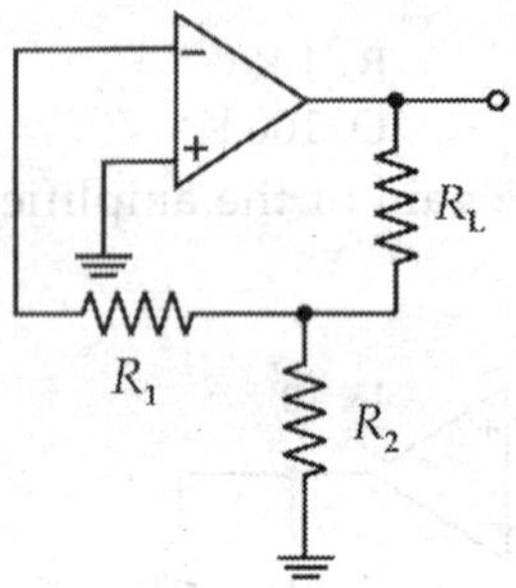

D. 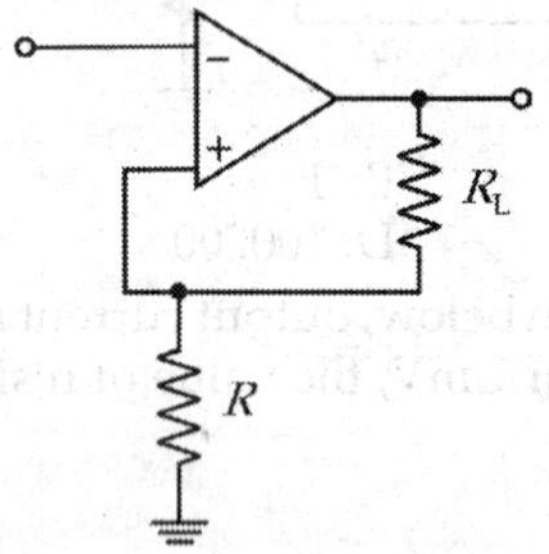

319. A 555 timer can be used as
A. an astable multivibrator only
B. a monostable multivibrator only
C. a frequency divider only
D. an astable multivibrator or a mono-stable multivibrator or a frequency divider

320. A constant voltage generator has
A. minimum efficiency
B. low internal resistance
C. high internal resistance
D. minimum current capacity

Ans. 308. B 309. B 310. A 311. B 312. C 313. D 314. A 315. C 316. B 317. D 318. C 319. D 320. B

321. Which of the following circuits can be used as a DC ammeter with high sensitivity?

A.
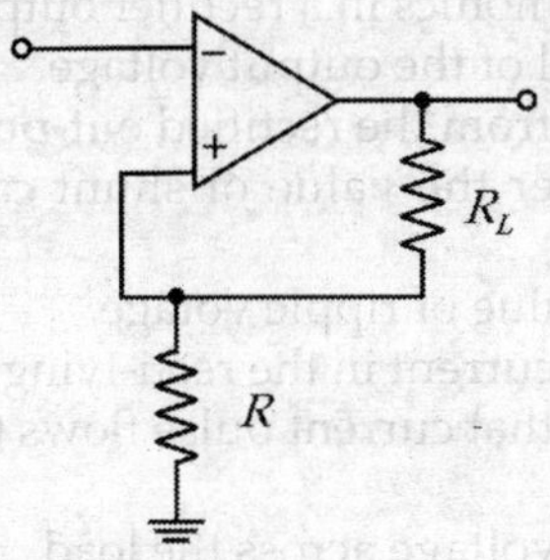

B.
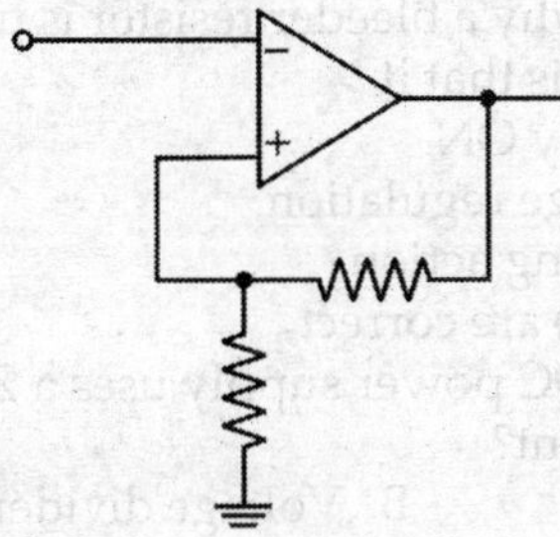

C.
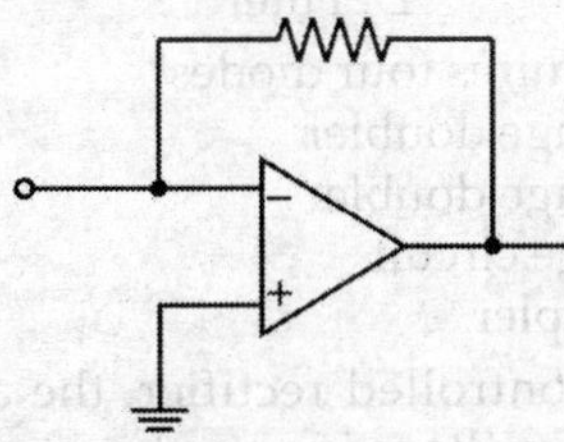

D.
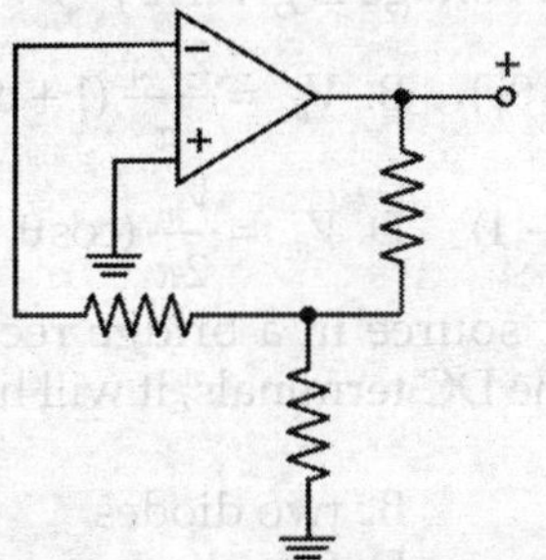

322. For the circuit shown below, the value of input current required for full scale deflection will be

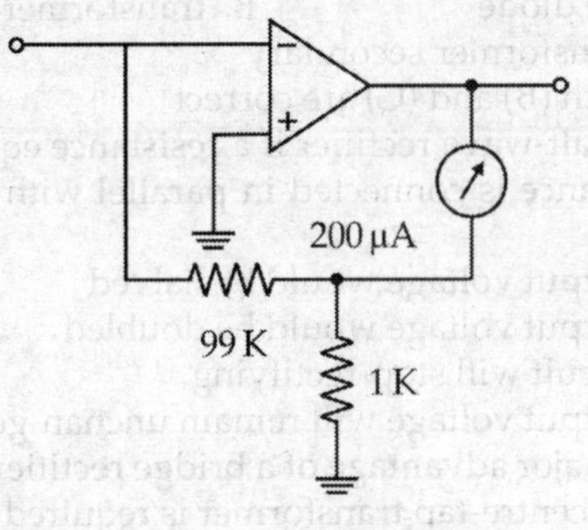

A. 0.5 µA B. 1 µA
C. 1.5 µA D. 2 µA

323. In a photostate machine, the drum is made of
A. selenium coating B. silicon coating
C. cadmium coating D. zircon coating

324. Which one of the following oscillators is used for generation of high frequencies?
A. RC phase shift B. Wien bridge
C. LC oscillator D. Blocking oscillator

325. The circuit shown below is a

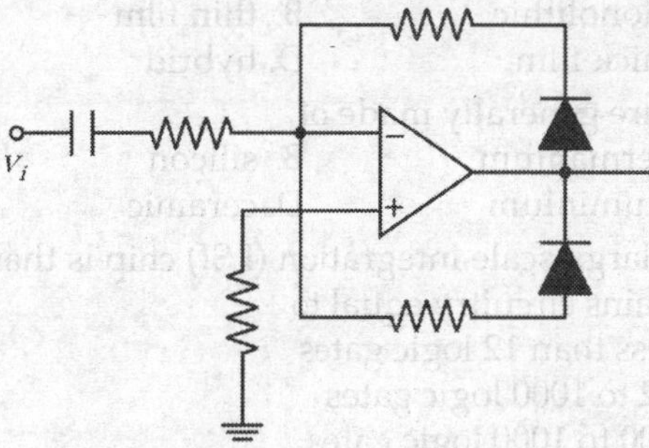

A. clamping circuit B. long amplifier
C. antilog amplifier D. rectifier circuit

326. For the above circuit, which of the following statements is not correct?
A. It is a half-wave rectifier circuit
B. The forward voltage drop of diode is effectively divided by the op amp open-loop gain
C. The op amp's offset voltage is amplified by DC open-loop gain
D. For low input signals, the amplifier input noise voltage is amplified by closed-loop gain

327. The best important application of an IC is in
A. computers B. simple circuits
C. nonlinear circuits D. none of these

328. The most popular form of IC package is
A. TO-5 B. DIL
C. flat pack D. all of these

329. It is not possible to produce ICs of power rating more than
A. 1 watt B. 100 watts
C. 10 watts D. 1000 watts

330. For fabrication of integrated circuit silicon is used
A. since silicon is mechanically strong
B. since silicon is more conductive semiconductor material
C. since silicon possesses characteristics which are best suited to IC manufacturing processes
D. since silicon is black in colour

331. The circuits in which the various separately manufactured components, such as transistors, diode capacitors, resistors, etc. are connected by plated conductors or printed boards are called
A. integrated circuits B. discrete circuits
C. linear circuits D. none of these

Ans. 321. C 322. D 323. A 324. A 325. D 326. D 327. A 328. B 329. C 330. C 331. B

332. An integrated circuits has extremely
A. small size B. small weight
C. heavy weight D. both A and B

333. To see the connections between the components of an IC, we generally require a
A. lense B. telescope
C. microphone D. microscope

334. The most commonly used ICs are
A. monolithic B. thin film
C. thick film D. hybrid

335. ICs are generally made of
A. germanium B. silicon
C. aluminium D. ceramic

336. The large-scale integration (LSI) chip is that which contains circuitry equal to
A. less than 12 logic gates
B. 12 to 1000 logic gates
C. 100 to 1000 logic gates
D. 1000 or more than 1000 logic gates

337. It is not possible to fabricate on an IC
A. inductor B. capacitor
C. resistor D. both A and B

338. An inverting amplifier has $R_f = 2$ M and $R_1 = 2$ K. Its scale factor is
A. 1000 B. −1000
C. 10^{-3} D. -10^{-3}

339. When in a negative sealer, both R_1 and R_f are reduced to zero, the circuit functions as
A. integrator B. subtracter
C. comparator D. unity follower

340. The ripple factor of a power supply is a measure of
A. its filter efficiency B. its voltage regulation
C. diode rating
D. purity of power output

341. The basic reason why a FW rectifier has twice the efficiency of a HW rectifier is that
A. it makes use of a transformer
B. its ripple factor is much less
C. it utilizes both half-cycles of the input
D. its output frequency is double the line frequency

342. The output of a half-wave rectifier is suitable only for
A. running car radios B. running AC motors
C. charging batteries
D. running tape-recorders

343. The ripple factor of a bridge rectifier is
A. 0.406 B. 0.812
C. 1.21 D. 1.11

344. The PIV of a half-wave rectifier circuit with a shunt capacitor filter is
A. $2V_{sm}$ B. V_{sm}
C. $V_{sm}/2$ D. $2V_{sm}$

345. The primary function of a rectifier filter is to
A. minimise ac input variations
B. suppress of harmonics in a rectifier output
C. stabilise dc level of the output voltage
D. remvoe ripples from the rectified out-put

346. In a rectifier, larger the value of shunt capacitor filter
A. larger the *p-p* value of ripple votage
B. larger the peak current in the recti-fying diode
C. longer the time that current pulse flows through the diode
D. smaller the DC voltage across the load

347. The main reason why a bleeder resistor is used in a DC power supply is that it
A. keeps the supply ON
B. improves voltage regulation
C. improves filtering action
D. both (B) and (C) are correct

348. Which stage of a DC power supply uses a Zener as the main component?
A. Rectifier B. Voltage divider
C. Regulator D. Filter

349. Which rectifier requires four diodes?
A. Half-wave voltage doubler
B. Full- wave voltage doubler
C. Full-wave bridge circuit
D. Voltage quadrupler

350. For a half-wave controlled rectifier, the average value of output DC voltage is given by
A. $V_{dc} = \frac{V_m}{2\pi}(1 - \cos\theta_1)$ B. $V_{dc} = \frac{2V_m}{\pi}(1 + \cos\theta_1)$
C. $V_{dc} = \frac{V_m}{\pi}(\cos\theta_1 - 1)$ D. $V_{dc} = \frac{V_m}{2\pi}(\cos\theta_1 + 1)$

351. If, by mistake, AC source in a bridge rectifier is connected across the DC terminals, it will burn out and hence short
A. one diodes B. two diodes
C. three diodes D. four diodes

352. The DC output polarity from a half wave rectifier can be reversed by reversing
A. the diode B. transformer primary
C. transformer secondary
D. both (B) and (C) are correct

353. In a half-wave rectifier if a resistance equal to load resistance is connected in parallel with the diode, then
A. output voltage would be halved
B. output voltage would be doubled
C. circuit will stop-rectifying
D. output voltage will remain unchan-ged

354. The major advantage of a bridge rectifier is that
A. no centre-tap transformer is required
B. the required peak inverse voltage of each diode is double of that for a full-wave rectifier

Ans. 332. D 333. D 334. A 335. B 336. C 337. D 338. B 339. D 340. D 341. C 342. B 343. C 344. A
345. D 346. B 347. D 348. C 349. B 350. D 351. D 352. A 353. C

C. peak inverse voltage of each diode is half of that for a full-wave rectifier
D. the output is more smooth.

355. The Zener diode shown in the circuit has a reverse breakdown voltage of 10 volts. The power dissipation in R_s would be

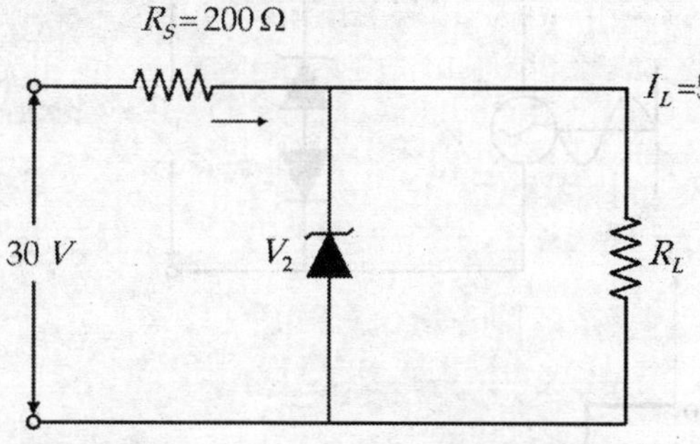

A. 0.5 W B. 1 W
C. 15 W D. 2 W

356. A bridge rectifier provides 50 mA current at 150 V: The transformer voltage speci-fication is
A. 220 V:167 V B. 120 V:150 V
C. 220V:236 V D. 120 V:110 V

357. A bridge rectifier provides 50 mA current at 150 V. The average current and PIVof each diode, respectively are
A. 79 mA, 167 V B. 25 mA, 236 V
C. 79 mA, 167 V D. 25 mA, 120 V

358. Which of the following circuits cannot be operated directly from the mains?
A. Half-wave rectifier
B. Centre-tap full-wave rectifier
C. Bridge rectifier D. Voltage doubler

359. For a filter, no load output voltage is 300 V and full-load output voltage is 280 V. The regulation of the circuit is
A. 9.6% B. 7.1%
C. 6.7% D. 4.2%

360. A typical value of filter capacitor for 50 Hz input is
B. 1,000 μF B. 50 μF
C. 1,000 pF D. 100 pF

361. A typical value of induction filter for 50 Hz input is
A. 10 H B. 100 mH
C. 10 mH D. 100 μH

362. If the ripple factor of the output wave of a rectifier is low, it means that
A. output voltage will have less ripple
B. output voltage will be low
C. filter circuits may not be required
D. none of these

363. The ripple factor of a waveform is defined as

$$\text{Ripple factor} = \frac{\text{RMS of the ripple}}{\text{Average output level}}$$

By the definition, the ripple factor of the waveform shown below will be

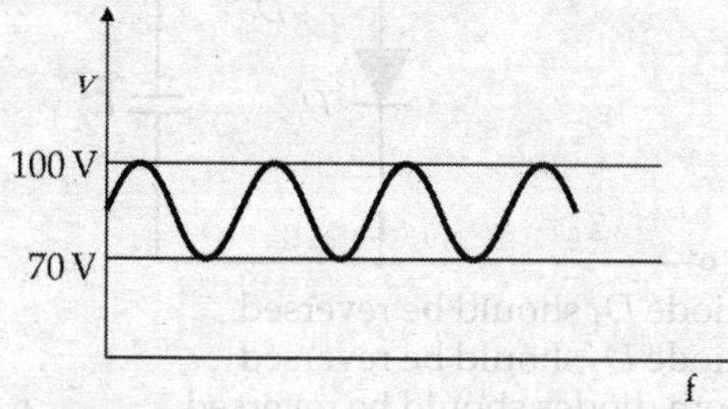

A. 15% B. 1.21%
C. 12.5% D. none of these

364. The resistance temperature coefficient of an ordinary junction diode is θ_D. The resistance temperature coefficient of a zener diode is
A. negative and much larger than θ_D
B. positive and much larger than θ_D
C. positive and much smaller than θ_D
D. positive and nearly equal to θ_D

365. In a rectifier circuit, the load connected is of low value. For proper filter operation, it is required that
A. a capacitor is to be included in the circuit
B. a bleeder resistance is to be placed in the circuit
C. an inductor, filter is to be included in the circuit
D. All of these

366. In the voltage regulator shown below, if the current through the load decreases

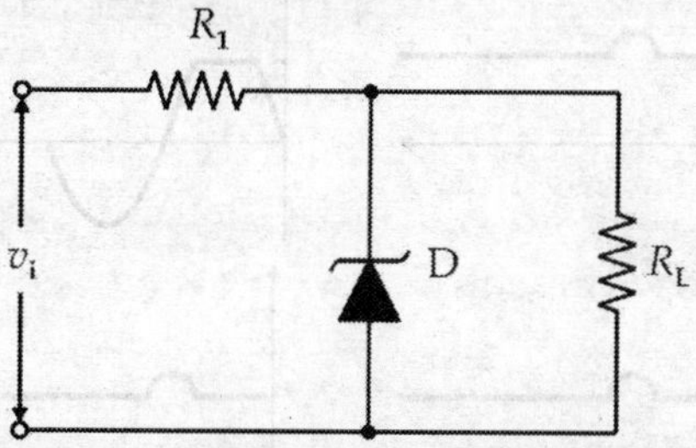

A. the current through R_1 will increase
B. the current through Ri_1 will decrease
C. zener diode current will increase
D. zener diode current will decrease

367. Ripple frequency of a six-phase half-wave rectifier for 230 V, 50 Hz input will be
A. 150 Hz B. 50 Hz
C. 300 Hz D. 100 Hz

368. The limitation of the voltage multiplying circuits is that
A. the output has high ripple content
B. high output voltage is difficult to obtain
C. high output current is difficult to obtain
D. the size of the capacitor becomes very large

369. In the voltage doubling circuit shown below, one of the diodes is connected wrongly. For correct operation

Ans. 354. C 355. D 356. A 357. B 358. B 359. B 360. B 361. A 362. A 363. C 364. D 365. C 366. C 367. C 368. C

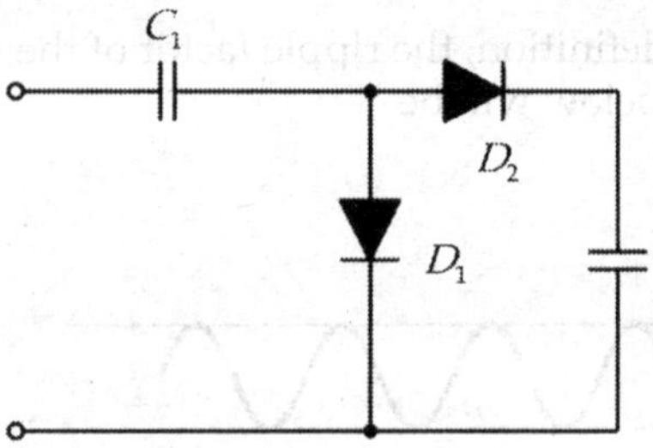

A. diode D_1 should be reversed
B. diode D_2 should be reversed
C. both diodes should be reversed
D. none of these

370. The outputs of the two circuits shown below are

A. 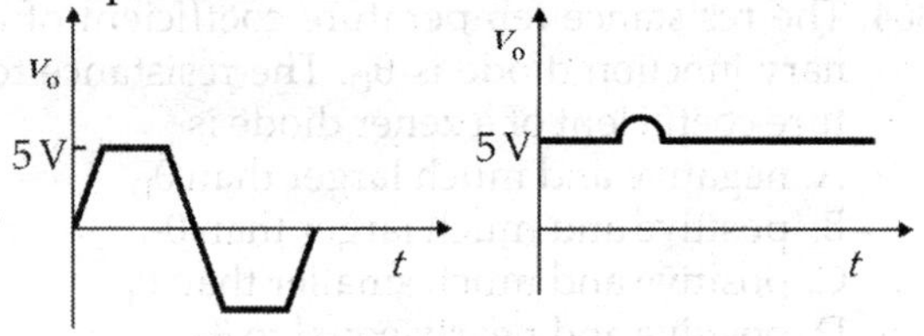

B. 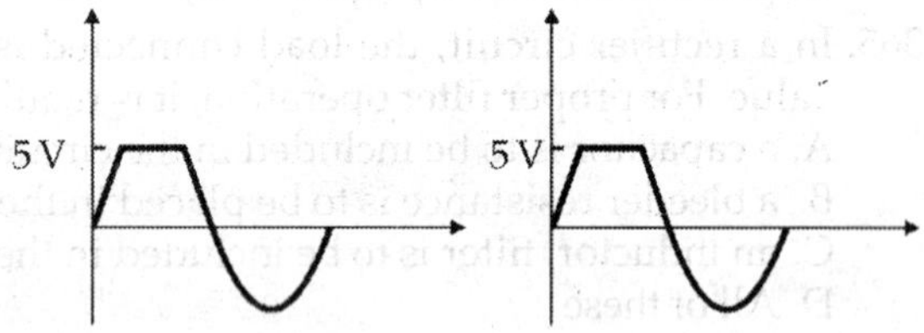

C. 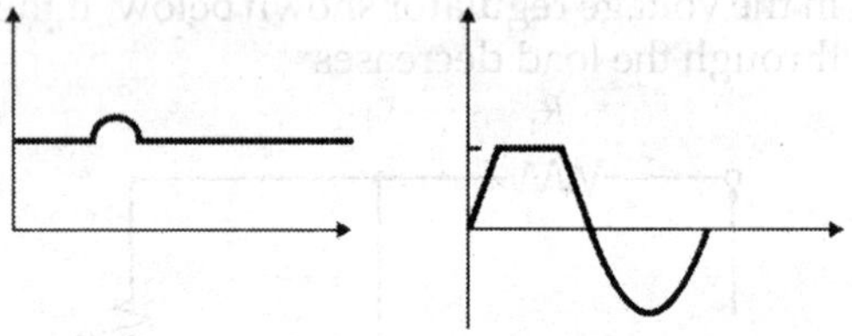

D. 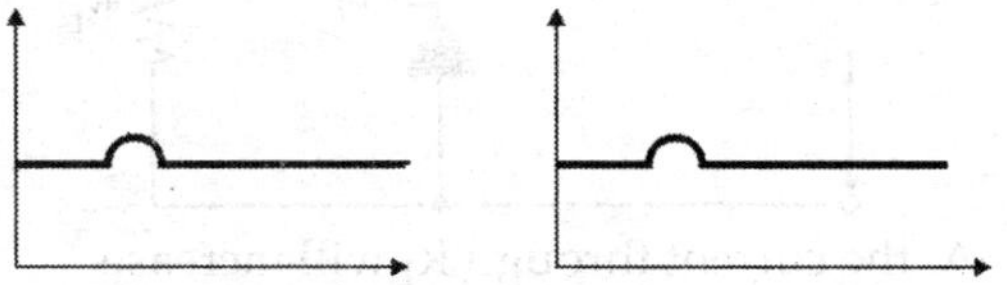

371. The function of bleeder resistance in filter circuits is
A. to maintain minimum current necessary for optimum inductor filter operation
B. to work as voltage divider in order to provide variable output from the supply
C. to provide discharge path to capaci-tors so that output becomes zero when the circuit has been deenergised
D. all of these

372. In a three-phase half-wave rectifier, each diode conducts for a duration of
A. 180° B. 30°
C. 60° D. 45°

373. As the number of phases in a multiphase rectifier are increased the output will
A. increase B. become more smooth
C. decrease
D. the diodes will require high peak inverse voltage

374. The output of the circuit shown for the given input will be

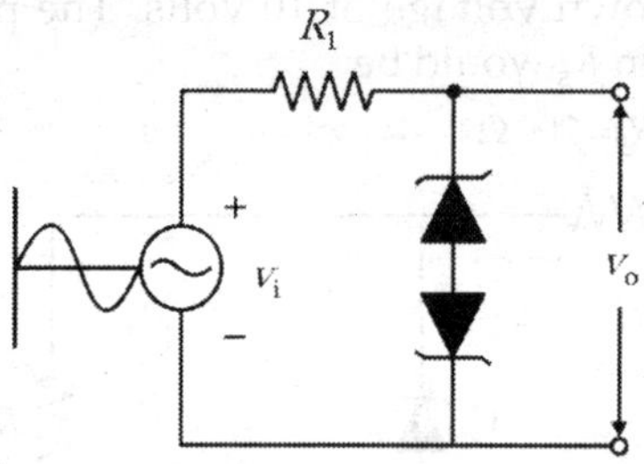

A. 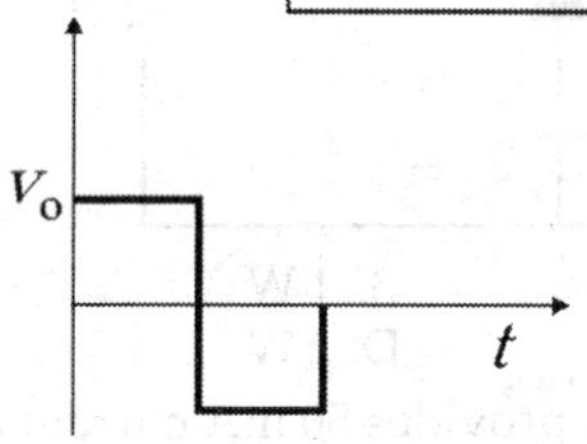

B. 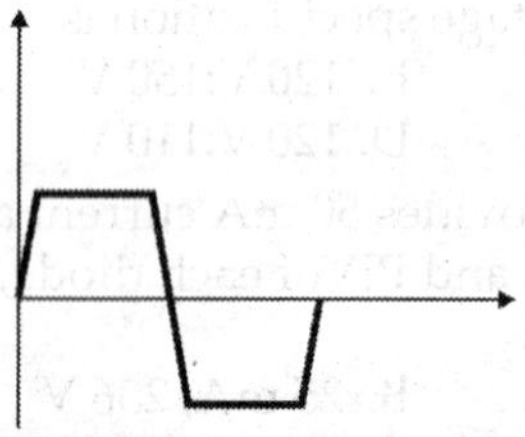

C. 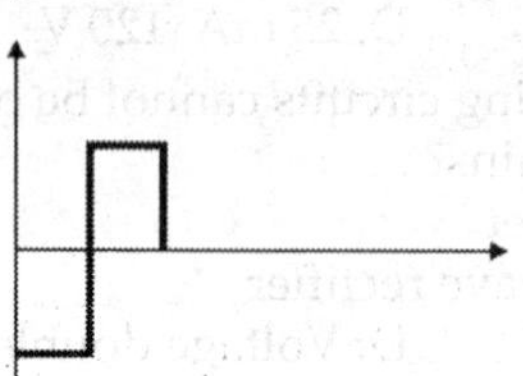

D. 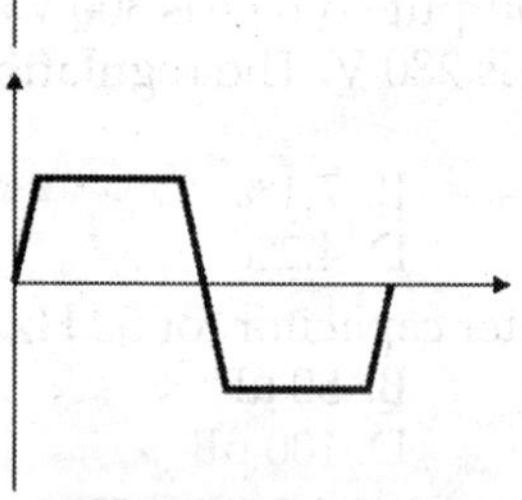

375. For a rectifier output of 150 kW at 500 V, an average current rating of 50 A for each diode is sufficient in
A. three-phase full-wave rectifier
B. six-phase half-wave rectifier
C. three-phase half-wave rectifier
D. none of these

376. The effect of DC saturation in a rectifier transformer is
A. to decrease the output
B. to increase the output
C. to decrease the AC components of the output
D. none of these

Ans. 369. B 370. B 371. D 372. C 373. B 374. B 375. B 376. A

377. The output waveform v_o in the circuit shown below is

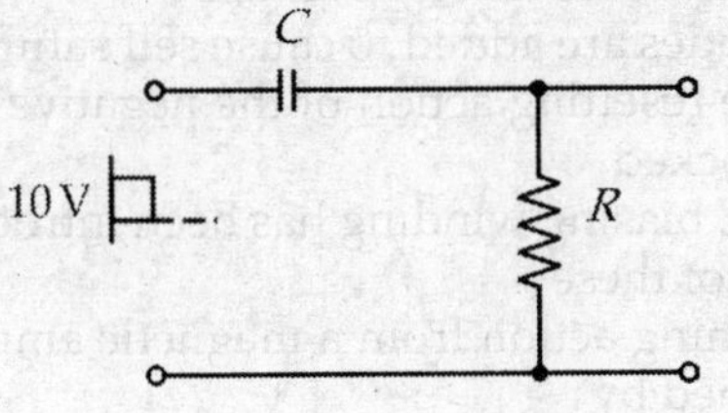

A.
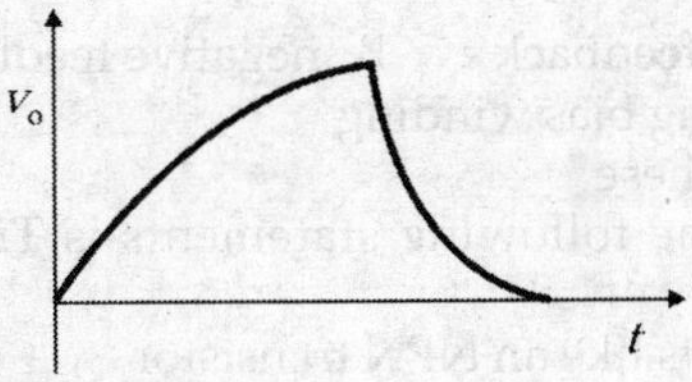

B.
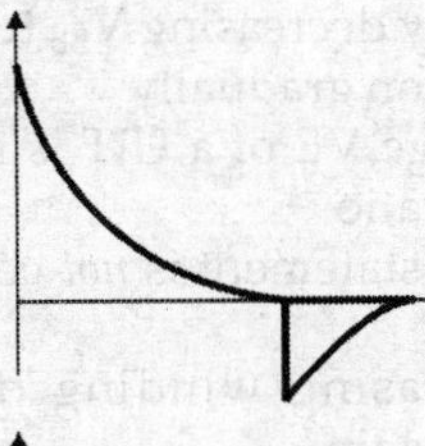

C.
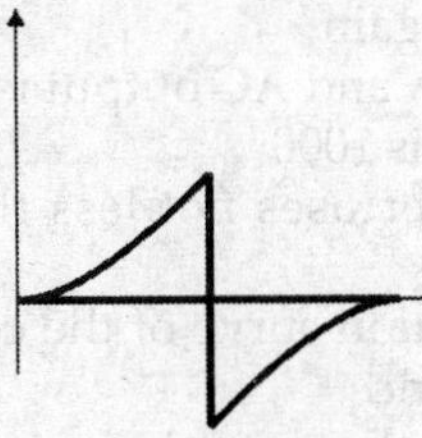

D.
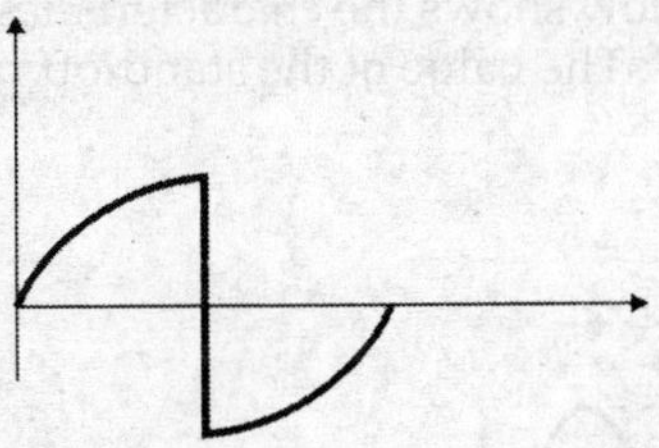

378. In a three-phase half-wave rectifier, the output voltage is equal to
A. the most positive input phase voltage at any instant
B. the difference of most positive and most negative input phases at any instant
C. the average value of the three phases
D. the difference of the two positive phase voltages

379. In a three-phase half-wave rectifier, if the input phase voltage is 200 V, the *PIV* required for each diode will be
A. 400 V B. 346 V
C. 370 V D. 200 V

380. The rectifier, which requires minimum amount of filtering is
A. half-wave rectifier B. full-wave rectifier
C. voltage doubler circuit
D. SCR half-wave rectifier

381. For the circuit shown below, the output for the given input shown is

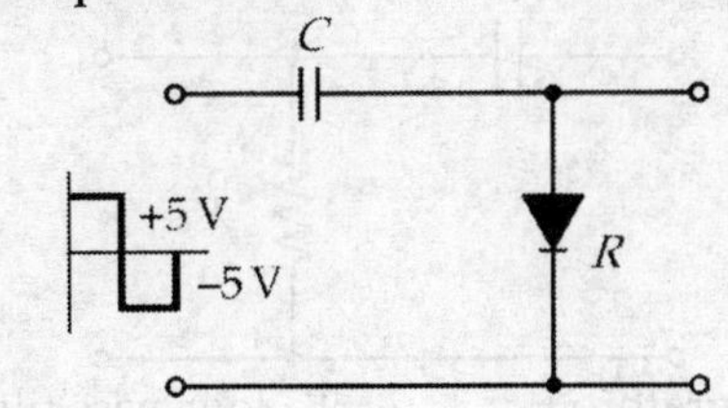

A.
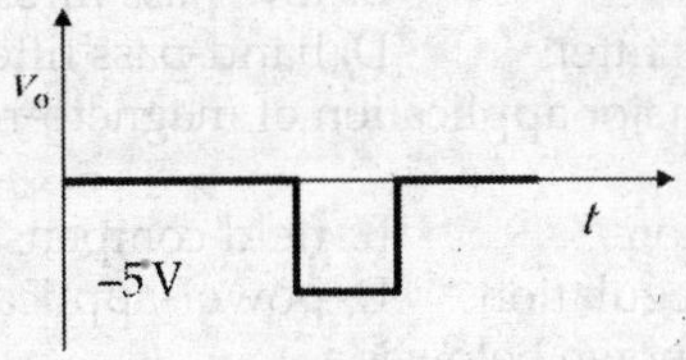

B.
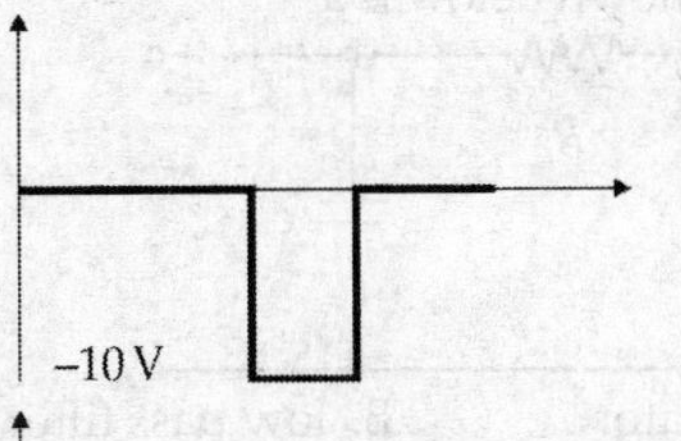

C.
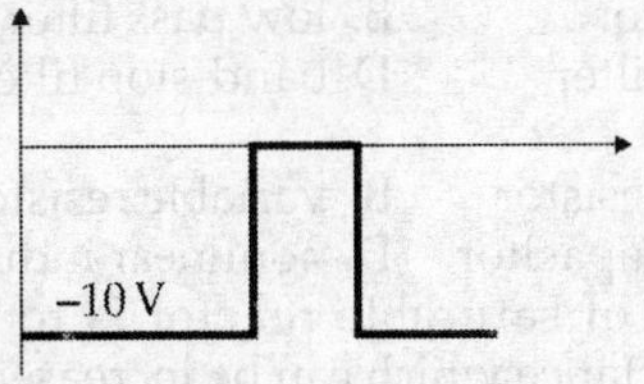

D.
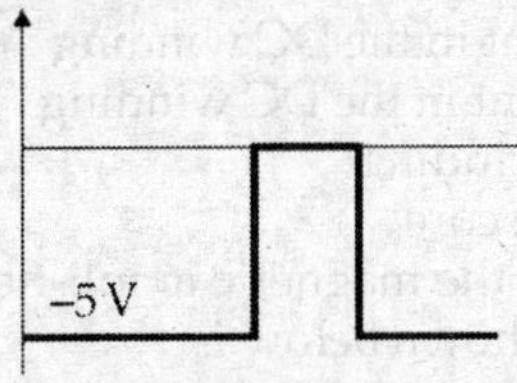

382. In a magneto-resistor,
A. magnetic field varies as resistance varies
B. resistance varies as magnetic field varies
C. magnetic field is created by varying a resistance
D. none of these

383. A capacitor of 100 μF is charged to 10 V through a resistance of 10 K. It will be fully charged in
A. 5 sec B. 0.1 sec
C. 1 sec D. 0.5 sec

384. In the above question, in one second the capacitor will charge to
B. 5 V B. 6.3 V
C. 10 V D. 7.9 V

385. If the output of a circuit is ramp pulses for a input of square pulses, the circuit is an/a
A. integrating circuit with low RC
B. integrating circuit with high RC
C. differentiating circuit with low RC
D. differentiating circuit with high RC

Ans. 377. B 378. A 379. C 380. D 381. B 382. B 383. A 384. B 385. A

386. The circuit shown below is an/a

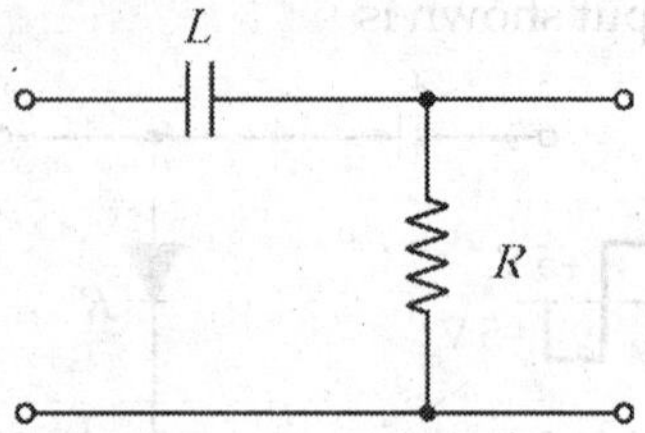

A. integrator B. low-pass filter
C. high-pass filter D. band-pass filter

387. One of the major application of magneto-resistor is in
A. rectification B. field compensation
C. voltage regulation D. power-application

388. The circuit shown below is a

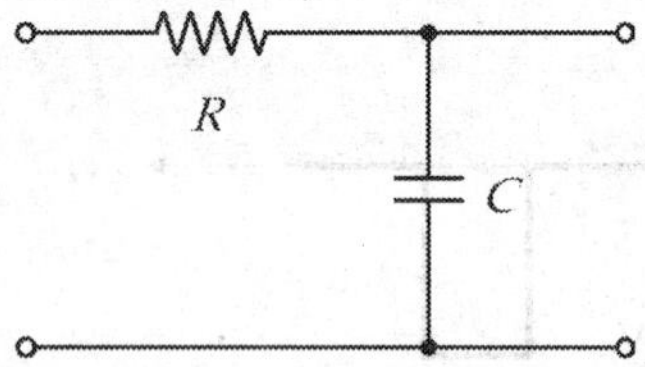

A. differentiator B. low pass filter
C. high-pass filter D. band-stop filter

389. Varactor is a
A. nonlinear resistor B. variable resistor
C. nonlinear capacitor D. nonlinear inductor

390. The function of saturable reactor is to provide variable inductance which can be increased by
A. increasing current in the DC winding
B. decreasing current in the DC winding
C. decreasing AC winding
D. by saturating the core

391. The current gain of the magnetic ampli-fier whose characteristics are shown below is

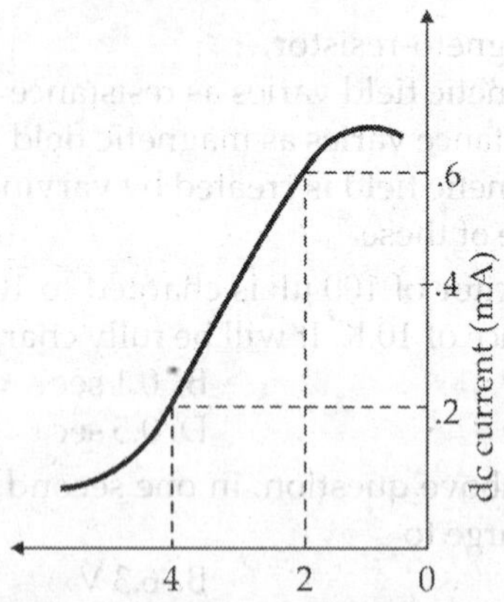

A. 100 B. 200
C. 300 D. 500

392. If the DC current is reversed in the control winding of a saturable reactor.
A. AC current will increase
B. AC current will decrease
C. same AC current will flow
D. the inductance will decrease

393. The gain of a magnetic amplifier is greater than that of a saturable reactor because
A. diodes are added to cause self saturation
B. the resetting action of the negative half-wave is blocked
C. AC biasing winding has been introduced
D. all of these

394. Switching action from a magnetic amplifier can be obtained by
A. positive feedback B. negative feedback
C. employing bias winding
D. none of these

395. Which of the following statements is True for a UJT?
A. A UJT acts like an NPN transistor
B. A UJT is turned off by decreasing V_{bb} to zero
C. A UJT can be turned on gradually
D. The triggering voltage VE of a UJT is independent of its stand-off ratio

396. Which of the following statement is *not* correct for a magnetic amplifier?
A. The addition of biasing winding does not change the available gain
B. If DC current is 1 mA and AC output is 1 A, the gain of the amplifier is 1000
C. A magnetic amplifier uses not less than four windings
D. A steeper magnetisation curve of the core indicates greater inductance

397. The figure below shows the characteristics of a *UJT* for V_{BB} = 10 V. The value of the stand-off ratio is

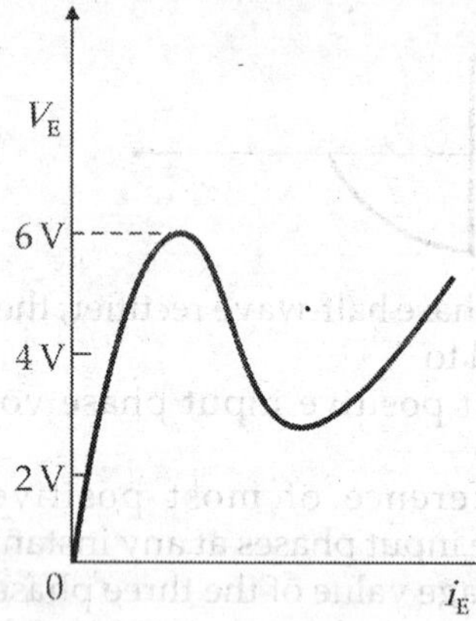

A. 0.8 B. 0.53
C. 0.60 D. 0.5

398. Which of the following statements is *not* true for a dielectric amplifier?
A. It is based on the principle of ferro-electricity
B. It uses a non-linear capacitor
C. Such amplifier does not exist
D. Its output is a modulated wave

399. In a full-wave rectifier, the diodes conduct for
A. one-half cycle B. full cycle
C. alternate half cycle D. none of these

Ans. 386. C 387. C 388. B 389. A 390. B 391. B 392. C 393. B 394. A 395. B 396. B 397. C 398. C 399. C

400. The maximum efficiency of a half-wave rectifier is
A. 33.33% B. 40.6%
C. 50% D. 68%

401. The ripple factor in case of a full-wave rectifier is
A. 1.21 B. 0.50
C. 0.48 D. 1.0

402. Peak inverse voltage for a full wave rectifier is, peak value of input voltage
A. E_m B. $2E_m$
C. $3E_m$ D. $4E_m$

403. The ripple frequency in case of a full-wave rectifier is
A. twice the supply frequency
B. equal to the supply frequency
C. thrice the supply frequency
D. none of these

404. For the sweep circuit given below, the input for the output shown should be

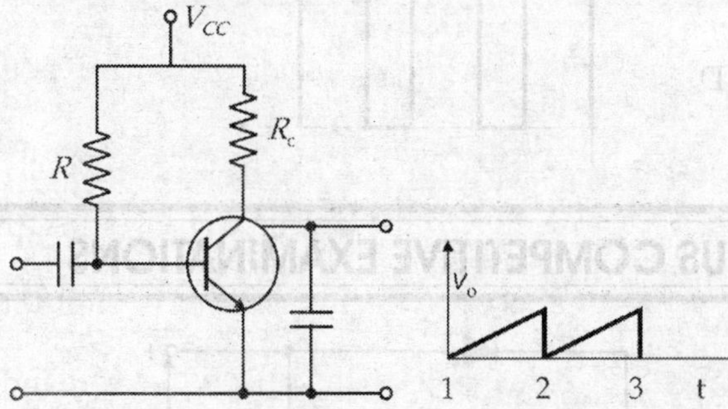

A.
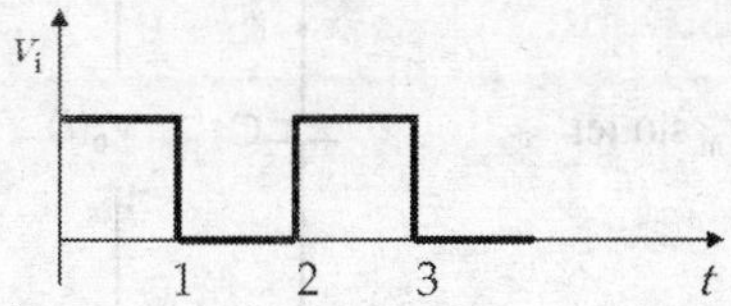

B.
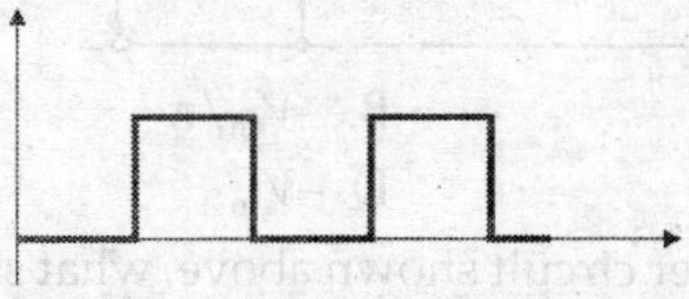

C.
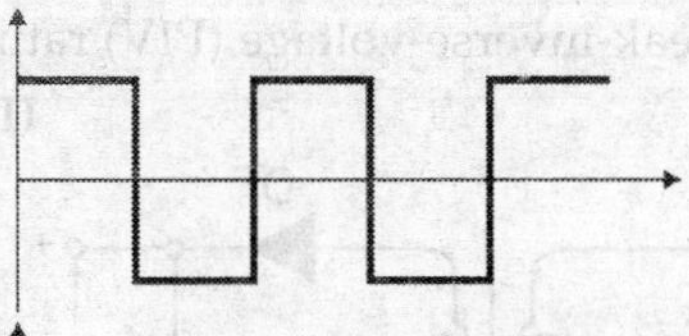

D.
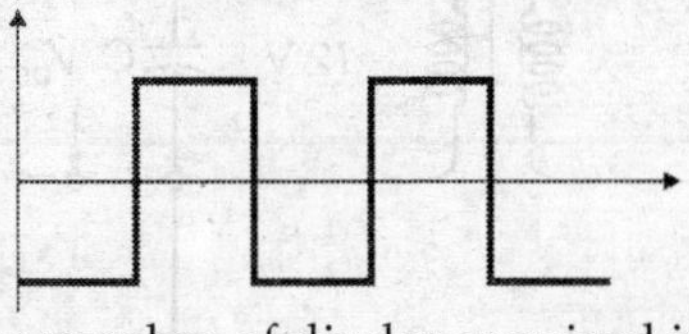

405. The number of diodes required in a bridge rectifier circuit is
A. one B. two
C. there D. four

406. The peak inverse voltage in case of a bridge rectifier for each diode is
A. E_m B. $2E_m$
B. $3E_m$ D. $4E_m$

407. The maximum efficiency of a full-wave rectifier circuit can be
A. 37.2% B. 40.6%
C. 53.9% D. 81.2%

408. The ripple factor of a full wave rectifier is
A. the same as that of half wave rectifier
B. higher than that of half wave rectifier
C. less than half the ripple factor of a half wave rectifier
D. more than half the ripple factor of a half wave rectifier

409. The output waveform of the circuit shown below for the given input (and large value of L and R) would be

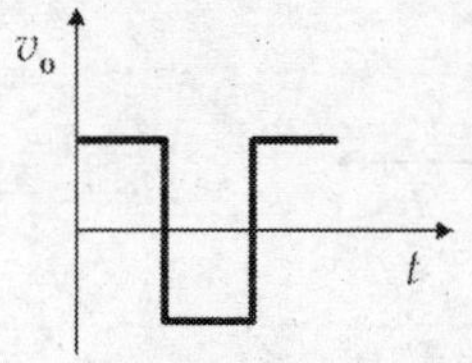

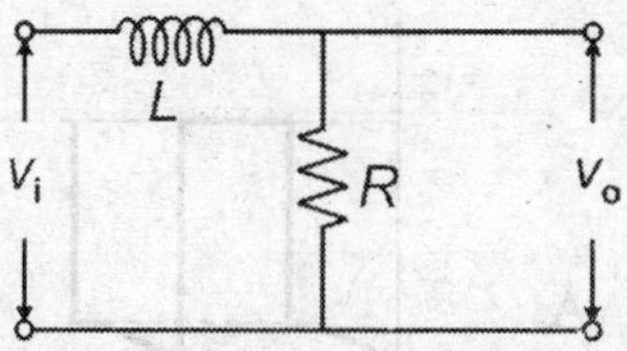

A.
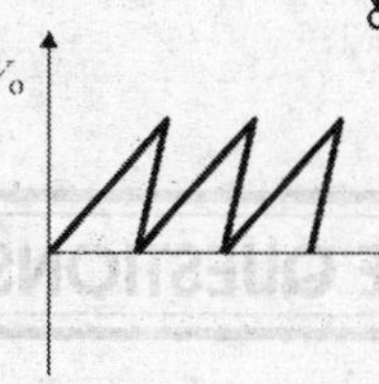

B.
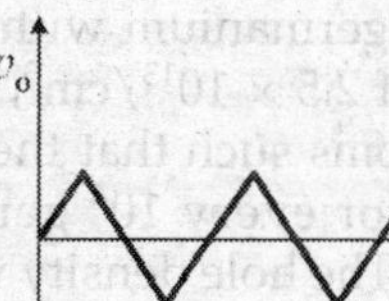

C.
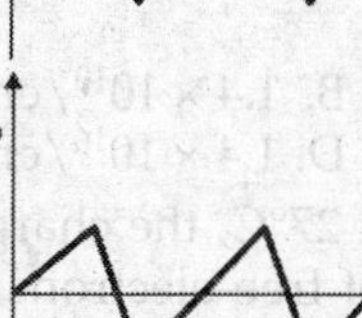

D.
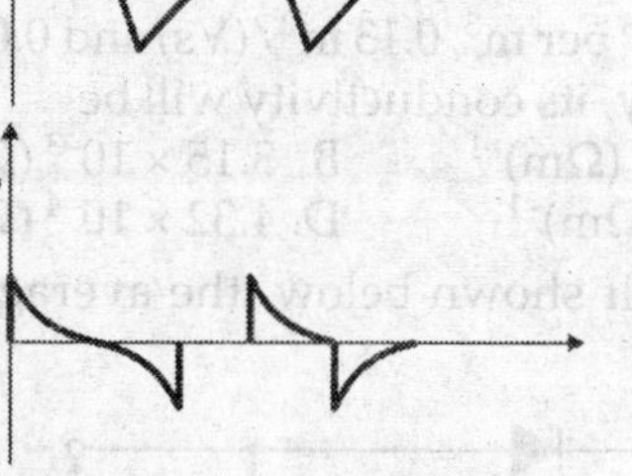

410. A transistor is a combination of two *pn* junction with their
A. P region connected together
B. N region connected together
C. N region connected to others P region
D. Both (A) and (B) are correct

411. For an n-pulse rectifier, the rms value of the AC current is related to the DC load/current as
A. $I_{rms} = I_d/n$ B. $I_{rms} = I_d/\sqrt{n}$
C. $I_{rms} = I_d$ D. $I_{rms} = \frac{2}{\pi} I_d$

Ans. 400. B 401. C 402. B 403. A 404. B 405. D 406. A 407. D 408. C 409. B 410. D 411. B

412. The output waveform of the circuit shown below for the given input would be

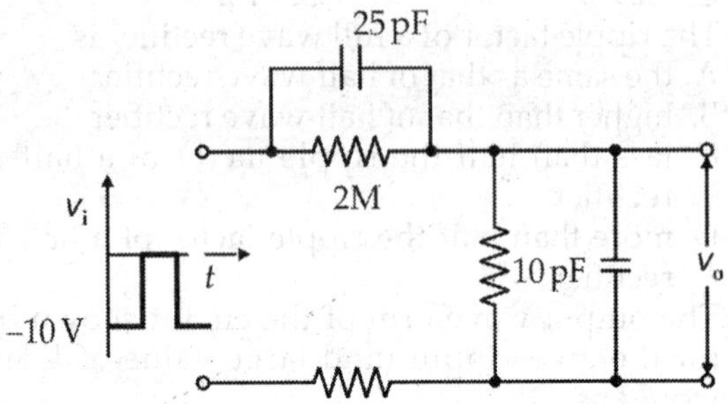

A.
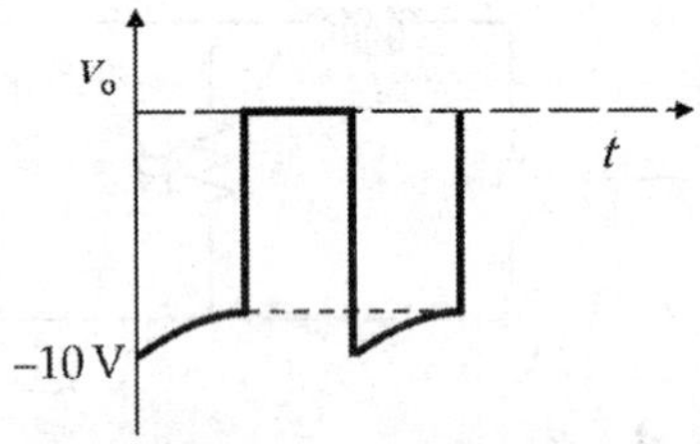

B.
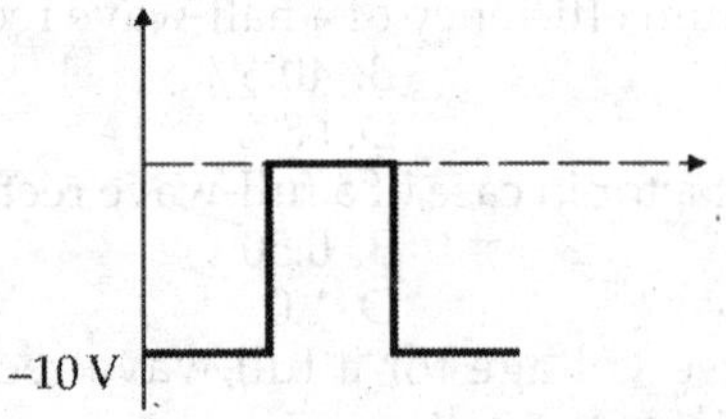

C.
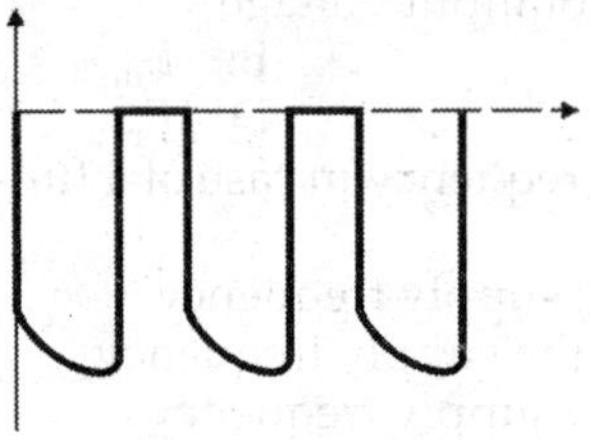

D.
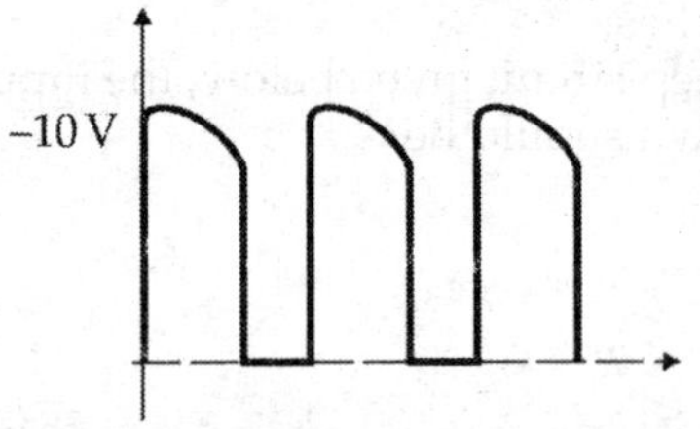

MULTIPLE CHOICE QUESTIONS FROM VARIOUS COMPETITIVE EXAMINATIONS

1. A specimen of intrinsic germanium with the density of charge carriers of $2.5 \times 10^{13}/\text{cm}^3$, is doped with donor impurity atoms such that there is one donor impurity atom for every 10^6 germanium atoms is $4.4 \times 10^{22}/\text{cm}^3$. The hole density would be **(IES 2001)**
   A. $4.4 \times 10^{16}/\text{cm}^3$ B. $1.4 \times 10^{10}/\text{cm}^3$
   C. $4.4 \times 10^{10}/\text{cm}^3$ D. $1.4 \times 10^{16}/\text{cm}^3$

2. If for intrinsic Silicon at 27°C, the charge concentration and mobilities of free-electrons and holes are 1.5×10^{16} per m³, 0.13 m²/(Vs) and 0.05 m²/(Vs) respec-tively, its conductivity will be **(IES 2001)**
   A. $2.4 \times 10^{-3}\ (\Omega\text{m})^{-1}$ B. $3.15 \times 10^{-3}\ (\Omega\text{m})^{-1}$
   C. $5 \times 10^{-4}\ (\Omega\text{m})^{-1}$ D. $4.32 \times 10^{-4}\ (\Omega\text{m})^{-1}$

3. In the circuit shown below, the average value of $V_0(t)$ will be **(IES 2002)**

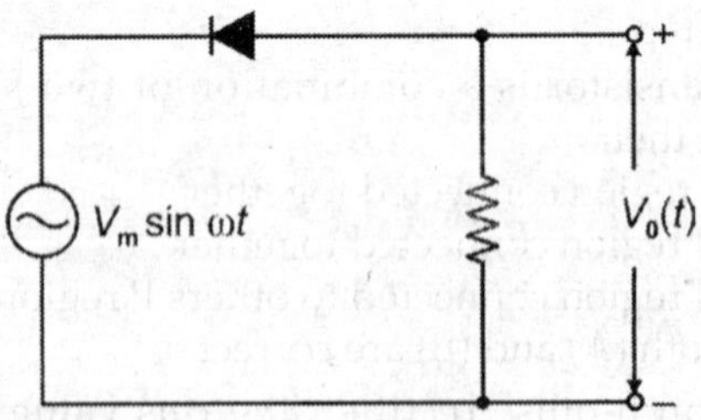

   A. 0 B. $-V_m/\pi$
   C. $-V_m/\sqrt{2}$ D. $-V_m$

4. In the circuit shown below, the average value of $V_0(t)$ will be **(IES 2002)**

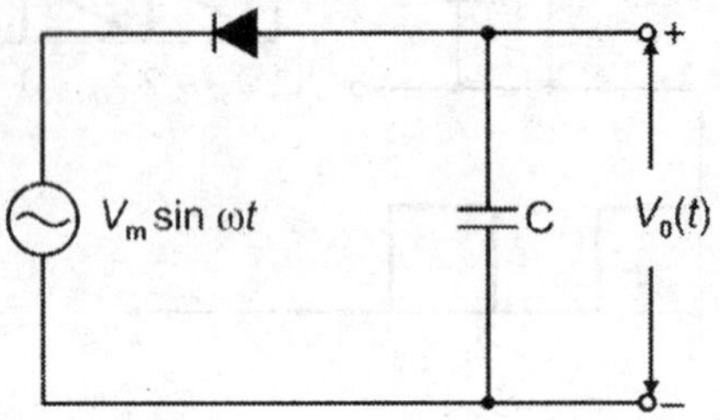

   A. 0 B. $-V_m/\pi$
   C. $-V_m/\sqrt{2}$ D. $-V_m$

5. In the rectifier circuit shown above, what should be minimum peak-inverse-voltage (PIV) rating of the diode? **(IES 2005)**

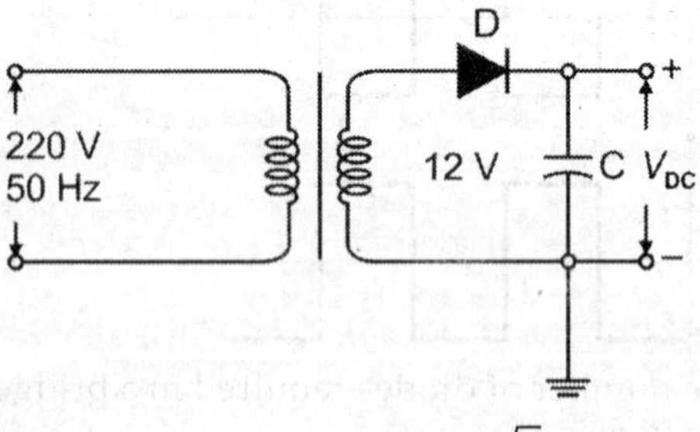

   A. 12 V B. $12\sqrt{2}$ V
   C. 24 V D. $24\sqrt{2}$ V

6. If n is the number of electrons per unit volume of the semiconductor and v_d is the drift velocity of the electrons, then the current flowing through a semiconductor is given by **(IES 2012)**

Ans. 412. A 1. B 2. D 3. B 4. D 5. D

A. $i = n/v_d$ B. $i = nv_d$
C. $i = v_d/n$ D. $i = nv_d^{1/2}$

7. A Zener diode regulator shown in the figure given below is to be designed to meet the following specifications:

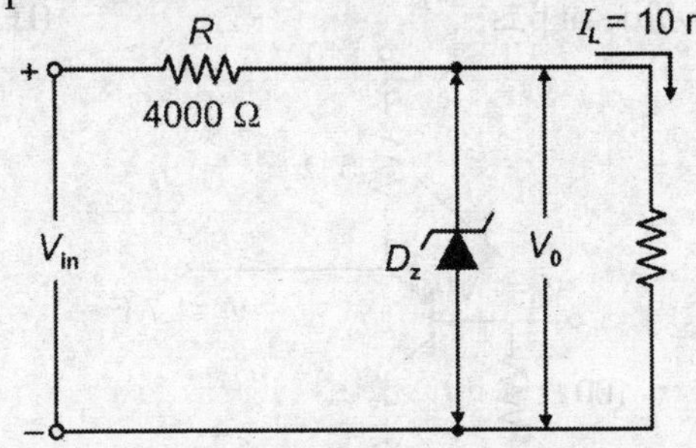

$I_L = 10$ mA, $V_0 = 10$ V, V_{in} varies from 30 V to 50 V. The Zener diode has $V_z = 10$ V and I_{zk} (knee current) = 1 mA. For satisfactory operation, which one of the following is correct? **(IES 2007)**

A. $R \leq 1800\ \Omega$ B. $2000\ \Omega \leq R \leq 2200\ \Omega$
C. $3700\ \Omega \leq R \leq 4000\ \Omega$ D. $R > 4000\ \Omega$

8. In the circuit given below, D_1 and D_2 are ideal. Which one of the following repre-sents the transfer characteristics of the circuit? **(IES 2007)**

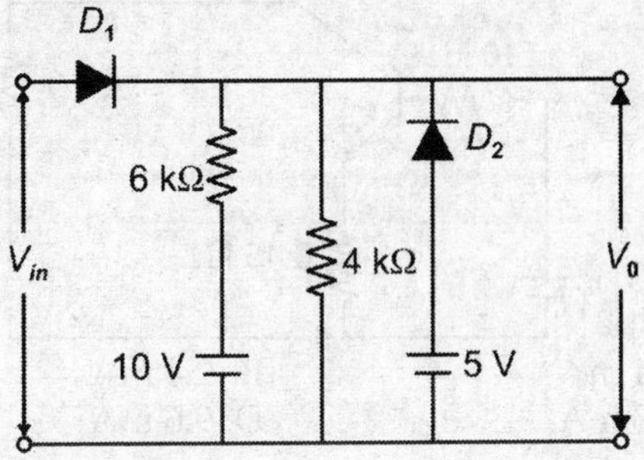

A.
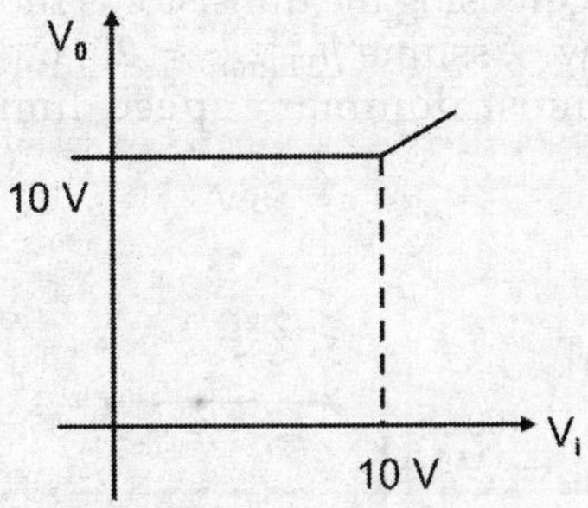

B.
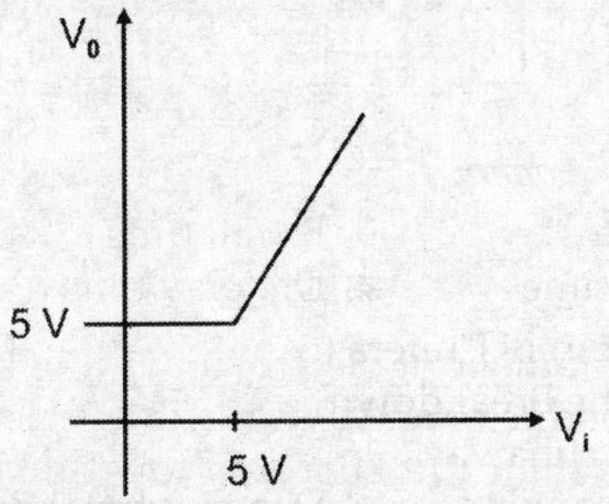

C.
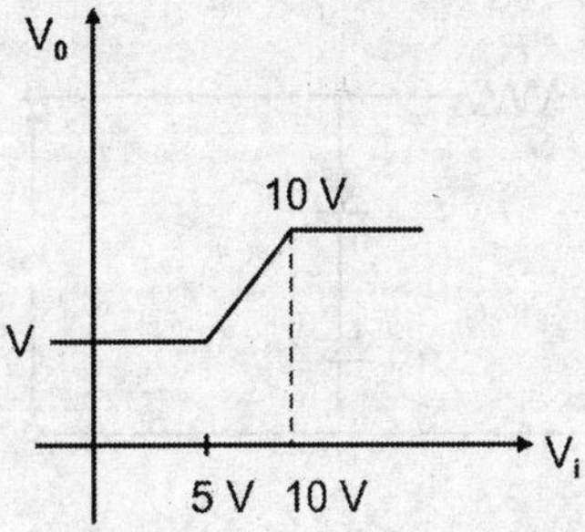

D.
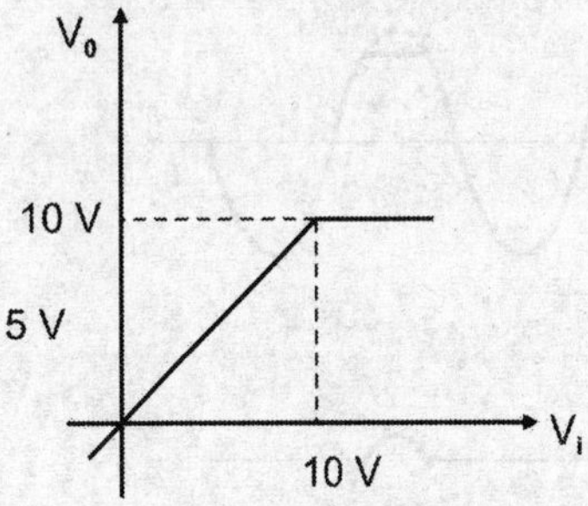

9. The forward resistance of the diode shown in the below circuit is 5 ohms, and the other parameters are same as those of an ideal diode. Then what is the DC component of the source current? **(IES 2009)**

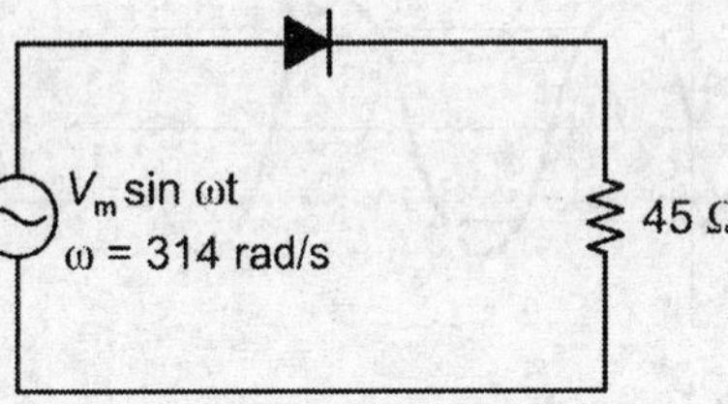

A. $\dfrac{V_m}{50\pi}$ B. $\dfrac{V_m}{50\pi\sqrt{2}}$

C. $\dfrac{V_m}{100\pi\sqrt{2}}$ D. $\dfrac{2V_m}{50\pi}$

10. The 6 V zener diode as shown in the circuit below, has zero zener resistance and a knee current of 5 mA. Then what is the minimum value of R so that the voltage across it does not fall below 6 V? **(IES 2009)**

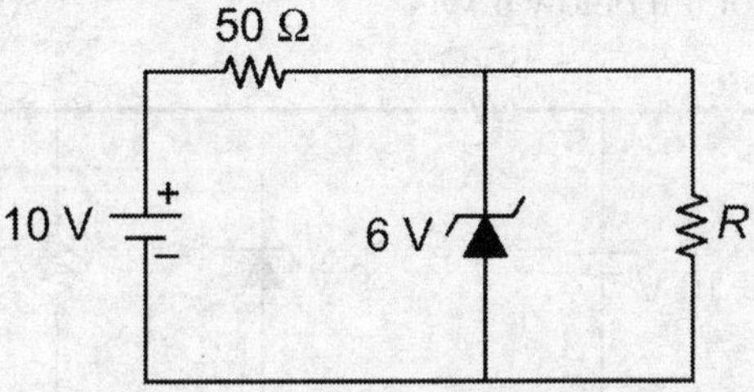

A. 1200 ohms B. 80 ohms
C. 50 ohms D. 40 ohms

11. For a sinusoidal input of peak value V_p, the output waveform V_0 will be **(IES 2010)**

Ans. 6. B 7. A 8. B 9. A 10. B

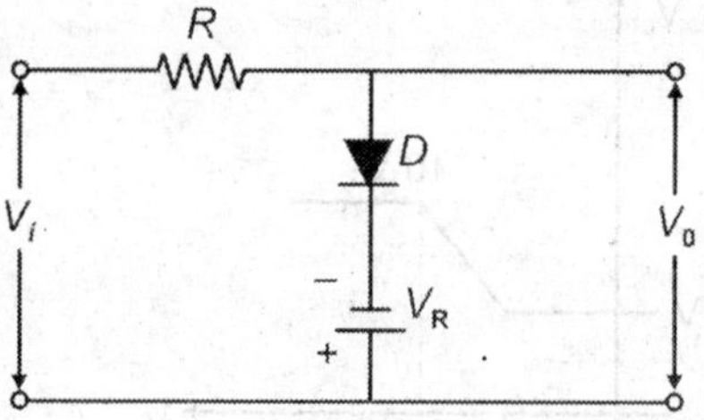

A. 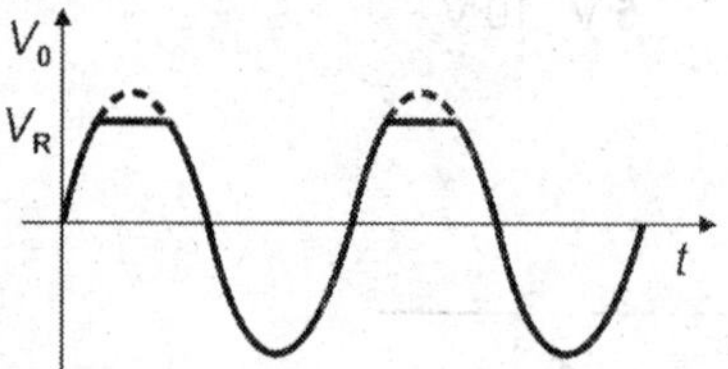

B. 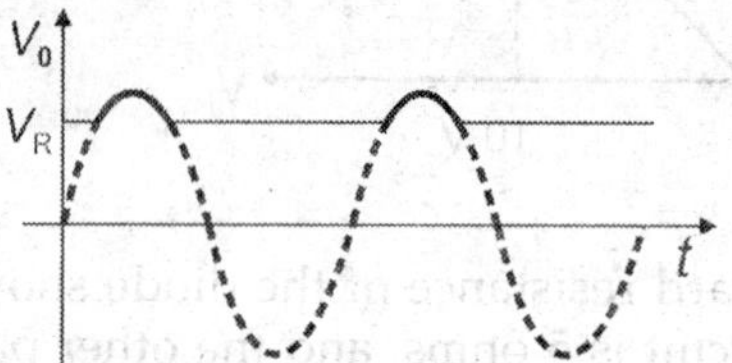

C. 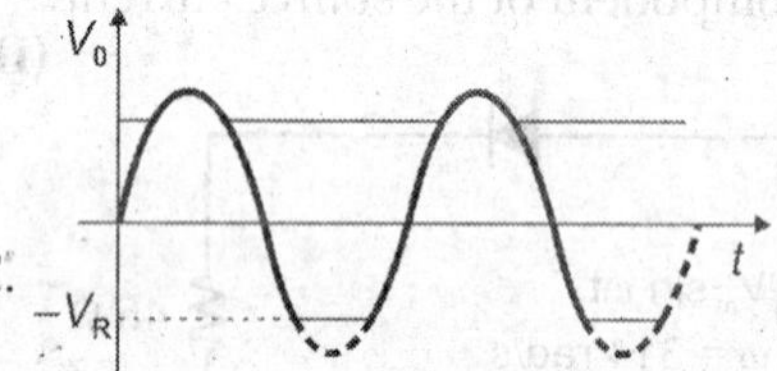

D. 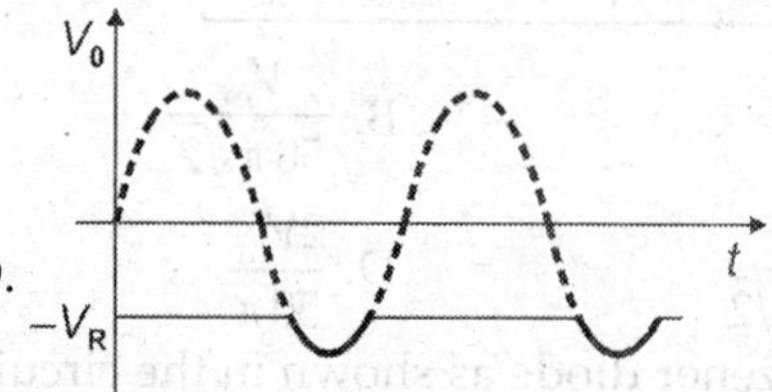

12. The 6 V Zener diode shown in figure has zero Zener resistance and a knee current of 5 mA. The minimum value of R, so that the voltage across it does not fall below 6 V is **(IES 2013)**

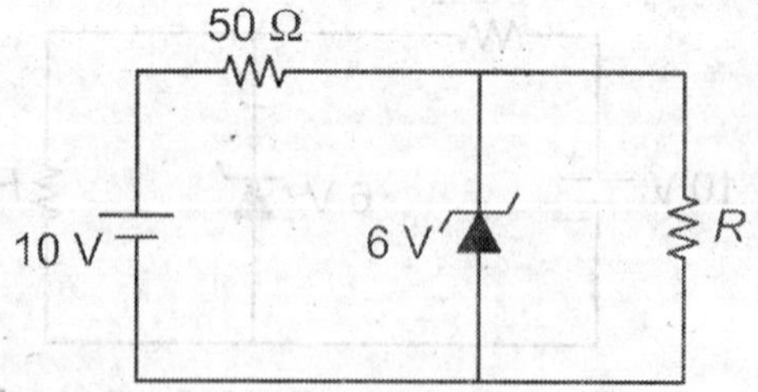

A. 1.2 kΩ
B. 80 Ω
C. 50 Ω
D. zero **(IES 2013)**

13. As compared to an LED, an LCD has the distinct advantage of **(IES 2012)**

A. extremely low power consumption
B. providing a silver display
C. being extremely thin
D. giving two types of display

14. A circuit using the BJT is shown in the below figure, the value of β is **(IES 2001)**

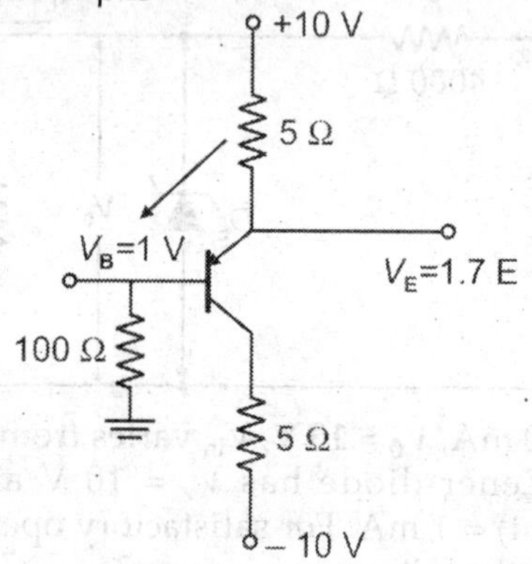

A. 120
B. 150
C. 165
D. 166

15. For the circuit shown in the below figure, by assuming β = 200 and V_{BE} = 0.7 V, the best approximation for the collector current I_c in the active region is **(IES 2001)**

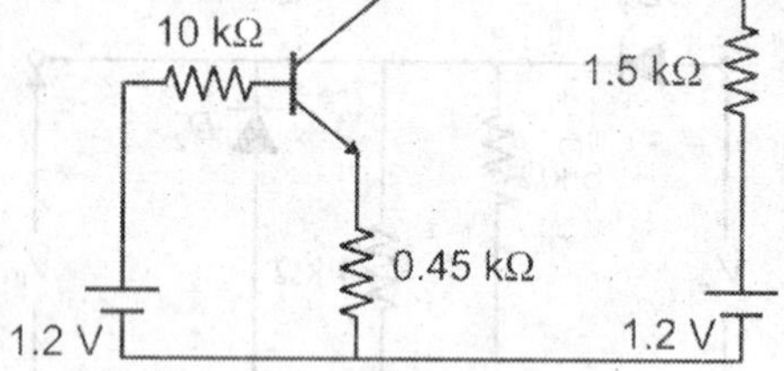

A. 1 mA
B. 2.4 mA
C. 3 mA
D. 9.6 mA

16. A switch circuit using the transistor is shown in the figure below. Assume $h_{FE(min)}$ = 20 and f_T = 100 MHz. The most dominant speed limitation is brought by **(IES 2001)**

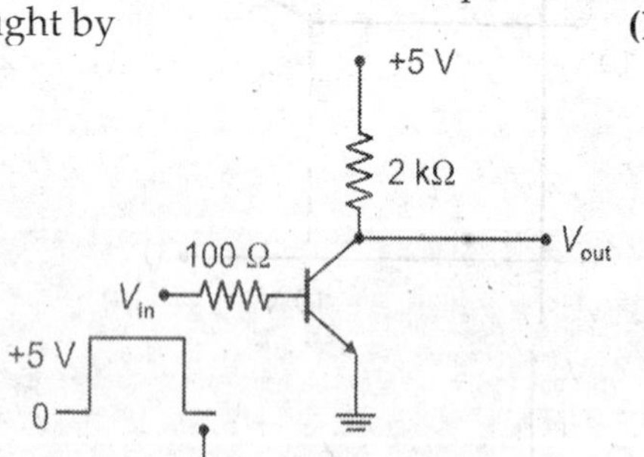

A. rise time
B. fall time
C. storage time
D. delay time

17. Early effect in BJT refers to **(IES 2002)**

A. avalanche breakdown
B. thermal runaway
C. base narrowing
D. zener breakdown

Ans. 11. D 12. B 13. A 14. C 15. B 16. B 17. C

18. In the circuit as shown below, the ratio of V_0 to $(V_2 - V_1)$ would approximately (neglecting constant due to V_{CC}) be **(IES 2003)**

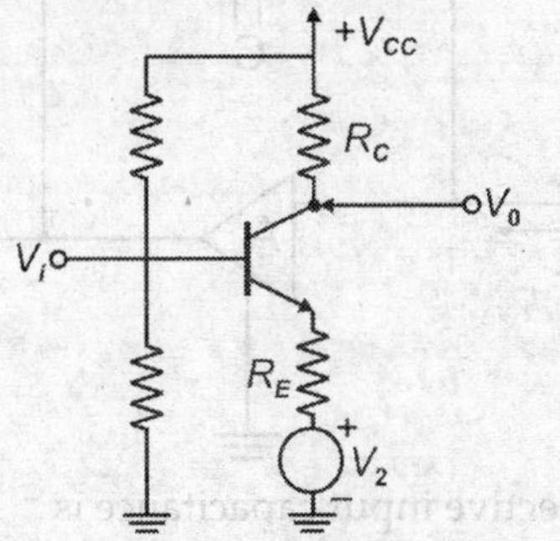

A. R_C/R_E
B. R_E/R_C
C. $-R_C/R_E$
D. $-R_E/R_C$

19. What is the thermal runaway in a bipolar junction transistor biased in the active region due to? **(IES 2006)**

A. Heating of the transistor emitter region
B. Changes in 'β' which increases with temperature
C. Base emitter voltage V_{BE} which decreases with rise in temperature
D. Increase in reverse collector-base saturation current due to rise in internal device temperature

20. If α_F and α_B denote the forward and inverted mode current gains of a BJT, which one of the following is correct? **(IES 2007)**

A. $\alpha_F = \alpha_R$
B. $\alpha_F < \alpha_R$
C. $\alpha_F \geq \alpha_R$
D. $\alpha_F >> \alpha_R$

21. Fixed biasing of CE configuration is shown in the figure given below.

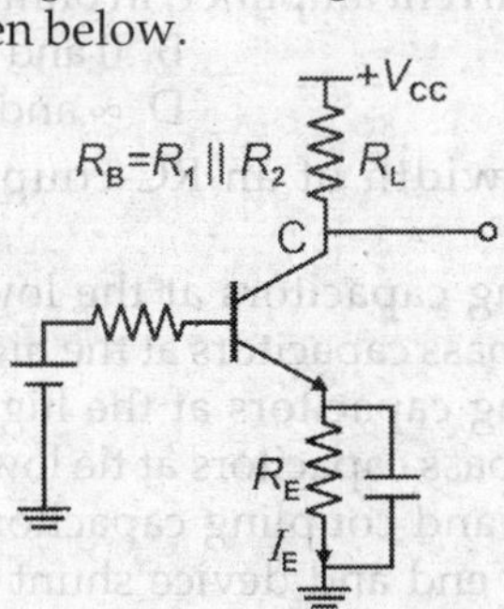

The current stabilisation factor for $R_B << \beta R_E$ is $S_i = 1 + \frac{R_B}{R_E}$.

For $R_B << R_E$, what is the voltage stabilisation factor S_V? **(IES 2007)**

A. $= \frac{R_E}{R_B + R_E}$
B. $= \frac{R_B}{R_E}$
C. $\approx \frac{1}{R_B}$
D. $\approx -\frac{R_E}{R_B + R_E}$

22. What is the transistor combination shown in the figure given below? **(IES 2007)**

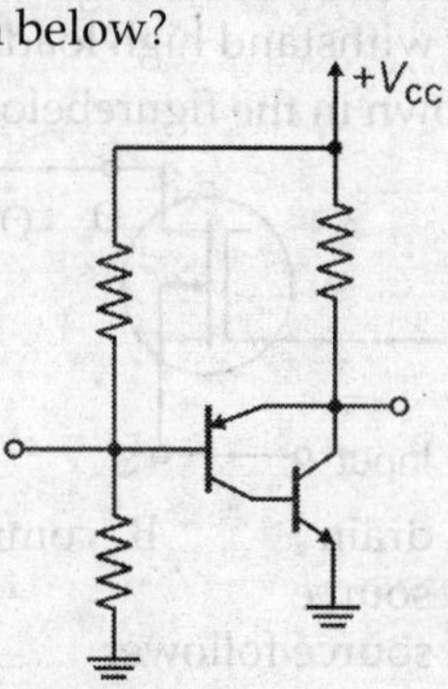

A. A Darlington pair
B. A complementary pair
C. If effectively acts as a single *pnp* tran-sistor
D. If effectively acts as a single *npn* tran-sistor

23. In the difference amplifier as shown below, the differential output **(IES 2010)**

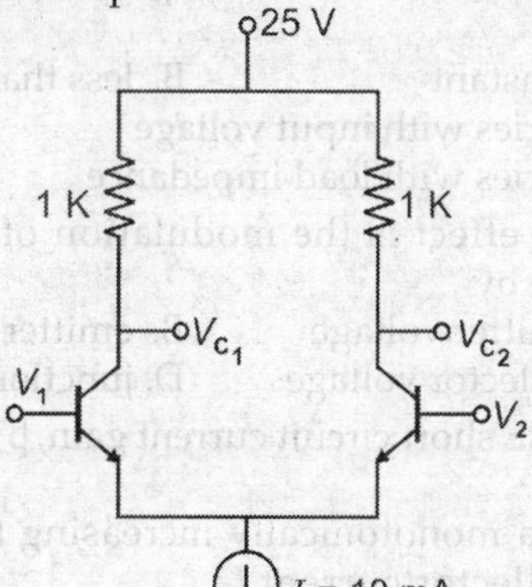

$(V_{c_2} - V_{c_1})$ for $V_1 = +5$ V, $V_2 = 0$ V is

A. 0 V
B. 5 V
C. 10 V
D. 15 V

24. The JEFT in the below circuit has an $I_{DSS} = 10$ mA, $V_P = 5$ V. The value of the resistance R_S for a drain current of $I_{DS} = 6.4$ mA is **(IES 2010)**

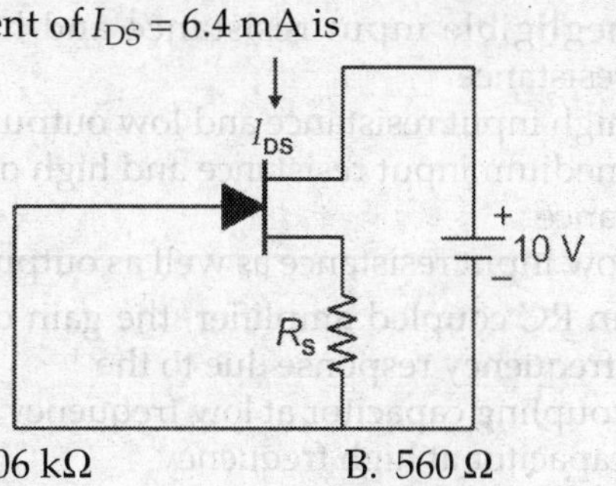

A. 1.06 kΩ
B. 560 Ω
C. 470 Ω
D. 156 Ω

25. An emitter follower regulator has the following disadvantage: **(IES 2011)**

A. It does not provide high gain
B. No provision exists for varying the output voltage

Ans.	18. C	19. D	20. D	21. C	22. B	23. D	24. D	25. B

C. Its output resistance is high
D. It cannot withstand high load current

26. The FET shown in the figurebelow is a **(IES 2010)**

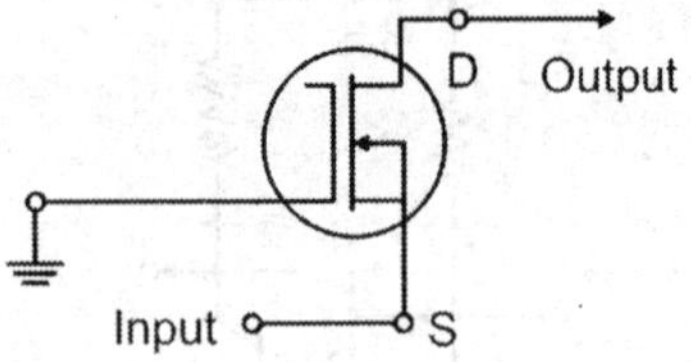

A. common drain
B. common gate
C. common source
D. common source follower

27. For a transistor, turn-off time is **(IES 2011)**
A. sum of storage time and fall time
B. maximum value of storage time
C. maximum value of fall time
D. sum of rise time and fall time

28. In a common collector amplifier the voltage gain is **(IES 2011)**
A. constant
B. less than 1
C. varies with input voltage
D. varies with load impedance

29. Early effect is the modulation of effective base width by **(IES 2012)**
A. emitter voltage
B. emitter current
C. collector voltage
D. junction temperature

30. The CE short circuit current gain, β of a transistor **(IES 2014)**
A. is a monotonically increasing function of the collector current I_C
B. is a monotonically decreasing function of I_C
C. for low values of I_C, it increases and reaches a maximum and then decrea-ses with further increase in I_C
D. is not a function of I_C

31. When a transistor is connected in common emitter mode it will have **(IES 2014)**
A. negligible input resistance and high output resistance
B. high input resistance and low output resistance
C. medium input resistance and high output resistance
D. low input resistance as well as output resistance

32. In an RC coupled amplifier, the gain decreases in the frequency response due to the **(IES 2001)**
A. coupling capacitor at low frequency and bypass capacitor at high frequency
B. coupling capacitor at high frequency and bypass capacitor at low frequency
C. coupling junction capacitance at low frequency and coupling capacitor at high frequency
D. device junction capacitor at high frequency and coupling capacitor at low frequency

33. An amplifier of gain *A* is bridged by a capacitance C as shown below.

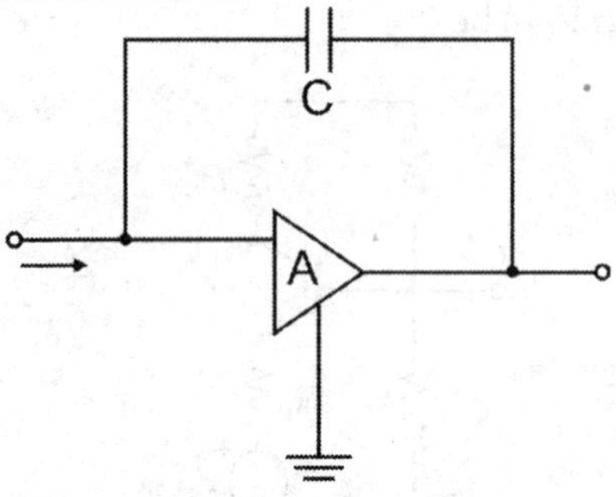

The effective input capacitance is **(IES 2002)**
A. C/A
B. $C(1-A)$
C. $C(1+A)$
D. CA

34. Two identical RC coupled amplifiers, each having a lower cut-off frequency f_l are cascaded with negligible loading. What is the lower cut-off frequency of the overall amplifier? **(IES 2004)**
A. $\frac{f_l}{\sqrt{\sqrt{2}-1}}$
B. $f_l\sqrt{\sqrt{2}-1}$
C. $f_u/2$
D. $2f_l$

35. What is the purpose of impedance matching between the output of previous stage and input of next stage in a cascaded amplifier? **(IES 2006)**
A. To achieve high efficiency
B. To achieve maximum power transfer
C. To achieve reduced distortion
D. To achieve reduced noise

36. The input resistance R_i and output resistance R_0 of an ideal current amplifier, in ohms are **(IES 2010)**
A. 0 and 0
B. 0 and ∞
C. ∞ and 0
D. ∞ and ∞

37. The bandwidth of an RC-coupled amplifier is limited by **(IES 2014)**
A. coupling capacitors at the low frequency end and bypass capacitors at the high frequency end
B. coupling capacitors at the high frequency end and bypass capacitors at tie low frequency end
C. bypass and coupling capacitors at the low frequency end and device shunt capacitors at the high frequency end
D. device shunt capacitors at the low frequency end and bypass as well as coupling capacitors at the high frequency end

38. The slew rate of an op-amp is 0.5 V/micro sec. The maximum frequency of a sinusoidal input of 2 V rms that can be handled without excessive distortion is **(IES 2001)**
A. 3 kHz
B. 30 kHz
C. 200 kHz
D. 2 MHz

39. An op-amp is used in the circuit as shown in the below figure. Current I_0 is **(IES 2001)**

Ans.	26. B	27. A	28. B	29. C	30. C	31. C	32. D	33. B	34. A	35. B	36. B	37. C	38. B

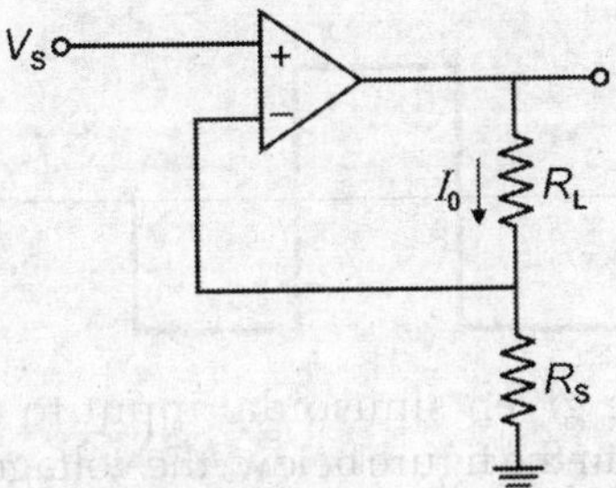

A. $V_S \times \dfrac{R_L}{R_S(R_L + R_S)}$ B. V_S/R_S

C. V_S/R_L D. $V_S\left(\dfrac{1}{R_s + R_S}\right)$

40. A circuit with op-amp is shown in the below figure. The voltage V_0 is **(IES 2001)**

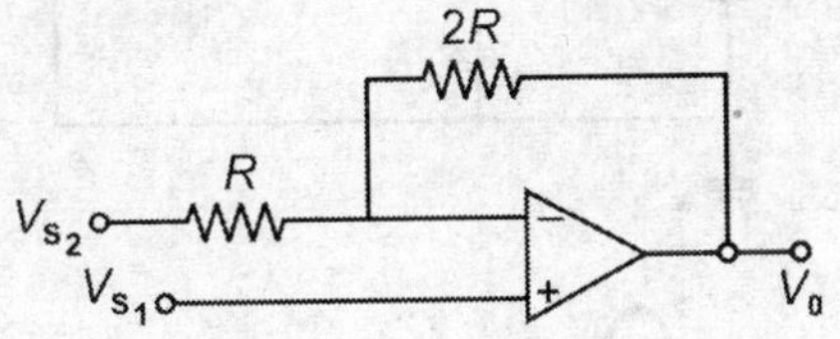

A. $3V_{s_1} - 6V_{s_2}$ B. $2V_{s_1} - 3V_{s_2}$
C. $2V_{s_1} - 2V_{s_2}$ D. $3V_{s_1} - 2V_{s_2}$

41. In the op-amp circuit shown below, $V_i > 0$ and $i = I_0e^{\alpha V}$. The output V_0 will be proportional to **(IES 2002)**

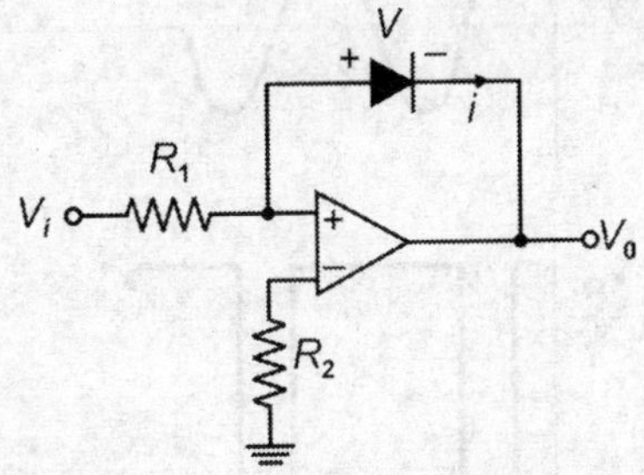

A. $\sqrt{V_i}$ B. V_i
C. e^{kv_i} D. $\ln(kV_i)$

42. In the inverting op-amp circuit shown below, the resistance R_g is chosen as $R_1 \| R_2$ **(IES 2002)**

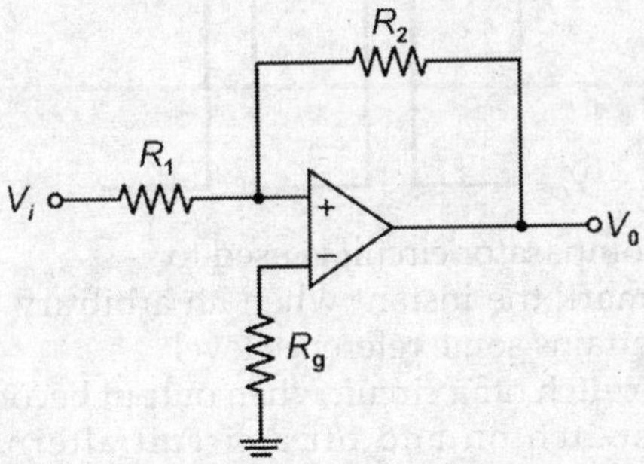

A. increase gain B. reduce offset voltage
C. reduce offset current D. increase CMRR

43. An op-amp has a differential gain of 10^3 and a CMRR of 100. The output of the op-amp with inputs of 120 μV and 80 μV will be **(IES 2003)**
A. 26 mV B. 41 mV
C. 100 mv D. 200 mV

44. In the op-amp circuit as shown below, the current i_L is **(IES 2003)**

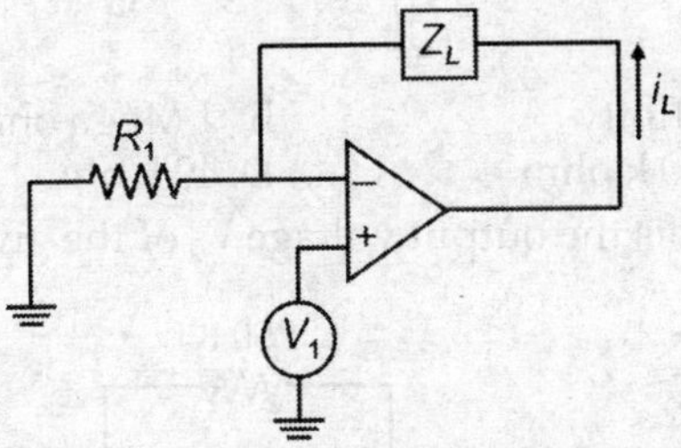

A. V_i/Z_L B. $V_i/Z_L \| R_1$
C. V_i/R_1 D. $V_i(R_1 + Z_L)$

45. Which one of the following is the correct expression for the current I_0? **(IES 2004)**

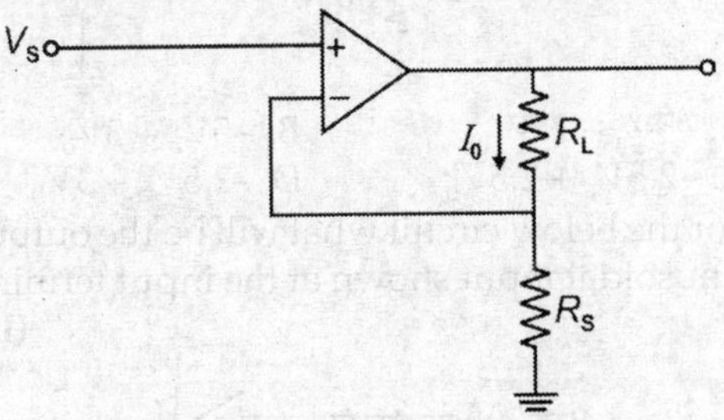

A. $\dfrac{V_S \cdot R_L}{R_S(R_L + R_S)}$ B. $\dfrac{V_S}{R_S}$

C. $\dfrac{V_S}{R_L}$ D. $V_S\left(\dfrac{1}{R_L} + \dfrac{1}{R_S}\right)$

46. Consider the following circuit:

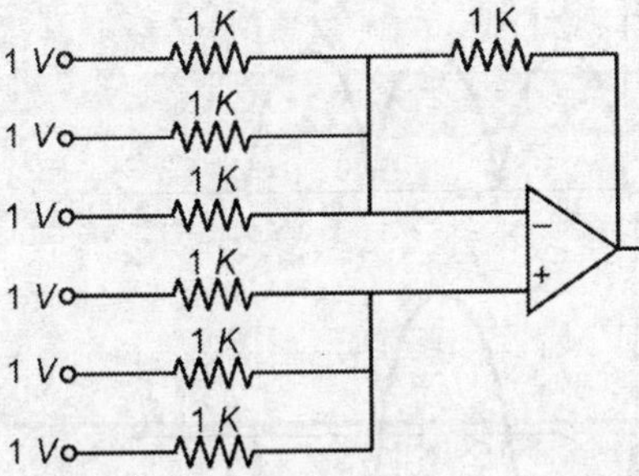

What is the output voltage V_0 in the above circuit? **(IES 2004)**

A. 9.5 V B. 3 V
C. 32.2 V D. 1 V

47. For the operational amplifier circuit shown in the figure below, what is the maximum possible value of R_1, if the voltage gain required is between −10 and −25? (The upper limit on R_F is a mΩ) **(IES 2007)**

Ans. 39. B 40. D 41. D 42. C 43. B 44. C 45. B 46. D

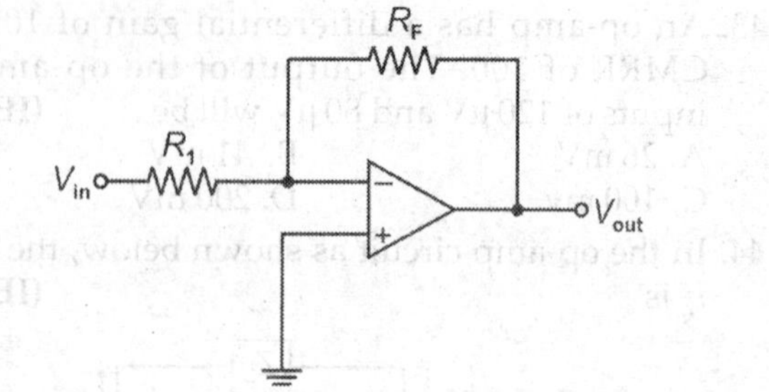

A. Infinity
B. 1 Mega ohm
C. 100 k ohm
D. 40 k ohm

48. What is the output voltage V_0 of the given circuit? **(IES 2008)**

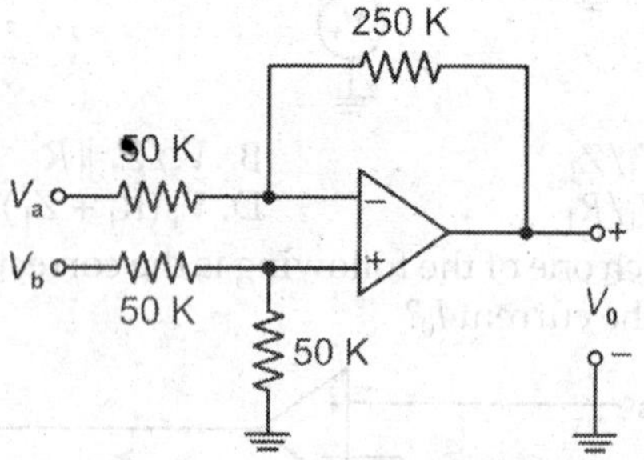

A. $-5V_a + 2.5V_b$
B. $-5V_a + 3V_b$
C. $-2.5V_a + 2.5V_b$
D. $-2.5V_a + 3V_b$

49. For the below circuit what will be the output for the sinusoidal input shown at the input terminal? **(IES 2009)**

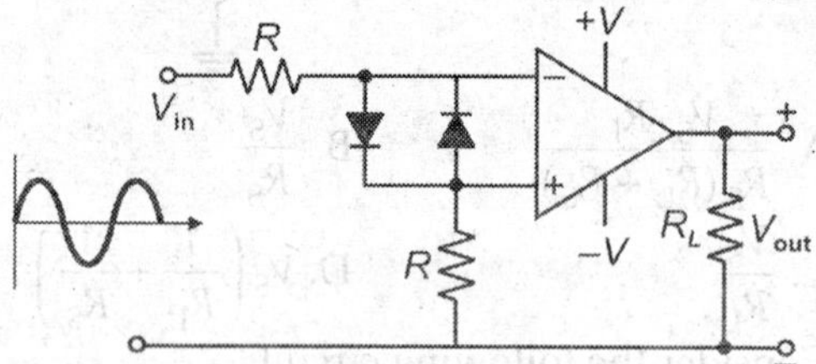

A. 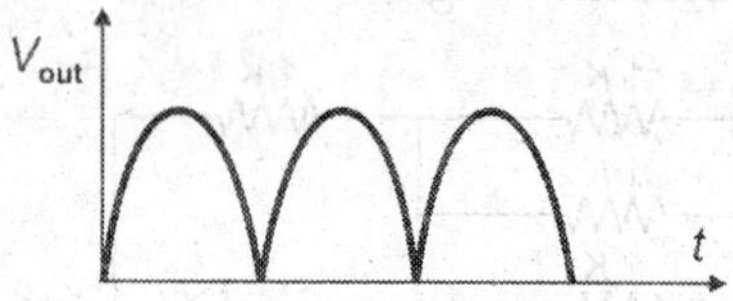

B. 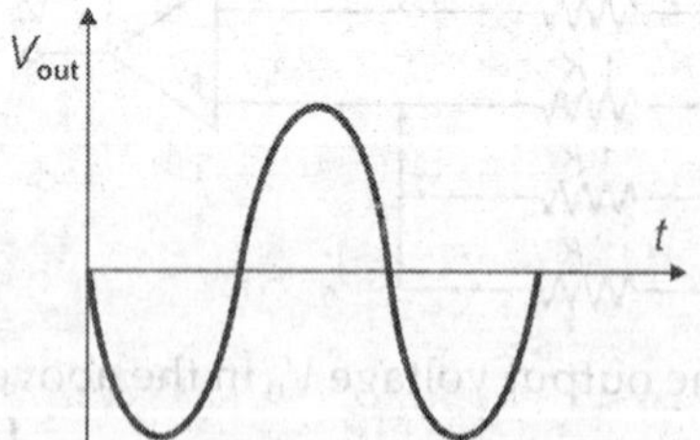

C. 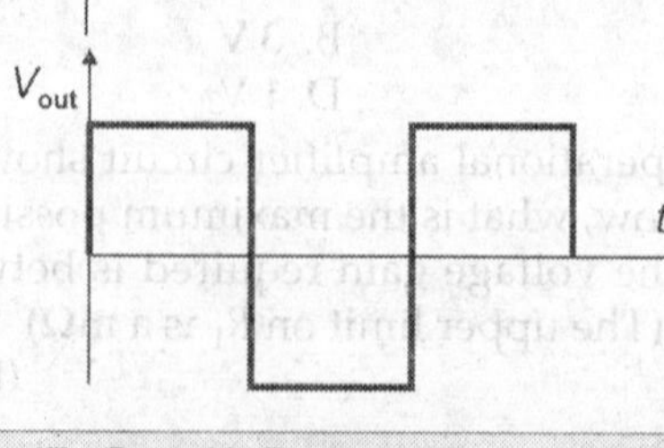

D. 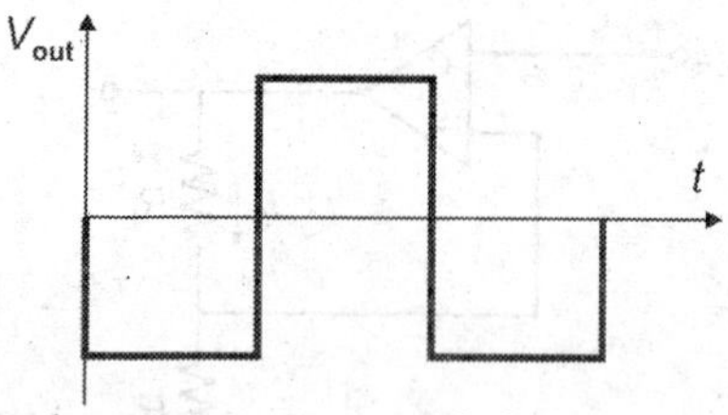

50. For the given sinusoidal input to the circuit as shown in the figure below, the voltage waveform at point P of the clamper circuit is **(IES 2010)**

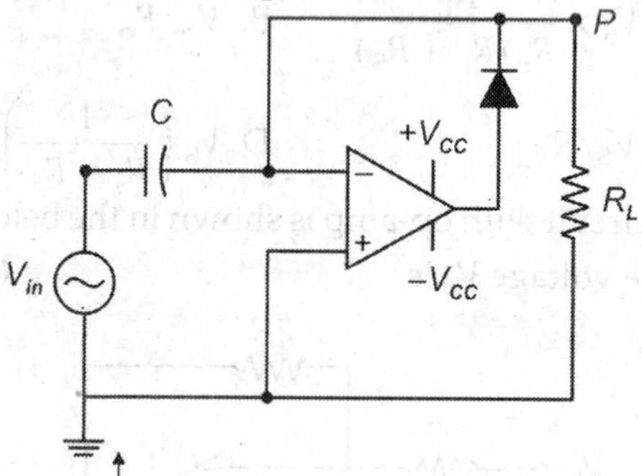

A. 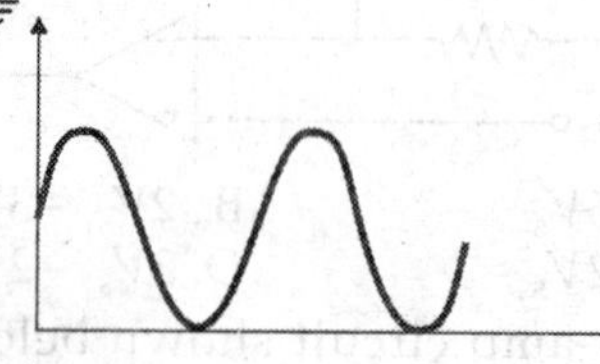

B. 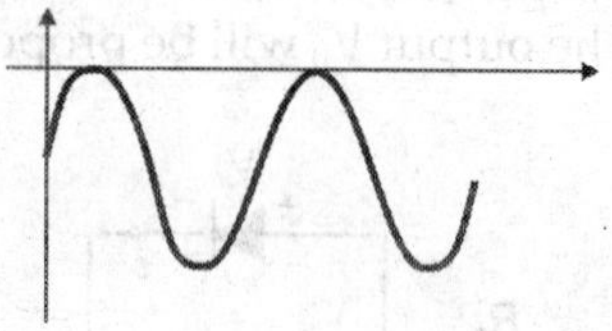

C. 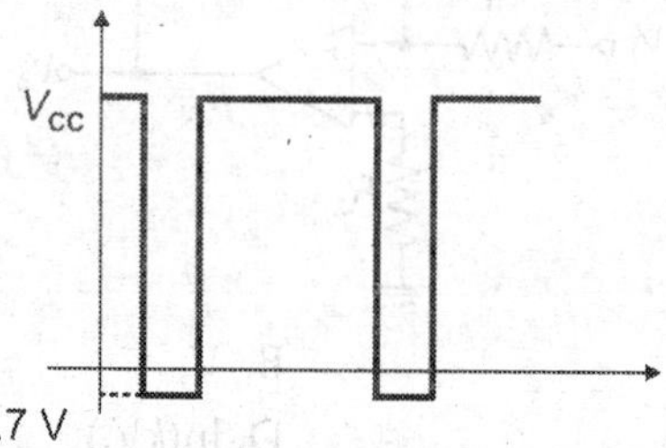

D. 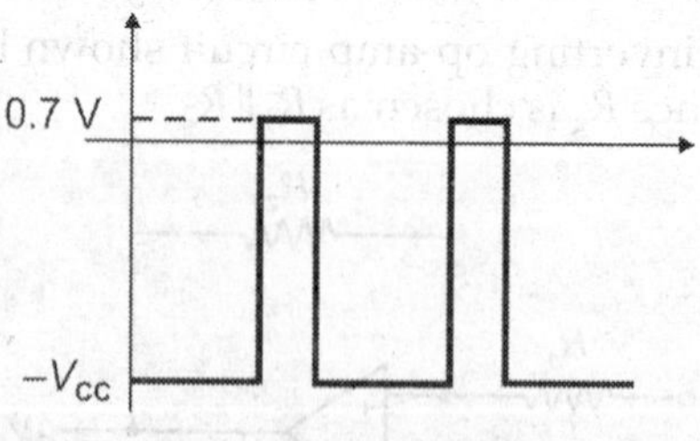

51. A comparator circuit is used to **(IES 2012)**
A. mark the instant when an arbitrary waveform attains some reference level
B. switch off a circuit when output becomes zero
C. switch on and off a circuit alternately at a particular rate
D. mark the instant when the input voltage becomes constant

Ans.	47. C	48. B	49. D	50. D	51. A

52. The circuit shown is (IES 2011)

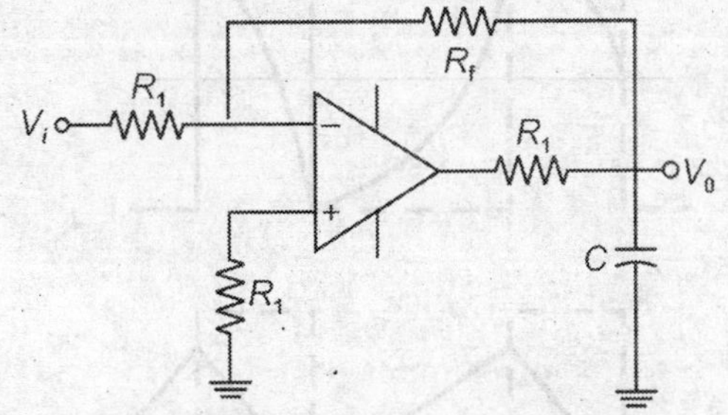

A. a low-pass filter
B. a clamper
C. a lag compensated inverting amplifier
D. a narrow band video amplifier

53. The circuit shown is (IES 2011)

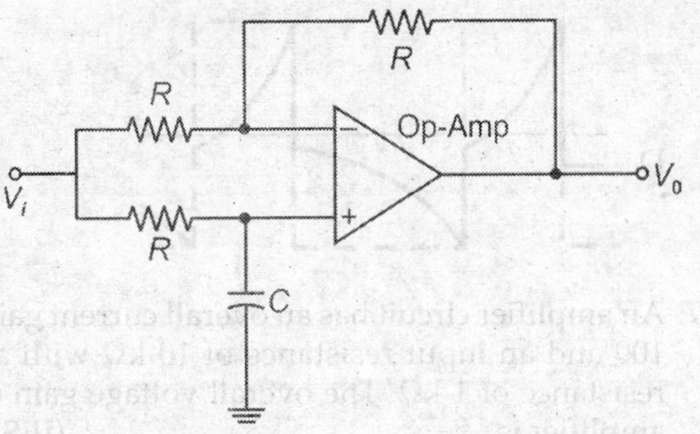

A. a low-pass filter
B. a high-pass filter
C. a comparator
D. an all-pass filter

54. An op-amp has a slew rate of 5 $V/\mu s$. The largest sine wave output voltage possible at a frequency of 1 MHz is (IES 2013)

A. 10 πV
B. 5 V
C. 5/π V
D. 5/2π V

55. The gain and distortion of an amplifier are respectively 150 and 5%. When used with a 10% negative feedback, the % distortion would be (IES 2014)

A. 5/16
B. 9/16
C. 6
D. 8

56. A two stage amplifier with negative feedback (IES 2014)

A. can become unstable for larger values of β
B. becomes unstable at high and very low frequencies if A is very large
C. becomes unstable when the pole fre-quencies become complex
D. is always stable

57. A sinusoidal waveform can be converted to a square waveform by using a (IES 2001)

A. two stage transistorised overdriven amplifier
B. two stage diode detector circuit
C. voltage comparator based on op-amp
D. regenerative voltage comparator circuit

58. An FET oscillator uses the given phase shift network as shown below. The minimum gain required for oscillation is (IES 2003)

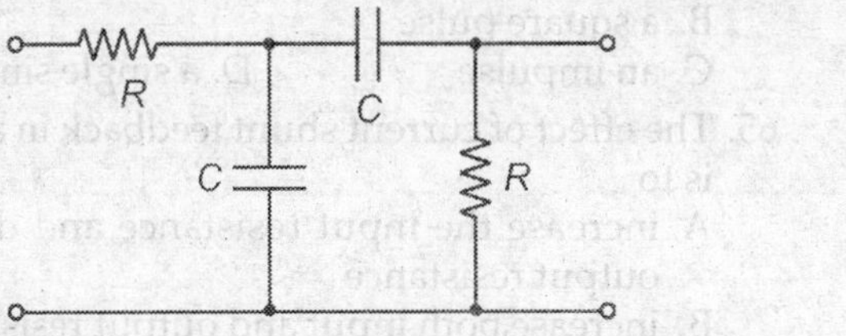

A. –29
B. 1
C. 3
D. 29

59. Negative feedback in an amplifier leads to which one of the following? (IES 2007)

A. Decrease in bandwidth
B. Increase in current gain
C. Increase in voltage gain
D. Decrease in voltage gain

60. Consider the following statements in respect of the Wien bridge oscillator shown in the figure below: (IES 2008)

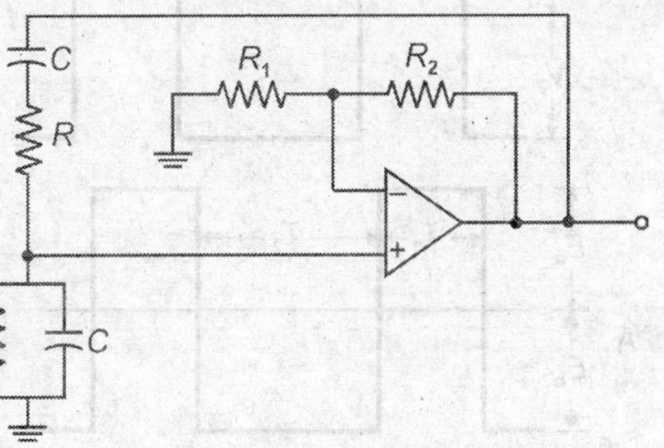

1. For $R = 1$ kilo ohm $C = \left(\frac{1}{2\pi}\right)\mu F$, $f = 1$ kHz

2. For $R = 3$ kilo ohm $C = \left(\frac{1}{18\pi}\right)\mu F$, $f = 3$ kHz

Which of these statements is/are correct?

A. 1 only
B. 2 only
C. Both 1 and 2
D. Neither 1 nor 2

61. An amplifier without feedback, when fed with a 1 V, 50 Hz input signal gives an output of 30 V, 50 Hz with a 5% 2nd order distortion. When 10% of the output is feedback, what is the 2nd order distortion? (IES 2009)

A. 0.375
B. 1.3 V
C. 0.75 V
D. 2 V

62. For a transconductance amplifier, input and output resistances are respectively (IES 2011)

A. ∞ and 0
B. 0 and ∞
C. 0 and 0
D. ∞ and ∞

63. The highest frequency stability is achieved by using an oscillator of the type (IES 2011)

A. Colpitts
B. Crystal controlled
C. Hartley
D. RC oscillator

Ans. 52. A 53. A 54. D 55. A 56. D 57. D 58. C 59. D 60. C 61. A 62. D 63. B

64. One shot multi-vibrator, with a pulse input gives an output **(IES 2011)**
A. a single triangular pulse
B. a square pulse
C. an impulse
D. a single sinusoid pulse

65. The effect of current shunt feedback in an amplifier is to **(IES 2013)**
A. increase the input resistance and decrease the output resistance
B. increase both input and output resistance
C. decrease both input and output resistance
D. decrease the input resistance and increase the output resistance

66. For low pass RC circuit, the input wave-form is as shown in the figure below. What will be the output waveform if the time constant of the circuit is equal to the time period of the input signal ($RC = T_1 + T_2$)? **(IES 2010)**

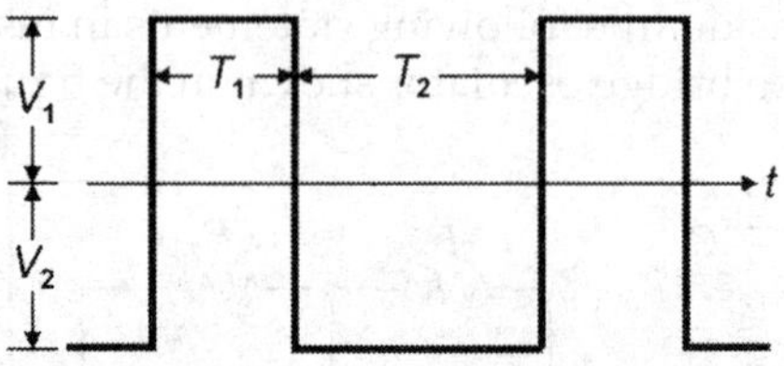

A.
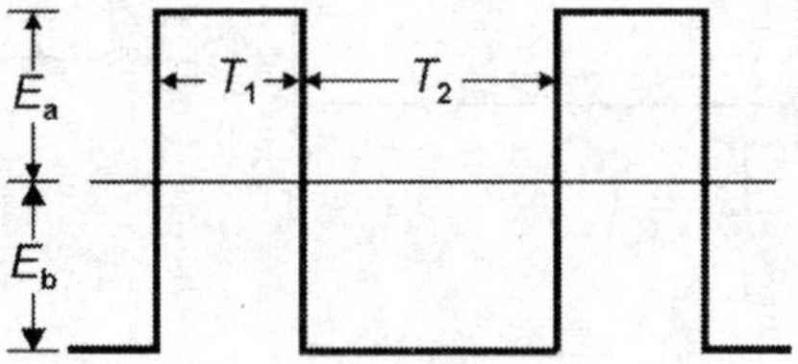

B.
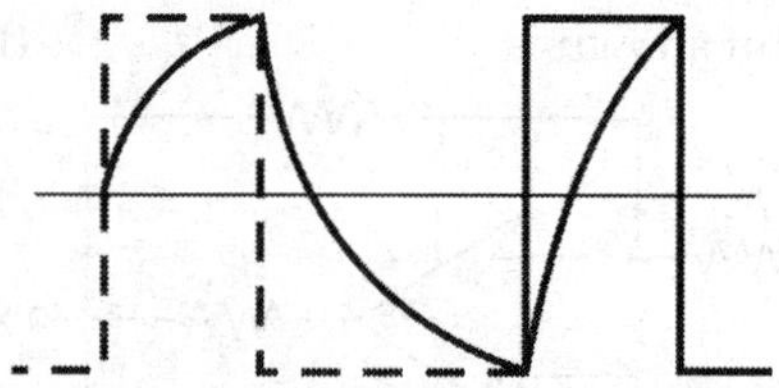

C.
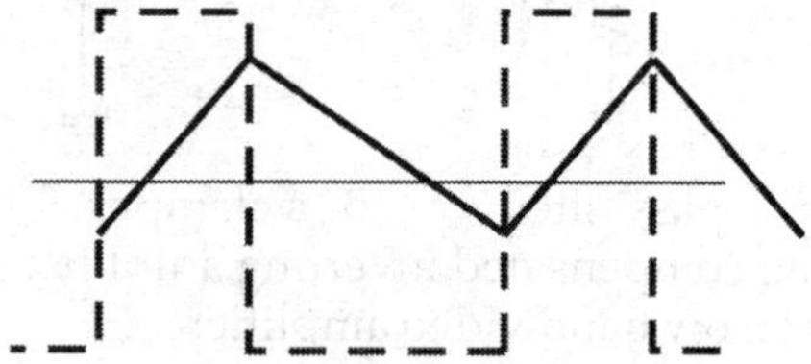

D.
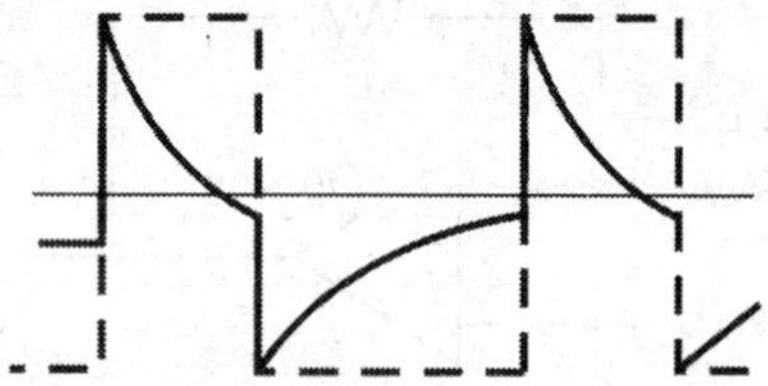

67. An amplifier circuit has an overall current gain of –100 and an input resistance of 10 kΩ with a load resistance of 1 kΩ. The overall voltage gain of the amplifier is **(IES 2013)**
A. 5 dB
B. 10 dB
C. 20 dB
D. 40 dB

Ans. 64. B 65. D 66. C 67. C

CHAPTER 16

Digital Techniques and Fundamentals of Computer

Introduction: Digital electronics find appliations in computers, integrated circuits and many digital devices and circuits. Digital circuits are used in many applications like display devices, digital clocks, digital computers, digital instruments etc. A small integrated circuit (chip) can perform tasks of large number of transistors, diode resistors, etc.

Digital circuit functions in two states, i.e. in a binary manner. Boolean algebra is being applied for the solution of logical problems.

Boolean algebra, originally intended for solving logic problems, also found its greatest use in the design of digital computers. The circuits in digital computers are the circuits which performs the logical processes. Any circuit that can be analyzed with Boolean algebra is known as a *logic circuit*. The output of a logic circuit is either a low (logical '0') or a high (logical 1) voltage level. Thus in a digital computer, we deal with a digital signal.

Digital logic: Postulates: The knowledge of number system is very important for the design of digital circuits and systems.

a. An operator '+' is defined such that if $c = a + b$ then $c \in X$, for every pair of elements $a, b \in X$. This operation replaces the 'union' (∪) operation of set theory and in switching theory, it is known as OR operation.

b. There exists as element '0' in X such that $a + 0 = a$, for every elements $a \in X$.

c. It holds commutative law true:
   i. $a + b = b + a$
   ii. $a \cdot b = b \cdot a$

d. It holds the distributive law true:
   $a \cdot (b + c) = a \cdot b + a \cdot c$

e. The complementary set $\overline{a}$, for every element $a \in X$, is defined as
   i. $a + \overline{a} = 1$
   ii. $a \cdot \overline{a} = 0$

f. There are at least two elements $a, b \in X$ such that $a \neq b$.

Based on these postulates, one can define now AND, OR and NOT operations. For example, the absorption law stated earlier, now can be written as

$A \cdot (A + B) = A$

$A + A \cdot B = A.$

De Morgan's theorem (inverse of Boolean function): It states that, "to obtain the inverse of any Boolean function invert all variables and replace all ORs by ANDs and all ANDs by ORs". The first De Morgan theorem says that a NOR gate $(\overline{A + B})$ is equivalent to an AND gate with NOT circuits in the inputs $(\overline{A} \cdot \overline{B})$. The second says that a NAND gate $(\overline{A} . \overline{B})$ is equivalent to an OR gate with NOT circuits in the input $(\overline{AB})$.

Exclusive NOR gate: The exclusive NOR gate, abbreviated as EX-NOR, operates exactly opposite to EX-OR gate (Fig. 16.1).

$$Y = A \oplus B = AB + \overline{A} \cdot \overline{B}$$

This circuit is also called an equality detector as its output is 1 only when the inputs are equal.

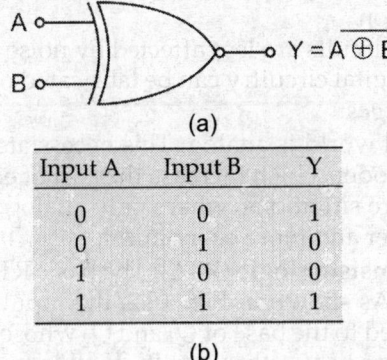

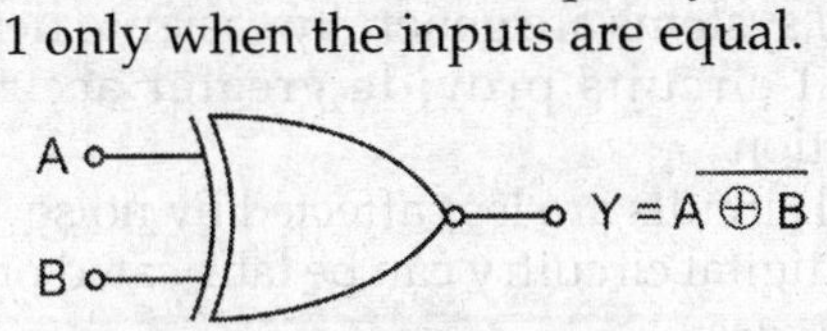

(a)

Input A	Input B	Y
0	0	1
0	1	0
1	0	0
1	1	1

(b)

Fig. 16.1: The exclusive-NOR gate

Boolean theorem: To understand the combinational logic system, these three fundamental operations (OR,

Table 16.1: Integrated circuits with their level of complexity

Complexity	Approximate no. of gates per chip	Typical products
Small scale integration (SSI)	Less than 12	Logic gates, flip flops
Medium scale integration (MSI)	12 to 99	Counters, multiplexers, adders
Large scale integration (LSI)	100 to 9999	8-bit microprocessors ROM, RAM
Very large scale integration (VLSI)	10,000 to 99,999	16-bit and 32 bit microprocessors, sophisticated computer peripherals
Ultra large scale integration	100,000 or more	64 bit microprocessors real-time image processing

AND and NOT), the various basic relations can be understood and proved as follows.

$1a.\ A + A = A$ $1b.\ A \cdot A = A$

$2a.\ A + 1 = 1$ $2b.\ A.0 = 0$

$3a.\ A + (A \cdot B) = A$ $3b.\ A \cdot (A + B) = A$

$4a.\ \bar{\bar{A}} = A$

$5a.\ A + (\bar{A} \cdot B) = A + B$ $5b.\ A \cdot (\bar{A} + B) = A \cdot B$

$6a.\ \overline{A + B} = \bar{A} \cdot \bar{B}$ $6b.\ \overline{A \cdot B} = \bar{A} + \bar{B}$

Relation (6) is known as *De Morgon's Law* and is helpful in solving Boolean expression.

Boolean function: It is defined as a set of Boolean variables, which are independent, and assigned by another Boolean variable which is dependent on the set of Boolean variables by some specific rule. In mathematical notion, we can write

$$z = f(a, b, c)$$

Encoder and decoder: A digital circuit can process numbers in binary form. The circuit that converts decimal to digital (binary) form is called *encoder* and the circuit that converts digital form to decimal form is called *decoder*.

Advantages:

i. Digital systems are generally easier to design

ii. Digital circuits provide greater accuracy and protection

iii. Digital circuits are less affected by noise

iv. More digital circuitry can be fabricated on IC chips

Disadvantages:

i. The real world is analog. This necessiates encoder and decoder which increase the cost of equipment

ii. There are situations, where only analog technique is simpler and more economical.

Resistor transistor logic (RTL): The basic RTL gate is a NOR gate. As shown in Fig. 16.2, the inputs A and B are connected to the base of Q_1 and Q_2 whose collectors are connected to V_{CC} through R_L. If all inputs are low, i.e. $V_i \leq V_{IL}$ (logical '0'), all the transistors remain off. Thus the common collector potential goes high so that $V_o \geq V_{OH}$ (logical '1'). If any one of the inputs goes high, the respective transistor becomes ON, the output

Logic gates: Table 16.2 provides the symbol, Boolean expression and the truth table of various logic gates.

Table 16.2

Logic	Symbol	Boolean expression	Truth table			Definition
AND		$X = A \cdot B$	A	B	X	The output is available only when all the inputs are available, i.e. the output is at logical '1' when all inputs are at logical '1'
			0	0	0	
			0	1	0	
			1	0	0	
			1	1	1	
OR		$X = A + B$	A	B	X	The output is available, if any one or more inputs are available
			0	0	0	
			0	1	1	
			1	0	1	
			1	1	1	
NOT		$X = \bar{A}$	A	X		The output is available when there is no input
			0	1		
			1	0		
NAND		$X = \overline{A \cdot B}$	A	B	X	The output is available to all states except when all the inputs are available (i.e. at logic '1')
			0	0	1	
			0	1	1	
			1	0	1	
			1	1	0	
NOR		$X = \overline{A + B}$	A	B	X	The output is available only when no input is available, i.e. all inputs are at logical '1'
			0	0	1	
			0	1	0	
			1	0	0	
			1	1	0	
EX. OR		$X = A\bar{B} + \bar{A}B$ $= A \times B$	A	B	X	The output is available to those states when the inputs are not identical
			0	0	0	
			0	1	1	
			1	0	1	
			1	1	0	
COINCIDENCE		$X = \bar{A}\bar{B} + AB$	A	B	X	The output is available to those states when the inputs are identical
			0	0	1	
			0	1	0	
			1	0	0	
			1	1	1	

becomes low, i.e. logical '0' value. If all the inputs are high, the output remains at low because of conduction of the transistors. Thus RTL gate behaves as NOR gate.

Diode transistor logic (DTL): The diode transistor logic behaves like a NAND gate. As shown in the circuit of Fig. 16.3, if all the inputs are kept high, the diodes D_3 and D_2 cannot pass any current. The series diode D_s now passes current through the base emitter junction of the transistor and holds it into saturation. The V_{CE} of this transistor during saturation being of the order of 0.1 V, the output potential remains at low $V_0 \leq V_{OL}$ i.e. logical 0.

If any one input drops to ground potential or $V_i \leq V_{IL}$, then the corresponding diode conduct bringing the potential at the input point of D_3 to about 0.7 V (forward voltage drop across the diode), which is not sufficient to drive current through D_s and base-emitter junction of the transistor. Thus the transister remains off and collector potential becomes high.

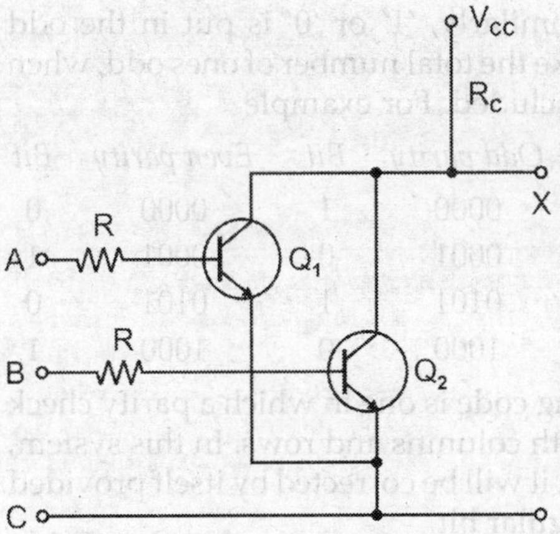

A	B	X
0	0	1
0	1	0
1	0	0
1	1	0

Fig 16.2

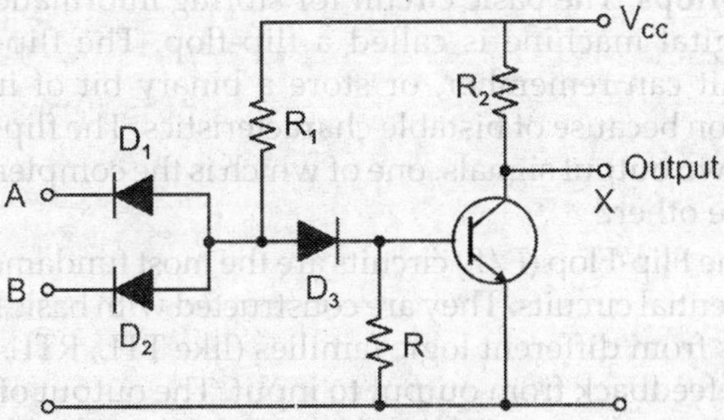

A	B	X
0	0	1
0	1	1
1	0	1
1	1	0

Fig. 16.3

Emitter coupled logic (ECL): As shown in Fig. 16.4, the basic ECL gate behaves as NOR gate. The emitters of logic transistors Q_1 and Q_2 are coupled to the emitter of a reference transistor Q_3. The common emitter resistance R_E is high enough to act as a constant-current source causing the overall loading remains more or less unaltered. So, there will be least disturbance on DC control voltage. Besides emitter coupling in the presence of suitable reference voltage prevents the transistor from entering into deep saturation. This results in a very fast switching speed (few *nano* seconds) but high power dissipation (about 50 mW).

If all the inputs are at logical '0' value, the transistors Q_1 and Q_2 are cut-off, resulting common collector potential rising towards V_{CC}. This drives the emitter follower Q_4 into conduction. The output from the emitter of Q_4 goes towards $+V_{CC}$ to give high output, i.e. logical '1'. If any of the input goes high, the respective input transistor conducts resulting the common collector potential going low. This causes Q_4 to turn-off and the output from the emitter of Q_4 goes low to give logical '0'.

Transistor transistor logical (TTL): In Fig. 16.5(a), the input transistor Q_1 has multiple emitters for multiple input connections. When any of the input goes low, Q_1 saturates making the base of Q_2 low and hence causing Q_2 as well as Q_3 to be cut-off. Thus, the output goes high with Q_4 acting as emitter-follower, supplying output current to the load. When all the inputs become high, Q_1 becomes off, forcing Q_2 and Q_3 to be saturated, so that the output goes low at logical '0'.

The function of the diode is to prevent Q_4 from turning ON, when Q_3 is saturated. The V_{BE} of Q_3 (approximately 0.7 V), and V_{CE} of Q_2 ($\approx$ 0·1 V), a total of about 0.8 V, is applied to the base of Q_4. Without the diode, Q_4 will turn ON.

MOS logic families: The advantages and disadvantages of various logic families are given in Table 16.3.

Self-checking/correcting codes: A parity bit is introduced in the side of a code to detect the error. It can follow either odd parity code or even parity code. This is an extra 1-bit depending on whether the total number of ones in the code is even or odd. In case of even parity code, '0' is put when the total number of

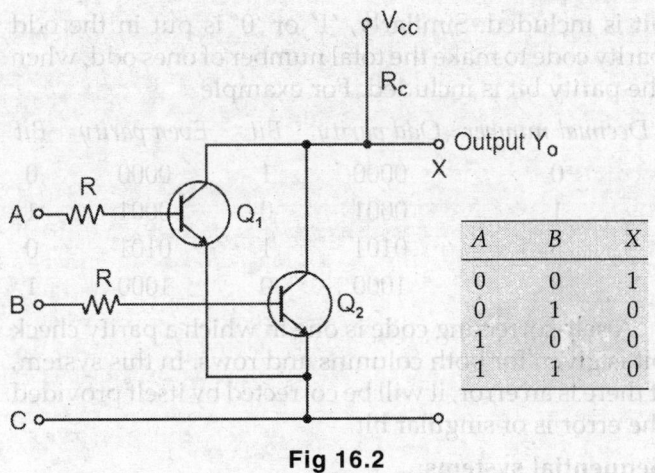

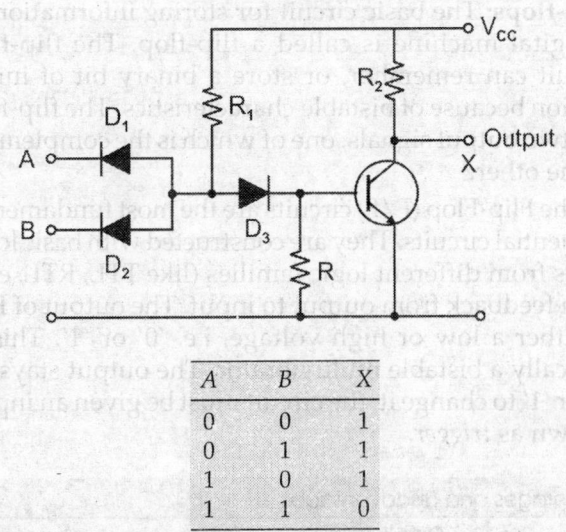

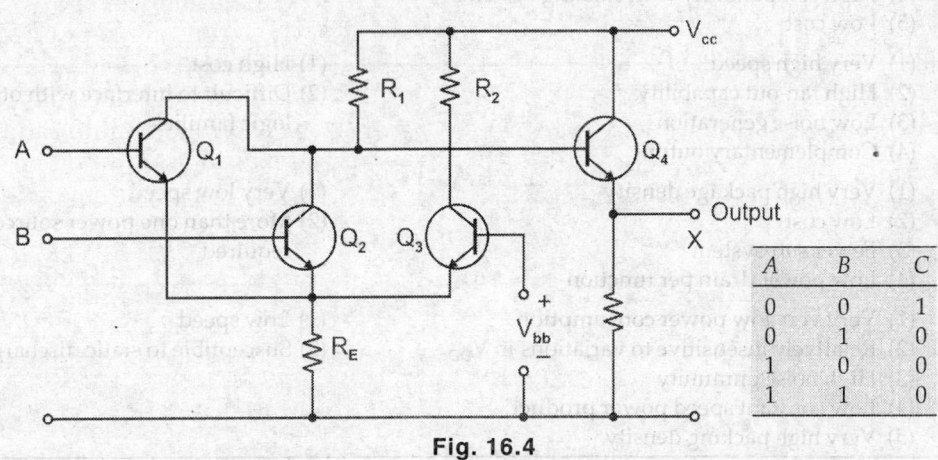

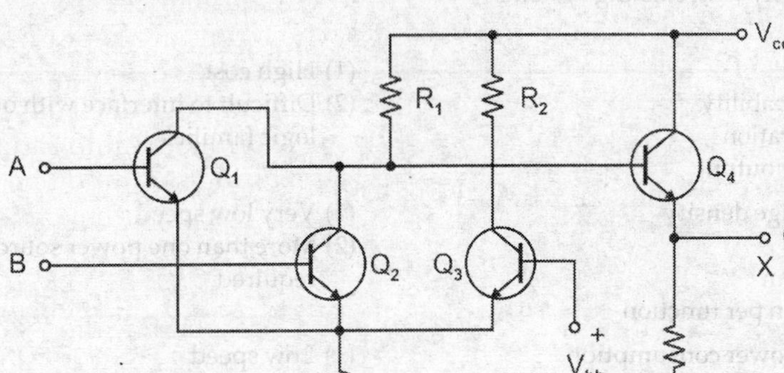

A	B	C
0	0	1
0	1	0
1	0	0
1	1	0

Fig. 16.4

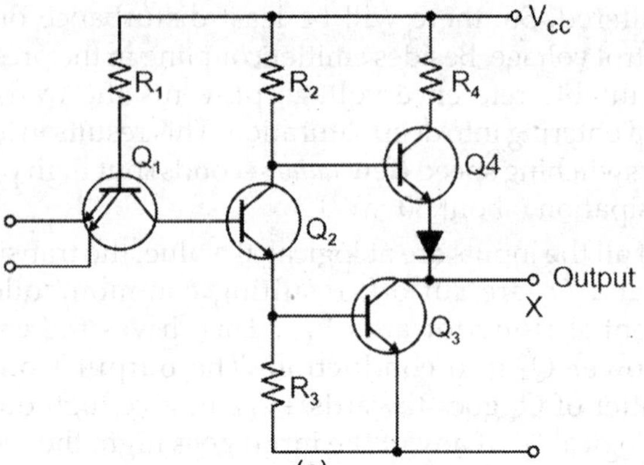

(a)

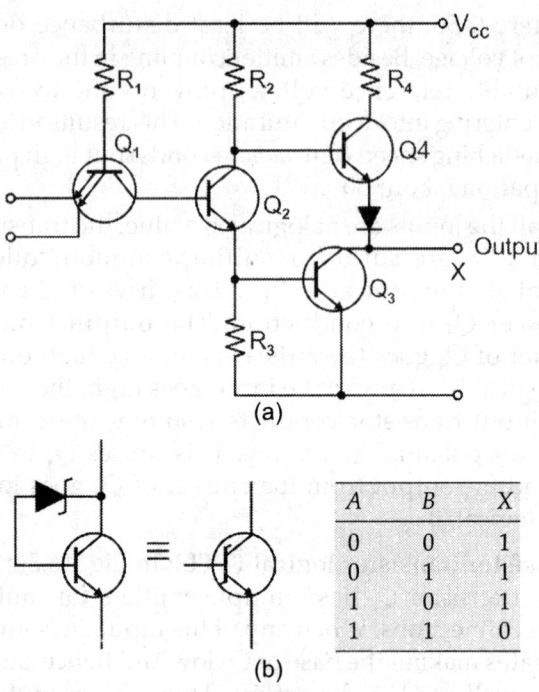

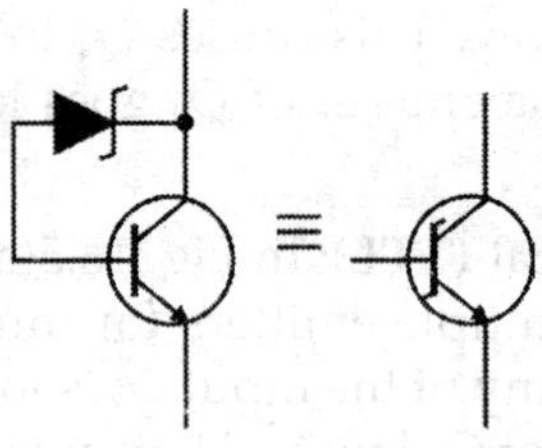

A	B	X
0	0	1
0	1	1
1	0	1
1	1	0

(b)

Fig. 16.5

ones are even and '1' is put when the total number of ones are odd, so to get the total number of ones in the code is even or odd. In case of even parity code, '0' is put when the total number of ones are even and '1' is put when the total number of ones are odd, so to get the total number of ones always even, when the parity bit is included. Similarly, '1' or '0' is put in the odd parity code to make the total number of ones odd, when the parity bit is included. For example

Decimal number	*Odd parity*	*Bit*	*Even parity*	*Bit*
0	0000	1	0000	0
1	0001	0	0001	1
5	0101	1	0101	0
8	1000	0	1000	1

A self-correcting code is one in which a parity check bit is given for both columns and rows. In this system, if there is an error, it will be corrected by itself provided the error is of singular bit.

Sequential systems:

Flip-flops: The basic circuit for storing information in a digital machine is called a flip-flop. The flip-flop circuit can remember, or store a binary bit of information because of bistable characteristics. The flip-flop has two output signals, one of which is the complement of the other.

The Flip-Flop (F/F) circuits are the most fundamental sequential circuits. They are constructed with basic logic gates from different logic families (like TTL, RTL, etc.) with feedback from output to input. The output of F/F is either a low or high voltage, i.e. '0' or '1'. This is basically a bistable multivibrator. The output stays on '0' or '1' to change it, the circuit must be given an input, known as *trigger*.

Table 16.3: Logic families with their advantages and disadvantages

Logic families	*Advantage*	*Disadvantage*
RTL	Low power dissipation	(1) Insufficient noise margin (2) Low speed
DTL	Low power dissipation	(1) Poor noise margin (2) Low speed
TTL	(1) Low power dissipation (2) High speed (3) High fan-out capability (4) High compatibility with existing systems (5) Low cost	(1) Tight V_{CC} tolerances (2) Susceptible to power transients
ECL	(1) Very high speed (2) High fan-out capability (3) Low noise generation (4) Complementary output	(1) High cost (2) Difficult to interface with other logic families
MOS	(1) Very high package density (2) Low cost (3) Fewer sub-system (4) Low power drain per function	(1) Very low speed (2) More than one power source required
CMOS	(1) Very very low power consumption (2) Relatively insensitive to variations in V_{DD}. (3) High noise immunity (4) Low constant speed power product (5) Very high packing density	(1) Low speed (2) Susceptible to static discharge damage

The RS flip-flop: The RS flip-flop is used to temporarily hold or store information until it is required. A single RS circuit will store one binary digit, i.e. 1 or 0. For storing a six digit binary number, six RS flip-flops will be required. The truth table of RS Flip-Flop is given below. The symbols R and S are input to the Flip-Flop which are known as 'Reset' and 'Set' inputs. Always Flip-Flop provides output in both true form; Q and in complement form $\overline{Q}$. The symbol Q_{n+1} indicates the next state while the inputs are applied. As shown in the Table, if R and S both are at logical '0' level, the output retains its last value. The realisation of F/F by NOR gates is also shown in Fig. 16.6.

R	S	$\overline{Q}_{n+1}$	1 Q_{n+1}
0	0	Qn	$\overline{Q}_n$
0	1	1	0
1	0	0	1
1	1	Not allowed	

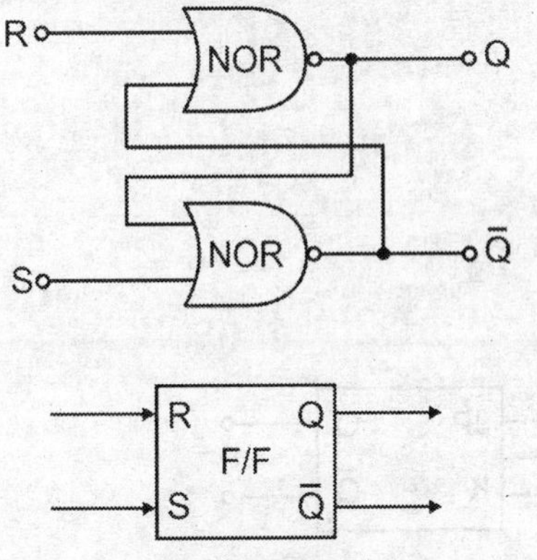

Fig. 16.6

JK flip-flop: JK flip-flop is the modified version of RS flip-flop with no valid or illegal output state. Here two inputs are interlocked so that they cannot be activated simultaneously. In RS F/F, the input $R = S = 1$ leads to an indeterminate output. This is inconvenient to use and thus this input is disallowed. In JK F/F, the circuit has been modified so that if a 1–1 input is applied, the circuit complements the previous outputs. The circuit is given in Fig. 16.7.

Toggled P or T flip-flop: These are basically JK flip-flops both J and K terminals connected together permanently. Thus such a flip-flop will have only one input terminal T as shown in Fig. 16.8. This F/F complements the output when a high signal (logical '1' input) is applied to terminal T. Its truth table can be written as follows:

T	Q_{n+1}
0	$\overline{Q}_n$
1	Q_n

Q	T	Q_{n+1}
0	0	0
0	1	1
1	0	1
1	1	0

Figure 16.8 shows how T F/F can be constructed using RS and JK F/F. If $Q = 1$, goes high, the lower AND gate will be enabled so that $\overline{R}$ input becomes '1'. The F/F resets resulting $Q = 0$, $\overline{Q} = 1$.

D flip-flop: It is a modification of clocked RS Flip-Flop. It transfers the data input at D to Q after one clock period. Figure 16.9 shows D flip-flop.

D	CP	S'	R'	Q_n	Q_{n+1}
0	1	1	0	0	0
0	1	1	0	1	0
1	1	0	1	0	1
1	1	0	1	1	1

Binary counters: The output pulse for a flip-flop is obtained for every two input pulses. In other words, a F/F halves the input frequency or clock pulses as shown in Fig. 16.10. The triggering is occured at the trailing edge of the clock pulse. Because the circuit can be thought of a frequency divider by two, any power of two can be counted by cascading the flip-flops.

Module-16 asynchronous counters: The F/Fs, when connected in cascade as shown in Fig. 16.11, can count upto

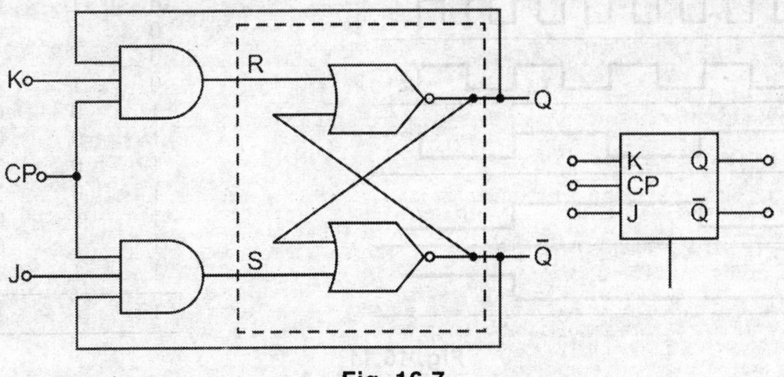

Fig. 16.7

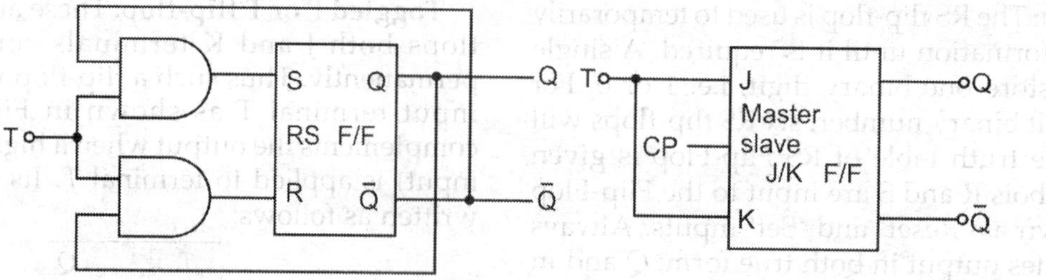

Fig. 16.8

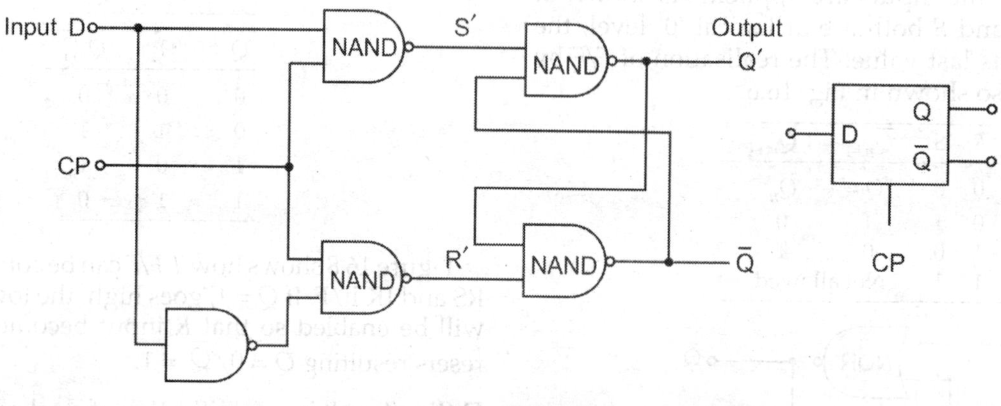

Fig. 16.9: D flip-flop

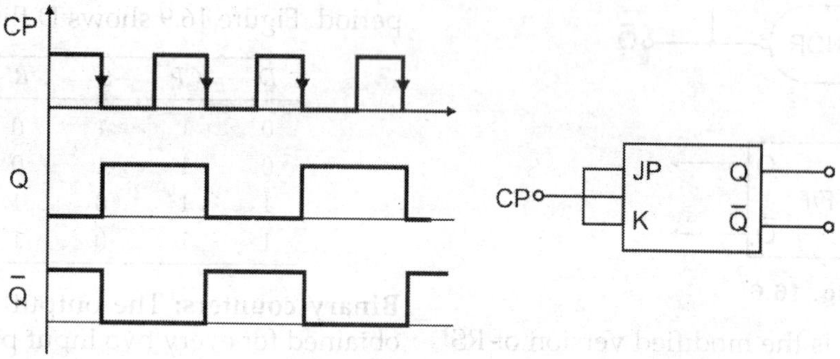

Fig. 16.10

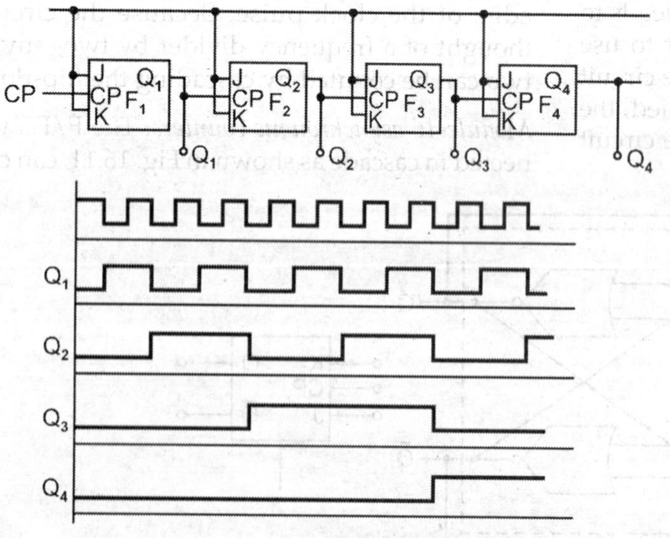

Fig. 16.11

Table 16.4

Q_4	Q_3	Q_2	Q_1
0	0	0	0
0	0	0	1
0	0	1	0
0	0	1	1
0	1	0	0
0	1	0	1
0	1	1	0
0	1	1	1
1	0	0	0
1	0	0	1
1	0	1	0
1	0	1	1
1	1	0	0
1	1	0	1
1	1	1	0
1	1	1	1

$2^4 = 16$ binary states, as the circuit gives one pulse for every 16 input clock pulses. This is known as *module-16* asynchronous ripple counter. The states of flip-flop follow every binary state from 0000 to 1111 and then returns to 0000.

If parallel outputs are taken out, $Q_4\ Q_3\ Q_2\ Q_1$ will follow binary sequence in Table 16.4.

D/A converters: D/A conversion is the process of converting digital input levels into an equivalent analog output voltage. This is most easily accomplished by the use of resistance networks as shown in Fig. 16.12. The basic problem in converting a digital signal into equivalent analog signal is to change the *n* digital voltage levels into one equivalent analog voltage. This can be accomplished by designing or current. The resistance network will change each of the digital levels into an equivalent binary weighted voltage (or current).

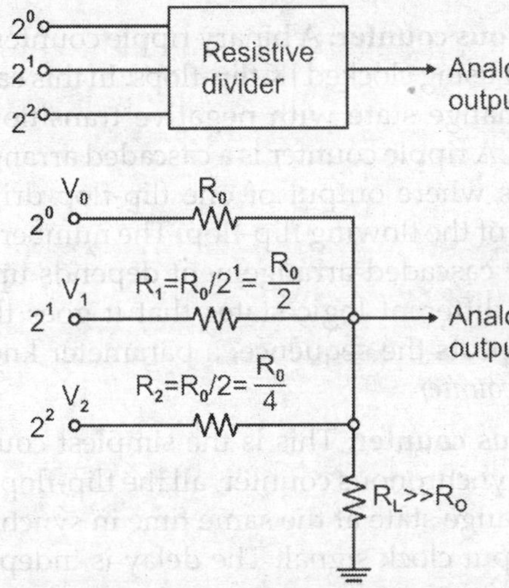

Digital 2^2	2^1	*Input* 2^0	*Analog output*
0	0	0	
0	0	1	
0	1	0	
0	1	1	
1	0	0	
1	0	1	
1	1	0	
1	1	1	

Fig. 16.12

The output analog voltage for binary inputs V_2, V_1 V_0 to the network will be

$$V = \frac{V_0/R_0 + V_1/(R_0/2) + V_2/(R_0/4)}{1/R_0 + 1/(R_0/2) + 1/(R_0/4)}$$

If '0' = 0V and '1' = +7V, the above truth table will hold true, while V_0, V_1 and V_2 are in binary form either 0 or 1.

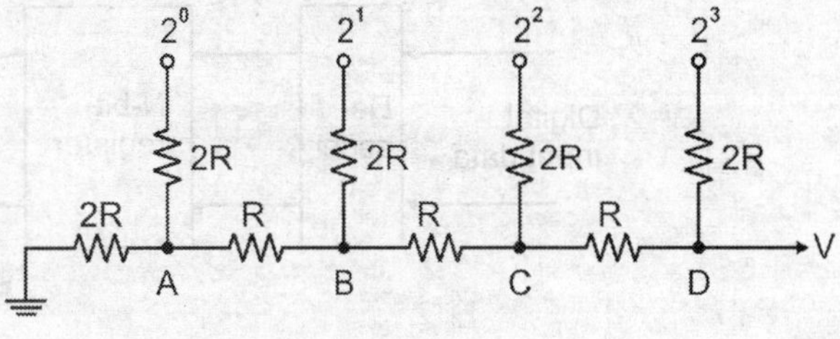

Fig. 16.13

Binary ladder: The binary ladder shown in Fig. 16.13 is a resistive network whose output voltage is properly weighted sum of the digital inputs.

The output voltage V_A, will depend on the bit position. Here, *V* is the voltage level of logic '1'.

The complete *D/A* converter consists of a binary ladder (usually), and a flip-flop register to hold the digital input information as shown in Fig. 16.13.

A/D converters: The process of converting the analog voltage into an equivalent digital signal is known as *A/D conversion.*

The simultaneous method of *A/D* conversion illustrated in Fig. 16.14 is based on the use of a number of comparator circuits, that will decide the number of bits of digital output signal. If the comparator is off, it will, give a 'low' output voltage and if it is on it will provide a 'high' output voltage. The comparator output levels for various ranges of input voltages are summarized in the following table.

Input voltage	*Comparator output*		
	C_1	C_2	C_3
0 to + V/4	Low	Low	Low
+ V/4 to + V/2	High	Low	Low
+ V/2 to 3V/4	High	High	Low
+ 3V/4 to + V	High	High	High

The three comparator outputs can then fed into a coding network to provide two bits. The bits of the coding network can be entered into a flip-flop registered for storage.

These levels are coded now to give the final digital output signal. This process is known as *simultaneous conversion*. The complete block diagram for such A/D counter is shown in Fig. 16.15.

Counter method: The counter method as illustrated in Fig. 16.16, provides a high resolution *A/D* conversion. In this method, the output of a binary counter is connected to the binary ladder or a *D/A* converter. A clock is applied to the input of the counter through a gate control. The output of the binary ladder is the staircase waveform. The waveform is applied to the one side of the comparator. The other side is connected to the analog input voltage. When a start signal appears, the gate allows the clock pulses to the counter. The counter advances its normal sequence. A corresponding

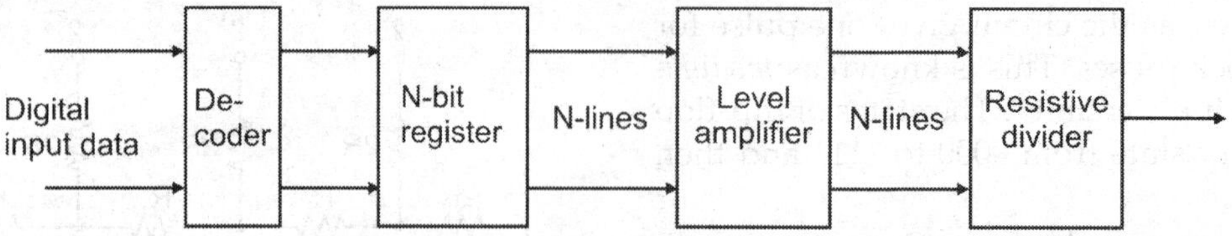

Fig. 16.14

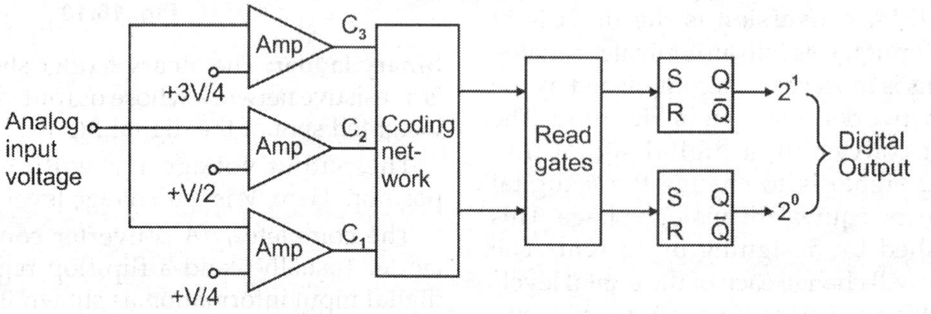

Fig. 16.15

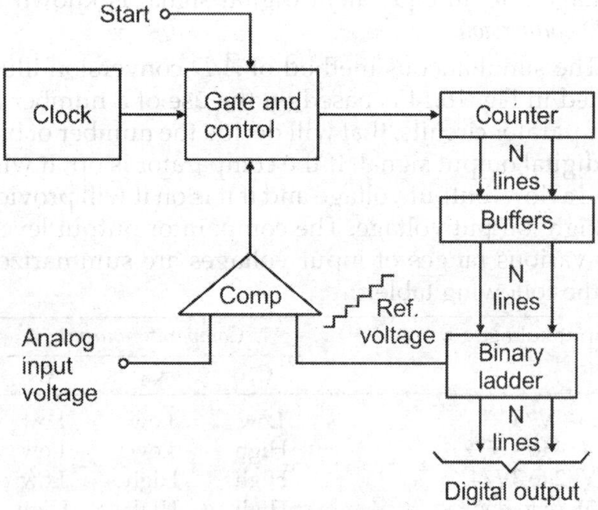

Fig. 16.16

reference voltage signal is formed by converting the counter output through binary ladder. When the analog output equals the reference voltage, the comparator closes the gate and the counter is stopped. The binary number spreads in the counter is the digital equivalent of analog input voltage. The counter-type A/D converter provides a good method for digitizing to a high resolution. This method is much simpler than the simultaneous method for high resolution, but the conversion time required is longer. The average conversion time is $2^n/2 = 2^{n-1}$ counts.

Counters and registers: Flip-flops are wired together that form circuits to do counting functions. Manufacturers make self contained counters in IC form. Counters and registers belong to the category of MSI sequential logic circuits. They have similar architecture as both counters as well as registers comprise a cascaded arrangement of more than one flip-flop with or without combinational logic devices.

Asynchronous counter: A binary ripple counter can be constructed using clocked JK flip-flops. In this each flip-flop will change state with negative transition at its clock input. A ripple counter is a cascaded arrangement of flip-flops where output of one flip-flop drives the clock input of the flowing flip-flop. The number of flip-flops in the cascaded arrangement depends upon the number of different logic states that it goes through before it repeats the sequence, a parameter known as *modulus of counter*.

Synchronous counter: This is the simplest counter to build. In a synchronous counter, all the flip-flops in the counter, change state at the same time in synchronism with the input clock signal. The delay is independent of the size of the counter.

Modulus of counter: The modulus of counter is the number of different logic states it goes through before it come back to the initial state to repeat the counter sequence. As n-bit counter that counts through all its natural states and does not skip any of the states has a modulus of 2^n. In general arrangement of flip-flop can be used to construct any counter with modulus given by

$$(2^{N-1} + 1) \leq \text{modulus} \leq 2^N$$

Shift register: A shift register is a digital device for storage and transfer data. The data to be stored could be the data appearing at the output of an encoding matrix before they are fed to the main digital system for processing or there might be data present at the output of a microprocessor before they are fed to the device circuitry of the output devices. The shift register thus forms an important link between the main digital system and input/output channels. Shift register can be classified as: (i) serial-in serial-out (SISO) shift

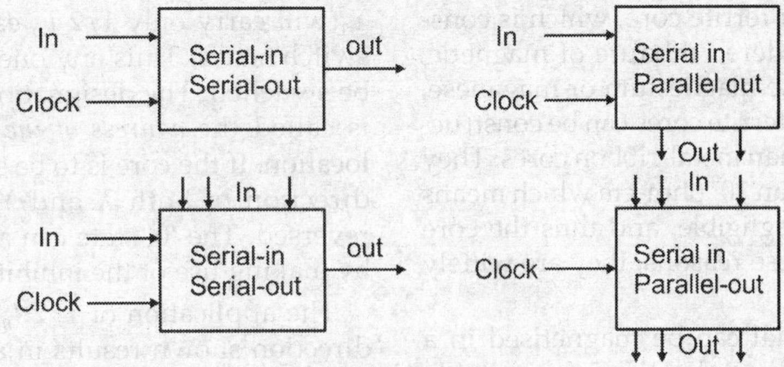

Fig. 16.17

registers, (ii) serial-in parallel-out (SIPO) shift registers, (iii) parallel-in serial-out (PISO) shift registers and (iv) parallel-in parallel-out (PIPO) shift registers. Figure 16.17 shows a circuit representation of above mentioned four types of shift registers.

Digital ICs: Digital ICs using MOS (Metal Oxide Semiconductor) technology were introduced around 1969, The MOS technology has several advantages over bipolar devices:

1. MOS devices require fewer manufacturing cycles.
2. MOS gates dissipate less power per gate.
3. MOS gates require less space on the silicon wafer.

Bipolar TTL ICs continue to dominate in SSI and MSI because of their high speed, wide acceptance and long history of reliable operations. Most LSI (Large Scale Integration) devices, including memories and microprocessors that require thousands of gates in a single IC, use MOS rather than bipolar technology.

MOS technology is divided into two types—PMOS which is based on P-doped silicon, and NMOS, which is based on n-doped silicon. These two types of implementation can be used individually or can be combined effectively to form a third category known as CMOS (Complementary MOS).

PMOS circuit speed is much slower (about one tenth) than bipolar devices, but has lower power dissipation per gate and reduced geometry to accommodate more complexity on a single chip.

NMOS is basically three times faster than PMOS. As a result NMOS has emerged the most popular process for microprocessor production. It is the process used for M 6800, Intel 8080, Fairchild F 8 and Zilog 80 etc.

CMOS combines a n-channel and a *p*-channel MOS transistor in each gate. It has the advantage of extremely low power dissipation when it is static.

Integrated circuit packages: It can be sub-divided into three basic groups, based on the size of integration of the package:

1. SSI (Small scale integration)
2. MSI (Medium scale integration)
3. LSI and VLSI (Large scale integration and Very large scale integration)

Small scale integration package: It contains fewer than 12 gates. Examples of SSI are NAND gates (7400, 7410, 7420, 7430), AND gates (7408, 7411. 7421), OR gates (7432), hex inverters (7404, 7405, 7406, 7407) etc.

Medium scale integrator: Depending on the technology used (bipolar or MOS) and the chip size, much more complicated IC package containing 12 to 100 gates have Full BCD encoder 7447 or BCD to 7-segment decoder-driver 7446, 7447, 7448 etc.

Large scale integration: When the circuit complexity crosses the limit of 100 logic gates, by using large scale integration techniques, highly completed circuits are manufactured on a single IC chip known as LSI or VLSI package. For example, TMS 2300, 2560 bit dynamic ROM or microprocessor chips Intel 8080, Intel 8085, CP 1600 etc.

The integrated circuit timer: The 555 monolithic timing circuit is highly stable controller capable of producing accurate time delays or oscillation. Additional terminals are provided for triggering or resetting, if desired. The circuit may be triggered at falling waveforms and the output structure can source or sink up to 200 mA or drive TTL circuits.

Magnetic memory: Magnetic tape, floppy discs, and hard discs are all capable of storing large quantities of digital data. A hard disc drive and a floppy disc drive are important components in all microcomputer and minicomputer systems. Large reels of magnetic tapes are economical and widely used mass storage components in large computer system.

Magnetic devices and memories: The ability to store information in two different states is available in ferromagnetic material. A large class of binary elements has been devised using the process, of magnetization. One of the most widely used magnetic element is magnetic core.

Magnetic cores: The typical core is of toroidal shape and of two types. The metal-ribbon core, which is constructed by winding a very thin metallic ribbon on a ceramic core form, and a ferrite core, which is constructed from a finely powdered mixture of magnetic, various bivalent metals like magnesium or maganese, and a binder material. The ferrite cores can be constructed in smaller dimension than metal-ribbon cores. They have resistivities greater than 10^5 ohm-cm which means eddy current losses are negligible, and thus the core heating is reduced. For these reasons, they are widely used in digital computers.

A magnetic core material can be magnetised in a particular direction by the application of a suitable magnetising force (a magnetic flux resulting from a current flow). The material remains magnetised in that direction after removal of the excitation. Application of magnetizing force of the opposite polarity will switch the material, and it will remain magnetised in the opposite direction after removal of the excitation. Thus, the information in two different states is stored. The arrangement and its symbol is shown in Fig. 16.18.

A pulse at the 1 input sets a '1' in the core, and a pulse at the 0 input sets a '0' in the core. During the advance pulse, a pulse appears at the output only if the core previously held at 1. By using cores, basic logic functions can also be implemented.

Coincident-current memory: Direct access memories with very fast access times can be constructed using either magnetic cores or transistors. The most popular method for constructing these memories is the coincident-current technique. For a coincident current memory, access time is the time required for one read/write cycle, i.e. to read one word from memory or to write one word into memory.

There are X drive wires (vertical) and the Y drive wires (horizontal) forming a matrix as shown in Fig. 16.19. One half select current $1/2\ I_m$ is applied to line X_1 and $1/2\ I_m$ is applied to line Y_1. The core threaded by both lines X_1 and Y_1 will have a total of $1/2\,I_m + 1/2\,I_m = I_m$ passing through it and it will switch state. The remaining cores which are threaded by X_1 or Y_1 will carry only $1/2\ I_m$ each and therefore will not switch states. Thus any one out of these 16 states can be switched. The designation as $X_1\ Y_1$ in this example is called the *address of the core* since it specifies its location. If the core is to be switched to be '0' state, the direction of both X and Y line currents should be reversed. The '0' state can also be written in any core by making use of the inhibit wire.

The application of $1/2\ I_m$ to the inhibit wire in the direction shown results in a complete cancellation of the Y line select current. Thus to wire a '0' in any core, it is only necessary to select the core in the same manner as if writing a '1' but at the same time apply an inhibit current to the inhibit wire (assuming all cores are initially reset).

To read information from the memory, we simply apply $-1/2\ I_m$ to the proper address and sense the outputs on the n sense lines. But this results in resetting all cores to the '0' state. Thus these memories are inherently DRO (destructive read-out) type since the data stored in memory is lost every time they are read out. But they can be transformed into NDRO (non-destructive read-out) type memories by the addition of external logic.

Semiconductor memory: Recent advances in semiconductor technology have provided a number of reliable and economical MSI and LSI memory circuits. The typical semiconductor memory consists of a rectangular array of memory cells, fabricated on a silicon wafer, and housed in a convenient package such as a DIP. The basic memory cell is typically a transistor flip-flop or a circuit capable of storing charge and is used to store 1-bit of information. Memories are usually classified as either bipolar, metal oxide semiconductor (MOS) or complementary metal oxide semiconductor (CMOS) according to the type of transistor used to construct the individual memory cells. In general, faster operation is obtained with a bipolar memory chip, but greater packing density and thus reduced size and cost,

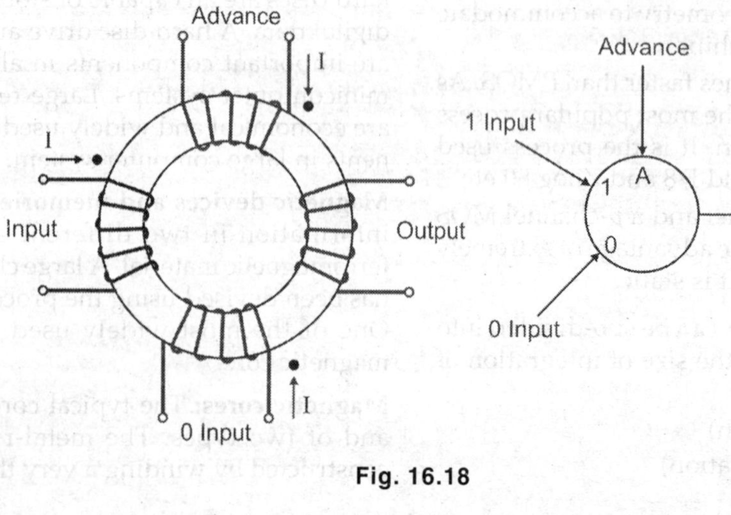

Fig. 16.18

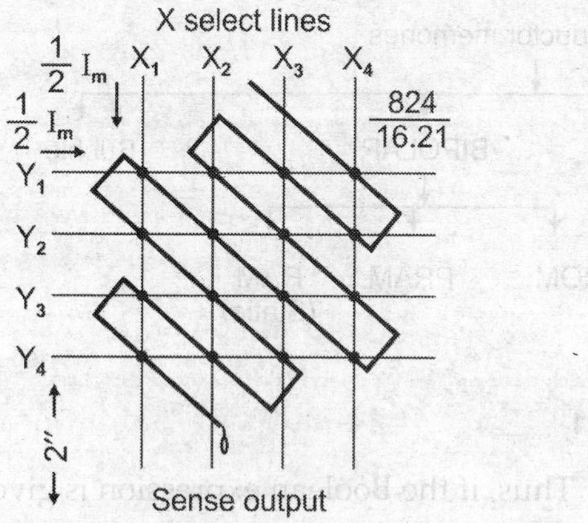

Fig. 16.19

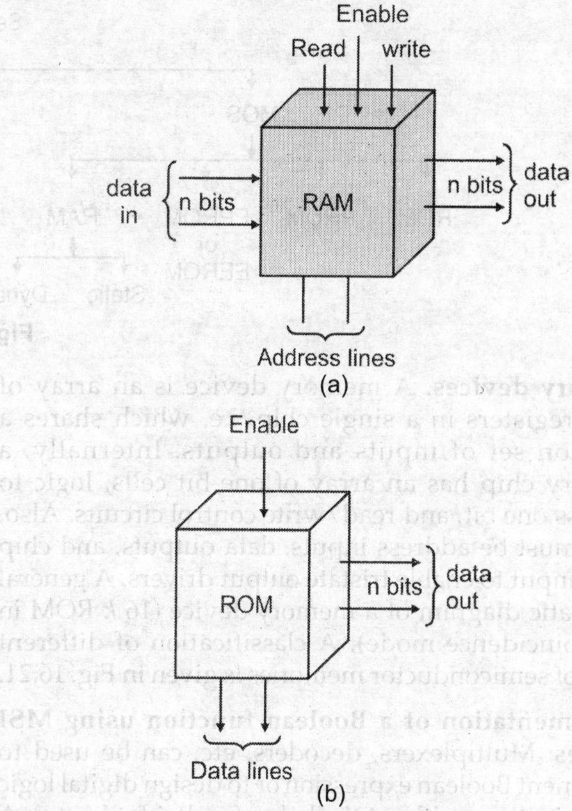

Fig. 16.20

as well as lower requirements, are characteristic of MOS and CMOS memory chips.

Characteristics: The two general (semiconductor) categories of memory RAM and ROM can be further divided. A DC power supply is required to energise any semiconductor memory chips. Once DC power is applied to a static RAM (SRAM), the SRAM retains stored information indefinitely, without any further action. The information (data) stored in ROM is fixed and will be retained permanently even if DC power is removed. Clearly a ROM is ideal for storing permanent instruction necessary for the start up and operation of a computer.

RAM: A block diagram of a typical RAM chip is shown in Fig. 16.20(a). An application in which data changes frequently calls for the use of a RAM. The logic circuitry associated with a RAM will allow a single bit of information to be stored in any of memory cells, this is the write operation.

ROM: A typical ROM chip is shown in Fig. 16.20(b). An application in which the data change dictates the use of a ROM. Since data is permanently stored in each cell, a loss of power does not cause a loss of data and thus ROM provides non-volatile data storage.

An application in which the data may change from time to time might call for the use of an EPROM.

MOS devices are somewhat simpler than bipolar devices; as a result, MOS memories can be constructed with more bits in a chip. They are less expensive than bipolar memories but slower than bipolar units.

Optical memory: Compact disc (CD) is being used for storing digital audio data since 1985. Different types of CDs flooding the market where binary data is optically coded the memory of CD is in the range of 650–750 MB. Its newer variety called digital versatile disc (DVD) can store data from 4.7 GB to 17.1 GB depending on configuration and make.

CDROM: CDROM or CD read only memory devices are mass produced in factory using a stamp press technology. CD ROM devices uses LASER technology to read data from it.

CD-R: CD-R or CD-recordable allows user to write data but once. CD-R drive has laser unit which uses higher intensity light wave for write operation than read.

CD-RW: CD-Read Write, previously known as CD-erasable gives user facility to write and erase data many time unlike CD-R, CD-RW uses an active layer of Ag-In-Sb-Te.

DVD: Digital versatile disc or digital video disc, popular as DVD resemble. Compact disc in dimension and look but contains much higher storage space. DVD uses smaller wavelength and lower numerical aperture of lens system to read smaller dimension land and pits.

Static memory: It is composed of cells capable of storing binary information indefinitely. For example, bipolar or *MOS* memory units remain set or reset as long as power is applied to the circuit. Magnetic core remains set or reset even if the power is removed. These basic cells are used to construct static memory. But 'dynamic memory' is one whose contents tend to decay over a period of time.

Magnetic drum storage: A magnetic drum is a cylindrical shaped drum, the surface of which has been coated with a magnetic material. The drum is rotated on its axis, and the read/write heads are used to record the information on the drum or read the information from the drum. Magnetic drums and disks provide larger storage capacities. But the disadvantage of using magnetic drums or disks is the increased access time.

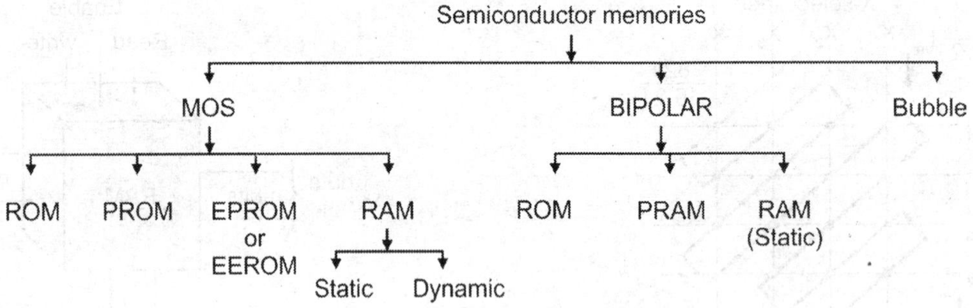

Fig. 16.21

Memory devices. A memory device is an array of *m*-bit registers in a single chip, i.e. which shares a common set of inputs and outputs. Internally, a memory chip has an array of one-bit cells, logic to address one bit, and read/write control circuits. Also, there must be address inputs, data outputs, and chip select input to enable tristate output drivers. A general schematic diagram of a memory device (16 *k* ROM in *X-Y* coincidence mode). A classification of different types of semiconductor memories is given in Fig. 16.21.

Implementation of a Boolean function using MSI devices: Multiplexers, decoders, etc, can be used to implement Boolean expression or to design digital logic system with or without the help of individual gates. A multiplexer selects one of the many inputs at a given time. For example, a four-to-one line multiplexer has two select lines to select one of the four inputs, as shown in Fig. 16.22. The input lines are selected by selecting their equivalent minterm coded by select-lines variables. This is clear from the table in Fig. 16.23, which shows a 4 × 1 line multiplexer.

S_1	S_0	*Minterm*	*Output selected*
0	0	$\bar{S}_1 \bar{S}_0$	I_0
0	1	$\bar{S}_1 S_0$	I_1
1	0	$S_1 \bar{S}_0$	I_2
1	1	$S_1 S_0$	I_3

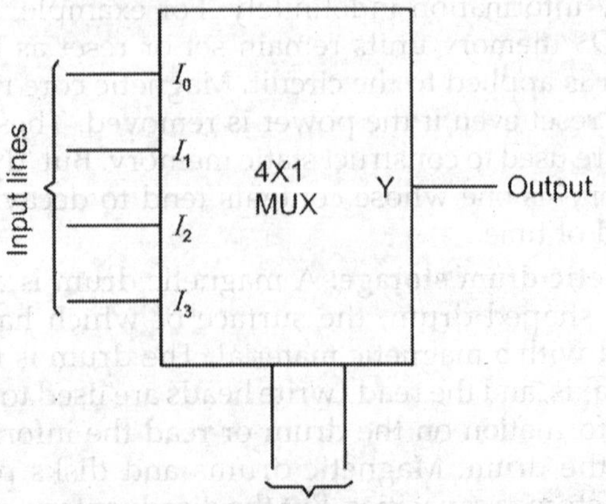

Fig. 16.22

Thus, if the Boolean expression is given as

$$F(A, B) = \bar{A}\bar{B} + A\bar{B} + AB$$
$$= m_0 + m_2 + m_3$$

A four-to-one line MUX chip can be used, as shown in Fig. 16.23, for the implementation.

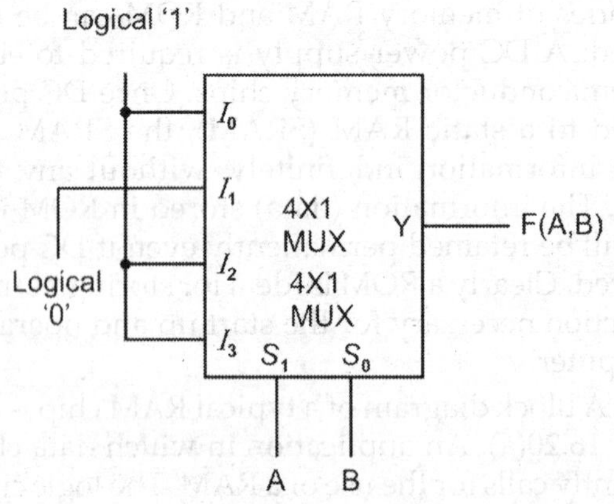

Fig. 16.23

Microcomputers: Computers were, first mechanical analog devices. These were later developed into electrical analog computers that could add, subtract, multiply and divide electrical quantities whose voltages were carefully made into an analog of the numbers being handled. But the advancements in digital circuits made analog computers obsolete, and they have been replaced almost entirely with digital units.

The logic gate was fabricated using a chip of silicon during the early 1960s. Very quickly, the number of gates per chip was increased and the term integrated circuit (IC) became part of nearly everyone's vocabulary. The arithmetic logic unit (ALU) developed in these early years could manipulate numbers just as the microcomputer chip of today was not then those listed, some contain less. The elements listed are described below:

Arithmetic logic unit (ALU): The arithmetic logic unit performs arithmetic and logic operations on data determined by signals from various control lines. The ALU can perform binary addition, subtraction, multi-

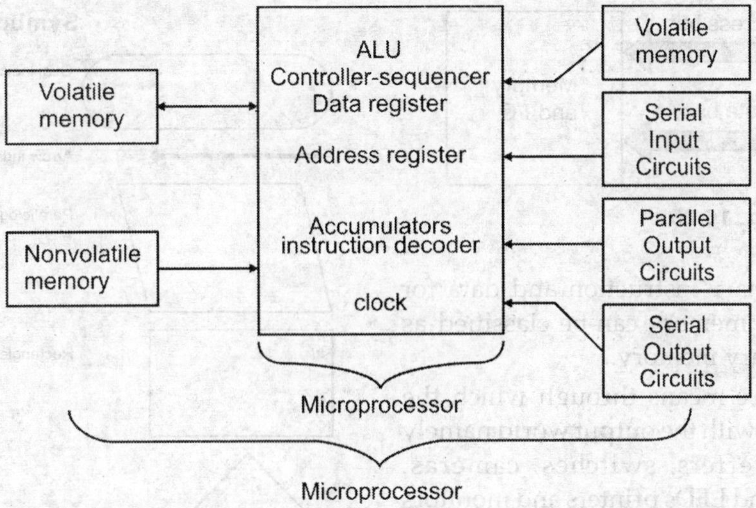

Fig. 16.24

plication and division, and also perform a variety of logical comparisons (Fig. 16.24).

Controller-sequencer: This section manages (controls) and supervises (sequences) the activities of the microprocessor. This is accomplished by producing a variety of control signals to various input pins to carry out an instruction.

Data register: The data register provides temporary storage for data received from the data bus. This may also be called the instruction register.

Address register: The address register is a temporary storage device to hold the location (address) of a data word in memory or I/O which is being used in a current operation.

Accumulators: The ALU is capable of only a few operations. The microprocessor gains additional power using storage registers like those described above. The ALU can operate on data from only one or two of the registers at a time. Some registers, which are more versatile than the others, are known as accumulators. During an arithmetic and logic operation the accumulator may perform a dual function. Before an operation, it holds an operand. After it holds the resulting sum, difference, or logical answer.

Instruction decoder: The instruction decoder is a circuit that analyzes the contents of the data register to determine what operation is to be performed. Once the operation code is recognized, the controller-sequencer can supply the appropriate preprogrammed signals for operation.

Clock: The precise timing of events in the microprocessor is controlled by a free-running oscillator. The clock input may be driven by a crystal, an LC-tuned circuit, or an external clock source.

Volatile memory: Data stored in volatile memory units will be lost when the power to the computer is turned off. Memory of this type is usually stored on IC circuits containing latch or flip-flop circuits. Volatile memories are used to hold data input for processing.

Nonvolatile memory: Data stored in nonvolatile memory will not be affected when the computer is turned off. This type of memory may be contained on a variety of magneic tapes or disks, or on special memory chips. Nonvolatile memories are used basically for program data storage.

Input-output (I/O): To become fully functional, the microprocessor must access memory and peripheral devices. This is accomplished by placing data on either an address bus or data bus, depending upon the function of the operation. The I/O data travels through logical circuits called ports.

Serial ports: These will transmit data one binary digit at a time. After one bit of information is sent, the next binary digit follows. This process is repeated until all of the information has been transmitted.

Parallel ports: The data is transmitted using two wires and is called parallel because each circuit is wired in parallel with respect to the next. This is continuous type of transmission and is capable of transmitting large amounts of serial data.

Microprocessors: The microprocessor is the heart of a microcomputer system. In fact, it form the central processing unit of any microcomputer and, has been rightly referred to as the computer on chip.

Any microcomputer based system essentially comprises three parts namely: (i) microprocessor, (ii) memory, (iii) peripheral I/O devices.

The three are interconnected by the data base, the address bus, and the control bus as shown in Fig. 16.25.

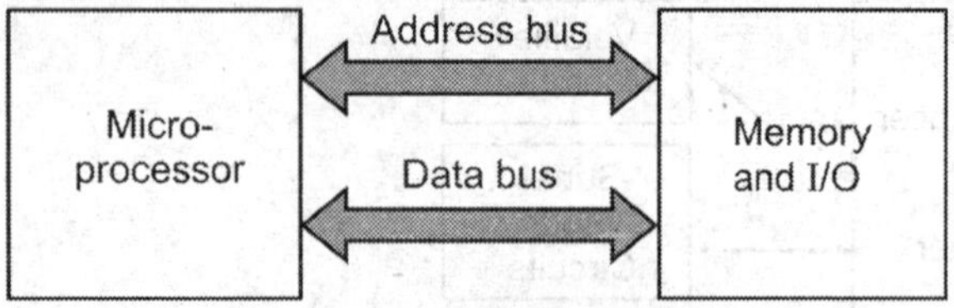

Fig. 16.25

Memory stores the binary instruction and data for the microprocessor. The memory can be classified as the primary and secondary memory.

Input/output devices are means through which the microprocessor interacts with the output world namely keyboards, A/D converters, switches, cameras, scanners, microphones and LEDs printers and monitors are examples of output devices.

A bus is basically a communication link between the processing and peripheral devices. It is a group of wires that carry information in the form of bits.

The address bus is unidirectional and is used by the CPU to send out the address of the memory location to be accessed. It is also used by the CPU to select a particular input or output port.

Data bus is bidirectional, that is, data flow occurs both to and fro from the microprocessor and peripherals. The control bus contain a number of individual lines carrying synchronizing signals.

Microprocessor based systems can be categorized as general purpose reprogrammable systems and embedded systems.

Intel 8085 microprocessor architecture: Intel 8085 is an 8-bit NMOS microprocessor. It is a 40 pin IC package fabricated on a single LSI chip. The intel 8085 uses a single +5 DC supply for its operation. Its clock speed is about 3 MHz. The clock cycle is 320 ns. It has 80 basic instructions and 246 opcodes. It consists of three main sections: (i) an arithmetic and logic unit, (ii) a timing and control unit, (iii) a set of registers.

Registers: Intel 8085 microprocessor has the following registers: (i) Accumulater (ACC), (ii) General purpose registers, (iii) Program counter (PC), (iv) Stack pointer (SP, (v) Instruction Register, (vi) Temporary Register.

Flags: Intel 8085 microprocessor contains five flip-flops to serve as status flags. Following are five status flags of Intel 8085: (i) Carry flags (CS), (ii) Parity flag (P), (iii) Auxiliary carry flag (AC), (iv) Zero flag (Z), (v) Sign flags (S).

There are some standard symbols shown in Fig. 16.26.

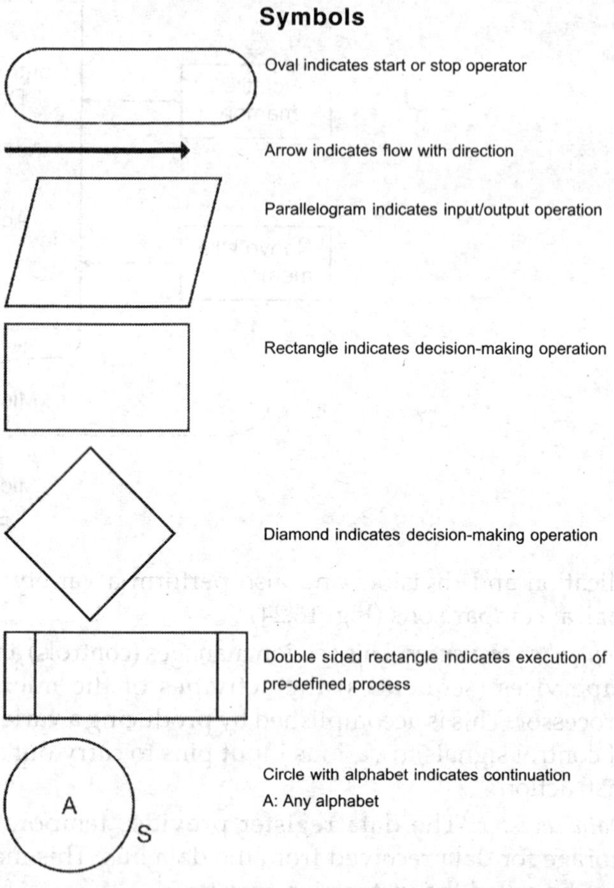

Fig. 16.26

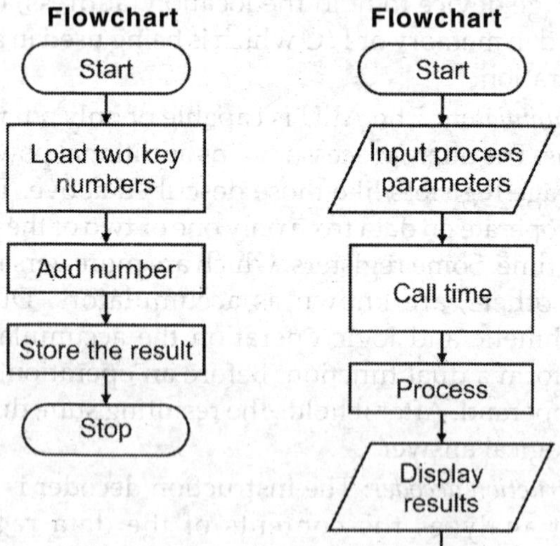

Assembly language program: Three tasks are involved in this program are shown in Flowchart. (i) Load two key numbers, (ii) Add number, (iii) Store the result in the memory.

Optical encoder: This provides a convenient means for the digitising shaft-position information or angular

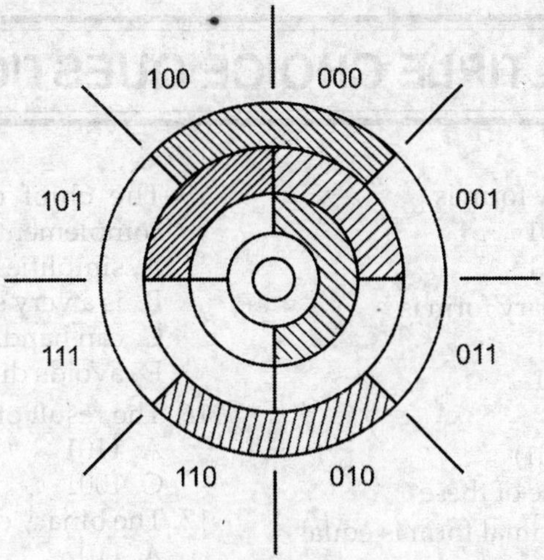

Fig. 16.27

position information. In Fig. 16.27, Gray code is used to code the optical encode wheel to eliminate the large-ambiguity error found in binary coded wheels. If the light areas on the wheel are transparent and the dark are opaque, the digital information can be obtained by placing light sources and photo-sensors on opposite sides of the disc. The output of photo-sensor is 'high' if light is sensed and 'low' if no light is sensed. Thus after amplification, the outputs received is 1 and 0 form is the translation of angular position into digital form.

MULTIPLE CHOICE QUESTIONS

1. The decimal number 29 in binary form is
   A. 11001 B. 11101
   C. 10001 D. 11111
2. The decimal fraction 0.375 in binary form is
   A. 0.0011 B. 0.011
   C. 0.0111 D. 0.111
3. The fraction 0.68_{10} is equal to
   A. $(0.010101)_2$ B. $(0.101)_2$
   C. $(0.10101)_2$ D. None of these
4. The binary fraction 0.0111 in decimal form is equal to
   A. 0.4375 B. 0.6225
   C. 0.8325 D. 0.1105
5. The binary fraction 0.01001 is equal to
   A. 0.225 B. 0.22125
   C. 0.28125 D. 0.132105
6. The digital system usually operate on .... system
   A. binary B. decimal
   C. octal D. hexadecimal
7. The binary system uses power of ...... for positional values.
   A. 2 B. 10
   C. 8 D. 16
8. After counting 0, 1, 10, 11, the next binary number is
   A. 12 B. 100
   C. 101 D. 110
9. The number 1000 is equivalent to decimal number
   A. one thousand B. eight
   C. four D. sixteen
10. In binary number, shifting the binary point one place to the right
    A. multiplies by 2 B. divides by 2
    C. decreases by 10 D. increases by 10
11. The binary addition 1 + 1 + 1 gives
    A. 111 B. 10
    C. 110 D. 11
12. The cumulative addition of the four binary bits (1 + 1 + 1 + 1) gives
    A. 1111 B. 111
    C. 100 D. 1001
13. The result of binary subtraction (100 – 011) is
    A. –111 B. 111
    C. 011 D. 001
14. The 2's complement of 1000_2 is
    A. 0111 B. 0101
    C. 1000 D. 0001
15. The chief reason why digital computers use complemental substraction is that it
    A. simplifies their circuitry
    B. is a very simple process
    C. can handle negative numbers easily
    D. avoids direct subtraction
16. The result of binary multiplication $111_2 \times 10_2$ is
    A. 1101 B. 0110
    C. 1001 D. 1110
17. The binary division $11000_2/100_2$ gives
    A. 110 B. 1100
    C. 11 D. 101
18. The number 12_8 is equivalent to decimal
    A. 12 B. 20
    C. 10 D. 4
19. The number 1000101_2 is equivalent to octal
    A. 54 B. 45
    C. 37 D. 25
20. The number 17_8 is equivalent to binary
    A. 11 B. 1110
    C. 10000 D. 1111
21. Which of the following is not an octal number?
    A. 19 B. 77
    C. 15 D. 101
22. Hexadecimal number system is used as a short-hand language for representing .... numbers.
    A. decimal B. binary
    C. octal D. large
23. The main advantage of a CMOS logic family over the TTL family is its
    A. much reduced power dissipation
    B. increased speed of operation
    C. very small physical size
    D. extremely low cost
24. BCD code is
    A. nonweighted
    B. the same thing as binary numbers
    C. a binary code D. an alphanumeric code
25. Which of the following 4-bit combinations is/are invalid in the BCD code?
    A. 1010 B. 0010
    C. 0101 D. 1000
26. Saturated logic circuits have inherently
    A. short saturation delay time
    B. low switching speed
    C. higher power dissipation
    D. lower noise immunity
27. In Excess-3 code, each coded number is .... than in BCD code.

Ans.												
1. B	2. B	3. C	4. A	5. C	6. A	7. A	8. B	9. B	10. A	11. D	12. C	13. D
14. C	15. A	16. D	17. A	18. C	19. B	20. D	21. A	22. B	23. A	24. C	25. A	26. A

A. four larger B. three smaller
C. three larger D. much larger

28. Boolean algebra is essentially based on
A. symbols B. logic
C. truth D. numbers

29. The first person who used Boolean algebra for the design of relay switching circuits was
A. Aristotle B. Boole
C. Shannon D. Ramanujam

30. Different variables used in Boolean algebra can have values of
A. 0 to 1 B. low or high
C. true or false D. ON or OFF

31. According to the algebra of logic, $(A + \overline{A})$ equals
A. A B. 1
C. 0 D. $\overline{AA}$

32. According to the absorptive laws of Boolean algebra, expression $(A + AB)$ equals
A. A B. B
C. AB D. $\overline{A}$

33. When we deorganize $\overline{AB}$, we get
A. $\overline{AB}$ B. $ABA + \overline{B}$
C. $\overline{A} + \overline{B}$ D. $A\overline{B}$

34. The dual of the statement $A + 1 = 1$ is
A. $A \cdot 1 = A$ B. $A \cdot 0 = 0$
C. $A + A = A$ D. $A \cdot A = 1$

35. The expression $\overline{ABC}$ can be simplified to
A. $\overline{A}, \overline{B}, \overline{C}$ B. $AB + BC + CA$
C. $AB = C$ D. $\overline{A} + \overline{B} + \overline{C}$

36. In a saturated bipolar circuit, transistors operate
A. in deep cut-off B. overactive region
C. in saturation D. just short of saturation

37. Octal coding involves grouping the bits in
A. 5's B. 7's
C. 4's D. 3's

38. Noise margin is expressed in
A. decibel B. watt
C. volt D. phon

39. DTL family employs
A. resistors and transistors
B. diode and resistor
C. diode and transistors
D. diodes, resistors and transistors

40. The chief advantages of Schottky TTL logic family is its least
A. power dissipation B. propagation delay
C. fan-in D. noise-immunity

41. The main advantage claimed for ECL, family of logic gates is its
A. very large fan-in
B. use of negative power supply voltage
C. extremely low propagation time
D. least power dissipation

42. Special feature of an I^2L logic circuit is that it
A. uses only high-value resistors
B. dissipate negligible power
C. is a bipolar saturated logic
D. uses no biasing and loading resistors

43. A unique advantageous feature of CMOS logic family is its
A. use of NMOS circuits B. speed
C. power dissipation in nanowatt range
D. dependence on frequency for power dissipation

44. CMOS circuits are extensively used for one-chip computers mainly because of their extremely
A. low power dissipation
B. large packing density
C. high noise immunity
D. low cost

45. The binary equivalent of A_{16} is
A. 1010 B. 1011
C. 1000 D. 1110

46. CMOS logic family uses only
A. MOSFETs and resistors
B. NMOS circuits
C. MOSFETs
D. bipolar transistors

47. Power is drawn by a CMOS circuit only when
A. its output is high B. its output is low
C. its switches logic levels
D. in static state

48. A logic gate is an electronic circuit which
A. makes logic decisions
B. allows electrons flow only in one direction
C. works on binary algebra
D. alternates between 0 and 1 values

49. In positive logic, logic state 1 corresponds to
A. positive voltage B. higher voltage level
C. zero voltage D. lower voltage level

50. In negative logic, the logic state 1 corresponds to
A. negative voltage B. zero voltage
C. more negative voltage
D. lower voltage level

51. The output of a 2-input OR gate is zero only when its
A. both inputs are 0 B. either input is 1
C. both inputs are 1 D. either input is 0

52. A NOR gate produces an output only when its two inputs are
A. high B. low
C. different D. same

Ans.													
	27. C	28. B	29. C	30. A	31. B	32. A	33. C	34. B	35. D	36. C	37. D	38. A	39. D
	40. B	41. C	42. D	43. C	44. B	45. A	46. C	47. C	48. A	49. B	50. D	51. A	52. C

53. An AND gate
A. implements logic addition
B. is equivalent to a series switching circuit
C. is an any-or-all gate
D. is equivalent to a parallel switching circuit

54. When an input electrical signal $A = 10100$ is applied to a NOT gate, its output signal is
A. 01011 B. 10101
C. 10100 D. 00101

55. The only function of a NOT gate is to
A. stop a signal B. recomplement a signal
C. invert an input signal
D. act as a universal gate

56. A NOR gate is ON only when all its inputs are
A. ON B. positive
C. high D. OFF

57. For getting an output from an XNOR gate, its both inputs must be
A. high
B. low
C. at the same high level
D. at the opposite logic levels

58. In a certain 2-input logic gate, when $A = 0, B = 0$, then $C = 1$ and when $A = 0, B = 1$, then again $C = 1$. It must be ..... gate.
A. XOR B. AND
C. NAND D. NOR

59. The logic symbol shown below represents

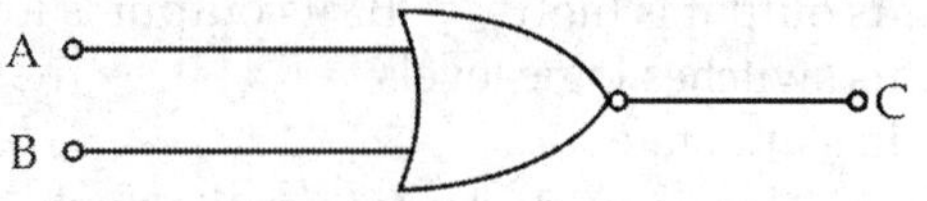

A. single-output AND gate
B. NAND gate
C. NAND gate used as NOT gate
D. NOR gate

60. The output from the logic gate shown below will be available when inputs ...... are present.

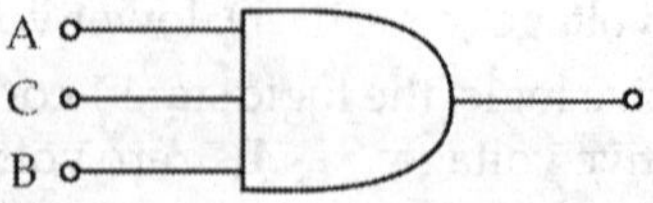

A. A and C B. B and C
C. A, B and C D. A and B

61. To get an output 1 from circuit shown below, the input must be ABC

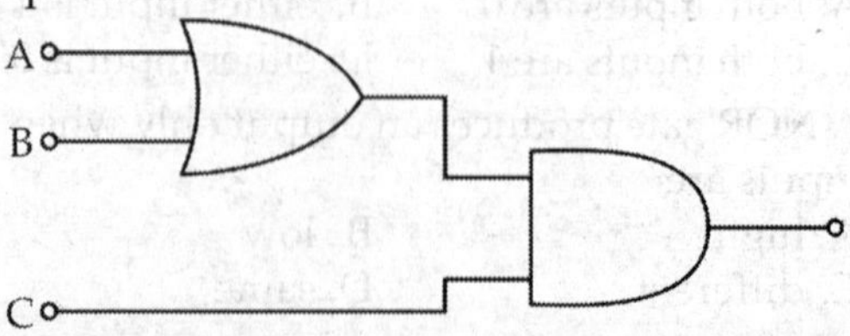

A. 010 B. 100
C. 101 D. 110

62. Which of the following logic gates shown below will have an output of 1?

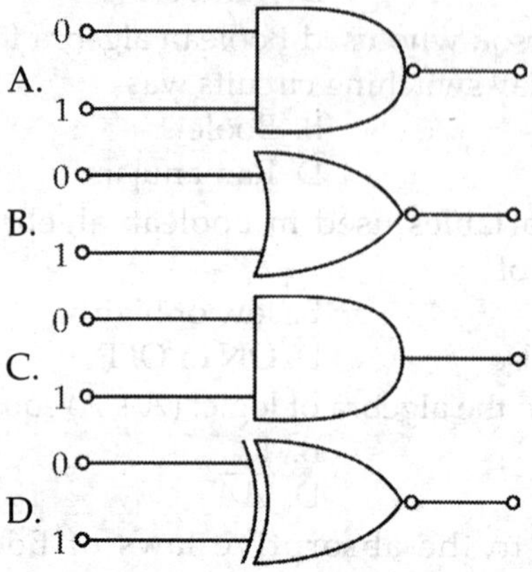

63. When an odd number is converted into the binary number, the least significant digit (LSD) is
A. 0 B. 1
C. 0 or 1 D. none of these

64. When the decimal number 16 is converted to the binary number, how many binary digits are needed?
A. 3 B. 4
C. 5 D. 6

65. When the decimal number 9 is converted to the binary number, the number of binary digits needed is
A. 3 B. 4
C. 5 D. 7

66. When the decimal number 9 is converted to the binary number, the number of binary digits needed is
A. 3 B. 4
C. 5 D. 6

67. How many binary numbers are created with 8 bits?
A. 256 B. 128
C. 64 D. 16

68. Number 84 in BCD is
A. 1000 0100 B. 0100 0100
C. 1000 1010 D. 1000 1100

69. Number 01000 1001 in BCD is the decimal number
A. 619 B. 419
C. 831 D. 413

70. What is the decimal equivalent of the binary number 110100?
A. 42 B. 52
C. 26 D. 62

71. The maximum count which a 6-bit binary word can represent is
A. 36 B. 64
C. 63 D. 65

Ans. 53. B 54. A 55. C 56. D 57. C 58. C 59. D 60. C 61. C 62. D 63. B 64. C 65. B 66. C 67. A 68. A 69. B 70. B 71. C

72. How many different numbers a 6-bit binary word can represent?
A. 63 B. 64
C. 124 D. 32

73. The highest decimal number that can be represented with 10 binary digit is
A. 1023 B. 1024
C. 512 D. none of these

74. How many fractional digits will 2^{-48} will have?
A. 24 B. 47
C. 49 D. 48

75. What is the binary equivalent of decimal 269?
A. 100001100 B. 100001010
C. 101001011 D. 100001101

76. What is the hexadecimal equivalent of binary number 1010 1111?
A. AF B. 9E
C. 8C D. none of these

77. What is the binary equivalent of the hexadecimal number B3?
A. 1011 0001 B. 1101 0011
C. 1011 0011 D. 1001 0101

78. What is the decimal equivalent of hexadecimal number 511?
A. FF1 B. 1FF
C. 3FF D. FF3

79. A computer sends a message to another computer using odd parity bit with ASCII code. The message 1100 1000 is received. The message
A. contains 1-bit error
B. does not contain 1-bit error
C. contains 3-bits error
D. may or may not contain 1-bit error

80. The octal equivalent of decimal 324.781 is
A. 504.771 B. 540.781
C. 215.234 D. 40.987

81. The Gray code for decimal 7 is
A. 0111 B. 1011
C. 0100 D. 0101

82. Which of the following code is known as the 8421 code?
A. Gray code B. Excess-3 code
C. ASCII code D. BCD code

83. The sum of weights in a self-complementing BCD code must be
A. 7 B. 9
C. 10 D. 8

84. The ASCII code is hexadecimal for number 5 is
A. 5 B. 25
C. 35 D. 65

85. The ASCII
A. is a subset of 8-bit EBCDIC
B. is used only in Western countries
C. is version II of the ASC standard
D. has 128 characters, including 32 control characters

86. Which number system is not a positional notation system?
A. Roman B. Binary
C. Decimal D. Hexadecimal

87. A 12 bit binary number has an accuracy equivalent to the decimal fraction?
A. 1/1024 B. 1/2048
C. 1/4096 D. 1/6400

88. $(r-1)$'s complement of an integer number is
A. the original number
B. the $(r-2)$'s complement of the number
C. the r's complement
D. none of these

89. The 10's complement of $(715)_8$ is
A. 63 B. 539
C. 285 D. 395

90. The 9's complement of 381 is
A. 372 B. 508
C. 618 D. 390

91. The result of the addition $37_{16} + 19_{16}$ is
A. 56_{16} B. 60_{16}
C. 59_{16} D. none of these

92. The result of the addition $A0_{16} + 6B_{16}$ is
A. $10\,B_{16}$ B. $16\,B_{16}$
C. $A0\,B_{16}$ D. $A6\,B_{16}$

93. The result of the subtraction FD_{16} and 88_{16} is
A. 75_{16} B. 65_{16}
C. $5E_{16}$ D. 10_{16}

94. The 1's complement of the binary number 1101101 is
A. 0000010 B. 0010010
C. 0010011 D. 1101110

95. The 2's complement of binary number 010111.1100 is
A. 101001.1100 B. 101000.0100
C. 010111.011 D. 101000.0011

96. The sign magnitude representation of binary number + 1101.011 is
A. 01101.011 B. 11101.011
C. 00010.100 D. 10010.100

97. The Excess-3 code is known as
A. cyclic redundancy code
B. weighted code
C. self complementing code
D. algebraic code

98. The sign magnitude representation of binary −1101.011 is

Ans.													
	72. B	73. A	74. D	75. D	76. A	77. C	78. B	79. B	80. A	81. C	82. C	83. B	84. C
	85. D	86. A	87. C	88. A	89. B	90. C	91. B	92. A	93. A	94. B	95. B	96. A	97. C

A. 01101.011 B. 10010.100
C. 11101.001 D. 10010.101

99. Consider n-bit one's representative of integer numbers. The range of integer values, N, that can be represented as
A. $-2^{n-1} < N < 2^{n-1}$
B. $-2^{n-1} < N < 2^{n-1} - 1$
C. $-2^{n-1} < N < 2^{n-1} + 1$
D. $-1 - 2^{n-1} < N < 2^{n-1}$

100. The number of two-input NAND gates required to produce the two-input OR function is
A. 3 B. 1
C. 2 D. 4

101. Which of the following is an unweighted code?
A. 8421 code B. Excess-3 code
C. 2421 code D. 5211 code

102. Cyclic codes are used in
A. data transfer
B. continuously varying signal representative
C. arithmatic and logical computation
D. none of these

103. ASCII and EBCDIC differ in
A. their efficiency in storing data
B. the random and sequential access method
C. the number of bytes used to store characters
D. their collecting sequences

104. The number of 1's present in the binary representation of $3 \times 512 + 7 \times 64 + 5 \times 8 + 3$ is
A. 7 B. 8
C. 9 D. 10

105. How many 1's are present in the binary representation of $15 \times 256 + 5 \times 16 + 3$?
A. 8 B. 9
C. 10 D. 11

106. What is the 2's complement representation of $-5/8$?
A. 0.1010 B. 1.0010
C. −0.1101 D. 1.0110

107. Number 85 in BCD code is
A. 1000–1100 B. 1101–1010
C. 1000–0101 D. 1101–1001

108. In order to get correct answer when two numbers are added in Excess-3 code and sum is less than 9, it is necessary to
A. subtract 0011 from the sum
B. add 0011 from the sum
C. subtract 0110 from the sum
D. add 0110 from the sum

109. The positive number 0.00001 is to be subtracted from 0.0011. The arithmatic unit of a computer would perform it by adding
A. 0.1111 to 0.0011 B. 0.1110 to 0.0011
C. 0.1100 to 0.0001 D. 0.1101 to 0.0001

110. The number 149 in octal code will be
A. 154 B. 178
C. 254 D. 225

111. The 2' complement of the number 0.0011 is
A. 1, 1100 B. 0, 1101
C. 1, 1101 D. 0, 1101

112. The quantity 837 in gray BCD code would be represented as
A. 1000 0011 0111 B. 1000 0010 0011
C. 1100 0100 0010 D. 1100 0010 0100

113. The quantity 837 in Excess-3 BCD code would be represented as
A. 1001 0011 0111 B. 1000 0001 1001
C. 1011 0110 1010 D. 1000 0011 0111

114. The advantage of self-correcting code is that it
A. is a weighted code B. has even parity
C. is easy to decode electronically
D. all of these

115. Which of the following codes is an unweighted code?
A. 8421 B. Excess-3 code
C. 2421 D. 63210

116. Decimal number 5 in level parity self-correcting code is
A. 00101 B. 01011
C. 01100 D. 10000

117. Which of the following codes is a weighted code?
A. Excess-3 code B. Shift-counter code
C. Gray code D. 5111 code

118. Decimal number 9 in 50 43210 code (binary code) is represented as
A. 00 01001 B. 10 10000
C. 10 01001 D. 11 00000

119. A six-bit alpha-numeric code is able to code
A. 36 characters B. 48 characters
C. 64 characters D. 128 characters

120. The number 34 is represented in Excess-3 BCD code is
A. 1.1001 1000 B. 0.1001 1001
C. 1.0110 0111 D. 0.0110 0111

121. The decimal fraction 0.375 in binary form is
A. 0.0011 B. 0.011
C. 0.0111 D. 0.111

122. The fraction 0.68_{10} is equal to
A. $(0.010101)_2$ B. $(0.101)_2$
C. $(0.10101)_2$ D. none of these

123. The decimal number 29 in binary form is
A. 11001 B. 11101
C. 10001 D. 11111

124. The binary fraction 0.0111 in decimal form is equal to
A. 0.4375 B. 0.6225
C. 0.8325 D. 0.1105

Ans. 98. C 99. A 100. A 101. B 102. B 103. A 104. C 105. A 106. D 107. C 108. A 109. B 110. D 111. C 112. C 113. D 114. C 115. B 116. B 117. D 118. A 119. C 120. C 121. B 122. C 123. B 124. A

125. The binary fraction 0.01001 is equal to
A. 0.225 B. 0.22125
C. 0.28125 D. 0.132105

126. The binary number 10.1100.110 in octal number is
A. 152.6 B. 154.6
C. 145.6 D. 174.6

127. Which of the following is *not* valid in case of binary product?
A. 0 × 0 B. 0 × 1 = 1
C. 1 × 1 = 1 D. all of these

128. 10 in BCD code is represented as
A. 10100 B. 11000
C. 010111 D. none of these

129. The ASC-II code is a
A. 5-bit code B. 7-bit code
C. 9-bit code D. 11-bit code

130. An 8-bit microprocessor uses 2's complement representation. How will the decimal number-41 appear?
A. 1101 10101 DOE B. 1101 1001 DEA
C. 1101 0000 DOH D. 1101 1111 DAH

131. The total number of inhibit required for a 16 K, 24-bit/word memory with a 3-D organisation is
A. 4 B. 8
C. 16 D. 24

132. In order to use NOR gate as an inverter
A. all its inputs are tied together
B. all its inputs are kept open
C. each input is used as an inverter input
D. None of these

133. Which of the following circuits exhibits memory?
A. Astable multivibrator
B. Bistable multivibrator
C. NAND gate D. Ex. OR gate

134. Which of the following is a sequential circuit?
A. AND gate B. NAND gate
C. Bistable multivibrator
D. Ex. OR gate

135. Which of the following statements is not true in regard to storage time of transistor?
A. The transistor is in active region
B. Both junction are forward-biased
C. The collector injects electrons or holes into the base region
D. It is the time taken by the carriers to leave the base region

136. The storage time of a transistor is part of the
A. turn-on time B. turn-off time
C. fall time D. rise time

137. If turn-on time of a transistor is decreased, its turn-on time
A. decreases. B. increases
C. is not affected D. infinite

138. The commutating capacitor reduces turn on time because
A. it acts as an open circuit
B. it allows maximum base current to flow
C. it aids in removing the excess carriers from the base region
D. None of these

139. An AND gate is a
A. sequential circuit B. combinational circuit
C. memory circuit D. relaxation circuit

140. The frequency of the square wave in Hz generated by an astable multivibrator is given by
A. 1/0.69 RC B. 1/1.38/ RC
C. 0.69/ RC D. 1.38/ RC

141. Which of the following statements is not true in regard to storage time of a diode?
A. During this time, the diode is reverse biased
B. The diode conducts during this time
C. It is due to the time taken by carriers to depart from the junction when, the diode is reverse-biased
D. During this time the diode current is zero

142. Storage time in transistors occurs when it is operating in
A. active region B. cut-off region
C. saturation region
D. either active or saturation region

143. Storage time of a transistor is the time taken for the collector current to fall to
A. 90 per cent from maximum value
B. 10 percent from maximum value
C. 10 percent to 90 percent from maximum value
D. 50 percent from maximum value

144. The commutation capacitor reduces turn-off time because
A. it acts an open circuit
B. it aids in removing the excess carriers from the base region
C. it decreases the fall time
D. All of these

145. The value of commutating capacitor is of the order of
A. 10 pF B. 50 pF
C. 0.5 μF D. 10 μF

146. Which of the following characteristics is not true for TTL logic?
A. Good speed.
B. Low power dissipation per gate
C. High cost D. None of these

147. Which of the following logic is the fastest?
A. TTL B. ECL
C. CMOS D. LSI

148. The ECL can be used to switch frequencies as high as

Ans. 125. C 126. B 127. B 128. D 129. B 130. C 131. D 132. A 133. B 134. C 135. A 135. B 137. B 138. B 139. B 140. B 141. D 142. C 143. A 144. B 145. B 146. C 147. B

A. 1 MHz B. 100 MHz
C. 500 MHz D. 1 GHz

149. Which of the following logics has the fan-out of more than 50?
A. TTL B. 5 nsec ECL
C. 8 nsec ECL D. CMOS

150. A transistor is operated as a non-saturated switch to eliminate
A. turn-on time B. turn-off time
C. storage time D. delay time

151. In Schottky-clamped TTL, the purpose of Schottky diode is
A. to increase propagation delay
B. to achieve efficient non-saturated switching
C. to improve noise margin
D. to decrease dissipation

152. The CMOS ICs can be operated at power supply of value
A. 5 V B. 5.2 V
C. 3-18 V D. 3-9 V

153. In medium-scale integration (MSI), a chip contains
A. less than 12 gates B. less than 100 gates
C. less than 1000 gates D. less than 5000 gates

154. In large-scale integration (LSI) a ship contains
A. More than 100 but less than 1000 gates
B. more than 500 but less than 1000 gates
C. more than 1000 gates D. more than 5000 gates

155. In grand-scale integration (GSI) a chip contains more than
A. 1500 gates B. 1000 gates
C. 5000 gates D. 10000 gates

156. Which of the following parameters is not specified for digital ICs?
A. Gate dissipation B. Propagation delay
C. Noise margin D. Band-width

157. The unit of speed-power product is
A. pico-watts/sec B. pico-watts
C. MHz D. pico-joules

158. Generally, the unit of propagation delay is
A. sec B. m-sec
C. p-sec D. n-sec

159. Noise margin of a digital IC is
A. the manimum extraneous voltage that causes a gate to change its state
B. the maximum extraneous voltage that causes a gate to change its state
C. the thermal noise voltage which causes a gate 10 change its state
D. None of these

160. Which of the following logics has excellent noise margin?
A. TTL B. ECL
C. CMOS D. All of these

161. Which of the following logics exhibit lowest power dissipation per gate?
A. TTL B. 4 *n*-sec-ECL
C. 8*n*-sec-ECL D. CMOS

162. Which of the following logics exhibit relatively highest power dissipation per gate?
A. TTL B. ECL
C. CMOS D. LSI

163. Current-mode logic (CML) is the same as
A. TTL B. ECL
C. CMOS D. LSL

164. The result of binary product $(101.1)_2$ multiply $(11.01)_2$ is given by
A. 1010.101 B. 1100.0011
C. 10001.001 D. 10001.111

165. A + B = Y is the Boolean expression for
A. OR gate B. NOR gate
C. AND gate D. NAND gate

166. The truth table given below is for

Input		Output
A	*B*	*Y*
0	0	0
0	1	0
1	0	0
1	1	1

A. AND gate B. NAND gate
C. OR gate D. NOR gate

167. In the figure below, *A*, *B* and *C* are inputs, and the output appears at *D*. This circuit is called a negative AND gate. The appropriate logic function is

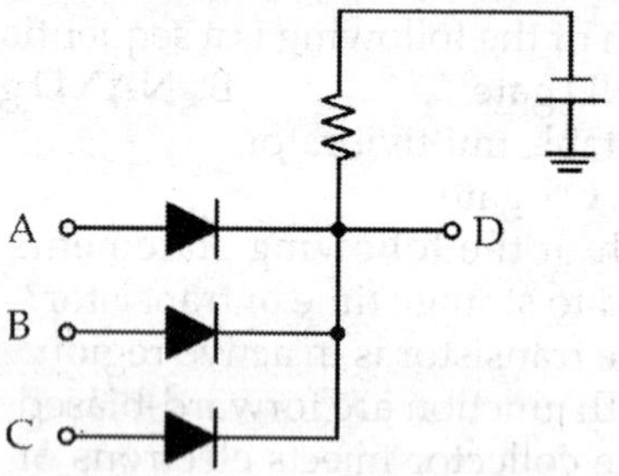

A. $A + B + C = D$ B. $AB + BC = D$
C. $ABC = D$ D. $AB = CD$

168. Which of the following statement is correct?
A. Decimal 9 is represented as 1011 in Excess-3 code
B. Decimal 9 is represented as 1001 in BCD code
C. Decimal 10 is represented as 1100 in Gray code
D. Decimal 10 is represented as 1001 in binary code

169. Identify *incorrect* relation if any
A. $A \times B = \overline{AB} + \overline{BA}$ B. $A + AB = A + B$

Ans. 148. B 149. D 150. C 151. B 152. C 153. B 154. A 155. B 156. D 157. D 158. D 159. B 160. C 161. D 162. B 163. B 164. D 165. B 166. A 167. C 168. B

C. $(A + B)(A + C) = A(A + B)$
D. $A + AB = A$

170. The product of binary numbers, $(-11101)_2 \times (100.1)_2$ is given by
A. –11110000.1 B. –10101010.10
C. –100000.01 D. –10000010.1

171. Which of the following is the minimum error code?
A. Octal codes B. Binary code
C. Gray code D. Excess-3 code

172. Any logic gate can be built using
A. NOR gates B. OR gates
C. AND gates D. All of these

173. According to Boolean algebra, $(1 + A + B + C)$ is equal to
A. $A + B + C$ B. ABC
C. $1 + ABC$ D. 1

174. A device with slowest switching speed is
A. LED B. LCD
C. Nixie tubes D. None of these.

175. Access in magnetic drum memory is
A. Completely random B. Sequential and cyclic
C. partly random and partly cyclic sequential
D. a cycle sequential

176. A decade counter requires
A. 2 flip flops B. 3 flip flops
C. 4flipflops D. 100 flip flops

177. Which of the following binary addition is incorrect?
A. 1001 + 1101 = 10110
B. 10101 + 10011 = 101000
C. 11111 + 11111 = 100000
D. 11111 + 10001 = 110000

178. Which code is suitable for input-output devices and analog-to-digital converters?
A. 8421 code B. Excess-3 code
C. Four bit BCD code D. Gray code

179. The switching circuit given in the figure can be expressed in binary logic notation as

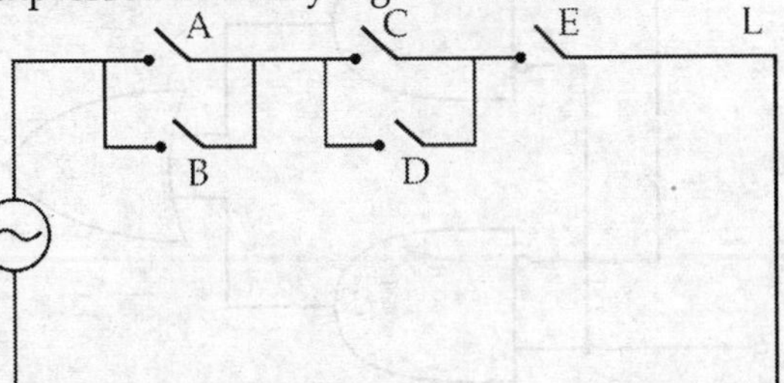

A. $L = (A + B)(C + D)E$
B. $L = AB + CD + E$
C. $L = E + (A + B)(C + D)$
D. $L = (AB + CD)E$

180. In Schottky TTL, a Schottky diode is used for
A. forming the gate
B. connecting the resistor
C. clamping of the base collector junction
D. none of these

181. Inter 8080 has
A. 16 pins B. 24 pins
C. 32 pins D. 40 pins

182. Which of the following hexagonal sum is equivalent to hexadecimal A8H?
A. 2 CH + 4 FH B. 5 FH + 1 AH
C. 3 BH + 6 DH D. 4 AH + 2 CH

183. If $\bar{x} \times \bar{y} = 0$, then which one of the following is true?
A. $\bar{x}y + \bar{y}x + xz = x\bar{y} + yz$
B. $\overline{xyz} + xyz = xy\bar{z} + \overline{xyz}$
C. $\bar{x}y + \bar{y}x = xy + \overline{xy}$
D. $\bar{x}yx = 1$

184. There are four Boolean variables x_1, x_2, x_3 and x_4. The following functions are defined on sets of them:
$f(x_3, x_2, x_1) = \Sigma(3, 4, 5)$
$g(x_4, x_3, x_2) = \Sigma(1, 6, 7)$
$h(x_4, x_3, x_2, x_1) = fg$
Then $h(x_4, x_3, x_2, x_1)$ is
A. zero B. $\Sigma(3, 12, 13)$
C. $\Sigma(3, 4, 5, 1, 6, 7)$ D. $\Sigma(3, 12, 15)$

185. In computer terminology 1 M byte memory means
A. 1000000 bytes B. 1000024 bytes
C. 1024000 bytes D. 1048576 bytes

186. Which of the following statement is true?
A. ICs are always linear
B. Digital circuits are linear circuits
C. AND gate is a logic circuit whose out-put is equal to its highest input
D. In a four-input AND circuit, all input must be UP for the output to be UP

187. The logic circuit given in the figure represents a

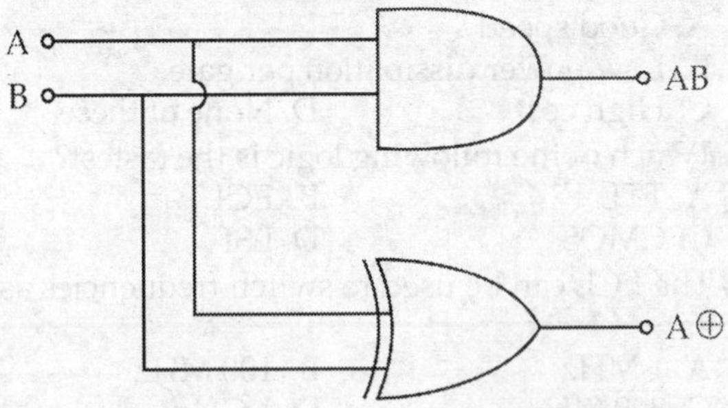

A. full adder B. half adder
C. half subtractor D. Boolean multiplier

188. According to De Morgan's second theorem a NAND gate is
A. always complementary to an AND gate
B. equivalent to a bubbled NAND gate
C. equivalent to bubbled AND gate
D. equivalent to bubbled OR gate

Ans.	169. C	170. D	171. C	172. A	173. D	174. B	175. C	176. C	177. C	178. D	179. A	180. C	181. A
	182. C	183. A	184. B	185. D	186. B	187. B	188. D						

189. Positive logic in a logic circuit is one in which
A. logic 0 and 1 are represented by 0 and positive voltage respectively
B. logic 0 and 1 are represented by nega-tive and positive voltage respectively
C. logic 0 and 1 voltage level is higher than logic 1 voltage level
D. logic 0 voltage level is lower than logic 1 voltage level

190. 7483 is a
A. TTL binary adder B. TTL clock
C. 8-bit adder-subtractor
D. none of these

191. PROM is
A. permanent read only memory
B. polarised read only memory
C. positive read only memory
D. programmable read only memory

192. In a 5 × 7 dot matrix format
A. 64 bits are required to store 64 alpha-numeric characters
B. 560 bits are required to store 64 alpha-numeric characters
C. 1120 bits are required to store 64 alphanumeric characters
D. 2240 bits are required to store 64 alphanumeric characters

193. $\overline{A+B} = Y$ is the Boolean expression for
A. OR gate B. NOR gate
C. AND gate D. NAND gate

194. According to Boolean algebra, (1 + A + B + C) is equal to
A. $A + B + C$ B. ABC
C. $1 + ABC$ D. 1

195. Which of the following characteristics is *not* true for TTL logic?
A. Good speed
B. Low power dissipation per gate
C. High cost D. None of these

196. Which of the following logic is the fastest?
A. TTL B. ECL
C. CMOS D. LSI

197. The ECL can be used to switch frequencies as high as
A. 1 MHz B. 100 MHz
C. 500 MHz D. 1 GHz

198. Which of the following logics has the fan-out of more than 50?
A. TTL B. 5 nsec ECL
C. 8-nsec ECL D. CMOS

199. A transistor is operated as a non-saturated switch to eliminate
A. turn-on time B. turn-off time
C. storage time D. delay time

200. Hamming code is
A. an error-indication code system
B. an error-correction code system
C. a programme conversion code system
D. none of these

201. Consider two normally open contacts in series for a normally closed relay. The logic operation performed by this circuit is
A. AND B. NAND
C. NOR D. none of these

202. In INHIBIT operation output
A. is 1 when both inputs are 0
B. is 0 when blocking input is 1
C. is 0 when blocking input is 0
D. is 1 when blocking input is either 0 or 1

203. In a digital control system, there are three inputs *a*, *b*, *c*. The output should be present when at least two inputs are present. The Boolean expression for the output is
A. $ab + bc + ac$ B. $abc + ab\bar{c} + \bar{a}bc + a\bar{b}c$
C. $ab\bar{c} + a\bar{b}c + \bar{a}bc$ D. $a\bar{b} + b\bar{c} + \bar{a}c$

204. A NAND circuit with positive logic will operate as
A. NOR with negative logic
B. AND with negative logic
C. OR with negative logic input
D. AND with negative logic output

205. Two binary signals *a*, *b* are to be compared. The output expression when the two signals are equal is given by
A. $a\bar{b} + a\bar{b}$ B. $ab + \overline{ab}$
C. $\overline{ab}$ D. ab

206. NAND operation is
A. $\overline{X + Y}$ B. $\bar{X} + \bar{Y}$
C. $\bar{X} \cdot \bar{Y}$ D. $(\bar{X} + \bar{Y})(X + Y)$

207. The circuit below represents an

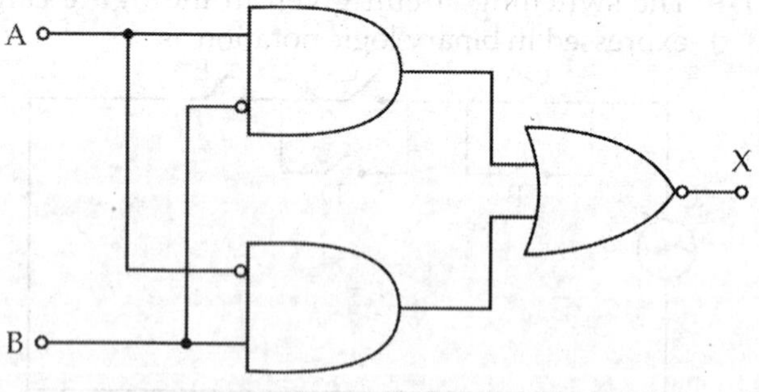

A. exclusive OR
B. comparator (outputs when both inputs are equal)
C. inhibit gate D. NAND gate

208. NOR operation is
A. $X + Y$ B. XY
C. $\bar{X} \cdot \bar{Y}$ D. $(\bar{X} + \bar{Y})(X + Y)$

Ans. 189. D 190. A 191. A 192. D 193. B 194. D 195. C 196. B 197. C 198. D 199. C 200. B 201. C 202. B 203. A 204. A 205. B 206. D 207. B 208. C

209. A lamp is controlled from two positions A and B (e.g. stair case circuit). The Boolean expression for this circuit is

A. $A\bar{B} + \bar{A}B$ B. $AB + A\bar{B}$
C. $AB + \bar{A}\bar{B}$ D. $A\bar{B} + \bar{A}\bar{B}$

210. The Boolean expression for the above circuit is

A. $X = AB + \overline{AB}$ B. $X = AB + A\bar{B}$
C. $X = (A + \bar{B})(A + \bar{B})$ D. $X = AB + \bar{A}\bar{B}$

211. The most widely used universal gates are

A. OR and AND gates
B. NOR and NAND gates
C. NOR and AND gates D. NAND and OR gates

212. A three-input NAND gate is to be used as an inverter. Which of the following measures will achieve this end better?

A. All inputs are connected together, making one input
B. The two inputs not used are connected to the logic "1" level
C. The two inputs not used are connected to ground ("0" level)
D. None of these

213. Which of the following logics has excellent noise margin?

A. TTL B. ECL
C. CMOS D. All of these

214. Which of the following logics exhibit lowest power dissipation per gate?

A. TTL B. 4n-sec-ECL
C. 8n-sec-ECL D. CMOs

215. Which of the following logics exhibit relatively highest power dissipation per gate?

A. TTL B. ECL
C. CMOS D. LSI

216. Current-mode logic CML is the same as

A. TTL B. ECL
C. CMOS D. LSL

217. Conversion of 13.8125 in digital form is

A. 1101.1011 B. 1011.1111
C. 1101.1101 D. 1011.1111

218. Which of the following ICs has only one NAND gate?

A. 7410 B. 7420
C. 7430 D. 7447

219. The number of 7400s required to make a two-point Ex OR gate is

A. 1 B. 2
C. 3 D. 4

220. Which of the following ICs allows wired AND connections?

A. 7410 B. 7990
C. 7447 D. 7440

221. Fan-out for the 74 series is

A. 4 B. 5
C. 8 D. 10

222. The circuit shown below is a

A. bistable multivibrator
B. astable multivibrator
C. monostable multivibrator
D. time-delay circuit

223. What is the maximum number (*in decimal notation*) that can be counted with above counter?

A. 999 B. 4096
C. 4095 D. 9999

224. In a full adder, there are

A. two binary number inputs and two outputs
B. three binary digit inputs and two binary outputs
C. three binary digit inputs and three binary digit outputs
D. none of these

225. Which binary adder is used as BCD adder, the sum is correct when it is

A. less than 9 B. greater than 9
C. less than 16 D. none of these

226. The gates required to build a half adder are

A. Ex OR gate and NOR gate
B. Ex OR gate and OR gate
C. Ex OR gate and AND gate
D. four NAND gates

227. Correction in decimal adder is required because

A. sums are represented wrongly in binary code
B. sums are represented wrongly in BCD code
C. the numbers to be added are not represented in BCD code
D. none of these

228. The voltage levels for a six-bit binary ladder are '0' = 0 V, '1' = +10 V. Its output for the input 101001 will be

A. 0.423 V B. 0.552 V
C. 0.641 V D. 0.923 V

229. Which of the following Boolean statements represent commutative law?

A. $(A + B) + CA + (B + C)$
B. $A - (B + C)(A - B) + (A - C)$
C. $A + BB + A$
D. $A + AA$

230. Which of the following is not functionally a complete set?

A. AND, OR B. NOR
C. NAND D. AND, OR, NOT

231. For what logic gate the output is complement of the input?

A. NOT B. AND
C. OR D. XOR

Ans.													
	209. A	210. D	211. B	212. B	213. C	214. D	215. B	216. B	217. C	218. C	219. D	220. D	221. D
	222. C	223. C	224. A	225. A	226. C	227. B	228. C	229. C	230. A	231. A			

232. In which of the following gates, the output is high if and only if all inputs are high?
A. NOT B. AND
C. OR D. XOR

233. In which of the following gates the output is 1 if and only if at least one input is 1?
A. NOT B. AND
C. OR D. NAND

234. In which of the following gates the output is 1 if and only if at least one input is 0?
A. NOT B. AND
C. OR D. NAND

235. In which of the following gates the output is 0 if and only if at least one input is 1?
A. NOT B. AND
C. NOR D. NAND

236. Let *X* and *Y* be the inputs and *Z* be the output of a NAND gate. The value of *Z* is given by
A. $X + Y$ B. X/Y
C. $X \cdot Y$ D. XY

237. The NAND can function as a NOT gate if
A. inputs are connected together
B. inputs are left open
C. one input is set to 0 D. one input is set to 1

238. What is the minimum number of two-input NAND gates used to perform the function of two-input OR gate?
A. One B. Two
C. Three D. Four

239. Which of the following gates are added to the inputs of the OR gate to convert it to the NAND gate?
A. NOT B. AND
C. OR D. XOR

240. What logic function is produced by adding inverters to the input of an AND gate?
A. NAND B. NOR
C. XOR D. OR

241. What logic function is produced by adding an inverter to the output of an AND gate?
A. NAND B. NOR
C. XOR D. OR

242. What logic function is produced by adding an inverter to each input and the output of an AND gate?
A. NAND B. NOR
C. OR D. XOR

243. Which of the following gates is known as *coincidence detector*?
A. AND B. OR
C. NOT D. NAND

244. An AND gate can be imagined as
A. switches connected in series
B. switches connected in parallel
C. MOS transistors connected in series
D. none of these

245. An OR gate can be imagined as
A. switches connected in series
B. switches connected in parallel
C. MOS transistors connected in series
D. none of these

246. Which of the following gates would be output 1 when one input is 1 and other input is 0?
A. OR gate B. AND gate
C. NAND gate
D. Both A and C are correct

247. Which of the following statements is correct?
A. The output of a NOR gate is high if all of its inputs are high
B. The output of a NOR gate is low if all of its inputs are low
C. The output of a NOR gate is high if all of its inputs are low
D. The output of a NOR gate is high if only one of its input is low

248. What will be the function performed if NOT gate is put at the end of OR gate?
A. NOR B. XOR
C. EQUIVALENCE D. OR

249. Which gate is known as universal gate?
A. NOT B. AND
C. NAND D. XOR

250. How many NAND gates are needed to perform the logic function $X \cdot Y$?
A. 1 B. 2
C. 4 D. 3

251. Which of the following is functionally complete set?
A. AND, OR B. AND, XOR
C. NOT, OR D. AND, OR, NOT

252. Which of the following represent Commutative law, Associative law, and Distributive law?
I. $A \cdot (B \cdot C) = (A \cdot B) \cdot C$
II. $A \cdot (B + C) = A \cdot B + A \cdot C$
III. $A + B = B + A$
A. I, III and II respectively
B. II, I and III respectively
C. III, II and I respectively
D. III, I and II respectively

253. Which of the following Boolean algebra expression is incorrect?
A. $A + 0 = A$ B. $A \cdot 1 = 1$
C. $A + \bar{A} = 1$ D. $A \cdot \bar{A} = 0$

Ans.	232. B	233. C	234. D	235. C	236. C	237. A	238. C	239. A	240. B	241. A	242. C	243. A	244. A
	245. B	246. D	247. C	248. A	249. C	250. B	251. A	252. D	253. B				

254. Which of the following Boolean algebra expression is incorrect?
A. $A + \overline{A} = A + B$ B. $A + AB = B$
C. $(A + B)(A + C) = A + BC$
D. $(A + \overline{B})(A + B) = A$

255. Which of the following Boolean algebra expression is incorrect?
A. $\overline{ABC} + BC + AC = C$
B. $(A + B)[\overline{A}(\overline{B} + \overline{C})] + \overline{BC} + \overline{AC} = 1$
C. $AB + \overline{A}C + BC = AB = \overline{A}C$
D. $AB + A\overline{C} = B$

256. Which of the following Boolean algebra expression is incorrect?
A. $AB + A(B + C) + B(B + C) = B + AC$
B. $[\overline{AB}(C + \overline{BD}) + \overline{A}B]C = B\overline{C}$
C. $\overline{AB(C + D)} + \overline{A} + B + \overline{C}D$
D. $(A + C)(ABC + ACD) = ABC + ACD$

257. How many lines the truth table for a four-input NOR gate would contain to cover all possible input combinations?
A. 4 B. 8
C. 12 D. 16

258. The number of two-input NAND gates required to produce the two-input OR function is
A. 1 B. 2
C. 3 D. 4

259. What logic function is obtained by adding an inverter to the output of an AND gate?
A. OR B. NAND
C. XOR D. NOR

260. What logic function is obtained by adding an inverter to the inputs of an AND gate?
A. OR B. NAND
C. XOR D. NOR

261. The clock signals are used in sequential logic circuits to
A. tell the time of the day
B. tell how much time has lapsed since the system was turned on
C. carry serial data signals
D. synchronize events in various parts of a system

262. How many bits are required to encode all twenty six letters, ten symbols, and ten numbers?
A. 5 B. 6
C. 7 D. 46

263. What table shows the logical state of a digital circuit output for every possible combination of logical states in the inputs?
A. Function table B. Truth table
C. Rounting table D. ASCII table

264. What table shows the electrical state of a digital circuit's output for every possible combination of electrical states in the inputs?
A. Function table B. Truth table
C. Rounting table D. ASCII table

265. How many truth tables can be made from one function table?
A. One B. Two
C. Three D. Any number

266. A positive AND gate is also a negative
A. NAND gate B. NOR gate
C. AND gate D. OR gate

267. The term sum-of-products in Boolean algebra means
A. The AND function of several OR functions
B. The OR function of several AND functions
C. The OR function of several OR functions
D. The AND function of several AND functions

268. The term product-of-sums in Boolean algebra means
A. The AND function of several OR functions
B. The OR function of several AND functions
C. The OR function of several OR functions
D. The AND function of several AND functions

269. A half-adder can be constructed from
A. two XNOR gates only
B. one XOR and one OR gate with their outputs connected in parallel
C. one XOR and one OR gate with their inputs connected in parallel
D. one XOR gate and one AND gate

270. The Boolean expression $F = xy + \overline{x}z$ can be expressed as a sum of minerms using
A. $F(x, y, z) = \Sigma\, 1, 3, 6, 7$
B. $F(x, y, z) = \Sigma\, 0, 2, 4, 5$
C. $F(x, y, z) = \Sigma\, 0, 3, 4, 6$ D. $F(x, y, z) = \Sigma\, 1, 2, 4, 5$

271. With x, y, z as input variables and F as output variable, the following K-amp represents

yz

x		1		1
	1		1	

A. the NOR of x, y, z
B. the NAND of x, y, z
C. the Exclusive-OR of x, y, z
D. the Exclusive-NOR of x, y, z

272. The content of 4-bit register is 1101. The register is shifted 6 times to the right with serial input being 101101. The final content of register will be
A. 1011 B. 0111
C. 0010 D. 10101

Ans. 254. B 255. D 256. B 257. D 258. C 259. B 260. D 261. D 262. B 263. B 264. A 265. B 266. D 267. B 268. A 269. C 270. A 271. D 272. B

273. The $Q(t)$ and $Q(t + 1)$ represent present and next states in a flip-flop as shown in the Table below. A and B are the inputs. The flip-flop is

$Q(t)$	$Q(t+1)$	A	B
0	0	0	x
0	1	1	x
1	0	x	1
1	1	x	0

A. as RSFF B. a JKFF
C. a DFF D. a TFF

274. The excess three code of the 5-bit Gray code number 11001 is
A. 10100 B. 00011
C. 11000 D. 10101

275. Decodes and multiplexes are examples of
A. combination circuits
B. combination and sequential circuits respectively
C. sequential circuits
D. sequential and combinational circuits respectively

276. In the following figure below, F represents

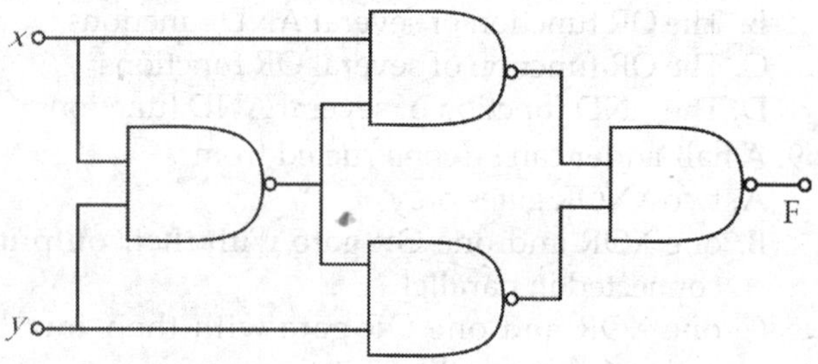

A. NAND of x, y B. NOR of x, y
C. Half-adder sum of x, y
D. Half-adder carry of x, y

277. Circuit hazards are present in
A. combinational circuits only
B. sequential circuits only
C. combinational and sequential circuits
D. none of these

278. Digital systems are the systems which
A. use digital information in the external world
B. use analog signals
C. manipulate information
D. none of these, handle information in digital form internally

279. Analog systems are different from digital systems because
A. they are transistors
B. they handle information in analog form
C. they handle information in digital form
D. none of these

280. Why are digital circuits easier to design than analog circuits?
A. Digital circuits do not control electricity precisely over a wide range
B. Digital circuits are made in the forms of ICs
C. All elements of digital circuits are from the same family
D. none of these

281. Digital gate can respond to an input signal in
A. about a second
B. about a hundredth of a second
C. a few billionth of a second
D. none of these

282. Analog methods are not used for handling extremely precise information because
A. they are very expensive
B. there are limits to how closely an analog signal can reproduce the information
C. analog information need never to be precise
D. none of these

283. Which of the following is the first integrated logic family?
A. RTL B. DTL
C. TTL D. MOS

284. Which of the following logic families has the highest noise immunity?
A. RTL B. DTL
C. TTL D. HTL

285. The worst-case noise margin in Trausis for logic circuit is approximately
A. 1 mV B. 400 mV
C. 100 mV D. 1 V

286. The worst-case noise margin occurs with
A. small fan-out B. full fan-out
C. no fan-out D. none of these

287. Which logic family is the fastest?
A. DTL B. TTL
C. ECL D. CMOS

288. Which logic family dissipates the minimum power?
A. DTL B. TTL
C. ECL D. CMOS

289. Which of the following logic has the maximum fan-out?
A. RTL B. ECL
C. NMOS D. CMOS

290. Which of the following logic families has the lowest power-delay product?
A. ECL B. TTL
C. IIL D. MOS

291. Which TTL sub-family has maximum speed?
A. Standard TTL
B. Schottky-clamped TTL
C. High speed TTL D. Low power TTL

292. What is the transition voltage for the voltage input of a CMOS operating from 10 V supply?

Ans. 273. B 274. A 275. A 276. A 277. C 278. D 279. B 280. A 281. C 282. B 283. A 284. D 285. B
286. B 287. C 288. D 289. D 290. C 291. B

A. 1 V B. 2 V
C. 5 V D. 12 V

293. The allowable supply range for LS TTL is
A. 2 to 5.6 V B. 3.5 to 6.2 V
C. 4.75 to 5.25 V D. 4 to 6 V

294. The 0 V reading on the output of an LS TTL gate indicates that the output circuit is
A. defective
B. short circuited to ground
C. either A or B is correct
D. none of these

295. The switching speed of ECL is very high because
A. it uses positive logic
B. it uses negative logic
C. it uses high speed transistors
D. its transistors remain unsaturated

296. What should be added so that an LS TTL gate can drive a CMOS gate operating from the supply?
A. Pull up transistor B. Pull-down transistor
C. A small transformer
D. A capacitor of high value

297. What should be added so that a CMOS gate drives TTL at the same supply voltage?
A. Pull up transistor B. Pull down transistor
C. A high value capacitor
D. A buffer

298. MOS logic gates have no current hogging problem because the gate terminal has
A. low input impedance
B. zero impedance
C. high impedance D. compensating effect

299. The time required for a gate or inverter to change its state is defined as
A. rise time B. decay time
C. propagation time D. charging time

300. The time required for a pulse to change from 10 to 90 percent of its maximum value is defined as
A. rise time B. decay time
C. propagation time D. operating speed

301. The time required for a pulse to decrease from 90 to 10 percent of its maximum value is known as
A. rise time B. decay time
C. binary level transition period
D. propagation delay

302. Which of the following statements is correct?
A. Propagation delay is the time required for a gate to change its state
B. The noise immunity is the amount of noise which can be applied to the input of a gate without causing the gate to change state
C. Fan-in of a gate is always equal to fan-out of the same gate
D. Operating speed is the maximum frequency at which digital data can be applied to a gate

303. Which of the following logic families has the maximum functional capacity?
A. SSI B. MSI
C. LSI D. VLSI

304. The functional capacity for SSI devices is
A. 1 to 11 gates B. 12 to 99 gates
C. 100 to 10000 gates D. more than 1000 gates

305. The functional capacity of LSI devices is
A. 1 to 11 gates B. 12 to 99 gates
C. 100 to 1000 gates D. more than 10000 gates

306. The functional capacity for VLSI devices is
A. 1 to 11 gates B. 12 to 99 gates
C. 100 to 1000 gates D. more than 10000 gates

307. The output 0 and 1 levels for ECL logic family is approximately
A. 0.1 and 5 V B. 0.9 and 1.75 V
C. 0.6 and 3.5 V D. 1.75 and −0.9 V

308. The output 0 and 1 level for TTL logic family is approximately
A. 0.1 and 5 V B. 0.6 and 3.5 V
C. 0.9 and 1.75 V D. −1.75 and −0.9 V

309. The output 0 and 1 level for CMOS logic family is approximately
A. 0.2 and 5 V B. 0.6 and 3.50 V
C. 30 percent and 70 percent of V_{cc}
D. −1.75 and −0.9 V

310. The propagation delay for standard TTL devices is approximately
A. 1 ns B. 10 ns
C. 25 ns D. 15 ns

311. Which of the following TTL sub-family is fastest?
A. Standard TTL B. High-speed TTL
C. Schottky TTL D. Low-speed TTL

312. The propagation delay for CMOS IC family is approximately
A. 10 ns B. 25 ns
C. 2 ns D. 50 ns

313. The propagation delay for ECL IC family is approximately
A. 2 ns B. 10 ns
C. 25 ns D. 50 ns

314. Which of the following logic family consumes the least amount of power?
A. I^2L B. ECL
C. TTL D. CMOS

315. Which of the following logic family is the fastest?
A. I^2L B. ECL
C. TTL D. CMOS

316. The digital IC which does not use bipolar transistor technology is
A. I^2L B. TTL
C. ECL D. MOSFET

Ans. 292. C 293. C 294. C 295. D 296. A 297. B 298. C 299. C 300. A 301. B 302. C 303. D 304. A 305. C 306. D 307. B 308. B 309. C 310. B 311. C 312. B 313. A 314. D 315. B 316. D

317. Which of the following logic families use bipolar transistor?
A. TTL B. CMOS
C. NMOS D. None of these

318. Which of the following input combinations is not allowed in an SR flip-flop?
A. S = 0, R = 0 B. S = 0, R = 1
C. S = 1, R = 0 D. S = 1, R = 1

319. When an inverter is placed between both inputs of an SR flip-flop, the resulting flip-flop is
A. JK flip-flop B. D flip-flop
C. T flip-flop
D. Master slave JK flip-flop

320. The functional difference between SR flip-flop and JK flip-flop is that
A. JK flip-flop is faster than SR flip-flop
B. JK flip-flop has a feedback path
C. JK flip-flop accepts both inputs 1
D. JK flip-flop does not require external clock

321. The master slave JK flip-flop is effectively a combination of
A. an SR flip-flop and a T flip-flop
B. an SR flip-flop and a D flip-flop
C. a T flip-flop and a D flip-flop
D. two T flip-flops

322. In a T flip-flop, the ratio of the frequency of an output pulse wave to the input wave pulse is
A. 1/2 B. 1
C. 2 D. 4

323. How many flip-flops are needed to divide the input frequency by 64?
A. 4 B. 5
C. 6 D. 8

324. For an input pulse train of clock period T, the delay produced by an n stage shift register is
A. $(n-1)T$ B. nT
C. $(n+1)T$ D. $2nT$

325. Which of the following flip-flops is free from race around problem?
A. T flip-flop B. SR slip-flop
C. Master slave JK flip-flop
D. All of these

326. Which of the following conditions must be met to avoid race around problem when t is propagation delay, t_p is pulse width and T is period of a pulse train
A. $t < t_p < T$ B. $2t_p < \Delta t < T$
C. $T > \Delta t > t_p$ D. $T > t_p > \Delta t$

327. An n stage ripple counter can count up to
A. 2^n B. $2^n - 1$
C. n D. 2^{n-1}

328. The number of flip-flops required in a decade counter is
A. 3 B. 4
C. 5 D. 10

329. The number of flip-flops required in a module N counter is
A. $\log_2(N) + 1$ B. $[\log_2(N+1)]$
C. $[\log_2(N)]$ D. $\log_2(N-1)$

330. What is the maximum counting speed of 4-bit binary counter which is composed of flip-flops with a propagation delay of 25 ns?
A. 1 MHz B. 10 MHz
C. 100 MHz D. 4 MHz

331. A shift register can be used for
A. parallel to serial conversion
B. serial to parallel conversion
C. digital delay line D. all of these

332. The dynamic race hazard problem occurs in
A. combinational circuits only
B. sequential circuits only
C. both combinational and sequential circuits
D. none of the combinational and sequential circuits

333. Which of the following devices selects one of several inputs and transmits it to a single output?
A. Decoder B. Multiplexer
C. Demultiplexer D. Counter

334. A demultiplexer is used to
A. perform serial to parallel data conversion
B. select data from several inputs and route it to a single output
C. route the data from a single input to one of the many outputs
D. perform parity checking

335. A multiplexer is also known as
A. coder B. decoder
C. data selector D. multivibrator

336. Which of the following logic circuits takes data from a single source and distribute it to one of the several output lines?
A. Multiplexer B. Data selector
C. Demultiplexer D. Parallel counter

337. A demultiplexer is also known as
A. data selector B. data distributor
C. multiplexer D. encoder

338. 74LS138 chip functions as
A. decoder/demultiplexer
B. encoder
C. multiplexer D. memory element

339. How many full adders are required to construct an m-bit parallel adder?
A. $m/2$ B. $m-1$
C. m D. $m+1$

340. The output of a sequential circuit depends on
A. present inputs only
B. past inputs only

Ans. 317. A 318. D 319. B 320. C 321. A 322. C 323. C 324. A 325. C 326. C 327. B 328. B 329. C
330. B 331. D 332. C 333. B 334. C 335. C 336. C 337. B 338. A 339. C

C. both present and past inputs
D. present outputs only

341. The logic performed by the circuit shown below is

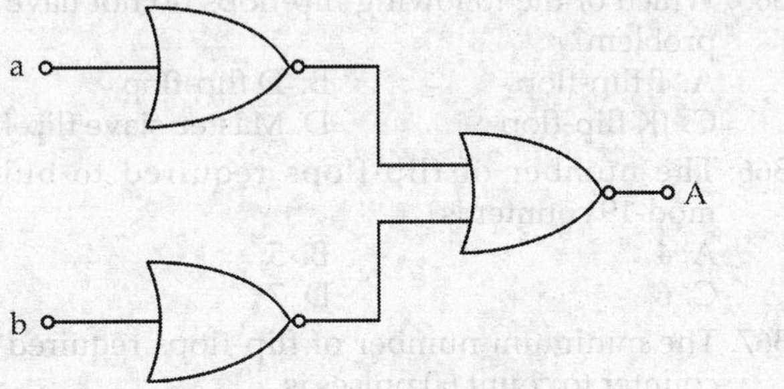

A. NAND B. AND
C. Ex OR D. none of these

342. Which of the following logic circuits accepts two binary digits on its inputs and produces two binary digits, a sum bit and a carry bit, on its outputs?
A. Full adder B. Half adder
C. Decoder D. Multiplexer

343. Consider an RS flip-flops with both inputs set to 0. If a monetary '1' is applied at the input *S*, then the output
A. *Q* will flip from 0 to 1
B. *Q* will flip from 0 and 1 and then back to 0
C. *Q* will flip from 1 to 0
D. *Q* will flip from 0 to 1 and then back to 1

344. Parallel adders are
A. combinational logic circuits
B. sequential logic circuits
C. both of these D. none of these

345. In which of the following adder circuits, the carry look ripple delay is eliminated?
A. Half adder B. Full adder
C. Parallel adder
D. Carry-look-ahead adder

346. Which of the following adders can add three or more numbers at a time?
A. Parallel adder
B. Carry-look-ahead adder
C. Carry-save-adder D. Full adder

347. A combinational logic circuit which generates a particular binary word or number is
A. decoder B. multiplexer
C. encoder D. demultiplexer

348. A logic circuit which is used to change a BCD number into an equivalent decimal number is
A. decoder B. encoder
C. multiplexer D. code converter

349. A combinational logic circuit which is used to send data coming from a single source to two or more separate destinations is called
A. decoder B. encoder
C. multiplexer D. demultiplexer

350. A combinational logic circuit which is used when it is desired to send data from two or more sources through a single transmission line is known as
A. encoder B. decoder
C. multiplexer D. demultiplexer

351. What is the largest number of data inputs which a data selector with two cotrol inputs can have?
A. Two B. Four
C. Eight D. Sixteen

352. Flip-flop outputs are always
A. complementary B. the same
C. independent of each other
D. same as inputs

353. Which of the following is the minimum error code?
A. Octal code B. Binary code
C. Gray code D. Excess-3-code

354. How many digits in binary notation are required for the decimal number 17?
A. 4 B. 6
C. 7 D. 5

355. To represent 1 out of 8 equiprobable events, the number of bits required are
A. 8 B. 4
C. 3 D. 2

356. When an even decimal number is converted into the binary number system, the least significant digit is
A. 1 B. 0
C. 1 or 0 D. none of these

357. Which of the following circuit converts a JK flip-flop to a T flip-flop?

A.
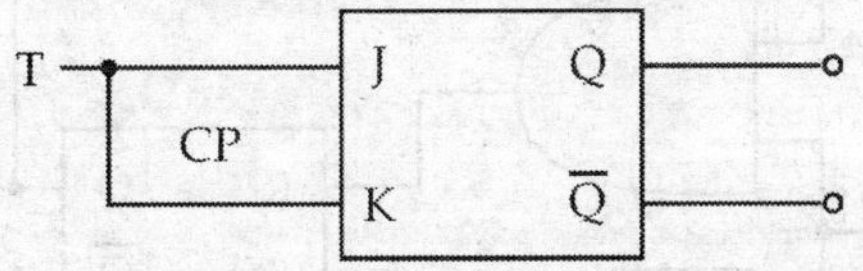

B.
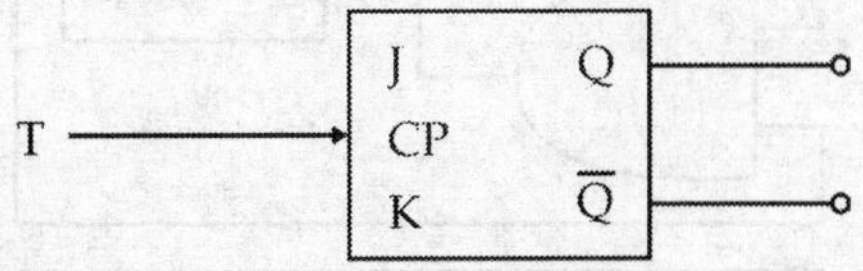

C.
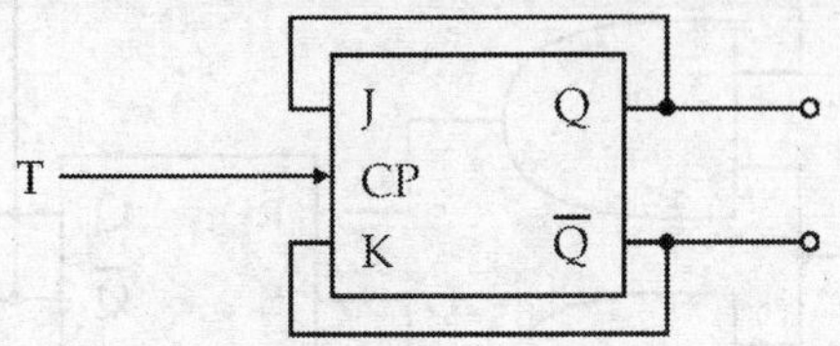

D.
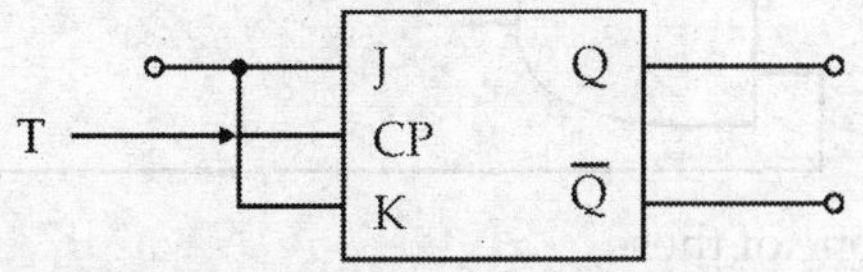

Ans. 340. C 341. B 342. B 343. A 344. A 345. D 346. C 347. B 348. A 349. D 350. C 351. B 352. A 353. C 354. D 355. D 356. B 357. A

358. Which of the following flip-flops is used as latch?
A. JK flip-flop B. RS flip-flop
C. D flip-flop D. T flip-flop

359. Which of the following can be used as an inverter?
A. AND gate B. OR gate
C. NOR gate D. None of these

360. Which of the following integrated circuit type provides Ex-OR operation?
A. 7404 B. 7486
C. 7470 D. 7476

361. Sixty capsules are to be filled in bottles automatically employing an electronic counter. The number of flip-flops in such a counter will be
A. 4 B. 5
C. 6 D. 7

362. How many 7490 ICs are to be cascaded to count up to 999?
A. 1 B. 2
C. 3 D. 4

363. Which of the following circuits given below converts an RS flip-flop to a T flip-flop?

A.
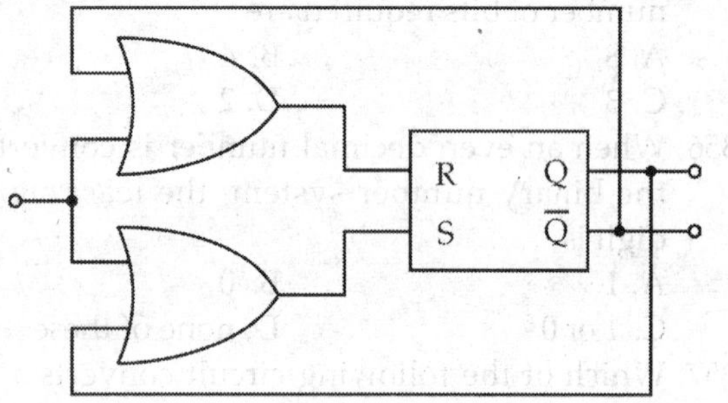

B.
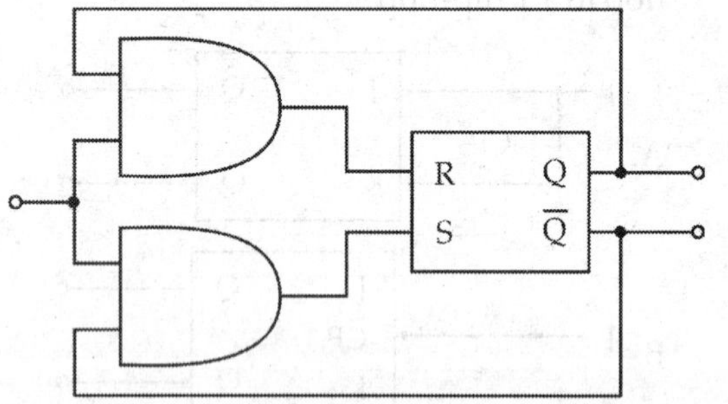

C.
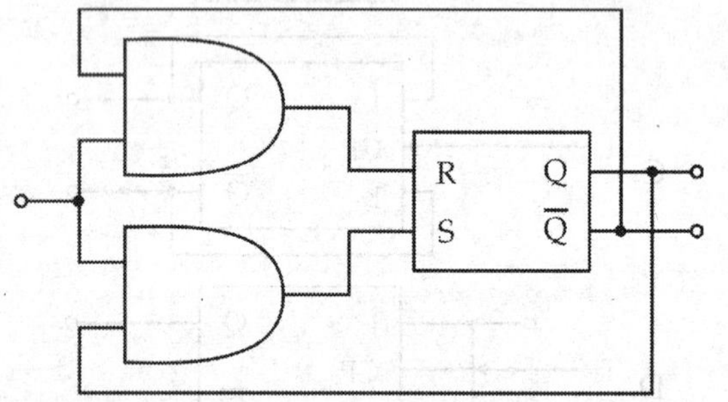

D. None of these

364. The output of a JK flip-flop toggles when
A. $J = 1,\ K = 0$ B. $J = 0,\ K = 1$
C. $J = 1,\ K = 1$ D. $J = 0,\ K = 0$

365. Which of the following flip-flops do not have race problem?
A. T flip-flop B. D flip-flop
C. JK flip-flop D. Master-slave flip-flop

366. The number of flip-flops required to build a mod-19 counter is
A. 4 B. 5
C. 6 D. 7

367. The minimum number of flip-flops required in a counter to count 60 pulses is
A. 4 B. 6
C. 8 D. 10

368. Flopy disc is a
A. flexible storage B. permanent storage
C. ROM D. fast storage

369. By placing an inverter between both input of an S-R flip-flop, the resulting flip-flop becomes
A. JK flip-flop B. D flip-flop
C. T flip-flop
D. Master slave JK flip-flop

370. If it is desired to clean up ragged looking pulses that have been distorted during transmission from one place to another, which of the following device will be appropriate?
A. Multiplexer B. D/A converter
C. JK flip-flop D. Schmitt trigger

371. The number of flip-flops required in a decade counter is
A. 2 B. 3
C. 4 D. 10

372. Three open collector TTL gates connected in AND configuration, drive the inputs of five TTL gates. What would be a suitable value for external pull-up resistor connected to the wired-AND circuit, if the following specifications are applied to TTL gates?
Low-level input current = 1.6 mA
Low-level output current = 16 mA
Low-level input voltage = 0.8 V
Low-level input voltage = 0.4 V
A. 600 Ω B. 300 Ω
C. 60 Ω D. 30 Ω

373. Consider the following logic circuit consisting of 2–4 decoder for each of decoder $f_0 = 1$ when $i_1 = 0$, $i_2 = 0$ is $f_1 = 1$ when $i_1 = 1$, $i_2 = 0$ and so on. Which one of the following gives value of $g(xyz)$?
A. $x\bar{y} + \bar{x}\bar{y}\cdot\bar{z}$ B. $x\cdot y\cdot z$
C. $[xy + \overline{xy}]\cdot z$ D. 1

374. A synchronous sequential circuit is to be designed which will produce an output '1' when previous two and present input represent an even number,

Ans. 358. D 359. C 360. B 361. C 362. C 363. C 364. C 365. D 366. B 367. B 368. A 369. B 370. D 371. C 372. B 373. D

with present input being least significant. The minimum number of states of the machine will be

A. 2 B. 3
C. 4 D. 5

375. For a periodic function, the spectral density and autocorrelation functions would be a
A. Fourier-transform pair
B. Laplace pair
C. Z-transform pair
D. Hilbert's transform pair

376. Programmable ROM has a decoder at the input and
A. both these blocks being fully programmable
B. only the former block being programmable
C. only the latter block being programmable
D. both these blocks being partially programmable

377. The circuit given in the figure represents a block commonly used in linear ICs. This is basically a

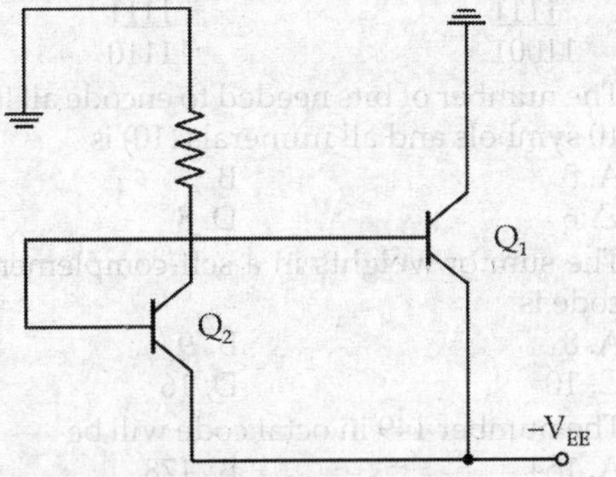

A. current amplifier
B. constant current source
C. constant voltage source
D. voltage amplifier

378. Which of the following are of the combinational logic types?
1. Read only memory
2. Random access memory
3. Shift register memory

Select the correct answer using the codes given below:

Codes
A. 1 and 2 B. 2 and 3
C. 1 and 3 D. 1, 2 and 3

379. The maximum time required to perform division using 8-bit binary operand with a clock of 500 MHz is
A. 5.2 μs B. 26 μs
C. 52 μs D. 104 μs

380. Which of the following binary addition is incorrect?
A. 1001 + 1101 = 10110
B. 10101 + 10011 = 101000
C. 11111 + 11111 = 100000
D. 11111 + 10001 = 110000

381. In the differentiating circuit given in the figure, the function of R_1 is to

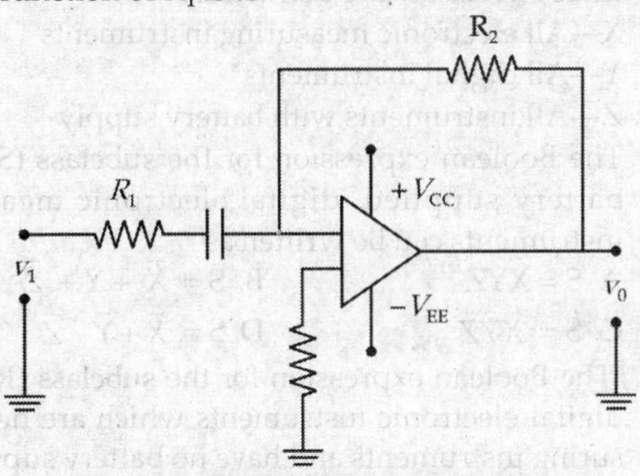

A. enable circuit to approach ideal differentiator
B. maintain high input impedance
C. eliminate high frequency noise spikes
D. prevent oscillations at high frequencies

382. Consider the following sizes related to floppy discs:
1. $3\frac{1}{4}$ inch diameter 2. $5\frac{1}{4}$ inch diameter
3. $3\frac{1}{2}$ inch diameter 4. $5\frac{1}{2}$ inch diameter

Of these sizes
A. 2 and 3 are correct B. 3 and 4 are correct
C. 1 and 4 are correct D. 1 and 2 are correct

383. Flag is a character for
A. identification of a word
B. occurrence of some condition
C. marking or tagging D. all of these

384. The circuit connections of OP-AMP given in figures (i) and (ii) represent

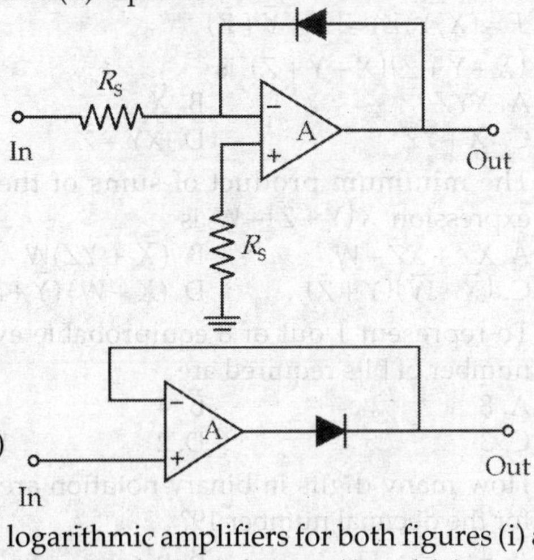

A. logarithmic amplifiers for both figures (i) and (ii)
B. detectors for both figures (i) and (ii)
C. detector for figure (i) and logarithmic amplifier for figure (ii)
D. logarithmic amplifier for figure (i) and detector for figure (ii)

Ans. 374. B 375. D 376. A 377. B 378. A 379. B 380. C 381. B 382. A 383. D 384. D

385. In the universal class of electronic instruments, three subclasses are defined as:
X—All electronic measuring instruments
Y—All digital instruments
Z—All instruments with battery supply
The Boolean expression for the subclass (S) of all battery supplied, digital electronic measuring instruments can be written as
A. $S = XYZ$ B. $S = \bar{X} + Y + \bar{Z}$
C. $S = \bar{X}\bar{Y}\bar{Z}$ D. $S = \bar{X} + \bar{Y} + \bar{Z}$

386. The Boolean expression for the subclass (R) of all digital electronic instruments which are not measuring instruments and have no battery supply is
A. $R = \bar{X} + \bar{Y} + \bar{Z}$ B. $R = XY + Z$
C. $R = \bar{X}\bar{Y}\bar{Z}$ D. $R = XY\bar{Z}$

387. The Boolean expression for the sub-class (Q) of all electronic instruments which are measuring instruments or are nondigital instruments with battery supply is
A. $Q = X(Y + Z)$ B. $Q = X + \bar{Y}Z$
C. $Q = X\bar{Y} + Z$ D. $Q = XY + \bar{Z}$

388. The numbers of terms in a logic function of three variables X, Y and Z is
A. 3 B. 4
C. 8 D. 16

389. The simplified form of Boolean function $F = (X, Y, Z) = (X + Y + XY)(X + Z)$, is
A. $X + Y + Z$ B. $X + YZ$
C. $XY + YZ$ D. XYZ

390. The simplified form of the Boolean function
$F = (X, Y, Z) = (X + \bar{Y} + \bar{Z})$
$(X + \bar{Y} + Z)(X + Y + Z)$ is
A. XYZ B. $\bar{X}$
C. $X + \bar{Y}Z$ D. $XY + \bar{Z}$

391. The minimum product of sums of the Boolean expression $X(Y + \bar{Z}) + W$ is
A. $XY + X\bar{Z} + W$ B. $(\bar{X} + YZ)W$
C. $(\bar{X} + \bar{W})(Y + \bar{Z})$ D. $(X + W)(\bar{Y} + \bar{Z} + W)$

392. To represent 1 out of 8 equiprobable events, the number of bits required are
A. 8 B. 4
C. 3 D. 2

393. How many digits in binary notation are required for the decimal number 19?
A. 4 B. 6
C. 7 D. 5

394. When an even decimal number is converted into the binary number system the least significant digit is
A. 1 B. 0
C. 1 or 0 D. None of these

395. Number 85 in BCD code is
A. 1000–1100 B. 1101–1010
C. 1000–0101 D. 1101–1001

396. 0110-0001-1001 in standard BCD is decimal number is
A. 615 B. 916
C. 619 D. 919

397. The decimal number 85 is encoded as 1100–0101 in
A. 8421 code B. 4421 code
C. 2421 code D. 2221 code

398. Which addition is correct?

A.	B.
0101	0101
1111	1111
11010	10100

C.	D.
0101	0101
1111	1111
11001	11110

399. The number of bits needed to encode all letters (26), 10 symbols and all numerals (10) is
A. 5 B. 7
C. 6 D. 8

400. The sum of weights in a self-complementing *BCD* code is
A. 8 B. 9
C. 10 D. 16

401. The number 149 in octal code will be
A. 154 B. 178
C. 254 D. 225

402. In which of the following codes a binary number $Y_1 Y_2 Y_3$ is converted by the circuit given below?

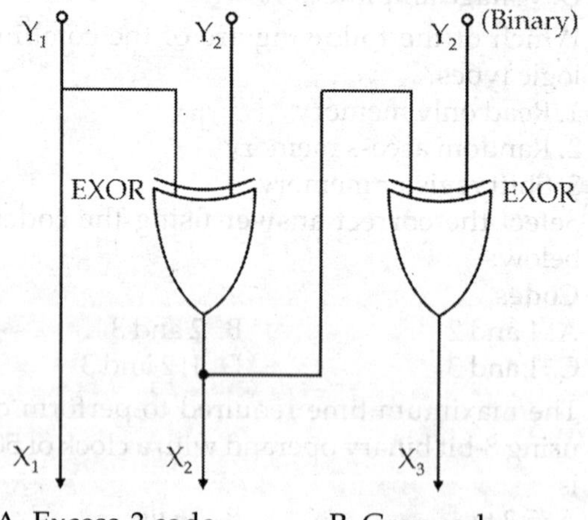

A. Excess-3 code B. Gray code
C. Decimal D. BCD code

403. The positive number 0.00001 is to be subtracted from 0.0011. The arithimatic unit of a computer would perform it by adding
A. 0.1111 to 0.0011 B. 0.1 110 to 0.0011
C. 0.1 100 to 0.0001 D. 0.1101 to 0.0001

Ans. 385. A 386. C 387. B 388. C 389. B 390. C 391. D 392. C 393. D 394. B 395. C 396. C 397. B
398. B 399. C 400. B 401. D 402. B 403. B

404. The 2' complement of the number 0.0011 is
A. 1.1100 B. 0.1101
C. 1.1101 D. 0.1101

405. The quantity 837 in Excess-3 BCD code would be represented as
A. 100100110111 B. 100000011001
C. 101101101010 D. 100000110111

406. The quantity 837 in gray BCD code would represented as
A. 10000110111 B. 100000100011
C. 110001000010 D. 110000100100

407. Which of the following statements is true for 2-out-of-5 code?
A. It is an unweighted code
B. It has even parity
C. It has exactly two I's in each code group
D. All of these

408. A six-bit alpha-numeric code is able to code
A. 36 characters B. 48 characters
C. 64 characters D. 128 characters

409. 2's complement of-7 is
A. 1.1001 B. 1.1000
C. 1.0111 D. 0.0111

410. The Boolean expressions for the circuits shown below is

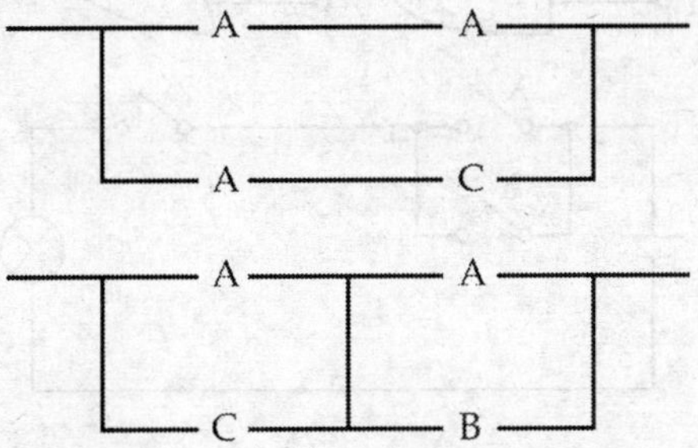

A. identical B. complementary
C. dual
D. different from each other

411. Logic table below gives the output for inputs A and B. The logic operation performed is

A	*B*	*C*
0	0	1
0	1	0
1	0	0
1	1	1

A. Negative of NOR B. Ex. OR
C. NAN D. Negative of Ex. OR

412. A NAND circuit with positive logic will operate as
A. NOR with negative logic
B. AND with negative logic
C. OR with negative logic input
D. AND with negative logic output

413. The Boolean expression for the network shown in the above question is
A. $A(B+C)$ B. $A+BC$
C. $A+BC+AC+AB$ D. $A(A+B+C)$

414. Consider two normally open contacts in series for a normally closed relay. The logic operation performed by this circuit is
A. AND B. NAND
C. NOR D. None of these

415. The frequency of h_1 and h_2 lighting up is half of the frequency of input pulses. Then, the logic element X is

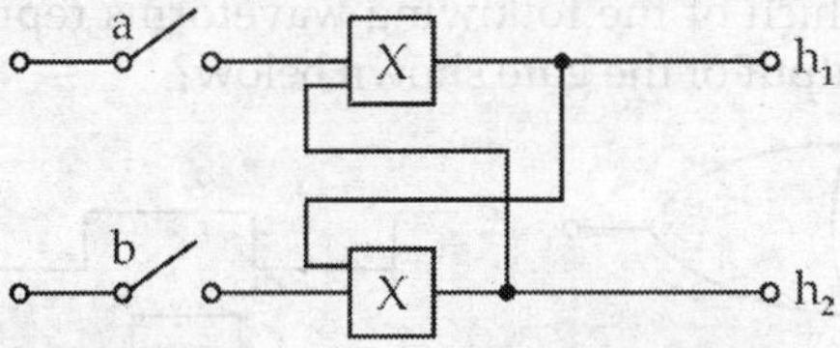

A. AND gate B. Ex. OR gate
C. Inverter D. NOR gate

416. In INHIBIT operation
A. output is 1 when both inputs are 0
B. output is 0 when blocking input is 1
C. output is 0 when blocking input is 0
D. output is 1 when blocking input is either 0 or l

417. In Boolean algebra, exclusive and coincidence operations are defined as below:
$X \oplus Y = X$ or Y but not both.
X·Y = either (X.Y) or $(\overline{X}\cdot\overline{Y})$
The two expressions $(X \oplus X + X.X)$ and $(X \oplus 1X\cdot 1)$
A. equal to 1 B. equal to 0
C. equal to X D. equal to $\overline{X}$

418. In a digital control system, there are three inputs *a*, *b*, *c*. The output should be present when at least two inputs are present. The Boolean expression for the output is
A. $ab + bc + ac$ B. $abc + ab\overline{c} + \overline{a}bc + a\overline{b}c$
C. $ab\overline{c} + a\overline{b}c + \overline{a}bc$ D. $a\overline{b} + b\overline{c} + \overline{a}c$

419. Two binary signals *a*, *b* are to be compared. The output expression when the two signals are equal is given by
A. $a\overline{b} + a\overline{b}$ B. $ab + \overline{a}\overline{b}$
C. $\overline{a}\overline{b}$ D. ab

420. A lamp is controlled from two positions *A* and *B* (e.g. stair case circuit) The Boolean expression for this circuit is
A. $A\overline{B} + \overline{A}B$ B. $AB + A\overline{B}$
C. $AB + \overline{A}\overline{B}$ D. $A\overline{B} + \overline{A}\overline{B}$

Ans. 404. C 405. C 406. D 407. D 408. C 409. A 410. C 411. D 412. A 413. B 414. C 415. D 416. B 417. A 418. A 419. B 420. A

421. The output Y of the circuit shown is

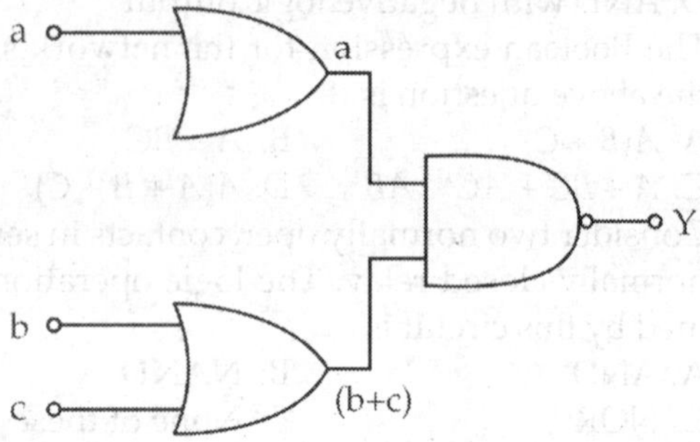

A. $\overline{a}+\overline{b}\,\overline{c}$ B. $a+bc$
C. $\overline{a}\,\overline{b}+\overline{a}\,\overline{c}$ D. $a+\overline{b}\,\overline{c}$

422. Which of the following waveforms represents the output of the gate shown below?

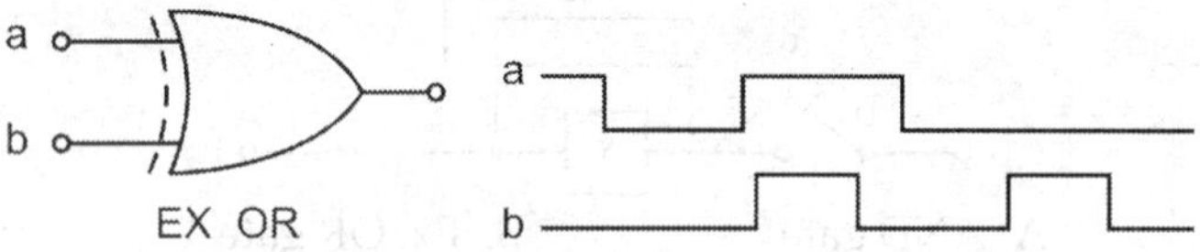

A. 
B. 
C. 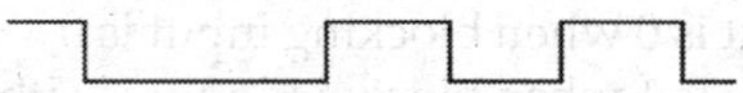
D. 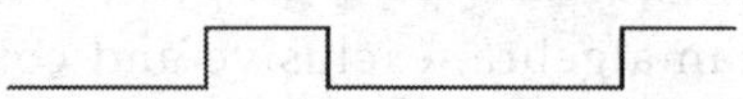

423. NAND operation is
A. $\overline{X+Y}$ B. $\overline{X}+\overline{Y}$
C. $\overline{X\cdot Y}$ D. $(\overline{Y}+\overline{Y})(X+Y)$

424. The circuit below represents a

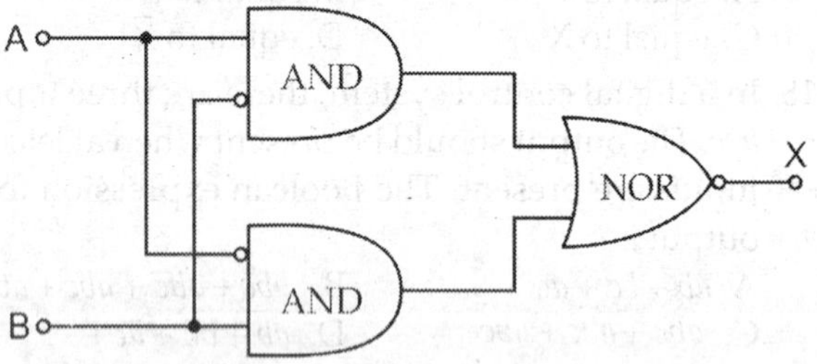

A. Exclusive OR
B. comparator (outputs when both inputs are equal)
C. inhibit gate D. NAND gate

425. The Boolean expression for the above circuit is
A. $X=AB+\overline{AB}$ B. $X=AB+A\overline{B}$
C. $X=\overline{(A+\overline{B})}+\overline{(A+\overline{B})}$ D. $X=AB+\overline{A}\,\overline{B}$

426. Which of the following statements is not correct?
A. $X+\overline{X}Y=X$ B. $X(\overline{X}+Y)=XY$
C. $XY+X\overline{Y}=X$ D. $ZX+ZXY=ZX+ZY$

427. The minimum Boolean expression for the circuit shown below is

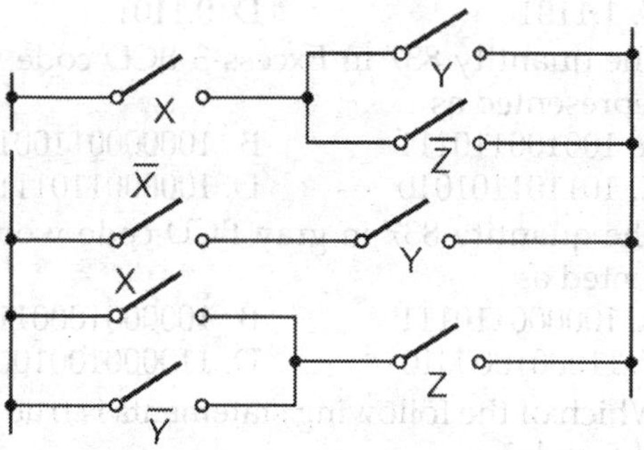

A. $XY+XZ+YZ$ B. $X+YZ$
C. $X+Y$ D. $X+Y+Z$

428. The minimum Boolean expression of $XYZ+\overline{X}Z+YZ+XY\overline{Z}$ when solved by complementary map method is
A. $\overline{XY}+XY+\overline{X}Y$ B. $\overline{XY}+XY+\overline{Y}X$
C. $(X+\overline{Y}+Z)(\overline{X}+Y)$ D. $(\overline{X}+\overline{Y})(X+Y+Z)$

429. Which of the following circuit is equivalent to the logic diagram shown below?

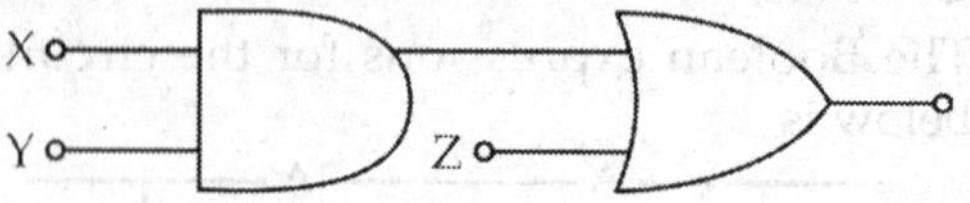

A. 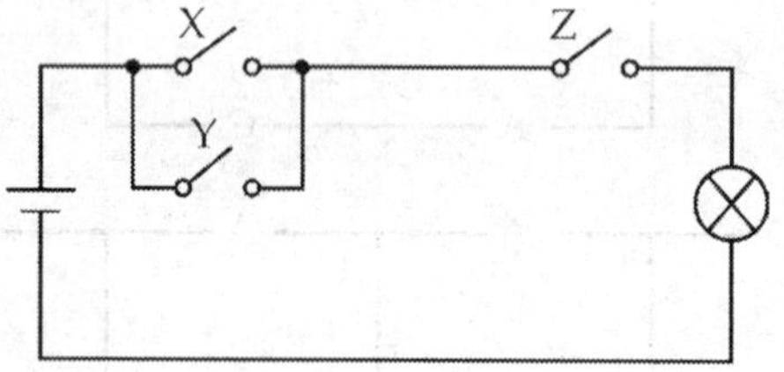

B. 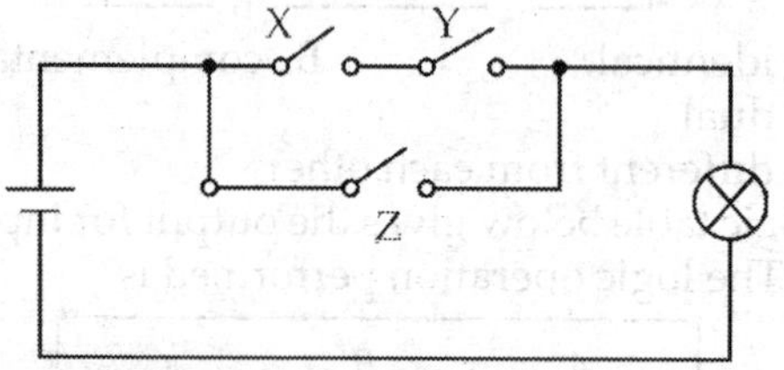

C. 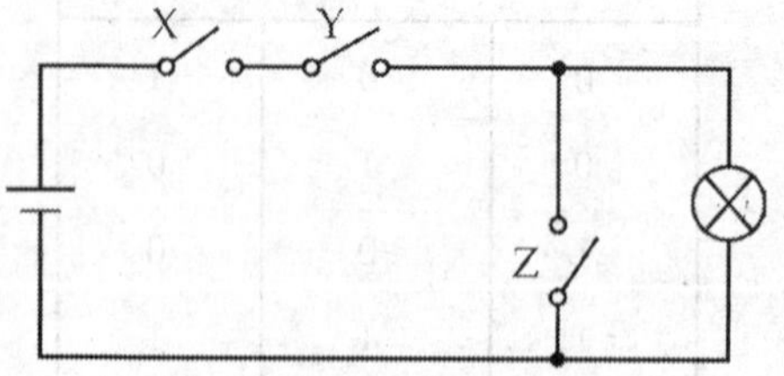

D. None of these

430. The most widely used universal gates are
A. OR and AND gates
B. NOR and NAND gates
C. NOR and AND gates
D. NAND and OR gates

Ans. 421. A 422. C 423. D 424. B 425. D 426. A 427. D 428. C 429. B 430. B

431. Which of the following circuit represents the shaded area of the accompanying Venn diagram?

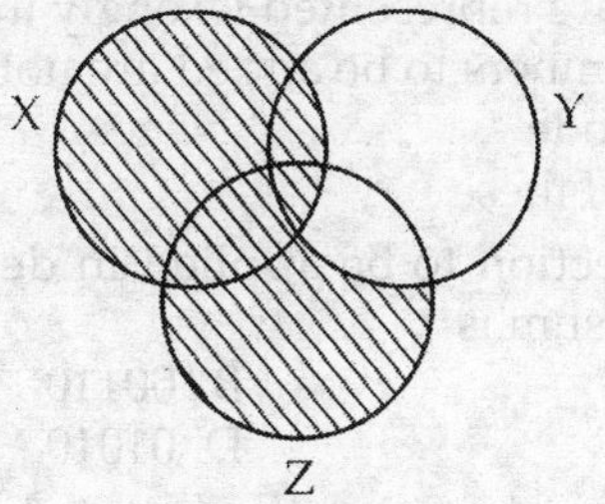

A. 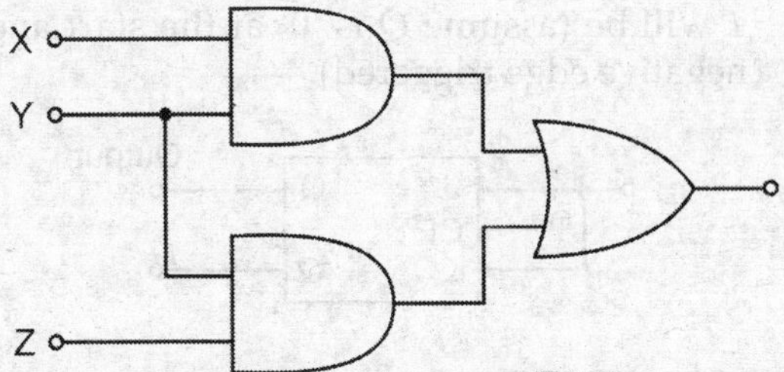

B. 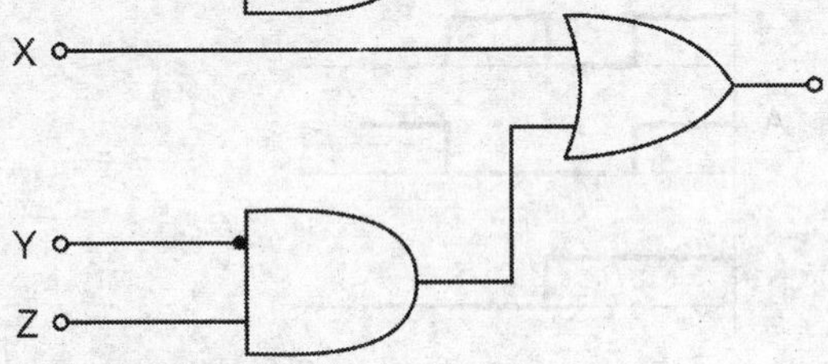

C. 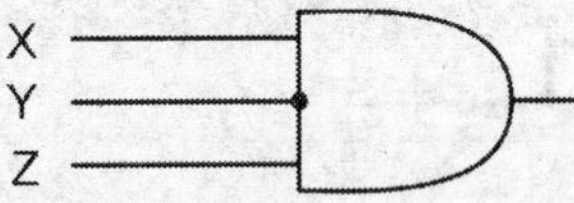

D. None of these

432. If the negative logic is used, the diode gate in the circuit below represents an

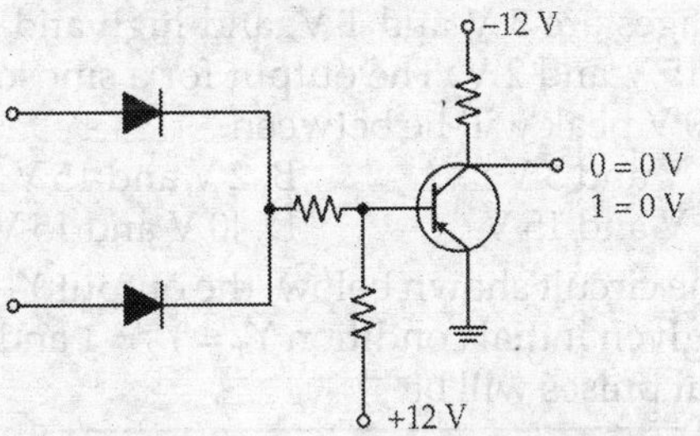

A. AND gate
B. NOR gate
C. OR gate
D. NAND gate

433. A three-input NAND gate is to be used as an inverter. Which of the following measures will achieve this end better?

A. All inputs are connected together, making one input
B. The two inputs not used are connected to the logic "1" level
C. The two inputs not used are connected to ground ("0" level)
D. None of these.

434. In positive logic, the transistor gate in the circuit below is an

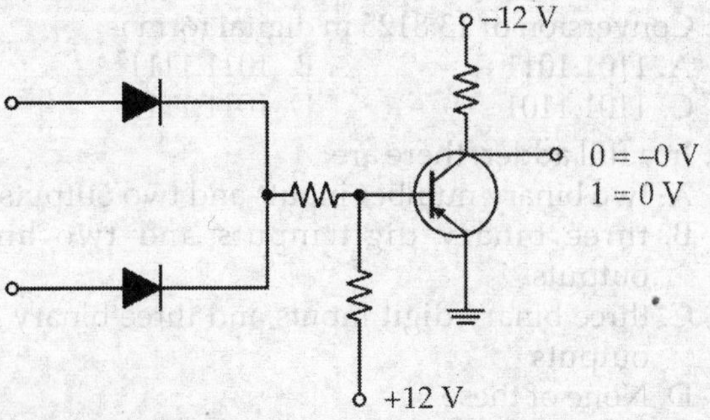

A. AND gate
B. NOR gate
C. NAND gate
D. OR gate

435. The circuit below (positive logic) is an

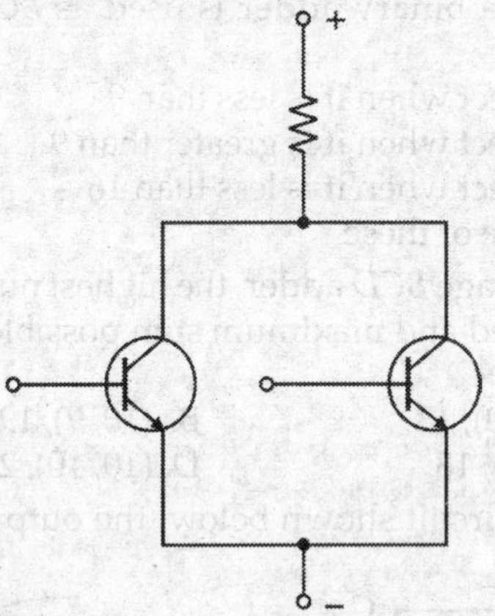

A. NOR gate
B. NAND gate
C. Flip-Flop
D. AND gate

436. What is the purpose of the standoff diode *D* in the basic *DTL* gate shown below?

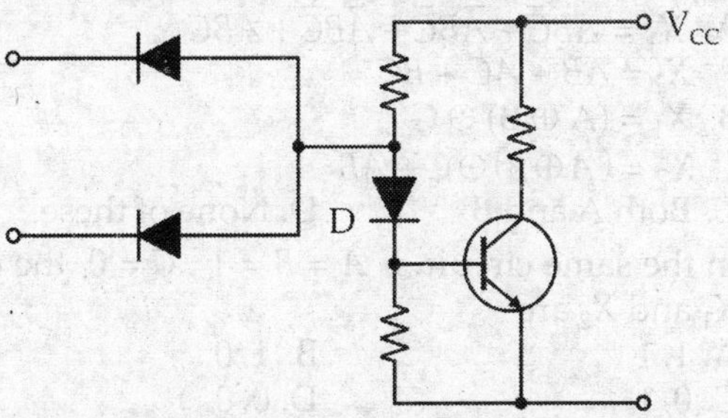

A. To protect the input against over voltages.
B. To raise the threshold level by 0.7 V
C. To provide facilities for wired-AND connections.
D. To provide temperature compensation

437. An inverter circuit using *NPN* transistor is not operated with positive logic. The transistor is replaced by a *PNP* transistor by mistake. The logic now performed will be

A. NOT operation with positive logic
B. NOT operation with negative logic

Ans. 431. B 432. A 433. B 434. B 435. A 436. B 437. B

C. NOT operation with proper battery connections
D. none of these

438. Conversion of 13.8125 in digital form is
A. 1101.1011 B. 1011.1111
C. 1101.1101 D. 1011.1111

439. In a full adder, there are
A. two binary number inputs and two outputs
B. three binary digit inputs and two binary outputs
C. three binary digit inputs and three binary digit outputs
D. None of these

440. In order to add 1111 + 1101, we require
A. one *FA*, one *HA* B. three-*FA*
C. three *FA* and one *HA* D. None of these

441. Which a binary adder is used as *BCD* adder, the sum is
A. correct when it is less than 9
B. correct when it is greater than 9
C. correct when it is less than 16
D. None of these

442. In a 4-stage *BCD* adder, the highest number that can be added and maximum sum possible respectively are
A. (9, 10), 18 B. (10, 9), 19
C. (9, 9), 18 D. (10, 10), 20

443. In the circuit shown below, the outputs X_1 and X_2 are

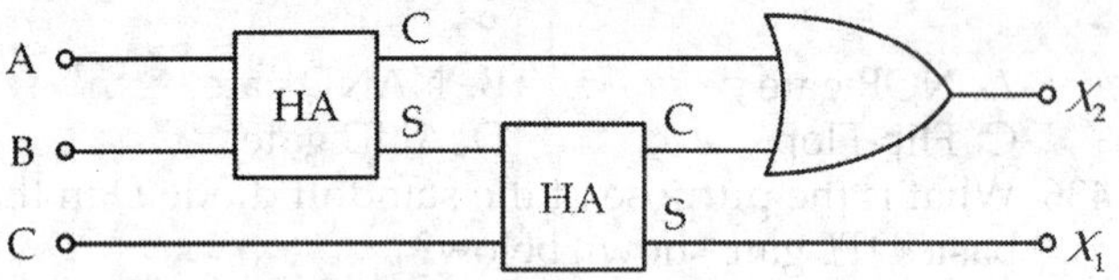

A. $X_1 = A\bar{B}\bar{C} + \bar{A}B\bar{C} + \bar{A}\bar{B}C + ABC$
$X_2 = AB + AC + BC$
B. $X_1 = (A \oplus B) \oplus C$
$X_2 = (A \oplus B) \oplus C + \overline{AB}$
C. Both A and B D. None of these

444. In the same circuit, if $A = B = 1$, $C = 0$, the output X_1 and X_2 are
A. 1, 1 B. 1, 0
C. 0, 1 D. 0, 0

445. The gates required to build an half adder are
A. Ex. OR gate and NOR gate
B. Ex. OR gate and OR gate
C. Ex. OR gate and AND gate
D. Four NAND gate

446. The advantage of serial transfer compared with parallel transfer for data transmission is that
A. it needs only one wire
B. it is faster
C. *BCD* is compatible
D. All of these

447. Correction in decimal adder is required because
A. sums are represented wrongly in binary code
B. sums are represented wrongly in *BCD* code
C. the numbers to be added are not represented in *BCD* code
D. none of these

448. The correction to be applied in decimal adder to incorrect sum is
A. 00101 B. 00110
C. 01101 D. 01010

449. The output of the circuit shown below for the input *T* will be (assume *Q* is '0' at the start and F/F is negative edge triggered).

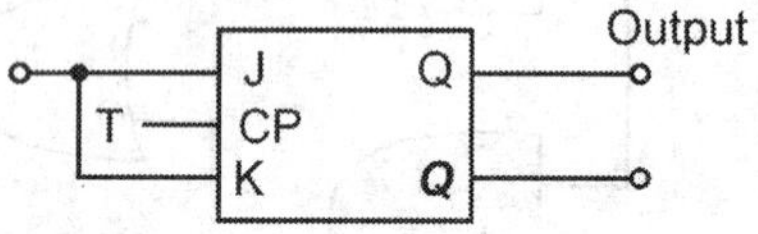

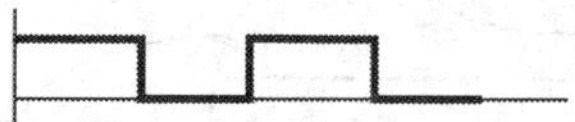

A. 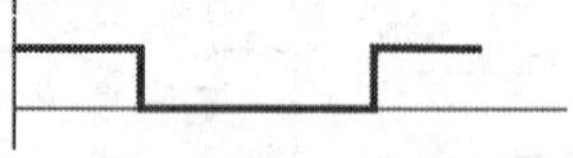

B. 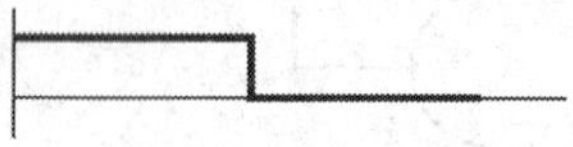

C. 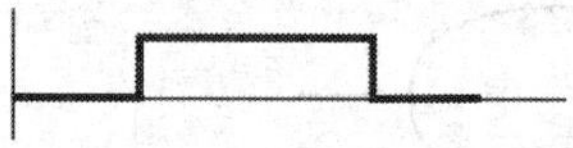

D. 

450. For a Schmitt trigger, the upper and lower trip voltages are 3 V and 1 V, and high and low states are 15 V and 2 V. The output for a sinosoidal input of 10 V peak will lie between.
A. 2 V and 3 V B. 2 V and 15 V
C. 1 V and 15 V D. 10 V and 15 V

451. In the circuit shown below, the output Y_1 and Y_2 for the given initial condition $Y_1 = Y_2 = 1$ and after four input pulses will be

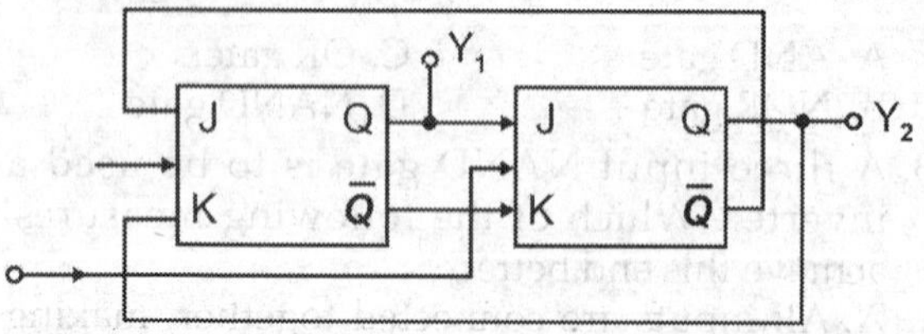

A. $Y_1 = 1$ $Y_2 = 0$ B. $Y_1 = 0, Y_2 = 1$
C. $Y_1, = 0, Y_2 = 0$ D. $Y_1 = 1, Y_2 = 1$

452. If the Schmitt trigger in the above question is driven by sawtooth wave of 10*V* peak, the output will be
A. 1 V and 3 V B. 2 V and 15 V
C. 3 V and 15 V D. 10 V and 15 V

Ans. 438. D 439. B 440. C 441. A 442. C 443. A 444. C 445. C 446. A 447. B 448. B 449. C 450. B 451. D 452. B

453. In a stable multivibrator, the base resistors of 100 K and capacitors of 1,000 pF are used. The frequency of the output will be
A. 1.2 kHz B. 5.5 kHz
C. 6.9 kHz D. 12.8 kHz

454. Schmitt trigger can be used as a
A. comparator
B. square-wave generator
C. flip-flop D. All of these

455. A ring counter is same as
A. up-down counter B. parallel counter
C. shift register D. None of these

456. The output of the circuit shown below will be of the frequency

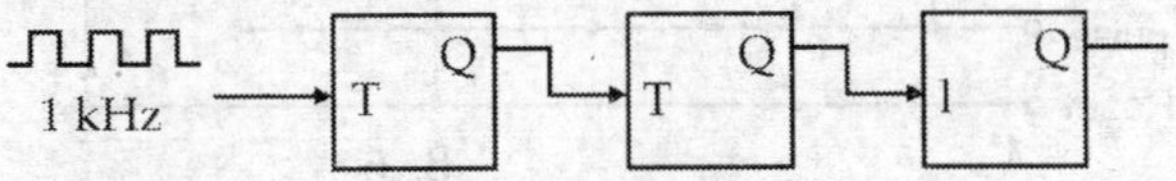

A. 125 Hz B. 250 Hz
C. 500 Hz D. 750 Hz

457. The counter circuit shown below will reset at

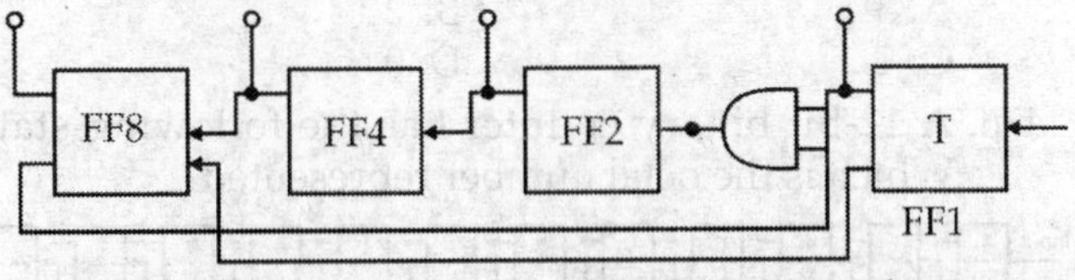

A. 8th input pulse B. 10th input-pulse
C. 12th input pulse D. 16th input pulse

458. In the same circuit, if the AND gate is removed and output of FF_1 is directly connected to FF_2 the output of the flip-flops at 10th pulse will be
A. 0000 B. 0001
C. 0010 D. 1001

459. Which of the following ICs can be used for frequency division by 5?
A. 7446 B. 7492
C. 74121 D. 7490

460. Which of the following ICs is a 4-bit latch?
A. 7475 B. 7446
C. 7400 D. 7490

461. The circuit shown below is a

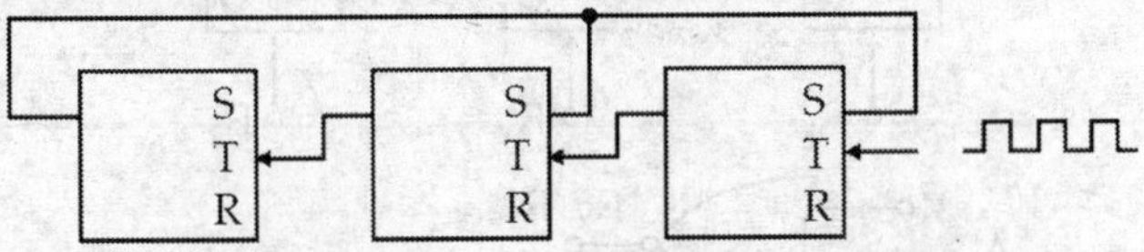

A. mod-4 counter B. mod-5 counter
C. mod-6 counter D. mod-7 counter.

462. The number of 7490s to obtain positive half cycle of 10 μsec is

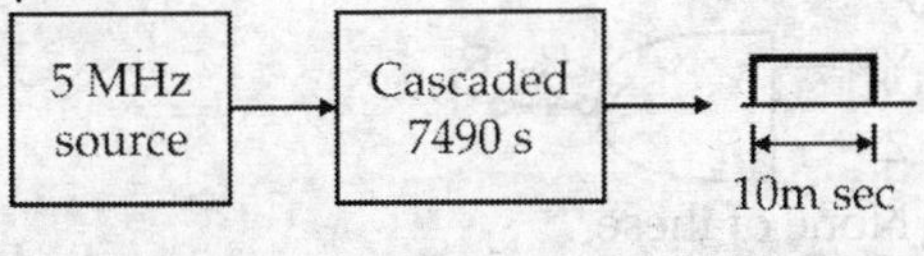

A. 3 B. 4
C. 5 D. 7

463. The output of the circuit for the given inputs '*a*' and '*b*' will be

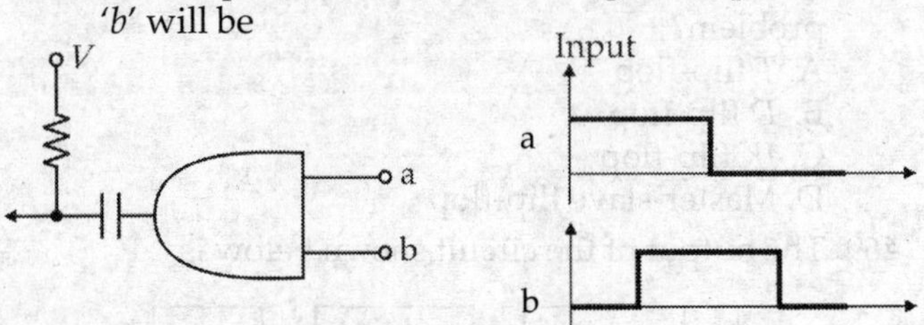

A.
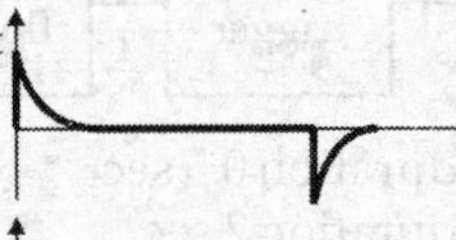

B.
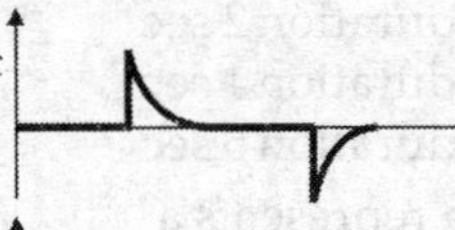

C.
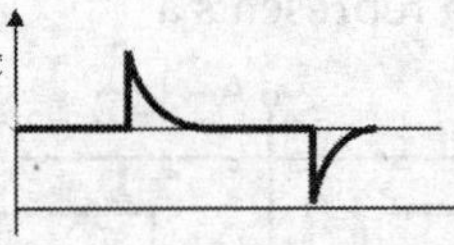

D.
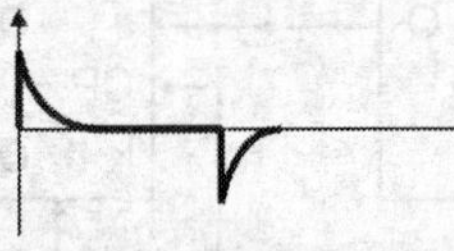

464. In the circuit shown below, the gate pulse is of 10 second duration and the frequency of the clock is 579 Hz. The output of the cascaded 7490 s will be

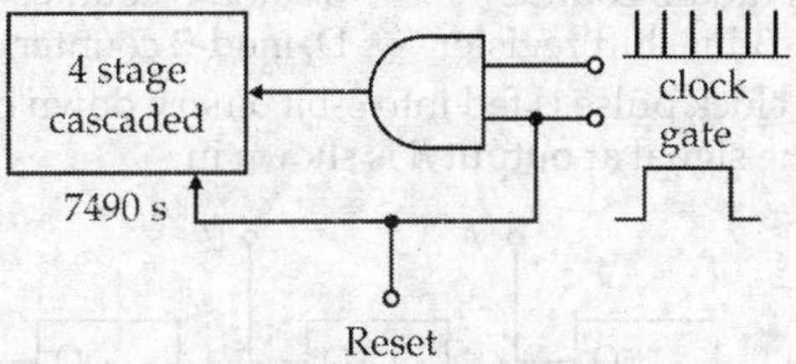

A. 1001 0111 0100 0000 B. 0101 0111 1001 0000
C. 0000 0101 1111 0000 D. 0000 0101 0111 1001

465. The output of a JK flip-flop toggles when
A. $J = 1, K = 0$ B. $J = 0, K = 1$
C. $J = l, K = I$ D. $J = 0, K = 0$

466. An electronic wrist-watch has a clock of frequency 32 KHz. To divide this frequency down to 1 Hz. it is necessary to have
A. two decade counters, one two-bit binary counter and a *T* flip-flop
B. three decade counters, one four-bit binary counter and one *T* flip-flop
C. one four-bit binary counter and *T* flip-flop and one decade counter
D. four four-bit binary Counter and decade counter

467. The number of flip-flops required to build a mod-19 counter is

Ans. 453. C 454. D 455. C 456. B 457. B 458. D 459. D 460. A 461. B 462. C 463. C 464. B 465. C 466. B

A. 4 B. 5
C. 6 D. 7

468. Which of the following flip-flops do not have race problem?
A. *T* flip-flop
B. *D* flip-flop
C. *JK* flip-flop
D. Master-slave flip-flop

469. The output of the circuit shown below is

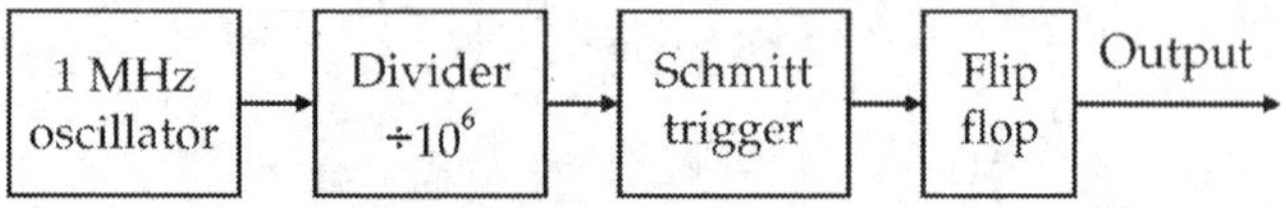

A. A pulse train of duration 0.5 sec
B. A pulse train of duration 2 sec
C. A pulse train of duration 1 sec
D. A pulse train of duration 5 sec

470. The circuit diagram represents a

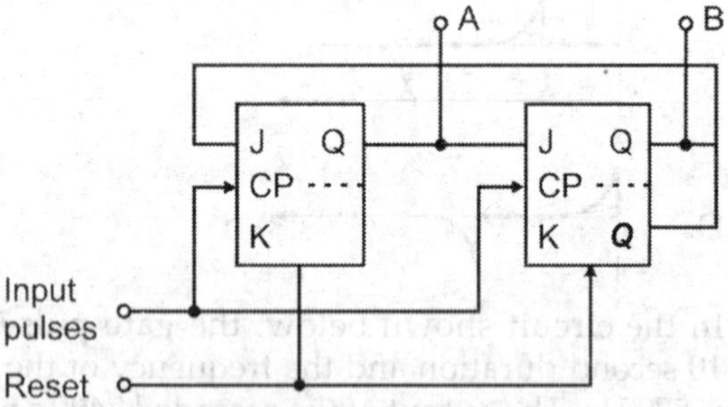

A. mod-3 counter B. mod-5 counter
C. 3-bit shift register D. mod-2 counter

471. A clock pulse is fed into 3-bit binary down counter. The signal at output *B* is shown in

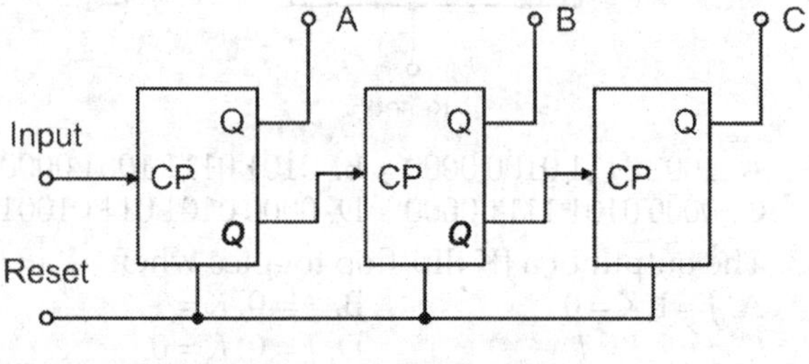

A.
B.

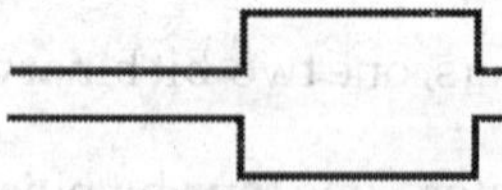

C.

D. None of these

472. In order to build a mode-6 counter using three flip-flops, the number of possible counter sequence is
A. 4 B. 8
C. 16 D. 28

473. The minimum number of flip-flops required in a counter to count 60 pulses is
A. 4 B. 6
C. 8 D. 10

474. The three-bit shift register is connected as shown in the figure below. After how many clock pulses (after reset to '0') the contents of the shift register are '000' again?

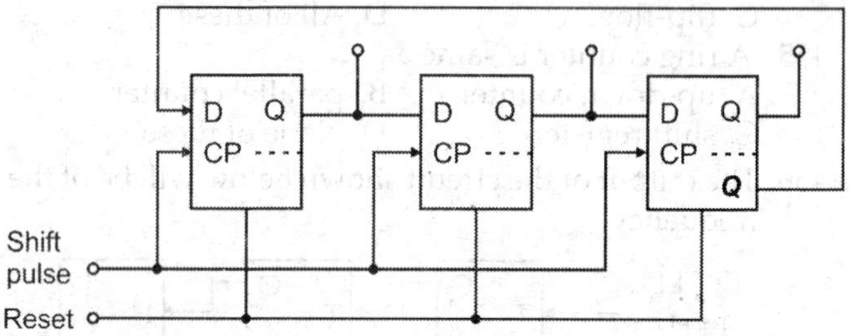

A. 3. B. 6
C. 8 D. 16

475. How many illegitimate states has a synchronous mod-6 counter?
A. 3 B. 2
C. 1 D. 0

476. A 12-bit binary counter has the following state. Which is the octal number represented?

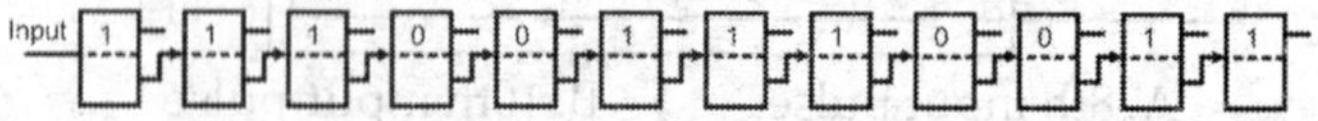

A. 677 B. 7163
C. 7183 D. 7128

477. What is the maximum number (in decimal notation) that can be counted with above counter?
A. 999 B. 4096
C. 4095 D. 9999

478. In another number system, e.g. 6, a "decade" counter has to recycle to "0" at the sixth count. Which of the connections indicated below will realise this resetting (a logic "0" at the *R* inputs resets the counters)?

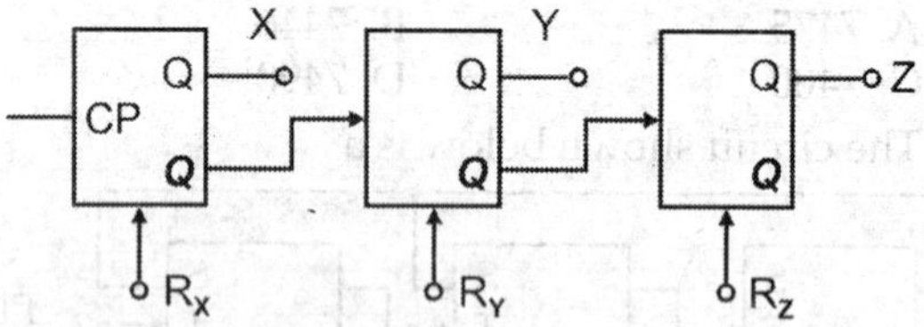

A.

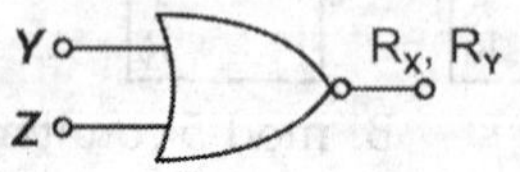

B.

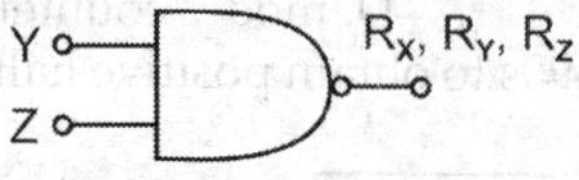

C.

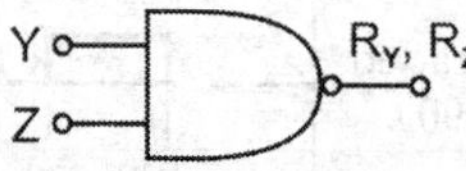

D. None of ihese

Ans. 467. B 468. D 469. B 470. A 471. A 472. D 473. B 474. B 475. B 476. B 477. C 478. C

479. The purpose of introducing feedback loop in a digital counter circuit is
A. it improve stability
B. to improve distortion
C. synchronise input and output pulses
D. to reduce the number of input pulses to reset the counter

480. For the inverter circuit shown below, I_c $_{(sat)}$ = 5 μA, $h_{fe\ (min)}$ = 30, $V_{CE\ (sat)}$ = 0.2V and $V_{BE\ (sat)}$ = 0.8V. The value of R_c should be

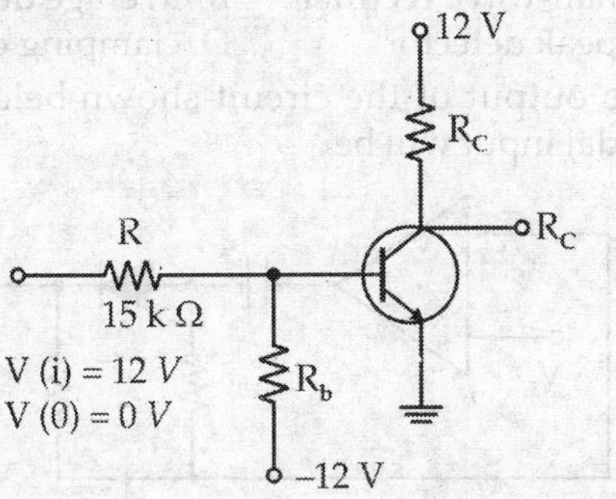

A. 10 K
B. 4.7 K
C. 6.8 K
D. 2.2 K

481. Which of the following logic gate can be used as an inverter?
A. AND gate
B. OR gate
C. NOR gate
D. None of these

482. Which of the following integrated circuit type provides Ex-OR operation?
A. 7404
B. 7486
C. 7470
D. 7476

483. The circuit given below is a

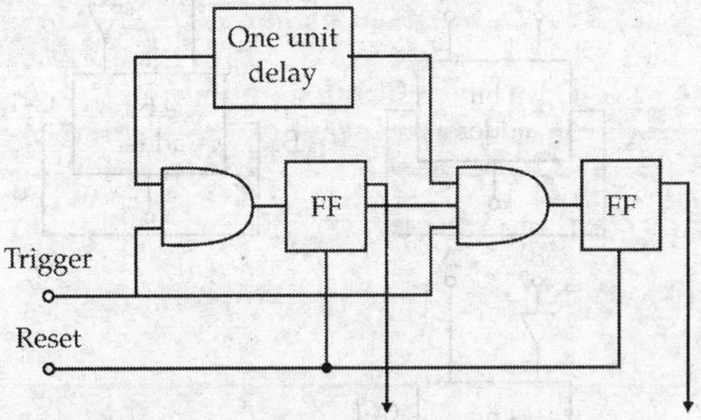

A. two-bit binary counter
B. analog-to-digital converter
C. two-bit series-to-parallel converter
D. two-bit parallel-to-series converter

484. For emittter-coupled logic, the switching speed is very high because
A. negative logic is used
B. the transistors are not saturated when conducting
C. emitter-coupled transistors are used
D. multi-emitter transistors are used

485. The circuit given below is a

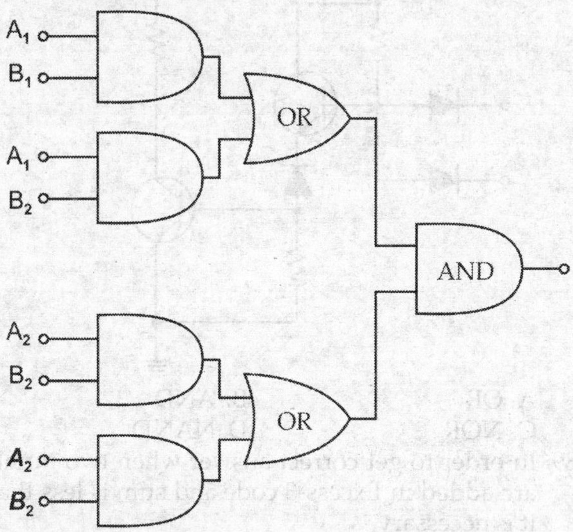

A. inverted Ex. OR
B. binary adder for two bit numbers
C. parity checker
D. binary comparator

486. The multiple-emitter transistor-logic circuit shown below represents a

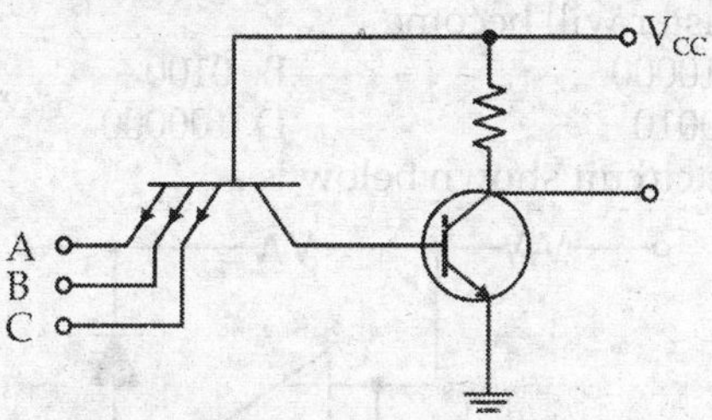

A. AND gate in positive logic
B. NOR gate in positive logic
C. NAND gate in positive logic
D. NAND gate in negative logic

487. The output of the circuit shown below is

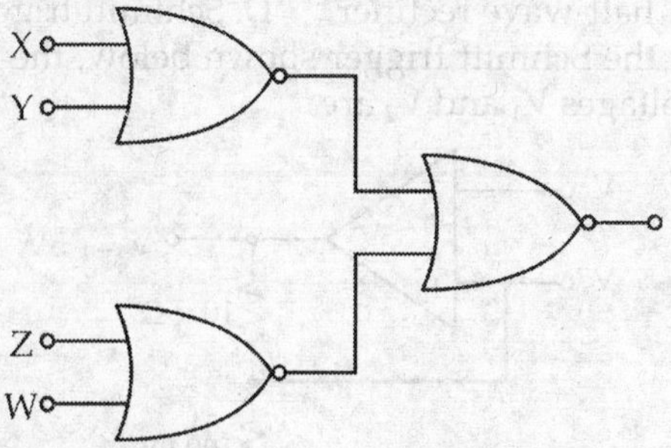

A. $XY + ZW$
B. $X\bar{Y} + ZW$
C. $(X + Y)(Z + W)$
D. $(X + \bar{Y})(\overline{ZW})$

488. The logic performed by the high-noise immunity logic circuit shown below is

Ans. 479. D 480. D 481. C 482. B 483. C 484. B 485. D 486. C 487. C

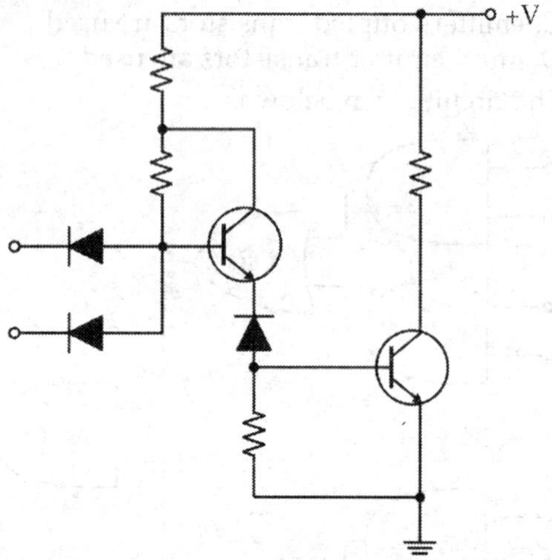

A. OR B. AND
C. NOR D. NAND

489. In order to get correct answer when two numbers are added in Excess-3 code and sum is less than 9, it is necessary
A. to subtract 0011 from the sum
B. to add 0011 from the sum
C. to subtract 0110 from the sum
D. to add 0110 from the sum

490. Number 1000 is stored in a shift register. If two right-shift pulses are applied, the number in the register will become
A. 10000 B. 0100
C. 0010 D. 100000

491. The circuit shown below is a

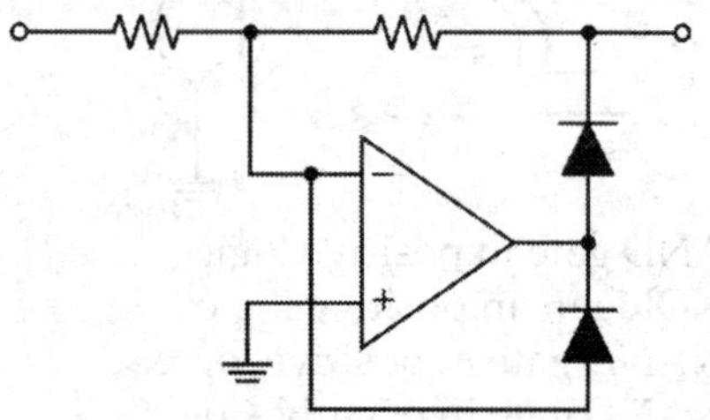

A. average detector B. clamping circuit
C. half-wave rectifier D. Schimitt trigger.

492. In the Schmitt trigger shown below, the threshold voltages V_1 and V_2 are

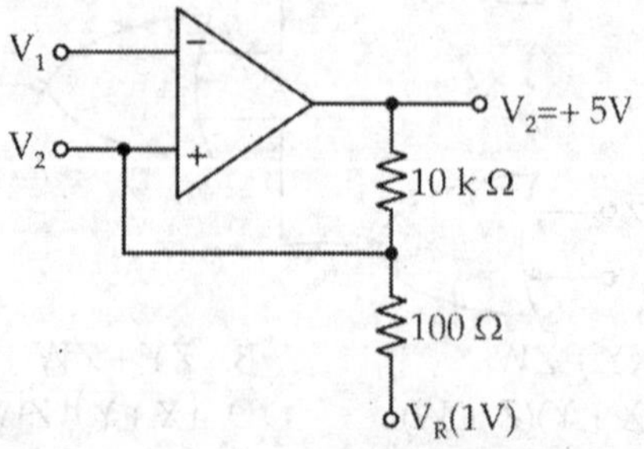

A. 2.5 V, 1.5 V B. 1.04 V, 0.94 V
C. 1.5 V, 1 V D. 1.85 V, 1.15 V

493. The circuit shown below is a

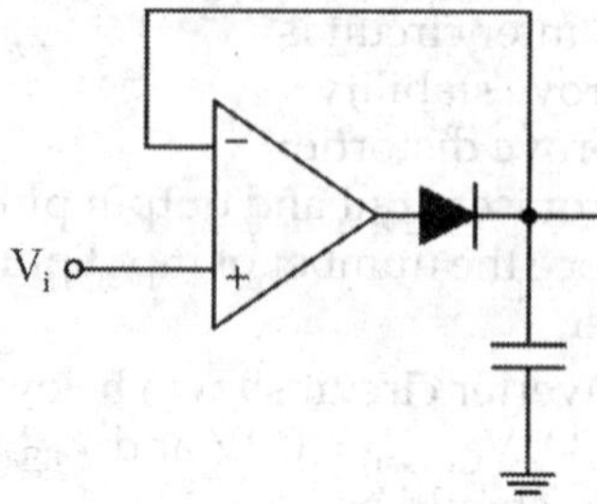

A. half-wave rectifier B. average detector
C. peak detector D. clamping circuit

494. The output of the circuit shown below for sinusoidal input will be

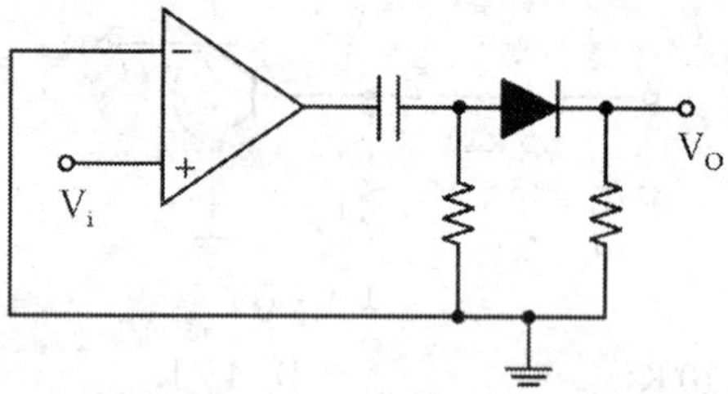

A. 

B. 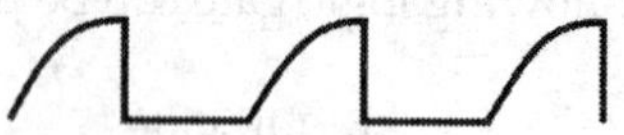

C. 

D. None of these

495. Which of the following circuit represents a subtracter to subtract *A* from *B*?

A. 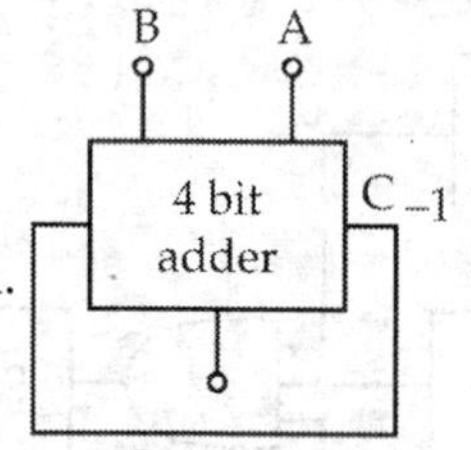

B. 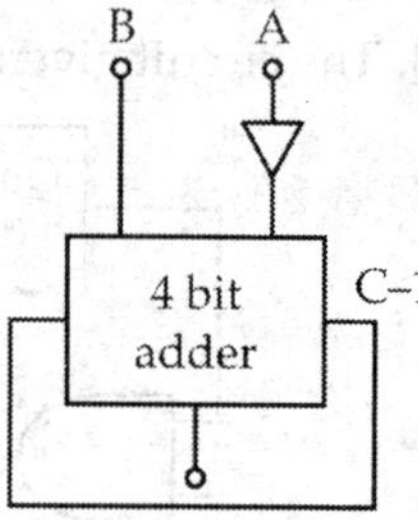

C. 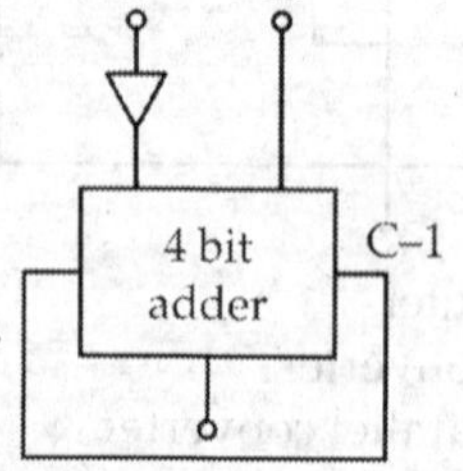

D. None of these

496. The function of a multiplexer is
A. to select 1 out of N input data sources and to transmit it to single channel
B. to transmit data on N lines
C. to perform serial-to-parallel conversion
D. to decode information

Ans. 488. D 489. A 490. C 491. C 492. B 493. C 494. C 495. B 496. A

497. In the 1-bit true-complement element shown below, the outputs for $L = 0$, $M = 0$ and $L = 0$ and $M = 1$ respectively are

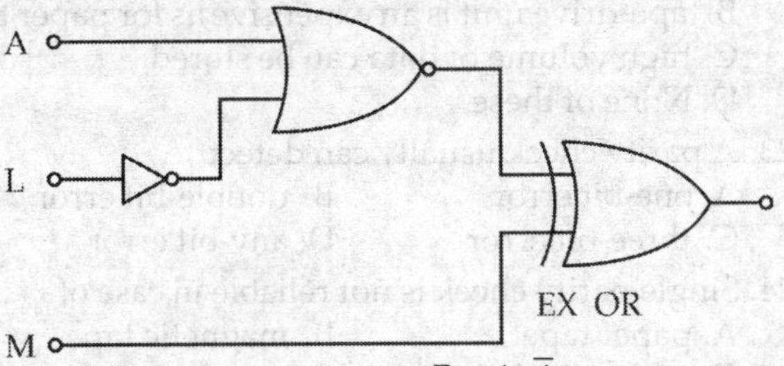

A. 1, 0 B. $A, \bar{A}$
C. $\bar{A}, A$ D. 0,1

498. Which of the following circuits can be used as parallel-to-serial converter?
A. Digital counter B. Decoder
C. Demultiplexer D. Multiplexer

499. The circuit shown below is a

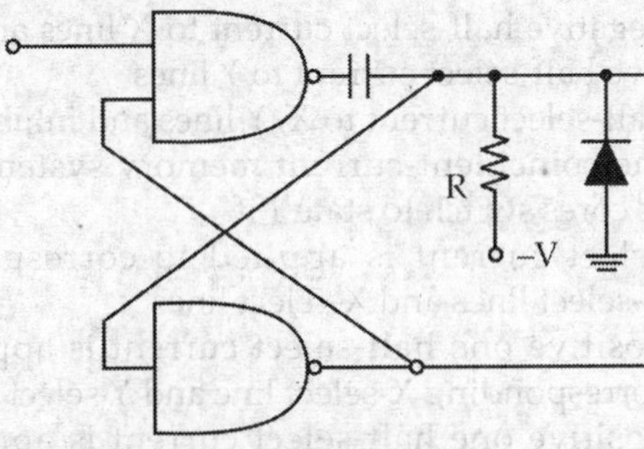

A. bistable multivibrator
B. astable multivibrator
C. monostable multivibrator
D. time-delay circuit

500. If an input 1011 drives a 74141 connected to a nixie tube, the output will be
A. 1011 B. 14
C. zero D. a blank

501. Which of the following *ICs* converts Excess-3 code to decimal code?
A. 74141 B. 7447
C. 7443 D. 7444

502. A 7447 decoder drives
A. Seven-segment display
B. nixie tube
C. relays D. diode matrix

503. Which of the following ICs has only one NAND gate?
A. 7410 B. 7420
C. 7430 D. 7447.

504. The number of 7400s required to make a two-input Ex.OR gate is
A. 1 B. 2
C. 3 D. 4

505. Which of the following ICs allows wired-AND connections?
A. 7410 B. 7990
C. 7447 D. 7440

506. Which of the following code is decoded to decimal values by the diode matrix shows below?

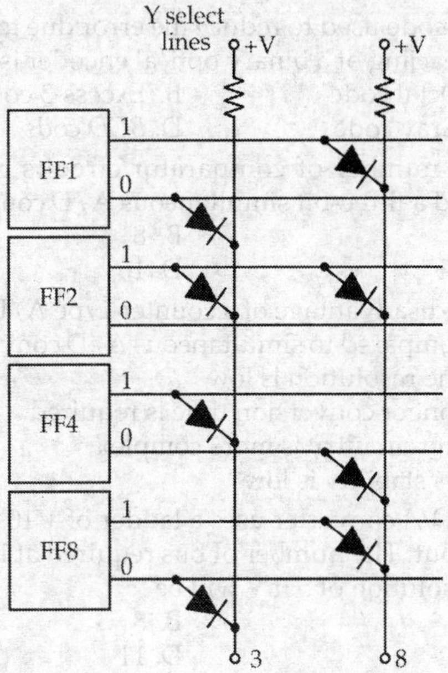

A. *BCD* code B. Gray code
C. Excess-3 code D. Binary code

507. Fan-out for the 74 series is
A. 4 B. 5
C. 8 D. 10

508. The output of the circuit shown below for a digital input of 1010 would be proportional to

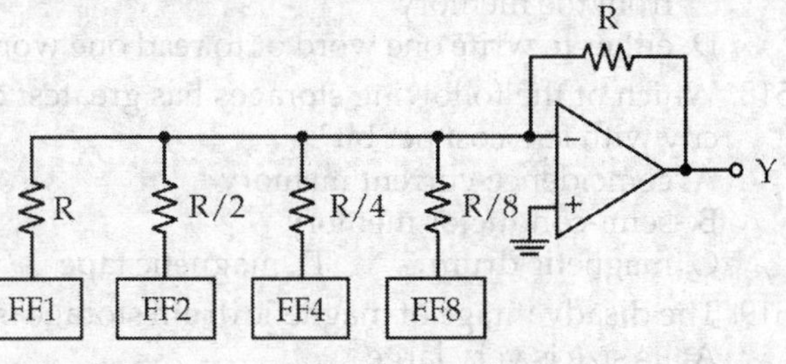

A. 15 C B. –10
C. $-10V/R$ D. $-15/R$

509. The voltage levels for a six-bit binary ladder are '0' = 0 V, '1' = +10 V. Its output for the input 101001 will be
A. 0.423 V B. 0.552 V
C. 0.641 V D. 0.923 V

510. The voltage levels for a four-input resistive divider are 0 = 0 V, 1 = +10 V, The analog output voltage for a digital input of 1001 will be
A. 3 V B. 6 V
C. 8 V D. 9 V

Ans. 497. C 498. D 499. C 500. D 501. C 502. A 503. C 504. A 505. D 506. C 507. D 508. B 509. C 510. B

511. The resolution of a 12-bit D/A converter using a binary ladder with + 10V as the full scale output will be
A. 2.44 mV B. 3.50 mV
C. 4.32 mV D. 5.12 mV

512. The code used to reduce the error due to ambiguity in reacjing of a binary optical encoder is
A. Octal code B. Excess-3 code
C. Gray code D. *BCD* cods

513. The number of comparator circuits required to build a three-bit simultaneous A/D converter is
A. 7 B. 8
C. 15 D. 16

514. The disadvantage of a counter-type A/D converter as compared to simultaneous A/D converter is that
A. the resolution is low
B. longer conversion time is required
C. the circuitry is more complex
D. its stability is low

515. An *D/A* converter uses a ladder of + 10 V full scale output. The number of bits required at its input for a resolution of 5 mV will be
A. 7 B. 8
C. 9 D. 11

516. The degree of resolution that can be obtained using an eight-bit optical encoder will be
A. 1.4° B. 2.8°
C. 3.2° D. 4.8°

517. Access time of a storage is the time required
A. to write one word into the memory
B. to read one word from the memory
C. to write one word into and to read one word from the memory
D. either to write one word or to read one word

518. Which of the following storages has greatest capacity with less cost per bit?
A. coincidence-current memory
B. semi-conductor memory
C. magnetic drum D. magnetic tape

519. The disadvantage of magnetic drum storage is that
A. its size is very large
B. only read out is possible
C. access time is high
D. access time is less

520. A magnetic drum of 8 inch diameter has 100 tracks and storage density of 200 bits/inch. Its storage capacity will be
A. 502400 bits B. 8402 bits
C. 1004800 bits D. 202400 bits

521. An DRO memory can be converted into NDRO memory by employing
A. decade counters B. storage registers
C. address registers D. inhibit lines

522. Which of the following is not true for magnetic tape as compared to paper tape?
A. reading and recoiding is much faster
B. tape-drive unit is an expensive as for paper tapes
C. high volume of data can be stored
D. None of these

523. A parity check usually can detect
A. one-bit error B. double-bit error
C. three-bit error D. any-bit error

524. Single parity check is not reliable in case of
A. paper tape B. magnetic tape
C. magnetic drum D. cores

525. In an *DRO* memory, the information in a core can be read by applying
A. negative half-select current to the proper *X* and *Y* lines
B. positive half-select current to the proper *X* and *Y* lines
C. negative half-select current to *X* lines and positive half-select current to *Y* lines
D. half-select current to *X*, *Y* lines and inhibit line

526. In the coincident-current memory system, magnetic cores switch to state 1 if
A. select current is applied to corresponding X-select lines and X-select lines
B. positive one half-select current is applied to corresponding X-select line and Y select line
C. positive one half-select current is applied to X-select line and negative half-select to Y-select line
D. positive select current to X-select line and negative select current to *Y*-select line

527. The output of the logic circuit using a magnetic core is

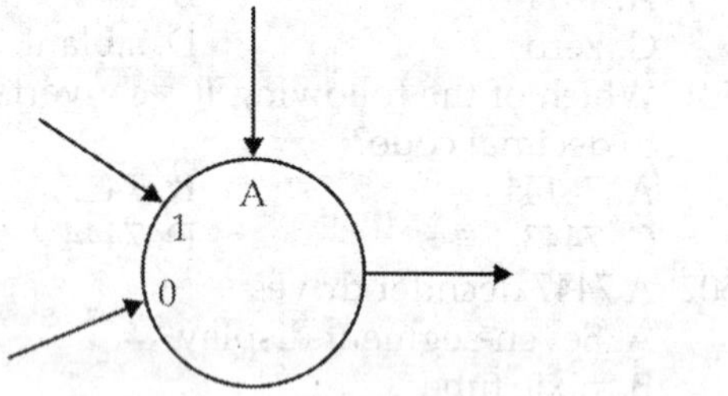

A. XY B. XY
C. XY D. $\bar{X} \cdot \bar{Y}$

528. The number of matrix planes required to store a 20-bit word in a memory array is
A. 2 B. 10
C. 20 D. 40

529. The function of inhibit line in a memory array is
A. to write 1 in a core
B. to read 1 from a core
C. to write 0 in a core
D. to stop writing 0 in a core

Ans. 511. A 512. C 513. A 514. B 515. D 516. A 517. D 518. C 519. C 520. A 521. B 522. B 523. A 524. B 525. A 526. B 527. A 528. C 529. C

530. In a memory system, there are 10 bits for an address word. If first five bits are used to designate an *X* line and second five for an *Y* line, the decimal address to 10110 00101 is
A. $X_{28}Y_4$ B. $X_{30}Y_3$
C. $X_{22}Y_5$ D. $X_{25}Y_{10}$

531. The maximum conversion time of a 10-bit counter type converter driven by a 1 MHz clock is
A. 1.42 μsec B. 10.54 μsec
C. 0.460 μsec D. 1.024 μsec

532. The average access time for a magnetic drum rotating at a speed of 50 revolutions per second is
A. 10 μsec B. 20 μsec
C. 30 μsec D. 50 μsec

533. The add-cycle time of a serial adder for 10-bit words and 1 MHz clock is
A. 10 μsec B. 12 μsec
C. 100 μsec D. 120 μsec

534. The multiplication time for 10-bit numbers with 1 MHz clock will be
A. 32 μsec B. 20 μsec
C. 21 μsec D. 22 μsec

535. The maximum time required to perform a division using 12-bit numbers and a clock of 2 MHz is
A. 12 μsec B. 19 μsec
C. 31 μsec D. 48 μsec

536. Dynamic memory cells are constructed using
A. transistors. B. flip-flops
C. MOSFETs D. FETs

537. Which of the following statements is true for dynamic memory?
A. It has low speed as compared to static memory.
B. It has low power dissipation
C. Its contents tend to decay over period of time
D. All of these

538. As compared to fixed-point machine, a floating-point machine has
A. a floating position for the binary point
B. no provision for powers off
C. provision for powers off
D. no position for the binary point

539. In a digital system, the maximum clock frequency that can be used with master/slave clocked flip-flops having total propagation delay of 200 μsec is
A. 20 MHz B. 50 MHz
C. 100 MHz D. 200 MHz

540. The function of strobe pulse in digital system is
A. to reset memory registers
B. to check the functioning of a logic gate
C. to avoid race problem
D. to shift data in registers

541. In a deck of cards, each card has twenty four 36-bit words. If the cards are read at a rate of 600 cards/min, the rate at which data are entering the system is
A. 1440 bits/sec B. 5184 bits/sec
C. 8640 bits/sec D. 14400 bits/sec

542. Data are recorded on a 2,400 ft reel of magnetic tape at a density of 556 characters per inch. If the record length is of 100 characters and 0.75 inch of record gap, the tape utilization factor is
A. 0.85 B. 0.67
C. 0.19 D. 0.08

543. In the above problem, the total storage capacity of the tape is
A. 3.1 × 10^6 characters B. 4.8 × 10^6 characters
C. 5.2 × 10^6 characters D. 6.4 × 10^6 characters

544. The disadvantage of an open-shift register is that
A. both shift-left and shift-right operations cannot be performed
B. the quantity stored is lost at every shift pulses
C. the register is reset when read-out is over
D. All of these

545. In small scale integration (557), chip contains
A. less than 5 gates B. less than 12 gates
C. less than 100 gates D. less than 500 gates.

546. The propagation delay of a digital IC is the difference between
A. initiating clock pulse and triggering of gates
B. the application of input and its presence at the output
C. the triggering of a flop-flop and change in its state
D. None of these

547. The speed of a digital IC indicates
A. the rate at which output changes.
B. how fast input triggers the output
C. how fast a flip-flop can change stages
D. None of these

548. The unit of the speed of a digital IC is
A. Hz B. KHz
C. MHz D. GHz

549. The speed-power product of a digital IC is the product of
A. speed and propagation delay
B. propagation delay and gate dissipation
C. speed and noise margin
D. speed and gate dissipation

550. In which of the following logics no resistors arc used?
A. TTL B. 4-n-sec ECL
C. 8-sec ECL D. CMOS

551. Which of the following MOSFETs is employed in the CMOS?
A. *p*-channel enhancement type
B. *n*-channel enhancement type

Ans. 530. C 531. D 532. A 533. B 534. D 535. B 536. C 537. C 538. C 539. B 540. C 541. C 542. C 543. A 544. C 545. B 546. B 547. C 548. C 549. B 550. D

C. both *n*-channel and *p*-channel type
D. *p*-channel depletion type

552. What is the function implemented by the following multiplexer chip?

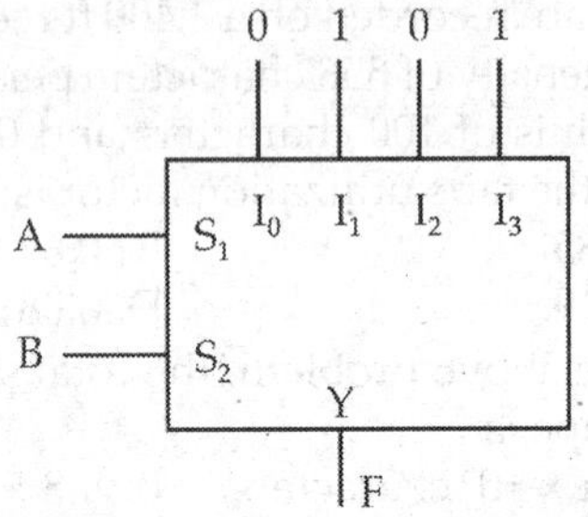

A. $F(A, B) = \overline{A}B$
B. $F(A, B) = A\overline{B} + AB$
C. $F(A, B) = A$
D. $F(A, B) = B$

553. What is the function implemented by the following multiplexer chip?

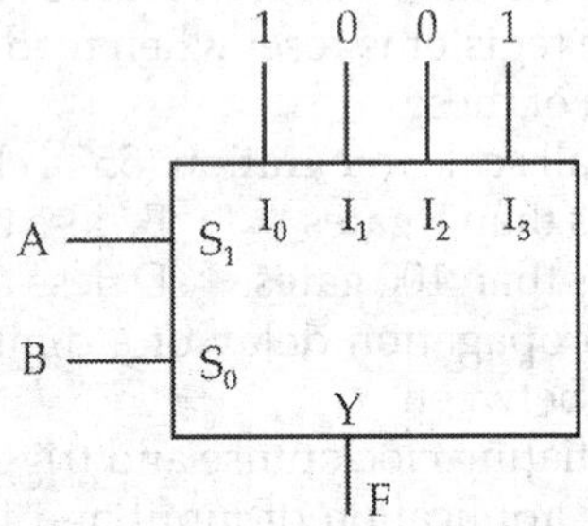

A. $F(A, B) = \overline{A} \oplus \overline{B}$
B. $F(A, B) = A \odot B$
C. $F(A, B) = \overline{A}B$
D. $F(A, B) = \overline{A}B + AB$

554. The microprocessor chip register unit ALU chip has
A. arithmatic functions only
B. basic control and arithmatic functions
C. instructions to execute data
D. memory functions

555. What is the function implemented by the following circuit

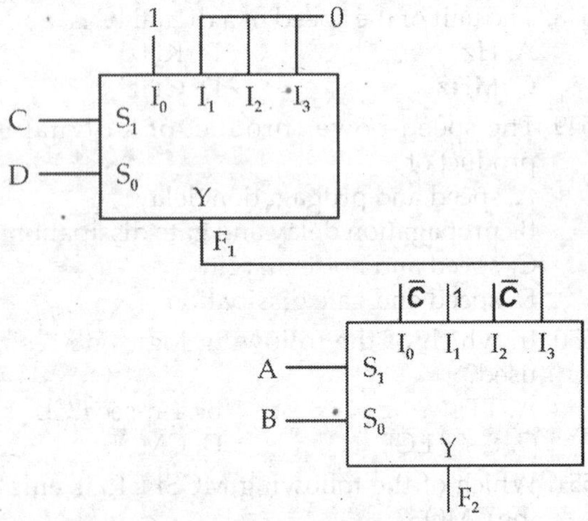

A. $F_2(A.B.C.D) = ABCD - \overline{A}BC\overline{D}$
B. $F_2(A.B.C.D) = CD + AB + BC$
C. $F_2(A.B.C.D) = \overline{A}B + \overline{B}\overline{C} + \overline{C}\overline{D}$
D. $F_2(A.B.C.D) = AB + BC + \overline{C}\overline{D}$

556. The function implemented by the following memory circuit is

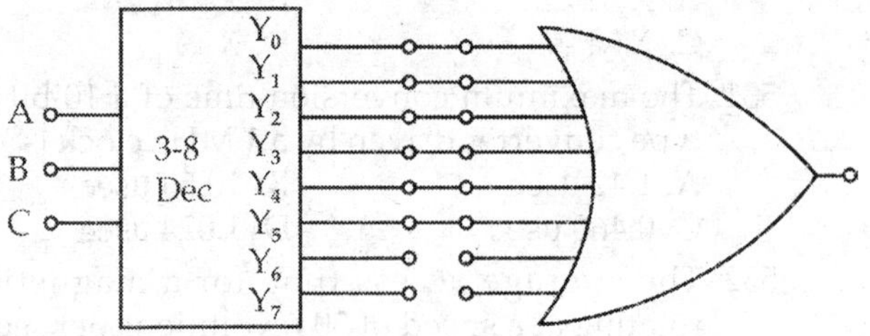

A. $F(A, B, C) = A + B$
B. $F(A, B, C) = \overline{\overline{A}} + B$
C. $F(A, B, C) = \overline{A} + \overline{B}$
D. $F(A, B, C) = \overline{A}B + B\overline{\overline{C}}$

557. The logic used in implementing Boolean expression in above question is
A. AND-OR-INVERTER
B. NAND-OR
C. NAND-NAND
D. AND-OR

558. For the op amp. the gain is 10,000. The exact output voltage will be
A. –5.11 V
B. –5.003V
C. –4.997V
D. –4.910.

559. The binary number 10.1 100.110 in octal numbers is
A. 152.6
B. 154.6
C. 145.6
D. 174.6

560. Which of the following is *not* valid in case of binary product?
A. 0×0
B. $0 \times 1 = 1$
C. $1 \times 1 = 1$
D. All of these

561. 10 in BCD code is represented as
A. 10100
B. 1100
C. 010111
D. None of these

562. Which of the following system is digital?
A. PPM
B. PFM
C. PWM
D. PCM

563. If it is required to multiply two binary numbers in a digital computer in the computer
A. a hardware multiplier is essential
B. it is adequate to have adder-subtractor unit and shift register
C. both a hardware multiplier and an adder subtractor unit are essential
D. a hardware divider is essential

564. Two point charges of 10^{-8} Coul, each are located at points (0,1,2) and (1,0,2). The electric field at point (2,2,0) will be
A. $(\overline{I}_x) + (\overline{I}_y) + (\overline{I}_z)$
B. $(3\overline{I}_x + 3\overline{I}_y + 3\overline{I}_z)$
C. $(3\overline{I}_x + 3\overline{I}_y - 4\overline{I}_z)$
D. $10/3[3\overline{I}_x + 3\overline{I}_y - 4\overline{I}_z]$

565. Loader in a computer is a
A. hardware
B. software
C. auxiliary
D. none of these

Ans. 551. C 552. D 553. B 554. B 555. C 556. C 557. D 558. C 559. B 560. B 561. D 562. D 563. B 564. D 565. A

Fundamentals of Computer

1. LSI usually contains
A. 10 gates B. 10 to 100 gates
C. more than 100 gates D. more than 10,000 gates
2. What is the state of the signal at X?
A. Don't know B. Floating
C. Changing their states
D. All high
3. The microprocessor contains *ROM* chip contains
A. control functions B. arithmatic functions
C. instructions to execute data
D. memory functions
4. The physical part or electronic circuitry of a computer system is called
A. hardware B. software
C. microprocessor D. peripheral device
5. The central processing unit of a small computer system, including ALU and control functions on a single LSI chip is called
A. hardware B. memory
C. microprocessor D. peripheral device
6. A device that works in conjunction with a computer but not as part of it is called
A. hardware B. memory
C. microprocessor D. peripheral device
7. A device in which computer information and data are stored and relieved at a later time is called
A. hardware B. memory
C. microprocessor D. peripheral device
8. The act of preparing a list or sequences of instructions for a computer for the solution to a problem is called
A. hardware B. memory design
C. programming D. controlling hardware
9. A single binary digit is called
A. byte B. bit
C. data D. logic
10. A fixed step by step procedure for finding a solution of a problem on computer is called
A. hardware B. logic
C. algorithm D. none of these
11. A bit or bits store for later use, indicating the status of CPU is called
A. flag B. data
C. logic D. word
12. A program structure that allows the repeated operation of a particular sequence of instructions until a specified termination is reached is called
A. sub-routine B machine
C. module D. loop
13. A program that can be used repeatedly throughout a major program is called
A. program module B. loop
C. sub-routine D. template
14. What do the symbol represents for?

A. Connector B. Terminator
C. Process D. Input/output
15. What do the symbol represents for?

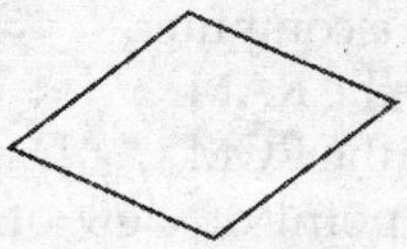

A. Flow connector B. Terminator
C. Decision D. Process
16. What do the symbol represents for?

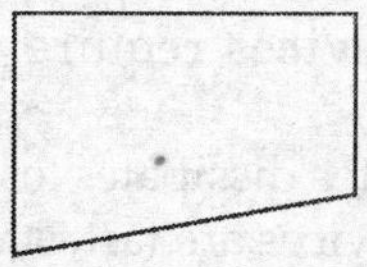

A. Magnetic Tape B. Process
C. Terminater D. Decision
17. What do the symbol represents for?

A. Connector B. Terminator
C. Decision D. Process
18. What do the symbol represents for?

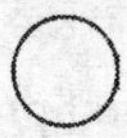

A. Process B. Terminates
C. Magnetic tape D. Connector
19. A block of successive memory locations that is accessible on a last-in, first-out basis is called
A. accumulator B. register
C. pointer D. stack
20. A programme that translates symbolically represented instructions into their binary equivalents is called
A. loader B. assembler
C. linker D. autoloader
21. To correct and eliminate errors in programmes the act of using an instruction, programme if action designed in software is called

Ans.	1. C	2. C	3. C	4. A	5. C	6. D	7. B	8. C	9. B	10. C	11. A	12. D	13. C
	14. D	15. C	16. A	17. B	18. D	19. D	20. B						

A. interrupting B. cross-assembling
C. documentation D. debugging

22. A system of letters, numbers and symbols adopted by computer manufacturer as an abbreviated form of instruction sets is called
A. modern B. monitor
C. mnemonic D. mask

23. The processing of causing an unplanned branching operation to occur, usually initiated by external system is called
A. debugging B. masking
C. interrupt D. iteration

24. A microprocessor contains
A. most of the control and arithmatic logic functions of a computer
B. most of the RAM
C. most of the ROM D. peripherial drivers

25. From the point of view of integration, a microprocessor chip employs
A. SSI B. MSI
C. LSI D. GSI

26. Which of the following statements is not correct?
A. MOS devices require fewer manufacturing cycles
B. MOS gates dissipates less power per gate
C. MOS devices are fastest devices
D. MOS gates require less space on the silicon wafer

27. The first practical microprocessor was introduced in
A. 1969 B. 1971
C. 1973 D. 1975

28. The first practical microprocessor was introduced by
A. Motorola
B. National semiconductors
C. Intel D. Fairchild

29. The number of first practical microprocessor was
A. Intel 4001 B. M6800
C. Intel 4004 D. F-8

30. The bit size of first practical microprocessor was
A. 2 bit B. 4 bit
C. 8 bit D. 4 bytes

31. How many pins Intel 8080 has?
A. 28 pins B. 36 pins
C. 40 pins D. 42 pins

32. Who manufactures 6800 μP?
A. Fairchild B. Intel
C. Zilog D. Motorola

33. What is the bit size of 6800 μF?
A. 4 bit B. 8 bit
C. 12 bit D. 16 bit

34. What is bit size of 8080 μP?
A. 4 bit B. 8 bit
C. 12 bit D. 16 bit

35. A byte is
A. a group of 2 bits B. a group of 4 bits
C. a group of 8 bits D. a group of 16 bits

36. A central processing unit of a computer contains.
A. control and arithmatic logic unit
B. control and RAM
C. control, clock, RAM and ROM
D. control, ALU clock, RAM and ROM

37. An instruction register is storage for
A. location of data in memory
B. location of instruction in memory
C. binary code for the operation to be performed
D. address of the next instruction to be executed

38. A program counter is a storage register for
A. location of data in memory
B. location of instruction in memory
C. binary code for the operation to be performed.
D. address of the next instruction to be executed

39. Stack pointer is a register which comes into use.
A. whenever a data is read from the memory
B. whenever a data is written into the memory
C. whenever the output variable is sent out of the CPU
D. Whenever an interrupt or high priority call comes from external devices

40. In microprocessor architecture, flag indicates.
A. the number of the microprocessor
B. the name of the manufacturer
C. the internal status of the CPU
D. The bit-size of the microprocessor

41. An address is the number used by the CPU to specify
A. a location in the memory
B. a location in flags
C. a location in accumulator
D. a location in stack pointer

42. Which of the following is not correct?
A. Bus is a group of wires
B. Boothstrap is a technique or device for loading first instruction
C. An instruction is a set of bits that defines a computer operation
D. An interrupt signal is required at the start of every programme

43. Which of the following is a correct statement?
A. Hardware is merely the preparation of printed circuit board for the connection of microprocessor μP
B. Software is to provide the required IC chips for μP

Ans.													
	21. D	22. C	23. C	24. A	25. C	26. C	27. B	28. C	29. A	30. B	31. C	32. D	33. B
	34. B	35. C	36. A	37. C	38. D	39. D	40. C	41. A	42. D				

C. Hardware is to develop programmes and to feed into the computer
D. Software is the permanent informational structure

44. Which of the following is the appropriate and correct statement?
A. Software is typing the given programme in card-punching machine
B. Software is to connect the RAMs and ROMs to the main µP
C. Hardware is the relatively permanent physical structure to perform an algorithm
D. Hardware is the technique how to install a computer

45. A microcomputer is a minimum combination of
A. µP and clock B. µP clock, ROMs
C. µP, clock, RAMs and ROMs
D. µP, clock, RAM, ROM, PIA and ACIA

46. *ACK* indicates reception of
A. correct data B. incorrect data
C. insufficient data D. sufficient data

47. Accumulation is a device which
A. stores a number B. adds two numbers
C. adds and stores previously stored number to another number
D. none of these

48. Address usually is
A. alphabetical B. numerical
C. alpha-numerical D. either of these

49. Algorithm is used
A. to bring itself into desired state by which its own action
B. to perform logarithmic operations
C. to describe a set of procedure by given result is obtained
D. none of these

50. American standard code for information interchange (ASCII) employs a coded character set consisting of
A. 7 bits B. 7 bits with parity check
C. 8 bit
D. 8 bit with parity check

51. In a train of binary signals, 1 baud is equal to
A. 1/4 bit/sec B. 1/2 bit/sec
C. 1 bit/sec D. 2 bits/sec

52. In a train of signals each of which can assume one out of eight different states, 1 baud is equal to
A. 2 bits/sec B. 3 bits/sec
C. 4 bits/sec D. 8 bits/sec

53. Benchmark Problem is used to evaluate the performance of
A. hardware B. software
C. both hardware and software
D. All of these

54. Bootstrap is a technique designed
A. to transfer control to specified closed sub-routine
B. to suspend the main programme for a while
C. to bring itself into a desired state by means of its own action
D. None of these

55. *Call* in computer systems means
A. to excute a part of main programme
B. to call a sub-progrmme to be executed
C. to transfer control to specified closed sub-routine
D. None of these

56. Carry-Look-Ahead is a type of adder in which
A. inputs to several stages are examined
B. carries are produced simultaneously
C. carries are produced before hand
D. None of these

57. CPU of a computer system does not contain
A. main storage B. arithmatic unit
C. special register group D. None of these

58. Which of the following operations is not performed by the CPU of a computer system?
A. Arithmatic operations
B. Providing timing signals
C. Control instruction processing
D. None of these

59. An assembler in a computer system prepares
A. machine-language programme from a symbolic language programme
B. object programme
C. assembles computer instructions and data in the machine
D. none of these

60. Which of the following storage elements can be programmed after packaging?
A. RAM B. ROM
C. PROM D. Silo memory

61. Propogation delay is the time required for
A. a change in level at the output of an element
B. output to appear when input is applied
C. the element of become operational
D. a signal to propagate in a circuit

62. In RAM, all the information can be obtained at the output with
A. unequal time delay
B. nearly same time delay
C. minimum time delay
D. maximum time delay

63. Scratch-pad memory is a
A. first-in first-out memory
B. local permanent memory
C. local temporary memory
D. last-in first-out memory

Ans. 43. D 44. C 45. D 46. A 47. C 48. B 49. C 50. D 51. C 52. B 53. D 54. C 55. C
56. C 57. D 58. B 59. A 60. C 61. A 62. B 63. C

64. Silo or FIFO Memory reads stored data in a
A. first-in first-out mode
B. first-in last-out mode
C. last-in first-out mode
D. last-in last-out mode

65. The disadvantage of static storage elements is that
A. information is to be refreshed frequently
B. information is lost when power is removed
C. information is stored forever
D. none of these

66. Volatile storage is a device in which data are lost
A. with time
B. when power is removed
C. When programme is over
D. none of these

67. In Push-down list, the item retrieved is the
A. most recently stored item
B. first item stored in last stack
C. oldest stored item D. none of these

68. In Push-up list, the item retrieved is the
A. most recently stored item
B. first item stored in last stack
C. oldest stored item D. none of these

69. Which of the following element does not belong to micro-computer system?
A. Microprocessor B. ROM
C. RAM D. all of these

70. The basic elements of a microprocessor are
A. ALU, memory
B. ALU, control unit
C. ALU, memory, I/O devices
D. ALU, control unit, memory

71. Byte is synonymous with
A. band B. character
C. bit D. word

72. Flopy disc is a
A. flexible storage B. permanent storage
C. ROM D. fast storage

73. Cycle time is the time required for the execution of a
A. item to be written and then read from the memory
B. item to be read from the memory
C. programme instruction
D. subroutine

74. Flag is a character for
A. identification of a word
B. occurance of some condition
C. marking or tagging D. all of these

75. Hamming code is
A. an error-indication code system
B. an error-correction code system
C. a programme conversion code system
D. none of these

76. The function of Interleave is to assign successive storage location numbers to
A. physically separated memory locations
B. physically adjacent memory locations
C. alternate memory location
D. none of these

77. Logic shift affects
A. odd positions B. all positions
C. even positions D. zero positions

78. Muemonic symbols are used
A. to denote address
B. to employ hamming code
C. to denote errors
D. to assist human memory

79. Modem is a
A. code changing system
B. programme tonversion system
C. modulator-demodulator system
D. demodulator-modulator system

80. An Input Port is
A. Tri-state buffer with register
B. Select logic
C. Input device D. R/W memory

81. Programmable Logic Array (PLA) uses
A. ROM matrices B. PROM matrices
C. RAM matrices D. Silo memory

82. To a great extent, computer in function is similar to
A. CPU B. Buffer
C. Assembler D. PLA

83. Which of the following is the redundancy check character?
A. Double precision B. Multiple precision
C. Parity bit D. CRC

84. Debug is synonmous to
A. erase B. exponent
C. emulate D. trouble shoot

85. Which of the following actions detect locations, and remove mistakes from a programme routine?
A. Erase B. Debug
C. Diagnose D. Emulate

86. Direct-memory access channel (DMA) facilitates data to move into and out of the system.
A. on first-come first-serve basis
B. with equal time delay
C. without sub-routine
D. without programme intervention

87. Double precision employs
A. double signal speed
B. two computer words to represent a number
C. erro-reduction code
D. None of these

Ans.												
64. A	65. B	66. B	67. A	68. C	69. D	70. B	71. B	72. A	73. C	74. D	75. B	76. A
77. C	78. D	79. C	80. A	81. A	82. C	83. D	84. D	85. B	86. D	87. B		

88. Dump means
A. erasing used data
B. storing used data in pushdown stack
C. copying data from internal stage to external storage
D. none of these

89. The disadvantage of dynamic storage elements is that
A. storage cells are to be refreshed frequently
B. information stored previously is lost
C. power has to be kept on
D. none of these

90. Every processor must necessarily have a
A. data bus B. control bus
C. data bus and address bus
D. data bus, a control bus and an address bus

91. A microprocessor with a 12-bit address bus will be able to access
A. 1 kilobytes of memory
B. 4 kilobytes of memory
C. 8 kilobytes of memory
D. 0.4 kilobytes of memory

92. Which of the following flag conditions are not available in 8085 processor?
A. Zero flag B. Parity flag
C. Overflow flag D. Auxiliary carry flag

93. The frequency of the driving network connected between pin 1 and 2 of a 8085 chip must be
A. twice of the desired clock frequency
B. equal to the desired clock frequency
C four times the desired clock frequency
D. None of the above

94. READY signal in 8085 is useful when the CPU communicates with
A. a slow peripheral device
B. a fast peripheral device
C. a DMA controller chip
D. aPPIchip

95. Which of the following task is *not* performed by an assembler
A. providing storage allocation
B. creating a table of lables etc.
C. doing assembly time arithmetic
D. translate a program written in high level language to machine code program

96. A microcomputer chip essentially contains
A. memory (ROMs and RAMs) and ALU
B. memory, CPU, and I/O lines
C. CPU only D. None of the above

97. A DAD H instruction is same as
A. shifting each bit one position to the left
B. shifting each bit one position to the right
C. shifting each bit one position to the left with a zero inserted in first position
D. shifting each bit one position to the right with a zero inserted in lsb position

98. Return from a subroutine is affected by
A. a jump instruction B. an RST instruction
C. a RESET instruction
D. a hardware interrupt signal

99. During a DMA transfer, the processor (check the incorrect statement)
A. continues its normal operations
B. suspends its normal operations
C. needs to initiate read (write) command
D. needs to check if the input/output device is ready for data transfer

100. In 8085, interrupts except TRAP are disabled (check the incorrect statement) by
A. DI instruction B. a system reset
C. acknowledgment of a previous interrupt
D. None of the above

101. Which of the following methods used in formatting a floppy disk is not IBM compatible
A. single density single side
B. double density single side
C. single density double side
D. All of the above

102. Which of the following control signal does not belong to a printer interface
A. DAV B. DACC
C. NEW LINE ACK D. TRACK 00

103. A USART chip provides
A. half duplex operation
B. full duplex operation
C. duplex operation
D. full duplex operation but cannot work in asynchronous mode.

104. A high on RESET OUT line signifies that
A. all the registers of the CPU are being reset
B. all the registers and counters are being icset
C. all the registers and counters are being reset and in addition this signal can be used to reset external support chips
D. processing can begin when this signal goes high

105. Normally a microprocessor cycles between
A. Fetch and Halt states
B. Fetch and Interrupt states
C. Fetch and Executes states
D. Halt and Executes states

106. Alter completing the execution, microprocessor returns to
A. Halt state B. Fetch state
C. Execute state D. Interrupt state

Ans. 88. C 89. A 90. D 91. A 92. C 93. A 94. A 95. D 96. B 97. C 98. C 99. A 100. D 101. B 102. D 103. B 104. C 105. C 106. B

107. The chip select access time for reading ROM contents is
 A. the delay between application of proper chip select signal and the stable output data
 B. the delay between the previous valid output data and the next change in address
 C. the time for which the output data remains valid when the device is no longer selected
 D. maximum time for which the valid address can be changed

108. The deselect to output float time is
 A. the delay between application of proper chip select signal and the stable output data
 B. the delay between the previous valid output data and the next change in address
 C. the time for which the output data remains valid when the device is no longer selected
 D. maximum time for which the valid address can be changed

109. The minimum time output data will be valid after address changes is called
 A. t_{DF} B. t_{RC}
 C. t_{OH} D. t_A

110. In writing mode of R/W memory, we can not begin writing until following time has elapsed
 A. t_{CW} B. t_{WCY}
 C. t_{DH} D. t_{AW}

111. The minimum time after write pulse application, the data should not change
 A. t_{CW} B. t_{WR}
 C. t_{DH} D. t_{AW}

112. The maximum allowable delay between address change and $1 \rightarrow 0$ transition of write pulse is
 A. t_{AW} B. t_{WR}
 C. t_{DH} D. t_{DW}

113. The maximum data/wire pulse overlays is
 A. t_{DH} B. t_{CW}
 C. t_{DH} D. t_{WR}

114. The maximum time from assertion of chip select to the end of write pulse is
 A. t_{DH} B. t_{AW}
 C. t_{WR} D. t_{CW}

115. The minimum time between writing operation is specified by
 A. t_{CW} B. t_{WCY}
 C. t_{WR} D. t_{DH}

116. A microprocessor, on arrival of RESET signal returns from HALT state to
 A. Execute B. Fetch
 C. Interrupt D. None of the above

117. What is the state of the bus at X?

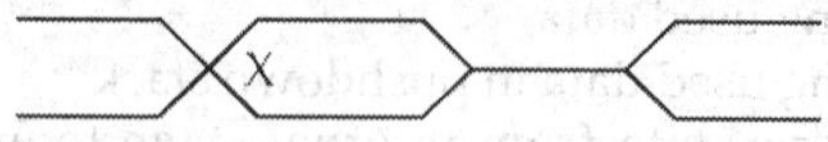

 A. Bus is High B. All wires are Low
 C. Some wires are floating
 D. Signals are changing their states

118. What is the state of the signals at X?

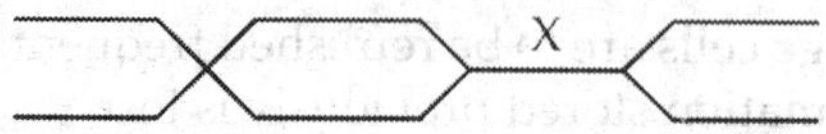

 A. Don't know B. Floating
 C. Changing their states D. All Low

119. What is the state of the signals at X?

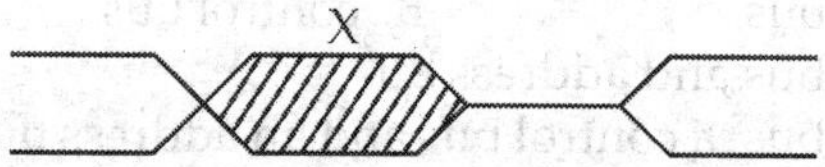

 A. Don't know B. Floating
 C. Changing their states D. All high

120. What is meant by the arrow shown in figure below?

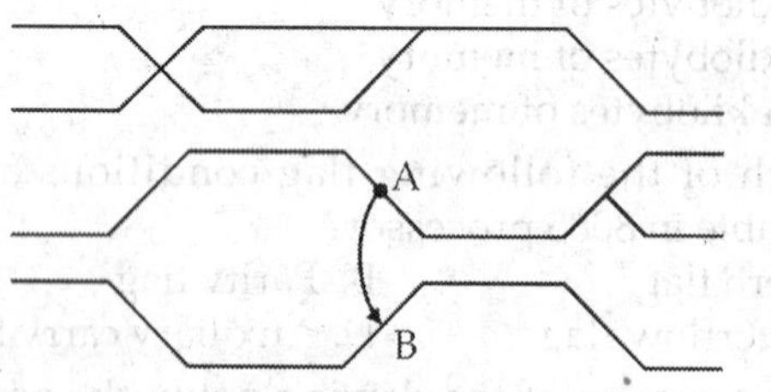

 A. A is high when B is low
 B. A causing don't know state
 C. High-low transition at A causes low high transition at B
 D. A causing floating signal at B

121. What is meant by the arrows shown in figure below?

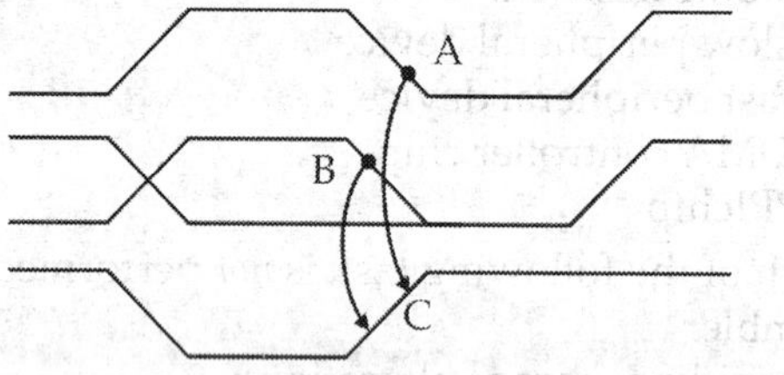

 A. Low transition of *B* causes *A* and *C*
 B. Low transition of *B* and low transition at *A* causes high transition at *C*
 C. Low transition of *B* or low transition at *A* causes high transition at *C*
 D. High transition at *C* causes low transition at *A* and *B*

122. The translator program that converts source code in high level language into machine code line by line is called
 A. assembler B. compiler
 C. loader D. interpreter

Ans. 107. A 108. C 109. C 110. D 111. C 112. A 113. C 114. D 115. B 116. B 117. D 118. B 119. B 120. C 121. B 122. D

123. The decimal equivalent of the hexadecimal number $(BAD)_{16}$ is
A. 111013 B. 5929
C. 3416 D. 2989

124. ACCESS in magnetic drum memory is
A. completely random
B. sequential and cyclic
C. partly random and partly cyclic sequential
D. a cyclic sequential

125. A decade counter requires
A. 2 flip-flops B. 3 flip-flops
C. 4 flip-flops D. 100 flip-flops

126. In RAM, all the information can be obtained at the output with
A. unequal time delay B. minimum time delay
C. nearly same time delay
D. maximum time delay

127. Cycle time is the time required for the execution of an
A. item to be written and then read from the memory
B. item to be read from the memory
C. programme instruction
D. subroutine

128. The output voltage v_0 in the given amplifier circuit is

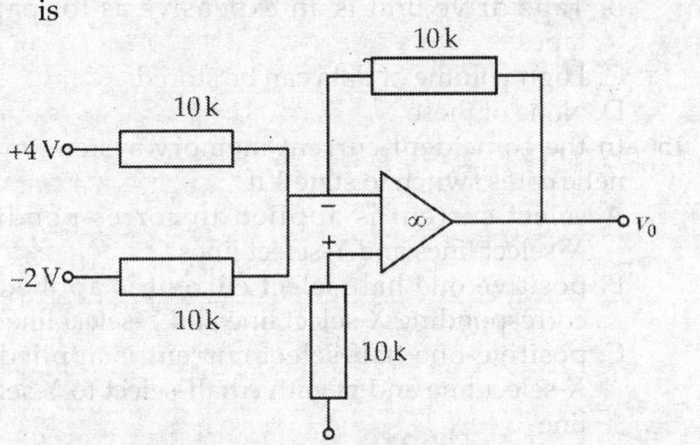

A. 4 V B. 6 V
C. 8 V D. 10 V

129. Logic shift affects
A. odd positions B. all positions
C. even positions D. zero positions

130. Direct-Memory Access Channel (DMA) facilitates data to move into and out of the system
A. on first-come first-serve basis
B. with equal time delay
C. without sub-routine
D. without programme intervention

131. *Dump* means
A. erasing used data
B. storing used data in pushdown stack
C. copying data from internal storage to external storage
D. none of these

132. An electronic wrist-watch has a clock of frequency 32 KHz. To divide this frequency down to 1 Hz, it is necessary to have
A. two decade counters, one two-bit binary counter and a T flip-flop
B. three decade counters, one four-bit binary counter and one T flip-flop
C. one four-bit binary counter and T flip-flop and one decade counter
D. four-bit binary counters and decade counter

133. The code used to reduce the error due to ambiguity in reading of a binary optical encoder is
A. Octal code B. Excess-3 code
C. Gray code D. BCD code

134. Which of the following ICs converts Excess-3 code to decimal code?
A. 74141 B. 7447
C. 7443 D. 7444

135. A ring counter is same as
A. up-down counter B. parallel counter
C. shift register D. none of these

136. The circuit shown below is a

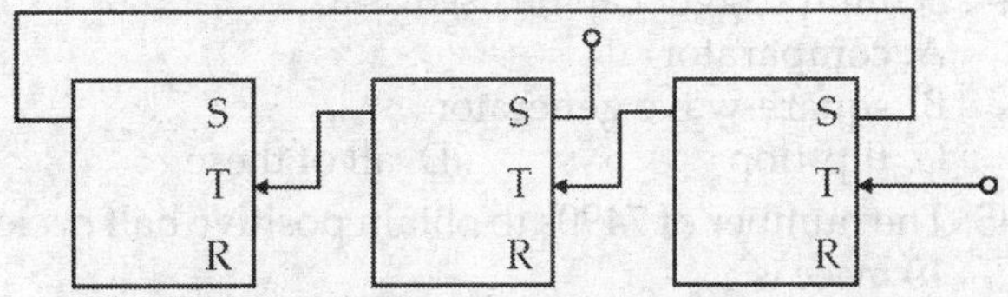

A. mod-4 counter B. mod-5 counter
C. mod-6 counter D. mod-7 counter

137. In order to built a mode-6 counter using three flip-flops, the number of possible counter sequence is
A. 4 B. 6
C. 16 D. 28

138. Which of the following circuits exhibits memory?
A. Astable multivibrator
B. Bistable multivibrator
C. NAND gate D. Ex OR gate

139. The storage time of a transistor is part of the
A. turn-on time B. turn-off time
C. fall time D. rise time

140. If turn-on time of a transistor is decreased, its turn-on time
A. decreases B. increases
C. is not affected D. infinite

141. A program counter is a storage register for
A. location of data in memory
B. location of instruction in memory
C. binary code for the operation to be performed
D. address of the next instruction to be executed

142. Stack pointer is a register which comes into use whenever
A. a data is read from the memory
B. a data is written into the memory

Ans. 123. D 124. C 125. C 126. C 127. C 128. D 129. C 130. D 131. C 132. B 133. C 134. C 135. C 136. B 137. D 138. B 139. B 140. B 141. D

C. the output variable is sent out of the CP
D. an interrupt or high priority call comes from external devices

143. A clock pulse is fed into 3-bit binary down counter. The signal at output *B* is shown in figure

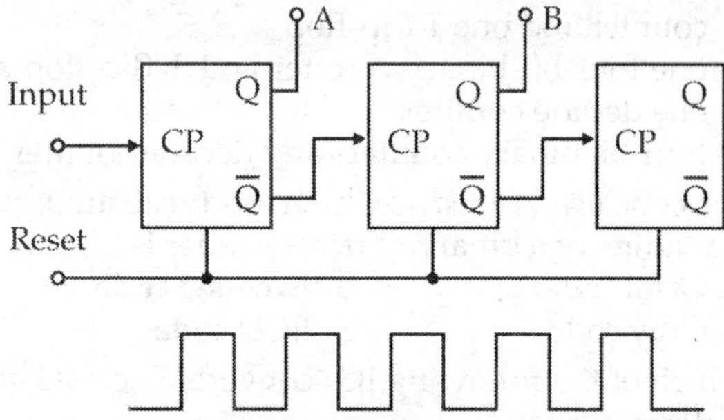

A. 

B. 

C. 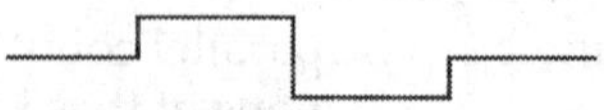

D. none of these

144. Schmitt trigger can be used as a
A. comparator
B. square-wave generator
C. flip-flop D. all of these

145. The number of 7490s to obtain positive half cycle of 10 msec is
A. 3 B. 4
C. 5 D. 7

146. Two scalers, one with scaling factor n_1 and other with scaling factor n_2 are connected in series n_1 before n_2. The maximum dividing speed of the combined scaler is determined by the speed of
A. the first scaler n_1 B. the second scaler n_2
C. both $(n_1 \times n_2)$ D. either n_1 or n_2

147. Access time of a storage is the time required to
A. write one word into the memory
B. read one word from the memory
C. write one word into and to read one word from the memory
D. either to write one word or to read one word

148. Which of the following storages has greatest capacity with less cost per bit?
A. Magnetic tape B. Magnetic drum
C. Coincidence-current memory
D. Semiconductor memory

149. The disadvantage of magnetic drum storage is that
A. its size is very large
B. only read out is possible
C. access time is high D. access time is less

150. A magnetic drum of 8 inch diameter has 100 tracks and storage density of 200 bits/inch. Its storage capacity will be
A. 502400 bits B. 8402 bits
C. 1004800 bits D. 202400 bits

151. A DRO memory can be converted into NDRO memory by employing
A. decade counters B. storage registers
C. address registers D. inhibit lines

152. A parity check usually can detect
A. one-bit error B. double-bit error
C. three-bit error D. any-bit error

153. In a DRO memory, the information in a core can be read by applying
A. negative half-select current to the proper *X* and *Y* lines
B. positive half-select current to the proper *X* and *Y* lines
C. negative half-select current to *X* lines and positive half-select current to *Y* lines
D. half-select current to *X*, *Y* lines and inhibit line

154. Single parity check is not reliable in case of
A. proper tape B. magnetic tape
C. magnetic drum D. cores

155. Which of the following is not true for magnetic tape as compared to proper tape?
A. Reading and recording is much faster
B. Tape drive unit is an expensive as for paper tapes
C. High volume of data can be stored
D. None of these

156. In the coincident-current memory system, magnetic cores switch to state 1 if
A. select current is applied to corres-ponding *X*-select lines and *Y*-select lines
B. positive one half-select current is applied to corresponding *X*-select line and *Y*-select line
C. positive-one half-select current is applied to *X*-select line and negative half-select to *Y*-select line
D. positive select current to *X*-select line and negative select current to *Y*-select line

157. The function of inhibit line in a memory array is to
A. write 1 in a core B. read 1 from a core
C. write 0 in a core D. stop writing 0 in a core

158. The average access time for a magnetic drum rotating at a speed of 50 revolutions per second is
A. 10 m sec B. 20 m sec
C. 30 m sec D. 50 m sec

159. The number of matrix planes required to store a 20-bit word in a memory array is
A. 2 B. 10
C. 20 D. 40

160. Data are recorded on a 2400 ft reel of magnetic tape at a density of 556 characters per inch. If the record length is of 100 characters and 0.75 inch of record gap, the tape utilisation factor is

Ans. 142. D 143. A 144. D 145. C 146. A 147. D 148. B 149. C 150. A 151. B 152. A 153. A 154. B 155. B 156. B 157. C 158. A 159. C

A. 0.85 B. 0.67
C. 0.19 D. 0.08

161. Dynamic memory cells are constructed using
A. transistors B. flip-flops
C. MOSFETs D. FETs

162. The maximum conversion time of a 10-bit counter type converter driven by a 1 MHz clock is
A. 1.42 μ sec B. 10.54 μ sec
C. 0.460 μ sec D. 1.024 μ sec

163. The add-cycle time of a serial adder for 10-bit words and 1 MHz clock is
A. 10 μ sec B. 12 μ sec
C. 100 μ sec D. 120 μ sec

164. Which of the following statements is true for dynamic memory?
A. It has low speed as compared to static memory
B. It has low power dissipation
C. Its contents tend to decay over period of time
D. All of these

165. The circuit diagram represents a

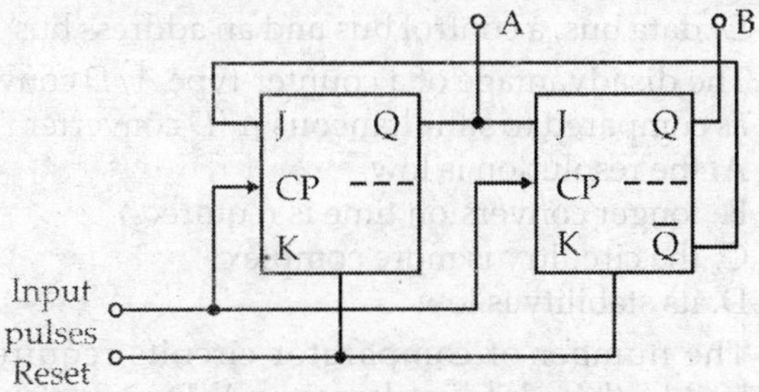

A. mod-3 counter B. mod-5 counter
C. 3-bit shift register D. mod-2 counter

166. The maximum time required to perform a division using 12-bit numbers and a clock of 2 MHz is
A. 12 μ sec B. 19 μ sec
C. 31 μ sec D. 48 μ sec

167. As compared to fixed-point machine, a floating-point machine has
A. a floating position for the binary point
B. no provision for powers of 2
C. provision for powers of 2
D. no position for the binary point

168. In a deck of cards, each card has twenty-four 36-bit words. If the cards are read at a rate of 600 cards/minute, the rate at which data are entering the system is
A. 1440 bits/sec B. 5184 bits/sec
C. 8640 bits/sec D. 14400 bits/sec

169. The disadvantage of an open-shift register is that
A. both shift-left and shift-right operations cannot be performed
B. the quantity stored is lost at every shift pulses
C. the register is reset when read-out is over
D. all of these

170. How many illegitimate states has a synchronous mod-6 counter?
A. 3 B. 2
C. 1 D. 0

171. Which of the following is the redundancy check character?
A. Double precision B. Multiple precision
C. Parity bit D. CRC

172. Which of the following storage elements can be programmed after packing?
A. RAM B. ROM
C. PROM D. Silo memory

173. A high on RESET OUT line signifies that
A. all the registers of the CPU are being reset
B. all the registers and counters are being reset
C. all the registers and counters are being reset and in addition this signal can be used to reset external support
D. processing can begin when this signal goes high

174. Scratch-pad memory is a
A. first-in first-out memory
B. local permanent memory
C. local temporary memory
D. last-in first-in memory

175. Silo or FIFO memory reads stored data in a
A. first-in first-out mode
B. first-in last-out mode
C. last-in first-out mode D. last-in last-out mode

176. The disadvantage of static storage elements is that
A. information is to be refreshed frequently
B. information is lost when power is removed
C. information is stored for ever
D. none of these

177. Volatile storage is a device in which data are lost
A. when programme is over
B. when power is removed
C. with time D. none of these

178. In push-down list, the item retrieved is the
A. most recently stored item
B. first item stored in last stack
C. oldest stored item D. none of these

179. In push-up list, the item retrieved is the
A. most recently stored item
B. first item stored in last stack
C. oldest stored item D. none of these

180. Which of the following element does not belong to microcomputer system?
A. Microprocessor B. ROM
C. RAM D. All of these

181. The disadvantage of dynamic storage elements is that
A. storage cells are to be refreshed frequently
B. information stored previously is lost

Ans.	160. C	161. C	162. D	163. B	164. C	165. A	166. B	167. C	168. C	169. C	170. B	171. D	172. C
	173. C	174. C	175. A	176. B	177. B	178. A	179. C	180. D	181. A				

C. power has to be kept on
D. none of these

182. The function of Interleave is to assign successive storage location numbers to
A. physically separated memory locations
B. physically adjacent memory locations
C. alternate memory locations
D. none of these

183. The commutating capacitor reduces turn on time because it
A. acts as an open circuit
B. allows maximum base current to flow
C. aids in removing the excess carriers from the base region
D. none of these

184. Semiconductor ROMs are sometimes preferred to semiconductor RAMs because
A. ROMs are cheaper than RAMs
B. ROMs are faster
C. ROMs do not require power supply for their operation
D. program stored in ROMs cannot be altered either by the power failure or by the user

185. Access in magnetic drum memory is
A. completely random B. sequential and cyclic
C. a cyclic sequential
D. partly random and partly cyclic sequential

186. Direct-memory-access channel facilitates data to move into and out of the system
A. without subroutine B. with equal time delay
C. without programmes intervention
D. on first come first serve basis

187. 2764 is 65536 bit EPROM organised as 8192 words of 8 bits each. It has
A. 10 address lines and 10 data lines
B. 12 address lines and 10 data lines
C. 13 address lines and 8 data lines
D. 15 address lines and 12 data lines

188. Fastest memory cell is
A. bubble memory B. semiconductor RAM
C. semiconductor memory
D. superconductor memory

189. All of the following are non-volatile memories except
A. ROMs B. Semiconductor RAM
C. PROMs D. EPROMs

190. A semiconductor read-only-memory basically is a
A. sequential circuit with flip-flop
B. sequential circuit with flip-flop and gates
C. set of flop-flip memory elements
D. combinational logic circuit

191. Number of comparators required to build a 5-bit analog to digital converter (ADC) is
A. 5 B. 11
C. 21 D. 31

192. Double precision constants in comparison with single precision constants have twice the
A. significant digits in decimal representation
B. number of binary digits for representation
C. range of magnitude D. accuracy

193. To a great extent, computer in function is similar to
A. CPU B. Buffer
C. Assembler D. PLA

194. Double Precision employs
A. double signal speed
B. two computer words to represent a number
C. error-reduction code D. none of these

195. After completing the execution, microprocessor returns to
A. halt state B. fetch state
C. execute state D. interrupt state

196. Every processor must necessarily have a
A. data bus
B. data bus and address bus
C. control bus
D. data bus, a control bus and an address bus

197. The disadvantage of a counter-type A/D converter as compared to simultaneous A/D converter is that
A. the resolution is low
B. longer conversion time is required
C. the circuitry is more complex
D. its stability is low

198. The number of comparator circuits required to build a three-bit simultaneous A/D converter is
A. 7 B. 8
C. 15 D. 16

199. The advantage of magnetic drum storage is
A. high access time B. large size
C. only read out is possible
D. low access time

200. The resolution of a 12-bit D/A converter using a binary adder with +10 V as the full scale output will be
A. 2.44 mV B. 3.50 mV
C. 4.32 mV D. 5.12 mV

201. The graph in the figure below represents a finite state machine. Which one of the following regular expressions describes the set strings recognised by the finite state machine?

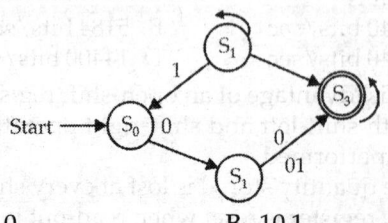

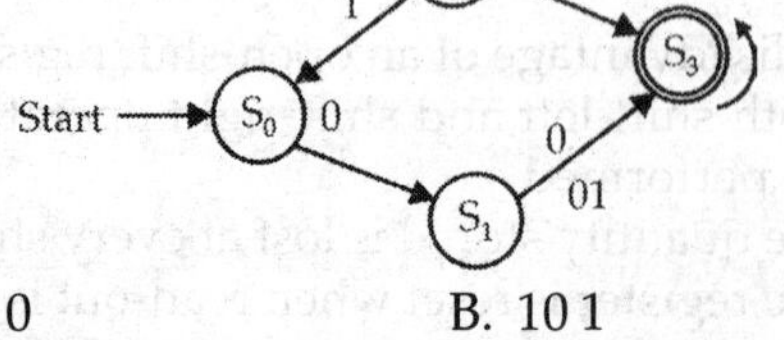

A. 10 0 B. 10 1
C. 10 1 + D. 0 (0 + 1) 0

Ans. 182. A 183. B 184. D 185. D 186. C 187. C 188. C 189. B 190. D 191. A 192. B 193. C 194. B 195. B 196. D 197. B 198. A 199. A 200. A 201. C

202. Consider a programme to compute XY

```
a = x
b = y
s = 0;
repeat, if b = 0, then exit,
else
begin;
s: = S + x;
b: = b - 1
end
until false;
```

What must be true for the programme (before the loop) to execute correctly?

A. $y > 0$ B. $y \geq 0$
C. $y < 0$ D. $y \leq x$

203. The function of strobe pulse in digital system is to
A. reset memory registers
B. check the functioning of a logic gate
C. avoid race problem D. shift data in registers

204. Condition $x \neq y$ is programmed as
A. IF(X·NT·Y)... B. IF(X·NE·Y)...
C. (IF X·NE·Y)... D. none of these

205. The FORTRAN expression
I. $10 \times 5 + 4 \times 4 + 3 \times 3 + 8 \times + 9$
and its equivalent form
II. 9 + P(8 + X(7 + X(3 + X (4 + 10 X))))
When translated into FORTRAN would respectively require which one of the following numbers of multiplications?
A. 15, 5 B. 15, 7
C. 19, 7 D. 10, 6

206. For reading a 50 × 50 matrix, the valid dimension statement is
A. A (50, 45) B. A (45, 50)
C. A (60, 60) D. none of these

207. Consider the following flow-chart: Assuming that M is positive, the value of T will be

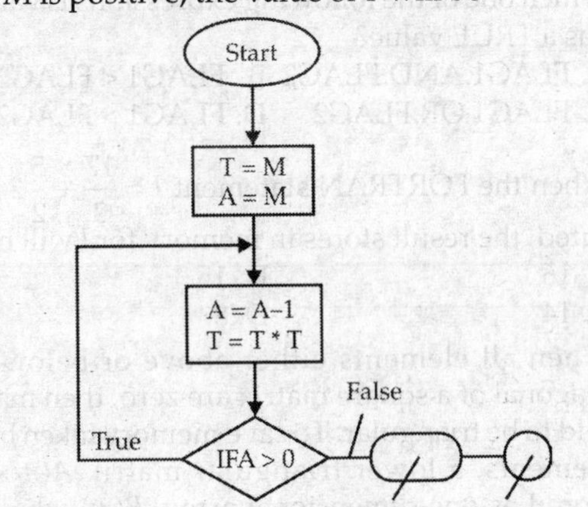

A. M^{2M} B. M^{2M}
C. M^{M^2} D. M^2

208. Which one of the following is a valid arithmetic expression?
A. X + Y sin (X) B. X + Y (sin (X)
C. X – Y (sin (X) D. X – Y sin (X)

209. Consider the following statements: A function subprogram is used because the
1. value of only variable is to be computed
2. values of more than one variable are to be computed
3. subprogram consists of more than one statement
4. name of the variable is the same as the name of the sub program

Of these statements
A. 1 and 3 are correct B. 2, 3 and 4 are correct
C. 1, 3 and 4 are correct D. 2 and 4 are correct

210. A subroutine can return
A. only one value B. only two values
C. only three values D. any number of values

211. If f is defined by function $f(x$: integer): integer; begin
if $x = 1$, then
f: = 0 else
f: = $xf(x - 1) + x$ end;
then the value of $f(4)$ is
A. 16 B. 29
C. 100 D. 124

212. If the loop notation is as shown in the given figure, then the series executed is

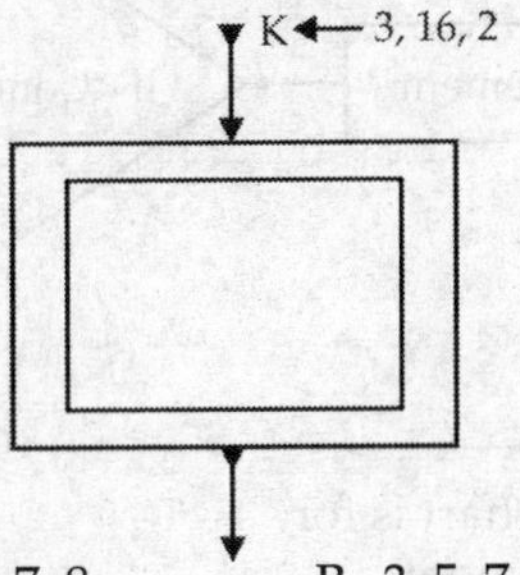

A. 3, 4, 5, 6, 7, 8 B. 3, 5, 7, 9, 11, 13, 15
C. 3, 6, 12, 8, 16 D. 3, 9, 11, 1, 3, 9

213. A FUNCTION subprogram is invoked in the main program by using
A. GO TO statement B. logical IF statement
C. CALL statement D. none of these

214. A FORTRAN line is segmented columnwise for specific purpose. Match List I (purpose) with List II (column numbers used) and select the correct answer using the codes given below the Lists:

List I	List II
a. Statement number	1. 6
b. Arithmatic statement	2. 1 to 80
c. Continuation mark	3. 7 to 72
d. Data	4. 1 to 5

Codes:

	a	b	c	d		a	b	c	d
A.	4	3	2	1	B.	4	3	1	2
C.	3	4	2	1	D.	3	4	1	2

Ans. 202. B 203. C 204. B 205. A 206. D 207. B 208. D 209. C 210. D 211. D 212. B 213. D 214. B

215. Given an algebraic or transcendental equation, exact solution of which is not obtainable mathematically, the computer
A. can be used to solve the equation using numerical methods
B. can solve the equation and provide an expression for the roots
C. can solve the equation only if the equation follows some fixed pattern
D. cannot solve the equation

216. Two coils 1 and 2 have self and mutual inductances of displacement 'x' metres as under: $L_{11} = 2 + 3x$, $L_{22} = 4 + 6x$, $L_{12} = 2 - 3x$. For constant $I_1 = 1$A and $I_2 = -0.5$A, the magnitude and direction of the developed electromagnetic force would be
A. 1.5 Newtons acting in the direction to increase x
B. 1.5 Newtons acting in the direction to decrease x
C. 3.75 Newtons acting in the direction to increase x
D. 4.5 Newtons acting in the direction to decrease x

217. If the following arithmetic statement is executed $IR = 10/3\,(4/6) + 7 - 25$, then the value of IR will be
A. +2 B. −3
C. +5 D. +3

218. Consider the flow chart given in the figure.

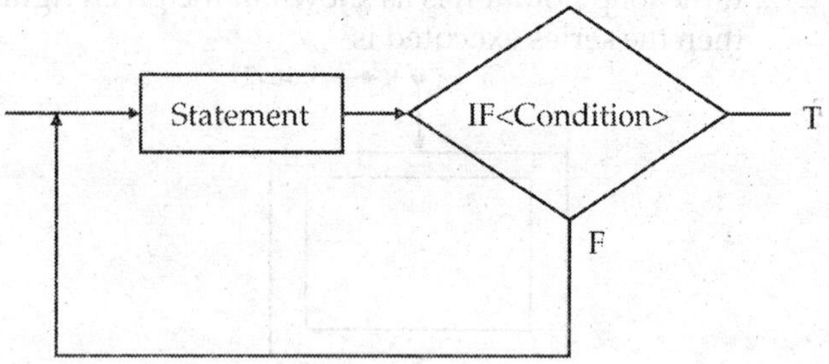

The flow chart is for
A. IF statement
B. While statement
C. Repeat until statement
D. DO statement

219. For a program written in BASIC language,

```
DIM X (5), TEM (5)
FOR I = 1 to 5
READ X(1)
T(1)= (5.0/9.0) (X(1) - 32) =
PRINT 1, X(I), T(I)
DATA 10, 25, 45, 95, 125, 65
END
```

The value of T greater than 40 is
A. 125 B. 95
C. 65 D. 45

220. What will be the output of the following program written in FORTRAN? (Input for A and B are T and F, respectively)

```
LOGICAL A, B, C, D
READ (5, 1) A, B
FORMAT (2LI)
C = A AND B
D = A OR B
WRITE (6) A, B, C, D
STOP
END
```

A. TTTF B. TFFT
C. FTFT D. FFTT

221. Consider the following statements convergence of the Newton-Raphson procedure:
1. It does not converge to a root when the second differential coefficient changes sign
2. It is preferred when the graph of (x) is nearly horizontal where it crosses the X-axis
3. It is used to solve algebraic and transcendental equations

Of these statements
A. 1 and 2 are correct B. 1, 2 and 3 are correct
C. 2 and 3 are correct D. 1 and 3 are correct

222. The series printed by the flowchart shown in the figure is

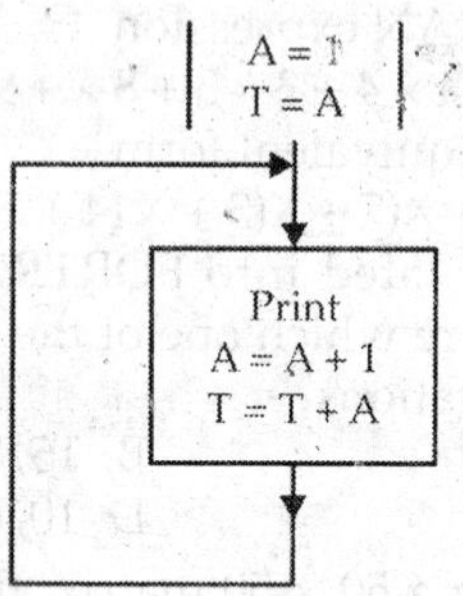

A. 1, 3, 6, 18, 36, ... B. 1, 3, 5, 7, ...
C. 1, 2, 4, 7, ... D. 1, 3, 6, 10, 15, ...

223. Two Boolean variables FLAG1 and FLAG2 are assigned values FALSE and TRUE respectively. Which one of the following expressions is valid and has a TRUE value?
A. FLAG1.AND.FLAG2 B. FLAG1 < FLAG2
C. FLAG1 OR.FLAG2 D. FLAG1 > FLAG2

224. When the FORTRAN statement $I = \frac{17}{2} + \frac{5}{2} 3$ is executed, the result stores in memory for I will be
A. 16 B. 15
C. 14 D. 8

225. When all elements either above or below main diagonal of a square matrix are zero, then matrix is said to be triangular. To save memory taken by zero elements, a lower triangular matrix $A(n \times n)$ is stored as one-dimensional array, $B(m)$ where m is the size of array. The value of m should be equal to
A. $n(n + 1)/2$ B. $n(n - 1)/2$
C. $(n^2 - 1)/2$ D. $(n^2 + 1)/2$

Ans. 215. A 216. C 217. D 218. C 219. C 220. B 221. B 222. D 223. C 224. C 225. A

226. Identify the slowest of the logic families listed below.
A. LSTTL B. TTL
C. ECL D. Low power TTL

227. One of the following logic families is particularly suited for implementing LSI and VLSI functions.
A. I^2L B. ECL
C. Schottky TTL D. CMOS

228. Of the logic families mentioned below, the one that consumes the least power is
A. low power TTL
B. low power Schottky TTL
C. CMOS D. ECL

229. Match the entries in List-I with those in List-II. Identify the correct matching sequence.

List-I	List-II
a. TTL	1. Maximum power consumption
b. ECL	2. Highest packing density
c. NMOS	3. Least power consumption
d. CMOS	4. Saturated logic

	a	b	c	d		a	b	c	d
A.	1	4	2	3	B.	1	4	3	2
C.	4	1	2	3	D.	4	1	3	2

230. The unused inputs of CMOS logic family devices should never be left open. They should
A. preferably be grounded
B. preferably be tied to $+V_{DD}$
C. be tied to either logic LOW or logic HIGH level or another used input
D. preferably be connected to one of the used inputs

231. Identify the bipolar logic family.
A. TTL B. ECL
C. I^2L D. All of them

232. In an *R-S* flip-flop with active-LOW inputs, for $R = S = 0$, *Q* output when the flip-flop is clocked is
A. indeterminate B. 0
C. 1
D. '1' if it was '0' and '0' if it was '1'

233. The basic sequential logic building blocks in which the output follows the data input as long as the ENABLE input is active is
A. *J-K* flip-flop B. *D* flip-flop
C. *T* flip-flop D. *D* latch

234. If the *J*-input of a *J-K* flip-flop is treated as an input and an inverter is wired between *J* and *K* inputs, the *J-K* flip-flop becomes
A. *R-S* flip-flop B. *D* flip-flop
C. *T* flip-flop D. *D* latch

235. One of the following is not a synchronous input with reference to flip-flops.
A. *J* input in a *J-K* flip-flop
B. *S* input in an *R-S* flip-flop
C. PRESET input in a *J-K* flip-flop
D. *D* input in a *D* flip-flop

236. For one of the following conditions, clocked *J-K* flip-flop can be used as a divide-by-2 circuit when the input signal is applied at clock-input.
A. $J = K = 1$ and flip-flop has active-HIGH inputs
B. $J = K = 0$ flip-flop has active-HIGH inputs
C. $J\ K = 10$ flip-flop has active-LOW inputs
D. $J = K = 1$ and flip-flop should be a negative edge-triggered

237. The *Q* output in a presettable, clearable *J-K* flip-flop with active-LOW asynchronous inputs is logic '1' at a certain time. The output immediately after the flip-flop is clocked for $J = 0, K = 1, Pr = 0, Cl = 1$ will
A. become '0' B. toggle
C. remain 1 D. become invalid

238. The characteristic equation of an *R-S* flip-flop with active-HIGH inputs is given by
A. $Q_{n+1} = S + R{\cdot}Q_n$ B. $Q_{n+1} = S' + R{\cdot}Q_n$
C. $Q_{n+1} = S + R'{\cdot}Q_n$ D. $Q_{n+1} = R{\cdot}Q'_n + S{\cdot}Q_n$

239. Mark the incorrect statement.
A. *D* flip-flop is same as *D* latch
B. When the PRESET input of flip-flop is active, it sets the flip-flop to logic '1' state irrespective of status of synchronous inputs
C. Both PRESET and CLEAR inputs should not be active simultaneously
D. A *J-K* flip-flop with active-LOW *J*, *K* inputs will function like a toggle flip-flop for $J = K = 0$

240. The minimum number of flip-flops needed to construct a BCD decade counter is
A. 4 B. 3
C. 10 D. none of these

241. A shift counter comprising five flip-flops with an inverse feedback from the output of the MSB flip-flop to the input of the LSB flip-flop is a
A. divide-by-32 counter B. divide-by-10 counter
C. divide by-5 counter D. five-bit shift register

242. A four-bit binary UP/DOWN counter is initially reset to 0000. The UP/DOWN mode select terminal designated as U/D on the pin connection diagram of the IC is tried to logic HIGH level. What would be the counter's output state at the end of the first clock pulse?
A. 0001 B. 1000
C. 1111 D. 0000

243. In any asynchronous counter
A. all flip-flops change state at the same time
B. only *D* flip-flops are used
C. the counter responds to negative going clock edges
D. each flip-flop output serves as clock input to the next flip-flop

Ans. 226. D 227. A 228. C 229. C 230. C 231. D 232. A 233. D 234. B 235. C 236. A 237. C 238. C 239. A 240. A 241. B 242. C 243. D

244. A five-bit counter
A. has a modulus of 5 B. has a modulus of 2
C. cannot have a modulus that is greater than 32
D. both C and D are true

245. All BCD counters
A. are decode counters because all decode counters are BCD counters
B. are not decode counters
C. have a modulus of 10
D. are constructed with only presettable *D* flip-flops

246. A 10 kHz clock signal having a duty cycle of 25% is used to clock a three-bit binary tripple counter. What will be the frequency and duty cycle of the true output of the MSB slip-flop?
A. 1.25 kHz, 25% B. 1.25 kHz, 50%
C. 3.33 kHz, 25% D. 3.33 kHz, 50%

247. A counter that has a modulus of 64 should use a minimum of
A. 6 flip-flops B. 6 *J-K* type flip-flops
C. 6 *D* flip-flops D. 64 flip-flops

248. A binary ripple counter is to be constructed using *J-K* flip-flops with each flip-flop having a propagation delay of 12 ns. The largest MOD counter that can be constructed using these flip-flops are still operate upto a clock frequency of 10 MHz is
A. MOD-16 B. MOD-64
C. MOD-256 D. MOD-8

249. A MOD-32 binary synchronous counter would require
A. 6 flip-flops and 3 AND gates
B. 5 flip-flops and 3 AND gates
C. 5 flip-flops D. none of these

250. Mark the false statement.
A. Ring counter is a synchronous counter
B. Johnson counter is a synchronous counter
C. The output of a ring counter is always a square wave
D. The decoding circuitry for a Johnson counter is simpler than that of a binary counter

251. The programmable logic device (PLD) having a programmable AND-array at the input and a programmable OR-array at the output is called a
A. Programmable logic array (PLA)
B. Programmable array logic (PAL)
C. Programmable gate array (PGA)
D. Application-specific integrated circuit (ASIC)

252. A programmable array logic (PAL) device has a
A. programmable AND-array at the input and a fixed OR-array at the output
B. fixed AND-array at the input and a programmable OR-array at the output
C. fixed AND-array at the input and a fixed OR-array at the output
D. none of these

253. A masked programmed version of a PAL device that is pin compatible to its PAL counterpart is known as
A. PLA device B. HAL device
C. PLS device D. PGA device

254. It is desired to implement a binary multiplier capable of multiplying two 3-bit binary numbers. Size of ROM would be
A. 64 × 4 B. 1024 × 6
C. 64 × 8 D. 128 × 6

255. A ROM is usually not preferred to implement those Boolean functions which
A. are very complex
B. can otherwise be implemented using a PLA
C. have a large number of don't care conditions
D. have large number of outputs

256. The architecture of a PLA device differs from that of a PROM device in the following.
A. The former has a fixed OR array while the latter has a programmable OR array
B. The former has a programmable AND array while the latter has a hard-wired AND array
C. The former has large number of AND gates in the AND array than the latter for a given number of variables
D. The latter has a fixed OR array while the former has a programmable OR array

257. A PROM device that can generate any eight desired four-variable Boolean functions will have
A. four 2-input AND gates in the AND array and eight 8-input OR gates in the OR array
B. eight 2-input AND gates in the AND array and eight 16-input OR gates in the OR array
C. sixteen 2-input AND gates in the AND array and eight 16-input OR gates in the OR array
D. sixteen 8-input AND gates in the AND array and eight 16-input OR gates in the OR array

258. Size of the ROM required to implement a 16-to-1 multiplexer would be
A. 515 K × 2 B. 1 M × 1
C. 1 M × 2 D. 256 × 2

259. A PLA-like architecture that can implement a full adder should at least have
A. eight 6-input programmable AND gates two 8-input programmable OR gates
B. seven 6-input programmable AND gates two 4-input programmable OR gates
C. eight 6-input programmable AND gates two 4-input programmable OR gates
D. sixteen 3-input programmable AND gates two 8-input programmable OR gates

260. A programmable interconnect device that has high initial resistance and can be permanently programmed to highly conducting path is known as
A. fuse B. floating-gate transistor
C. anti-fuse D. PGA device

Ans. 244. C 245. A 246. B 247. A 248. C 249. B 250. C 251. A 252. A 253. B 254. A 255. B 256. B 257. D 258. B 259. B 260. C

MULTIPLE CHOICE QUESTIONS FROM VARIOUS COMPETITIVE EXAMINATIONS

1. In Boolean algebra, **(IES 2001)**
   If $F = (A + B)(\bar{A} + C)$, then
   A. $F = AB + \bar{A}C$ B. $F = AB + \overline{AB}$
   C. $F = AC + \bar{A}B$ D. $F = AA + \bar{A}B$

2. The Boolean expression for the output Y in the logic circuit is **(IES 2002)**

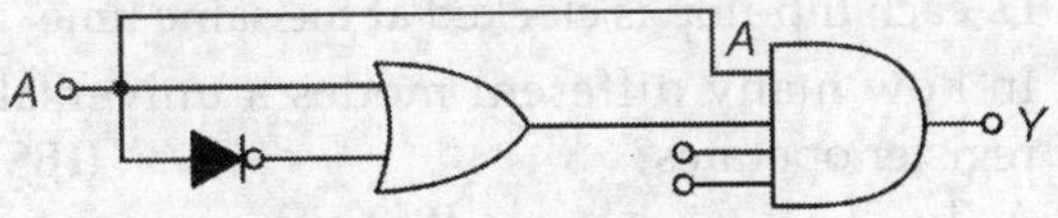

   A. $A\bar{B}C$ B. ABC
   C. $\bar{A}BC$ D. $\bar{A}\bar{B}\bar{C}$

3. The Boolean expression for the shaded area in the Venn diagram shown is **(IES 2011)**

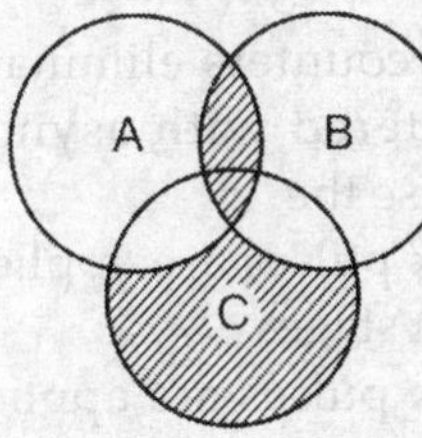

   A. $AB + \bar{A}BC$ B. $C + \bar{A}B$
   C. $A\bar{B}C + \bar{A}BC$ D. $AB + \bar{A}\bar{B}C$

4. What is the output of the gate circuit shown in the below figure? **(IES 2007)**

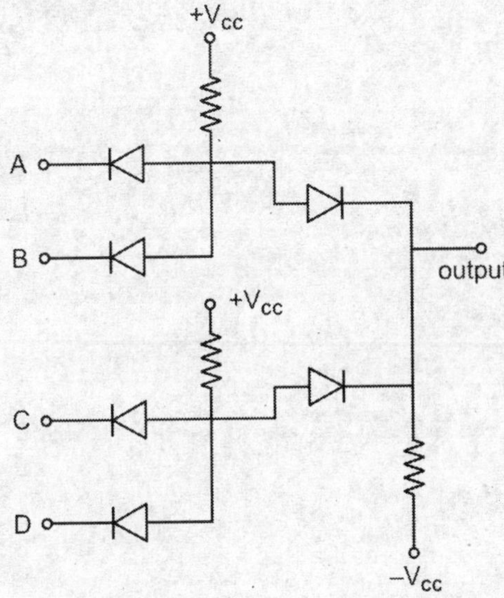

   A. $(A + B)(C + D)$ B. $AB + CD$
   C. $\overline{AB + CD}$ D. $\overline{(A + B)(C + D)}$

5. The simplified form of a logic function **(IES 2003)**
   $Y = \overline{\overline{(A\bar{B})}\cdot\overline{(\bar{A}B)}}$ is
   A. $A + B$ B. AB
   C. $\bar{A} + \bar{B}$ D. $\bar{A}B + A\bar{B}$

6. The Boolean expression $\bar{Y}\bar{Z} + \bar{X}\bar{Z} + \bar{X}\bar{Y}$ is logically equivalent to **(IES 2005)**
   A. $YZ + \bar{X}$
   B. $YZX + \bar{X}\bar{Y}\bar{Z}$
   C. $YZ + XZ + XY$
   D. $X\bar{Y}\bar{Z} + \bar{X}\bar{Y}\bar{Z} + \bar{X}Y\bar{Z} + \bar{X}\bar{Y}Z$

7. What is the simplified form of the Boolean expression **(IES 2008)**
   $T = (X + Y)(X + \bar{Y})(\bar{X} + Y)$?
   A. $\bar{X}\bar{Y}$ B. $\bar{X}Y$
   C. XY D. $X\bar{Y}$

8. The Boolean expression
   $F = \overline{A + \bar{B} + C} + \overline{\bar{A} + \bar{B} + C} + \overline{A + \bar{B} + \bar{C}} + ABC$
   reduces to **(IES 2003)**
   A. A B. B
   C. C D. $A + B + C$

9. In the given circuit, the output Y equals which one of the following? **(IES 2008)**

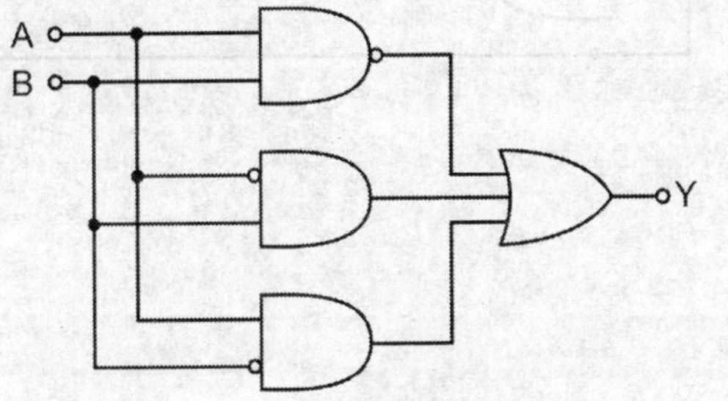

   A. $A + B$ B. $\bar{A}B + A\bar{B}$
   C. AB D. $\bar{A} + \bar{B}$

10. Which one of the following is the correct output (f) of the below circuit? **(IES 2009)**

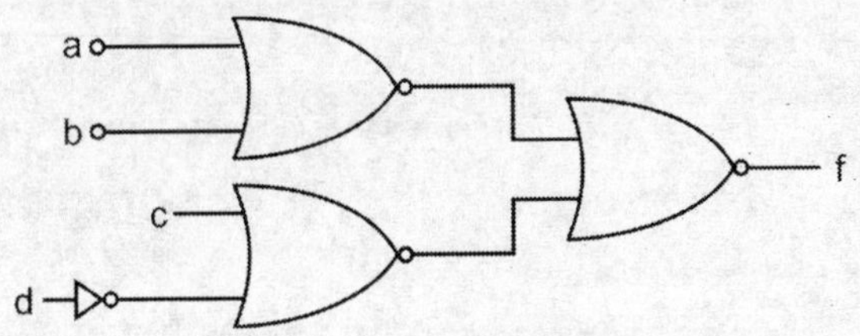

   A. $(a + b)(c + \bar{d})$ B. $(\bar{a} + \bar{b})(c + \bar{d})$
   C. $(a + \bar{b})(c + \bar{d})$ D. $(a + b)(\bar{a} + \bar{d})$

Ans. 1. C 2. B 3. D 4. B 5. D 6. D 7. C 8. B 9. D 10. A

11. The AND function can be realised by using only n number of NOR gates. What is n equal to? **(IES 2008)**
 A. 2 B. 3
 C. 4 D. 5

12. For logic circuit shown, the required inputs A, B and C to make the output X = 1 are respectively **(IES 2011)**

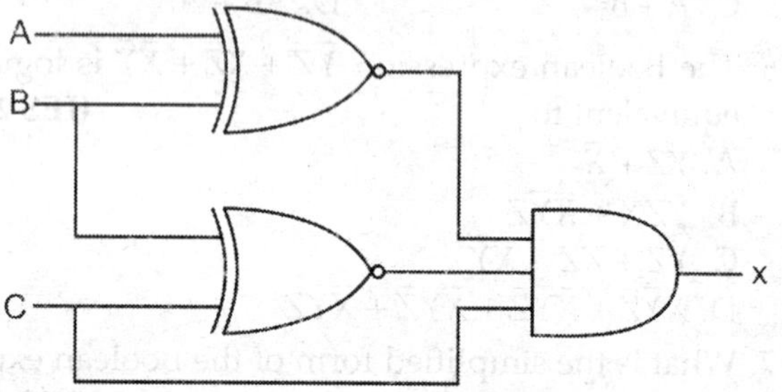

 A. 1, 0 and 1 B. 0, 0 and 1
 C. 1, 1 and 1 D. 0, 1 and 1

13. What is represented by the digital circuit given below? **(IES 2006)**

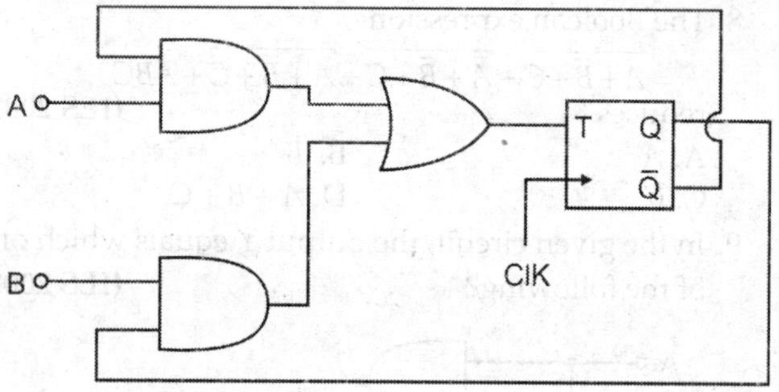

 A. An SR flip-flop with $A = S$ and $B = R$
 B. A JK flip-flop with $A = K$ and $B = J$
 C. A JK flip-flop with $A = J$ and $B = K$
 D. An SR flip-flop with $A = R$ and $B = S$

14. For a bi-directional synchronous counter **(IES 2012)**
 A. each flip-flop divides the frequency of its clock input by 2
 B. each flip-flop output is used as the clock input to the next flip-flop
 C. no decoding logic is required
 D. each flip-flop is clocked at the same time

15. In how many different modes a universal shift register operates? **(IES 2010)**
 A. 2 B. 3
 C. 4 D. 5

16. A divide-by 6 counter is obtained using **(IES 2012)**
 A. 6-bit ripple counter B. 6-bit ring counter
 C. 3-bit ripple counter
 D. 3-bit twisted-ring counter

17. Synchronous counters eliminate the delay problems encountered with asynchronous (ripple) counter because the **(IES 2013)**
 A. input clock pulses are applied only to the first and the last stages
 B. input clock pulses are applied only to the last stage
 C. input clock pulses are not used to activate any of the counter stages
 D. input clock pulses are applied simultaneously

Ans. 11. B 12. C 13. C 14. D 15. C 16. D 17. D

CHAPTER 17

Communication Systems

The function of a communication system is to transfer information from one place to the other through some communication link or medium. Modern communication system in use include telephony and telegraphy, ratio broadcasting, mobile communication, satellite communication and so on. Modern communication radio through personal computers have increased the speed of communication.

Basic components of a communication system: Figure 17.1 represents the block diagram of the communication system. Mainly basic components of a communication system are (i) transmitter, (ii) communication channel, (iii) receiver.

The signal is given as input to the transmitter which then transmits the information through communication channel. The message sent is received by the receiver at the receiving end and is sent to places (destination) for use.

Transmitter is an electronic device that converts the message into signals which are easy to transmit through communication channel.

Communication channel is used to send the message from one place to another. Signals are communicated from one place to another using electromagnetic field waves. Signals are converted into electric and magnetic fields that propagate easily over long distances in a minute of a second. At the transmitter end, one antenna transmits the signal through free space in the form of electromagnetic waves using a high-frequency carrier wave (called *modulations*). The receiver picks-up the signal through another antenna and then separates the signal from the carrier (demodulation).

Receiver converts the signal received back to its original form. The receiver is a complex circuit.

Noise is undesirable electronic energy which enters the communication channel because of a number of reasons.

Modulation is a process of using high-frequency carrier wave for sending information which is comparatively of low frequency to distant places. The baseband signal containing the information modulates the carrier frequency in three ways, i.e. by (i) modulating the amplitude, (ii) modulating the frequency, (iii) modulating the phase.

The information signal is said to be impressed upon the carrier wave, which is high frequency sine wave. The information signal is called the modulating signal or wave. The characteristics of carrier wave in terms of amplitude frequency and phase can be changed by the modulating signal. Accordingly, the modulation methods are: (a) amplitude modulation, (b) frequency modulation, (c) phase modulation.

Figure 17.2 shows the wave shape of an amplitude modulated carrier signal. The information signal, i.e. base band, is a sine wave of low frequency. The carrier wave is generated by transmitter electronic circuitry. The information signal is made to ride over the carrier signal through a process called *modulation*.

The information signals are generally low frequency signals. They are used to modulate a carrier of higher frequency. The modulated electromagnetic wave travel through the free space at a high frequency. A receiver picks-up this modulated electromagnetic wave. This is

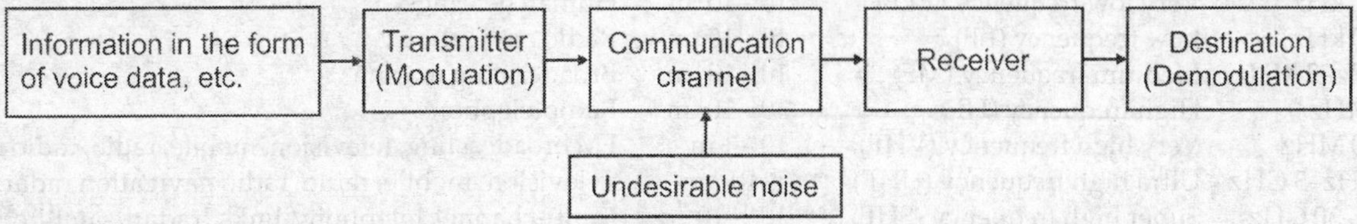

Fig. 17.1: Block diagram representation of communication system

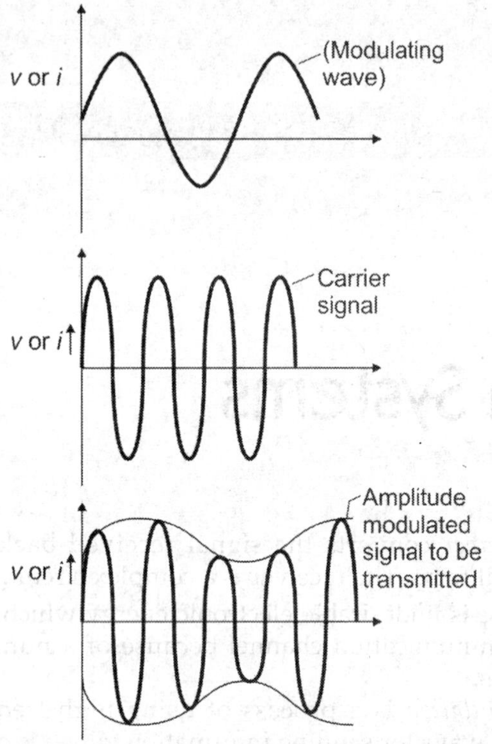

Fig. 17.2: Process of amplitude modulation

then converted into electrical signals. The carrier and the information signals are separated out through a process called demodulation and thus, the original message is recovered at the user end.

Antenna is a device which can radiate and receive electromagnetic waves. The electromagnetic waves will propagate through the space and receive by another antenna at the receiving end.

A communication system using modulation technique is represented in block diagram is shown in Fig. 17.3.

Need of modulation: Modulation is a technique of transmitting a low frequency information signal using a carrier wave of higher frequency.

Advantages:

i. Reduction in the height of antenna used.
ii. Increase in the range of communication.
iii. Avoidance of mixing up of information signals.
iv. Improvement of quality of reception of the information sent.
v. Allowing multiplexing of signal, making possible transmission of more signals simultaneously using the same channel.
vi. Allowing adjustment of bandwidth.

Types of modulation: The process of modulation creates a higher frequency modulated wave containing the information signal. There are three types of modulation used in communication:

i. Amplitude modulation (AM)
ii. Frequency modulation (FM)
iii. Pulse modulation (PM).

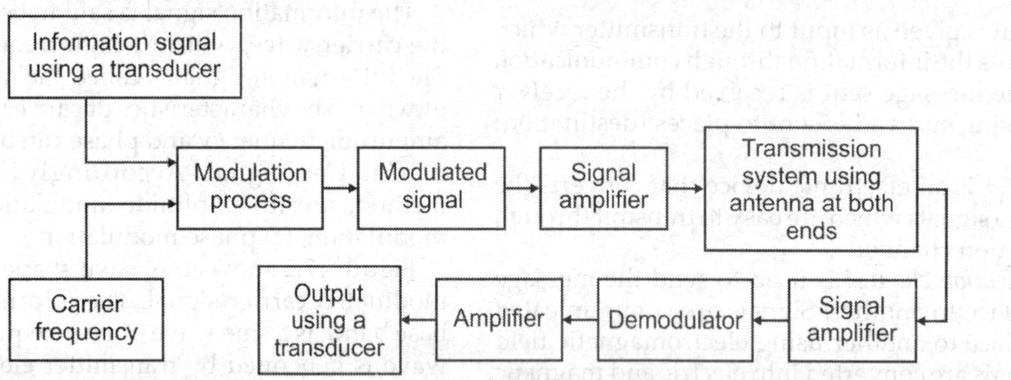

Fig. 17.3: Block diagram of a long-distance communication system using modulation technique

Table 17.1: Radio frequency spectrum

S. no.	*Frequency range*		*Wavelength*	*Typical uses*
1.	30–300 Hz	Extremely low frequency (ELF)	10^7–10^6 m	Long distance point to point communication
2.	30–3000 Hz	Voice frequency (VF)	10^6–10^5 m	Defence communication
3.	3–3030 kHz	Very low frequency (VLF)	10^5–10^4 m	Human hearings
4.	30–300 kHz	Low frequency (LF)	10^4–10^3 m	Radio navigation
5.	300 kHz–3 MHz	Medium frequency (MF)	10^3–10^2 m	Broadcasting marine
6.	3–30 MHz	High frequency (HF)	10^2–10^1 m	Radiotelephony
7.	30–300 MHz	Very high frequency (VHF)	10^1–1 m	FM broadcasting, television, mobile, radio, radio navigation
8.	300 MHz–3 GHz	Ultra high frequency (UHF)	1–10^{-1} m	Television, mobile, radio, radio navigation radar
9.	3 GHz–30 GHz	Super high frequency (SHF)	10^{-1}–10^{-2} m	Multi-channel telephony links, radar, satellite communication
10.	30 GHz–300 GHz	Extra high frequency (EHF)	10^{-2}–10^{-3} m	Satellite communication

Comparison of amplitude modulation and frequency modulation: Both AM and FM are suitable for long-distance communication. However, FM is considered as a superior method than AM. The advantages of FM over AM are as follows:

i. Transmitting efficiency is higher
ii. Rejects interfering signals
iii. Immune to external noise.

Antenna: An antenna is a structure, generally metallic and sometimes very complex, designed to provide an efficient coupling between space and output of a transmitter or the input to a receiver. Like a transmission line, an antenna is a device with distributed constants so that current, voltage and impedance, all vary from one point to the next one along it. This factor must be taken into account when considering important antenna properties, such as impedance, gain and shape of radiation pattern.

***Terms and definitions related to antenna*:** Important terms used in connection with antennas and their radiation pattern:

i. *Antenna gain*: Certain types of antennas focus their radiation pattern in a specific direction is compared to omnidirectional antenna.

ii. *Directive gain*: It is defined as the ratio of power density in a particular direction of one antenna to the power density that would be radiated by an omnidirectional antenna (isotropic antenna).

iii. *Directivity and power gain (ERP)*: Power gain is a comparison of the output power of an antenna in a certain direction that of an isotropic antenna. The gain of the antenna is a power ratio comparison between an omnidirectional and unidirectional radiator. The ratio can be expressed as

$$A\,(\text{dB}) = 10 \log_{10} (P_2/P_1)$$

where

A (dB) = antenna gain in decibels
P_1 = power of unidirectional antenna
P_2 = power of reference antenna.

***Radiation measurement and field intensity*:** The voltage induced in a receiving antenna are very small, generally in microvolt range.

***Field strength*:** The field intensity of an antenna's radiation at a given point in space, is equal to an amount of voltage induced in a wire antenna 1 m long, located at that given point. The field strength or the induced voltage is affected by a number of conditions such as the time of day, atmospheric condition and distance.

***Antenna resistance*:** Antenna resistance is the ratio of the power radiated by the antenna to the square of the current at the feed point.

***Antenna losses and efficiency*:** Antenna losses can be caused by ground resistance, corona effects, imperfect dielectric near antenna, energy loss due to eddy currents induced into nearly metallic objects and I^2R losses in the antenna itself. These losses can be given by the equation as

$$P_{\text{in}} = P_d + P_{\text{rad}}$$

where

P_{in} = power delivered to the feed point
P_d = power lost
P_{rad} = power actually radiated

Bandwidth: It refers to the range of frequencies, the antenna will radiate effectively.

Beamwidth: The beamwidth of an antenna is described as the angles created by comparing the half-power points (3 dB) on the main radiation lobe to its maximum power point.

Polarization: Polarization of an antenna refers to the direction in space of the *E*-field portion of the electromagnetic wave being radiated by the transmitting system.

Impedance matching with stubs: When the characteristic impedance of a transmission line is not equal to the impedance of an antenna quarter or half-wave stubs can be utilized as matching transformers. These stubs are generally constructed from a low-loss metallic material of predetermined length and are connected as shown in Fig. 17.4(a) where a quarter-wave ($\lambda/4$) stub acting as a matching transformer between a coaxial feedline and an end fed half-wave ($\lambda/2$) antenna.

Another commonly used method of impedance matching, especially where cost may be a factor, is the delta (Δ) match which is quick and inexpensive. This

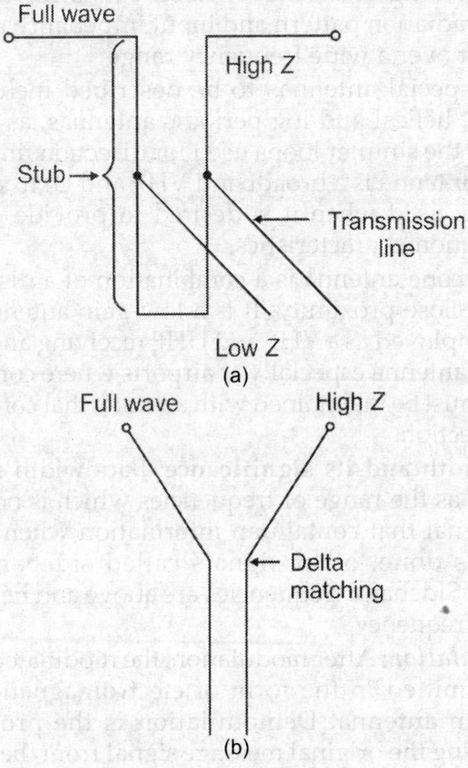

Fig. 17.4: (a) Stubs (b) Delta matching

method is accomplished by spreading the ends of the feed line as shown in Fig. 17.4(b) and adjusting the spacing until optimum performance.

UHF and microwave antennas: Transmitting and receiving antennas designed for use in UHF (0.3–3 GHz) and microwave (1–100 GHz) regions tend to be directive. This condition results from a combination of factors. A number of UHF and microwave applications, such as radar, are in the direction finding and measuring field, so that the need for directional antennas is widespread.

The VHF region, spanning the 30–300 MHz frequency range, is an overlap region. One of the most commonly seen VHF antennas used around the world is the Yagi Uda, most often used as a TV receiving antenna.

Antennas with parabolic reflectors: The parabola is a plane curve defined as the locus of a point which moves so that its distance from another point plus its distance from a straight line is constant. These geometric properties yield an excellent microwave or light reflector.

Properties of paraboloid reflectors: The directional pattern of antennas using a paraboloid reflector has a very sharp main lobe, surrounded by a number of minor lobes which are much smaller.

Wideband and special purpose antennas: It is often desirable to have antenna capable of operating over a wide frequency range. In TV reception, the requirement for wideband properties is magnified by the fact that it is desirable to use the same antenna for a group of neighbouring channels. A need exists for antenna whose radiation pattern and input impedance remains constant over a wide frequency range.

The special antennas to be described include the discone, helical and log periodic antennas, as well as some of the simpler loops used for direction finding. A helical antenna is a broadband VHF and UHF antenna which is used when it is desired to provide circular polarization characteristics.

A discone antenna is a combination of a disc and a cone in close proximity. It is a low-gain antenna. It is often employed as a VHF and UHF receiving and transmitting antenna especially at airports where communication must be maintained with aircrafts that come from any direction.

Bandwidth and its significance: Bandwidth may be defined as the range of frequencies which is occupied by a signal that contain an information when modulation is done, other signals called sidebands are created. Sideband frequencies are above and below the carrier frequency.

Demodulation: After modulation, the modulated signal is transmitted in the form of electromagnetic wave using an antenna. Demodulation is the process of recovering the original message signal from the modulated carrier signal or wave.

Demodulator is an electronic circuit used to recover the information contained in the modulated signal received by the receiver. Different kinds of circuits are used as demodulators depending upon whether the carrier signal is amplitude modulated or frequency modulated or phase modulated.

Multiplexing: Multiplexing is a technique or process by which a large number of information signal can be communicated through a single communication channel. The communication channel could be a cable connection or a radio link. This way communication will be less complex and inexpensive.

Multiplexing of information is done in case of telephone communication, telemetry radio and TV stereo broadcasting and satellite communication. There are two types of multiplexing, namely (i) frequency division multiplexing (FDM), and (ii) time division multiplexing (TDM).

In FDM, multiple signals are adjusted within the bandwidth of one communication channel and different signals are assigned different frequencies within the given bandwidth. In TDM, multiple signals are transmitted using different time slots.

Communication receivers: In wireless communication system using electromagnetic waves, the transmitted signal becomes very weak when it reaches the receiver like a radio receiver or telephone receiver set. The signal gets distorted due to interference from a large number of other radio signals on its way. These interferring signals are called *noise signals* or *unwanted signals*. The function of a radio receiver is to have high level of sensitivity and selectivity so as to reproduce the original modulating signal communication receivers are of two types:

i. Tuned radio frequency receiver (TRF receiver)
ii. Superheterodyne receiver.

Tuned radio frequency receiver (TRF): TRF receiver in block diagram is shown in Fig. 17.5. Electromagnetic waves are received by the antenna and voltage is induced in it. Due to demodulation, the carrier signal is bypassed and the original information signal is recovered. This signal is amplified to the necessary power level using audio and power amplifiers before connecting to the speaker.

TRF receivers have some problems with respect of selectivity and variation of bandwidth over the tuning range. These problem were overcome in an improved type of receiver called *super heterodyne receiver.*

Superheterodyne receivers: A superheterodyne receiver uses frequency mixing technique to convert all incoming signals into a fixed lower intermediate frequency which can be more conveniently processed than the radio carrier frequency. Block diagram of a superheterodyne receiver has been shown in Fig. 17.6. An automatic gain control (AGC) controls the gains of RF and IF amplifiers to maintain constant output if the

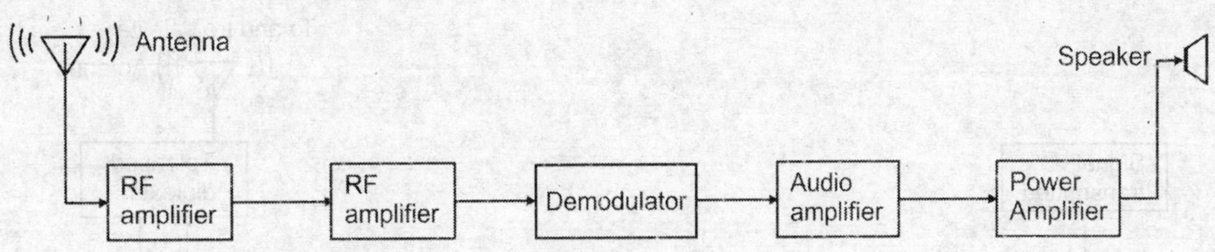

Fig. 17.5

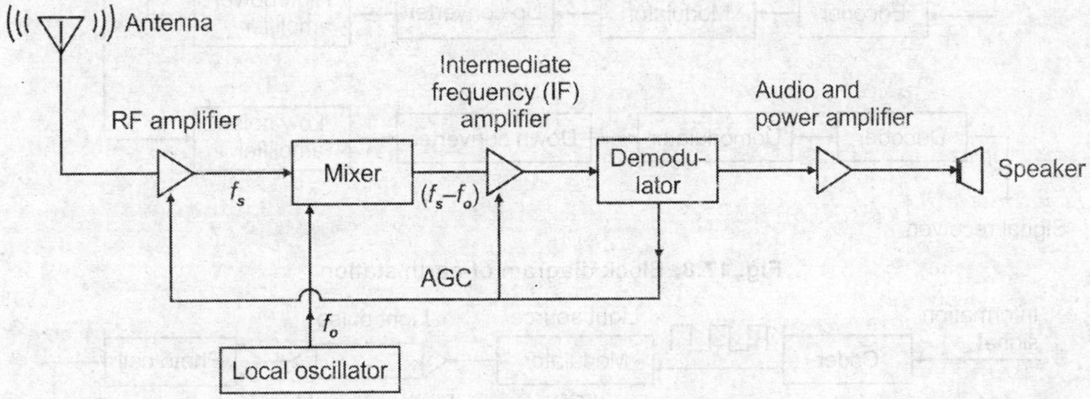

Fig. 17.6: Block diagram of a receiver

strength of the signal received at the input stage is fluctuating.

Microwave communication: The frequency spectrum used for radio communication has become overcrowded due to larger number of users. Increased number of users have crowded the airspace. Advancement of technology has enabled us to extend radio communication into higher frequencies. Microwave frequency spectrum is used for long-distance telephone communication. Radio detection and ranging (radar) communication also use microwave frequency. Radar is used to detect presence of distant objects like an aeroplane or a ship, etc.. The radar system transmits high frequency signal which gets reflected from the object to be detected. Television station also use microwave to transmit TV signals. In space communication, satellites, microwaves are used.

It is interesting to note that microwave has other applications like in heating, microwave ovens used in kitchen, in hospitals for heating of injured muscles (diathermy machines), etc.

Satellite communication: Satellite communication is meant for long distance communication. Almost one-third of the earth except the polar region is visible from a geostationary satellite. Communication link can be established to very distant remote areas. Mobile communication can be provided by mounting an earth station on a vehicle within a short period of time. Figure 17.7 shows a block diagram of a satellite transponder which works as a relay station.

The satellite transponder receives uplink transmission information and retransmits the signal through

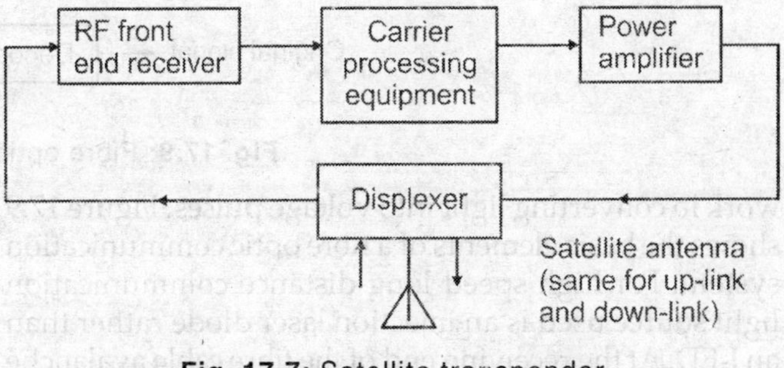

Fig. 17.7: Satellite transponder

its downlink mechanism. The diplexer facilitates simultaneous reception and transmission through the signal antenna. The power amplifier with sufficient gain is used to increase the power level of transmission signal so that information sent reaches the earth station. A block diagram on earth station is shown in Fig. 17.8.

The signal received from the satellite through the same antenna is of different frequency which goes through the demodulation process and through the decoder and the information signal is received in its original form.

Fibre-optic communication: In fibre-optic communication, information can be put on a light beam and transmitted through space or through a special cable called *fibre-optic cable.* Fibre-optic cable is essentially a narrow light pipe that is used to carry light beam from one end to the other. The information signal, which may be a voice signal, is converted into digital using A to D converter. The other light emitting source is a laser. The signal is then amplified and transformed into original digital signal or data. The most commonly used receiver is a photodiode. A photodiode can be made to

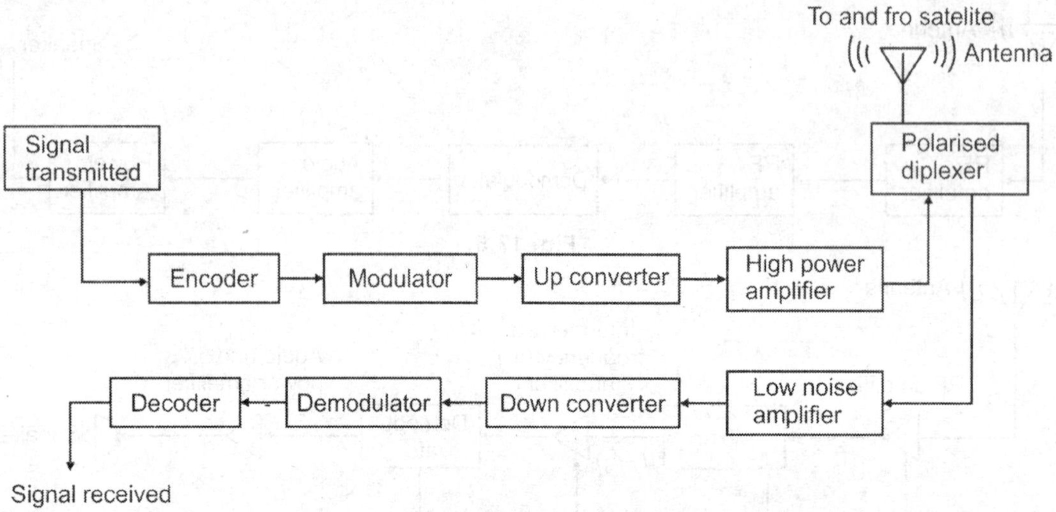

Fig. 17.8: Block diagram of earth station

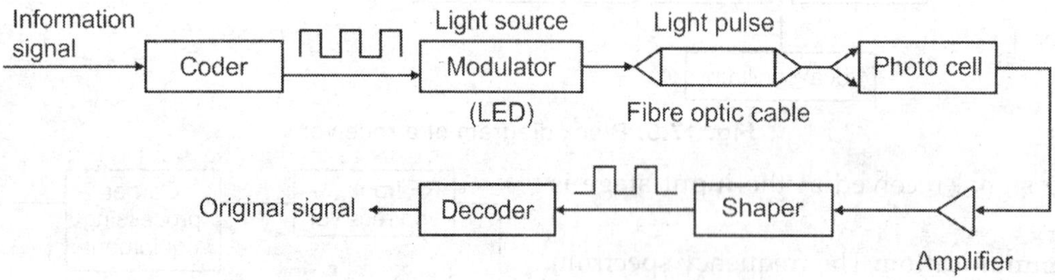

Fig. 17.9: Fibre optic communication system

work in converting light into voltage pulses. Figure 17.9 shows the basic elements of a fibre optic communication system. For high speed long-distance communication light source used is an injection laser diode rather than an LED. At the receiving end of the fibre cable avalanche photodiodes are used.

Mobile communication: Mobile telephone system involve complex communicating networks and computer controlled switching centres. Mobile telephone stations are small handsets which can be carried in a pocket. The basic concept of cellular telephone is that each radio coverage area is divided into hexagon-shaped cells that fit together to form a honey-comb pattern as shown in Fig. 17.10(a). The concept of frequency reuse is used to increase the capacity of a mobile telephone channel. A geographical area is divided into small cells and using a low power transmitter with low antenna at base station in each cell, a cellular communication network is developed.

Figure 17.10(b) shows a cell with its base transmitter located at the centre with omnidirectional antenna. Omnidirectional antennas radiate and receive signals equally in all directions.

Instead of having high-powered fixed base station located at the centre of a city at very high altitude, a number of low-powered duplicates distributed over the covered area using a cellular telephone concept are used with advantage.

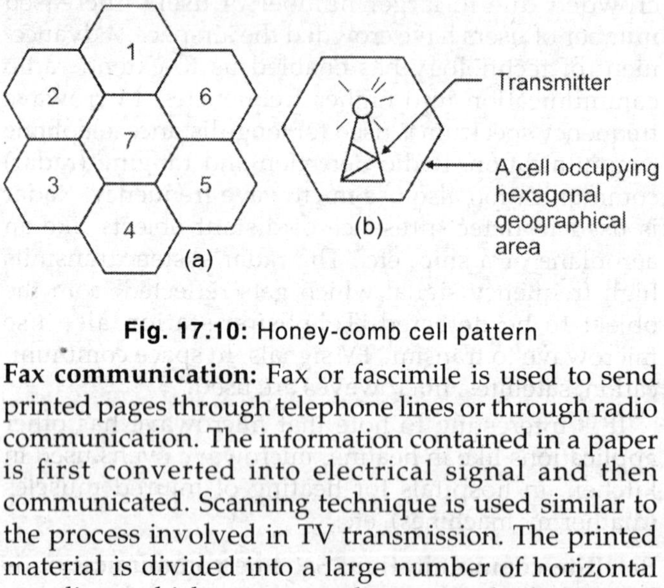

Fig. 17.10: Honey-comb cell pattern

Fax communication: Fax or fascimile is used to send printed pages through telephone lines or through radio communication. The information contained in a paper is first converted into electrical signal and then communicated. Scanning technique is used similar to the process involved in TV transmission. The printed material is divided into a large number of horizontal scan lines which are scanned at a very high rate and transmitted and received serially and simultaneously.

The fax machine at the receiving end produces the document and provides a printout through its printer. The fax machine first demodulates the signal and the processes to recover the original data. The data is decompressed and the printer provides a printout. At both ends, the fax machine can transmit as well as receive printed information. Fax machines are provided with built-in telephone and printer.

Television (TV): A television system transmits a picture or a scene along with sound to a distant place. The scene or the picture signal is generated by a TV camera and the sound signal by a microphone. The sound may be a stereo and the picture may be coloured. This way a TV signal is a complex one. The picture signal is amplitude modulated. The sound signal is frequency modulated. Two carrier frequencies are separated by a wide margin to avoid mixing of two. The two modulated frequencies are transmitted through a common antenna. For stereo sound, signal multiplexing technique is used. For colour transmission of coloured pictures, FDM is used.

TV transmitter: The picture information is converted into electrical signal by considering the light intensity and colour details of a picture scene. The scene to be transmitted is focussed by the lense of the TV camera upon a light sensitive imaging device. Using a method of scanning, the optical information is converted into an electrical signal in a sequential manner line by line so as to cover the entire scene. Scanning is done at a very fast rate and is repeated a very large number of times. Use of synchronization technique of scanning by the TV camera and its reproduction at the TV receiver is necessary. This creates an illusion, an impression on the eye of simultaneously viewing the whole scene at a time.

A colour TV camera is used to develop signal voltage proportional to the intensity of light of three primary or basic colours, namely red, green and blue. By mixing these colours in right proportion, any colour can be created. This focus the basis of colour television. By mixing any two or three of the basic colours, other colours can be created. The colour TV camera develops voltage proportional to the intensity of the primary colours.

A simple diagram of black and white (monochromic) TV transmitter is shown in Fig. 17.11. A signal from the video camera is amplified using a video amplifier. Synchronizing pulses are added before amplitude modulation using an AM modulating amplifier. Synchronization is necessary between TV camera and TV picture tube. For this, synchronizing pulses are transmitted along the carrier is modulated by modulating signal received from the modulating amplifier. This signal is mixed with audio signal and transmitted through the antenna. The sound signal is picked-up using a microphone which is amplified and then frequency modulated using a carrier frequency. Frequency modulated sound signal is combined with amplitude modulated picture signal before transmission as shown in Fig. 17.11.

TV receiver: Television transmitter antenna radiates electromagnetic waves from its antenna. At the receiving antenna, intercepts the radiated RF signals. Figure 17.12 shows a block diagram of TV receiver. The signal that carries picture is amplified using a video amplifier and then connected to TV picture tube. In the picture tube, the electrical signal is converted to picture on its illuminated screen. The frequency modulated signal is demodulated after amplification. The audio output is connected to a loudspeaker where electrical signal is converted to sound. It may be noted that the block diagram represents a black and white TV receiver. A colour receiver is similar to a black and white receiver. However, a colour receiver circuit will have a colour subsystem to accept and process colour signals. A colour TV camera have three colours trick up tubes and a colour picture tube will have three colour guns, red, green, and blue. The electronic guns produce electron beams to illuminate corresponding colour

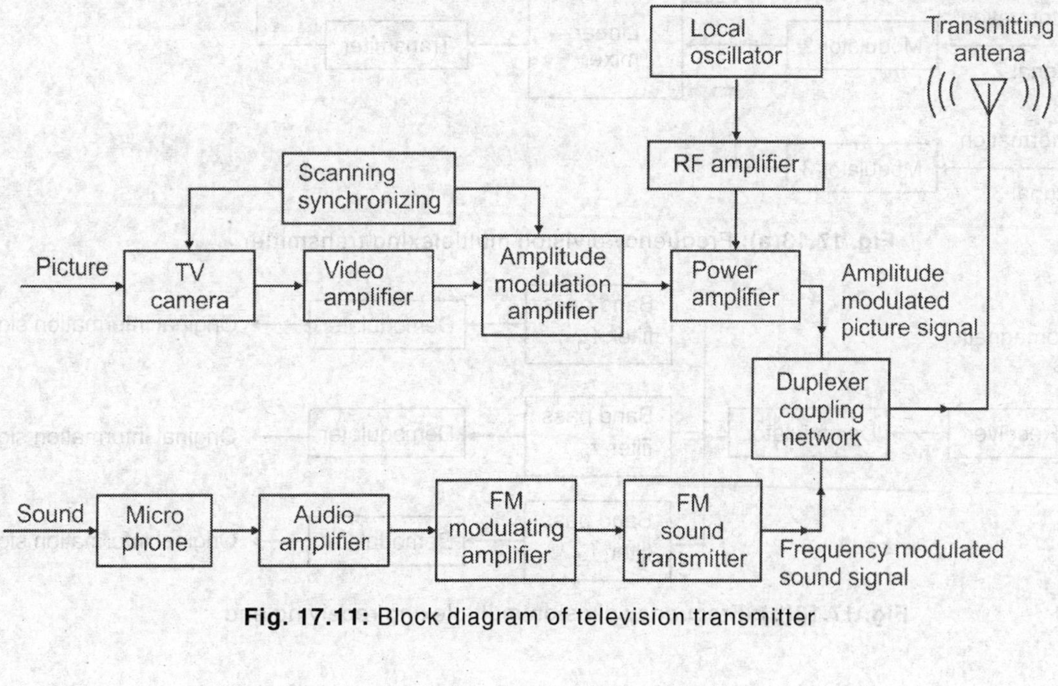

Fig. 17.11: Block diagram of television transmitter

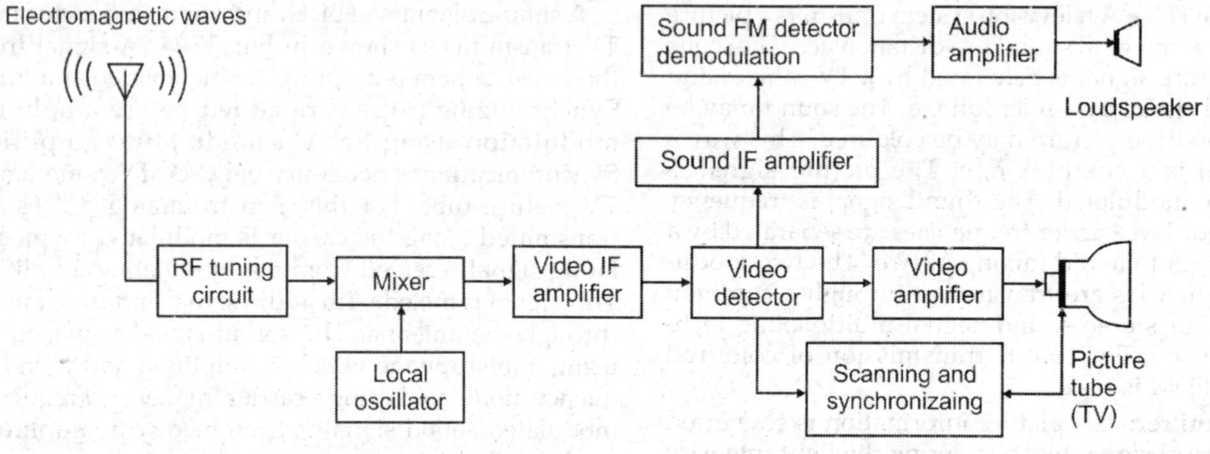

Fig. 17.12

phosphor on the screen. The TV viewer's eye integrate coloured information and their illuminance to visualise the actual picture.

Need of synchronization: When a TV camera starts scanning line 1, perpendicular, the receiver, that is the picture tube should also starts scanning line 1. If this does not happen, the top line of the camera will appear as a middle line at the receiver which is not desirable. The transmission reception synchronization is essential to avoid any distortion of the picture received on the TV screen.

Frequency division multiplexing (FDM): Figure 17.13 shows the representation of frequency division multiplexing, in which different signals modulate carriers are used. Each carrier has different frequency within a given range. The modulated carriers are then mixed, summed up in a linear mixer. All the carriers are combined into a composite signal and this composite signal is transmitted using a transmitter over the single communication channel using an antenna.

At the receiving end, an antenna receives the electromagnetic signal and a demodulator demodulates it and gives out the composite signal. Using band pass filters and demodulators each input information signal is recovered as shown. Earlier, FDM is used in telemetry, telephone systems, stereo FM, etc.

Time division multiplexing (TDM): By using TDM multiple signals can be transmitted over a single channel by time sharing mechanism. The whole of the bandwidth available can be utilised by each signal but

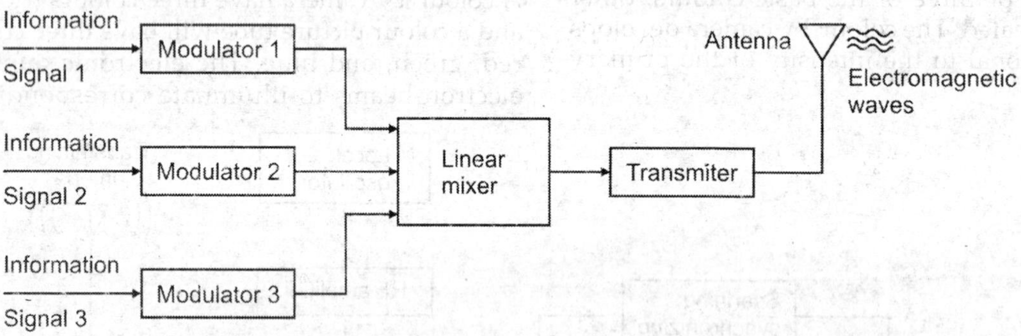

Fig. 17.13(a): Frequency division multiplexing transmitter

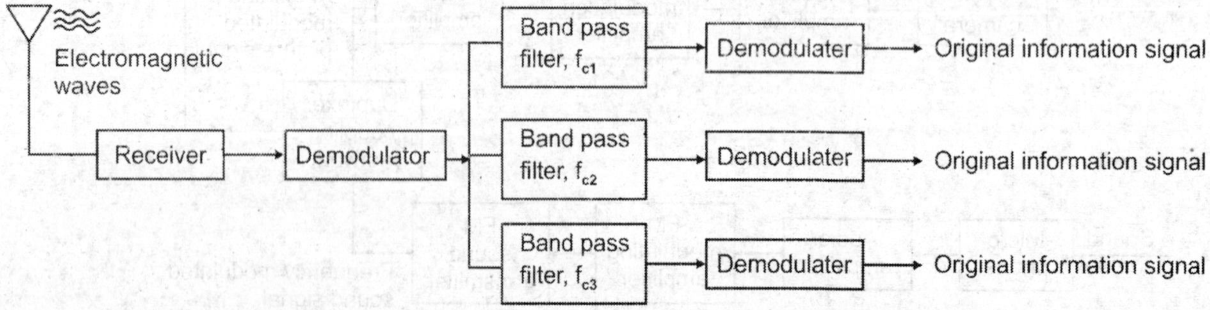

Fig. 17.13(b): Frequency division multiplexing receiving end

on time sharing basis. For the multiple signals to be transmitted, each signal is allowed to use the channel for a short period of time turn by turn. Some mechanism of switching off each of the signals in a sequence for a very brief period of time is made as shown in Fig. 17.14 using electronic switching circuits. Time division multiplexing is possible for both analog and digital signals.

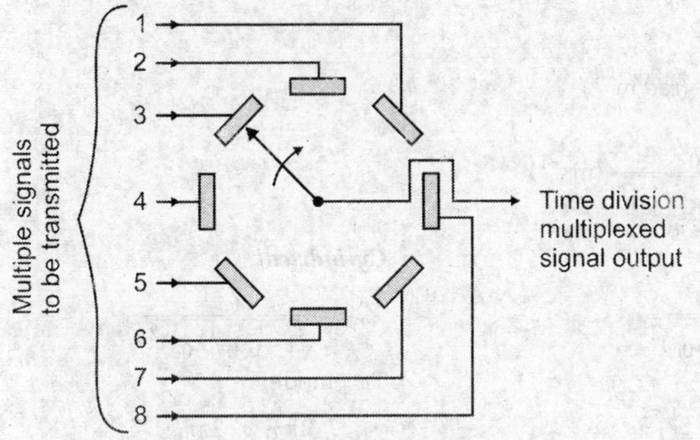

Fig. 17.14: Simple rotary switch

Analog signals, like voice signals, can also be transmitted using TDM system. The analog signal is sampled at a very high rate using electronic switching technique. Pulse code modulation is used in telephone system to transmit voice signals, data exchange through digital computers and video data through TDM. Pulse code modulation signals are retrieved using demultiplexing technique and conversion from digital to analog signal by D to A converter.

Waveguides: The waves on transmission lines are transverse electromagnetic (TEM) waves with electric and magnetic fields entirely transverse to the direction of propagation. Transmission system which can convey electromagnetic waves only in higher-order modes are usually called *waveguides* or *simply guides.*

***TE mode wave in waveguides*:** Consider an infinite-parallel-plane transmission line as shown in Fig. 17.15. This is two-conductor line which is capable of guiding energy in a transverse electromagnetic (TEM) mode with E in the z-direction.

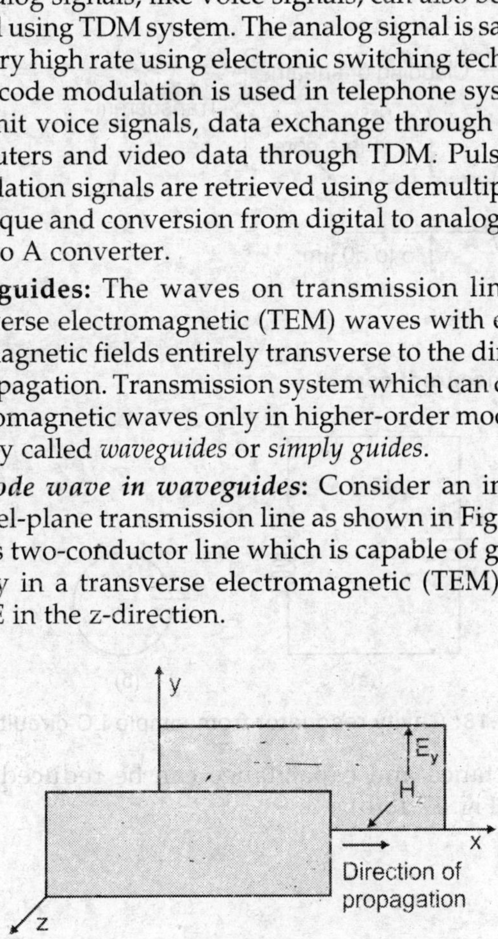

Fig. 17.15

In higher-order mode, where the electric field is everywhere in the *y*-direction, with the transmission in the *x*-direction, i.e. the electric field has only an E_y component. Since E_y is transverse to the direction of transmission, this mode is designated as transverse electric (TE) mode.

Hollow rectangular waveguide: Certain properties of an infinite parallel-plane transmission line and of a hollow rectangular guide are obtained by considering that the higher-mode wave consists of two plane TEM component waves.

TE_{10} mode: For the anode $m = 1$, and $n = 0$, the three components E_y, H_x and H_z that are not zero (Fig. 17.16). The six field components for the TE_{10} mode are then, $E_x = 0$.

TE mode requirement

$$E_y = \frac{\gamma Z_{yz} H_0}{k^2} \cdot \frac{\pi}{z_1} \sin \frac{\pi z}{z_1} e^{-\gamma x}$$

$$E_z = 0$$

$$H_x = H_0 \cos \frac{\pi z}{z_1} e^{-\gamma x}$$

$$H_y = 0$$

$$H_z = \frac{\gamma H_0}{k^2} \cdot \frac{\pi}{z_1} \sin \frac{\pi z}{z_1} e^{-\gamma x}$$

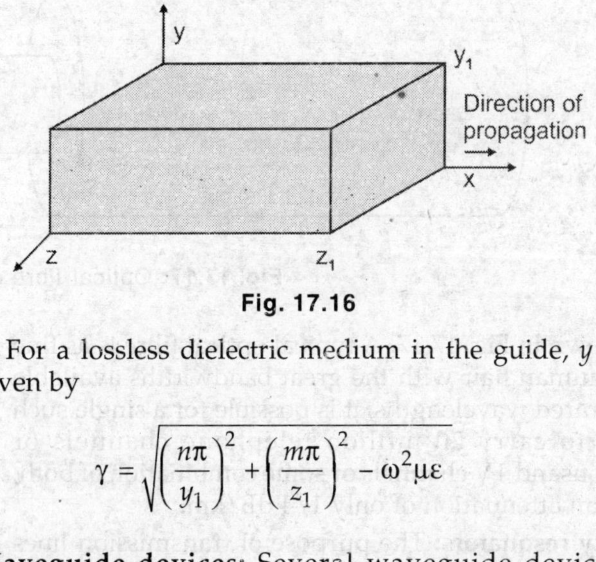

Fig. 17.16

For a lossless dielectric medium in the guide, *y* is given by

$$\gamma = \sqrt{\left(\frac{n\pi}{y_1}\right)^2 + \left(\frac{m\pi}{z_1}\right)^2 - \omega^2 \mu \varepsilon}$$

Waveguide devices: Several waveguide devices namely terminations, power dividers, and guide to line transitions are available.

Fibre optics: At or near optical wavelength the dielectric cylinder guide can be physically small or threadlike in diameter. Such guides, called optical fibres, consists typically of a transparent core fibre of glass of index of refraction η_1 surrounded by a transparent glass sheath or cladding, of slightly lower index η_2, with both enclosed in an opaque protective jacket

Table 17.2: Rectangular and circular cylindrical waveguide parameters

Symbol	Name	Equation (rectangular or cylindrical)	
α	Attenuation constant		
	Below cut-off frequency	$\alpha = \frac{2\pi}{\lambda_0}\sqrt{\left(\frac{\lambda_0}{\lambda_{0c}}\right)^2 - 1}$ $(N_p\,m^{-1})$	
	Above cut-off frequency	$\alpha = \frac{R_e Z_c \int \lvert H_{t_1}\rvert^2 dl}{2R_e Z_{yz}\int \lvert H_{t_2}\rvert^2 ds}$ $(N_p\,m^{-1})$	
β^+	Phase constant	$\beta = \sqrt{\left(\frac{2\pi}{\lambda_0}\right)^2 - k^2}$ (rad m^{-1})	
v_p	Phase velocity	$v_p = \frac{\omega}{\beta} = \frac{v_0}{\sqrt{1-(\lambda_0/\lambda_{0c})^2}}$ (ms^{-1})	
		Rectangular	*Cylindrical*
$Z_{yz}, Z_{r\phi}$	Transverse wave impedance	$Z_{yz} = \frac{Z_d}{\sqrt{1-(\lambda_0/\lambda_{0c})^2}}$ TE_{mn} mode only	$Z_{r\phi} = \frac{Z_d}{\sqrt{1-(\lambda_0/\lambda_{0c})^2}}$ TE_{nr} mode
λ_{0c}	Cut-off wavelength	$\lambda_{0c} = \frac{2}{\sqrt{(n/y_1)^2 + (m/z_1)^2}}$ TE_{mn} or TM_{mn} mode	$\lambda_{0c} = \frac{2\pi r_0}{k_{nr}}$ or $\frac{2\pi r_0}{k_{nr}}$ TE_{nr} $(k'_{nr}) \cdot TM_{nr}$ (k_{na})

$+y$ = propagation constant = $\alpha + j\beta$

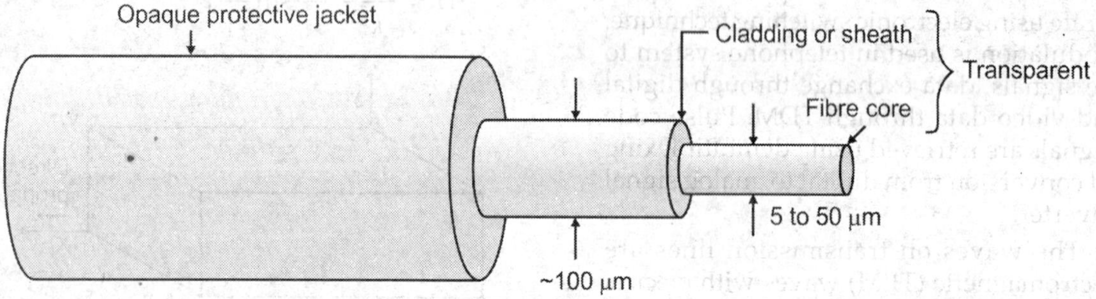

Fig. 17.17: Optical-fibre guide with transparent core

as shown in Fig. 17.17. A typical optical fibre is as fine as a human hair with the great bandwidths available at infrared wavelengths, it is possible for a single such fibre to carry 20 million telephone channels or 20 thousand TV channels or some combination of both, with an attenuation of only 1/4 dB/km.

Cavity resonators: The purpose of transmission lines and waveguides is to transmit electromagnetic energy efficiently from one point to another. A resonator, on the other hand, is an energy storage device. At low frequencies, a parallel-connected capacitor and inductor as shown in Fig. 17.18(a) form a resonant circuit. To make this combination resonate at shorter wavelength, the inductance and capacitance can be reduced as shown in Fig. 17.18(b).

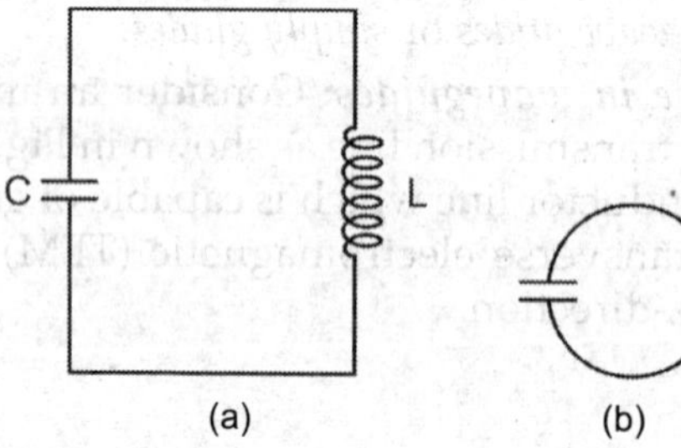

Fig. 17.18: Cavity resonator from simple LC circuit

MULTIPLE CHOICE QUESTIONS

1. The Nyquist sampling rate for a signal $g(t) = A \sin^2 (2\pi f_m)$ is
   A. $4 f_m$ B. $3 f_m$
   C. $2 f_m$ D. f_1
2. Consider a binary PCM transmission of video signals with $f_s = 10$ MHz and $S_x = 0.1\ \Omega$ (where f_s is the sampling rate and S_x is the signal power): The signalling rate required to get the quantisation noise power at least 50 dB below the signal power is
   A. 200 Mbps B. 100 Mbps
   C. 50 Mbps D. 20 Mbps
3. The output bit rate in the first level multiplexer in the CCITT hierarchy is 2.048 Mbps. Assuming no bit robbing, the number of framing plus signalling bits per second is
   A. 128000 B. 16000
   C. 8000 D. 16
4. A source output power is 1.0 W. It is connected to a channel having 3 dB loss. The channel output is dBm is
   A. 30 B. 27
   C. 15 D. none of these
5. The envelope of the output of a bandpass filter excited by white noise at the input follows
   A. Gaussian distribution
   B. uniform distribution
   C. Rayleigh distribution
   D. exponential distribution
6. Signal-to-quantisation noise ratio in A-law companding for voice PCM
   A. remains essentially unchanged irrespective of talkers
   B. changes with talkers arbitrarily
   C. increases linearly with the voice level of the talker
   D. cannot be determined
7. The number of line components in the power spectrum of a bipolar signal is
   A. 3 B. 2
   C. 1 D. nil
8. The rows of the parity submatrix in the generator matrix of a (7, 4) block code is given as (101), (111), (011) and (110). The code for a typical 4-bit message (1101) is
   A. 1101001 B. 1101100
   C. 1011010 D. none of these
9. The output of the filter matches to a rectangular pulse when the input is a
   A. rectangular B. sine function
   C. sawtooth D. triangular pulse
10. The joint probability $P(B/A)$ of two events A and B is equal to
    A. $P(AB)/P(B)$ B. $P(AB)$
    C. $P(AB)/P(A)$ D. none of these
11. If two events are statistically independent, then
    A. $P(AB) = P(A)(B)$ B. $P(AB) = P(A)P(B/A)$
    C. $P(AB) = P(B)P(A/B)$ D. none of these
12. If two coins are tossed simultaneously, the probability function $P(X)$ of getting no head will be
    A. zero B. 1/4
    C. 1/2 D. 1
13. In the above question, the probability function of getting one head will be
    A. zero B. 1/4
    C. 1/2 D. 1
14. In the same question, the probability function of getting two heads will be
    A. 1/4 B. 1/2
    C. 1 D. 2
15. Two events X and Y are statically independent and mutually exclusive. The probability $P(XY)$ is
    A. zero B. 1
    C. infinity D. none of these
16. The standard deviation is a measure of
    A. deviation from average value
    B. variation in the observed values
    C. spread of observed values
    D. none of these
17. An ergodic process is
    A. always stationary B. need not be stationary
    C. stationary but not random
    D. none of these
18. A stationary random process is
    A. not always ergodic B. always ergodic
    C. always ergodic but non-random
    D. none of these
19. The autocorrelation function is a measure of
    A. spread of a random process
    B. inter-dependence of two random processes
    C. regularity of the random process
    D. none of these
20. Which of the following conditions is necessary for two events to be mutually exclusive?
    A. Both events occur simultaneously
    B. The event occur at different times.
    C. One event precludes the other event
    D. None of these
21. To events X and Y are statistically independent and mutually exclusive. The probability $P(XY)$ is

Ans. 1. A 2. B 3. C 4. B 5. A 6. C 7. C 8. B 9. D 10. A 11. A 12. B 13. C 14. A 15. A 16. C 17. A 18. A 19. C 20. C

A. zero B. one
C. infinity ✓ D. not possible

22. The total area under probability distribution curve is
 A. one B. infinite
 C. dependent upon the distribution function
 D. zero

23. In a binary PCM system, if the number of bits per sample were increased from n to $(n + 1)$, then the signal to quantisation noise ratio will increase by
 A. a factor of $(n + 1)/n$ B. a factor of $(n + 1)^2/n^2$
 C. $6n$ dB
 D. a factor which is independent of n

24. The gain function of a system may be written as $H(j\omega)$ $(1 + 2a \cos \omega T)$ $H_0(j\omega)$ where $H_0(j\omega)$ corresponds to a time function $h_0(t)$. The impulse response of the system may be written as
 A. $(1 + 2a) h_0(t)$
 B. $(1 + a) h_0(1) + ah_0 (t - T) + ah_0 (t + T)$
 C. $h_0(1) + ah_0 (t - T) + ah_0 (t + T)$
 D. $h_0(1) + 2a \cos T$

25. The two fundamental physical limitations faced by an engineer in designing a communication system would include
 A. bandwidth and interference
 B. interference and distortion
 C. distortion and noise D. noise and bandwidth

26. Entropy is basically a measure of
 A. rate of information B. average information
 C. probability of information
 D. channel capacity for transmission information

27. The AC power of a random signal is represented by
 A. standard deviation of the signal
 B. variance of the signal
 C. mean square value of the signal
 D. average value of the signal

28. The *bit* is a unit of
 A. entropy B. channel capacity
 C. information D. rate of information

29. One *bit* is equivalent to choice of one event from a set of
 A. 16 equiprobable events
 B. 10 equiprobable events
 C. 8 equiprobable events
 D. 4 equiprobable events

30. One *bit* is equal to
 A. 1.14 bits B. 1.17 bits
 C. 3.32 bits D. 10 bits

31. Entropy essentially is a measure of
 A. rate of information B. average information
 C. disorder in information
 D. probability of information

32. The entropy of two equiprobable messages is
 A. zero B. 1/2
 C. 1 D. 2

33. The maximum value of entropy is
 A. 1 B. 2
 C. 4 D. infinite

34. Redundancy of communication systems
 A. reduces efficiency of communication
 B. helps to correct errors
 C. helps to detect errors D. all of these

35. The efficiency of a binary system which selects one message out of 13 equiprobable messages will be
 A. 55.5% B. 67.5%
 C. 92.5% D. 98.5%

36. The efficiency of a decimal system which selects one message out of 13 equiprobable messages will be
 A. 55.5% B. 67.5%
 C. 92.5% D. none of these

37. The probabilities of symbols A, B, C produced by a source are 1/2, 1/4 and 1/8, respectively. If all the symbols are independent, the probabilities of a message $X = BACA$ will be
 A. 1/6 B. 1/32
 C. 1/64 D. 1/128

38. In the above question, the total information in the message X will be
 A. 2-bits B. 5-bits
 C. 7-bits D. 14-bits

39. In a PCM system, there are four quantised levels of probabilities 1/2, 1/4, 1/8 and 1/8, respectively. The entropy will be equal to
 A. 0.50 bits/sec B. 1/25 bits/sec
 C. 700 bits/message D. 2.25 bits/message

40. In the above question, if the input is a 50 Hz continuous signal sampled at Nyquist rate, the symbol rate will be
 A. 25 symbols/sec B. 50 symbols/sec
 C. 175 symbols/sec D. 200 symbols/sec

41. In the above question, the information rate is
 A. 50 bits/sec B. 125 bits/sec
 C. 175 bits/sec D. 225 bits/sec

42. The entropy of English text (letters plus space in-between) is
 A. 1.6 bits/symbol B. 3.2 bits/symbol
 C. 4.8 bits/symbol D. 6.4 bits/symbol

43. The occurrence of turning up of head when a coin is tossed has the information of
 A. 0-bit B. 1-bit
 C. 2-bit D. 4-bit

44. Essentially channel capacity is a measure of
 A. maximum rate of information
 B. amount of information a channel can take
 C. entropy rate D. none of these

Ans.	21. A	22. A	23. A	24. B	25. D	26. B	27. B	28. C	29. B	30. C	31. B	32. C	33. A
	34. D	35. C	36. C	37. D	38. C	39. C	40. C	41. C	42. C	43. B	44. A

45. The amount of information in a continuous signal is
A. zero B. 1
C. 2 D. infinite

46. Which of the following measures can help in transmitting a continuous signal band limited to f_m Hz over a channel of bandwidth less than f_m Hz?
A. Average rate of information is increased
B. Enough signal power is transmitted
C. Amount of signal information is increased
D. None of these

47. The entropy of the fact that the letter '*q*' is followed by letter '*u*' in English language is
A. zero B. 50%
C. 100% D. infinity

48. The auto-correlation function of an ergodic random process is given by
A. $\underset{T\to\infty}{\text{Lt}} \frac{1}{T}\int_0^T x(t)\cdot x(t-\tau)\,dt$
B. $\underset{T\to\infty}{\text{Lt}} \frac{1}{T}\int_{-T}^T x(t)\cdot x(t-\tau)\,dt$
C. $\underset{T\to\infty}{\text{Lt}} \frac{1}{2T}\int_0^T x(t)\cdot x(t-\tau)\,dt$
D. $\underset{T\to\infty}{\text{Lt}} \frac{1}{2T}\int_{-T}^T x(t)\cdot x(t-\tau)\,dt$

49. The auto-correlation function for the function $x(t) = E \sin \omega t$ is given by
A. $E^2\omega\tau$ B. $E^2 \cos \tau$
C. $\frac{1}{2}E^2 \cos \omega\tau$ D. $\frac{3}{2}E^2 \cos^2 \omega\tau$

50. If η is the positive frequency power density, the power density spectrum of white noise $\delta(\omega)$ is equal to
A. $1/\eta$ B. $\eta/2$
C. η D. 2η

51. The auto-correlation function of the white noise is a
A. step function B. impulse function
C. Gaussian function D. constant

52. In the study of power of the aperiodic signal, it is important to note that different harmonic frequency
A. interact to produce power
B. do not interact to produce power
C. few of them interact to reduce power
D. none of these

53. The power associated with a periodic signal is equal to
A. power contributed by the DC term
B. power contributed by all frequency terms
C. sum of A and B D. none of these

54. The pulse width controls the speed of the frequency composition for a
A. periodic pulse B. aperiodic pulse
C. in both A and B D. none of these

55. A periodic pulse signal of pulse width, τ can be reconstructed with close resemblance with the original signal by using harmonics of frequencies below
A. $2/\tau$ B. $3/\tau$
C. $5/\tau$ D. $6/\tau$

56. In the frequency spectrum of an aperiodic signal, the spacing among components is
A. large B. small
C. absent D. none of these

57. If the frequency spectrum of signal is defined, its time-domain function is
A. simultaneously determined
B. unique
C. both A and B D. none of these

58. Once a waveform is established, it can be described in
A. time domain B. frequency domain
C. both time and frequency domains
D. none of these

59. In an aperiodic signal, the larger is the off-time as compared to on-time, the more signal is
A. periodic B. distorted
C. non-sinusoidal D. none of these

60. A periodic function contains
A. all positive frequencies which are harmonically related to the fundamental
B. all negative frequencies
C. both A and B D. none of these

61. The frequency of any given component in the frequency spectrum of a periodic pulse depends only on
A. time period T B. pulse width τ
C. both A and B D. none of these

62. The analysis of periodic waveform is usually performed in
A. time domain B. frequency domain
C. both A and B D. none of these

63. The faithful reproduction of physically reliable pulse, the minimum number of required harmonics is
A. 5 B. 10
C. 20 D. 40

64. In a rectangular pulse, if the width τ remains constant and the period becomes greater than 2τ then bandwidth
A. increases B. decreases
C. remains constant D. none of these

65. In a tape recorder, the beginning section is recorded with a signal. The tape is continued with no signal

Ans. 45. D 46. B 47. C 48. D 49. C 50. B 51. B 52. B 53. C 54. C 55. C 56. C 57. C 58. C 59. A 60. C 61. A 62. B 63. B 64. C

applied for a time equal to ten lengths of the recorded signal. The overall signal is
A. periodic B. aperiodic
C. both A and B D. none of these

66. In the Fourier representation of a periodic signal, the associated power is equal to
A. power of the DC term
B. sum of the powers contributed by DC term plus harmonic components
C. power of the first ten harmonics and DC terms
D. none of these

67. In a pulse of width τ, the harmonics for which the amplitude coefficient is zero occurs at
A. $n\tau$ B. n/τ
C. $1/n\pi\tau$ D. $1/2\tau$

68. As the periodic length of a pulse increases, the harmonic density in its amplitude frequency spectrum
A. increases B. decreases
C. remains constant D. none of these

69. An image uses 512 × 512 picture elements. Each of the picture elements can take any of the 8 distinguishable intensity levels. The maximum entropy in the above image will be
A. 2097152 bits B. 786432 bits
C. 648 bits D. 144 bits

70. The information capacity (bits/sec) of a channel with bandwidth W and transmission time T is given by
A. $C \propto WT$ B. $C = W/T$
C. $C = T/W$ D. $C = W^2T$

71. The rate at which information can be passed through a telecommunication channel depends on the
A. carry frequency B. bandwidth
C. transmission loss D. transmitter power

72. A source has two symbols x_1 and x_2 with probability will be (ε is a number such that $0 < \varepsilon < 1$)
A. P_1, P_2, P_3, P_4 B. P_1, P_1, P_2, P_2
C. $(1-\varepsilon)P_1, (1-\varepsilon)P_2, \varepsilon P_1, \varepsilon P_2$
D. $\frac{P_1}{4}, \frac{P_2}{4}, \frac{P_1}{4}, \frac{P_2}{4}$

73. Consider the following communication system:
1. Telephony
2. Radio communication
3. Microwave communication
4. Optical communication
The correct sequence of these systems from the point of view of increasing order of base band channels each one of them can accommodate is
A. 2, 4, 3, 1 B. 3, 4, 1, 2
C. 1, 2, 3, 4 D. 4, 2, 1, 3

74. A TDM link has 20 signal channels and each channel is sampled 8000 times/sec. Each sample is represented by seven binary bits and contain an additional bit for synchronisation. The total bit rate for the TDM link is
A. 1180 k bits/sec B. 1280 k bits/sec
C. 1180 M bits/sec D. 1280 M bits/sec

75. The configuration used by a line communication system capable of transmitting a data at the rate of 1000 MB/s is
A. open line system B. optical fibre system
C. coaxial cable system D. twisted wire system

76. Which one of the following functions qualifies to be an autocorrelation function of a real-valued time signal?
A. $\sin(\omega\tau)$ B. $\exp(-\tau)$
C. $\delta(\tau)$ D. $u(\tau)$

77. The autocorrelation function of white noise is
A. a delta function B. a constant
C. Gaussian
D. $\exp(-|\tau|)$ with usual notation

78. A source generates four messages. The entropy of the source will be maximum when
A. all probabilities are equal
B. one of the probabilities equals 1.0 and others are zero
C. the probabilities are 1/2, 1/4 and 1/8
D. two of the probabilities are 1/2 each and others are zero

79. The spectral density of white noise is
A. exponential B. uniform
C. Poisson D. Gaussian

80. A source generates four equiprobable symbols. If the source coding is adopted, the average length of code for 100 per cent efficiency is
A. 6 bits per symbol B. 4 bits per symbol
C. 3 bits per symbol D. 2 bits per symbol

81. The probability density function of the envelope of narrow band noise is
A. uniform B. Gaussian
C. Rayleigh D. Rician

82. A message signal band limited to 5 kHz is sampled at the minimum rate as dictated by the sampling theorem. The number of quantisation levels is 64. If the samples are encoded in binary form, the transmission rate is
A. 60 kbps B. 50 kbps
C. 32 kbps D. 10 kbps

83. The Fourier transform of a Gaussian time pulse is
A. uniform B. a pair of impulses
C. Gaussian D. Rayleigh

84. The channel capacity under the Gaussian noise environment for a discrete memory less channel with a bandwidth of 4 MHz and SNR of 31 is
A. 20 mbps B. 4 mbps
C. 8 kbps D. 4 kbps

Ans.												
65. B	66. B	67. B	68. A	69. B	70. D	71. B	72. C	73. D	74. C	75. B	76. C	77. A
78. A	79. B	80. B	81. B	82. A	83. C	84. A						

85. Which of the following statements is not correct for power-density spectrum?
A. It retains information of magnitude of frequency spectrum
B. A given signal has a unique power-density spectrum
C. Many signals cannot have the same power density
D. It is an even function

86. The normalised energy of the signal $f(t) = \exp(-et)\,u(t)$ volt is
A. infinite B. 1/6 volt^2
C. 1/9 volt^2 D. zero

87. If on doubling the modulating frequency, the modulation index is halved, and the modulating voltage remains constant, then the system is
A. amplitude modulated
B. frequency modulated
C. phase modulated D. pulse modulated

88. The fidelity of a receiver relates to variation of the output with the modulating frequency when the output load is
A. resistive B. inductive
C. capacitive
D. a combination of all the above three

89. In an AM wave with 100 percent modulation, each sideband carriers .... of the total transmitted power.
A. one-half B. one-sixth
C. one-third D. two-third

90. Given a carrier frequency of 100 kHz and a modulation frequency of 5 kHz, the bandwidth of AM transmission is
A. 5 kHz B. 200 kHz
C. 10 kHz D. 20 kHz

91. When modulation of an AM wave is decreased
A. percentage carrier power is decreased
B. percentage carrier power is increased
C. total transmitted power is increased
D. percentage sideband power is unaffected

92. In AM transmission, power content of the carrier is maximum when m equals
A. 0 B. 1
C. 0.8 D. 0.5

93. In AM transmission with $M = 1$, suppression of carrier cuts power dissipation by a factor of
A. 6 B. 2
C. 3 D. 4

94. As compared to DSB-FC 100% modulation transmission, power saving in SSB-SC system is
A. 94.4% B. 50%
C. 100% D. 83.3%

95. In an AM transmission with 100% modulation, 66.7% of power is saved when ...... is/are suppressed.
A. carrier B. carrier and USB
C. carrier and LSB D. USB and LSB

96. In an AM system, full information can be conveyed by transmitting only
A. the carrier B. the upper sideband
C. the lower sideband D. any one sideband

97. In measurement of modulation index of an AM wave using CRO, the waveform shown in the figure below is observed. The value modulation index is

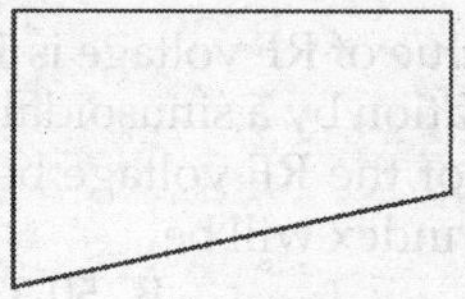

A. 13.5% B. 25.2%
C. 42.9% D. 67.7%

98. In the same question, the power in each sideband will be
A. 0.765 W B. 1.53 W
C. 3.05 W D. 6.10 W

99. In the AM wave shown in the figure below, the modulation index is

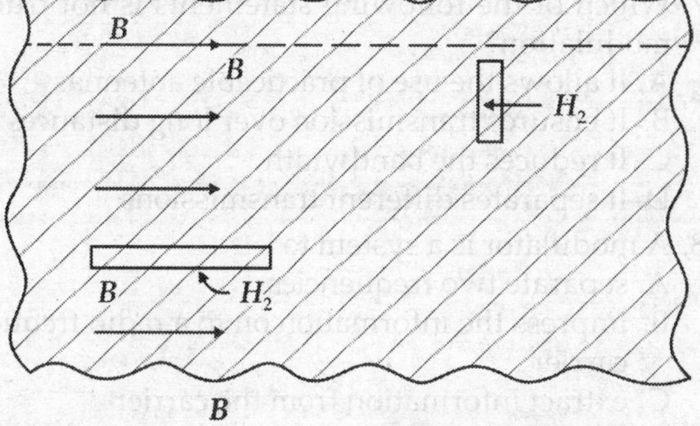

A. zero B. less than 100%
C. equal to 100% D. greater than 100%

100. In AM, if modulation index is more than 100% then
A. power of the wave increases
B. efficiency of transmission increases
C. the wave gets distorted
D. bandwidth increases

101. An AM wave has modulated index of 100%. If the carrier is suppressed, the percentage power saving will be
A. 52% B. 66.6%
C. 75% D. 95.5%

102. If the modulation index of an AM wave is changed from 0 to 1, the transmitted power
A. increases by 50% B. increases by 75%
C. increases by 100% D. remains unaffected

103. A carrier is simultaneously modulated by sine wave with modulation indices of 30% and 40% respectively. The overall modulation index will be

Ans. 85. C 86. A 87. B 88. A 89. B 90. C 91. B 92. A 93. C 94. D 95. A 96. D 97. C 98. A 99. D 100. C 101. B 102. A

A. 50%
B. 70%
C. 100%
D. indefinite as modulation by two waves simultaneously is not possible

104. The antenna current of an AM transmitter is 8 A when only the carrier is sent. When the carrier is sinusoidally modulated with depth of 50%, the antenna current becomes
A. 40.3%
B. 60.5%
C. 70.5%
D. 100%

105. The RMS value of RF voltage is 100 V. After amplitude modulation by a sinusoidal audio voltage, the RMS value of the RF voltage becomes 110 V. The modulation index will be
A. 40.5%
B. 50.3%
C. 64.8%
D. 100%

106. In measurement of modulation index for an AM wave using oscilloscope, the modulated signal and the modulating signal are applied respectively to
A. *Y*-input, *X*-input of the oscilloscope
B. *X*-input, *Y*-input
C. *Y*-inputs of a dual-beam oscilloscope
D. *Y*-input alternatively

107. Which of the following statements is not true for modulation?
A. It allows the use of practicable antennas
B. It ensures transmission over long distances
C. It reduces the bandwidth
D. It separates different transmissions

108. A modulator is a system to
A. separate two frequencies
B. impress the information on to a radio frequency carrier
C. extract information from the carrier
D. amplify the audio frequency signal

109. In AM transmission, the frequency which is not transmitted is
A. upper side frequency
B. lower side frequency
C. audio frequency
D. carrier frequency

110. The AM broadcast band is given by
A. 100 kHz to 30 kHz
B. 500 kHz to 1500 kHz
C. 3 to 30 MHz
D. 30 to 300 MHz

111. In an AM wave with 100% modulation, the carrier is suppressed. The percentage of power saving will be
A. 100%
B. 50%
C. 25%
D. 66.7%

112. A 400 W carrier AM modulated to a depth of 75%. The total power of the modulated wave would be
A. 210 W
B. 512 W
C. 450 W
D. 620 W

113. In high-power AM transmission, modulation is done at
A. buffer stage
B. RF power stage
C. oscillator stage
D. IF stage

114. An audio signal 15 sin 2π 1500t amplitude modulates 60 sin $2\pi \times 10^3$. The modulation index will be
A. 20%
B. 25%
C. 50%
D. 100%

115. In the above question, the two sidebands are of frequencies
A. 92.5 kHz, 102.5 kHz
B. 95 kHz, 105 kHz
C. 98.5 kHz, 101.5 kHz
D. 1.5 kHz, 100 kHz

116. In the same question, the amplitude of the two sidebands respectively will be
A. 7.5 V, 7.5 V
B. 15 V, 15 V
C. 15 V, 7.5 V
D. 7.5 V, 15 V

117. In the same question, the total bandwidth required to transmit the AM wave is
A. 1.5 kHz
B. 3 kHz
C. 100 kHz
D. 101.5 kHz

118. In AM, pilot-carrier transmission has
A. two sidebands
B. carrier, one sideband and part of other sideband
C. two sideband and a trace of carrier
D. carrier and part of one sideband

119. The number of AM broadcast stations that can be accommodated in a 100 kHz bandwidth for the highest modulating frequency of 5 kHz will be
A. 5
B. 10
C. 20
D. 100

120. For an AM wave having a power of 800 W at its carrier frequency and modulation index of 90%, the power content of each sideband is
A. 81 W
B. 162 W
C. 200 W
D. 400 W

121. The highest harmonic generated in human voice is
A. 1 kHz
B. 3 kHz
C. 5 kHz
D. 10 kHz

122. Which of the following is not an advantage of SSB over double-side band full carrier?
A. Noise in the signal is less
B. More channel space is available
C. Less power is required for the same signal strength
D. None of these

123. In a demodulator of a typical superhet broadcast receiver (medium and short-wave), the frequency of the input will be approximately
A. 10 MHz
B. 10 kHz
C. 500 kHz
D. 1000 Hz

124. A certain transmitter radiates 9 kW of power unmodulated and 10.125 kW when the carrier is sinusoidally modulated. The depth modulation will be
A. 40%
B. 50%
C. 60%
D. 65%

Ans. 103. A 104. C 105. C 106. A 107. C 108. B 109. C 110. B 111. D 112. B 113. B 114. B 115. C 116. A 117. B 118. C 119. B 120. B 121. B 122. D 123. C 124. B

125. In a radio receiver, which of the following stages does not need alignment?
A. TRF stage B. IF stage
C. Antenna input stage D. Audio stage

126. In a radio receiver, the oscillator frequency required to tune 1000 kHz signal would be
A. 455 kHz B. 1455 kHz
C. 645 kHz D. 1000 kHz

127. Carrier wave is generated by crystal oscillator because
A. it provides high power output
B. no buffer stage is needed
C. tuned circuit is not required
D. it has good stability

128. The function of buffer amplifiers in transmitters is
A. to amplify audio signals
B. to multiply frequency
C. to isolate RF stages D. none of these

129. The efficiency of Class C amplifier is high because
A. the grid is positive for one-half of the cycle
B. the plate current flows when plate voltage is low
C. the plate-tank circuit has sinusoidal output
D. a resonant circuit is used as load impedance

130. In a modulation system, if modulating frequency is doubled, the modulation index also becomes double. The system is
A. FM B. AM
C. PM D. Both FM and PM

131. The disadvantage of FM over AM is that
A. noise is very high for high frequency signals
B. larger bandwidth is required
C. high modulating power is required
D. high output power is required

132. The modulation index of a phase-modulated wave is
A. proportional to modulating frequency
B. proportional to reciprocal of modulating frequency
C. same as in frequency modulation
D. proportional to the phase of modulating signal

133. Frequency shift keying is used mostly in
A. radio transmission B. telegraphy
C. all of above D. none of these

134. In delta modulation which of the following drawbacks are existing?
1. Slope overload 2. Serration noise
3. Granular noise
A. 1 and 2 only B. 2 and 3 only
C. 1 and 3 only D. 1, 2 and 3

135. Which one of the following statements regarding the following signal
$x(t) = 5 \sin (2\pi \times 10^3 t) \sin (2\pi \times 10^6 t)$ is incorrect?
A. The upper side band frequency is 1001000
B. The lower side band frequency is 999000
C. The carrier amplitude is 5
D. $x(t)$ is a DSB-SC signal

136. To balance out the local oscillator noise at the input to the intermediate frequency or to IF amplifier in superheterodyne receivers, the microwave circuit used is
A. balanced amplifier B. balanced isolator
C. balanced phase shifter
D. balanced microwave mixer

137. An AM broadcast radio transmitter radiates 10 kW power, when the modulation percentage is 60, the power of the carrier is
A. 5.00 kW B. 7.69 kW
C. 8.47 kW D. 9.17 kW

138. If carrier modulated by a digital bit stream had one of the possible phases of 0, 90, 180 and 270 degrees, then the modulation is called
A. BPSK B. QPSK
C. QAM D. MSK

139. The most common detector used in an AM radio broadcast receiver is
A. envelope detector B. coherent detector
C. discriminator D. ratio detector

140. In FM transmission, amplitude of the modulating signal determines
A. rate of frequency variations
B. amount of frequency shift
C. total balance of transmission
D. distance of broadcast

141. In FM, when frequency deviation is doubled
A. modulation is doubled
B. modulation index is decreased
C. modulation is halved
D. carrier swing is halved

142. In an FM broadcast in VHF band, channel width is
A. 75 B. 25
C. 88 D. 200

143. Modern FM receivers use deemphasis circuit for
A. reducing high frequency noise
B. reducing the amplitude of high frequencies in the audio signal
C. increasing the amplitude of higher modulating frequencies
D. making demodulation easy

144. One of the serious disadvantages of FM transmission is its
A. high static noise B. expensive equipment
C. limited line-of-sight range
D. adjacent channel interference

145. FM broadcast band lies in
A. VHF band B. UHF band
C. SHF band D. HF band

Ans. 125. D 126. B 127. D 128. C 129. B 130. C 131. B 132. C 133. B 134. B 135. C 136. D 137. B 138. B 139. A 140. B 141. A 142. D 143. B 144. C 145. A

146. If a periodic saw-tooth wave frequency modulates the carrier, maximum frequency in FM wave will occur
A. in the beginning of the saw-tooth wave
B. at the end of the saw-tooth wave
C. in the middle of the saw-tooth wave
D. frequency will remain the same

147. In FM, if the modulating voltage remains constant, then lowering of modulating frequency would lead to
A. increase in amplitude of distant side-bands
B. decrease in amplitude of distant side-bands
C. no change in amplitude of distant side-bands
D. none of these

148. Which of the following statements is not true for FM?
A. The carrier never becomes zero
B. The *J*-coefficients occasionally are negative
C. The total power remains constant with respect to modulation index
D. The total bandwidth increases with increase in modulation index

149. In FM, the noise can be further decreased by
A. decreasing deviation B. increasing deviation
C. keeping deviation constant
D. none of these

150. Which of the following statements is not a reason for high signal-to-noise ratio in case of FM?
A. There is less noise at frequencies at which FM is used
B. Amplitude limiters are incorporated in FM receivers
C. Interference from other FM transmitter is very less
D. None of these

151. The percent modulation of the audio portion of a TV signal is 80%. If the maximum frequency deviation allowed is 25 kHz for audio, the actual frequency deviation will be
A. 10 kHz B. 20 kHz
C. 26 kHz D. 40 kHz

152. In the above question, the carrier swing will be
A. 20 kHz B. 30 kHz
C. 40 kHz D. 50 kHz

153. An FM wave is represented by $e = 12 \sin (6 \times 10^8 t + 5 \sin 1250t)$. Its carrier frequency is
A. 60 MHz B. 99.5 MHz
C. 125 kHz D. 199 MHz

154. In the same question, the modulating index is
A. 5 B. 6
C. 10 D. 12

155. In the same question, the maximum deviation is
A. 99.5 MHz B. 60 MHz
C. 995 Hz D. 199 Hz

156. In the same question, the power dissipated by the FM wave in a 10 ohm resistor will be
A. 2.1 W B. 3.6 W
C. 4.5 W D. 10 W

157. In an FM wave, the modulating frequency is 2 Hz and the maximum frequency deviation is 10 kHz. If number of significant pair of side-band is 8, the theoretical bandwidth required is
A. 8 kHz B. 16 kHz
C. 32 kHz D. 64 kHz

158. In the above question, the practical bandwidth is
A. 8 kHz B. 16 kHz
C. 24 kHz D. 32 kHz

159. The frequency deviation of an FM wave with modulation index of 7 and a practical bandwidth of 160 kHz is
A. 10 kHz B. 35 kHz
C. 70 kHz D. 140 kHz

160. For an FM wave of carrier 100 MHz, modulating frequency 10 kHz and the maximum frequency deviation of 1 MHz, the bandwidth required is
A. 20 kHz B. 1 MHz
C. 2 MHz D. 10 MHz

161. In the above question, if the modulating signal amplitude is doubled, the bandwidth required will be
A. 1 MHz B. 4 MHz
C. 10 MHz D. of the same value

162. In the same question, if the modulation frequency is doubled, the bandwidth required will be
A. 1 MHz B. 2 MHz
C. 4 MHz D. of the same value

163. An increase in the modulation index leads to increase in bandwidth in case of
A. AM B. FM
C. PM D. both FM and PM

164. Frequency and phase modulation differ in
A. different definitions of the modulation indices
B. their actual waveform
C. compatibility towards each other
D. all of these

165. Which of the following circuits does not have similar characteristics?
A. Amplitude discriminator
B. Diode detector
C. Phase discriminator D. Ratio discriminator

166. Which of the following microphones does not have a flat frequency response?
A. Carbon microphone
B. Condenser microphone
C. Ribbon microphone D. Crystal microphone

167. Which of the following type of speakers is used in a telephone receiver?

Ans. 146. B 147. A 148. A 149. B 150. D 151. B 152. C 153. B 154. A 155. C 156. B 157. C 158. C 159. C 160. C 161. B 162. A 163. B 164. A 165. B 166. A

A. Coaxial type B. Moving coil type
C. Fixed coil type D. Tweeter type

168. The most common method of erasing the magnetic tape to a tape recorder is to apply
A. signal of frequency more than 20 kHz
B. small DC voltage
C. signal of frequency less than 20 kHz
D. none of these

169. The treble response of a tape recorder
A. extends if the tape speed is increased
B. extends if the tape speed is decreased
C. is not affected by the tape speed
D. none of these

170. The modulation index of wide-band FM signal is
A. less than 1 B. equal to 1
C. greater than 1 D. none of these

171. The approximate bandwidth of a wide-band FM signal of modulating frequency f_m and maximum frequency deviation of δ is equal to
A. f_m B. $2f_m$
C. δ D. 2δ

172. The modulation index of a narrow-band FM signal is
A. less than 1 B. nearly equal to 1
C. more than 1 D. none of these

173. The bandwidth of a narrow-band FM signal is equal to
A. f_m B. $2f_m$
C. δ D. 2δ

174. From bandwidth point of view, narrow-band FM is equivalent to
A. AM B. PM
C. SSB
D. DSB-suppressed carrier

175. Which of the following circuits is an indirect method of generating FM?
A. Varactor diode modulator
B. Armstrong modulator
C. Reactance tube modulator
D. None of these

176. The frequency range used in frequency modulator is
A. 30 MHz to 300 MHz B. 88 MHz to 108 MHz
C. 3 MHz to 30 MHz D. 1 MHz to 3 MHz

177. Which of the following statements is not correct for Varactor diode modulator?
A. It generates FM
B. The diode is always reverse-biased
C. The junction capacitance vary with the modulating signal
D. There is no need of an oscillator circuit

178. According to Schwartz's criterion for developing a graph between bandwidth of an FM signal and its modulator index is that any frequency component should be considered insignificant if its signal strength is less than
A. 0.1% of unmodulated carrier
B. 1% of unmodulated carrier
C. 5% of unmodulated carrier
D. 10% of unmodulated carrier

179. Which of the following statements is not true for an FM system?
A. It requires less modulating power
B. It provides better S/N ratio
C. It requires less bandwidth
D. None of these

180. An FM station requires a bandwidth of 150 kHz for the signal and 26 kHz guard band. If a 6 MHz band is used, the number of such FM stations that can be accommodated is
A. 5 B. 10
C. 15 D. 25

181. The figure given below shows part of an FM transmitter. The carrier frequency at point X will be

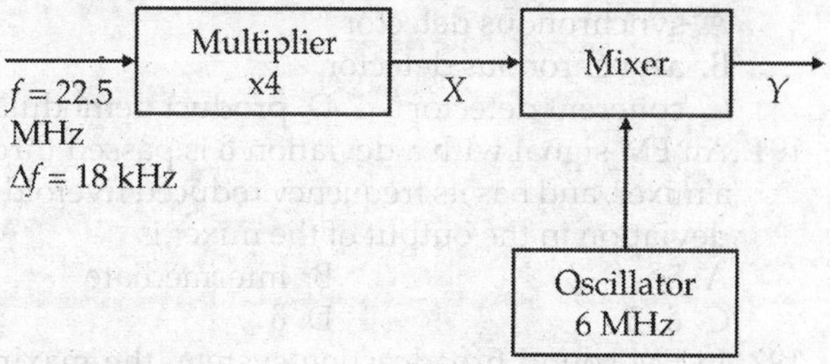

A. 5.6 MHz B. 18 kHz
C. 72 kHz D. 180 MHz

182. A 2 kHz audio signal frequency modulates 50 MHz causing a frequency deviation of 2.5 kHz. The bandwidth of the FM signal is
A. 2 kHz B. 4 kHz
C. 6 kHz D. 8 kHz

183. In the above question, the frequency deviation at point X is
A. 6 kHz B. 18 kHz
C. 72 kHz D. 180 kHz

184. In the same question, the frequency of the signal at point Y is
A. 90 MHz B. 84 MHz
C. 432 MHz
D. either 84 MHz or 96 MHz

185. In the same question, the frequency deviation is
A. 12 kHz B. 72 kHz
C. 432 kHz
D. either 12 kHz or 432 kHz

186. As compared to a phase discriminator, a ratio detector provides
A. double the output obtainable from a phase discriminator

Ans. 167. C 168. A 169. A 170. C 171. D 172. B 173. B 174. A 175. B 176. B 177. D 178. B 179. C
180. C 181. C 182. B 183. C 184. D 185. A

B. higher stabilisation against signal variations
C. all the advantages of a phase discriminator
D. lower linearity

187. The best way for signal coupling of two stages of a transmitter located physically far apart is
A. link coupling B. impedance coupling
C. single tuned transformer coupling
D. double tuned transformer coupling

188. Frequency frogging is used in carrier system to
A. conserve frequencies B. reduce distortion
C. reduce cross talk D. none of these

189. A comparison of frequency division and time division multiplexing systems shows that
A. FDM requires a lower bandwidth, but TDM has greater noise immunity
B. FDM has greater noise immunity and requires lower bandwidth than TDM
C. FDM requires channel synchronisation, while TDM has greater noise immunity
D. FDM requires more multiplexing while TDM requires band pass filter

190. The envelope detector is a/an
A. synchronous detector
B. asynchronous detector
C. coherent detector D. product demodulator

191. An FM signal with a deviation δ is passed through a mixer, and has its frequency reduced fivefold. The deviation in the output of the mixer is
A. 5δ B. intermediate
C. $\delta/5$ D. δ

192. In FM sound broadcasting system, the maximum frequency deviation is usually
A. 15 kHz B. 75 kHz
C. 200 kHz D. 5.2 MHz

193. In the spectrum of a FM wave, the
A. carrier frequency cannot disappear
B. carrier frequency disappears when on the modulation index is large
C. amplitude of any side-band depends on the modulation index
D. total number of side-bands depends on the modulation index

194. In FM reception, amplitude disturbances due to static change
A. reduce the signal frequencies
B. increase the signal frequencies
C. disturb the signal frequencies
D. do not affect the signal frequencies

195. The FM modulation index is given by
A. maximum frequency/minimum frequency
B. minimum frequency/maximum frequency
C. modulating frequency/maximum frequency deviation
D. maximum frequency deviation/modulating frequency

196. Consider the following statements:
The fundamental characteristic of an FM wave is that Δf (maximum deviation of instantaneous frequency from carrier frequency) is
1. proportional to the peak amplitude of the information signal
2. independent of its frequency
3. proportional to the RMS amplitude of the information signal

Of these statements
A. 1 alone is correct B. 2 and 3 are correct
C. 3 alone is correct D. 1 and 2 are correct

197. The frequency deviation for audio signal is
A. ± 3 kHz B. ± 10 kHz
C. ± 20 kHz D. ± 100 kHz

198. The main purpose of modulation is to
A. combine two waves of different frequencies
B. achieve wave-shaping of the carrier wave
C. transmit low-frequency information over long distances efficiently
D. produce sidebands

199. Domodulation
A. is performed at the transmitting station
B. removes sidebands
C. rectifies modulated signal
D. is opposite of modulation

200. In amplitude modulation
A. carrier frequency is changed
B. carrier amplitude is changed
C. three sidebands are produced
D. fidelity is improved

201. 100% modulation is produced in AM when carrier
A. frequency equals signal frequency
B. frequency exceeds signal frequency
C. amplitude equals signal amplitude
D. amplitude exceeds signal amplitude

202. For a given carrier wave, maximum undistorted power is transmitted when value of modulation is
A. 1 B. 0.8
C. 0.5 D. 0

203. Spread spectrum modulation technique is used in communication because
A. it is immune to interference
B. occupies large bandwidth
C. improves SNR
D. incorporates secrecy

204. In QPSK modulation scheme, to resolve the phase ambiguities, one should use
A. staggered QPSK modulation scheme
B. differential encoding
C. nothing needs to be used since QPSK modulation has got well defined phase transitions
D. filtering

Ans. 186. D 187. A 188. C 189. C 190. B 191. D 192. B 193. C 194. D 195. D 196. A 197. C 198. C 199. D 200. B 201. C 202. A 203. C 204. B

205. Filtering of BPSK/QPSK signal
A. improves the overall bit energy to noise spectral density ratio
B. causes signal distortion and inter-symbol interference
C. reduces overall RF power requirements
D. reduces complexity of demodulation

206. A square-law demodulator of AM signals suffers from
A. diagonal clipping B. positive peak clipping
C. second harmonic distortion
D. negative peak clipping

207. Which modulation scheme is most immune to distortion over non-linear channels?
A. VSB B. SSB
C. FM D. AM

208. The VSB signal is produced from the DSB signal by employing
A. simpler filters B. balance modulator
C. ring modulator D. phase-shift circuit

209. In AM, if one of the sidebands is suppressed, which of the following statements is not true?
A. Saving in bandwidth B. Saving in power
C. The equipment becomes more complex
D. Part of message is lost

210. Which of the following will carry the same information as the AM wave itself?
A. DSB B. SSB
C. VSB D. All of these

211. If carrier e_c and modulating signal e_m is applied to the input of a non-linear device, the output will be
A. e_c, e_m B. harmonics of e_c, e_m
C. $(e_c - e_m)$ and $(e_c + e_m)$ D. all of these

212. Essentially, balanced modulation produces
A. DSB B. SSB
C. BSB D. AM

213. The SSB can be obtained from balanced modulator by connecting at its output a
A. buffer B. clipper
C. filter D. adder

214. The filter required to obtain SSB from DSB signal is
A. low-pass filter B. high-pass filter
C. band-pass filter D. band-stop filter

215. In the phase-shift method of producing SSB, the number of balanced modulators and phase shift circuits required respectively are
A. 1, 1 B. 2, 1
C. 2, 2 D. 1, 2

216. In the phase-shift method to obtain USB
A. the two LSBs are out of phase
B. the two LSBs are in phase
C. the two USBs are out of phase
D. all side-bands are in phase

217. The SSB demodulator is known as
A. balanced modulation
B. product demodulation
C. amplitude discrimination
D. none of these

218. The product demodulator essentially is a
A. balanced modulator B. mixer
C. amplifier D. oscillator

219. The output of a product demodulator consists of frequencies
A. ω_m only B. ω_m and ω_c
C. $(\omega_c - \omega_m)$ and $(\omega_c + \omega_m)$ D. ω_m and $2\omega_c + \omega_m$

220. In order to obtain modulating signal from product demodulator
A. a low-pass filter is to be connected at the output
B. a high-pass filter is to be connected
C. a band-pass filter is to be connected
D. no filtering is required

221. High frequency in a superhet receiver
A. reduces tracking problem
B. reduces adjacent channel rejection
C. improves selectivity D. none of these

222. In superhet broadcast receiver, the frequency of local oscillator is
A. higher than the incoming signal
B. lower than the incoming signal
C. equal to the incoming signal
D. none of these

223. The function of buffer amplifier in transmitters is to provide
A. amplification of RF signal
B. frequency stability of the oscillator
C. impedance matching D. none of these

224. The type of modulator amplifier used in AM transmitter is
A. class A B. class B
C. class AB
D. class C with negative feedback

225. During the heterodyne process in a receiver, the modulation of the signal
A. remains unaffected B. decreases
C. increases D. is eliminated

226. In a typical AM receiver circuit, the oscillator frequency is
A. same as signal frequency
B. always equal to 455 Hz
C. lower than the signal frequency by 455 KHz
D. higher than the signal frequency by 455 kHz

227. A radio receiver is tuned to 710 kHz and its oscillator frequency is 885 kHz. The image frequency would be
A. 1420 kHz B. 1770 kHz
C. 1050 kHz D. 1060 kHz

Ans. 205. C 206. C 207. C 208. A 209. D 210. D 211. D 212. A 213. C 214. C 215. C 216. A 217. B 218. B 219. D 220. A 221. D 222. A 223. A 224. D 225. A 226. D 227. D

228. The image frequency of a station of frequency 1200 kHz is
A. 910 kHz B. 1290 kHz
C. 2110 kHz D. 2400 kHz

229. A transmitter radiates a total power of 10 kW. The carrier is modulated to a depth of 60%. The power in the carrier will be
A. 4.28 kW B. 8.47 kW
C. 16.52 kW D. 32.63 kW

230. The function of a pre-emphasis circuit in a communication system is to boost the
A. higher audio frequencies
B. lower audio frequencies
C. whole audio band D. modulated wave

231. MODEM stands for
A. modulator at transmitting side and detector at the receiving side
B. which deals with analog signals and shows digital information
C. analog to digital at transmitting side and digital to analog at a receiving side
D. a device which deals with digital signals only

232. For signal amplitude, modulated to a depth of 100% by a sinusoidal signal, power is
A. same as the power of unmodulated carrier
B. twice as the power of unmodulated carrier
C. 3/2 times the power of unmodulated carrier
D. 2/3 times the power of unmodulated carrier

233. A 1000 kHz carrier is simultaneously modulated with 300 Hz and 2 kHz audio sine waves. Which of the following frequencies will not be present in the output?
A. 998 kHz B. 999.7 kHz
C. 100.3 kHz D. 700 kHz

234. Line of sight transmission is used in
A. radio frequencies below 1 MHz
B. VHF only
C. sound signals only
D. VHF and UHF

235. A carrier simultaneously modulated by two sine waves with modulation indices of 0.3 and 0.4. If the modulated power is 10 kW, then what is the total modulated power?
A. 12.5 kW B. 10 kW
C. 10.125 kW D. 10.5 kW

236. The saving in power in a DSBSC system modulated at 80% is
A. nil B. 80%
C. 75.75% D. 50%

237. The DC voltage on a plate modulated class C amplifier is 8 kV and the peak modulating voltage is 4 kV. The peak RF voltage delivered to the load is
A. 32 kV B. 12 kV
C. 0.5 kV D. 24 kV

238. In a DM (delta modulation) system, the granular (idling) noise occurs when the
A. modulation signal increases rapidly
B. pulse rate decreases
C. modulating signal remains constant
D. pulse amplitude decreases

239. The most commonly used system of modulation for telegraphy is
A. single tone modulation
B. on-off keying
C. frequency shift keying
D. pulse code modulation

240. In PC for 132 standard quantising levels, the maximum error will be
A. 1/66 of the total amplitude range
B. 1/132 of the total amplitude range
C. 1/264 of the total amplitude range
D. 1/528 of the total amplitude range

241. A 3A modulation is sometimes used to
A. reduce the bandwidth required for transmission
B. reduce the power that must be transmitted
C. simplify the frequency stability problem in reception
D. allow the receiver to have a frequency synthesizer

242. The IF stage of a receiver employs
A. capacitive coupling B. impedance coupling
C. double-tuned transformer coupling
D. single-tuned transformer coupling

243. Which one of the following pairs is not correctly matched?
A. DSB-SC modulation: Balanced modulator
B. PCM: Pre-emphasis
C. FM: Reactance modulator
D. SSB modulation: Waver's method

244. Which one of the following systems offers the best trade off between bandwidth and S/N ratio
A. PAM B. PDM
C. PPM D. PCM

245. Which one of the following systems gives the highest figure-of-merit (a measure of the noise performance)?
A. WBFM B. NBFM
C. AM D. SSB

246. The types of modulation used generally in TV transmission of video and audio signals respectively are
A. FM, AM B. FM, FM
C. AM, AM D. AM, FM

247. If the video carrier is 62.25 MHz, the audio carrier will be
A. 62.25 MHz B. 56.75 MHz
C. 67.75 MHz D. 68.25 MHz

Ans.	228. C	229. B	230. A	231. B	232. C	233. D	234. D	235. A	236. C	237. D	238. C	239. C	240. C
	241. C	242. C	243. B	244. D	245. D	246. D	247. C						

248. Which of the following Pulse-time modulation does not exist in practice?
A. PWM B. PAM
C. PPM D. PFM

249. As compared to PPM, which of the following statements is not true for PWM?
A. PWM will still work if synchronisation between transmitter and receiver fails
B. The PWM transmitter should be able to handle maximum width pulse
C. The demodulation of PWM is more complex
D. The pulse amplitude in PWM remains constant

250. PDM is generated by employing
A. Schmitt trigger
B. free running multivibrator
C. JK flip-flop
D. monostable multivibrator

251. The monostable multivibrator is used for generating PWM because it is a/an
A. integrator
B. voltage-to-time converter
C. voltage-to-frequency converter
D. pulse generator

252. Which of the following statements is not true for PPM?
A. It requires constant transmitter power output
B. It depends upon transmitter-receiver synchronisation
C. It can be generated from PWM
D. The pulse amplitude does not remain constant

253. The PPM can be converted into PWM by employing
A. monostable multivibrator
B. bistable multivibrator
C. astable multivibrator D. integrator

254. As compared to PPM, the disadvantage of PDM is that it requires
A. powerful transmitter
B. pulses of larger widths
C. more samples per second
D. none of these

255. Which of the following technique is different from others?
A. PDM B. PWM
C. PCM D. PPM

256. In PCM, the number of pulses 'p' in a code-group for 'q' quantising levels is given by
A. $\log_2 q$ B. $\log_0 q$
C. $\log_{10} q$ D. $\log_e p$

257. In PCM, the number of quantising level is 16. The number of pulses in a code group will be
A. 16 B. 8
C. 4 D. 3

258. In the above question, the number of code-groups is
A. 7 B. 8
C. 15 D. 16

259. The bandwidth required for transmitting 4 kHz signal using PCM with 128 quantising levels is
A. 8 kHz B. 16 kHz
C. 20 kHz D. 128 kHz

260. In a PCM system, there are 256 quantising levels, and pulse allocation width of 0.625 μsec. Many messages each of band-width 4 kHz are simultaneously transmitted on a signal channel. The number of pulses per code-groups is
A. 3 B. 4
C. 8 D. 16

261. In the above question, the time period of each code-group is
A. 0.625 μsec B. 2.5 μsec
C. 5 μsec D. 10 μsec

262. In the same question, the time period between two samples is
A. 5 μsec B. 10 μsec
C. 52.5 μsec D. 125 μsec

263. In the same question, the number of messages that can be transmitted is
A. 10 B. 15
C. 25 D. 50

264. The standard value of sampling rate is
A. 4000 samples/sec B. 8000 samples/sec
C. 10000 samples/sec D. 16000 samples/sec

265. In an FM demodulator
A. the capacitors charged to the amplitude of FM wave
B. the frequency deviations are converted into voltage
C. simple diode is employed
D. none of these

266. Which of the following statements is not an advantage of phase discriminator?
A. Alignment is much easier
B. Linearity is better
C. It relies less on frequency response
D. It provides amplitude limiting

267. An FM has advantages over AM, all forms of pulse-time modulation (PTM) has over
A. PDM B. PAM
C. PCM D. FM

268. As compared to PCM, which of the following is not an advantage of other pulse modulation systems?
A. PCM circuitry is complex
B. PCM requires large bandwidth
C. PCM is not noise immune
D. None of these

Ans. 248. D 249. C 250. D 251. B 252. D 253. B 254. A 255. C 256. A 257. C 258. D 259. C 260. C 261. C 262. D 263. C 264. B 265. B 266. D 267. B 268. C

269. Which of the following statements are not true for commanding process?
   A. The signal is compressed at the transmitter and expanded at the receiver
   B. Small amplitude signals are expanded by the amplifier
   C. It can be used in PCM, PDM and PAM
   D. It reduces quantising noise

270. Which of the following events will not happen when quantising noise is decreased in PCM?
   A. Increase in the number of standard levels
   B. Increase in bandwidth
   C. Decrease in channel noise
   D. Decrease in randomness due to difference in digits sent and actual signal at an instant

271. To generate PCM, the signal is sampled and converted into
   A. PWM B. PPM
   C. PAM D. PDM

272. If the PCM is to be modulated for transmission, the most common type of modulation employed is
   A. AM B. FM
   C. PM D. PAM

273. To demodulate PCM, it is first converted into
   A. PWM B. PPM
   C. PAM D. PDM

274. In which of the following contexts, the PWM and PPM are not similar to FM?
   A. Use of amplitude limiter to reduce noise
   B. The ability of trade bandwidth for improved signal-to-noise ratio
   C. Limited range of transmission
   D. None of these

275. In PCM, for 128 standard quantising levels, the maximum error will be
   A. 1/128 of the total amplitude range
   B. 1/256 of the total amplitude range
   C. 1/64 of the total amplitude range
   D. 1/4 of the total amplitude range

276. For transmission of normal speech signal, the PCM channel needs a bandwidth of
   A. 64 kHz B. 8 kHz
   C. 4 kHz D. none of these

277. The main advantage of PCM system is lower
   A. bandwidth B. power
   C. noise D. none of these

278. A signal of 4 kHz is to be transmitted using sampling techniques. The time between two samples should be
   A. 62.5 µsec B. 125 µsec
   C. 250 µsec D. 400 µsec

279. A TDM system
   A. needs lower bandwidth
   B. gives lower signal-to-noise ratio
   C. uses simple circuits as compared to FDM
   D. all of these

280. In TDM, if number of signals being multiplexed are band limited in f_m, the bandwidth requirement is
   A. $\le nf_m$ B. $\ge nf_m$
   C. $\le 2nf_m$ D. $\ge 2nf_m$

281. Which of the following statement is not true for TDM?
   A. Pulse modulated signals can be sent during wide spaces of narrow pulses
   B. The transmitter and receiver are to be synchronised
   C. Analog signals can also be combined
   D. None of these

282. Low speed TDM is produced by
   A. electronic switching circuits
   B. delay lines
   C. rotating mechanical switches
   D. pulse oscillators

283. In high speed TDM, the channels are separated in the receiver employing
   A. OR gate B. AND gate
   C. NAND gate D. NOR gate

284. The overall bandwidth of a radar using PCM should be at least equal to
   A. reciprocal of pulse width
   B. pulse width
   C. half of the pulse width
   D. none of these

285. In systems like position radar or PDM
   A. only direction of pulse is necessary
   B. correct shape of the pulse is necessary
   C. either A or B D. none of these

286. In systems where proper shape of pulse received is required, the system bandwidth should be at least equal to
   A. ten times the pulse width
   B. reciprocal of five times the width
   C. reciprocal of two times the width
   D. none of these

287. The system response in pulse system is measured in terms of
   A. rise-time of the pulse B. fall-time of the pulse
   C. both A and B D. none of these

288. The rise-time of a pulse is defined as the time required for the amplitude of the leading edge from
   A. 0 to 90 per cent B. 10 to 100 per cent
   C. 10 to 90 per cent D. 0 to 10 per cent

289. The harmonic content of a periodic wave-form without sharp discontinuities is
   A. large B. small
   C. moderate D. none of these

Ans. 269. C 170. C 271. C 272. A 273. C 274. C 275. B 276. A 277. C 278. B 279. C 280. B 281. C 282. C 283. B 284. A 285. B 286. B 287. C 288. C 289. B

290. The waveforms without sharp discontinuities contain at the most
A. three harmonics B. five harmonics
C. seven harmonics D. none of these

291. The most rapid changes that the vocal chords can generate is controlled by
A. first harmonic B. third harmonic
C. fifth harmonic D. none of these

292. The highest harmonic generated in human voice is
A. 1 kHz B. 3 kHz
C. 5 kHz D. 10 kHz

293. The pulse length controls the spread of frequency composition for a pulse which is
A. periodic B. aperiodic
C. periodic or aperiodic D. none of these

294. In a pulse system, the system bandwidth is determined by
A. presence of the pulse B. number of pulses used
C. the application for which the pulse is used
D. none of these

295. A matched filter receiver for binary signals leads to
A. complete elimination of noise
B. maximum output signal-to-noise ratio
C. no distortion of signal
D. perfect input impedance matching

296. The function of a noise limiter is communication receivers is that at
A. eliminates noise as the signal enters the receiver
B. creates phase difference between the signal and the noise
C. reduces the effect of noise pulses
D. limits the noise bandwidth

297. If the value of a resistor is halved, the thermal noise power generated would become
A. doubled B. halved
C. unchanged D. one-fourth

298. As far as the effect of thermal-agitational noise is concerned, a noisy resistor can be considered as a noise-free-resistor in
A. series with a constant-current generator
B. parallel with a constant current generator
C. parallel with a constant voltage generator
D. none of these

299. Leak type bias is used in a plate modulated class C amplifier to
A. increase the bandwidth
B. prevent excessive grid current
C. prevent tuned circuit damping
D. prevent over modulation

300. In ionosphere the incidence angle is 60. The maximum usable frequency will be
A. $2f_c$ B. f_c
C. $2/\sqrt{3}\, f_c$ D. $\sqrt{3}\, f_c$

301. In PCM, if the transmission path is very long
A. repeater stations are used
B. pulse width may be increased
C. pulse amplitude is increased
D. pulse spacing is reduced

302. If large amounts of power are required, a push pull amplifier is designed for operation as
A. class-A B. class-B
C. class-AB D. class-C

303. A signal having uniformly distributed amplitude in the interval $(-V, +V)$ is to be encoded using PCM with uniform quantisation. The signal to quantizing noise ratio is determined by the
A. dynamic range of the signal
B. sampling rate
C. number of quantizing levels
D. power spectrum of signal

304. In PCM, the biggest advantage as compared to AM is
A. larger bandwidth B. larger noise
C. inability to handle analog signals
D. incompatibility with time division multiplex system

305. In a PCM, the amplitude levels are transmitted in a 7 unit code. The sampling is done at the rate of 10 Hz. The bandwidth should be
A. 5 kHz B. 35 kHz
C. 70 kHz D. 5 MHz

306. A RF carrier of 12 kV at 1 MHz is amplitude modulated to a 1 kHz signal to 6 kV peak. The modulation pattern is observed on a CRO calibrated suitably. The voltage indicated will be
A. 18 kV B. 1.001 kV
C. 18 V D. 36 kV

307. Pulse code modulation (PCM) employing 4-bit code is used to transmit a data signal having frequency components from DC to 2 kHz. The minimum bandwidth of the carrier channel should be
A. 2 kHz B. 4 kHz
C. 8 kHz D. 16 kHz

308. In a PCM system, the number of quantization levels are 16 and the maximum signal frequency is 4 kHz, the bit transmission rate is
A. 64 k bits/sec B. 16 k bits/sec
C. 32 k bits/sec D. 32 bits/sec

309. The signal-to-quantisation noise ratio in a PCM system depends upon
A. sampling rate
B. number of quantisation levels
C. message-signal bandwidth
D. size of the transmission system

Ans. 290. A 291. B 292. B 293. C 294. C 295. B 296. C 297. C 298. B 299. B 300. A 301. A 302. B 303. B 304. A 305. B 306. D 307. B 308. B 309. B

310. Under identical conditions, FM and PM are indistinguishable for a single modulating frequency. Now, if the modulating frequency is increased,
A. PM modulation index will remain constant, whereas FM modulation index will increase
B. PM modulation index will remain constant, whereas FM modulation index will decrease
C. PM modulation index will decrease and FM modulation index will increase
D. both PM as well as FM modulation indices will increase

311. Which of the following pulse communication system is inherently immune to noise?
A. PPM B. PCM
C. PWM D. PAM

312. Companding is used
A. to overcome quantising noise in PCM
B. in PWM receivers to reduce impulse noise
C. to protect small signals in PC from quantising noise
D. none of these

313. In PCM system, if the quantization levels are increased from 2 to 8, the relative bandwidth requirement will
A. remain same B. be doubled
C. be trippled D. become four times

314. A GSO spacecraft for each coverage should have a single antenna beam width as
A. 4.5° B. 17.3°
C. 120° D. 180°

315. In the amplitude frequency spectrum, radio waves lie between
A. audio range and infrared region
B. infrared and ultraviolet
C. ultraviolet and gamma rays
D. gamma rays and X-rays

316. Which of the following has the least wavelength?
A. X-rays B. Ultraviolet
C. Infrared D. UHF

317. The radiation pattern of Marconi antenna is

A. 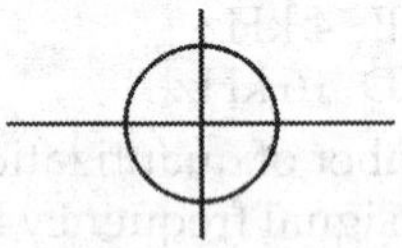 B. 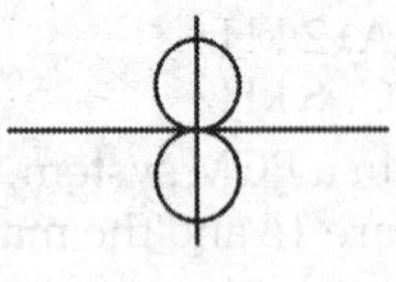

C. 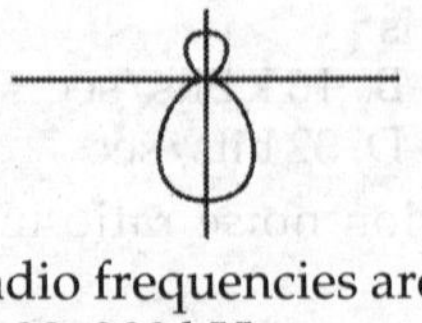 D. 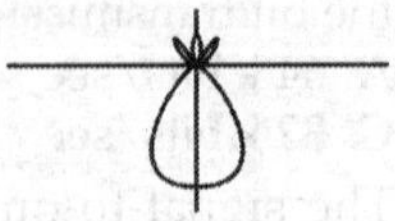

318. Radio frequencies are of the range
A. 30–300 kHz B. 300–3000 kHz
C. 3000–30000 kHz D. 30000–300000 kHz

319. The frequency of normal sound wave determines its
A. pitch B. quality
C. loudness D. echo

320. The waveform of sound wave determines its
A. loudness B. pitch
C. quality D. echo

321. Audio frequency range used in telephone is
A. 20–20000 kHz B. 30–30000 kHz
C. 300–3000 kHz D. 30–300 MHz

322. The sound intensity level due to radio playing loudly is of the order
A. 10 dB B. 50 dB
C. 90 dB D. 160 dB

323. The threshold of audibility of sound intensity for human ear is
A. zero dB B. –10 dB
C. –20 dB D. –50 dB

324. Loss of hearing may occur for sound intensities more than
A. 20 dB B. 50 dB
C. 100 dB D. 160 dB

325. UHF band is given by
A. 3 MHz to 30 MHz B. 30 MHz to 300 MHz
C. 300 to 3000 MHz D. 300 kHz to 30 MHz

326. Long wave transmission corresponds to frequency range of
A. 5–15 MHz B. 15–500 kHz
C. 3–30 MHz D. 30–70 MHz

327. The refractive index of the ionosphere for radio wave is
A. greater than 1 B. equal to 1
C. less than 1 D. equal to 390

328. The *D* layer of the ionosphere during day time
A. exists B. disappears
C. splits into two parts D. becomes thicker

329. Which of the following layers persists at night?
A. *D* layer B. *E* layer
C. F_1 layer D. F_2 layer

330. Which is the least important layer in regard to high frequency propagation?
A. *D* layer B. *E* layer
C. F_1 layer D. F_2 layer

331. Which of the following statements is not true for F_2 layer?
A. It is highly ionised layer
B. Its air density is very low
C. Its maximum height increases during night
D. Its average thickness is about 200 kms

332. The maximum usable frequency for sky wave propagation lies in the range
A. 10–500 kHz B. 8–35 MHz
C. 30–300 MHz D. 100–500 MHz

Ans. 310. B 311. B 312. C 313. D 314. B 315. A 316. A 317. B 318. C 319. A 320. C 321. C 322. C 323. B 324. D 325. C 326. B 327. C 328. A 329. D 330. A 331. C 332. B

333. The critical frequency for a layer is
A. lowest frequency that can return back to earth at any angle of beaming
B. highest frequency that can return back to earth at any angle of beaming
C. lowest frequency that can return back to earth at beaming angle of 90°
D. highest frequency that can return back to earth at beaming angle of 90°

334. In case of surface wave, to prevent short-circuiting of the electric component, the wave should be
A. vertically polarized
B. horizontally polarised
C. circularly polarised
D. of high power

335. In case of surface wave, the field strength at a point is directly proportional to
A. frequency of the wave
B. distance of the point from the antenna
C. height of the antenna
D. current of the antenna

336. Which of the following statement is not true with regard to VLF propagation?
A. The degree of tilt of wave-front is small
B. It propagates for small distances
C. Rapid fading does not occur
D. It is used for ship communication

337. Which of the following statement is not true for ground-wave propagation?
A. The wave gradually tilts more and more due to diffraction
B. The wave gradually gets attenuated by greater short-circuiting of electric field component
C. The propagation is beyond the horizon
D. The range is independent of the frequency

338. Which of the following modes permits beyond the horizon propagation?
A. Tropospheric scatter B. Surface wave
C. Satellite communication
D. None of these

339. For critical frequency of 10 MHz, the maximum usable frequency for angle of incidence of 60° will be
A. 10 MHz B. 20 MHz
C. 11.7 MHz D. 45 MHz

340. If a vertical antenna is used in conjunction with a loop antenna, the field pattern would be

A. 
B. 
C. 
D. 

341. Skip distance is
A. independent of transmitting power
B. dependent upon transmitting power
C. independent of frequency
D. independent of the state of ionisation

342. Under which of the following situations, fading will not occur?
A. The frequency of the signal is high
B. A single sky wave is present
C. Many sky waves arrive by different paths
D. None of these

343. The height of the transmitting antenna is 225 m above ground level. Its radio horizon will be
A. 60 km B. 76 km
C. 120 km D. 225 km

344. In the above question, if the receiving antenna is 16 m above the ground, then the total radio horizon will be
A. 16 m B. 60 km
C. 76 km D. 20 km

345. Virtual height of an ionosphere layer is
A. more than the height of a wave actually penetrates
B. less than the height a wave actually penetrates
C. same as the height a wave actually penetrates
D. none of these

346. For satellite communication, the frequency should be
A. less than the critical frequency of ionosphere
B. equal to the critical frequency of ionosphere
C. more than the critical frequency of ionosphere
D. none of these

347. VHF wave propagation exists with the help of
A. ionosphere B. ground
C. troposphere D. none of these

348. For an electromagnetic wave, the electric field component always meets a perfectly ground surface at an angle of
A. 0° B. 45°
C. 90° D. 180°

349. The frequency used for HF communication between two fixed points is generally higher in
A. winter B. summer
C. rainy season D. same in all the seasons

350. Which of the following frequencies cannot be used for reliable beyond the horizon communication without repeaters?
A. 20 MHz B. 400 MHz
C. 900 MHz D. 10 GHz

351. The utility of VLF waves in communication is due to their
A. low power requirements
B. these waves are not used at all
C. simpler antennas D. high reliability

Ans. 333. C 334. A 335. B 336. B 337. D 338. D 339. B 340. D 341. A 342. D 343. A 344. C 345. A
346. C 347. C 348. C 349. B 350. D 351. D

352. Troposphere scatter occurs most usefully in the frequency range of
A. 3 MHz–30 MHz B. 30 MHz–70 MHz
C. 400 MHz–5 GHz D. 30 MHz–300 MHz

353. In tropospheric scatter links, it is necessary to have
A. high transmitting powers
B. parabolic antenna reflectors
C. low-noise receivers D. all of these

354. Which one of the following effects occur in tropospheric scatter propagation?
A. Faraday effect B. Fading
C. Super refraction D. Atmospheric storms

355. In which of the following frequency range tropospheric scatter is used?
A. HF B. VHF
C. UHF D. VLF

356. Propagation by means of space waves occur in
A. HF range B. VHF range
C. UHF range D. VLF range

357. The radiated field of a short dipole is directly proportional to
A. inverse of wavelength
B. length of the antenna
C. antenna current D. all of these

358. In order to convert a transmission line into a radiating antenna for coupling to free space, it is necessary that
A. the line is short-circuited
B. the line is open-circuited
C. the receiving ends are bent outwardly at right angles
D. none of these

359. In the above question, the antenna will be of the form

A.
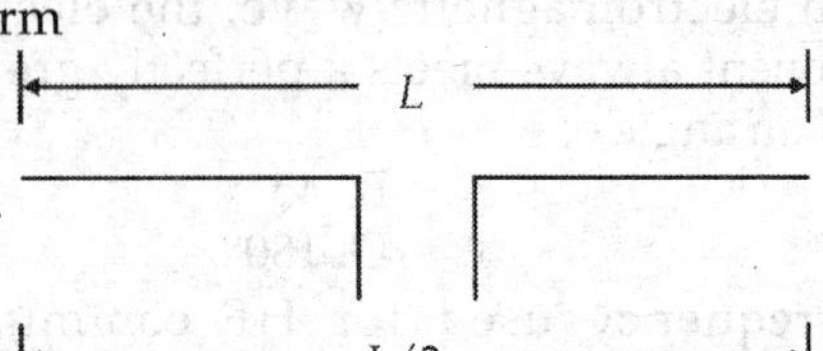

B.
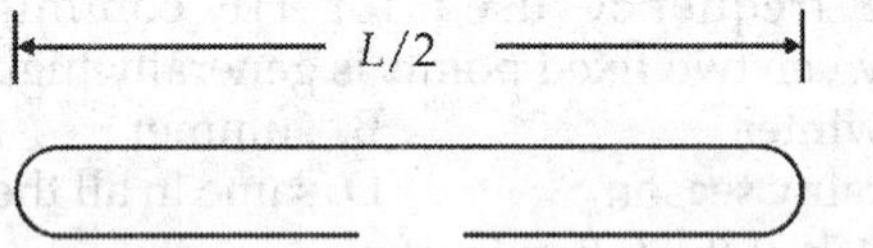

C.
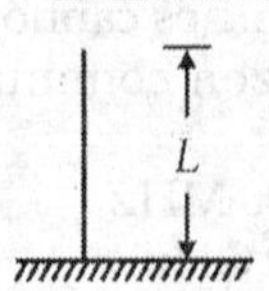

D.
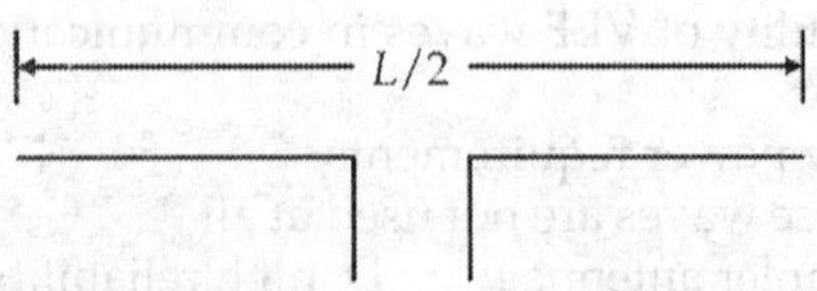

360. A Marconi antenna is required to transmit a 100 MHz signal. If the velocity factor for the antenna is 0.90, its optimum length l is
A. 0.675 m B. 1.35 m
C. 2.7 m D. 5.4 m

361. A beam antenna consists of a dipole, one reflector and one director. The velocity factor is 0.90 and the antenna is to be cut for a frequency of 150 MHz. The length of the dipole will be
A. 0.4 m B. 0.5 m
C. 0.9 m D. 1.2 m

362. In the above question, the length of the reflector is
A. 0.52 m B. 0.66 m
C. 0.82 m D. 0.945 m

363. In the above question, the length of the director is
A. 0.523 m B. 0.855 m
C. 0.18 m D. 0.28 m

364. In the above question, the spacing between the dipole and the reflector is
A. 0.9 m B. 0.12 m
C. 0.18 m D. 0.28 m

365. An antenna is constructed using antenna traps. The higher optimum frequency of operation will be
A. 37.5 MHz B. 7.2 MHz
C. 1.50 MHz D. 300 MHz

366. In the above question, the lower optimum frequency of operation is
A. 25 MHz B. 50 MHz
C. 100 MHz D. 200 MHz

367. In the set of figures given below, the radiation pattern of a $\lambda/2$ resonant antenna is shown in

1.
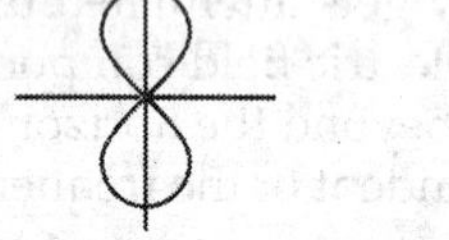
2.

3.
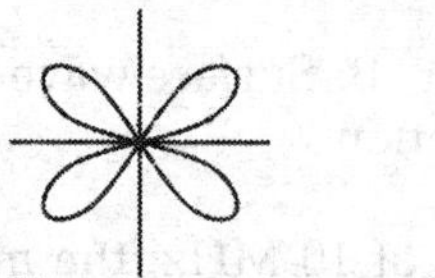
4.
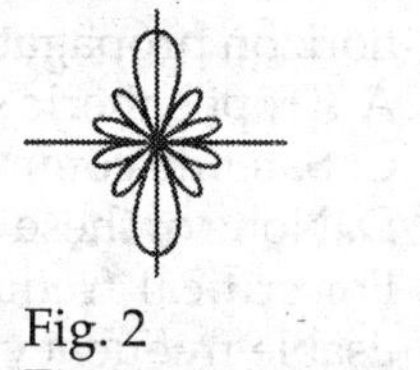

A. Fig. 1 B. Fig. 2
C. Fig. 3 D. Fig. 4

368. The radiation pattern of a folded dipole antenna is of the form

A.
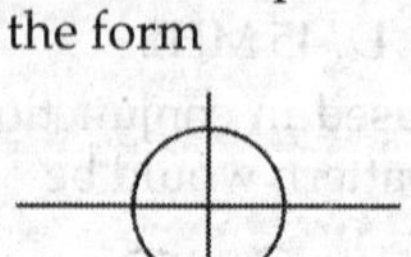
B.
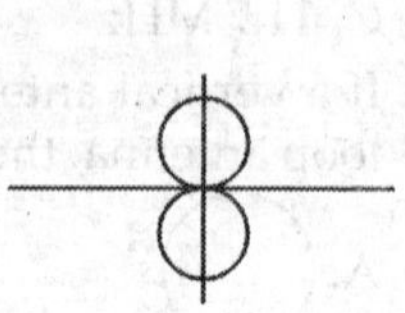
C.
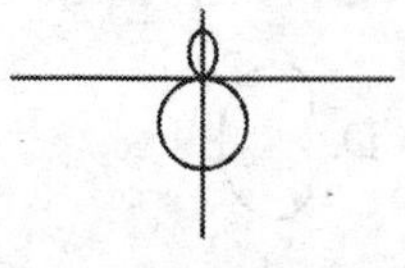
D.

Ans. 352. C 353. D 354. B 355. C 356. C 357. D 358. C 359. C 360. A 361. C 362. D 363. B 364. C 365. B 366. B 367. D 368. C

369. An antenna is to be used to receive a line-of-sight wave transmitted from an antenna located at a distance of 50 miles from this installation which is 800 ft in height. The minimum necessary height of the receiving antenna is
A. 12.5 ft B. 22 ft
C. 50 ft D. 100 ft

370. In a loop-antenna, the unbalanced current

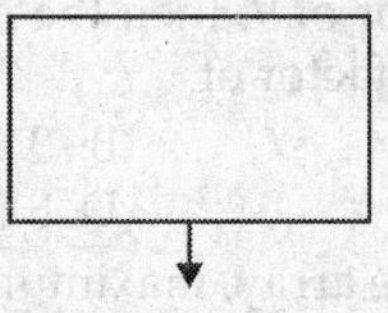

A. leads to an elliptical pattern
B. is independent of loop orientation
C. does not let true null to occur
D. none of these

371. The typical squelch circuit
A. cuts-off an audio-amplifier when the carrier is absent
B. eliminates the RF interference when the signal is weak
C. cuts-off an IF amplifier when the AGC is maximum
D. cuts-off an IF amplifier when the AGC is minimum

372. The dominant mode in a rectangular waveguide is
A. TE_{01} B. TM_{01}
C. TE_{10} D. TM_{11}

373. A transmission line of characteristic impedance 300 ohms is terminated by a load of (300 – j300) ohms. The transmission coefficient is
A. $1.265 \angle -18.43°$ B. $1.01 \angle -10°$
C. $1.41 \angle 66.68°$ D. $1.09 \angle 66.68°$

374. Two isotropic antennas are separated by a distance of two wavelengths. If both the antennas are with current of equal phase and magnitude, the number of lobes in the radiation pattern in the horizontal plane is

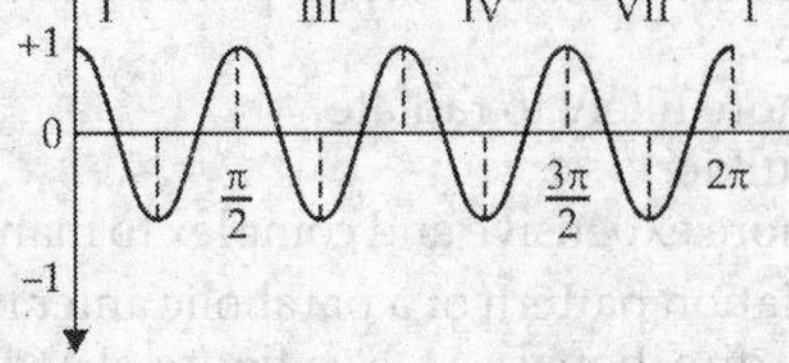

A. 2 B. 4
C. 6 D. 8

375. What is the numerical aperture of an optical fiber when its critical angle is 30°?
A. 0.5 B. 0.704
C. 0.866 D. 0.2

376. If the antenna diameter in a radar i increased by factor of 4, the maximum range will be increased by a factor of
A. $\sqrt{2}$ B. 2
C. 4 D. 8

377. Of the target cross-section is changing, the best system for accurate tackling is
A. lobe switching B. sequential lobing
C. conical scanning D. mono-pulse

378. A differential amplifier has a differential gain of 20000. CMRR = 80 dB. The common mode gain is given by
A. 2 B. 1
C. 1/2 D. 0

379. The hysteresis phenomenon in Schmitt trigger has, which of the following features?
1. It is due to negative feedback
2. In the inverted mode, circuit triggers at a higher voltage for increasing signals than for decreasing signals
3. Noise signal voltages greater than V_H are cut-off where V_H is the Hysteresis voltage
4. Slowly varying signals are converted to output waveform with abrupt changes at two specified levels of input signals

Of these four statements, the true statements are
A. 2 and 3 only B. 2 and 4 only
C. 2, 3 and 4 only D. 3 and 4 only

380. The transmission loss for a 3 GHz microwave system over a certain distance is 130 dB, if the frequency is now doubled, then the transmission loss will be
A. 136 dB B. 133 dB
C. 127 dB D. 139 dB

381. The demarcation between Fresnel region and Fraunhofer region fairly approximated by minimum distance between antennas for pattern measurement is (given that R = distance between antennas, D = largest aperture dimension and λ = wavelength)
A. $R = \lambda/2D^2$ B. $R = \lambda^2/2D^2$
C. $R = 2D^2/\lambda$ D. $R = 2D^2/\lambda^2$

382. Zoning is used with a dielectric antenna in order to
A. reduce the bulk of the lens
B. increase the bandwidth of the lens
C. permit pin-point focussing
D. correct the curvature of the wavefront from a horn that is too short

383. A parabolic dish has a diameter of 10 m. The maximum possible (ideal) gain of the antenna at λ = 314 cm will be
A. 30 dB B. 40 dB
C. 50 dB D. 60 dB

Ans. 369. C 370. D 371. A 372. C 373. A 374. D 375. A 376. C 377. D 378. A 379. B 380. B 381. C 382. A 383. D

384. With solid dielectrics such as those used in twisted pair and coaxial cables, the shunt conductance will be proportional to the
A. frequency and dielectric constant but independent of the power factor
B. frequency and power factor but independent of the dielectric constant
C. power factor and dielectric constant but independent of the frequency
D. frequency, dielectric constant and power factor

385. Two identical one-directional radiating elements are spaced 'd' apart excited by equal current strengths having a phase difference of ϕ. If this array is to be used for unambiguous direction finding, what should be the values of d and ϕ?
A. $d = \lambda/2, \phi = 0°$ B. $d = \lambda, \phi = 0°$
C. $d = \lambda/2, \phi = 180°$ D. $d = \lambda/2, \phi = 90°$

386. The antenna most commonly used for TV broadcasting in the UHF band is
A. turnstile antenna B. dipole antenna
C. yagi antenna D. rhombic antenna

387. A half-wave dipole operated at 30 MHz and has a maximum directive gain of 1.5. A plane wave carrying π watts of power per square metre is
A. $(\pi)^2$ watts B. 25 watts
C. 3.25 watts D. 37.5 watts

388. A short dipole and a half-wave dipole have radiation resistance of 0.4 Ω and 72 Ω, respectively. If the former requires 10 A rms current for certain total output power radiated, what would be the current required (approximately) by the latter for the same output?
A. 0.25 A B. 0.50 A
C. 0.75 A D. 1.0 A

389. When EM waves travel along a metallic waveguide in which there is a discontinuity due to some lossy material filled in a small length and other end is termined with a matched load, the discontinuity will cause
A. progressive waves throughout the line
B. standing waves on the generator side and progressive waves on the matched load side
C. progressive waves on the generator side and standing waves on the load side
D. standing waves throughout the line

390. In antenna measurements using two aperture antennas of dimensions D_1 and D_2, minimum separation between the two should be (λ is free-space wavelength of radiation uses)
A. $(D_1 + D_2)/(2\lambda)$ B. $(D_{12} + D_{22})/(8\lambda)$
C. $(D_{12} + D_{22})k/(\lambda)$ D. $(D_{12} \cdot D_{22}/2\lambda)$

391. Noise temperature (T) and gain (G) are two important parameters of satellite antennas. Which of the following antennas has the largest G/T ratio?
A. Parabolic reflector
B. Cassegranian antenna
C. Pyramidal horn antenna
D. Dipole antenna

392. If the antenna diameter in a radar system is increased by a factor of 9, then the maximum range will increase by a factor of
A. $\sqrt{3}$ B. 3
C. 9 D. 81

393. In a broadside array, maximum radiation occurs
A. at 90° to the line of array
B. at 45° to the line of the array
C. along the line of the array
D. at 60° to line of the array

394. For a transmission line, the propagation constant for a TEM wave travelling in it, is given by (where the symbols have the usual meanings)
A. $[(R + j\omega L)(G + j\omega e)]$
B. $[(R + j\omega L)(G + j\omega c)]^{1/2}$
C. $[(R - j\omega L)(G + j\omega c)]^{1/2}$
D. $[(R - j\omega L)(G + j\omega^2 c)]^{1/3}$

395. In a turnstile antenna, the crossed dipoles are excited with voltage such that the phase shift between the voltage is
A. zero B. 45°
C. 90° D. 180°

396. The electric field at a nequipotential surface is
A. unity B. zero
C. always parallel to the surface
D. always perpendicular to the surface

397. Which one of the following is the correct angular aperture for a paraboloidal reflector antenna for which the aperture number is 0.25.
A. 45° B. 90°
C. 120° D. 180°

398. A disadvantage of microstrips with respect to stripline circuits is that the former
A. do not lend themselves to printed-circuit techniques
B. are more likely to radiate
C. are bulkier
D. are more expensive and complex to manufacture

399. The radiation pattern of a parabolic antenna is
A. omni-directional B. a figure of eight
C. highly directional D. multi-directional

400. For a wire radiator of length equal to the wavelength of the transmitting signal, the number of loops in the radiation pattern will be
A. 1 B. 2
C. 3 D. 4

Ans. 384. D 385. D 386. C 387. B 388. C 389. D 390. C 391. B 392. C 393. A 394. B 395. C 396. D 397. C 398. B 399. C 400. D

401. Both end-fire and broad-band arrays are
A. non-linear B. non-resonant
C. of wide bandwidth
D. not suitable for short-wave transmission

402. Which of the following is not an advantage of folded dipole over one-wire dipole?
A. Input impedance is four times
B. Only half of the total current flows in each arm
C. Its radiation pattern is different
D. None of these

403. The number of parasitic elements in Yagi antenna is
A. 1 B. 2
C. 3 D. 4

404. The antenna employed in television receivers is a
A. half-wave dipole B. yagi antenna
C. rhombic antenna D. horn antenna

405. Which of the following statements is not true for rhombic antenna?
A. It is a resonant antenna
B. It operates over most of 3 to 30 MHz range
C. Its length need not be an integral number of half-wavelength
D. It is broadband antenna

406. Which of the following statements is not true for non-resonant antenna?
A. It is open-circuited at the far end
B. It has standing wave patterns
C. There is no power reflected
D. Its radiation pattern is unidirectional

407. In a dipole array, the overall radiation pattern does not depend upon
A. directional patterns of the elements
B. spacing between the elements
C. phase difference between the various feed points
D. none of these

408. Which of the following statements is not true for a broad-band array?
A. The dipoles are of equal size
B. The dipoles are equally placed along a straight line
C. The dipoles are fed in the same phase from the same source
D. There is phase difference between dipole currents

409. A one-half wavelength antenna has radiation resistance equal to
A. 700 ohms B. 300 ohms
C. 73 ohms D. 47 ohms

410. A half-wave dipole antenna is required to radiate 250 MHz signal. If the velocity factor of the antenna elements is 0.85, its length should be
A. 0.3 m B. 0.51 m
C. 0.6 m D. 1.2 m

411. The intended frequency of operation of dipole antenna cut-off a length of 3.5 is
A. 21.45 MHz B. 42.9 MHz
C. 85.8 MHz D. 100.5 MHz

412. Which of the following statements is not true for beam-width of an antenna?
Beam-width is the angle within which the
A. power radiated is above one-half of the power that exists at the maximum point
B. voltage developed remains 70.7% of the voltage at the maximum point
C. power becomes 3 dB than the maximum power
D. none of these

413. The maximum gain of an antenna using an evenly illuminated 6 ft paraboloid reflector used at 6 MHz will be
A. 1008 B. 8050
C. 950 D. 428

414. The folded dipole antenna is quite conveniently connected to
A. flat ribbon-type transmission line
B. coaxial line
C. two-wire line D. shielded line

415. The front-to-back ratio of an antenna which puts out 3 kW in its most optimum direction and 500 W in the opposite directions is
A. $10 \log_{10} 3$ dB B. $20 \log_{10} 6$ dB
C. $10 \log_{10} 6$ dB D. $20 \log_{10} 3$ dB

416. The input impedance of a folded dipole antenna is approximately
A. 12 ohm B. 220 ohm
C. 280 ohm D. 377 ohm

417. The characteristic impedance of twin transmission line used with TV receivers is
A. 98 ohm B. 127 ohm
C. 300 ohm D. 500 ohm

418. Directive gain of elementary doublet relative to isotropic radiator is
A. 1.5 times B. 2 times
C. 1.64 times D. 3.5 times

419. If the diameter of the parabolic reflector of a micro-wave antenna is doubled, its gain will increase by
A. 0 dB B. 2 dB
C. 4 dB D. 6 dB

420. Top loading is used with an antenna in order to increase its
A. beam-width B. input capacitor
C. effective length D. all of these

421. If length of the antenna is increased, the directive gain
A. increases B. decreases
C. remains the same D. becomes infinite

Ans. 401. C 402. C 403. B 404. B 405. A 406. A 407. D 408. D 409. C 410. B 411. B 412. D 413. B
414. A 415. C 416. C 417. C 418. A 419. D 420. C 421. A

422. As compared to resonant antenna, the directive gain of a non-resonant antenna of the same length is
A. greater B. equal
C. smaller D. about 6 dB more

423. Which of the following characteristics is shown by microwave antenna?
A. High gain B. Low input impedance
C. High directivity D. None of these

424. Which of the following antenna is best suited for low frequency transmission?
A. End-fire array B. Broad-side array
C. Unipolar antenna D. None of these

425. Which of the following antennas can be used in direction finding?
A. Loop antenna B. Long-wire antenna
C. Broad-side array D. None of these

426. The function of Cassegrain feed used with parabolic reflector is to
A. increase the beam-width
B. increase gain of the system
C. allow the feed to be placed at a convenient point
D. all of these

427. A Rhombic antenna is a
A. resonant antenna B. non-resonant antenna
C. directional HF antenna
D. none of these

428. A Yagi Uda array does not have
A. high gain B. high bandwidth
C. parasitic reflector D. parasitic director

429. Which of the following antennas is the standard reference antenna for the directiveness?
A. Elementary doublet B. Half-wave dipole
C. Isotropic antenna D. Rhombic antenna

430. An aerial array consists of 10 vertical aerials in a straight line spaced half a wavelength apart and equally energised in phase. The angular width of the forward beam in horizontal plane will be about
A. 32° B. 27°
C. 23° D. 18°

431. Maximum voltage will be induced in a loop antenna if
A. it is placed parallel to the incoming wave
B. it is placed at right angle to the incoming wave
C. it is placed at 45° to the incoming wave
D. its width is greater than $\lambda/2$

432. When loop antenna is rotated 360°, the induced voltage will experience
A. two maxima
B. two minima
C. two maxima and two nulls
D. one maxima and one null

433. The differential loops signal voltage in a loop antenna depends upon
A. the wavelength of incoming signal
B. the angle of the loop orientation
C. the loop area D. all of these

434. In the aural-null direction-finder, null indication is used because
A. the unbalanced current in minimised
B. in the null position a small rotation null produces large signal
C. there are two nulls available
D. it is easier to keep the loop at right angles to the incoming wave

435. The function of Adcock antenna is to eliminate
A. night error
B. unbalanced current
C. there are two nulls available
D. it is easier to keep the loop at right angles to the incoming wave

436. The disadvantage of Adcock antenna is that
A. it cannot eliminate night error
B. its output voltage is very small
C. it is not accurate
D. it is not the same as loop antenna

437. A common transmitter is switched from one loop antenna to another and the loops are positioned at right angles to each other. A receiver with the plane of either loop would indicate no difference in the output if it is placed on a line making an angle of
A. 0° B. 45°
C. 90° D. 180°

438. The aural-range beacon system employs
A. crossed vertical dipoles
B. sense vertical antenna
C. interlocking code letters
D. none of these

439. When an aeroplane is directly above the 'cone of silence', it means that the plane is
A. directly over the beacon antenna
B. very far from the range station
C. on the course
D. on the reciprocal course

440. As an aeroplane approaches a beacon, the aural signal
A. disappears B. increases in value
C. increases in volume and then disappears
D. is in a continuous tone

441. A klystron is used as
A. the main power source in an X-band radar
B. an amplifier in the radars at RF stage
C. a local oscillator in the receiver
D. a modulating signal generator for the magnetron

Ans. 422. A 423. B 424. C 425. A 426. C 427. B 428. A 429. C 430. C 431. A 432. C 433. D 434. B
435. A 436. B 437. B 438. C 439. B 440. B 441. C

442. The function of de-emphasis circuit in an FM receiver is to
A. control the amplitude of FM wave
B. restore the various frequency components to their original levels
C. reduce the high frequency component
D. none of these

443. The function of an amplitude limiter in an FM receiver is to
A. eliminate any change in amplitude of received FM signal
B. reduce the amplitude of the signal to suit IF amplifier
C. to amplify low frequency signals
D. none of these

444. Which of the following stages is not present in AM receiver as compared to FM receiver?
A. IF amplifier B. Amplitude limiter
C. Demodulator D. AF amplifier

445. Typical bandwidth of an FM receiver is
A. 20 kHz B. 200 kHz
C. 1 kHz D. 20 kHz

446. Which of the following codes is not a telegraphic code?
A. Five-unit code B. Gray code
C. 7.5 unit code D. Cable code

447. Which of the following codes requires minimum average number of signal elements to transmit one character?
A. 5-unit code B. 7.5 unit code
C. Cable code D. Morse code

448. The fundamental frequency of signalling for a telegraphic speed of 100 bauds is
A. 25 Hz B. 50 Hz
C. 100 Hz D. 200 Hz

449. In a telegraph relay, the polarising flux increases the
A. sensitivity of the relay
B. speed of operation
C. capacity of the relay D. all of these

450. Which of the following circuits transmits two messages simultaneously in one direction?
A. Duplex B. Diplex
C. Simplex D. Quadruplex

451. One of the main functions of the RF amplifiers in a superhetrodye receiver is to
A. provide improved tracking
B. permit better adjacent channel rejection
C. increase the tuning range of the receiver
D. improve the reflection of the image frequency

452. An A3J transmitter operating at 16 MHz has a frequency stability of 1 part per million. If its transmission is reproduced by a receiver whose frequency stability is 8 parts per million, the maximum frequency error that the output of this receiver could have in reproducing this transmission will be
A. 20 Hz B. 48 Hz
C. 72 Hz D. 144 Hz

453. Which of the following statement about receivers is false?
A. Double conversion is used to improve image rejection
B. Double conversion is used to improve selectivity
C. Variable sensitivity is used to improve selectivity
D. Variable sensitivity is used to eliminate selective fading

454. Synchronous Time Division Multiplexing (STDM) used in digital communication systems has 24 slots. If the maximum frequency of speech frequency of talkers is 8 kHz, the revolutions per minute of the STDM switch will be
A. 192000 B. 480000
C. 960000 D. 140000

455. Radar is a device, with the help of which we can
A. perform mathematical calculations automatically
B. listen more melodious music
C. detect the pressure of an aircraft and its location in the vicinity of a particular range
D. cure the damaged tissues in the human body

456. In RF communication, EIR means
A. equally important radiated power
B. equi isotropic rotational phase
C. equivalent isometric radiation of power
D. equivalent isotropic radiated power

457. A high PRF would
A. increase the maximum range
B. decrease the maximum range
C. affect the range ambiguity
D. increase the efficiency of the radar system

458. With a large ratio of the serial dish diameter to the wavelength in a radar system, there will be
A. decreased capture area
B. easy target acquisition
C. poor target discrimination
D. small maximum range

459. 'A'-scope displays
A. both the target position and range
B. the target position but not range
C. neither range nor position but only velocity
D. the target range but not position

460. Circular polarization in radar system is used for
A. detecting moving targets
B. reducing clutter due to rain drops
C. range increase during radio interference

Ans. 442. B 443. A 444. B 445. B 446. B 447. C 448. B 449. A 450. B 451. D 452. D 453. D 454. A 455. C 456. B 457. B 458. B 459. D 460. B

D. determining height as well as bearing of the target

461. If the target cross-section is changing, the best system for accurate tracking is
A. monopulse B. conical scanning
C. sequential lobing D. lobe switching

462. For a radio direction finder, the antenna pattern is
A. omnidirectional B. fan-beam type
C. cardioid D. pencil beam

463. Compared to other type of radars, the phased array radar does not have the following advantage.
A. Very fast scanning B. Circuit simplicity
C. Ability to track and scan simultaneously
D. Ability to track many targets simultaneously

464. Which of the following statements is true for a reflex klystron?
A. Maximum efficiency and maximum output voltage cannot be obtained together
B. Maximum efficiency and maximum output voltage can be obtained together
C. Maximum power can be obtained
D. None of these

465. The disadvantage of a two-cavity klystron is that
A. its efficiency is low B. its output power is low
C. it can be used only at one operating frequency
D. none of these

466. The efficiency of an ideal klystron is about
A. 25% B. 50%
C. 75% D. 100%

467. The purpose of duplexer is
A. to convert TDM to FDM
B. to provide same antenna both for transmission and reception
C. to convert pulses transmission to CW transmission
D. none of these

468. The klystron tube can be used as
A. microwave oscillator B. microwave amplifier
C. microwave mixer D. all of these

469. The reflex klystron can be used as
A. microwave oscillator B. microwave amplifier
C. microwave mixer D. all of these

470. In magnetron, the diode plate current is influenced by
A. magnetic field B. electric field
C. plate voltage D. none of these

471. In klystron, the velocity modulation takes place in
A. input cavity resonator
B. output cavity resonator
C. drift tube D. collector

472. The advantage of a tunnel-diode mixer is that it has
A. low noise figure B. high conversion gain
C. low local oscillation power requirements
D. all of these

473. An LSA diode is similar to
A. tunnel diode
B. large-scale integrated diode
C. gun diode D. IMPATT diode

474. Which of the following is not employed by a magnetron?
A. Grid B. Magnetic field
C. Anode D. Cavity resonators

475. Which of the following statements is true for a klystron?
A. The butcher grids modulate the velocity of electrons
B. Bunches of electrons transfer energy to the cavity resonator
C. The cavity resonator is connected to the catcher grid
D. All of these

476. Most audio communication utilises frequencies in the range
A. 100 Hz–1 kHz B. 100 Hz–3 kHz
C. 100 Hz–30 kHz D. none of these

477. For a very high degree of fidelity, the voice communication should have a bandwidth of
A. 1 kHz B. 3 kHz
C. 10 kHz D. 20 kHz

478. In order to represent an aperiodic signal pulse without sharp discontinuities of the rectangular pulse, the bandwidth necessary to allow passage of all the frequency information necessary to produce the pulse is equal to
A. reciprocal of the longest pulse received
B. twice the reciprocal of the longest pulse received
C. twice the reciprocal of the shortest pulse received
D. none of these

479. Most of human voice energy is concentrated in the vicinity of
A. 100 Hz–1 kHz B. 100 Hz–3 kHz
C. 100 Hz–5 kHz D. none of these

480. For intelligible voice communication, minimum bandwidth required is
A. 900 Hz B. 1 kHz
C. 3 kHz D. 6 kHz

481. In human voice communication, if the bandwidth is less than 900 Hz, the sound produced in the receiver will be
A. clear B. mushy
C. highly fidel D. none of these

482. The frequency of sound produced by forcing air between the tongue and teeth (like 's' sound) has highest harmonic of
A. 1 kHz B. 3 kHz
C. 10 kHz D. 20 kHz

Ans. 461. A 462. C 463. B 464. A 465. C 466. B 467. B 468. C 469. A 470. A 471. A 472. D 473. C 474. A 475. D 476. B 477. C 478. C 479. A 480. A 481. B 482. C

483. In a perfect electronic music system, the bandwidth necessary for high fidelity should be
A. 3 kHz B. 10 kHz
C. 20 kHz D. 40 kHz

484. Major bottleneck in the design of high fidelity system is
A. noise B. high gain amplifier
C. transducer D. none of these

485. The microwave range is
A. 30 MHz–300 MHz B. 300 MHz–3 GHz
C. 3 GHz–30 GHz D. 30 GHz–300 GHz

486. Which of the following statements are not true for microwave links?
A. Propagation is by space wave
B. Typical repeater spacing is more than 100 km
C. The repeaters send signals of different frequencies in two directions
D. Frequency modulation is employed

487. Typical operating frequencies for microwave links are close to
A. 10 MHz–100 MHz B. 500 MHz–1 GHz
C. 4 GHz–6 GHz D. 10 GHz–50 GHz

488. Vacuum tubes cannot be used at microwave frequencies because of
A. small series inductive reactances
B. large shunt capacitive reactances
C. large transit time
D. high noise temperature

489. Transit time in microwave tubes cannot be reduced by
A. bringing electrodes closer together
B. increasing anode current
C. employing multiple or coaxial leads
D. increasing anode voltage

490. The reflex klystron is often preferred to the two-cavity klystron because it is
A. more efficient
B. has higher number of modes
C. able to give higher output power
D. easier to tune

491. The function of squelch circuit in radio receiver is to
A. reduce radio frequency interference in strong signals
B. increase the output of audio amplifier for weak signals
C. cut-off audio amplifier between stations
D. reduce AVC bias for strong signals

492. The tracking between RF amplifier and local oscillator stages of a receiver is obtained by
A. variable tuning capacitors
B. variable running inductors
C. tracking circuits
D. a ganged tuning arrangement

493. The IF stage of a receiver employs
A. impedance coupling B. capacitive coupling
C. double-tuned transformer coupling
D. single-tuned transformer coupling

494. The passband of the tuned circuits of a radio receiver should be equal to
A. 455 kHz B. 1000 kHz
C. 20 kHz D. 5 kHz

495. Automatic gain control is used to
A. maintain the tuning correct
B. reduce the volume of loud passage of music
C. maintain the same volume of output, when stations of different strength are achieved
D. increase the amplification at high frequencies

496. The AGC voltage in a radio receiver is proportional to
A. the amount of modulation
B. the amplitude of the audio signal
C. the amplitude of the IF carrier
D. none of these

497. The output of a diode detector contains
A. modulating signal B. DC voltage
C. RF ripple D. all of these

498. Which of the following statements is not true in case of AVC in an AM receiver?
A. The gain control has to be readjusted for changing from one station to another
B. It helps to smooth out the rapid fading
C. It prevents the overloading of the final IF stage
D. none of these

499. The function of padders in receivers is to improve
A. tracking
B. image frequency rejection
C. noise reduction D. sensitivity

500. The function of AM detector circuit is
A. to rectify the input signal
B. to discard the carrier
C. to provide audio signal
D. all of these

501. The intermediate frequency of a superhet receiver is 450 kHz. If it is tuned to 1200 kHz, the image frequency will be
A. 750 kHz B. 900 kHz
C. 1650 kHz D. 2100 kHz

502. In a broadcast receiver, the local oscillator is tuned to a frequency higher than the incoming frequency to facilitate
A. image frequency rejection
B. easier tracking
C. adequate frequency coverage
D. noise reduction

503. Public broadcasting employs double sideband system because

Ans. 483. C 484. C 485. C 486. B 487. C 488. C 489. C 490. D 491. C 492. D 493. C 494. C 495. C 496. C 497. D 498. A 499. A 500. D 501. D 502. C

A. it requires less transmitting power
B. it requires smaller bandwidth
C. the circuits are simple and less expensive
D. all of these

504. In double side-band suppressed carrier system, detection requires expensive circuitry because
A. synchronous detection is required
B. it is difficult to generate local carrier of the receiver
C. received signal is of low power
D. none of these

505. The advantage of SSB—suppressed carrier system is that it provides
A. higher efficiency of transmission
B. better quality of communication
C. simpler and inexpensive circuitry
D. none of these

506. The bandwidth requirement for VSB system is
A. less than the bandwidth for SSB system
B. same as the bandwidth for SSB system
C. more than the bandwidth for SSB system
D. double the bandwidth for SSB system

507. Which of the following microphones provides high directivity?
A. Carbon microphone
B. Condenser microphone
C. Ribbon microphone D. Crystal microphone

508. Telephone receiver employs
A. carbon microphone
B. condenser microphone
C. ribbon microphone D. crystal microphone

509. Which of the following microphones does not require polarising current?
A. Carbon microphone
B. Condenser microphone
C. Crystal microphone D. None of these

510. In which of the following microphones, background noise is present?
A. Carbon microphone
B. Condenser microphone
C. Ribbon microphone D. Crystal microphone

511. The disadvantage of a ribbon microphone is that
A. it tends to accentuate low frequency
B. it tends to accentuate high frequency
C. the speaking voice has to be placed near the microphone
D. it provides low directional characteristics

512. The function of baffles in speaker system is to
A. allow high frequencies to pass
B. avoid cancellation of compression and ratification of air
C. provide stability to the speaker
D. produce echo effect

513. In a tri-axial speaker, the low frequency diaphragm will be
A. largest in size B. smallest in size
C. medium in size D. independent of size

514. The condition for no distortion in a transmission line is
A. LC = GR B. LG = CR
C. LR = GC D. none of these

515. Loading in a transmission line leads to an increase in
A. inductance of the line
B. capacitance of the line
C. resistance of the line D. conductance of the line

516. If the reflection coefficient of a line is zero, the line is
A. infinite line B. open-circuited
C. short-circuited D. none of these

517. Along a transmission line with standing waves, the distance between a voltage maximum and adjacent current minimum is
A. $\lambda/8$ B. $\lambda/4$
C. $\lambda/2$ D. λ

518. Impedance matching over wide frequency range can be obtained by using
A. double stub B. single stub
C. quarter-wave transformer
D. none of these

519. By stub matching, standing waves are eliminated on
A. source side of stub B. load side of stub
C. both sides of stub D. none of these

520. If the frequency increases, the characteristic impedance of a transmission line
A. decreases B. increases
C. remains constant
D. first increases and then decreases

521. The input impedance of a line of infinite length is equal to
A. zero B. Z_0
C. $Z_0/2$ D. infinity

522. The input impedance of a short-circuited loss-less $\lambda/4$ line is equal to
A. zero B. Z_0
C. $Z_0/2$ D. infinity

523. As compared to short-circuited stubs, open-circuited stubs are not preferred because the latter are
A. of different characteristic impedance
B. of tendency to radiate
C. difficult to connect D. all of these

524. Which one of the following systems is best for transmission line load matching over a range of frequencies?

Ans. 503. C 504. A 505. A 506. B 507. C 508. A 509. C 510. A 511. A 512. B 513. A 514. B 515. A
516. A 517. C 518. A 519. C 520. A 521. B 522. D 523. B

A. Broad-band directional coupler
B. Single-stub with adjustable position
C. Single-stub D. Double-stub

525. Which of the following can provide impedance inversion?
A. A quarter-wave line B. A short-circuited stub
C. An open-circuited stub
D. A half-wave line

526. In a travelling wave tube
A. bunches of electrons are formed
B. the gain of the tube is independent of length
C. the gain is governed by relative velocity of the energy wave and the electron beam
D. all of these

527. Which of the following statements is not true for Gunn diode?
A. It does not have a junction
B. It cannot rectify
C. It has negative resistance characteristics
D. It cannot be used as CW oscillator

528. In Gunn diodes electrons are transferred from
A. high to low mobility energy bands
B. low to high mobility energy bands
C. valley to domain formation
D. domain to valley formation

529. For a directional coupler, the power is in the ratio of
A. 40 dB B. 30 dB
C. 20 dB D. 10 dB

530. In an optical fibre, the refractive index of the cladding material should be
A. nearly unity B. very low
C. less than that of the core
D. more than that of the core

531. Microwave link repeaters are typically 50 km apart
A. because of atmospheric attenuation
B. because of output tube power limitations
C. because of earth's curvature
D. to ensure that the applied DC voltage is not excessive

532. Any two-port network having a 6 dB loss will give an output
A. power which is one-quarter of the input power
B. power which is one-half of the input power
C. voltage which is 0.707 of the input voltage
D. power which is 0.707 of the input power

533. To couple a coaxial line to a parallel wire line, it is best to use a
A. slotted line B. balun
C. directional couplet D. $\lambda/4$ transformer

534. Consider the following statement:
SWR for a transmission line is infinity when the load is
1. a short-circuit 2. a complex impedance
3. an open-circuit 4. a pure reactance
Of these statements
A. 1, 3 and 4 are correct B. 1, 2 and 3 are correct
C. 2, 3 and 4 are correct D. 1, 2 and 4 are correct

535. Microwave link repeaters are typically 50 km apart in TV transmission because of
A. atmospheric attenuation
B. output power tube limitations
C. microwave transmission is through surface wave which attenuates faster
D. earth's curvature

536. If the peak transmitted power in a radar system is increased by a factor of 81, then the maximum range will be increased by a factor of
A. 3 B. 9
C. 27 D. 81

537. A radar have a maximum range of 120 km. The maximum allowable pulse repetition frequency (in pulse per second) for unambiguous reception will be
A. 1250 B. 330
C. 2500 D. 8330

538. The radiation pattern of the antenna used in VHF omnidirectional range (VOR) is

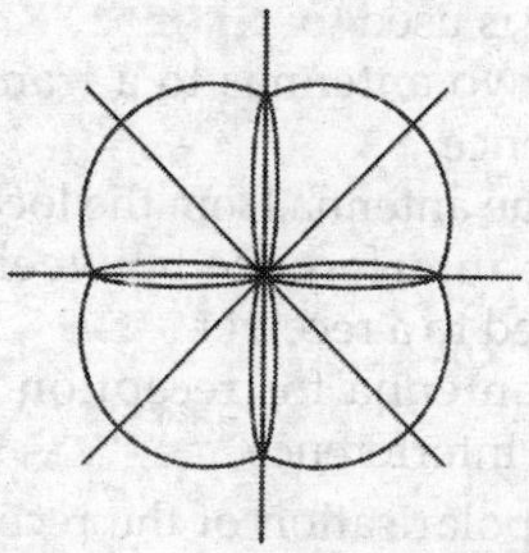

A. pencil beam B. omnidirectional
C. cardioid D. elliptical

539. The following figure shows the conductor pattern of a microstrip or stripline version of a microwave filter. This represents a

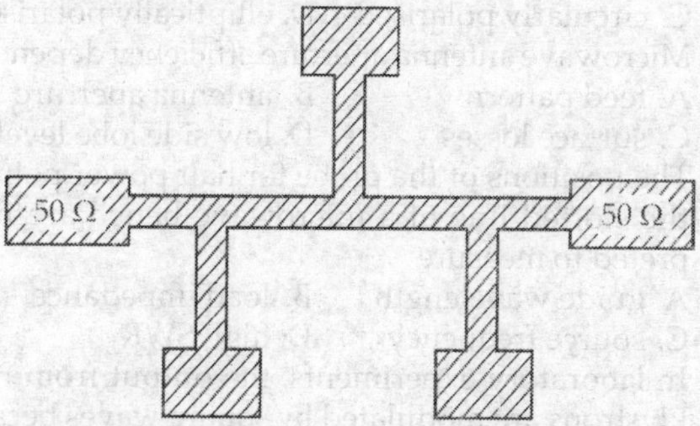

A. band pass filter B. high pass filter
C. low pass filter
D. channel dropping filter

540. A microwave junction is supposed to be matched at all ports if in the S matrix

Ans. 524. D 525. A 526. A 527. A 528. A 529. C 530. C 531. C 532. A 533. B 534. A 535. D 536. A 537. A 538. A 539. D

A. all the diagonal elements are zero
B. all the diagonal elements are equal but not zero
C. all the diagonal elements are complex
D. is Hermitian

541. Consider the following statements:
Hyperbolic system to navigation is associated with
1. time of arrival of the echo pulse from the time of transmission of the signal to measure distance
2. time of arrival of two signals from two transmitters at the receiver located in the craft
3. doppler shift in the frequency
4. the master station operating independently and the distant slave stations being synchronised with the master station by ground waves

Of these statements
A. 1, 2 and 3 are correct B. 1 and 2 are correct
C. 1 and 3 are correct D. 2 and 4 are correct

542. Waveguides are pressurised above normal atmospheric pressure for
A. increasing their power handling capacity
B. improving the conductivity of their walls
C. preventing higher order modes from propagating
D. varying the wave impedance

543. A duplexer is used to
A. couple two antennas to a transmitter without interference
B. isolate the antenna from the local oscillator
C. prevent interference between two antennas connected to a receiver
D. use an antenna for reception of transmission without interference

544. When the polarisation of the receiving antenna is unknown, to ensure that it receives at least half the power (except in one particular situation), the transmitted wave should be
A. horizontally polarised
B. vertically polarised
C. circularly polarised D. elliptically polarised

545. Microwave antenna aperture efficiency depends on
A. feed pattern B. antenna aperture
C. surface losses D. low side lobe level

546. The positions of the probe for half power points in the slotted line of a microwave bench are interpreted to measure
A. guide wavelength B. load impedance
C. source frequency D. high SWR

547. In laboratory experiments, the output from reflex klystrons are modulated by square waves because
A. it is easy to generate a square wave
B. crystal diode operates in the square law region of the I-V characteristics
C. it prevents frequency modulation
D. detector circuit is less complicated

548. The guide wavelength is measured by short-circuiting a waveguide and shifting the tunable probe along the slotted line to locate the voltage minima. If the shorting plate is replaced by a matched load, then
A. it would improve the accuracy of the measurement
B. the guide wavelength cannot be measured
C. it will give less accurate result, as no reflected wave is present
D. it will change the value of the guide wavelength

549. A generator of 50 ohm internal impedance and operating at 1 GHz feeds a 75 Ω load via a coaxial line of characteristic impedance 50 ohm. The Voltage Standing Wave Ratio on the feed line is
A. 0.5 B. 1.5
C. 2.5 D. 1.75

550. A loss-less 50 ohm transmission line is terminated in (A) 25 ohm and (B) 100 ohm loads. Which one of the following statement would be correct if the voltage standing wave patterns measured in the two cases are compared?
A. The two patterns will be identical in all respects and cannot be distinguished
B. The two patterns will have identical locations of maximum/minima but the VSWR will be higher in the case of A
C. The two patterns will have identical locations of maxima/minima but the VSWR will be higher in the case of B
D. The two patterns will be identical except for a relative spatial shift of quarter wavelength in the two cases

551. A 50 Ω lossless transmission line is terminated in 100 Ω load and is excited by a 300 MHz source of internal resistance of 50 Ω. What should be the length of the transmission line for maximum power transfer?
A. 5.0 m B. 1.25 m
C. 2.5 m D. 10.0 m

552. A quarterwave transformer, made of air-filled coaxial line, matches two transmission lines of characteristic impedance 50 ohms and 72 ohms, respectively. If the inner conductor of the coaxial line is made 10 mm in diameter, what should be the diameter of the outer conductor (approximately)?
A. 16 mm B. 20 mm
C. 27 mm D. 32 mm

553. The input impedance of short-circuited loss-less line of strength less than a quarter wavelength is
A. purely resistive B. purely inductive
C. purely capacitive D. complex

554. Rain-drop attenuation in most microwave communication links is caused due to

Ans. 540. A 541. D 542. C 543. D 544. B 545. B 546. D 547. B 548. B 549. B 550. D 551. B 552. C 553. B

A. scattering of microwaves by water drops of specific size
B. scattering of microwaves by a collection of droplets acting as a single body
C. absorption of microwaves by water and consequent heating of the liquid
D. absorption of microwaves by water-vapour in the atmosphere

555. A pair of loss-less transmission lines termined by resistance $R_1 = 2R_0$ (R_0 being the characteristic resistance) is excited by a source of emf E_s and internal resistance $R_s = 2R_0$. The build up of the voltage at the middle of the line will be as shown in

A.
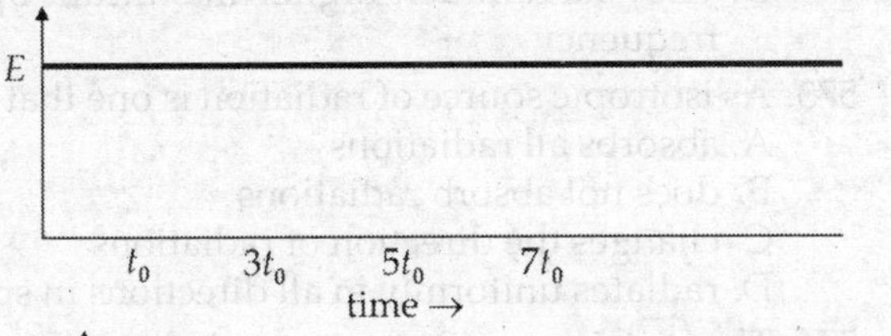

B.
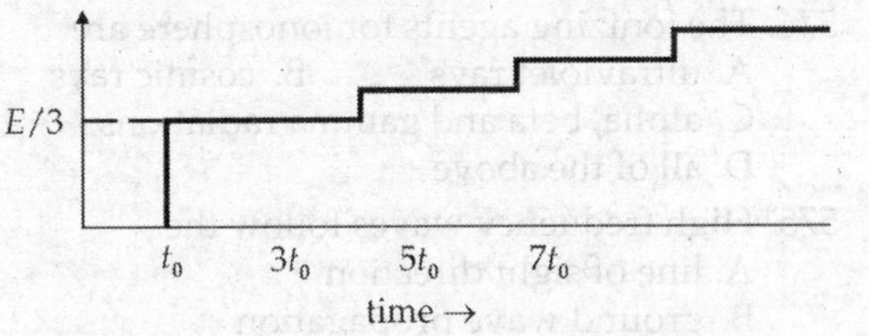

C.
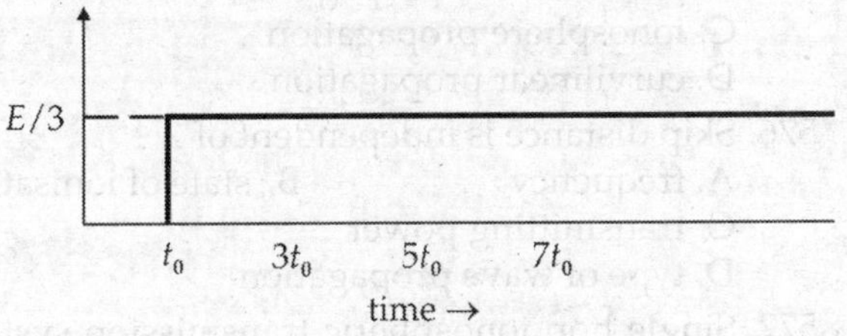

D.
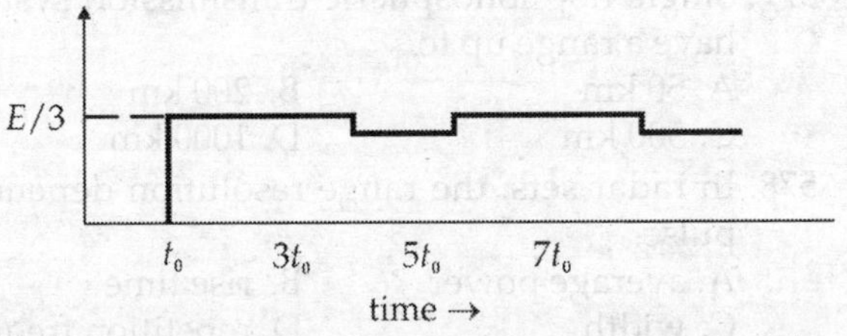

556. In a travelling-wave tube, the phase velocity of the axial component of the off-field on the slow-wave structure is kept
A. equal to the velocity of the electrons
B. slightly less than the velocity of electrons
C. slightly more than the velocity of electrons
D. equal to the velocity of light in free space

557. In both the sequential-lobing and conical-scan tracking techniques of radar, the measurements of angular errors in two orthogonal coordinates (azimuth and elevation) require a minimum of
A. three pulses be processed
B. two pulses be processed
C. four pulses be processed
D. only one pulse be processed

558. In parametric amplifiers used in microwave communication systems, the amplification is limited by
A. type of biasing B. a maximum limit of 10
C. pump energy D. frequency of operation

559. In a peak measurement with the arrangement given in the figure, the oscilloscope measures the PRF to be 1 kHz and the pole width as 2 μs. If the peak power to be measured is 50 kW and the power sensor has a rating of 10 mW average, then for protecting the sensor, the minimum value of the attenuator R should be

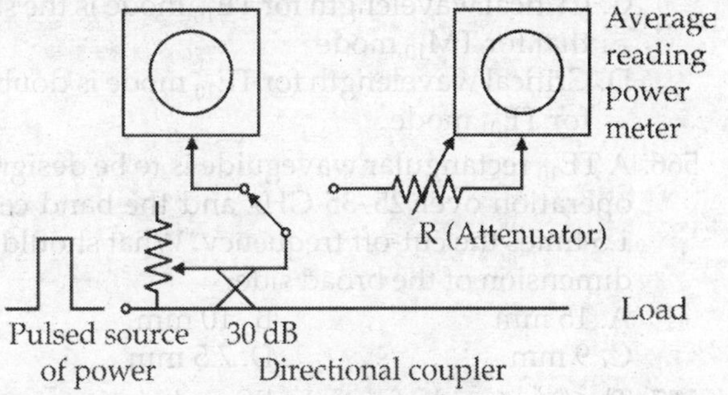

A. 3 dB B. 10 dB
C. 15 dB D. 20 dB

560. The pulse repetition frequency in a radar system is determined primarily from
A. transmitter power B. signal/noise ratio
C. the minimum range at which targets are expected
D. the maximum range at which targets are expected

561. For a given bandwidth of the receiver in a radar system, high discrimination between targets is achieved, when the
A. PRF is high
B. receiver sensitivity is high
C. pulse width is increased
D. diameter of the antenna aperture is increased

562. If the minimum range is to be doubled in a radar, the peak power has to be increased by a factor of
A. two B. four
C. eight D. sixteen

563. The instrument landing system has which of the following components?
1. Goniometer 2. Marker beacons
3. Ad cock antenna 4. Localiser
5. Glide slop
Select the correct answer using the codes given below:
Codes
A. 2, 4 and 5 B. 1, 2, 4 and 5
C. 1, 2, 3 and 5 D. 2, 3, 4 and 5

Ans. 554. C 555. B 556. A 557. D 558. A 559. D 560. D 561. A 562. D 563. C

564. In a rectangular waveguide, there is one half-wave variation of electric field across the narrow dimension and two half-wave variations of electric field across the wider dimension. What is the dominant mode?
A. TM_{12} B. TE_{12}
C. TE_{21} D. TM_{12}

565. Which one of the following statements is true, or cm wave propagation through a rectangular waveguide of cross-section $2a \times a$?
A. Critical wavelength for TE_{10} mode is the same as that for TE_{01} mode
B. Critical wavelength for TE_{10} mode is half that for TM_{01} mode
C. Critical wavelength for TE_{10} mode is the same as that for TM_{10} mode
D. Critical wavelength for TE_{10} mode is double that for TE_{01} mode

566. A TE_{10} rectangular waveguide is to be designed for operation over 25–35 GHz and the band centre is 1.5 times the cut-off frequency. What should be the dimension of the broad side?
A. 15 mm B. 10 mm
C. 9 mm D. 7.5 mm

567. The front end of an amplifier chain in a manufacture communication system is kept immersed in liquid nitrogen, to
A. dissipate heat generated by amplifier
B. expand its frequency response
C. improve its noise figure
D. reduce the distortion by the amplifier

568. Short-term fading in microwave communication links can be overcome by
A. increasing the transmitted power
B. changing the antenna
C. changing the modulation scheme
D. diversity reception and transmission

569. To increase the radar range of ground and surface targets to see well beyond the normal radar horizon, the electromagnetic wave propagation adopted is
A. ionospheric scatter B. troposcatter
C. ground wave propagation
D. duct

570. Which of the following statements is not true for waveguide?
A. They are superior to coaxial cables in microwave range
B. They can pass several signals simultaneously
C. Signals in different modes are propagated to separate them at the receiving point
D. Signals have to be of the same frequency

571. Which of the following statements is not true for waveguides?
A. The waveguide itself is virtually lossless
B. The propagation of energy is through the dielectric filling the guide
C. The propagation of energy takes place through the walls of the waveguide
D. The losses n waveguide depends upon the frequency of the wave

572. Which of the following statements is not an advantage of waveguides compared to coaxial lines?
A. Waveguides are simple to manufacture
B. Power-handling ability is higher
C. The velocity of propagation is more than that in free space
D. They have much higher maximum operating frequency

573. As isotropic source of radiation is one that
A. absorbs all radiations
B. does not absorb radiations
C. changes the direction of radiations
D. radiates uniformly in all directions in space

574. The ionizing agents for ionosphere are
A. ultraviolet rays B. cosmic rays
C. alpha, beta and gamma radiations
D. all of the above

575. High frequency waves follow the
A. line of sight direction
B. ground wave propagation
C. ionosphere propagation
D. curvilinear propagation

576. Skip distance is independent of
A. frequency B. state of ionisation
C. transmitting power
D. type of wave propagation

577. Single hop ionospheric transmission systems can have a range up to
A. 50 km B. 200 km
C. 500 km D. 1000 km

578. In radar sets, the range resolution depends upon pulse
A. average power B. rise time
C. width D. repetition frequency

579. In case of communication involving microwave communication system repeaters are used approximately at a distance of 20 to 25 km. These repeaters are used to
A. amplify the signal B. multiplex the signal
C. filter out the noise at each repeater place
D. keep the signal at the line of sight and to provide direction

580. The frequencies in ultra high frequency range are propagated by means of
A. ground wave B. space waves
C. sky waves D. surface waves

Ans. 564. C 565. C 566. D 567. C 568. D 569. D 570. D 571. C 572. C 573. D 574. D 575. C 576. C 577. D 578. D 579. A 580. B

581. When a transmission line is terminated by its characteristic impedance
A. it behaves as an infinite line
B. standing wave pattern is obtained
C. zero power reaches the load
D. power gets reflected to the generator

582. The perfect 4-port circulators are cascaded, as shown in the figure, to form a 6-port circulator.

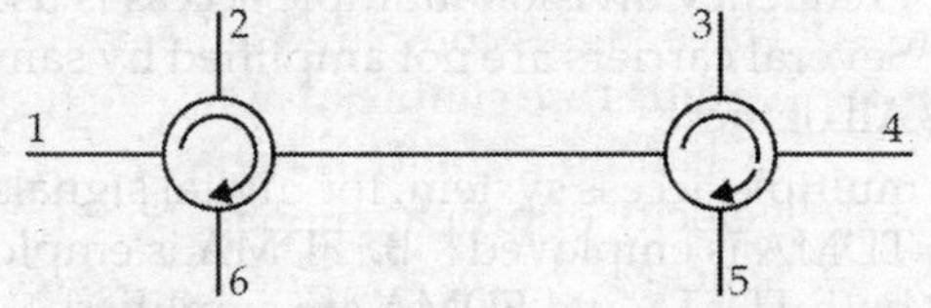

Transmitter (T), receiver (R), antenna (A) and match load (L) are connected to these ports. Which one of the following interconnection schemes is correct:

	T	R	A	L
A.	4	5	6	1, 2, 3
B.	3	5	4	1, 2, 4
C.	2	3	4	1, 5, 6
D.	1	3	2	4, 5, 6

583. The advantage offered by *optical fibers* over coaxial cables for submarine cable transmission include
A. longer repeater spacing
B. smaller cable size and greater flexibility
C. high transmission capacity
D. all of these

584. The aperture effect occurs, when the signal is
A. undersampled
B. sampled at the Nyquist type
C. oversampled
D. sampled using pulses of non-zero width

585. An optical fiber has a core with refractive index 1.52 and a cladding with a refractive index 1.45. Its numerical aperture is
A. 0.07 B. 0.20
C. 0.46 D. 0.95

586. The input power to a fiber-optic cable is 1 mW. The cable loss is 20 dB. What is the output power?
A. 0.10 mW B. 0.05 mW
C. 0.01 mW D. 0.001 mW

587. Which one is not a potential application of fiber optics?
A. Sensor B. Power transmission
C. Image transmission D. Signal transmission

588. Which of the following fibers gives highest modal dispersion?
A. Graded-index single-mode
B. Graded-index multimode
C. Step-index single-mode
D. Step-index multimode

589. Operating wavelength of GaAIAs LEDs and laser diodes is about
A. 850 nm B. 665 nm
C. 1300 nm D. 1550 nm

590. Si detectors are usable at wavelength of
A. 800–900 nm B. 1300 nm
C.1500 nm D. all of these

591. In a fiber connector, the role of ferrules is to
A. relieve strain on the cable
B. allow adjustment of attenuation
C. prevent back reflections
D. hold the fiber precisely in place

592. The function of an optical isolator is to
A. limit output power from a laser
B. prevent reflected light from fiber into a laser
C. reduce power level to match a receiver's dynamic range
D. prevent excess optical power from damaging the fiber

593. A computer has the following negative numbers stored in binary form as shown. Which one of these numbers is wrongly stored?
A. –37 as 11011011 B. –89 as 10100111
C. –48 as 11101000 D. –32 as 11100000

594. After an operation code has been fetched, which one of the following is the next process that takes place?
A. Reading the operands
B. Displaying the result
C. Decoding the opcode
D. Clearing the accumulator

595. The quality factor Q of a cavity resonator at a given frequency for higher mode orders
A. increases B. decreases
C. remains unaffected D. none of these

596. A geostationary communication satellite transponder receives signals from an earth station at a frequency f and sends back to the earth at
A. f B. F
C. $f/2$ D. 3860 MHz

597. The maximum eclipse duration for a geostationary spacecraft is
A. 72 minutes once in a calendar year
B. 10 minutes every midnight
C. 72 minutes twice a year
D. full local night

598. A sidereal day is
A. 23 hours 59 minutes
B. 23 hours 56 minutes and 4 seconds
C. 24 hours and 1.008665 minutes
D. exactly equal to the Mean Solar Day of 24 hours

599. The skip distance does not depend upon
A. frequency of the signal transmitted

Ans. 581. A 582. C 583. D 584. D 585. C 586. D 587. B 588. C 589. A 590. A 591. B 592. C 593. C 594. C 595. A 596. B 597. A 598. D

B. angle of beaming of the wave
C. type of wave propagation
D. the power transmitted

600. Geostationary satellites are placed in equatorial orbits at a height of approximately
A. 900 km B. 1600 km
C. 2000 km D. 36000 km

601. Which of the following has longest wavelength?
A. HF B. VHF
C. UHF D. SHF

602. In the above question, the display on the CRO will be of the form

A. 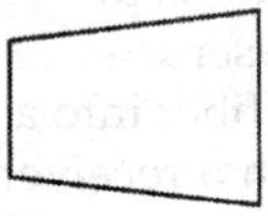 B. 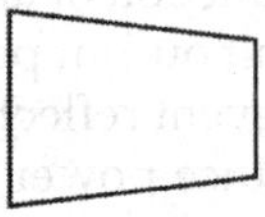

C. 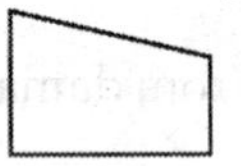 D. 

603. In how much time, a geostationary satellite makes one revolution around the earth?
A. 12 hours 40 minutes B. 23 hours 56 minutes
C. 36 hours 30 minutes
D. The satellite is stationary and thus no revolution

604. The source of primary power for a satellite is
A. nickel radium cells B. solar cells
C. inverters D. load batteries

605. The purpose of satellite repeater is to
A. repeat the satellite signals many times
B. translate signal to new frequency band
C. receive signal from the satellite
D. none of these

606. Before transmitting to the satellite, the wide-band satellite signal is amplified by
A. klystron B. travelling wave tube
C. tunnel-diode amplifier
D. waveguide

607. The complete bandwidth of a typical multipurpose satellite is
A. 5 MHz B. 50 MHz
C. 500 MHz D. 5 GHz

608. The carrier-to-noise ratio for a satellite depends upon
A. effective isotropically radiated power
B. free-space path losses
C. bandwidth D. all of these

609. The signal is sent back to earth by satellite by means of
A. chicken-mesh antenna
B. horn antenna
C. yagi antenna D. duplexer

610. The purpose of steerable spot-beam paraboloid aerials is to
A. increase EIRP B. reduce bandwidth
C. increase receiver gain to noise temperature ratio
D. none of these

611. Which of the following statement is not correct for multiple access system in a satellite earth station?
A. Access to same repeater sub-systems and same RF channel is possible
B. Frequency division multiple access is used
C. Several carriers are not amplified by same TWT
D. All of these

612. In multiple access system, for digital signals
A. TDMA is employed B. FDMA is employed
C. both TDMA and FDMA are amplified
D. none of these

613. In a communication satellite, the telephone channels are assembled in
A. TDM B. FDM
C. FM D. AM

614. Which of the following operations are carried out by INSAT?
A. Telecasting B. Parabolic antenna
C. Chicken-mesh antenna
D. All of these

615. The Satellite Instructional Television Experiment (SITE) conducted in India employed
A. INSAT B. ATS-6 satellite
C. APPLE D. Bhaskara

616. Which of the following systems is an international system?
A. INSAT B. INTELSAT
C. ATS-6 D. All of these

617. Which of the following is India's first operational satellite?
A. SITE B. INSAT
C. APPLE D. Aryabhatta

618. To telecast programmes directly to TV receivers, the additional circuit required with the receiver is
A. modulation converter
B. front-end converter
C. frequency converter D. all of these

619. Satellites used for international communication are known as
A. cossat B. domsat
C. marisat D. intelsat

620. In a satellite system, circular polarization is to be obtained. The antenna used is
A. parabolic antenna B. home-antenna
C. log-periodic antenna D. helical antenna

621. When the largest cross-section is changing, the best system for accurate tracking is
A. lobe switching B. sequential lobing
C. conical scanning D. monopulse

Ans. 599. C 600. D 601. A 602. B 603. C 604. B 605. C 606. B 607. C 608. D 609. B 610. A 611. C 612. A 613. B 614. D 615. B 616. B 617. A 618. B 619. D 620. D 621. D

622. In a satellite communication, circular polarization is to be obtained, which of the following antenna will be preferred?
A. Parabolic antenna B. Horn antenna
C. Log periodic antenna D. Helical antenna

623. Earth-coverage satellite antennas from synchronous altitudes have a beam-width of
A. 1.73° B. 7.3°
C. 17.3° D. 34.6°

624. The full-duplex round-trip delay through a synchronous satellite is approximately
A. 300 msec B. 500 msec
C. 600 msec D. 800 msec

625. The code division multiple access technique is not used in satellite communication because of
A. wastage of baseband spectrum
B. wastage of power
C. increase in delay
D. complexity and unreliability of operation

626. Basic transmission loss in decibels for a synchronous satellite signal for uplink frequency of 6 GHz would be
A. 184 B. 200.5
C. 92 D. 60

627. Without a spectrum analyser, it is not possible to determine
A. carrier frequency B. antenna pattern
C. pulse width
D. spurious signal strength and its location

628. The transmission path loss L is the free space between a communication satellite operating at a frequency f and at a distance d from the earth station is
A. directly proportional to fd
B. directly proportional to f^2d^2
C. inversely proportional to f
D. directly proportional to f^3d^3

629. Consider the following statement:
Laser is acronym for light amplification by stimulated emission of radiation. In comparison to LED, it has
(i) higher emission efficiency
(ii) no tuning arrangements
(iii) narrow spectral width
(iv) provision for confinement
Of these statements
A. (i), (ii) and (iv) are correct
B. (i), (ii) and (iii) are correct
C. (i) and (iii) are correct
D. (i), (iii) and (iv) are correct

630. A satellite has an EIRP of 47 dBw and the ground station figure of merit is 31 dB/K. The free space loss is 206 dB. Boltzmann's constant is –228 dB/K. Carrier to noise density at the ground receiver station is
A. –168 dB B. –159 dB
C. +100 dB D. +22 dB

631. The value of the integral $\int_{-\infty}^{\infty} f(t)\delta(t)$ is
A. $f(t)$ B. $f(0)$
C. unity D. zero

632. The Fourier transform of the function $f(t) = 1$ is
A. unity B. $\pi\,\delta(\omega)$
C. $2\pi\,\delta(\omega)$ D. $\delta(\omega)$

633. If $F(\omega) = \frac{1}{2\pi}\int_{-\infty}^{\infty} f(t)e^{-j\omega t}dt$, the corresponding function $f(t)$ is given by
A. $\frac{1}{2\pi}\int_{-\infty}^{\infty} f(\omega)e^{-j\omega t}d\omega$ B. $\int_{-\infty}^{\infty} f(\omega)e^{-j\omega t}d\omega$
C. $\frac{1}{2\pi}\int_{-\infty}^{\infty} f(\omega)e^{j\omega t}d\omega$ D. none of these

634. A time function is defined as $f(t) = 0,\ t < 0.1 = e^{-xt}$, $t \geq 0 < 0$. Its Fourier transform is
A. $\frac{1}{2\pi}(\alpha + j\omega)$ B. $\frac{1}{2\pi}(\alpha - j\omega)$
C. $(\alpha + j\omega)/2\pi$ D. $(\alpha - j\omega)/2\pi$

635. The Laplace transforms of an impulse time function is
A. zero B. infinity
C. 1 D. $\delta(\omega)$

636. The Laplace transform of a unit step function is
A. $1/s$ B. 1
C. infinity D. zero

637. The convolution of a function $f(t)$ with a unit impulse function is
A. $f(-t)$ B. $f(t)$
C. $F(\omega)$ D. $F(-\omega)$

638. The convolution of two functions $f_1(t)$ and $f_2(t)$ in frequency domain is
A. $\frac{1}{2\pi} f_1(t)\, f_2(t)$ B. $\frac{1}{2\pi} F_1(\omega)\, F_2(\omega)$
C. $F_1(\omega){\cdot}F_2(\omega)$ D. $F_1(\omega)\, F_2(\omega)$

639. If an impulse function is applied to a network with transfer function $H(f)$, then the output $V_0(t)$ will correspond to
A. $1/H(f)$
B. inverse transform of $H(f)$
C. $H(f)$ D. $\delta(f)$

640. A shift of ω_0 in the frequency domain for a function is equivalent to multiplication of the function by
A. $e^{j\omega_0 t}$ B. $e^{-j\omega_0 t}$
C. $\omega_0/2\pi$ D. none of these

Ans. 622. D 623. C 624. A 625. B 626. B 627. ? 628. B 629. D 630. B 631. B 632. C 633. C 634. A 635. C 636. A 637. B 638. C 639. B 640. A

641. The Fourier transform of the gate function of amplitude A and duration is $(-1/2 \le t \le 1/2)$

A. $A\dfrac{T}{2}\dfrac{\sin\omega\tau}{\omega\tau}$ B. $A\tau\dfrac{\sin(\omega\tau)}{\omega\tau}$

C. $A\tau\dfrac{\sin(\omega\tau/2)}{(\omega\tau/2)}$ D. $\dfrac{A\tau}{2}\dfrac{\sin(\omega\tau/2)}{(\omega\tau/2)}$

642. The spectral density of auto-correlation function $e^{-2\alpha/T}$ is

A. $\dfrac{2\alpha}{\omega^2+2\alpha^2}$ B. $\dfrac{4\alpha}{\omega+4\alpha}$

C. $\dfrac{4\alpha^2}{\omega^2+4\alpha^2}$ D. $\dfrac{4\alpha}{\omega^2+4\alpha^2}$

643. The response to unit step for RC low-pass filter for $t \ge 0$ is

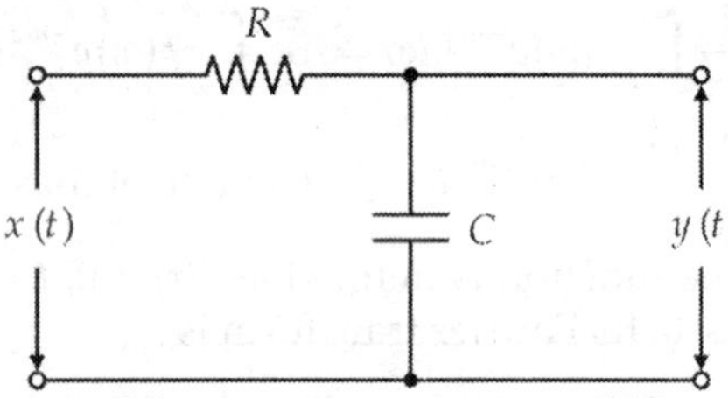

A. $1+e^{-t/RC}$ B. $1-e^{-t/RC}$

C. $1+e^{-tR/C}$ D. $1-e^{-tR/C}$

644. The integral $\int_{-\infty}^{\infty} e^{j\omega t}$ is equal to

A. $u(\omega)$ B. $r(\omega)$

C. $\delta(\omega)$ D. 1

645. The signum function is defined as

A. $\text{sgn}(t) = -1 \quad t>0$
$\quad = 1 \quad t<0$
B. $\text{sgn}(t) = 0 \quad t>0$
$\quad = -1 \quad t<0$

C. $\text{sgn}(t) = 1 \quad t>0$
$\quad = -1 \quad t<0$
D. $\text{sgn}(t) = 1 \quad t>0$
$\quad = 0 \quad t<0$

646. The signum function can be written as

A. $u(t)+1$ B. $2u(t)-1$

C. $1-u(t)$ D. $u(t)/2$

647. The fourier transform of signum function is given by

A. $j\omega$ B. $\delta(\omega)$

C. $\dfrac{2}{j\omega}$ D. $\pi\delta(\omega)t\dfrac{1}{j\omega}$

648. $f(\omega)$ in fourier transform of $f(t)\cdot e^{-j\omega_0 t}$ will be

A. $F(\omega-\omega_0)$ B. $f(\omega_0-\omega)$

C. $F(\omega_0)$ D. $F(\omega)$

649. If $F(\omega)$ is fourier transform of $f(t)$, the fourier transform of $f(t)\cos\omega_0 t$ will be

A. $\frac{1}{2}[F(\omega+\omega_0)-F(\omega-\omega_0)]$

B. $j/2[F(\omega+\omega_0)+F(\omega-\omega_0)]$

C. $\frac{1}{2}[F(\omega+\omega_0)+F(\omega-\omega_0)]$

D. $j/2[F(\omega+\omega_0)-F(\omega-\omega_0)]$

650. If $F(\omega)$ is fourier transform of $f(t)$, the fourier transform of $f(t-t_0)$ will be

A. $F(\omega)e^{-j\omega_0 t}$ B. $F(\omega)e^{+j\omega_0 t}$

C. $F(\omega)e^{+j\omega t}$ D. $F(\omega_0)e^{-j\omega t}$

651. The implication of the above question is that

A. there is a frequency translation

B. there is no phase shift

C. the magnitude of fourier transform has not changed

D. all of these

652. Another implication of the same question is that there is a

A. change in phase spectrum

B. change in frequency spectrum

C. frequency translation

D. none of these

653. The energy-density spectrum $\psi(\omega)$ is given by

A. $|F(\omega)|$ B. $1/|F(\omega)|$

C. $|F(\omega)|^2$ D. $1/|F(\omega)|^2$

654. The energy-density spectrum of a signal represents

A. relative contribution of energy by positive components

B. relative contribution of energy by negative components

C. energy per unit bandwidth

D. all of these

655. The unit of energy density spectrum is

A. J B. J/Hz

C. Hz D. Hz/J

656. If in the frequency spectrum of a signal, the maximum component is of frequency f_m, the energy of the signal is

A. $\psi(\omega)$ B. $\psi(\omega)\cdot f_m$

C. $|\psi(\omega)\cdot|^2 f_m$ D. $2\psi(\omega)\cdot f_m$

657. The energy of a signal of energy-density spectrum $\psi(\omega)$ is given by

A. $\int_{-\infty}^{\infty}\psi^2(\omega)d\omega$ B. $\dfrac{1}{2\pi}\int_{-\infty}^{\infty}\psi^2(\omega)d\omega$

C. $\dfrac{1}{2\pi}\int_{0}^{\infty}\psi^2(\omega)d\omega$ D. none of these

658. If $\delta(\omega)$ denotes the power density spectrum, then the average power of a signal is given by

A. $\dfrac{1}{2\pi}\int_{-\infty}^{\infty}\delta(\omega)d\omega$ B. $\dfrac{1}{2\pi}\int_{-\infty}^{\infty}\delta^2(\omega)d\omega$

C. $\dfrac{1}{\pi}\int_{0}^{\infty}\delta(\omega)d\omega$ D. $\dfrac{1}{\pi}\int_{0}^{\infty}\delta^2(\omega)d\omega$

659. The power-density spectrum is a/an

A. off function B. even function

C. linear function D. square function

660. Which of the following statements is not correct for power-density spectrum?

Ans. 641. C 642. D 643. B 644. B 645. C 646. B 647. C 648. A 649. C 650. A 651. C 652. A 653. C 654. D 655. B 656. D 657. B 658. C 659. B

A. It retains information of magnitude of frequency spectrum
B. A given signal has a unique power-density spectrum
C. Many signals cannot have the same power-density
D. It is an even function

661. Two signals have same frequency spectrum but different phase functions. Their power-density spectrums will be
A. differing in phase B. differing by $e^{\omega_0 t}$
C. differing by ω_0 D. identical

662. The power-density spectrum represents
A. power per unit frequency
B. power per unit bandwidth
C. power per unit wavelength
D. none of these

663. The unit of power-density spectrum is
A. W B. Hz/W
C. W/Hz D. W·Hz

664. Which of the following statements is not correct for a band-limited signal, $f(t)$?
A. There is no spectral components above a frequency f_{max}
B. $F(\omega) = 0$ for $f > f_{max}$
C. It is represented by samples taken uniformly at a distance of $1/2f_{max}$
D. None of these

665. A band-limited signal when sampled is represented as $f_s(t)$. Its frequency spectrum will show no overlaps if the sampling time is
A. $\leq 1/f_{max}$ B. $\leq 1/2f_{max}$
C. $\geq 1/f_{max}$ D. $\geq 2f_{max}$

666. According to sampling theorem
A. the signal should be sampled at least twice each cycle of its lowest frequency
B. twice each cycle of its highest frequency
C. guard time should be as large as possible
D. Nyquist rate should be as slow as possible

667. In sampling of a signal, guard time is provided when the sampling time is
A. $< 1/2f_{max}$ B. $= 1/2f_{max}$
C. $> 1/2f_{max}$ D. all of these

668. In order to get the original signal from the sampled signal, it is necessary to use
A. low-pass filters B. high-pass filters
C. band-pass filters D. band-stop filters

669. The Nyquist rate of a signal is
A. f_{max} samples/sec B. $2f_{max}$ samples/sec
C. $1/f_{max}$ samples/sec D. $1/2f_{max}$ samples/sec

670. In sampling of a signal, overlapping of frequencies components will occur if sampling time is
A. $< 1/2f_{max}$ B. $= 1/2f_{max}$
C. $> 1/2f_{max}$ D. none of these

671. If sampling time is less than the Nyquist interval
A. simple filters can be used to obtain original signal
B. bandwidth increases
C. channel capacity increases
D. guard time becomes less

672. In communications, the sampling technique leads to
A. better efficiency
B. highest speed of communication
C. less costly equipment
D. all of these

673. A signal of maximum frequency of 8 kHz is sampled at Nyquist rate. The time intervals between the two successive samples will be
A. 62.5 μsec B. 125 μsec
C. 1250 μsec D. zero

674. The effect of noise in a communication system is most adverse with reference to
A. source B. encoder
C. channel D. receiver

675. Which of the following requirements is necessary for fast communication?
A. High transmitter power
B. Higher channel capacity
C. Large bandwidth D. High S/N ratio

676. Two cards are drawn from a standard deck of cards. If the first card is replaced before the second drawing, the probability of drawing two aces is
A. $\left(\frac{4}{52}\right)^2$ B. $\left(\frac{4}{52}\right)\times\left(\frac{3}{51}\right)$
C. $2\left(\frac{4}{52}\right)$ D. none of these

677. A random variation has the following probability density function
$p(x) = 3e^{-6|x|}, -\infty < x < \alpha$. The PDF for $x \leq 0$ is
A. 0 B. $\frac{1}{2}e^{-6x}$
C. $\frac{1}{2}e^{6x}$ D. $\frac{1}{2}e^{-3/x}$

678. In the above problem, the expected value of the process is given by
A. 0 B. 1/2
C. –1 D. 1

679. In the same problem, the second moment will be
A. 0 B. 1/2
C. 1/4 D. 1/8

680. The output of a network $z(t)$ is related to the input (t) by the relationship $z = y^2$ with $p(y) = e^{-y}$, $(0 \leq y \leq \infty)$. The expectation $E(z)$ will be equal to
A. 0 B. 1
C. 2 D. none of these

Ans. 660. C 661. B 662. B 663. C 664. DD 665. B 666. B 667. A 668. A 669. C 670. C 671. A 672. B 673. A 674. C 675. C 676. A 677. C 678. A 679. D 680. C

681. A band-limited signal has no spectral components above the frequency of 100 kHz. The signal can be uniquely determined by its values at uniform intervals of duration less than
A. 2 μsec B. 5 μsec
C. 10 μsec D. 20 μsec

682. For nearly distortion-less received signal in pulse modulation, it is required that the speed should be
A. less than the signal frequency
B. less than twice the signal frequency
C. more than twice the signal frequency
D. equal to the signal frequency

683. A communication channel has band-width of 5 Hz and if signal-to-noise ratio is 5, the corresponding channel capacity will be
A. 18000 bits/sec B. 4000 bits/sec
C. 1500 bits/sec D. 1000 bits/sec

684. According to Shanon's theorem, the output from any source of rate R can be coded and transmitted over a channel of capacity C with the condition that
A. $C < R$ B. $C < R^2$
C. $C > R$ D. $C > R^2$

685. The capacity of a channel is
A. the maximum rate of information transmission
B. the bandwidth required for information
C. volume of information it can take
D. the number of digits used in coding

686. There is going to be snow-fall in the month of June. This statement has
A. no information content
B. very high information content
C. very low information content
D. no truth in it

687. Indicate the false statement related to white noise.
A. It is a thermal noise
B. Its frequency spectrum contains all frequency components but in unequal proportion
C. Theoretically it has infinite power
D. It is a Gaussian noise signal

688. The probability density function of thermal noise is
A. Binomial B. Poisson
C. Gaussian D. none of these

689. If for a function $x(t)$, $|R(t_0)|x^{-2} = 0$, then $x(t + t_0)$ and $x(t)$ are
A. correlated and statistically independent
B. uncorrected but not necessarily statistically independent
C. uncorrected but statistically dependent
D. corrected but not necessarily statistically dependent

690. For a Gaussian process, auto-correlation also implies
A. statistical independence
B. statistical dependence
C. ergodic process D. none of these

691. The auto-correlation function of a signal $R(t)$ at $t_0 = 0$ is equal to
A. average power of the signal
B. average value of the signal
C. zero D. infinite

692. The auto-correlation function is a/an
A. even function in t_0 B. odd function in t_0
C. square function in t_0
D. exponential function in t_0

693. The value of auto-correlation function at t_0, $R(\theta)$ is
A. $\leq R(t_0)$ B. $\geq R(t_0)$
C. $= R(t_0)$
D. of any value with respect to $(R + t_0)$

694. Indicate the false statement concerning auto-correlation function.
A. It is an average function
B. It is a non-unique description of signal
C. It is a measure of both time variation and statistical dependence
D. It is defined only for periodic signals

695. The auto-correlation function of a function $x(t)$ is defined as
A. $R(t) = \frac{1}{T}\int_0^T x(t - t_0)x(t + t_0)dt$
B. $R(t) = \frac{1}{T}\int_0^T x(t)\,x(t + t_0)dt$
C. $R(t_0) = \frac{1}{T}\int_0^T x(t)\,x(t + t_0)dt$
D. $R(t) = \frac{1}{T}\int_0^T x(t)\,x(t - t_0)dt$

696. If two coins are tossed simultaneously, the probability density function $P(x)$ of getting no head will be
A. zero B. 1/4
C. 1/2 D. 1

697. In the above question, the probability density function of getting one head will be
A. zero B. 1/4
C. 1/2 D. 1

698. In the above question, the probability density function of getting two heads will be
A. 1/4 B. 1/2
C. 1 D. 2

699. In the above question
A. $\Sigma P(x) = 0$ B. $\Sigma P(x) = 1$
C. $\Sigma P(x) = 2$ D. $\Sigma P(x) = 4$

700. Ensemble averages of many random signals are average of sample function values taken
A. over a long time B. at one instant
C. during the average time period of signals
D. none of these

Ans. 681. B 682. C 683. A 684. C 685. A 686. B 687. B 688. C 689. B 690. A 691. A 692. A 693. B 694. D 695. C 696. B 697. C 698. A 699. B 700. B

701. An ergodic process in communication is present if many random signals have
A. identical time averages
B. identical ensemble averages
C. identical time and ensemble averages
D. none of these

702. If the random process is ergodic, then which of the following conditions shows that it is an even function of T?
A. $R(T) \neq R(-T)$ B. $R(T) = R(-T)$
C. $R(T) > R(-T)$ D. $R(T) < R(-T)$

703. The information content of the source producing four messages of probabilities 1/2, 1/4, 1/8, 1/8 is
A. 1 B. 2
C. $1\frac{3}{4}$ D. $1\frac{1}{5}$

704. In a TV receiver, there are 525 lines/frame, about 500 identifiable dots on each line and 30 frames per second. Also assume that the human eye can distinguish 10 graduation of light intensity.
The number of picture-dots per second will be
A. 8.215×10^6 B. 5.225×10^6
C. 7.875×10^6 D. 4.115×10^6

705. In the above problem, the quantity of information contained in a given picture-dot of a television would be
A. 3.58 bits/picture element
B. 3.32 bits/picture element
C. 5.0 bits/picture element
D. 2.46 bits/picture element

706. In the same problem, the information rate would be
A. 2.62×10^7 bits/sec B. 3.18×10^7 bits/sec
C. 8.15×10^7 bits/sec D. 3.32×10^7 bits/sec

707. In the voltage standing ratio of a line is infinity, the line is terminated in a
A. pure reactance B. complex impedance
C. pure resistance D. short-circuit

708. A balun is a
A. quatar-wave transformer
B. balance-to-unbalance transformer
C. double stub D. single stub

709. Indicate the false statement.
A. Single stub is useful only at one frequency
B. Double stub generally is a coaxial line
C. The length cannot be changed independently
D. The two stubs are placed $3\lambda/8$ distance apart

710. A square wave repeats at a frequency of 100 Hz. Its fundamental frequency will be
A. 50 Hz B. 100 Hz
C. 1 kHz D. 0.5 kHz

711. For periodic time function, the frequency spectrum is
A. periodic B. continuous
C. discrete D. none of these

712. The frequency spectrum indicates the relative
A. frequency of each of the frequency components
B. phase of each of the frequency components
C. magnitude of each of frequency components
D. none of these

713. A matched filter has a voltage-transfer characteristics
A. identical with the signal time function
B. identical with the signal frequency spectrum
C. absolutely linear D. none of these

714. In the frequency spectrum of a signal
A. only frequencies of the components are given
B. only amplitude of the components are given
C. both A and B are correct
D. none of these

715. If signal is completely known, it represents
A. either time or frequency domain
B. both in time and frequency domains
C. both frequency and amplitude
D. none of these

716. When a signal is displayed on the oscilloscope, we see it in
A. frequency domain B. time domain
C. both time and frequency domains
D. none of these

717. When a signal is seen on a frequency analyser, we see it in
A. frequency domain B. time domain
C. both time and frequency domains
D. none of these

718. The total collection of sample points covering all possible outcome/values of phenomenon is called
A. set point B. sample point
C. sample space D. none of these

719. Each value is the sample space of a phenomenon is called
A. sample point B. outcome point
C. set point D. none of these

720. A telephone system is a
A. point-to-multipoint system
B. local area network
C. switched system D. point-to-point system

721. What type of transmission can readily accommodate terminals sending signals at different rates?
A. Packet switching
B. Time-division multiplexing
C. Frequency-division multiplexing
D. Standard digital telephone hierarchy

722. If the periodic time of a pulse of width 1 sec is extended for 24 hours or 86400 sec, its fundamental frequency will be
A. 1 Hz B. 86400 Hz
C. 1.16×10^5 Hz D. 2.32×10^{10} Hz

Ans. 701. C 702. B 703. C 704. C 705. B 706. A 707. D 708. B 709. C 710. B 711. D 712. C 713. B 714. C 715. A 716. B 717. A 718. C 719. A 720. C 721. A 722. C

723. The pulse length controls the spread of frequency composition for a pulse which is
A. periodic B. aperiodic
C. periodic or aperiodic D. none of these

724. In a pulse system, the system bandwidth is determined by
A. presence of the pulse B. number of pulses used
C. the application for which the pulse is used
D. none of these

725. Which of the following equations is incorrect for a conducting medium?
A. $\nabla + \overline{H} = \frac{\partial \overline{D}}{\partial t} + \overline{J}$ B. $\nabla + \overline{E} = \frac{\partial \overline{B}}{\partial t}$
C. $\nabla + \overline{D} = \sigma$ D. $\nabla + \overline{B} = 0$

726. If a charge density σ exists within σ conductor at certain time, it would be reduced to $1/e$ times, this value in
A. σ/e seconds B. σ/ε seconds
C. ε/σ seconds D. ρ/σ seconds

727. The function $f(x - v_0 t)$ represents
A. a wave motion in forward direction
B. a wave motion in reverse direction
C. a stationary wave D. none of these

728. Which of the following is an incorrect Maxwell's equation?
A. $\nabla \times \overline{H} = \frac{\partial \overline{D}}{\partial t} + \sigma \overline{E}$ B. $\nabla \times \overline{E} = \frac{\partial \overline{H}}{\partial t}$
C. $\nabla \times \overline{D} = \sigma$ D. $\nabla \times \overline{B} = 0$

729. The characteristic impedance of free space is
A. 397 ohms B. 377 ohms
C. 387 ohms D. 367 ohms

730. The electric intensity for free space is given by
A. D/ε_0 B. ε_0/σ
C. $\varepsilon_0 D$ D. σ/ε_0

731. Maxwell's divergence equation for the electric field is represented by
A. $\nabla \cdot \overline{E} = \frac{\rho}{\varepsilon_0}$ B. $\nabla \times \overline{E} = -\frac{\rho}{\varepsilon_0}$
C. $\nabla \cdot \overline{E} = -\frac{\rho}{2\pi\varepsilon_0}$ D. none of these

732. The Poisson's equation can be represented by
A. $\nabla^2 V = -\frac{\rho}{\varepsilon_0}$ B. $\nabla^2 V = \frac{\rho}{\varepsilon_0}$
C. $\nabla \cdot E = \frac{\rho}{\varepsilon_0}$ D. $E = -\nabla V$

733. A sheet charge of uniform density ρ_{so} coulombs/m^2 extends over the entire XY plane. Gauss's law in differential form for a polar be written as
A. $\nabla \cdot \boldsymbol{D} = \rho_{so}\delta$ B. $\nabla \cdot \boldsymbol{D} = \rho_{so}\delta(z - z_0)$
C. $\nabla \cdot \boldsymbol{D} = \rho_{so}\delta(z_0 - z)$ D. none of these

734. In cylindrical coordinates $F = \left(\frac{k}{r^2}\right)[\cos\phi\, a_r + \sin a_\phi]$ can be written as
A. magnetic field B. static electric field
C. electromagnetic field D. none of these

735. Due to the law of conservation of charge for a closed surface, the eqn. $[I_c]_s + [I_d]_s$
A. may not necessarily be equal to zero
B. should not be equal to zero
C. should be equal to zero
D. none of these

736. Poynting vector for an electromagnetic wave is equal to
A. $\boldsymbol{H} \cdot \boldsymbol{E}$ B. $\boldsymbol{H} \times \boldsymbol{E}$
C. $\boldsymbol{E} \times \boldsymbol{H}$ D. $\boldsymbol{E} \cdot \boldsymbol{H}$

737. Poynting vector can be represented by the equation
A. $\boldsymbol{E} \times \boldsymbol{B}/\mu_0$ B. $\boldsymbol{EB}/\mu_0$
C. $\mu_0/\boldsymbol{EB}$ D. none of these

738. A plane wave starting in air incidents normally on a loss-free dielectric medium of relative permittivity is 4. The reflected portion of the wave is to be eliminated by placing a quarter-wavelength plane between the air and the dielectric. The required relative permittivity of the material of the plane is
A. 2 B. 3
C. 4 D. 5

739. When an electromagnetic wave undergoes reflections from a plane of conducting units, it acquires
A. group velocity greater than the velocity of light
B. group velocity less than the velocity of light
C. phase velocity greater than the velocity of light
D. phase velocity less than the velocity of light

740. An arbitrarily eliptically-polarised wave can be broken up into
A. two circularly polarised components rotating in the same direction
B. two oppositely rotating circularly polarised components
C. two stationary circularly polarised components
D. none of these

741. When a wave travelling in air enters into a wave guide
A. the phase velocity will increase
B. the group velocity will increase
C. the phase velocity will decrease
D. none of these

742. A combination of left-handed circular polarisation and right-handed polarisation leads to
A. circular polarisation B. elliptical polarisation
C. parabolic polarisation
D. none of these

Ans. 723. C 724. C 725. C 726. C 727. A 728. B 729. B 730. D 731. A 732. A 733. C 734. B 735. C 736. C 737. A 738. A 739. C 740. C 741. A 742. B

743. The type of time variation ($\delta/\delta t$) implied with reference to the field quantities in Maxwell's equation is
A. sinusoidal B. non-sinusoidal
C. both sinusoidal and non-sinusoidal
D. exponential

744. For a lossy transmission line short-circuited at the receiving end, the input impedance is given by (Z_0 is the characteristic impedance, γ is the propagation constant and l is the length of the line)
A. $Z_0 \cot h\gamma l$ B. $Z_0 \cot \gamma l$
C. $Z_0 \tan h\gamma l$ D. $Z_0 \tan \gamma l$

745. A short grounded vertical antenna has a length L which is 0.05λ at frequency f. If its radiation resistance at f is R ohms, then its radiation resistance at a frequency $2f$ will be
A. $R/2$ B. R ohms
C. $2R$ ohms D. $4R$ ohms

746. The mouth diameter of two parabolic reflectors are in the ratio of 2:1. Their power gains with respect to a resonant half-wave dipole will be in the ratio of
A. 2:1 B. 1:4
C. 4:1 D. 1:2

747. The beamwidth between nulls of 140 cm parabolic, reflector used at 6 GHz is
A. 2.5 degrees B. 5 degrees
C. 10 degrees D. 7.5 degrees

748. Five isotropic elements placed along the x-axis are fed by in-phase currents of equal amplitudes. The spacing between successive elements is $\lambda/4$. Consider the following statements about the directivity pattern of the array:
1. The array will have a maximum in the direction of Y-axis
2. The field strength in the $+X$-axis direction will be zero
3. The directivity pattern will have minor lobes

Of the statements given above
A. 1, 2 and 3 are correct B. 1 and 2 are correct
C. 2 and 3 are correct D. 1 and 3 are correct

749. Two isotropic radiators A and B are placed along the X-axis at the pisitions $(-\lambda/8, 0)$ and $(+\lambda/8, 0)$, respectively. The elements are fed by current of equal amplitude with the phase of B leading that of A by 90°. The array pattern will have maxima
A. in the direction of the $+X$-axis
B. an angle of +45° with respect to the Y-axis
C. in the direction of $+Y$-axis
D. in the direction of $-X$-axis

750. The dominate mode in a waveguide is characterised by
A. longest cut-off wavelength
B. shortest cut-off wavelength
C. infinite attenuation D. zero attenuation

751. The frequency of resonance of water vapour in the atmosphere is
A. 10.225 GHz B. 12.850 GHz
C. 17.500 GHz D. 60.000 GHz

752. A signal propagated in a waveguide has a full wave of electric intensity change between the two further walls and no component of the electric field in the direction of propagation. The mode is
A. TE_{11} B. TE_{10}
C. TM_{22} D. TE_{20}

753. When a particular mode is excited in a waveguide, there appears an extra electric component in the direction of propagation. The resulting mode is
A. longitudinal electric
B. transverse electromagnetic
C. transverse magnetic D. transverse electric

754. For a uniform metallic waveguide, the scalar wave equation is given by $\Delta_{12}\psi + k_{02}\psi = 0$. When this equation is solved for the cut-off wave number kc, one will obtain
A. infinite number of discrete eigen values
B. continuous spectrum of eigen values
C. a finite number of discrete eigen values
D. eigen values whose number depends on the boundary conditions

755. In a cylindrical cavity resonator, the two modes which are degenerate would include
A. TE_{111} and TM_{111} B. TE_{011} and TM_{011}
C. TE_{022} and TM_{111} D. TE_{111} and TM_{011}

756. Consider the following statements regarding a linearly polarised wave travelling in a dielectric medium which is incident at the planar interface with a rare medium and is totally internally reflected:
The transmitted wave in the second medium will be
1. growing 2. non-uniform
3. linearly polarised 4. attenuated
5. circularly polarised 6. elliptically polarised

Of these statements
A. 2, 4 and 6 are correct B. 1, 2 and 4 are correct
C. 3, 4 and 5 are correct D. 2, 4 and 5 are correct

757. The main components of atmosphere responsible for absorption of EM waves are
A. nitrogen and oxygen
B. nitrogen and hydrogen
C. oxygen and water vapour
D. nitrogen and water vapour

758. What are the direction of $\overline{E}$ and $\overline{H}$ in TEM mode transmission lines with respect to the direction of propagation?
A. Both $\overline{E}$ and $\overline{H}$ are entirely transverse to the direction of propagation

Ans. 743. C 744. C 745. D 746. ? 747. B 748. B 749. D 750. A 751. C 752. D 753. C 754. B 755. D 756. B 757. C 758. A

B. E is entirely transverse to and E has a component in the direction of propagation
C. H is entirely transverse to and E has a component in the direction of propagation
D. Both E and H have components in the direction of propagation

759. In the ionosphere, the layer F_2
A. disappears in the day time
B. gets merged with F_1 layer at night
C. disappears in the night time
D. is lower than F_1 layer

760. When an electromagnetic wave makes a normal incidence on a perfect conductor
A. no energy is absorbed in the conductor
B. electrical field component forms standing wave pattern
C. magnetic field component forms standing wave pattern
D. all of these

761. When an electromagnetic wave undergoes reflections from a plane of conducting units, it acquires
A. group velocity greater than the velocity of light
B. group velocity less than the velocity of light
C. phase velocity greater than the velocity of light
D. phase velocity less than the velocity of light

762. An arbitrarily elliptically-polarised wave can be broken up into
A. two circularly polarised components rotating in the same direction
B. two oppositively rotating circularly polarised components
C. two stationary circularly polarised components
D. none of these

763. An electromagnetic wave incidents on a perfect insulator at an angle. If the wave is not entirely parallel polarised,
A. there will be no reflection
B. the reflected wave will be entirely polarised perpendicular to the plane of incidence
C. the reflected wave will be entirely polarised parallel to the plane of incidence
D. all of these

764. Van de Graaff generator is used
A. for generating microwave frequencies
B. in radar system for bearings
C. at a device for producing high speed neutrons
D. for generating high DC voltages

765. Electromagnetic waves may be diffracted
A. by reflection from the ground
B. when encountered with a spherical wavefront
C. around the edge of a sharp obstacle
D. while passing through a large slot

766. The absorption of radiowaves by the atmosphere depends on
A. polarisation of the waves
B. polarisation of the atmosphere
C. distance from the transmitter
D. frequency

767. As far as high frequency propagation is concerned, which of the following layer is least important?
A. D-layer B. E-layer
C. F_1 layer D. F_2 layer

768. The electric and magnetic fields travel through a medium of perfect insulator. The modified Maxwell's equations are
A. $\nabla \cdot D = 0$, rest 3 not changed
B. $\nabla \cdot D = 0$, $\nabla \times E$, rest 2 not changed
C. $\nabla \cdot D = 0, \nabla \times H = \delta D / \delta t$, rest 2 not changed
D. none of these

769. The following wave does not exist in waveguides.
A. TM waves B. TE wave
C. TEM waves D. TE and TM waves

770. E and H are related by the equation
A. $\frac{E}{H} = \sqrt{\varepsilon_0 \mu_0}$ B. $\frac{E}{H} = \sqrt{\frac{\mu_0}{\varepsilon_0}}$
C. $\frac{H}{E} = \sqrt{\mu_0 \varepsilon_0}$ D. $\frac{H}{E} = \sqrt{\frac{\mu_0}{\varepsilon_0}}$

771. Maxwell's wave equations are
A. $\nabla \times E = \frac{-\partial B}{\partial t}$
B. $\nabla \times E = \frac{-\partial B}{\partial t}, \nabla \times H = \frac{-\partial D}{\partial t}$
C. $\nabla D = 0, \nabla^2 E = \mu_0 \in_0 \frac{\partial^2 E}{\partial t^2}$
D. $\nabla^2 E = \mu_0 \in_0 \frac{\partial^2 E}{\partial t^2}, \nabla^2 B_0 \in \frac{\partial^2 B}{\partial t^2}$

772. VSWR is defined as
A. $\frac{E_{min}}{E_{max}}$ B. $\frac{E_{max}}{E_{min}}$
C. $E_{max} \times E_{min}$ D. $\frac{E_{max} - E_{min}}{E_{max} + E_{min}}$

773. Maxwell's equation in free space is
A. $\nabla \cdot B = 0$ B. $\nabla \cdot B = \rho$
C. $\nabla \cdot B = J$ D. $\nabla \cdot B = \sigma J$

774. The Poisson's equation can be represented by
A. $\nabla^2 V = -\frac{\rho}{\varepsilon_0}$ B. $\nabla^2 V = \frac{\rho}{\varepsilon_0}$
C. $\nabla \cdot E = \frac{\rho}{\varepsilon_0}$ D. $E = -\nabla V$

775. The wave propagation in plasma is known as
A. parametric dispersion

Ans. 759. B 760. D 761. C 762. B 763. B 764. B 765. C 766. D 767. A 768. C 769. C 770. B 771. D 772. B 773. A 774. B 775. A

B. solution dispersion
C. spherical dispersion D. none of these

776. The waves of same frequency have opposite phases when the phase angle between them is
A. 0° B. 90°
C. 360° D. π rad

777. VSWR is minimum when
A. line is terminated with full load
B. line is terminated with load equal to the characteristics impedance of the load
C. line is short circuited D. line is open circuited

778. The average power flow per unit area in a uniform plane wave in an electric field of maximum voltage E_0 and impedance Z_0 is
A. $\frac{E_0^2}{Z_0}$ B. $I_0 E_0$
C. $\frac{1}{2} \cdot \frac{E_0^2}{Z_0}$ D. $I_0^2 \cdot Z_0$

779. When ρ is the coefficient of reflection, the value of VSWR will be
A. $\frac{1+|\rho|}{1-|\rho|}$ B. $\frac{1-|\rho|}{1+|\rho|}$
C. $\frac{1+|\rho|^2}{1-|\rho|^2}$ D. $\frac{1-|\rho|^2}{1+|\rho|^2}$

780. In electromagnetic waves, polarization
A. is caused by reflection
B. is due to the transverse nature of the sky waves
C. results from the longitudinal nature of the waves
D. is always vertical in an isotropic medium

781. For propagation of electromagnetic wave in a good conductor, the value of
A. $\alpha = \sqrt{\left(\frac{\omega\mu\sigma}{2}\right)}, \beta = 180°$
B. $\alpha = \sqrt{2\omega\mu\sigma} \cdot \beta = 90°$
C. $\alpha = \beta = \sqrt{\varepsilon\mu}$ D. $\alpha = \beta = \sqrt{\left(\frac{\varepsilon\mu\sigma}{2}\right)}$

782. An electromagnetic wave incident on a perfect dielectric is
A. fully transmitted B. fully reflected
C. partially transmitted and partially reflected
D. none of these

783. A dominant wave is characterised by
A. lower cut-off wavelength
B. highest cut-off wavelength
C. no attenuation D. infinite attenuation

784. A range of microwave frequencies more easily passed by the atmosphere than other is called a
A. resonance in the atmosphere
B. gyro-frequency range
C. critical frequency D. window

785. The relation coefficient is generally a
A. real number B. fraction
C. imaginary number D. complex number

786. Circular polarised waves result when
A. phases are the same
B. magnitudes are same and phase difference is 90°
C. magnitudes are same and phase difference is zero
D. magnitudes are the same

787. In a uniform plane wave E and H related by
A. $\frac{E}{H} = \sqrt{\left(\frac{\mu}{\varepsilon}\right)}$ B. $\frac{E}{H} = \sqrt{\left(\frac{\varepsilon}{\mu}\right)}$
C. $\frac{E}{H} = 1$ D. none of these

788. The intrinsic impedance of free space is given by
A. $Z_0 = \mu_0 \varepsilon_0$ B. $Z_0 = \sqrt{\left(\frac{\mu_0}{\varepsilon_0}\right)}$
C. $Z_0 = \frac{\mu_0}{\varepsilon_0}$ D. $Z_0 = \sqrt{\mu_0 \varepsilon_0}$

789. The propagation coefficient for a transmission line is $(0.0005 + j\pi/10)\ m^{-1}$. The attenuation in nepers and dB, for a 50 m line is
A. 0.0005 N, 0.0043 dB B. 0.025 N, 0.217 dB
C. 0.025 dB, 0.217 N D. 0.025 N, 0.0043 dB

790. An electromagnetic wave travelling in x-direction has E_x and E_y components at 45° phase apart. The wave can be termed as
A. linearly polarised B. circularly polarised
C. elliptically polarised
D. horizontally polarised

791. High frequency waves are
A. absorbed by the F_2 layer
B. reflected by the D layer
C. capable of use for long-distance communications on the moon
D. affected by the solar cycle

792. For a uniform plane wave travelling in Z-direction in free space, if $E_x 2\cos(\omega t - \beta z)$ volts/ampere, then Hy is given by
A. $Hy = \frac{2}{377}\cos(\omega t - \beta z)$ amperes/m
B. $Hy = 2 \times 377\cos(\omega t - \beta z)$ ampere/m
C. $Hy = 4 \times 377\cos(\omega t - \beta z)$ ampere/m
D. $Hy = \frac{2}{377}\cos(\omega t - \beta z)$ volts/m

793. Maxwell's divergence equation for the electric field is represented by

Ans. 776. D 777. B 778. C 779. A 780. B 781. D 782. C 783. A 784. D 785. D 786. B 787. A 788. B 789. B 790. C 791. D 792. B

A. $\nabla \cdot E = \frac{\rho}{\varepsilon_0}$ B. $\nabla \times E = \frac{\rho}{\varepsilon_0}$
C. $\nabla \times E = \frac{\rho}{2\pi\varepsilon_0}$ D. none of these

794. When electric field is parallel to the plane of incidence, the electromagnetic wave is said to be
A. linearly polarized B. circularly polarized
C. elliptically polarized D. parallelly polarized

795. Polarization in electromagnetic wave is caused by
A. reflection B. refraction
C. transverse nature of electromagnetic waves
D. longitudinal nature of electromagnetic waves

796. The wave propagation occurs without attenuation in case of a
A. perfect dielectric B. semiconductor
C. conductor D. all of the above

797. The direction of which of the following vector gives the direction of propagation of electromagnetic waves?
A. Vector H B. Vector E
C. Vector $(E \cdot H)$ D. Vector $(E \times H)$

798. The absorption of radiowaves by the atmosphere depends on
A. the polarization of atmosphere
B. the polarization of the waves
C. their frequency
D. their distance from the transmitter

799. A plane wave starting in air incidents normally on a loss-free dielectric medium of relative permittivity 4. The reflected portion of the wave is to be eliminated by placing a quarter-wavelength plate between the air and the dielectric. The required relative permittivity of the material of the plate is
A. 2 B. 3
C. 4 D. 5

800. The critical angle for electromagnetic wave passing from glass to air $(\varepsilon/\varepsilon_0 = 0)$ will be nearly
A. 10.5° B. 19.5°
C. 26.5° D. 38.5°

801. When VSWR is 4, reflection coefficient is
A. 1/2 B. 1
C. 0 D. 1/4

802. Mark the false statement regarding the boundary conditions between different media.
A. The tangential component of E is continuous at the surface. That is, it is the same just outside the surface as it is just inside the surface
B. The tangential component of H in continuous across a surface except at the surface of a perfector conductor
C. The normal component of D in continuous, if there is no surface charge density otherwise D is discontinuous by an amount equal to the surface charge density
D. The normal component of B is zero at the surface of discontinuity

803. Polarisation of electromagnetic waves is due to
A. transverse nature of the waves
B. longitudinal nature of the waves
C. reflection from another medium
D. none of these

804. Propagation of frequencies in UHF range takes place by means of
A. surface waves B. space waves
C. ground waves D. sky waves

805. When a wave travelling in air enters into a wave guide
A. the phase velocity will increase
B. the group velocity will increase
C. the phase velocity will decrease
D. none of these

806. Distilled water is a dielectric having $\epsilon_\rho = 81$ and $\mu_r = 1$. If a wave is incident from water into a water air interface, the critical angle will be nearly
A. 6° B. 12°
C. 15° D. 22°

807. For different loads, the range of the values for VSWR is
A. 0 to 0.5 B. 0.5 to 1
C. 0 to 1 D. 1 to ∞

808. In a horizontally polarized wave
A. electric component is perpendicular to the ground
B. electric component is parallel to the ground
C. magnetic component is parallel to the ground
D. none of these

809. If the magnetic field vector H as only Z component given by $H_z = 3x \cos\beta + 6y \sin\gamma$. Also the field is invariant with time. The current
A. $J = 6k \sin\gamma$ B. $J = -6k \sin\gamma$
C. $J = -6k \cos\gamma$ D. $J = 6k \cos\gamma$

810. The function $f(x - v_0 t)$ represents
A. a wave motion in forward direction
B. a wave motion in reverse direction
C. a stationary wave D. none of these

811. A prominent wave should have
A. no phase shift
B. no attenuation
C. highest cut-off frequency
D. lowest cut-off frequency

812. A dominant wave is characterised by
A. highest cut-off wavelength
B. lowest cut-off wavelength
C. no attenuation D. all of the above

813. All of the following are Maxwell's equations except
A. div $D = 0$ B. div $B = 0$
C. curl $H = J + \frac{\partial D}{\partial t}$ D. curl $H = J + \frac{\partial B}{\partial t}$

Ans.	793. A	794. D	795. C	796. A	797. D	798. C	799. A	800. B	801. A	802. D	803. A	804. B	805. A
	806. A	807. A	808. B	809. B	810. A	811. D	812. B	813. A					

814. Assuming the velocity of EM waves reaching an antenna as 3×10^8 m/sec, the length of a half-wave dipole antenna required to receive a 5 MHz radio signal is
A. 7.5 m B. 15 m
C. 30 m D. 60 m

815. When an electromagnetic wave makes a normal incidence on a perfect conductor
A. no energy is absorbed in the conductor
B. electrical field component forms standing wave pattern
C. magnetic field component forms standing wave pattern
D. all of these

816. An electromagnetic wave incidents on a perfect insulator at an angle. If the wave is not entirely parallels polarised, then
A. there will be no reflection
B. the reflected wave will be entirely polarised perpendicular to the plane of incidence
C. the reflected wave will be entirely polarised parallel to the plane of incidence
D. all of these

817. Polarisation of electromagnetic waves is due to
A. transverse nature of the wave
B. longitudinal nature of the waves
C. reflection from another medium
D. none of these

818. Refraction of electromagnetic waves occurs, when they
A. are polarised at right angles to the direction of propagation
B. face a perfect conductor
C. enter a medium of different dielectric constant
D. all of these

819. Which one of the following cannot be used to remove the unwanted sideband in SSA?
A. Phase shift method B. Balanced modulator
C. Third method D. Filter system

820. Propagation of frequencies in UHF range takes place by means of
A. surface waves B. space waves
C. ground waves D. sky waves

821. The range of microwave frequencies, which easily pass through the atmosphere is called
A. critical frequency range
B. window frequency range
C. gyro frequency range
D. resonant frequency range

822. A wave of frequency 6 GHz is propagated in a parallel plane waveguide with plane separation of 3 cm. The dominant mode of the wave will
A. not propagate B. propagate
C. be stand still D. reflect

823. In the above problem, the group velocity will be
A. 1.66×10^8 m/sec B. 2.12×10^8 m/sec
C. 3.2×10^8 m/sec D. 5.20×10^8 m/sec

824. Which of the following statements is not correct for a TEM wave?
A. Electromagnetic fields are entirely transversed
B. The velocity of the wave is greater than the velocity of light
C. The cut-off frequency is zero
D. Characteristic impedance is equal to 37 ohms

825. An electromagnetic wave in a waveguide has a full wave of electric intensity change between the two further walls, and no component of the electric field in the direction of propagation. The mode of the wave is
A. $TE_{1,0}$ B. $TE_{1,1}$
C. $TE_{2,0}$ D. $TM_{2,2}$

826. The advantage of waveguides over co-axial line is that
A. these are simpler to manufacture
B. power losses are lower
C. higher operating frequencies are possible
D. all of these

827. The function of conducting walls in a waveguide is to
A. allow conduction of energy
B. reduce power losses
C. confine energy within the waveguide
D. allow higher frequencies transmitted

828. Waveguides are generally superior to co-axial lines in the range
A. 300 MHz–3 GHz B. 3 GHz–10 GHz
C. 3 GHz–100 GHz D. 100 GHz–300 Hz

829. In actual practice, waveguide acts as a
A. low-pass filter B. high-pass filter
C. band-pass filter D. none of these

830. Which of the following statements is not for waveguides?
A. The waveguide itself is virtually lossless
B. The propagation of energy is through the dielectric filling of the waveguide
C. The propagation of energy takes place through the walls of the waveguide
D. The losses in waveguide depends upon the frequency of the wave

831. Waveguides are not used for frequencies below
A. 1 GHz B. 10 GHz
C. 100 GHz D. 500 MHz

832. Waveguide of convenient dimensions are employed in the range of
A. 500 MHz–1 GHz B. 3–100 GHz
C. 50–500 GHz D. none of these

Ans. 814. C 815. D 816. B 817. A 818. C 819. C 820. B 821. B 822. B 823. A 824. B 825. C 826. D 827. C 828. C 829. B 830. C 831. A 832. B

833. Which of the following statement is not true for waveguides?
A. They are superior to coaxial cables in microwave range
B. They can pass several signals simultaneously
C. Signals in different modes are propagated to separate them at the receiving point
D. Signals have to be of the same frequency

834. Which of the following statements is not an advantage of waveguides as compared to coaxial lines?
A. Waveguides are simply to manufacture
B. Power-handling ability is higher
C. The velocity of propagation is more than that in free space
D. They have much higher maximum operating frequency

835. Which of the following statements is not true for strip-line in comparison with waveguide?
A. It can be directly connected to semiconductor microwave devices
B. Its size is much smaller
C. Its bandwidth is smaller
D. Losses are less

836. Strip-line is analogous to
A. flattened coaxial line B. parallel-wire line
C. waveguide D. none of these

837. Microstrip is analogous to
A. flattened coaxial line B. parallel-wire line
C. waveguide D. none of these

838. Typical material thickness in microstrip is
A. 0.001 to 0.01 mm B. 0.1 to 1.5 mm
C. 2 to 2.5 mm D. 3 to 5 mm

839. Which of the following statements is not true for microstrip as compared to strip-line?
A. It is simpler in construction
B. It can be print-circuited
C. Its radiation losses are less
D. None of these

840. Which of the following statements is not true for FDM?
A. It is used for analog signals
B. Each channel has one frequency band for its exclusive use
C. The modulated carriers travel on one line simultaneously
D. One channel is transmitted at a time

841. Which of the following statements is not true for decoder?
A. It selects the desired signal from other signals
B. It demodulates the signal
C. It transforms the signal into a form suitable for decoding process
D. It amplifies the signal

842. Which of the following is not uniformly distributed over all frequencies?
A. White noise B. Flicker noise
C. Short noise D. Thermal noise

843. At high frequencies, which of the following types of noise becomes most important?
A. Flicker noise B. Short noise
C. Thermal noise D. Transit time noise

844. Indicate the false statement.
A. Pentodes are generally noiser than triodes in the same connection
B. Thermal noise is independent of the frequency at which it is measured
C. Partition noise voltage is independent of the bandwidth
D. All of these

845. Satellite tracking stations are located in remote areas in order to minimise the effect of
A. solar noise B. man-made noise
C. cosmic noise D. thermal noise

846. DSB-SC and SSB-SC systems are identical from the point of view of
A. bandwidth requirements
B. efficiency of transmission
C. noise improvements D. none of these

847. The signals contaminated with large noise are demodulated by
A. envelope detector
B. synchronous detector
C. envelope detector followed by low-pass filter
D. envelope detector followed by high-pass filter

848. The harmonic content of a periodic wave-form without sharp discontinuities is
A. large B. small
C. moderate D. none of these

849. The waveforms without sharp discontinuities contain at the most
A. three harmonics B. five harmonics
C. seven harmonics D. none of these

850. The most rapid changes that the vocal chords can generate is controlled by
A. first harmonic B. third harmonic
C. fifth harmonic D. none of these

851. A periodic pulse becomes an aperiodic signal if its periodic time approaches
A. zero B. infinity
C. ten times the pulse width
D. none of these

852. To eliminate ghosts in the picture
A. use a long transmission line
B. connect a booster
C. change the antenna orientation of location
D. twist the transmission line

Ans.													
833. D	834. C	835. C	836. A	837. B	838. B	839. C	840. D	841. D	842. B	843. D	844. C	845. B
846. C	847. B	848. B	849. A	850. B	851. B	852. C						

853. A communication channel disturbed by additive white Gaussian noise has a bandwidth of 4 kHz and SNR of 15. The highest transmission rate that such a channel can support (in K-bits/sec) is
A. 16
B. 1.6
C. 3.2
D. 60

854. A discrete zero memory information source has 40 symbols and each symbol is equally likely. The minimum number of bits required to code the source with uniform length code and the entropy of the source are respectively
A. 5, 5.03
B. 6, 5.83
C. 5, 6.64
D. 6, 5.32

855. If each stage had a gain of 10 dB, and noise figure of 10 dB, then the overall noise figure of a two-stage cascade amplifier will be
A. 10
B. 1.09
C. 1.0
D. 10.9

856. Which of the following are useful in comparing the noise performance of receivers?
1. Input noise voltage
2. Equivalent noise resistance
3. Noise temperature
4. Noise figure
Select the correct answer using the codes given below
Codes:
A. 1, 2 and 3
B. 1, 3 and 4
C. 2, 3 and 4
D. 1, 2 and 4

857. Quantising noise is produced in
A. FDM
B. PCM
C. all modulation system
D. all pulse modulation systems

858. The dimension specified by the manufacture for the TV screen is
A. width
B. height
C. diagonal
D. none of these

859. The ratio of width to height (aspect ratio) of TV screen is
A. 3:4
B. 4:3
C. 1:1
D. 2:1

860. The bandwidth of video signal in India TV sets is
A. 4.5 MHz
B. 5.5 MHz
C. 6.5 MHz
D. 7.5 MHz

861. Sound in TV is usually modulated using
A. AM
B. FM
C. PCM
D. PM

862. The RF amplifier in a TV receiver is of
A. cascade configuration
B. cascode configuration
C. differential configuration
D. common-base

863. Which of the following stages furnish maximum gain?
A. RF amplifier
B. IF amplifier
C. Video amplifier
D. None of these

864. Purpose of DC restorer is to correct
A. average brightness
B. sync pulses
C. flyback duration
D. none of these

865. Rolling picture can be set right by the knob
A. synch hold
B. vertical linearity
C. contrast
D. all of these

866. The amplitudes of the components in the frequency domain of a signal are
A. peak amplitudes
B. RMS amplitudes
C. average amplitudes
D. none of these

867. In TV system equalizing pulses are sent during
A. horizontal blanking
B. vertical blanking
C. serrations
D. horizontal retrace

868. In broadcast television, the number of scanning per frame is chosen to be an odd number to
A. reduce the required bandwidth
B. improve the picture resolution
C. reduce peak power requirement
D. make interlacing easier

869. In TV signals, the colour burst is used to
A. maintain the colour sequence
B. ensure the I and Q phase correctly
C. synchronise colours
D. interlace each horizontal line

870. The resolution of a TV picture is determined by
A. video amplification factor
B. video bandwidth
C. the number of frames scanned
D. the output of the video detector

871. A correlation receiver consists of a/an
A. multiplier and an integrator
B. adder and integrator
C. multiplier only
D. integrator only

872. In push button keypad dialling in telephony, touching a button generates a tone which is a
A. combination of two frequencies
B. single frequency
C. chirp tone
D. sequence of pulses

873. The attenuation in a telephone cable is 1 dB per km and the transmitted voltage is 1 V. The voltage received 1000 km away is
A. 10^{-100} V
B. 10^{-50} V
C. 10^{-3} V
D. 10^{-4} V

874. Bandwidth of colour television transmission is of the order of
A. 100 kHz
B. 1 MHz
C. 6 MHz
D. 10 MHz

875. The sound signal in TV broadcast is modulated in
A. SSB
B. VSB
C. FM
D. DSB-SC

Ans. 853. A 854. D 855. D 856. D 857. B 858. C 859. B 860. C 861. B 862. B 863. A 864. A 865. C
866. A 867. A 868. D 869. B 870. C 871. A 872. A 873. B 874. C 875. C

876. The channel required for FM telemetry is
A. the same as that required for AM telemetry
B. smaller than that required for AM telemetry
C. 100 times that requires for AM telemetry
D. 10 times that required for AM telemetry

877. For transmission line load matching over a range of frequencies, it is best to use a
A. balun B. double stub
C. broad band directional coupler
D. single stub of adjustable position

878. In television system, VSB, modulation is preferred to SSB in order to
A. save on bandwidth B. simplify generation
C. enable demodulation using an envelope detector
D. save on signal power

879. In TV transmission, picture signal is amplitude modulated and sound signal is frequency modulated. This is done because
A. it is not possible to frequency modulate the picture signal
B. bandwidth requirement is minimised
C. sound signal is more susceptible to noise than picture signal
D. synchronisation of picture frames becomes easier

880. In television transmission, the type of modulation used for video is
A. DSB B. SSB
C. VSB D. FM

881. The number of still pictures displayed taken in one second in case of movies is
A. 16 B. 24
C. 2 D. 30

882. The number of still picture frames taken in one second in case of television is
A. 16 B. 24
C. 25 D. 30

883. The function of interlacing in television is to
A. avoid flicker
B. provide vertical sync pulse position
C. increase frame frequency
D. avoid picture noise

884. The retrace period in field scanning as percentage of total time is
A. 7% B. 10%
C. 12% D. 20%

885. The horizontal retrace in a TV receiver is also called
A. equaliser B. blanker
C. burst D. flyback

886. Which of the following pulses are used to keep the horizontal oscillator of a TV receiver in synchronisation during vertical blanking period?
A. Sync pulse B. Serration pulse
C. Equalising pulse D. Blanking pulse

887. The aspect ratio in a TV receiver is
A. 3:2 B. 4:2
C. 4:3 D. 3:1

888. The function of diplexer bridge in a TV transmitter is to
A. prevent the loading of several transmitters by video transmitter
B. increase the efficiency of transmission
C. increase the bandwidth
D. increase power output

889. In horizontal scanning in TV receiver, the beam starts from
A. left top corner and moves to right
B. centre of top of the frame and moves to right
C. centre of the bottom of the frame and moves to right
D. right top corner of the frame and moves to right

890. The vertical scanning is from
A. top middle to bottom middle
B. bottom middle to top middle
C. top right corner to bottom left corner
D. bottom right corner to top left corner

891. Photomultiplier boosts the video signal by virtue of
A. its high gain B. negative modulation
C. secondary emission D. its low gain

892. The time for sweeping a picture frame is
A. 1/60 sec B. 1/150 sec
C. 1/25 sec D. 1/30 sec

893. In case of Indian TV system, the difference between audio and video carriers is
A. 4.5 MHz B. 5.5 MHz
C. 6 MHz D. 6.5 MHz

894. The intermediate frequency for the video stage is
A. 5.5 MHz B. 45.75 MHz
C. 56.75 MHz D. 61.25 MHz

895. The intermediate frequency for the audio stage is
A. 5.5 MHz B. 45.75 MHz
C. 56.75 MHz D. 62.25 MHz

896. The sync pulses are transmitted during
A. horizontal scanning retrace time
B. vertical scanning retrace time
C. both horizontal and vertical retrace times
D. equalising pulses

897. The horizontal sync frequency is
A. 50 Hz B. 13125 Hz
C. 15625 Hz D. 15750 Hz

898. The vertical sync frequency is
A. 50 Hz B. 60 Hz
C. 100 Hz D. 120 Hz

899. The blanking pulse is interposed during
A. the retrace period of vertical line
B. the retrace period of horizontal line

Ans. 876. ? 877. D 878. A 879. B 880. C 881. B 882. C 883. A 884. A 885. D 886. C 887. C 888. A
889. B 890. B 891. C 892. B 893. B 894. B 895. A 896. C 897. C 898. A

C. the retrace interval between each line
D. equalising pulse

900. Which of the following signals are not present beyond the black region of the composite signal?
A. Sync pulses B. Blanking pulses
C. Equalising pulses D. Video signal

901. Horizontal pulses are separated from vertical pulses by
A. sync separator B. integrator
C. differentiator D. sweep circuit

902. Vertical pulses are separated from horizontal pulses by
A. sync separator B. integrator
C. differentiator D. sweep circuit

903. Sync separation essentially is a
A. clipper B. differentiator
C. clamper D. integrator

904. The function of the equalising pulses is to
A. smooth out differences between vertical sync signals
B. smooth out differences between horizontal sync signals
C. maintain horizontal synchronisation over three line period
D. none of these

905. The spacing of the equalising pulses is at
A. half-line intervals B. one-line intervals
C. two-line intervals D. three-line intervals

906. Which of the following pulses are of largest width?
A. Horizontal pulses B. Vertical pulses
C. Equalising pulses D. None of these

907. Which of the following functions is not performed by video detector?
A. It demodulates the composite video signal
B. It separates sync signals
C. It separates audio and sound IF signals
D. It heterodynes the two IF signals

908. The video detector used in TV receiver is a
A. diode detector B. ratio detector
C. phase detector D. balanced demodulator

909. The horizontal sync and deflection currents are similar to the vertical circuits except the addition of
A. automatic voltage control
B. automatic frequency control
C. automatic phase control
D. automatic gain control

910. Which of the following statements is not true for automatic frequency control?
A. It contains a phase detector
B. It compares the phase difference between horizontal deflection voltage and sync pulses
C. Its output is AC
D. It changes the bias of sweep oscillator

911. The oscillator used for deflection or electron beam in the picture tube is a
A. RC oscillator B. tunnel diode oscillator
C. blocking oscillator D. crystal oscillator

912. In the picture tube, the electron beam is
A. electrostatically deflected
B. magnetically deflected
C. both electrostatically and magnetically deflected
D. none of these

913. The high voltage power supply for anode of a TV picture tube is obtained
A. during trace period of vertical sweep signal
B. during trace period of horizontal sweep signal
C. from a step-up transformer connected to the mains
D. none of these

914. The function of iron trap magnet in a TV picture tube is to
A. deflect away ionised gas particles
B. trap electrons returning from screen
C. trap secondary electrons
D. none of these

915. The TV camera used to telecast movies is
A. telecine B. image orthogon
C. vidicon D. plumbicon

916. The colour TV system adopted in India is
A. NTSC system B. PAL system
C. CCTV system D. Indian TV system

917. The three primary colours are
A. red, orange and blue B. red, green and yellow
C. red, blue and green D. green, blue and yellow

918. As compared to black and white TV system, compatibility in colour TV system requires
A. smaller bandwidth B. same bandwidth
C. larger bandwidth
D. three times the bandwidth for black and white system

919. The function of interleaving in colour TV system is to
A. mix colour information with black and white signals
B. extent black and white information from colour signals
C. superimpose black and white signals
D. produce *i* and *Q* signals

920. Which of the following systems is most noise immune?
A. SSB B. PCM
C. PDM D. FM

921. Omni directional range (ODR) operates by
A. comparing the frequencies of two signals
B. comparing the phase between two signals
C. measuring time taken by a pulse to reach a station
D. interlocking code letter

Ans.	899. C	900. D	901. C	902. B	903. A	904. A	905. A	906. B	907. B	908. A	909. B	910. C	911. C
	912. B	913. B	914. A	915. A	916. B	917. C	918. B	919. A	920. B	921. B			

922. In a loop-antenna, the unbalanced current
A. leads to an elliptical pattern
B. is independent of loop orientation
C. does not let true null to occur
D. all of these

923. Which of the following navigational aids operates by measuring the time difference between signals transmitted from master-slave transmitters.
A. Omni-range B. Loran
C. Tacon
D. Ground-controlled apparatus

924. Which of the following types of presentation for a radar system displays elevation and azimuth?
A. *PPI* presentation B. A-scan presentation
C. B-scan presentation
D. C-scan presentation

925. For direct reception of TV signal from a communication satellite, the type of antenna required is
A. horn antenna B. parabolic antenna
C. chicken-mesh antenna
D. loop antenna

926. Which of the following measures cannot be used to reduce noise?
A. By increasing the channel bandwidth
B. By increasing the transmitted power
C. By reducing the signalling rate
D. By using redundancy

927. The advantage of guard time between pulses is that
A. it increases the efficiency of transmission
B. it suppresses the cross talk
C. message can be reconstructed with particular filters
D. none of these

928. The term noise temperature is used for
A. white-noise source only
B. thermal-noise source only
C. non-thermal-noise source only
D. all of these

929. The noise figure of a receiver connected to an antenna whose resistance is 50 ohm has an equivalent noise of 30 ohm is
A. 1.6 dB B. 2.04 dB
C. 4.2 dB D. 6.05 dB

930. For DSB-suppressed carrier, the improvement in S/N ratio at the output of the demodulator as compared to its input S/N ratio is by a factor of
A. 2 B. 2/3
C. 1 D. 3

931. For SSB-suppressed carrier, the improvement in S/N ratio at the output of the demodulator as compared to its S/N ratio at the input is by a factor of
A. 3 B. 2
C. 2/3 D. 1

932. For AM, the improvement in S/N ratio at the output of the envelope detector as compared to its S/N ratio at the input is by a factor of
A. 3 B. 2
C. 2/3 D. 1

933. The ratio of S/N ratios at the detector output of FM and AM signals depends upon
A. bandwidth B. carrier frequency
C. modulating frequency
D. all of these

934. In FM, if the bandwidth is doubled, the S/N ratio improves by a factor of
A. 6 B. 6 dB
C. 8 dB
D. the statement is not true

935. In which of the following systems, S/N ratio can be improved by increasing bandwidth?
A. AM
B. DSB-suppressed carrier
C. SSB D. FM

936. At FM demodulator output
A. noise power density decreases with frequency
B. noise power density increases with frequency
C. signal power density increased with frequency
D. none of these

937. The emphasis circuits are used for improving S/N ratio at
A. lower frequencies B. middle frequencies
C. higher frequencies D. complete frequencies

938. If the bandwidth in a PCM system is B, the S/N ratio at the output and input of the receiver is proportional to
A. B B. B^2
C. $2B$ D. $1/B$

939. Which of the following statement is not true for PCM?
A. The noise is removed by regenerating pulses at each repeater station
B. The noise accumulates throughout the transmission path
C. Noise-free signal is transmitted at each repeater station
D. Only noise on the link between repeater station is to be considered

940. Quantising noise is produced in
A. all pulse modulation system
B. all modulation system
C. PCM D. PDM

941. Quantising noise can be reduced by increasing
A. bandwidth
B. sampling rate
C. number of standard quantum levels
D. all of these

Ans. 922. D 923. B 924. D 925. C 926. A 927. C 928. B 929. B 930. A 931. D 932. C 933. D 934. B 935. B 936. B 937. C 938. C 939. B 940. C 941. C

942. The thermal noise is due to random
A. motion of atoms and molecules
B. motion of free electrons
C. vibrations of atoms about their mean position, inside the conducting medium
D. none of these

943. Indicate which one of the following type of noise does not occur in transistors?
A. Shot noise B. Partition noise
C. Resistance noise
D. None of these

944. Noise temperature is
A. same as physical temperature of the body
B. a measure of the available noise power for non-thermal sources
C. a measure of noise figure of a receiver
D. none of these

945. The maximum frequency at which digital data can be applied to gate is defined as
A. operating speed B. propagation speed
C. binary level transition period
D. charging time

MULTIPLE CHOICE QUESTIONS FROM VARIOUS COMPETITIVE EXAMINATIONS

1. Power spectral density of a signal is **(IES 2003)**
A. complex, even and non-negative
B. real, even and non-negative
C. real, even and negative
D. complex, odd and negative

2. In a communication system, a process for which statistical averages and time averages are equal, is called **(IES 2003)**
A. Stationary B. Ergodic
C. Gaussian D. BIBO stable

3. A microphone with inbuilt amplifier has an output noise level in a very quiet room of about 30 μV rms. The maximum output level is about 300 mV rms before severe distortion occurs. What dynamic range does this represent in dB? **(IES 2005)**
A. 4 dB B. 36 dB
C. 40 dB D. 80 dB

4. A system has a receiver noise resistance of 50 Ω. It is connected to an antenna with an output resistance of 50 Ω. The noise figure of the system is **(IES 2007)**
A. 1 B. 2
C. 50 D. 101

5. In an AM signal when the modulation index is one, the maximum power P_t (where P_c is the carrier power) is equal to **(IES 2001)**
A. P_c B. $1.5P_c$
C. $2P_C$ D. $2.5P_c$

6. An AM superheterodyne receiver with IF of 455 kHz is tuned to the carrier frequency of 1000 kHz. The image frequency is **(IES 2001)**
A. 545 kHz B. 1 MHz
C. 1455 kHz D. 1910 kHz

7. If the modulation index of an AM wave is changed from 0 to 1, the transmitted power **(IES 2001)**
A. increases by 50% B. increases by 75%
C. increases by 100% D. remains unaffected

8. For an AM signal, the bandwidth is 10 kHz and the highest frequency component present is 705 kHz. The carrier frequency used for this AM signal is **(IES 2003)**
A. 695 kHz B. 700 kHz
C. 705 kHz D. 710 kHz

9. If large amount of information is to be transmitted in a small amount of time, we require **(IES 2003)**
A. low-frequency signals
B. narrow-band signals
C. wide-band signals D. high-frequency signals

10. A modulator is a device to **(IES 2003)**
A. separate two frequencies
B. impress the information on a radio frequency carrier
C. extract information from the carrier
D. amplify the audio frequency signal

11. The detection-gain for coherent DSB demodulator is **(IES 2003)**
A. 2 B. 4
C. 8 D. 16

12. For an AM signal, the bandwidth is 10 kHz and the highest frequency component present is 705 kHz. What is the carrier frequency used for this AM signal? **(IES 2008)**
A. 695 kHz B. 700 kHz
C. 705 kHz D. 710 kHz

13. The antenna current of an AM transmitter is 8 A when only carrier is sent, but it increases to 8.93 A when the carrier is modulated. Then what is the percentage modulation of the wave? **(IES 2009)**
A. 43% B. 70.15%
C. 57% D. 100%

14. Which one of the following is considered as an AM signal? **(IES 2009)**

Ans. 942. B 943. B 944. B 945. A 1. C 2. B 3. D 4. B 5. B 6. D 7. A 8. B 9. C 10. B 11. A 12. B 13. B

A. Binary phase shift keying (BPSK)
B. Differential phase shift keying (DPSK)
C. Differential encoded PSK
D. Quadrature PSK

15. An AM modulator has output
$S(t) = 20 \cos (300\pi t) + 6 \cos (320\pi t) + 6 \cos (280\pi t)$
When what is the modulation index of the wave? **(IES 2009)**
A. More than 100% B. 0.93
C. 0.3 D. 0.6

16. How much power will an AM transmitter, rated at 50 kW, radiate if it is modulated to 100%? **(IES 2009)**
A. 25 kW B. 50 kW
C. 75 kW D. 100 kW

17. For which one of the following modulated signals, the original message, up to scaling factor can be recovered using envelope detection? **(IES 2009)**
A. $20 \cos (200\pi t) + 30\, m(t) \cos (200\pi t)$
B. $20 \cos (200\pi t) + 16\, m(t) \cos (200\pi t)$
C. $10\, m(t) \cos (400\pi t)$ D. $10 \cos m(t) \cos (400\pi t)$

18. What is the assigned bandwidth of each of the channels in the AM broadcast band? **(IES 2010)**
A. 5 kHz B. 10 kHz
C. 15 kHz D. 200 kHz

19. Which one of the following modulation techniques has got maximum SNR? **(IES 2010)**
A. AM-SSB B. AM-JSB
C. FM D. AM-SC

20. Diode demodulator will ignore **(IES 2011)**
A. the amplitude modulation
B. the frequency modulation
C. PCM D. PWM

21. In double sideband suppressed carrier modulation, the modulated wave undergoes phase reversal, whenever **(IES 2011)**
A. modulating signal's amplitude decreases
B. modulating signal's amplitude increases
C. modulating signal crosses zero
D. carrier signal crosses zero

22. Bandwidth occupied by 100 MHz carrier, AM modulated by signal frequency of 10 kHz is **(IES 2012)**
A. 100 MHz B. 20 kHz
C. 10 kHz D. 110 MHz

23. RF carrier 10 kV at 1 MHz is amplitude modulated and modulation index is 0.6. Peak voltage of the signal is **(IES 2013)**
A. 600 kV B. 1200 kV
C. 6 kV D. 10 kV

24. A carrier wave of frequency 2.5 GHz amplitude is modulated with two modulating frequencies equal to 1 kHz and 2 kHz. The modulated wave will have the total bandwidth **(IES 2013)**
A. 6 kHz B. 2 kHz
C. 4 kHz D. 3 kHz

25. In an amplitude modulated (AM) wave with 100% modulation (m_a), the carrier is suppressed. The percentage of power saving will be **(IES 2013)**
A. 100% B. 50%
C. 25% D. 66.7%

26. A 1000 W carrier is amplitude-modulated and has a side-band power of 300 W. The depth of modulation is **(IES 2013)**
A. 0.255 B. 0.545
C. 0.775 D. 0.95

27. A 100 V carrier peak changes from 160 V to 40 V by a modulating signal. The modulation factor is **(IES 2014)**
A. 0.3 B. 0.5
C. 0.6 D. 0.7

28. A broadcast AM transmitter radiates 50 kW of carrier power. The radiation power at 85% of modulation is **(IES 2014)**
A. 68.1 kW B. 60.8 kW
C. 61.8 kW D. 62.0 kW

29. In a typical AM receiver circuit, the oscillator frequency is **(IES 2014)**
A. same as signal frequency
B. always equal to 455 Hz
C. lower than the signal frequency by 455 Hz
D. higher than the signal frequency by 455 kHz

30. The waveform $A \cos (\omega_1 t + k \cos \omega_2 t)$ is **(IES 2002)**
A. amplitude modulated
B. frequency modulated
C. phase modulated
D. frequency as well as phase modulated

31. An FM wave uses a 2–5 V, 500 Hz modulating frequency and has a modulation index of 50. The deviation is **(IES 2002)**
A. 500 Hz B. 100 Hz
C. 1250 Hz D. 25000 Hz

32. An angle modulated signal is described by the equation
$x_c(t) = 10 \cos [2\pi f_c t + 10 \sin (4000\pi t) + 5 \sin 2000\pi t]$
What is the bandwidth of this modulated signal? **(IES 2009)**
A. 6 kHz B. 45 kHz
C. 54 kHz D. 63 kHz

Ans.													
	14. D	15. D	16. B	17. B	18. B	19. C	20. B	21. C	22. B	23. C	24. C	25. D	26. C
	27. C	28. A	29. D	30. D	31. D	32. C							

33. Which one of the following factors is limited in a case of FM? **(IES 2009)**
A. Maximum frequency deviation
B. Maximum permissible modulation index
C. Signal to noise voltage ratio
D. Minimum permissible modulation index

34. Time division multiplexing is used when the data to be transmitted is **(IES 2001)**
A. slow changing B. of small bandwidth
C. slow changing and has a small bandwidth
D. fast changing and has a wide bandwidth

35. Which one of the following PCM schemes is depicted in the below figure? **(IES 2001)**

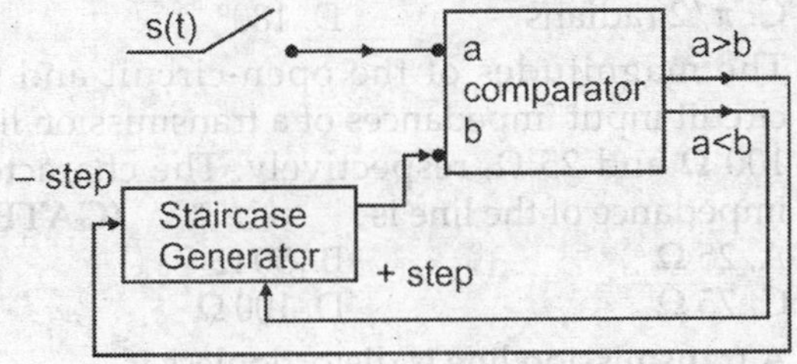

A. Adaptive DM B. Differential PCM
C. Companding D. Delta modulation

36. "Slope overload" occurs in delta modulation when the **(IES 2001)**
A. frequency of the clock pulses is too low
B. rate of change of analog waveform is too large
C. step size is too small
D. analog signal varies very slowly with time

37. Which one of the following statements is correct? In TDM, non-essential frequency components of the modulating signal are removed by **(IES 2004)**
A. sampler B. attenuator
C. pre-alias filter D. modulator

38. For a 10-bit PCM system, the signal to quantization noise ratio is 62 dB. If the number of bits is increased by 2, then how would the signal to quantization noise ratio change? **(IES 2007)**
A. Increase by 6 dB B. Decrease by 6 dB
C. Increase by 12 dB D. Decrease by 12 dB

39. The demodulation of a delta modulated signal is achieved by **(IES 2011)**
A. integration B. differentiation
C. sampling D. band-pass filtering

40. PPM signal is **(IES 2011)**
A. differentiation of PWM
B. integration of PWM
C. differentiation of PAM
D. not related to PWM or PAM

41. Consider a binary Hamming code of block length 31 and rate equal to (26/31). Its minimum distance is **(IES 2002)**
A. 3 B. 5
C. 26 D. 31

42. In which type of the switchings given below, entire capacity of a dedicated link is used? **(IES 2003)**
A. Virtual circuit packed switching
B. Data gram packet switching
C. Circuit switching D. message switching

43. The channel capacity of a noiseless channel is equal to **(IES 2005)**
A. rate at which information is transmitted
B. signalling speed
C. bandwidth
D. bandwidth-SNR product

44. A conventional telephone line with 3 kHz bandwidth and having 30 dB signal-to-noise ratios can carry information at a rate of **(IES 2010)**
A. 30 kbps B. 15 kbps
C. 3 kbps D. 16 mpbs

45. In TV transmission, picture signal is amplitude modulated and sound signal is frequency modulated. This is done because **(IES 2003)**
A. it is not possible to frequency modulate the picture signal
B. bandwidth requirement is minimised
C. sound signal is more susceptible to noise than picture signal
D. synchronization of picture frames become easier

46. The power gain of an antenna using paraboloid reflector is directly proportional to **(IES 2013)**
A. mouth diameter B. wavelength
C. aperture ratio
D. square of aperture ratio

47. A transmission line of real characteristic impedance is terminated with an unknown load. The measured value of VSWR on the line is equal to 2 and a voltage minimum point is found to be at the load. The load impedance is then **(GATE 1987)**
A. complex B. purely capacitive
C. purely resistive D. purely inductive

48. A two-wire transmission line of characteristic impedance Z_0 is connected to a load of impedance Z_L ($Z_L \neq Z_0$). Impedance matching cannot be achieved with **(GATE 1988)**
A. a quarter-wavelength transformer
B. a half-wavelength transformer
C. an open-circuited parallel stub
D. a short-circuited parallel stub

Ans. 33. A 34. C 35. A 36. B 37. C 38. C 39. A 40. A 41. A 42. C 43. A 44. A 45. B
46. D 47. C 48. B

49. A 50 ohm lossless transmission line has a pure reactance of (j 100) ohms as its load. The VSWR in the line is **(GATE 1989)**
A. 1/2 (half) B. 2 (two)
C. 4 (four) D. ∞ (infinity)

50. The input impedance of a short circuited lossless transmission line quarter wave long is **(GATE 1991)**
A. purely reactive B. purely resistive
C. infinite
D. dependent on the characteristic impedance of the line

51. A transmission line whose characteristic impedance is a real **(GATE 1992)**
A. must be a lossless line
B. may not be a lossless line
C. may not be a distortionless line
D. none of these

52. Consider a transmission line of characteristic impedance 50 ohm. Let it be terminated at one end by (+j50) ohm. The VSWR produced by it in the transmission line will be **(GATE 1993)**
A. +1 B. 0
C. ∞ D. +j

53. A lossless transmission line having 50 Ω characteristic impedance and length λ/4 is short circuited at one end and connected to an ideal voltage source of 1 V at the other end. The current drawn from the voltage sources is **(GATE 1996)**
A. 0 B. 0.02 A
C. ∞ D. none of these

54. The capacitance per unit length and the characteristic impedance of a lossless transmission line are C and Z_0, respectively. The velocity of a travelling wave on the transmission line is **(GATE 1996)**
A. Z_0C B. $1/(Z_0C)$
C. Z_0/C D. C/Z_0

55. All transmission line section in figure, have a characteristic impedance $R_0 + j0$. The input impedance Z_{in} equals **(GATE 1998)**

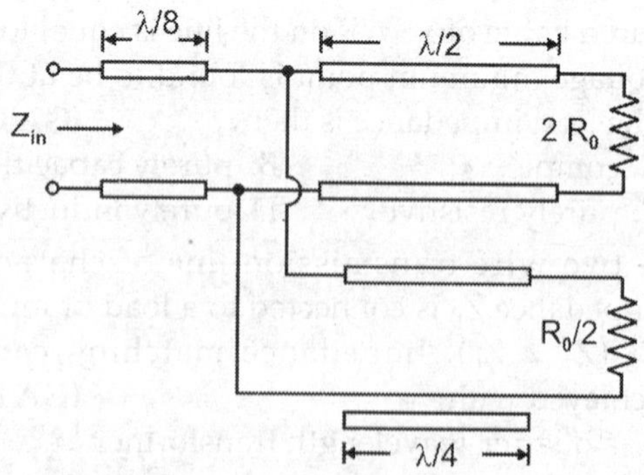

A. $2/3R_0$ B. R_0
C. $3/2R_0$ D. $2R_0$

56. A transmission line of 50 Ω characteristic impedance is terminated with a 100 Ω resistance. The minimum impedance measured on the line is equal to **(GATE 1997)**
A. 0 Ω B. 25 Ω
C. 50 Ω D. 100 Ω

57. In a twin-wire transmission line in air, the adjacent voltage maxima are at 12.5 cm and 27.5 cm. The operating frequency is **(GATE 1999)**
A. 300 MHz B. 1 GHz
C. 2 GHz D. 6.28 GHz

58. In air, a lossless transmission line of length 50 cm with L = 10 μH/m, C = 40 pF/m is operated at 25 MHz. Its electrical path length is **(GATE 1999)**
A. 0.5 metres B. λ metres
C. π/2 radians D. 180°

59. The magnitudes of the open-circuit and short-circuit input impedances of a transmission line are 100 Ω and 25 Ω, respectively. The characteristic impedance of the line is **(GATE 2000)**
A. 25 Ω B. 50 Ω
C. 75 Ω D. 100 Ω

60. A transmission line is distortionless if **(GATE 2001)**
A. $RL = 1/GC$ B. $RL = GC$
C. $LG = RC$ D. $RG = LC$

61. A uniform plane electromagnetic wave incident normally on a plane surface of a dielectric material is reflected with a VSWR of 3. What is the percentage of incident power that is reflected? **(GATE 2001)**
A. 10% B. 25%
C. 50% D. 75%

62. The VSWR can have any value between **(GATE 2002)**
A. 0 and 1 B. −1 and +1
C. 0 and ∞ D. 1 and ∞

63. A short-circuited stub is shunt connected to a transmission line as shown in the figure. If Z_0 = 50 Ω, the admittance Y seen at the junction of the stub and the transmission line is **(GATE 2003)**

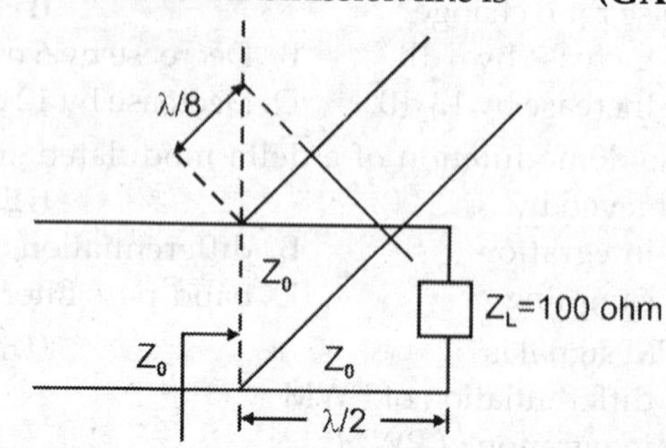

A. (0.01 − j0.02) mho B. (0.02 − j0.01) mho
C. (0.04 − j0.02) mho D. (0.02 + j0) mho

Ans.	49. D	50. C	51. A	52. C	53. A	54. B	55. B	56. B	57. B	58. C	59. B	60. C	61. B
	62. D	63. A											

64. In an impedance Smith chart, a clockwise movement along a constant resistanc4e circle gives rise to (GATE 2002)
A. a decrease in the value of reactance
B. an increase in the value of reactance
C. no change in the reactance value
D. no change in the impednce value

65. Consider a 300 Ω quarter-wave long (at 1 GHz) transmission line as shown in the figure. It is connected to a 10V, 50 Ω source at one end and is left open circuited at the other end. The magnitude of the voltage at the open circuit end of the line is (GATE 2004)

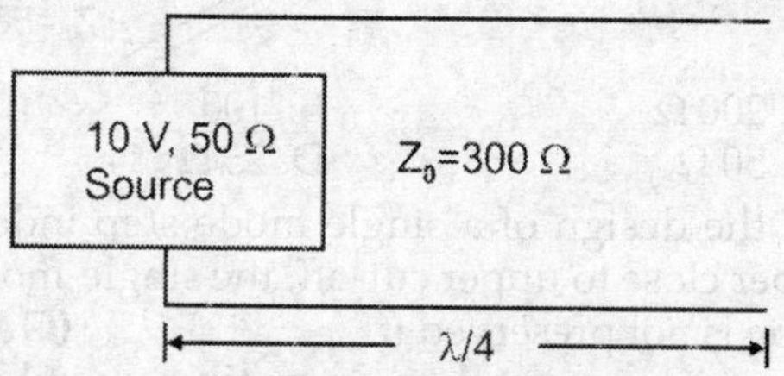

A. 10 V B. 5 V
C. 60 V D. 60/7 V

66. Consider an impedance $Z = R + jX$ marked with point P in an impedance Smith chart as shown in the figure. The movement from point P along a constant resistance circle in the clockwise direction by an angle 45° is equivalent to (GATE 2004)

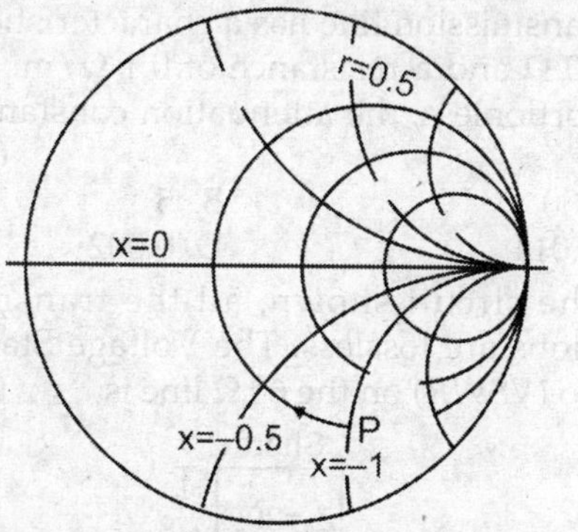

A. addint an inductance in series Z
B. adding a capacitance in series Z
C. adding an inductance in shunt across Z
D. adding a capacitance in shunt across Z

67. A plane electromagnetic wave propagation in free space is incident normally on a large slab of lossless, non-magnetic, dielectric material with $\epsilon > \epsilon_0$. Maxima and minima are observed when the electric field is measured in front of the slab. The maximum electric field is found to be 5 times the minimum field. The intrinsic impedance of the medium should be (GATE 2004)
A. 120πΩ B. 60πΩ
C. 600πΩ D. 24πΩ

68. A lossless transmission line is terminated in a load which reflects a part of the incident power. The measured VSWR is 2. The percentage of the power that is reflected back is (GATE 2004)
A. 57.73 B. 33.33
C. 0.11 D. 11.11

69. Characteristic impedance of a transmission line is 50 Ω. Input impedance of the open-circuited line is Z_{OC} = 100 + j150 Ω. When the transmission line is short-circuited, then value of the input impedance will be (GATE 2005)
A. 50 Ω B. 100 + j150 Ω
C. 7.69 + j11.54 Ω D. 7.69 – j11.54 Ω

70. Many circles are drawn in a Smith chart used for transmission line calculations. The circles shown in the figure represent (GATE 2005)

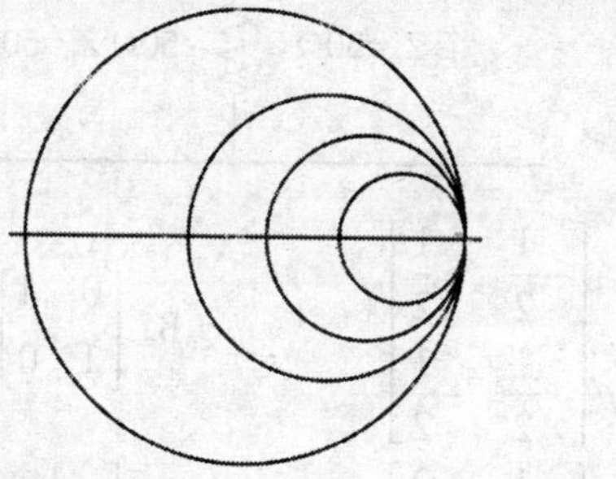

A. unit circles
B. constant resistance circles
C. constant reactance circles
D. constant reflection coefficient circles

71. The reflection coefficient of the above figure is given by (GATE 2005)
A. –0.6 B. –1
C. 0.6 D. 0

72. Voltage standing wave pattern in a lossless transmission line with characteristic impedance 50 Ω and a resistive load is shown in the figure. The value of the load resistance is (GATE 2005)

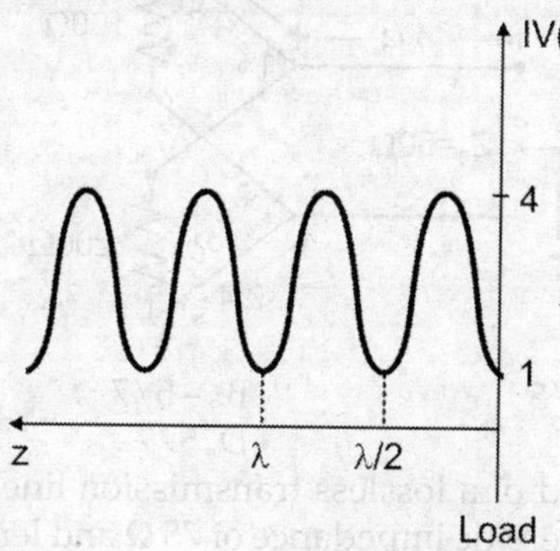

A. 50 Ω B. 200 Ω
C. 12.5 Ω D. 0 Ω

Common data for questions 73 and 74: A 30 Volt battery with zero source resistance is connected to a coaxial line of characteristic impedance of 50 ohms at t = 0 second and

Ans. 64. B 65. C 66. A 67. D 68. D 69. D 70. B 71. A 72. C

terminated in an unknown resistive load. The line length is such that it takes 400 μs, the voltage at the load end is found to be 40 Volts.

73. The load resistance is **(GATE 2006)**
A. 25 Ohms B. 50 Ohms
C. 75 Ohms D. 100 Ohms

74. The steady-state current through the load resistance is **(GATE 2006)**
A. 1.2 Amps B. 0.3 Amps
C. 0.6 Amps D. 0.4 Amps

75. A load of 50 Ω is connected in shunt in a 2-wire transmission line of Z_0 = 50 Ω as shown in the figure. The 2-port scattering parameter matrix (S-matrix) of the shunt element is **(GATE 2007)**

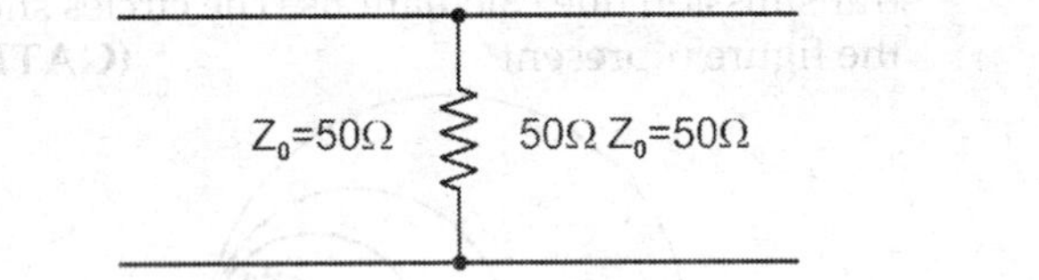

A. $\begin{bmatrix} -\frac{1}{2} & \frac{1}{2} \\ \frac{1}{2} & -\frac{1}{2} \end{bmatrix}$ B. $\begin{bmatrix} 0 & 1 \\ 1 & 0 \end{bmatrix}$

C. $\begin{bmatrix} -\frac{1}{3} & \frac{2}{3} \\ \frac{2}{3} & -\frac{1}{3} \end{bmatrix}$ D. $\begin{bmatrix} \frac{1}{4} & -\frac{3}{3} \\ -\frac{3}{4} & \frac{1}{4} \end{bmatrix}$

76. The parallel branches of a 2-wire transmission line are terminated in 100 Ω and 200 Ω resistors as shown in the figure. The characteristic impedance of the line is Z_0 = 50 Ω and each section has a length of λ/4. The voltage reflection coefficient Γ at the input is **(GATE 2007)**

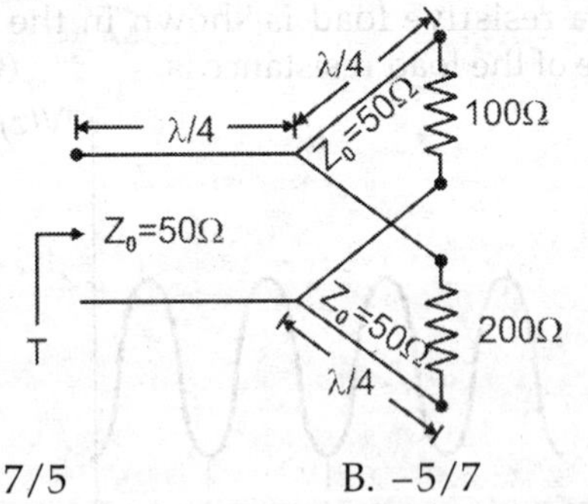

A. $-j7/5$ B. $-5/7$
C. $j5/7$ D. $5/7$

77. One end of a lossless transmission line having the characteristic impedance of 75 Ω and length of 1 cm is short-circuited. At 3 GHz, the input impedance at the other end of the transmission line is **(GATE 2008)**
A. 0 B. resistive
C. capacitive D. inductive

78. A transmission line terminates in two branches, each of length λ/4 as shown. The branches are terminated by 50 Ω loads. The lines are lossless and have the characteristic impedances shown. Determine the impedance Z_i as seen by the source. **(GATE 2009)**

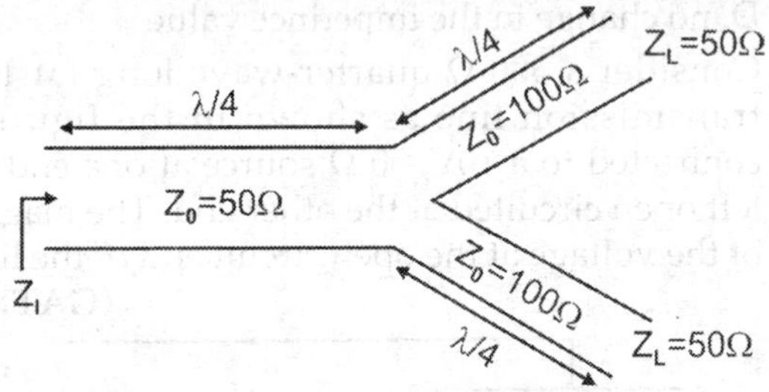

A. 200 Ω B. 100
C. 50 Ω D. 25 Ω

79. In the design of a single mode step index optical fiber close to upper cut-off, the single-mode operation is not preserved if **(GATE 2008)**
A. radius as well as operating wavelength are halved
B. radius as well as operating wavelength are doubled
C. radius is halved and operating wavelength is doubled
D. radius is doubled and operating wavelength is halved

80. A transmission line has a characteristic impedance of 50 Ω and a resistance of 0.1 Ω/m. If the line is distortionless, the attenuation constant (in Np/m) is **(GATE 2010)**
A. 500 B. 5
C. 0.014 D. 0.002

81. In the circuit shown, all the transmission line sections are lossless. The Voltage Standing Wave Ratio (VSWR) on the 60 Ω line is **(GATE 2010)**

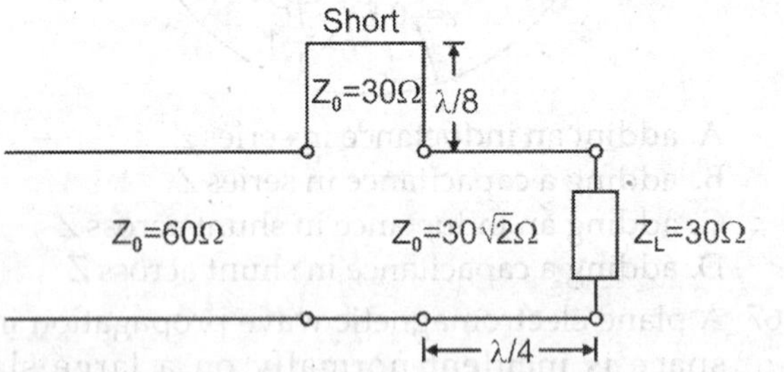

A. 1.00 B. 1.64
C. 2.50 D. 3.00

82. A transmission line of characteristic impedance 50 Ω is terminated by a 50 Ω load. When excited by a sinusoidal voltage source at 10 GHZ, the phase difference between two points spaced 2 mm apart on the line is found to be π/4 radians. The phase velocity of the wave along the line is **(GATE 2011)**

Ans. 73. D 74. B 75. B 76. D 77. D 78. D 79. D 80. D 81. B

A. 0.8×10^8 m/s B. 1.2×10^8 m/s
C. 1.6×10^8 m/s D. 3×10^8 m/s

83. A transmission line of characteristic impedance 50 Ω is terminated in a load impedance Z_L. The VSWR of the line is measured as 5 and the first of the voltage maxima in the line is observed at a distance of $\lambda/4$ from the load. The value of Z_L is **(GATE 2011)**

A. 10 Ω B. 250 Ω
C. $(19.23 + j46.15)$ Ω D. $(19.23 - j46.15)$ Ω

84. A transmission line with a characteristic impedance of 100 Ω is used to match a 50 Ω section to 1200 Ω section. If the matching is to be done both at 429 MHz and 1 GHz, the length of the transmission line can be approximately **(GATE 2012)**

A. 82.5 cm B. 1.05 m
C. 1.58 m D. 1.75 m

85. The return loss of a device is found to be 20 dB. The voltage standing wave ratio (VSWR) and magnitude of reflection coefficient are respectively **(GATE 2013)**

A. 1.22 and 0.1 B. 0.81 and 0.1
C. –1.22 and 0.1 D. 2.44 and 0.2

86. For a parallel plate transmission line, let v be the speed of propagation and Z be the characteristic impedance. Neglecting fringe effects, a reduction of the spacing between the plates by a factor of two result in **(GATE 2014)**

A. halving of v and no change in Z
B. no change in v and halving of Z
C. no change in both v and Z
D. halving of both v and Z

87. In the following figure, the transmitter T_x sends a wideband modulated RF signal via a coaxial cable to the receiver Rx. The output impedance Z_T or T_x, the characteristic impedance Z_0 of the cable and the input impedance Z_R of R_x are all real. **(GATE 2014)**

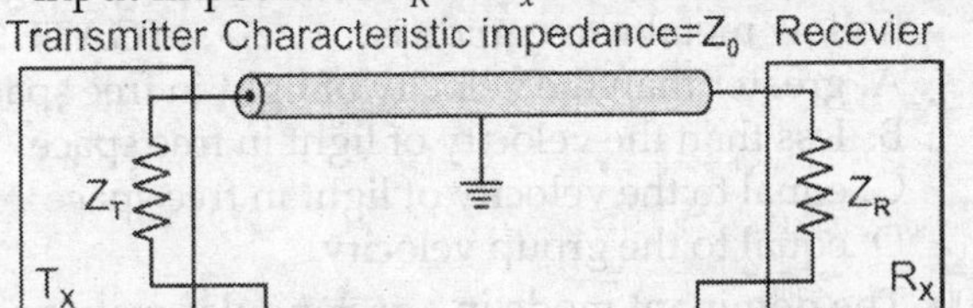

Which one of the following statements is true about the distortion of the received signal due to impedance mismatch?

A. The signal gets distorted if $Z_R \neq Z_0$, irrespective of the value of Z_T
B. The signal gets distorted if $Z_T \neq Z_0$, irrespective of the value of Z_R
C. Signal distortion implies impedance mismatch at both ends: $Z_T \neq Z_0$ and $Z_R \neq Z_0$
D. Impedance mismatches do not result in signal distortion but reduce power transfer efficiency

88. A coaxial cable is made of two brass conductors. The spacing between the conductors is filled with Teflon (ε_r = 2.1, and tanδ = 0). Which one of the following circuits can represent the lumped element model of a small piece of this cable having length Δz? **(GATE 2015)**

A. 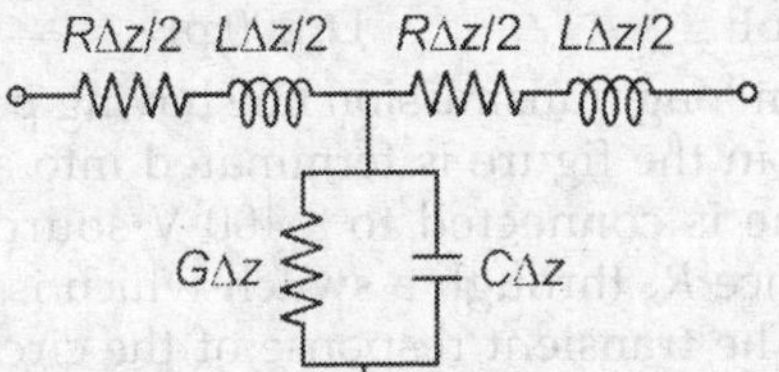

B. 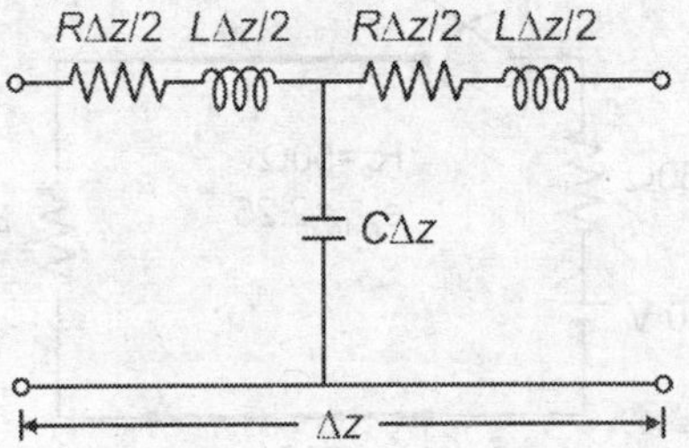

C. 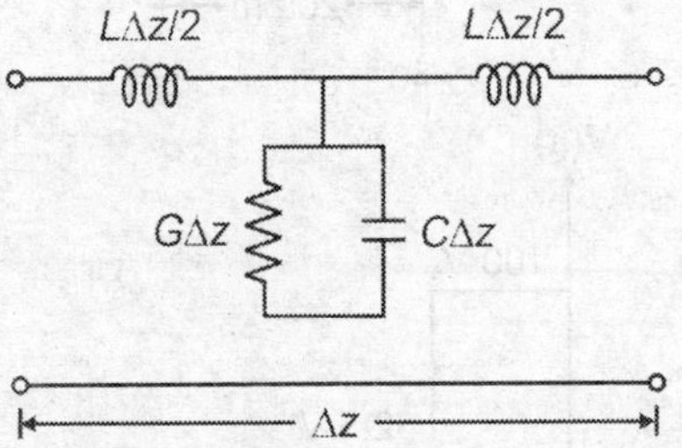

D. 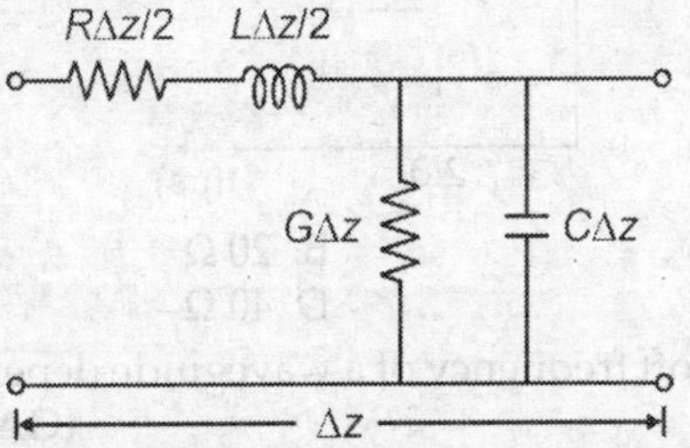

89. Consider the 3 m long lossless air-filled transmission line shown in the figure. It has a characteristic impedance of 120 πΩ is terminated by a short circuit, and is excited with a frequency of 37.5 MHz. What is the nature of the input impedance (Z_{in})? **(GATE 2015)**

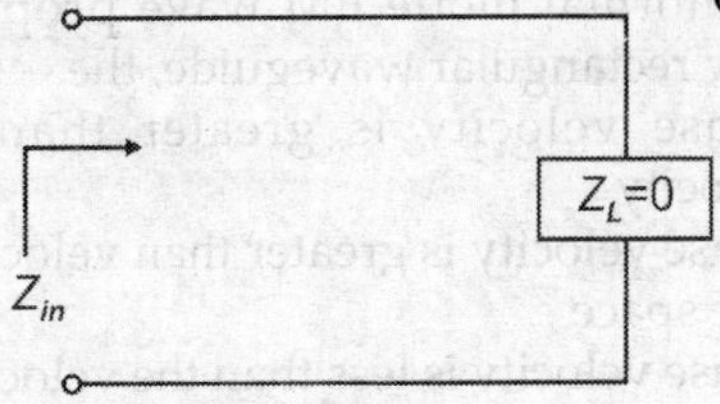

A. Open B. Short
C. Inductive D. Capacitive

Ans. 82. C 83. A 84. C 85. A 86. B 87. C 88. B 89. D

90. A coaxial capacitor of inner radius 1 mm and outer radius 5 mm has a capacitance as per unit length of 172 pF/m. If the ratio of outer radius to inner is doubled, the capacitance per unit length (in pF/m) is **(GATE 2015)**
A. 50 pF B. 100 pF
C. 120 pF D. 200 pF

91. A 200 m long transmission line having parameters shown in the figure is terminated into a load R_L. The line is connected to a 400 V source having resistance R_S through a switch which is closed at $t = 0$. The transient response of the circuit at the input of the line ($z = 0$) is also drawn in the figure. The value of R_L (in Ω) is ...... **(GATE 2015)**

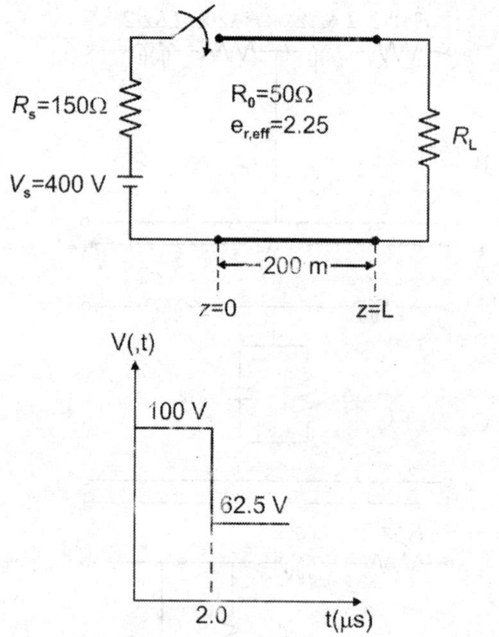

A. 10 Ω B. 20 Ω
C. 30 Ω D. 40 Ω

92. The cutoff frequency of a waveguide depends upon the **(GATE 1987)**
A. dimensions of waveguide
B. dielectric property of the medium in the waveguide
C. characteristic impedance of the waveguide
D. transverse and axial components of the fields

93. For a normal mode EM wave propagating in a hollow rectangular waveguide, the **(GATE 1988)**
A. phase velocity is greater than the group velocity
B. phase velocity is greater than velocity of light in free space
C. phase velocity is less than the velocity of light in free space
D. phase velocity may be either greater than or less than the group velocity

94. Choose the correct statement.
For a wave propagating in an air filled rectangular waveguide **(GATE 1990)**
A. guided wavelength is never less than the free space wavelength
B. wave impedance is never less than the free space impedance
C. phase velocity is never less than the free space velocity
D. TEM mode is possible if the dimensions of the wave guide are properly choosen

95. A rectangular air filled waveguide has a cross-section of 4 cm × 10 cm. The minimum frequency which can propagate in the waveguide is **(GATE 1998)**
A. 1.5 GHz B. 2.0 GHz
C. 2.5 GHz D. 3.0 GHz

96. Assuming perfect conductors of a transmission line, pure TEM propagation is not possible in **(GATE 1999)**
A. coaxial cable
B. air-filled cylindrical waveguide
C. parallel twin-wire line in air
D. semi-infinite parallel plate waveguide

97. A TEM wave is incident normally upon a perfect conductor. The E and H fields at the boundary will be, respectively **(GATE 2000)**
A. minimum and minimum
B. maximum and maximum
C. minimum and maximum
D. maximum and minimum

98. A rectangular waveguide has dimensions 1 cm × 0.5 cm. Its cutoff frequency is **(GATE 2000)**
A. 5 GHz B. 10 GHz
C. 15 GHz D. 12 GHz

99. The phase velocity of waves propagating in a hollow metal waveguide is **(GATE 2001)**
A. greater than the velocity of light in free space
B. less than the velocity of light in free space
C. equal to the velocity of light in free space
D. equal to the group velocity

100. The dominant mode in a rectangular waveguide is TE_{10} because this mode has **(GATE 2001)**
A. no attenuation B. no cutoff
C. no magnetic field component
D. the highest cutoff wavelength

101. The phase velocity for the TE_{10}-mode in an air-filled rectangular waveguide is **(GATE 2002)**
A. less than c B. equal to c
C. greater than c D. none of the above
Note: c is the velocity of plane waves in free space.

102. A rectangular metal waveguide filled with a dielectric material of relative permittivity $\epsilon_r = 4$ has the

Ans. 90. C 91. C 92. A 93. A 94. A 95. A 96. B 97. A 98. C 99. A 100. D 101. C

inside dimensions 3.0 cm × 1.2 cm. The cutoff frequency for the dominant mode is **(GATE 2003)**

A. 2.5 GHz B. 5.0 GHz
C. 10.0 GHz D. 12.5 GHz

103. The phase velocity of an electromagnetic wave propagating in a hollow metallic rectangular waveguide in the TE_{10} mode is **(GATE 2004)**

A. equal to its group velocity
B. less than the velocity of light in free space
C. equal to the velocity of light in free space
D. greater than the velocity of light in free space

104. Which one of the following does represent the electric field lines for the TE_{02} mode in the cross-section of a hollow rectangular metallic waveguide? **(GATE 2005)**

A.
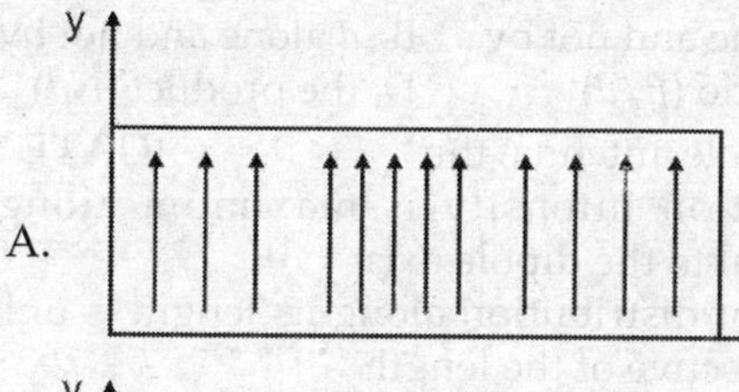

B.

C.
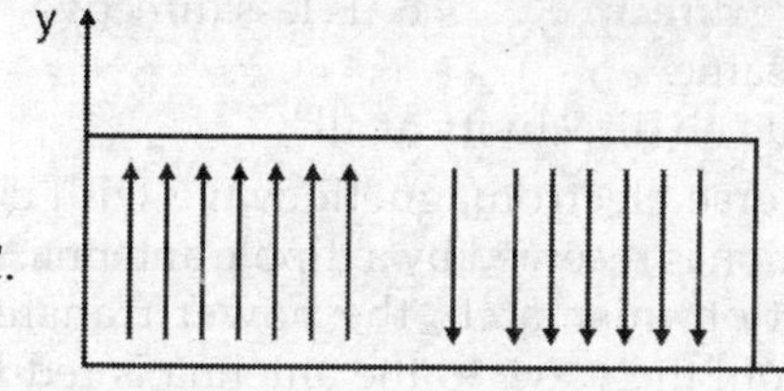

D.
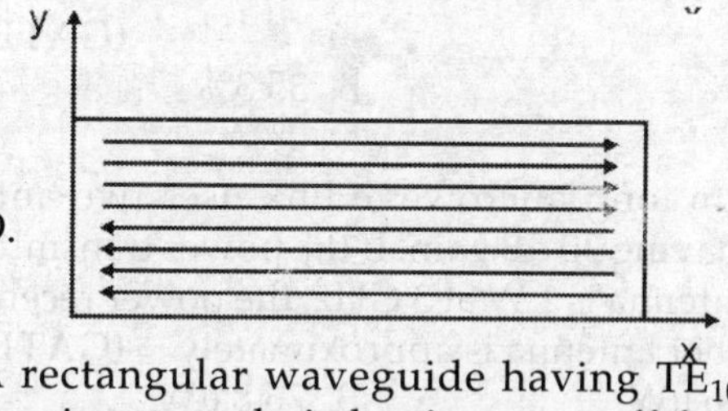

105. A rectangular waveguide having TE_{10} mode as dominant mode is having a cutoff frequency of 18 GHz for the TE_{30} mode. The inner broad-wall dimension of the rectangular waveguide is **(GATE 2006)**

A. 5/3 cms B. 5 cms
C. 5/2 cms D. 10 cms

106. In a microwave test bench, why is the microwave signal amplitude modulated at 1 kHz? **(GATE 2006)**

A. To increase the sensitivity of measurement
B. To transmit the signal to a far-off place
C. To study amplitude modulation
D. Because crystal detector fails at microwave frequencies

107. An air-filled rectangular waveguide has inner dimensions of 3 cm × 2 cm. The wave impedance of the TE_{20} mode of propagation in the waveguide at a frequency of 30 GHz is (free space impedance $\eta_0 = 377\ \Omega$) **(GATE 2007)**

A. 308 Ω B. 355 Ω
C. 400 Ω D. 461 Ω

108. The *E* field in a rectangular waveguide of inner dimensions $a \times b$ is given by

$$E = \frac{\omega\mu}{h^2}\left(\frac{\pi}{a}\right)H_0 \sin\left(\frac{2\pi x}{a}\right)\sin(\omega t - \beta z)\hat{y}$$

where H_0 is a constant, and *a* and *b* are the dimensions along the *x*-axis and the *y*-axis, respectively. The mode of propagation in the waveguide is **(GATE 2007)**

A. TE_{20} B. TM_{11}
C. TM_{20} D. TE_{10}

109. A rectangular waveguide of internal dimensions (a = 4 cm and b = 3 cm) is to be operated in TE_{11} mode. The minimum operating frequency is **(GATE 2008)**

A. 6.25 GHz B. 6.0 GHz
C. 5.0 GHz D. 3.75 GHz

110. Which of the following statements is true regarding the fundamental mode of the metallic waveguides shown? **(GATE 2009)**

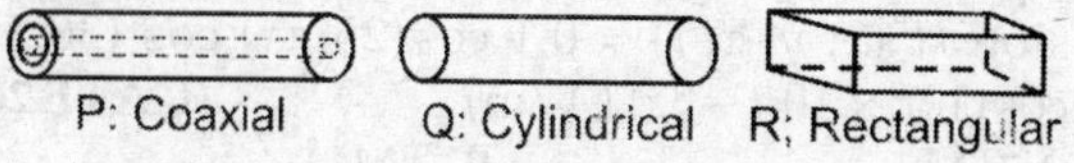

A. Only *P* has no cutoff frequency
B. Only *Q* has no cutoff frequency
C. Only *R* has no cutoff frequency
D. All three have cutoff frequencies

111. The magnetic field along the propagation direction inside a rectangular waveguide with the cross-section shown in figure is $H_z = 3\cos(2.094 \times 10^2 x)\cos(2.618 \times 10^2 y)\cos(6.283 \times 10^{10} t - \beta z)$ **(GATE 2012)**

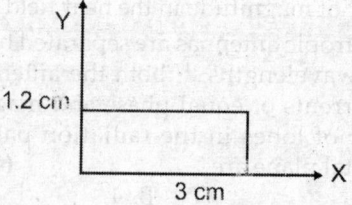

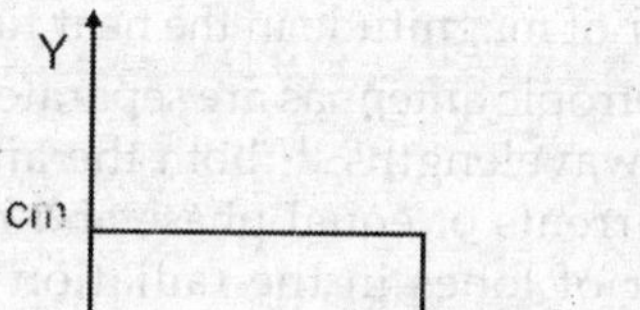

The phase velocity v_p of the wave inside the waveguide satisfies

A. $v_p > c$ B. $v_p = c$
C. $0 < v_p < c$ D. $v_p = 0$

Ans. 102. A 103. D 104. D 105. C 106. D 107. C 108. A 109. A 110. A 111. A

112. The modes in a rectangular waveguide are denoted by TE_{mn}/TM_{mn} where m and n are the eigen numbers along the larger and smaller dimensions of the waveguide respectively. Which one of the following statement is true? **(GATE 2011)**
A. The TM_{10} mode of the waveguide does not exist
B. The TE_{10} mode of the waveguide does not exist
C. The TM_{10} and the TE_{10} modes both exist and have the same cutoff frequencies
D. The TM_{10} and TM_{01} modes both exist and have the same cutoff frequencies

113. A two-port network has scattering parameters given by $[S]=\begin{bmatrix} s_{11} & s_{12} \\ s_{21} & s_{22} \end{bmatrix}$. If the port-2 of the two-port is short circuited, the s_{11} parameter for the resultant one-port network is **(GATE 2014)**
A. $\dfrac{s_{11} - s_{11}s_{22} + s_{12}s_{21}}{1+s_{22}}$ B. $\dfrac{s_{11} + s_{11}s_{22} - s_{12}s_{21}}{1+s_{22}}$
C. $\dfrac{s_{11} + s_{11}s_{22} + s_{12}s_{21}}{1-s_{22}}$ D. $\dfrac{s_{11} - s_{11}s_{22} + s_{12}s_{21}}{1-s_{22}}$

114. Which one of the following field patterns represents a TEM wave travelling in the positive x direction? **(GATE 2014)**
A. $E = +8\hat{y}, H = -4\hat{z}$ B. $E = -2\hat{y}, H = -3\hat{z}$
C. $E = +2\hat{z}, H = -2\hat{y}$ D. $E = +3\hat{y}, H = -4\hat{z}$

115. The longitudinal component of the magnetic field inside an air-filled rectangular waveguide made of a perfect electric conductor is given by the expression $H_z(x, y, z, t) = 0.1 \cos(25\pi x) \cos(30.3\pi y) \cos(12\pi \times 10^9 t - \beta z)$ (A/m) **(GATE 2015)**
A. TM_{12} B. TM_{21}
C. TE_{21} D. TE_{12}

116. The electric field E and the magnetic field H of a short dipole antenna satisfy the condition **(GATE 1988)**
A. the r component of E is equal to zero
B. both r and θ components of H are equal to zero
C. the θ component of E dominates the r component in the far-field region
D. the θ and ϕ components of H are of the same order of magnitude in the near field region

117. Two isotropic antennas are separated by a distance of two wavelengths. If both the antennas are fed with currents of equal phase and magnitude, the number of lobes in the radiation pattern in the horizontal plane are **(GATE 1990)**
A. 2 B. 4
C. 6 D. 8

118. In a board side array of 20 isotropic radiators, equally spaced at distance of $\lambda/2$, the beam width between first nulls is **(GATE 1991)**
A. 51.3 degree B. 11.46 degree
C. 22.9 degree D. 102.6 degree

119. Two dissimilar antennas having their maximum directivities equal **(GATE 1992)**
A. must have their beam widths also equal
B. cannot have their beam widths equal because they are dissimilar antenna
C. may not necessarily have their maximum power gains equal
D. must have their effective aperture areas (capture areas) also equal

120. The beam width between firs nulls of a uniform linear array of N equally spaced (element spacing = d), equally excited antennas is determined by **(GATE 1992)**
A. N alone and not by d B. d alone and not by N
C. the ratio (n/d) D. the product (Nd)

121. For a dipole antenna, the **(GATE 1994)**
A. radiation intensity is maximum along the normal to the dipole axis
B. current distribution along its length is uniform irrespective of the length
C. effective length equals its physical length
D. input impedance is independent of the location of the feed-point

122. An antenna, when radiating, has a highly directional radiation pattern. When the antenna is receiving, its radiation pattern **(GATE 1995)**
A. is more directive B. is less directive
C. is the same
D. exhibits no directivity at all

123. A transverse electromagnetic wave with circular polarization is received by a dipole antenna. Due to polarization mismatch, the power transfer efficiency from the wave to the antenna is reduced to about **(GATE 1996)**
A. 50% B. 35.3%
C. 25% D. 0%

124. A 1 km long microwave link uses two antennas each having 30 dB gain. If the power transmitted by one antenna is 1 W at 3 GHz, the power received by the other antenna is approximately **(GATE 1996)**
A. 98.6 μW B. 76.8 μW
C. 63.4 μW D. 55.2 μW

125. A parabolic dish antenna has conical beam 2° wide. The directivity of the antenna is approximately **(GATE 1997)**
A. 20 dB B. 30 dB
C. 40 dB D. 50 dB

126. The vector $\boldsymbol{H}$ in the far field of an antenna satisfies **(GATE 1998)**
A. $\nabla \cdot \boldsymbol{H} = 0$ and $\nabla \times \boldsymbol{H} = 0$
B. $\nabla \cdot \boldsymbol{H} \neq 0$ and $\nabla \times \boldsymbol{H} \neq 0$

Ans. 112. A 113. B 114. B 115. C 116. C 117. D 118. B 119. A 120. D 121. A 122. C 123. D 124. C 125. C

C. $\nabla \cdot H = 0$ and $\nabla \times H \neq 0$
D. $\nabla \cdot H \neq 0$ and $\nabla \times H = 0$

127. The radiation resistance of a circular loop of one turn is 0.01 Ω. The radiation resistance of five turns of such a loop will be **(GATE 1998)**
A. 0.002 Ω B. 0.01 Ω
C. 0.05 Ω D. 0.25 Ω

128. An antenna in free space receives 2 μW of power when the incident electric field is 20 mV/m rms. The effective aperture of the antenna is **(GATE 1998)**
A. 0.005 m^2 B. 0.05 m^2
C. 1.885 m^2 D. 3.77 m^2

129. The far field of an antenna varies with distance r as **(GATE 1998)**
A. $1/r$ B. $1/r^2$
C. $1/r^3$ D. $1/\sqrt{r}$

130. A transmitting antenna radiates 251 W isotropically. A receiving antenna, located 100 m away from the transmitting antenna, has an effective aperture of 500 cm^2. The total power received by the antenna is **(GATE 1999)**
A. 10 μW B. 1 μW
C. 20 μW D. 100 μW

131. The frequency range for satellite communication is **(GATE 2000)**
A. 1 kHz to 100 kHz B. 100 kHz to 10 kHz
C. 10 MHz to 30 MHz D. 1 GHz to 30 GHz

132. If the diameter of a $\lambda/2$ dipole antenna is increased from $\lambda/100$ to $\lambda/50$, then its **(GATE 2000)**
A. bandwidth increases B. bandwidth decreases
C. gain increases D. gain decreases

133. For an 8 feet (2.4 m) parabolic dish antenna operating at 4 GHz, the minimum distance required for far field measurement is closest to **(GATE 2000)**
A. 7.5 cm B. 15 cm
C. 15 m D. 150 m

134. A medium wave radio transmitter operating at a wavelength of 492 m has a tower antenna of height 124 m. What is the radiation resistance of the antenna? **(GATE 2001)**
A. 25 Ω B. 36.5 Ω
C. 50 Ω D. 73 Ω

135. In a uniform linear array, four isotropic radiating elements are spaces $\lambda/4$ apart. The progressive phase shift between the elements required for forming the main beam at 60° off the end-fire is **(GATE 2001)**
A. $-\pi$ radians B. $-\pi/2$ radians
C. $-\pi/4$ radians D. $-\pi/8$ radians

136. The line-of-sight communication requires the transmit and receive antennas to face each other. If the transmit antenna is vertically polarized, for best reception the receiver antenna should be **(GATE 2002)**
A. horizontally polarized
B. vertically polarized
C. at 45° with respect to horizontal polarization
D. at 45° with respect to vertical polarization

137. Two identical antennas are placed in the $\phi = \pi/2$ plane as shown in the figure. The elements have equal amplitude excitation with 180° polarity difference, operating at wavelength λ. The correct value of the magnitude of the far-zone resultant electric field strength normalized with that of a single element, both computed for $\phi = 0$, is **(GATE 2003)**

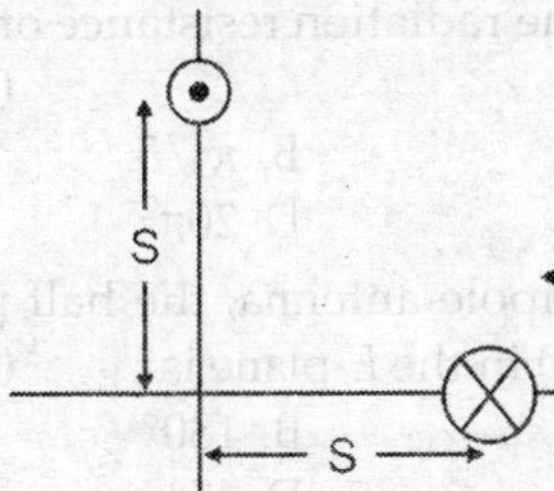

A. $2\cos\left(\frac{2\pi s}{\lambda}\right)$ B. $2\sin\left(\frac{2\pi s}{\lambda}\right)$
C. $2\cos\left(\frac{\pi s}{\lambda}\right)$ D. $2\sin\left(\frac{\pi s}{\lambda}\right)$

138. A person with a receiver is 5 km away from the transmitter. What is the distance that this person must move further to detect a 3 dB decrease in signal strength? **(GATE 2002)**
A. 942 m B. 2070 m
C. 4978 m D. 5320 m

139. Two identical and parallel dipole antennas are kept apart by a distance of $\lambda/4$ in the H-plane. They are fed with equal currents but the right most antenna has a phase shift of +90°. The radiation pattern is given as **(GATE 2005)**

A. 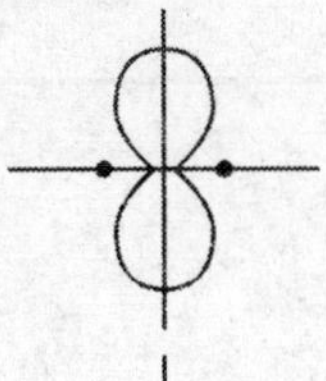 B. 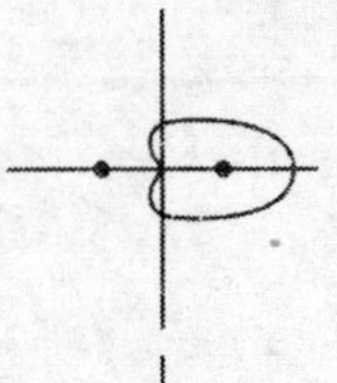

C. 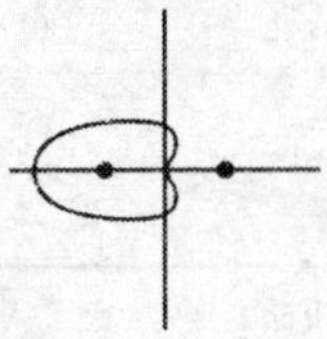 D. 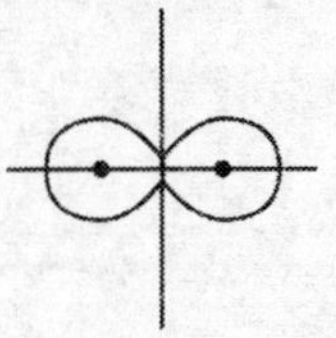

Ans. 126. C 127. D 128. C 129. A 130. D 131. D 132. B 133. D 134. B 135. C 136. B 137. D 138. B 139. A

140. Consider a lossless antenna with a directive gain of +6 dB. If 1 mW of power is fed to it, the total power radiated by the antenna will be (GATE 2004)
A. 4 mW B. 1 mW
C. 7 mW D. 1/4 mW

141. A transmission line is feeding 1 Watt of power to a horn antenna having a gain of 10 dB. The antenna is matched to the transmission line. The total power radiated (GATE 2006)
A. 10 watts B. 1 watt
C. 0.1 watt D. 0.01 watt

142. A mast antenna consisting of a 50 meter long vertical conductor operates over a perfectly conducting ground plane. It is base-fed at a frequency of 600 kHz. The radiation resistance of the antenna in Ohms is (GATE 2006)
A. $2\pi^2/5$ B. $\pi^2/5$
C. $4\lambda^2/5$ D. $20\pi^2$

143. For a Hertz dipole antenna, the half power beam width (HPBW) in the *E*-plane is (GATE 2008)
A. 360° B. 180°
C. 90° D. 45°

144. At 20 GHz, the gain of a parabolic dish antenna of 1 meter diameter and 70% efficiency is (GATE 2008)
A. 15 dB B. 25 dB
C. 35 dB D. 45 dB

145. A $\lambda/2$ dipole is kept horizontally at a height of $\lambda_0/2$ above a perfectly conducting infinite ground plane. The radiation pattern in the plane of the dipole (*E* plane) looks approximately as (GATE 2007)

A.
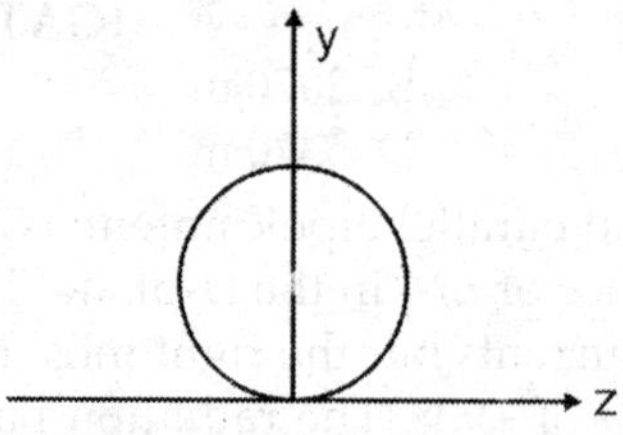

B.
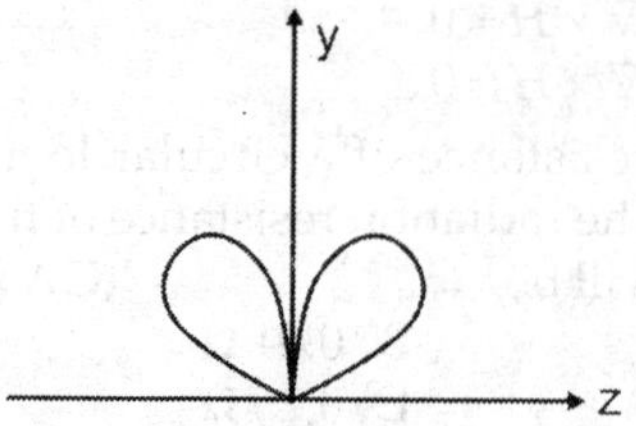

C.
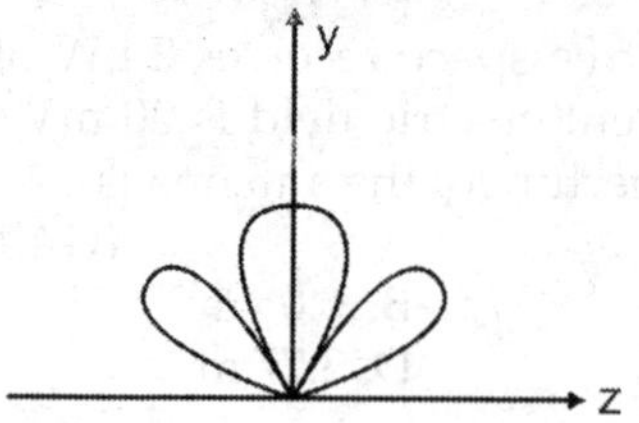

D.
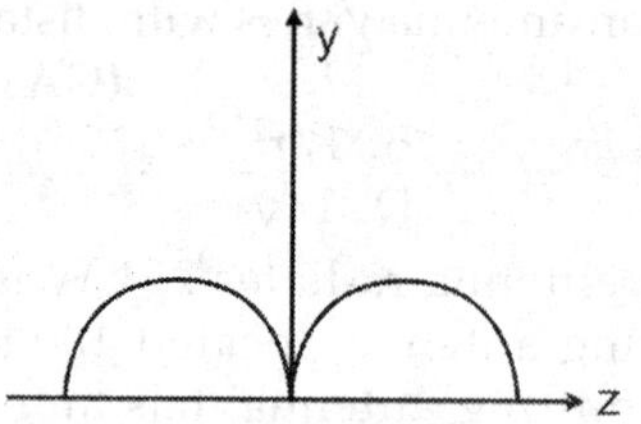

146. The radiation pattern of an antenna in spherical coordinates is given by $F(\theta) = \cos^4\theta$, $0 \leq \theta \leq \pi/2$. The directivity of the antenna is (GATE 2012)
A. 10 dB B. 12.6 dB
C. 11.5 dB D. 18 dB

147. Match column A with column B. (GATE 2014)

Column A	Column B
1. Point electromagnetic source	P. Highly directional
2. Dish antenna	Q. End fire
3. Yagi Uda antenna	R. Isotropic

P Q R P Q R
A. 1 2 3 B. 2 3 1
C. 2 1 3 D. 3 2 1

148. The directivity of an antenna array can be increased by adding more antenna elements, as a larger number of elements (GATE 2015)
A. improves the radiation efficiency
B. increases the effective area of the antenna
C. results in a better impedance matching
D. allows more power to be transmitted by the antenna

Ans. 140. B 141. B 142. A 143. C 144. D 145. B 146. A 147. B 148. B

CHAPTER **18**

Computer and Microprocessor

The computer: Before the launch of microprocessor, a computer was designed as shown in Fig. 18.1. The computer is divided into four components namely: (i) Memory (ii) Input devices (iii) Output devices and (iv) CPU (central processing unit). The CPU contains registers to store data, the ALU which performs arithmatic and logic operations on data, the instruction decodes, counters circuits and control lines. The CPU reads instructions from memory and perform the operation specified. Its job also includes communication with input and devices so as to accept or send data.

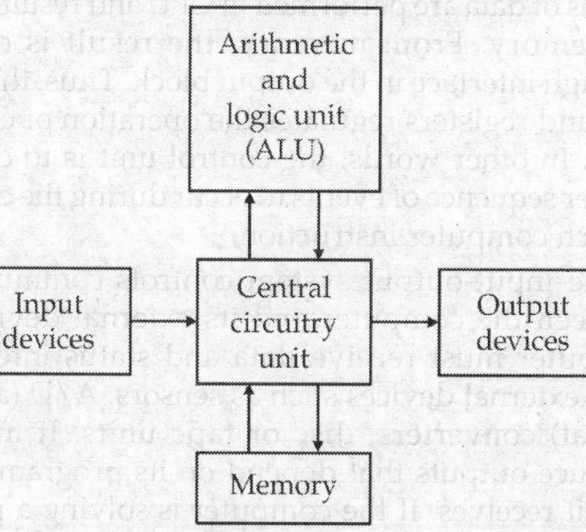

Fig. 18.1

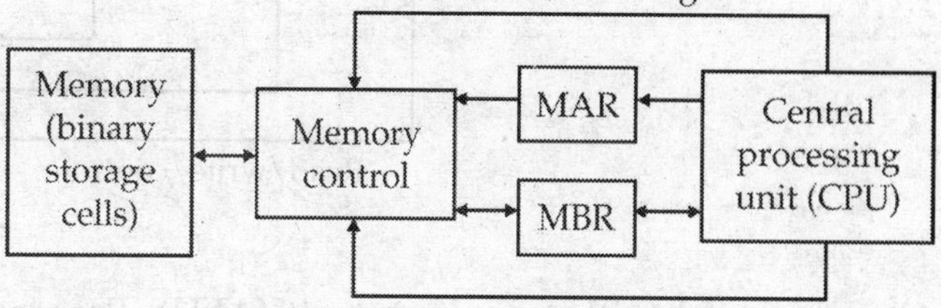

Fig. 18.2: Control signal

Basic organisation: As shown in Fig. 18.2, the memory unit has an assembly of binary cells, a control section and a set of registers. There are two registers, namely, MAR and MBR, and two signal paths. The memory address register (MAR) holds the address of the word in memory to be accessed. The operation to be performed, namely, reading from or writing into the memory is specified by the read/write signal. The data read from the memory in the case write operation are held in the memory buffer register (MBR). Memory access operation is initiated by the presence of the control signal. It is the task of CPU (central processing unit) to load and clear these registers appropriately. The memory control unit decodes the address contained in MAR and routes the control signal to the desired location in the memory.

Instruction word: An instruction word is a word having two sections—the *op-code* section and *address* section. In some cases, another section comes in between to show the number of the register [general purpose register (GPR)] to be used in CPU for carrying out the operations, as in the case of SMAC—B computer. This is shown in Fig. 18.3.

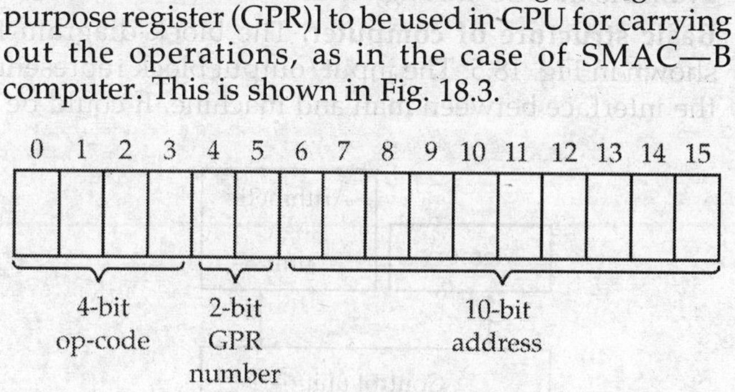

Fig. 18.3

Op-code represents the code for the operation to be performed and 'Address Code' indicates the location of the data. Sometimes it is also known as *operand.*

CPU organisation: The control and processing unit shown in Fig. 18.4, enables a computer to execute the instructions. It consists of a number of registers,

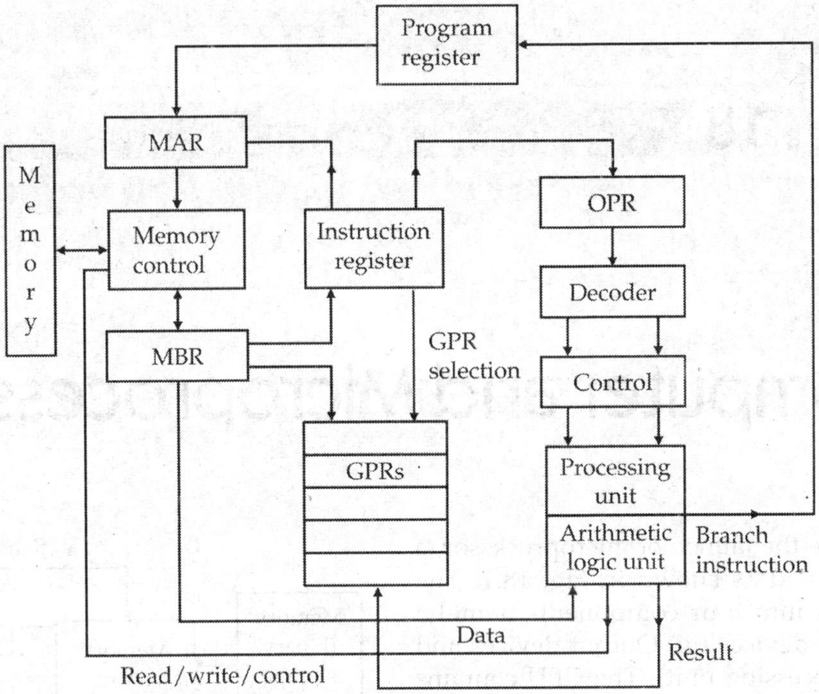

Fig. 18.4

decoders and arithmetic logic unit (ALU). Program register holds the address (in memory) of the instruction to be executed next. Instruction register contains the instruction word being executed. The OP-code part of the instruction word is stored in OP-code register (OPR). Instruction register also sends signal for GPR selection. The selected GPR store intermediate results obtained during arithmetic operation. According to the op-code decoder, control signal generator controls the ALU processing unit. The processing unit also sends result to be stored in GPR. It also provides control signal for memory control so that the desired data could be available in ALU through MBR.

Basic structure of computer: The block diagram is shown in Fig. 18.5. The input/output block represents the interface between man and machine. It could be a tele-type unit or a punched paper tape or punched unit record-cards or a magnetic type. The input data are taken into the system and stored in the memory according to the appropriate signal generated by control block. The computations and other manipulations of data are performed in CPU and result is stored in memory. From memory, the result is obtained through interface in the output block. Thus, the control unit and registers regulates the operation of computer itself. In other words, the control unit is to cause the proper sequence of events to occur during the execution of each computer instruction.

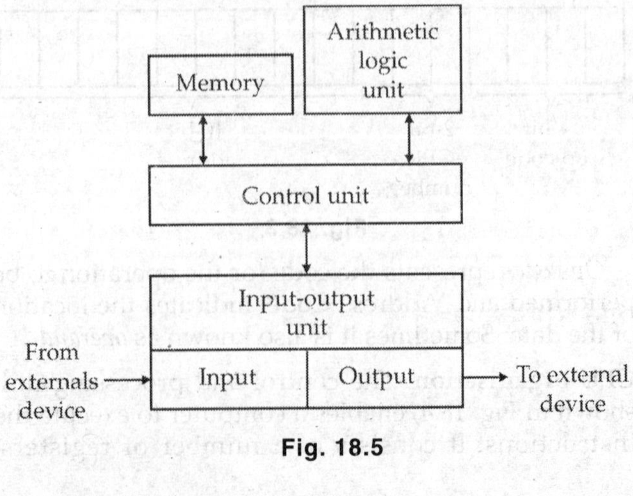

Fig. 18.5

The input-output system controls communication between the computer and its external devices. The computer must receive data and status information from external devices such as sensors, A/D (analog to digital) converters, disc or tape units. It must also produce outputs that depend on its program and the data it receives. If the computer is solving a problem, the output is usually a printout of the result. In this case, it must issue commands to a printer to cause the results to be printed for the user to read. If the function of the computer is to control a physical process, however, its output will be electric pulses that are translated into COMMANDS (to open a valve, reduce the air flow, etc.) and regulate a process.

The huge advantage of microprocessor based control system is that these procedures can be absolutely standardised and not subject to error or variation from operator to operator. Another advantage is the greatest reliability and flexible architecture. The separately available CPUs, RAMs, ROMs, PROMs, PIAs and other interfaces give the designer full freedom and flexibility in design. Also, the other advantage is the uniformity of reporting format usually of microprocessors.

The 8-bit microprocessor has undoubtedly caught on it fits many of the controller and medium sized data handing jobs that formerly went by default to mini-computer designs. But the 8-bit word length can be a handicap for large systems, where big bytes of memory must be processed or in high performing data-acquisition systems, where speed and high resolution are needed. This is where the 16-bit microprocessor comes in, its 16-bit words reach external memory locations two bytes at a time, while its long words can easily accommodate the 10-, 12- and 14-bit converter resolution that is standard for most of the systems.

8085 Microprocessor: The 8085 is an 8-bit general purpose microprocessor, capable of addressing 64 k of memory. The device has 40 pins, requires a +5 V single power supply and can operate with 3 MHz single phase clock. All the pins can be classified into six groups:

1. Address bus
2. Data bus
3. Control and status signals
4. Power supply and frequency signals
5. Externally initiated signals
6. Signal input/output ports.

Application of microprocessors: The typical applications of microprocessors are as follows:

1. Microprocessor based control of firing circuit of thyristors.
2. Microprocessor based process controllers such as motor controllers (dc and ac) temperature monitoring, level controllers, controller for reversing drives, and pulse width controllers, etc.
3. Musical instruments such as synthesizers.
4. Medical instruments such as stress ecocardiogram, ultrasound and blood analysers.
5. Instruments used in the defence such as RADAR and missiles, etc.
6. Computerised eye testing equipments.
7. High performance industrial drives such as rolling mills and paper mills.

Basics of microprocessors and microcomputers: A microcomputer is a bus-oriented control unit using several LSI (large-scale integration) chips. Figure 18.6 shows the basic blocks in a microcomputer structure.

Microprocessor (μP): The heart of the microcomputer is the microprocessor, which performs calculations and controls various functions. This central chip is also known as the *central processing unit* (CPU) or *microprocessing unit* (MPU). The microprocessor is the most important development in the electronics industry. This chip was introduced to meet the need for a universal large-scale integrated (LSI) circuit. Before the microprocessor, the LSI chip became more dedicated to a particular application.

The microprocessor has the ability to perform a wide variety of functions. It carries out an operation by

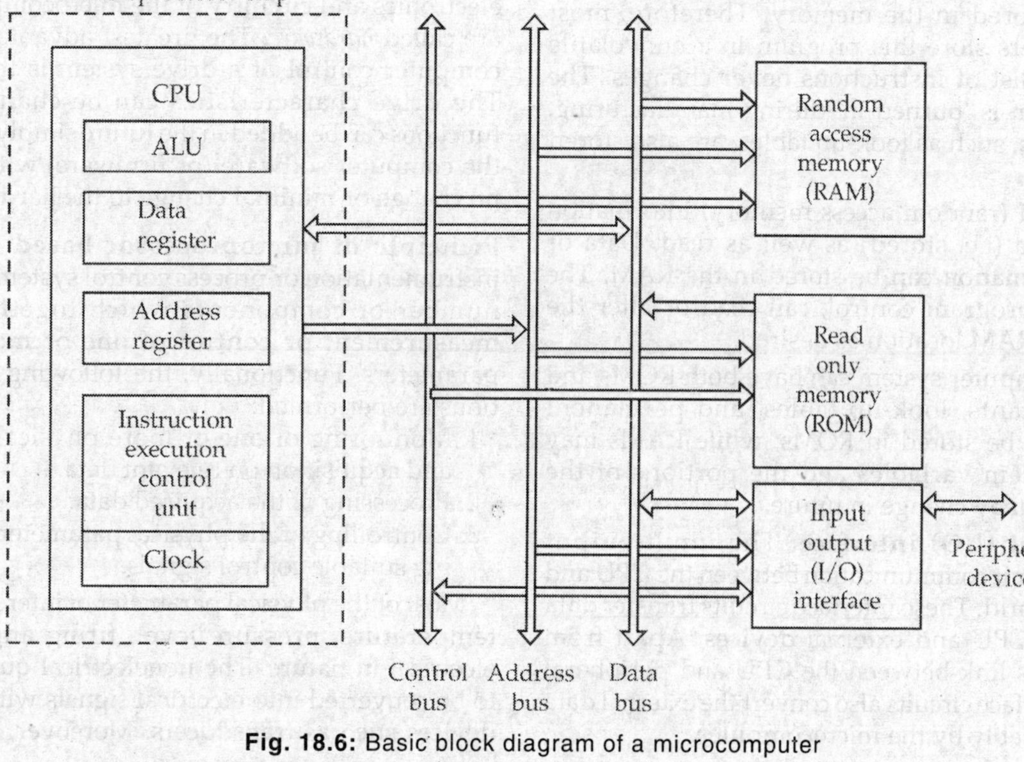

Fig. 18.6: Basic block diagram of a microcomputer

executing as sequence of instructions called a software program that are stored in a memory connected to the microprocessor.

The various components within the microprocessor are shown in the simplified block diagram of a microprocessor (μP) chip in Fig. 18.6. The arithmetic logic unit (ALU) is the processing unit where all logical and arithmetic operations such as addition, subtraction, and bit manipulation are performed. The data registers are used for intermediate data manipulation and storage, thereby cutting down the number of transfers from the memory. They also receive and transfer data into and out of the CPU. The address registers are used for storing memory addresses and are used in conjunction with the data registers for data transfer from the memory of the input/output interface. The control unit controls and supervises the correct execution of instructions. The clock, or CPU timebase generator, is contained in the control unit. The clock frequency determines the basic operating speed of the microprocessor.

Memory ROM and RAM: The memory is a unit where instructions and data are stored. The memory contains a large number of locations or cells. The size of the memory is equal to the number of locations in the memory.

In the ROM (read only memory) instructions are stored permanently. They can only be read and cannot be destroyed. Most microcomputers are used for only one purpose. This dedicated use means that only a limited number of programs (i.e. sets of instructions) need to be stored in the memory. Therefore, most microcomputers store this program in a nonvolatile ROM, as the list of instructions never changes. The actual program is 'burned in' during manufacturing. Data constants, such as look-up tables, are also stored in ROMs.

In the RAM (random access memory) information can be written (i.e. stored) as well as read. Data or variable information can be stored in the RAM. The CPU, under program control, can read or alter the contents of a RAM location as desired.

A microcomputer system can have both ROMs and RAMs. Constants, look-up tables, and permanent programs can be stored in ROMs, while RAMs may contain problem variables and the portions of the program that may change in future.

Input-Output (I/O) interface: The input-output interfaces permit communication between the CPU and the outside world. These interface circuits transfer data between the CPU and external devices. Apart from providing this link between the CPU and peripheral units, the interface circuits also convert the external data into a form usably by the microcomputer.

Data bus, address bus, and control bus: Microcomputers are bus-oriented control units. Information is communicated among various parts of the system through the data bus, the address bus, and the control bus. In a microcomputer system, using 8-bit microprocessors, there are typically 8 lines for the data bus, 16 lines for the address bus, and 6 lines for the control bus.

Program or computer software: The microprocessor can carry out various tasks only if it is provided with a set of instructions, called a *program*. The microprocessor, however, understands only binary codes and operates on binary instructions of 'ones' and 'zeros'. Writing a program in binary codes is very tedious and prone to error. A programmer, therefore, uses a convenient language to write the instructions. The most basic languages for a microprocessor is called *assembly language* in which symbolic (or *mnemonic*) codes are used. Every microprocessor has its own assembly language. In this symbolic coding system, an instruction is represented by a group of three or four letters chosen so as to suggest the function of the instruction. The symbolic coding of instructions cannot be understood by the microprocessor, which acts only on binary information. A program, called assembler program, converts the symbolic program or the source program into a *binary program*, also called the object program, which the microprocessor can execute.

If the program is changeable and re-loadable, it is called *software*. The program that is not changeable by the user and is held in ROM is known as *firmware*. The electronics and circuitry of the microcomputer system are called *hardware*. The greatest advantage of microcomputer control of a drive system is the flexibility. The drive characteristics can be changed or new functions can be added in the future simply by changing the computer software (or firmware) with practically no change or minimal change to the hardware.

Principle of microprocessor based system: An instrumentation or process control system comprises a number of components which together perform measurement or control of one or more physical parameters. Functionally, the following three operations are performed:

1. Monitoring of one or more physical parameters and acquisition on relevant data.
2. Processing of the acquired data.
3. Controlling of the physical parameters by generating suitable control signals.

Most of the physical parameter of interest, i.e. speed, temperature, pressure level, firing angle are non-electrical in nature. The nonelectrical quantities have to be converted into electrical signals with the help of devices such as transducers. Moreover, the electrical signals generated by the transducers are usually analog in form. Therefore, these analog electrical signals have to be converted into digital form. Extra hardware known as *analog to digital converters* (ADC) have to be used for this purpose. In many situations, process control variables are monitored by using opto-isolators, so that the noise and transients of an industrial environment do not disturb the processing elements.

On the other hand, analog signals have to be generated to control various physical parameters. The hardware which is used to convert digital data into equivalent analog form is known as the *digital to analog converter* (DAC). In many situations, relays have to be used to control process variables of high voltage or currents.

To analyze an instrumentation, three major elements are required, namely:

1. Input devices
2. Signal processing devices
3. Output devices

The function of signal processing devices is performed by MPU and necessary input and output devices are to be interfaced to it. Block diagram of a typical process control system is shown in Fig. 18.7.

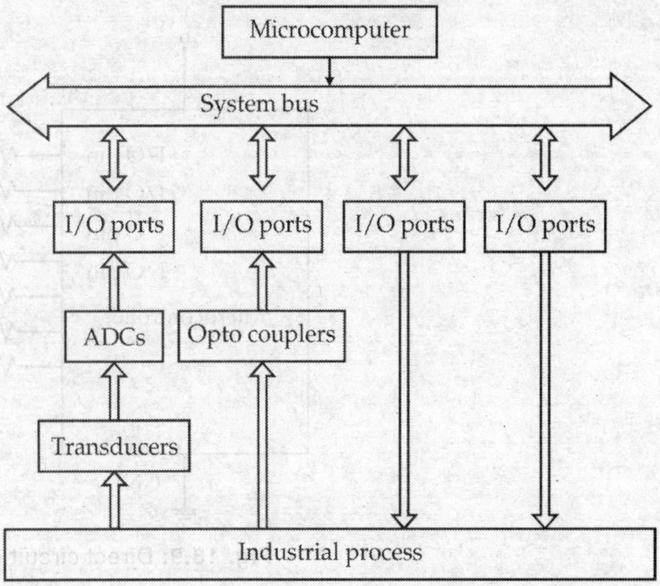

Fig. 18.7: Basic block diagram of process control system

Interfacing seven-segment displays: Seven-segment displays commonly contains LED segments arranged as a figure-of-eight pattern, with one common led (anode or cathode) and seven individual leads for each segment. When the common lead is the anode, it is referred to as common anode. Figure 18.8 show one of the possible configurations of interfacing a GC display with the microcontroller. The IC CD 4511 is a BCD to seven-segment decoder/driver. The microcontroller feeds the BCD equivalent of the digit to the display to the 4511 IC.

Seven-segment displays can also be connected directly without the use of a BCD to seven-segment decoder. In this case the seven-segment code of the digit is generated by the microcontroller program itself. Figure 18.9 shows the direct circuit connection for CA display.

If more than one display is to be used, the displays are time multiplexed. The human eye cannot detect the blinking display if each display is relit every 10 ms or so. The 10 ms time is divided by the number of displays used to find the interval between updating each display. In the case of GC display, the display is selected by driving the common cathode to logic LOW, in the case of CA display, the display is selected by driving the

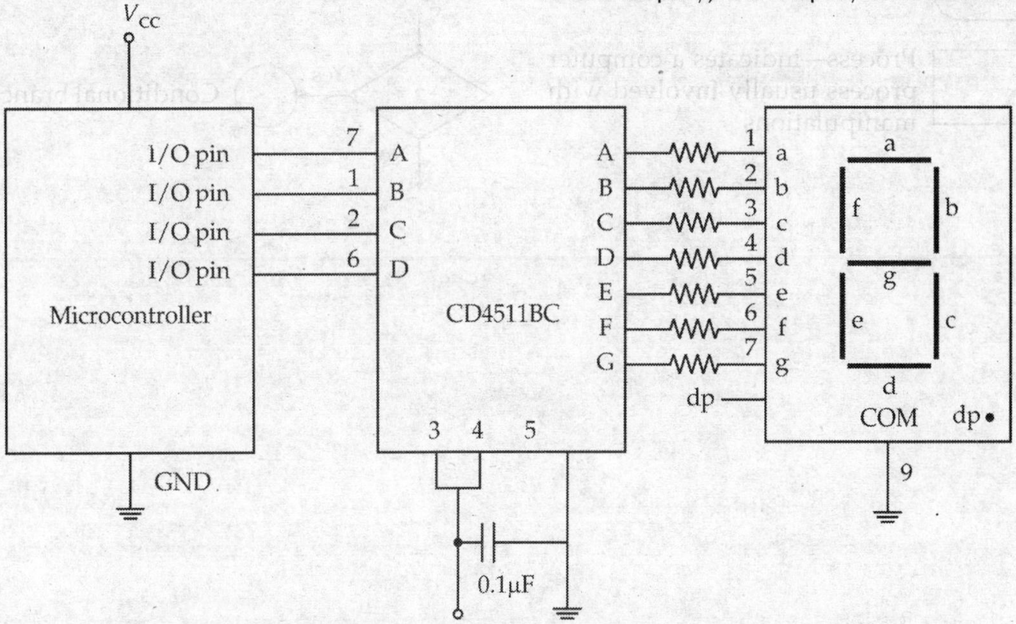

Fig. 18.8: Possible configuration of interfacing a GC display with a microcontroller

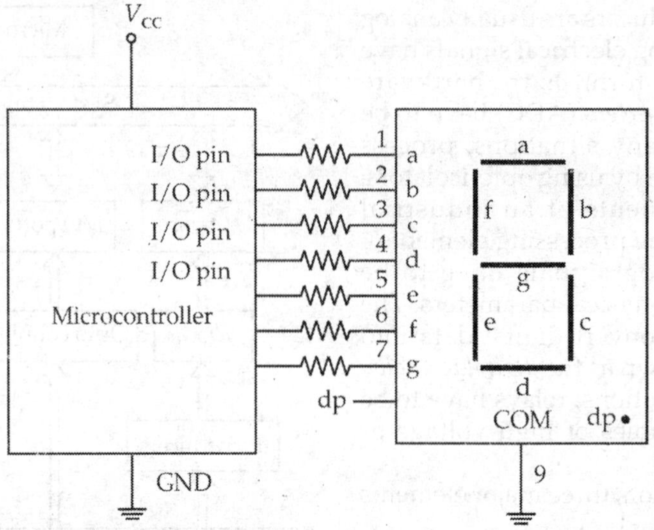

Fig. 18.9: Direct circuit connection for CA display

There are some standard symbols, which are shown below.

Flow line indicates sequence direction

Input/output—indicates the reception and transmission of information into and out of the system

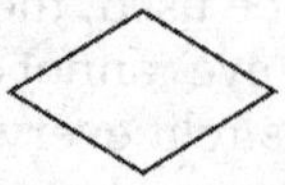
Decision—indicates a point of logical comparison. YES or NO usually is the answer of comparison

Connector

Terminator

Process—indicates a computer process usually involved with manipulations

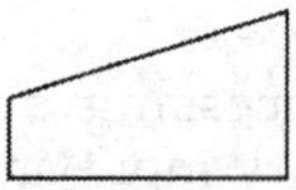
Magnetic tape I/O

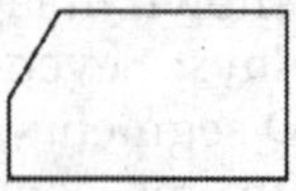
Card I/O

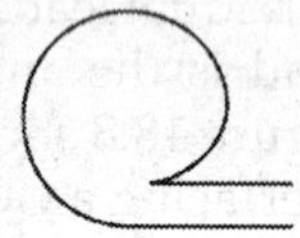
Manual I/O

Punch tape

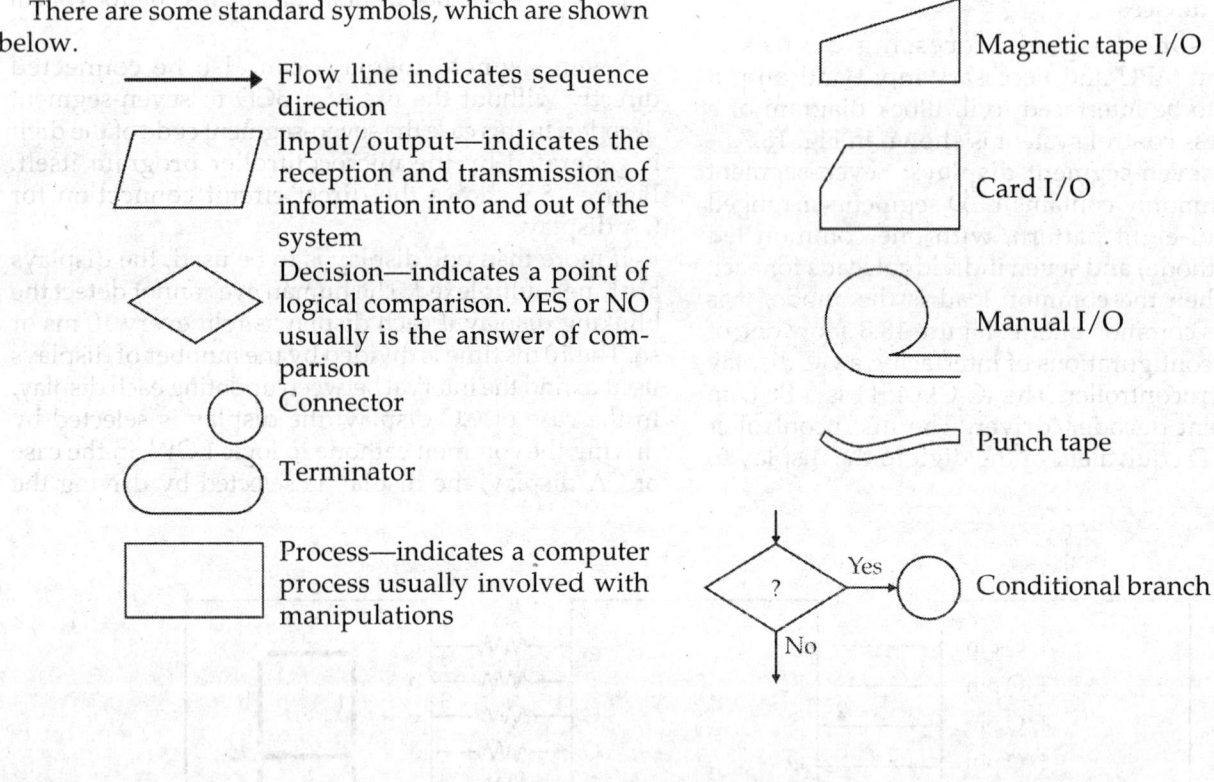

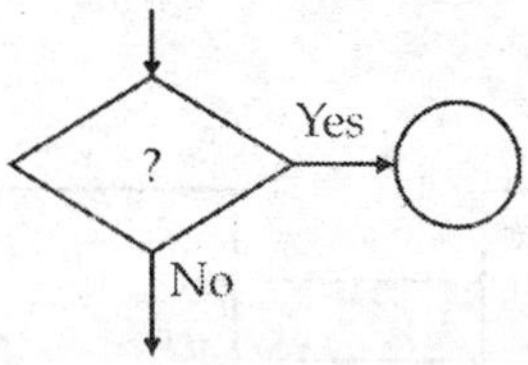

Conditional branch

MULTIPLE CHOICE QUESTIONS

1. Where does a computer add and compare data?
   A. Hard disk B. Floppy disk
   C. CPU chip D. Memory chip
2. Which generations of computers is covered by the period 1965–77?
   A. First B. Second
   C. Third D. Fourth
3. Computer cannot do anything without a
   A. chip B. memory
   C. output device D. program
4. The word 'computer' usually refers to the central processor unit plus
   A. external memory B. internal memory
   C. input devices D. output devices
5. A computer is a box full of electronic
   A. switching devices B. chips
   C. circuits D. registers
6. Which of the following belongs to the first generation of computer?
   A. ENLAC B. IBM 1401
   C. IBM 8090 D. UNIVAC
7. What hardware was used in the first generation computers?
   A. Transistors B. Valves
   C. VLSI D. ICs
8. Processors of all computers either micro, mini or mainframe must have
   A. ALU B. primary storage
   C. control unit D. all of these
9. A collection of eight bits is called
   A. byte B. word
   C. record D. hill
10. The basic components of a modern digital computer are
    A. input device B. output device
    C. central processor D. all of these
11. Which of the following is a part of the central processing unit?
    A. Printer B. Keyboard
    C. Mouse D. Arithmetic Logic Unit
12. Control unit of a digital computer is often called the
    A. clock B. nerve center
    C. ICs D. all of these
13. A typical modern computer uses
    A. LSI chips B. vacuum tubes
    C. valves D. all of these
14. An integrated circuit (IC) is
    A. fabricated on a tiny silicon chip
    B. a complicated circuit
    C. much costlier than a single transistor
    D. an integrating device
15. A computer program consists of
    A. system flowchart B. program flowchart
    C. algorithms written in computer's language
    D. discrete logical steps
16. The term "memory" applies to which one of the following?
    A. Logic B. Storage
    C. Control D. Input device
17. Which device can produce the final product of machine processing into a form usable by humans?
    A. Storage B. Input device
    C. Output device D. Control
18. Which can read data and convert them into a form that a computer can use?
    A. Logic B. Storage
    C. Control D. Input device
19. Which kind of hardware is used the most in the input phase of a computer-based information system?
    A. Keyboard B. Printer
    C. Monitor D. Hard disk
20. Which kind of device allows the user to add components and capabilities to a computer system?
    A. System boards B. Storage devices
    C. Input devices D. Expansion slots
21. Which kind of storage device can be carried around?
    A. Floppy disk B. Hard disk
    C. System cabinet D. Hard disk drive
22. Which of the following persons probably has the least amount of technical knowledge?
    A. User B. Computer operator
    C. System analyst D. Programmer
22. Which is the most powerful type of computers?
    A. Microcomputer B. Minicomputer
    C. Mainframe computer D. Supercomputer
24. Which is not true for primary storage?
    A. Information must be transferred to primary storage
    B. It is relatively more expansive
    C. It allows very fast access to data
    D. All of these
25. Which is not a factor when categorising a computer?
    A. Speed of the output device
    B. Amount of main memory the CPU can use

Ans.												
1. C	2. C	3. D	4. B	5. A	6. D	7. B	8. D	9. A	10. D	11. D	12. B	13. A
14. A	15. C	16. A	17. C	18. D	19. A	20. D	21. A	22. A	23. D	24. D		

C. Cost of the system
D. Where it was purchased

26. The most common input device used today is
A. motherboard B. track ball
C. scanner D. keyboard

27. What is the control unit's function in the CPU?
A. To transfer data to primary storage
B. To store program instructions
C. To perform logic operations
D. To decode program instructions

28. The ascending order of a data Hierarchy is
A. bit - bytes - field - record - file - data-base
B. bit - bytes - record - field - file - data-base
C. bytes - bit - field - record - file - data-base
D. bytes - bit - record - field - file - data-base

29. Which is not an input device?
A. CRT (cathode ray tube)
B. Optical scanners
C. Voice recognition devices
D. COM (Computer Output Microfilm)

30. The central processing unit (CPU) consists of
A. input, output and processing
B. control unit, primary storage, and secondary storage
C. control unit, arithmetic-logic unit and primary storage
D. control unit, processing, primary storage

31. Where was the India's first computer installed and when?
A. Institute of Social Science, Agra, 1955
B. Indian Institute of Statistics, Delhi, 1957
C. Indian Statistical Institute, Calcutta, 1955
D. Indian Institute of Science, Bangalore, 1971

32. The first electronic digital computer contained
A. electronic valves B. vacuum tubes
C. transistors
D. semiconductor memory

33. The first computer made available for commercial use was
A. Mark-I B. ENIAC
C. EDSAC D. UNIVAC

34. The first computer used to store a program is
A. EDSAC B. ENIAC
C. EDVAC D. ACE

35. Who is called the "grandfather" of the computer?
A. Blaise Pascal B. Charles Babbage
C. Joseph Jacquard D. Dr Herman Hollerith

36. The first mechanical computer designed by Charles Babbage was called
A. abacus B. processor
C. calculator D. analytical engine

37. The fifth generation digital computer will be
A. extremely low cost B. very expensive
C. versatility D. artificial intelligence

38. Hard disks and diskettes are
A. direct access storage devices
B. sequential access storage devices
C. rarely used with microcomputers
D. both A and C are correct

39. The transistorized computer circuits were introduced in the
A. first generation B. second generation
C. third generation D. fourth generation

40. The third generation of computers covers the period
A. 1971–1982 B. 1982–1994
C. 1959–1964 D. 1965–1971

41. When did the ELSA consortium 'Gang of Nine' come into being in USA?
A. 1988 B. 1986
C. 1984 D. 1982

42. A six-digit card field used for postal ZIP codes is defined as a/an
A. letter field B. alphabetic field
C. numeric field D. alphanumeric field

43. The decreased cost and increased performance of computer hardware were the distinguishing features of which generation of computers?
A. First and second B. Second and third
C. Both A and B are correct
D. All generations

44. The first machine to successfully perform a long service or arithmetic and logical operations was
A. EDSAC B. UNIVAC-I
C. Mark-I D. ENLAC

45. Which is true for the digital computer?
A. It is less accurate than the analog computer
B. It represents the decimal numbers through a string of binary digits
C. It is used primarily in scientific applications
D. Both A and B are correct

46. Primary storage is .... as compared to secondary storage.
A. slow and inexpensive B. fast and inexpensive
C. fast and expensive D. slow and expensive

47. Which of the following is an example of nonvolatile memory?
A. ROM B. VLSI
C. LSI D. RAM

48. The most popular secondary storage today is
A. magnetic tape B. floppy disk
C. mass storage D. semiconductor

49. Which is widely used in academic testing?
A. MICR B. POS
C. OCR D. OMR

50. Linkage between the CPU and the users is provided by
A. storage B. control unit
C. peripheral devices D. software

51. EBCDIC can code up to how many different characters?

Ans.													
	25. D	26. D	27. D	28. A	29. D	30. C	31. C	32. A	33. D	34. A	35. B	36. D	37. D
	38. A	39. B	40. D	41. A	42. C	43. D	44. C	45. B	46. C	47. A	48. B	49. D	50. C

A. 8 B. 16
C. 32 D. 256

52. Magnetic tape can serve as
A. input media B. output media
C. secondary storage media
D. all of these

53. Which is considered a direct entry input device?
A. Optical scanner B. Mouse
C. Light pen D. All of these

54. What is required when more than one person uses a central computer at the same time?
A. Light pen B. Mouse
C. Digitizer D. Terminal

55. Which kind of terminals is entirely dependent for all its capabilities on the computer system to which it is connected?
A. Smart terminal B. Dumb terminal
C. Intelligent terminal D. Microcomputer

56. Which is a unit of measurement used with computer systems?
A. Byte B. Kilobyte
C. Megabyte D. All of these

57. The microcomputers today have hard disk because it
A. can be moved easily from one computer to another
B. has very high storage capacity
C. is expensive D. all of these

58. Who was the father of Punched Card Processing?
A. J Presper Eckert B. Charles Babbage
C. Dr Herman Hollerith D. Blaise Pascal

59. The two main components of the CPU are
A. control unit and registers
B. control unit and ALU
C. registers and main memory
D. ALU and bus

60. Which is used for manufacturing chips?
A. Bus B. Control unit
C. Semiconductors
D. Both A and B are correct

61. Large computer system typically use
A. line printers B. ink-jet printers
C. dot-matrix printers D. daisy wheel printers

62. The computer code for the interchange of information between terminals is
A. ASCII B. BCD
C. EBCDIC D. all of these

63. The data recording area between the blank gaps on magnetic tape is called a/an
A. record B. block
C. field D. database

64. A hard copy would be prepared on a
A. line printer B. dot matrix printer
C. typewriter terminal D. all of these

65. Which is a secondary memory device?
A. CPU B. ALU
C. Floppy disk D. Mouse

66. Two kinds of main memory are
A. ROM and RAM B. direct and sequential
C. floppy disk and hard disk
D. primary and secondary

67. RAM is used as a short memory because it
A. is volatile B. is very expensive
C. has small capacity D. is programmable

68. A hybrid computer resembles
A. digital computer B. analog computer
C. both a digital and an analog computer
D. none of these

69. Binary number need more places for counting because
A. 0's and 1's can be added in front of them
B. 0's and 1's have to be properly placed
C. they are always big numbers
D. binary base is small

70. The CPU chip used in a computer is partially made out of
A. copper B. iron
C. gold and silver D. silica

71. The silicon chips used for data processing are called
A. RAM chips B. ROM chips
C. Microprocessor D. PROM chips

72. The hardware in which data may be stored for a computer system is called
A. registers B. bus
C. control unit D. memory

73. Which is the computer memory that does
A. ROM B. RAM
C. PROM D. all of these

74. The computer memory used for temporary storage of data and program is called
A. ROM B. RAM
C. EROM D. EPROM

75. Which computer memory is essentially empty?
A. RAM B. ROM
C. EPROM D. PROM

76. The most common binary code in use today is the 8-bit ASCII code. What do the letters ASCII stands for?
A. American Standard Code for International Interchange
B. American Standard Code for Information Interchange
C. American Standard Code for Intelligence Interchange
D. American Scientific Code for Information Interchange

77. Which of the following is not an output device of a computer?

Ans.													
	51. D	52. D	53. D	54. D	55. B	56. D	57. B	58. C	59. B	60. C	61. A	62. A	63. B
	64. D	65. C	66. A	67. A	68. C	69. D	70. D	71. C	72. D	73. A	74. B	75. B	76. B

A. Printer B. Keyboard
C. VDU D. CRT screen

78. A monitor looks like a TV set but it does not
A. receive TV signals B. give a clear picture
C. give a steady picture D. display graphics

79. The metal disks which are permanently housed in sealed and contamination free containers are called
A. hard disk B. floppy disk
C. winchester disk D. flexible disk

80. A half byte is known as
A. data B. bit
C. half byte D. nibble

81. Everything in computer is controlled by its
A. RAM B. ROM
C. CPU D. storage devices

82. An error in computer data is called
A. chip B. bug
C. bit D. byte

83. Which of the following memory is capable of operating at electronics speed?
A. Semiconductor memory
B. Magnetic disks
C. Magnetic drums D. Magnetic tapes

84. Which of the following is not used as secondary storage?
A. Semiconductor memory
B. Magnetic disks
C. Magnetic drums D. Magnetic tapes

85. A computer consists of
A. a central processing unit
B. input and output units
C. a memory D. all of these

86. The heart of any computer is the
A. CPU B. Memory
C. I/O unit D. Disks

87. Which of the following is not an advantage of stored programs?
A. Reliability
B. Reduction in operational costs
C. Ability of the computer to operate at electronic speeds
D. The computers becoming general purpose

88. Which of the following is not true of immediate processing?
A. It is often used in realtime applications
B. It can occur with either sequential or direct-access files
C. It can be used in an airline-reservation system
D. Transactions are processed shortly after a real-world event occurs

89. Which is the product of data processing?
A. Data B. Information
C. Software D. A computer

90. Software instructions intended to satisfy a user's specific processing needs are called
A. systems software B. a microcomputer
C. documentation D. applications software

91. People typically interface with a computer-based system when
A. information must be output
B. information must be reviewed
C. data must be input D. all of these

92. Which of the following statements best describes the batch method of input?
A. Data is processed as soon as it is input
B. Data is input at the time it is collected
C. Data is collected in the form of source, documents, placed into groups and then input to the computer
D. Source documents are not used

93. Which of the following terms applies to communication between separate computer systems?
A. Computer literacy B. Power supply
C. Applications software
D. Connectivity

94. The following typically happens in the output phase of a computer-based information system.
A. Data is put into the computer for processing
B. Information is produced in hardcopy and/or softcopy form
C. Mathematical calculations are performed
D. The computer is turned-off

95. Which of the following best describes a computer-based information system?
A. Processing data B. Input data
C. A system in which a computer is used to turn data into information
D. Performing complex mathematical calculations

96. One thousand bytes represents a
A. megabyte B. gigabyte
C. kilobyte D. all of these

97. Super computers are primarily useful for
A. input-output intensive processing
B. data-retrieval operations
C. mathematical-intensive scientific applications
D. all of these

98. A transaction file is a type of
A. master file B. data file
C. access method D. all of these

99. Which of the following is a computer code?
A. EPROM B. FAT
C. EBCDIC D. All of these

100. First generation computers are characterised by
A. vacuum tube and magnetic drum
B. magnetic tape and transistors
C. microcomputers D. all of these

Ans.	77. B	78. A	79. C	80. D	81. C	82. B	83. A	84. A	85. D	86. A	87. B	88. B	89. B
	90. D	91. D	92. C	93. D	94. B	95. C	96. C	97. C	98. B	99. C	100. A		

101. The Octal equivalent of hexadecimal number 3DE is
A. 1736 B. 3176
C. 1037 D. all of these

102. The binary equivalent of the Octal number 13.54 is
A. 1011.1011 B. 1101.1110
C. 1001.1110 D. all of these

103. A Winchester disk is a
A. disk stack B. removable disk
C. flexible disk D. all of these

104. An application's program that helps the user to change any number and immediately see the result of that change is
A. desktop publishing program
B. database
C. spreadsheet D. all of these

105. Which of the following holds the ROM, CPU, RAM and expansion cards?
A. Hard disk B. Cache memory
C. Motherboard D. All of these

106. The instructions for starting the computer are housed on
A. random-access memory
B. read-only memory chips
C. CD-ROM D. all of these

107. Which of the following is responsible for coordinating various operations using timing signals?
A. Control unit B. Memory unit
C. Arithmetic-logic unit D. Input/Output unit

108. Which of the following is the fastest?
A. CPU B. Video terminal
C. Magnetic tapes and disks
D. Sensors, mechanical controllers

109. The least significant bit of the binary number which is equivalent to any odd decimal number is
A. 0 B. 1
C. 1 or 0 D. 3

110. The decimal equivalent of octal number 56 is
A. 46 B. 66
C. 53 D. 49

111. The decimal equivalent of binary number 0.0111 is
A. 4.375 B. 0.4375
C. 0.5375 D. −0.4375

112. The binary code of $(21\ 125)_{10}$ is
A. 10101.001 B. 10100.001
C. 10101.010 D. 10100.100

113. The binary code of $(73)_{10}$ is
A. 1010001 B. 1000100
C. 1100101 D. 1001001

114. The range of numbers which can be stored in an eight bit register (with left bit of sign) is
A. −128 to +127 B. −128 to +128
C. −99999999 to +99999999
D. can't be determined without knowing the details of the computers

115. Which of the following is true of the digital computer?
A. It represents the decimal numbers through a string of binary digits
B. It is used primarily in scientific applications
C. It is less accurate than the analog computer
D. All of these

116. Which of the following is not true of primary storage?
A. It represents the decimal numbers through a string of binary digits
B. It stores operating-system programs
C. It stores data while they are being processed by the CPU
D. It stores the bulk of data used by a computer application

117. Compared with secondary storage, primary storage is
A. slow and inexpensive B. fast and inexpensive
C. fast and expensive D. slow and expensive

118. The CPU can perform read or write operations at any point in time is
A. ROM B. PROM
C. EPROM D. RAM

119. The technique of placing software or programs in a ROM semiconductor chip is called
A. PROM B. EPROM
C. firm ware D. microprocessor

120. The advantages of magnetic tape include all of the following except
A. low cost B. compact and portable
C. direct-access storage medium
D. highly reliable

121. Currently the most popular form of secondary storage is
A. magnetic tape B. semiconductor
C. magnetic core D. disk

122. Which of the following is an example of nonvolatile memory?
A. ROM B. RAM
C. LSI D. VLSI

123. Which can be programmed one time by either the manufacturer or the computer user? Once programmed, it cannot be modified.
A. RAM B. ROM
C. LSI D. VLSI

124. Which of the following is not true of a magnetic disk?
A. Users can easily update records by writing over the old data
B. It provides only sequential access to stored data
C. It is expensive relative to magnetic tape
D. It does not provide an automatic audit trail

Ans. 101. A 102. A 103. A 104. C 105. C 106. B 107. A 108. A 109. A 110. B 111. B 112. A 113. D 114. A 115. A 116. D 117. C 118. D 119. C 120. C 121. D 122. A 123. C 124. B

125. Which of the following is not true of punched cards as data-entry media?
A. They can be used as turn-around documents
B. They are inexpensive
C. Input is slow compared with other media
D. They are easily damaged

126. The primary advantage of key-to-tape data-entry systems is that
A. a large percentage of editing can be performed at the time of data entry
B. key verification is easily performed
C. the tape is reusable
D. keying errors can be detected as they occur

127. Linkage between the CPU and the users is provided by
A. peripheral devices B. storages
C. control unit D. software

128. Which of the following is widely used in academic testing?
A. MICR B. POS
C. OCR D. OMR

129. Which printer is a non-impact printer that can produce very high quality, letter-perfect printing?
A. Dot-matrix printer B. Daisy-wheel printer
C. Electrostatic printer D. Laser printer

130. The POS data-entry system is used most extensively by the
A. banking industry B. grocery industry
C. railroad industry
D. word-processing industry

131. A disadvantage of the laser printer is that
A. it is quieter than an impact printer
B. the output is of a lower quality
C. it is very slow D. none of these

132. Data-entry can be performed with all of the following except
A. OCR B. OMR
C. COM
D. Voice-recognition systems

133. Magnetic tape can serve as
A. input media B. output media
C. secondary-storage media
D. all of these

134. The advantages of COM are its .... and ....
A. compact size; readability
B. compact size; speed
C. readability; speed D. low-cost; readability

135. EBCDIC can code up to how many different characters?
A. 8 B. 16
C. 32 D. 256

136. The hexadecimal number system has a base of
A. 2 B. 4
C. 8 D. 16

137. The parity bit is added for which purposes?
A. Coding B. Indexing
C. Error-detection D. Controlling

138. Which of the following fields in a student file can be used as a primary key?
A. Class
B. Social security number
C. GPA D. Major

139. The two basic types of record-access methods are
A. sequential and random
B. sequential and indexed
C. on-line and realtime D. direct and immediate

140. Which file organisation is allowed by a direct-access storage device?
A. Direct only
B. Sequential and direct only
C. Indexed and direct only
D. Sequential, indexed and direct

141. Sequential file organisation is most appropriate for which of the following applications?
A. Grocery-store checkout
B. Bank checking accounts
C. Payroll D. Airline reservations

142. Which of the following is not a practical data-processing approach?
A. Batch-sequential B. Batch-direct
C. Immediate-direct D. Immediate-sequential

143. Which of the following file organisations is most efficient for a file with a high degree of file activity?
A. Sequential B. ISAM
C. VSAM D. B-Tree index

144. Which of the following is not one of the three primary functions that online direct-access systems can serve?
A. Inquiry B. Backup
C. Update D. Programming

145. Which of the following is considered a direct-entry input device?
A. Optical scanner B. Mouse
C. Light pen D. All of these

146. Which of the following types of input media is used much less now, than in the 1960s?
A. Hard disk B. Punched cards
C. Magnetic tape D. Floppy disk

147. Which of the following is required when more than one person uses a central computer at the same time?
A. Terminal B. Light pen
C. Digitizer D. Mouse

148. Which of the following typically uses a keyboard for input?
A. Desktop terminal B. Point-of-sale terminal
C. Financial transaction terminal
D. All of these

Ans. 125. B 126. C 127. A 128. D 129. D 130. B 131. D 132. C 133. D 134. B 135. D 136. D 137. C 138. B 139. A 140. D 141. C 142. D 143. A 144. B 145. D 146. B 147. A 148. D

149. Which of the following types of terminals is entirely dependent for all its capabilities on the computer system to which it is connected?
A. Smart terminal B. Dumb terminal
C. Microcomputer D. Intelligent terminal
150. Which of the following is an advantage(s) of magnetic input media?
A. High speed
B. Flexibility in accessing data
C. Low cost D. All of these
151. Which of the following is used only for data entry and storage, and never for processing?
A. Mouse B. Dumb terminal
C. Microcomputer
D. Dedicated data entry system
152. Which of the following are often used to ensure that data has been accurately input to the computer?
A. Keyboards B. Light pens
C. Digitizers D. Input controls
153. Which of the following is not a direct entry input device?
A. Keyboard B. Light pen
C. Digitizer D. Input controls
154. Which of the following would you most likely use at home as well as in the office connected to a central computer?
A. Dumb terminal B. Point-of-sale terminal
C. Financial transaction terminal
D. Microcomputer
155. Which of the following is a unit of measurement used with computer systems?
A. Byte B. Megabyte
C. Gigabyte D. All of these
156. Why do so many microcomputers today have hard disks?
A. It can be moved easily from one computer to another
B. It is inexpensive
C. It has very high storage capacity
D. They are a sequential access storage device
157. Hard disks and diskettes are
A. sequential access storage devices
B. direct access storage devices
C. rarely used with microcomputers
D. capable of storing terabytes of data and information
158. Which of the following terms is not used to refer to the recording density of a disk?
A. Mega-density B. Single-density
C. Double-density D. Quad-density
159. Which of the following is the most appropriate unit for measuring the storage capacity of a hard disk?
A. Byte B. Megabyte
C. Bit D. Terabyte
160. Which of the following statements is false?
A. Secondary storage is nonvolatile
B. Primary storage is volatile
C. Secondary storage contains data for immediate processing
D. When the computer is turned-off, data and instructions stored in primary storage are erased
161. Which of the following storage and retrieval methods would be well suited to your processing requirements if you only need to retrieve records one at time and there is no fixed pattern to the requests for data and records?
A. Direct B. Sequential
C. Indexed sectors D. Indexed direct
162. Which of the following is not used for storage purpose with mainframe computers
A. Removable disks B. Fixed disks
C. Mass storage systems D. None of these
163. Which of the following would you disregard when determining the storage capacity of a hard disk?
A. Track density
B. Height of the hard disk drive
C. Recording density D. Number of platters
164. Which of the following holds data and processing instructions temporarily until the CPU needs it?
A. ROM B. Control unit
C. Main memory D. Coprocessor chips
165. Which of the following is used for manufacturing chips?
A. Control bus B. Control unit
C. Parity unit D. Semiconductors
166. Which of the following is the most characteristics of ROM?
A. It is measured in megabytes
B. It is volatile
C. It performs mathematical calculations
D. Instructions are stored there permanently
167. Which of the following hardware components is the most volatile?
A. ROM B. RAM
C. PROM D. EPROM
168. Which of the following is used to check for error in RAM chips?
A. ROM chip B. Microprocessor chip
C. Parity chip D. EPROM chip
169. Which of the following are used to quickly accept, store, and transfer data and instructions that are being used immediately by the CPU?
A. Microprocessors B. Registers
C. ROM chips D. Data buses
170. Why is the width of a data bus so important to the processing speed of a computer?
A. The narrower it is, the greater the computer's processing speed

Ans. 149. B 150. D 151. D 152. D 153. D 154. D 155. D 156. C 157. B 158. A 159. B 160. C 161. D 162. D 163. B 164. C 165. D 166. D 167. B 168. C 169. B

B. The wider it is, the more data that can fit into main memory
C. The wider it is, the greater the computer's processing speed
D. The wider it is, the slower the computer's processing speed

171. Which of the following terms is most closely related to main memory?
A. Nonvolatile B. Permanent
C. Control unit D. Temporary

172. Which of the following affects processing power?
A. Data bus capacity B. Addressing scheme
C. Clock speed D. All of these

173. Which of the following is the principal difference between a monochrome monitor and an RGB monitor?
A. Number of electron guns
B. Resolution
C. Size D. Cost

174. Which of the following can be the output by a computer?
A. Graphics B. Voice
C. Text D. All of these

175. Output hardware is often categorised according to whether it
A. is expensive B. can fit on a desktop
C. produces hardcopy or softcopy
D. requires a large amount of electricity to work

176. Large computer systems typically use
A. dot-matrix printers B. daisy-wheel printers
C. ink-jet printers D. line printers

177. Which of the following printers are not to be used if your objective is to print on multicarbon forms?
A. Daisy-wheel B. Dot-matrix
C. Laser D. Thimble

178. Which of the following is not a part of CRT?
A. Phosphor screen B. Shadow mask
C. Electron gun D. Gas plasma

179. Which of the following does not affect the resolution of a video display image?
A. Bandwidth B. Faster scan rate
C. Vertical and horizontal lines of resolution
D. Screen size

180. To produce high-quality graphics (hardcopy) in colour, you would want to use a/an
A. RGB monitor B. plotter
C. inkjet printer D. laser printer

181. The terminal device that functions as a cash register, computer terminal, and OCR reader is the
A. video display terminal
B. OCR register terminal
C. data collection terminal
D. POS terminal

182. The technique designed to support the effective access of micro-filmed data is
A. Microfiche retrieval B. COM
C. micrographics D. all of these

183. An impact printer that is used as interchangeable, rotating printing unit for hardcopy output is the
A. thermal printer B. wire-matrix printer
C. drum printer D. daisy-wheel printer

184. Which of the following printing devices provides an output composed of a series of dots?
A. Wire-matrix printer B. Band printer
C. WANG image printer
D. Both A and B are correct

185. Which of the following terminals output most closely resembles the output produced by a plotter?
A. Graphics terminal B. POS terminal
C. Hardcopy terminal D. All of these

186. Which of the following impact printers print fastest?
A. Band printer B. Chain printer
C. Drum printer D. Wire-matrix printer

187. Which of the following statements is true?
A. All hardcopy terminals use punched paper tapes
B. Intelligent terminals provide hardcopy outputs only
C. Microfiche are always produced directly from printed outputs
D. None of these

188. Softcopy outputs available from the IBM 3279 Color Display terminal may include
A. graphic displays of data
B. three-dimensional line drawings
C. line-by-line outputs of character data
D. all of these

189. The terminal device often used in checking charge cards that offers both a limited keyboard input and visual output is the
A. intelligent terminal B. POS terminal
C. video display terminal
D. audio response unit

190. Which of the following printing devices can generate printed outputs composed of two or more types of printed characters?
A. IBM 300 printing subsystem
B. Twin-heat daisy-wheel printer
C. WANG image printer
D. All of these

191. The computer code for the interchange of information between terminals is
A. ASCII B. BCD
C. EBCDIC D. HOLLERITH

192. A temporary storage area, attached to the CPU, for I/O operations is a

Ans. 170. C 171. D 172. D 173. A 174. D 175. C 176. D 177. C 178. B 179. D 180. B 181. D 182. C 183. D 184. D 185. A 186. B 187. D 188. D 189. C 190. D 191. A

A. channel B. buffer
C. register D. core

193. In virtual storage, program segments stored on disk during processing are called
A. sections B. partitions
C. pages D. sectors

194. The EBCDIC code for the character X, with odd parity is
A. 0 1110 0110 B. 1 1110 0111
C. 1 1110 0110 D. 0 1110 0111

195. A characteristic of the ASCII code is that
A. its limitation is to a maximum of 96 character configuration
B. its use of the zone code 1010 and 1100
C. its independence from the Hollerith code
D. all of these

196. The comparison of data inside the arithmetic logic unit is referred to as a
A. question B. data operation
C. conditional question D. logical operation

197. The ASCII code for the character J is
A. 1101 0001 B. 1101 1010
C. 1010 1010 D. 1010 0001

198. A computer-controlled device for training exercise that duplicates the work environment is a
A. simulator B. duplicator
C. trainer D. COM device

199. An advantage of overlapped processing activities is
A. more effective use of the CPU
B. increased processing activities for the entire system
C. better coordination of I/O activities
D. all of these

200. First-generation computer systems used
A. transistors B. vacuum tubes
C. magnetic cores D. silicon chips

201. A complete set of programs for one specific data processing application is called
A. utility application B. canned program
C. program package
D. both B and C are correct

202. A characteristic generally associated with a fixed disk device is
A. simultaneous movement of all read/write heads
B. one read/write head per track
C. slower access times than removable disks
D. all of these

203. The number of records contained within a block of data on magnetic tape is defined by the
A. record per block factor
B. record contains clause
C. block definition D. blocking factor

204. The reflective strip noting of the physical end of a tape is the
A. load-point marker B. tapemark
C. end-of-reel marker D. tape end marker

205. The Josephson tunnelling device illustrates principles associated with the advanced storage technique
A. cryogenics B. CCD
C. EBAM D. holographing

206. A mass storage system is a direct access device capable of storing within its array of cartridges
A. over 500 megabytes of data
B. almost 16 billion characters
C. 2000 megabytes of data
D. 472 billion characters

207. One disadvantage of a direct access file is
A. the delay in computing the storage address
B. duplication of address locations
C. unused, but available, storage locations
D. all of these

208. The recently introduced disk concept for reducing time lost from rotational delay is
A. fixed block addressing
B. cylinder, track, sector data
C. staggered addressing
D. graduated block identification

209. The optical technique that uses laser is
A. photodigital storage B. EBAM storage
C. CCD storage D. cryogenic storage

210. A billionth of a second is defined as a
A. millisecond B. microsecond
C. nanosecond D. picosecond

211. Which of the following is not a part of the CPU?
A. Storage unit
B. Arithmetic and logic unit
C. Program unit D. Control unit

212. The largest of the four conceptual areas of storage is likely to be the
A. program area B. output area
C. working area D. input area

213. In EDCDIC, a character is represented by
A. one bit B. two bits
C. four bits D. eight bits

214. Currently, the most popular type of DASD is
A. magnetic tape strips B. magnetic drums
C. magnetic cards D. magnetic disks

215. The most economical way to process data is
A. batch processing B. transaction processing
C. realtime processing D. distributed processing

216. Transaction processing was made possible by the development of
A. direct access storage B. magnetic tape
C. cash register input D. minicomputers

Ans. 192. B 193. C 194. B 195. C 196. D 197. C 198. A 199. D 200. B 201. C 202. B 203. D 204. C 205. A 206. D 207. D 208. C 209. A 210. C 211. C 212. A 213. D 214. D 215. A 216. A

217. The central computer in a distributed processing system is called the
A. father B. monitor
C. multiplexer D. host

218. The amount of a cheque is recorded in magnetic ink, using an
A. encoder B. embosser
C. inscriber D. imprinter

219. A serial card reader would be expected to be
A. faster than a parallel reader
B. more expensive than a parallel reader
C. both A and B are correct
D. neither A nor B

220. MICR has made possible a
A. cashless society B. checkless society
C. creditless society D. none of these

221. The company that pioneered the development of key driven input devices, with output on magnetic tape or disk rather than punched cards.
A. Mohawk Data Science
B. Sperry Univac
C. Boroughs D. National Cash Register

222. In mark sensing, data is used
A. optically B. electrically
C. magnetically D. mechanically

223. Some card readers detect errors by
A. accumulating control totals on certain data
B. comparing data with acceptable limits
C. reading the card twice
D. none of these

224. Optical character readers can read
A. machine printed data only
B. machine printed letters and numbers, and hand printed numbers only
C. both machine and hand printed letters, but not handwriting
D. machine

225. The word 'serial' describes
A. an expensive and reader that reads one column at a time
B. a character-by-character transfer of data between an input/output or auxiliary storage unit and a channel
C. both A and B are correct
D. neither A nor B

226. During the past decade, the most improvements probably have been made in
A. input and output units
B. auxiliary storage units
C. data communication facilities
D. main storage

227. A slow input unit such as a card reader would be attached to the CPU by means of a
A. selector channel B. buffer channel
C. byte multiplexer channel
D. block multiplexer channel

228. The input bottleneck has been relieved by
A. key-to-tape machines
B. combination keypunch and key verifier machines
C. MICR D. all of these

229. The data appearing first in a punched card (in the left portion) usually is
A. descriptive data B. coded data
C. quantitative data D. none of these

230. A floppy disk can contain approximately
A. 2,500 bytes B. 25,000 bytes
C. 1,440,000 bytes D. 2,500,000 bytes

231. A turnaround document frequently is used by
A. utilities B. supermarkets
C. manufacturing firms D. wholesalers

232. EFTS makes use of
A. MICR B. port-a-punch
C. mark sensing D. none of these

233. The least expensive OCR units can read
A. hand printed number
B. machine printed numbers
C. marks D. handwriting

234. The most common means of preparing printed computer output is the
A. plotter B. console typewriter
C. IBM printing subsystem
D. line printer

235. The IBM 3800 printing subsystem prints upto
A. 1,360 lines per minute
B. 13,600 lines per minute
C. 1,36,000 lines per minute
D. 1,360,000 lines per minute

236. Plotter's print
A. with ball point pens B. with ink pens
C. electrostatically D. all of these

237. A byte is comprised of
A. 1-bit B. 4-bits
C. 8-bits D. 32-bits

238. In the language of the number profession, one thousand positions of main storage is represented with the letter
A. K B. L
C. M D. C

239. The IBM 3850 mass storage facility uses
A. magnetic disks B. magnetic drums
C. magnetic tape strips D. magnetic cards

240. Each cartridge of the IBM 3850 mass storage facility has a capacity of
A. 5 million bytes B. 50 million bytes
C. 500 million bytes D. 5 billion bytes

Ans. 217. D 218. A 219. D 220. A 221. A 222. B 223. C 224. D 225. C 226. C 227. C 228. C 229. B
230. C 231. A 232. D 233. C 234. D 235. B 236. D 237. C 238. A 239. A 240. B

241. The standard size reel of magnetic tape is
A. 120 feet in length B. 240 feet in length
C. 1200 feet in length D. 2400 feet in length

242. The IBM 305 RAMAC was the first computer system with
A. auxiliary storage B. magnetic tape units
C. disk units D. magnetic core storage

243. The RAMAC address 12345 identifies
A. disk 12 B. disk 23
C. disk 34 D. disk 5

244. The RAMAC had one access arm per
A. disk stack B. disk
C. track D. sector

245. When a disk stack is serviced by a comb-like access mechanism, there is one read/write head for each
A. disk B. disk surface
C. track D. cylinder

246. The modern disk pack, called the data module, offers increased access speed by employing a read/write head for each
A. disk B. track
C. sector D. disk surface

247. The model base would most likely be organised as
A. sequential B. partitioned
C. indexed sequential D. direct

248. A hashing scheme is used with
A. sequential file organisation
B. direct file organisation
C. indexed sequential file organisation
D. partitioned file organisation

249. DASD is required for
A. transaction processing
B. realtime processing
C. time sharing D. all of these

250. DASD is most valuable as a/an
A. master file medium B. input medium
C. historical storage medium
D. data communication medium

251. A 24-bit operand permits as many as
A. 1,600 addresses B. 16,000 addresses
C. 160,000 addresses D. none of these

252. If a virtual storage segment contains 20,000 positions, the relative addresses will be
A. 0–19,999 B. 1–20,000
C. 20,000–39,999 D. 21,000–40,000

253. DASD is good for
A. historical storage B. master files
C. both A and B are correct
D. neither A nor B

254. Magnetic tape is good for
A. historical storage B. computer input
C. both A and B are correct
D. neither A nor B

255. If 40 character records are being written on tape at 800 bpi, what is the blocking factor that will reduce the proportion of blank tape area to one-half?
A. 1 B. 4
C. 8 D. 10

256. If 80 character records are being written on tape at 800 bpi, what is the blocking factor that will reduce the proportion of blank tape area to one-third?
A. 1 B. 4
C. 10 D. 15

257. The storage within the addressing scheme of the computer but exceeding the physical capacity is called
A. address space B. overlay storage
C. secondary storage D. scratch pad memory

258. File specification books are created primarily for the use of
A. system analysts B. programmers
C. operators D. managers

259. A logic gate is an electronic circuit which
A. makes logic decisions
B. allows electron flow only in one direction
C. works on binary algebra
D. alternates between 0 and 1 values

260. In positive logic, logic state 1 corresponds to
A. positive voltage B. higher voltage level
C. zero voltage D. lower voltage level

261. In negative logic, the logic state 1 corresponds to
A. negative voltage B. zero voltage
C. more negative voltage
D. lower voltage level

262. Truth table of a logic function
A. summarises its output values
B. tabulates all its input conditions
C. displays all its input/output possibilities
D. is not based on logic algebra

263. The output of a 2 input OR gate is 0 only when its
A. both inputs are 0 B. either input is 1
C. both inputs are 1 D. either input is 0

264. An AND gate
A. implements logic addition
B. is equivalent to a series switching circuit
C. is an any-or-all gate
D. is equivalent to a parallel switching circuit

265. When an input electrical signal A = 10100 is applied to a NOT gate, its output signal is
A. 01011 B. 10101
C. 10100 D. 00101

266. The NOT symbol at the output of an OR gate converts it into which gate?
A. AND B. NAND
C. XOR D. NOR

Ans.	241. D	242. C	243. A	244. A	245. B	246. B	247. B	248. B	249. D	250. A	251. D	252. A	253. B
	254. C	255. D	256. C	257. A	258. B	259. A	260. B	261. D	262. C	263. A	264. B	265. A	266. D

267. The only function of a NOT gate is to
A. stop a signal
B. act as universal gate
C. invert an output signal
D. recomplement a signal

268. Most of the errors blamed on computers are actually due to
A. programming errors B. hardware fatigue
C. data entry errors D. defects in floppy disks

269. A dot-matrix printer
A. is an input device B. is an output device
C. cannot print alphabet D. has a speed of 500 cps

270. A computer will function only if it
A. is given input data B. has control unit
C. has a program in its memory
D. has software package

271. Multiple choice examination answer sheets are evaluated by which reader?
A. Optical mark B. Optical character
C. Magnetic tape D. Magnetic ink character

272. Two new types of semiconductor memories are
A. magnetic disks
B. charge-coupled devices
C. magnetic bubble memory
D. both B and C are correct

273. A computer programmer
A. does all the thinking for a computer
B. can enter input data quickly
C. can operate all types of computer equipment
D. can draw only flow charts

274. Which of the following storage is volatile?
A. Floppy disk
B. Semiconductor memory
C. Bubble memory D. Core memory

275. Which access method is used for obtaining a record from a cassette tape?
A. Direct B. Sequential
C. Random D. None of these

276. The availability of low-price mini-computer/microcomputer is largely responsible for the current interest in
A. realtime processing B. batch processing
C. distributed processing
D. transaction processing

277. A set of programs that handle firm's data base responsibilities is called a
A. Data Base Management System (DBMS)
B. Data Base Processing System (DBPS)
C. Data Management System (DMS)
D. all of these

278. A hard copy would be prepared on a
A. typewriter terminal B. line printer
C. plotter D. all of these

279. The two kinds of main memory are
A. primary and secondary
B. random and sequential
C. ROM and RAM D. central and peripheral

280. Following is a way to access secondary memory
A. random access memory
B. action method
C. transfer method D. density method

281. Which of the following is a secondary memory device?
A. Disk
B. Central processing unit
C. Keyboard D. Arithmetic logic unit

282. Programmable Read-Only Memory (PROM) of a digital computer (mark the wrong statement)
A. is permanent
B. contains program instructions
C. is programmed during manufacture
D. is volatile

283. As compared to the secondary memory, the primary memory of a computer is
A. large B. cheap
C. fast D. slow

284. Computer memory consists of which microchips?
A. RAM B. ROM
C. PROM D. All of these

285. Backing storage is so named because it
A. lags behind the main memory
B. is slow and backward
C. backs up the computer's main memory
D. is always kept at the back of the CPU

286. In a computer, RAM is used as a short memory because it
A. is volatile B. has small capacity
C. is very expensive D. is programmable

287. Main memory contains
A. data B. instructions
C. both A and B are correct
D. neither A nor B

288. A hybrid computer resembles
A. analog computer B. digital computer
C. both analog and digital computer
D. none of these

289. Which of the following is not connected with backing or back-up storage?
A. Diskette B. Hard disk
C. Dot-matrix printer D. Cassette

290. The function of CPU is to
A. provide external storage of text
B. communicate with the operator
C. read, interpret and process the information and instruction
D. provide a hard copy

Ans.													
	267. C	268. C	269. B	270. C	271. A	272. D	273. A	274. B	275. C	276. A	277. D	278. A	279. C
	280. A	281. A	282. C	283. C	284. D	285. C	286. A	287. C	288. C	289. C	290. C		

291. The term used to define all input and output devices in a computer system is
A. monitor B. software
C. shared resources D. hardware

292. The essential features of a number system are
A. radix B. bits
C. set of distinct counting digits
D. both A and B are correct

293. The radix of a number system
A. is variable
B. has nothing to do with digit position value
C. equals the number of its distinct counting digits
D. is always an even number

294. The binary system uses powers of four positional values
A. 2 B. 10
C. 8 D. 16

295. Hexadecimal numbers are a mixture of
A. binary and decimal numbers
B. letters and decimal digits
C. binary and octal numbers
D. octal and decimal numbers

296. Which of the following is not an octal number?
A. 29 B. 75
C. 18 D. 102

297. The number 1000 is equivalent to decimal number
A. one thousand B. eight
C. four D. sixteen

298. Octal numbers are used
A. in computer hardware
B. when binary numbers are too long
C. external to the computer
D. in preference to 'hex' numbers

299. What is a microprocessor?
A. Same as a microcomputer
B. A small piece of equipment
C. A small device that controls other equipments
D. A way of doing something fast

300. A processor is popularly called computer's brain. What are its main functions?
A. To control peripheral devices
B. To obey the instructions
C. To read instructions D. All of these

301. Which are the following major components of a central processing unit?
A. Primary memory B. ALU
C. Control unit D. All of these

302. Which are the three major components of a central processing unit?
A. Primary memory B. ALU
C. Control unit D. All of these

303. What are the two main sections of a CPU chip?
A. Primary memory B. Control section
C. Arithmetic-logic section
D. Both B and C are correct

304. The CPU chip used in a computer is partially made out of
A. carbon B. copper
C. aluminium D. silica

305. The computer hardware made of silica that can store a vast amount of data is called
A. disk B. chip
C. magnetic tape D. qwerty

306. The silicon chips which are used for data processing are called
A. memory chips B. RAM chips
C. microprocessors D. ROM chips

307. The microelectronics is the technology of
A. microprocessors B. microcomputers
C. chips D. automatic processing

308. The silicon chips presently used in computers are made from
A. germanium B. gallium-arsenide
C. copper D. ordinary sand

309. A microchip is usually put into a plastic case which protects it and makes it easier to handle. Next, the chip is connected to the pins in the case with the help of tiny wires. What are these wires made of?
A. Silver B. Copper
C. Aluminium D. Gold

310. A small square of rectangular piece of silica on which several layers of an integrated circuit are etched or imprinted is called
A. VLSI B. micro
C. wafer D. chip

311. The metal-oxide semiconductor (MOS) transistor is the real forerunner of the modern chip. In which year was it invented?
A. 1958 B. 1959
C. 1960 D. 1962

312. The first microprocessors produced by Intel Corpn. and Texas Instruments were used primarily to control small
A. microwaves B. washing machines
C. calculators D. personal computers

313. The ALU and control unit of most of the microcomputers are combined and manufactured on a single silicon chip. What is it called?
A. Monochip B. Microprocessor (MP)
C. Microprocessor unit (MPU)
D. Either B or C

314. The processor of a computer system is called its
A. nerve centre B. eye
C. brain D. ear

315. When did Intel Corpn. develop the Intel 4004 chip?
A. 1970 B. 1972
C. 1967 D. 1969

316. Jack S Kilby is credited with the invention of the Integrated Circuit (IC) which consists of a number

Ans. 291. D 292. D 293. C 294. A 295. B 296. A 297. B 298. C 299. C 300. A 301. D 302. D 303. D
304. D 305. D 306. C 307. C 308. D 309. D 310. D 311. C 312. C 313. D 314. C 315. D

of discrete components integrated into one tiny piece of semiconductor material. This invention was patented by the company where the young engineer Kilby was working. Can you name the company?
A. Texas Instruments Company
B. Digital Equipment Corporation
C. Intel Corpn D. Hewlett-Packard Ltd

317. What is the generation of the computers which are built with VLSI technology and microprocessors?
A. Third B. Fourth
C. Second D. First

318. When did Intel announce its 16-bit 80286 chip?
A. 1980 B. 1982
C. 1984 D. 1986

319. When did Intel announce its 80386 processor?
A. 1983 B. 1984
C. 1985 D. 1986

320. When did Motorola come out with its enhanced 68030 processor?
A. 1983 B. 1984
C. 1985 D. 1986

321. When did Intel come out with its RISC i860 chip?
A. 1989 B. 1986
C. 1987 D. 1985

322. When did Motorola announce its newest microprocessor chip 68040?
A. 1990 B. 1984
C. 1988 D. 1987

323. What is the usual expected basic life of any processor architecture?
A. 5 years B. 7 years
C. 9 years D. 11 years

324. What are the two major types of computer chips?
A. Microprocessor chip B. Primary memory chip
C. External memory chip
D. Both B and C are correct

325. What is the name of the chip which has more than one processor on it?
A. Parallel chip B. Multi-processor chip
C. Transputer D. Parallel processor

326. Computer scientists are exploring the possibility of culturing bacteria to create living chips which have digital electronic properties. What are these chips called?
A. Bacterial chips B. Biochips
C. Natural chips D. Neural chips

327. One of the Intel's co-founders had made the statement that "transistor densities on a single chip would continue to double every 18 months". The above statement goes under the name of which law?
A. Moore's Law B. Murphy's Law
C. Engelbart's Law D. Amdahl's Law

328. The hardware in which data may be stored for a computer system is called
A. register B. memory
C. chip D. peripheral

329. Which is the computer memory that does not forget?
A. RAM B. ROM
C. DRAM D. SRAM

330. The computer memory used for temporary storage of data and program is called
A. ROM B. Sector
C. RAM D. EPROM

331. The computer memory holds data and
A. bytes B. program
C. registers D. words

332. Which computer memory is used for storing programs and data currently being processed by the CPU?
A. Mass memory B. Internal memory
C. Non-volatile memory D. PROM

333. What is meant by the term RAM?
A. Memory which can only be read
B. Memory which can be read and written to
C. Memory which is used for permanent storage
D. Memory which can only be written to

334. Which memory of the computer is used for storing word processing or spread-sheet programs?
A. RAM B. ROM
C. EROM D. EPROM

335. Which computer memory is essentially empty?
A. ROM B. PROM
C. EPROM D. RAM

336. The bubbles in bubble memory pack are created with the help of
A. laser beam B. magnetic field
C. electric beam D. X-ray

337. Which term is not used to describe the memory component of a computer?
A. ROM B. PROM
C. RAM D. SAM

338. Which computer company invented the RAM chip?
A. Motorola B. Intel
C. Toshiba D. IBM

339. What are the alternative names for the internal storage of a computer?
A. Real storage B. Primary memory
C. Main memory
D. Both A and B are correct

340. Which of the following is used for serial access storage only?
A. RAM B. Magnetic tape
C. Magnetic disk D. Core memory

341. Which of the following is not an alternative name for primary memory?

Ans. 316. A 317. B 318. B 319. C 320. D 321. A 322. A 323. B 324. D 325. C 326. B 327. A 328. B 329. B 330. C 331. B 332. B 333. B 334. A 335. A 336. B 337. D 338. B 339. D 340. B

A. Main memory B. Primary storage
C. Internal storage D. Mass storage

342. What do the present-day digital computers use for their internal memory?
A. Magnetic cores
B. Integrated circuits (ICs)
C. Electrical switches D. Magnetic disks

343. Both silicon and gallium arsenide are used for primary storage devices. Can you tell how many times gallium arsenide devices are faster than the silicon devices?
A. 3 B. 5
C. 7 D. 10

344. In modern computers, bipolar semiconductor chips are often used in the arithmetic-logic unit. Can you tell what material is used for the slower and less expensive primary storage section?
A. Gallium arsenide (GaAs)
B. Metal-oxide semiconductor (MOS)
C. Silicon D. Germanium

345. Which type of memory chips are likely to be used in the primary storage of the future generation of computers?
A. Selenium chips B. Optical chips
C. Biochips D. GaAs chips

346. Which of the following chips can be re-programmed with special electric pulses?
A. EPROM B. PROM
C. ROM D. EEPROM

347. The storage location in the internal storage of a CPU are called
A. contents B. addresses
C. locations D. intersections

348. What is the name given to the specialised storage element in the processor unit of a computer which is used as a 'scratch pad' during processing operations?
A. Storage register B. Primary memory
C. Cache memory D. Accumulator

349. How many bits can be stored in the 8 K RAM?
A. 8000 B. 8192
C. 4000 D. 4096

350. A computer has a 1024 K memory. What does the letter K stand for?
A. Kilometre B. Thousand
C. 1024 D. Core

351. A microcomputer has primary memory of 640 K. What is the exact number of bytes contained in this memory?
A. 640000 B. 655360
C. 640 × 1024 D. Either B or C

352. If, in a computer, 16-bits are used to specify addresses in a RAM, the number of addresses will be
A. 216 B. 65536
C. 64 K D. any of these

353. What was the amount of memory required by the earliest operating system called DOS 1.0?
A. 4 K B. 8 K
C. 16 K D. 32 K

354. What is the name of the memory device which for the first time made large internal storage possible?
A. Core memory B. Electronic valve
C. Delay lines D. Cathode ray tube

355. There is only one key in the keyboard of a computer which is without any marking and is also the widest key of the keyboard. What is the name of this key?
A. Shift key B. Enter key
C. Backspace key D. Space bar

356. What is the function of the 'home key' in a microcomputer qwerty keyboard?
A. To capitalise letters
B. To reposition the cursor
C. To interrupt executions
D. To cancel the entry

357. In a microcomputer keyboard if you discover a mistake after inputting data with the enter key, which key will have to be pressed to retype the data?
A. Control key B. Insert key
C. Backspace key D. Tet key

358. A hand-held terminal used for inputting data in a computer has no
A. keys B. memory
C. leads D. stored data

359. What does OCR stand for?
A. Outsized Character Reader
B. Optical Character Recognition
C. Operational Character Reader
D. Only Character Reader

360. Which of the following organisations uses magnetic ink character recognition extensively?
A. Banks B. Electricity Boards
C. Housing departments
D. Oil and Natural Gas Commission

361. What is the name of the reading device which makes use of photosensors on the source documents and then convert them directly into computer usable input. Which of the following is not a character reader?
A. OCR B. MICR
C. OMR D. LCD

362. Print out the item amongst the following.
A. Computer mouse B. Touchpad
C. Light pen D. Printer

363. The barcode which is used on all types of items from books to candybars is read by a scanning device directly into the computer. What is the name of this scanning device?

Ans.	341. D	342. B	343. B	344. B	345. C	346. D	347. B	348. C	349. B	350. C	351. B	352. D	353. B
	354. A	355. D	356. B	357. C	358. C	359. B	360. B	361. B	362. D				

A. Laser scanner B. Wand
C. OCR D. MICR

364. Which is not necessary when using barcodes in supermarkets?
A. Point-of-sale (POS) terminal
B. Check digit on the bar code
C. Price on the shelf D. Price on the goods

365. Bar codes store information using
A. punch holes B. thick and thin lines
C. magnetised spots D. bits

366. The point of sale (POS) terminals are used for reading
A. kimball tags B. bar codes
C. printed information
D. both B and C are correct

367. Which one of the following input device is user-programmable?
A. Dumb terminal B. Smart terminal
C. VDT D. Intelligent terminal

368. What is the name of the screen symbol that shows the placement of the next character?
A. Mouse B. Cursor
C. Track ball D. Graphic tablet

369. Which one of the following is termed a combination input-output device?
A. VDT B. Keyboard
C. Printer D. Laser jet

370. A printed document reader which scans the characters and then encodes them in a computer-compatible code for direct transmission to the computer is called
A. Optical Character Reader
B. Kurzweil Data Entry Machine (KDEM)
C. MICR D. laser scanner

371. The computer that we use are digital whereas we live in an analog word which means that we have to translate analog data into digital data. What is the name of the circuit which helps us in this conversion?
A. D/A converter B. A/D converter
C. Voice recognition D. Adapter

372. What is the name of the screen pointing device that rolls on bearings and has one more button on its top?
A. Track ball B. Mouse
C. Light pen D. Joy stick

373. Who built the first computer mouse?
A. Douglas Engelbart B. William English
C. Daniel Cougar D. Robert Zawacki

374. A mouse is a simple palm-sized device that moves a pointer on the screen as you move it on the desk. When was the first computer mouse developed?
A. 1960 B. 1962
C. 1964 D. 1966

375. Who invented the computer input device called mouse?
A. Vannevar Bush B. Douglas Engelbart
C. Ted Nelson D. Wolfgang Haken

376. What is a Kimball tag?
A. An address label B. A gift token
C. A security key D. An input medium

377. Small punched cards are used to give details on the style, colour and size etc. of an item, usually clothing. What are these cards called?
A. Bar codes
B. Point-of-sale (POS) devices
C. Kimball tag D. Badges

378. Whereas a computer mouse moves over the table surface, the trackball is
A. stationary B. difficult to move
C. dragged D. moved in small steps

379. A trackball is manipulated by
A. palm B. foot
C. fingers D. sound

380. Touchpads are particularly popular with microcomputer users who cannot
A. read B. type
C. see D. hear

381. Voice recognition systems for programming a computer are used by persons who are handicapped?
A. visually B. educationally
C. socially D. physically

382. What is the general range of speed in revolutions per minute (rpm) at which floppy disks rotate?
A. 2400–4700 B. 390–600
C. 150–250 D. 300–600

383. Which type of electric motor is used to activate the read/write head of a floppy diskette drive?
A. Single-phase electric motor
B. Reluctance motor
C. DC motor D. Stepper motor

384. What is the size of the diskettes used by IBMs newer PS/2 personal computers?
A. 5¼″ B. 8″
C. 3½″ D. 2½″

385. Which of the following is not an output device of a computer?
A. Printer B. Keyboard
C. VDU D. CRT screen

386. The equipment with which the computer talks to its users is called a
A. word processor B. peripheral
C. software D. diskette

387. The monitor of a computer is connected to it by a
A. cable B. wire
C. bus D. modem

Ans. 363. D 364. A 365. C 366. D 367. D 368. B 369. A 370. B 371. B 372. B 373. B 374. C 375. B 376. D 377. C 378. A 379. A 380. B 381. D 382. C 383. D 384. C 385. B 386. B 387. A

388. In a computer system, which device is functionally opposite to a keyboard?
A. Mouse B. Trackball
C. Printer D. Joystick

389. A monitor looks like a TV set but it does not
A. receive TV signals B. give a steady picture
C. display graphics D. give a clear picture

390. What is the name of the hardware/software boundary that permits communication between people and computer?
A. Interface B. Keyboard
C. VDT D. Monitor

391. Which of the following is classified as an impact printer?
A. Jet printer B. Daisywheel printer
C. Thermal printer D. Laser printer

392. A kind of serial dot matrix printer that forms characters with magnetically charged ink sprayed dots is called
A. laser printer B. inkjet printer
C. drum printer D. chain printer

393. Which printer is very commonly used for desk-top publishing?
A. Laser printer B. Bank printer
C. Daisywheel printer D. Dot-matrix printer

394. What is the general name of the device which produces hardcopy graphics?
A. COM B. Plotter
C. Printer D. Microfilm

395. Which one of the following is a character-at-a-time printer?
A. Chain printer B. Bank printer
C. Drum printer D. Daisywheel printer

396. Which of the following printer can be classified as a page-at-a-time printer?
A. Laser printer B. Dot-matrix printer
C. Thermal printer D. Inkjet printer

397. Dot-matrix is a type of
A. tape B. printer
C. disk D. bus

398. Which of the printer used in conjunction with computers uses dry ink powder?
A. Daisywheel printer B. Line printer
C. Laser printer D. Laser writer

399. What is the name of printer which prints all the A's in a line before all the B's?
A. Thermal printer B. Electrostatic printer
C. Line printer D. Inkjet printer

400. Line printers are those which prints complete in one go. What is the name of the printer which is opposite to the line printer?
A. Band printer B. Chain printer
C. Drum printer D. Serial printer

401. Which of the following is a serial printer?
A. Daisywheel printer B. Chain printer
C. Drum printer D. Line printer

402. Which of the following is non-impact printer?
A. Drum printer B. Line printer
C. Chain printer D. Laser printer

403. Which computer company introduced the printed LaserJet in 1984?
A. Hewlett-Packard Inc B. Mitsubishi Electronics
C. Ashton-Tate Corpn D. Mitgart India Ltd

404. Which of the following printers cannot print graphics?
A. Inkjet B. Daisywheel
C. Laser D. Dot-matrix

405. With the popularity of PCs, CRT screens have become the most common output device for communication between humans and computers. What is the number of colours displayed by the so-called monochrome CRT?
A. 1 B. 2
C. 3 D. 4

406. A pixel on the screen of a VDT appears light or dark when the corresponding bit in the computer memory is turned ON or OFF. What this technique is known as?
A. Bit mapping B. Colour graphics
C. Pixel modulation D. Pixel processing

407. What is the name of the computer terminal which gives paper printout?
A. Display screen B. Soft copy terminal
C. Hard copy terminal D. Plotter

408. A flat-bed plotter uses a pen which moves in two directions across a piece of paper fixed on a flat-bed. Can you tell who controls the movements of this pen?
A. Microfilm B. Microfiche
C. Film card D. COM

409. Regarding a VDU, which statement is more correct?
A. It is an output device B. It is an input device
C. It is a peripheral device
D. It is a hardware item

410. Computers are being increasingly used for putting together different notes and pieces of music in order to make new sounds. What is this process called?
A. Music creation B. Music synthesis
C. Speech synthesis D. Compumusic

411. What does a 'dumb' terminal essentially consists of?
A. Keyboard and microprocessor
B. Microprocessor and VDU
C. Keyboard and VDU D. Keyboard and printer

Ans. 388. C 389. A 390. A 391. B 392. B 393. A 394. B 395. D 396. A 397. B 398. C 399. C 400. D
401. A 402. D 403. A 404. B 405. B 406. A 407. C 408. C 409. C 410. B 411. C

412. Which type of display is the latest to be used for portable computer?
A. LED display B. LCD display
C. Plasma display
D. Electroluminescent display

413. What is the brand name of the popular floppy disk manufactured by Allied Electronics and Magnetics Ltd?
A. Dycill B. Verbatim
C. Amkette D. Maxwell

414. Because of their reduced bulk, good readability and flicker-free screen, plasma displays are well-suited for
A. personal computers B. portable computers
C. laptop computers D. minicomputers

415. Main problem with LCDs is that they are very difficult to read
A. directly B. in bright light
C. in dull light
D. both B and C are correct

416. What is the name of the record-like object which is coated with a magnetic material and is used for storing data in a computer?
A. Tape B. Track
C. Disk D. Format

417. Floppy disks which are made from flexible plastic material are also called
A. hard disks B. high-density disks
C. diskettes D. templates

418. Which part of the diskette should never be touched?
A. Hub B. Hole in the centre
C. Over slot D. Corner

419. If you see a diskette with a piece of foil covering its notch, it is said to be
A. write-protected B. copy-protected
C. write-enabled D. foil-covered

420. What is the alternative name for a diskette?
A. Winchester disk B. Flexible disk
C. Hard disk D. Floppy disk

421. Before anything can be stored on a disk, it has to be
A. cleansed B. formatted
C. initialized D. either A or B

422. What are the concentric circles on floppy disks known as?
A. Tracks B. Cylinders
C. Sectors D. Segments

423. What is the normal speed of the motor in revolutions per minute (rpm) used for a floppy diskette drive?
A. 1000–1200 B. 600–900
C. 300–360 D. 200–300

424. Disk units that use unremovable platters are called
A. unremovable disks B. hard disks
C. sealed disks D. high-density disks

425. Reel-to-teel tape drives are commonly used with
A. minicomputers B. microcomputers
C. laptop computers
D. mainframe computers

426. As compared to diskettes, the hard disks are
A. more expensive B. more portable
C. less rigid D. slowly accessed

427. The secondary storage devices can only store data but they cannot perform
A. arithmetic operations B. logic operations
C. fetch operations D. either of these

428. The metal disks which are permanently housed in sealed and contamination free containers are called
A. hard disks B. Winchester disks
C. floppy disks D. aluminium disks

429. Which is the most popular medium for direct-access seondary storage of a computer?
A. Magnetic tape B. Magnetic disk
C. RAM D. ROM

430. The data on the tracks on a magnetic disk are written as
A. up or down B. tiny magnetic spots
C. 0 or 1 D. high or low voltage

431. The data on the magnetic disks are stored in a number of concentric circles called tracks. Unlike the grooves on a phonogram, these tracks are
A. less crowded B. invisible
C. thicker D. finer

432. What is the most common speed in rpm at which the hard disks usually rotate?
A. 2400 B. 3600
C. 4700 D. 600

433. Before a disk drive can access any sector record, a computer program has to provide the record's disk address. What information does this address specify?
A. Track number B. Sector number
C. Surface number D. All of these

434. Most disk drives have a single read/write head for each disk surface. What is the advantage of using multiple heads on each movable access arm?
A. Reduced seek time B. Less disk speed
C. Reduced search time D. Less latency time

435. What is meant of quad-density (QD) diskette?
A. It is double-sided disk
B. It is double density disk
C. It has double the number of tracks per inch
D. All of these

Ans. 412. D 413. C 414. C 415. D 416. C 417. C 418. C 419. A 420. D 421. B 422. A 423. D 424. B
425. D 426. A 427. D 428. B 429. B 430. B 431. B 432. B 433. D 434. A 435. D

436. The space between the read/write ceramic head of a disk drive and the disk platter called fly-height is of the order of
A. 0.005 mm B. 0.0025 mm
C. 0.006 mm D. 0.1 mm

437. When the read/write head of a disk drive accidentally touched the recording surface, the disk is said to suffer from a/an
A. accident B. head crash
C. dent D. injury

438. Which of the following are likely to jam the read/write head of a disk pack?
A. Dust B. Smoke particle
C. Finger print smudge D. All of these

439. What is the shape of the tracks on the recording surface of a disk platter?
A. Spiralling B. Elliptical
C. Concentric D. Wavy

440. What is the name of the window which exposes the diskette surface to be drive's read/write head?
A. Index window B. Access window
C. Alignment notch D. None of these

441. In a disk pack containing 6 disks, there are 200 tracks on each disk surface, each track is divided into 50 sectors and each sector contains 512 characters of one byte each. What is the total storage capacity of disk pack in bytes?
A. 512 × 50 × 200 × 12 B. 512 × 50 × 200 × 10
C. 512 × 50 × 200 × 6 D. 512 × 50 × 200/10

442. The time taken for the read/write head to move to the correct track on the magnetic disk is called
A. search delay B. latency delay
C. seek time D. approach time

443. What is the approximate speed with which a magnetic tape moves through a tape drive?
A. 50 km/h B. 100 km/h
C. 20 km/h D. 200 km/h

444. What is the number of read-write heads in the drive for a 9-track magnetic tape?
A. 9 B. 6
C. 18 D. 12

445. What is the usual number of tracks on a 1/2 inch wide magnetic tape?
A. 7 B. 9
C. 18 D. Either B or C

446. What is the name of the visible silver rectangle which separates the take-up portion of the magnetic tape from the data recording portion?
A. Load maker B. Load marker
C. Tape separator D. Tape marker

447. Why do magnetic tape drives for mini computers use vacuum columns?
A. For immediate starting and stopping of the tape
B. For quick acceleration and deceleration of the tape
C. To avoid stress in the tape
D. To provide extra tape length

448. What is the size of the optical compact disk which is used for recording high quality music?
A. 4.7 inch B. 3½ inch
C. 5½ inch D. 8 inch

449. What is the name of the secondary storage device which uses a silvery groove-less surface and is encoded by a laser beam in the form of microscopic pits?
A. Laser disk B. Compact disk
C. Photo disk D. Video disk

450. As a secondary storage medium, what is the most important advantage of a video disk?
A. Compactness B. Potential capacity
C. Durability D. Cost effectiveness

451. What is the name of the memory card which is conceptually related to the smart card but is similar to the video disk?
A. Laser card B. Master card
C. Visa D. Optical card

452. To reduce their paper load, many large companies are turning to electronic record systems consisting of computers, digital scanners, OCRs, does an IMB, 30 cm optical disk contain?
A. 25,000 B. 5,000
C. 10,000 D. 2,000

453. What is the latest write-once optical storage media?
A. Digital paper B. CD-ROM disk
C. Worm disk D. Magneto-optical disk

454. Which optical phenomenon is utilised in the operation of the latest write once optical storage medium called digital paper?
A. Polarisation B. Interference
C. Internal reflection D. Diffraction

455. Compact disk (CD) audio technology has been adapted to develop CDROM optical storage disks for use with PCs. What are the special advantages of such disk?
A. Enormous storage density
B. Very low storage cost
C. Faster access time D. All of these

456. The Encyclopaedia Britannica comes in thirty-three volumes and contains 450 million characters. How many optical disks would be required to store the entire contents of the encyclopaedia?
A. 1 B. 2
C. 3 D. 4

457. The wire-once, read-many (WORM) disks have huge storage capacity, hence they are highly desirable for archival application. If a person types

Ans. 436. B 437. B 438. D 439. C 440. B 441. B 442. C 443. C 444. A 445. D 446. B 447. C 448. A 449. D 450. B 451. A 452. A 453. A 454. B 455. D 456. A

90 words per minute for 8 hours everyday, how many years will he take to fill a single 12 inch WORM disk?
A. 5 years B. 10 years
C. 50 years D. 25 years

458. What is the name of the storage device which is used to compensate for the difference in rates of flow of data from one device to another?
A. Cache B. Concentrator
C. Buffer D. I/O device

459. What is the fundamental unit of storage in digital computers?
A. Bit B. Byte
C. Code D. Baud

460. A common example of a data storage medium is
A. disk B. modem
C. cluge D. track

461. A half byte is known as
A. bug B. nybble
C. data D. bit

462. A single binary integer consists of '1' or a '0' and is called a 'bit'. How many bits taken together to form a byte?
A. 8 B. 5
C. 4 D. 2

463. As compared to a magnetic tape, the main advantage of magnetic disk is its
A. rigidity B. direct access
C. low cost D. high packing density

464. Can you give an approximate idea of the capacity of a present day optical disk?
A. 500 KB B. 125 MB
C. 1000 MB D. 250,000 MB

465. A double side magnetic disk pack has six disks. Can you tell how many surfaces of this pack are normally used?
A. 10 B. 12
C. 8 D. 6

466. What is the number of bits needed for an address in a 4 K memory?
A. 6 B. 8
C. 12 D. 16

467. What is the highest address possible if 16-bits are used for each address?
A. 65536 B. 32768
C. 16384 D. 131072

468. In the mid-1950s, the most popular computer was IBM-650. What was the type of storage used by this computer?
A. Semiconductor storage
B. Rotating drum
C. Magnetizable cores D. Bubble memory

469. The floppy diskettes store data on circular concentric tracks sub-divided into sectors which contain bytes of storage. In addition to the data, a certain specific information called formatting information is also stored on the tracks. The function of the formatting information is to tell the computer
A. where to look on the diskette for any given data
B. which track this data is on
C. which sector it is reading
D. all of these

470. By using molecular computing technology, it would be possible to use polyethylene molecules as tape for mass storage. Can you give an idea of the size of a 64 KB molecular device including tape, reel, drive and read/write apparatus?
A. 100 nm across B. 50 Hm across
C. 10 nm across D. 10 Hm across

471. What is the storage capacity of a 5.25 inch double sided, double density disk having 40 tracks with each track having 9 sectors and each sector storing 512 bytes?
A. $40 \times 9 \times 512 \times 2$ KB B. $40 \times 9 \times 512/1024$ KB
C. $40 \times 9 \times 512 \times 2/1024$ Bytes
D. $40 \times 9 \times 512 \times 2/1024$ KB

472. The arranging of data in a logical sequence is called
A. classifying B. sorting
C. summarising D. reproducing

473. Usually, which type of computers use the 8-bit code called EBCDIC?
A. Minicomputers B. Microcomputers
C. Mainframe computers
D. Supercomputers

474. For what purpose is the 7-bit ASCII code widely used?
A. For data communication work
B. In IBM mainframe models
C. For coding external memory
D. In large machines produced by non-IBM vendors

475. A byte consists of one group of 8-bits put together. What is the largest decimal number which can be represented by a byte?
A. 127 B. 256
C. 255 D. 128

476. One byte can be used to encode any integer between '0' and inclusive
A. 16 B. 256
C. 128 D. 255

477. Which of the following code used in present day computing was developed by IBM Corporation?
A. ASCII B. Hollerith Code
C. Baudot Code D. EBCDIC Code

Ans. 457. C 458. C 459. A 460. A 461. B 462. A 463. B 464. D 465. A 466. C 467. A 468. B 469. D 470. C 471. D 472. B 473. C 474. A 475. C 476. D 477. D

478. Bytes can be combined to form 'words' of memory which may contain 16-, 24-, 32- or 64-bits that a computer is designed to transfer and manipulate as a unit. How many different values can a 32-bit word encode?
A. 65,536 B. 56,653
C. Over 4 billion D. Either B or C

479. A code used for standardising the storage and transfer of information amongst various computing devices is called
A. CRT B. CPU
C. ASCII D. Dvorak

480. What is the number of bit patterns provided by a 7-bit code?
A. 256 B. 128
C. 64 D. 512

481. Both computer instructions and memory addresses are represented by
A. character codes B. binary codes
C. binary word D. parity bit

482. How many alphanumeric characters are there in BLT 1987?
A. 0 B. 3
C. 4 D. 7

483. A logic is an electronic circuit which
A. makes logic decisions
B. allow electron flow only in the one direction
C. works on binary algebra
D. alternates between 0 and 1 values

484. The output of a 2 input OR gate is 0 only when its
A. both inputs are 0 B. either input is 1
C. both inputs are 1 D. either input is 0

485. An AND gate
A. implements logic addition
B. is equivalent to a series switching circuit
C. is an any or all gate
D. is equivalent to a parallel switching circuit

486. A NOR gate is ON only when all its inputs are
A. ON B. positive
C. high D. OFF

487. What is the number of binary digits which can be added by a full-adder?
A. 2 B. 3
C. 4 D. 8

488. What is the name of the logic circuit which can add two binary digits?
A. Full adder B. Half adder
C. Parallel adder D. Flat register

489. What is needed to assemble a logic network for adding three digits?
A. Two half adders B. OR gate
C. AND and NOT gates
D. Both A and B are correct

490. How many full adders are needed to add 4-bit numbers with a parallel adder?
A. 8 B. 4
C. 2 D. 16

491. All modern computers operate on
A. diskettes B. information
C. data D. words

492. Everything in a computer is controlled by its
A. random access memory
B. input devices
C. central processing unit
D. external storage devices

493. A certain computer hardware component does not process the data itself but acts as a central nervous system for other data manipulating components. What is it called?
A. ALU B. RAM
C. Control unit D. Secondary storage

494. In a computer, all calculations are performed and all comparisons are made in-unit
A. control B. arithmetic-logic
C. central processing D. primary storage

495. A computer system consisting of its processor, memory and I/O devices accepts data, processes it and produces the output results. Can you tell in which component is the raw data fed?
A. Mass memory B. Main memory
C. Arithmetic unit D. Logic unit

496. The act of retrieving existing data from memory is called
A. read-out B. read from
C. read D. all of these

497. The ALU of a computer responds to the commands coming from
A. primary memory B. control section
C. external memory D. cache memory

498. A computer performs its arithmetic in
A. RAM B. registers
C. ROM D. control unit

499. Presently, the most expensive component in a computer system is its
A. hardware B. peripherals
C. software D. external storage

500. A smart or intelligent device is so called because it contains within it a
A. computer B. microcomputer
C. programmable microprocessor
D. sensor

501. The ALU of a central processing unit does the essential maths works for the computer. What does the control unit do?
A. Communicates its results
B. Activates the output device
C. Monitors the flow of information
D. Controls the printer

Ans. 478. C 479. C 480. B 481. B 482. D 483. A 484. A 485. B 486. D 487. B 488. B 489. D 490. B 491. C 492. C 493. C 494. A 495. B 496. D 497. B 498. B 499. C 500. C 501. C

502. What name is given to the collection of commands that tells a computer how to perform a specific task?
A. Information B. Program
C. File D. Data processing

503. How many types of arithmetic operations does the ALU of all computer perform?
A. 3 B. 2
C. 5 D. 8

504. What are the three decision-making operations performed by the ALU of a computer?
A. Greater than B. Less than
C. Equal to D. All of these

505. Where have the program and data to be located before the ALU and control unit of a computer can operate on it?
A. Internal memory B. Secondary memory
C. Microprocessor D. Magnetic tapes

506. The microprocessor of a computer cannot operate on any information if that information is not in its
A. secondary storage B. main storage
C. ALU D. logic unit

507. Who coordinates the sequencing of events within the central processor of a computer?
A. Logic unit B. Arithmetic unit
C. Register D. Control unit

508. The CPU of a computer takes instruction from the memory and executes them. What is this process called?
A. Load cycle B. Time sequence
C. Fetch-execute cycle D. Clock cycle

509. The data between the computer and peripherals always passes via an input/output
A. port B. A/D converter
C. D/A converter D. buffer memory

510. Can you tell what passes into and out from the computer via its ports?
A. Data B. Bytes
C. Graphics D. Pictures

511. Which part of the computer is used for calculating and comparing?
A. Disk unit B. Control unit
C. ALU D. Modem

512. How does a computer solve a problem?
A. By human like thinking
B. By first understanding the problem
C. By recalling the answer from memory
D. By following the pre-written programs

513. Preparing a magnetic disk for data storage is called
A. booting B. formatting
C. debugging D. commissioning

514. Computers are extremely fast and have fantastic memories. However, the only thing they can remember is
A. alphanumeric B. series of 1's and 0's
C. Boolean algebra D. logic theorems

515. Earlier computer used to breakdown every few hours. But present day microprocessor based computers run faultlessly for many
A. days B. months
C. years D. decades

516. A computer performs operations by comparing data items and then, depending on the results, follows predetermined course of action. What are such operations called?
A. Sequential B. Logical
C. Digital D. Physical

517. If the processor of a computer does not have a direct and unassisted access to data items, these items are said to be
A. offline B. remote
C. detached D. disconnected

518. In which area of the primary storage section are the intermediate processing results held temporarily?
A. Input storage area B. Program storage area
C. Output storage area
D. Working storage space

519. Sometimes data are gathered over a period of time and collected into a group before entering them into a computer for processing. What type of processing is it called?
A. Interactive processing
B. Sequential processing
C. Batch processing D. Group processing

520. Which of the following tasks is not suited for sequential processing by a computer?
A. Preparing of mailing labels
B. Processing payroll cheques
C. Accounting for credit card purchases
D. Inventory control

521. What is the name of the program which allows the processor to alternate between processing a user's on-going activity and controlling the printing process?
A. Utility program B. Application program
C. Spooler program D. Driver program

522. Multi-user systems provide cost savings for small business because they use a single processing unit to link several
A. personal computers B. workstations
C. dumb terminals D. mainframes

523. The interleaved execution of two or more different and independent programs by the CPU of a computer is called
A. multiprocessing B. multiprogramming
C. multitasking D. either B or C

Ans. 502. B 503. C 504. D 505. A 506. B 507. D 508. C 509. A 510. B 511. C 512. D 513. B 514. B 515. D 516. B 517. A 518. D 519. C 520. D 521. C 522. C 523. D

524. Execution of instructions from different and independent programs by a computer at the same instant in time is called
A. multiprogramming B. multiprocessing
C. concurrent programming
D. multitasking

525. A multiprogramming system is one that can
A. compute many programs simultaneously
B. share hardware resources with many programs simultaneously
C. run very fast
D. use many operating systems

526. In some computer applications, the data is processed almost simultaneously as it is generated. What is this processing called?
A. Simultaneous processing
B. Real time processing
C. Instantaneous processing
D. Quick processing

527. The main purpose of time sharing techniques used in computers is to make the best use of the
A. CPU B. peripherals
C. secondary storage D. floppy disk

528. What type of computer programming is used for airplane ticket reservation different users linked to the computer?
A. By satellite B. By direct cables
C. By telephone line D. Either of these

529. In time sharing a computer is used by many people at the same time. How are different users linked to the computer?
A. By satellite B. By direct cable
C. By telephone lines D. Either of these

530. Which of the following is not an example of time shared real time processing?
A. Hotel reservation B. Airline reservation
C. Seat booking D. Batch processing

531. The newly inaugurated computer centre at Jamia Millia as a part of the Indo-US research project in Particle Physics uses Magnum multi-RISC computer. What is the range of speed in MIPS at which this computer operates?
A. 10 to 25 B. 25 to 50
C. 50 to 100 D. 25 to 100

532. Whenever a peripheral is being controlled by the CPU, it is said to be online. If additionally, the computer files are updated as soon as any change takes place, the system is called
A. quick processing system
B. batch processing system
C. real time system
D. remote processing system

533. The hardware components of a modern digital computer are connected to each other (logically and physically) by a number of parallel wires. What are those wires called?
A. Cable B. Bus
C. Line D. Conductor

534. The mechanical diskette drive in which you insert your diskette is connected to the computer's-bus
A. data B. communication
C. address D. parallel

535. The word length of a computer is measured in
A. bytes B. millimetres
C. metres D. bits

536. What is the common method used by low-megaflop computers for high speed data transmission?
A. Print spooling B. Cycle stealing
C. Looping D. Multiplexing

537. Often computer users have information in two different files which they want to combine in one common file. When the two files are combined one after the other and end to end, it is called
A. sorting B. merging
C. searching D. concatenating

538. Before the arithmetic logic unit (ALU) of a computer can perform arithmetic and logic operations, it must first transfer the data from primary memory to
A. read only memory B. control unit
C. ALU registers D. secondary memory

539. The processing capability of a microcomputer can be enhanced with the help of
A. expansion slots B. expansion cards
C. secondary memory D. buffer memory

540. The CPU of a computer transfers print output to a temporary disk memory at high speed and then gets back of processing another job without waiting for the output to go to the printer. In this way, the CPU does not remain idle due to its own high speed as compared to the low speed of the printer. What is the name of this memory?
A. External memory B. I/O memory
C. ROM D. Buffer memory

541. The temporary or volatile memory of a computer is erased when the computer is turned-off. Accordingly, the computer user before turning off the machine must transfer anything in the memory which is important to
A. control unit B. ALU
C. CPU D. permanent memory

542. A regular TV set can be hooked up to a computer so as to receive computer signal instead of a television program. This hooking up is achieved with the help of
A. modem B. adapter
C. cable D. aerial

Ans. 524. B 525. B 526. B 527. A 528. B 529. D 530. D 531. D 532. C 533. B 534. B 535. D 536. B
537. D 538. C 539. B 540. D 541. D 542. B

543. A group of wires running parallel to each other between two connected computer parts and which transfer information is called a
A. track B. bus
C. cable D. bundle

544. A piece of computer hardware that is physically placed between two devices each of which manages data in a different way is called
A. modem B. interface
C. cluge D. data bus

545. A computer program that converts an entire program into machine language at one time is called a/an
A. interpreter B. simulator
C. compiler D. commander

546. A computer program that translates one program instruction at a time into machine language is called a/an
A. interpreter B. CPU
C. compiler D. simulator

547. What is the responsibility of the logical unit in the CPU of a computer?
A. To interpret instructions
B. To compare numbers
C. To control flow of information
D. To do maths work

548. When any part of memory can be read equally quickly, it is called
A. random access B. sequential access
C. virtual access D. indirect access

549. The CPU of a computer operates at a much higher rate than peripherals. How is the operating speed difference between the two reduced?
A. By using interrupt line
B. With the help of buffer memory
C. By using matching peripherals
D. By using parallel input/output ports

550. The larger the RAM of a computer, the faster its processing speed is, since it eliminates
A. need of ROM B. frequent disk I/Os
C. need for external memory
D. need for wider data path

551. Which of the following performs modulation and demodulation?
A. Fiber optic B. Satellite
C. Coaxial cable D. Modem

552. The process of converting analog signals into digital signals so that they can be processed by a receiving computer is referred to as
A. modulation B. demodulation
C. synchronising D. none of these

553. Which of the following communication modes support two-way traffic but in only one direction at a time?
A. Simplex B. Half-duplex
C. Three-quarters duplex
D. All of these

554. Which of the following might be used by a company to satisfy its growing communications needs?
A. Front-end processor B. Multiplexer
C. Controller D. None of these

555. Which of the following is considered a broadband communications channel?
A. Coaxial cable B. Fiber optic cable
C. Microwave circuits D. None of these

556. Which of the following is not a transmission medium?
A. Coaxial cable B. Fiber optic cable
C. Microwave circuits D. Satellite systems

557. Which of the following does not allow multiple users or devices to share one communication line?
A. Doubleplexer B. Multiplexer
C. Concentrator D. Controller

558. Which of the following is an advantage of using fiber optics data transmission?
A. Resistance to data theft
B. Fast data transmission rate
C. Low noise level D. None of these

559. Which of the following is required to communicate between two computers?
A. Communications software
B. Communications hardware
C. Protocol D. None of these

560. Which of the following types of channels moves data relatively slowly?
A. Wideband channel B. Voiceband channel
C. Narrowband channel
D. Broadband channel

561. A subdivision of main storage created by operational software is referred to as a
A. compartment B. time-shared program
C. divided core D. partition

562. An anticipated result from multi-programming operations is
A. reduced computer idle time
B. the handling of more jobs
C. better scheduling of work
D. all of these

563. Most data communications involving telegraph lines use
A. simplex lines B. widespread channels
C. narrowband channels
D. dialled service

564. A distributed data processing configuration in which all activities must pass through a centrally located computer is called a
A. ring network B. spider network
C. hierarchical network D. data control network

Ans. 543. B 544. B 545. C 546. A 547. B 548. A 549. B 550. B 551. D 552. B 553. B 554. D 555. D 556. C 557. A 558. D 559. D 560. C 561. D 562. D 563. B 564. B

565. A communications device that combines transmissions from several I/O devices into one line is a
A. concentrator B. modifier
C. multiplexer D. full-duplex line

566. A characteristic of a multiprogramming system is
A. simultaneous execution of program instructions from two applications
B. concurrent processing of two or more programs
C. multiple CPUs D. all of these

567. Which of the following communications lines is best suited to interactive processing applications?
A. Narrowband channels
B. Simplex lines
C. Full-duplex lines D. Mixedband channels

568. A remote batch-processing operation in which data is solely input to a central computer would require a
A. telegraph line B. simplex line
C. mixedband channel D. all of these

569. A required characteristic of an online real-time system is
A. no delay in processing
B. offline batch processing
C. more than one CPU D. all of these

570. The systematic access of small computers in a distributed data processing system is referred to as
A. dialled service B. multiplexing
C. polling D. conversational mode

571. A teleprocessing system may consist of
A. computer centre systems
B. communications system
C. user systems D. all of these

572. To make possible the efficient online servicing of many teleprocessing system users on large computer systems, designers are developing
A. communication systems
B. multiprogramming systems
C. virtual storage systems
D. all of these

573. A bond is always equivalent to
A. a byte B. a bit
C. 100 bits D. none of these

574. Communication circuits that transmit data in both directions but not at the same time are operating in a/an
A. simplex mode B. half-duplex mode
C. full-duplex mode D. asynchronous mode

575. An example of a medium speed, switched communications service is
A. Series 100 B. Dataphone 50
C. DDD D. all of these

576. Microprogramming is
A. assembly language programming
B. programming of minicomputers
C. control unit programming
D. macroprogramming of microcomputers

577. Operating system function may include
A. input/output control B. virtual storage
C. multiprogramming D. all of these

578. A front-end processor is
A. a user computer system
B. a processor in a large-scale computer that executes operating system instructions
C. a minicomputer that relieves mainframe computers at a computer centre of communications control functions
D. preliminary processor of batch jobs

579. Data communication monitors available on the software market include
A. ENVIRON/1 B. TOTAL
C. BPL D. Telenet

580. An example of an analog communication method is
A. laser beam B. microwave
C. voice grade telephone line
D. all of these

581. Terminals are used to
A. collect data from the physical system
B. provide information for the manager
C. communicate management decisions to the physical system
D. all of these

582. A hard copy would be prepared on a
A. typewriter terminal B. line printer
C. plotter D. all of these

583. If a firm wanted to transmit data from 1000 punched cards to a remote computer, they would be a/an
A. POS terminal B. intelligent terminal
C. batch processing terminal
D. data collection terminal

584. An example of digital, rather than analog communication is
A. DDA B. DDS
C. WATS D. DDT

585. Terminals are required for
A. realtime, batch processing and timesharing
B. realtime, timesharing, and distributed processing
C. realtime, distributed processing, and manager inquiry
D. realtime, timesharing, and message switching

586. The channel in the data communication model can be
A. postal mail service B. telephone lines
C. radio signals D. all of these

587. A data terminal serves as a/an
A. effector B. sensor

Ans. 565. C 566. B 567. C 568. B 569. A 570. C 571. D 572. D 573. D 574. B 575. C 576. C 577. D 578. C 579. A 580. D 581. A 582. B 583. C 584. B 585. D 586. D

C. both A and B are correct
D. neither A nor B

588. A WATS arrangement is
A. always less expensive than flat-rate service
B. less expensive than flat-rate service only when the number of calls is large and the duration of each is short
C. less expensive than flat-rate service only when the number of calls is small and the duration of each is long
D. never less expensive than flat-rate service

589. An ROP would be attached to a
A. simplex channel B. duplex channel
C. half duplex channel D. full duplex channel

590. Typewriter terminals can print computer-generated data at a rate of
A. 10 characters per second
B. 120 characters per second
C. 120 characters per minute
D. 1200 characters per minute

591. A batch processing terminal would not include a
A. CPU B. card reader
C. card puncher D. line printer

592. Terminals are required for
A. batch processing B. realtime processing
C. both A and B are correct
D. neither A nor B

593. The term "remote job entry" relates to
A. batch processing B. real time processing
C. transaction processing
D. distributed processing

594. "Store and forward" applies to
A. timesharing B. distributed processing
C. both A and B are correct
D. neither A nor B

595. The OCR reading unit attached to a POS terminal is called a
A. light pen B. wand
C. cursor D. none of these

596. UPC has been vigorously opposed by
A. the federal government
B. independent supermarket operators
C. both A and B are correct
D. neither A nor B

597. When UPC is used, the price of the item is located
A. on the item
B. on the item and on the shelf
C. in computer storage
D. on the shelf and in computer storage

598. The dialogue techniques for terminal use do not include
A. questions and answers
B. open-ended questions
C. form fillings D. menu display

599. What is the name of the software package that allows people to send electronic mail along a network of computers and workstations?
A. Memory resident package
B. Project management package
C. Data communication package
D. Electronic mail package

600. Transmission of computerised data from one location to another is called
A. data transfer B. data flow
C. data communication D. data management

601. The coming together of three technologies, i.e. microelectronics, computing and communications has users in
A. business revolution B. information explosion
C. information technology
D. educational upgradation

602. The word telematics is a combination of
A. computer B. telecommunication
C. informatics
D. both B and C are correct

603. What is the name of the device that connects two computers by means of a telephone line?
A. Tape B. Modem
C. Bus D. Cable

604. Which of the following data transmission media has the largest terrestrial range without the use of repeaters or other devices?
A. Hardwiring B. Microwave
C. Satellite D. Laser

605. What function does a serial interface perform in data communication?
A. Converts serial data into audio signals
B. Converts analog signals into digital signals
C. Converts parallel data into a stream of bits
D. Decodes incoming signals into computer data

606. What are the data transmission channels available for carrying data from one location to another?
A. Narrowband B. Voiceband
C. Broadband D. All of these

607. Which of the following is a voiceband channel?
A. Telephone line B. Telegraph line
C. Coaxial cable D. Microwave systems

608. What does the acronym ISDN stand for?
A. Indian Standard Digital Network
B. Integrated Services Digital Network
C. Intelligent Services Digital Network
D. Integrated Services Data Network

609. In a PC to telephone hookup for long distance communication, modem is connected between the telephone line and
A. PC B. synchronous port
C. crossover cable D. asynchronous port

Ans. 587. C 588. B 589. A 590. B 591. A 592. B 593. A 594. D 595. B 596. D 597. D 598. B 599. C 600. C 601. C 602. D 603. B 604. C 605. C 606. D 607. A 608. B 609. D

610. Identify the old term amongst the following group.
A. Coaxial cable B. Optical fibre
C. Twisted pair wire D. Microwaves

611. Many data communication networks have been established which provide a wealth of on-demand information services to people at home. What is the name of the system which provides an interactive, graphics-rich service that permits user to select what they want?
A. Teletex system B. Fax system
C. Videotex system D. Microwave system

612. Sales persons and other employees of the company who spend much of their time away from their offices but keep in touch with their company's microcomputers or mainframe computers over telephone lines are called
A. field workers B. telecommuters
C. teleprocessors D. company directors

613. Business meetings and conference can be held by linking distantly located people through a computer network. Not only the participants exchange information but are able to see each other. What is it called?
A. Telemeeting B. Telemailing
C. Teleconferencing D. Teletalking

614. For connecting modem, a computer must be equipped with a port that conform to the RS-232 standard of the Electronic Industries Association of America. What do the letter 'RS' stand for?
A. Recognised standard B. Random sequence
C. Recommended standard
D. Registered source

615. A smart modem can dial, hang up and answer incoming calls automatically. Can you tell who provides the appropriate instruction to the modem for this purpose?
A. Communications software
B. Error detecting protocols
C. Link access procedure (LAP)
D. Telecommunications

616. When two computers communicate with each other, they send information back and forth. If they are separated by a reasonable distance, they can send and receive the information through a direct cable connection which is called a null-modem connection. Presently what is the maximum distance in metres permitted in this null-modem connection?
A. 50 B. 100
C. 30 D. 150

617. Communication between computers is almost always
A. serial B. parallel
C. series parallel D. direct

618. What is the name given to the exchange of control signals which is necessary for establishing a connection between a modem and a computer at one end of a line and another modem and computer at the other end?
A. Handshaking B. Modem options
C. Protocol D. Duplexing

619. When you connect to an online information service, you are asked to provide some kind of identification such as your name, an account number and a password. What is the name given to this brief dialogue between you and the information system?
A. Security procedure B. Safeguard procedure
C. Identification procedure
D. Log-on procedure

620. If communication software can be called the "traffic con" of a microcommunication system, then what should the modem be called?
A. Park B. Bridge
C. Interface D. Link

621. Which of the following items is not used in Local Area Networks (LANs)?
A. Computer B. Modem
C. Printer D. Cable

622. What is the name of the device that converts computer output into a form that can be transmitted over a telephone line?
A. Teleport B. Modem
C. Multiplexer D. Concentrator

623. Computers cannot communicate with each other directly over telephone lines because they use digital pulses whereas telephone lines use analog sound frequencies. What is the name of the device which permits digital to analog conversion at the start of a long distance transmission?
A. Interface B. Modem
C. Attenuation D. Teleprocessor

624. Modem is used in data transmission. When was it invented and in which country?
A. 1960, USA B. 1965, Germany
C. 1950, USA D. 1962, Japan

625. Nowadays computers all over the world can talk to each other. Which is one of the special accessories essential for this purpose?
A. Keyboard B. Modem
C. Scanner D. Fax

626. A modem is connected in between a telephone line and a
A. network B. computer
C. communication adapter
D. serial port

627. Who invented the modem?
A. Wang Laboratories Ltd
B. AT&T Information Systems, USA

Ans. 610. D 611. B 612. B 613. C 614. C 615. A 616. C 617. A 618. A 619. D 620. B 621. B 622. B 623. B 624. C 625. B 626. C 627. B

C. Apple Computers Inc
D. Digital Equipment Corpn

628. A network which is used for sharing data, software and hardware among several users owning micro-computers is called
A. WAN B. MAN
C. LAN D. VAN

629. Different computers are connected to a LAN by a cable and a/an
A. modem B. interface card
C. special wires D. telephone lines

630. When a group of computers is connected together in a small area without the help of telephone lines, it is called
A. Remote Communication Network (RCN)
B. Local Area Network (LAN)
C. Wide Area Network (WAN)
D. Value Added Network (VAN)

631. A communication network which is used by large organisations over regional, national or global area is called
A. LAN B. WAN
C. MAN D. VAN

632. Working of the WAN generally involved
A. telephone lines B. microwaves
C. satellites D. all of these

633. Many large organisations with their offices in different countries of the world connect their computers through telecommunication satellites and telephone lines. Such a communication network is called
A. LAN B. WAN
C. ECONET D. ETHERNET

634. What is the name of the network topology in which there are bidirectional links between each possible node?
A. Ring B. Star
C. Tree D. Mesh

635. What is the commonly used unit for measuring the speed of data transmission?
A. Bytes per second B. Baud
C. Bits per second D. Either B or C

636. A 2400-character text file has to be transmitted by using a 1200 band modem. Can you tell how long will it take?
A. 2 seconds B. 20 seconds
C. 120 seconds D. 12 seconds

637. What are the most commonly used transmission speeds in BPS used in data communication?
A. 300 B. 1200
C. 2400 D. 4800

638. Usually, it takes 10-bits to represent one character. How many characters can be transmitted at a speed of 1200 BPS?
A. 10 B. 12
C. 120 D. 1200

639. What is the name of the computer based EMMS that provides a common forum where users can check in at their convenience, post message, actively exchange ideas and participate in ongoing discussions?
A. E-mail
B. Bulletin Board System (BBS)
C. Teleconferencing D. Videoconferencing

640. Messages from one computer terminal can be sent to another by using data networks. The message to be sent is converted to an electronic digital signal, transmitted via a cable, telephone or satellite and then converted back again at the receiving end. What is this system of sending messages called?
A. Paperless office B. Electronic mail
C. Global network D. Electronic newspaper

641. Videotex is a combination of
A. television B. communication
C. computer technology
D. all of these

642. We can receive data either through our television aerial or down our telephone lines and display this data on our television screen. What is the general name given to this process?
A. Viewdata B. Teletext
C. Telesoftware D. Videotex

643. The transfer of data from a CPU to peripheral devices of a computer is achieved through
A. modems B. computer ports
C. interfaces D. buffer memory

644. After coding a document into a digital signal, it can be sent by telephone, telex or satellite to the receiver where the signal is decoded and an exact copy of the original document is made. What is it called?
A. Telex B. Word processor
C. Facsimile D. Electronic mail

645. What is the usual number of bits transmitted simultaneously in parallel data transmission used by microcomputers?
A. 6 B. 9
C. 8 D. 7

646. Which transmission mode is used for data communication along telephone lines?
A. Parallel B. Serial
C. Synchronous D. Asynchronous

647. What is the minimum number of wires required for sending data over a serial communication links?
A. 2 B. 1
C. 4 D. 3

648. The term 'duplex' refers to the ability of the data receiving stations to echo back a confirming

Ans. 628. B 629. B 630. B 631. B 632. D 633. B 634. D 635. B 636. B 637. D 638. C 639. B 640. B 641. D 642. D 643. B 644. C 645. B 646. B 647. A

message to the sender. In full duplex data transmission, both the sender and the receiver
A. cannot talk at once
B. can receive and send data simultaneously
C. can send or receive data one at a time
D. can do one way data transmission only

649. Which data communication method is used for sending data in both directions at the same time?
A. Super duplex B. Simplex
C. Half duplex D. Full duplex

650. Which of the following does not describe one or more characteristics of a computer?
A. Electronic B. External storage
C. Stored program
D. Program modification at execution

651. The CPU (Central Processing Unit) consist of
A. input, output and processing
B. control unit, primary storage and secondary storage
C. control unit, arithmetic-logic unit and primary storage
D. input, processing and storage

652. Which of the following is not an advantage of stored programs?
A. Reliability
B. Reduction in operational costs
C. Ability of the computer to operate at electronic speeds
D. The computer becoming general-purpose

653. All of the following are examples of input devices except
A. COM (Computer Output Microfilm)
B. CRT (Cathode Ray Tube)
C. Optical scanners
D. Voice-recognition devices

654. Which of the following is not true of primary storage?
A. It is a part of the CPU
B. It allows very fast access to data
C. It is relatively more expensive
D. All of these are true

655. What is the control unit's function in the CPU?
A. To decode program instructions
B. To transfer data to primary storage
C. To perform logical operations
D. To store program instructions

656. The ascending order of a data hierarchy is
A. bit-type-record-field-file-database
B. byte-bit-field-record-file-database
C. byte-bit-record-file-field-database
D. bit-byte-field-record-file-database

657. Which of the following is not true of immediate processing?
A. It is often used in real time applications
B. It can occur with either sequential or direct-access files
C. It can be used in an airline-reservation system
D. Transactions are processed shortly after a real-world event occurs

658. Electronic spread sheets are most useful in a situation where relatively ...... data must be input, and (but) ..... calculations are required.
A. little; simple B. large; simple
C. large; complex D. little; complex

659. Which of the following statements is true?
A. The smart approach of using computers is to write programs
B. Knowledge of the system development life-cycle is not important to operators who use computers without programming
C. Hands-on exposure to the computer is not helpful to those who write programs
D. Personal computers have been an important contributing factor in the movement toward using computers without programming

660. Which is the product of data processing?
A. Data B. Information
C. Software D. A computer

661. The most common input device used today is the
A. motherboard B. central processing unit
C. keyboard D. system unit

662. Software instructions intended to satisfy a user's specific processing needs are called
A. systems software B. a microcomputer
C. documentation D. applications software

663. Which of the following is not a factor when categorising a computer?
A. Amount of main memory, the CPU can use
B. Capacity of the storage devices
C. Cost of the system
D. When it was purchased

664. Which of the following is the most powerful type of computer?
A. Supermicro B. Superconductor
C. Microcomputer D. Supercomputer

665. Which of the following terms is related to a monitor?
A. Screen B. Monochrome monitor
C. RGB monitor D. Video display

666. Which of the following terms applies to communication between separate computer systems?
A. Computer literacy B. Power supply
C. Connectivity D. Applications software

667. People typically interface with a computer-based system when
A. information must be output

Ans. 648. B 649. D 650. B 651. C 652. B 653. A 654. D 655. A 656. D 657. B 658. D 659. D 660. B 661. C 662. D 663. D 664. D 665. D 666. C

B. information must be reviewed
C. data must be input D. all of these

668. Which of the following typically happens in the output phase of a computer-based information system?
A. Data is put into the computer for processing
B. Information is produced in hardcopy and/or softcopy form
C. Mathematical calculations are performed
D. The computer is turned-off

669. Which of the following best describes a computer-based information system?
A. A system in which a computer is used to turn data into information
B. Performing complex mathematical calculations
C. Processing data D. Inputting data

670. Which of the following is an example processing activities?
A. Classifying B. Summarising
C. Performing calculations
D. All of these

671. Which of the following pieces of hardware is used the most in the input phase of a computer-based information system?
A. Printer B. Diskette
C. Monitor D. Keyboard

672. Which of the following statements best described the batch method of input?
A. Data is processed as soon as it is input
B. Data is input at the time it is collected
C. Data is collected in the form of source documents, placed into groups, and then input to the computer
D. Source documents are not used

673. Which of the following might occur when an organisation uses online processing?
A. Data is acted on immediately
B. Master files are updated immediately
C. Output is produced without delay
D. All related files are updated

674. The principal advantages of the centralised approach for organising a computer facility is
A. cost-effectiveness
B. processing activities are easier to coordinate
C. processing activities are easier to control
D. all of these

675. Which of the following is not a type of computer facility?
A. Decentralised B. Centralised
C. Re-distributed D. Distributed

676. Data system management has long-term viability as a separate business function because
A. it requires much technical knowledge
B. it requires large investments
C. specialists in data systems cannot be integrated into a marketing for manufacturing organisation
D. an integrated database accessible to all requires independent management

677. To be information, data must be
A. factual B. relevant
C. news D. all of these

678. A teleprocessing system may be a
A. card processing system
B. data communication system
C. computer system D. all of these

679. Data management systems may be implemented as
A. system software B. application software
C. computer programs D. all of these

680. A data system for calculating measures used in statistical inference is an example of a
A. teleprocessing system
B. data management system
C. computing system D. all of these

681. Strategic planning is
A. planning operation B. supervising
C. problem identification
D. all of these

682. An example of providing supervisory information by the use of teleprocessing technique is the use of a
A. computer to prepare customer invoices
B. computer to control a machine
C. phone system to report job status
D. all of these

683. For the purpose of defining data needs, a responsibility area is
A. marketing B. administration
C. personnel D. all of these

684. Data system for planning are often called
A. decision analysis systems
B. planning analysis systems
C. decision support systems
D. all of these

685. Computing systems can be effective in generating strategic information because
A. the future can be predicted from the pattern of the past
B. they require managers to clarify their thinking about their plans and future possibilities
C. they give management access to a large database
D. all of these

686. Coded entries which are used to gain access to a computer system are called
A. entry codes B. passwords
C. security commands D. codewords

687. Electronic images of people speaking that are used in computer security operations are referred to as

Ans. 667. D 668. B 669. A 670. D 671. D 672. C 673. B 674. D 675. C 676. D 677. D 678. D 679. D 680. C 681. C 682. C 683. C 684. C 685. B 686. B

A. voiceprints B. password images
C. vocal passwords D. electronic prints

688. A factor which might cause an individual to consider using a computer in criminal activities is
A. the computer's access to large sums of money
B. the speed with which the crime can be accomplished
C. EFTS (Electronic Funds Transfer Systems)
D. all of these

689. The repeated access of a particular flight number from an airline reservation system is an example of
A. frequency B. repetitive processing
C. updating D. volume

690. A factor which would strongly influence a business person to adopt a computer is its
A. accuracy B. reliability
C. speed D. all of these

691. The total number of messages handled by a computerised telephone system on a daily basis is an example of
A. frequency B. updating
C. volume D. all of these

692. Which of the following statements is true?
A. The installation of a computer is favourably received by all employees
B. Some form of training is necessary for employees who work with computers
C. Computers are portrayed solely as society's benefactor
D. A businessperson is only interested in the computer's accuracy

693. The status of all data handled by a DP centre is determined by a
A. data number B. control number
C. reference number D. item number

694. The individual within the operations group who ordinarily uses a variety of keyboard devices is the
A. data clerk B. keypunch operator
C. data entry clerk D. computer operator

695. The daily processing of corrections to customer accounts best examplifies the processing mode of
A. batch processing B. real-time processing
C. time-sharing D. offline processing

696. Which of the following terms could be used to describe the concurrent processing of computer programs, via CRTs, on one computer system?
A. Time-sharing B. Online processing
C. Interactive processing
D. All of these

697. The unit of hardware, an operator uses to monitor computer processing is the
A. card reader B. CPU
C. line printer D. console

698. The individual who catalogs storage media is the
A. data clerk B. control clerk
C. tape librarian D. all of these

699. The data processing job expected to further decrease in the 1990s is that of
A. key puncher B. data entry clerk
C. computer operator D. programmer

700. Which of the following statements is true?
A. Analysis usually work alone and sometimes as part of a team
B. Most systems projects are completed in 6 to 12 weeks
C. An analyst's primary concern is the development of software
D. Analysts evaluate data flow through an organisation

701. The computer device primarily used to provide hardcopy is the
A. CRT B. line printer
C. computer console D. card reader

702. The term interchangeable with control clerk is
A. data clerk B. data control clerk
C. tape librarian D. data entry clerk

703. In computer terminology, information means
A. raw data B. alphanumeric data
C. data is more useful or intelligible form
D. program

704. Data processing is
A. the same thing as data collection
B. similar to computer programming
C. mostly associated with commercial work
D. skin to data coding

705. Which one of the following can read data and convert them to a form that a computer can use?
A. Logic B. Storage
C. Control D. Input device

706. Which one of the following can produce the final product of machine processing in a form usable by humans?
A. Logic B. Storage
C. Control D. Output device

707. The term "memory" applies to which one of the following?
A. Logic B. Storage
C. Control D. Input device

708. List of detailed instructions that directs a computer is called
A. logic B. storage
C. memory D. program

709. A computer program consists of
A. a completed flowchart
B. algorithms written in computer's language
C. algorithms D. discrete logical steps

Ans. 687. C 688. D 689. A 690. D 691. C 692. B 693. B 694. C 695. A 696. D 697. D 698. C 699. A 700. D 701. B 702. A 703. C 704. C 705. D 706. D 707. B 708. D 709. B

710. In computer terminology a compiler means
A. a person who computes source programs
B. the same things as a programmer
C. key punch operator
D. a program which translates source program into object program

711. In which language source program is?
A. English B. Symbolic
C. High-level D. Machine

712. A program written in machine language is called
A. assembler B. object
C. computer D. machine

713. A computer programmer
A. does all the thinking for a computer
B. can enter input data quickly
C. can operate all types of computer equipment
D. can draw only flowchart

714. Most of the errors blamed on computers are actually due to
A. programming errors B. hardware fatigue
C. defects in floppy disks
D. data entry errors

715. An integrated circuit (IC) is
A. a complicated circuit B. an integrating device
C. much costlier than a single transistor
D. fabricated on a tiny silicon chip

716. What is the most significant difference between a simple desk calculator and computer? (Note that some of the following may not even be true).
A. The computer is an electronic machine while the desk calculator may or may not be electronic
B. The computer is useful in business applications while the desk calculator is not
C. The computer can print its result while the desk calculator can only show it on a display
D. The computer is controlled by a program stored in its memory while calculator requires step-by-step manual control

717. A visual display unit or terminal
A. is, by definition, a dumb terminal
B. can possess either graphic or alpha-numeric capabilities, but not both
C. must be located at the site of the CPU
D. is the most popular input device used today in direct-access processing

718. The language in which computer programs are usually written differ from the language that the computer directly 'obeys' or 'understands'.
A. The card reader B. The programmer
C. A computer program
D. The computer operator

719. The basic components of a modern digital computer are
A. central processor B. input device
C. output device D. all of these

720. Arithmetic and Logic Unit (ALU) are called the ...... of a computer.
A. heart B. master despatcher
C. primary memory D. all of these

721. A typical modern computer uses
A. magnetic cores for secondary storage
B. LSI chips
C. magnetic tapes for primary memory
D. more than 10,000 vacuum tubes

722. An ADU is used as
A. input device B. output device
C. voice data entry device
D. both A and B are correct

723. Computer memory
A. performs all calculations
B. is better than human memory
C. receives input data D. is extremely limited

724. The central processor of a modern digital computer consists of
A. control unit B. primary memory
C. all of these D. none of these

725. Control unit of a digital computer is often called
A. nerve centre B. master despatcher
C. clock D. none of these

726. A CPU's processing power is measured in
A. IPS B. CIPS
C. MIPS D. nanoseconds

727. Which of the following is a part of the Central Processing Unit?
A. Keyboard B. Printer
C. Tape D. Arithmetic Logic Unit

728. The retrieval of information from the computer is defined as
A. collection of data
B. data retrieval operations
C. output D. data output collection

729. A term associated with the comparison of processing speeds of different computer systems is
A. EFTS B. MPG
C. MIPS D. CPS

730. A computer enthusiast is
A. user friendly B. a hacker
C. a computerist D. none of these

731. Which of the following require large computer memory?
A. Imaging B. Graphics
C. Voice D. All of these

732. What is the term which represents the use of links between information of all sorts whether text, graphics, video or audio-based?
A. Hypertext B. Hypermedia
C. Hyper Card D. Wild Card

Ans. 710. D 711. C 712. B 713. A 714. D 715. D 716. D 717. D 718. C 719. D 720. A 721. B 722. D 723. D 724. C 725. A 726. C 727. D 728. C 729. C 730. B 731. D 732. B

733. What hardware was used by first generation computers?
A. Transistors B. ICs
C. Valves D. SSI

734. General purpose computers are those that can be adopted to countless uses simply by changing its
A. keyboard B. printer
C. program D. display screen

735. Processors of all computers, whether micro, mini or mainframe, must have
A. primary storage B. ALU
C. control unit D. all of these

736. Modern computers are very reliable but they are not
A. fast B. powerful
C. infallible D. cheap

737. When did the ELSA consortium 'Gang of Nine' come into being in USA?
A. 1988 B. 1986
C. 1984 D. 1982

738. According to you, which of the following statement is correct?
A. Generally, computers do not make mistakes
B. Computers eliminate jobs
C. Computers can think
D. Maths is necessary to understand computers

739. From amongst the following, pick out the item that does not belong to computer.
A. Mouse B. OCR
C. MICR D. Plotter

740. Which one of the following words has both a common meaning and a computer meaning?
A. Mode B. Bus
C. Quick D. Efficiency

741. A dumb terminal can do nothing more than communicate data to and from a CPU of a computer. How does a 'smart' terminal differ from a dumb terminal?
A. It has a primary memory
B. It has a cache memory
C. It has a microprocessor
D. It has an input device

742. What is the name of the hardware and software package that is ready for use as soon as it is installed?
A. Hands-on-system
B. Quick implementation system
C. Turnkey computer system
D. Operational system

743. Which is the quantity of the number of children in a family?
A. Analog B. Digital
C. Hybrid D. Hyperbolic

744. Most of the inexpensive personal computers do not have any disk or diskette drive. What is the name of such computers?
A. Home computers
B. Diskless computers
C. Dedicated computers
D. General-purpose computers

745. If a computer is on but does not respond to a system reset, what is they said to be?
A. Dead B. Off
C. Hang D. Insensitive

746. A menu-driven operating system is one which allows you to pick-up from the menu of choices it displays on the screen. What is the name given to the images which are used in such image oriented menus?
A. Icon B. Figure
C. Symbol D. Model

747. What is the name given to the weapons which use computerised guidance system?
A. Guided weapons B. Smart weapons
C. Dumb weapons D. Starwars weapons

748. A computer has no more sense than a light
A. bulb B. pen
C. switch D. pad

749. One of the most impressive features of a modern digital computer is its speed. A fast computer is able to do more calculations in one minute than a person using a pencil and paper can do in
A. 10 years B. 20 years
C. lifetimes D. 50 years

750. Which of the following does not obtain a microprocessor?
A. Robot B. Microwave oven
C. Washing machine D. Ball pen

751. What is meant by computer literacy?
A. Ability to write computer programs
B. Knowing what a computer can and cannot do
C. Knowing computer related vocabulary
D. Ability to assemble computers

752. The digital computer was developed primarily in
A. USSR B. Japan
C. USA D. UK

753. The pieces of equipment which are attached to the CPU of a computer and which it can access are called
A. output devices B. control units
C. peripherals D. ALU

754. A computer is a box full of electronic
A. chips B. switching devices
C. circuits D. components

755. The analog computer measures dimensions and its circuits use the differential and integral equations

Ans. 733. C 734. C 735. D 736. C 737. A 738. C 739. D 740. B 741. C 742. C 743. B 744. A 745. C 746. A 747. B 748. C 749. C 750. D 751. B 752. C 753. C 754. B

of continuous variables. The digital computer counts units and its circuits use
A. logic gates B. discrete switches
C. Boolean algebra D. Bayes' theorem

756. The word 'computer' usually refers to the central processor units plus
A. keyboard B. external memory
C. internal memory D. peripheral devices

757. If a computer had no decision-making function, what will it be reduced to?
A. Electrical machine B. Adding machine
C. Counting machine D. Slider rule

758. What is the name given to the molecular-scale computer?
A. Supercomputer B. Nanocomputer
C. Femtocomputer D. Microcomputer

759. A new technology which provides the ability to create an artificial world and have people interact with, it is called
A. televirtuality B. virtual reality
C. alternative reality D. 3D reality

760. Which generation of computer is covered by the period 1964–77?
A. First B. Second
C. Third D. Fourth

761. Where does a computer add, compare and shuffle its data?
A. Memory chip B. CPU chip
C. Floppy disk D. Hard disk

762. Which American computer company is called Big Blue?
A. Microsoft Corpn B. Compaq Corpn
C. IBM D. Tandy Svenson

763. A collection of wires that connects several devices is called
A. link B. bus
C. bidirectional wires D. cables

764. The register which holds the address of the location to or from which data are to be transferred is known as
A. index register B. instruction register
C. memory address register
D. memory data register

765. The ALU of a computer normally contains a number of high speed storage elements called
A. semiconductor memory
B. registers
C. hard disk D. magnetic disk

766. Memories in which any location can be reached in a fixed (and short) amount of time after specifying its address is called
A. sequential-access memory
B. random-access memory
C. secondary memory D. mass storage

767. A device or peripheral equipment which is not in direct communication with CPU of a computer is known as
A. online device B. offline device
C. channel D. intelligent terminal

768. A storage device used to compensate the difference in flow of data, is known as
A. auxiliary storage B. main storage
C. buffer D. core memory

769. Which of the following registers hold the information before it goes to the decoder?
A. Control register B. Data register
C. Accumulator D. Address register

770. How many units in a single bus structure communicate at a time?
A. One B. Two
C. Three D. Four

771. A single bus structure is primarily found in
A. main frames B. super computers
C. high performance machines
D. mini- and micro-computers

772. The principal advantage of the centralised approach of organising a computer facility is
A. cost-effectiveness
B. processing activities are easier to co-ordinate
C. processing activities are easier to control
D. all of these

773. Data System Management has long-term viability as a separate business function because
A. it requires much technical knowledge
B. it requires large investments
C. specialists in data systems cannot be integrated into a marketing or manufacturing organisation
D. an integrated database accessible to all require independent management

774. Which of the following is not used in the storage phase of a computer-based information system?
A. Magnetic B. Keyboard
C. Diskette D. Hard disk

775. Which of the following is not type of computer facility?
A. Decentralised B. Centralised
C. De-distributed D. Distributed

776. Data systems for planning are often called
A. decision analysis systems
B. planning analysis systems
C. decision support systems
D. all of these

777. Computing systems can be effective in generating strategic information because
A. the future can be predicted from the pattern of the best
B. they require managers to clarify their thinking about their plans and future possibilities

Ans. 755. C 756. C 757. B 758. B 759. B 760. C 761. B 762. C 763. D 764. C 765. B 766. B 767. B 768. A 769. B 770. B 771. D 772. D 773. D 774. B 775. C 776. C 777. B

C. they give management access to a large database
D. all of these

778. For the purposes of defining data needs, a responsibility area is
A. marketing B. administration
C. personnel D. all of these

779. The total number of messages handled by a computerised telephone system on a daily basis is an example of
A. frequency B. updating
C. volume D. all of these

780. A factor which would strongly influence a businessperson to adopt a computer is its
A. accuracy B. reliability
C. speed D. all of these

781. A factor which might cause an individual to consider using a computer in criminal activities is
A. the computer's access to large sums of memory
B. the speed with which the crime can be accomplished
C. EFTS (Electronic Funds Transfer System)
D. all of these

782. An example of providing supervisory information by the use of teleprocessing technique is the use of
A. computer to prepare customer invoices
B. computer to control a machine
C. phone system to report job status
D. all of these

783. Coded entries which are used to gain access to a computer system are called
A. entry codes B. passwords
C. security commands D. codewords

784. The computer device primarily used to provide hardcopy is the
A. CRT B. line printer
C. computer console D. card reader

785. In computer terminology, a compiler means
A. a person who computes source programs
B. the same thing as programmer
C. key punch operator
D. a program which translates source program into object program

786. CD-ROM is a
A. semiconductor memory
B. memory register
C. magnetic memory D. none of these

787. Which of the following is the user programmed semiconductor memory?
A. SRAM B. DRAM
C. EPROM D. All of these

788. Which of the following is not a primary?
A. Magnetic tape B. Magnetic disk
C. Optical disk D. None of these

789. Average access time in the magnetic disk is of the order of
A. 50 millisecond B. 50 microsecond
C. 50 second D. all of these

790. A name or number used to identify a storage location is called a/an
A. byte B. record
C. address D. all of these

791. The access method used for cassette tape is
A. direct B. random
C. sequential D. all of these

792. Which of the following is used for input and output both?
A. Graph plotter B. Teletype terminal
C. Line printer D. All of these

793. Which of the following is a secondary memory device?
A. Keyboard B. Disk
C. ALU D. All of these

794. By word-processing, we understand
A. processing string of only words
B. processing string of numbers and special symbols
C. string manipulation D. all of these

795. The difference between memory and storage is that memory is ...... and storage is ....
A. temporary; permanent
B. permanent; temporary
C. slow; fast D. all of these

796. A floppy disk contains
A. circular tracks only B. sectors only
C. both circular tracks and sectors
D. all of these

797. The octal equivalent of 111010 is
A. 81 B. 72
C. 71 D. none of these

798. The first electronics computer in the world was
A. UNIVAC B. EDVAC
C. ENIAC D. all of these

799. Microfilm reader is a
A. OCR device B. COM device
C. MICR device D. none of these

800. The register which contains the instruction that is to be executed is known as
A. index register
B. instruction register
C. memory address register
D. memory data register

801. If some device requires urgent service, normal execution of programs may sometimes be preempted using
A. an interrupt signal
B. a request to memory modules
C. DMA D. all of these

Ans. 778. C 779. C 780. D 781. D 782. C 783. B 784. B 785. D 786. D 787. C 788. D 789. A 790. C 791. C 792. B 793. B 794. C 795. A 796. C 797. B 798. C 799. B 800. A 801. A

802. A memory bus is mainly used for communication between
A. processor and memory
B. processor and I/O devices
C. I/O devices and memory
D. input device and output device

803. The input unit of a computer
A. feeds data to the CPU or memory
B. retrieves data from CPU
C. directs all other units
D. all of these

804. Which of the following is used as storage locations both in the ALU and the control section of a computer?
A. Accumulator B. Register
C. Adder D. Decoder

805. Offline device is a/an
A. device which is not connected to CPU
B. device which is connected to CPU
C. direct access storage device
D. I/O device

806. Which of the following is a set of general purpose internal registers?
A. Stack B. Scratch pad
C. Accumulator D. Status register

807. The unit of a computer system that executes program, communications with and often controls the operation of other sub-systems of the computer is known as
A. CPU B. Control unit
C. I/O unit D. Peripheral unit

808. In which addressing mode, the effective address of the operand is generated by adding a constant value to the contents of a register?
A. Absolute mode B. Indirect mode
C. Immediate mode D. Index mode

809. Which of the following registers is loaded with the contents of the memory location pointed by the PC?
A. Memory Address Register
B. Memory Data Register
C. Instruction Register
D. Program Counter

810. Which of the following registers is used to keep track of address of the memory location where the next instruction is located?
A. Memory Address Register
B. Memory Data Register
C. Instruction Register D. Program Counter

811. What is the minimum number of operations required for a microprocessor with 8 data pins, to read a 32-bits word?
A. 1 B. 2
C. 4 D. 8

812. Microprocessors can be used to make
A. computers B. digital system
C. calculations D. all of these

813. The magnetic storage chips used to provide non-volatile direct access storage of data and that have no moving parts are known as
A. magnetic core memory
B. magnetic tape memory
C. magnetic disk memory
D. magnetic bubble memory

814. A beam of light used to record and retrieve data on optical disks is known as
A. polarized light
B. unpolarized concentric light
C. laser
D. coloured light

815. A microprocessor with 12 address lines is capable of addressing
A. 1024 locations B. 2048 locations
C. 4096 locations D. 64 K locations

816. The stack pointer in the 8085 microprocessor is a
A. 16-bits register that points to stack memory locations
B. 16-bits accumulator
C. memory location in the stack
D. flag register used for the stack

817. How many address lines are needed to address each memory location in a 2048 × 4 memory chip?
A. 10 B. 11
C. 8 D. 12

818. CD-TV is basically a PC without keyboard or floppy disk drive but with a remote control and a CD-ROM drive. Who has announced this system?
A. Commodore B. Sony
C. Philips D. Toshiba

819. The process of starting or restarting a computer system by loading instructions from a secondary storage device into the computer memory is called
A. Duping B. Booting
C. Padding D. all of these

820. Which chips are erasable by ultraviolet rays after removing them from the main circuit?
A. EPROM chips B. EEPROM chips
C. PROM chips D. All of these

821. Serial access memories are useful in applications where
A. data consists of numbers
B. short access time is required
C. each stored word is processed differently
D. data naturally needs to flow in and out in serial form

822. How many input lines are needed to construct 1024 bits coincident core plate?
A. 8 B. 16
C. 32 D. 64

Ans. 802. A 803. A 804. B 805. A 806. B 807. A 808. D 809. C 810. D 811. C 912. D 813. D 814. C 815. C 816. A 817. B 818. A 819. C 820. A 821. D 822. D

823. How many types of storage loops exist in magnetic bubble memory?
A. 8 B. 4
C. 16 D. 2
824. The disadvantage of dynamic RAM over static RAM is
A. higher power consumption
B. variable speed
C. need to refresh the capacitor charge every once in two milliseconds
D. higher bit density
825. The minimum number of MOS transistors required to make a dynamic RAM cell is
A. 1 B. 2
C. 3 D. 4
826. Which of the following memories must be refreshed many times per second?
A. Static RAM B. Dynamic RAM
C. EPROM D. ROM
827. EE-ROM is
A. electrically erasable B. easily erasable
C. non-erasable D. effective erasable
828. EPROM consist of
A. bipolar transistors B. UJTs
C. MOSFETs D. diodes
829. The memory which is ultraviolet light erasable and electrically programmable is
A. ROM B. PROM
C. RAM D. EPROM
830. Which memory is nonvolatile and may be written only once?
A. RAM B. EPROM
C. ROM D. PROM
831. The memory which is programmed at the time it is manufactured?
A. ROM B. RAM
C. PROM D. EPROM
832. Which memory is volatile?
A. RAM B. ROM
C. EPROM D. PROM
833. Which of the following memory medium is not used as main memory system?
A. Magnetic core
B. Semiconductor
C. Magnetic tape
D. Both semiconductor and magnetic tape
834. Which of the following statements is wrong?
A. An EPROM can be programmed, erased and reprogrammed by the user with an EPROM programming instrument
B. Magnetic tape is nonvolatile
C. Magnetic core and semiconductor memories are used as mass memory medium
D. RAM is a type of volatile memory
835. For a memory system, the cycle time is
A. same as the access time
B. longer than the access time
C. shorter than the access time
D. submultiple of the access time
836. The control unit of a microprocessor
A. stores data in the memory
B. accepts input data from a keyboard
C. performs arithmetic/logic functions
D. none of these
837. The data bus in 8080A/8085 microprocessor is a group of
A. eight bidirectional lines that are used to transfer 8-bits between the microprocessor and its I/O and memory
B. sixteen bidirectional lines that are used for data transfer between the microprocessor and memory
C. eight unidirectional lines that are used for I/O devices
D. eight lines used to transfer data among the registers
838. Registers which are partially visible to users and used to hold conditional codes (bits set by the CPU hardware as the result of operations), are known as
A. PC
B. flags
C. general purpose registers
D. memory address registers
839. In a genetic microprocessor, instruction cycle time is
A. shorter than machine cycle time
B. larger than machine cycle time
C. exactly double the machine cycle time
D. exactly the same as the machine cycle time
840. One of the main feature that distinguish microprocessors from microcomputers is
A. words are usually larger in microprocessors
B. words are shorter in microprocessors
C. microprocessors does not contain I/O devices
D. exactly the same as the machine cycle time
841. The flow and timing of data from the microprocessor is regulated by
A. control pins B. address pins
C. data pins D. power pins
842. Which of the following is true about parity bit?
A. Internal code used for storing alpha-numeric information
B. The means used by the computer to check for errors during internal manipulation of data
C. It is used in self-complementing code
D. All of these

Ans. 823. D 824. C 825. A 826. B 827. A 828. C 829. D 830. D 831. A 832. A 833. C 834. C 835. B 836. D 837. A 838. B 839. B 840. C 841. A 842. B

843. The cyclic codes are used in
A. arithmetic and logical computations
B. continuously varying signal representation
C. data transfer
D. all of these

844. Which of the following is true about Hamming code?
A. It is distancè 10 code B. It is weighted code
C. It is minimum distance 3 code
D. All of these

845. Which of the following is true about 2-out-of-5-code?
A. It is weighted code B. It is unweighted code
C. It has odd parity D. All of these

846. An interpreter is a program that
A. places programs into memory and prepares them for execution
B. automates the translation of assembly language into machine language
C. accepts a program written in a high level language and produces an object program
D. appears to execute a source program as if it were machine language

847. A compiler is a program that
A. places programs into memory and prepares them for execution
B. automates the translation of assembly language into machine language
C. accepts a program written in a high level language and produces an object program
D. appears to execute a source program as if it were machine language

848. A loader is a program that
A. places programs into memory and prepares them for execution
B. automates the translation of assembly language into machine language
C. accepts a program written in a high level language and produces an object program
D. appears to execute a source program as if it were machine language

849. Virtual memory
A. is a method of memory allocation by which the program is subdivided into equal portions, or pages and core is subdivided into equal portions or blocks
B. consists of these addresses that may be generated by a processor during execution of a computation
C. is a method of allocating processing time
D. allows multiple programs to reside in separate areas of core at the time

850. Paging
A. is a method of memory allocation by which the program is subdivided into equal portions, or pages and core is subdivided into equal portions or blocks
B. consists of those addresses that may be generated by a processor during execution of a computation
C. is a method of allocating processing time
D. allows multiple programs to reside in separate areas of core at the time

851. Multiprogramming
A. is a method of memory allocation by which the program is subdivided into equal portions, or pages and core is subdivided into equal portions or blocks
B. consists of those addresses that may be generated by a processor during execution of a computation
C. is a method of allocating processing time
D. allows multiple programs to reside in separate areas of core at the time

852. The Memory Buffer Register (MBR)
A. is a hardware memory device which denotes the location of the current instruction being executed
B. is a group of electrical circuits (hardware), that performs the intent of instructions fetched from memory
C. contains the address of the memory location that is to be read from or stored into
D. contains a copy of the designated memory location specified by the MAR after a "read" or the new contents of the memory prior to a "write"

853. The memory addresses register
A. is a hardware memory device which denotes the location of the current instruction being executed
B. is a group of electrical circuits (hardware), that performs the intent of instructions fetched from memory
C. contains the address of the memory location that is to be read from or stored into
D. contains a copy of the designated memory location specified by the MAR after a "read" or the new contents of the memory prior to a "write"

854. The instruction register
A. is a hardware memory device which denotes the location of the current instruction being executed
B. is a group of electrical circuits (hardware), that performs the intent of instructions fetched from memory

Ans. 843. B 844. C 845. B 846. D 847. C 848. A 849. B 850. A 851. D 852. D 853. D 854. B

C. contains the address of the memory location that is to be read from or stored into
D. contains a copy of the designated memory location specified by the MAR after a "read" or the new contents of the memory prior to a "write"

855. Advantage(s) of using assembly language rather than machine language is (are)
A. mnemonic and easy to read
B. address any symbolic, not absolute
C. introduction of data to program is easier
D. all of these

856. The storage-to-storage instructions
A. have both their operands in the main store
B. which perform an operation on a register operand and an operand which is located in the main store, generally leaving the result in the register, except in the case of store operation when it is also written into the specified storage location
C. which perform indicated operations on two fast registers of the machine and leave the result in one of the registers
D. all of these

857. Which of the following are (is) language processor(s)?
A. Assemblies B. Compilers
C. Interpreters D. All of these

858. The specific tasks storage manager performs are
A. allocation/deallocation of storage to programs
B. protection of the storage area allocated to a program from legal access by other programs in the system
C. the status of each program
D. both A and B are correct

859. The microprocessor must perform
A. recognise macro definitions and macro calls
B. save the macro definitions
C. expand macros calls and substitute arguments
D. all of these

860. In which way(s) a macroprocessor for assembly language can be implemented?
A. Independent two-pass processor
B. Independent one-pass processor
C. Processor incorporated into pass 1
D. All of these

861. Which of the following functions is (are) performed by the loader?
A. Allocate space in memory for the programs and resolve symbolic references between object decks
B. Adjust all address dependent locations, such as address constants to correspond to the allocated space
C. Physically place the machine instructions and data into memory
D. All of these

862. When a computer is first turned-on or restarted, a special type of absolute loader is executed, called a
A. "compile and go" loader
B. boot loader
C. bootstrap loader D. relating loader

863. Terminal Table
A. contains all constants in the program
B. is a permanent table of decision rules in the form of patterns for matching with the uniform symbol table to discover syntactic structure
C. consists of a full or partial list of the tokens as they appear in the program. Created by Lexical analysis are used for syntax analysis and interpretation
D. is a permanent table which lists all key words and special symbols of the language in symbolic form

864. Assembly code data base
A. contains all constants in the program which is created by the code generation phase and is input to the assembly phase
B. is a permanent table of decision rules in the form of patterns for matching with the uniform symbol table to discover syntactic structure
C. consists of a full or partial list of the tokens as they appear in the program created by lexical analysis are used for syntax analysis and interpretation
D. is a permanent table which lists all key words and special symbols of the language in symbolic form

865. The function(s) of the syntax phase is/are to
A. recognise the major constructs of the language and to cell the appropriate action routines that will generate the intermediate form or matrix for these constructs
B. build a literal table and an identifier table
C. build a uniform symbol table
D. parse the source program into the basic elements or tokens of the language

866. The function(s) of the Storage Assignment is (are)
A. to assign storage to all variables referenced in the source program
B. to assign storage to all temporary locations that are necessary for intermediate results
C. to assign storage to literals, and to ensure that the storage is allocated and appropriate locations are initialised
D. all of these

867. A non-relocating program is one which
A. cannot be made to execute in any area of storage other than the one designated for it at the time of its coding or translation

Ans. 855. D 856. A 857. D 858. D 859. D 860. D 861. D 862. C 863. D 864. A 865. A 866. D 867. A

B. consists of a program and relevant information for its relocation
C. can itself perform the relocation of its address-sensitive portions
D. all of these

868. A Relocating Program is one which
A. cannot be made to execute in any area of storage other than the one designated for it at the time of its coding or translation
B. consists of a program and relevant information for its relocation
C. can itself perform the relocation of its address-sensitive portions
D. all of these

869. A self-relocating program is one which
A. cannot be made to execute in any area of storage other than the one designated for it at the time of its coding or translation
B. consists of a program and relevant information for its relocation
C. can itself perform the relocation of its address-sensitive portions
D. all of these

870. The function(s) of Scheduler is (are)
A. selects which ready process is to be run next
B. specifies the time slice
C. waiting for an event to occur before continuing execution
D. both A and B are correct

871. A hardware device that is capable of executing a sequence of instructions, is known as
A. CPU B. ALU
C. CU D. Processor

872. The function(s) performed by the Paging Software is (are)
A. implementation of the access environment for all programs in the system
B. management of the physical address space
C. sharing and protection
D. all of these

873. The file label is a descriptor containing all the information which would be required to
A. connect the file to a program
B. guard against the destruction of the file, inadvertently or deliberately, by another program
C. facilitate easy creation, storage and access of files
D. both A and B are correct

874. The function(s) of file system is (are)
A. to provide complete file naming freedom to the users and to permit controlled sharing of files
B. to provide for long and short term storage of files with appropriate economic trade offs
C. to provide security against loss of information due to system failure
D. all of these

875. While running DOS on a PC, which command would be used to duplicate the entire diskette?
A. COPY B. DISKCOPY
C. CHKDSK D. TYPE

876. Which of the following filename extension suggests that the file is a backup copy of another file?
A. TXT B. COM
C. BAS D. BAK

877. With MS-DOS which command will divide the surface of the blank floppy disk into sectors and assign a unique address to each one?
A. FORMAT B. FAT
C. VER D. CHKDSK

878. If you do not know which version of MS-DOS you are working with, which command will you use after having booted your operating system?
A. FORMAT B. DIR
C. VER D. DISK

879. The higher versions of the operating systems are so written that programs designated for earlier versions can still be run. What is it called?
A. Upgradability B. Upward mobility
C. Universality
D. Upward compatibility

880. What is the name given to the process of initialising a microcomputer with its operating system?
A. Cold booting B. Booting
C. Warm booting D. Boot recording

881. Can you name the major Operating System used in computers?
A. MS DOS B. OS/2
C. UNIX D. All of these

882. What is the name of operating system which was originally designed by scientists and engineers for use by scientists and engineers?
A. XENIX B. UNIX
C. OS/2 D. MS DOS

883. Which are the most important features of Microsoft Windows program?
A. Windows B. Pull-down menus
C. Icons D. All of these

884. What is the name given to the organised collection of software that controls the overall operation of a computer?
A. Working system B. Peripheral system
C. Operating system D. Controlling system

885. What is the name given to the values that are automatically provided by software to reduce keystrokes and improve a computer user's productivity?
A. Defined values B. Fixed values
C. Default values D. Special values

886. The powerful text editor called PC-Write can be used by anybody by paying a small fee. Such programs are called

Ans. 868. B 869. B 870. D 871. D 872. D 873. D 874. D 875. B 876. D 877. A 878. C 879. D 880. B
881. D 882. B 883. D 884. C 885. C

A. software B. shareware
C. firmware D. mindware

887. What is the name of the operating system that reads and reacts in terms of actual time?
A. Batch system B. Quick response system
C. Real time system D. Time sharing system

888. What is the name of the technique in which the operating system of a computer executes several programs concurrently by switching back and forth between them?
A. Partitioning B. Multitasking
C. Windowing D. Paging

889. What is the name given to the software which can be legally compiled and often used for free?
A. Shareware program
B. Public domain program
C. Firmware program D. Mindware

890. All the time a computer is switched on, its operating system software has to stay in
A. main storage B. primary storage
C. floppy disk D. disk drive

891. The primary job of the operating system of a computer is to
A. command resources B. manage resources
C. provide utilities D. be user friendly

892. The operating system of a computer serves as a software interface between the user and
A. hardware B. peripheral
C. memory D. screen

893. The term "operating system" means
A. a set of programs which controls computer working
B. the way a computer operator works
C. conversion if high level language into machine code
D. the way a floppy disk drive operates

894. Which of the following software types is used to simplify using systems software?
A. Spreadsheet
B. Operational environment
C. Timesharing D. Multitasking

895. Which of the following is not applications software?
A. Word processing B. Spreadsheet
C. UNIX D. Desktop publishing

896. Which of the following types of software should you use if you often need to create, edit, and print documents?
A. Work processing B. Spreadsheet
C. UNIX D. Desktop publishing

897. Which of the following will determine your choice of systems software for your computer?
A. Is the application software you want to use compatible with it
B. Is it expensive
C. Is it compatible with your hardware
D. Both A and B are correct

898. Which of the following might be used to convert high-level language instructions into machine language?
A. System software
B. Applications software
C. An operating environment
D. An interpreter

899. Multiprogramming was made possible by
A. input/output units that operate independently of the CPU
B. operating systems
C. both A and B are correct
D. neither A nor B

900. Which of the following systems software does the job of merging the records from two files into one?
A. Security software B. Utility program
C. Networking software
D. Documentation system

901. What is the name of the system which deals with the running of the actual computer and not with the programming problems?
A. Operating system B. Systems program
C. Object program D. Source program

902. Systems software is a program that directs the overall operation of the computer, facilitates its use and interacts with the users. What are the different types of this software?
A. Operating system B. Languages
C. Utilities D. All of these

903. What is the name given to all the programs inside the computer which makes it usable?
A. Application software B. System software
C. Firmware D. Shareware

904. A translator which reads an entire program written in a high level language and converts it into machine language code is
A. assembler B. translator
C. compiler D. system software

905. Which of the following is helpful in evaluating applications software that will best suit your needs?
A. Recommendations by other users
B. Computer magazines
C. Objective software reviews
D. All of these

906. If you want to execute more than one program at a time, the systems software you are using must be capable of
A. word processing B. virtual memory
C. compiling D. multitasking

Ans. 886. B 887. C 888. C 889. B 890. B 891. B 892. A 893. A 894. C 895. C 896. A 897. D 898. D 899. C 900. B 901. B 902. D 903. B 904. C 905. D 906. D

907. Which of the following types of software you must have in main memory in order to use your keyboard?
A. Work processing B. Systems
C. Spreadsheet D. Applications

908. A front-end processor is usually used in
A. multiprogramming B. virtual storage
C. timesharing D. multiprocessing

909. The problem of thrashing is affected significantly by
A. program structure B. program size
C. primary-storage size D. all of these

910. Two basic types of operating systems are
A. sequential and direct B. batch and timesharing
C. sequential and realtime
D. batch and interactive

911. Remote computing services involve the use of timesharing and
A. multiprocessing B. interactive processing
C. batch processing D. realtime processing

912. Under Virtual Storage
A. a single program is processed by two or more CPUs
B. two or more programs are stored concurrently in primary storage
C. only the active pages of a program are stored in primary storage
D. interprogram interference may occur

913. An example of system development program is
A. operating systems B. performance monitors
C. data-base management systems
D. language translators

914. Which of the following is not a part of the operating system?
A. Supervisor B. Performance monitor
C. Job-control program
D. Input/output control program

915. Which of the following is not an advantage of multiprogramming?
A. Increased throughout
B. Shorter response time
C. Decreased operating-system overhead
D. Ability to assign priorities to jobs

916. In which addressing mode, the contents of a register specified in the instruction are first decremented, and then these contents are used as the effective address of the operands?
A. Index addressing B. Indirect addressing
C. Auto increment D. Auto decrement

917. In which addressing mode, the effective address of the operand is generally by adding a constant value to the contents of register?
A. Absolute mode B. Indirect mode
C. Immediate mode D. Index mode

918. In what module, multiple instances of execution will yield the same result even if one instance has not terminated before the next one has begun?
A. Reenterable module B. Non-reusable module
C. Serially reusable D. Recursive module

919. Which of the following system program forgoes the production of object code to generate absolute machine code and load it into the physical main storage location from which it will be executed immediately upon completion of the assembly?
A. Two pass assembler
B. Load-and-go assembler
C. Macroprocessor D. Compiler

920. Which of the following scheduling objectives should be applied to the following? The system should admit jobs to create a mix that will keep most devices busy.
A. To be fair
B. To balance resource utilisation
C. To obey priorities D. To be predictable

921. Which policy replace a page if it is not in the favoured subset of a process's page?
A. FIFO B. LRU
C. LFU D. Working set

922. In which of the storage placement strategies a program is placed in the smallest available hole in the main memory?
A. Best fit B. First fit
C. Worst fit D. Buddy

923. Which of the following statements is false?
A. Protection in segmentation systems is more natural than it is in paging systems
B. Sharing is more natural in segmentation systems than in paging system
C. Associative memory speeds up the dynamic address translation
D. Every cell of the associative memory in searched sequentially

924. Banker's algorithm for resource allocation deals with
A. deadlock prevention B. deadlock avoidance
C. deadlock recovery D. mutual exclusion

925. A compiler for a high-level language that runs on one machine and produces code for a different machine is called
A. optimising compiler B. one pass compiler
C. cross compiler D. multipass compiler

926. What problem is solved by Dijkstra's banker's algorithm?
A. Mutual exclusion B. Deadlock recovery
C. Deadlock avoidance D. Cache coherence

927. A system program that combines the separately compiled modules of a program into a form suitable for execution is

Ans. 907. B 908. C 909. A 910. D 911. C 912. C 913. D 914. B 915. C 916. D 917. D 918. A 919. B 920. B 921. D 922. A 923. D 924. B 925. C 926. B

A. assembler B. linking loader
C. cross compiler D. load and go

928. Which of the following operating systems use write through catches?
A. UNIX B. DOS
C. ULTRIX D. XENIX

929. Moving process from main memory to disk is called
A. scheduling B. caching
C. swapping D. spooling

930. Producer consumer problem can be solved using
A. seamphores B. event counters
C. monitors D. all of these

931. Situations where two or more processes are reading or writing some shared data and the final result depends on who runs precisely when, are called
A. race conditions B. critical sections
C. mutual exclusions D. message passing

932. Process is
A. program in high level language kept on disk
B. contents of main memory
C. a program in execution
D. a job in secondary memory

933. In which addressing mode, the effective address of the operand is the contents of a register specified in the instruction and after accessing the operand, the contents of this register is incremented to point to the next item in the list?
A. Index addressing B. Indirect addressing
C. Auto increment D. Auto decrement

934. Which of the following modules does not incorporate initialisation of values changes by the module?
A. None reusable module
B. Serially reusable module
C. Re-enterable module D. All of these

935. Which of the following statements is false?
A. A process scheduling algorithm is preemptive if the CPU can be forcibly removed from a process
B. Time sharing systems generally use preemptive CPU scheduling
C. Response time are more predictable in preemptive systems than in non-preemptive systems
D. Real time systems generally use non-preemptive CPU scheduling

936. Non-modifiable procedures are called
A. serially usable procedures
B. concurrent procedures
C. topdown procedures D. reentrant procedures

937. An instruction in a programming language that is replaced by a sequence of instructions prior to assembly or compiling is known as
A. procedure name B. macro
C. label D. literal

938. The main function of the dispatcher (the portion of the process scheduler) is
A. swapping a process to the disk
B. assigning ready process to the CPU
C. suspending some of the processes when the CPU load is high
D. bring processes from the disk to the main memory

939. In order to allow only one process to enter its critical section, binary semaphore are initialised to
A. 0 B. 1
C. 2 D. 3

940. A relationship between processes such that each has some part (critical section) which must not be executed while the critical section of another is being executed, is known as
A. semaphore B. mutual exclusion
C. multiprogramming D. multitasking

941. The process of transferring data intended for a peripheral device into a disk (or intermediate store), so that it can be transferred to peripheral at a more convenient time or in bulk, is known as
A. multiprogramming B. spooling
C. caching D. virtual programming

942. Which of the following statements is false?
A. A small page size cause large page tables
B. Internal fragmentation is increased with small pages
C. A large page size causes instructions and data that will not be referenced brought into primary storage
D. I/O transfers are more efficient with large pages

943. In which addressing mode, the operand is given explicitly in the instruction?
A. Absolute mode B. Immediate mode
C. Indirect mode D. Index mode

944. Which of the following program is not a utility?
A. Debugger B. Editor
C. Spooler D. All of these

945. Job Control Language (JCL) statements are used to
A. read the input from the slow-speed card reader to the high-speed magnetic disk
B. specify, to the operating system, the beginning and end of a job in a batch
C. allocate the CPU to a job
D. all of these

946. Operating System is a collection of
A. hardware components
B. input-output devices
C. software routines D. all of these

947. Indicate which of the following is not true about Nassi-Shneiderman charts?
A. These charts are types of graphical design tool

Ans. 927. B 928. B 929. C 930. D 931. A 932. C 933. C 934. A 935. D 936. D 937. B 938. B 939. B 940. B 941. B 942. B 943. B 944. C 945. B 946. C

B. These charts cannot represent CASE constructs
C. These charts can represent three fundamental control structures
D. All of these

948. Indicate which of the following is not true about a data flow diagram (DFD)?
A. It is graphical representation of the flow of data through the system
B. It is used to analyse any system or software at any level of abstraction
C. It is a very important tool, used by system analysts and designers
D. None of these

949. Which of the following checks, cannot be carried out on the input data to a system?
A. Consistency check B. Syntax check
C. Range check D. All of these

950. Which of the following is not true about documentation?
A. Documentation of a system should be as clear and direct as possible
B. Documentation increases the maintenance time and cost
C. Documentation gives better understanding of the problem
D. All of these

951. C is
A. a/an assembly language
B. third generation high-level language
C. machine language
D. all of these

952. Indicate, which of the following is not true about an interpreter?
A. Interpreter generates an object program from the source program
B. Interpreter is a kind of translator
C. Interpreter analyses each source statement every time it is to be executed
D. All of these

953. Indicate which of the following is not true about 4GL?
A. 4GL does not support a high-level of screen interaction
B. Many database management system packages support 4GLs
C. A 4GL is a software tool which is written, possibly, in some third generation language
D. All of these

954. An algorithm is best described as a
A. computer language
B. step by step procedure for solving a problem
C. branch of mathematics
D. all of these

955. A sequence of instructions, in a computer language to get the desired result, is known as
A. Algorithm B. Decision Table
C. Program D. all of these

956. A characteristic of an online real-time system is
A. more than one CPU
B. no delay in processing
C. offline batch processing
D. all of these

957. Indicate, which of the following is the worst type of module coupling?
A. Content coupling B. No coupling
C. Control coupling D. All of these

958. Indicate whether the statement LDA B is a statement in
A. machine language B. assembly language
C. high level language D. all of these

959. Indicate which describes the term "software"?
A. Systems programs only
B. Application programs only
C. Both A and B are correct
D. All of these

960. Bug means
A. documenting programs using an efficient documentation tool
B. a difficult syntax error in a program
C. a logical D. all of these

961. A development strategy whereby the executive control modules of a system are coded and tested first, is known as
A. bottom-up development
B. top-down development
C. left-right development
D. all of these

962. Which of the following is true for testing and debugging?
A. Testing checks for logical error in the programs, while debugging is a process of correcting those errors in the program
B. Testing detects the syntax errors in the program while debugging corrects those errors in the program
C. Testing and debugging indicate the same thing
D. All of these

963. If special forms are needed for printing the output, the programmer specifies these forms through
A. JCL B. IPL
C. utility programs D. load modules

964. Under multiprogramming, turnaround time for short jobs is usually ...... and that for long jobs is slightly ......
A. lengthened; shortened
B. shortened; lengthened
C. shortened; shortened
D. lengthened; lengthened

Ans. 947. B 948. D 949. B 950. B 951. B 952. A 953. A 954. B 955. C 956. B 957. A 958. B 959. C 960. C 961. B 962. B 963. A 964. B

965. Which of the following type of systems software is used on microcomputers?
A. MS-DOS B. PC-DOS
C. Unix D. All of these

966. Which of the following are loaded into main memory when the computer is booted?
A. Internal command instructions
B. External command instructions
C. Utility programs
D. Word processing instructions

967. What is the name of the operating system for the laptop computer called MacLite?
A. Windows B. DOS
C. MS-DOS D. OZ

968. What is the operating system used by Macintosh computers?
A. System 7.0 B. AU/X
C. Unix D. Either A or B

969. Most of the microcomputer's operating systems like Apple DOS, MS DOS and PC DOS, etc. are called disk operating systems because they are
A. memory resident
B. initially stored on disk
C. available on magnetic tapes
D. partly in primary memory and partly on disk

970. When did IBM release the first version of disk operating system DOS version 1.0?
A. 1981 B. 1982
C. 1983 D. 1984

971. IBM released its first PC in 1981. Can you name the operating system which was most popular at that time
A. MS-DOS B. PC-DOS
C. OS/360 D. CP/M

972. Page fault frequency in an operating system is reduced when the
A. size of pages is reduced
B. processes tend to be I/O-bound
C. processes tend to be CPU-bound
D. locality of reference is applicable to the process

973. Spooling is most beneficial in a multi-programming environment where
A. most jobs are CPU-bound
B. most jobs are I/O-bound
C. jobs are evenly divided as I/O-bound and CPU-bound
D. there is limited primary memory and need for secondary memory

974. Software that measures, monitors, analyzes and controls real-world events is called
A. system software B. real-time software
C. scientific software D. business software

975. Object modules generated by assemblers that contain unresolved external references are resolved for two or more object modules by a/an
A. operating system B. loader
C. linker D. compiler

976. Which of the following is false about disk when compared to main memory?
A. Non-volatile
B. Longer storage capacity
C. Lower price per bit D. Faster

977. To avoid race condition, the maximum number of processes that may be simultaneously inside the critical section is
A. hundred B. one
C. two D. three

978. Daisy chain is a device for
A. connecting a number of controllers to a device
B. connecting a number of devices to a controller
C. interconnecting a number of devices to a number of controllers
D. all of these

979. Scissoring enables
A. a part of data to be displayed
B. entire data to be displayed
C. full data display on full area of screen
D. no data to be displayed

980. Information in a memory that is no longer valid or wanted is known as
A. non-volatile B. volatile
C. surplus D. garbage

981. The state transition initiated by the user process itself in an operating system is
A. block B. dispatch
C. wake up D. time run out

982. Which of the following terms refers to be the degree to which data in a database system are accurate and correct?
A. Data security B. Data validity
C. Data independence D. Data integrity

983. The total to prepare a disk drive mechanism for a block of data to be read from it is
A. latency
B. latency plus transmission time
C. latency plus seek time
D. latency plus seek time plus transmission time

984. Special software to create a job queue is called a
A. drive B. spooler
C. interpreter D. linkage editor

985. The most common security failure is
A. carelessness by users
B. depending on passwords
C. too much emphasis on preventing physical access
D. insufficient technology used to prevent breaches

Ans. 965. D 966. A 967. D 968. D 969. B 970. A 971. D 972. D 973. C 974. B 975. C 976. C 977. B
978. B 979. A 980. D 981. A 982. D 983. C 984. B 985. A

986. A public key encryption system
A. allows anyone to decode the transmission
B. allows only the correct sender to decode the data
C. allows only the correct receiver to decode the data
D. does not encode the data before transmitting it

987. Supervisor state is
A. never used
B. entered by programs when they enter the processor
C. required to perform any I/O
D. only allowed to the operating system

988. A high paging rate
A. may cause a high I/O
B. keep the system running well
C. is a symptom of too much processor activity
D. always creates a slow system

989. The paging rate
A. should never be greater than 100 per second
B. is greater for large programs
C. is the number of I/O interrupts each second
D. increases as the number of page faults increase

990. Thrashing
A. is a natural consequence of virtual memory systems
B. can always be avoided by swapping
C. always occurs on large computers
D. can be caused by poor paging algorithms

991. A page fault
A. is an error in a specific page
B. occurs when a program accesses a page of memory
C. is an access to a page not correctly in memory
D. is a reference to a page belonging to another program

992. In memory systems, boundary registers
A. are used for temporary program variable storage
B. are only necessary with fixed partitions
C. track page boundaries
D. track the beginning and ending of programs

993. Relocatable programs
A. cannot be used with fixed partitions
B. can be loaded almost anywhere in memory
C. do not need a linker
D. can be loaded only at one specific location

994. The FIFO algorithm
A. executes first the job that last entered the queue
B. executes first the job that first entered the queue
C. executes first the job that has been in the queue the longest
D. executes first the job with the least processor needs

995. Round-robin scheduling
A. allows interactive tasks quicker access to the processor
B. is quite complex to implement
C. gives each task the same chance at the processor
D. allows processor-bound tasks more time in the processor

996. Interprocess Communication
A. is required for all processes
B. is usually done via disk drives
C. is never necessary
D. allows processes to synchronise activity

997. Fork is
A. the dispatching of a task
B. the creation of a new job
C. the creation of a new process
D. increasing the priority of a task

998. A process is another name for
A. a job B. a task
C. paging
D. the operating system dispatcher

999. The computational technique used to compute the disk storage address of individual records is called
A. bubble memory B. key fielding
C. dynamic reallocation D. hashing

1000. Capacity planning
A. requires detailed system performance information
B. is independent of the operating system
C. does not depend on the monitoring tools available
D. is not needed in small installations

1001. Assemblers language
A. is usually the primary user interface
B. requires fixed-format commands
C. is a form of machine language
D. is quite different from the SLC interpreter

1002. Which of the following is/are the advantage(s) of modular programming?
A. The program is much easier to change
B. Modules can be reused in other programs
C. Easy debugging D. Easy to compile

1003. Which of the following can be accessed by transfer vector approach of linking?
A. External data segments
B. Data located in other procedures
C. External subroutines
D. All of these

1004. The Linker
A. is the same as the loader
B. is required to create a load module
C. uses source code as input
D. is always used before programs are executed

Ans. 986. C 987. D 988. A 989. D 990. D 991. C 992. D 993. B 994. B 995. C 996. D 997. C 998. B 999. D 1000. B 1001. C 1002. A 1003. C 1004. B

1005. The user interface
A. is relatively unimportant
B. is slanted toward novice users
C. supports both novice and experienced users
D. is always used before programs are executed

1006. Memory Management is
A. not used in modern operating system
B. replaced with virtual memory on current systems
C. not used on multiprogramming systems
D. critical for even the simplest operating systems

1007. The practice of "bundling" refers to
A. selling computers alone
B. selling peripheral devices with computer
C. selling software to run on computers
D. giving away software with a computer purchase

1008. The primary purpose of an operating system is to
A. make computer easier to use
B. keep system programmers employed
C. make the most efficient use of the hardware
D. allow people to use the computers

1009. Multiprogramming Systems
A. are easier to develop than single programming system
B. execute each job faster
C. execute more jobs in the same time period
D. are used only on large mainframe computers

1010. Which of the following capabilities is required for a system program to execute more than one program at a time?
A. Word processing B. Compiling
C. Virtual memory D. Multitasking

1011. A Critical Region is
A. a program segment that has not been proved bug-free
B. a program segment that causes unexpected system crashes
C. a program segment where shared resources are accessed
D. one which is enclosed by a pair of *P* and *V* operations on semaphores

1012. Which of the following addressing modes, facilities access to an operand whose location is defined relative to the beginning of the data structure in which it appears?
A. Assending B. Sorting
C. Index D. Indirect

1013. The register or main memory location which contains the effective address of the operand is known as
A. pointer B. indexed register
C. special location D. scratch pad

1014. System programs such as compilers are designed so that they are
A. reenterable B. non-reusable
C. serially usable D. recursive

1015. Round robin scheduling is essentially the preemptive version of
A. FIFO B. FCFS
C. FILO D. Longest time first

1016. If the number of bits in a virtual address of a program is 12 and the page size of 0.5 K bytes, the number of pages in the virtual address space is
A. 16 B. 32
C. 64 D. 128

1017. In which of the storage placement strategies a program is placed in the largest available hole in the main memory?
A. Best fit B. First fit
C. Worst fit D. Buddy

1018. For how many processes which are sharing common data, the Dekker's algorithm implements mutual exclusion?
A. 1 B. 2
C. 3 D. 4

1019. A disk scheduling algorithm in an operating system causes the disk arm to move back and forth across the disk surface in order to service all requests in its path. This is a
A. first come first served
B. shortest seek time first (SSTF)
C. scan D. FIFO

1020. The strategy of allowing processes that are logically runnable to be temporarily suspended is called
A. preemptive scheduling
B. nonpre-emptive scheduling
C. shortest job first D. first come first served

1021. Part of a program where the shared memory is accessed and which should be executed indivisibly, is called
A. semaphores B. directory
C. critical section D. mutual exclusion

1022. The technique, for sharing the time of a computer among several jobs which switches jobs so rapidly such that each job appears to have the computer to itself
A. time sharing B. time out
C. time domain D. FIFO

1023. The operating system manages
A. memory B. processor
C. disks and I/O devices
D. all of these

1024. A form of code that uses more than one process and processor, possibly of different type, and that may on occasions have more than one process or processor active at the same time, is known as

Ans. 1005. B 1006. B 1007. D 1008. A 1009. C 1010. D 1011. D 1012. A 1013. A 1014. A 1015. A 1016. D 1017. C 1018. C 1019. C 1020. A 1021. C 1022. A 1023. D

A. multiprogramming B. multithreading
C. broadcasting D. time sharing

1025. Which of the following is a block device?
A. Mouse B. Printer
C. Terminals D. Disk

1026. Which of the following statements is false?
A. The technique of storage compaction involves moving all occupied areas of storage to one end or other of main storage
B. Compaction does not involve relocation of programs
C. Compaction is also known as garbage collection
D. The system must stop everything while it performs the compaction

1027. Thrashing can be avoided if
A. the pages, belonging to the working set of the programs, are in main memory
B. the speed of CPU is increased
C. the speed of I/O processor is increased
D. all of these

1028. Operating System
A. links a program with the subroutines in references
B. provides a layered, user-friendly interface
C. enables the programmer to draw a flowchart
D. all of these

1029. The principles of structured programming forbid use of
A. WHILE-DO B. GO TO
C. IF-THEN-ELSE D. DO-WHILE

1030. Which of the following is necessary to work on a computer?
A. Compiler B. Operating system
C. Assembly D. Interpreter of these

1031. Dividing a project into segments and smaller units in order to simplify the analysis, design and programming efforts is known as
A. modular approach B. to-down approach
C. bottom-up approach D. left-right approach

1032. The errors that can be pointed out by the compiler are
A. syntax errors B. semantic errors
C. logical errors D. internal errors

1033. A translator is best described as
A. an application software
B. a hardware component
C. a system software D. all of these

1034. The part of machine level instruction, which tells the central processor what has to be done, is
A. operating code B. address
C. locator D. flip-flop

1035. Which of the following instruction steps would be written within the diamond-shaped box of a flowchart?
A. $S = B - C$ B. IS $A < 10$
C. PRINT A D. DATA X, 4, Z

1036. Which of the following is not true about the description of a decision table?
A. A decision table is easy to modify
B. A decision table is directly understood by the computer
C. A decision table is easy to understand
D. All of these

1037. The memory allocation scheme subject of external fragmentation is
A. segmentation B. swapping
C. pure demand paging
D. multiple contiguous fixed partitions

1038. The advantage of a command processor running only built-in commands is
A. flexibility to the users in running lists of commands by simply collecting them in named batch command files
B. command set being common across different hardware configurations
C. uses can create system programs and run them as commands
D. the processing is much faster than would otherwise be when user defined commands are used

1039. Which of the following is not true about the memory management?
A. Virtual memory is used only in multi-user systems
B. Segmentation suffers from external fragmentation
C. Paging suffers from internal fragmentation
D. Segmented memory can be paged

1040. In a multiprogramming system, a set of processes is deadlocked if each process in set is waiting for an event to occur that can be initiated only by another process in the set. Which of the following is not one of the four conditions that are necessary for deadlock to occur?
A. Non-preemption B. Process suspension
C. Partial assignment of resources
D. Circular wait

1041. Block or buffer caches are used to
A. improve disk performance
B. handle interrupts
C. increase the capacity of main memory
D. speed up main memory read operation

1042. The CPU, after receiving an interrupts from an I/O device
A. halts for a predetermined time
B. hands over control of address bus and data bus to the interrupting device
C. branches off to the interrupt service routine immediately
D. branches off to the interrupt service routine after completion of the current instruction

Ans. 1024. B 1025. D 1026. B 1027. A 1028. B 1029. B 1030. B 1031. A 1032. A 1033. C 1034. A 1035. B 1036. B 1037. A 1038. B 1039. A 1040. B 1041. A 1042. D

1043. Which of the following is not a characteristic of a daisy chaining priority control scheme?
A. Priority is programmable
B. It is relatively easy to add more devices to the chain
C. The failure of one device may affect other device on the chain
D. The number of control lines is independent of the number of devices on the chain

1044. The working set theory of programming behaviour of processes running within an operating system involves
A. the collection of pages that a process accesses
B. disk scheduling mechanisms
C. coalescing holes in memory
D. assigning the CPU to processes

1045. Trojan-Horse Programs
A. are legitimate programs that allow unauthorised access
B. do not usually work
C. are hidden programs that do not show upon the system
D. are usually immediately discovered

1046. Page Stealing
A. is a sign of an efficient system
B. is taking page frames from other working sets
C. should be the tuning goal
D. is taking larger disk spaces for pages paged out

1047. In virtual memory systems, dynamic address translation
A. is the hardware necessary to implement paging
B. stores pages at a specific location on disk
C. it useless when swapping is used
D. is part of the operating system paging algorithm

1048. The Garbage Collector
A. prevents fragmentation from occurring
B. is mostly used with fixed partitions
C. collects fragmented areas of memory
D. is critical for efficient virtual memory systems

1049. The Dispatcher
A. actually schedules the tasks into the processor
B. puts tasks in I/O wait
C. is always small and simple
D. never changes task priorities

1050. The SJF algorithm executes first the job
A. that last entered the queue
B. that first entered the queue
C. that has been in the queue the longest
D. with the least processor needs

1051. Semaphores
A. synchronise critical resources to prevent deadlock
B. synchronise critical resources to prevent contention
C. are used to do I/O
D. are used for memory management

1052. Fragmentation of the file system
A. occurs only if the file system is used improperly
B. can always be prevented
C. can be temporarily removed by compaction
D. is a characteristic of all file systems

1053. The Command Interpreter
A. is usually the primary user interface
B. requires fixed format commands
C. is menu drive
D. is quite different from the SCL interpreter

1054. Relocation bits used by relocating loader are specified (generated) by
A. relocating loader itself
B. assembler or translator
C. linker D. macroprocessor

1055. Device Independence
A. allows the computer to run without I/O devices
B. makes all devices look the same to the operating system
C. allows programs to be written more easily
D. allows tape drivers to be substituted for disk drives

1056. User-Friendly Systems are
A. required for object-oriented programming
B. easy to develop
C. common among traditional mainframe operating systems
D. becoming more common

1057. In which addressing mode, the address of the location of the operand is given explicitly as a part of the instruction?
A. Absolute mode B. Immediate mode
C. Index mode D. Modulus mode

1058. Which of the following translator program converts assembly language program to object program?
A. Assembler B. Compiler
C. Microprocessor D. Linker

1059. In which of the following page replacement policies, Balady's anomaly occurs?
A. FIFO B. LRU
C. LFU D. NRU

1060. Four necessary conditions for deadlock to exist are: mutual exclusion, no-preemption, circular wait and
A. hold and wait B. deadlock avoidance
C. race around condition
D. buffer overflow

1061. A system program that sets up an executable program in main memory ready for execution is
A. assembler B. linker
C. loader D. compiler

1062. The principle of *locality of reference* justifies the use of
A. reenterable B. non-reusable
C. virtual memory D. cache memory

Ans. 1043. A 1044. A 1045. A 1046. B 1047. A 1048. C 1049. A 1050. D 1051. A 1052. C 1053. A 1054. C 1055. C 1056. D 1057. A 1058. A 1059. A 1060. A 1061. C 1062. D

1063. Block caches or buffer caches are used to
A. improve disk performance
B. handle interrupts
C. increase the capacity of the main memory
D. speed up main memory read operation

1064. To avoid the race condition, the number of processes that may be simultaneously inside their critical section is
A. 8 B. 1
C. 16 D. 0

1065. What is the initial value of the semaphore to allow only one of the many processes to enter their critical section?
A. 8 B. 1
C. 16 D. 0

1066. Which technique stores a program on disk and then transfer the program into main storage as and when they are needed, is known as
A. spooling B. swapping
C. thrashing D. all of these

1067. Page-map table is
A. a data file B. a directory
C. used for address translation
D. all of these

1068. Which of the following rules out the use of GO TO?
A. Flowchart B. Hipo-diagrams
C. Nassi-Shneiderman diagram
D. All of these

1069. Which of the following is a phase of the compilation process?
A. Lexical analysis B. Code generation
C. Both A and B are correct
D. Static analysis

1070. A computer cannot "boot" if it does not have the
A. compiler B. loader
C. operating system D. assembler

1071. Which of the following is a measure to test how good or bad a modular design is?
A. Module strength B. Module coupling
C. Static analysis D. All of these

1072. A flowchart that uses predefined symbols to describe data flow in a system is known as
A. program flowchart B. system flowchart
C. data flow diagram D. all of these

1073. Which of the following is true about pseudocode?
A. A machine language B. An assembly language
C. A high-level language
D. None of these

1074. A program that converts a high-level language program to a set of instructions that can run on a computer is called a/an
A. compiler B. debugger
C. editor D. all of these

1075. Which of the following statements is not true about the FORTRAN language?
A. FORTRAN is a high level language
B. A FORTRAN program, written for the IBM-PC is totally different from a FORTRAN program written for execution on the SUN machine
C. FORTRAN is extensively used to write programs for performing scientific computations
D. All of these

1076. The initial value of the semaphore that allows only one of the many processes to enter their critical sections is
A. 8 B. 1
C. 16 D. 0

1077. Which of the following statements is not true?
A. Time sharing is an example of multi-programming
B. JCL is used only to communicate between systems programmers
C. A batch file contains a series of operating system commands
D. The primary function of operating systems is to make the computer hardware easily usable

1078. What scheduling algorithm allows processes that are logical runnable to be temporarily suspended?
A. Preemptive scheduling
B. Non-preemptive scheduling
C. FIFO D. FCFS

1079. The term 'polling' in a computer means a process by which a computer system
A. detects/corrects errors
B. multiplexes the inputs and updates the memory accordingly
C. decides correct anternative by analysing several ones
D. enquires to see if a terminal has any transaction to send

1080. In a magnetic disk, data is recorded in a set of concentric tracks which are subdivided into
A. periods B. sectors
C. zones D. groups

1081. Which of the following is true for machine language?
A. Repeated execution of program segments
B. Depicting flow of data in a system
C. A sequence of instructions which, when followed properly, solves a problem
D. The language which communicates with the computer using only the binary digits 1 and 0

1082. Data Encryption
A. is mostly used by public networks
B. is mostly used by financial networks
C. cannot be used by private installations
D. is not necessary since data cannot be intercepted

Ans. 1063. A 1064. B 1065. B 1066. B 1067. C 1068. C 1069. C 1070. A 1071. A 1072. B 1073. D 1074. A 1075. B 1076. B 1077. C 1078. A 1079. D 1080. B 1081. D 1082. B

1083. Seeks Analysis is
A. used for analysing paging problems
B. used for analysing device busy problems
C. used for analysing control-unit busy problems
D. only shown on real-time displays

1084. Swapping
A. works best with many small partitions
B. allows many programs to use memory simultaneously
C. allows each program in turn to use the memory
D. does not work with overlaying

1085. Feedback queues
A. are very easy to implement
B. dispatch tasks according to execution characteristics
C. are used to favour real-time tasks
D. require manual intervention to implement properly

1086. A task is a blocked state
A. is executable B. is running
C. must still be placed in the run queues
D. is waiting for same temporarily unavailable resources

1087. Backups should be done
A. daily for most installations
B. weekly for most installations
C. as several image copies, followed by an incremental
D. as several incrementals, followed by an image copy

1088. Which of the following statement is true?
A. The LRU algorithm pages out pages that have been used recently
B. Thrashing is natural consequence of virtual memory systems
C. Seek analysis is used for analysing control-unit busy problems
D. All of these

1089. The details of all external symbols and relocation formation (relocation list or map) is provided to linker by
A. macroprocessor B. translator
C. loader D. editor

1090. Scheduling is
A. allowing jobs to use the processor
B. unrelated to performance consideration
C. not required in uniprocessor systems
D. the same regardless of the purpose of the system

1091. Real-time Systems are
A. primarily used on mainframe computers
B. used for monitoring events as they occur
C. used for program analysis
D. used for real-time interactive users

1092. Access time is the highest in the case of
A. floppy disk B. cache
C. swapping devices D. magnetic disk

1093. Which of the following refers to the associative memory?
A. The address of the data is generated by the CPU
B. The address of the data is supplied by the users
C. There is no need for an address, i.e. the data is used as an address
D. The data is accessed sequentially

1094. Link encryption
A. is more secure than end-to-end encryption
B. is less secure than end-to-end encryption
C. cannot be used in a public network
D. is used only to debug

1095. Which of the following is characteristic of an operating system?
A. Resource management
B. Memory management
C. Error recovery D. All of these

1096. Files can have
A. read access B. copy access
C. write access D. all of these

1097. The most common systems security method is
A. passwords B. encryption
C. firewall D. all of these

1098. Poor response times are caused by
A. processor busy B. high I/O rate
C. high paging rates D. any of these

1099. The LRU algorithm
A. pages out pages that have been used recently
B. pages out pages that have not been used recently
C. pages out pages that have been least used recently
D. pages out the first page in a given area

1100. A linker
A. creates a load module
B. is not necessary with variable partitions
C. must be run after the loader
D. is not needed with a good compiler

1101. Global locks
A. synchronise access to local resources
B. synchronise access to global resources
C. are used to avoid local locks
D. prevent access to global resources

1102. The dynamic allocation of storage areas with VSAM files is accomplished by
A. hashing B. control splits
C. overflow areas D. relative recording

1103. File organisation component of a VSAM file is
A. relative record data set
B. keyed sequential data set
C. entry sequential data set
D. all of these

Ans. 1083. B 1084. C 1085. B 1086. D 1087. D 1088. C 1089. B 1090. A 1091. B 1092. D 1093. C 1094. B 1095. D 1096. D 1097. A 1098. D 1099. C 1100. A 1101. B 1102. B 1103. D

1104. The file structure that redefines its first record at a base of zero uses the term
A. relative organisation B. key fielding
C. dynamic reallocation D. hashing

1105. File record length
A. should always be fixed
B. should always be variable
C. depends upon the size of the file
D. should be chosen to match the data characteristics

1106. An incremental backup
A. uses more tapes B. saves all files
C. should be done each month
D. saves only files that have recently changed

1107. A partitioned data set is mostly used for
A. a program or source library
B. storing program data
C. storing backup information
D. storing ISAM files

1108. A file is sometimes called
A. collection of input data
B. temporary place to store data
C. data set D. program

1109. In MS-DOS, relocatable object files and load modules have extensions
A. .OBJ and .COM or .EXE respectively
B. .COM and .OBJ respectively
C. .EXE and .OBJ respectively
D. .DAS and .EXE respectively

1110. Resolution of externally defined symbols is performed by
A. linker B. loader
C. compiler D. assembler

1111. System generation
A. is always quite simple
B. is always very difficult
C. varies in difficulty between systems
D. requires extensive tools to be understandable

1112. Object code
A. is ready to execute
B. is the output of compilers, but not assemblers
C. must be "loaded" before execution
D. must be rewritten before execution

1113. Virtual memory is
A. simple to implement
B. used in all major commercial operating systems
C. less efficient in utilisation of memory
D. useful when fast I/O devices are not available

1114. System maintenance
A. is usually not necessary
B. is necessary on all systems, regardless or how good
C. is not required if the system is well written
D. always requires several programs

1115. What type of control pins are needed in a microprocessor to regulate traffic on the bus, in order to prevent two devices from trying to use it at the same time?
A. Bus control B. Interrupts
C. Bus arbitration D. Status

1116. The process of fetching and executing instructions, one at a time, in the order of increased addresses is known as
A. instruction execution
B. straight line sequencing
C. instruction fetch D. random sequencing

1117. Non-volatility is an important advantage of
A. CCDs B. magnetic bubbles
C. magnetic tapes and disks
D. both B and C are correct

1118. Which was the world's first microprocessor that used Intel 80386 microprocessor chip?
A. IBM PS/2 B. HP-9830
C. DeskPro-386 D. IBM-360

1119. Who developed the personal computer called NEXT which was the first to incorporate a drive for optical storage disk, voice recognition and object Phillips oriented languages?
A. Niklaus Wirth B. Steve Jobs
C. Kahn D. Vannevar Bush

1120. When did IBM come out with one of the first RISC-based workstation, the RTPG?
A. 1980 B. 1982
C. 1986 D. 1988

1121. When was the world's first laptop computer introduced in the market and by whom?
A. Hewlett-Packard, 1980
B. Epson, 1981
C. Laplink Travelling Software Inc, 1982
D. Tandy Model-200, 1985

1122. What was the name of the first personal computer electronic spreadsheet software package which became a smash hit as soon as it was introduced in 1978?
A. WordStar B. VisiCalc
C. Lotus 1-2-3 D. Excel

1123. When was the first IBM microprocessor called IBM PC with 16-bits microprocessor introduced?
A. 1979 B. 1981
C. 1982 D. 1984

1124. In which year were chips used inside the computer for the first time?
A. 1964 B. 1975
C. 1977 D. 1981

1125. The first microprocessor built by the Intel Corpn was called
A. 8008 B. 8080
C. 4004 D. 8800

Ans. 1104. A 1105. D 1106. D 1107. A 1108. C 1109. A 1110. A 1111. C 1112. C 1113. B 1114. B 1115. C 1116. B 1117. D 1118. B 1119. B 1120. C 1121. B 1122. B 1123. B 1124. B 1125. C

1126. What was the name of the first commercially available processor chip?
A. Intel 8008 B. Intel 8080
C. Intel 4004 D. Motorola 6809

1127. When was the first minicomputer built?
A. 1965 B. 1962
C. 1971 D. 1966

1128. The first digital computer built with IC chips was known as
A. IBM 7090 B. Apple-1
C. IBM System/360 D. VAX-780

1129. When was the first solid-state computer produced and by which country?
A. 1964, Japan B. 1960, USA
C. 1965, UK D. 1960, Germany

1130. Third generation computers are those which are built with integrated circuits. What was the name of the first third generation computer and when was it introduced?
A. IBM-1620, 1964 B. IBM-360, 1965
C. CDC-6600, 1962 D. PDP-1401

1131. Which one of the following is the first second-generation computer?
A. IBM 7090 B. IBM 801
C. IBM 7070 D. IBM 650

1132. Which is the first compiler-level language developed by a team of IBM programmers led by John Backus and unveiled in 1957?
A. BASIC B. PL/1
C. FORTRAN D. APL

1133. What is the name of the British gentleman who was the first to put forward in 1952 the idea for the integrated circuit?
A. Jack S Kilby B. GW Dummer
C. William Shockley D. John Bardeen

1134. Which of the following computers was the first to use a stored program?
A. ENIAC B. EDVAC
C. EDSAC D. ACE

1135. The magnetic tape consists of a plastic tape with a surface coating of magnetic material. When was it invented?
A. 1968 B. 1964
C. 1956 D. 1946

1136. The first large scale electronic computer which became operational in 1946 and contained approximately 18000 vacuum tubes and could perform 300 multiplications per second was known as
A. ILLIAC B. ENIAC
C. JOHNNIAC D. EDSAC

1137. Who built the world's first electronic calculator using telephone relays, light bulbs and batteries?
A. Claude Shannon B. Konrad Zuse
C. George Stibitz D. Howard H Aiken

1138. In 1943, the first general-purpose, all electronic deciphering computer using 1500 vacuum tubes was built. What was it called?
A. Mark-I B. Whirlwind
C. Colossus D. ENIAC

1139. A modern digital computer has
A. extremely high speed B. large memory
C. almost unlimited accuracy
D. all of these

1140. Where was India's first computer installed and when?
A. Indian Institute of Technology, Delhi, 1977
B. Indian Institute of Science, Bangalore, 1971
C. Indian Iron & Steel Co Ltd., 1968
D. Indian Statistical Institute, Calcutta, 1955

1141. Which of the following terms is the most closely related to main memory?
A. Nonvolatile B. Permanent
C. Control unit D. Temporary

1142. Why is the width of the data bus so important to the processing speed of a computer?
A. The narrower it is, the greater the computer's processing speed
B. The wider it is, the more data that can fit into main memory
C. The wider it is, the greater the computer's processing speed
D. The wider it is, the slower the computer's processing speed

1143. Which of the following is used to check for errors in RAM chips?
A. ROM chips B. Microprocessor chip
C. Parity chip D. EPROM chip

1144. Which of the following are used to quickly accept, store, and transfer data and instructions that are being used immediately by the CPU?
A. Microprocessors B. Registers
C. ROM chips D. Data buses

1145. Which of the following affects processing power?
A. Clock speed B. Addressing scheme
C. Data bus capacity D. All of these

1146. Which of the following is used for manufacturing chips?
A. Control bus B. Control unit
C. Parity unit D. Semiconductor

1147. Which of the following are that two main components of the CPU?
A. Control unit and registers
B. Registers and main memory
C. Control unit and ALU
D. ALU and bus

1148. Which of the following holds data and processing instructions temporarily until the CPU needs it?

Ans. 1126. C 1127. A 1128. C 1129. B 1130. B 1131. A 1132. C 1133. B 1134. C 1135. D 1136. B 1137. C 1138. B 1139. D 1140. D 1141. D 1142. C 1143. C 1144. B 1145. D 1146. D 1147. C

A. ROM B. Control unit
C. Main memory D. Coprocessor chips

1149. Which of the following typically uses a keyboard for input?
A. Desktop terminal B. Point-of-sale terminal
C. Financial transaction terminal
D. All of these

1150. Which of the following is an advantage of magnetic input media?
A. Low cost
B. Flexibility in accessing data
C. High speed D. All of these

1151. Which of the following are often used to ensure that data has been accurately input to the computer?
A. Digitizers B. Light pens
C. Keyboards D. Input control

1152. Which of the following is considered a direct-entry input device?
A. Optical scanner B. Mouse
C. Light pen D. All of these

1153. Which of the following file organisations is most efficient for a file with a high degree of file activity?
A. Sequential B. ISAM
C. VSAM D. B-Tree index

1154. Which of the following is required when more than one person uses a central computer at the same time?
A. Terminal B. Light pen
C. Digitizer D. Mouse

1155. Two basic types of record-access methods are
A. sequential and random
B. sequential and indexed
C. direct and immediate
D. online and realtime

1156. Which of the following fields in a student file can be used as a primary key?
A. Class
B. Social security number
C. GPA D. Major

1157. The parity bit is added for which purposes?
A. Coding B. Indexing
C. Error-detection D. Controlling

1158. Data entry can be performed with all of the following except
A. OCR B. OMR
C. COM
D. Voice-recognition systems

1159. The advantages of COM are its .... and ....
A. compact size; readability
B. compact size; speed
C. readability; speed D. low-cost; readability

1160. Which of the following is not used for storage purpose with mainframe computers?
A. Removable disks B. Fixed disks
C. Mass storage systems diskettes
D. Diskettes

1161. Which of the following is used to ensure the high quality of computer output?
A. Computer output microfilm
B. Output controls
C. Voice output systems
D. Liquid crystal display

1162. Which of the following technologies will you likely see in laptop computers?
A. Voice output systems
B. Output controls
C. Computer crystal microfilm
D. Liquid crystal display

1163. Some software tools used in developing computer programs are
A. text editors
B. compilers and assemblies
C. operating systems, and debugging programs
D. all of these

1164. Software specifications tell exactly ...... a program will do but not ...... it will do it.
A. what, how B. when, why
C. if, then D. none of these

1165. Ways to make your programs readable and easily understood are
A. use of comments and self-documenting code
B. pretty printing and constants
C. avoidance of tricky code
D. all of these

1166. What is the first step in developing any software program?
A. System design B. System study
C. Coding D. Think

1167. Describe Line Status of the Process Program File in terms of Big-O, if *N* refers to the number of lines in the Program File.
A. $O(N)$ B. $O(I)$
C. $O(\log_2 N)$ D. $O(N)^2$

1168. A syntax error is an error
A. in the use of the programming language, and will most likely be found at compile time, if it has not been detected by visual inspection of the program
B. in how the program works, and will most likely be detected at run time, by causing the program to crash or to produce wrong results
C. that occur during the execution of a program. Its are often result of the programmer's making too many assumptions
D. none of these

1169. A logical error is an error
A. in the use of the programming language, and will most likely be found at compile time, if it

Ans. 1148. C 1149. D 1150. D 1151. D 1152. D 1153. A 1154. A 1155. A 1156. B 1157. C 1158. C 1159. B 1160. D 1161. B 1162. D 1163. D 1164. A 1165. D 1166. D 1167. B 1168. A

has not been detected by visual inspection of the program
B. in how the program works, and will most likely be detected at run time, by causing the program to crash or to produce wrong results
C. that occur during the execution of a program. Its are often result of the programmer's making too many assumptions
D. none of these

1170. Run Time Error is an error
A. in the use of the programming language, and will most likely be found at compile time, if it has not been detected by visual inspection of the program
B. in how the program works, and will most likely be detected at run time, by causing the program to crash or to produce wrong results
C. that occur during the execution of a program. Its are often result of the programmer's making too many assumptions
D. none of these

1171. Data Encapsulation means that
A. the separation of the representation of data from the applications that use the data at a logical level
B. the logical picture of a data type, plus the specifications of the operations required to create and manipulate objects of this data type
C. a collection of data elements whose organisation is characterised by accessing operations that are used to store and retrieve the individual data elements
D. none of these

1172. The structured data types that are built into Pascal are
A. arrays and records B. variant records
C. sets and files D. all of these

1173. Which of the following can be described as a pointer?
A. Stack B. Lîst
C. Lîst Info D. Stâck Info

1174. Which of the following can be described as an array?
A. Stack B. Lîst
C. Lîst Info D. Stâck Info

1175. Which of the following can be described as a record?
A. Stack B. Lîst
C. Lîst Info D. Info

1176. Which of the following can be described as a pointer?
A. Queue B. Queue.Bar
C. Queue.Rêar D. Queuê.Font.Info

1177. Which of the following can be described as an integer
A. Queue B. Queue.Bar
C. Queue.Rêar D. Queuê.Font.Info

1178. Which of the following can be described as a record?
A. Queue.Front.Nêxt B. Queue.Rear
C. Queue.Front.Next D. Queue.Front.Info

1179. Memory is
A. a device that performs a sequence of operations specified by instructions in memory
B. the device where information is stored
C. a sequence of instructions
D. typically characterised by interactive processing and time-slicing of the CPU's time to allow quick response to each user

1180. A Program is
A. a device that performs a sequence of operations specified by instructions in memory
B. the device where information is stored
C. a sequence of instructions
D. typically characterised by interactive processing and time of the CPU's time to allow quick response to each user

1181. A Processor is
A. a device that performs a sequence of operations specified by instruction in memory
B. the device where information is stored
C. a sequence of instructions
D. typically characterised by interactive processing and time of the CPU's time to allow quick response to each user

1182. An Assembler is a program that
A. places programs into memory and prepared them for execution
B. automates the translation of assembly language into machine language
C. accepts a program written in a high level language and produces an object program
D. appears to execute a source program as if it were machine language

1183. Advances in computer hardware and software are generally classified into generations. We are currently in which generation?
A. Second B. Third
C. Fourth D. Fifth

1184. The first machine to successfully perform a long series of arithmetic and logical operations was
A. ENIAC B. Mark I
C. Analytic engine D. UNIVAC-I

1185. Which of the following was (were) not used in first-generation computers?
A. Vacuum tubes B. Cards
C. Magnetic core D. Punched paper tape

Ans. 1169. B 1170. C 1171. A 1172. D 1173. A 1174. C 1175. B 1176. B 1177. D 1178. A 1179. B 1180. C 1181. A 1182. B 1183. C 1184. B 1185. C

1186. Which of the following was not associated with second-generation computers?
A. High-level procedural language
B. Operating system
C. Magnetic core D. None of these

1187. Which of the following is true of the ENIAC?
A. It was developed by Charles Babbage
B. It was the first stored-program electronic digital computer
C. It was an electromechanical computer
D. It was vacuum tubes in place of electromagnetic relays

1188. The third generation of computers covers the period
A. 1959–1964 B. 1965–1971
C. 1971–1981 D. 1981–till date

1189. In the third generation of computers
A. high-level procedural languages were first used
B. an operating system was first developed
C. distributed data processing first became popular
D. online, realtime systems first became popular

1190. All of the following became popular during the fourth generation of computers except
A. minicomputers B. semiconductors
C. CRT terminals D. personal computers

1191. The decreased cost and increased performance of computer hardware were the distinguishing features of which generation of computers?
A. First B. Second
C. Third D. None of these

1192. Which of the following is not true of future computers?
A. Increased use of CAD/CAM techniques
B. Faster turnaround time
C. Developments in artificial intelligence systems
D. None of these

1193. The card type that requires you to return a portion of it with payment is the
A. dual card B. document card
C. stub card D. either B or C is correct

1194. A five-digit card field used for postal ZIP codes is defined as a/an
A. alphabetic field B. numeric field
C. alphameric field D. letter field

1195. The gang punching of cards is performed on the
A. collator B. reproducer
C. interpreter D. keypunch

1196. The BCD code for the character W is composed of
A. A, 4-, 3- and 1-bits B. A 4-, 3- and 2-bits
C. A and 5-bits D. A and 6-bits

1197. The field name assigned to card field should
A. indicate the type of data items the field will contain
B. have the same number of characters as the length of the field
C. begin with the same character as the data in the field
D. all of these

1198. A characteristic of card systems is
A. slowness in processing data
B. using cards as records of transactions
C. needing a larger DP staff
D. all of these

1199. The subdivision of fields is
A. always done to give the programmer greater flexibility
B. dependent on the programming language used
C. never accomplished on fields containing numeric data
D. all of these

1200. The checking operation performed on input data is called the
A. validation of data B. verification of data
C. vilification of data D. control of data

1201. The EAM device that does not use a control panel is the
A. collator B. reproducer
C. interpreter D. sorter

1202. The card type designed to accept the manual entry of data onto its surface is the
A. mark-sense card B. document card
C. dual card
D. both A and C are correct

1203. Because it can function in more ways than a line printer, the IBM 6670 is sometimes referred to as a/an
A. plotter B. mag card printer
C. intelligent printer D. photostatic copier

1204. An I/O device which provides photographic outputs for printing galleys is the
A. camera printer B. automatic typesetter
C. radix printer D. all of these

1205. An IBM System/38 represents the computer class of
A. small-scale computer
B. medium-scale computer
C. large-scale computer
D. supercomputer

1206. A term used interchangeably with diskette is
A. disk cartridge B. disk pack
C. floppy disk D. none of these

1207. A characteristic normally associated with a large-scale computer system is
A. real-time processing
B. a complex array of I/O devices
C. a CPU of 1000 K D. all of these

Ans. 1186. D 1187. D 1188. B 1189. D 1190. A 1191. D 1192. D 1193. C 1194. B 1195. B 1196. B 1197. A 1198. D 1199. B 1200. A 1201. D 1202. D 1203. C 1204. B 1205. A 1206. C 1207. D

1208. A software package generally used within a distributed word processing system is
A. ATMS B. DWPS
C. RPG D. DBMS

1209. An advantage of a distributed word processing network is
A. increased number of reports handled
B. less delay in interoffice communications
C. greater corporate control over outputs
D. all of these

1210. ROM is composed of
A. magnetic cores B. microprocessors
C. photoelectric cells D. floppy disk

1211. A peripheral device used in a word processing system is
A. floppy disk B. magnetic card reader
C. CRT D. all of these

1212. Multiprogramming was made possible by
A. input/output units that operate independently of the CPU
B. operating system
C. both A and B are correct
D. neither A nor B

1213. The person contributing the idea of the stored program was
A. John von Neumann B. Charles Babbage
C. Howard Aiken D. Thomas J Watson Sr

1214. Transistorised computer circuits were introduced in the
A. first generation B. second generation
C. third generation D. fourth generation

1215. The first firm to mass-market a microcomputer as a personal computer was
A. Data General Corporation
B. Sperry Univac
C. Radio Shack D. IBM

1216. The availability of low-price minicomputers is largely responsible for current interest is
A. realtime processing B. batch processing
C. distributed processing
D. transaction processing

1217. In a punched card, data is processed by a
A. keypunch machine, sorter and posting machine
B. sorter, posting machine and billing machine
C. accounting machine, posting machine and billing machine
D. accounting machine, keypunch machine and sorter

1218. The "Father of Punched Card Processing" was
A. J Presper Eckert B. Charles Babbage
C. Blaise Pascal D. Dr Herman Hollerith

1219. The punched card used in the IBM System/3 contains
A. 80 columns B. 90 columns
C. 96 columns D. 126 columns

1220. Thick-film ICs use which technique?
A. Screen printing B. Cathode sputtering
C. Monolithic D. Hybrid

1221. An IC contains 50 gates each of which consists of 6 components. It is an example of
A. VLSI B. LSI
C. MSI D. SSI

1222. First integrated circuit chip was developed by
A. CV Raman B. WH Brittain
C. JS Kilby D. Robert Noyce

1223. Most important advantage of an IC is its
A. easy replacement in case of circuit failure
B. extremely high reliability
C. reduced cost
D. low power consumption

1224. As compared to discrete circuit, an IC has the advantage of
A. small size B. smaller weight
C. higher reliability D. all of these

1225. An integrated circuit is
A. a complicated circuit B. an integrating device
C. much costlier than a single transistor
D. fabricated on a tiny silicon chip

1226. A modern fast mainframe digital computer can perform up to operations second
A. 1000 (1 K) B. 100,000
C. 1,000,000 (1 M) D. 1,000,000,000 (1 G)

1227. A modern electronic computer is a machine that is meant for
A. doing quick mathematical calculations
B. input, storage, manipulation and outputting of data
C. electronic data processing
D. performing repetitive tasks accurately

1228. The operation of a digital computer is based on which principle?
A. Counting B. Measuring
C. Electronic D. Logical

1229. The proper definition of a modern digital computer is
A. an electronic automated machine that can solve problems involving words and numbers
B. a more sophisticated and modified electronic pocket calculator
C. any machine that can perform mathematical operations
D. a machine that works on binary code

1230. A modern digital computer has
A. extremely high speed
B. almost unlimited accuracy
C. large memory D. all of these

1231. The very small and cheap computer built into many home devices is called .... computer.

Ans. 1208. A 1209. B 1210. A 1211. D 1212. C 1213. A 1214. B 1215. C 1216. C 1217. D 1218. D 1219. C 1220. A 1221. C 1222. C 1223. B 1224. D 1225. D 1226. B 1227. D 1228. A 1229. A 1230. D

A. mainframe B. mini
C. micro D. analog

1232. A digital computer scores over an analog computer in terms of
A. speed B. accuracy
C. cost D. memory

1233. The main distinguishing features of fifth generation digital computers will be
A. liberal use of microprocessors
B. artificial intelligence
C. extremely low cost D. versatility

1234. IBM 7000 digital computer
A. belongs to second generation
B. uses VLSI
C. employs semiconductor memory
D. has modular construction

1235. The first mechanical computer designed by Charles Babbage was called
A. analytical engine B. abacus
C. calculator D. processor

1236. The modern abacus which uses small beads strung on counting frames is still being used in some oriental countries like China and Japan. How would you classify this calculator?
A. Analog B. Hybrid
C. Digital D. Electrical

1237. The counting board in which numbers are represented as beads strung on wires is commonly called abacus. By what name is it known in Japan?
A. Suanpan B. Soroban
C. Napier's beads D. Suharto

1238. Who developed a mechanical device in the 17th century that could add, subtract, multiply, divide and find square roots?
A. Napier B. Babbage
C. Pascal D. Leibnitz

1239. What is the name of the earliest calculating machine which was based on concepts found in modern computers but was unfortunately never built?
A. Babbage's difference engine
B. Pascal's adder
C. Leibnitz's multiplier D. Differential analyser

1240. Who is called the "grandfather" of the computer?
A. Herman Hollerith B. Blaise Pascal
C. Charles Babbage D. Joseph Jacquard

1241. In 1830, Charles Babbage designed a machine called the Analytical Engine which he showed at the Paris Exhibition. In which year was it exhibited?
A. 1835 B. 1860
C. 1855 D. 1970

1242. Charles Babbage, who was Professor of Mathematics at Cambridge University in England, was an eccentric and quick-tempered man. He planned and dreamt of many machines which he could not complete. One such machine was known as "Babbage Folly". Can you tell the name of that machine?
A. Differential Analyzer B. Difference Engine
C. Analytical Engine D. Mark-I

1243. In 1994, an electromechanical computer was built having thousands of relays which sounded "like a roomful of old ladies knitting away with steel needles" as they clicked open and closed. What was the name of this computer?
A. Colossus B. Mark I
C. Whirlwind D. EDSAC

1244. ENIAC (Electronic Numerical Integrator and Calculator) had huge advantage over Mark I because it used electronic valves in place of the electromagnetic switches. In the beginning, ENIAC was used for calculating the path of artillery shells. For which other war weapon design was it utilised?
A. Hydrogen bomb B. Atom bomb
C. Submarines D. Fighter aircraft

1245. What was the total number of UNIVAC-I sold eventually and by which company?
A. 20, British Tabulating Machine Co (BTM)
B. 80, International Business Machines (IBM)
C. 48, Remington Rand
D. 40, International Computers Ltd (ICL)

1246. Who is credited with the development of "blueprints" for the first digital computer?
A. Blaise Pascal
B. William Seward Burroughs
C. Charles Babbage D. Herman Hollerith

1247. The first practical commercial typewriter was invented in 1867 in the United States by
A. Christopher Latham Sholes
B. Carlos Glidden
C. Samuel Soule D. all of these

1248. Punched-card equipment was first introduced in Britain in 1904 by a small company. What was the name of that company?
A. Tabulator Ltd
B. Tabulator Machine Company
C. International Business Machines (IBM)
D. Accounting and Tabulating Corpn of Great Britain

1249. When was punched-card equipment used for the first time to process the British census?
A. 1904 B. 1907
C. 1911 D. 1914

1250. The first major electronic analog computer called the differential analyser was built as Massachusetts Institute of Technology (MIT) in the late 1920s. Can you tell who directed the development of this computer?

Ans. 1231. C 1232. C 1233. B 1234. D 1235. A 1236. C 1237. B 1238. D 1239. A 1240. C 1241. C 1242. C 1243. B 1244. A 1245. C 1246. C 1247. D 1248. A 1249. C

A. Clifford Berry B. Thomas J Watson
C. AM Turing D. Vannevar Bush

1251. Which was the first company in the world to build computer for sale?
A. International Business Machines
B. Remington Rand Corporation
C. English Electric Computers Ltd
D. Sperry Univac

1252. In which year was the first analog computer built by an American named Vannevar Bush?
A. 1920 B. 1950
C. 1930 D. 1944

1253. The first special-purpose electronic computer which contained about 300 vacuum tubes and became operational in early 1940s was called?
A. Colossus B. ENLAC
C. Atanasoff-Berry Computer (ABC)
D. EDVAC

1254. Mark the wrong statement regarding Atanasoff-Berry Computer. It was the first computer to
A. perform its operations in binary
B. incorporate electronic logic circuits
C. use vaccum tubes to store information
D. work as a general purpose computer

1255. Which was the first electronic computer that belonged to the first generation of computers?
A. Mark-1 computer
B. Atanasoff-Berry computer
C. ENIAC D. UNIVAC-I

1256. In 1943, the first general-purpose, all electronic deciphering computer using 1500 vacuum tubes was built. What was it called?
A. Mark-I B. Whirlwind
C. Colossus D. ENIAC

1257. Who invented the first electromechanical computer Mark-I?
A. John W Mauchly B. JP Eckert
C. Howard Aiken D. Clifford Berry

1258. Who amongst the following is credited with the building of the first operational stored program computer?
A. Maurice V Wilkes B. HH Goldstine
C. AW Burks D. Grace Murray Hopper

1259. Who built the world's first electronic calculator using telephone relays, light bulbs and batteries?
A. Claude Shannon B. Konrad Zuse
C. George Stibitz D. Howard H Aiken

1260. The first large scale electronic computer which became operational in 1946 and contained approximately 1800 vacuum tubes and could perform 300 multiplications per second was known as
A. ILLIAC B. ENIAC
C. JOHNIAC D. EDSAC

1261. The magnetic tape consists of a plastic tape with a surface coating of magnetic material. When was it invented?
A. 1956 B. 1964
C. 1968 D. 1946

1262. The first electronic computer using vacuum tubes was built in 1946 by JP Eckert and JW Mauchly at the University of Pennsylvania. It could multiply two ten-digit numbers in three thousand of a second compared to roughly three seconds for Howard Mark-I computer. What was the name of that computer?
A. UNIVAC B. ENIAC
C. MANIAC D. JOHNIAC

1263. Which of the following computers was the first to use a stored program?
A. ENIAC B. EDVAC
C. EDSAC D. ACE

1264. What was the name of the first stored program electronic computer which was built in 1949 at Cambridge University in England?
A. EDVAC B. EDSAC
C. UNIVAC D. MARK-I

1265. In early electronic computer, vacuum tubes were used as the storage media. Later on, they were replaced by small doughnut-shaped pieces of magnetic material called magnetic core. Can you tell which of the following electronic stored-program computers was the first to use this core memory?
A. IBM 701 B. Whirlwind
C. Colossus D. MANIAC

1266. The first computer made available for commercial use was
A. Mark-I B. ENIAC
C. UNIVAC D. EDSAC

1267. Which computer was the first to use to magnetic drum for memory?
A. IBM-650 B. IBM-7090
C. IBM-701 D. PDP-1

1268. What is the name of the British gentleman who was the first to put forward in 1952 the idea for the integrated circuit?
A. Jack S Kilby B. GW Dummer
C. William Shockley D. John Bardeen

1269. What did the first electronic digital computer contains?
A. Transistors B. Electronic valves
C. Core memory
D. Semiconductor memory

1270. The first large commercial magnetic disk storage device was introduced in 1956 by
A. Texas Instruments B. IBM
C. Digital Equipment Corporation
D. Hewlett-Packard

Ans. 1250. D 1251. B 1252. C 1253. C 1254. D 1255. B 1256. B 1257. C 1258. B 1259. C 1260. B 1261. B 1262. B 1263. C 1264. B 1265. B 1266. C 1267. A 1268. B 1269. B 1270. B

1271. Which is the first compiler-level language developed by a team of IBM programmers led by John Backus and unveiled in 1957?
A. BASIC B. PL/1
C. FORTRAN D. APL

1272. Which was the world's first minicomputer and when was it introduced?
A. PDP-1, 1958 B. IBM Systems/36, 1959
C. PDP-II, 1960 D. VAX 11/780, 1962

1273. Which one of the following is the first second-generation computer?
A. IBM 7090 B. IBM 701
C. IBM 7070 D. IBM 650

1274. Which was the first personal computer aimed squarely at small businesses that had floppy drive whose motor and circuits were run by a program called CP/M (computer program for microcomputers)?
A. IMSAI B. Apple-1
C. Apple Macintosh D. Altair-8800

1275. The first disk system which had fifty 24-inch magnetic disks capable of storing about 5 million characters of information was called
A. RAMAC B. RAM
C. ROM D. ROMA

1276. A microprocessor integrates the arithmetic logic and control circuitry of a computer into one chip. The first microprocessor was built by a group of engineers at the Intel Corpn. Can you tell who headed this group?
A. Jack S Kilby B. Stan Mazor
C. Marcian E (Ted) Hoff
D. Seymour Cray

1277. Which was the first computer to use transistors instead of vacuum tubes?
A. IBM-650 B. Burroughs E-101
C. Datamatic-1000 D. IBM-1401

1278. The first minicomputers were developed and built in 1965 by
A. IBM B. Data General
C. Digital Equipment Corporation (DEC)
D. Hewlett Packard

1279. Name the two persons who were the first to develop a model of the microprocessor chip.
A. Marcian Ted Hoff B. Victor Poor
C. Harry Pyle
D. Both B and C are correct

1280. What was the name of the first commercially available microprocessor chip?
A. Intel 8008 B. Intel 8080
C. Intel 4004 D. Motorola 6809

1281. What is the name of the personal computer/home computer which was the first to become available in the world and which company introduced it?
A. Altair 880, Micro Instrumentation and Telemetre Systems (MITS)
B. ZX 80, Sinclair Research
C. HX-20, Epson D. BBC Micro, Arcon

1282. The first microprocessor built by the Intel Corpn was called
A. 8008 B. 8080
C. 4004 D. 8800

1283. Which bank was the first to introduce ATMs to the world and when?
A. Standard Chartered Bank, 1972
B. Bank of America, 1968
C. Citibank, 1970 D. Hongkong Bank, 1970

1284. Which microprocessor chip was the first to be used in personal computer?
A. Intel 4004 B. Intel 8088
C. Intel 8080 D. Z 80

1285. Who marketed the world's first pocket calculator called 'Executive' which made world headlines in 1972?
A. Charles Tandy B. Sir Clive Sinclair
C. Steve Wozniak D. Mike Markkula

1286. The first microprocessor-based personal computer (PC) was advertised in March 1974 in an American magazine for amateur radio buffs. What was the name of that personal computer?
A. Scelbi-8H B. Altair 8800
C. Apple 1 D. TRS-80

1287. In which year were chips used inside the computer for the first time?
A. 1972 B. 1975
C. 1977 D. 1981

1288. Can you name the first guided weapon in the world which used a programmable digital computer?
A. Sting Ray Torpedo B. Mk 46 Torpedo
C. Air-Launched Cruise Missile (ALCM)
D. Tomahawk Missile

1289. Texas Instruments Corporation was the first to introduce speech synthesizer contained in a single integrated circuit (IC) chip. It was the basis for the "Speak and Spell" game used for teaching children to read and write the words stored in the chip's memory. In which year was it produced?
A. 1960 B. 1978
C. 1984 D. 1974

1290. What was the name of the first personal computer electronic spreadsheet software package which became a smash hit as soon as it was introduced in 1978?
A. WordStar B. VisiCale
C. Lotus 1-2-3 D. Excel

1291. Who developed the first personal computer spreadsheet package called VisiCale?
A. Niklaus Wirth B. Dan Bricklin

Ans. 1271. C 1272. A 1273. A 1274. A 1275. A 1276. C 1277. D 1278. C 1279. D 1280. C 1281. A 1282. C 1283. C 1284. C 1285. B 1286. A 1287. B 1288. A 1289. B 1290. B

C. Bob Frankston
D. Both B and C are correct

1292. Which gentleman in the computer industry became the world's youngest billionaire at the age of 31?
A. John Akers B. William Gates
C. Carlo De Benedetti D. Mac Jeffery

1293. When was the first IBM microcomputer called IBM PC with 16-bit microprocessor introduced?
A. 1970 B. 1981
C. 1982 D. 1984

1294. The first IBM PC was released in the market in 1981. Which microchip did it use?
A. Intel 8088 B. Intel 4004
C. Z 8000 D. Motorola 6502

1295. When was the world's first laptop computer introduced in the market and by whom?
A. Hewlett-Packard, 1980
B. Epson, 1981
C. Laplink Travelling Software Inc 1982
D. Tandy Model-200, 1985

1296. First CAD system called Auto CAD for the PCs was announced in the year 1982. Can you tell who announced it?
A. Intel Corpn B. Rolta India Ltd
C. Autodesk Inc D. AT&T Corpn

1297. The first movie with terrific computer animation and graphics was released in 1982. What was its name?
A. Star Wars B. Tron
C. Forbidden Planet D. Dark Star

1298. Which company announced the first IBM PC clone, the MPC?
A. Columbia Data Products
B. Compaq Computer Inc
C. Carona Systems D. Honeywell Bull Inc

1299. When did IBM announce its IBM XT microcomputer?
A. 1982 B. 1983
C. 1984 D. 1985

1300. Who introduced the world's first file server LAN called Netware and when?
A. Lotus Development Corpn, 1981
B. Zenith Computers Ltd, 1984
C. Novel Inc, 1983 D. Wang Labs Inc, 1982

1301. What is the title of the first book written by a computer and published by Warner Books in mid-1984?
A. The Moon is a Harsh Mistress
B. The Multiple Man
C. The Nine Billion Names of God
D. The Policeman's Beard is Half Constructed

1302. When did Hewlett-Packard Inc introduce its first HP-110 laptop computer?
A. 1984 B. 1982
C. 1980 D. 1986

1303. What is the designation of IBM's first optical storage device specifically designed for its PS/2 series of computers?
A. IBM-2717 B. IBM-3363
C. IBM-3131 D. IBM-4040

1304. Which was the first supercomputer purchased by India for medium range weather forecasting?
A. PARAM B. Cray XMP-14
C. Medha-930 D. CDC Cyber 930-11

1305. When did Intel introduce 32-bit 386DX CPU for the first time?
A. 1987 B. 1985
C. 1983 D. 1988

1306. When did IBM come out with one of the first RISC-based workstations, the RT-PG?
A. 1982 B. 1984
C. 1986 D. 1988

1307. Who developed the personal computer called NEXT which was the first to incorporate a drive for optical storage disk, voice recognition and object-oriented languages?
A. Niklaus Wirth B. Steve Jobs
C. Phillipe Kahn D. Vannevar Bush

1308. Which was the world's first microcomputer that used Intel 80386 microprocessor chip?
A. IBM PS/2 B. HP-9830
C. DeskPro-386 D. IBM-360

1309. CPU of a computer system does not contain
A. main storage B. arithmatic unit
C. special register group D. none of these

1310. A program that can be used repeatedly throughout a major program is called
A. program module B. loop
C. sub-routine D. template

1311. Modem is a
A. code changing system
B. programme conversion system
C. modulator-demodulator system
D. demodulator-modulator system

1312. An Input Port is
A. tri-state buffer with register
B. select logic
C. input device D. R/W memory

1313. Programmable Logic Array (PLA) uses
A. ROM matrices B. PROM matrices
C. RAM matrices D. Silo memory

1314. Mnemonic symbols are used to
A. denote address B. employ hamming code
C. denote errors D. assist human memory

Ans. 1291. D 1292. B 1293. B 1294. A 1295. B 1296. C 1297. B 1298. A 1299. B 1300. C 1301. D 1302. A 1303. B 1304. B 1305. B 1306. C 1307. B 1308. C 1309. D 1310. C 1311. C 1312. A 1313. A 1314. D

1315. Propagation delay is the time required for
A. a change in level at the output of an element
B. output to appear when input is applied
C. the element to become operational
D. a signal to propagate in a circuit

1316. An assembler in a computer system prepares
A. machine-language programme from a symbolic language programme
B. object programme
C. assembles computer instructions and data in the machine
D. none of these

1317. READY signal in 8085 is useful when the CPU communicates with a
A. slow peripheral device
B. fast peripheral device
C. DMA controller chip
D. PPI chip

1318. Which of the following ICs can be used for frequency division by 5?
A. 7446 B. 7492
C. 74121 D. 7490

1319. For the generation of a sinusoidal function on an analog computer, with usual available modules, the required number of OP-AMPs is
A. 1 B. 2
C. 3 D. 4

1320. Assuming an 8-bit word length, –5 in 2's complement form is
A. 11111011 B. 10000101
C. 00000101 D. 00001011

1321. The number of locations that can be selected using 12 address lines is
A. 1024 B. 2048
C. 4096 D. 8192

1322. Which of the following ICs is a 4-bit latch?
A. 7475 B. 7446
C. 7400 D. 7490

1323. The purpose of introducing feedback loop in a digital counter circuit is to
A. improve stability B. improve distortion
C. synchronise input and output pulses
D. reduce the number of input pulses to reset the counter

1324. During a DMA transfer, the processor (check the incorrect statement)
A. continues its normal operations
B. suspends its normal operations
C. needs to initiate read (write) command
D. needs to check if the input/output device is ready for data transfer

1325. In 8085, interrupts except TRAP are disabled (check the incorrect statement) by
A. DI instruction B. system reset
C. acknowledgement of a previous interrupt
D. none of these

1326. The basic elements of a microprocessor are
A. ALU, memory B. ALU, control unit
C. ALU, memory, I/O devices
D. ALU, control unit, memory

1327. A system of letters, numbers and symbols adopted by computer manufacturer as an abbreviated form of instruction sets is called
A. modern B. monitor
C. mnemonic D. mask

1328. The processing of causing an unplanned branching operation to occur, usually initiated by external system is called
A. debugging B. masking
C. interrupt D. iteration

1329. A microprocessor contains
A. most of the control and arithmatic logic functions of a computer
B. most of the RAM
C. most of the ROM D. peripheral drivers

1330. Which of the following statements is not correct?
A. MOS devices require fewer manufacturing cycles
B. MOS gates dissipate less power per gate
C. MOS devices are fastest devices
D. MOS gates require less space on the silicon wafer

1331. From the point of view of integration, a microprocessor chip employs
A. SSI B. MSI
C. LSI D. GSI

1332. The first practical microprocessor was introduced in
A. 1969 B. 1971
C. 1973 D. 1975

1333. The first practical microprocessor was introduced by
A. Motorola
B. National Semiconductors
C. Intel D. Fairchild

1334. The number of first practical microprocessor was
A. Intel 4001 B. M 6800
C. Intel 4004 D. F-8

1335. The bit size of first practical microprocessor was
A. 2-bit B. 4-bit
C. 8-bit D. 4 bytes

1336. How many pins intel 8080 has?
A. 28 B. 36
C. 40 D. 42

Ans. 1315. A 1316. A 1317. A 1318. D 1319. C 1320. A 1321. C 1322. A 1323. D 1324. A 1325. D 1326. B 1327. C 1328. C 1329. A 1330. C 1331. C 1332. B 1333. C 1334. A 1335. B 1336. C

1337. Who manufactured 6800 μP?
A. Fairchild B. Intel
C. Zilog D. Motorola

1338. A byte is a group of
A. 2-bits B. 4-bits
C. 8-bits D. 16-bits

1339. A Central processing unit of a computer contains
A. control and arithmetic logic unit
B. control and RAM
C. control, clock, RAM and ROM
D. control, ALU, clock, RAM and ROM

1340. An instruction register is storage for
A. location of data in memory
B. location of instruction in memory
C. binary code for the operation to be performed
D. address of the next instruction to be executed

1341. Which of the following operations is not performed by the CPU of a computer system?
A. Arithmetic operations
B. Providing timing signals
C. Control instruction processing
D. None of these

1342. A microprocessor with a 12-bit address bus will be able to access
A. 1 kilobytes of memory
B. 4 kilobytes of memory
C. 8 kilobytes of memory
D. 0.4 kilobytes of memory

1343. Which of the following flag conditions are not available in 8085 processor?
A. Zero flag B. Parity flag
C. Overflow flag D. Auxiliary carry flag

1344. A microcomputer chip essentially contains
A. memory (ROMs and RAMs) and ALU
B. memory, CPU, and I/O lines
C. CPU only D. none of these

1345. The frequency of the driving network connected between pins 1 and 2 of a 8085 chip must be
A. twice of the desired clock frequency
B. equal to the desired clock frequency
C. four times the desired clock frequency
D. none of these

1346. A bit or bits store for later use, indicating the status of CPU is called
A. flag B. data
C. logic D. word

1347. In microprocessor architecture, flag indicates the
A. number of the microprocessors
B. name of the manufacturer
C. internal status of the CPU
D. bit-size of the microprocessor

1348. A microcomputer is a minimum combination of
A. μP and clock
B. μP clock, ROMs
C. μP, clock, RAMs and ROMs
D. μP, clock, RAM, ROM, PIA and ACIA

1349. Accumulation is a device which
A. stores a number B. adds two numbers
C. adds and stores previously stored number to another number
D. none of these

1350. Address usually is
A. alphabetical B. numerical
C. alpha-numerical D. either of these

1351. Algorithm is used to
A. bring itself into desired state by its own action
B. perform logarithmic operations
C. describe a set of procedure by given result is obtained
D. none of these

1352. American Standard Code for Information Interchange (ASCII) employs a coded character set consisting of
A. 7-bits B. 8-bits
C. 7-bits with parity check
D. 8-bits with parity check

1353. In a train of binary signals, 1 band is equal to
A. 1/4 bit/sec B. 1/2 bit/sec
C. 1 bit/sec D. 2 bits/sec

1354. In a train of signals each of which can assume one out of eight different states, 1 band is equal to
A. 2 bits/sec B. 3 bits/sec
C. 4 bits/sec D. 8 bits/sec

1355. In medium-scale integration (MSI), a chip contains less than
A. 12 gates B. 100 gates
C. 1000 gates D. 5000 gates

1356. In large-scale integration (LSI) a chip contains more than
A. 100 but less than 1000 gates
B. 500 but less than 100 gates
C. 1000 gates D. 5000 gates

1357. In grand-scale integration (GSI) a chip contains more than
A. 1500 gates B. 1000 gates
C. 5000 gates D. 10000 gates

1358. Which of the following parameters is not specified for digital ICs?
A. Gate dissipation B. Propagation delay
C. Noise margin D. Bandwidth

1359. Generally, the unit of propagation delay is
A. sec B. m-sec
C. p-sec D. n-sec

1360. Noise margin of a digital IC is
A. the minimum extraneous voltage that causes a gate to change its state
B. the maximum extraneous voltage that causes a gate to change its state

Ans. 1337. D 1338. C 1339. A 1340. C 1341. B 1342. A 1343. C 1344. B 1345. A 1346. A 1347. C 1348. D 1349. C 1350. B 1351. C 1352. D 1353. C 1354. B 1355. B 1356. A 1357. B 1358. D 1359. D 1360. B

C. the thermal noise voltage which causes a gate to change its state
D. none of these

1361. In small scale integration (SSI), chip contains less than
A. 5 gates B. 12 gates
C. 100 gates D. 500 gates

1362. The propagation delay of a digital IC is the difference between
A. initiating clock pulse and triggering of gates
B. the application of input and its presence at the output
C. the triggering of a flip-flop and change in its state
D. none of these

1363. The speed of a digital IC indicates
A. the rate at which output changes
B. how fast input triggers the output
C. how fast a flip-flop can change stages
D. none of these

1364. The unit of the speed of a digital IC is
A. Hz B. KHz
C. MHz D. GHz

1365. The speed-power product of a digital IC is the product of
A. speed and propagation delay
B. propagation delay and gate dissipation
C. speed and noise margin
D. speed and gate dissipation

1366. Which of the following MOSFETs is employed in the CMOS?
A. *p*-channel enhancement type
B. *n*-channel enhancement type
C. both *n*-channel and *p*-channel type
D. *p*-channel depletion type

1367. The microprocessor chip register unit ALU chip has
A. arithmetic functions only
B. basic control and arithmetic functions
C. instructions to execute data
D. memory functions

1368. Which of the following system is digital?
A. PPM B. PFM
C. PWM D. PCM

1369. If it is required to multiply two binary numbers in a digital computer
A. a hardware multiplier is essential
B. it is adequate to have adder-subtractor unit and shift register
C. both a hardware multiplier and an adder subtractor unit are essential
D. a hardware divider is essential

1370. Which code is suitable for input-output devices and analog-to-digital converters?
A. 8421 code B. Excess-3 code
C. 4-bit BCD code D. Gray code

1371. LSI usually contains
A. 10 gates B. 10 to 100 gates
C. more than 100 gates
D. more than 10000 gates

1372. The microprocessor contains ROM chip contains
A. control functions B. arithmetic functions
C. instructions to execute data
D. memory functions

1373. The address bus of Intel 8085 is 16-bit wide and hence the memory which can be addressed by this address bus is
A. 2 k bytes B. 4 k bytes
C. 16 k bytes D. 64 k bytes

1374. The physical part or electronic circuitry of a computer system is called
A. hardware B. software
C. microprocessor D. peripheral device

1375. The central processing unit of a small computer system, including ALU and control functions on a single LSI chip is called
A. hardware B. memory
C. microprocessor D. peripheral device

1376. A device that works in conjunction with a computer but not as part of it is called
A. hardware B. memory
C. microprocessor D. peripheral device

1377. A device in which computer information and data are stored and relieved at a later time is called
A. hardware B. memory
C. microprocessor D. peripheral device

1378. A single binary digit is called
A. byte B. bit
C. data D. logic

1379. A fixed step by step procedure for finding the solution of a problem on computer is called
A. hardware B. logic
C. algorithm D. none of these

1380. A bit or bits stored for later use, indicating the status of CPU is called
A. flag B. data
C. logi D. word

1381. The 8085 A instruction that takes maximum number of machine cycles for execution is
A. STA Addr B. MVI A, data
C. Out Port D. CALL Addr

1382. The contents of Stack Pointer (SP) can be accessed by the instruction
A. SP HL B. PC HL
C. XT HL D. none of these

Ans. 1361. B 1362. B 1363. C 1364. C 1365. B 1366. C 1367. B 1368. D 1369. B 1370. D 1371. C 1372. C 1373. D 1374. A 1375. C 1376. D 1377. B 1378. B 1379. C 1380. A 1381. D 1382. D

1383. The address lines required to address 2048 K byte of memory are
A. 11 B. 18
C. 20 D. 21

1384. Intel 8085 A supports
A. isolated I/O and memory mapped I/O
B. only memory mapped I/O
C. only isolated I/O
D. none of these

1385. For the entire block of data transfer, the preferred DMA mode of operation is
A. cycle stealing DMA B. transparent DMA
C. burst DMA D. None of these

1386. The port of 8255 A that can be split-up into two nibbles is
A. port A B. port B
C. port C D. all of these

1387. IC 8253 programmable interval timers are programmed with
A. binary data only B. BCD data only
C. binary and BCD data both
D. none of these

1388. The number of additional flags in 8086 as compared to 8085 A are
A. 1 B. 2
C. 3 D. 4

1389. A program structure that allows the repeated operation of a particular sequence of instructions until a specified termination is reached is called
A. subroutine B. machine
C. module D. loop

1390. A programme that can be used repeatedly throughout a major program is called
A. program module B. loop
C. sub-routine D. template

1391. A programme that translates symbolically represented instructions into their binary equivalents is called
A. loader B. assembler
C. linker D. autoloader-5

1392. To correct and eliminate errors in programmes, the act of using an instruction, programme or action designed in software is called
A. interrupting B. cross-assembling
C. documentation D. debugging

1393. Bootstrap is a technique designed to
A. transfer control to specified closed subroutine
B. suspend the main programme for a while
C. bring itself into a desired state by means of its own action
D. none of these

1394. *Call* in computer system means
A. to execute a part of main programme
B. to call a sub-programme to be executed
C. to transfer control to specified closed subroutine
D. none of these

1395. *Carry-Look-Ahead* is a type of adder in which
A. inputs to several stages are examined
B. carries are produced simultaneously
C. carries are produced before hand
D. none of these

1396. Which of the following task is not performed by an assembler?
A. Providing storage allocation
B. Creating a table of lables, etc.
C. Doing assembly time arithmetic
D. Translate a program written in high level language to machine code program

1397. A program structure that allows the repeated operation of a particular sequence of instructions until a specified termination is reached is called
A. sub-routine B. machine
C. module D. loop

1398. Which of the following actions detect locations, and remove mistakes from a programme routine?
A. Erase B. Debug
C. Diagnose D. Emulate

1399. A multiprocessor is also referred to as the
A. chip that does some calculations for the computer
B. computer on a chip
C. chip that is responsible for data transfer
D. none of these

1400. A microprocessor is called an *n*-bit microprocessor depending on
A. size of external data bus
B. size of internal data bus
C. register length
D. none of these

1401. Two operands can be checked for equality using
A. OR-operation
B. AND-operation
C. EX-OR operation
D. none of these

1402. Setting contents of a register to zero can be efficiently done by
A. MOV Immediate instruction using zero as immediate data
B. AND Immediate instruction using zero as immediate data
C. XORing register with itself
D. none of these

Ans. 1383. B 1384. A 1385. C 1386. C 1387. C 1388. D 1389. D 1390. C 1391. B 1392. D 1393. C 1394. C 1395. C 1396. D 1397. D 1398. B 1399. B 1400. B 1401. C 1402. C

1403. Stack memory is used to
A. provide additional memory to the base memory
B. save return address of a subroutine
C. save the status of the microprocessor
D. none of these

1404. 8085 microprocessor is a
A. zero address microprocessor
B. one address microprocessor
C. two address microprocessor
D. none of these

1405. SHIFT LEFT instruction causes all bits shifted one position to the left with right-most bit set to zero. The effect is to
A. multiply by 2 B. divide by 2
C. SET the most significant bit
D. none of these

1406. The following signal is used when a microprocessor wants to address the memory
A. I_0/M' B. Status signals
C. ALE D. HOLD and HLDA

1407. The following signal is used when a peripheral device requests the microprocessor to have a DMA operation
A. I_0/M' B. READY
C. HOLD and HLDA D. RD′ and WR′

1408. Program counter is used to
A. store address of the next instruction to be executed
B. store temporary data to be used in arithmetic operations
C. store the status of the microprocessor
D. none of these

1409. The synchronization between microprocessor and memory is done by
A. ALE signal B. HOLD signal
C. READY signal D. none of these

1410. As compared to 16-bit microprocessors, 8-bit microprocessors are limited in
A. data handling capability
B. directly addressable memory
C. speed D. all of above

1411. The set of commands which give directions to the assembler during the assembly process but are not translated into machine instructions are called
A. mnemonics B. directives
C. identifiers D. operands

1412. RISC is characterized by
A. hard-wired control design with no micro codes
B. fixed instruction format
C. executing most of the instructions in a single clock cycle
D. several general-purpose registers and large cache memories that support very fast access to data

1413. The theoretical dividing line between Reduced Instruction Set Computing (RISC) chips and Complex Instruction Set Computing (CISC) chips is
A. number of pins in the chip
B. number of address and data lines
C. instruction execution rate to be one instruction per clock cycle
D. none of these

1414. It is a type of microprocessor instruction in which the contents of the source are copied into the destination register without modifying the contents of the source.
A. Arithmetic instructions
B. Data transfer instructions
C. Control transfer instructions
D. Machine control instructions

1415. With reference to 8085, ANAR/M is
A. a logic instruction B. control instruction
C. data transfer instruction
D. an arithmetic instruction

1416. Identify the true statement.
A. Stack pointer is a register that tells about occurrence or non-occurrence of different conditions like carry, overflow, etc.
B. Program counter stores the code of the instruction currently being executed
C. Microprocessors use a stack because it is faster to move data using PUSH and POP instructions than to move data using MOVE instructions
D. *STA address* is a logic instruction

1417. The largest number that can be processed by a microprocessor in a single operation is determined by the size of its
A. external data bus B. internal data bus
C. address bus D. control bus

1418. Microprocessors are programmed using
A. assembly language B. Pascal
C. high-level language such as C, C++, Pascal
D. assembly language or some suitable high level language

1419. With reference to microcontrollers and microprocessors,
A. they are the names of similar devices
B. microcontroller is a complete functional unit and has a processor, some memory, I/O ports and some peripheral devices on the same chip
C. microcontroller is that part of microprocessor, which controls its operations
D. none of these

1420. Peripheral devices that can possibly be found on a microcontroller chip include the following.

Ans. 1403. B 1404. A 1405. A 1406. C 1407. C 1408. A 1409. C 1410. D 1411. B 1412. C 1413. C 1414. B 1415. A 1416. C 1417. B 1418. D 1419. B

A. Computer/timer B. USB port
C. Serial communication interface
D. All of these

1421. Identify the asynchronous communication interface.
A. UART
B. Inter-integrated circuit bus
C. Controller area network
D. All of these

1422. Identify the synchronous communication interface.
A. UART
B. Inter-integrated circuit bus
C. Serial peripheral interface
D. Controller area network

1423. Identify the serial interface commonly used to integrate intelligent sensors and actuators in modern automobiles.
A. UART
B. Local interconnect network bus
C. Controller area network
D. Both B and C

1424. A microcontroller uses an eight-bit counter/timer and is clocked by 1 MHz signal. When used to measure width of a pulse, the rising and falling edges of the pulse are recorded at the count values of FE and FF (in hex). The pulse width is
A. 1.0 μs B. 10 μs
C. 100 ns D. none of these

1425. Minimum pulse width that a microcontroller can be programmed to measure depends upon
A. number of bits in the counter/timer
B. clock frequency
C. both A and B D. none of these

1426. Minimum time period that a microcontroller can measure depends upon
A. clock frequency B. size of counter/timer
C. both A and B D. none of these

1427. The processor architecture that uses separate memory for program instructions and data is
A. Von Neumann architecture
B. Pipeline architecture
C. Harvard architecture
D. none of these

1428. The processor architecture that uses same memory for program instructions and data is
A. Von Neumann architecture
B. Harvard architecture
C. Accumulator-based architecture
D. none of these

1429. Identify the power saving mode or modes that usually come with 8051 microcontroller.
A. Stop clock mode B. Power down mode
C. Idle mode D. All of these

1430. Identify eight-bit microcontrollers.
A. 68HC11 family B. 80C51 family
C. 16X84 family D. All of these

1431. Identify the peripheral function that is provided in microcontrollers by a pulse width modulator followed by a filter.
A. A/D converter B. D/A converter
C. Timer D. Counter

1432. The processor architecture that allows different stages of execution of an instruction to overlap and perform with parallelism is
A. Pipeline architecture
B. Accumulator-based architecture
C. Register-based architecture
D. Stack-based architecture

1433. A microcontroller is to be programmed to generate a square waveform with minimum and maximum time period of 100 ns and 10 ms. Minimum clock frequency required would be
A. 100 Hz B. 1 MHz
C. 10 MHz D. none of these

1434. Memory devices are used for
A. temporary storage of data
B. semi-permanent storage of data
C. permanent storage of data
D. all of these

1435. The primary memory of a computer system
A. holds the instructions of the program to be executed
B. stores data to be processed
C. stores the intermediate results of any calculation during processing of data
D. all of these

1436. Identify the characteristic feature or features of random access memory.
A. It is a serial access memory
B. Data can be read from or written into any of the memory locations regardless of order in which they are arranged
C. All memory locations can be accessed with the same speed
D. Both B and C

1437. A 64 K RAM chip can store
A. 64000 bytes B. 65536 bytes
C. 65536 bits D. 64000 bits

1438. Three main building blocks of RAM are
A. array of memory cells, address decoder, and read/write control logic
B. array of memory cells, input buffers, and output buffers
C. array of memory cells, address decoder, and buffers
D. address decoder, address encoder, and memory cells

Ans. 1420. D 1421. D 1422. C 1423. D 1424. A 1425. B 1426. A 1427. C 1428. A 1429. D 1430. D 1431. B 1432. A 1433. C 1434. D 1435. D 1436. D 1437. B 1438. A

1439. Which one of the following types of RAM needs periodic refreshing.
A. DRAM B. Asynchronous SRAM
C. Synchronous SRAM
D. All types of RAM need periodic refreshing

1440. Type of RAM in which data is stored in the form of change on a capacitor is
A. asynchronous SRAM B. synchronous SRAM
C. DRAM D. all of these

1441. SRAM devices are made using
A. bipolar, MOS, or BiMOS technologies
B. bipolar technology
C. MOS technology D. BiMOs technology

1442. A 64 × 8 SRAM will use
A. 6-line to 64-line address decoder
B. 8-line to 64-line address decoder
C. 64 memory cells D. both B and C

1443. The memory device that communicates directly with the CPU and has relatively much higher speed than the main memory is
A. Cache memory B. Hard disk
C. DVD ROM D. Primary memory

1444. With reference to level 1 and level 2 cache memories, which one of the following is true?
A. Level 1 cache has relatively higher storage capacity
B. Level 2 cache is a part of microprocessor chip itself
C. Level 2 cache has relatively lower storage capacity and is mounted external to the microprocessor
D. Level 1 cache has relatively lower storage capacity and is part of microprocessor chip

1445. Identify the volatile memory.
A. SRAM B. Flash memory
C. EPROM D. EEPROM

1446. With reference to a 2 K bit ROM organized as 256 × 8 array of memory cells, which one of the following is true.
A. It uses 200 memory cells
B. It uses 256 rows of eight memory cells each
C. It uses 2048 memory cells and 8-line to 256-line address decoder
D. Both B and C are true

1447. With reference to erasable PROM devices, identify the true statement.
A. In the case of EEPROM, selective erasure is not possible
B. UV EPROM erasure process is very fast
C. In the case of UV EPROM, selective erasure is not possible
D. EEPROM has relatively much higher density as compared to UV EPROM

1448. The memory device has both high-density, high-speed access and in-circuit electrical erasability feature is
A. EEPROM B. UV EPROM
C. Cache memory D. Flash memory

1449. Two 1 K × 8 ROM chips can be used to configure
A. one 2 K × 8 ROM B. one 1 K × 16 ROM
C. both A and B D. none of these

1450. Identify the serial port.
A. IEEE-488 B. USB
C. IEEE-1284 D. All of these

1451. Serial ports are used to connect the following to the computer.
A. Mouse B. Keyboard
C. Modem D. All of these

1452. Identify the sequential access secondary storage device.
A. Magnetic tape B. Hard disk
C. Floppy disk D. CD-ROM

1453. Identify the optical secondary storage device.
A. CD-ROM B. DVD-ROM
C. CD-R D. All of these

1454. For a certain two-input logic gate, the output is '1' for like inputs and '0' for unlike inputs. The logic gate is ......
A. EX-OR B. NAND
C. NOR D. EX-NOR

1455. In general, a logic gate whose all output entries are logic '0' except one output entry that is logic '1' is ......
A. either AND or NOR B. either NAND or OR
C. either EX-OR or EX-NOR
D. either AND or OR

1456. One of the following two-input logic gates can be used to implement an inverter circuit ......
A. AND B. EX-NOR
C. OR D. none of these

1457. A logic gate with four inputs can have ......
A. 16 possible input combinations
B. 4 possible input combinations
C. 8 possible input combinations
D. none of these

1458. Minimum number of two-input NAND gates required to construct a two-input OR gate is ......
A. one B. two
C. three D. four

1459. Minimum number of two-input AND gates needed to construct one-input AND gate is ......
A. three B. four
C. two D. five

1460. Minimum number of two-input NAND gates needed to construct one four-input NAND gate is ......
A. three B. four
C. two D. five

Ans. 1439. A 1440. C 1441. A 1442. A 1443. A 1444. D 1445. A 1446. D 1447. C 1448. D 1449. C 1450. B 1451. D 1452. A 1453. D 1454. D 1455. A 1456. B 1457. A 1458. C 1459. A 1460. D

CHAPTER 19

Electrical Aircraft, Automotive Electric and Electronic Systems

Energy for the operation of most electrically operated equipment in an aircraft is supplied by a generator which may be of direct current (DC) or alternating current (AC) type. In this chapter, we will elaborate generators, serving as the source of primary DC supply for the aircraft electrical installation. A review of relevant fundamental current-generating principles are discussed here.

Fundamental principles: A generator is a machine that converts mechanical energy into electrical energy by the process of electromagnetic induction.

Generator classification: Generators are classified according to the method by which their magnetic circuits are energised.

i. Permanent magnet generators
ii. Separately excited generators
iii. Self-excited generators

In aircraft DC power supply systems, self-excited shunt-wound generators are employed and the following details are therefore related only to this type:

a. *Fixed winding arrangement*: This has an arrangement of the fixed windings of a basic four-pole machine suitable for use as self-excited generator.
b. *Self-excited shunt-wound generators*: Self-excited shunt wound generators are used in aircraft DC power supply systems.

Voltage regulation: The efficient operation of aircraft equipment requiring DC depends on the fundamental requirement that the generator voltage at the distribution bus bar system be maintained constant under all conditions of load at varying speeds. It is necessary to provide a device that will regulate the output voltage of a generator.

Different types of regulators:

a. *Vibrating contact regulator*: Vibrating contact regulators are used in several types of small aircraft employing comparatively low DC output generators.
b. *Voltage regulator*: They are designed for use with twin-generator systems.
c. *Carbon pile generator*: This has a granular surface and the contact resistance between two carbon faces that are held together depends not only on the actual area of contact but also on the pressure with which two faces are held together.

Power supplies—Batteries: The batteries selected for use in aircraft employ secondary cells which are either of:

i. Lead-acid secondary cell
ii. Nickel-cadmium secondary cell

Ground power supplies: Electrical power is required for the starting of engine operation of certain services during turn-round serving periods at airports, i.e. lighting, and for the testing of electrical systems during routine maintenance checks.

DC system: A basic system for supply of DC is shown in Fig. 19.1 and from this, it will also be noted how in addition to the ground power supply, the battery may be connected to the main bus bar by selecting the 'flight' position of the switch.

AC system: In aircraft, electrical power system is of AC type. It is essential for the ground supply system of the installation to include a section through which an external source of AC power may be supplied. The circuit arrangement for appropriate system vary between aircraft types but in order to gain some understanding of the circuit requirements and operation. Generally, we may consider the current shown in Fig. 19.2. When ground power is coupled to the phase, a three phase supply is fed to the main contacts of the ground power breaker to a ground power transformer/rectifier unit (TRU) and to a phase sequence protection unit.

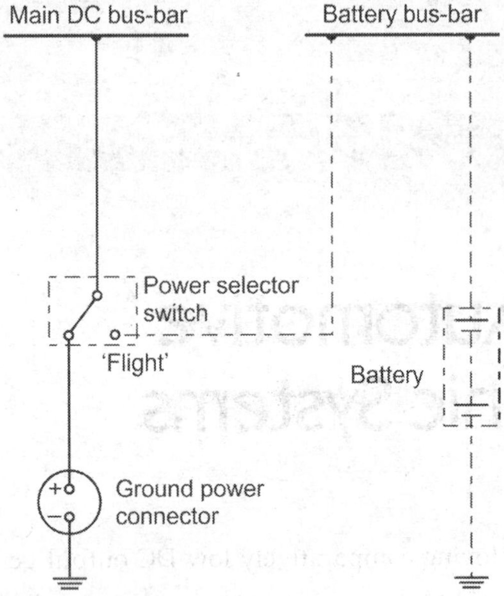

Fig. 19.1: Basic ground power supply system

Measuring instruments warning indicators and lights: For monitoring the operating conditions of various supply and utilization system, it is necessary for measuring instruments and working devices in the form of indicators and light to be included in the system. The number of indicating devices required and the type employed depend on the type of aircrafts and overall nature of electrical installation having the layout as shown in Fig. 19.3 which is generally represents systems monitoring requirements and usefully serve as basis for the study of appropriate indicating devices.

Ammeter and voltmeter: These instruments are provided in DC and AC power generating systems and in most instances are of the permanent magnet and moving coil types.

Instrument transformers: Transformers are used in conjunction with ac measuring instrument for scaling down large current and voltage to a suitable level for handling by standardised types of instruments. They are of two types:

i. Current transformers (CT)
ii. Potential transformers (PT)

Their application is shown in Fig. 19.4.

Frequency meters: These instruments form part of the metering system required for main ac power generating system and in some aircraft, they may also be employed in secondary ac generating system utilizing inverters. The dial presentation and circuit diagram of a typical meter are shown in Fig. 19.5. The indicating element, which is used in a mutual inductance circuit is of the standard electrodynamometer pattern consisting essentially of a moving coil and a fixed field coil.

Warning and indicator lights: Warning and indicator lights are used to alert the flight crew to conditions affecting the operation of aircraft system. The lights may be divided into different categories namely: (i) Warning lights (ii) Caution lights (iii) Indicating lights.

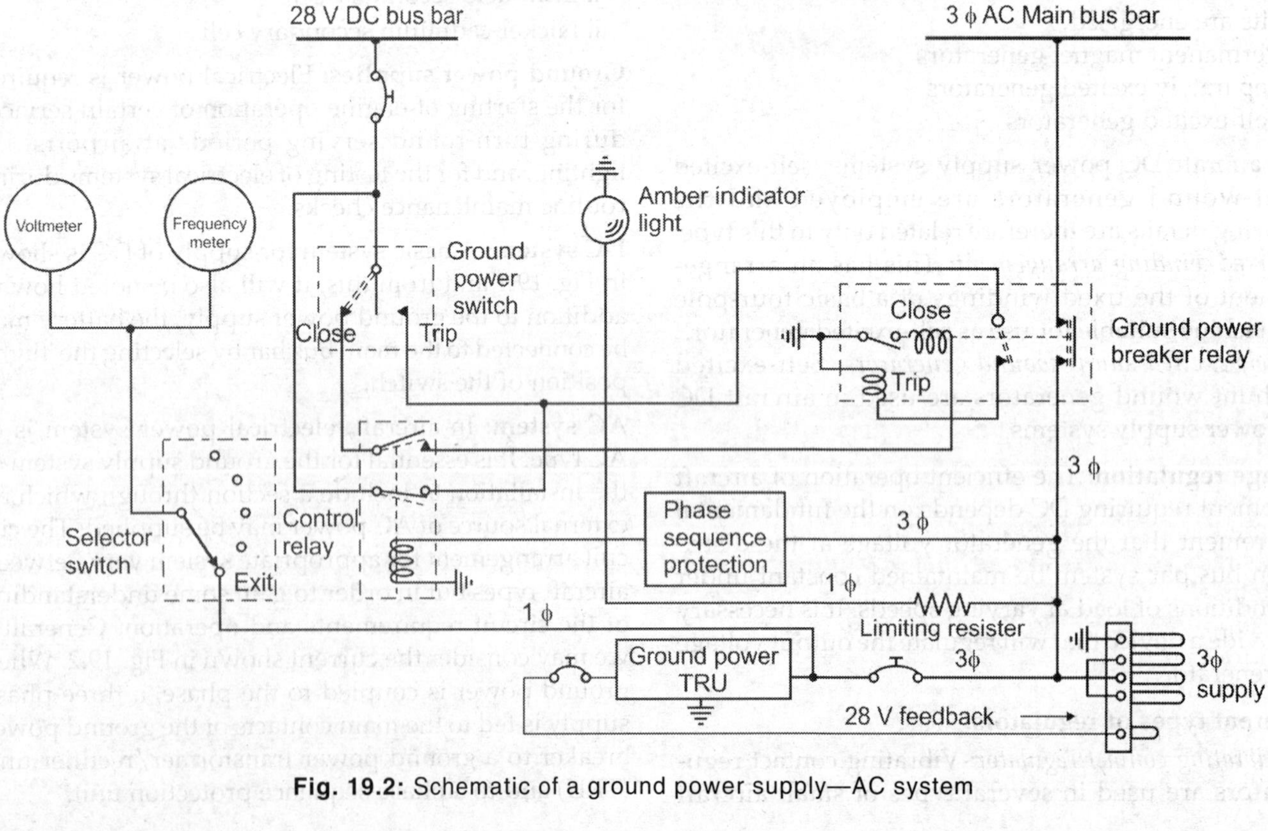

Fig. 19.2: Schematic of a ground power supply—AC system

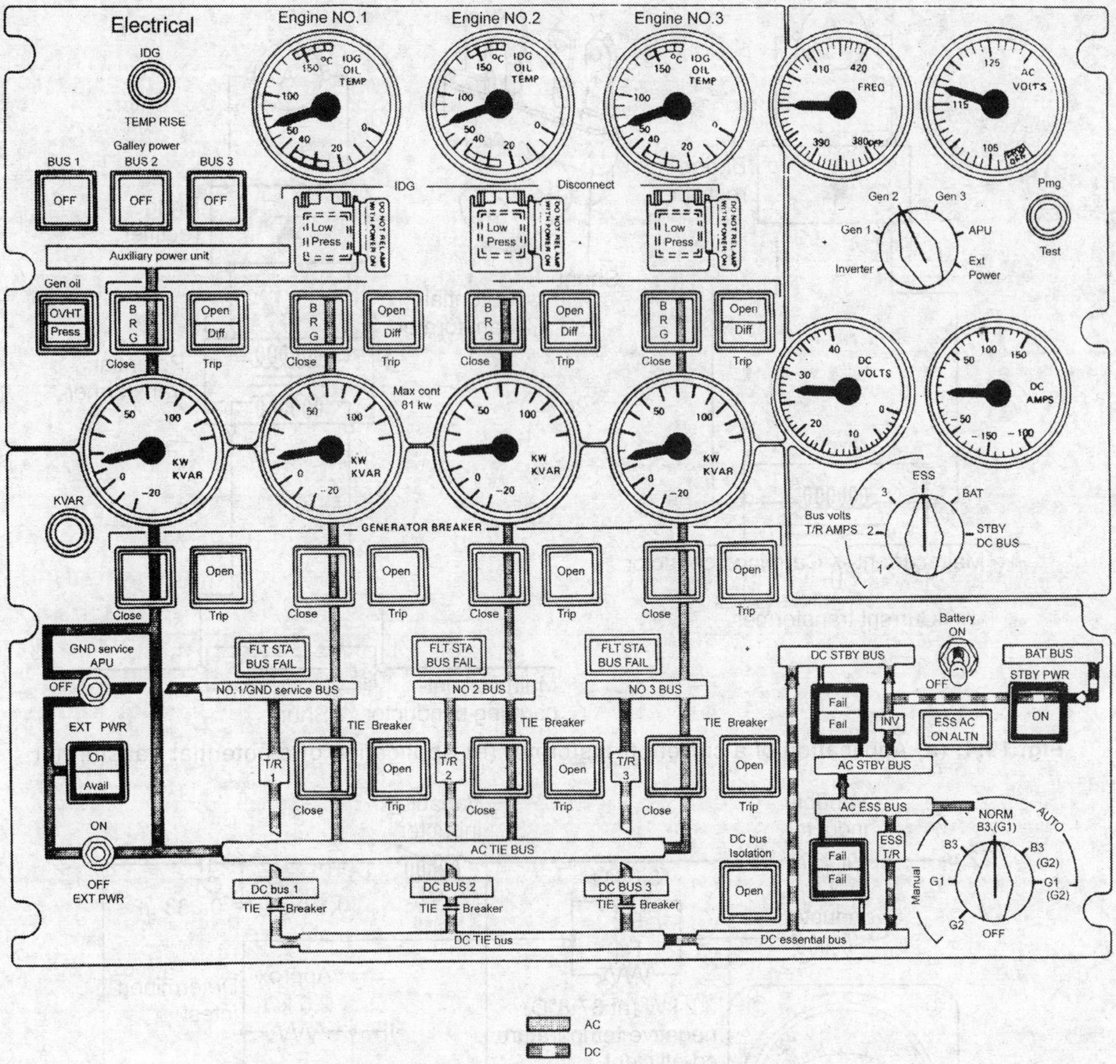

Fig. 19.3: Electrical system control panel

i. *Warning lights*: These are designed to alert the flight crew of unsafe conditions and accordingly coloured rod.

ii. *Caution lights*: These are amber in color to indicate abnormal but not necessarily dangerous condition requiring caution, i.e. hydraulic system pressure running low.

iii. These lights, which are either green or blue, are provided to indicate that a system is operable or has assumed a safe condition, i.e. a landing gear down and locked.

Magnetic indicators: In many aircraft systems, components require electrical control. For example, in a fuel system, electric actuators position values which permit the supply of fuel from the main tanks to engines and also for cross-feeding the fuel supply. In its simplest form by a magnetic indicator is of two position type comprising a ball pivoted on its axis and spring returned to the off-position. Figure 19.6 shows a development of the basic indicator.

Power distribution: To facilitate the power available at the appropriate generating source to be made available at the terminal of the power consuming form of distribution throughout an aircraft is essential. The precise manner in which this is arranged is governed principally by the consumers and the location of consumer components. In a small light aircraft, electrical power requirements may be limited to a few consumer services and components situated within a small area and the power may be distributed via only a few yards of cable, some terminal blocks, circuit breakers or fuses (Fig. 19.7).

Circuit protection devices: In the event of a short circuit, an overload or other fault condition occurring in the circuit formed by cables and components of an electrical system, it is possible for extensive damage and failure to result. The devices normally employed are fuses, circuit breakers and current limiters. In addition, other devices are provided to serve as protection against such fault condition as reverse

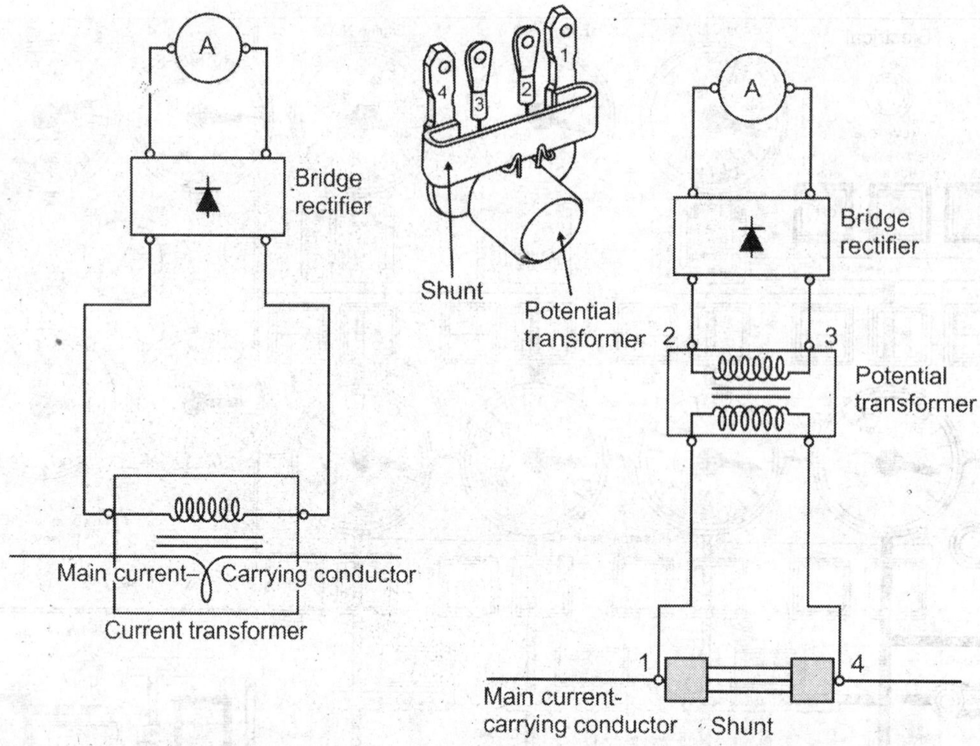

Fig. 19.4: (a) Application of a current transformer (b) Application of a potential transformer

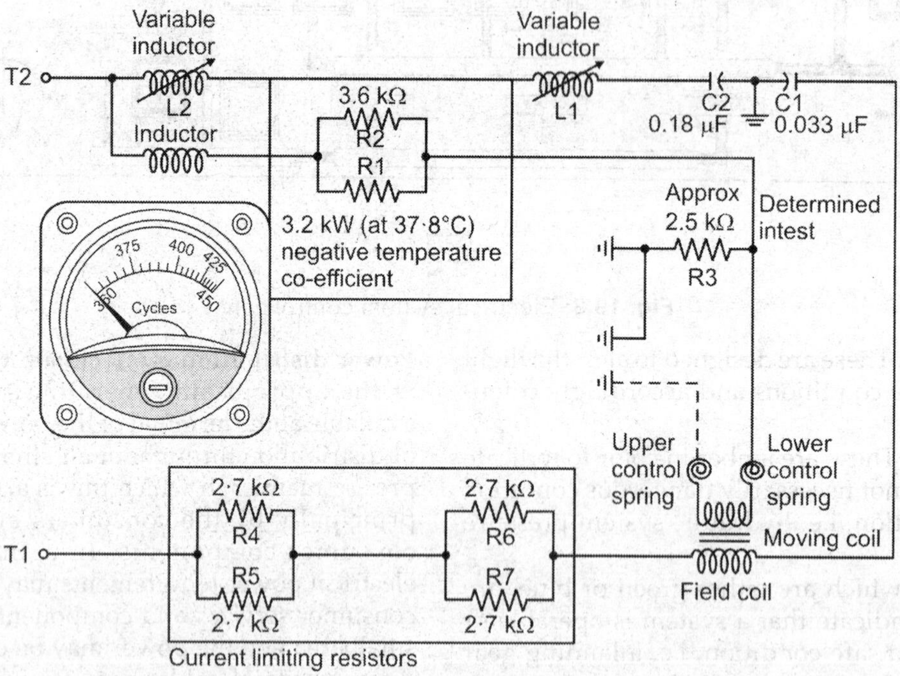

Fig. 19.5: Circuit arrangements of a frequency meter

current over-voltages, under-voltages, over-frequency, under-frequency, phase imbalance etc.

Fuses: A fuse is a thermal device designed primarily to protect the cables of a circuit against the flow of short circuit and overload currents.

Current limiters: Current limiters are designed to limit the current to some pre-determined amperage value. They are also thermal devices but unlike ordinary fuses, they have high m.p. so their time/current characteristics permit them to carry a considerable overload current before rupturing.

For this reason their application is confined to the protection of heavy-duty power distribution circuits. A typical current limiter named as airfuse is illustrated

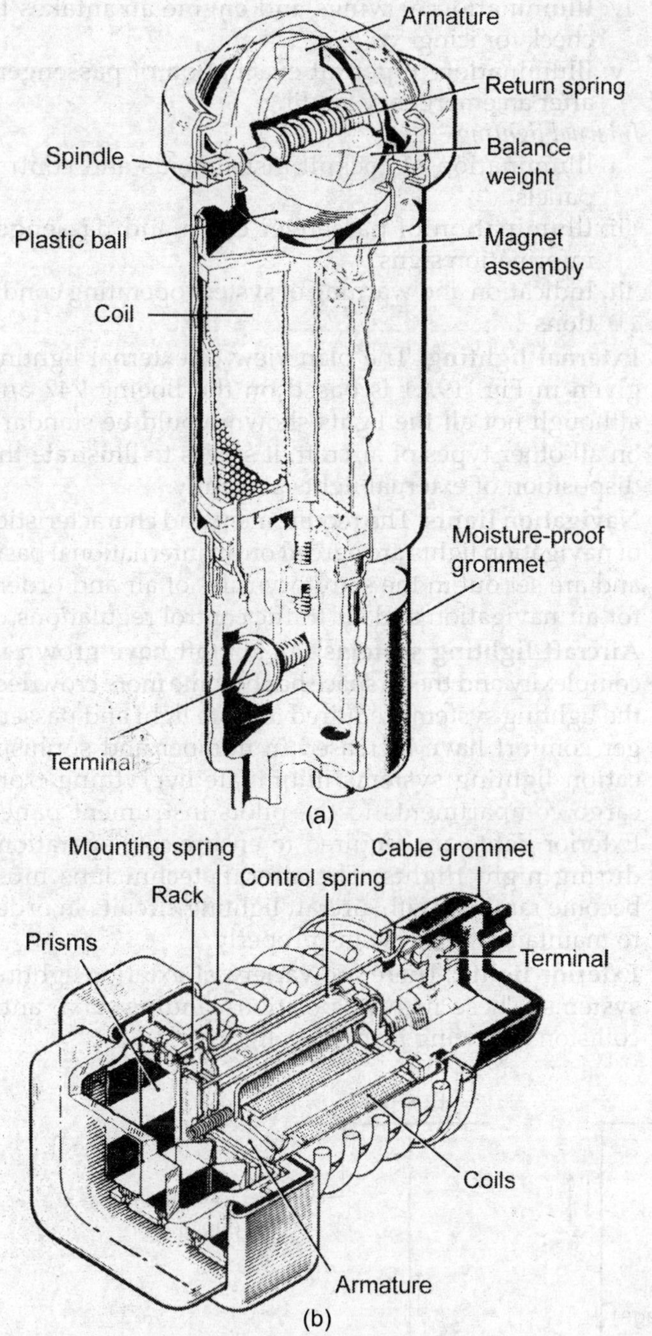

Fig. 19.6: Magnetic indicators

in Fig. 19.8. It incorporates a fusible element which is, in effect, a simple strip of tinned copper, drilled and shaped at each end to form log type connection with the central portion, 'waisted' to the required width to form the fusing area.

Limiting resistors: These device provide another form of protection particularly in DC circuits in which the initial current surge is very high. Figure 19.9 illustrates the application of a limiting resistor to a turbine engine shorter motor circuit incorporating a time switch the initial current flow may be as high as 1500 A.

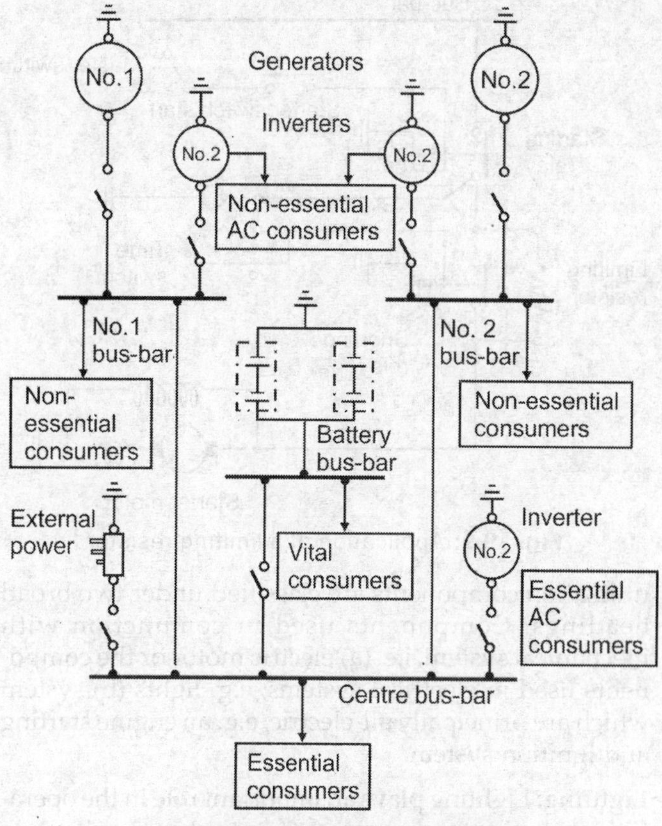

Fig. 19.7: Split bus-bar systems

Fig. 19.8: Typical current limiter (airfuse)

Protection against reverse current: A reverse current cut-out relay is used principally in a DC generating system either as a separate unit or as a part of voltage regulator. The circuit arrangement, as applied to the generating system is a typical of several types of small aircraft as shown in Fig. 19.10. The relay consists of two coils wound on a core and a spring controlled armature and contact assembly.

Power utilization components: Utilization can extend over very wide areas depending on the size and type of aircraft, and whether systems are employed which require full or only partial use of electrical power. Four

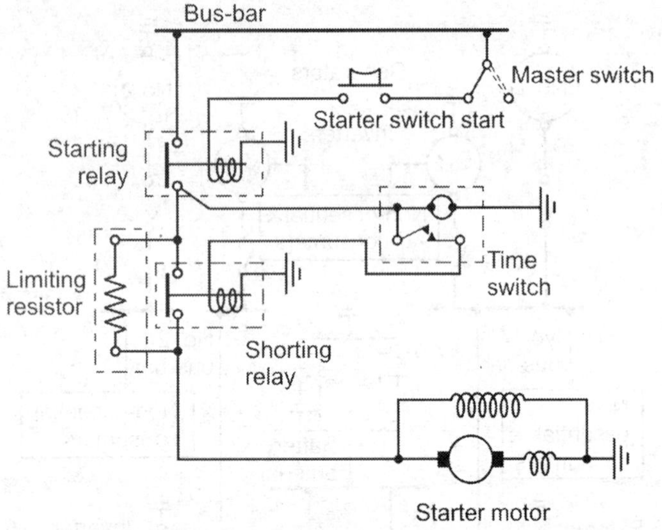

Fig. 19.9: Application of a limiting resistor

utilization components are classified under two broad headings. Components used in conjunction with mechanical system, i.e. (a) electric motor or the components used in electrical systems, e.g. lights (b) system which are principally all-electric, e.g. an engine starting and ignition system.

Lighting: Lighting plays an important role in the operation of an aircraft and many of its system, having two groups: (a) External lightening, (b) Internal lightening.

External lightning

i. The marking of an aircraft's position by means of navigation lights
ii. Position marking by means of flashing lights
iii. Forward illumination for landing and taxing
iv. Illumination of wings and engine air intakes to check for icing
v. Illumination to permit evacuation of passengers after an emergency landing

Internal lighting

i. Illumination of cockpit instruments and control panels
ii. Illumination of passenger cabin and passenger information signs
iii. Indication and warning of system operating conditions

External lighting: The plan view of external lighting given in Fig. 19.11 is based on the Boeing 747 and although not all the lights shown would be standard on all other types of aircraft, it serves to illustrate the disposition of external lights generally.

Navigation lights: The requirement and characteristics of navigation lights are agreed on an international basis and are set out in the statutory rules of air and orders for air navigation and air traffic control regulations.

Aircraft lighting systems: As aircraft have grown in complexity and the airspace has become more crowded, the lighting systems required for safe light and passenger comfort have increased in number and sophistication lighting systems illuminate everything from cargo compartments to the pilots instrument panel. Exterior lights are required to ensure safe operations during night flights. The aircraft technicians must become familiar with aircraft lighting circuits in order to maintain these systems properly.

Exterior lights: There are variety of exterior lighting systems. These include position, landing, taxi, anti-collision and wing inspection lights.

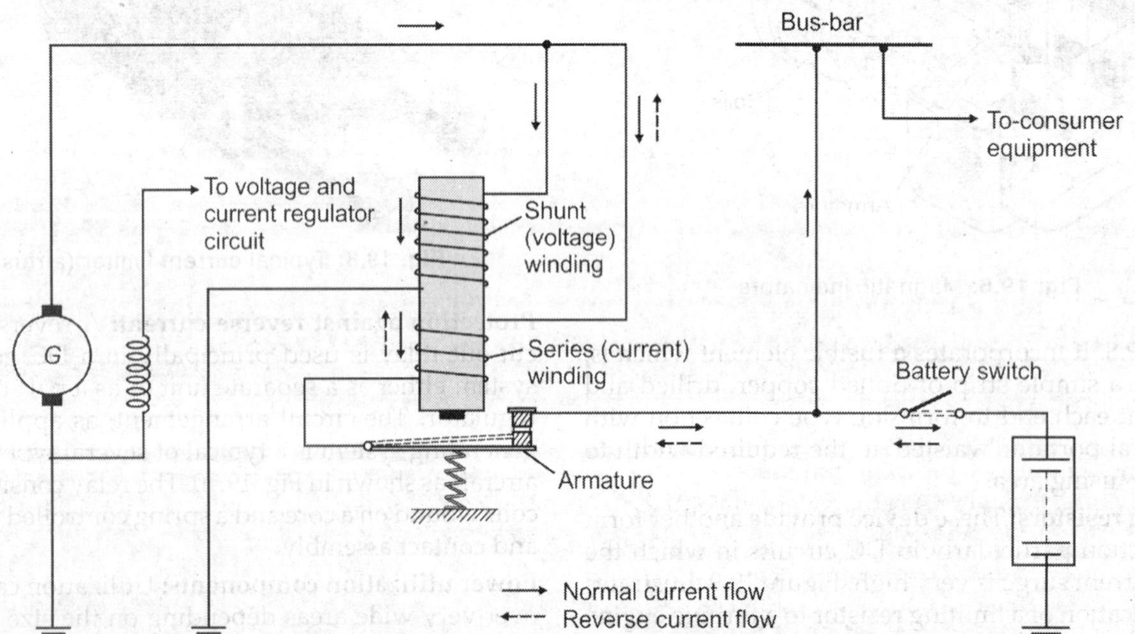

Fig. 19.10: Reverse current cut-out operation

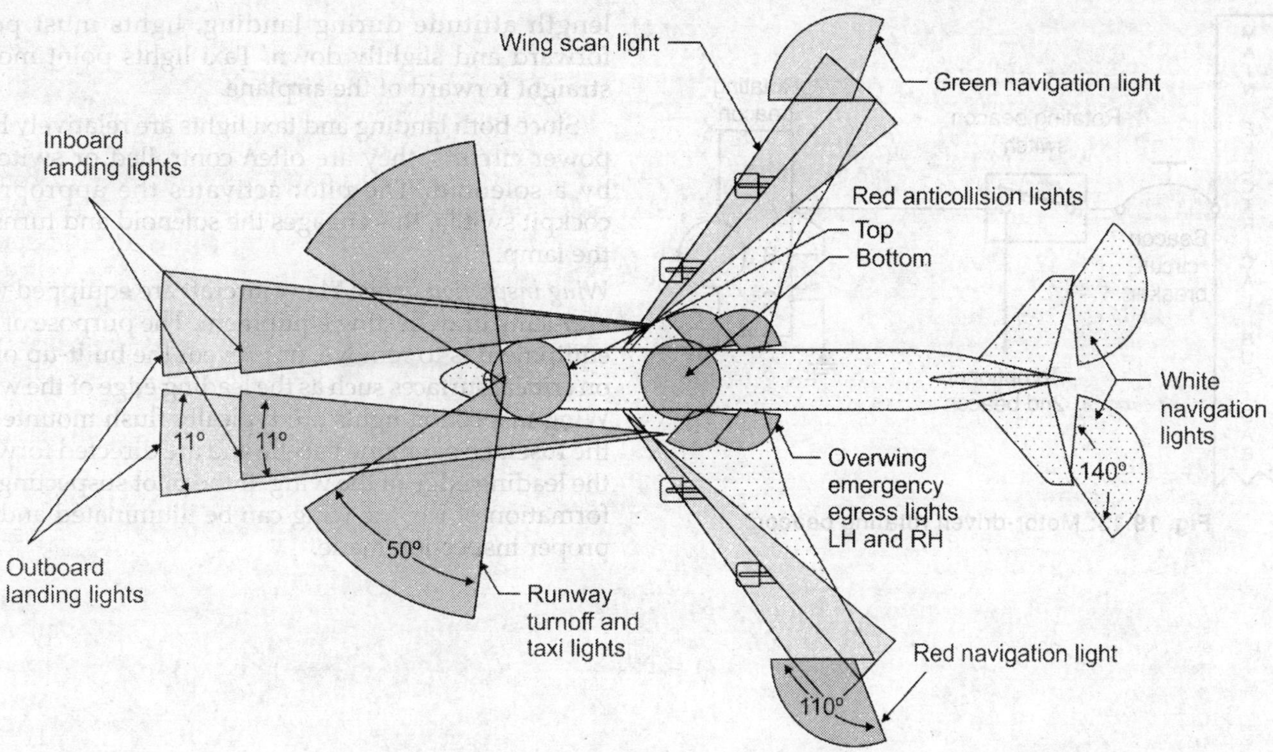

Fig. 19.11: Disposition of external lighting

Position lights: Position lights are used to indicate the position of an aircraft during night operations. If pilots can identify the position of another aircraft from its position lights, they may safely navigate around that aircraft, hence, position lights are often referred to as navigation lights. One or more position lights must be located on each wing tip and the tail of the aircraft. The right wing tip must have a green colored light, the left has a red light, and the tail must have a white light. These lights are required on any aircraft certified for night flight. The actual light bulb is covered with a clear glass, the color is achieved by placing a red or green lens over the bulb. Most navigation lights operate in one mode only, however, some older model aircraft have dimmer or flasher circuits incorporated with position lights. A typical position light circuit is shown in Fig. 19.12.

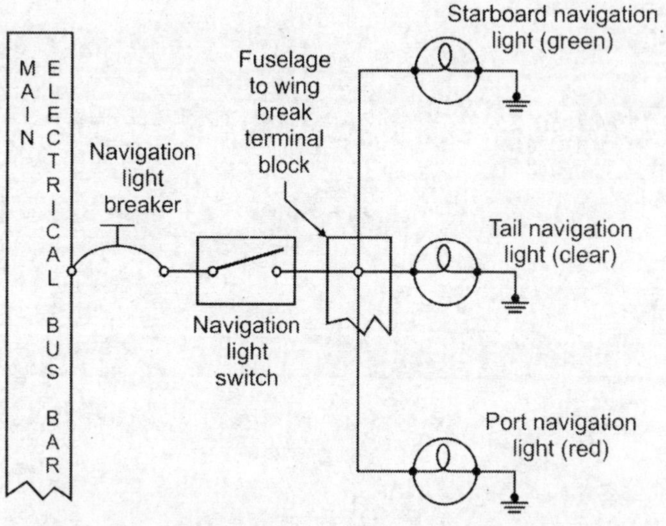

Fig. 19.12: Typical position light circuit for an aircraft

Anticollision lights: Anticollision lights are found in two basic styles and the difference typically comes as a function of age. Older aircraft were originally equipped with rotating beacons, either on top of the vertical stabilizer or, on the top or bottom of the fuselage. The rotating beacon system typically contained a stationary light bulb and a rotating reflector covered by a red glass lens as shown in Fig. 19.13.

Newer systems utilize solid state electronics to create a flashing- or strobe-type anticollision light. The strobe-type flashing anticollision light has an extremely bright flash produced by a xenon tube which requires approximately 400 volts. The high voltage is produced by the strobe power supply which uses a capacitor charging system to achieve the high voltage. Typically, two white wing tip strobe lamps and one red tail strobe are used on modern aircraft.

Landing and taxi lights: In order to provide night visibility for landing and taxing, two or more sealed beam lamps are mounted on the aircraft facing forward. These lights can be mounted in the leading edge of the wing, in the cowling or on landing gear struts. Some aircraft employ a retractable light which extends from the wing during use. Since the airplane is in a nose-

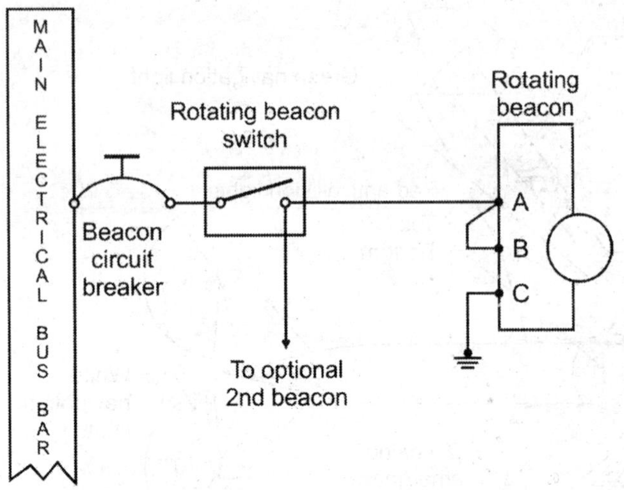

Fig. 19.13: Motor-driven rotating beacon

length attitude during landing, lights must point forward and slightly down. Taxi lights point mostly straight forward of the airplane.

Since both landing and taxi lights are relatively high power circuits, they are often controlled or switched by a solenoid. The pilot activates the appropriate cockpit switch, this engages the solenoid and turns on the lamp.

Wing inspection lights: Many aircraft are equipped with anti-icing or deflecting equipment. The purpose of this equipment is to remove, or prevent the built-up of ice on critical surfaces such as the leading edge of the wing. Wing inspection lights are typically flush mounted in the fuselage or engine nacelle and are directed forward the leading edge of the wing. If the pilot suspecting the formation of ice, the wing can be illuminated and the proper inspection made.

MULTIPLE CHOICE QUESTIONS

1. An electron has
   A. a positive (+) charge B. a negative (–) charge
   C. a neutral charge D. no charge
2. An insulator has
   A. eight or more free electrons
   B. more than four electrons in its outer orbit
   C. exactly four electrons in its outer orbrt
   D. fewer than four electrons in its outer orbit
3. An ampere is a unit of
   A. electrical pressure B. electrical resistance
   C. of static electricity
   D. the amount of current flow
4. A volt is a unit of
   A. electrical pressure B. electrical resistance
   C. static electricity
   D. the amount of current flow
5. An ohm is a unit of
   A. electrical pressure B. electrical resistance
   C. static electricity
   D. the amount of current flow
6. As temperature increases
   A. the resistance of a conductor decreases
   B. the resistance of a conductor increases
   C. the resistance of a conductor remains the same.
   D. None of these
7. Piezoelectric crystals can create a voltage whenever
   A. pressure is put on the crystals
   B. the crystal are heated
   C. the crystals are cooled
   D. the crystals are put into water
8. Which material releases electrons when exposed to light?
   A. Thermoelectric B. Selenium or cesium
   C. Piezoelectric D. Chemical
9. An electric circuit consists of
   A. a battery only
   B. a power source, an electrical load, and connecting wires
   C. a battery an electrical and a power-side wire only
   D. None of these
10. A open circuit is one
    A. in which no current flows
    B. created by a blown fuse
    C. in which the electrical unit does not operate
    D. Either of these
11. A short circuit is
    A. a type of normal circuit
    B. copper-copper type of defective circuit
    C. copper-to-steel type of defective circuit
    D. None of these
12. What is a ground circuit?
    A. A type of normal circuit
    B. Copper-to-copper type of circuit
    C. Copper-to-steel type of defective circuit
    D. Both A and C could be correct
13. The major parts of any circuit include are
    A. a power source, a power lead, and a return lead
    B. a power source, a return lead and a ground lead
    C. a battery a light and a ground return lead
    D. Any of these
14. The four basic automotive circuits include
    A. lighting, accessories, starter and battery
    B. ignition, starter, battery, and radio
    C. charging, starting, battery, and ignition
    D. starting, charging, ignition and lighting accessories
15. What should be disconnected to facilitate safely working on the electrical system of any car?
    A. The power-side battery lead from the battery
    B. The ground (return path) lead from the battery
    C. The ignition wire
    D. The power lead from the starter
16. Which statement is correct?
    A. A "short" causes an "open" by blowing a fuse
    B. A "short" is an electrical connection that is from "copper to copper"
    C. A "short to ground" is an electrical connection that is from "copper to steel"
    D. All of these
17. Ohm's law states that it
    A. takes 1 ohm of pressure to push 1A through 1V
    B. It takes 1V of pressure to push 1A through 1 ohm of resistance
    C. It takes 1 A of pressure io push 1 ohm through 1 V
    D. None of these
18. A watt is
    A. amperes times ohms
    B. amperes times volts
    C. ohms times volts
    D. amperes divided by volts
19. In a series circuit
    A. all the current fows through every resistor
    B. some of the current flows through every resistor
    C. none of the current flows through every resistor
    D. only about one-half of the current flows through any single resistor

Ans. 1. B 2. B 3. D 4. A 5. B 6. B 7. A 8. B 9. B 10. D 11. ? 12. C 13. A 14. D 15. B 16. B 17. B 18. B 19. A

20. The higher the resistance
    A. the smaller the battery has to be
    B. the greater the voltage drop
    C. the lower the voltage drop
    D. the voltage remains the same
21. In a parallel circuit
    A. only a part of the total current flows in each leg of the circuit
    B. all of the current flows through all of the resistors
    C. one-half of all the current flows through all of the resistors
    D. the voltage will increase as the current flows through each resistor
22. How many amperes will flow through the circuit shown in below figure?

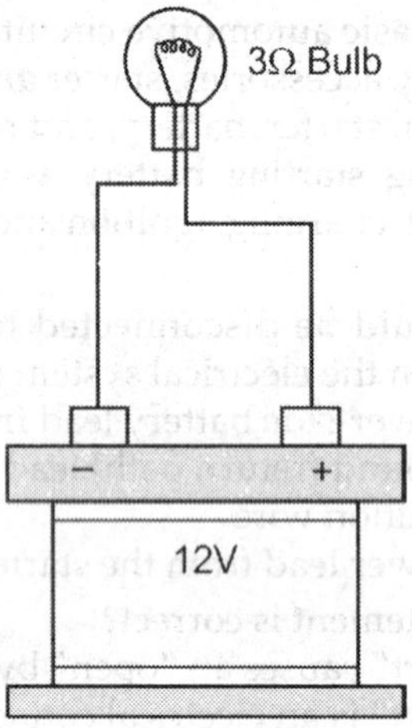

    A. 12 V B. 4 V
    C. 4 A D. 3 V
23. How much will the voltage be dropped by the 3 ohm resistor in the above figure?
    A. 3 V B. 4 V
    C. 12 V D. 0 V
24. Electromagnetism is measured in units of
    A. ampere-turns B. farads
    C. volt-turns D. coulombs
25. Electromagnetism is used in almost every automotive circuit except
    A. lights and horn
    B. horn and cigarette lighter
    C. radio and horn
    D. lights and cigarette lighter
26. A coil of wire wrapped around a soft-iron core. If the wires are connected to a power source, it would result in
    A. coil B. electromagnet
    C. solenoid D. electric switch
27. In the above figure, what is the total current through the entire circuit?
    A. 6 A B. 12 A
    C. 2 A D. 22 A
28. What is the total resistance of the circuit shown in figure below?

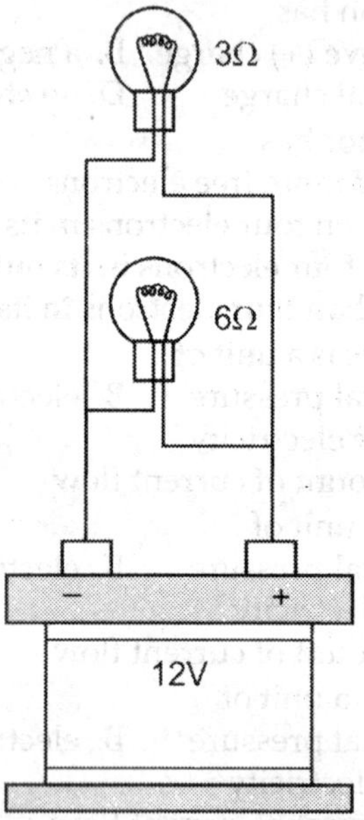

    A. 2 ohms B. 4 ohms
    C. 18 ohms D. 12 ohms
29. How much current flows through the 3-ohm resistor in circuit shown in the above figure?
    A. IV B. 1 ohm
    C. 12 A. D. 4 A
30. Capacitance is measured in units of
    A. ohms B. coulombs
    C. amperes D. farads
31. A capacitor can
    A. prevent high-voltage surges
    B. reduce radio interference
    C. prevent ignition-point arcing
    D. all of these
32. The major difference between a solenoid and a relay is
    A. a relay uses a movable core
    B. a solenoid uses movable core
    C. a solenoid uses a movable arm
    D. none of these
33. A natural magnetic material is
    A. lodes tone B. permalloy
    C. alnico
    D. neodymium-iron-boron
34. The name of the electromagnetic switch that can conduct a high current flow is
    A. a relay B. a solenoid
    C. an ignition switch D. a permalloy switch

Ans. 20. B 21. A 22. C 23. C 24. A 25. D 26. B 27. A 28. A 29. D 30. D 31. D 32. B 33. A 34. B

35. A semiconductor is a material with
    A. fewer than four electrons in the outer orbit of its atoms
    B. more than four electrons in the outer orbit of its atoms.
    C. exactly four electrons in the outer orbit of its atoms
    D. other factors beside the number of electrons.
36. Two types of semiconductors are
    A. holes and blocks B. p and n
    C. white and black D. None of these
37. The arrow of a semiconductor
    A. points toward the negative
    B. points away from the negative
    C. is attached to the emitter
    D. both A and C are correct
38. A transistor is controlled by the polarity and current at
    A. the collector B. the emitter
    C. the base
    D. both the collector and the emitter
39. A transistor can
    A. switch ON and OFF
    B. amplify
    C. throttle D. All of these
40. What happens when a transistor or diode "blows"?
    A. The N-type materials becomes P
    B. The P-type materials becomes N
    C Both A and B
    D. The junction between the N-and the P-type material is destroyed
41. Zener diodes are used
    A. in voltage regulators
    B. to control radio static
    C. to improve radio reception
    D. both A and C
42. The voltage required to turn a transistor on is
    A. 0.7 V for silicon transistors
    B. 0.7 V for germanium transistors
    C. 0.3 V for silicon transistors
    D. None of these
43. The higher the gauge number
    A. the smaller the wire
    B. the same the wire size
    C. the larger the wire D. the thicker the wire
44. Metric wire size is measured in units of
    A. metres B. cubic centimeters
    C. liters D. square millimeters
45. The metal in fuses is made from
    A. tin B. lead
    C. copper D. silver
46. Jumper cables should be made from
    A. copper at least 8 gauge
    B. aluminium or copper at least 6 gauge
    C. copper at least 4 gauge
    D. None of these
47. 18-gauge wire is illustrated in the metric system on a wiring diagram as
    A. 0.5 mm^2 B. 0.8 mm^2
    C. 1.0 mm^2 D. 2.0 mm^2
48. Most automotive "primary" wire is made from
    A. solid copper B. stranded copper
    C. solid aluminium D. stranded aluminium
49. All fuses should be tested with
    A. a fuse tester B. anohmmetei
    C. a voltmeter D. a test light
50. Blades fuses are
    A. all the same amperage rating
    B. colour-coded according to amperage rating
    C. colour-coded according to time delay
    D. None of these
51. Bulb numbers mean
    A. the trade number B. amperage required
    C. candlepower D. none of these
52. Low sealed-beam headlights
    A. contain only a low beam
    B. contain both a low beam and a high beam
    C. use a two-wire connector
    D. none of these
53. Technician *A* says that the brake lights operate the high-intensity filament of the taillight bulb. Technician *B* says that the current from the brake lights goes to the turn signal switch before going to the brake lights. Which technician is correct?
    A. A only B. B only
    C. Both A and B D. Neither A or B
54. A slow turn signal on one side only is most likely caused by
    A. a defective flasher unit
    B. the wrong flasher unit
    C. a defective (open) bulb on the affected side
    D. a poor ground connection on the affected side
55. The headlight switch controls
    A. the headlights, taillights, and dash lights
    B. the headlights, taillights, and side marker lights
    C. the dash lights, side marker lights, taillights and headlights
    D. All of these
56. Illuminated entry operates
    A. only when the doors are locked
    B. the headlights
    C. the trunk light and keyhole light
    D. by grounding the door jam switch
57. The major difference between the flasher units commonly used for directional (turn) signals and four-way hazard flashers is
    A. the size of the flasher unit
    B. four-way flasher units are thermoelectric

Ans.													
	35. C	36. B	37. D	38. C	39. D	40. D	41. A	42. A	43. A	44. D	45. A	46. C	47. B
	48. B	49. D	50. B	51. A	52. B	53. C	54. D	55. D	56. A				

C. two-way directional signal units are constant-rate flasher units
D. directional signal units are thermoelectric and four-way units are electromagnetic

58. A defective directional (turn) signal flasher unit could cause
A. the turn signals to flash slowly in both directions
B. the turn signals not to operate in both directions
C. the hazard flashers and turn signals not to operate
D. Both A and B

59. Technician *A* says that thermoelectric gauges require an instrument voltage regulator. Technician *B* says that if the power lead of the fuel gauge of an electro-magnetic gauge is grounded, the dash units reads empty. Which technician is correct?
A. A only B. B only
C. Both A and B D. Neither A or B

60. The oil pressure light 'lights' when oil pressure drops between 3 and 7 psi by
A. opening the circuit B. shorting the circuit
C. grounding the circuit
D. conducting current to the dash light by oil

61. The "hot" light comes on at about 250F by
A. opening the circuit B. shorting the circuit
C. grounding the circuit
D. conducting electricity to the dash light by steam

62. The brightest electronic dash display is called
A. LED B. VTF
C. LCD D. analog

63. Technician *A* says that LCD displays may be slow to work at low temperatures. Technician *B* says that LED displays re-quire very low electrical power. Which technician is correct?
A. A only B. B only
C. Both A and B D. Neither A nor B

64. Technician A says that some electronic dashes use an analog speedometer. Technician B says that all electronic dash fuel gauges use a numerical display. Which technician is correct?
A. A only B. B only
C. Both A and B D. Neither A nor B

65. Which device do digital odometers use to retain the miles travelled?
A. An EEPROM B. A stepper motor
C. NVRAM D. Either A or (C)

66. When servicing an electronic dash
A. if the "WOW" does not display, the cluster must be replaced
B. individual gauges can be replaced if one is defective
C. the sensors used for an electronic dash are not replaceable
D. static electricity discharge can damage any electronic instrument cluster

67. Technicial A says that all halogen bulbs must not be touched with bare hands. Technician B says that it is normal for a VTF display to dim to 75% brightness whenever the headlights are "on". Which technician is correct?
A. A only B. B only
C. Both A and B are correct
D. Neither A nor B

68. Technician *A* says that a defective high-speed relay prevents high-speed blower operation yet low-speed operates nor-mally. Technician *B* says that a defective (open) blower motor resistor can prevent low-speed blower operation, yet the high-speed mode operates normally. Which technician is correct?
A. A only
B. B only
C. Both A and B are correct
D. Neither A nor B

69. Technician *A* says that all automotive electric horn circuits use a relay. Technician *B* says that wiper motors are usually series wound electric motors. Which technician is correct?
A. A only B. B only
C. Both A and B are correct
D. Neither A nor B

70. PM motors as used in power windows and power seats can be reversed by
A. sending current to a reversed field coil
B. using a relay
C. reversing the polarity of the current to the motor
D. using a relay and a two-way clutch

71. Technician A says that a misadjusted brake switch could cause the cruise (speed) control to be inoperative. Technician *B* says that a defective low-speed switch could cause the cruise (speed) control to be inoperative. Which technician is correct?
A. A only B. B only
C. Both A and B are correct
D. Neither A nor B

72. Technician *A* says that either a defective circuit breaker or a defective ground connection can cause all power windows to fail to operate. Technician *B* says that if one control wire is disconnected, all windows will fail to operate, Which technician is correct?
A. A only B. B only
C. Both A and B are correct
D. Neither A nor B

73. Six way power seats
A. can use one or three motors
B. must use six seperate motors
C. can use as many as eight motors for six-way operation
D. can use permanent magnet (PM) motors only.

Ans.	57. D	58. D	59. C	60. C	61. C	62. B	63. A	64. A	65. C	66. D	67. C	68. C	69. D
	70. C	71. C	72. A	73. A									

74. When checking the operation of a rear-window defogger with a voltmeter
A. the voltmeter should indicate decreasing voltage when the grid is tested across the width of the glass
B. the voltmeter must be set to 110 V on the ac scale.
C. the voltmeter should read battery voltage anywhere along the grid
D. the voltmeter must be set on a low-voltage scale

75. Technician *A* says that a radio can receive AM signals but not FM signals if the antenna is defective. Technician B says that the speakers used with a stereo system must have matching impedance. Which technician is correct?
A. A only B. B only
C. Both A and B are correct
D. Neither A nor B

76. Technician *A* says that most power sun-roofs and power mirrors use PM motors. Technician B says that PM motors must be properly grounded to operate correctly. Which technician is correct?
A. A only B. B only
C. Both A and B are correct
D. Neither A nor B

77. A meter used to measure amperes is called
A. an amp meter B. a coulomb meter
C. an ampmeter D. an ammeter

78. A voltmeter should be connected to the circuit being tested
A. in series B. in parallel
C. only when no power is flowing
D. None of these

79. An ohmmeter
A. contains its own power source (battery)
B. actually "measures" voltage drop.
C. must be used in a component or circuit not carrying current
D. All of these

80. A high reading on an ohmmeter (needle toward the left side of the dial) indicates
A. a short circuit or component
B. an open circuit or component (no current flows)
C. a high-resistance circuit or component
D. that a component or circuit is defective

81. A high-impedance meter is a meter that
A. measures a high amount of current flow
B. measures a high amount of resistance
C. measures a high voltage
D. has a high internal resistance

82. When using a test light
A. one lead must be grounded to a good vehicle ground
B. both leads must be grounded to a good vehicle ground
C. the device will also "light" with the high-voltage spark plug current
D. None of these

83. VOM can mean
A. volt-ohmmeter
B. volt-ohm-milli-ammeter
C. volts-only meter D. Either A or B

84. A jumper wire can be used during testing
A. to provide electrical power directly to an electrical component
B. to jump-start another car
C. A and B only D. All of these

85. A voltage drop is commonly used to test
A. the starter ampere draw
B. the condition of the starter brushes
C. the condition of the cables and connections
D. the spark plug wires

86. If cranking voltage is below 9.6 V
A. normally everything is okay
B. perform a voltage-drop test to check cables and connections
C. the alternator is possibly defective
D. none of these

87. The correct specification for a voltage-drop test is
A. above 9.6 V B. more than 0.2 V
C. 0.2 V maximum
D. below basic battery voltage

88. The correct specification for a battery-drain test using a voltmeter is
A. below battery voltage
B. 13.5 to 15.0V
C. 0.2 V maximum D. above 9.6 V

89. If the ohmmeter reads 1000 ohm during a battery drain test, the technician should
A. perform the general voltmeter test to determine the problem
B. perform a voltage-drop test
C. perform a battery dram test using a test light
D. look elsewhere for the problem because the i test indicates no battery drain

90. If the charging voltage is above 15.0 V there is
A. a possible alternator problem
B. a possible voltage regular problem
C. a possible starter problem
D. a possible spark plug problem

91. After removing the surface charge, a 12 V battery indicating 12.6 Vmeans that the battery is
A. 100% charged B. 60% charged
C. 25% charged D. overcharged

92. If a voltmeter indicates 14.4 V while the engine is running at 2000 RPM this means that
A. the battery is being overcharged
B. the alternator and voltage regulator are okay

Ans. 74. A 75. D 76. A 77. D 78. B 79. D 80. C 81. D 82. A 83. D 84. C 85. C 86. B 87. C 88. A 89. D 90. B 91. A 92. B

C. the starter and alternator are okay
D. the battery is defective

93. Voltage-drop testing should be used to determine the condition of
A. the starter B. the battery
C. the cables and connections
D. the alternator and the voltage regulator

94. To find a short circuit
A. one ohmmeter lead can be connected to the circuit side of the fuse holder and the other lead to a good ground
B. one ohmmeter lead can be connected to the power side of the fuse holder and the other lead to positive (+) post of the battery.
C. a Gauss gauge can be used
D. Both *A* and *C* are correct

95. The positive (+) plate of an automotive battery is
A. PbO_2 B. Pb
C. H_2SO_4 D. $PbSO_4$

96. The negative (-) plate of an automotive battery is Pb
A. PbO_2 B. Pb
C. H_2SO_4 D. $PbSO_4$

97. The electrolyte of an automotive battery is
A. PbO_2 B. Pb
C. H_2SO_4 D. $PbSO_4$

98. A maintenance-free battery
A. does not contain water
B. uses lead-antimony plates
C. uses lead-calcium plates
D. does not contain electrolyte

99. Each cell of an automotive battery
A. contains one more negative plate than positive plate
B. contains one more positive plate than negative plate
C. produces 2.1 *V* when fully charged
D. both A and C

100. A fully charged 12 *V* automotive battery should indicate
A. 12.6 V or higher
B. a specific gravity of 1.265 or higher
C. 12.0 V D. Both A and B

101. Deep cycling means
A. overcharging the battery
B. the battery is fully discharged and then recharged
C. the battery is overfilled with water
D. the battery is overfilled with acid

102. During discharge, the positive and negative plates become
A. sulfuricacid B. lead peroxide
C. porous lead D. lead sulfate

103. "Reserve capacity" for batteries means
A. the number of hoi ars the battery can supply 25 A and story above 10.5 V
B. the number of minutes the battery can supply 25 A and stay above 10.5 V
C. the number of minutes the battery can supply 50 A and stay above 9.6 V
D. the number of minutes the battery can supply 150 A and stay above 9.6 V

104. As the battery temperature decreases, the specific gravity
A. increases B. decreases
C. stays the same D. None of these

105. Which of the following conditions indicate that a battery uses excessive water?
A. the charging voltage may be too high
B. the battery plates could be sulfated
C. Both A and B D. None of these

106. A fully charged 12-V battery has
A. 12.0V B. 12.2V
C. 12.4V D. 12.6 V or higher

107. The maximum allowable difference between the highest and lowest hydrometer (specific gravity) reading is
A. 0.010 B. 0.020
C. 0.050 D. 0.500

108. A battery loadtest tells the
A. condition of the starter
B. condition of the battery cables
C. condition of the battery
D. state of charge of the starter

109. A battery high-rate discharge (load, capacity) test is being performed on a 12 V battery. Technician *A* says that a good battery should have a voltage reading below 9.6 V while under load. Technician *B* says that the battery should be discharged (loaded) to twice its cold cranking rating. Which technician is correct?
A. A only B. B only
C. Both A and B D. Neither A nor B

110. When charging maintenance-free (lead-calcium) battery
A. the initial charging rate should be about 35 A for 30 minutes
B. the battery may not accept a charge for several hours
C. the battery temperature should not exceed
D. All of these

111. Whenever jump starting
A. the last connection should be positive (+) post of the dead battery
B. the last connection should be on the engine block of the dead vehicle

Ans. 93. C 94. D 95. A 96. B 97. C 98. C 99. D 100. D 101. B 102. D 103. B 104. A 105. C 106. D 107. C 108. C 109. D 110. D 111. B

C. the alternator must be disconnected on both vehicle
D. the vehicles should touch, if possible, to be assured of a good ground connection

112. The charge indicator on some batteries
A. lights if the battery voltage is 12.6 V or higher
B. is a built-in hydrometer on one cell only
C. can indicate if the electrolyte level is too low
D. Both B and C are correct

113. When performing a capacity test (load test) on a battery, what is the proper load to apply?
A. 15 *A* for 45 sec.
B. three times the ampere-hour rating for 15 seconds
C. four times the ampere-hour rating for 30 seconds
D. 10 times the ampere-hour rating for 15 seconds

114. Technician *A* says that a battery hydro-meter test can be used to find a defective battery. Technician *B* says that a battery load test can be used to determine a discharged or defective battery. Which tech-nician is correct?
A. A only B. B only
C. Both A and B D. Neither A nor B

115. Starter brushes
A. are constructed of a combination of copper and carbon
B. consist of one hot brush and one ground brush
C. consist of two hot brushes and one ground brush
D. Both A and B are correct

116. Starter motors operate on the principle that
A. the filed coils rotate in the opposite direction from the armature
B. opposite magnetic poles repel
C. like magnetic poles repel
D. the armature rotates from a strong magnetic field toward a weaker magnetic field

117. Series motors
A. produce electrical power
B. produce maximum power at zero RPM
C. produce maximum power at 5000 RPM
D. use a shunt coil

118. Automotive starters are
A. series-wound B. compound-wound
C. shunt-wound
D. both A and B are correct

119. Starters that do not have a solenoid mounted on starter
A. are called positive engagement starters
B. are called gear-reduction starters
C. use a drive coil in the field housing to move a metal pole shoe
D. Both A and (C) are correct

120. The instant the ignition switch is turned to the 'start position',
A. both pull-in and hold-in windings are energised
B. the hold-in winding is energised
C. the pull-in winding is energised
D. the starter motor starts to rotate before energis-ing the starter pinion

121. A solenoid
A. is an electromagnetic switch
B. uses a moveable-arm electrical contact
C. uses a movable core
D. Both A and (C) are correct

122. Technician *A* says that a discharged batte-ry (lower than normal battery voltage) is harmful to the starter motor. Technician *B* says that a discharged battery or dirty (coroded) battery cable can cause solenoid clicking. Which technician is correct?
A. A only
B. B only
C. Both A and B are correct
D. Neither A not B

123. Technician *A* says that a battery should be at least 75% charged for accurate starting system test results. Technician *B* says that the starter drive can be replaced without removing the starter from the engine. Which technician is correct?
A. A only B. B only
C. Both A and B are correct
D. Neither A nor B

124. Slow cranking by the starter can be caused by ail except the following
A. a low or discharged battery
B. corroded or dirty battery cables
C. engine mechanical problem
D. an open neutral safety switch

125. A defective or misadjusted neutral safety switch will be indicated by what symptom?
A. slow cranking especially when warm
B. starter whine
C. "non crank" (no starter noise at all)
D. solenoid chatter or clicking

126. Technician *A* says that GM V-8 engines should not be require more than 250 A to crank the engine. Technician *B* says that the overrunning clutch is part of a replaceable starter drive unit. Which technician is correct?
A. A only B. B only
C. Both A and B are correct
D. Neither A nor B

127. To check the pull-in windings of a GM solenoid, connect the battery as follows
A. positive (+) to S terminal, negative (–) to ground
B. positive (+) to BAT terminal, negative (–) to S terminal

Ans. 112. D 113. C 114. C 115. A 116. D 117. B 118. D 119. D 120. A 121. D 122. C 123. A 124. D 125. C 126. C

C. positive (+) to S terminal, negative (–) to M terminal
D. negative (–) to S terminal, positive (+) to M terminal

128. To check the hold-in-windings of a *GM* solenoid connect the battery as follows
A. positive (+) to S terminal, negative (–) to ground
B. positive (+) to BAT terminal, negative (–) to S terminal
C. positive (+) to S terminal, negative (–) to M terminal
D. negative (–) to S terminal, positive (+) to M terminal

129. Field coils should be tested using
A. a 110 V test light B. an ohmmeter
C. a growler D. Either A or B

130. If the starter pinion clearance with the engine flywheel is too great
A. the starter will produce a high-pitched whine during cranking
B. the starter will produce a high-pitches whine after the engine starts
C. the starter drive will not rotate
D. the solenoid will not operate

131. To decrease the starter pinion clearance
A. install a shim under the inboard bolt
B. remove a long shim under both mounting bolts
C. install long shim under both mounting bolts
D. replace the starter drive

132. Technician A says that the diodes regulate the alternator output voltage. Technician B says that degenerators require a current limiter to protect the generator. Which technician is correct?
A. A only B. B only
C. Both A and B are correct
D. Neither A nor B

133. In and alternator, the magnetic field is created in the
A. stator B. diodes
C. rotor D. bearings

134. The voltage regulator controls the current
A. through the alternator brushes
B. through the alternator field
C. through the rotor D. All of these

135. Technician *A* says that a wye-wound stator in an alternator can produce greater amperage output at low speed than can a delta-wound stator. Technician *B* says that excessive amperage output can damage the alternator. Which technician is correct?
A. A only B. B only
C. Both A and B are correct
D. Neither A nor B

136. Alternator output voltage
A. is controlled by the voltage regulator
B. is controlled by the diode
C. can be too high due to a poor ground on the voltage regulator
D. Both A and C are correct

137. Mechanical voltage regulators
A. use A field circuits only
B. use B field circuits only
C. use either A or B field circuits
D. none of these

138. Field circuit types are classified as follows
A. A means internally grounded
B. B means internally grounded
C. all electronic regulators are A circuits
D. Both A and C are correct

139. Alternator brushes must carry current approximately
A. 1 to 3 A B. 2 to 5 A
C. 10 to 36 A D. 36 to 108 A

140. Using an alternator for an extended period in an attempt to charge a defective battery can cause
A. burned diodes B. an overheated stator
C. burned rotor slip rings
D. an overheated voltage regulator

141. An average charging circuit voltage On a 12 V system is
A. 7.5 to 8.5 V B. 11.5 to 13.3 V
C. 14.9 to 16.1 V D. 13.5 to 15.0 V

142. Technician *A* says that by full fielding the alternator, you are bypassing the voltage regulator. Technician *B* says that voltage regulators control the alternator output by controlling the field current. Which technician is correct?
A. A only B. B only
C. Both A and B D. Neither A nor B

143. Technician *A* says that a voltage-drop test of the charging circuit can be performed only if current is flowing through the circuit being tested. Technician *B* says that if the ground for the voltage regulator is corroded, the charging voltage could be too high. Which technician is correct?
A. A only B. B only
C. Both A and B are correct
D. Neither A nor B

144. Clock circuits are built into automatic computers to
A. display time of day
B. tell the driver when to change oil
C. wake up the computer at the correct time
D. None of these

145. An alternator output test
A. reads the ammeter and the voltmeter at 2000 RPM

Ans. 127. C 128. A 129. D 130. A 131. B 132. B 133. C 134. C 135. A 136. ? 137. ? 138. ? 139. B 140. B 141. D 142. C 143. C 144. D

B. must be within 10% of the specifications
C. must include the ignition amperage in the reading
D. All of these

146. To obtain maximum output from a GM alternator with an external voltage regulator
A. ground the field
B. supply the battery voltage to the field terminal
C. connect a resistor to the output terminal
D. ground the diodes

147. When checking an alternator rotor, if an ohmmeter shows zero ohm between the slip rings and the rotor shaft, the rotor is
A. okay-normal
B. defective-shorted rotor windings
C. defective-grounded rotor windings
D. defective-open rotor windings

148. When checking an alternator stator, if an ohmmeter shows low ohms between all three stator connections, the stator is
A. okay-normal B. okay-open
C. defective-grounded D. defective-shorted

149. When checking an alternator diode trio, if an ohmmeter reads low ohms in both directions on one leg, the diode trio is
A. okay-normal B. defective-shorted
C. okay-open D. defective-grounded

150. When checking an alternator rectifier bridge, if an ohmmeter reads 5 ohm on all connections, the part is
A. okay-normal B. defective-open
C. defective-grounded D. defective-shorted

151. Noisy alternators can be caused by
A. defective bearing(s) B. a shorted diode
C. loose mounting brackets
D. All of these

152. Cam dwell or can angle means
A. the length of time the points are closed
B. the angle at which the rubbign block contacts the cam
C. the difference between the lobes on the cam
D. the number of degrees of distributor cam rotation that the points are closed

153. Excessive point gap results in
A. retarded timing B. reduced dwell
C. Both A and B D. None of these

154. Ignition coils reach saturation in approximately
A. 10 sec B. 1 sec
C. 0.1 sec D. 0.01 sec

155. With the ignition switch in the "start" or "crank" position, the engine starts but as soon as the switch is released to run position, the engine stalls. The probable cause is
A. a defective bypass circuit
B. a defective primary ignition resistor
C. a defective coil D. a defective condenser

156. The maximum coil output is approximately
A. 20,000 V and 5 A B. 40,000 V and 0.080 A
C. 25,000 V and 20m A D. 80,000 V and 60m A

157. Technician *A* says that dwell must only be adjusted at the specified idle RPM. Technician *B* says that if the dwell is adjusted 2°, the timing will be changed by 2°. Which technician is correct?
A. A only B. B only
C. Both A and B are correct
D. Neither A nor B

158. Coil energy is measured in units of
A. volts B. amperes
C. watt-sec D. coulombs

159. As ignition points wear
A. the point gap increases
B. the points remain the same
C. the dwell decreases
D. the point gap decreases

160. The spark occurs when
A. the points open B. the points close
C. the ballast resistor is "on"
D. the ignition switch is "on"

161. The items that increase the voltage requi-red to fire a spark plug include
A. a rich mixture, a wide plug gap
B. dirty spark plug wires, a lean fuel mixture
C. a narrow spark plug gap a rich mixture
D. a rich mixture; a wide spark plug gap

162. The pulse generator
A. fires the spark plugs directly
B. signals the electronic control unit (module)
C. replaces the electronic control unit (module)
D. fires the distributor rotor

163. Technician *A* says that dwell cannot be adjusted on cars equipped with pointless electronic ignition. Technician *B* says that dwell cannot be determined at all on pointless electronic ignition. Which technician is correct?
A. A only B. B only
C. Both A and B are correct
D. Neither A nor B

164. Most direct-fire ignition systems
A. use one coil to fire all cylinders at the same time.
B. fire two cylinders at the same time
C. fire four cylinders at the same time
D. use two coils for a six-cylinder engine

165. The ford thick-film-integration (TFI) module
A. is mounted on the inner fender
B. is mounted on the distributor
C. does not use a pickup coil
D. Both B and C

Ans. 145. D 146. B 147. C 148. A 149. B 150. D 151. D 152. D 153. B 154. D 155. B 156. B 157. B 158. C 159. D 160. A 161. B 162. B 163. A 164. B 165. B

166. Technician *A* says that a defective pickup coil can cause a "no start" situation. Tech-nician B says that a defective pickup coil will cause the module to fail. Which tech-nician is correct?
A. A only B. B only
C. Both A and B are correct
D. Neither A nor B

167. Ballast resistors are
A. used on some GM HEI units and all Chrysler electronic ignitions
B. used on most older-style Ford and Chrysler electronic ignition systems
C. not used on electronic ignition systems
D. only used on direct-fire ignition systems

168. Chrysler electronic ignition systems (EIS) use
A. either a three-four-pin module
B. either a four-or five-pin module
C. either a five or six-pin module
D. either a six or seven-pin module

169. GMHEI systems
A. all use integral coil (ignition coil in distributor)
B. do not use an ignition coil
C. all use a four-pin module
D. None of these

170. Electronic ignition systems
A. use special 80,000 V ignition coils
B. fire the spark plug by charging, then discharging the ignition coil
C. have lower spark energy than that of point-type systems
D. require unleaded fuel to burn the spark plugs correctly

171. Hall-effect ignition systems
A. use a Hall-effect switch h instead of a pick-up coil to trigger the module (electronic control unit)
B. use a blade shutter to open and close a special set of contact points.
C. must have the shutter blades grounded for proper engine operation
D. both A and B are correct

172. When should the ignition timing be adjusted?
A. Before adjusting the ignition contact points
B. After adjusting the ignition contact points
C. Either before or after adjusting the contact points
D. None of these

173. Technician A says that the vacuum advances the ignition timing with increasing rpm. Technician B says that the vacuum and mechanical advance can be tested using a timing light. Which technician is correct?
A. A only B. B only
C. Both A and B are correct
D. Neither A nor B

174. Technician *A* says that retarded timing can cause ping. Technician *B* says that point gap increases with wear. Which technician is correct?
A. A only B. B only
C. Both A and B are correct
D. Neither A nor B

175. Ignition timing can be adjusted using
A. a timing light B. a magnetic probe
C. a vacuum gauge D. All of these

176. Technician A says that a mechanical advance mechanism advances the timing by changing the dwell (point gap). Technician B says that a defective vacuum advance unit usually causes a decrease in fuel mileage. Which technician is correct?
A. A only B. B only
C. Both A and B are correct
D. Neither A nor B

177. Technician A says the increasing dwell 1° advances the timing 10°. Technician B says that increasing point gap advances the timing. Which technician is correct?
A. A only B. B only
C. Both A and B are correct
D. Neither A nor B

178. Technician A says that over advanced ignition timing can cause slow, jerky cranking. Technician B says that timing changes dwell a degree for a degree. Which technician is correct?
A. A only B. B only
C. Both A and B are correct
D. Neither A nor B

179. Technician A says that timing should be adjusted at the specified rpm. Technician B says that a defective vacuum advance reduces power and performance. Which technician is correct?
A. A only B. B only
C. Both A and B are coirect
D. Neither A nor B

180. Magnetic offset timing can be employed
A. on all engines B. on many engines
C. for average timing
D. on all Ford-built engines

181. Technician A says that the direction the rotor moves when rotated by hand while in the engine is the direction of rotation. Technician B says that a defective vacuum advance unit is most noticeable by the reduction of fuel economy. Which technician is correct?
A. A only B. B only
C. Both A and B are correct
D. Neither A nor B

182. Spark plug wire
A. increases EMI B. increases RFI
C. is made of metal of original equipment
D. should measure 10,000 ohm or less per foot of length

Ans. 166. A 167. B 168. B 169. D 170. B 171. D 172. B 173. B 174. D 175. D 176. D 177. B 178. A 179. A 180. B 181. C 182. D

183. 8-mm spark plug wire
A. uses 8-rrtm deep boots
B. fits 8-mm spark plugs
C. uses 8-mm diameter metallic wire
D. is 8-mm in outside diameter

184. Cross-fire means that
A. a spark occurs between two spark plug wires
B. current from one spurk plug wire is induced into another spark plug wire
C. two sparks occur al the same instant from the ignition coil
D. distributor cap must be replaced

185. Spark plug wires should be tested
A. with an ohmmeter B. with a voltmeter
C. visually D. Both A and (C)

186. Distributor caps should be checked for
A. chipped or cracked center carbon inserts
B. cracks or carbon tracks
C. a corroded tower(s) D. All of these

187. Dirt on the distributor cap and spark plug wires can cause
A. rapid spark plug wear
B. a higher voltage to be required to fire the spark plugs
C. a lower voltage to be required to fire the spark plugs
D. a rapid centre carbon insert wear

188. A "no start" or poor engine operation during damp or wet weather condition could be an indication of
A. defective distributor cap
B. a defective spark plug wire
C. dirty spark plug wires and/or distributor cap
D. All of these

189. Silicon grease should be installed on both sides of the rubber seal inside a GM/HEI distributor cap with integral coil to
A. lubricate the rotor
B. lubricate and cool the coil
C. help seal out moisture
D. help heat escape from coil

190. Spark plug numbering and lettering identifies
A. the heat range B. reach
C. electrode type D. All of these

191. For American-made spark plugs and many import brands, the heat range code number indicates
A. the higher the number, the hotter the plug
B. the higher the number, the colder the plug
C. that resistor plugs are colder than non-resistor types
D. that resistor plugs are hotter than non-resistor types

192. A spark plug that could correctly replace an AC R46TSX is
A. AC45XLS B. AC R45 TX
C. ACR46TS6 D. ACR44NSX

193. A spark plug that has a white centre insulator indicates that
A. the engine may have vacuum leak
B. the ignition timing may be overadvanced
C. the ignition timing may be retarded
D. Both A and B are correct

194. A spark plug that has a black center insulator indicates that
A. the engine may have a vacuum leak
B. the ignition timing may be overadvanced
C. the choke in the carburetor may be stuck closed
D. Both A and B are correct

195. Resistor spark plugs are used
A. with unleaded fuel only
B. to reduce electrode wear
C. to help control RFI D. Both *A* and *B* only

196. Most spark plug brands use the same code letter for
A. reach B. resistor
C. diameter thread D. electrode type

197. Copper-core spark plugs
A. should be used only in engines run on unleaded fuel.
B. will not last as long as standard plugs.
C. are more likely to foul during rich-mixture conditions
D. None of these

198. Technican A says that a pickup coil (pulse generator) can be tested with an ohm-meter. Technician B says that ignition coils can be tested with an ohmmeter. Which technician is correct?
A. A only B. B only
C. Both A and B are correct
D. Neither A nor B

199. Technician A says that a defective spark plug wire can cause an engine miss. Technican B says that a defective pickup coil wire can cause an engine miss. Which technician is correct?
A. A only B. B only
C. Both A and B are correct
D. Neither A nor B

200. Which component sends a pulse signal to an electronic ignition module?
A. Ballast resistor B. Pickup coil
C. Ignition coil D. Condenser

201. Typical primary coil resistance specifications usually range from
A. 100 to 450 ohms B. 500 to 1500 ohms
C. 1 to 3 ohms D. 6000 to 30,000 ohms

202. Technician A says that an engine will not start and run without a good ground on the points and condenser (breaker plate). Technician B says that

Ans. 183. D 184. B 185. D 186. D 187. B 188. D 189. C 190. D 191. A 192. C 193. D 194. C 195. D 196. B 197. D 198. C 199. C 200. B 201. C

one wire of any pickup coil must be grounded. Which technician is correct?
A. A only B. B only
C. Both A and B are correct
D. Neither A nor B

203. Technican A says that a GM HEI distributor rotor can burn through and cause an engine miss during acceleration. Technician B says that a defective pickup coil can cause an engine miss during acceleration. Which technician is correct?
A. A only B. B only
C. Both A and B are correct
D. Neither A nor B

204. The secondary ignition circuit can be tested using
A. an ohmmeter B. a test light
C. an ammeter D. Both A and B

205. Electronic ignition modules can be tested using
A. special electronic ignition module testers
B. a soldering gun for some modules
C. a test light and jumper wires for some modules
D. All of these

206. The scratch test tests operation of
A. the pickup coil B. ignition coil
C. ignition switch and ignition coil
D. spark plug wires

207. The height of firing lines can be observed on
A. display (parade) only
B. raster (stacked) only
C. superimposed only D. all settings

208. The length of the spark line should be
A. 5 to 15 kV B. 1.0 to 1.5 ms
C. 5 or more 'bumps' D. 1.5 to 2.0 ms

209. Too short spark line on all cylinders could indicate
A. a worn distributor cap and/or rotor
B. an open coil wire (if equipped)
C. too lean a fuel mixture
D. All of these

210. A high firing line without a downward spike below the zero line indicates
A. a defective spark plug wire with good insulation
B. a defective spark plug wire with poor insulation
C. a weak ignition coil
D. a rich fuel mixture in the affected cylinder

211. The "roller coaster" height of firing lines usually indicates
A. an engine mechanical problem
B. a distributor cap/rotor problem
C. a fuel mixture problem
D. a spark plug wire problem

212. Some electronic ignition systems show
A. no firing lines on the scope
B. a current-limiting hump
C. no point-close point
D. no point-open point

213. To test the rotor gap voltage
A. a spark plug wire must be removed
B. a spark plug wire must be grounded
C. the rotor must be removed from the engine
D. the engine must not be running

214. Dwell on a scope pattern is read at
A. the point-closed point
B. the point-open point
C. the coil section D. the spark line section

215. Technician A says that the dwell on the scope should increase as the engine speed is increased on some engines. Technician B says that all firing lines should be within 3 kV of each other. Which technician is correct?
A. A only B. B only
C. Both A and B are correct
D. Neither A nor B

216. Technician A says that if one cylinder drops in rpm much more than the remaining cylinders during a power balance test, that the cylinder is weak. Technician B says that rotor gap voltage must be tested with a spark plug wire grounded. Which technician is correct?
A. A only B. B only
C. Both A and B are correct
D. Neither A nor B

217. The type of memory that disappears or is lost when the ignition is turned off is called
A. RAM B. ROM
C. PROM D. EPROM

218. The coolant sensor
A. is a thermistor
B. operates the "hot" light on the dash.
C. has a high resistance when hot and low resistance when cold.
D. operates the thermostat.

219. Computer inputs include
A. BARO, TPS, M/C solenoid
B. MAP, O_2, TPS
C. air management, M/C solenoid, VAC
D. vehicle speed sensor, air management, TPS

220. Computer output include
A. air management, M/C solenoid, idle speed control
B. TPS, O_2 and ignition timing control
C. BARO, MAP, VAC
D. vehicle speed, TPS, M/C solenoid

221. The oxygen sensor
A. supplies a high voltage (1 V) when the exhaust is rich
B. supplies a high voltage (1 V) when the exhaust is lean
C. tells the computer if the engine is bruning oil.
D. tells the computer the attitude of the engine based on the oxygen content of the exhaust

Ans. 202. A 203. C 204. D 205. D 206. C 207. A 208. B 209. D 210. B 211. C 212. B 213. B 214. A 215. C 216. B 217. A 218. A 219. B 220. A 221. A

222. When the computer is sensing all engine factors and making engine fuel mixture and/or ignition timing changes based on the sensors, it is called
A. stoichiometric B. open loop
C. closed loop D. clear flood

223. Technician A says that the throttle position sensor is a variable resistor that sends a high voltage (about 5 V) back to the computer if the throttle is wide open (WOT). Technician B says that the oxygen sensor must be about 600 F before the voltage can be used by the computer. Which technician is correct?
A. A only B. B only
C. Both A and B are correct
D. Neither A nor B

224. Technician A says that a computer-equipped vehicle should never be jump started. Technician B says that a computer-equipped vehicle should never jump start another vehicle. Which technician is correct?
A. A only B. B only
C. Both A and B are correct
D. Neither A nor B

225. Computer fuel injection systems
A. use a higher-pressure fuel pump
B. have the computer control the pulse width.
C. have the computer operate the fuel injector units
D. All of these

226. Technician A says that if no trouble codes are present and the engine runs poorly, the coolant sensor should be checked first. Technician B says to check all non-computer units first. Which technician is correct?
A. A only B. B only
C. Both A and B are correct
D. Neither A nor B

227. All troubles codes
A. tell the technician exactly what part requires replacement
B. tell the technician what circuit may have problem
C. tell the technician exactly which component or wire is defective
D. None of these

228. In the GM system performance check
A. a high dwell reading indicates a rich mixture being driven leaner
B. a high dwell reading indicates a lean mixture being driven richer
C. a fixed dwell means that the computer is in a closed loop
D. both B and C are correct

229. Ignition timing checks on computer-equipped cars
A. are the same as noncomputer engines
B. are not adjustable
C. require special equipment such as magnetic offset timing units
D. None of these

230. Oxygen sensors
A. are ruined if leaded fuel is used
B. are ruined if some silicone sealers are used on some parts of the engine
C. must be replaced every 5000 miles for proper operation
D. Both A and B are correct

231. Coolant sensors
A. warn the driver when coolant temperatures are too high
B. provide an output from the computer
C. can be tested using an ohmmeter
D. should be replaced every year for best performance

232. Throttle position sensors
A. must be adjusted correctly
B. if not correct, can cause engine ping.
C. if not correct, can cause lack of power
D. All of these

233. Scan tools
A. will always find the computer problem
B. scan the various input and output readings
C. are required tools to adjust or service any computer-equipped car
D. can be used to check previous gas mileage history on all computer engines.

234. Technician A says that if there are no trouble codes, the computer system and engine must be functioning correctly. Technician B says that many engine prob-lems, such as ignition module problems, often are not part of the computer engine controls diagnostics. Which technician is correct?
A. A only B. B only
C. Both A and B are correct
D. Neither A nor B

235. Technician A says that the engine control computer and body computer are not connected. Technician B says that accessories must never be connected to a computer power supply or ground connection. Which technician is correct?
A. A only B. B only
C. Both A and B are correct
D. Neither A nor B

236. The heating, ventilation, and airconditioning system controls operated by the body computer include
A. the blower motor speed
B. the hot water valve
C. the blend door position
D. All of these

Ans. 222. C 223. C 224. B 225. D 226. B 227. B 228. A 229. D 230. D 231. C 232. D 233. B 234. A 235. C 236. D

237. A "wake-up" signal is sent by
A. the ignition switch B. the door switch
C. the seat switch
D. the accelerator pedal switch

238. The process of sampling information from more than one source is called
A. sampling B. multiplexing
C. using EEPROMS D. clock circuits

239. Mpst body computers
A. communicate with engine control computers
B. use replaceable PROMS
C. have built-in diagnostics
D. All of these

240. Ignition timing control is affected by
A. the temperature the lower the temperature, the greater the spark advance
B. the MAP the lower the engine vacuum, the less the spark advance
C. the knock sensor, the greater the engine knock, the less spark advance
D. All of these

241. The working voltage of a capacitor to which a.c. of pulsating d.c. is applied should be
A. the same as or greater than the applied voltage
B. at least 50 percent greater than the applied voltage
C. 1.41 times the applied voltage
D. 0.707 times applied voltage

242. A circuit contains 10 ohms of resistance, 20 ohms of inductive reactance, and 30 ohms of capacitive reactance. The circuit is
A. inductive B. in resonance
C. resisstive D. capacitive

243. The opposition offered by a coil to the flow of alternating current is known as
A. conductivity B. impedance
C. reluctance D. inductive reactance

244. An increase in the inductive reactance of a circuit will be due to increase in
A. inductance and frequency
B. capacitance and voltage
C. Resistance and voltage
D. Resistance and capacitive reactance

245. The resistive force in a dc electrical circuit is measured in ohms and referred to as
A. resistance B. capacitance
C. reactance D. inductance

246. When the capacitive reactance in an a.c. electrical circuit is equal to the inductive reactance, the circuit is
A. in correct voltage phase angle
B. in correct current phase angle
C. out of phase D. resonant

247. In an alternating current circuit, the effective voltage
A. is equal to the maximum instantaneous voltage
B. is greater than the maximum instantaneous voltage
C. may be greater than the maximum instantaneous voltage
D. is less than the maximum instantaneous voltage

248. The amount of electricity a capacitor can store is directly proportional to
A. the distance between the plates and inversely proportional to the plate area.
B. the distance between the plates and inverserly proportional to the plate area.
C. the plate area and inversely propor-tional to the distance betwee the plates
D. the distance between the plates and is not affected by the plate area

249. A transformer with a step-up ratio of 5 to 1 has a primary voltage of 24 V and a secondary amperage of 0.20 A The pri-mary amperage will be
A. 1 A B. 4.8 A
C. 0.40 A
D. cannot be determined from the infor-mation given

250. The phase relationship between the current and voltage in an inductive circuit is
A. The current lags the voltage by 0°
B. The current lags the voltage by 90°
C. The current leads the voltage by 90°
D. The current leads the voltage by 0°

251. Current flow is measured in
A. amperes B. volts
C. watts D. electron flow

252. Unless otherwise specified, any values given for current or voltage in an alternating cuijent circuit are assumed to be
A. average values B. instantaneous values
C. effective values D. maximum values

253. The devices which will require the most electrical power during operation? (Note: 1 hp = 746 W).
A. A 12 V motor requiring 8 A
B. Four 30-W lamps in a 12 V parallel circuit
C. Two lights requiring 3 A each in a 24 V parallel system
D. A 1/10-horsepower, 24 V motor which is 75 percent efficient

254. The number of amperes that will be required by a 24-V, 1/3- horsepower electric motor, when operating at its rated load is (Note: 1 hp = 746 W)
A. 10.4 B, 13.8
C. 7.9 D. 25.6

Ans. 237. B 238. B 239. D 240. A 241. B 242. D 243. D 244. A 245. A 246. D 247. D 248. C 249. A 250. A 251. A 252. C 253. C 254. A

255. A unit in a 28 V aircraft electrical system has a resistance of 10 ohms. The power it will use is
A. 280 W B. 7.84 W
C. 78.4 W D. 28 W

256. A 12 V electric motor has 1,000 W input and 1 hp. output. Maintaining the same efficiency, the input power that will a 24 V, 1-hp electric motor require is (Note: 1 hp = 746 W).
A. 1,000 W B. 2,000 W
C. 500 W
D. Cannot be determined from the infor-mation given

257. A 28 V generator required to supply to a circuit containing five lamps in parallel, three of which have a resistance of 6 ohms each and two of which havp a resistance of 5 ohms. The number of amperes required is:
A. 1.11 A B. 1 A
C. 0.9 A D. 25.23 A

258. The rate of work done which equal to 1 hp is
A. 33,000 ft. ib. per minute
B. 746 ft. ib. per second
C. 3,300 ft. ib. per minute
D. 55 ft, ib. per second

259. The wattage rating of a carbon resistor is deter mined by
A. a gold band B. a silver band
C. the size of the resistor
D. a red band

260. The potential difference between two conductors which are insulated from each other is measured in
A. ohms. B. volts
C. amperes D. coulombs.

261. The ratio of the true power to the apparent power in an a.c. electrical circuit is called the power factor. If the true power and the power factor of a circuit are known, the apparent power can be deter-mined by
A. multiplying the true power times 100 times the power factor
B. multiplying the power factor times 100 times the power
C. dividing the true power times 100 by the power factor
D. dividing the power factor times 100 by the true power.

262. A 24-V source is required to furnish 48 W to a parallel circuit consisting of four resistors of equal value. The voltage drop across each resistor is
A. 12V B. 6V
C. 3V D. 24V

263. When calculating power in a reactive or inductive a.c. circuit in the true power is
A. more than the apparent power
B. more than the apparent power in a reactive circuit and less than the apparent power in an inductive circuit
C. less than the apparent power in a reactive circuit and more than the apparent power an inductive circuit
D. less than the apparent power

264. The power furnished in watts by the generator of the circuit in figure will be

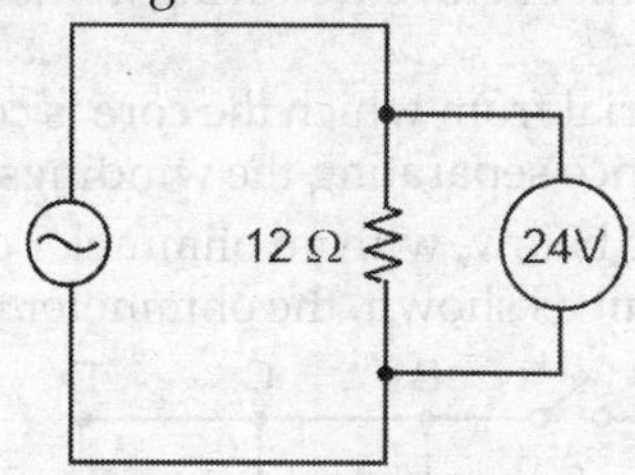

A. 288 W B. 24 W
C. 48 W D. 12 W

265. In figure shown if resistor R_5 is disconnected at the junction of R_3 as shown. The ohmmeter reading will be

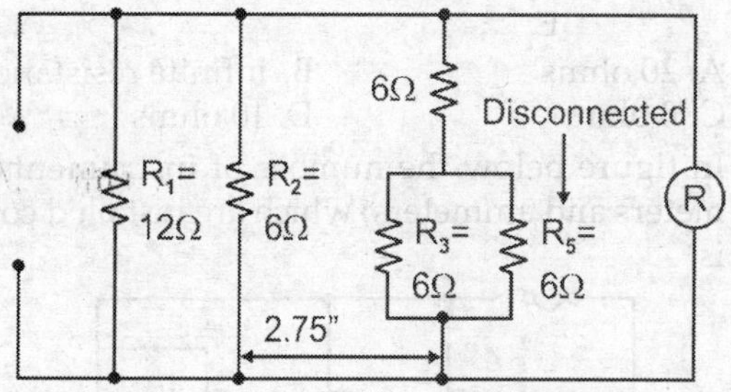

A. 9 ohms B. 2.76 ohms
C. 3 ohms D. 12 ohms

266. Which of the following electrical measuring instru-ments is most likely to obtain its own source of electrical power?
A. Wattmeter B. Ammeter
C. Voltmeter D. Ohmmeter

267. In figure below, resistor R_3 is disconnected at terminal D, the ohmmeter reading will be

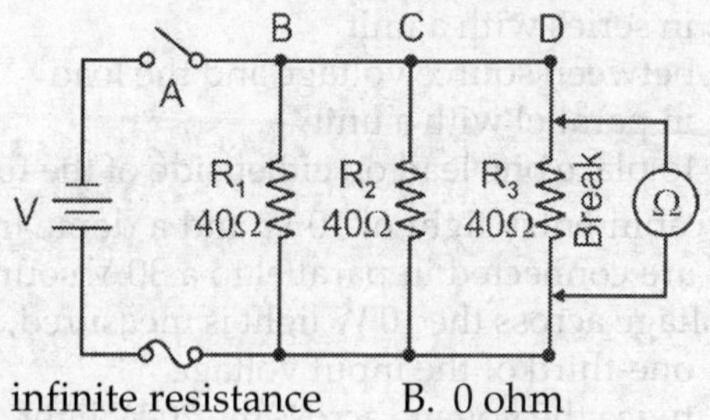

A. infinite resistance B. 0 ohm
C. 10 ohms D. 20 ohms

268. The D' Arsonval type meter movement used in an ammeter, voltmeter, or ohm-meter measures
A. current flow through the movement

Ans. 255. C 256. A 257. D 258. A 259. C 260. B 261. C 262. D 263. D 264. C 265. C 266. D 267. A 268. A

B. potential difference across the movement
C. amount of resistance in series with the movement
D. electrical power consumed by the movement

269. The secondary voltage of a transformer depends upon the efficiency of the transformer and the ratio of the number of turns in the primary winding to the
A. number of turns in the secondary winding
B. amount of current flowing in the primary winding
C. material from which the core is constructed
D. distance separating the windings

270. In figure below, with an ohmmeler co-nnected into the circuit as shown, the ohmmeter reading will be

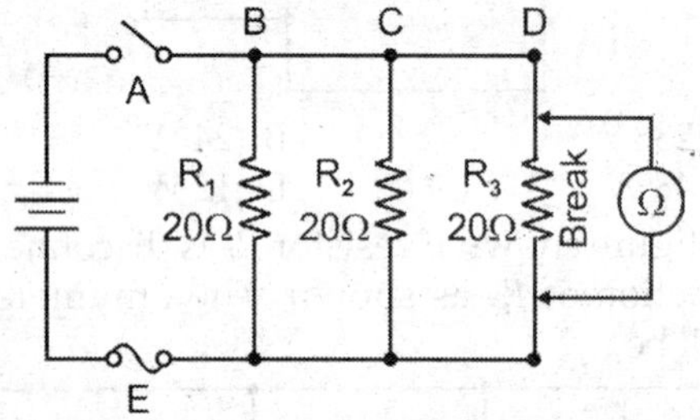

A. 20 ohms B. infinite resistance
C. 0 ohm D. 10 ohms

271. In figure below the number of instruments (voltmeters and ammeters) which are installed correctly is

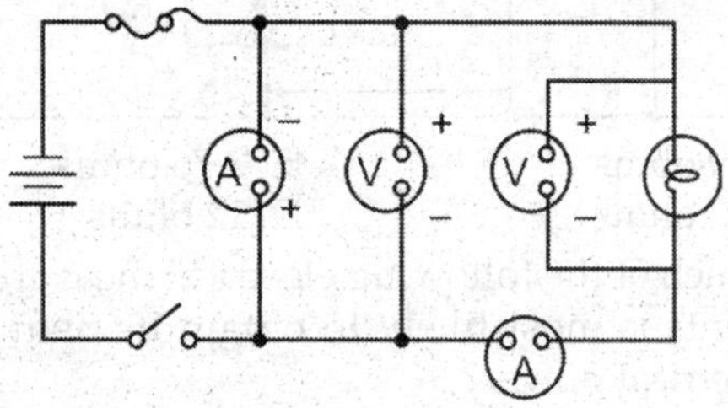

A. Three B. One
C. Two D. Four

272. The correct way to connect a test voltmeter in a circuit is
A. in series with a unit
B. between source voltage and the load
C. in parallel with a unit
D. to place one lead on either side of the fuse

273. A cabin-entry light of 10 W and a dome light of 20 W are connected in parallel to a 30 V source. If the voltage across the 10 W light is measured, it will be
A. one-third of the input voltage
B. twice the voltage across the 20 W light
C equal to the voltage across the 20 W light
D. half the voltage across the 20 W light

274. The device used to measure the very high insulation resistance of electric cables is
A. High-resistance voltmeter
B. Moving iron-vane meter
C. Megger D. Multimeter

275. Before trouble-shooting, an electrical circuit with a continuity light, must be
A. connected to the aircraft battery
B. connected to the aircraft generator
C. isolated
D. connected to an external source of power

276. A 14-ohm resistor is to be installed in a series circuit carrying 0.05 A. The power to be dissipated by the resistor will be
A. at least 0.70 milliwatt B. at least 35 milliwatts
C. less than 0.035 watt
D. less than 0.70 milliwatt

277. The maximum number of electrical wire terminals that can be installed on one stud is
A. Four terminals per stud
B. Three terminals per stud
C. Two terminals per stud
D. As many terminals as you ean stack on and still have the required number of threads showing through the nut

278. In figure, the measured voltage of the series circuit between terminals *A* and *B* will be

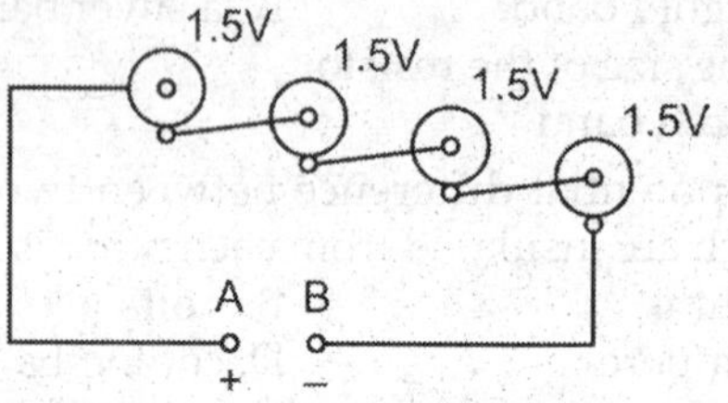

A. 1.5 V B. 3.0 V
C. 4.5 V D. 6.0 V

279. The efficiency of power in an ac circuit is expressed by
A. volt-amperes B. true power
C. power factor D. apparent power

280. The current in a 60-W. 120 V electric light bulb is
A. 0.5 A B. 2 A
C. 1/3 A D. 1/4 A

281. The device/system which will require the most electrical power is
A. Four 30-W lamps arranged in 12 V parallel circuit
B. A 12 V landing gear retraction motor which requires 8 A when operating the landing gear.
C. A 1/10 horsepower, 24 V motor which is 75 percent efficient.
D. 24 V anticollision light circuit consisting of two light assemblies which require 3 A each during operation

Ans. 269. A 270. D 271. C 272. C 273. C 274. C 275. C 276. B 277. A 278. D 279. C 280. A 281. D

282. The unit used to express electrical power is
A. Colomb B. Volt
C. Watt D. Ampere

283. What is the operating resistance of a 30-W light bulb designed for a 28-V system?
A. 30 ohms B. 1.07 ohms
C. 26 ohms D. 0.93 ohm

284. Which of the following statements is correct when made in reference to a parallel circuit?
A. The current is equal in all portions of the circuit
B. The current in amperes in the product of the EMF in volts times the tota, resistance of the circuit in ohms
C. The total current is equal to the sum of the currents through the individual branches of the circuit
D. The cdfrent in amperes can be found by dividing the EMF in volts by the sum of the resistors in ohms

285. Diodes are used in electrical power circuits primarily as
A. current eliminators B. circuit cutout switches
C. rectifiers
D. power transducer relays

286. Three resistors of 3 ohms, 5 ohms, and 22 ohms are connected in series in a 28 V circuit. The current will flow through the 3-ohm resistor is
A. 9.3 A B. 1.05 A
C. 1.03 A D. 0.93 A

287. A good conductor of electricity is a material
A. through or along which electrons move freely
B. whose protons are all on the outside
C. that contains few electrons
D. through or along which protons move freely

288. A circuit has an applied voltage of 30 V and load consisting of a 10 ohm resistor in series with a 20 ohm resistor. The voltage drop across the 10 ohm resistor will be
A. 15V B. 10V
C. 20V D. 30V

289. In figure, the total current flowing in the wire between points C and D in will be

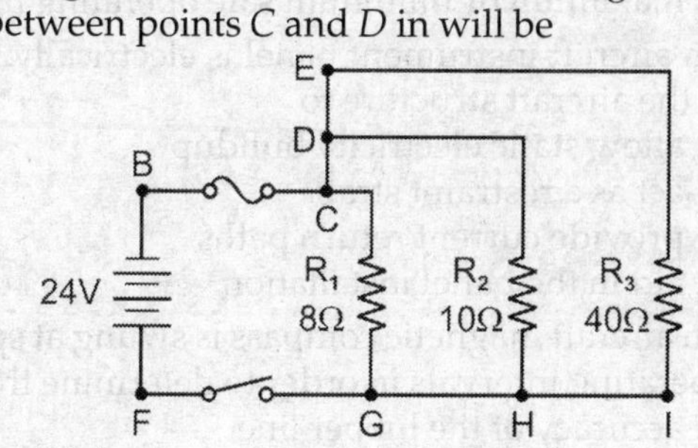

A. 6.0 A B. 2.4 A
C. 3.0 A D. 0.6 A

290. In above figure, the voltage across the 8 ohm resistor will be
A. 2.4 V B. 12 V
C. 20.4 V D. 24 V

291. In the following figure, the total resistance of the circuit across battery will be

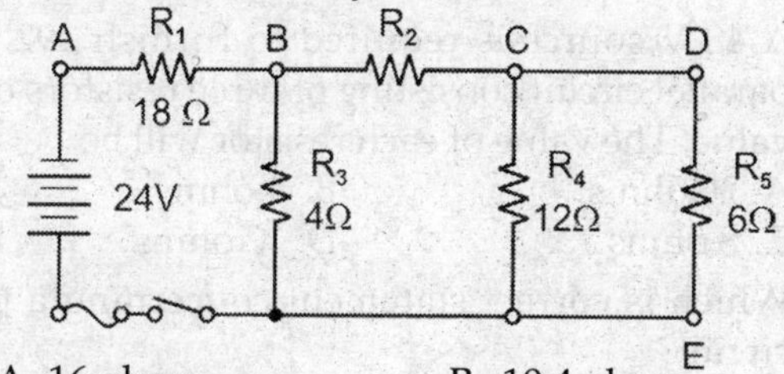

A. 16 ohms B. 10.4 ohms
C. 2.6 ohm D. 21.1 ohms

292. Which of the following is correct in reference to electrical resistance?
A. Two electrical devices will have the same combined resistance if they are connected in series as they will have if connected in parallel
B. If one of three bulbs in a parallel lighting circuit is removed, the total resistance of the circuit will become greater
C. An electrical device that has a high resistance will use more power than one with a low resistance with the same applied voltage
D. A 5 ohm resistor in a 12 volt circuit will use less current than a 10-ohm resistor in a 24 volt circuit

293. An electric cabin heater draws 25 A at 110 V. The current will that flow if the voltage is reduced to 85 will be
A. 19.3 A B. 44.0 A
C. 4.4 A D. 1.93 A

294. In figure, the total current L, flow in the circuit is

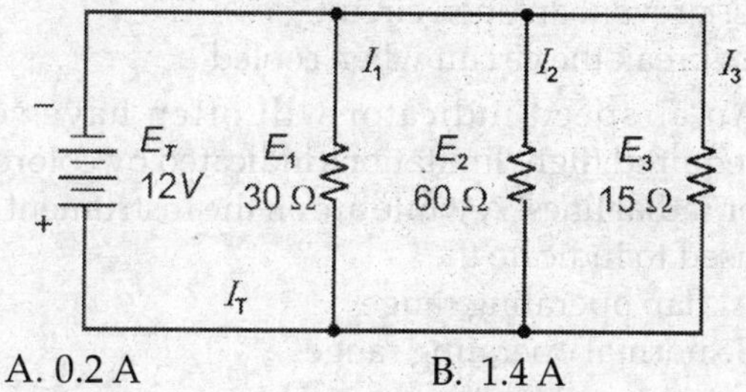

A. 0.2 A B. 1.4 A
C. 0.4 A D. 0.8 A

295. In figure, the total resistance of the circuit is

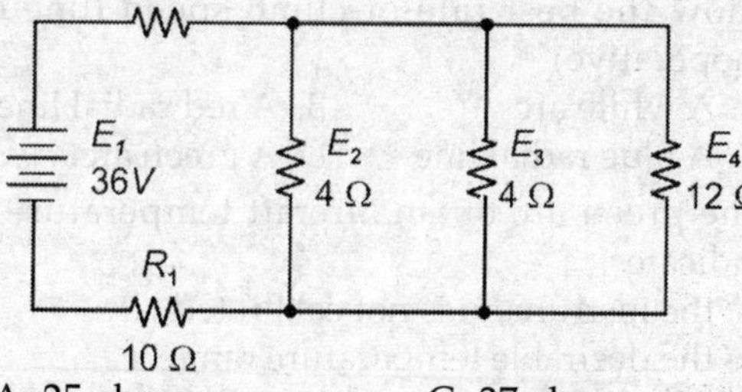

A. 25 ohms C. 37 ohms
B. 35 ohms D. 17 ohms

Ans. 282. C 283. C 284. C 285. C 286. D 287. A 288. B 289. C 290. D 291. D 292. B 293. A 294. B 295. D

296. Through which of the following will magnetic lines of force pass the most readily?
A. Copper B. Iron
C. Aluminium D. Titanium

297. A 48 V source is required to furnish 192 W to a parallel circuit consisting of three resistors of equal value. The value of each resistor will be
A. 36 ohms B. 4 ohms
C. 8 ohms D. 12 ohms

298. Which is correct statement concerning a parallel circuit?
A. Total resistance will be smaller than the smallest resistor
B. Total resistance will decrease when one of the resistances is removed
C. Total voltage drop is the same as the total resistance
D. Total amperage remains the same, regardless of the resistance

299. The voltage drop in a conductor of known resistance is dependent on
A. the voltage of the circuit
B. the amount and thickness of wire insulation
C. only the resistance of the conductor and does not change with a change in either voltage or amperage
D. the amperage of the circuit

300. An electric motor malfunctions causing it to over heat, which will cause an incorporated thermal switch to
A. close the circuit B. break the circuit
C. prevent an open circuit
D. break the circuit when cooled

301. An airspeed indicator will often have several required flight limitations indicated by colored arcs or radial lines. A white arc on the instrument face is used to indicate the
A. flap operating range
B. normal operating range
C. caution range
D. permissible aerobatics range

302. How would an airspeed indicator be marked to show the best rate-of-climb speed (one engine inoperative)?
A. A white arc B. A red radial line.
C. A blue radial line D. A green arc

303. The green arc on an aircraft temperature guage indicates.
A. the instrument is not calibrated
B. the desirable temperature range
C. a low, unsafe temperature range
D. a high, unsafe temperature range

304. What marking on an instrument is used to indicate whether the glass has slipped?
A. Yellow arc B. White index mark
C. Green radial line D. Red radial line

305. What color paint is used to indicate if the cover glass has slipped?
A. Red B. White
C. Yellow D. Green

306. How is a flangeless instrument case mounted in an instrument panel?
A. By four machine screws which extend through the instrument panel
B. By an expanding-type clamp secured to the back of the panel and tightened by a screw from the front of the instrument panel
C. By a metal shell separate from and located behind the instrument panel
D. By press fit into the instrument panel and held in place by friction

307. Why are most electrical instruments mounted in iron or steel cases?
A. To avoid damage to the instrument during maintenance
B. To facilitate removal or installation
C. To prevent interference from outside magnetic fields
D. To reduce heat buildup in the instrument

308. When installing an instrument in an aircraft, who is responsible for making sure it is properly marked?
A. An authorized inspector
B. The aircraft owner
C. The instrument installer
D. The instrument manufacturer

309. A red radial line on the face of an engine instrument indicates.
A. normal operating range
B. caution range
C. operation is permitted under certain conditions
D. maximum or minimum safe operating limits

310. An aircraft instrument panel is electrically bonded to the aircraft structure to
A. allow static electricity buildup
B. act as a restraint strap
C. provide current return paths
D. aid in the panel installation

311. An aircraft magnetic compass is swung at specified operating intervals in order to determine the
A. accuracy of the lubber line
B. compass precession
C. compass variation D. compass deviation

312. The operating mechanism of most hydraulic pressure gauges is

Ans. 296. B 297. A 298. A 299. D 300. B 301. A 302. C 303. B 304. B 305. B 306. B 307. C 308. C 309. D 310. C 311. D

A. a Bourdon tube B. an airtight bellows
C. an airtight diaphragm
D. an evacuated bellows filled with an inert gas to which suitable arms, levers, and gears are attached

313. What is the fixed reference marker attached to the compass bowl of a magnetic compass called?
A. Reederline B. Lubber line
C. Card line D. Pole line

314. Magnetic compass bowls are filled with a liquid to
A. retard precession of the float
B. reduce deviation errors
C. counteract temperature and attitude changes
D. dampen the oscillation of the float

315. Instrument static system leakage can be detected by Observing the rate of change in indication of the.
A. airspeed indicator after suction has been applied to the static system to cause a prescribed equivalent airspeed to be indicated
B. altimeter after pressure has been applied to the static system to cause a prescribed equivalent altitude to be indicated
C. airspeed indicator after pressure has been applied to the static system to cause a prescribed equivalent air-speed to be indicated
D. altimeter after suction has been applied to the static system to cause a prescribed equivalent altitude to be indicated

316. A turn-and-bank instrument indicates
A. the longitudinal attitude of the aircraft during climb
B. trim and serves as an emergency source of bank information in case the attitude gyro fails
C. the longitudinal attitude of the aircraft during descent
D. the need for corrections in pitch and bank anytime the aircraft deviates from preselected attitude

317. Thermocouple leads.
A. are designed for a specific installation and may not be altered
B. may be adjusted in length to fit any installation
C. may be installed with either lead to either post of the indicator
D. may be repaired using solderless connectors

318. Turbine engine exhaust gas temperatures are measured by
A. using iron/constantan thermocouples
B. using electrical resistance thermometers
C. using chromel/alumel thermo-couples
D. using ratiometer electrical resistance thermometers

319. Fuel flow transmitters are designed to transmit data
A. mechanically B. electrically
C visually D. utilizing fluid power

320. Who is authorized to repair an aircraft instrument?
A. A certificated mechanic with airframe and powerplant ratings
B. A certificated repair station approved for that class instrument
C. An appropriately rated airframe repair station
D. A certificated mechanic holding an inspection authorizaiton

321. What is true primary purpose of an auto-pilot?
A. To relieve the pilot of control of the aircraft during long periods of flight
B. To provide a secondary system of aircraft during long periods of flight
C. To fly a more precise course for the piloi
D. To obtain the navigational aid necessary for extended overwater flights

322. Which of the following provides manual meneuverability of the aircraft while the autopilot is engaged?
A. Servo-amplifier B. Attitude indicator
C. Directional gyro indicator
D. Flight controller

323. In autopilot, which signal nullifies the input signal to the ailerons?
A. Displacement signal B. Course signal
C. Rate signal D. Followup signal

324. In which control element of an autopilot system is an attitude indicator?
A. Command B. Sensing
C. Computer D. Input

325. What is the operating principle of the sensing device used in an autopilot system?
A. The reaction of the force 90° away from the applied force in the direction of gyro rotation
B. The relative motion between a gyro and its supporting system
C. The rate of change of motion between the gyro gimbal rings and the aircraft
D. The interaction of the applied force and the rigidity of the gyro

326. What is the purpose of the position transmitter in an autopilot system?
A. To transmit a rate-change signal proportional to the rate of recovery
B. To develop and transmit a negative feedback signal
C. To develop a signal proportional to the amount of flight deviation
D. To transmit a signal to the controller when the aircraft on course

Ans. 312. A 313. B 314. D 315. D 316. B 317. A 318. C 319. B 320. B 321. A 322. D 323. D 324. B 325. B 326. B

327. What component of an autopilot system applies torque to the control surfaces of an aircraft?
A. Servo B. Controller
C. Gyro D. Computer

328. What is the main purpose of a servomotor in an autopilot system?
A. Correct for displacement of the aircraft about its axis
B. Change mechanical energy to electrical energy
C. Move the control surface as commanded
D. Drive the control surface back to the streamlined position

329. What component is the sensing device in electromechanical autopilot system?
A. Servo B. Turn and bank
C. Gyro D. Controller

330. A fully integrated autopilot controls the aircraft around how many axes?
A. One B. Two
C. Three D. Four

331. The ELT (emergency locator transmitter) battery
A. replacement date must be marked on the outside of the transmitter
B. must be a dry-cell type
C. must be replaced annually
D. replacement date must be computed from the date of installation

332. What is the most common power source for an ELT (emergency locator transmitter)?
A. A self-contained battery
B. A direct unfused connection to the aircraft electrical power system
C. A special interconnection to the air-craft battery which allows current to flow in only one direction
D. An isolating transformer connected to the normal aircraft electrical power system

333. When an antenna is installed, it should be fastened
A. to the primary structure at the aircraft's directional pivotal point
B. to the primary structure at the approxi-mate intersection of the three aircraft axes
C. with a reinforcing doubler on each side of the aircraft skin
D. so that loads imposed are transmitted to the aircraft structure

334. After an automatic direction finding antenna has been installed
A. the antenna must be grounded
B. the loop must be calibrated
C. the extra length of wire between the loop and receiver must be removed
D. the transceiver must be compensated

335. Doublers are used when antennas are installed
A. to eliminate antenna vibration
B. to reduce aircraft flutter
C. to prevent oil canning of the skin
D. to reinstate the structural strength of the aircraft skin

336. One antenna can be used for the radio range and standard broadcast bands in light aircraft because the
A. two ranges are close together
B. antenna in omnidirectional
C. antenna length may be electronically adjusted
D. quadrantal error is minimised

337. A gasket or sealant is used between the antenna mast and fuselage skin.
A. to prevent the entry of moisturre
B. so the attachment studs may be drawn tighter
C. For aircraft pressurisation only
D. to prevent abrasion betwene the antenna mast and fuselage skin

338. The preferred location of a VOR antenna on light-aircraft is
A. om the bottom of the fuselage and as far forward as possible
B. any convenient location on the top of the fuselage
C. on top of the cabin with the apex on the V pointing forward
D. on top of the vertical stabilizer

339. The purpose of a localiser is to
A. locate lost airplanes
B. set the airplane on the proper approach angle to the runway
C. indicate the distance the airplane is from the end of the runway
D. align the airplane with the center of the runway

340. An antenna is a special type of electrical circuit designed to radiate and receive
A. Electromagnetic energy
B. Audible signals C. Visual signals
D. Subharmonic frequencies

341. A DME antenna should be located in a position on the aircraft that will
A. not be blanked by the wing when the aircraft is banked
B. allow an antenna functional check to be made on the ground without a DME test set
C. permit interruptions in DME operation
D. eliminate the possibility of the DME locking on a station

342. When bending coaxial cable, the bend radius should be at least
A. 5 times the diameter of the cable
B. 10 times the diameter of the cable

Ans. 327. A 328. C 329. C 330. C 331. A 332. A 333. D 334. B 335. D 336. A 337. A 338. C 339. D 340. A 341. A 342. B

C. 15 times the diameter of the cable
D. 20 times the diameter of the cable

343. When installing a DME antenna, it should be aligned with the
A. angle of decalage
B. null position
C. angle of incidence
D. center line on the airplane

344. The addition of avionics and associated antenna systems forward of the CG limit will affect
A. empty weight and useful load
B. C.G. limits and useful load
C. useful load and maximum gross weight
D. maximum gross weight and datum

345. Which system on aircraft uses to communicate with ground control?
A. VOR receiver B. ADR
C. VHP transceiver D. HF transmitter

346. The probe of a capacitance-type fuel level gauge is essentially a
A. float-actuated variable condenser
B. condenser with fuel and air action gas one plane
C. condenser with ruei and air acting as a dielectric
D. float-actuated variable resistor

347. The capacitance type (electronic type) fuel quantity indicator
A. has no moving parts in the tank
B. has two tubes separated by a mica dielectric in the tank
C. is not accurate when fuel temperature exceeds 100°R
D. measures the amount of the fuel in each tank and reads in gallons

348. The type of remote-reading fuel quantity indicating system which has several probes installed in each fuel tank is:
A. Mechanical B. Electromechanical.
C. Electronic D. Direct reading

349. The aircraft fuel quantity indicating systems which incorporates a signal amplifier is:
A. Electronic B. Sight glass
C. Mechanical D. Electrical

350. A drip guage can measure
A. the amount of fuel in the tank
B. fuel selector valve leakage
C. system leakage with the system shut down
D. fuel pump diaphram leakage

351. The electronic-type fuel quantity indicating system consists of a bridge circuit
A. an amplifier, an indicator, and a tank unit
B. a tank, an amplifier, and an indicator
C. a tank unit, a tank, and an amplifier
D. an indicator, a tank unit, and a tank

352. A probe or a series of probes is used in
A. magnesyn B. Selsyn
C. Capacitor D. Synchro

353. Why is the electronic-type fluid quantity indicating system is more accurate in measuring fuel level?
A. It measures in gallons and converts to pounds
B. Only one probe an one indicator are necessary for multiple tank configurations
C. Aircraft attitude has no effect on fluid quantity indication
D. It measures by weight instead of gallons

354. One advantage of electrical and electronic fuel quantity indicating systems is that
A. the indicators are calibrated in gallons therefore, no conversion is necessary
B. only one transmitter and one indicator are needed regardless of the number of tanks
C. several fuel tank levels can be read on one indicator
D. once calibrated/no further adjustment or calibration is required

355. The dielectric (nonconducting material) of the condenser in a capacitor-type fuel quantity indicating system is:
A. Outer shell of the condenser
B. Wire coil external to the tank
C. Fuel in the tank
D. Fuel and air (vapour) above the fuel

356. Temperature affects fuel weight as:
A. Cold fuel is heavier per gallon
B. Warm fuel is heavier per gallon
C. Temperature has no effect
D. Temperature affects fuel weight at seal level but has no effect at cruising altitude

357. One advantage of electrical and electronic fuel quantity indicating systems is that the indicator
A. can be located at any distance from the tank
B. indicates the fueljquantity for one tank only
C. has no movable devices
D. is always calibrated in gallons, therefore, no conversion is necessary

358. When fuel quantity is measured in pounds instead of gallons, the measurement will be more accurate because fuel volume.
A. varies with temperature change
B. remains unchanged when temperature decreases
C. increases when temperature decreases
D. remains unchanged when temperature increases

359. An electrical-type fuel quantity system con-sists of an indicator in the cockpit and a
A. float-operated transmitter installed in the tank
B. glass or plastic tube placed on the same level as the tank

Ans. 343. D 344. A 345. C 346. C 347. A 348. C 349. A 350. A 351. A 352. C 353. D 354. D 355. D 356. A 357. A 358. A 359. A

C. Float resting on the surface of the tank
D. float-operated receiver installed in the tank

360. In an electronic-type fuel quantity indicating system, the tank transmitter is
A. an electric condenser
B. an electric amplifier
C. a variable resistor D. a variable capacitor

361. The unit which would be adjusted to change the fuel pressure warning limits is
A. Fuel flowmeter bypass valve
B. Engine fuel pump bypass valve
C. Pressure-sensitive mechanism
D. Fuel pressure relief valve

362. The unit which is generally used to actuate the fuel pressure warning system is:
A. Fuel flowmeter
B. Pressure sensitive mechanism
C. Engine fuel pump bypass valve
D. Fuel pressure guage

363. The method used on turbine-powered aircraft to determine when the condition of the fuel is approaching the danger of forming ice crystals is:
A. Fuel pressure warning
B. Fuel pressure guage
C. Fuel strainer pressure gauge
D. Fuel temperature indicator

364. The first positive indication that a change-over from one fuel tank to another is needed is given by:
A. Fuel pressure warning
B. Fuel pressure guage
C. Fuel flowmeter D. Fuel quantity indicator

365. The fuel pressure is taken for the pressure warning signal on most aircraft engines at
A. outlet side of the boost pump
B. fuel pressure line of the carburetor
C. between the fuel pump and the strainer
D. upstream of the fuel shutoff valve

366. In some aircraft with several fuel tanks, the possible danger of allowing the fuel supply in one tank to become exhausted before the selector valve is switched to another tank is prevented by the installa-tion of
A. A fuel pressure warning signal system
B. a fuel pressure relief valve
C. an engine fuel pump bypass valve
D. a fuel flowmeter bypass valve

367. The primary purpose of a ftiel tank pump is to provide a
A. positive system of maintaining the designed minimum fuel supply for safe operation
B. means of visually checking the amount of fuel actually in the tank prior to flight
C. place where water and dirt accumu-lations in the tank can collect and be drained
D. reserve supply of fuel to enable the aircraft to land safely in the event of fuel exhaustion

368. Integral fuel tanks used in many large aircraft because they
A. reduce fire hazards B. facilitate servicing
C. minimize leakage D. reduce weight

369. Aircraft defueling should be accomplished.
A. as soon as possible after engine shut-down
B. with the aircraft's communication equipment on and in contact with the tower in case of fire
C. in a hangar where activities can be controlled
D. in the open air for good ventilation

370. Integral fuel tanks are
A. constructed of nonmetallic material
B. readily removed from the aircraft
C. located within the cabin space of an aircraft
D. formed by the aircraft structure

371. The precautions which must be observed if a gravity-feed fuel system is permitted to supply fuel to an engine from more than one tank at a time is:
A. The tank airspaces must be interconnected
B. The fuel outlet ports of each tank must have the same cross-sectional area
C. Each tank must have a valve in its outlet that automatically shuts off the line when the tank is empty
D. The tanks must always be serviced with identical amounts of fuel

372. The purpose of the baffle plate in a fuel tank is:
A. To provide an expansion space for the fuel
B. To resist fuel surging within the fuel tank
C. To provide internal structural integrity
D. To prevent fuel overflow during refueling

373. Fuel-boost pumps are operated.
A. for providing a positive flow of fuel to the engine
B. during take off only
C. primarily for fuel transfer
D. automatically from fuel pressure

374. The function of flapper valves used in fuel tanks is:
A. to reduce pressure
B. to prevent a negative pressure
C. to serve as variable restrictors
D. to act as check valves

375. Centrifugal-type boost pumps used in systems of aircraft operating at thigh altitude because they
A. save space by having pumps submerged in this tank
B. are positive displacement pumps
C. supply fuel under pressure to engine-driven pump
D. permit air to circulate around the motor

376. It is necessary to vent all aircraft fuel tanks because it
A. ensures a positive head pressure for a submerged boost pump

Ans. 360. A 361. C 362. B 363. D 364. A 365. B 366. A 367. C 368. D 369. ? 370. D 371. A 372. B 373. A 374. D 375. C

B. exhausts fuel vapours
C. limits pressure differential between the tank and atmosphere
D. eliminates vapour lock

377. The type of fuel-booster pump which requires a pressure relief valve is:
A. Wobble B. Concentric
C. Sliding vane D. Centrifugal

378. To prevent vapour lock in fuel lines at high altitude some aircraft are equipped with
A. vapour separators
B. direct-injection-type carburetors
C. booster pumps D. vapour resistors

379. A fuel temperature indicator is located in the fuel tanks on some turbine powered airplanes to tell when fuel may be
A. getting cold enough to form hard ice
B. in danger of forming ice crystals
C. getting too cold to burn
D. about to form rime ice

380. The type of fuel-boost pump that separates air and vapour from the fuel before it enters the line to the carburetor is the
A. diaphragm-type pump
B. gear-type pump
C. centrifugal-type pump
D. sliding vane-type pump

381. Some electric motors have two sets to field windings wound in opposite directions with the result that the
A. speed of the motor can be more closely controlled
B. power output of the motor can be more closely controlled
C. motor can be operated at any speed within its rated range without a change in power output
D. motor can be operated in either direction

382. The purpose of a growler test is to determine the presence of
A. an out-of-round commutator
B. a broken field lead
C. a shorted armature D. None of these

383. What happens if a short circuit occurs between the positive armature lead and the field lead of a shunt generator, which has the voltage regulator located in the positive side of the field circuit?
A. The generator voltage will drop to zero
B. The generator will only produce residual voltage
C. The reverse-current cut-out relay will open and remain open until the fault is corrected
D. The generator voltage will increase

384. The principal advantage of the series-wound d.c. motor is:
A. High starting torque
B. Suitable for constant speed use
C. Low starting torque
D. Speed slightly higher when unloaded

385. If a generator is equipped with a vibrator-type voltage regulator, the actual time the voltage regulator points remain open
A. depends on the load carried by the generator
B. is controlled by the current limiter point clearance
C. is controlled by the reverse-current cutout relay point clearance
D. is increased when the external load is greater than the generator output

386. One cause of generator brush arcing is:
A. Seating brushes with No. 000 sandpaper
B. Excessive spring tension
C. Low spring tension D. Carbon dust particles

387. When can be a.c. generators operated in parallel?
A. Ampere and frequency must be equal
B. Wattage and voltage must be equal
C. Frequency and voltage must be equal
D. Ampere and voltage must be equal

388. The starting current of a series-wound dc motor, in passing through both the field and armature windings, produces a
A. low starting torque B. high starting torque
C. speed slightly higher when unloaded
D. force for constant speed

389. The motor which would be most likely to have an armature brake is:
A. Starter motor B. Inverter drive motor
C. Landing light retraction motor
D. Anticollision beacon operating motor

390. Why must aluminum wire be stripped very carefully?
A. High resistance will develop in stripping nicks
B. Low resistance will develop in stripping nicks
C. Stripping nicks will cause short circuits in wire runs
D. Individual strands will break easily after being nicked

391. The purpose of a commutator in a dc motor is:
A. To allow the transfer of current by converting it
B. To reverse the current in the field coils at the proper time in order to maintain the current flow in the same direction in all conductors under a given pole
C. To provide a means of transferring mechanical energy
D. To reverse the current in the coil just at the time the coil becomes parallel to the lines of force

Ans. 376. C 377. D 378. C 379. B 380. C 381. D 382. C 383. D 384. A 385. A 386. C 387. C 388. B 389. C 390. D 391. D

392. The purpose of an ammeter in a battery charging system is for indicating the
A. amperage available for use
B. total amperes being used in the airplane
C. rate of current used to charge the battery
D. electrical potential of battery
393. Which of the following is *not* one of the purposes of interpoles in a generator?
A. Counteract field distortion
B. Reduce field strength
C. Overcome armature reaction
D. Reduce arcing at the brushes
394. What is the procedure for testing of generator or motor armature windings for opens?
A. Place armature in a growler and connect a 110V test light on adjacent segments; lights should light
B. Check adjacent segments on commutator with an ohmmeter on the high resistance scale
C. Use a 12/24V test light between the armature core segments and the shaft
D. Use a 110V test light between the armature core segments and the shaft
395. The nominal current rating of an aircraft switch is normally stamped on the
A. switch housing B. face plate
C. inside of the switch D. nonmetallic housing
396. The depth to which the mica insulation between the commutator bars of a dc generator undercut is:
A. One-half the width of the mica
B. Equal to twice the width of the mica
C. Equal to the width of the mica
D. Never undercut
397. How does a voltage regulator control generator output?
A. Introducing a resistance in generator-to-battery lead in the event of overload
B. Shorting out field coil in the event of overload
C. Varying current flow to generator field coil
D. Motorising generator to oppose its action
398. Aircraft operation at night must be equipped with position lights that meet the minimum requirements specified by the
A. safety control regulations
B. air safety regulations
C. Federal Aviation Regulations
D. air transportation board
399. The type of d.c. generator is not used as an airplane generator is:
A. externally grounded B. internally grounded
C. series wound D. compound wound
400. The type of ammeter used to measure radio frequency alternating current is:
A. half-wave bridge type
B. full-wave bridge type
C. thermocouple type D. emitter-base type
401. The most accurate type of frequency measuring instrument is:
A. Integrated circuit chip having a clock circuit
B. Electrodynamometers using electromagnetic fields
C. Electromagnets using one permanent magnet
D. Repulsion-type, moving-vane meter
402. During ground operation, how is a starter-generator normally cooled?
A. Ram air B. Engine bleed air
C. An integral fan
D. The environmental system cooled air
403. During flight operation, how a starter-generator is cooled?
A. The environmental system cooled air
B. An integral fan and ram air
C. Engine bleed air
D. An external motor-driven fan
404. The function of a rectifier is:
A. To change direct current into alternating current
B. To step up voltage
C. To change alternating current into direct current
D. To reduce voltage
405. The type of instrument used for mea-suring very high values of resistance is:
A. Megohmmeter B. Shunt-type ohmmeter
C. Thermocouple D. Multimeter
406. A diode to be checked for an open circuit or a short circuit. It should be
A. in the circuit
B. checked with a milliamp ammeter
C. disconnected from the circuit
D. checked from positive to negative only
407. The type of rectifier used to change alternator output to direct current is:
A. Carbonpile
B. Single phase, half wave
C. Three phase, full wave, solid-state
D. Sodium-filled, semiconductor
408. It is necessary to observe caution when handling a high-voltage capacitor in an electrical circuit because
A. a capacitor may emit toxic gases if not properly ventilated
B. the polarity of the plates may be reversed by improper attachment of an ohmmeter
C. a capacitor may lose its ability to hold charge if intentionally discharged
D. a capacitor may retain its charge after power is removed
409. The type of capacitor which can be checked with an ohmmeter to determine its condition is
A. extremely low-capacity capacitor

Ans. 392. C 393. B 394. A 395. A 396. C 397. C 398. C 399. C 400. C 401. A 402. C 403. B 404. C 405. A 406. C 407. C 408. D

B. sodium filled capacitor
C. dielectric capacitor
D. electrolytic capacitor

410. If a transformer winding has some of its turns shorted together, it can be determined by observing that
A. the output voltage will be high
B. the transformer will not function
C. the transformer will get hot in normal operation
D. None of these

411. The two general types of a.c. motors used in aircraft systems are
A. induction and synchronous
B. shaded pole and Universal
C. ac series and capacitor-start
D. theostat-series and condenser-start

412. The splices can be arranged it several are to be located in an electrical wire bundle as:
A. staggered along the length of the boundle
B. it is not permissible to splice aircraft wiring
C. enclosed in a conduit D. grouped together

413. The minimum bend radii for an electrical wire bundle is:
A. ten times the outside diameter of the bundle
B. five times the outside diameter of the bundle
C. fifteen times the outside diameter of the bundle
D. twenty times a diameter of the largest wire in the bundle

414. The voltage output of an alternator may be regulated by controlling the
A. speed of the alternator
B. voltage output of the dc exciter
C. resistance in the rotor windings
D. exciter frequency

415. If several long lengths of electrical cable are to be installed in rigid conduit, the possibility of damage to the cable as it is pulled through the conduit will be reduced by
A. dusting the cable with powdered graphite
B. dusting the cable with powdered soap-stone
C. blowing powdered graphite into the conduit runs before installation of the cables
D. the application of a light coat of oil or grease

416. Grounding is electrically connecting a conductive object to the primary structure so as to
A. prevent current return paths
B. prevent stability of radio transmission and reception
C. allow accumulation of static charge
D. prevent development of radio frequency potential

417. The components used to bond non-continuous stainless steel aircraft components are
A. Printed circuits B. Stainless steel jumpers
C. Copper jumpers D. Aluminum jumpers

418. The rating of aircraft fuse capacity is in
A. volts B. ohms
C. amperes D. microfarads

419. The inadvertent operation of a switch can be prevented by
A. mounting a suitable guard over the switch
B. installing a derated spring-loaded toggle switch
C. installing circuit breakers as a switch
D. installing a low amperage fuse across the contacts

420. It is important to prevent oil canning in electrical junction box sides because it is necessary to
A. aid in the installation of terminal strips
B. prevent internal short circuits
C. provide space for relay switches
D. provide space for wire ties and clA

421. During inspection or the terminal strips of an aircraft electrical system, it should be determined that
A. only locknuts have been used for terminal attachment to the studs
B. the terminal studes are anchored against rotation
C. the heat rise at any single terminal stud does not exceed 120°F above ambient temperatures
D. Only plain nuts and lockwashers have been used for terminal attachment to the studs

422. What protection to wires and cables does conduit provide when used in aircraft installations?
A. Electromagnetic B. Thermal
C. Mechanical D. Structural

423. Which of the following should be avoided in conduit installation?
A. Support the conduit to prevent chafing against the structure
B. Provide drainholes at the lowest point in a conduit run
C. Locate conduit to provide a footstep or handhold for personnel
D. Drilling burrs should be carefully removed

424. Electrical wiring installed in aircraft without special enclosing means is called
A. master wiring B. open wiring
C. stranded wiring
D. non-conductive wiring

425. If the (+) terminal of a voltmeter is connected to the (–) terminal of the source voltage and the (–) terminal of the meter is connected to the (+) terminal of the source voltage, the voltmeter will read
A. high voltage B. correctly
C. low voltage D. backwards

Ans. 409. D 410. C 411. A 412. A 413. A 414. B 415. B 416. D 417. C 418. C 419. A 420. B 421. B 422. C 423. C 424. B 425. D

426. The nominal rating of electrical switches refers to continuous
A. voltage rating with the contacts open
B. current rating with the contacts open
C. voltage rating with the contacts closed
D. current rating with the contacts closed

427. The material used in the construction of aircraft electrical junction boxes in fire zone is
A. fireproof aluminum B. asbestos
C. cadmium-plated steel
D. stainless steel

428. The conduit which is normally used to minimise radio interference is:
A. Flexible brass B. Flexible aluminum
C. Rigid steel D. Rigid aluminum

429. What are the primary considerations when selecting electric cable size?
A. Current-carrying capacity and allowable voltage drop
B. The voltage and amperage of the load it must carry
C. The cable location and normal opera-ting temperature
D. The system voltage and cable length

430. How are electric circuits protected from overheating?
A. Thermocouples B. Shunts
C. Fuses D. Solenoids

431. The routing of coaxial cables differs from the routing of electrical wiring as
A. coaxial cable are routed parallel with stringers or ribs
B. coaxial cable are routed at right angles to stringers or ribs
C. coaxial cables must not be clamped
D. coaxial cables are routed as directly as possible

432. The aluminum electrical cable sizes which would be selected to replace a No. 10 copper electrical cable is
A. No. 4 B. No. 6
C. No. 8 D. No. 10

433. Which of the following statements relating to electric wiring is true?
A. When attaching a terminal to the end of an electric cable, it should be determined that the strength of the cable-to-terminal joint is at least twice the tensile strength of the cable
B. When splicing an electric cable in a location subjected to extreme vibration, it is generally recommended that solder spices be used
C. When attaching a terminal to the end of an electric cable, it should be determined that the strength of the cable-to-terminal joint is at least equal to the tensile strength of the cable itself
D. All electric cable splices should be covered with soft insulating tubing (spaghetti) for mechanical protection against external abrasion

434. For which value are bonding connections tested?
A. resistance value B. amperage value
C. reactance D. voltage value

435. A voltmeter is connected in
A. series with the load B. parallel with the load
C. series with the source
D. series-parallel with the source

436. If it is necessary to use an electrical connector where it may exposed to moisture, the mechanic should.
A. coat the connector with grease
B. use a special moisture-proof type
C. wrap the connector with waxed paper
D. spray the connector with varnish or zinc chromate

437. The aircraft components are bonded because it is necessary
A. to allow electrical charges to move through the aircraft structure without causing sparks
B. to prevent electrical charges from moving through the aircraft structure
C. to maintain the electrostatic charge of the aircraft equal to that of the surrounding atmosphere
D. to allow the electrostatic charge of the aircraft to dissipate before it contacts the ground after flight

438. The advantage of a circuit breaker when compared to a fuse is
A. never needs replacing
B. tesponds rester to overload
C. always eliminates the need of a switch
D. resettable and reusable

439. The advantage of a current limiter is
A. it breaks circuit quickly
B. it will take overload for a short period
C. it is easily replaced D. it can be reset easily

440. Where electric cables must pass through holes in bulkheads, formers, ribs, firewalls, etc. the wires should be protected from chafing by
A. wrapping with tape
B. using a rubber grommet
C. several coats of varnish
D. Wrapping with plastic

441. In aircraft electrical systems, automatic reset circuit breakers
A. should not be used as circuit protective devices
B. are useful where only temporary overloads are normally encountered
C. need not be made accessible to crew members in flight
D. must be used in all circuits essential to safe operation of the aircraft

Ans. 426. D 427. D 428. A 429. A 430. C 431. D 432. B 433. C 434. A 435. B 436. B 437. A 438. D 439. D 440. B 441. A

442. The throw of a single-pole, double-throw switch (SPOT) indicates the
A. number of circuits each pole can complete through the switch
B. method of actuating the switch (push-pull, laterally, vertically, etc.)
C. number of terminals at which current can enter or leave the switch
D. number of places at which the operating device (toggle, plunger, etc) will come to rest and at the same time open or close a circuit

443. An important factor in selecting aircraft fuses is .
A. The current exceeds a predetermined value
B. The voltage rating should be lower than the maximum circuit voltage
C. The inner strip of metal is made of an alloy of tin and bismuth.
D. Capacity matches the needs of the circuit

444. The function of circuit breaker in the instrument lighting system is to protect the
A. lights from too much current
B. wiring from too much current
C. wiring from too much voltage
D. lights from too much voltage

445. An ohmmeter is used to check the conti-nuity of a circuit. The ohmmeter would read infinity if
A. the resistance in the circuit has become practi-cally zero
B. a separation in the wire has caused a short circuit
C. the circuit is shorted to ground
D. the circuit is open

446. A voltmeter and an external shunt ammeter are to be installed. The ammeter should be placed in
A. series with the shunt, the shunt in series with circuit, and the voltmeter in parallel with the circuit
B. parallel with the shunt, the shunt in series with circuit, and the voltmeter in series with the circuit
C. series with the shunt, the shunt in series with circuit, and the voltmeter in series with the circuit
D. parallel with the shunt, the shunt in series with circuit, and the voltmeter in parallel with the circuit

447. The range switch on a multimeter is set on 300mA. How much current is the meter in-dicating if the needle is pointing to 1.85 on a scale of zero to three?
A. 0.185 A B. 1.85 A
C. 1.85 mA D. 18.5 mA

448. Before using an ohmmeter to check for open or short circuits, it is necessary to
A. activate the circuit to be checked
B. never short the ohmmeter leads together
C. isolate the circuit to be checked
D. set the ohmmeter to infinity

449. The precaution to be followed a multimeter is:
A. Do not store near a power source
B. Do not store near a permanent magnet
C. Keep lead probes from touching
D. select function switch in other than ohm position

450. When a multimeter is usdd to measure resistance, a range is selected which will
A. read on lower half of scale
B. prevent pointer from fluctuating
C. allow meter to be adjusted to zero
D. read on upper half of scale

451. Capacitors connected in series can be combined in the same manner as
A. resistance is series B. conductors in series
C. farads in parallel D. resistance in parallel

452. In a D'Arsonval ammeter, the amount of current required to turn the meter pointer to full-scale defection depends on the
A. magnet strength and number of turns of wire in the moving coil
B. strength of the balance springs and the size of wire in the moving coil
C. strength of the balance springs
D. magnet strength and zero adjustment

453. In a parallel d.c. circuit the current flow in any one resistor is equal to
A. the current flow in any other part of the circuit
B. the sum of the currents in the other parallel resistors
C. the total current flow
D. the total current flow minus the sum of the currents in the other parallel resistors

454. Which of the following principles is not true in reference to a parallel electrical circuit?
A. The total resistance is less than the smallest individual resistor
B. The total resistance will decrease as additional parallel resistors are added
C. The total resistance is equal to the sum of the individual resistors
D. The total resistance will increase if one or more of the parallel resistors is removed

455. What is an advantage of using a.c. electrical power in aircraft?
A. AC electrical motors can be reversed while DC motors cannot
B. Self-induction due to voltage change contributes to the effective power, thus causing the power, output to be 1.707 times the power input
C. Greater ease in stepping the voltage up or down
D. The effective voltage is 1.41 times the maximum instantaneous voltage; therefore, less power input is required

Ans. 442. A 443. D 444. B 445. D 446. D 447. A 448. C 449. D 450. C 451. D 452. A 453. D 454. C 455. C

456. Aircraft position lights consist of at least three lights. What are their colour and location?
A. White in front, red in the rear, and green mid-way on the aircraft center-line
B. Red on the left, green on the right, and white on the rear
C. Green in front, red in the rear, and while mid-way on the aircraft center-line
D. Red on the right, green on the left, and white in the rear

457. The iron cores of most induction coils are laminated in order to
A. reduce the core reluctance
B. increase the core permeability
C. reduce the effects of eddy currents
D. reduce the production of weak areas and strong areas on the core faces

458. Certain transport aircraft use AC electrical power for all normal operation and battery furnished DC electrical power for stand by emergency use. In aircraft of this type that operate no d.c. generators, the batteries are kept charged by
A. inverters which use the aircraft's ac generators as a source of power
B. a.c. current directly from the aircraft's generators
C. alternators which use the aircraft's ac generators as a source of power
D. rectifiers which use the aircraft's a.c. generators as a source of power

459. The voltage in an ac transformer secondary that contains twice as many laps as the primary will be
A. greater and the amperage less than in the primary
B. greater and the amperage greater than in the primary
C. less and the amperage greater than in the primary
D. less and the amperage less than in the primary

460. If the positive field lead between a generator and a generator control panel breaks and is shorted while the engine is running, a voltmeter connected to generator output would indicate
A. zero voltage B. residual voltage
C. normal voltage
D. slightly below normal voltage

461. A method used for restoring generator field residual magnetism is
A. Flash the fields
B. Demagnetise the commutator
C. Reseat the brushes
D. Energise the armature

462. One of the chief advantages of alternating current is that it can be transmitted at a high voltage with a low power loss. The voltage can then be changed to any desired value of
A. dc by means of inverters
B. dc by means of transformers
C. ac by means of inverters
D. ac by means of transformers

463. Which of the following must be accomplished when installing an anticollision light?
A. Connect the light to the primary electrical bus
B. Install a switch independent of the position light switch
C. Use shielded electrical cable to assure fail-safe operation
D. Connect the anticollision light to the aircraft position light switch

464. The inductor-type inverter output voltage is controlled by the
A. number of poles and the speed of the motor
B. dc stator field current
C. voltage regulator D. ac armature coils

465. While using an ohmmeter to check the continuity of a generator field coil, the coil should
A. be removed from the generator housing
B. show high resistance when the meter prods are connected to be terminals of the coil
C. be heated to operating temperature
D. show very low resistance if it is a series field coil

466. The strength of the core of an electromagnet depends upon the material from which it is constructed the which of the two other factors?
A. The number of turns of wire in the coil and the applied voltage
B. The size (cross section) of the wire and the amount of current (amperes) passing through the coil
C. The number of turns of wire in the coil and the amount of current (amperes) passing through the coil
D. The size (cross section) and the number of turns of wire in the coil and the applied voltage

467. How does a voltage regulator control generator voltage?
A. By changing the resistance in the generator output circuit
B. By changing the residual magnetism of the generator
C. By changing the current in the generator output circuit
D. By changing the resistance of the generator field circuit

468. The overvoltage control automatically protects the generator system when excessive voltage is present by

Ans. 456. B 457. C 458. D 459. A 460. B 461. A 462. D 463. B 464. B 465. D 466. C 467. D

A. opening the shunt field circuit
B. opening and resetting the field control relay
C. breaking a circuit to the trip coil of the field control relay
D. closing the generator switch circuit

469. When d.c. generators are operated in parallel to supply power for a single load, their controls include an equaliser circuit to assure that all generators share the load equally. The equaliser circuit operates by
A. switching all new electrical loads to the low generator to maintain an equal load division among the generators
B. increasing the output of the low generator to equal the output of the high generator
C. decreasing the output of the high generator to equal the output of the low generator
D. increasing the output of the low generator and decreasing the output of the high generator until they are equal

470. Which of the following is considered to be an intermittent duty circuit?
A. Anticollision light circuit
B. Instrument panel light circuit
C. Landing light circuit D. navigation light circuit

471. The most common method of regulating the voltage output of a compound d.c. generator is to vary the
A. current flowing through the shunt field coils
B. total effective field strength by changing the reluctance of the magnetic circuit
C. resistance of the series field circuit
D. number of rotating conductors being affected by the field flux

472. Upon completion of the landing gear extension cycle, the green light illumi-nated and the red light remained lit. The probable cause is short in the
A. down limit switch B. gear safety switch
C. up limit switch D. nose gear down switch

473. The instrument panel voltmeter indicates 10V to 15V in a 24-V system with a properly adjusted voltage regulator. The trouble is most likely in the
A. reverse current cutout
B. generator circuit
C. Battery circuit
D. inverter

474. The direction of rotation of a dc electric motor can be changed by
A. Interchanging the wires which connect the motor to the external power source
B. Reversing the electrical connections to either the field or armature windings
C. Rotating the brush assembly 90°
D. Removing the starting winding

475. Aircraft which operate only ac generators (alternators) as a primary source of electrical power normally provide current suitable for battery charging through the use of a/an
A. stepdown transformer and a rectifier
B. network of condensers and choke coils to filter the alternating current with negligible power loss
C. inverter and a voltage-dropping resistor
D. dynamotor with a half-wave dc output

476. On installations requiring alternating current from the battery-generator system, it is necessary to have.
A. a transformer B. an inverter
C. two or more generators
D. a variable resistor between the battery and generator

477. What is a relay?
A. A magnetically operated switch
B. A device which increases voltage
C. A device which converts electrical energy to heat energy
D. Any conductor which receives electrical energy and passes it on with little or no resistance

478. The purpose of a rectifier in an electrical system is to change.
A. the frequency of alternating current
B. the voltage of alternating current
C. the voltage and amperage of alternating current
D. alternating current to direct current

479. The ratio of turns between the primary coil winding and the secondary coil winding of a transformer designed to triple its input voltage would be as primary will have
A. one-third as many turns as its secondary
B. one-half as many turns as its secondary
C. twice as many turns as its secondary
D. three times as many turns as its secondary

480. In an a.c. circuit with no phase lead or lag
A. Real power is zero
B. Reactive power is maximum
C. Real power is greater than apparent power
D. Real power equals apparent power

481. The rating of generators is in
A. Watts at rated voltage
B. Farads at rated voltage
C. Amperes at rated voltage
D. The imppdance at rated voltage

482. A shunt woijnd d.c. generator is connected as:
A. one field is shunted across the other
B. both fields are shunted across the armature
C. the field and armature are shunted with a capacitor

Ans. 468. A 469. D 470. C 471. A 472. C 473. A 474. B 475. A 476. B 477. A 478. D 479. A 480. D 481. C 482. B

D. the armature and fields are shunted by a variable resistor

483. The poles of a generator are laminated to
A. reduce hysteresis losses
B. reduce flux losses
C. increase flux concentration
D. reduce eddy current losses

484. The frequency of an alternator depends upon
A. voltage B. RPM
C. current D. wattage rating

485. Where is the generator rating usually found stamped on?
A. Firewall B. Generator
C. Engine D. Cowling

486. Residual voltage is a result of
A. magnetism in the field windings
B. current flow in the field coils
C. magnetism in the field shoes
D. magnetism in the armature

487. Three resistors of 3 ohm, 10 ohm, and 15 ohm are connected in parallel in a 30 V circuit. The current will that flow through the 3-ohm resistor is:
A. 30 A B. 10 A
C. 6 A D. 2 A

488. A 24 V source is required to furnish 48 W to a parallel circuit consisting of two resistors of equal value. The value of each resistor is
A. 20 ohm B. 24 ohm
C. 12 ohm D. 10 ohm

489. Which of the following is not one of the three methods by which an electromotive force can be induced in a conductor using the principle of electromagnetic induction?
A. Varying the speed of a conductor moving parallel to the lines of force of a stationary magnet
B. Moving a magnet so that its lines of force cut across a stationary conductor
C. Moving a conductor to that it cuts across the lines offeree of a stationary magnet
D. Varying the strength of stationary magnet so that its lines of force cut across a stationary conductor

490. A 24 V source is required to furnish 48 W to a series circuit consisting of two resistors of equal value. The value of each resistor is
A. 2 ohm B. 6 ohm
C. 12 ohm D. 24 ohm

491. The unit of power used in d.c. electrical circuits is
A. Ampere B. volt
C. Joule D. Watt

492. An electric cabin heater draws 20 A at 110 V. The current that will flow if the voltage is reduced to 90 V is:
A. 1.64 A B. 5.5 A
C. 16.4 A D. 14.5 A

493. The resistance of a conductor can be decreased by
A. increasing the length or decrease the cross-sectional area
B. decreasing the length or increase the cross-sectional area
C. fecreasing the length or the cross-sectional area
D. increasing the length or the cross-sectional area

494. If a 12 V circuit furnish 3 A to a parallel circuit consisting of three equal resistors. The value of each resistor will be
A. 12 ohm
B. 4 ohm
C. 1.33 ohm
D. cannot be determined from the information given

495. A simple circuit has three parallel resistors of 5 ohm, 10 ohm, and 17 ohm. If the measured voltage across the 17 ohm resistor is 25 V. The current that will flow through the 5 ohm resistor will be
A. 8.97 A B. 1.47 A
C. 5 A
D. cannot be figured from the information given

496. In a circuit having an increasing resistance with a constant voltage, the current flow will
A. be directly proportional
B. increase
C. decrease D. be constant

497. In a parallel circuit, most current flow will be through the branch with the
A. least voltage B. greatest voltage
C. least resistance D. greatest resistance

498. An electric cabin heater draws 15 A at 110 V. If the voltage is reduce to 95V, the current will be
A. 13.0A B. 1.30A
C. 7.3 A D. 73.0A

499. The purpose of skid detectors is:
A. To reduce brake drag
B. To aid in effective braking
C. To reduce hydraulic pressure
D. To indicate the tires are skidding

500. The rotor in an autosyn remote indicating system employs
A. an electromagnet B. a permanent magnet
C. an electromagnet and a permanent magnet
D. neither an electromagnet or a permanent magnet

501. What is the basic difference between an autosyn and a magnesyn indicating system?
A. Rotor B. Transmitter
C. Receiver D. Winding

502. The rotor in a magnesyn remote indica-ting system employs

Ans. 483. D 484. B 485. B 486. C 487. B 488. B 489. A 490. B 491. B 492. C 493. B 494. A 495. C 496. C 497. C 498. A 499. B 500. A 501. A

A. a permanent magnet
B. an electromagnet
C. an electromagnet and a permanent magnet
D. neither an electromagnet nor a permanent magnet

503. One check necessary for proper operation of a pitot/static tube heater after replacement is
A. Ammeter reading B. Voltmeter reading
C. Visual inspection of all connections
D. Continuity check or system

504. The type of instrument indicating system that is used in a fuel flow system in large airplanes is usually
A. a direct reading system
B. an autosyn system
C. a synchro resolver system
D. a servomechanism system

505. The current required to operate an aircraft autosyn fuel-flow indicating system is
A. direct current B. pulsating current
C. alternating current D. pulsating voltage

506. Fuel-flow transmitters are designed to transmit data
A. mechanically B. electrically
C. visually D. fluidly

507. The fuel-flow transmitter converts the fuel flow into an electrical signal which represents the rate of fuel flow in pounds per hour. It then transmits the signal to
A. a receiver device on the instrument panel
B. the fuel quantity guages
C. the control valve on the fuel control regulator
D. the bypass valve solenoid on the fuel control regulator

508. The unit which most accurately indicates fuel consumption of an internal combustion engine is:
A. Fuel quantity gauge B. Fuel totalizer
C. Electronic fuel quantity indicator
D. Fuel flowmeter

509. A ratiometer-type oil temperature indicator moves off-scale on the high side of the dial as soon as the master switch is turned. The most probable trouble is
A. a short in the power circuit
B. an open in the bulb circuit
C. an open in the power circuit
D. a short in the bulb circuit

510. Which mechanism drives the fuel-flow indicator needle?
A. A friction clutch on the motor shaft
B. A mechanical gear train
C. A direct coupling to the motor shaft
D. A magnetic linkage

511. The function of an engine analyser
A. to measure the fuel/air ratio being burned in the cylinders
B. to monitor the temperature of exhaust gases
C. to measure specific fuel consumption
D. to detect ignition system faults

512. The function of a manifold pressure gauge is
A. to maintain constant pi assure in the intake manifold
B. to indicate differential pressure bet-ween the intake manifold and atmos-pheric pressure
C. to indicate variations of atmospheric pressure at different attitudes
D. to indicate absolute pressure in the intake manifold

513. The function of an exhaust gas analyser is to indicate the
A. brake specific fuel consumption
B. fuel/air ratio being burned in the cylinders
C. temperature of the exhaust gases in the exhaust manifold
D. grade of fuel being used

514. The types of electric motors commonly used in electric tachometers are:
A. direct current, series-wound motors
B. synchronous motors
C. direct current, shunt-wound motors
D. direct current, compound-wound motors

515. The hot and cold junctions in an engine cylinder temperature indicating system are located at
A. both junctions are located at the instrument
B. both junctions are located at the cylinder
C. the hot junction is located at the cylinder and the cold junction is located at the instrument
D. the cold junction is located at the cylinder and the hot junction is located at the instrument

516. The indicator of a tachometer system is responsive to change in
A. current flow B. voltage polarity
C. frequency D. voltage value

517. A thermocouple-type temperature indicating instrument system
A. is a balanced-type, variable resistor circuit
B. requires no eternal power source
C. usually contains a balancing circuit in the instrument case to prevent fluctuations of the system voltage from affecting the temperature reading
D. will not indicate a true reading if the system voltage varies beyond the range for which it is calibrated

518. In a thermocouple-type cylinder head temperature measuring system

Ans. 502. A 503. A 504. B 505. C 506. B 507. A 508. D 509. B 510. C 511. D 512. D 513. B 514. B 515. C 516. C 517. B

A. the resistance required for cylinder head temperature indicators is measured in farads
B. the voltage output of a thermocouple system is determined by the temperature difference between the two ends of the thermocouple
C. if a resistor is installed in a thermocouple lead, it is placed in the positive lead
D. when the master switch is turned on, a thermocouple indicator will move offscale to the low side

519. The basic meter used to indicate cylinder head lemperaiure in most aircrafts is
A. Iron-vane meter B. Elecirodynamometer
C. Galvanometer
D. Thermocouple-type meter

520. Which of the following is a primary engine instrument?
A. Tachometer B. Torque meter
C. Fuel flowmeter D. Airspeed indicator

521. A complete break in the line between the manifold pressure gauge and the induction system will be indicated by the gauge registering
A. prevailing atmospheric pressure
B. higher than normal for conditions prevailing
C. lower than normal for conditions prevailing
D. None of these

522. Engine oil temperature gauges indicate the temperature of the oil
A. entering the engine B. entering the oil cooler
C. in the oil storage tank
D. in the return lines to the oil storage tank

523. Helicopters require a minimum of two synchronous tachometer systems because
A. one indicates engine RPM and the other tail rotor RPM
B. one indicates main rotor RPM and the other main rotor RPM
C. one indicates engine RPM and the other main rotor RPM
D. only helicopters with turbine engines employing a dual compressor require two systems

524. The thermocouple leads are inadvertently crossed at installation. What would the cylinder temperature gauge pointer indicate?
A. normal temperature for prevailing condition
B. oscillating pointer
C. moves off-scale on the zero side of the meter
D. moves off-scale on the higher side of the meter

525. Which instrument is used to check heat sensitive elements and heat sensitive bulb resistance?
A. Wheatstone-bridge meter
B. standing wave ratio (SWR) meter
C. thermocouple-type meter
D. megohmmeter

526. Cylinder head temperatures are measured by the use of a thermocouple circuit which measures the
A. resistance in a metal gasket
B. difference in the resistance between two dissimilar metals used in the circuit between the hot and cold junctions
C. difference in the voltage between two dissimilar metal gaskets

527. Which statement is true in case of thermocouple leads?
A. These may be adjusted in length to fit any installation
B. These may be installed with either lead to either post of the indicator
C. These are designed for a specific installation and may not be altered
D. These may be repaired using solderless connectors

528. The unit in a tachometer system which sends information to the indicator is
A. The three-phase a.c. generator
B. The two-phase a.c. generator
C. The synchronous motor
D. The miniature dc motor

529. The device used to convert alternating current, which has been induced into the loops of the rotating armature of a dc generator, to direct current is:
A. An alternator B. A rectifier
C. A commutator D. An inverter

530. A direct current series motor mounted within an aircraft draws more amperes during start than when it is running under its rated load. The most logical conclusion that may be draws is
A. the starting winding is shorted
B. the brushes are floating at operating RPM because of weak brush springs
C. the condition is normal for this type of motor
D. hysteresis losses have become excessive through armature bushing (or bearing) wear

531. The stationary field strength in a direct-current generator is varied
A. by the reverse-current relay
B. because of generator speed
C. because of the number of rotating armature loops available
D. according to the load requirements

532. The type of electric motor generally used with a direct-cranking engine starter is
A. direct current, shunt-wound motor
B. direct current, series-wound motor
C. direct current, compound-wound motor
D. synchronous motor

533. The output frequency of an ac generator (alternator) depends upon

Ans. 518. B 519. C 520. A 521. A 522. A 523. C 524. C 525. A 526. B 527. C 528. A 529. C 530. C 531. D 532. B

A. the speed of rotation and the strength of the field
B. the strength of the field and the number of field poles
C. the speed of rotation, the strength of the field, and the number of field poles
D. the speed of rotation and the number of field poles

534. There is a high surge of current require when a dc electric motor is first started. As the speed of the motor increases
A. the counter emf decreases proportionally
B. the applied emf increases proportionally
C. the net counter emf increased until its value is greater than the applied emf
D. the counter emf builds up and opposes the applied emf, thus reducing the current flow through the armature

535. Alternators are often driven by a constant speed drive mechanisms to permit a nearly constant
A. voltage output B. amperage output
C. number of cycles per second
D. total power output

536. Commutators or slip rings are polished using
A. fine ernery cloth B. very fine sandpaper
C. crocus cloth or fine oilstone
D. aluminum oxide or garnet paper

537. If the generator is malfunctioning, its voltage can be reduced to residual by actuating the
A. rheostat B. master solenoid
C. overvoltage circuit breaker
D. master switch

538. The points in a vibrator-type voltage regulator stick in the closed position while the generator is operating. The probable result will be
A. generator output voltage will decrease
B. generator output voltage will not affected
C. generator output voltage will increase
D. the reverse-current cutout relay will remove the generator from the line

539. A constant-speed drive is used to control the speed of some aircraft engine-driven generators because
A. the voltage output of the generator will remain within limits
B. uncontrolled surges of current to the electrical system are eliminated
C. both voltage and amperage output can be controlled directly
D. the frequency of the alternating current output will remain constant

540. A short circuit occurs between the positive field lead and the positive armature lead of a generator with the engine operating at cruising RPM. What will be the outcome?
A. The reverse-current cutout relay will not close
B. A high generator voltage
C. Failure of the generator to produce any voltage
D. The generator will only produce residual voltage

541. Aircraft that operate more than one generator connected to a common electrical system must be provided with
A. automatic generator switches that operate to isolate any generator whose output is less than 80 percent of its share of the load
B. an automatic device that will isolate non-essential loads from the system if one of the generators fails
C. A generator switch arrangement that will prevent any one generator from being connected to the system unless the other generators are operating
D. Individual generator switches that can be operated from the cockpit during flight

542. The most effective method of regulating aircraft direct current generator output is to vary, according to the load requirements, the
A. strength of the stationary field
B. generator speed
C. effective resistance in the load circuit
D. number of rotating armature loops in use

543. Aircraft engine starter motors are generally of the type
A. compound B. series
C. differential compound
D. shunt (parallel)

544. As the generator load is increased (within its reted capacity), the voltage will
A. decrease and the amperage output will increase
B. increase arid the amperage output will increase
C. remain constant and the amperage output will increase
D. remain constant and the amperage output will decrease

545. As the flux density in the field of a dc generator increases and the current flow to the system increases.
A. A force required to turn the generator decreases
B. the generator voltage decreases
C. the generator amperage decreases
D. the force required to turn the generator increases

546. The automatic ignition relight switch is activated on a gas turbine engine by
A. a sensing switch located in the tailpipe
B. a decrease in tailpipe temperature
C. a drop in the compressor discharge pressure
D. a drop in fuel flow

Ans. 533. D 534. D 535. C 536. B 537. D 538. C 539. D 540. B 541. D 542. A 543. B 544. C 545. D 546. C

547. The rotor windings of an aircraft alternator usually are excited by
A. a constant ac voltage from the battery
B. alternating current from a permanent condenser
C. a constant ac voltage
D. a variable direct current

548. The precaution usually taken to prevent electrolyte from freezing in a lead acid battery is
A. place the aircraft in hangar
B. remove the battery and place it in a warm area
C. keep the battery fully charged
D. drain the electrolyte

549. A 140 ampere-hour battery will deliver 15 A for a period of
A. 15.0 hours B. 1.40 hours
C. 9.33 hours D. 14.0 hours

550. The basic advantage of using ac for electrical power for a large aircraft is
A. AC systems operate at higher voltage than dc systems and therefore use less current and can use smaller and lighter weight, wiring
B. AC systems operate at lower voltage than dc systems and therefore use less current and can use smaller and lighter weight wiling
C. AC systems operate at higher voltage than dc systems and therefore use more current and can use smaller and lighter weight wiring
D. AC systems operate at lower voltage than dc system is and therefore are more current and can use smaller and lighter weight wiring

551. Two types of a.c. motors that are used to produce a relatively high starting torque are
A. shaded pole and shunt field
B. shunt field and single phase
C. three-phase induction and repulsion
D. single-phase induction and rotating field

552. The frequency of most aircraft alternating current is
A. 50 Hertz B. 120 Hertz
C. 208 Hertz D. 400 Hertz

553. Which of the following aircraft circuits does *not* contain a fuse?
A. Generator circuit B. Starter circuit
C. Exterior lighting circuit
D. Airconditioning circuit

554. The maximum number of terminals that may be connected to any one terminal stud in an aircraft electrical system is:
A. Four B. One
C. two D. three

555. The maximum number of bonding jumper wires that may be attached to one terminal grounded to a flat surface is:
A. Four B. Five
C. Two D. Three

556. When installing an electrical switch, under which of the following conditions should the switch be derated from its nominal current rating?
A. Conductive circuits B. Capacitive circuits
C. Low rush-in circuits
D. Direct-current motor circuits

557. The resistance of the current return path through the aircraft is always considered negligible, provided the
A. voltage drop across the circuit is checked
B. circuit resistance is checked
C. generator is properly grounded
D. structure is adequately bonded

558. Parallel electrical wires should be twisted on installation when
A. it is desired to keep the bundle rigid
B. they are not tied in bundles
C. it is necessary to reduce the wire diameter
D. they are used in the vicinity of a mag-netic compass

559. Current flow through the coil of a solenoid-operated electrical switch is
A. Continually, as long as the aircraft's electrical system master switch is on
B. Continually, as long as the control circuit is complete
C. Only for a short time period following movement of the control switch
D. Only until the movable points contact the stationary points

560. The lubricant that may be used to aid in pulling electrical wires or cables through conduits is
A. lightweight, vegetable-base grease
B. powdered graphite
C. soapstone talc D. rubber lubricant

561. Bonding jumpers should be designed and installed in such a manner that they
A. are not subject to flexing by relative motion of airframe or engine components
B. limit the relative motion of the parts to which they are attached by acting as a secondary stop
C. provide a low electrical resistance in the ground circuit
D. prevent buildup of a static electrical charge between the airframe and the surrounding atmosphere

562. On a turbine engine, with the starter-generator circuit energised, the engine would not crank. What is the probable cause?
A. Overvoltage relay is defective
B. Throttle ignition switch is defective
C. Igniter relay is defective
D. Starter relay is defective

Ans. 547. D 548. C 549. C 550. A 551. C 552. D 553. B 554. A 555. A 556. D 557. D 558. D 559. D 560. C 561. C 562. D

563. Arcing at the brushes and the burning of the commutator of a motor valve be caused by
A. weak brush springs
B. excessive brush spring tension
C. smooth commutator D. low-mica

564. Switches used to control engine electrical circuits should be installed
A. upside down to prevent debris from shorting the terminals
B. so the toggle will move in the same direction as the desired motion of the unit controlled
C. under a guard
D. so the ON position is reached by a forward or upward position

565. When installing electrical wiring parallel to a fuel line, the wiring should be
A. in a metal conduit B. in a vinyl sleeve.
C. above the fuel line D. below the fuel line

566. The speed of an eight-pole ac generator to produce 400 Hertz a.c. will be
A. 400 RPM B. 1,200 RPM
C. 6,000 RPM D. 12,000 RPM

567. The number of basic types of circuit breakers used in powerplant installation electrical systems is:
A. One B. Two
C. Three D. Four

568. Electrical switches are rated according to the
A. voltage and the current they can control
B. resistance rating of the switch and the wiring
C. temperature and capacitance rating
D. resistance and the temperature rating

569. Electrical circuit protection devices are installed primarily to protect the
A. relays B. switches
C. units D. wiring

570. Most electrical terminal strips in an aircraft are the barrier type and are made of a
A. strong paper-base phenolic compound
B. petrolatum and zinc dust compound
C. layered aluminum impregnated with compound
D. compressed vinyl impregnated with aluminum

571. When two or more electrical terminals are installed on a single lug of a terminal strip, the process is called
A. stripping B. stepping
C. stacking D. fanning

572. Aircraft electrical wire size is measured according to the
A. Military Specification system
B. American Wire Gage system
C. Society of Automotive Engineers system
D. Military Standards system

573. The majority of the wire used in aircraft electrical systems is stranded copper wire that meets the specifications of
A. MIL-W-7042 B. MIL-W-5606
C. MIL-W-7072 D. MIL-W-5086

574. The basic unit of measure for capacitance is:
A. Henery B. Gauss
C. Ohm D. Farad

575. If the voltage applied across the plates of a capacitor (condenser) is too high, then
A. the plates will becomes saturated and not accept an electrical charge
B. the dielectric will break down and arcing will occur between the plates
C. the induced counter emf will cause the capacitor to act as a resistor in the circuit
D. it will probably not damage the capacitor because there is no physical connection between the plates; however, it may cause some secondary problems in other parts of the circuit

576. In an alternating current circuit, the effective voltage is
A. equal to the maximum instantaneous voltage
B. less than the maximum instantaneous voltage
C. greater than the maximum instantaneous voltage
D. none of these

577. The amount of electric charge a capacitor can store is directly proportional to
A. the plate area and inversely proportional to the distance between the plates
B. the distance between the plates and inversely proportional to the plate area
C. the plate area and is not affected by the distance between the plates
D. the distance between the plates and is not affected by the plate area

578. When capacitors are connected in parallel, the total capacitance is equal to
A. the sum of the reciprocals of the capacitances
B. the sum of the capacitances divided by the number of capacitors
C. the reciprocal of the sum of the reciprocals of the capacitances
D. the sum of the capacitances

579. The result of operating an engine in extremely high temperatures using a lubricant recommended by the manufacturer for a much lower temperature is:
A. the oil pressure will be higher than normal
B. the oil pressure gauge will fluctuate excessively
C. the oil temperature and oil pressure will be higher than normal
D. the oil pressure will be lower than normal

Ans. 563. D 564. D 565. C 566. C 567. C 568. A 569. D 570. A 571. C 572. B 573. D 574. D 575. B 576. B 577. A 578. D 579. D

580. The tail rotor of a helicopter could fail to compensate accurately for the torque of the main rotor if the
A. power transmission is erratic.
B. engine power is reduced.
C. roor is inaccurately rigged
D. engine power is applied suddenly to the main rotor

581. The vertical flight of a helicopter is controlled by
A. increasing or decreasing collective pitch
B. tilting the rotor disc
C. cyclic pitck changes
D. increasing or decreasing the RPM of the mam rotor

582. A decrease in pitch angle of the tail rotor blades on a helicopter
A. causes the tail to pivot in the opposite direction of torque rotation around the main rotor axis
B. is produced by depressing the left anti-torque pedal
C. causes the tail to pivot in direction of torque rotation around the main rotor axis
D. is required to counteract main rotor torque produced by takeoff RPM

583. How is tracking of helicopter rotor blades normally done?
A. On a special fixture before they are installed due to their flexibility
B. While the blades are rotating at a specified RPM and collective pitch angle
C. By triangulation while the blades are at rest and in a specified location in reference to the fuselage centerline
D. By direct measurement from the fuselage structure to specified blade station when the blade is parallel to the fuselage center-line

584. A helicopter in forward flight, cruise configuration changes direction by
A. varying the pitch of the main rotor blades
B. changing tail rotor RPM
C. tilting the main rotor disk in the desired direction
D. tilting the tail rotor

585. A hovering helicopter equipped with a tail rotor maintains directional control by
A. changing the tail rotor RPM
B. tilting the main rotor disk in the desired direction
C. a conventional rudder and rudder control system
D. varying the pitch of the tail roior blades

586. A single-rotor helicopter is in horizontal flight. The angle of attack of the advancing blade is
A. more than the retreating blade
B. equal to the retreating blade
C the same at any point around the rotor disk
D. less than the retreating blade

587. Main rotor blades that do not cone by the same amount during rotation, are said to be out of
A. balance B. lateral cyclic pitch
C. collective pitch D. track

588. The torque force is associated with single-rotor helicopter compensated as:
A. A tail rotor with a variable pitch mechanism that is atuated by pilot controls
B. A twist in the mam rotor blade chord
C. A vertical flat plate that is acted upon by the main rotor down wash
D. A double set of planetary gears in the main transmission

589. The purpose of the free-wheeling unit in a helicopter drive system is:
A. If disconnects the rotor whenever the engine stops or slows beiow the equivalent or rotor RPM
B. It releases the rotor break for starting
C. It relieves bending stress on the rotor blades during starting
D. it allows the engine to be over-revived for landing

590. Lateral dihedral, a rigging consideration on most airplanes of conventional design, contributes most to stability of the airplane about its.
A. longuitudinal axis B. vertical axis.
C. lateral axis D. transverse axis

591. As the result of an accident, it is necessary to replace the fuselage fittings for the wing attachment. Prior to welding, the fittings in place, a great deal of care should be exercised in locating them in their proper position because the
A. angle of attach will be reestablished at this time
B. angle of incidence is one of the factors to be reesablished at time
C. main spar attachment fittings must line up with fuselage fittings; otherwise, the spar location in the wing will have to be changed
D. aspect ratio of the wing will be controlled by the fuselage fittings

592. A pilot reported that an airplane flew left wing heavy. The condition is corrected by
A. washing-in the right wing, the washing-out the left wing, or both
B. washing-in the left wing, or washing-out the right wing, or both
C. washing-in the right wmg only
D. washing-out the left wmg only

593. If the vertical fin of a single-engine, propeller-driven airplane is rigged properly, it will generally be parallel to
A. the longitudinal axis but not the vertical axis

Ans. 580. C 581. A 582. C 583. B 584. C 585. D 586. D 587. D 588. A 589. A 590. A 591. B 592. B

B. the vertical axis but not the longitudinal axis
C. neither the longitudinal nor the vertical axis
D. both the longitudinal and vertical axes

594. If an airplane has good longitudinal stability, it should have a minimum tendency to
A. roll B. pitch
C. yaw D. stall

595. Airplanes of conventional design have a horizontal stabiliser located left of the wings and usually set at a different angle of incidence from the wing, this difference in angle is called
A. the plane-form ratio B. the angle of slugger
C. the angle of decalage
D. the longitudinal dihedral angle

596. If the angle of attack of an airfoil increases, the centre of pressure will
A. move toward the trailing edge
B. remain stationary because both lift and drag components increase proportionally to increased angle of attack
C. remain stationary because of no change in the incidence angle
D. move toward the leading edge

597. The angle of incidence is that acute angle formed by
A. the angular difference between the setting of the main airfoil and the auxiliary airfoil (horizontal stabiliser) in reference to the longitudinal axis of the aircraft
B. the angular difference between the wing settings of the two wings of a biplane
C. a line parallel to the wing chord and a line parallel to the longitudinal axis of the aircraft
D. a line parallel to the wing from root to tip and a line parallel to the lateral axis of the aircraft

598. An airplane's centre of lift is usually located at its centre of gravity because
A. the airplane will have a tail-heavy tendency
B. the airplane will have a nose-heavy tendency
C. it is impossible to predict the exact location of either, it is considered next best to having them both fall at the same point
D. stability about the longitudinal axis improves

599. An airplane is controlled directionally about its vertical axis by
A. the rubber B. the elevator(s)
C. the ailerons
D. a combination of two of the above.

600. The elevators of a conventional airplane are used to provide rotation about the
A. longitudinal axis B. lateral axis
C. directional axis D. vertical axis

601. Flaps increase the effective lift of an airfoil by
A. increasing the camber of the airfoil
B. introducing drag aft of the center of pressure
C. reducing the profile drag
D. increasing the angle of attack of the airfoil

602. When the right wing of a monoplane is improperly rigged to a greater angle of incidence than designated in the manufacturer's specifications, it will result in
A. the airplane to be off balance both laterally and directionally
B. the airplane to pitch and roll about the lateral axis
C. the airplane to be out of lateral balance only
D. the right wing to have both an increased lift and a decreased drag

603. The chord of a wing is measured from
A. wingtip to wingtip
B. wing attachment point to the wingtip
C. leading edge to trailing edge
D. maximum upper camber to the base line

604. The physical factors which are involved in the aspect ratio of airplane wings are:
A. Thickness and chord B. Span and chord
C. Dihedral and angle to attack
D. Sweepback and lateral axis

605. An airplane that has a tendency to gradually increase a pitching moment that has been set into motion has
A. poor longitudinal stability
B. good lateral stability
C. poor lateral stability
D. good longitudinal stability

606. Differential control on an aileron system implies that
A. the down travel is more than the up travel
B. the up travel is more than the down travel
C. they travel up or down equally, but bell cranks are used
D. one aileron on one wing travels further up than the aileron on the opposite wing to adjust for wash-in and wash-out

607. Generally it is necessary to jack an aircraft indoors for weighing because
A. aircraft may be placed in a level position
B. aircraft empty weight can be determined
C. the weighing scales are stabilised
D. None of these

608. Before jacking an aircraft, it is necessary to
A. remove all optional equipment
B. install critical stress panels or plates
C. determine that the fuel tanks are, empty
D. make sure the aircraft is leveled laterally

609. An aircraft can be operated with a 100-hour inspection overdue when
A. it is necessary to reach a place at which the inspection can be done

Ans. 593. B 594. B 595. D 596. D 597. C 598. B 599. A 600. B 601. A 602. A 603. C 604. B 605. A 606. B 607. C 608. B 609. A

B. the aircraft is operated under a continuous airworthiness program
C. When the aircraft is operated under a progressive inspection program
D. the aircraft is equipped with an hour meter

610. The maximum time a 100-hour inspection may be extended to
A. 10 hr B. 2 hr
C. 5 hr D. 15 hr

611. Which statement is correct when an aircraft has not been approved for return to service after an annual inspection because of several items requiring minor repair?
A. Only the person who performed the annual inspection may approve the aircraft for return o service
B. An appropriately rated mechanic may repair the defects and approve the aircraft for return to service
C. An appropriately rated mechanic may repair the defects, but an IA must approve the aircraft for return to service
D. An authorized repair station may repair the defects, but an appropriately rated mechanic must approve the aircraft for return to service

612. An aircraft that is due an annual inspection may be operated
A. by a certificated pilot, but no passengers may be carried
B. if a special permit has been issued for the aircraft
C. for the purpose of performing maintenance
D. for a period of time not to exceed 10 hr

613. A 100-hour inspection of an aircraft is authorised to be conducted by:
A. Any machanic working for a repair station
B. Certificated mechanic with airframe and power plant ratings
C. Certificated repairman
D. Certificated mechanic with power plant rating

614. If the piston return spring broke in a brake master cylinder, the effect would be:
A. the brakes would become spongy
B. the brake travel would become excessive
C. the brakes would drag
D. the brake linkage would seize

615. Spongy brakes are usually due to
A. air in the system
B. both internal and external leakage
C. internal leakage D. external leakage

616. A pilot reports that the right brake on an aircraft grabs when the brake pedal is depressed in a normal manner. The probale cause is that the right
A. master cylinder piston rod is bent
B. wheel actuating cylinder is leaking fluid on the brake lining
C. master cylinder piston return spring is weak
D. brake drum contains considerable brake lining dust

617. Most aircraft tire manufacturers recommend that the tubes in newly installed tries be first inflated, fully deflated, and then reinflated to the correct pressure. Why?
A. To allow the tube to position itself correctly inside the tire
B. To eliminate all the air between the tube and the inside of the tire
C. To ensure that the wheel rim has been properly assembled
D. To test the entire assembly for leaks

618. The function of metering pins in (air/oil) shock struts serve is to
A. lock the struts in the up position
B. lock the struts in the down position
C. retard the how of oil as the struts are compressed
D. meter the proper amount of air in the struts

619. Tire and wheel manufacturers often recommend that the tires on split rim wheels be deflated before removing the wheel from the axle. Why?
A. To relieve the strain on the wheel retaining nut and axle threads
B. As a safety precaution in case the bolts that hold the wheel halves together have been damaged or weakened
C. So that the person removing the wheel will be sure to remember to jack the wheel
D. To remove the static load imposed upon the wheel bearings by the inflated tire

620. The braking action of single-disc breaks is accomplished by compressing a rotating brake disc between two opposite brake lining. Equal pressure on both sides of the rotating disc is assured by
A. hydraulically interconnecting the brake linings located on opposite sides of the rotating disc
B. keeping the brake clearances closely adjusted
C. keying the disc loosely in the wheel so that it can move from side to side
D. letting the brake linings automatically equalise because of the greater rate of wear on the lining with the most pressure

621. It is determined that spongy brake action is not caused by air in the brake system. The next most likely cause is
A. worn brake lining
B. internal leakage in the master cylinder
C. deteriorated flexible hoses
D. unevenly adjusted brakes

622. A stripe or mark extending from the rim of a wheel into the tire

Ans. 610. A 611. B 612. A 613. B 614. C 615. A 616. B 617. A 618. C 619. B 620. C 621. C

A. is a slippage mark
B. is a balance mark
C. indicates the tire is a high-pressure type
D. indicates the wheel is a high-speed type

623. All air has been purged from the brake system. This can be determined by
A. operating a hydraulic unit and watching the system pressure gauge for smooth, full-scale deflection
B. the increased pedal movement for brake application
C. submerging the bleeder hose in hydraulic fluid and watching until all air bubbles have escaped
D. observing the reservoir sight gauge to determine if there is any fluid movement

624. Over-inflated aircraft tires may cause damage to the
A. brake linings B. wheel hub
C. brake drum D. wheel flange

625. The function of debooster valves used in brake systems is:
A. To ensure rapid application and release of the brakes
B. To reduce brake pressure and maintain static pressure
C. To reduce the pressure and release the brakes rapidly
D. To increase pressure and release the brakes rapidly

626. If an out-of-tolerance to-in condition of main landing gear wheels exists and it is determined not to be the result of bent or twisted components, it would be necessary to
A. reposition the axle in the oleo trunnion
B. change the oleo assembly
C. insert, remove, or change location of washers or spacers at the center pivotal point of the scissor torque links
D. place shims or spacers behind the bearing of the out of tolerance wheel or wheels

627. To prevent rapid deterioration, aircraft tires should be stored in an area that is
A. dry and warm B. damp and cool
C. dry and cool D. damp and warm

628. A raised H on the stem of an air valve core denotes that the valve core is a
A. hydraulic type B. hard-core type
C. high-pressure type D. heavy-duty type

629. The primary purpose for balancing aircraft wheel assemblies is:
A. To prevent heavy spots and reduce vibration
B. To distribute the aircraft weight properly
C. To reduce excessive wear and turbulence
D. To improve the braking action

630. On all aircraft equipped with retractable landing gear, some means must be provided to
A. retract and extend the landing gear if the normal operating mechanism fails
B. extend the landing gear if the normal operating mechanism fails
C. prevent extension of the landing gear at airspeeds greater than that determined structurally safe
D. prevent the throttle from being reduced below a safe power setting while the landing gear is retracted

631. The type of hydraulic fluid to be used in a shock strut depends upon the
A. ambient temperature at the time of servicing
B. amount of air pressure in the shock strut
C. type of seals used in the shock strut
D. type of fluid in the landing gear actuating systems

632. An automatic damping action occurs at the steer damper if for any reason the flow of high-pressure fluid is removed from the
A. outlet of the steer damper
B. inlet of the steer damper
C. replenishing check valve
D. tension steering valve

633. The purpose of a brake master cylinder is:
A. To constantly maintain the correct volume of fluid in the hydraulic lines and brake components
B. To transform mechanical motion to hydraulic pressure
C. To provide a means for building up, varying, and releasing the hydraulic pressure required for brake operation
D. To per form all of the above functions

634. Aircraft brakes requiring a large volume of fluid to operate the brakes generally
A. do not use brake pressure relief valves
B. use independent master cylinder systems
C. do not use brake system accumulators
D. use power brake control valves

635. The purpose of a sequence valve in a hydraulic retractable landing gear system is to ensure the operation at the proper time of the
A. main gear safety switches
B. landing gear doors
C. nose gear safety switches
D. main gear downlocks

636. The pressure source for power brakes is
A. the main hydraulic system
B. the power brake reservoir
C. a master cylinder
D. pressure being applied to the rudder pedals

Ans. 622. A 623. C 624. D 625. C 626. C 627. C 628. C 629. A 630. B 631. C 632. B 633. D 634. D 635. B 636. A

637. The type of valve used in the brake actuating line to isolate the emergency brake system from the normal power brake control valve system is a/an
A. bypass valve B. orifice check valve
C. brake pressure relief valve
D. shuttle valve

638. When tightening the packing nut on an air/oil shock strut installed on an aircraft, the
A. packing gland should be replaced
B. aircraft should be jacked
C. fluid should be removed from the strut
D. Torque should not exceed 800 ft lb

639. The purpose of a relief valve in a brake system is to:
A. compensate for a drop in pressure
B. reduce pressure for brake application
C. prevent the tire from skidding
D. compensate for thermal expansion

640. An aircraft equipped with multiple-disc brakes was reported to have sluggish and jerky brake operation. The action should taken is
A. reduce the brake clearance
B. disassemble and clean the automatic adjuster
C. replace retractor springs
D. bleed the brakes

641. If an aircraft shock strut (air/oil type) bottoms upon initial landing contact, but functions correctly during taxi, the most probable cause is
A. low fluid B. low air charge
C. a restricted material pin orifice
D. reversed metering pin

642. If an airplane equipped with master cylinders and single-disc brakes has excessive brake pedal travel, but the brakes are hard and effective, the probable cause is
A. the master cylinder one-way cup is leaking
B. worn brake linings
C. the pusfy rod between the brake pedal and the master cylinder is too long
D. worn brake disc causing excessive clearance between the notches on the perimeter of the disc and the splines or keys on the wheel

643. The correct inflation pressure for an aircraft tire can be obtained from
A. tire pressure charts based on gross landing weight
B. the information stamped on the wheel
C. the aircraft's logbook D. the operator's manual

644. What should be checked when shock struts bottom during a landing?
A. Air pressure in the struts
B. Struts for correct alignment
C. Packing seals for binding
D. Fluid level

645. Failure of the landing gear to retract completely may be caused by
A. a leaking accumulator air valve
B. the wrong grade of fluid
C. insufficient fluid in the system
D. a leak in the return line

646. The left brake is dragging excessively on an airplane on which no recent brake service work has been performed. The most probable cause is
A. foreign particles stuck in the master cylinder compensating port
B. excessively worn brake linings
C. too much play in the linkage interconnecting the pedals and the master cylinder
D. low fluid supply in the brake system reservoir

647. A tire has been subjected to a temperature high enough to melt one of the fusible plugs in the wheel. The tire should be
A. be scrapped
B. removed from the wheel and inspected before returning it to service
C. recapped before returning it to service
D. returned to service if an external inspection reveals no carcass or tread damage

648. To avoid nylon tire spotting on an aircraft that is to remain idle for a period longer than 3 days, the aircraft
A. should be moved every 60 hr
B. should be moved every 36 hr
C. should be blocked up so that no weight is on the tires or moved every 48 hr
D. none of these

649. The best safeguards against heat buildup in aircraft tires are
A. proper tire inflation, minimum braking, and fast taxi speeds
B. short ground rolls, slow taxi speeds, minimum braking and proper tire inflation
C. minimum braking, proper tire inflation, and long ground rolls
D. fast taxi speeds, short ground rolls, and coordinated turns

650. The fusible plugs installed in some aircraft wheels will
A. facilitate servicing of the wheel assembly
B. indicate tire tread separation
C. eliminate the need to check air pressure
D. melt at a specified elevated temperature

651. Excessive wear in the shoulder area of an aircraft tire is an indication of
A. over-inflation B. excessive toe-in
C. under-inflation
D. fast cornering during taxi

Ans. 637. D 638. B 639. D 640. B 641. A 642. B 643. D 644. D 645. C 646. A 647. A 648. C 649. B 650. D 651. C

652. Excessive wear in the center of the tread of an aircraft tire is an indication of
A. under-inflation B. excessive toe-out
C. excessive braking during landing roll and taxi
D. over-inflation

653. When an empty shock strut is filled, care should be taken to extend and compress the strut completely at least two times to
A. thoroughly lubricate the piston rod
B. show if there are any fluid leaks
C. force out any excess fluid
D. ensure proper packing ring seating and removal of air bubbles

654. In shock struts, chevron seals are used to
A. prevent air from escaping
B. absorb bottoming effect
C. prevent oil from escaping
D. serve as a bearing surface

655. On most aircraft, the old level of an air and oil shock strut is checked by
A. removing the oil filler plug and inserting a gage
B. measuring the overall length of the strut
C. releasing the air and seeing that the oil is to the level of the filler plug
D. checking the air pressure

656. An electric motor used to raise and lower a landing gear is a
A. shunt field series wound motor
B. split field shunt wound motor
C. continuous duty motor
D. split field series wound motor

657. When installing a chevron-type seal in an aircraft hvdraulic cylinder, the open side of the seal should face
A. opposite the direction of fluid pressure
B. up or forward when the unit is installed in a horizontal position
C. the direction of fluid pressure
D. down or aft when the unit is installed in a vertical position

658. The device which in a hydraulic system with a constant delivery pump allows circulation of the fluid when no demands are on the system is
A. Pressure relief valve B. Shuttle valve
C. Pressure regulator D. Debooster valve

659. A fully-charged hydraulic accumulator provides.
A. air pressure to the various hydraulic components
B. a source for additional hydraulic power when heavy demands are placed on the system
C. positive fluid flow to the pump inlet
D. additional hydraulic fluid

660. The condition which would most likely cause excessive fluctuation of the pressure guage when the hydraulic pump is operating is
A. Bourdon tube in the guage is broken
B. Accumulator air pressure low
C. Inadequate supply of fluid
D. System relief valve sticking closed

661. A filter incorporating specially treated cellulose paper is
A. sediment trap B. finger strainer
C. cuno filter D. micronic filter

662. The purpose of an orifice check valve is to
A. relieve pressure to a sensitive component
B. restrict flow in one direction
C. relieve pressure in one direction and prevent ilow in the other direction
D. relieve pressure in locked mechanisms

663. To prevent external and internal leakage in aircraft hydraulic units, the most commonly used type of seal is the
A. cup seal B. O-ring seal
C. rubber seal D. chevron seal

664. The component which allows free fluid flow in one direction and no fluid flow in the other direction is:
A. check valve B. selector valve
C. metering piston D. master cylinder

665. The valve used in a hydraulic system that directs pressurized fluid to one end of an actuating cylinder and simultaneously directs return fluid to the reservoir from the other end is:
A. sequence B. shuttle
C. check D. selector

666. The function of the absolute pressure regulator perform in the pneumatic power system is to regulate
A. the compressor outlet air pressure to stabilize the system pressure
B. the pneumatic system pressure to protect the moisture separator from internal explosion
C. the compressor inlet air to prevent excessive speed variation and/or over-speeding of the compressor
D. the compressor inlet air to provide a stabilized source of air for the compressor

667. One of the distinguishing characteristics of an open center selector valve used in hydraulic system is that
A. this type valve has only three ports
B. fluid flows through the valve in the OFF position
C. fluid flows in three directions in the ON position
D. a limited amount of fluid flows in one direction and no fluid flows in the opposite direction

668. The type of packings which should be used in hydraulic components to be installed in a system containing skydrol is

Ans. 652. D 653. D 654. C 655. C 656. D 657. C 658. C 659. B 660. C 661. D 662. B 663. B 664. A 665. D 666. D 667. B

A. AN packings made of natural rubber
B. AN packings made of either natural rubber or neoprene
C. packing materials made for ester-base fluids
D. AN packings made of neoprene

669. An aircraft pneumatic system, which incorporates an engine-driven multistage reciprocating compressor, also requires
A. an oil separator B. a surge chamber
C. a vacuum relief valve D. a moisture separator

670. The removal of air from an aircraft hydraulic system generally presents no problem because
A. the air will bleed out if the system is allowed to remain inoperative for a period of time
B. the air will be removed by operating the various hydraulic components through several cycles
C. the high working pressure in the system will physically unite the air and fluid which will result in the air being dissipated in the form of heat
D. each hydraulic actuator has a bleeder valve located on or adjacent to it for this purpose

671. Pneumatic systems employ
A. hand pumps B. relief valves
C. freon D. diluter valves

672. The component in the hydraulic system that is used to direct the flow of fluid is the
A. check valve B. orifice check valve
C. relief valve D. selector valve

673. The type of selector valve which is one of the most commonly used in hydraulic systems to provide for simultaneous flow of fluid into and out of a connected actuating unit is:
A. three-port, three-way valve
B. four-port, closed center valve
C. three-port, four-way valve
D. two-port, open center valve

674. A flexible sealing element subject to motion is a
A. washer B. compound
C. packing D. gasket

675. An advantage of hydraulic systems is that they
A. requires maximum effort to inspect
B. complicated to install
C. maximum maintenance requirements
D. Minimum maintenance requirements

676. The internal resistance of a fluid which ends to prevent it from flowing is called
A. volatility B. flashpoint
C. viscosity D. acidity

677. The viscosity of hydraulic fluid is defined as the:
A. increase in volume of a fluid due to temperature change
B. temperature at which a fluid gives off vapour insufficient quantity to ignite momentarily
C. fluid's ability to resist oxidation and deterioration for long periods
D. The internal resistance of a fluid which tends to prevent it from flowing

678. The colour with which Petroleum base hydraulic fluid is dyed is
A. purple B. blue
C. green D. red

679. The properties and characteristics of petroleum base hydraulic fluid are improved by
A. filters
B. lower operating temperatures
C. chemical analysis D. additives

680. A characteristic of synthetic base hydraulic fluid is
A. low moisture retention
B. high viscosity
C. high flash point D. low flash point

681. Which of the following statements about fluids is correct?
A. Any fluid will completely fill its container
B. All fluids are considered to be highly compressible
C. All fluids readily transmit pressure
D. Only liquids are considered to be fluids

682. What are three types of hydraulic fluids currently being used in civil aircraft?
A. Mineral base, mixed mineral and vegetable base, and vegetable base
B. Mineral base, vegetable base, and phosphate ester base
C. Mineral base, phosphate ester base and mixed mineral and phosphate ester base
D. Mineral base, phosphate ester base, the alcohol base

683. An aircraft's constant pressure hydraulic system cycles more frequently than usual and no unusual fluid leakage can be detected. The most probable cause is
A. excessively high relief valve setting
B. moisture absorbed in the fluid
C. low accumulator air preload
D. an obstruction in the reservoir vent line

684. The valve installed in a hydraulic system which will have the highest pressure setting is:
A. Pressure regulator valve
B. Main relief valve
C. Thermal relief valve
D. Pump unloading valve

685. If two actuating cylinders which have the same cross-sectional area but different lengths of stroke are connected to the same source of hydraulic pressure, they will exert

Ans. 668. C 669. D 670. A 671. B 672. D 673. B 674. ? 675. D 676. C 677. D 678. D 679. D 680. C
681. C 682. B 683. C 684. B

A. different amounts of force but will move at the same rate of speed
B. equal amounts of force but will move at different rates of speed
C. equal amounts of force and will move at the same rate of speed
D. different amounts of force and will move at different rates of speed

686. The purpose of a hydraulic pressure regulator is to:
A. prevent the system pressure from rising above a predetermined amount due to thermal expansion
B. direct the entire pump output to the essential units in case of emergency
C. boost the pressure in the portions of the system that require higher pressure
D. relieve the pump of its load when no actuating units are being operated

687. Severe kickback of the emergency hydraulic hand pump handle during the normal intake stroke will indicate the
A. hand pump inport check valve is sticking open
B. main system relief valve is set too high
C. accumulator still contains its normal air charge and the emergency hand pump cannot override it
D. hand pump outport check valve is sticking open

688. Before removing the filler cap of a pressurized hydraulic reservoir
A. relieve the hydraulic system pressure
B. actuate several components in the system
C. relieve the air pressure
D. disconnect all electrical power

689. The primary function of the hydraulic system accumulator is to
A. act as an emergency supply of fluid for the hand pump
B. ensure a continuous supply of fluid to the engine driven pump
C. relieve the pump of the constant load when no units in the system are operating
D. supplement the pump when it is under peak load

690. Chattering of the hydraulic pump during engine runup is an indication that
A. the pressure gauge snubber is inoperative
B. the main system relief value is sticking open
C. the ball-check valve has not been installed at the pump outlet
D. there is an air leak in the pump inlet line

691. A hydraulic motor converts fluid pressure to
A. linear motion B. rotary motion
C. angular motion D. vertical motion

692. The main purpose of a pressurised reservoir in a hydraulic system is to present
A. tank collapse
B. hydraulic pump cavitation
C. hydraulic fluid from foaming
D. normal accumulator pressure from saving

693. An emergency supply of fluid is often retained in the main hydraulic system reservoir by the use of a standpipe located in the
A. reservoir filler neck extension
B. inlet from the main hydraulic system
C. outlet to the emergency pump
D. outlet to the main system pump

694. The phosphate ester base hydraulic fluid is removed from aircraft tires by washing with
A. soap and water B. petroleum solvent
C. acetone D. alcohol

695. The material used to remove petroleum base hydraulic fluid from aircraft tires is
A. acetone B. soap and water
C. petroleum solvent D. alcohol

696. Hydraulic system thermal relief valves are set to open at
A. a lower pressure than the system relief valve
B. a higher pressure than the system relief valve
C. a lower pressure than the system pressure regulator
D. the same pressure required to open the system relief valve

697. Chatter in a hydraulic system is caused by
A. excessive system pressure
B. insufficient system pressure
C. air in the system
D. improper adjustment of the pressure regulator

698. If the hydraulic system pressure is normal while the engine-driven pump is running, but there is no pressure after the engine has been shut off, it indicates
A. the system relief valve setting is too high
B. no air pressure in the accumulator
C. the pressure regulator is set too high
D. the pressure of air in the system

699. The section of a turbojet engine which provides air for the pressurisation and airconditioning systems of a jet aircraft is
A. exhaust B. compressor
C. combustion D. intake

700. The component of an air-cycle cooling system, which undergoes a pressure and temperature drop of air during operation is
A. water separator
B. expansion turbine
C. primary heat exchanger
D. refrigeration bypass valve

Ans. 685. C 686. D 687. D 688. C 689. D 690. D 691. B 692. B 693. D 694. A 695. B 696. B 697. C 698. B 699. B 700. B

701. In a freon vapour-cycle cooling system, coolant air for the condenser is obtained from
A. turbine engine compressor
B. ambient air
C. subcooler air
D. pressurised cabin air

702. Ventilating air is used on a combustion heater because it
A. provides combustion air for ground blower
B. carries heat to the places where needed
C. keeps overheat thermoswitch cool
D. provides oxygen required to support the flame

703. Turbine engine air used for airconditioning and pressurisation is generally called
A. compressed air
B. ram air
C. conditioned air
D. bleed air

704. When the cabin pressure regulator is operating in the differential mode, reference pressure is vented to the atmosphere by the
A. dump valve
B. relief valve
C. isobaric metering valve
D. differential metering valve

705. In the combustion air system, what prevents too much air from entering the heaters as air pressure increases?
A. Either a combustion air relief valve or a differential pressure regulator
B. Only a differential pressure regulator can be used
C. Only a combustion air relief valve can be used
D. Both a combustion air relief valve and a differential pressure regulator in every system

706. When the cabin pressure regulator is operating in the isobaric range, cabin pressure is maintained constant by
A. the movement of the regulator bellows
B. limiting the flow of air to the cockpit
C. the action of the cabin pressure safety valve
D. limiting the air from the reference chamber

707. The parameter which controls the operation of the cabin pressure regulator is:
A. cabin air pressure
B. ram air pressure
C. bleed air pressure
D. compression air pressure

708. The basic air-cycle cooling system consists of
A. a source of compressed air, heat exchangers, and a turbine
B. heaters, coolers, and compressors
C. ram air source, compressors, and engine bleeds
D. heat exchangers and evaporators

709. The purpose of the dump valve in a pressurized aircraft is to relieve
A. all positive pressure from the cabin
B. a negative pressure differentia
C. the load on the compressors
D. pressure in excess of the maximum differential

710. The component in a vapour cycle aircooling system which would most likely be at fault if a system would not take a freon charge is
A. evaporator
B. expansion valve
C. condenser
D. receiver-dryer

711. Frost or ice buildup on a vapour-cycle, air-cooling system evaporator would most likely be caused by
A. too much freon in the system
B. moisture in the evaporator
C. inadequate airflow through the evaporator
D. high humidity

712. The test used to determine the serviceability of an oxygen cylinder is pressure test with
A. high pressure nitrogen
B. compressed air
C. oxygen
D. water

713. The type of oxygen system which uses the rebreather bag principle is
A. pressure demand
B. diluter demand
C. continuous flow
D. demand

714. The cabin pressure of a pressurised aircraft is usually controlled by a/an
A. valve that stops the pressurisation pump when a pressure equivalent to the maximum safe cabin altitude has been reached
B. pressure-sensitive switch that causes the pressurisation pump to turn on or off as required
C. automatic outflow valve that dumps all the pressure in excess of the amount for which it is set
D. pressure-sensitive valve that controls/the output pressure of the pressurisation pump

715. A good practice concerning the inspection of heating and exhaust systems of aircraft utilizing a jacket around the engine exhaust as a heat source is
A. all exhaust and heating system components should be replaced at each engine overhaul period
B. supplement physical inspections with periodic operational carbon monoxide detection tests
C. all exhaust system components should be removed periodically, and their condition determined by the magnetic-particle inspection method
D. All exhaust system components should be removed and replaced at each 100 hour inspection period

716. What will result if auxiliary (ambient) ventila-tion is selected during pressurised flight while at cruising altitude?
A. An increase in cabin pressure
B. Cabin compressor overspeed
C. Increased cabin altitude
D. Increased conditioned air efficiency

Ans. 701. B 702. B 703. D 704. D 705. A 706. D 707. A 708. A 709. B 710. B 711. C 712. D 713. C 714. C 715. B 716. C

717. The cabin pressure control setting has a direct influence upon the
A. outflow valve opening
B. cabin supercharger compression ratio
C. pneumatic system pressure
D. turbocompressor speed

718. The function of the evaporator in a freon cooling system is to
A. liquefy freon in the line between the compressor and the condenser
B. lower the temperature of the cabin air
C. transfer heat from the freon gas to ambient air
D. evaporate water and other impurities from the liquid freon

719. The purpose of a mixing valve in an airconditioning system is to
A. dehumidify cabin air by mixing it with dry air
B. control the supply of hot, cool, and cold air
C. distribute conditioned air evenly to all parts of the cabin
D. combine emergency ram air with conditioned air

720. The component of a pressurisation system which prevents the cabin altitude from becoming higher than airplane altitude is:
A. cabin rate of descent control
B. negative pressure relief valve
C. supercharger overspeed valve
D. compression ratio limit switch

721. If the cabin rate of climb is too great, the controls should be adjusted to cause the
A. outflow valve to close slower
B. cabin compressor speed to increase
C. outflow valve to close faster
D. cabin compressor speed to decrease

722. The position of the thermostatic expansion valve in a vapour-cycle cooling system is determined by temperature and pressure of the
A. freon entering the evaporator
B. air in the outlet of the condenser
C. air in outlet of the thermostatic expansion valve
D. freon in the outlet of the evaporator

723. A loss of pressurisation through a disengaged cabin air compressor is prevented by
A. firewall shutoff valve
B. supercharger disconnect mechanism
C. cabin pressure outflow valve
D. deliery air duct theck valve

724. When servicing an airconditioning system that has lost all of its freon charge, it is necessary to
A. evacuate the system, purge it of air and moisture, evacuate it again, and add the oil and freon
B. evacuate the system, purge it of air and moisture, and add oil and freon
C. purge the system of air and moisture, and add oil and freon
D. add oil and freon only, since attempting to evacuate or purge the system will add to the danger of contamination

725. The function of the pressurisation system outflow valve is to provide for
A. air outflow only
B. air outflow, pressure relief, and cabin pressure warning
C. manual dumping of cabin pressure
D. air outflow, pressure relief, and vaccum felief

726. One prupose of a jet pump in a pressurisation and airconditioning system is to:
A. produce a high pressure for operation of the outflow valve
B. circulate fluids through the fluids through the freon system
C. conctol the speed of the engine-driven cabin compressor
D. control the valume of the exhaust air in the radio rack cooling duct

727. The operation of an aircraft combustion heater is usually controlled by a thermostat circuit which
A. alternately turns the fuel on and off, a process known as cycling
B. meters the amount of fuel continuously entering the heater and therefore regulates the heater's BTU output
C. regulates the voltage applied to the heater's ignition transformer
D. operates a duct damper and therefore allows only the required part of the heater's output to circulate in the cabin ducts; the remainder is dumped over-board

728. The air-cycle cooling system produces cold air by
A. routing conditioned air through the cooling fan
B. passing heated air through a compressor
C. passing air through cooling coils that contain a refrigerant
D. exracting heat energy across an expansion turbine

729. Maximum tapper contact between crank-shaft and propeller hub is determined by using a
A. telescoping gauge
B. bearing blue colour transfer
C. micrometer
D. surface guage

730. An airplane is cruising at 8,000 ft., and the cabin altitude is stabilised at 3,000 ft. using manual controls after failure of the automatic controls. If the airplane climbs 500 ft and the manual control setting is not changed, the cabin altitude will

Ans. 717. A 718. B 719. B 720. B 721. C 722. D 723. D 724. B 725. A 726. D 727. A 728. D 729. B

A. remain at 3,000 ft B. climb to 7,500 ft
C. descend to 2,500 ft D. climb to 3,500 ft

731. When checking a freon system, a steady stream of bubbles in the sight gauge indicates
A. here is a little too much charge
B. the charge is extremely high
C. the proper amount of air is in the system
D. the charge is low

732. Which of the following articles of safety gear must be worn when working with Freon 12?
A. Particle mask B. Earplugs
C. Rubber-soled safety shoes
D. Goggles

733. An aircraft fuselage is subjected to five major stresses. Pressurisation would be classified as a
A. tension stress B. compression stress
C. torsion stress D. shear stress

734. The purpose of evacuating freon airconditioner system is to
A. assure that all check valves are sealed
B. remove all stale worn-out liquid
C. check for leaks D. remove moisture

735. The operating cabin pressurisation ranges of a pressurised aircraft are
A. isobaric, differential, and maximum differential
B. differential, unpressurised, and isobaric
C. ambient, unpressurised, and isobaric
D. unpressurised, differential, an ambient

736. A pressurisation controller uses which of the following?
A. Bleed air pressure, outside air tempe-rature, and cabin rate of climb
B. Barometric pressure, cabin altitude, and cain rate of change
C. Cabin rate of claimb, bleed air volume, and cabin pressure
D. Outside air temperature, cabin rate of climb, and cabin pressure

737. What is the condition of the refrigerant as it enters the condenser of a vapour-cycle air-cooling system?
A. High-pressure liquid B. Low-pressure liquid
C. High-pressure vapour
D. Low-pressure vapour

738. What unit in a vapour-cycle air-cooling system serves as a reservoir for the refrigerant?
A. Receiver-dryer B. Evaporator
C. Compressor D. Condenser

739. What is the condition of the refrigerant as it enters the evaporator of a vapour-cycle air-cooling system?
A. High-pressure liquid
B. Low-pressure vapour
C. low-pressure liquid
D. High-pressure vapour

740. What drives the airconditioning compressor on most small aircraft?
A. An electric motor B. Ram air
C. The aircraft engine through a belt drive
D. A hydraulic motor

741. What is the condition of the refrigerant as it j leaves the evaporator of a vapour-cycle air-cooling system?
A. High-pressure liquid
B. Low-pressure liquid
C. Low-pressure vapour
D. High-pressure vapour

742. What is the condition of the refrigerant as itl leaves thj condenser of a vapour-cycle air-cooling system?
A. Low-pressure liquid B. High-pressure liquid
C. High-pressure vapour
D. Low-pressure vapour

743. When purging a freon airconditioning system it is important to release the charge at a slow rate. What is the reason or the slow-rate discharge?
A. Prevent the large amount offeron from contami-nating the surrounding atmos-phere
B. Prevent excessive loss of refrigerant oil
C. Prevent excessive chattering of the expansion valve
D. Prevent condensation from forming and contaminating the system

744. What is used in some oxygen systems to change high cylinder pressure to low system pressure?
A. Pressure reducer valve
B. Pressure relief valve
C. Calibrated fixed orifice
D. Diluter-demand regulator

745. If the pressure reducer fails, what prevents high-pressure oxygen from entering the system down-steam?
A. Check valve B. Cylinder control valve
C. Pressure relief valve
D. Manifold control valve

746. Oxygen systems is unpressurised aircraft are generally of
A. the continuous-flow and pressure-demand types
B. the continuous-flow type only
C. the pressure-demand type only
D. the portable-bottle type only

747. A contaminated oxygen system is normally purged with
A. castile soap and water
B. oxygen
C. compressed air D. nitrogen

748. How do you determine the amount of oxygen in a portable, high-pressure cylinder?

Ans. 730. D 731. D 732. D 733. A 734. D 735. B 736. B 737. C 738. A 739. C 740. C 741. C 742. A 743. B 744. A 745. C 746. A 747. B

A. Weigh the cylinder and its contents
B. Read the pressure guage mounted on the cylinder
C. Measure the pressure at the mask
D. Check the flow indicator during use

749. The maximum takeoff weight for a transport category aircraft is 105% of the maxi-mum landing weight. The system required is
A. fuel jettison B. fuel injection
C. crossfeed bypass D. fuel transfer

750. How is fuel jettisoning usually accomplished?
A. Through a common mainfold and outlet in each wing
B. By gravity flow into the outboard wing tanks and over board through a common outlet in each wing
C. By pump pressure into the crossfeed manifold and overboard through the vent lines
D. Through individual outlets for each tank

751. One of the requirements of fuel jettisoning systems is the
A. system and its operation must be free from fire hazards
B. time required to jettison fuel must not exceed 10 minutes
C. fuel must be discharged from outlets at the tail of the aircraft
D. fuel must be discharged from outlets at the wingtips

752. A fuel jettisoning system comprises
A. filters, switched, valves, dump chute, and chute operating mechanisms
B. lines, valves, dump chutes, and chute operating mechanisms
C. tanks, filters, valves, lines, dump chutes, and chute operating mechanisms
D. flowmeters, filters, valves, lines, dump chutes, and chute operating mechanisms

753. The system employed to maintain lateral stability when jettisoning fuel is
A. two separate independent systems
B. two interconnected systems
C. crossfeed system D. equaliser system

754. A fuel jettison system is required if the maximum takeoff weight exceeds the maximum landing weight on
A. transport category aircraft only
B. aircraft that weight over 12,500 IB only
C. both transport category and general aviation aircraft
D. general aviation aircraft only

755. The procedure which must be followed when defueling aircraft with sweptback wings is defuel
A. all the tanks at one time
B. the fuselage tanks last
C. the inboard wing tanks first
D. the outboard wing tanks first

756. Oil leakage around the rear cone of a hydromatic propeller usually indicates a defective
A. piston gasket B. blade-barrel packing
C. spider-shaft oil seal D. dome-barrel oil seal

757. The corssfeed systems are used in multi-engine aircraft fuel systems to
A. permit dumping of excess fuel.
B. reduce number of fuel lines.
C. maintain a balanced fuel conditions
D. reduce refueling time.

758. The purpose of the crossfeed system is to
A. allow the feeding of any engine from any tank
B. allow removal of residual fuel
C. allow the feeding of fuel from tank for defueling
D. provide automatic refueling of a tank to any desired level

759. Fuel system components must be bonded and grounded so as to
A. drain off static charges
B. prevent stray currents
C. identify the components
D. stabilise the units

760. One precaution that must be observed during fueling operations is
A. all outside electrical sources must be disconnected from the aircraft.
B. fuel to be used must be appropriately identified
C. all electrical switches must be in OFF position
D. fuel must be filtered through a chamois

761. Before fueling an aircraft by using the pressure fueling method, an important precaution that should be observed is the:
A. truck pump pressure must be correct for that refueling system
B. truck pump pressure must be adjusted for minimum filter pressure
C. hose must be connected before grounding
D. aircraft's electrical system rnust be on to indicate quantity gauge readings

762. The flight safety related advantage of a pressure fueling system is that it
A. eliminates aircraft skin damage
B. increase the chances of static electricity ligniting fuel vapours
C. reduces the chances for fuel contamination
D. reduces the time required for fueling

763. Aircraft pressure fueling systems instructional procedures are normally placarded on the

Ans. 748. B 749. A 750. A 751. A 752. B 753. A 754. B 755. D 756. C 757. C 758. A 759. A 760. B 761. A 762. C

A. flightcrew checklist
B. fuel control panel access door
C. lower wing surface adjacent to the access door
D. ground crew checklist

764. The pressure fueling method is more practical for large aircraft than the over-the-wing method because it
A. reduces training time for fuel servicing personnel
B. saves time by providing a single point to fuel the entire aircraft
C. saves fuel by reducing the chances for overfills
D. It reduces the possibility of structural damage to bladder-type fuel cells

765. Most integral fuel tanks are repaired by
A. welding B. brazing
C. soldering D. riveting

766. The recommended practice for cleaning a fuel tank before welding is
A. immerse in a 5-percent solution of nitric acid
B. purge the tank with air
C. wash inside of tank with clean water
D. clean the tank interior with steam

767. An aircraft's integral fuel tank is
A. removable from the aircraft
B. usually located in the bottom of the fuselage
C. a part of the aircraft structure
D. a self-sealing tank

768. The gas used for purging an aircraft fuel tank is:
A. hydrogen B. carbon dioxide
C. oxygen D. carbon monoxide

769. The main fuel strainer located at the lowest point in the fuel system because it
A. eliminates the need for fuel tank sumps
B. traps any small amount of water that may be present in the fuel system
C. provides a drain for residual fuel
D. filters and traps all micro-organisms that may be present in the fuel system

770. Entrained water in aviation turbine fuel is a hazard because of its susceptibility to freezing as it passes through the filters. A common method of preventing this hazard is
A deicing fluid in the fuel
B. micromesh fuel strainers
C. high-velocity fuel pumps
D. anti-icing fuel additives

771. One of the first items to be checked when troubleshooting a rough running fuel-injected engine is
A. excessive fuel pressure
B. inadequate fuel flow
C. clogged air screen D. clogged fuel nozzles

772. The presence of fuel stains around a fuel nozzle indicates
A. too much fuel pressure
B. excessive airflow across the venturi
C. clogged fuel nozzle D. loose mainfold

773. The method which would be used to check for internal leakage of a fuel valve without removing the valve from the aircraft is
A. place the valve in the OFF position, drain the strainer bow 1, and with boost pump on, watch to see if fuel flows to the strainer bowl
B. remove fuel cap(s), turn boost pump(s) on and watch for bubbling in the tanks
C. disconnect valve inlet line and observe for fuel leakage
D. apply regulated air pressure on the downstream side of the fuel pump and listen for air passing through the valve

774. Jet fuels are more susceptible to water contamination than aviation gasoline because
A. jet fuel has a higher viscosity that gasoline
B. condensation is greater because of the irapici burn-off of jet fuels
C. jet fuel is lighter than gasoline, therefore, water is more easily suspended
D. processing and handling of jet fuel is less stringent than gasoline

775. When installing a rigid fuel line, 1/2 inch in diameter, at what intervals should be line be supported?
A. 24 inches B. 36 inches
C. 12 inches D. 16 inches

776. Some aircrafts are protected against air-frame icing by heating the leading edges of the air-foils and intake ducts. This type of anti-ice system usually operated during flight is
A. continuously while the aircraft is in flight
B. lia symmetric cycles during icing conditions to remove ice as it accumulates
C. at all times while the outside air temperature is below freezing
D. whenever icing conditions are first en-countered to expected to occur

777. The purpose of the oil separator in the pneumatic deicing system is to
A. protect the deicer boots from oil deterioration
B. remove oil from air exhausted from the deicer boots
C. prevent an accumulation of oil in the vaccum system
D. remove oil from the vaccum pump

778. The heat sehsors are located on most aircraft with electrically heated wind-shields as:
A. imbedded in the glass
B. attached to the glass
C. around the glass D. attached to the frame

Ans. 763. B 764. B 765. D 766. D 767. C 768. B 769. B 770. D 771. D 772. C 773. A 774. A 775. C 776. D 777. A 778. A

779. What are three possible sources of hot air for the operation of a wing thermal anti-icing system?
A. Turbo-compressors, air storage tank, vacuum pump
B. Engine bleed air, vaccum pump, com-pressed air tank
C. Engine bleed air, combustion heaters, augmentor tubes
D. Combustion heaters, augmentor tubes, ex-haust gases

780. What does breakdown of arcing in an electrically heated windshield panel usually indicate?
A. Temperature sensing elements
B. Thermal overheat switches
C. Conductive coating D. Autotransformers

781. Carburetor icing may be eliminated by
A. alcohol spray and heated induction air
B. alcohol spray and heating induction duct
C. ethylene glycol spray and heated induction air
D. electrically heating air intake and ethyleneglycol spray

782. The mixture that may be used as a deicing fluid to remove frost from an aircraft surface is
A. ethylene glycol and isopropyl alcohol
B. mild soap and water
C. methyl ethyl ketone and ethylene glycol
D. naphtha and isopropyl alcohol

783. Wet snow deposits from an aircraft are removed by:
A. a brush or a squegee B. hot air
C. warm water D. a chamois or a mop

784. Ice formation on a pitot tube is prevented by a/an:
A. electric heating element built into the pitot head
B. ribbon heater installed around the pitot head
C. blanket-type heater installed on the pitot head
D. gasket heater installed at the base of the pitot head

785. The icing condition which may occur in warm weather when there is no visible moisture present is:
A. glaze ice B. rime ice
C. carburetor ice D. wing leading edge ice

786. The ice in a turbine engine, if the compressor is immobile because of ice, can be melted by
A. deicing fluid B. hot water
C. anti-icing fluid D. hot air

787. The temperature sensing element used in an electrically heated windshield is
A. Resistor B. Thermistor
C. Capacitor D. Both A and B

788. The area of an aircraft which carbon monoxide detector is loaded is
A. Surface combustion heater compartment
B. Cockpit and/or cabin
C. Auxiliary power unit compartment
D. Engine and/or naçelle

789. When a visual smoke detector is activated
A. a warning bell within the indicator alarms automatically
B. a lamp within the indicator illuminates automatically
C. a lamp within the indicator extinguishes automatically
D. the test lamp illuminates and an alarm is provided automatically

790. The three types of fire extinguishing agents for aircraft interior fires are
A. water, methyl bromide, and carbon dioxide
B. water, dry chemical, and methyl bromide
C. water, carbon dioxide, and dry chemical
D. water, dry chemical, and chlorobro-momethane

791. When air samples contain carbon monoxide, portable carbon monoxide detectors containing yellow silica gel will turn
A. blue B. green
C. pink D. red

792. Smoke detectors which use a measurement of light transmissibility in the air are known as
A. electromechanical devices
B. photoelectrical devices
C. visual devices
D, electromeasuring devices

793. A contaminated carbon monoxide portable test unit would be returned to service by
A. heating the indicating element to 300°F to reactivate the chemical
B. installing a new indicating element
C. evacuating the indicating element with conductor
D. taking no action, unit is self-reactivating

794. The fire-detection system which measures temperature rise compared to a reference temperature is called
A. Fenwal continuous loop
B. thermal switch
C. Lindberg continuous element
D. thermocouple

795. A carbon dioxide hand-held fire extin-guisher may be used on an electrical fire if the
A. siphon tube is rigid
B. horn is nonmetallic
C. handle is insulated
D. horn is nonmagnetic

796. The proper fire-extinguishing agent to use on an aircraft brake fire is
A. carbon tetrachloride B. dry chemical
C. water D. carbon dioxide

Ans. 779. C 780. C 781. A 782. A 783. A 784. A 785. A 786. D 787. D 788. B 789. B 790. C 791. B 792. B 793. B 794. D 795. B 796. B

797. The most desirable agent for use in hand-type fire extinguishers installed in aircraft is
A. water
B. carbon dioxide
C. carbon tetrachloride
D. bromotnfluoromethane

798. Which of the following statements is currect concerning the operation of a photoelectric smoke detector?
A. A photoelectric smoke detector measures the amount of smoke under a specific set of conditions
B. A photoelectric smoke detector measures the amount of light available under a specific set of conditions
C. A photoelectric smoke detector will warn only when smoke is present
D. A photoelectric smoke detector is not affected by dust, soot, or other contaminants because it senses the difference between these and smoke

799. The thermocouple fire-warning system is activated by a
A. slowly overheated engine
B. certain temperature
C. more resistance drop
D. rate or temperature rise

800. Built-in-aircraft fire-extinguishing systems are ordinarily charged with
A. carbon monoxide and nitrogen
B. freon and nitrogen
C. carbon tetrachloride
D. sodium bicarbonate

801. The type of fire-detection system would also signal an overheat condition is:
A. Thermocouple system
B. Continuous loop system
C. Pressure-sensitive rcsponder system
D. thermograph system

802. On a periodic check of fire-extinguisher containers, the pressure was not between minimum and maximum limits. The pro-cedure should be followed is:
A. Release pressure if it is above limits
B. Replace the extinguisher container
C. increase pressure if it is below limits
D. Leave it alone, as pressure will change with temperature change

803. The fire-extinguisting agent which is the safest to use from the point of toxicity and corrosion hazards is
A. carbon dioxide
B. methyl bromide
C. chlorobromethane
D. all of these

804. In aircraft using gas turbines, the cycle used is
A. simple
B. regeneration
C. reheating
D. reheating with regeneration

Ans. 797. B 798. C 799. D 800. B 801. C 802. B 803. A 804. A

MULTIPLE CHOICE QUESTIONS FROM VARIOUS COMPETITIVE EXAMINATIONS

2015

1. A 2000/200 V, 50 Hz single-phase transformer has an exciting current of 0.5 A and a core-loss of 600 W. When HV side is energized by the rated voltage and frequency, the magnetizing current is
A. 0.1 A B. 0.2 A
C. 0.3 A D. 0.4 A

2. A 250 kVA, 11000 V/400 V and 50 Hz single-phase transformer has 80 turns on the secondary, what is the maximum value of flux?
A. 2475 mWb B. 0.2 mWb
C. 22.5 mWb D. 55.2 mWb

3. In a normal operation of 400 V, 50 Hz transformer, the total iron loss is 2500 W. When the supply voltage and frequency are reduced to 200 V and 25 Hz respectively, the corresponding loss is 850 W. The eddy-current loss at normal voltage and frequency is
A. 400 W B. 800 W
C. 1600 W D. 200 W

4. A magnetic circuit has 150 turns-coil, the cross-sectional area 5×10^{-4} m^2 and the length of the magnetic circuit 25×10^{-2} m. What are the values of magnetic field intensity and relative permeability when the current is 2A and total flux is 0.3×10^{-3} Wb?
A. 1200 AT/m and 397.9
B. 300 AT/m and 500×10^{-6}
C. 300 AT/m and 397.9
D. 1200 AT/m and 500×10^{-6}

5. A 240 V/120 V, 12 kVA auto-transformer has 96.2% efficiency at full-load and unity power-factor. The full-load efficiency at 0.85 pf lagging when connected to a load across 360 V would be
A. 89.5% B. 92.8%
C. 98.5% D. 88.9%

6. The percentage resistance and reactance of a transformer are 2% and 4%, respectively. The approximate regulation on full load at 0.8 pf lag is
A. 12% B. 8%
C. 6% D. 4%

7. A three-phase transformer has 420 and 36 turns on the primary and secondary windings, respectively. The supply voltage is 3300 V. The secondary line voltage on no-load when the windings are connected in star-delta is nearly
A. 22260 V B. 1908 V
C. 164 V D. 490 V

8. A 4 kVA transformer has iron-loss of 200 W and full load copper loss of 200 W. The maximum efficiency at unity power factor will be
A. 90.9% B. 85.6%
C. 80.6% D. 70.9%

9. A uniformly distributed winding on the stator has three full-pitched coils, each coil having N turns and each turn carrying a current I. The mmf produced by this winding is
A. sinusoidal in waveform with an amplitude 3 NI
B. sinusoidal in waveform with an amplitude 3 NI/2
C. trapezoidal in waveform with an amplitude 3 NI
D. trapezoidal in waveform with an amplitude 3 NI/2

10. In DC machines, the field-flux axis and armature-mmf axis are respectively along
A. direct axis and indirect axis
B. direct axis and inter-polar axis
C. quadrature axis and direct axis
D. quadrature axis and inter-polar axis

11. A shunt motor supplied at 250 V, runs at 900 rpm and the armature current drawn is 30 A. The resistance of the armature circuit is 0.4 Ω. The resistance required in series with the armature to reduce the speed to 600 rpm when the armature current is 20 A will be
A. 3.17 Ω B. 2.17 Ω
C. 5.17 Ω D. 4.17 Ω

12. The internal characteristics of a DC generator is plotted between the
A. field current and voltage generated at no load
B. armature current and voltage generated on load
C. armature current and voltage generated after armature reaction
D. field current and voltage generated on load

13. The drawbacks of 'Armature Resistance Control' method of speed control of a DC motor are
1. A large amount of power is lost in the external resistance R
2. This method gives the speed below normal values
3. For a specified value of R, speed reduction is not constant. It varies with motor load

Which of the above statements are correct?
A. 1, 2 and 3 B. 1 and 2 only
C. 1 and 3 only D. 2 and 3 only

14. A DC motor running at 2000 rpm has a hysteresis loss of 500 W and eddy-current loss of 200 W. The

Ans.	1. D	2. C	3. C	4. B	5. C	6. D	7. C	8. A	9. D	10. B	11. D	12. C	13. A

flux is maintained constant but the speed is reduced to 1000 rpm. At the reduced speed, the total iron-loss would be
A. 500 W B. 400 W
C. 300 W D. 200 W

15. The speed control of dc shunt motor in both directions can be obtained by
A. Armature resistance control method
B. Ward-Leonard method
C. Field diverter method
D. Armature voltage control method

16. A 10 hp, 240 V dc shunt motor, having armature-circuit resistance of 0.5 Ω and a full-load current of 40 A, is started by a starter, such that sections of required resistances in series with the armature-circuit should limit the starting current to 150% of the full-load current. The steady-state emf developed by the machine at full-load when the arm of the starter is moved to the next step is
A. 120 V B. 100 V
C. 80 V D. 60 V

17. A synchronous motor operates at rated voltage and frequency and has a load torque angle of 30°. If both the terminal voltage and frequency are reduced by 10%, then
A. the load torque angle remains the same
B. the load torque angle increases
C. the load torque angle decreases
D. nothing can be said about the torque angle

18. The nature of armature mmf for a zero power factor lagging load in a three-phase alternator is
A. magnetizing B. de-magnetizing
C. neutral D. cross-magnetizing

19. The flux/pole in a synchronous motor with stator not connected to supply is Φ_1 and when connected to supply it is $\frac{3}{4}\Phi_1$. The no-load current drawn from the supply under this condition would be
A. lagging the supply voltage
B. leading the supply voltage
C. in phase with the supply voltage
D. zero

20. A cylindrical-rotor generator with internal voltage 2.0 pu and X_S = 1.0 pu is connected by a line of reactance 0.5 pu to a round-rotor synchronous motor of synchronous reactance 1.2 pu and excitation voltage 1.35 pu. When 0.5 pu power is supplied by the generator, the electrical angular difference between the rotors would be
A. 25° B. 30°
C. 60° D. 120°

21. The synchronizing power for one mechanical degree of displacement for a three-phase, 20000 kVA, 6600 V, 50 Hz, 12-pole machine having X_S = 1.65 Ω and negligible resistance is
A. 1024.6 kW B. 921.9 kW
C. 782.6 kW D. 1182.6 kW

22. For obtaining very quick breaking of a three-phase, wound-rotor induction motor running on load
A. a large external resistance has to be inserted in the rotor-circuit
B. a large external resistance has to be inserted in the stator-circuit
C. interchange any two terminals of the stator supply
D. interchange any two terminals of the rotor to the slip-rings

23. Consider the following statements:
As a three-phase induction motor is loaded from no-load to rated load,
1. There is an improvement in the power factor
2. The torque increases almost in proportion to slip
3. The air-gap flux falls sharply
Which of the above statements are correct?
A. 1, 2 and 3 B. 1 and 2 only
C. 1 and 3 only D. 2 and 3 only

24. A small three-phase induction motor has a short-circuit current 5 times of full-load current and full-load slip 5%. If starting resistance starter is used to reduce the impressed voltage to 60% of normal voltage, the starting torque obtained in terms of full load torque would be
A. 30% B. 45%
C. 55% D. 80%

25. The frequency of rotor emf of an 8-pole induction motor is 2 Hz. If the supply frequency is 50 Hz, then the motor speed is
A. 1500 rpm B. 750 rpm
C. 375 rpm D. 720 rpm

26. For a given applied voltage and current, the speed of a universal motor will be
A. higher in dc excitation than in ac excitation
B. higher in ac excitation than in dc excitation
C. same in both dc and ac excitation
D. dangerously high in dc excitation

27. Two single-phase transformers A and B with equal turn's ratio have reactances of $j3$ Ω and $j9$ Ω referred to secondary. When operated in parallel, the load-sharing of 100 kW at 0.8 pf lag between A and B transformers would respectively be
A. 75 kW and 25 kW B. 60 kW and 40 kW
C. 20 kW and 80 kW D. 25 kW and 75 kW

28. Compared to turbines in the conventional coal-fired thermal stations, nuclear power plant turbines use steam at

Ans. 14. C 15. B 16. C 17. A 18. B 19. B 20. B 21. B 22. C 23. B 24. B 25. D 26. B 27. A

A. higher pressure and higher temperature
B. lower pressure and lower temperature
C. higher pressure and lower temperature
D. lower pressure and higher temperature

29. In an L-section filter, a bleeder resistance connected across the load
A. provides good regulation for all values of load
B. ensures lower PIV of the diodes
C. ensures lower values of capacitance in the filter
D. reduces ripple content

30. The power output from a hydroelectric power plant depends on
A. head, type of dam and discharge
B. head, discharge and efficiency of the system
C. type of draft tube, type of turbine and efficiency of the system
D. type of dam discharge and type of catchment area

31. Incremental fuel costs of two units A and B of a power station are

$$dF_A/dP_A = 0.4P_A + 400$$
$$dF_B/dP_B = 0.48P_B + 320$$

For the minimum cost of generation of a total load of 900 MW, the generation allocation for A and B units respectively are
A. 200 MW and 700 MW
B. 300 MW and 600 MW
C. 400 MW and 500 MW
D. 500 MW and 400 MW

32. A power station has a maximum demand of 2500 kW and number of kWh generated per year is 45×10^5. The load factor is
A. 10.25% B. 20.5%
C. 41% D. 82%

33. A long overhead lossless power transmission line is terminated with its characteristic impedance. While the line is in operation
A. a resonance of reactive powers occurs in the line
B. the line becomes purely inductive
C. the line becomes purely capacitive
D. there is no reflected wave on the line

34. For exact compensation of voltage drop in the feeder, the booster must
A. be earthed B. work on line voltage
C. work on its linear portion of V-I characteristics
D. work on its non-linear portion of V-I characteristics

35. A 100 MVA, 50 Hz turbo-generator operates at no load at 3000 rpm. A load of 25 MW is suddenly applied to the machine and the steam valve to the turbine commences to open after 0.6 seconds due to the governor time-lag. Assuming the inertia constant $H = 5.0$ kW-s per kVA of the generator rating, the frequency to which the generated voltage drops before the steam-flow commences to meet the new load is
A. 49.0 Hz B. 50.15 Hz
C. 49.24 Hz D. 49.82 Hz

36. In case of single line to ground fault
A. all sequence networks are connected in parallel
B. all sequence networks are connected in series
C. positive and negative sequence networks are connected in parallel
D. zero and negative sequence networks are connected in series

37. What is the region of operation of a three-phase inverter employing sinusoidal PWM when the peak-to-peak values of both the carrier and the modulating waves are made equal?
A. Linear modulation B. Over modulation
C. Boundary of linear modulation and over modulation
D. Six-step operation

38. Which of the following circuit breakers has the lowest voltage range?
A. SF_6 circuit breaker
B. Air-blast circuit breaker
C. Tank type of circuit breaker
D. Air-break circuit breaker

39. Which of the following will be provided to reduce the harmonics on the ac side of an HVDC transmission line?
A. Synchronous motors in over excited condition
B. Shunt capacitor
C. Static compensator D. Shunt filters

40. Modified McMurray full-bridge inverter works on
A. load commutation B. current commutation
C. voltage commutation
D. complementary commutation

41. When a pnp transistor is properly biased, the holes from the emitter
A. diffuse through the base into the collector region
B. recombine with the electrons in the base region
C. recombine with the electrons in the emitter region
D. diffuse through the emitter to collector

42. In a p-n junction diode under reverse bias, the magnitude of electric field is maximum at
A. the edge of the depletion region on the P side
B. the edge of the depletion region on the N side
C. the centre of the depletion region on the N side
D. the p-n junction

Ans. 28. B 29. A 30. B 31. C 32. B 33. D 34. C 35. C 36. B 37. C 38. D 39. D 40. D 41. A 42. D

43. A silicon diode is preferred to a germanium diode because of its
A. higher reverse current
B. lower reverse current and higher reverse breakdown voltage
C. higher reverse current and lower reverse breakdown voltage
D. none of the above

44. Compared to an ordinary semiconductor diode, a Schottky diode has
A. higher reverse saturation current and zero cut-in voltage
B. higher reverse saturation current and higher cut-in voltage
C. higher reverse saturation current and lower cut-in voltage
D. lower reverse saturation current and lower cut-in voltage

45. Which of the following is called 'hot carrier diode'?
A. PIN diode B. LED
C. Photo diode D. Schottky diode

46. In a voltage-series-feedback amplifier with open loop gain A_v and the feedback factor β, the input resistance becomes
A. $\dfrac{R_i}{(1+\beta A_v)}$ B. $R_i\sqrt{(1+\beta A_v)}$
C. $R_i(1+\beta A_v)$ D. $\dfrac{R_i}{\sqrt{(1+\beta A_v)}}$

47. Ripple-rejection ratio of voltage regulator is the ratio of
A. output voltage to input ripple voltage
B. output power to input power of regulator
C. input power to output power of regulator
D. input ripple voltage to output ripple voltage

48. Variation in β in a BJT can cause a fixed bias circuit to go
A. into active mode of operation from saturation mode
B. out of active mode
C. out of saturation
D. into cut-off mode from active mode of operation

49. In a McMurray inverter, diodes are connected in inverse parallel to thyristors to
1. Protect the thyristor
2. Make the turn-off of the thyristor successful
3. Make the turn-on of the thyristor successful
4. Provide path to the reactive component of the load current
Which of the above statements are correct?
A. 1 and 3 B. 1 and 4
C. 2 and 4 D. 2 and 3

50. Following pulse is applied to an SR flip-flop.

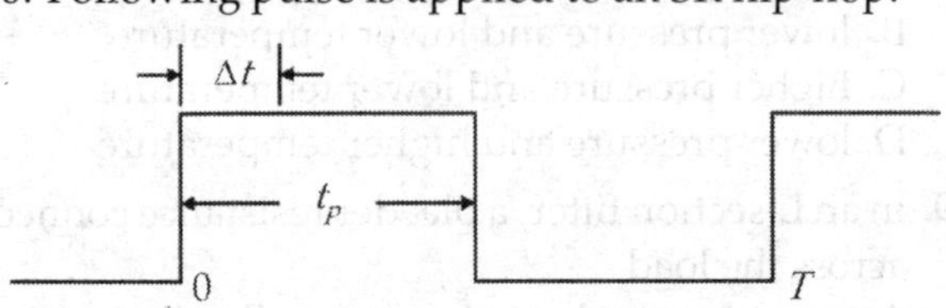

If Δt is propagation delay, which of the following relations should hold to avoid the race condition?
A. $t_p < \Delta t < T$ B. $\Delta t < t_p < T$
C. $2\Delta t < t_p$ D. $2\Delta t < t_p < T$

51. For a four variable K-Map, if each cell is assigned one integer value in range 0–15 then which is the cells adjacent to the cell corresponding to decimal value 7?
A. 3, 5, 6 and 8 B. 3, 5, 10 and 11
C. 3, 5, 6 and 15 D. 4, 6, 8 and 15

52. In order to generate a square wave from a sinusoidal input signal, one can use
1. Schmitt trigger circuit
2. Clippers and amplifiers
3. Monostable multivibrator
Which of the above statements are correct?
A. 1, 2 and 3 B. 1 and 2 only
C. 1 and 3 only D. 2 and 3 only

53. An analog voltage in the range of 0–8 V is divided in eight equal intervals for conversion to 3-bit digital output. The maximum quantization error is
A. 0 V B. 0.5 V
C. 1.0 V D. 2.0 V

54. The logic function $A + BC$ is the simplified form of which of the following?
A. $AB + BC$ B. $\bar{A}B + A\bar{B}C$
C. $\overline{ABC}$ D. $(A + B)(A + C)$

55. Program counter in a digital computer
A. counts the numbers of programs run in the machine
B. counts the number of times a subroutine is called
C. counts the number of times the loops are executed
D. points the memory address of the current or the next instruction to be executed

56. Each cell of a static RAM contains
A. 4 MOS transistors
B. 4 MOS transistors and 1 capacitor
C. 2 MOS transistors
D. 4 MOS transistors and 2 capacitors

57. If an *npn* silicon transistor is operated at $V_{CE} = 5$ V and $I_C = 100$ μA and has a current gain of 100 in the

Ans. 43. B 44. C 45. D 46. C 47. D 48. B 49. C 50. A 51. C 52. B 53. C 54. D 55. D 56. A

CE connection, then the input resistance of this circuit will be

A. 250 kΩ B. 450 kΩ

C. 4500 Ω D. 2500 Ω

58. A 4-bit synchronous counter has flip-flops having propagation delay of 50 ns each and AND gates having propagation delay of 20 ns each. The maximum frequency of clock pulses can be

A. 20 MHz B. 50 MHz

C. 14.3 MHz D. 5 MHz

59. A self-starting counter is one that can start

A. the sequence from initial count and continues its sequence

B. the sequence from any state among the sequence and continues its normal count sequence

C. from any state but eventually reaches the required count sequence

D. none of the above

60. When a program is being executed in an 8085 microprocessor, its program counter contains

A. the memory address as the instruction that is to be executed next

B. the memory address of the instruction that is being currently matched

C. the total number of instructions in the program being executed

D. the number of instructions in the current program that have already been executed

61. A 'DADH' instruction is the same as shifting each bit by one position to the

A. left B. right

C. left with a zero inserted in LSB position

D. right with a zero inserted in LSB position

62. Which one of the following statements is correct about 8086?

A. It is 46 PM IC and uses 5 V dc supply

B. It uses 20 lines for data bus

C. It is multiplexes status signals with address bus

D. It is manufactured using CMOS technology

63. During which T-state, contents of OP code from memory are loaded into IR (instruction register)?

A. T_1 OP code Fetch B. T_2 OP code Fetch

C. T_3 OP code Fetch D. T_4 OP code Fetch

64. In 8085 microprocessor, the addrss for 'TRAP' interrupt is

A. 0024H B. 002CH

C. 0034H D. 003CH

65. Assuming LSB is at position 0 and MSB at position 7, which bit positions are not used (undefined) in Flag Register of an 8085 microprocessor?

A. 1, 3, 5 B. 2, 3, 5

C. 1, 2, 5 D. 1, 3, 4

66. At the beginning of a fetch cycle, the contents of the program counter are

A. incremented by one

B. transferred to address bus

C. transferred to memory address register

D. transferred to memory data register

67. What will be the contents of DE and HL register pairs respectively after the execution of the following instructions?

LX1H, 2500H
LX1D, 0200H
DAD D
XCHG

A. 0200H, 2500H B. 0200H, 2700H

C. 2500H, 0200H D. 2700H, 0200H

68. Direction flag is used with

A. String instructions B. Stack instructions

C. Arithmetic instructions

D. Branch instructions

69. XCHG instruction of 8085 exchanges the content of

A. top of stack with contents of register pair

B. BC and DE register pairs

C. HL and DE register pairs

D. none of the above

70. To interface a slow memory, wait states are added by

A. extending the time of the chip select logic

B. causing READY signal to go low

C. causing READY signal to go high

D. by increasing the clock frequency

71. A ripple counter with *n* flip-flops can function as a

A. $n:1$ counter B. $n/2:1$ counter

C. $2n:1$ counter D. $2^n:1$ counter

72. The relation among IC (Instruction Cycle), FC (Fetch Cycle) and EC (Execute Cycle) is

A. IC = FC – EC B. IC = FC + EC

C. IC = FC + 2EC D. EC = IC + EC

73. Each instruction in an assembly language program has the following fields:

1. Label field 2. Mnemonic field

3. Operand field 4. Comment field

What is the correct sequence of these fields?

A. 1, 2, 3 and 4 B. 2, 1, 4 and 3

C. 1, 3, 2 and 4 D. 2, 4, 1 and 3

74. A memory system has a total of 8 memory chips, each with 12 address lines and 4 data lines. The total size of the memory system is

A. 32 k bytes B. 48 k bytes

C. 64 k bytes D. 6 k bytes

75. Among the given instructions, the one which affects maximum number of flags is

Ans.	57. D	58. C	59. C	60. A	61. C	62. C	63. C	64. C	65. A	66. C	67. D	68. A	69. C
	70. B	71. D	72. B	73. A	74. A								

A. RAL B. POP PSW
C. XRA A D. DCR A

76. The internal characteristic of a dc generator is plotted between the
A. armature current and voltage generated after armature reaction
B. field current and voltage generated at no load
C. field current and voltage generated on load
D. armature current and voltage generated at the output terminals

77. In 8085 microprocessor with memory mapped I/O, which one of the following is correct?
A. I/O devices have 16-bit addresses
B. I/O devices are accessed during IN and OUT instructions
C. There can be a maximum of 256 input and 256 output devices
D. Logic operations cannot be performed

78. Race-around condition occurs in
A. Multiplexer B. ROM
C. Flip-flops D. Voltage regulator

79. The DMA transfer technique where one word data is transferred at a time is called
A. Cycle stealing B. Memory stealing
C. Hand shaking D. Interleaving

80. In a single-phase to single-phase cycloconverter, the magnitudes of harmonic components are quite large. How can they be reduced?
A. By using a chopper circuit
B. By using an RC oscillator
C. By using a three-phase input supply
D. By adding an alternator to the input

81. Which one of the following is used for serial I/O transfer in 8085 based system?
A. 8251 B. 8255
C. 8259 D. 8279

82. An analog voltage is in the range of 0 to 8 V. It is divided in 8 equal intervals for conversion to 3-bit digital output. The maximum quantization error is
A. 0 V B. 0.5 V
C. 1 V D. 2 V

83. Digital modulation techniques are used in satellite communication system, because
A. they are easier to handle
B. large bandwidth utilization is possible
C. they have a spectral efficiency
D. they are less prone to interference

84. In FM modulation, when the modulation index increases, the transmitted power is
A. increased B. decreased
C. unchanged D. not related

85. In TV, video signals are transmitted through
A. amplitude modulation
B. frequency modulation
C. either amplitude or frequency modulation
D. neither amplitude nor frequency modulation

86. If the number of bits per sample in a PCM system is increased from 8 to 16, then the bandwidth will be increased by
A. 2 times B. 4 times
C. 8 times D. 16 times

87. In a PCM system, a five bit-encoder is used. If the difference between two consecutive levels is 1 V, then the range of the encoder is
A. 0–31 V B. 0–30 V
C. 1–32 V D. 1–30 V

88. Consider the following types of modulation:
1. Amplitude modulation
2. Frequency modulation
3. Pulse modulation 4. Phase modulation
Which of the above modulation are used for telecasting TV programs?
A. 2 and 3 B. 3 and 4
C. 1 and 2 D. 1 and 4

89. In a step up operation, what is the condition to be satisfied for controllable power transfer?
A. $0 < V_S < E$ B. $0 < V_S > E$
C. $V_S < E$ D. $V_S > E$

90. A Schottky diode is
A. a majority carrier device
B. a minority carrier device
C. a fast recovery diode
D. both majority and minority carrier diode

2016

1. Compared to the salient-pole hydroelectric generators, the steam and the gas-turbine generators have cylindrical rotors for
A. better air-circulation in the machine
B. reducing the eddy-current losses in the rotor
C. accommodating larger number of terms in the field winding
D. providing higher mechanical strength against the centrifugal stress

2. Consider the following losses for short circuit test on a transformer:
1. Copper loss 2. Copper and iron losses
3. Eddy current and hysteresis losses
4. Friction and windage losses
Which of the above is/are correct?
A. 1 only B. 2 only
C. 3 only D. 2, 3 and 4

3. A 2000 V/200 V, 20 kVA, two winding, single phase transformer is reconnected as a step up auto-trans-

Ans.	75. B	76. A	77. A	78. C	79. A	80. C	81. A	82. C	83. D	84. C	85. A	86. A	87. A
	88. C	89. A	90. A	1. D	2. A								

former having 200 V/2200 V ratings. Then the power rating for the auto transformer will be
A. 160 kVA B. 180 kVA
C. 200 kVA D. 220 kVA

4. The regulation of a transformer in which ohmic loss is 1% of the output and reactance drop is 5% of the voltage, when operating at 0.8 power factor lagging, is
A. 3.8% B. 4.8%
C. 5.2% D. 5.8%

5. In a power transformer, the core loss is 50 W at 40 Hz, and 100 W at 60 Hz, under the condition of same maximum flux density in both cases. The core loss at 50 Hz will be
A. 64 W B. 73 W
C. 82 W D. 91 W

6. Consider the following advantages of a distributed winding in a rotating machine:
1. Better utilization of core as a number of evenly placed small slots are used
2. Improved waveform as harmonic emf are reduced
3. Diminished armature reaction and efficient cooling
Which of the above advantages are correct?
A. 1 and 2 only B. 1 and 3 only
C. 2 and 3 only D. 1, 2 and 3

7. The breadth factor for 3rd harmonic emf of a three-phase, four-pole, synchronous machine having 36 stator slots is
A. 0.47 B. 0.53
C. 0.67 D. 0.73

8. Consider the following factors for a DC machine:
1. Interpole 2. Armature resistance
3. Reduction in field current
Which of the above factors are responsible for decrease in terminal voltage of a shunt generator?
A. 1 and 2 only B. 2 and 3 only
C. 1 and 3 only D. 1, 2 and 3

9. A dc motor develops an electromagnetic torque of 150 N-m in a certain operating condition. From this operating condition, a 10% reduction in field flux and 50% increase in armature current is made. What will be new value of electromagnetic torque?
A. 225 N-m B. 202.5 N-m
C. 22.5 N-m D. 20.25 N-m

10. A dc machine, having a symmetrical closed-circuit armature winding and a sinusoidal air-gap flux-density distribution, will have a sinusoidal voltage induced in the individual coils. The resultant brush-to-brush voltage will have a waveform
A. sinusoidal with the negative-half reversed
B. unidirectional and constant without any ripples
C. unidirectional and constant with ripples superimposed
D. sinusoidal positive-half and zero negative-half, in each cycle

11. A three-phase induction motor operating at a slip of 5% develops 20 kW rotor power output. What is the corresponding rotor copper loss in this operating condition?
A. 750 W B. 900 W
C. 1050 W D. 1200 W

12. What are the signs of load angle in an alternator during generator and motor operations, respectively?
A. Negative, negative B. Positive, negative
C. Negative, positive D. Positive, positive

13. In an alternator, the armature winding is kept stationary while the field winding is kept rotating for the following reasons:
1. Armature handles very large current and high voltage
2. Armature fabrication, involving deep slots to accommodate large coils, is easy if armature is kept stationary
3. It is easier to cool the stator than the rotor
Which of the above reasons are correct?
A. 1 and 2 only B. 1 and 3 only
C. 2 and 3 only D. 1, 2 and 3

14. Increasing the air-gap of a squirrel-cage induction motor would result in
A. increase in no-load speed
B. increase in full-load power-factor
C. increase in magnetizing current
D. maximum available torque

15. A cumulative compound DC motor runs at 1500 rpm on full load. If its series field is short circuited, its speed
A. becomes zero B. remains same
C. increases D. decreases

16. If the capacitor of a capacitor-start single-phase motor fails to open when the motor picks-up speed,
A. the motor will stop
B. the auxiliary winding will be damaged
C. the capacitor will be damaged
D. the winding will be damaged

17. For a three-phase induction motor, what fraction/multiple of supply voltage is required for a direct-on-line starting method such that starting current is limited to 5 times the full-load current and motor develops 1.5 times full-load torque at starting time?
A. 1.632 B. 1.226
C. 0.816 D. 0.456

Ans. 3. D 4. A 5. B 6. D 7. D 8. C 9. B 10. C 11. C 12. B 13. D 14. C 15. C 16. C 17. C

18. What is the material of slip-rings in an induction machine?
A. Carbon B. Nickel
C. Phosphor bronze D. Manganese

19. The stator loss of a three-phase induction motor is 2 kW. If the motor is running with a slip of 4% and power input of 90 kW, then what is the rotor mechanical power developed?
A. 84.48 kW B. 86.35 kW
C. 89.72 kW D. 90.52 kW

20. In a single-phase capacitor-start induction motor, the direction of rotation
A. can be changed by reversing the main winding terminals
B. cannot be changed
C. is dependent on the size of the capacitor
D. can be changed only in large capacity motors

21. Air pollution due to smoke around a thermal power station can be reduced by installing
A. induced draft fan B. super heater
C. economizer
D. electrostatic precipitator

22. The load curve is useful in deciding
1. The operating schedule of generating units
2. The total installed capacity
Which of the above statements is/are correct?
A. 1 only B. 2 only
C. both 1 and 2 D. neither 1 nor 2

23. The maximum demand on a steam power station is 480 MW. If the annual load factor is 40%, then the total energy generated annually is
A. 19819.2 × 10^5 kWh B. 18819.2 × 10^5 kWh
C. 17819.2 × 10^5 kWh D. 16819.2 × 10^5 kWh

24. To equalize the sending and receiving end voltages, impedance is connected at the receiving end of a transmission line having the following *ABCD* parameters:

$$A = D = 0.9\angle 0°, B = 200\angle 90°\ \Omega$$

The impedance so connected would be
A. 1000 $\angle 0°$ Ω B. 1000 $\angle 90°$ Ω
C. 2000 $\angle 90°$ Ω D. 2000 $\angle 0°$ Ω

25. The maximum efficiency in the transmission of bulk ac power will be achieved when the power factor of the load is
A. slightly less than unity lagging
B. slightly less than unity leading
C. unity
D. considerably less than unity

26. A speed of a dc motor is
A. directly proportional to back emf and inversely proportional to flux
B. inversely proportional to back emf and directly proportional to flux
C. directly proportional to back emf as well as to flux
D. inversely proportional to back emf as well as to flux

27. When the sending end voltage and current are numerically equal to the receiving end voltage and current respectively, then the line is called a
A. tuned line B. transposed line
C. long line D. short line

28. If V_m is the peak value of an applied voltage in a half wave rectifier with a large capacitor across the load, then the peak inverse voltage will be
A. $0.5V_m$ B. V_m
C. $1.5V_m$ D. $2.0V_m$

29. A 100 MVA generator operates on full-load of 50 Hz frequency. The load is suddenly reduced to 50 MW. The steam valve begins to close only after 0.4 s and if the valve of the inertia constant *H* is 5 s, then the frequency at 0.4 s is nearly
A. 38 Hz B. 44 Hz
C. 51 Hz D. 62 Hz

30. A 25 MVA, 33 kV transformer has a pu impedance of 0.9. The pu impedance at a new base 50 MVA at 11 kV would be
A. 10.4 B. 12.2
C. 14.4 D. 16.2

31. Symmetrical components are used in power system for the analysis of
A. balanced three-phase fault
B. unbalanced three-phase fault
C. normal power system under steady conditions
D. stability of system under disturbance

32. For *V*-curves for a synchronous motor, the graph is drawn between
A. terminal voltage and load factor
B. power factor and field current
C. field current and armature current
D. armature current and power factor

33. Critical clearing angle is related to
A. stability study of power system
B. power flow study of power system
C. regulation of transmission line
D. power factor improvement of the system

34. A 2-pole, 50 Hz, 11 kV, 100 MW alternator has a moment of inertia of 10,000 kg·m². The value of inertia constant, *H* is
A. 3.9 s B. 4.3 s
C. 4.6 s D. 4.9 s

Ans.													
18. C	19. A	20. A	21. D	22. C	23. D	24. C	25. A	26. A	27. A	28. B	29. C	30. D
31. B	32. B	33. A	34. D									

35. Stability of a power system can be improved by
1. Using series compensators
2. Using parallel transmission lines
3. Reducing voltage of transmission

Which of the above statements are correct?
A. 1 only B. 2 only
C. 2 and 3 D. 1 and 2

36. Equal-area criterion is employed to determine the
A. steady-state stability B. transient stability
C. reactive power limit
D. rating of a circular breaker

37. Consider the following advantages with respect to HVDC transmission:
1. Long distance transmission
2. Low cost of transmission
3. Higher efficiency

Which of the above advantages are correct?
A. 1 and 2 only B. 1 and 3 only
C. 2 and 3 only D. 1, 2 and 3

38. The three sequence voltages at the point of fault in a power system are found to be equal. The nature of the fault is
A. L–G B. L–L–L
C. L–L D. L–L–G

39. A distance relay with inherent directional property is known as
A. Buchholtz relay B. admittance relay
C. directional over current relay
D. directional switch relay

40. Consider the following circuit breakers for 220 kV substations:
1. Air 2. SF_6
3. Vacuum

Which of the above circuit breakers can be used in an indoor substation?
A. 1, 2 and 3 B. 1 only
C. 2 only D. 3 only

41. A semiconductor differs from a conductor in that it has
A. only one path for the free electrons in the valence band
B. only one path for holes in the conduction band
C. two paths followed by free electrons and holes, one an ordinary path in the conduction band and the other one an extraordinary path in the valence band, respectively
D. two paths followed by free electrons and holes, one an extraordinary path in the conduction band and the other one an ordinary path in valence band, respectively

42. Which of the following circuits is used for converting a sine wave into a square wave?
A. Monostable multivibrator
B. Bistable multivibrator
C. Schmitt trigger circuit
D. Darlington complementary pair

43. What is the type of breakdown that occurs in a Zener diode having breakdown voltage (bV)?
A. Avalanche breakdown only
B. Zener breakdown only
C. Avalanche breakdown where breakdown voltage is below 6 V and Zener breakdown otherwise
D. Zener breakdown where breakdown voltage is below 6 V and Avalanche breakdown otherwise

44. Consider the following statements:
A power supply uses bridge rectifier with a capacitor input filter. If one of the diodes is defective, then
1. The DC load voltage will be lower than its expected value
2. Ripple frequency will be lower than its expected value
3. The surge current will increase considerably

Which of the above statements are correct?
A. 1 and 2 only B. 1 and 3 only
C. 2 and 3 only D. 1, 2 and 3

45. The lowest frequency of ac components in the outputs of half-wave and full-wave rectifiers are, respectively (where ω is the input frequency)
A. 0.5ω and ω B. ω and 2ω
C. 2ω and ω D. ω and 3ω

46. A half-wave rectifier circuit using ideal diode has an input voltage of 20 sin ωt volt. Then average and rms values of output voltage respectively, are
A. $\frac{10}{\pi}$ V and 5 V B. $\frac{20}{\pi}$ V and 10 V
C. $\frac{20}{\pi}$ V and 5 V D. $\frac{10}{\pi}$ V and 10 V

47. For a BJT, $I_C = 5$ mA, $I_B = 50$ μA and $I_{CBO} = 0.5$ μA, then the value of β is
A. 99 B. 91
C. 79 D. 61

48. Which of the following conditions must be satisfied for a transistor to be in saturation?
1. Its collector to base junction should be under forward bias
2. Its collector to base junction should be under reverse bias
3. Its emitter to base junction should be under reverse bias
4. Its emitter to base junction should be under forward bias

Which of the above conditions are correct?
A. 1 and 3 B. 2 and 3
C. 2 and 4 D. 1 and 4

Ans. 35. D 36. B 37. D 38. D 39. B 40. C 41. C 42. C 43. D 44. D 45. B 46. B 47. A 48. D

49. In an amplifier with a gain of 1000 without feedback and cut-off frequencies at 2 kHz and 20 kHz, negative feedback of 1% is employed. The cut-off frequencies with feedback would be
A. 220 Hz and 22 kHz B. 182 Hz and 220 kHz
C. 220 kHz and 200 kHz
D. 182 Hz and 22 kHz

50. Consider the following circuits:
1. Oscillator 2. Emitter follower
3. Power amplifier
Which of the above circuits employ feedback?
A. 1 and 2 only B. 2 and 3 only
C. 1 and 3 only D. 1, 2 and 3

51. Three identical amplifiers each having a voltage gain of 50 are cascaded. The open loop voltage gain of the combined amplifier is
A. 71 dB B. 82 dB
C. 91 dB D. 102 dB

52. A clamper circuit
1. Adds or subtracts a dc voltage to or from a waveform
2. Does not change the shape of the waveform
Which of the above statements is/are correct?
A. 1 only B. 2 only
C. Both 1 and 2 D. Neither 1 nor 2

53. The operational amplifier circuit shown in figure having a voltage gain of unity has

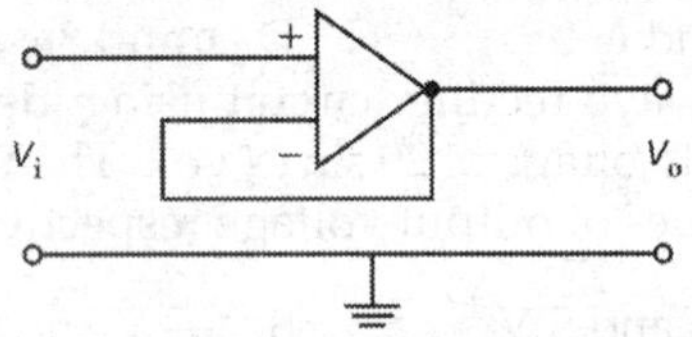

A. high input impedance and high output impedance
B. high input impedance and low output impedance
C. low input impedance and low output impedance
D. low input impedance and high output impedance

54. Consider the following statements:
1. Race-around condition occurs in a JK flip-flop when the inputs are 1, 1
2. A flip-flop is used to store one bit of information
3. A transparent latch consists of D-type flip-flops
4. Master-slave configuration is used in a flip-flop to store two bits of information
Which of the above statements are correct?
A. 1, 2 and 3 only B. 1, 2 and 4 only
C. 3 and 4 only D. 1, 2, 3 and 4

55. An operational amplifier has a slew rate of 2 V/μ sec. If the peak output is 12 V, what will be the power bandwidth?
A. 36.5 kHz B. 26.5 kHz
C. 22.5 kHz D. 12.5 kHz

56. A voltage follower is used as
1. An isolation amplifier
2. A buffer amplifier
Which of the above statements is/are correct?
A. 1 only B. 2 only
C. Both 1 and 2 D. Neither 1 nor 2

57. If a, b, c are 3-input variables, then Boolean function $y = ab + bc + ca$ represents
1. A 3-input majority gate
2. A 3-input minority gate
3. Carry output of a full adder
4. Product circuit for a, b and c
Which of the above statements are correct?
A. 2 and 3 B. 2 and 4
C. 1 and 3 D. 1 and 4

58. In a 2-input CMOS logic gate, one input is left floating, i.e. connected neither to ground nor to a signal. What will be the state of that input?
A. 1 B. 0
C. Same as that of the other input
D. Indeterminate (neither 1 nor 0)

59. The expression for MOD number for a ripple counter with N flip-flops is
A. N B. 2^N
C. 2^{N-1} D. $2^N - 1$

60. Why a ROM does not have data inputs?
A. It does not have a WRITE operation
B. Data inputs are integrated with data outputs
C. Data inputs are integrated with address inputs
D. ROM is sequentially accessed

61. Consider the following statements:
1. RAM is a non-volatile memory whereas ROM is a volatile memory
2. RAM is a volatile memory whereas ROM is a non-volatile memory
3. Both RAM and ROM are volatile memories but in ROM data is not when power is switched off
Which of the above statements are correct?
A. 1 only B. 2 only
C. 3 only D. None of the above

62. Consider the following instructions:
1. LOCK 2. STD
3. HLT 4. CLI
Which of the above are machine control instructions?
A. 1 and 4 B. 1 and 3
C. 2 and 3 D. 2 and 4

Ans. 49. B 50. A 51. D 52. C 53. B 54. A 55. B 56. C 57. C 58. D 59. B 60. A 61. B 62. B

63. What is the assembler directive statement used to reserve an array of 100 words in memory and initialize all 100 words with 0000 and give it a name STORAGE?
A. STORAGE DW 100 B. STORAGE DB 100
C. STORAGE DW 100 DUP (?)
D. STORAGE DW 100 DUP (0)

64. Consider the following statements:
1. Auxiliary carry flag is used only by DAA and DAS instructions
2. Zero flag is set to 1 if the two operands compared are equal
3. All conditional jumps are long-type jumps
Which of the above statements are correct?
A. 1, 2 and 3 B. 1 and 2 only
C. 1 and 3 only D. 2 and 3 only

65. If a three-phase slip ring induction motor is fed from the rotor side with stator winding short circuited, then frequency of currents flowing in the short circuited stator is
A. slip × frequency B. supply frequency
C. frequency corresponding to rotor speed
D. zero

66. The reversing of a 3ϕ induction motor is achieved by
A. Y–Δ starter B. DOL starter
C. auto transformer
D. interchanging any two of the supply line

67. Consider the following interrupts for 8085 microprocessor:
1. INTR 2. RST 5.5
3. RST 6.5 4. RST 7.5
5. TRAP
If the interrupt is to be vectored to any memory location then which of the above interrupts is/are correct?
A. 1 and 2 only B. 1, 2, 3 and 4
C. 5 only D. 1 only

68. The instruction JNC 16-bit refers to Jump to 16-bit address if
A. sign flag is set B. CY flag is reset
C. zero flag is set D. parity flag is reset

69. Consider the symbol shown below:

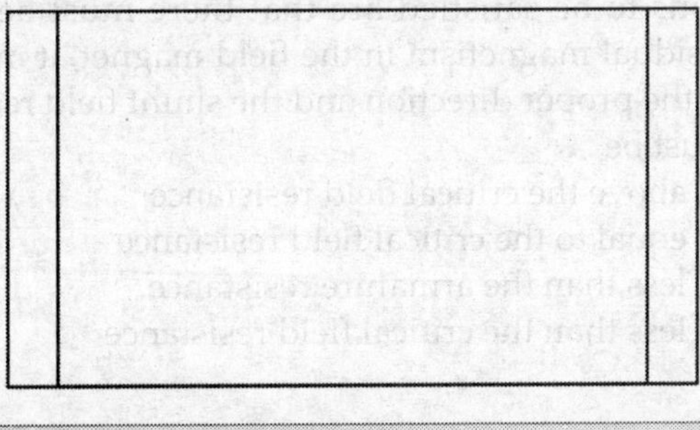

What function does the above symbol represent in a program flow chart?
A. A process B. Decision-making
C. A subroutine D. Continuation

70. Which one of the following statements is correct regarding the instruction CMP A?
A. Compare accumulator with register A
B. Compare accumulator with memory
C. Compare accumulator with register H
D. This instruction does not exist

71. The instruction RET executes with the following series of machine cycle
A. fetch, read, write B. fetch, write, write
C. fetch, read, read D. fetch, read

72. Consider the following circuits:
1. Full adder 2. Half adder
3. JK flip-flop 4. Counter
Which of the above circuits are classified as sequential logic circuits?
A. 1 and 2 B. 3 and 4
C. 2 and 3 D. 1 and 4

73. When a peripheral is connected to the Microprocessor in Input Output mode, the data transfer takes place between
A. any register and I/O device
B. memory and I/O device
C. accumulator and I/O device
D. HL register and I/O device

74. While execution of IN/OUT instruction takes place, the 8-bit address of the port is placed on
A. lower address bus
B. higher address bus
C. data bus
D. lower as well as higher order address bus

75. The port C of 8255 can be configured to work in
A. mode 0, mode 1, mode 2 and BSR
B. mode 0, mode 1 and mode 2
C. mode 2 and BSR
D. BSR mode only

76. Consider the following statements:
1. Semiconductor memories are organized as linear array of memory locations
2. To address a memory location out of N memory locations, at least log N bits of address are required
3. 8086 can address 1,048,576 addresses
4. Memory for an 8086 is set up as two banks to make it possible to read or write a word with one machine cycle
Which of the above statements are correct?
A. 1, 2 and 3 only B. 1, 2 and 4 only
C. 3 and 4 only D. 1, 2, 3 and 4

Ans. 63. D 64. B 65. A 66. D 67. D 68. B 69. C 70. A 71. C 72. B 73. C 74. D 75. A 76. C

77. The sticker over the EPROM window protects the chip from
A. infrared light from sunlight
B. UV light from fluorescent lights and sunlight
C. magnetic field D. electrostatic field

78. A 2400/240 V, 200 kVA, single-phase transformer has a core loss of 1.8 kW at rated voltage. Its equivalent resistance is 1.1%, then the transfer efficiency at 0.9 power factor and on full load is
A. 95.60% B. 96.71%
C. 97.82% D. 98.93%

79. The 8259A Programmable Interrupt Controller in cascade mode can handle interrupts of
A. 8 priority levels B. 16 priority levels
C. 32 priority levels D. 64 priority levels

80. 8259A Programmable Interrupt Controller uses the following initialization commands:
1. ICW_1 2. ICW_2
3. ICW_3 4. ICW_4
If 8259A is to be used in cascaded and fully nested mode, the ICW_1 bits D_0 and D_1 are
A. 0 and 0 B. 1 and 0
C. 0 and 1 D. 1 and 1

81. The induced emf in the armature conductor of a DC machine is
A. sinusoidal B. trapezoidal
C. rectangular unidirectional
D. triangular

82. If a carrier of 100% modulated AM is suppressed before transmission, the power saving is nearly
A. 50% B. 67%
C. 100% D. 125%

83. An FM signal is represented by $v = 12 \sin (6 \times 10^8 t + 5 \sin 1250t)$. The carrier frequency f_c and frequency deviation δ, respectively, are
A. 191 MHz and 665 Hz B. 95.5 MHz and 995 Hz
C. 191 MHz and 995 Hz D. 95.5 MHz and 665 Hz

84. When the modulating frequency is doubled the modulation index is halved and the modulating voltage remains constant. This happens when the modulating system is
A. AM B. PM
C. FM D. delta modulation

85. $v = A \sin (\omega_c t + m \sin \omega_m t)$ is the expression for
A. amplitude modulated signal
B. frequency modulated signal
C. phase modulated signal
D. carrier signal used for modulation

86. The four basic elements in a PLL are loop filter, loop amplifier, VCO and
A. up converter B. down converter
C. phase detector D. frequency multiplier

87. In a frequency modulated (FM) system, when the audio frequency is 500 Hz and audio frequency voltage is 2.4 V, the frequency deviation δ is 4.8 kHz. If the audio frequency voltage is now increased to 7.2 V then what is the new value of deviation?
A. 0.6 kHz B. 3.6 kHz
C. 12.4 kHz D. 14.4 kHz

88. Modulation is used to
1. Separate different transmission
2. Reduce the bandwidth requirement
3. Allow the use of practicable antennas
4. Ensure that intelligence may be transmitted over long distances
Which of the above statements are correct?
A. 1, 2 and 3 only B. 1, 3 and 4 only
C. 2 and 4 only D. 1, 2, 3 and 4

89. Carson's rule is (with symbols having their standard meaning)
A. B = 2 DW B. B = 2 (D + 1) W
C. $B = \sqrt{2}$ (D + 1) W D. $B = \sqrt{2}$ DW

90. Consider the following features of FM vis-a-vis AM:
1. Better noise immunity is provided
2. Lower bandwidth is required
3. The transmitted power is better utilized
4. Less modulating power is required
Which of the above are advantages of FM over AM?
A. 1, 2 and 3 only B. 1, 3 and 4 only
C. 2 and 4 only D. 1, 2, 3 and 4

91. The ideal characteristic of a stabilizer is
A. constant output voltage with low internal resistance
B. constant output current with low internal resistance
C. constant output voltage with high internal resistance
D. constant internal resistance with variable output voltage

92. For a dc shunt generator to self excite, the conditions to be satisfied are that there must be some residual magnetism in the field magnet, it must be in the proper direction and the shunt field resistant must be
A. above the critical field resistance
B. equal to the critical field resistance
C. less than the armature resistance
D. less than the critical field resistance

Ans. 77. B 78. C 79. D 80. C 81. A 82. B 83. B 84. C 85. C 86. C 87. D 88. B 89. B 90. B 91. A 92. D

Multiple Choice Questions from Various Competitive Examinations Including GATE and IES with Explanations

2017

1. When the temperature of a ferromagnetic material exceeds the Curie temperature, it behaves similar to a
A. Diamagnetic material
B. Ferrimagnetic material
C. Paramagnetic material
D. Antiferromagnetic material

2. Photoconductivity is a characteristic of semiconductors. When light falls on certain semiconductors, it
A. Sets free electrons from some of the atoms, increasing the conductivity
B. Ejects electrons into space
C. Establishes a potential difference creating a source of EMF
D. Produces heat raising the temperature.

3. The resistivity of intrinsic germanium at 30°C is 0.46Ω·m. What is the intrinsic carrier density n_i at 30°C, taking the electron mobility as 0.38 $m^2/V \cdot s$ and hole mobility μ_n as $0.38^2/V \cdot s$ and hole mobility μ_p as 0.18 $m^2/V \cdot s$.
A. $2.4 \times 10^{19}/m^3$ B. $4.2 \times 10^{19}/m^3$
C. $2.4 \times 10^{10}/m^3$ D. $4.2 \times 10^{18}/m^3$

Explanation: $\rho = \dfrac{1}{ne \cdot \mu_n + p \cdot e\mu_p}$

$$= \frac{1}{n_i[\mu_n + \mu_p]e}$$

[$\because n = \rho = n_1$ for an intrinsic semiconductor]

$$n_1 = \frac{1}{\rho[\mu_n + \mu_p]e}$$

$$= \frac{1}{0.46[0.38 + 0.18] \times 1.602 \times 10^{-19}}$$

$$n_i = 2.4 \times 10^{19}/m^2$$

4. For intrinsic gallium arsenide, conductivity at room temperature is 10^{-6} $(W \cdot m)^{-1}$, the electron and hole mobilities are, respectively 0.85 and 0.04 $m^2/V\text{-}s$. The intrinsic carrier concentration n at room temperature is
A. $7.0 \times 10^{12}\ m^{-3}$ B. $0.7 \times 10^{12}\ m^{-3}$
C. $7.0 \times 10^{-12}\ m^{-3}$ D. $0.7 \times 10^{-12}\ m^3$

Explanation: $n_i = \dfrac{\sigma}{[\mu_n + \mu_p]e}$

$$= \frac{10^{-6}}{0.89 \times 1.602 \times 10^{-19}}$$

$$= 7.0137 \times 10^{12}\ m^{-3}$$

5. Consider the following statements:
1. The critical magnetic field of a superconductor is maximum at absolute zero
2. Transition temperature of a superconductor is sensitive to its structure
3. The critical magnetic field of a superconductor is zero at its critical temperature
4. Superconductors show very high conductivity below the critical temperature

Which of the above statements are correct?
A. 1, 2 and 3 only B. 1, 2 and 4 only
C. 2, 3 and 4 only D. 1, 3 and 4 only

6. Which of the following statements are correct regarding dot product of vectors?
1. Dot product is less than or equal to the product of magnitudes of two vectors
2. When two vectors are perpendicular to each other, then their dot product is non-zero
3. Dot product of two vectors is positive or negative depending whether the angle between the vectors is less than or greater than π/2
4. Dot product is equal to the product of one vector and the projection of the vector on the first one

Select the correct answer using the codes given below:
A. 1, 2 and 3 only B. 1, 3 and 4 only
C. 1, 2 and 4 only D. 2, 3 and 4 only

7. Susceptibility of a diamagnetic material is
1. Negative 2. Positive
3. Dependent on the temperature
4. Independent of the temperature

Select the correct answer using the codes given below:
A. 1 and 3 only B. 2 and 3 only
C. 1 and 4 only D. 2 and 4 only

8. Consider the following statements:
1. The susceptibility χ of diamagnetic materials is small and negative
2. The susceptibility of para and anti-ferromagnetic materials is small but positive
3. The susceptibility has a finite value for free space or air.

Which of the above statements are correct?
A. 1 and 2 only B. 1 and 3 only
C. 2 and 3 only D. 1, 2 and 3

9. Eddy current losses in transformer cores can be reduced by the use of
1. Solid cores 2. Laminated cores
3. Ferrites

Select the correct answer using the codes given below:

A. 2 and 3 only B. 1 and 2 only
C. 1 and 3 only D. 1, 2 and 3

Explanation: Eddy current losses can be reduced by having thin laminated construction

$\left(\because R=\frac{\rho l}{a}\right)$ if a reduces $\Rightarrow R$ increases and I decreases

Ferrite material will possess more magnetic properties than conducting properties. Therefore *electron conduction* will be less and hence Eddy losses also less.

10. The phenomenon of magnetostriction occurs when a ferromagnetic substance is magnetized resulting in
A. Heating
B. Small changes in its dimensions
C. Small changes in its crystal structure
D. Some change in its mechanical properties

11. Which type of defect causes F-centres in a crystal?
A. Stoichiometric defect
B. Metal excess defect due to anion vacancies
C. Metal excess defect due to extra cations
D. Frenkel defect

Hint: Schottky defects

12. Consider the following statements:
1. Superconductors exhibit normal conductivity behaviour above a transition temperature T_c
2. Superconductors lose their superconducting nature in an external magnetic field, provided the external magnetic field is above a critical value
3. High T_c superconductors have T_c values in the range 1 to K.

Which of the above statements are correct?
A. 1 and 2 only B. 1 and 3 only
C. 2 and 3 only D. 1, 2 and 3

Hint: $T > T_c \Rightarrow$ Loss of superconductivity
$H > H_c \Rightarrow$ Loss of superconductivity

13. Superconductivity is a material property associated with
A. Changing shape by stretching
B. Stretching without breaking
C. A loss of thermal resistance
D. A loss of electrical resistance

14. An atom in a crystal vibrates at a frequency, determined by
1. Crystal heat current
2. Crystal temperature
3. The stiffness of the bonds with neighbour atoms

Select the correct answer using the codes given below:
A. 1 only B. 2 only
C. 3 only D. 1, 2 and 3

15. Consider the following statement
1. Nano means 10^{-9} so that nano materials have an order of dimension higher than the size of atom and come in the form of rods, tubes, spheres or even thin sheets/films
2. Nano materials have enhanced or changed structural property
3. Nano elements lend themselves to mechanical processing like rolling, twisting, positioning
4. Nano elements show important electrical, magnetic and optical characteristics that are useful in electrical industry

Which of the above statements are correct?
A. 1, 2 and 3 only B. 1, 2, 3 and 4
C. 3 and 4 only D. 1, 2 and 4 only

16. The voltage and current waveforms for an element are shown in the figure.

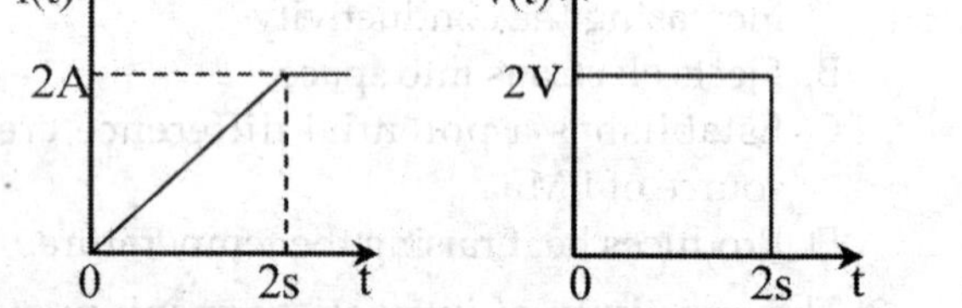

The circuit element and its value are
A. Capacitor, 2 F B. Inductor, 2 H
C. Capacitor, 0.5 F D. Inductor, 0.5 H

Explanation: By comparing $v(t)$ and $i(t)$ we can say that

$$v(t) = K\frac{di(t)}{dt}$$

So it is inductor, but $\frac{di(t)}{dt} = 1$

$$v(t) = L\frac{di(t)}{dt}$$

$$\Rightarrow \quad 2 = L1 \Rightarrow L = 2H$$

17. In a connected graph, the total number of branches is b and the total number of nodes is n. Then the number of links L of a co-tree is
A. $b - n$ B. $b - n - 1$
C. $b + n - 1$ D. $b - n + 1$

Explanation: Number of links $= [(\text{No. of branch}) - (\text{No. of nodes})] - 1 = b - n + 1$.

18. For the circuit shown, Thevenin's open circuit voltage V_{oc} and Thevenin's equivalent resistance R_{eq} at terminals A – B are, respectively

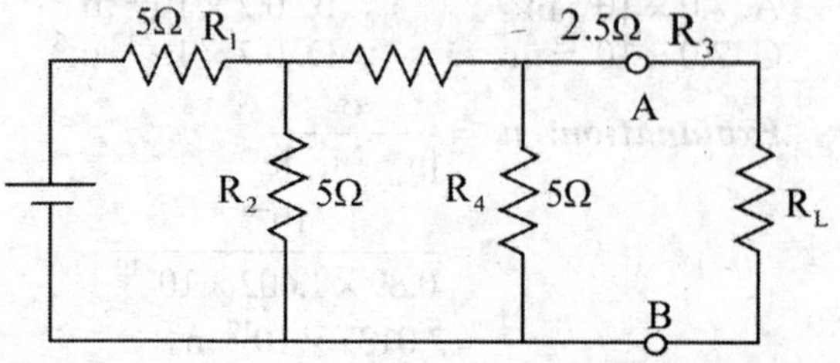

A. 6.25 V and 2.5 Ω B. 12.5 V and 5 Ω
C. 6.25 V and 5 Ω D. 12.5 V and 2.5 Ω

Explanation: To find V_{OC}, remove R_L and measure V_{OC}

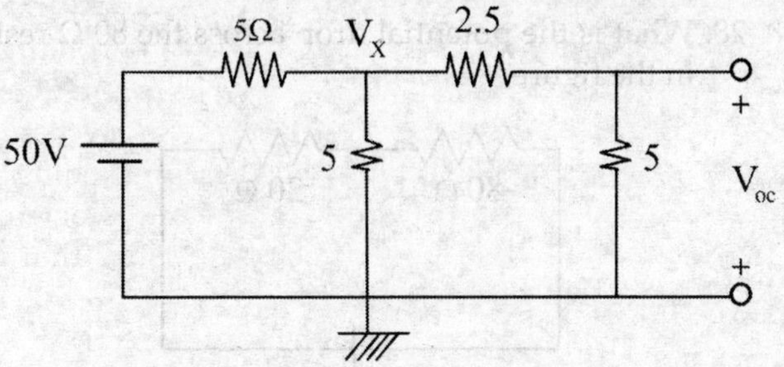

$$\frac{V_x - 50}{5} + \frac{V_x}{5} + \frac{V_x}{7.5} = 0 \Rightarrow V_x = \frac{75}{4} V$$

$$V_{oc} = \frac{5}{2.5+5} V_x = \frac{5}{7.5} \times \frac{75}{4} = 12.5\,V$$

$\rightarrow R_{th} = 5 \,||\, [2.5 + (5 \,||\, 5)] = 5 \,||$

$[2.5 + 2.5] = 5 \,||\, 5 = 2.5\,\Omega$

19. What is the current through the 8Ω resistance connected across terminals, M and N in the circuit?

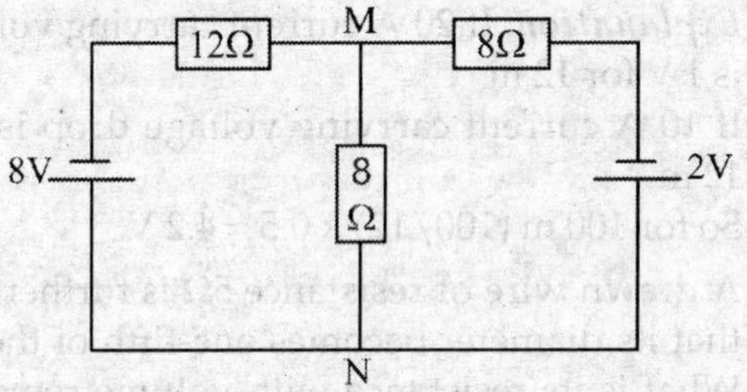

A. 0.34 A from M to N B. 0.29 A from M to N
C. 0.29 A from N to M D. 0.34 A from N to M

Explanation:

$$\frac{V_x + 8}{12} + \frac{V_x}{8} + \frac{V_x + 2}{8} = 0 \Rightarrow V_x = -11/4$$

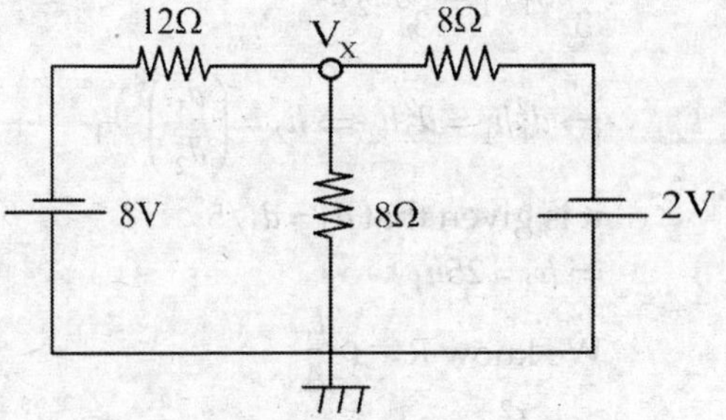

$$I_{MN} = \frac{-11}{32};\, I_{NM} = \frac{11}{32} = 0.34A$$

20. Two resistors of 5Ω and 10Ω and an inductor L are connected in series across a 50 cosωt voltage source. If the power consumed by the 5Ω resistor is 10W, the power factor of the circuit is

A. 1.0 B. 0.8
C. 0.6 D. 0.4

Explanation:

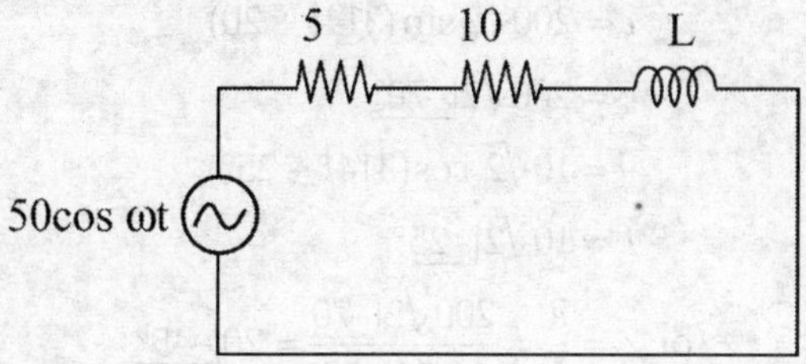

$$P_{5\Omega} = 10\text{ W} \Rightarrow I_{rms}{}^2 R = 10$$

$$\Rightarrow |I_{rms}| = \sqrt{\frac{10}{5}} = \sqrt{2}, |V_{rms}| = \frac{50}{\sqrt{2}}$$

$$\rightarrow |Z| = \left|\frac{V_{rms}}{I_{rms}}\right| = \frac{50}{2} = 25\,\Omega,\ R = |Z| \cos\phi$$

$$\Rightarrow \cos\phi = \frac{R}{|Z|} = \frac{5+10}{25} = 0.6$$

21. What is the value of resistance *R* which will allow maximum power dissipation in the circuit?

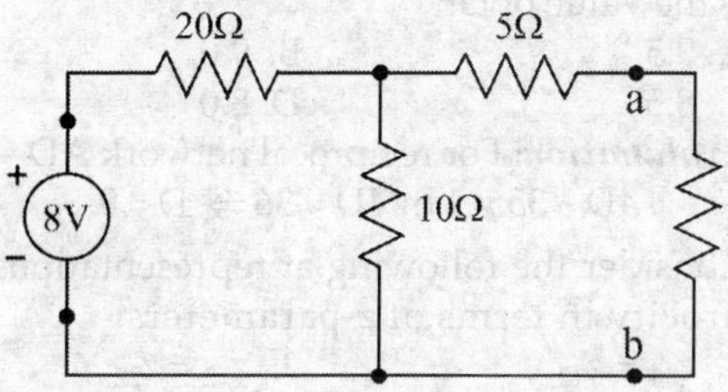

A. 11.66 Ω B. 10.33 Ω
C. 8.33 Ω D. 7.66 Ω

22. How fast can the output of an OP Amp change by 10 V, if its slew rate is 1 V/μs?

A. 5 μs B. 10 μs
C. 15 μs D. 20 μs

Explanation:

dv_0 = change of output voltage = 10 V

d = slew rate = 1 V/μs = 10^6 V/s

dt = change of time

$$= \left.\frac{dv}{dt}\right|_{max} \Rightarrow dt = \frac{dv}{s} = \frac{10}{10^6} = 10\,\mu s$$

23. A two-element series circuit is connected across an AC source given by $e = 200\sqrt{2}\sin(314t + 20)$V. The current is then found to be $= 10\sqrt{2}\cos(314t - 25)A$. The parameters of the circuit are

A. $R = 20\,\Omega$ and $C = 160\,\mu F$
B. $R = 14.14\,\Omega$ and $C = 225\,\mu F$

C. L = 45 mH and C = 225 μF
D. l = 45 mH and C = 160 μF
Explanation:

$$e = 200\sqrt{2}\sin(314t + 20)$$
$$\bar{e} = 200\sqrt{2}\angle -70°$$
$$i = 10\sqrt{2}\cos(314t - 25°)$$
$$\bar{I} = 10\sqrt{2}\angle -25°$$
$$\text{or } \bar{Z} = \frac{\bar{e}}{\bar{I}} = \frac{200\sqrt{2}\angle -70}{10\sqrt{2}\angle -25} = 20\angle -45°$$
$$= 20[\cos 45 - j\sin 45] = 20\frac{1}{\sqrt{2}} - j20\frac{1}{\sqrt{2}}$$
$$= 10\sqrt{2} - j10\sqrt{2}$$
$$Z = R - j\frac{1}{\omega C};\ R = 10\sqrt{2} = 14.14$$
$$\frac{1}{\omega C} = 10\sqrt{2}$$
$$\Rightarrow C = \frac{1}{\omega \times 10\sqrt{2}} = \frac{1}{314\ 10\sqrt{2}} = 225\ \mu F$$

24. For a two-part reciprocal network, the three transmission parameters are A = 4, B = 7 and C = 5. What is the value of D?
A. 9.5 B. 9.0
C. 8.5 D. 8.0
Explanation: For reciprocal network AD – BC = 1
4D – 35 = 1 or 4D = 36 ⇒ D = 9

25. Consider the following at representations of reciprocity in terms of z-parameters:
1. $z_{11} = z_{12}$ 2. $z_{12} = z_{22}$
3. $z_{12} = z_{21}$
Which of the above representations is/are correct?
A. 1 only B. 2 only
C. 3 only D. 1, 2 and 3
Explanation: $Z_{12} = Z_{21}$

26. A parallel plate capacitor is made of two circular plates separated by a distance of 5 mm and with a dielectric with dielectric constant of 2.2 between them. When the electric field in the dielectric is 3×10^4 V/m, the charge density of the positive plate will be nearly
A. $58.5 \times 10^4\ C/m^2$ B. $29.5 \times 10^4\ C/m^2$
C. $23.4 \times 10^{-4}\ C/m^2$ D. $58.5 \times 10^{-4}\ C/m^2$

Explanation. $Q = CV = \frac{\epsilon A}{d} \times \frac{E}{d}$

$$\Rightarrow \frac{Q}{A} = \frac{\epsilon \times E}{d^2} = \frac{1}{36\pi} \times \frac{10^{-9} \times 2.2 \times 3 \times 10^4}{25 \times 10^{-6}}$$
$$= 23.34 \times 10^{-4}\ mC/m^2$$

27. A three phase star connected load is operating at a power factor angle φ, with φ being the angle between
A. line voltage and line current
B. phase voltage and phase current
C. line voltage and phase current
D. phase voltage and line current

28. What is the potential drop across the 80 Ω resistor in the figure?

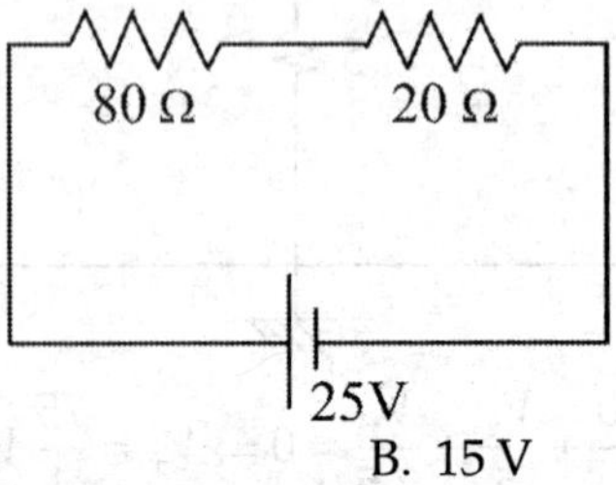

A. 20 V B. 15 V
C. 10 V D. 5 V
Explanation:
$$V_{80\Omega} = \frac{80}{80 + 20} \times 25 = \frac{80}{100} \times 25 = \frac{80}{4} = 20\ V$$

29. When 7/0.029 VIR cable is carrying 20A, a drop of 1 V occurs every 12 m. The voltage drop in a 100 m run of this cable when it is carrying 10A is nearly
A. 4.2 V B. 3.2 V
C. 1.2 V D. 0.42 V
Explanation: If 20 A current carrying voltage drop is 1 V for 12 m
If 10 A current carrying voltage drop is 0.5 V for 12 m
So for 100 m (100/12) × 0.5 = 4.2 V

30. A drawn wire of resistance 5Ω is further drawn so that its diameter becomes one-fifth of the original. What is its resistance with volume remaining the same?
A. 25 Ω B. 125 Ω
C. 625 Ω D. 3125 Ω
Explanation: Since volume remains same, we can write
$$\frac{4}{3}\pi r_1^2 h_1 = \frac{4}{3}\pi r_2^2 h_2$$
$$\Rightarrow d_1^2 h_1 = d_2^2 h_2 \Rightarrow h_2 = \left(\frac{d_1}{d_2}\right)^2 h_1$$
it is given that $d_2 = d_1/5$
$\Rightarrow h_2 = 25h_1$
We know $R = \rho\frac{L}{A}$
$$= \rho\frac{L^2}{AL} = \frac{\rho}{V}L^2 \Rightarrow R\alpha l^2$$
$$\frac{R_1}{R_2} = \left(\frac{l_1}{l_2}\right)^2$$
$$\Rightarrow R_2 = R_1\left(\frac{l_1}{l_2}\right)^2 = 5(25)^2 = 625\Omega$$

31. The three non-inductive loads of 5 kW, 3 kW and 2 kW are connected in a star network between R, Y and B phases and neutral. The line voltage is 400 V. The current in the neutral wire is nearly

A. 11 A B. 14 A
C. 17 A D. 21 A

Explanation: $P_{RN} = V_R \cdot I_R$

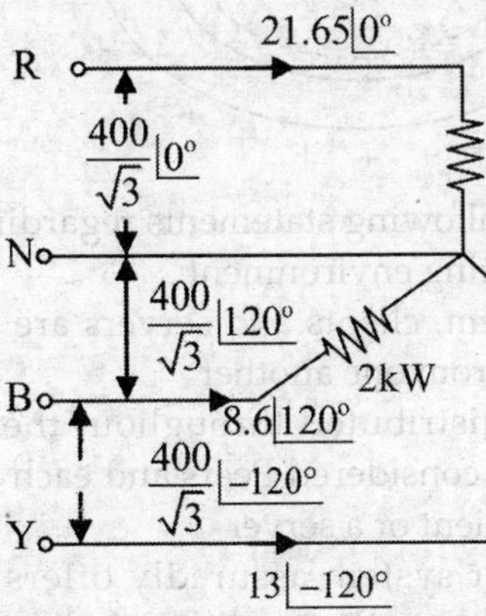

$$I_{RN} = \frac{P_R}{V_R} = \frac{5\times10^3}{400/\sqrt{3}} = 21.65\angle 0^\circ$$

$$I_{BN} = \frac{P_B}{V_B} = \frac{2\times10^3}{400/\sqrt{3}} = 8.6\angle +120^\circ$$

$$I_{YB} = \frac{P_Y}{V_Y} = \frac{3\times10^3}{400/\sqrt{3}} = 13\angle -120^\circ$$

$$I_N = I_{RN} + I_{BN} + I_{YB}$$

$$= 21.65\angle 0^\circ + 8.6\angle 120^\circ + 13\angle -120^\circ = 11.45\angle 19.11^\circ$$

32. Consider the following statements:
If a high Q parallel resonant circuit is loaded with a resistance
1. The circuit impedance reduces
2. The resonant frequency remains the same
3. The bandwidth reduces
Which of the above statements is/are correct?

A. 3 only B. 2 only
C. 1 only D. 1, 2 and 3

33. Kirchoff's current law is applicable to
1. Closed loops in a circuit
2. Junction in a circuit
3. Magnetic circuits
Which of the above is/are correct?

A. 1 only B. 2 only
C. 3 only D. 1, 2 and 3

34. Which of the following are satisfied in a non-linear network?
1. Associative 2. Superposition
3. Homogeneity 4. Bilaterality
Select the correct answer using the codes given below:

A. 1 and 3 only B. 1 and 4 only
C. 2 and 3 only D. 2 and 4 only

35. $\nabla \times \bar{H} = \sigma E + \varepsilon\left(\frac{\partial E}{\partial t}\right)$ is

A. Biot–Savart law B. Gauss's law
C. modified Faraday's law
D. modified Ampere's law

36. Consider the following statements:
1. Network theorems are not derivable from Kirchhoff's law
2. To get the Norton current, one has to short the current source
3. Thevenin's theorem is suitable for a circuit involving voltage sources and series connections
Which of the above statements is/are correct?

A. 1, 2 and 3 B. 1 only
C. 2 only D. 3 only

37. What are the Thevenin's equivalent voltage V_{TH} and resistance R_{TH} between the terminals A and B of the circuit?

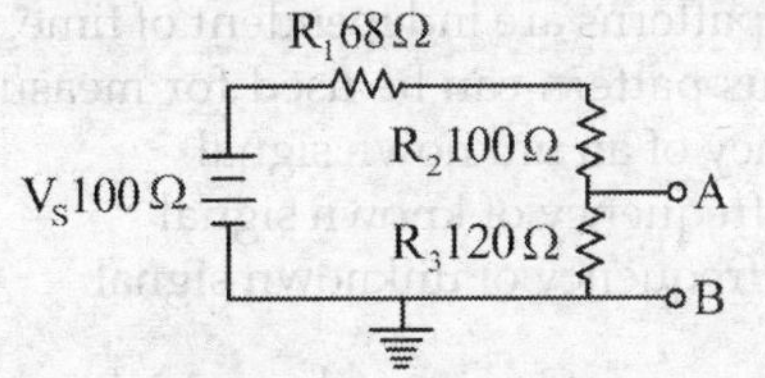

A. 4.16 V and 120 Ω B. 41.67 V and 120 Ω
C. 4.16 V and 70 Ω D. 41.67 V and 70 Ω

Explanation:

$$V_{th} = \frac{100}{68+100+120}120 = 41.67\text{ V}$$

$$R_{th} = 168 \times 120/288 = 70\ \Omega$$

38. What is the current through the 5 Ω resistance in the circuit shown?

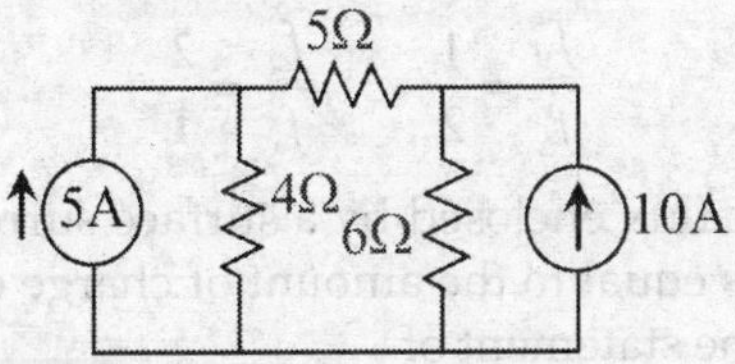

A. 5.33 A B. 4.66 A
C. 2.66 A D. 1.33 A

Explanation: Using source transformation

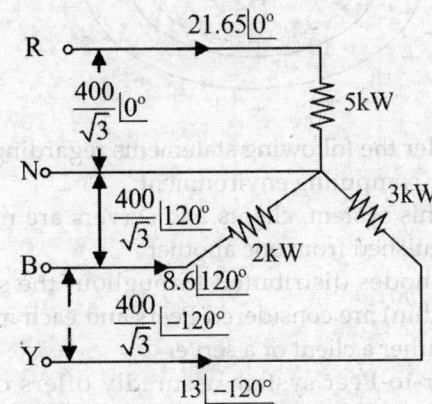

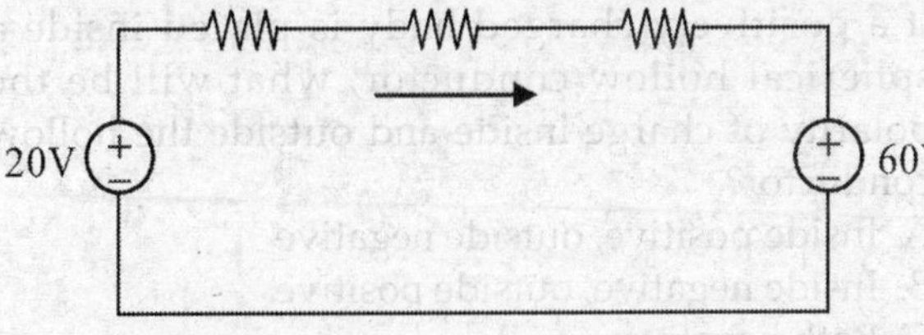

$$I = \frac{60-20}{15} = \frac{40}{15} = 2.66\text{ A}$$

39. Consider the following statements with regard to Lissajous pattern on a CRO:
   1. It is stationary pattern on the CRO
   2. It is used for precise measurement of frequency of a voltage signal
   3. The ratio between frequencies of vertical and longitudinal voltage signals should be an integer to have a steady Lissajous pattern

   Which of the above statements is/are correct?
   A. 1 only B. 2 only
   C. 3 only D. 1, 2 and 3

Explanation: Lissajous pattern is a stationary pattern.

Example:

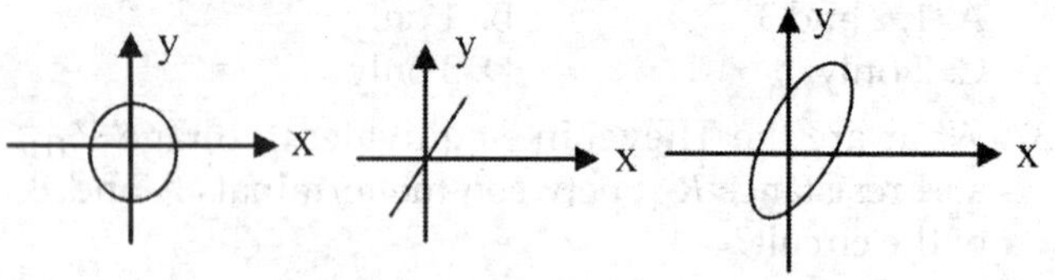

All the patterns are independent of time.

Lissajous pattern can be used for measurement of frequency of an unknown signal.

If f_x = frequency of known signal

f_y = frequency of unknown signal

then $\frac{f_y}{f_x} = \frac{\text{no. of horizontal tangencies}}{\text{no. of vertical tangencies}}$ (an integer)

Example:

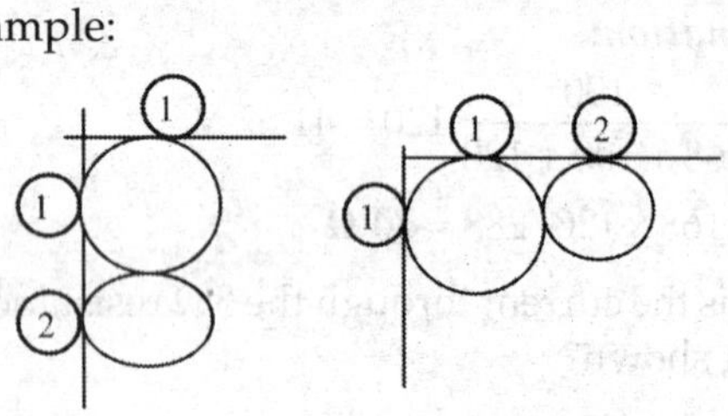

$\frac{f_y}{f_x} = \frac{1}{2}$ $\frac{f_y}{f_x} = \frac{2}{1}$

40. "Electric flux enclosed by a surface surrounding a charge is equal to the amount of charge enclosed." This is the statement of
   A. Faraday's law
   B. Lenz's law
   C. Modified Ampere's law
   D. Gauss's law

41. It a positively charged body is placed inside a spherical hollow conductor, what will be the polarity of charge inside and outside the hollow conductor?
   A. Inside positive, outside negative
   B. Inside negative, outside positive
   C. Both negative
   D. Both positive

Explanation:

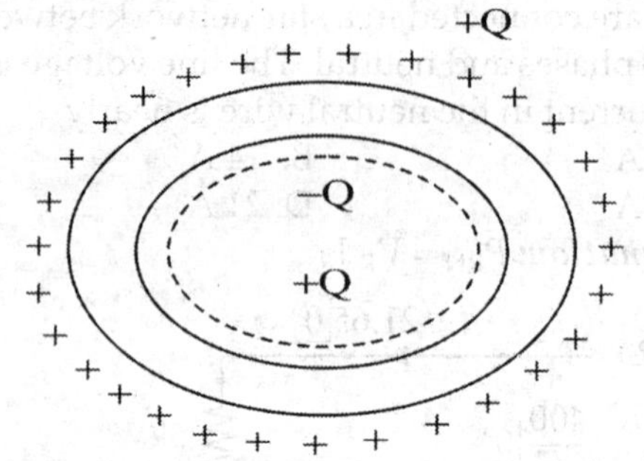

42. Consider the following statements regarding Peer-to-Peer computing environment:
   1. In this system, clients and servers are not distinguished from one another
   2. All nodes distributed throughout the system (within) are considered Peers and each may act as either a client or a server
   3. Peer-to-Peer system assuredly offers certain advantages over the traditional client server sstem
   4. Peer-to-Peer system is just a replica of the file server system

   Which of the above statements are correct?
   A. 1, 2, 3 and 4 B. 1, 2 and 3 only
   C. 1 and 4 only D. 1, 2 and 4 only

43. The Laplace transform of the below function is
   A. $F(s) = 8s(1 - e^{-s})$ B. $F(s) = 8/s(1 + e^{-s})$
   C. $F(s) = 8s(1 + e^{-s})$ D. $F(s) = 8/s(1 - e^{-s})$

44. The number of complex additions and multiplications in direct DFT are, respectively
   A. $N(N-1)$ and N^2 B. $N(N+1)$ and N^2
   C. $N(N+1)^2$ and N D. N and N^2

45. The Fourier transform of a unit rectangular pulse shown in the figure is

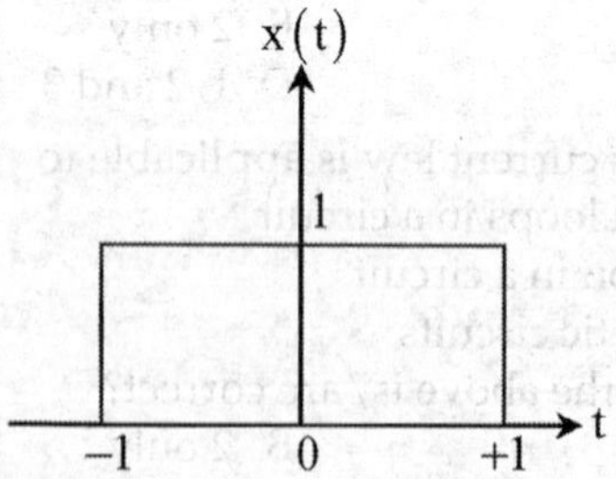

   A. $\omega \sin \omega$ B. $2 \sin \omega$
   C. $\omega / \sin \omega$ D. $\cos \omega / 2\omega$

46. The number of complex additions and multiplications in FFT are, respectively
   A. $N/2 \log_2 N$ and $N \log_2 N$
   B. $N \log_2 N$ and $N/2 \log_2 N$
   C. $N/2 \log_2 N$ and $\log_2 N$
   D. $\log_2 N$ and $N/2 \log_2 N$

47. A power system has two synchronous generators having governor turbine characteristic
$P_1 = 50\,(50 - f)$
$P_2 = 100\,(51 - f)$
where f represents the system frequency. Assuming a lossless operation of the complete power system, what is the system frequency for a total load of 800 MW?

A. 55.33 Hz B. 50 Hz
C. 45.33 Hz D. 40 Hz

Explanation: $P_1 = 50\,(50 - f)$, $P_2 = 100\,(51 - f)$
Total load, $P = 800$ MW

$$P = P_1 + P_2$$
$$800 = 50\,(50 - f) + 100\,(51 - f)$$
$$= 2500 - 50f + 5100 - 100f$$
$$800 = 7600 - 150f$$
$$\Rightarrow \quad 150f = 7600 - 800$$
$$f = 45.33 \text{ Hz}$$

48. Two networks are connected in cascade in the figure. The equivalent ABCD constants are obtained for the combined network having $C = 0.1\angle 90°$. What is the value of Z_2?

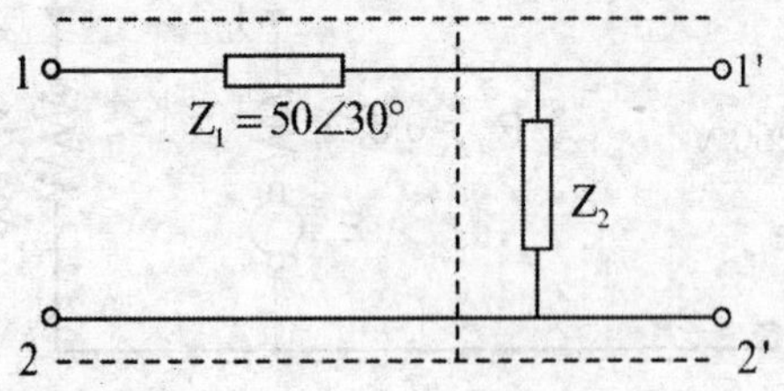

A. $500\angle -60°$ B. $0.10j$
C. $-10j$ D. $50\angle -60°$

Explanation:

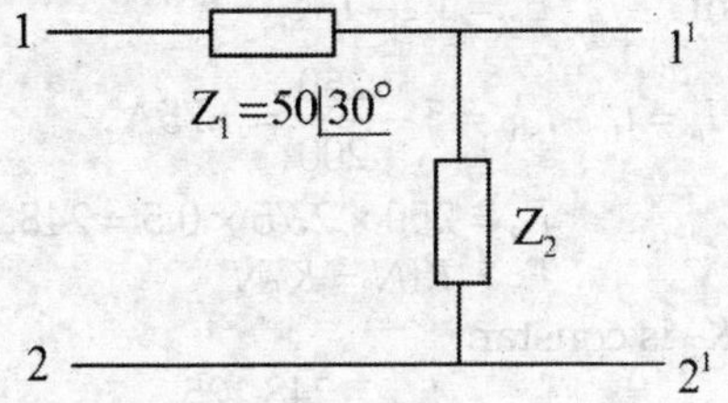

$$Z_2 = -10j\,\Omega$$

49. Which one of the following does not have an effect on corona?

A. Spacing between conductors
B. Conductor size
C. Line voltage
D. Length of conductor

Hint: Corona is independent on length of conductor.

50. Consider the following statements regarding corona:
1. It causes radio interference
2. It attenuates lightening surges
3. It causes power loss
4. It is more prevalent in the middle conductor of a transmission line employing flat conductor configurations

Which of the above statements are correct?

A. 1, 2 and 3 only B. 1, 2 and 4 only
C. 1, 2, 3 and 4 D. 3 and 4 only

51. The loss formula coefficient matrix for a two plant system is given by

$$B = \begin{bmatrix} 0.001 & -0.0001 \\ -0.0001 & 0.0013 \end{bmatrix} M\omega^{-1}$$

The economic schedule for a certain load is given as $P_1 = 150$ MW and $P_2 = 275$ MW.

What is the penalty factor for plant 1 for this condition?

A. 1.324 B. 1.515
C. 1.575 D. 1.721

Explanation: We have formula for Penalty factor

$$L_1 = \frac{1}{1 - \dfrac{\partial p_{loss}}{\partial P_{GI}}}$$

$$P_{loss} = H_{11}P_{G1}^2 + B_{22}P_{G2}^2 + 2P_{G1}P_{G2}B_{12}$$

$$\frac{\partial p_{loss}}{\partial P_{G1}} = 2 \cdot B_{11}P_{G1} + 2P_{G2} \cdot B_{12}$$

$$\text{Given } B = \begin{bmatrix} 0.001 & -0.0001 \\ -0.0001 & 0.0013 \end{bmatrix} = \begin{bmatrix} B_{11} & B_{12} \\ B_{21} & B_{22} \end{bmatrix}$$

$$P_{GI} = 150 \text{ Mw}; \; P_{G2} = 275 \text{ Mw}$$

substituting all the values, we get

$$\frac{\partial p_{loss}}{\partial P_{G1}} = (2 \times 0.001 \times 150) + [2 \times 275 \times (-0.0001)]$$
$$= 0.245$$

Penalty factor or plant-1

$$L_1 = \frac{1}{1 - 0.245} = 1.324$$

52. A large dc motor is required to control the speed of the blower from a 3-phase ac source. The suitable AC to DC converter is 3-phase

A. fully controlled bridge converter
B. fully controlled bridge converter with free-wheeling diode
C. half controlled bridge converter
D. converter pair in sequence control

Hint: For motoring operation, half controlled bridge is suitable

2018

1. What is phasor sum of currents $I_1 = (10a - a^2)$ and $I_2 = -j\,10$ for two complex operators which are individually defined by $a^3 = 1$ and $j^2 = 1$?
A. 17.32∠90° B. 7.32∠90°
C. 17.32∠0° D. 7.32∠0°
Explanation: We know that

$$1 + a + a^2 = 0$$
$$\therefore \quad a^2 = -(1 + a) \qquad \text{(i)}$$

Given, $I_1 = 10(a - a^2)$

Using equation (i)

$$I_1 = 10(a + (1 + a))$$
$$= 10(2a + 1)$$
$$\therefore \quad I = I_1 + I_2$$
$$= 10(2 \times 1 \angle 120° + 1) - j10$$
$$= 7.32\angle 90°$$

2. A 220 V DC compound generator connected in long-shunt mode has the following parameters: $R_a = 0.1\Omega$, $R_{sh} = 80\Omega$, $R_{series} = 0.05\Omega$. For a load of 150A at rated terminal voltage, the induced emf of the generator should nearly by
A. 233 V B. 243 V
C. 251 V D. 262 V
Explanation: DC compound generator is long shunt mode.

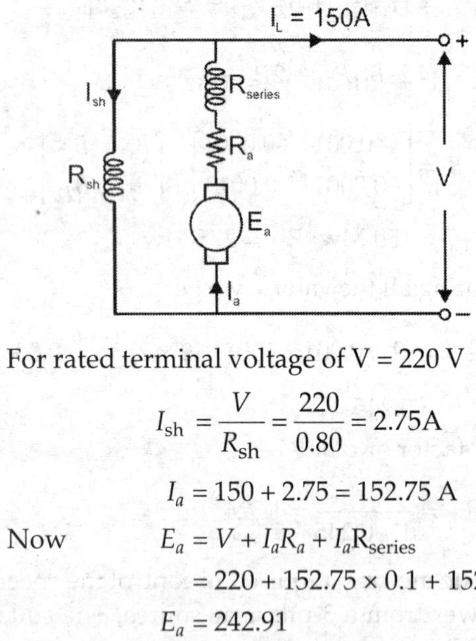

For rated terminal voltage of V = 220 V

$$I_{sh} = \frac{V}{R_{sh}} = \frac{220}{0.80} = 2.75\text{A}$$
$$I_a = 150 + 2.75 = 152.75 \text{ A}$$

Now $E_a = V + I_aR_a + I_aR_{series}$

$$= 220 + 152.75 \times 0.1 + 152.75 \times 0.05$$
$$E_a = 242.91$$

The nearest option is $E_a = 243$ V

3. The impulse response of an LTI system is given by $5u(t)$. If the input to the system is given by e^{-t} then the output of the system is
A. $5(1 - e^{-t})u(t)$ B. $(1 - 5e^{-t})u(t)$
C. $5 - e^{-t}u(t)$ D. $5u(t) - e^{-t}$
Explanation: Impulse response is given as

$$h(t) = 5u(t)$$
$$\frac{C(s)}{R(s)} = H(s) = \frac{5}{s}$$

When input is $r(t) = e^{-t}$

$$R(s) = \frac{1}{(s+1)}$$
$$C(s) = H(s)\cdot R(s)$$
$$= \frac{5}{s(s+1)} = \frac{5}{s} - \frac{5}{s+1}$$

taking inverse laplace transform,

$$C(t) = 5u(t) - 5e^{-t}U(t)$$
$$C(t) = 5(1 - e^{-t})u(t)$$

4. A DC shunt motor has the following characteristics, $R_a = 0.5\ \Omega$, $R_f = 200\ \Omega$, base speed = 1000 rpm, rated voltage = 250 V. On no load it draws a current = 5A. At what speed will this run while delivering a torque of 150 N/m?
A. 881 rpm B. 920 rpm
C. 950 rpm D. 990 rpm
Explanation: DC shunt motor circuit

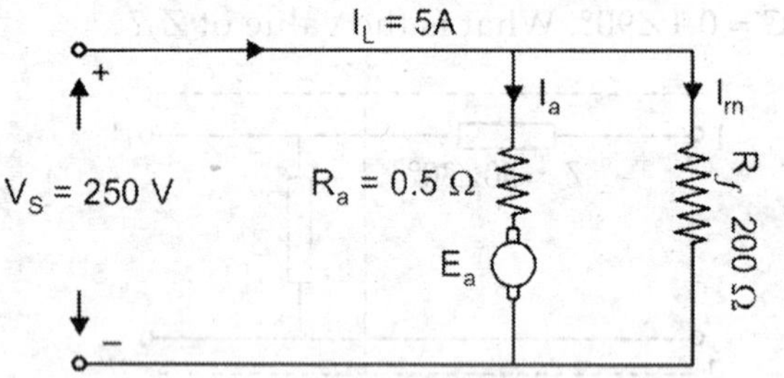

At no load $I_a = 5\text{A}, N_o = 1000$ rpm

$$= \frac{1000 \times 2\pi}{60} = 104.72 \text{ rad/sec}$$

Back emf $E_a = V_S - I_aR_a$

where $I_a = I_L - I_{sh} = 5 - \frac{250}{200} = 3.75\text{A}$

$$\therefore \quad E_a = 250 \times 2.75 \times 0.5 = 248.125 \text{ V}$$

Since, $E_a = Kf N = K_T N$

where K_T is constant.

$$\therefore \quad K_T = \frac{E_a}{N} = \frac{248.125}{104.72} = 2.369$$

At load torque, $\tau = 150$ N·m

Torque, $\tau = K_T I_a$ (as flux is constant)

$$\therefore \quad I_a = t/K_T = 150/2.369 = 63.318 \text{ A}$$

Now, back emf

$$E_a = V_S - I_aR_a = 250 - 63.18 \times 0.5$$

As $E \propto N$

$\therefore \quad \frac{E_a}{E'_a} = \frac{N_0}{N}$ for constant flux which is true for shunt machine

$$\therefore \quad \frac{28.125}{218.341} = \frac{1000}{N} \text{ or } N = 879.96 \text{ rpm} \approx 881 \text{ rpm}$$

5.

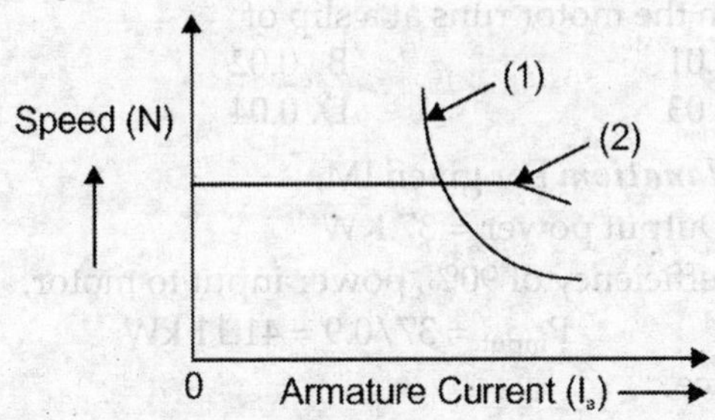

The The figure shows plots of speed (*N*) Vs. armature current (I_a) of a dc motor for two different operating conditions. Which one of the following features is relevant?

A. (1) represents stronger shunt field, and (2) represents stronger series field of a compound motor

B. (1) represents stronger series field, and (2) represents stronger shunt field of a compound motor

C. (1) represents only shunt excitation, and (2) represents only series excitation

D. (1) represents only series excitation, and (2) represents only shunt excitation

Explanation:

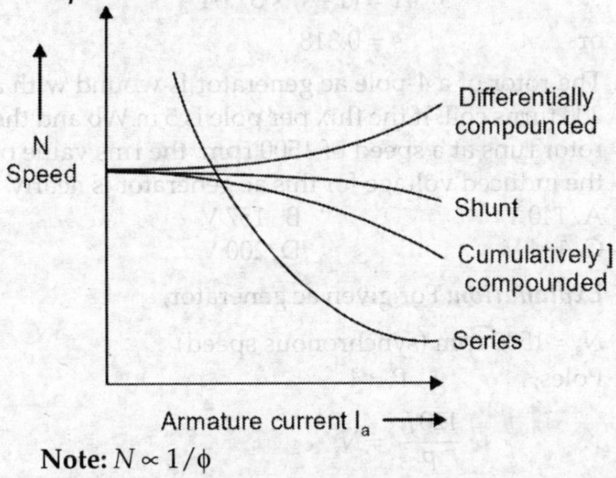

Note: $N \propto 1/\phi$

So in series field winding, as I_a increases ϕ increases, hence speed (N) decreases.

In shunt field winding, voltage applied across it, remains constant (almost), so ϕ remains constant. Thus speed is generally constant. The small drop at high I_a, is due to reduced back emf as $E \propto N$.

In comutatively compounded dc motor,

$$\phi_{total} = \phi_{series} + \phi_{shunt}$$

while in differentially compounded motor

$$\phi_{total} = \phi_{series} - \phi_{shunt}$$

6. The Laplace transform of $f(t) = t^n e^{-\omega t} u(t)$ is

A. $\dfrac{(n+1)!}{(s+\alpha)^{n+1}}$ B. $\dfrac{n!}{(s+\alpha)^n}$

C. $\dfrac{(n-1)!}{(s+\alpha)^{n+1}}$ D. $\dfrac{n!}{(s+\alpha)^{n+1}}$

Explanation: We know that $L\{t^n\} = \dfrac{|\overline{n+1}}{s^{n+1}}$

and $u(t) = \begin{cases} 1, & t \geq 0 \\ 0, & t < 0 \end{cases}$

So $L\{t^n \cdot u(t)\} = \dfrac{|\overline{n+1}}{s^{n+1}}$

Now by using 1st shifting property

$$L\{e^{-\omega t} \cdot t^n u(t)\} = \frac{|\overline{n+1}}{(s+\alpha)^{n+1}} = \frac{|\underline{n!}}{(s+\alpha)^{n+1}}$$

7. A 230V, 50 Hz, 4-pole, single-phase induction motor is rotating clockwise (forward) direction at a speed of 1425 rpm. If the rotor resistance at standstill is 7.8Ω, then the effective rotor resistance in the backward branch of the equivalent circuit will be

A. 2.0 Ω B. 4.0 Ω

C. 78 Ω D. 156 Ω

Explanation: For single phase induction motor (rotor speed) $N_r = 1425$ rpm

$$N_S \text{ (synchronous speed)} = \frac{120f}{P} = \frac{120 \times 50}{4}$$

$$= 1500 \text{ rpm}$$

Hence, $\quad \text{slips} = \dfrac{N_S - N_f}{N_S} = \dfrac{1500 \times 1425}{1500}$

Standstill rotor resistance, $R_2 = 7.8\Omega$ (given)

Effective rotor resistance for backward branch

$$= \frac{R_2}{2(2-s)}$$

$$= \frac{7.8}{2(2-0.05)} = 2\Omega$$

8. A 400 V, 50 Hz, 30 hp, three phase induction motor is drawing 50A current at 0.8 power factor lagging. The stator and rotor copper losses are 1.5 kW and 900 W respectively. The friction and windage losses are 1050 W and the core losses are 1200 W. The air gap power of the motor will be, nearly

A. 15 kW B. 20 kW

C. 25 kW D. 30 kW

Explanation: For given three phase induction motor

$$P_{input} = \sqrt{3}\text{ VIcos}\phi$$

$$= \sqrt{3} \times 400 \times 50 \times 0.8 = 27712.81 \text{ W}$$

or $\quad P_{input} = 27.713$ kW

Air gap power

$$P_g = P_{inpput} - \text{Stator Cu loss} - \text{Core loss}$$

$$= 27.713 - 41.5 - 1.2 = 25.013 \text{ kW}$$

Nearest option is 25 kW.

9. When the value of slip of an induction motor approaches zero, the effective resistance
A. is very low and the motor is under no-load
B. of the rotor circuit is very high and the motor is under no-load
C. is zero
D. of the rotor circuit is infinity and the motor is equivalent to short-circuited two-winding transformer

Explanation: For induction motor,
Effective resistance of rotor is given as R_2/s
where R_{20} = standstill rotor resistance
s = slip
Under no load, slip ≈ 0
As the load on motor increases, rotor speed decreases, hence slip increases.
Therefore,
As s approaches to zero, $R_2/s \approx \infty$ and motor is under no load.
By observing the rotor equivalent circuit at line frequency.

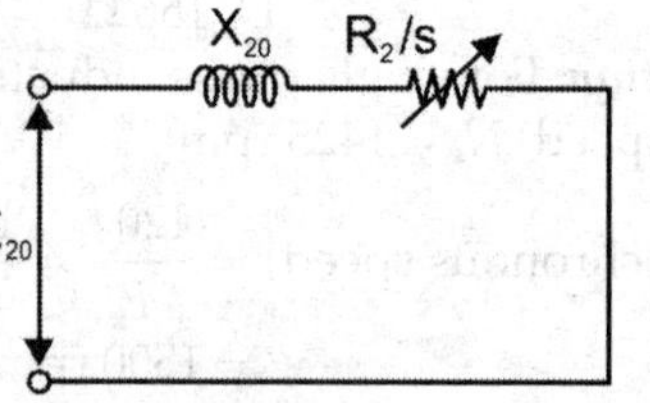

As slip approaches zero, $R_{20}/s \approx \infty$ and secondary i.e. rotor winding acts as open circuit.

10. A 4-pole, 50 Hz, 3-phase induction motor with a rotor resistance of 0.25 Ω develops a maximum torque of 25 N·m at 1400 rpm. The rotor reactance x_2 and slip at maximum torque $s_{max,T}$ respectively would be
A. 2.0 and 1/15
B. 3.75 and 1/12
C. 2.0 and 1/12
D. 3.75 and 1/15

Explanation: For given 3 – ϕ IM

$$N_s = \frac{120f}{P} = \frac{120 \times 50}{4} = 1500 \text{ rpm}$$

Given that, at maximum torque condition,
N_r (rotor speed) = 1400 rpm

$$s_m = \frac{N_s - N_r}{N_s} = \frac{1500 - 1400}{1500} = \frac{1}{5}$$

Also at maximum torque condition

$$R_2 = s_m X_2$$

$$\therefore \quad 0.25 = \frac{1}{15} X_2$$

or $X_2 = 15 \times 0.25 = 3.75\ \Omega$

or $X_2 = 2.75\ \Omega$

11. A 3-phase, 37 kW induction motor has an efficiency of 90% when delivering full load. At this load the stator copper losses and rotor copper losses are equal and are equal to stator iron losses. The mechanical losses are one-third of no-load losses. Then the motor runs at a slip of
A. 0.01
B. 0.02
C. 0.03
D. 0.04

Explanation: For given IM,
Output power = 37 KW
For efficiency of 90%, power input to motor,

$$P_{input} = 37/0.9 = 41.11 \text{ kW}$$

Hence,
total tosses = 41.11 – 37 = 4.111 kW
Now, let stator copper loss = P, then
Rotor copper loss = P
Stator iron loss = P and Mechanical loss = $P/3$
Total losses $P_{Loss} = P + P + P + P/3 = 10P/3$
Therefore, $10P/3 = 4.11$
or $P = 1.23$ kW
Now, air gap power;

$$P_g = P_{input} - \text{stator iron loss} - \text{stator copper loss}$$
$$= 41.11 - 1.23 - 1.23 = 38.665 \text{ kW}$$

Mechanical power developed

$$P_{nd} = P_{output} + \text{mechanical loss}$$
$$= 37 + (1.23/3) = 37.41 \text{ kW}$$

Also, $P_{md} = (1 - s)Pg$

$\therefore$ $37.41 = (1 - s) \times 37.64$

or $s = 0.318$

12. The rotor of a 4-pole ac generator is wound with a 200 turns coil. If the flux per pole is 5 m Wb and the rotor runs at a speed of 1500 rpm, the rms value of the induced voltage for this ac generator is nearly
A. 140 V
B. 157 V
C. 164 V
D. 200 V

Explanation: For given ac generator,
N_s = 1500 rpm (synchronous speed)
Poles, P = 4

$$\therefore \quad \frac{120f}{P} = N_s$$

or $\frac{120 \times f}{4} = 1500$

or $f = 50$ Hz

Given that,
flux per pole, $\phi = 5$ mWb $= 5 \times 10^{-3}$ Wb
Number of turns in rotor, $T = 200$
Induced emf is given y

$$E = 4.44\, f\phi T$$
$$= 4.44 \times 50 \times 5 \times 10{-}3 \times 200$$

or $E = 222$ V

Nearest option is 200 V

13. A 3-MVA, 6-pole, 50 Hz, 3-phase synchronous generator is connected to an infinite bus of 3300 V;

and it is run at 1000 rpm. The synchronous reactance of the machine is 0.915Ω per phase. The synchronizing torque for 1° mechanical displacement of the rotor is

A. 7500 N.m B. 7000 N.m
C. 6000 N·m D. 4500 N.m

Explanation: As we know that

$$E_f = V_t + I_a Z_s$$

$$\Rightarrow \quad E_f = V_t + jI_a X_S$$

Here, $$V_t = \frac{3300}{\sqrt{3}}$$

$$I_a = \frac{3\times10^6}{\sqrt{3}\times3300} = 524.8$$

$$X_s = 0.915 \text{ W/phase}$$

$$\therefore \quad E_f = 1964.8\angle 14.14°$$

Now, synchronous torque

$$\tau = \frac{P_{\text{syn}}}{\omega_m} = \frac{3E_f v_t}{X_a}\cdot\frac{1}{\omega_m}\cdot\cos\delta\cdot\propto$$

$$= \frac{3\times1964.8\times3300}{0.915\times\sqrt{3}}\times\frac{1}{\left(\frac{2\pi\times1000}{60}\right)}$$

$$\times\cos 14.14\times3\times\frac{\pi}{180}\text{ N·m}$$

$$\cong 6000 \text{ N·m}$$

14. The second-harmonic component of the power P versus load angle δ characteristic of a synchronous machine, operating at a terminal voltage V_t and having the d- and q-axis reactance per phase of X_d and X_q, respectively, is

A. $\frac{V_t^2}{2}\cdot\frac{X_d X_q}{X_d+X_q}\sin 2\delta$ B. $\frac{V_t^2}{2}\cdot\left(\frac{1}{X_q}-\frac{1}{X_d}\right)\sin 2\delta$

C. $\frac{V_t^2}{2}\cdot\frac{X_d X_q}{X_d+X_q}\cos 2\delta$ D. $\frac{V_t^2}{2}\cdot\left(\frac{1}{X_q}-\frac{1}{X_d}\right)\cos 2\delta$

Explanation: The real power for a salient pole machine is given by

$$P_{1\phi} = \frac{V_t E_t}{X_d}\sin\delta + \frac{V_t^2}{2}\left(\frac{1}{X_q}-\frac{1}{X_d}\right)\sin 2\delta$$

The second harmonic component is

$$\frac{V_t^2}{2}\left(\frac{1}{X_q}-\frac{1}{X_d}\right)\sin 2\delta$$

15. The term synchronous condenser refers to

A. A synchronous motor with a capacitor connected across the stator terminal to improve the power factor

B. A synchronous motor operating at full-load with leading power factor

C. An over-excited synchronous motor partially supplying mechanical load and also improving the power factor of the system to which it is connected

D. An over-excited synchronous motor operating at no-load with leading power factor used in large power stations for improvement of power factor

Explanation: Synchronous condenser is an over excited synchronous motor at no load.

Over excited means, $E_f \cos\delta > V$

By phasor diagram

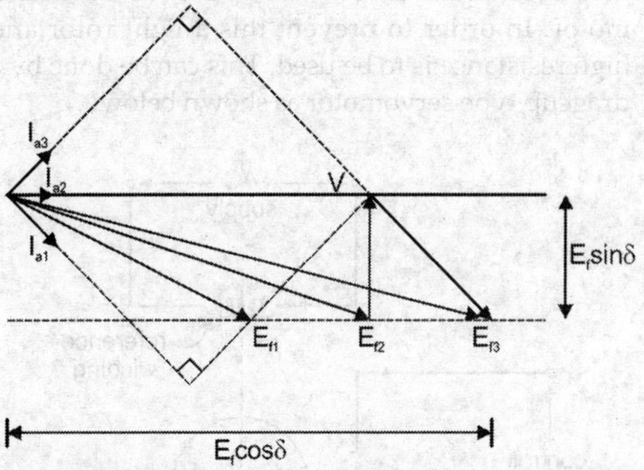

When $E_f \cos\delta > V$, pf is leading, as for current I_{a3}

Synchronous condenser is used in large power stations. It is not cost effective for small power stations.

16. Which of the following operating aspects necessitate the computation of regulation of an alternator?

1. When load is thrown off
2. For designing of an automatic voltage control equipment
3. For determination of steady-state and transient stability
4. For parallel operation of alternators

A. 1, 2 and 3 only B. 1, 2 and 4 only
C. 1, 3 and 4 only D. 1, 2, 3 and 4

Explanation: Need for voltage regulation computation.

1. To know the voltage drop when load is thrown off.
2. For designing automatic voltage-control equipment.
3. For parallel operation.
4. For determination of steady-state and transient stability

17. A 2-phase AC servomotor has a tendency to run as a single-phase induction motor, if the voltage across the control winding becomes zero. To prevent this

A. Rotor having high mass and moment of inertia is to be used

B. Drag-cup type of light rotor and high resistance is to be used
C. A low resistance rotor is to be used
D. The number of turns in the control winding is to be kept lesser than in the main reference winding

Explanation: A two phase ac servomotor has reference winding supplied by constant AC voltage supply and control winding supplied by variable control voltage obtained from servo amplifier. If the voltage across the control winding becomes zero then servo motor acts as a 1 – φ induction motor. In order to prevent this a light rotor and high resistance is to be used. This can be done by a dragcup type servomotor as shown below:

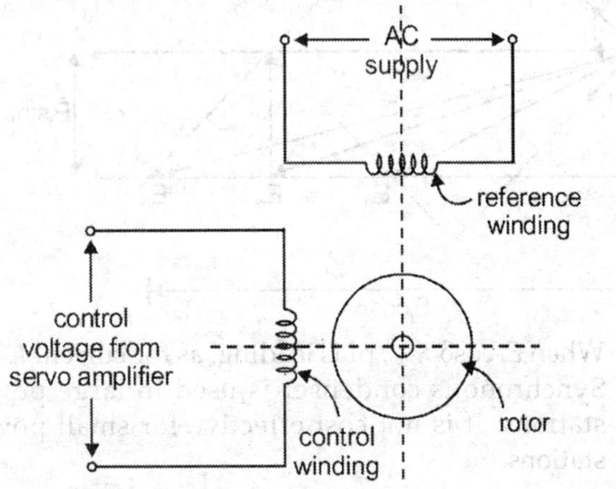

18. A single-stack, 8-phase (stator), multiple-step motor has 6-rotor teeth. The poles are excited one at a time. If excitation frequency is 120 Hz, the speed of the motor is
A. 3 rps B. 5 rps
C. 10 rps D. 15 rps

Explanation: Given that,
stator poles, $N_s = 8$
rotor teeth, $N_r = 6$
As, step size,

$$\alpha = \frac{N_S - N_f}{N_S N_r} \times 360 = \frac{8-6}{8 \times 6} \times 360 = 15°$$

Shaft speed is given by

$$n = \frac{\alpha f}{360} = \frac{15 \times 120}{360}$$

or $n = 5$ rps

19. A lossy capacitor C_x, rated for operation of 5 kV, 50 Hz is represented by an equivalent circuit with an ideal capacitor C_p in parallel with a resistor R_p. C_p is 0.102 μF; and R_p = 1.25 MΩ. The power loss, and tanδ, of this lossy capacitor when operating at the rated voltage are, respectively
A. 20 W and 0.04 B. 10 W and 0.04
C. 20 W and 0.025 D. 10 W and 0.025

Explanation: Lossy capacitor

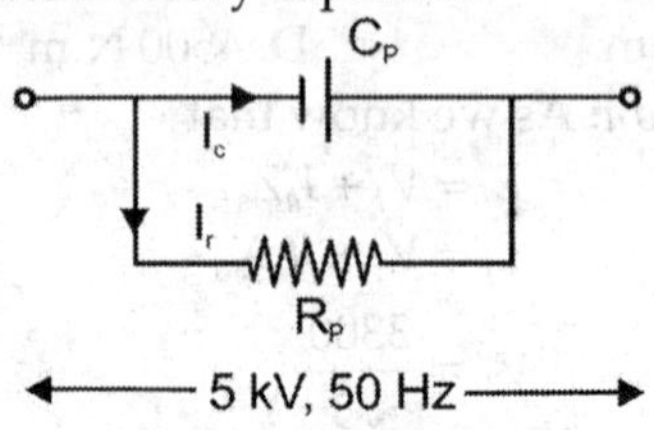

where $C_P = 0.102\ \mu F$
$R_P = 1.25\ M\Omega$

tanδ is given by

$$\tan\delta = \frac{I_r}{I_c} = \frac{\dfrac{5 \times 10^3}{1.25 \times 10^6}}{\dfrac{5 \times 10^3}{\left(\dfrac{1}{2\pi \times 50 \times 0.102 \times 10^{-6}}\right)}}$$

$= 0.025$

Power loss,

$P = V^2 C\omega \tan\delta$
$= (5 \times 10^3)^2 0.102 \times 10^{-6} \times 2\pi \times 50 \times 0.025$
$= 20.028$ W

20. An extra high voltage transmission line of length 300 km can be approximated by a lossless line having propagation constant β = 0.00127 rad/km. The percentage ratio of line length to wavelength will nearly be
A. 24% B. 19%
C. 12% D. 6%

Explanation: line length = 300 km
Propagation constant; β = 0.00127rad/km
The wavelength for one complete sinusoidal variation will be,

$$\lambda = \frac{2\pi}{\beta} = 4947.4$$

$$\lambda = \frac{2\pi}{0.00127} = 0.0606 \text{ or } 6.06\%$$

21. At what power factor will a lossless line with a reactance of 0.6 pu exhibit zero regulation given that the sending-end voltage is 1.0 pu?
A. 0.800 lag B. 0.800 lead
C. 0.954 lead D. Unity p.f.

Explanation: The zero voltage regulation is always obtained at leading power factor so, let the current be $1\underline{|\phi}$ receiving end voltage, $V_r = 1\underline{|0°}$.
Sending end voltage for zero voltage regulation, will be, $V_s = 1\underline{|\delta°}$.

$$V_s = V_r + I_r jX_s$$

$$\therefore\ 1\underline{|\delta} = 1\underline{|0°} + (1\underline{|\phi})(0.6\underline{|90°})$$

$$1\angle\delta = 1\angle 0° + 0.6\angle(90+\phi)$$
$$1\angle\delta = 1 + 0.6\cos(90+\phi) + j0.6\sin(90+\phi)$$
$$1\angle\delta = 1 - 0.6\sin\phi + j0.6\cos\phi \quad (1)$$
$$\therefore \quad 1 = (1-0.6\sin\phi)^2 + (0.6\cos\phi)^2$$
$$1 = 1 + 0.36 - 1.2\sin\phi$$
$$\sin\phi = \frac{0.36}{1.2} = 0.3$$
$$\cos\phi = \sqrt{1-\sin^2\phi} = \sqrt{1-(0.3)^2}$$
$$\cos\phi = 0.954 \text{ lead}$$

22. The dielectric loss in the insulation of a lossy underground cable, due to leakage current is (using standard notations)
A. $\omega CV^2\cos\delta$ B. $\omega CV\tan\delta$
C. $\omega CV^2\tan\delta$ D. $\omega CV\sin\delta$

Explanation: The dielectric loss in the insulation of a lossy underground cable due to leakage correct is

$$\text{dielectric loss} = \frac{V^2}{X_c}\tan\delta$$
$$= \frac{V^2}{(1/\omega c)}\tan\delta$$
$$= V^2\omega c\tan\delta$$

23.

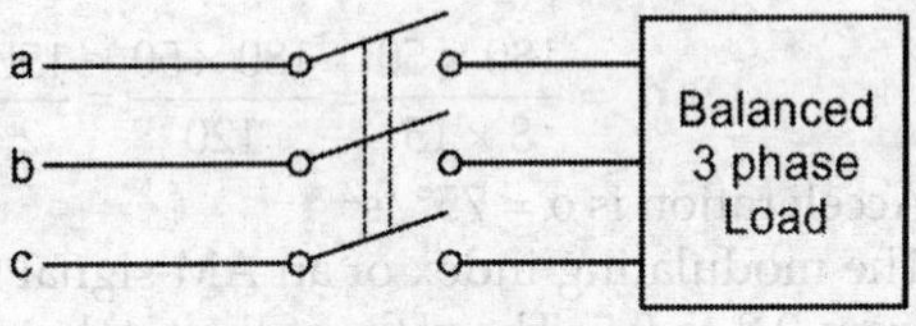

Triple-pole switch

A balanced 3-phase load is supplied from a 3-phase supply. The contact in line c of the triplepole switch contactor fails to connect when switched on. If the line-currents in lines *a* and *b* record 25A each, then the positive-sequence component of the current is
A. $14.4\angle+30°$A B. $25.0\angle-30°$A
C. $14.4\angle-30°$A D. $25.0\angle+30°$A

Explanation: The current I_a and I_b will be such that,

$$I_a + I_b = 0 \text{ or } I_a = -I_b$$
$$I_a = 25\angle 0° \quad I_b = 25\angle 180°$$
$$\alpha = 1\angle 120°$$

The positive sequence current will be

$$I_{a_1} = \frac{1}{3}(I_a + \alpha I_b + \alpha^2 I_c)$$
$$= \frac{1}{3}\, 25\angle 0° + 25\angle 300°$$
$$I_{a_1} = 14.4\angle -30°$$

24. A 3-phase, 100 MVA, 11 kV generator has the following p.u. constants. The generator neutral is solidly grounded. $X_1 = X_2 = 3X_0 = 0.15\Omega$. The ratio of the fault current due to threephase dead-short-circuit to that due to L-G fault would be nearly
A. 0.33 B. 0.56
C. 0.78 D. 1.0

Explanation: $X_1 = X_2 = 3X_0 = 0.15\Omega$

$$E_a = \text{phase voltage generated} = \frac{11}{\sqrt{3}} \text{ kV}$$

For $3-\phi$ deal short circuit, fault current

$$(I_f)_1 = \frac{E_a}{jX_1} \quad \text{(i)}$$

For L-G fault, fault current

$$(I_f)_2 = \frac{3E_a}{jX_1 + jX_2 + jX_0} \quad \text{(ii)}$$
$$\Rightarrow \frac{I_{f_1}}{I_{f_2}} = \frac{j(X_1 + X_2 + X_0)}{3jX_1}$$
$$\frac{I_{f_1}}{I_{f_2}} = \frac{0.15 + 0.15 + 0.05}{3\times 0.15} = 0.78$$

25. In a circuit-breaker, the arc is produced due to
1. Thermal emission
2. High temperature of air
3. Field emission

Which of the above statements are correct?
A. 1, 2 and 3 B. 2 and 3 only
C. 1 and 2 only D. 1 and 3 only

Hint: In a circuit breaker Arc is produced
1. Electronic (emission/field emission): As field is emitting high amount of electrons between the contacts.
2. Thermionic emission: This heat will continue to liberate the electrons which further increases the heat. Hence statement 1,3 are correct.

26. An alternator is
A. A polyphase synchronous machine operated with DC exciter
B. A polyphase synchronous machine operated with AC exciter
C. A three-phase induction machine with prime mover
D. Any AC generator

Hint: An alternator is a polyphase synchronous machine in which excitation is provided by rotor winding connected to dc supply.

27. The stability of a system, when subjected to a disturbance, is assessable by which of the following methods?
1. Swing curve 2. Equal-area criterion
3. Power-angle diagram 4. Power-circle diagram
A. 1, 2 and 4 only B. 1, 3 and 4 only
C. 2, 3 and 4 only D. 1, 2 and 3 only

Explanation: The stability of a system when subjected to a disturbance is assesable by
1. swing curve 2. equal area criterion
3. power angle diagram
From the power-circle diagram Liariour types of losses, torques etc can be found in a induction machine. Hence statements 1,2,3 are correct and statements 4 is false.

28. The line reactances of a power network are as follows :

Line No.	From Bus	To Bus	Reactance
1	0	1	0.2 p.u.
2	1	2	0.4 p.u.

The bus impedance matrix with '0' as ref-bus is

A. $\begin{bmatrix} 0.2 & 0.4 \\ 0.4 & 0.6 \end{bmatrix}$ B. $\begin{bmatrix} 0.2 & 0.2 \\ 0.4 & 0.6 \end{bmatrix}$

C. $\begin{bmatrix} 0.2 & 0.2 \\ 0.2 & 0.6 \end{bmatrix}$ D. $\begin{bmatrix} 0.2 & 0.2 \\ 0.2 & 0.4 \end{bmatrix}$

Explanation: First considering only line 1 ⇒

$Z_{Bus} = [j0.2]$

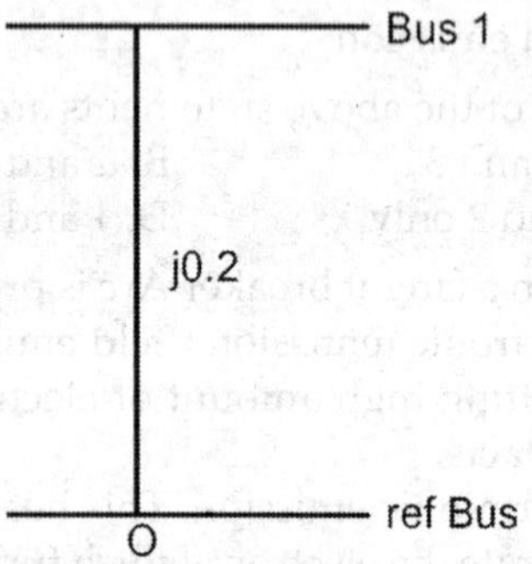

Now, if we add a new line of impedance j 0.4, between Bus 1 and Bus 2 then,

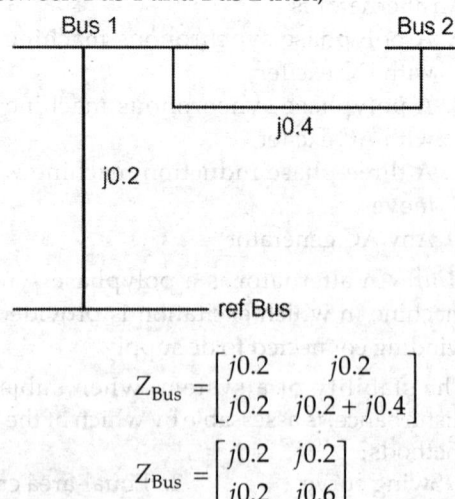

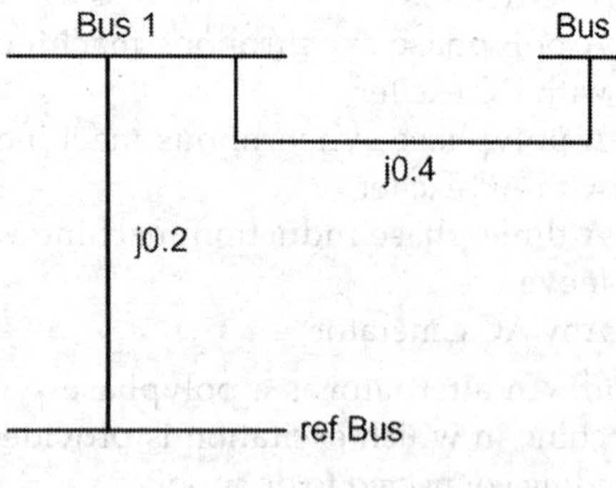

$$Z_{Bus} = \begin{bmatrix} j0.2 & j0.2 \\ j0.2 & j0.2 + j0.4 \end{bmatrix}$$

$$Z_{Bus} = \begin{bmatrix} j0.2 & j0.2 \\ j0.2 & j0.6 \end{bmatrix}$$

29. A 40 MVA, 11 kV, 3-phase, 50 Hz, 4-pole turbo-alternator has an inertia constant of 15 sec. An input of 20 MW developed 15 MW of output power (Neglecting losses). Then the acceleration is

A. $60°/s^2$ B. $65°/s^2$
C. $70°/s^2$ D. $75°/s^2$

Explanation: Given data:

$S = 40$ MVA (rating of the machine)
$V = 11$ kv
$f = 50$ Hz
$P = 4$ (no of poles)
$H = 15$ sec (inertia constant)
$P_i = 20$ MW (input power)
$P_o = 15$ MW (output power)

Acceleration power $P_a = (P_i - P_o)$

$P_a = (20 - 15)$ MW

$P_a = 5$ MW

We know $P_a = M\dfrac{d^2\delta}{dt^2}$

$$P_a = M\alpha \quad \left[\alpha = \frac{d^2\delta}{dt^2}\right]$$

$$P_a = \left(\frac{SH}{180f}\right)\alpha$$

$$5M = \left(\frac{40M \times 15}{180 \times 50}\right)\alpha$$

$$\alpha = \frac{180 \times 50}{8 \times 15} = \frac{180 \times 50}{120} = \frac{150}{2} = 75°$$

Acceleration is $\alpha = 75°/s^2$

30. The modulating index of an AM-signal is reduced from 0.8 to 0.5. The ratio of the total power in the new modulated signal to that of the original signal will nearly be

A. 0.39 B. 0.63
C. 0.85 D. 1.25

Explanation:
Total power in modulated signal

$$P_T = A_C\left(1 + \frac{\mu^2}{2}\right)$$

For $\mu = 0.8\, P_1 = A_c\left(1 + \left(\frac{0.8}{2}\right)^2\right)$ → original modulated signal

For $\mu = 0.5\, P_2 = A_c\left(1 + \left(\frac{0.5}{2}\right)^2\right)$ → new modulated signal

$$\frac{P_2}{P_1} = \frac{2 + (0.5)^2}{2 + (0.8)^2} = 0.85$$

31. An 8-bit DAC uses a ladder network. The fullscale output voltage of the converter is +10V. The resolution expressed in percentage and in volts is, respectively

A. 0.25% and 30 mV B. 0.39% and 30 mV
C. 0.25% and 39 mV D. 0.39% and 39 mV

Explanation: Number of bits of DAC (n) = 8

Percentage resolution of DAC is given by

$$\% \text{ Resolution} = \left(\frac{100}{2^n - 1}\right) = 0.392\%$$

$$\text{Resolution in volts} = \frac{\text{Full scale output}}{2^n - 1}$$

$$= 0.039 \text{ V} = 39 \text{ mV}$$

32.

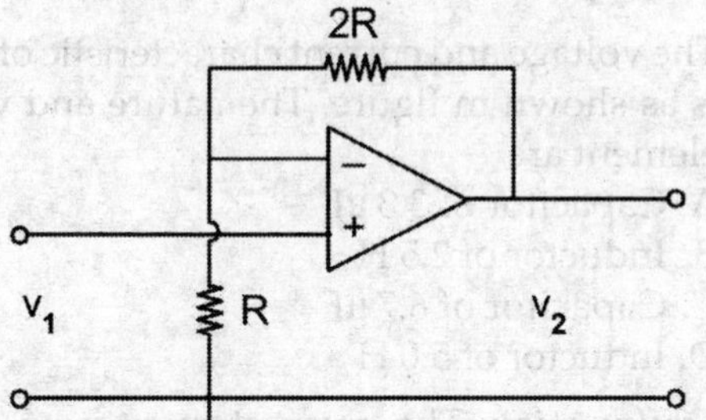

An ideal operational amplifier is connected as shown in figure. What is the output voltage V_2?

A. $3V_1$ B. $2V_1$

C. $1V_1$ D. $3V_1$

Explanation:

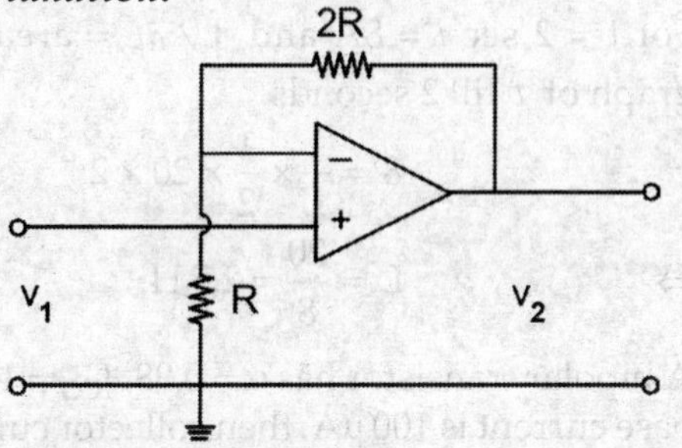

This can be redrawn as

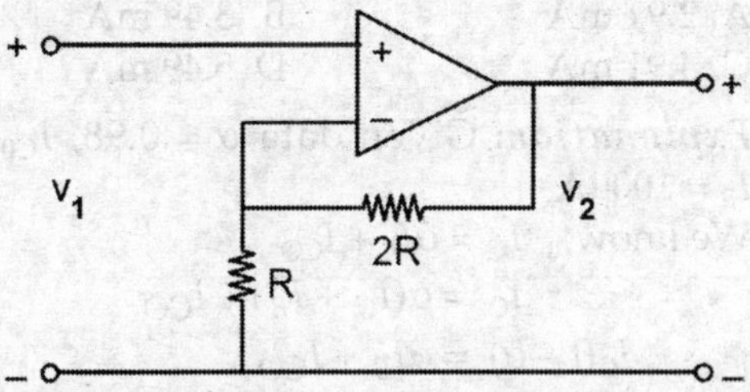

by virtual node concept the voltage at –ve terminal is V_1.

Applying nodal analysis at V_1 we get

$$\frac{V_1}{R} + \frac{V_1 - V_2}{2R} = 0$$

$$\frac{V_1}{R} + \frac{V_1}{2R} = \frac{V_2}{2R}$$

$$V_1\left(\frac{1}{R} + \frac{1}{2R}\right) = \frac{V_2}{2R}$$

$$V_1\left(\frac{2+1}{2R}\right) = \frac{V_2}{2R}$$

$$\frac{3V_1}{2R} = \frac{V_2}{2R}$$

$$V_2 = 3V_1$$

33. The Truth table for the function

$f(ABCD) = \Sigma m\,(0, 1, 3, 4, 8, 9)$ is

A	B	C	f
0	0	0	W
0	0	1	X
0	1	0	Y
0	1	1	0
1	0	0	Z
1	0	1	0
1	1	0	0
1	1	1	0

where W, X, Y, Z are given by (d is the complement of D)

A. D, d, 1, 1 B. 1, d, D, 1

C. 1, 1, D, d D. 1, D, d, 1

34. For a state model $\dot{X} = AX$, where $A = \begin{bmatrix} 1 & 0 \\ 1 & 1 \end{bmatrix}$, the state transition matrix is

A. $\begin{bmatrix} te^{-t} & 0 \\ e^t & e^t \end{bmatrix}$ B. $\begin{bmatrix} 0 & t \\ e^t & te^t \end{bmatrix}$

C. $\begin{bmatrix} e^t & 0 \\ te^t & e^t \end{bmatrix}$ D. $\begin{bmatrix} t & 0 \\ t^2 & e^t \end{bmatrix}$

Explanation: Given

$$A = \begin{bmatrix} 1 & 0 \\ 1 & 1 \end{bmatrix}$$

The state transition matrix is given by

$$\phi(t) = L^{-1}[sI - A]^{-1}$$

$$[sI - A] = \begin{bmatrix} s-1 & 0 \\ -1 & s-1 \end{bmatrix}$$

$$[sI - A]^{-1} = \frac{1}{|sI - A|}\text{Adj}[sI - A]$$

$$= \frac{1}{(s-1)^2}\begin{bmatrix} s-1 & 0 \\ 1 & s-1 \end{bmatrix}$$

$$= \begin{bmatrix} \frac{1}{s-1} & 0 \\ \frac{1}{(s-1)^2} & \frac{1}{s-1} \end{bmatrix}$$

$$L^{-1}[sI - A]^{-1} = \begin{bmatrix} e^t & 0 \\ te^t & e^t \end{bmatrix}$$

35. A second order system with a zero at –2 has its poles located at –3 + j4 and –3 – j4 in the s-plane. The undamped natural frequency and the damping factor of the system respectively are

A. 3 rad/s and 0.80 B. 5 rad/s and 0.80

C. 3 rad/s and 0.60 D. 5 rad/s and 0.60

Explanation:

$$TF = \frac{s+2}{(s+3-j4)(s+3+j4)}$$

$$= \frac{s+2}{(s+3)^2-(j4)^2} = \frac{s+2}{s^2+6s+9+16}$$

$$TF = \frac{s+2}{s^2+6s+25}$$

Therefore

$$\omega_n^2 = 25 \Rightarrow 7\ \omega_n = 5$$

$$2\xi w_n = 6 \Rightarrow 2\xi \times 5 = 6$$

or $\xi = 0.6$

36. An ideal transformer is having 150 turns primary and 750 turns secondary. The primary coil is connected to a 240 V, 50 Hz source. The secondary winding supplies a load of 4A at lagging power factor of 0.8. What is the power supplied by the transformer to the load?

A. 4200 W B. 3840 W
C. 2100 W D. 1920 W

Explanation: Primary turns, $N_P = 150$
Secondary turns, $N_S = 750$
Primary voltage = 240 (say VP)

Since $\frac{V_P}{V_S} = \frac{N_P}{N_S}$

where V_S = secondary voltage

$$\therefore \quad V_S = \frac{V_P \times N_S}{N_P} = \frac{240 \times 750}{150} = 1200\text{ V}$$

for load current of 4A at pf = 0.8 lagging

$$P_{output} = VI\cos\phi$$

$$= 1200 \times 4 \times 0.8 = 3840\text{ W}$$

37. In an induction motor for a fixed speed at constant frequency

A. Both line current and torque are proportional to voltage
B. Both line current and torque are proportional to the square of voltage
C. Line current is proportional to voltage and torque is proportional to the square of voltage
D. Line current is constant and torque is proportional to voltage

Explanation: In induction motor

$$I_2 = \frac{sE_{20}}{R_{20}+jsX_{20}}$$

$$= s\left(\frac{Te_2}{Te_1}\right)E_1 \times \frac{1}{R_{20}+jsX_{20}} \quad \left(\because E_{20} = \frac{Te_2}{Te_1}E_1\right)$$

or $I_2 \propto E_t$

$$\tau_d = \frac{kE_1^2 sR_2}{R_2^2 + s^2 X_{20}^2}$$

where $K = \frac{3}{2\pi n_s}\left(\frac{Te_2}{Te_1}\right)^2$ = constant for fixed speed

$$\therefore \quad t_d \propto E_1^2$$

Note: At constant frequency, value of X remains constant.

38.

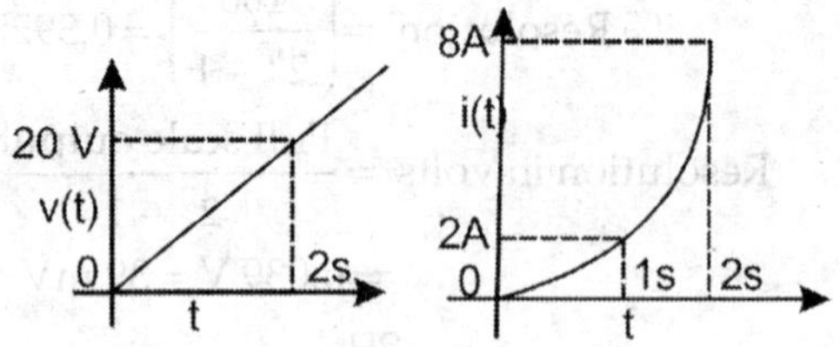

The voltage and current characteristic of an element is as shown in figure. The nature and value of the element are

A. Capacitor of 3.3 μF
B. Inductor of 2.5 H
C. Capacitor of 6.7 μF
D. Inductor of 5.0 H

Explanation: The given element is an inductor as the graph of current is an integration of that of current.

The voltage in an inductor is given as

$$i = \frac{1}{L}\int V\,dt$$

For t = 2 sec i = 8A and $\int V\,dt$ = area under the graph of v till 2 seconds.

$$\therefore \quad 8 = \frac{1}{L} \times \frac{1}{2} \times 20 \times 2$$

$$\Rightarrow \quad L = \frac{20}{8} = 2.5\text{ H}$$

39. A bipolar transistor has $\alpha = 0.98$, $I_{CO} = 10\ \mu A$. If the base current is 100 μA, then collector current would be

A. 2.91 mA B. 3.49 mA
C. 4.91 mA D. 5.49 mA

Explanation: Given data α = 0.98, I_{CO} = 10μA , I_B = 100μA

We know, $I_C = \alpha I_E + I_{CO}$

$$I_C = \alpha(I_C + I_B) + I_{CO}$$

$$I_C(t-\alpha) = \alpha I_B + I_{CO}$$

$$I_C = \left(\frac{\alpha}{1-\alpha}\right)I_B + \frac{I_{CO}}{(1-\alpha)}$$

$$= \left(\frac{0.98}{1-0.98}\right)100\mu + \left(\frac{10\mu}{1-0.98}\right)$$

$$= \frac{0.98}{0.02} \times 100\mu + \frac{10\mu}{0.02}$$

$$= 10\mu\left(\frac{9.8+1}{0.02}\right)$$

$$= 10\mu \times \frac{10.8}{0.02}$$

$$= 10\mu \times \frac{1080}{2}$$

$$= 5400 \times 10^{-8} = 5.49\text{ mA}$$

Answer Key to Multiple Choice Questions from Various Competitive Examinations Including GATE and IES

2017

1. C	2. A	3. A	4. A	5. D	6. B	7. C	8. A	9. A	10. B	11. B	12. A	13. D	14. D
15. B	16. B	17. D	18. D	19. D	20. C	21. A	22. B	23. B	24. B	25. C	26. C	27. B and D	
28. A	29. A	30. C	31. A	32. D	33. B	34. B	35. D	36. D	37. D	38. C	39. D	40. D	41. B
42. A	43. D	44. A	45. B	46. A	47. C	48. C	49. D	50. C	51. A	52. C			

2018

1. B	2. B	3. A	4. A	5. D	6. D	7. A	8. C	9. B	10. D	11. C	12. D	13. C	14. B
15. D	16. D	17. B	18. B	19. C	20. D	21. C	22. C	23. C	24. C	25. D	26. C	27. A	28. D
29. D	30. A	31. D	32. A	33. D	34. C	35. D	36. B	37. C	38. B	39. D			

Electrical Symbols

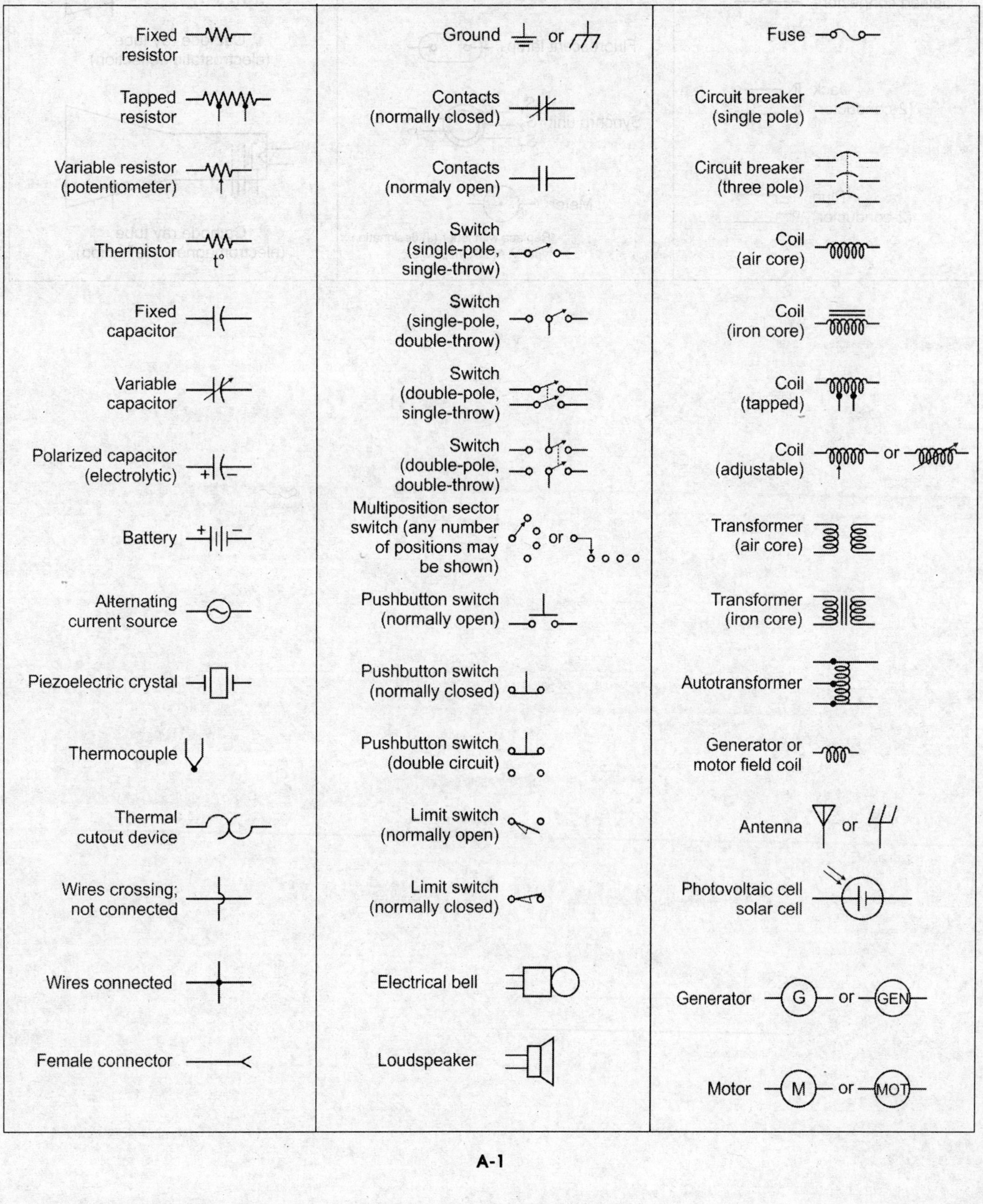
Fixed resistor
Tapped resistor
Variable resistor (potentiometer)
Thermistor
t°
Fixed capacitor
Variable capacitor
Polarized capacitor (electrolytic)
+
–
Battery
+
–
Alternating current source
Piezoelectric crystal
Thermocouple
Thermal cutout device
Wires crossing; not connected
Wires connected
Female connector
Ground
or
Contacts (normally closed)
Contacts (normaly open)
Switch (single-pole, single-throw)
Switch (single-pole, double-throw)
Switch (double-pole, single-throw)
Switch (double-pole, double-throw)
Multiposition sector switch (any number of positions may be shown)
or
Pushbutton switch (normally open)
Pushbutton switch (normally closed)
Pushbutton switch (double circuit)
Limit switch (normally open)
Limit switch (normally closed)
Electrical bell
Loudspeaker
Fuse
Circuit breaker (single pole)
Circuit breaker (three pole)
Coil (air core)
Coil (iron core)
Coil (tapped)
Coil (adjustable)
or
Transformer (air core)
Transformer (iron core)
Autotransformer
Generator or motor field coil
Antenna
or
Photovoltaic cell solar cell
Generator
G
or
GEN
Motor
M
or
MOT

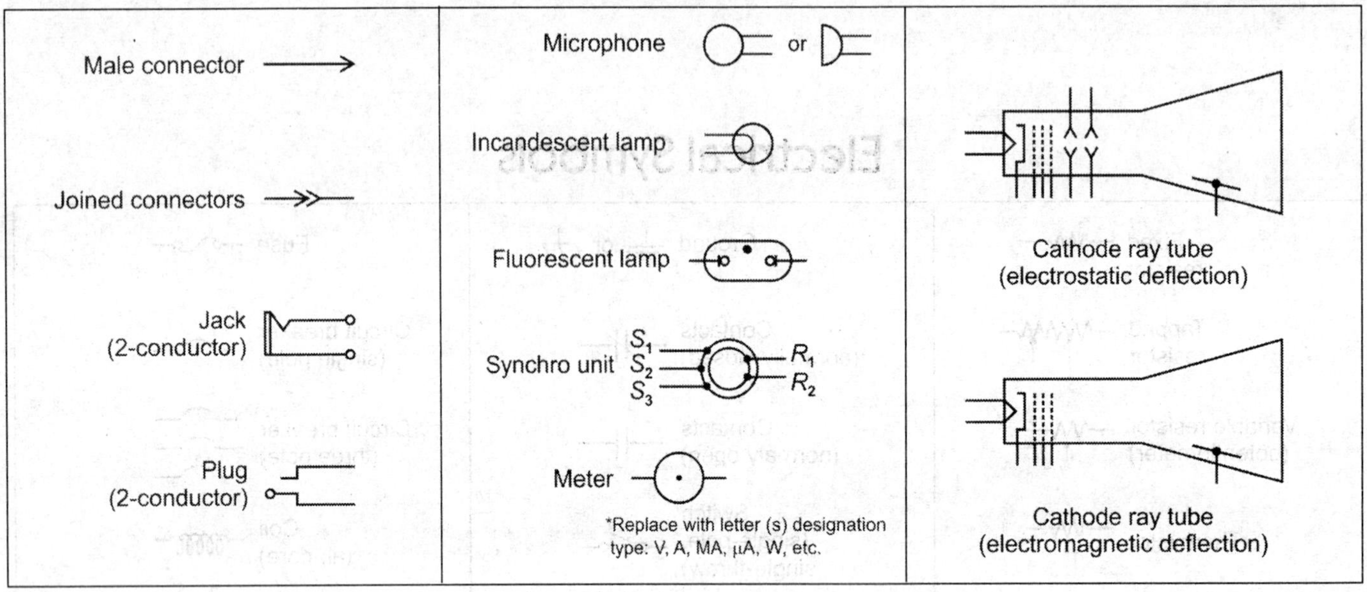
Male connector
Joined connectors
Jack
(2-conductor)
Plug
(2-conductor)
Microphone
or
Incandescent lamp
Fluorescent lamp
Synchro unit
S_1
S_2
S_3
R_1
R_2
Meter
*Replace with letter (s) designation type: V, A, MA, μA, W, etc.
Cathode ray tube
(electrostatic deflection)
Cathode ray tube
(electromagnetic deflection)

Electronic Symbols

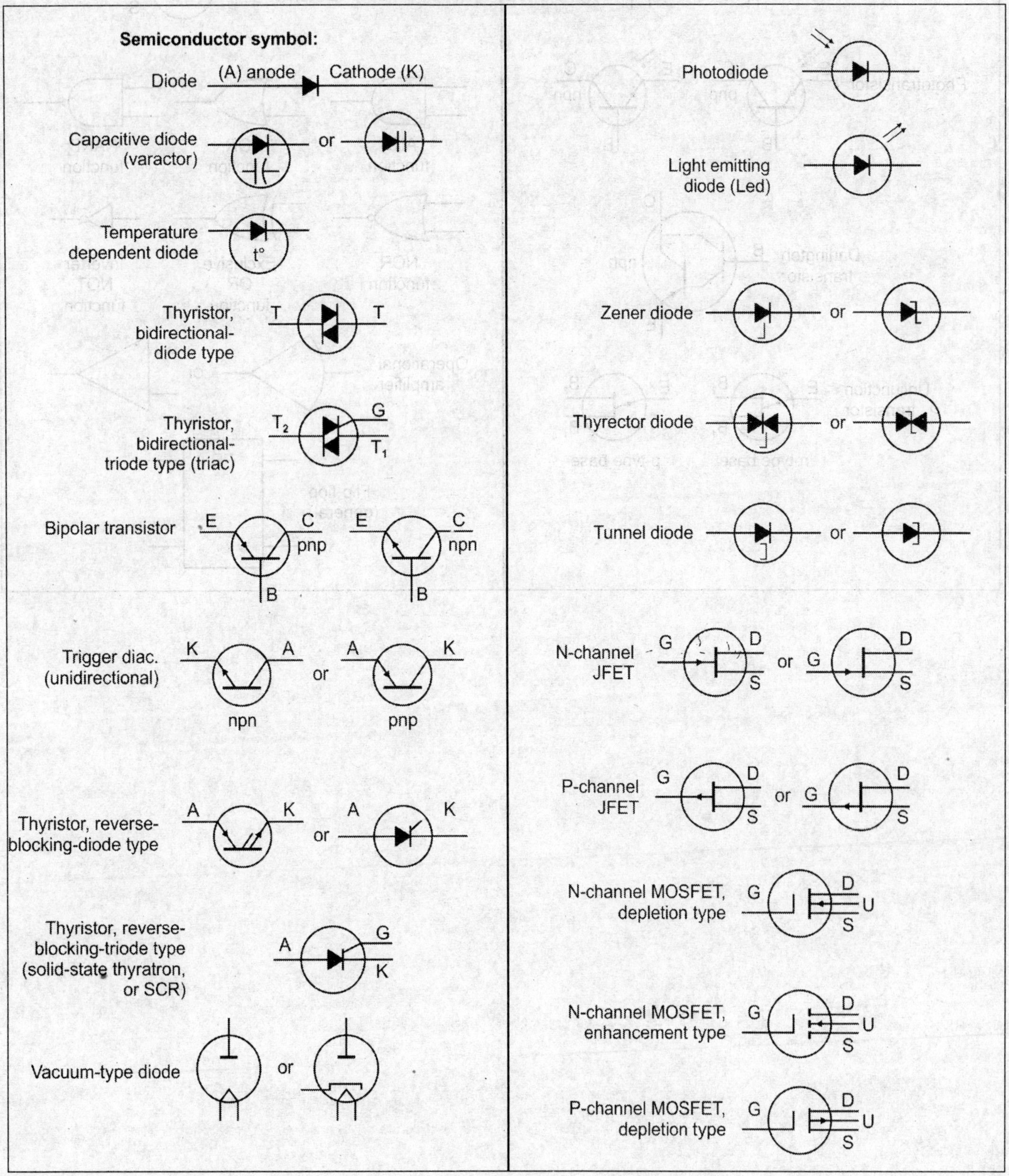
Semiconductor symbol:
Diode
(A) anode
Cathode (K)
Capacitive diode (varactor)
or
Temperature dependent diode
t°
Thyristor, bidirectional-diode type
T
T
Thyristor, bidirectional-triode type (triac)
T2
G
T1
Bipolar transistor
E
C
pnp
B
E
C
npn
B
Trigger diac. (unidirectional)
K
A
npn
or
A
K
pnp
Thyristor, reverse-blocking-diode type
A
K
or
A
K
Thyristor, reverse-blocking-triode type (solid-state thyratron, or SCR)
A
G
K
Vacuum-type diode
or
Photodiode
Light emitting diode (Led)
Zener diode
or
Thyrector diode
or
Tunnel diode
or
N-channel JFET
G
D
S
or
G
D
S
P-channel JFET
G
D
S
or
G
D
S
N-channel MOSFET, depletion type
G
D
U
S
N-channel MOSFET, enhancement type
G
D
U
S
P-channel MOSFET, depletion type
G
D
U
S

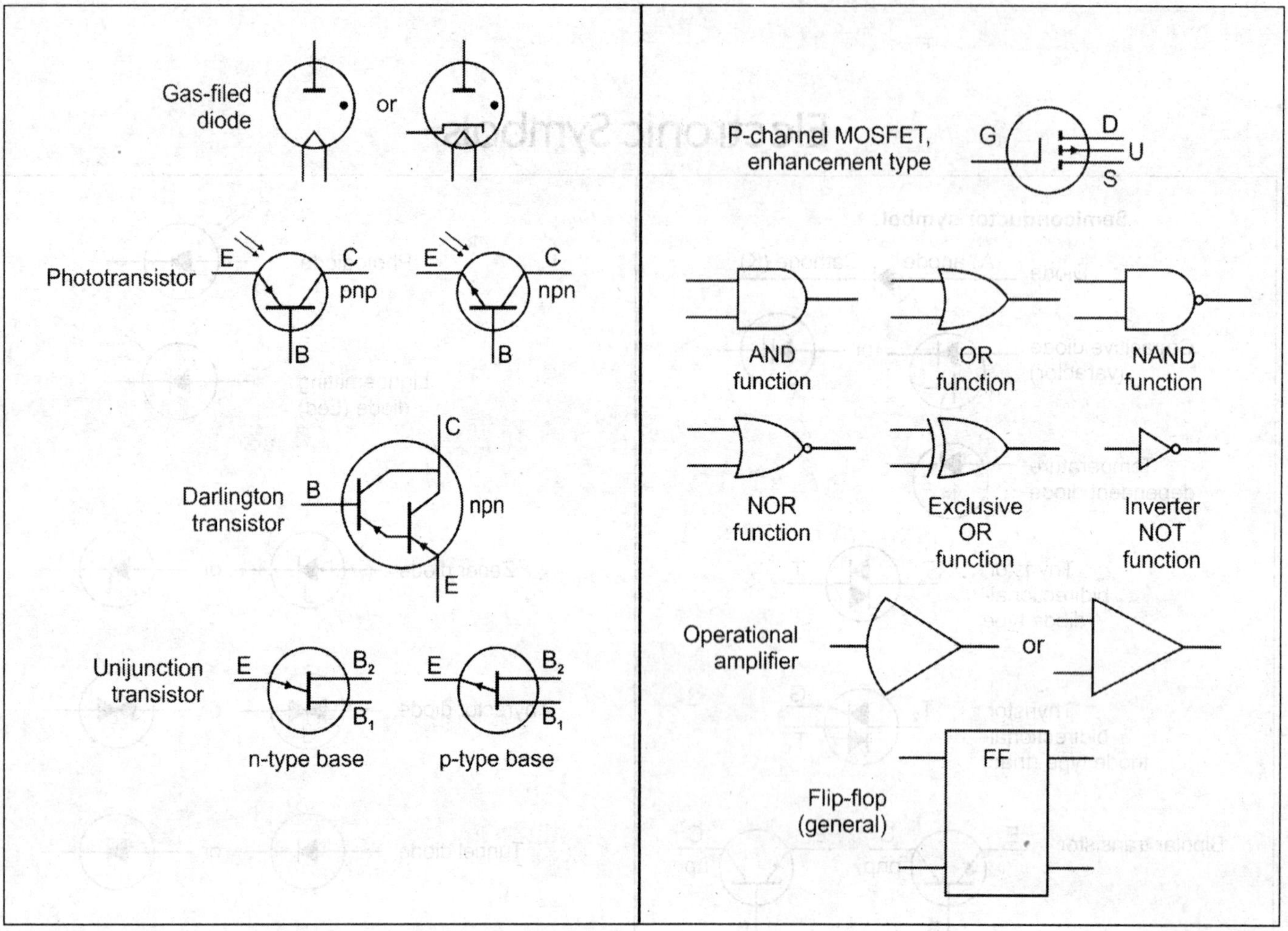
Gas-filed diode
or
P-channel MOSFET, enhancement type
G
D
U
S
Phototransistor
E
C
pnp
B
E
C
npn
B
AND function
OR function
NAND function
NOR function
Exclusive OR function
Inverter NOT function
Darlington transistor
B
C
npn
E
Operational amplifier
or
Unijunction transistor
E
B_2
B_1
n-type base
E
B_2
B_1
p-type base
FF
Flip-flop (general)